Innovative Renewable Energy

Series editor

Ali Sayigh
World Renewable Energy Congress, Brighton, UK

The primary objective of this book series is to highlight the best-implemented worldwide policies, projects and research dealing with renewable energy and the environment. The books are developed in partnership with the World Renewable Energy Network (WREN). WREN is one of the most effective organizations in supporting and enhancing the utilisation and implementation of renewable energy sources that are both environmentally safe and economically sustainable. Contributors to books in this series come from a worldwide network of agencies, laboratories, institutions, companies and individuals, all working together towards an international diffusion of renewable energy technologies and applications. With contributions from most countries in the world, books in this series promote the communication and technical education of scientists, engineers, technicians and managers in this field and address the energy needs of both developing and developed countries.

Each book in the series contains contributions from WREN members and cover the most-up-to-date research developments, government policies, business models, best practices, and innovations from countries all over the globe. Additionally, the series publishes a collection of best papers presented during the annual and bi-annual World Renewable Energy Congress and Forum each year.

More information about this series at http://www.springer.com/series/15925

Ali Sayigh
Editor

Renewable Energy and Sustainable Buildings

Selected Papers from the World Renewable Energy Congress WREC 2018

Springer

Editor
Ali Sayigh
World Renewable Energy Congress
Brighton, UK

ISSN 2522-8927 ISSN 2522-8935 (electronic)
Innovative Renewable Energy
ISBN 978-3-030-18487-2 ISBN 978-3-030-18488-9 (eBook)
https://doi.org/10.1007/978-3-030-18488-9

This Springer imprint is published by the registered company Springer Nature Switzerland AG
The registered company address is: Gewerbestrasse 11, 6330 Cham, Switzerland

Contents

Introduction

At the Congress, 21 plenary papers, 8 keynote papers and 140 technical papers were presented. The opening address was given by H E Dr. Altwaijri, director general of the ISESCO and WREN honorary chairman, representing 54 countries, who presented his excellent talk 'Leading Role of ISESCO in the field of Renewable Energy and the Promotion of the Concept of Green and Sustainable Cities in the Islamic World'.

The message from Professor Ali Sayigh to the participants at the opening was:

On behalf of the World Renewable Energy Network and Congress, I welcome you all to our 28th anniversary Congress at the University of Kingston, London, UK. Once again, it gives me very great pleasure to welcome our permanent member of WREN Council, H E Dr. Altwaijri, our honorary chairman, and the rest of my distinguished colleagues from more than 60 countries.

WREN, with your continued support, has managed to change the equation of 'energy from fossil-fuel' to 'renewable energy fuel'.

Recently, the second largest exporter of oil, the Kingdom of Saudi Arabia, embraced renewable energy on a very large scale. Their present power capacity is 77 GW, and they are planning projects of renewable energy to generate 200 GW by 2030.

In Europe, the EU has set an energy consumption target of 32% from renewables. However, this figure is considered by many to be too low since renewables already nearly meet this amount and bearing in mind that the target did not include biomass.

Meanwhile, China has already reached its 2020 carbon emission reduction goal 2 years ahead of its target. India is another country which contributes on a large scale to global pollution but now aims to install 100 GW of renewables by 2022. Many countries in the world are now embracing renewable energy not only because it is clean and cost-effective but also because it reduces energy imports and creates local jobs.

While the outlook for renewables is in many places very encouraging, there was a feeling of dismay at the recent withdrawal by the US from the Paris Agreement on Climate Change to be effective in 2020. Nevertheless, many States in the US will continue to implement their own renewable energy policies and place the US among the leading renewable energy producers.

Now, it is up to you, the participants in this Congress, to continue to spread and develop all forms of renewable energy and to commit yourselves to the very urgent reduction of the climate change effect. It goes without saying that this goal includes the improvement of energy efficiency and energy conservation through recycling and sustainable development. I urge you to do this not primarily for the sake of renewable energy industries but more importantly for the future welfare of our fragile environment.

Our thanks to all the sponsors and to the staff and students of Kingston University for their generous support and help. My thanks and gratitude to the Technical Committee and our publishers, Springer (USA) and EDP (France), who are attending the Congress in person.

Special thanks to ISESCO and their Director General, Dr. Altwaijri, who is WREN honorary chairman and main supporter of WREN during the last 18 years. My thanks to Dr. Doukas and the European Commission in supporting WREN Congresses over the last 4 years. Last, but not least, our thanks to the Institute of Engineering and Technology, in particular to Mrs. Jones in sponsoring WREN Congresses over the last 5 years.

I wish you a great success in your research and utilization of green energy and seize this opportunity to network and learn from each other while you are in London. Wishing you a very enjoyable time.

One of our pioneers and invited speakers to WREC-18, Prof. David Elliott, Open University, has written a summary of the event which was published by Physics World (IoP); the link is https://physicsworld.com/a/specialists-gather-at-world-renewable-energy-congress/.

This could be an introduction to WREC-18:

'World Renewable Energy Congress: 18 in the UK'

The World Renewable Energy Congress (WREC) is a long-running biannual gathering of academics and practitioners, linked via the World Renewable Energy Network (WREN) . Although UK-based, its biannual sessions, and other regular sub-gatherings, are held all over the world. This year, however, for its 18th session, WREC returned (for a second time) to Kingston University in suburban Surrey. Big hitters included David Renné, president of the International Solar Energy Society, who offered an inspiring *Overview of pathways to 100% Renewable Energy*, and Rainer Hinrichs-Rahlwes, vice president, European Renewable Energies Federation (EREF), and board member of the German Renewable Energy Federation (BEE), looking critically at *Europe's 2030 Targets*: the EU is trying, he said, but needs to do better.

Wherever it is held, WREC always attracts significant international participation, including from **Africa**, with, this year, a range of technical papers being presented, for example, on mini-grids and rural electrification. At the more general policy level, David Elliott, Open University, and his colleague Terry Cook relayed some ideas from their new book (https://www.palgrave.com/gb/book/9783319747866) by focussing on the impact of market-based private investment approaches to renewable development in Africa. The new 'trade not aid' approach being adopted by the EU aims to stimulate private sector investment to create new markets, something that China was already good at, although often with less emphasis on the social and environmental aspects.

There were also several good presentations on developments in **Asia**, including one on the state of play with offshore wind in Japan. It was reported that were or had been ten projects, 40MW in all, including some 2MW floating systems and some now at 5 and 7MW. But though the potential was very large (100s of GW for both offshore and onshore wind), there were problems with developing wind power in Japan. It has cumbersome EIA procedures (they can take 4 years) and three separate regional electricity supply bodies in the East and six in the West, making grid integration of outputs from onshore wind, which is mainly in the North, hard. Floating offshore projects might however circumvent some of the bureaucratic issues and supply power direct to load centres like Tokyo.

Integration is clearly an increasingly important issue, with storage being part of that. Controversially, WREN Director Donald Swift-Hook argued that **energy storage** was mostly irrelevant and that renewables were best seen as direct (fossil) fuel savers. So, if they were available, and especially at increasingly low cost, then just use them. Other times, use gas! Turning surplus renewables into gas (PtG) would evidently be too expensive. So post-generation storage of renewable outputs was basically not needed or viable: 'Commercial storage of energy on a power system works by arbitrage, buying cheap electricity [typically in the middle of the night] and selling it when electricity is dear [during the day-time or evening]. If fuel saving renewables are stored, the round-trip losses from putting their electricity into store and taking it out again wastes some of the fuel already saved. There is no difference in fuel costs around the clock, so storing electricity from renewable fuel savers cannot be arbitraged or commercially justified'.

Reinhard Haas, from the Energy Economics Group at the Technical University of Vienna, also looked at the storage issue, in the context of dealing with surplus outputs from renewables, and their short- and long-term variations. Batteries were OK for short-term balancing, and they reduced the strain on the grid. But pumped hydro was needed for longer-term storage—though its cost would rise as new sites became scarce. However, power-to-gas (PtG hydrogen/methane) would get cheaper. *But* there are diminishing returns—each storage success makes it less economically attractive to store more! Or as Haas put it: 'Every additional storage unit makes this one and every other less cost-effective'! That's the so-called principle of self-cannibalism in energy economics, which, as far as I can see, means that the cheaper anything gets, the less profit can be made from going for more of it. But that's surely only if the market remains the same size—if it's expanding, so will the opportunity to make profits from meeting it. However, Haas says that while local batteries might do well for short-term storage, 'in a dynamic market framework the costs of all centralized long-term storage technologies will finally be too high to become competitive', and he sees competition with demand response options and demand-side management challenging storage as well.

So, decentralized batteries apart, his final conclusion is quite grim, almost as grim as Swift Hook's, 'with respect to all centralized long-term storage technologies, the future perspectives will be much less promising than currently indicated in several papers: new long term hydro storages will not become economically attractive in the next decades [and] for PtG-technologies it will become very hard to compete in electricity markets, despite a high technological learning potential', though Haas does say that 'for hydrogen and methane there might be prospects in the transport sector'.

There were, as ever, many papers on specific technical and operational issues and developments (over 170 in all), with the **PV solar** field heavily represented, including one from Oman, on the impacts of dust, and a nice historical one on the use of PV by the US Vanguard satellite—a lot to take in but with some overall very positive messages. For example, it was suggested that PV may surpass 1TW globally by 2022. It was also suggested that new luminescent wavelength downshifting materials could boost PV productivity significantly. That would make PV more valuable in less sunny areas—the paper on that came from Ireland. Arguably, it might then be clearly even more preferable to biomass in land use terms. It was claimed that given the low-energy conversion efficiency of bio-photosynthesis compared to PV cells, PV could already generate 40–80 times more power output/acre than biomass crops. Of course, the difference would be less if it was biomass *waste* that was being used, an issue also well covered at WREC. However, overall PV does seem to be romping ahead in many areas, both at the large and small scales, with, for example, PV being used in refugee camps and in Afghanistan's reconstruction and hybrid PV thermal playing a role in desalination.

WREC was held in the midst of the sweltering UK summer, and interestingly, there were some timely papers on **cooling and ventilation**, including, significantly, some from China, where, with a rapidly growing more affluent urban population, this is becoming a major issue. Much of the Middle East and Africa also has increasing air con demand, and this will rise everywhere as climate change increases. Fortunately, some new technologies are emerging which may help, including systems using high heat capacity phase-changing materials. As some papers from the UAE noted, they can also be used with concentrating solar power systems. So maybe high temperatures are not always a problem—although I still would not like to be trapped on the 100th floor of a Dubai high rise when and if the currently still mostly fossil-derived power supply fails and there are no lifts or air con. But it was good to hear that some progress is being made on greening buildings and power there!

However, as everywhere, there is a lot more to do. As Prof. Ali Sayigh, the indefatigable chair of WREC, put it, 'many countries in the world are now embracing renewable energy not only because it is clean and cost effective but also because it reduces energy imports and creates local jobs…Now it is up to you, the participants in this Congress, to continue to spread and develop all forms of renewable energy, and to commit yourselves to the very urgent reduction of the climate change effect'.

Chapter 1
Leading Role of ISESCO in the Field of Renewable Energy and Promotion of the Concept of Green and Sustainable Cities in the Islamic World

H. E. Abdulaziz Othman Altwaijri

Honarary Chairman of World Renewable Energy Network (WREN) and Director General of ISESCO Excellences Ladies and gentlemen,

It is my pleasure to talk to you at the opening of the World Renewable Energy Congress, WREC-18′, hosted by University of Kingston, UK, and supported by ISESCO.

Allow me, first, to express my gratitude to His Excellency Prof. Ali Sayigh, Chairman of the World Renewable Energy Network (WREN), for his kind invitation to take part in this Congress, for which I wish every success in achieving its goals.

Excellences,
Ladies and gentlemen,

The choice of theme for the 18th session of the Congress clearly mirrors the major interests of WREN, which are shared by ISESCO. The close systematic connection between renewable energy and sustainable buildings deeply highlights the daunting equation, whose treatment is the best solution for overcoming the dangers of climate change, today's primary concern of the international community.

This correlation between renewable energy and sustainable buildings is the cornerstone of the advanced.

Concept and modern sense of eco-friendly green cities, which act as incubators for development and a lever of progress and prosperity, preserve the environment and contribute to promoting comprehensive sustainable development in all its fields. As such, all the components of this equation harmoniously come together to attribute a vital role to renewable energies in the protection of clean environment, the natural incubator of comprehensive sustainable development.

H. E. Abdulaziz Othman Altwaijri (✉)
Islamic Educational, Scientific and Cultural Organization (ISESCO), Rabat, Morocco

A. Sayigh (ed.), *Renewable Energy and Sustainable Buildings*, Innovative Renewable Energy, https://doi.org/10.1007/978-3-030-18488-9_1

While the 17th session of the Congress, hosted by Murdoch University in Australia, in February 2017, focused on the theme "**Full Transition to Renewable Energy**", the theme of this session reaffirms, in many ways, our collective interest in new and outstanding initiatives in the fields of solar and wind energy, as well as in taking cognizance of modern technologies in the fields of biomass, energy conversion systems and biofuel, and testifies to the Congress's progress and WREN's remarkable action. Moreover, this leap from "**Full Transition to Renewable Energy**" to "**Renewable energy and Sustainable Buildings as the Best Solution to Climate Change"** unequivocally affirms the qualitative evolution of renewable energy concepts and their relation to sustainable buildings and green cities in terms of concept and methodology. Thus, the continuous advancement in this scientific field helped us lay the foundations of a new concept: **green culture**. Following in the footsteps of renewable energy scientists and ecologists who established the concept of **green economy**, from which the concept of **green cities** was derived, **green culture** will be similarly tantamount to the term of **peace culture** adopted in the literature of both UNESCO and ISESCO.

If the world today is in dire need for deepening the concept of peace culture to counter the rising trends of hatred, racism, extremism and terrorism, and halt armed conflicts arising in many regions, humanity, now and in the near and distant future, needs to adopt the concept of green culture to be the driving force for creating sustainable green cities based on renewable energy and sustainable buildings.

Excellences,
Ladies and gentlemen,

Against this backdrop, given ISESCO's avid interest in the promotion of sustainable energy use, and pursuant to the "**Strategy of Developing Renewable Energy in the Islamic World"** adopted by the 3'd Islamic Conference of Environment Ministers, held in Rabat, Kingdom of Morocco, in 2008, the Organization plays a leading role in promoting the concept of green and sustainable cities in the Islamic world. In this regard, and as part of the Three-Year Action Plan for 2019–2021, the new concept of transition to green economy has been methodically inculcated in various fields such as energy, transport, agriculture and industry. ISECO has also launched "Islamic Green cities Excellence Award" as part of the "**Kingdom of Saudi Arabia Award for Environmental Management in the Islamic World**" which was approved by the Islamic Conference of Environment Ministers.

In addition, ISESCO will continue, under its upcoming action plan, to undertake its role in this vital field through the implementation of a number of activities on capacity building, and promoting innovation, while focusing on technology transfer in the field of renewable energies. ISESCO will also continue co-holding regional and international conferences on renewable energies, in cooperation with WREN, in order to promote the use of these energies in its 54 Member States.

In this regard, and in furtherance of this trend, ISESCO will continue holding experts meetings and training sessions on renewable energies. It will also promote cooperation and expand the prospects of joint Islamic action with its partners in the fields of renewable energies in order to deepen the understanding of these rigorous scientific concepts and promote interest in vital issues that fall within this wide field.

Given its prominent status in the sphere of joint Islamic action within and outside the Islamic world, ISESCO is fully aware of the importance of the transition to green economy in its member States, and has accordingly exerted remarkable efforts to develop and implement strategies and action plans on green cities. ISESCO maintained its leading role in this field and will continue to do so through the implementation of activities on capacity building, promoting innovation, and enhancing transfer of green technology to various fields such as renewable energies, water management, solid waste management, and biodiversity.

In this regard, a number of activities were carried out in 2018 pertaining to sustainable green cities, including:

- National training session on sustainable management of solid waste, Bamako, Republic of Mali.
- Funding the adoption of water projects to implement the nationally set contributions and achieve sustainable development goals: challenges, prospects and opportunities, Nouakchott, Islamic Republic of Mauritania.
- Regional expert meeting to implement ISESCO's Strategy on Green Cities in Member States, Baku, Republic of Azerbaijan.
- 1st International Symposium on research works and innovative technology in the water sector to achieve SDG 6, Amman, Hashemite Kingdom of Jordan.

Other activities scheduled later this year include:

- Regional workshop on presenting biodiversity, cooperation with Al-Farabi University, Almaty, Republic Of Kazakhstan.
- Two regional workshops on preparedness for the dangers of Tsunami and disaster risk reduction, November, ISESCO headquarters. These two activities aim to enhance cities' resilience to climate change and natural hazards. We should also not forget the World Renewable Energy Congress, which is annually supported and sponsored by ISESCO.

Excellences,
Ladies and gentlemen,

The lofty goals that the World Renewable Energy Congress strives to achieve are at the core of the United Nations' Development Goals, which makes the recommendations issued by this Congress worthy of Consideration by all world states. For building sustainable green cities on the foundations of renewable energies is the strategic goal to counter the creeping climate change and address the environmental problems arising therefrom.

In order to activate the outcomes of this international event, and taking into consideration the international Dimensions of renewable energies and sustainable buildings in green cities, ISESCO invites the Congress to adopt the concept of green culture and accordingly feature the theme "**spreading the green culture to promote the culture of peace",** as one of the topics to be addressed at its 19 ft. session.

Wishing this Congress every success in achieving its goals, I thank you for your attention.

Chapter 2
Energy Policies at Crossroads: Will Europe's 2030 Targets and Framework be in Line with the Paris Climate Agreement?

Rainer Hinrichs-Rahlwes

2.1 Introduction

The European Union and its Member States have been among the pioneers of renewable energy and frontrunners of climate protection. Denmark, Germany, Spain, Portugal and Austria were among the first countries to successfully accelerate the deployment of renewable energies. Dedicated policies, most of them using feed-in tariffs or feed-in premiums, triggered a significant boost of new renewable energy installations in the power sector resulting in drastic cost reductions, whereas less support was aiming at transport, heating and cooling, industry and other end-use sectors. In these sectors, renewables are developing much slower and will need more supportive policies and frameworks including for sector coupling and system integration, if the rapid decarbonisation of the energy sector is a serious objective of all those who have committed to the Paris Climate Agreement.

Europe has an ambitious and widely effective policy and regulatory framework for renewable energy, energy efficiency and greenhouse gas reduction up to 2020. The 2030 targets and policies were agreed before the end of 2018, and a new 2050 strategy is being developed to bring the EU in line with the climate commitments. Unfortunately, too many Member States have tried to reduce policy ambition and to water down directives and regulations supporting renewables and efficiency while at the same time preventing effective carbon pricing and continuing to directly and indirectly support fossil and nuclear energy. Some Member States, e.g. France, the UK, Poland and their allies, are (ab)using the national prerogative to define their own energy mix (which is granted by the European Treaties) as a pretext to continue supporting fossil and nuclear energy—irrespective of the fact that renewables

R. Hinrichs-Rahlwes (✉)
European Renewable Energies Federation (EREF), Brussels, Belgium

German Renewable Energy Federation (BEE), Berlin, Germany

A. Sayigh (ed.), *Renewable Energy and Sustainable Buildings*, Innovative Renewable Energy, https://doi.org/10.1007/978-3-030-18488-9_2

(in particular wind and solar) are meanwhile the cheapest sources of energy, even in distorted markets.

At the same time, investment in renewables is significantly decreasing in the EU, whereas other countries, among them more and more developing countries, are significantly extending their quest for a renewable energy transition, with China and some other countries including the USA (despite major interventions from the president in favour of fossil fuels) leading the way in an accelerating race for clean energy for all.

On this background, the European Union and its Member States are developing targets and frameworks for 2030. Until summer, compromise has been reached on four legislative proposals, whereas four other pieces of legislation were only finalised in early 2019. Compromise results had to be reached before the elections in May 2019 for a new European Parliament would put an end to pending cases for at least half a year. It is encouraging to see that the compromises are better than the original proposals from the European Commission and certainly more ambitious than some backwards leaning Member States would have them.

2.2 Global Development of Renewable Energy

Renewables have become mainstream, and **targets and policies** for accelerating their development and deployment are implemented worldwide. Most policies and targets are aiming at the power sector, whereas policies for renewable heating and cooling and renewables in the transport sector are less frequently found. The International Renewable Energy Agency (IRENA) [1] estimates that 10.3 million **jobs** exist in renewables worldwide in 2016, up 5% from 2015. 4.2 million people had jobs in solar energy, 3 million in bioenergy, 1.8 million in hydropower, 1.2 million in wind power and 100,000 in geothermal energy.

Investment in renewable energy was up again in 2017, after some slowdown in the years before. The Global Policy Network for the 21st Century (REN21) in its Renewables 2018 Global Status Report[1], based on data from the Bloomberg New Energy Finance (BNEF), estimates that 280 billion USD (or 320 including large hydropower) were invested in renewables in 2017, up 2.2% from 2016. Investment in renewables was roughly three times as much as in new fossil or nuclear capacity. In 2017, the balance of investment in renewable energy shifted further towards developing country. Whereas developed countries invested 103 billion USD in 2017 (down from 151 in 2014), developing countries increased investment from 133 billion USD in 2014 to 177 in 2017, with China providing the largest share with 127 billion USD in 2017 (up from 85 in 2014). The Chinese figures are even more impressive when compared to the European Union. China increased investment in renewables from 48 billion USD in 2011 via 85 in 2014 to 127 in 2017, while the EU's investment decreased—after an all-time high in 2011 of more than 129 billion

[1] Facts and figures in this chapter are from REN21 [2], if not quoted differently.

USD—via 86 in 2014 to below 41 billion in 2017, which is less than China's investment in 2010.

By end of 2016 (more recent figures are not available), renewable energy provided **18.2% of the world's final energy consumption** (up from 15% in 2005), out of which 10.4 percentage points came from modern renewables and 7.8 from traditional biomass. The use of traditional biomass is slowly decreasing, while modern renewables have been growing at an average annual rate of 5.4% since 2005—still too slow for achieving the objectives of the Paris Agreement but going in the right direction. Growth of renewables has been strongest in the **electricity sector**, with a 26.4% share of global power production in 2017.[2] Although the power sector represents only 20% of global energy consumption, this development is encouraging, because it shows that—based on well-designed and effective policies driven by strong political will—the necessary fundamental transformation of the energy system is possible. **Heating and cooling** and **transport** need to be addressed by decision-makers rather sooner than later.

Growth of modern renewables is a valuable indicator of a country's or a region's or a community's willingness to further the development of sustainable renewable energy and related fundamental transformation of the energy sector. In the last 10 years, **global wind capacity** increased from 94 GW in 2007 to 539 GW in 2017. 52 GW was added in 2017, with China leading with 19.7 GW newly installed, followed by the USA and Germany as distant second and third with 7 GW and 6.1 GW, respectively, followed by the UK (4.3 GW) and India (4.1 GW). China also is the lonely champion of **overall renewable power** capacity with 334 GW, followed by the European Union with 320 GW and the USA a distant third with 161 GW and Germany with 106 GW (which is included in the EU's 320 GW), India 61 GW, Japan 57 GW and the UK with 38 GW (also included in the EU's 320 GW). It goes without saying that China is also the lonely leader of the **global solar PV markets** with more than 131 GW of installed capacity, out of which 53.1 GW was added in 2017. The USA with an overall capacity of 51 GW (10.6 GW added in 2017) and Japan with 50 GW (7 GW added in 2017) are second and third, closely followed by Germany with 42 GW (with only 1.7 GW added in 2017). In terms of new capacity, India was third with 9.1 GW added and nearly doubling capacity in 1 year. China's dominance is even stronger in solar thermal heating with a market share of more than 71%, leaving 3.9% for the USA as a distant second, followed by Turkey (3.3%) and Germany (3%).

It is obvious that the **old champions** in Europe (Germany, Denmark, Spain) have found **serious competitors** (China, but also the USA, India and a growing number of developing and emerging economies) in the quest for global leadership of the energy transition.

The impressive growth in the champion countries should not overshadow some **remarkable and ambitious efforts in smaller and poorer countries**. **On a per capita base**, Iceland and Denmark (and not China and USA) are the global champions

[2] IEA [3] has slightly lower figures (25%), but is basically showing a similarly strong and increasing growth of renewables globally.

of installed renewable power capacity (excluding large hydro). On a per capita base, Germany and Japan are the champions of installed PV capacity, instead of China and the USA on mere capacity base, followed third by Belgium, Italy and Australia. On a per capita base, Denmark, Ireland and Sweden are the champions of installed wind capacity, followed by Germany and Portugal—leaving GW champions China and the USA far behind and leaving Germany on the fourth rank only. In solar thermal for heating, the lonely champion on capacity basis, China, is not even among the top 5 when it comes to per capita ranking, where Barbados, Austria and Cyprus are closely followed by Israel and Greece. A similar difference can be observed, when it comes to annual investment in new capacity. Whereas, in 2017, China, the USA, Japan, India and Germany were leading new investment in renewables, the top 5 look completely different when **assessed per unit GDP**. The Marshall Islands are leading this competition, followed by Rwanda, Solomon Islands, Guinea Bissau and Serbia.

The global energy transition is underway, much faster than expected only a few years ago, and it will gain more momentum in the years ahead—and there certainly is no alternative to significantly increasing renewables, if we are serious about limiting global warming to a just tolerable level of 2 °C (or only 1.5 °C) above pre-industrialised times—in other words, if we are serious about our commitments under the historic Paris Climate Agreement.

2.3 Europe's Policies and Frameworks for Renewables

The European Union and many of its Member States have been early champions of greenhouse gas reduction and policies for renewable energy. The Kyoto Protocol would not have been agreed on and ratified by the necessary majority of countries without the proactive role the European negotiators. And it is also safe to assume that the Paris Agreement would not have been possible without a strong role of European countries—but certainly the willingness of China to agree on an ambitious deal was another major condition, as well as the constructive role of the USA with the Obama administration in charge.

The EU is trying to maintain high ambition in programmatic objectives and visions. Not only did the EU present decarbonisation scenarios in 2011, which were all in line with ambitious greenhouse gas reduction efforts and all with very high or dominant shares of renewable energy by 2050. When taking office in 2015, the (now outgoing) European Commission's President, Jean-Claude Juncker, underlined the ambition to **make Europe the global number 1 in renewables**. Under his leadership, the EU developed a strategy for an ambitious **Energy Union**: "The goal of a resilient Energy Union with an ambitious climate policy at its core is to give EU consumers—households and businesses—secure, sustainable, competitive and affordable energy. Achieving this goal will require a fundamental transformation of Europe's energy system" [4].

Europe's positive role in climate negotiations so far is founded on domestic policy objectives—**and on an effective energy and climate framework up to 2020**—

underpinned by continued efforts to maintain a momentum for full implementation of the Paris Agreement in the international political arena. Unfortunately, commitments of European leaders have become less credible, because of a significant slowdown of implementation of agreed policies for renewables and efficiency. European investment in renewables fell dramatically in the last few years to the low levels of more than 10 years ago. The energy transition is slowing down, and policy discussions about a new framework up to 2030 have not yielded really ambitious objectives and the strong policies needed for implementing the Paris Climate Agreement. The public consultation[3] on a **strategy for long-term EU greenhouse gas emission reductions** was an opportunity to increase existing targets and develop new momentum for a more ambitious climate and energy policy—but doubts remain.

2.3.1 Towards 2020: Good Framework to Be Implemented

Up to 2020, the EU is aiming at share of at least 20% renewables in gross final energy consumption (GFEC), at least 20% efficiency increase and at least 20% greenhouse gas reduction (or 30%, if other relevant countries commit to similar objectives).[4] These objectives are underpinned by **differentiated binding national targets** for each of the 28 Member States and guided by an **indicative trajectory**. Target reaching is facilitated by National Action Plans, biannual reporting obligations, a set of **voluntary cooperation mechanisms between** Member States and with third countries for target achievement and infringement procedures in case of missing the binding 2020 targets. The national reports are evaluated in a biannual report from the European Commission—meant to point out, if the Union as a whole and each Member State is on track towards the 2020 targets or if additional efforts are necessary.

The most recent Renewable Energy Progress Report [5] shows that target achievement in 2020 cannot be taken for granted. Figure 2.1 shows the growth of renewable energy in GFEC from 8.5% in 2004 to 16.4% in 2015. More recent figures [6] indicate a share of about 17% in 2016. Figure 2.1 also shows the indicative trajectory up to 2020, which is becoming steeper at the end. And it shows—a little bit above the trajectory—the projected shares of renewables in the National Renewable Action Plans. The European Commission points to the **risk of not reaching the 20% share in 2020**: "…as the trajectory becomes steeper in the years ahead, efforts to keep on track will need to intensify… ". The need becomes more evident when it comes to the different sectors. Whereas renewables are growing slowly but smoothly in the electricity sector, particularly wind and solar, growth is

[3] https://ec.europa.eu/clima/consultations/strategy-long-term-eu-greenhouse-gas-emissions-reductions_en—consultation closing on 8 October 2018

[4] Although the Paris Agreement could certainly be understood as such a commitment, the EU never officially increased the 2020 target from 20 to 30%—among other reasons with reference to the ongoing 2030 negotiations.

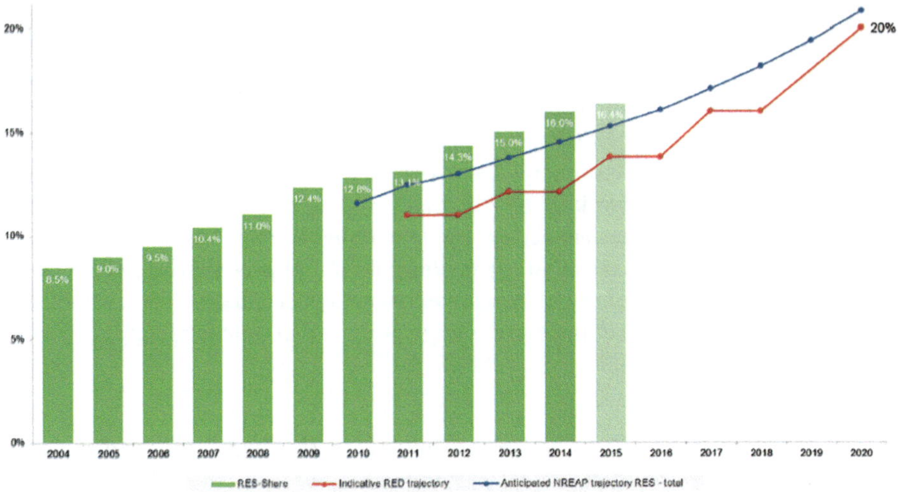

Fig. 2.1 Renewable energy shares in the EU vs. Renewable Energy Directive (RED) and National Renewable Energy Action Plan (NREAP) trajectories (based on EUROSTAT, Oka-Institut)

slower in the heating and cooling sector with stagnating deployment since 2013 and in the transport sector with only 6% achieved so far and no relevant increase since 2011. Experts and industry have voiced their concerns that the 2020 targets may not be reached, including by some of the old frontrunner countries including Germany and Spain. In 2016, 11 Member States had already reached their 2020 targets, and 9 were far away with little chance to catch up until 2020 [6].

2.3.2 Towards 2030: Business as Usual or Implementing the Paris Agreement

Due to long investment cycles in the energy sector and to long lead times for political decisions, the discussion about a climate and energy policy framework beyond 2020 gained momentum in 2014. First evaluations of the 2020 framework showed that more efforts may be needed to achieve the targets, and some Member States advocated for a "more market-based" framework towards 2030. Poland and other Eastern European countries were opposing any new targets after 2020. The UK, France and others were advocating for a European greenhouse gas reduction target only, opposing specific targets for renewables and energy efficiency and national targets as in the 2020 framework. Their instrument of choice was a "**new governance model**", which would require national climate and energy plans but would not allow the European Commission to reject them for inadequacy. Others, including Germany, Austria and Sweden, underlined their **willingness to develop strong targets** for greenhouse gas reduction, efficiency increase and renewable energy.

They were not very outspoken, however, in terms of defending national binding targets beyond 2020.

2.3.2.1 European Council Conclusions

On this background, the European Council (the heads of state and government of all Member States) in October 2014 agreed on a weak position for a 2030 climate and energy framework. They agreed on a domestic greenhouse gas reduction target of at least 40% by 2030 and at least 27% renewables and 27% efficiency increase. The targets were criticised by NGOs for being insufficient for meeting climate targets and by the renewables industry for being hardly more than business as usual and not providing investment security. The Council strongly agreed on **not supporting national binding targets** and establishing a robust governance model instead, which was still to be defined.

Among the positive elements of the Council Conclusions was the commitment to **fully implement the 2020** targets and to develop a **new energy market design** which should be fit for renewables including a multitude of decentralised producers and consumers.

2.3.2.2 Clean-Energy-for-All-Europeans Package

Two years later, on 30 November 2016, the European Commission opened the legislative process for a 2030 climate and energy framework. They presented the Clean-Energy-for-All-Europeans Package (also called Winter Package) with eight legislative pieces to implement and specify the 2014 Council Conclusions. As required by the European Treaties, the package was submitted to the European Parliament and to the Council. Both institutions need to agree on the Regulations and Directives, before they can enter into force. Given upcoming elections for the European Parliament in May 2019, the process had to be formally concluded by early 2019—more than 8000 pages to be read, discussed and agreed.

The Winter Package [7] consisted of proposals for recasts of the Renewable Energy Directive (RED), the Energy Efficiency Directive (EED) and the Energy Performance of Buildings Directive (EPBD), for a Governance Regulation (GovR), for a Risk Preparedness Regulation (RPR), for an ACER Regulation and for Market Design (MD) Directive and Regulation. They all needed to be voted in Council and Parliament.

Taking stock of the wide range of proposals, renewable energy and environmental stakeholders saw a **mixed picture**. On the downside, the **low level of ambition** of the 2014 Council Conclusions had been kept, as well as the **lack of national targets and a meaningful governance proposal**. In addition, the proposal required mandatory technology neutral auctions or tenders as the main support scheme for renewable energy, and it widely abolished priority access and dispatch for renewables. Only exceptionally and for small installations below 500 kW (250 kW as of 2026), feed-in policies and technology-specific support or exceptions from balancing responsibility should be possible after 2020. Member State should be obliged to increasingly open their auctions to competitors from other Member States. No effective proposals for carbon pricing beyond ailing Emissions Trading System (ETS) or phase-out plans for fossil and nuclear power were included in the package.

On the positive side, the obligation to fully implement the 2020 targets and ruling out retroactive and retrospective changes of support systems were welcomed by the renewables sector. Suggestions for supporting energy communities, consumers and self-consumption were a good basis for further discussions. The intention of creating flexible markets fit for a multitude of decentralised producers and consumers and allowing renewable energies to participate in all market segments are on the positive side, with some need for clarification and improvement. Also, language was welcomed on simplified and transparent procedures and on limiting the use of capacity markets, which more often than not result in significant lock-in of dirty and inflexible fossil capacity.

The renewables industry and environmental NGOs worked hard to support those in the European Parliament who were willing to improve the Winter Package and to work with those national governments which were open for arguments. Apart from biomass-related questions, where environmental NGOs asked for very strong sustainability criteria and for limiting and phasing out first-generation biofuels, and irrespective of some hesitation on the side of a few of the renewables associations, there was a wide and strong agreement on the key points which needed improvement.

The 27% target for renewables in 2030 was based on outdated and much too high cost assumptions for renewables and too low costs for other energy. Many associations reminded policymakers that already in 2011, the renewables industry had suggested a target of 45% in 2030. The European Wind Association, WindEurope, in the beginning only asked for "more than 27%", whereas some others, including important members of the European Parliament, suggested to go for "at least 35%", which was eventually adopted as the Parliaments position and supported by renewables industry and environmental groups as "a bare minimum" of what is needed. Most stakeholders, as a next priority, strongly advocated for continuation of binding national targets, or at least strong benchmarks. A common rule book for renewables support, including the right for Member States to apply technology-specific auctions and tenders and to maintain priority dispatch and access for renewables, as well as exceptions from balancing responsibility, at least for existing installations, was another major request. For the Market Design Initiative, flexibility should be the overarching principle allowing for market participation of renewable energy in all segments.

2.3.2.3 Compromise in Trilogue Negotiations

When this paper was drafted, **four of the eight legislative proposals had been voted and politically agreed between Parliament and Council**. The first to pass was the EPBD, next the GovR. RED and EED were controversial until the 2018 summer break, particularly because of the target discussion. ACER, RPR, MDR and MDD were decided in early 2019.

For EPBD and EED, compromise was found for all major issues. The indicative EU **energy efficiency target for 2030 was increased from 27% to at least 32.5% with an upward review clause for 2023**. A similar increase and review were agreed for the **EU-binding 2030 renewables target of at least 32%** [8, 9]. Given the strong reluctance of some Member States to allow for any target higher than 27%, this is a positive development, but with a view to the necessity of adjusting policies to the commitments under the Paris Agreement, it is only a first step [10].

Apart from the agreement about higher targets, other important points were settled. In contrast to the original proposal, Member States' right to decide whether they chose technology-neutral or **technology-specific support schemes remains in place**. Opening of support schemes for other Member States remains an important objective, but it is not mandatory. For giving project developers, manufacturers and investors a reasonable perspective, Member States are required to regularly publish plans for new renewable capacity with a **5-year visibility**. Strong language against retroactive changes of support schemes and for transparent and effective administrative procedures and fees was included.

A highlight of the agreement is a support for **community energy** projects and a basic right for every citizen to produce, consume and sell the energy produced from installations of less than 30 kWp on their own premises, including in third-party ownership and in cooperation with others. Member States must not impose burdensome or discriminate procedures and fees or taxes on **self-consumption**. Consumer organisations and the solar industry celebrated this as a breakthrough for more decentralised renewable energy production and strengthening the role of prosumers and potentially ruling out our reducing significantly existing fees and taxes which have prevented a strong development for more decentralised and community-owned projects.

The institutions agreed on a **14% target for the 2030 share of renewables in transport**. This objective had been controversially debated, because environmental groups campaigned for exclusion of first-generation biofuels and for stronger sustainability criteria, whereas others suggested more pragmatic solutions. Eventually, it was agreed to cap the maximum share of first-generation biofuels to 7% with a flexibility for Member States to choose a lower cap and thus reduce the 14% target. Member may incentivise waste-based and innovative biofuels by multiplying their shares for target achievement, which also applies to renewables in shipping, aviation and rail transport and with up to 3.5 for electric vehicles.

Agreement includes an indicative obligation for Member States to increase the share of **renewables in heating and cooling** by 1.3 percentage points every year—with a 40% flexibility for waste heat. Clarifications about guarantees of origin will help power purchase agreements to be more easily established and implemented, which is an important element for direct corporate sourcing of renewable energy by smaller and larger companies.

2.3.3 Markets Fit for Renewables: Creating a Level Playing Field

After summer break and after finalizing the main text of this paper, the remaining four legislative pieces of the summer break were negotiated. Whereas the RPR passed without much controversy, the ACER regulation was more difficult to agree on, because Member States have a strong position on not weakening national authorities in favour of a European institution (ACER). Compromise was found by agreeing on indicative cooperation mandates instead of increased competency for ACER.

The most important and controversial dossiers were the **Market Design Regulation and Directive** (together **MDI**), which will be the basis for future EU's energy markets. Results have a strong impact on whether or not or how fast energy markets will be a **level playing** field fit for increasing shares of variable and more decentralised production and consumption including for private and industrial self-consumption.

Although carbon pricing is not a direct part of the MDI, there are elements which are designed to complement the existing Emissions Trading System. Weak or strong criteria for limiting **capacity markets** to the absolute minimum and only after having considered cross-border capacities based on transparent criteria will impact future decisions of Member States to continue or phase out highly emitting power plants. This is why stakeholders are advocating for strictly limiting capacity markets and why they are strongly defending or trying to strengthen a proposal from the European Commission for **Energy Performance Standards (EPS)**, which would rule out participation of fossil power plants in capacity markets, if they are emitting more than 550 g of CO_2. Poland and other countries with a high share of coal in their energy mix are against any such EPS, whereas NGOs have asked for reducing the value to 350 g. Compromise was found around the original 550 g becoming effective later than suggested.

A major step towards a level playing field in energy markets would be removing and phasing out direct and indirect subsidies for unsustainable energy sources. But again, this is not part of the MDI. Nevertheless, the closer the compromise agreement comes to removing market distortions, the more effectively can it accelerate the energy transition. The objective of MDI is to facilitate **participation of all energy sources on a level playing field**, which would translate into transforming all energy market segments so that renewables can **compete on equal footing**.

Renewable energies will have to be allowed and enabled to **participate in all sectors of the energy market**—futures, day ahead, intraday, balancing, demand response, auxiliary services, etc. This requires **shifting energy markets from futures to day ahead and from day ahead to intraday** and (close to) real time. Reducing gate closure times to a maximum of 15 min, which was a potential compromise within the MDI, would significantly improve opportunities for wind and solar power in these markets. Real-time trading would be the next (and parallel) step. Improved forecasts, smarter grids and digitalisation will help opening markets for renewables. Integrating demand response would be another step towards making flexibility the key driver of energy markets.

Until markets are fit for renewables and no longer distorted by direct and indirect subsidies for unsustainable energies, **protective measure for renewable energy will need to be continued**, e.g. protection for small installations through priority access and dispatch, where appropriate, and by exemptions from balancing responsibilities. Forging compromise in the spirit of creating a really competitive flexible market fit for renewables could considerably accelerate phasing out fossil and nuclear energy and pave the way for a fully renewable-based energy system.

2.4 The Way Forward: Towards 100% Renewable Energy

The European Union's ambition to create a resilient Energy Union and to become the world's number one in renewable energy are pointing in the right direction. Recent proposal from the EU Commissioner for Energy and Climate Action, Miguel Arias Cañete, to increase Europe's greenhouse gas reduction target to at least 45% by 2030 is another point to be acknowledged. What is missing is **decisive action instead of mainly lip services** from European institutions and Member States. If it were not the Paris Agreement that required higher emission reduction and more clean energy, this summer's record temperatures are another incentive to speed up efforts to reduce global warming. The upward revision of Europe's renewables and efficiency targets in 2023 is a good step, but there is **no reason to wait for 5 years**. Evidence is overwhelming.

Europe's **2050 decarbonisation strategy** could be another opportunity to increase ambition for the energy transition. Already in 2011, despite outdated cost assumptions and methodology, the European Commission presented roadmaps and scenarios showing that effective decarbonisation by 2050 requires high or dominating shares of renewable energy in all sectors with little or no additional system costs.

Today, renewables are the lowest cost options for new capacity in an increasing number of markets, including where **direct and indirect subsidies for fossil and nuclear** are still supporting dirty sources. Phasing them out could **significantly accelerate the global shift towards renewables**, as could higher efficiency standards. Technologies for renewable energy production, smart grids, storage and demand response are available and becoming cheaper at an unprecedented pace.

Given the dramatic cost decreases in renewable energy, particularly in wind and solar photovoltaics, there should be no doubt that increased development and deployment of renewable energy[5] is the safest and cheapest way towards leaving a liveable planet to future generations. We no longer have to choose whether protecting the environment and limiting global warming is more important than growing the economy or vice versa. Today, **it is obvious that the energy transition will create and maintain tens of millions of jobs, avoid societal costs and costs of imported energy sources, grow GDP, generate value and increase energy supply security**.

All that is needed is clear and unambiguous political will to implement phase-out strategies for the old system and smart policies for accelerating the transformation towards a 100% renewable energy system. The sooner and the clearer decision-makers embrace the clean energy future, the easier can they make sure not to leave anybody behind.

References

1. International Renewable Energy Agency (IRENA) (2018) Renewable Energy Statistics 2018, Abu Dhabi. http://www.irena.org/publications/2018/Jul/Renewable-Energy-Statistics-2018
2. Renewable Energy Policy Network for the 21st Century (REN21) (2018) Renewables 2018 Global Status Report, Paris. http://www.ren21.net/wp-content/uploads/2018/06/17-8652_GSR2018_FullReport_web_final_.pdf
3. International Energy Agency (IEA) (2018) Global Energy & CO_2 Status Report 2017, March 2018, Paris. http://www.iea.org/publications/freepublications/publication/GECO2017.pdf
4. European Commission (2015) Energy Union Package. A Framework Strategy for a Resilient Energy Union with a Forward-Looking Climate Change Policy, Brussels. https://ec.europa.eu/energy/sites/ener/files/publication/FOR%20WEB%20energyunion_with%20_annex_en.pdf
5. European Commission (2017) Renewable Energy Progress Report, COM(2017) 57 final, Brussels. https://ec.europa.eu/transparency/regdoc/rep/1/2017/EN/COM-2017-57-F1-EN-MAIN-PART-1.PDF
6. Eurobserver (2018) The State of Renewable Energies in Europe, 2017 Edition, Brussels. https://www.eurobserv-er.org/category/barometer-2017/
7. European Commission (2016) Clean Energy for all Europeans, Legislative Package, Brussels. https://ec.europa.eu/energy/en/topics/energy-strategy-and-energy-union/clean-energy-all-europeans
8. Council of the European Union (2018) Interinstitutional File 2016/0382 (COD), 10308/18, Proposal for a Directive of the European Parliament and of the Council on the promotion of the use of energy from renewable sources—analysis of the final compromise text with a view to agreement, Brussels
9. Council of the European Union (2018) Interinstitutional File 2016/0307 (COD), 10307/18, Proposal for a REGULATION OF THE EUROPEAN PARLIAMENT AND OF THE COUNCIL on the Governance of the Energy Union, amending Directive 94/22/EC, Directive 98/70/EC, Directive 2009/31/EC, Regulation (EC) No 663/2009, Regulation (EC) No 715/2009, Directive 2009/73/EC, Council Directive 2009/119/EC, Directive 2010/31/EU,

[5] IRENA and others recently calculated that a sixfold increase of global investment in renewables would be necessary to likely limiting global warming to a maximum of 2 °C or even 1.5 °C.

Directive 2012/27/EU, Directive 2013/30/EU and Council Directive (EU) 2015/652 and repealing Regulation (EU) No 525/2013—analysis of the final compromise text with a view to agreement, Brussels
10. European Parliament (2018) MEPs set ambitious targets for cleaner, more efficient energy use, Brussels. http://www.europarl.europa.eu/news/en/press-room/20180112IPR91629/meps-set-ambitious-targets-for-cleaner-more-efficient-energy-use

Chapter 3
Field Test, Dynamic Simulation and Fatigue Analysis of a Small Wind Turbine Operating in a Highly Turbulent Environment

Tebogo Pooe, Mitsumasa Iino, and Abdulrhman Elawady

3.1 Introduction

Highly turbulent environments are generally not good for fatigue as they increase the number of loading cycles and dynamics of a small turbine. The higher rotational speeds of small wind turbine blades make them exposed to a high number of stress cycles. The lower moment of inertia of micro wind turbines makes them more susceptible to changes in the ambient wind turbulence. The smaller mass of SWT also makes them have much higher natural frequencies of vibration for the blades and other components. Usually, a wind turbine is expected to have a fatigue life of at least 20 years or more. Unlike ferrous metals, composite materials (e.g. GFRP) for the most part do not have a fatigue endurance limit, and their SN curves tend to be a continuous straight line. If the turbulence of a site is severe enough, it can cause premature fatigue failure if such a site was not taken into account in the design phase of the wind turbine. It is therefore necessary to be able to quantify how much turbulence of a site affects the power output and fatigue life of a wind turbine. If accurate simulations of wind turbines can be done using software, it can save much time and resources in validating the dynamic performance of a wind turbine.

3.2 Objectives

– To quantify the effect of varying turbulence intensity wind profiles on the fatigue/lifetime DEL of a small wind turbine
– To carry out field test and analyse the turbulence intensity effect on the power output of a small wind turbine

T. Pooe (✉) · M. Iino · A. Elawady
Ashikaga University, Tochigi, Japan

A. Sayigh (ed.), *Renewable Energy and Sustainable Buildings*, Innovative Renewable Energy, https://doi.org/10.1007/978-3-030-18488-9_3

– To verify using FAST simulation the effect of turbulence intensity on the SWT's power output

3.3 Methodology

3.3.1 *Field Test*

The SWT field test data collected was over a continuous period of 3 months. The details of the turbine used for the test are shown in Table 3.1. The wind turbine system data logger takes raw data samples at a set frequency of 1 Hz in voltage form of various parameters such as turbine RPM, hub-height wind speed and direction, battery current, battery voltage, shaft rotational speed, etc. The raw voltages are then converted into engineering values through their respective conversion factors and algebraic calculations. The wind turbine is connected to a 24 V battery charge controller system which manages the torque-speed relationship of the generator.

3.3.2 *Power Output FAST Simulation*

NREL FAST (fatigue, aerodynamics, structural and turbulence) v. 7 was used to create a dynamic simulation of the current SWT in this research. A FAST model was previously created by Iino et al. [1] for this 1-m-diameter turbine under study. The source code was modified to make it a five-bladed FAST model (original FAST code is for two- or three-bladed turbines). The design parameters for the FAST model (airfoil tables, dimensions, etc.) were received from the original manufacturer. The model was verified with the original manufacturer.

The range of TI simulated was between 5 and 45%. The random number generator was used in FAST to generate random time series wind. Each FAST data point shown in Figs. 3.4 and 3.5 was the average of a minimum of 30 simulations (each simulation generated a different random time series; however the Gaussian mean was still the same). The simulation time period was for 1 min. The field test data also used 1 min averages. A diagram showing how the turbulence grid was set up using NREL TurbSim is shown in Fig. 3.1.

Table 3.1 Small wind turbine specifications

Rated power	135 W (at 10 m/s)
Max power output	500 W
No. of blades	5
Cut-in wind speed	2 m/s
Cut-out wind speed	16 m/s
Rotor diameter	1000 mm
Maker and model	AURA 1000

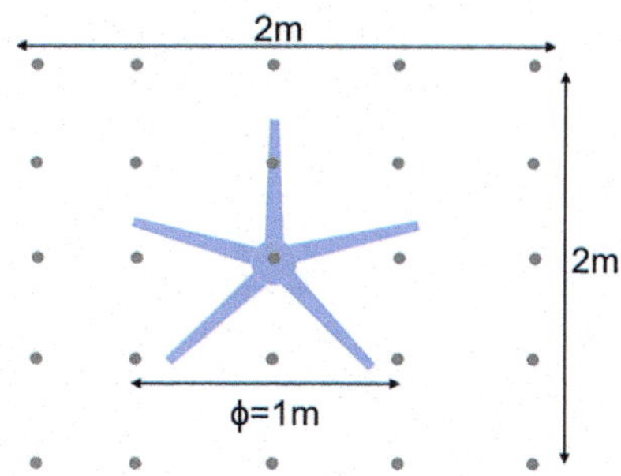

Fig. 3.1 TurbosSim turbulence grid layout

From Fig. 3.1, the inflow turbulence was modelled as a 5 × 5 grid. The grid area was larger than the turbine diameter, so it could be fully covered. The grid size was deemed satisfactory considering the small size of the rotor diameter (1 m). The IEC Kaimal model was used in all the turbulence simulations.

3.3.3 *Fatigue Aeroelastic Simulations*

IEC 61400-2 specifies that for power production under the NTM, either the simple load or aeroelastic analysis is necessary for fatigue computations. Previous research has shown the limitations of the simple load method [1]. For this research, aeroelastic modelling will be done for fatigue analysis of this SWT's composite blades. NREL FAST, TurbSim and Crunch were used for the aeroelastic modelling and fatigue analysis of this micro wind turbine [2–4].

The subsequent calculations and fatigue analysis were for the blades of the SWT, which are made out of GFRP composite material. Equations 3.1–3.3 show the equations used to compute the fatigue and DEL of the micro turbine. The sum average of ten simulations was used in the rainflow-counting algorithm. For the rainflow counting, incomplete cycles were assigned a value of 0.5. For lifetime DEL, the Rayleigh distribution was used to extrapolate the cycles for the blades over a time period of 20 years. IEC 61400 DLC (damage load condition) that was assessed for this research is DLC 1.1, i.e. power production.

$$\mathrm{DEL}_{\mathrm{agg}}^{\mathrm{ST}} = \left\{ \frac{\sum_j \sum_i \left(n_{ji} \left(L_{ji}^R \right)^m \right)}{\sum_j n^{\mathrm{life,\,eq}}} \right\}^{1/m} \tag{3.1}$$

$$\mathrm{DEL}^{\mathrm{life}} = \left\{ \frac{\sum_j \sum_i \left(n_{ji}^{\mathrm{life}} \left(L_{ji}^R \right)^m \right)}{n^{\mathrm{life,\,eq}}} \right\}^{1/m} \tag{3.2}$$

$$n^{\mathrm{life,\,eq}} = \sum_j f_j^{\mathrm{life}} n_j^{\mathrm{STeq}} \tag{3.3}$$

where $n^{\text{life,eq}}$ = lifetime equivalent count, n_{ji}^{life} = damage count for cycle i and time series j for the design life of the blade, L_{ji}^{R} = load range for cycle i and time series j,m = Whöler exponent, DEL^{life} = lifetime damage equivalent load, f_{j}^{life} = lifetime damage count factor for time series j,n_{j}^{STeq} = equivalent counts for time series j and $\text{DEL}_{\text{agg}}^{\text{ST}}$ = short-term average damage rate. Regarding the subscripts, i = fatigue cycle index and j =time series index (for full details of the fatigue calculation algorithms, please refer to NREL MLife manual [3]).

A Whöler exponent of $m = 10$ was used for the blades of this SWT. Previous research was done on the composite blade of this particular SWT by running it to failure at different load cycle ranges and recording the number of cycles to failure; the SN curve obtained was found to be best represented by a value of $m = 10$ [5]. This factor is typical for a GFRP composite material.

The purpose of this research was to see the effect of varying TI on fatigue. Three wind speed profiles with varying turbulence intensities were used in the aeroelastic simulation: NTM Class C, NTM Class A and high turbulence value similar to the current site (hereby referred to as "site turbulence"). For the lifetime damage predictions, an IEC 61400-2 Class IV wind profile with a Rayleigh distribution and average wind speed of 6 m/s was chosen. For all three wind profiles analysed, the shape factor, scaling factor and average wind speed were kept constant so that the only changing variable was the turbulence intensity at each particular wind speed (see Figs. 3.2 and 3.3). From Fig. 3.3, the k factor was 2 (Rayleigh distribution), and the lambda was chosen to ensure an average wind velocity of 6 m/s.

For the rainflow count cycles, the in-plane and out-of-plane load ranges were separated into 300 different load ranges and counted from there. Equations 3.2 and 3.3 were then used to compute the DEL. The short-term DEL was also calculated for each wind speed using Eq. 3.1.

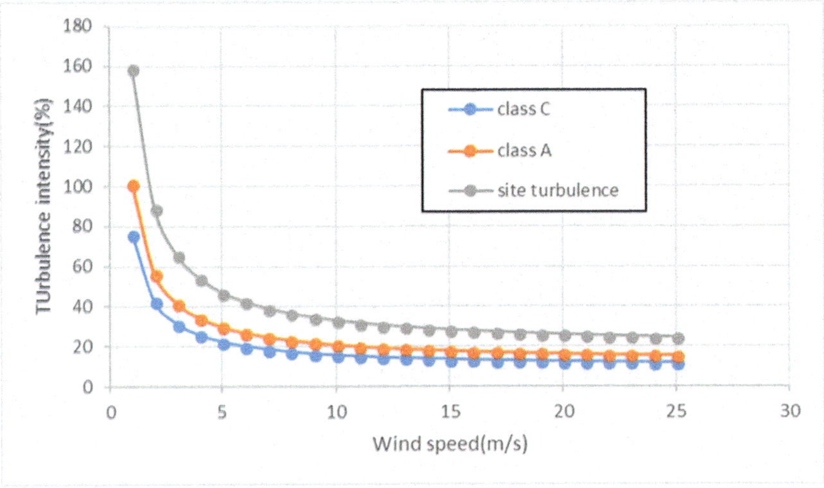

Fig. 3.2 Wind speed vs TI for various wind profiles

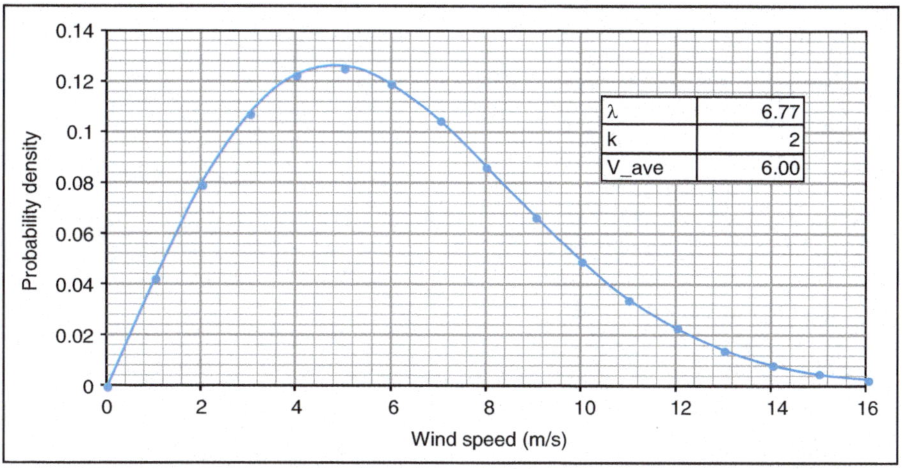

Fig. 3.3 Wind speed probability distribution

For the rainflow counting and Crunch post-processing of FAST time series simulation data, the 1st 10 s of data was removed from the rainflow counting as the simulation initial conditions are still stabilizing in the software (this is also recommended in IEC 61400-2 in the fatigue damage calculations section). The random number generator in TurbSim was used so that each simulation could have a different random wind speed profile (while keeping the same TI and V_{avg} for each wind speed bin) to ensure that an average value reflecting various random profiles could be aggregated.

From Fig. 3.2, at a wind speed of 6 m/s, the TI for the three wind profiles are as follows: Class C, 20.2%; Class A, 26.93%; and site turbulence, 42.08%. The site TI was more than double that of Class C. This was the case for all the wind speed values for this research. The root in-plane and out-of-plane bending moments were analysed for this research.

3.4 Results

3.4.1 Field Test and Simulation for Power Output

A comparison between the FAST and field test data is now made. The data was split into lower wind speeds (5–6.5 m/s) and higher wind speeds (7–8 m/s).

From Fig. 3.4, it can be seen there is generally a good agreement between the simulation and the field test. The maximum error between the simulation and field test occurred at the wind speed of 5 m/s, and this error was 20%. Towards the higher turbulence intensities, some of the errors were below 1% with almost exact correlation between the simulation and the field test results. From 5 to 20% TI, there is very

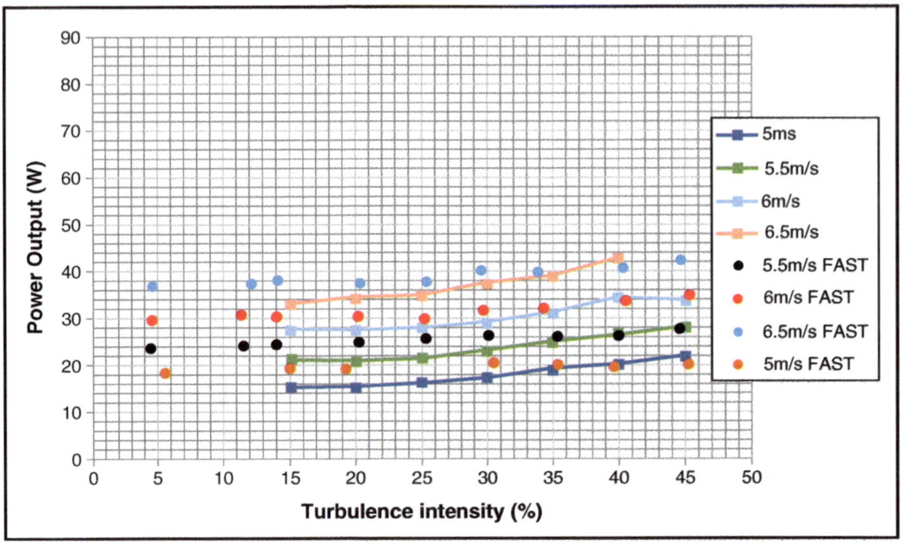

Fig. 3.4 Lower wind speed field test vs simulation

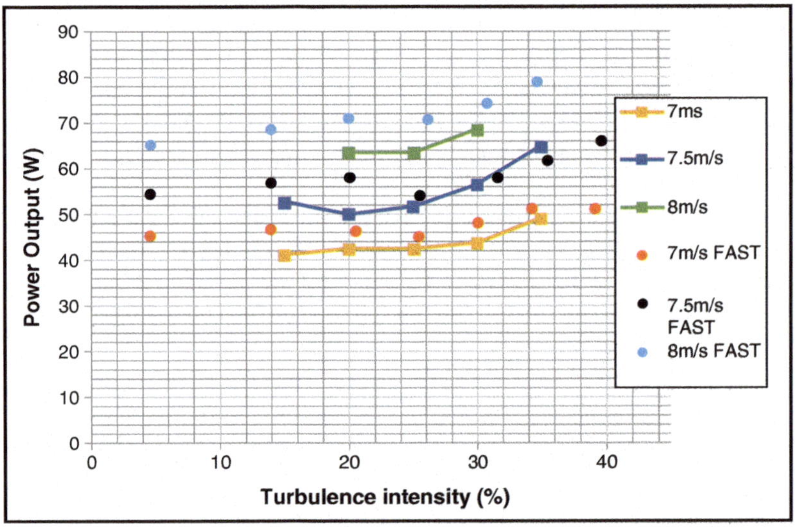

Fig. 3.5 Higher wind speed field test vs simulation

minimal increase in the power output. At these levels, the turbulence energy is not sufficient to cause a major increase in the power output.

From Fig. 3.5, similar to the field results, the simulation shows that there was only a slight increase in power from 5 to 25% TI for the turbine. Above 20/25% TI, the simulation output power gradient increased similar to the field data. The correlation between simulation and field test at 8 m/s was not as good as other wind speeds due to a lower number of observations (thus lower statistical significance). The maximum error between simulation and field test occurred at 8 m/s at TI = 20%, and the error value was 8%. Both the simulation and field tests show an increase in output power with increasing wind speed for the majority of the cases.

Some of the differences between the field test and the simulation are due to simulation yawing not being taken into account, only a single anemometer being used in the field placed at a distance from the wind turbine, cup anemometer dynamics/inertia not compensated for in simulation input, and the anemometer that only took samples at a frequency of 1 Hz.

3.4.2 Fatigue Load Computations

The results from the aeroelastic simulation and subsequent fatigue life calculations will now be discussed.

Figures 3.6 and 3.7 show the DEL for each wind speed and the lifetime DEL for the composite blade. The DEL for the MOOP increased for all three wind profiles for all wind speeds. Comparing Figs. 3.6 and 3.9, it can be seen that the in-plane moment DEL loads were smaller than the MOOP at wind speed bins far below the cut-out of 16 m/s. Up to 10 ms, the DEL of the site condition was more than double that of the NTM Class C wind.

A time series analysis is shown in Fig. 3.8 of the root MOOP for Class C wind and the site turbulence (i.e. the lowest and highest turbulence intensity profiles). The wind profile was the same, with only the intensity of the change of amplitudes

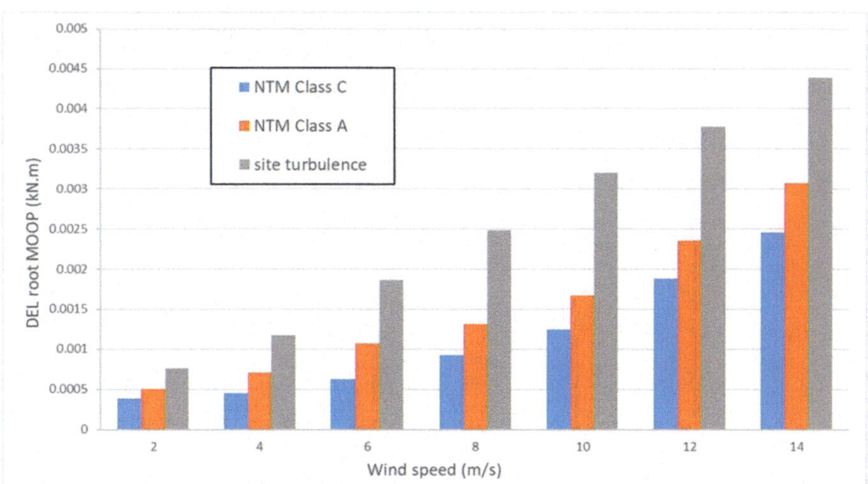

Fig. 3.6 Blade root MOOP DEL for each wind profile

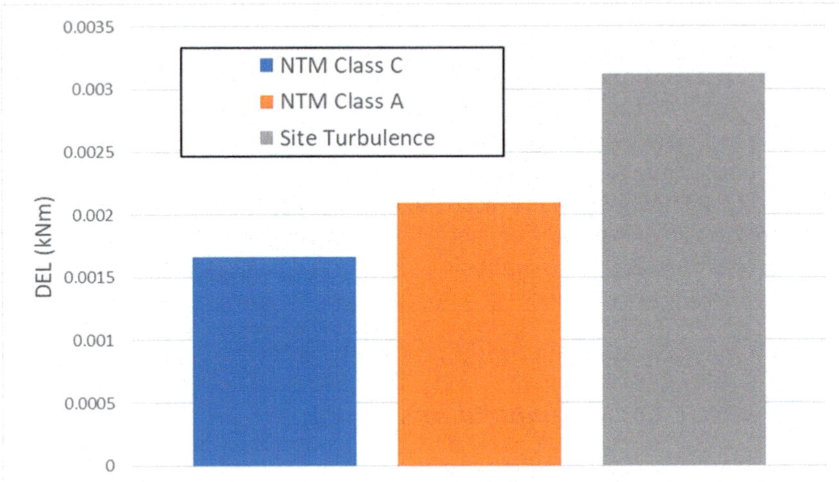

Fig. 3.7 Blade lifetime DEL for each wind profile (20-year life)

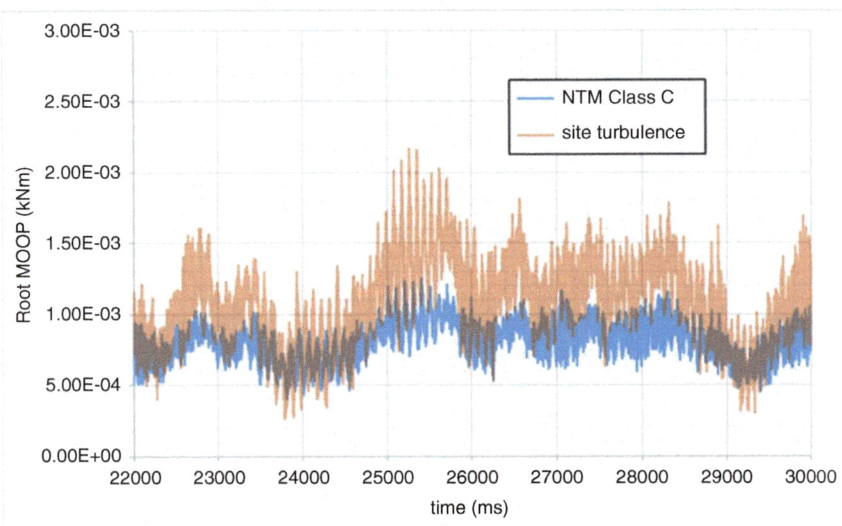

Fig. 3.8 Blade root MOOP time series

varying in the simulation. It can be seen from this time series that the site turbulence causes the MOOP to have higher fatigue fluctuation ranges. The average value of the MOOP is also higher. A highly turbulent site is not good for long-term fatigue loading of a wind turbine. From this figure, it can also be seen that the frequency of vibrations is high for this particular SWT, and the flap first mode, flap second mode and first mode edgewise natural frequencies are 122, 430 and 414 Hz, respectively [1].

Figure 3.9 shows the results for the root MIP,it can be seen that in up to 8 m/s, there is no correlation between the wind speed/turbulence intensity and the DEL of the SWT. The increase from 10 m/s was found to be due to an increase in turbine cut-out (the cut-out speed is 16 m/s). An example of this cut-out behaviour is shown in Fig. 3.10, which shows the root MIP during a high wind speed condition. At $T = 50$ s (5×10^3 ms), the turbine cuts out, and the root MIP has a sudden increase. This sudden stop of the turbine rotor is what causes the root MIP to increase and

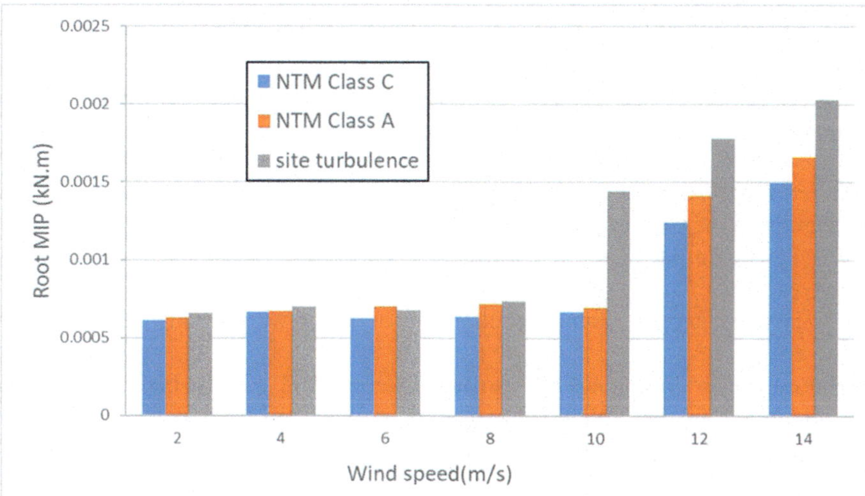

Fig. 3.9 Blade root MIP DEL for each wind profile

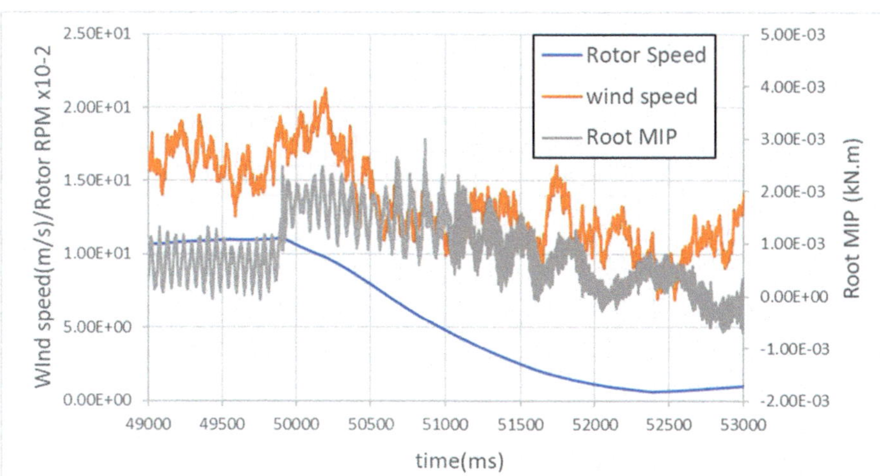

Fig. 3.10 Turbine cut-out condition blade root MIP characteristics

thus the MIP DEL to increase (the cut-out corresponds to a rotor RPM of 1150 RPM). Analysis of the simulation time series results for the root MIP showed that the cut-out condition was frequent at these higher wind speeds. The results agree with the physics of a wind turbine as the MIP is mainly affected by gravitational load of the blade mass, rotor torque as well as inertial forces associated with the rotation of the blade.

3.5 Conclusions

- Increasing turbulence intensity has a negative effect on the root MOOP fatigue life of small wind turbines by increasing the DEL.
- The turbulence intensity had no impact on the root MIP fatigue DEL, except for increasing the turbine cut-out.
- There is an increase in power output with turbulence intensity.
- The power output increase with turbulence is more pronounced at TI values above 25%.
- The increase in power is more pronounced at higher wind speeds, i.e. from 7 m/s and above.
- There is generally a good agreement between FAST and the field test, and the simulation can save much time and resources when evaluating various field conditions and potential scenarios.

References

1. Iino M et al (2015) Examination of simplified load model on fatigue analysis of small wind turbine blade using aeroelastic analysis. Bull JSME 81:823
2. Jonkman JM, Buhl ML Jr (2005) FAST User's Guide, Technical report NREL/EL
3. Hayman GJ, Buhl M Jr (2012) MLife User's Guide, version 1.00, Technical report, NREL/TP
4. Jonkman BJ (2012) Turbsim Users Guide, v1.50 NREL (National Renewable Energy Laboratory), Technical report NREL/TP-500-46918
5. Tokuyama H et al (2014) Proposal of an evaluation method of equivalent fatigue load for small wind turbine blade. Bull JSME 80:816

Chapter 4
Evaluation of Performance Losses and Degradation of Aged Crystalline Si Photovoltaic Modules Installed in Minas Gerais (Brazil)

Antonia Sônia A. C. Diniz, Denio A. Cassini, Michele C. C. de Oliveira, Vanessa F. C. de Lins, Marcelo Machado Viana, Daniel Sena Braga, and Lawrence L. Kazmerski

4.1 Introduction

"How long does the PV system perform well and within specifications?" is a very important question for PV technology investors and consumers in order to ensure their investment return and technology buy-in. The understanding of ageing process of a PV module is critical for predicting and warranting their performance, for identifying and correcting any problems, and for addressing the fundamental considerations for the bankability and component integrity of photovoltaic (PV) systems.

The economics of the PV systems can only be competitive if the modules operate reliably with assured longevity for more than 20–25 years (i.e., the typical warranty periods), with power degradation smaller than 20–25%, respectively, for these time frames (typically designated at "less than 1% per year") [1]. The degradation modes in PV modules dictate the symptoms of failure and degradation mechanisms and the relative progression of these symptoms. Consequently, a single failure mechanism (e.g., encapsulant breakdown) can lead to more than one observable failure modes (e.g., delamination, corrosion, discoloration, etc.) [1]. The degradation modes observed in field-aged modules suggest that the various modes of degradation that

A. S. A. C. Diniz (✉) · D. A. Cassini · D. S. Braga
GREEN-IPUC, Pontificia Universidade Catolica Minas Gerais—PUC Minas, Belo Horizonte, Brazil
e-mail: asacd@pucminas.br

M. C. C. de Oliveira · V. F. C. de Lins · M. M. Viana
Universidade Federal de Minas Gerais—UFMG, Belo Horizonte, Brazil

L. L. Kazmerski
Renewable and Sustainable Energy Institute (RASEI), University of Colorado Boulder, Boulder, CO, USA

National Renewable Energy Laboratory, Golden, CO, USA

A. Sayigh (ed.), *Renewable Energy and Sustainable Buildings*, Innovative Renewable Energy, https://doi.org/10.1007/978-3-030-18488-9_4

are ultimately responsible for performance loss and failure can be of five types: (i) degradation of packaging (glass, encapsulant, backsheet) materials, (ii) loss of adhesion, (iii) degradation of cell/module interconnects, (iv) degradation caused by moisture intrusion, and (v) degradation of the semiconductor device [1].

Photodegradation studies of the PV modules is of economic and social (technology acceptance and public confidence) interest since the reduction of the cost of producing solar energy is associated with the increase in the PV module service life, which does depend significantly on the materials used in the solar cell encapsulation [2–12]. A primary concern is to identify and understand the mechanisms of degradation of the solar module encapsulant.

Jordan and Kurtz [13] reviewed the degradation rates from the field testing studies performed during the last 40 years. They found that the degradation rates observed in the photovoltaic modules/systems deployed after year 2000 have significantly reduced failure rates over those before the year 2000, indicating a substantial improvement in design, materials, manufacturing, and durability of PV modules [13].

Wohlgemuth reported 20 key results of a study for crystalline silicon PV modules relating to field failure and warranty return rates based on various reports in the literature [14, 15]. Their analysis shows that less than 0.1% of annual field failure rate on 10-year-old qualified PV modules, 0.005% for PV modules up to 5 years, 0.13% for PV modules manufactured between 1994 and 2005, and 0.01% annual return rate for PV modules manufactured between 2005 and 2008 [14, 15].

The investigation and diagnosis of PV systems that have been installed and operating for decades or more have been providing valuable information for the entire PV value chain (from material supplies through installers/operators) for addressing critical reliability issues [2, 16]. The installation of these distributed systems in rural in Brazil's tropical climate conditions provides a special opportunity to get insight into two crystalline Si module technologies from two different manufacturers produced and then installed at approximately the same time. These technologies, **Case A** (polycrystalline Si-cell-based modules, sometimes termed "multicrystalline") and **Case B** (monocrystalline Si-cell-based modules), were in operation for similar time frames (15 years)—but their ageing characteristics are much different. This paper addresses "Why?". What are the reasons for the fact that the **Case A** modules had power degradation losses averaging 0.4–0.5%/year, well within the warrantied specifications, while **Case B** modules had average power losses in the range 2.3–3.7%/year over the similar installation time?

This paper presents the results of careful investigations and classifications of these performance losses of photovoltaic modules and the identification of major degradation mechanisms. The approach starts with the visual inspection of these modules for indications of discoloration, corrosion, and delamination or other defects, proceeds with the nonintrusive electrical characterization (*I*–*V* parameters) and subsequent thermal mapping of the modules, and concludes with chemical and compositional diagnostic characterization of the key components uncovering the bases for the various degradation modes and mechanisms.

The technical contributions and scientific understanding provided by these studies are important. But the correlation of the effects of the environmental conditions on the PV performance also contributes to decision-making for future investments in PV systems and selection of appropriate technology for a better performance within these specific tropical climate zones.

4.2 Methodology and Framework

This study covers an analysis from representative PV module samples selected from the approximately 1000 stand-alone PV systems that are operated by the Energetic Company of Minas Gerais (CEMIG). These PV systems were installed in the northern and eastern region of the state of Minas Gerais, classified as tropical zone according to the Köppen-Geiger classification (Fig. 4.1), with dry winters and rainy summers [17]. The average annual maximum temperature in the regions where the photovoltaic systems were installed varies between 28 °C and 32 °C [18]. Figure 4.1 also shows the solar resource map for the state of Minas Gerais, indicating the very good insolation levels for PV utilization [18].

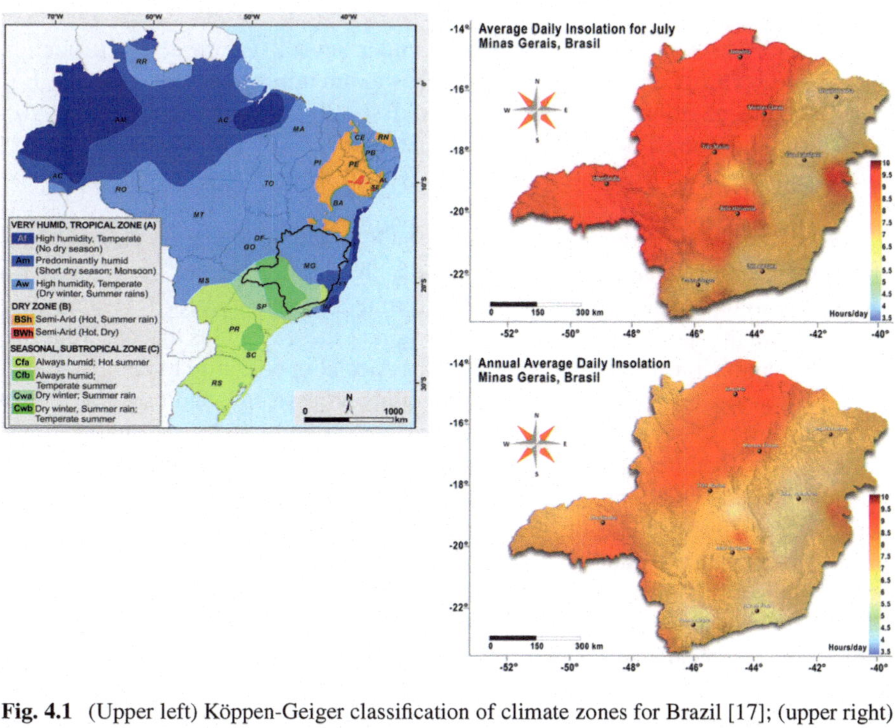

Fig. 4.1 (Upper left) Köppen-Geiger classification of climate zones for Brazil [17]; (upper right) average daily insolation for the state of Minas Gerais (hours of sunlight/day) for the month of July; (lower right) annual average daily insolation [18]

Fig. 4.2 School PV system, typical of the distributed installations for the modules from this study

These PV systems were implemented under several PV rural electrification demonstration programs, such as US-Brazil program initiated in 1995, PRODEEM in 1997, and Luz Solar in 1999, with the oldest ~20 years in operation [19, 20]. The PV modules used in this study are located at 16° 10′ 13″ S latitude and 42° 17′ 25″ W longitude. Figure 4.2 shows one typical school PV system of Luz Solar program.

As noted, these photovoltaic systems in such long field operation contain valuable information to map, analyze, and classify the performance of the PV generators. They were installed at about the same time and operated for the same time period within the same climatic zone. The origin of these modules that are from two different manufacturing sources presented the special opportunity to compare them for performance, technology type, and materials/processing—and especially their operation under essentially the same tropical conditions and same applications. These systems had *not* had a rigorous evaluation for their reliability and operation performance that could have documented any module ageing issues or identifying major degradation mechanisms previously during their time in the field.

The investigation of the performance of these PV systems was initiated with a review of the available history of PV systems' operations and possible failures from CEMIG database. A sample of these PV systems was inspected and selected in the field during technical visits, and the representative PV modules (some having severe degradation issues) were removed from the field to be analyzed at in our laboratories of GREEN Solar-IPUC in Belo Horizonte. These modules were inspected following the visual inspection procedures developed by Freitas Souza [19] based on the NREL spreadsheet methodology. The manufacturers are not identified specifically in this paper, but rather referred to **Case A** (polycrystalline Si) and **Case B** (monocrystalline Si).

Case A: Module 1 is ***polycrystalline Si***. Module 1 was selected because it was typical of all the modules in this Case A, having technology characteristics that are typified by (i) no visible discoloration (the encapsulant EVA); (ii) it was a different Si cell technology and manufacturer; and (iii) it was installed at about the same time, at the same location, and operated for the same period as the monocrystalline Si devices. **Case B:** Modules 2, 3, and 4 are ***monocrystalline Si***. Several modules were selected because they visually represented different degrees of discoloration/delamination/corrosion and/or the *IV* characteristics were in various degrees of degradation. Figures 4.3 and 4.4 show photos of these **Case A** and **Case B** modules selected for diagnosis in this study.

Following the detailed visual inspection procedure, the modules had their *I–V* characteristics recorded using the AAA commercial flash simulator at "Laboratorio Fotovoltaico" from the University of São Paulo (IEE-USP). The specific electrical parameters monitored were (1) open-circuit voltage (V_{oc}), (2) short-circuit current (I_{sc}), (3) fill factor (FF), (4) efficiency, (5) current and voltage at maximum power point (I_m and V_m), (6) maximum power (P_m), and (7) series resistance (R_s).

The P_m (watts) of the tested module ($P_m$test) was determined by the flash test P_m that was obtained from the module data sheet. The degradation rate (**DR** in %) and the mean *annual* degradation rate (%/year) were calculated using Eqs. (4.1) and (4.2):

$$\mathbf{DR} = \left[\left(P_{\mathrm{m}}\mathrm{ref} - P_{\mathrm{m}}\mathrm{tested}\right) / P_{\mathrm{m}}\mathrm{ref}\right] \times 100 \tag{4.1}$$

And, the annual average

$$\mathbf{DR} = \left(\mathbf{DR}\right) / N \tag{4.2}$$

where *N* is the number of years in the field.

These photovoltaic modules were also subjected separately to thermographic (IR) imaging analysis to identify possible hot spots (using a commercial infrared camera: NEC/Sane), capable of time-resolved imaging. For these IR studies, the modules were installed outside, positioned at the azimuthal deviation and the ideal inclination angle for the test site and the minimum irradiation of 700 W/m^2 [21, 22]. The modules were short-circuited, and after 15 min in this condition, the thermal maps were recorded for the entire module. Note that the temperatures reached during these tests are typically below the 50–70 °C that the modules could encounter during their field operation.

Attenuated total reflectance Fourier transform infrared (ATR-FTIR) spectroscopy was performed using Bruker equipment (Model Alpha) equipped with a single reflection *diamond* ATR with a scanning range of 400–4000 cm^{-1}. This technique provided the identification of the chemical state of the glass and encapsulants (EVA). In addition, surface analytical techniques (Auger electron microscopy and

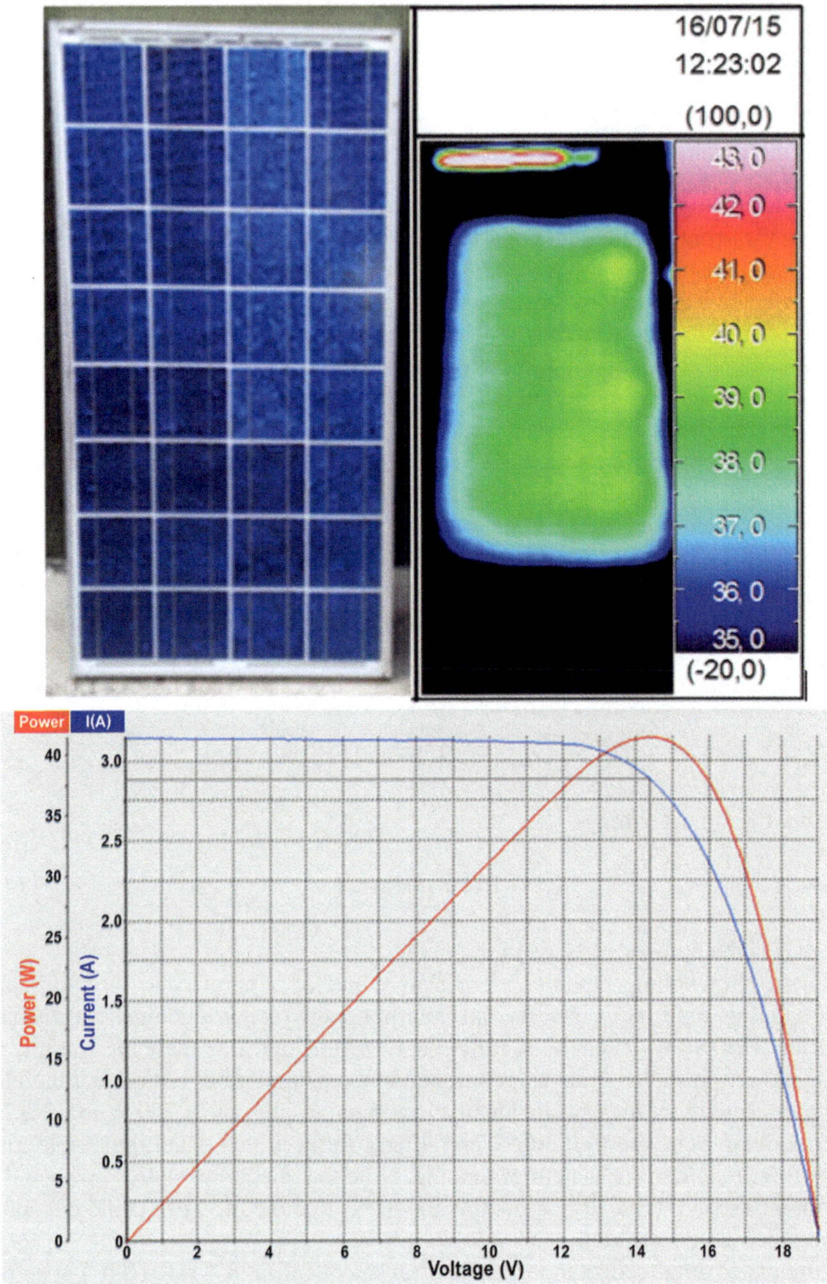

Fig. 4.3 Case A module (upper right). Thermal mapping showing uniform response with no hot spot (upper right). Current voltage and power characteristics (lower) with data provided in Table 4.1

(a)

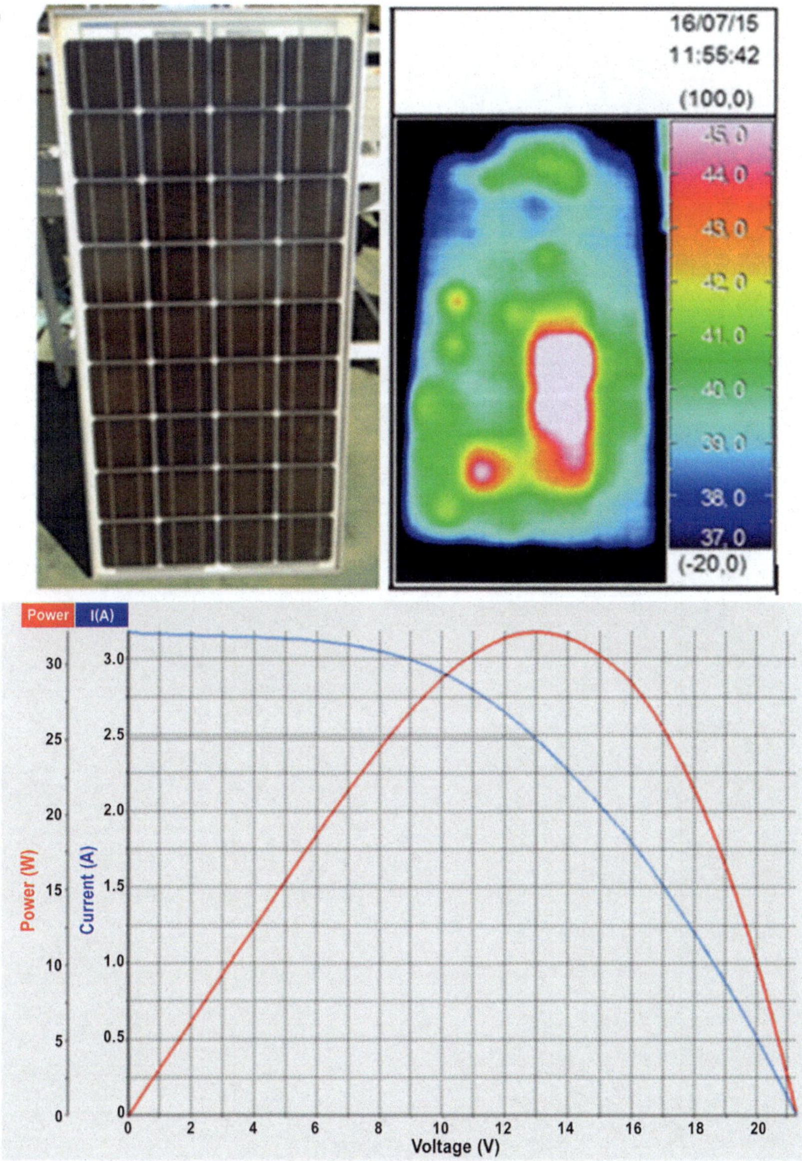

Fig. 4.4 Case B modules: Degraded modules (**a**) module 2, (**b**) module 3, and (**c**) module 4. Photo of each module is shown in the upper left of each and the thermal scan in the upper right. The current voltage and power characteristics are shown at the bottom of each

(b)

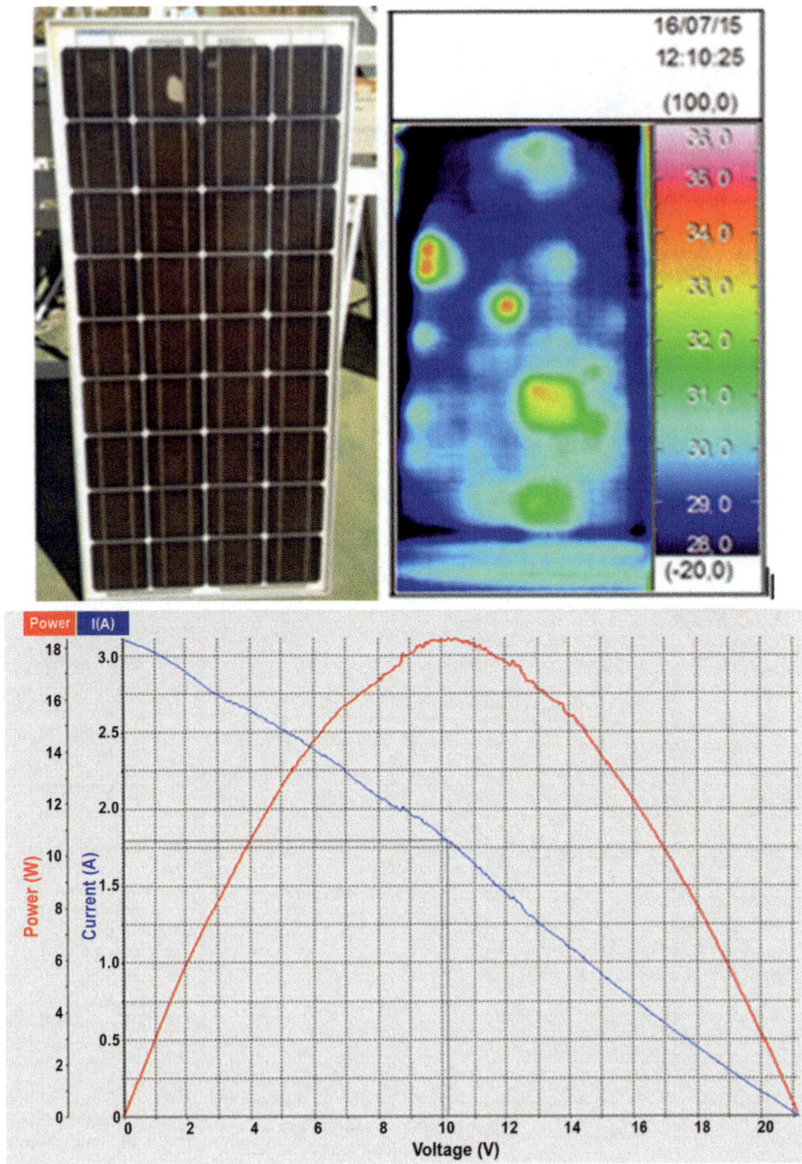

Fig. 4.4 (continued)

(c)

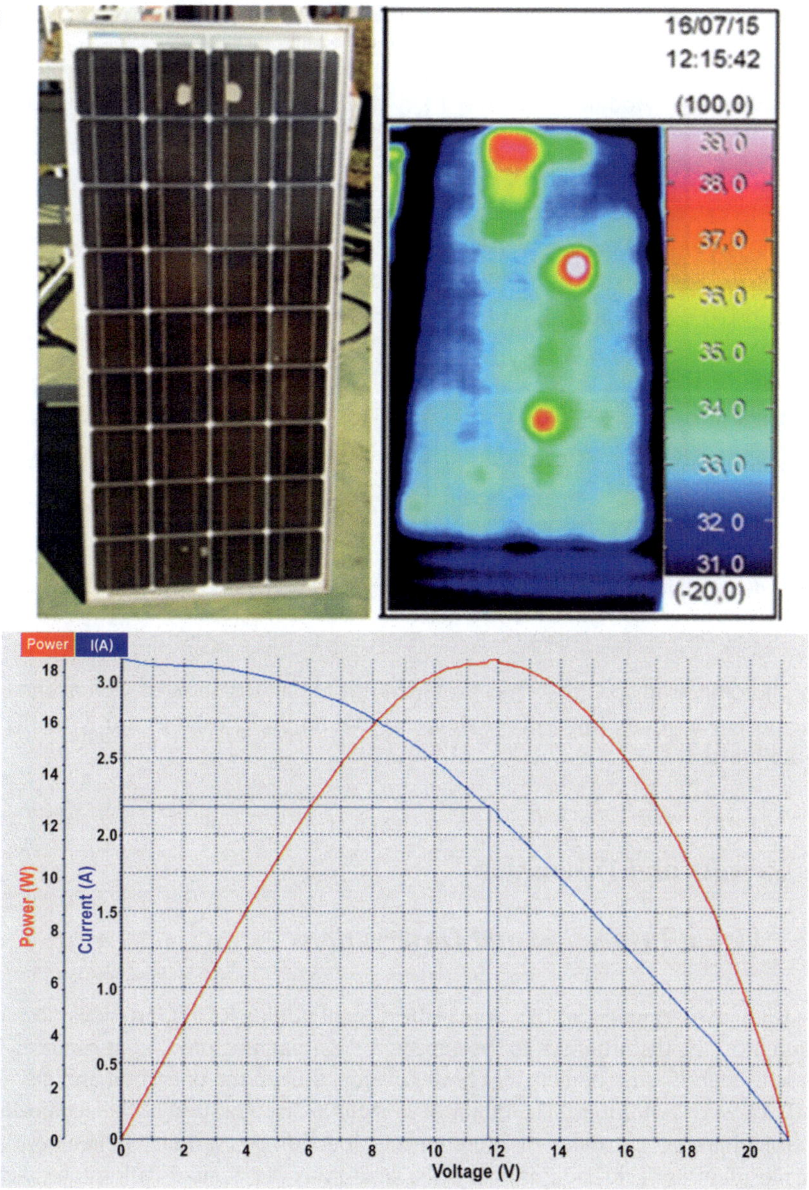

Fig. 4.4 (continued)

Table 4.1 Electrical parameters measured for **Case A** and **Case B** modules, comparing to manufacturers' specification sheets for these delivered modules

Parameters	CASE A Module Specification Sheet	CASE A Module 1	CASE B Module Specification Sheet	CASE B Module 2	CASE B Module 3	CASE B Module 4
Isc (A)	3.25	3.15	3.50	3.18	3.14	3.19
Im (A)	3.02	2.89	3.17	2.47	1.80	2.18
Voc (V)	18.90	19.06	21.70	21.34	21.08	21.34
Vm (V)	15.00	14.36	17.40	13.00	10.18	11.71
Pm (W)	45.30 ±5%	41.46	55.00 ±5%	32.17	18.28	25.56
Rs (Ohms)		0.99		2.67	6.72	3.95
Total Change in Pm from Specification (%)		-8.47%		-45.51%	-66.76%	-53.52%
Average annual degradation (%)		0.44%		2.30%	3.70%	2.97%

Total change in the P_m for the field-operated modules over the operation periods and the calculated annual P_m degradation (%) are also indicated

X-ray photoelectron spectroscopy) provided initial investigation of the encapsulant for the presence of any lower-level elements that might be used as ultraviolet (UV) light inhibitors.

4.3 Results and Discussion

4.3.1 Visual Inspection and Observations

The visual assessment (in the spreadsheet methodology [9]) provides the first evaluation of any disturbances to the expected ("as-manufactured") appearance. The intent is to identify any defects that have developed from the operation and environmental exposure conditions. The major ones include discoloration of the encapsulant, backsheet chalking, corrosion of interconnects, junction box failures, delamination of the encapsulant or backsheet, broken cells, snail trails (associated with microcracks), broken/cracked module glass, moisture ingression, broken contacts/ribbons, and other features that may indicate fires or lighting strikes, for example. The inspection can also note the buildup of soiling or growth of biofilms, especially in tropical areas, on the module surface or along the interfaces with the frames.

For the modules of this study, the two cases have significantly different observations from the visual inspection. **Case A** modules showed few signs of any defects—and no discoloration of the encapsulant to the human eye. Additionally, now significant defects such as delamination or cell contact corrosion, indicating that the encapsulant retained integrity over the 15 years of operation. **Case B** presented a noticeable contrast: The modules had various degrees of discoloration (with the majority showing browning) and significant encapsulant delamination. Metal cell interconnects also displayed various levels of corrosion. Figures 4.3 and 4.4 compare the encapsulant integrity of **Case A** with the observations of **Case B**.

If no discoloration of the encapsulant is observed from normal human eye inspection, it does not mean that some changes in this material have not occurred. TalizhMani et al. [23] had noted that initial discoloration can be occurring despite that it is not observable under normal inspection conditions. This study employed their UV light methodology (380–395 nm wavelengths) to determine if there were any initial stages of changes in the chemistry of the EVA encapsulation [18]. This UV inspection did uncover some modules with slight (initial) discoloration of several **Case A** modules investigated. But it is unlikely that at this stage of chemical change, the efficiency of the module was significantly affected. Certainly, this is confirmed by the *IV* measurements that show these 15-year-old modules from **Case A** were well within the manufacturer's specifications (discussed in the following sections).

4.3.2 Electrical and Thermal Characterization

Following and correlating with the visual inspection results, the electrical and thermal characteristics of the two module types are contrasting. Figure 4.3 shows a representative *I–V* curve from the **Case A** samples. The characteristic maintains the relative shape of its specifications at manufacture—with a fill factor of about 0.63 (slightly below that of the modules delivered). Moreover, the maximum power delivered has only decreased by about 8.5% over the operational lifetime.

Figure 4.4a–c shows three **Case B** PV modules, chosen because they typified various degrees of EVA discoloration and overall degradation. Corresponding to the discoloration, corrosion, and delamination of these components, the *I–V* characteristics show losses in the major parameters, especially noticeable in the V_m and the FF.

The electrical characteristics of all four modules (**Cases A and B**) are summarized in Table 4.1, comparing the values to reference modules for each type—taken from the manufacturers' reference module data sheets. Although the measured *I–V* characteristics of the modules at the time of delivery were no longer available, all modules met initial specifications. The significant changes in P_m (decrease), FF (decrease), and r_s (increase) are documented for the aged modules.

Module 1 had a measured degradation rate of about 0.5%/year (representative of other **Case A** modules), within the warranty for this period of time—and typical of other reports in the literature [7–9]. The degradation rate in PV modules 2–4 (**Case B**) were considerably higher (>2%/year) and outside the warranty specifications.

(These PV modules were in the field—tropical environment—for more than 15 years and were manufactured before the year 2000.)

Case A results correlate with the observations from the visible, with the module not showing signs of discoloration, corrosion, or delamination (see Fig. 4.3). In fact, this was typical of all modules inspected of this type and age also well under the 1%/year acceptable degradation. In contrast, Case B results indicate significant degradation associated with EVA discoloration, delamination, and corrosion (Fig. 4.4).

Figures 4.3 and 4.4 also present thermographic images of the modules. **Case A** modules typically show fairly uniform temperature mapping—with no significant occurrence of unusual heated regions. However, thermal imaging for Case B modules identified hot spots within the modules, which not only restricts the performance of the entire string but also causes PV module degradation in overextended operating time periods [1]. Such hot spots are major issues for system reliability and service lifetime. Due to delamination, the incident sunlight is not able to reach the solar cells resulting in performance degradation. In addition, these defects inhibit uniform dissipation of heat, resulting in higher operating cell temperatures (hot spots), which further degrade the performance [1–5].

4.3.3 Chemical Analysis

Since UV radiation with high temperature is a documented cause of polymer breakdown, we postulated that the differences between the evident differences between the two cases could be due to either differences in the glass or the EVA itself. PV manufacturers have used additives to both to absorb the UV, preventing this radiation from breaking the organic bonds underlying the material degradation.

4.3.3.1 Module Glass

Cerium (cerium oxide) had been used in module glass to absorb the deleterious UV radiation before reaching the EVA. Because of the differences in the discoloration between the polycrystalline and the monocrystalline modules, we first conjectured that this could be due to the UV absorber in the glass. To identify the presence of the cerium element (Ce) in the module glass, by attenuated total reflectance Fourier transform infrared (ATR-FTIR) analysis was performed in attenuated total reflectance (ATR) mode.

Multiple module glass samples from both **Case A** and **Case B** were analyzed. The FTIR did not detect the presence of cerium in either Case. XPS and EDS analysis showed similar results—indicating no cerium or its oxide present within the detection limits of the techniques. Additionally, Carvalho de Oliveira reported that the composition of these glasses appeared to be identical [2]. Thus, cerium-containing glass was eliminated in differentiating the degradation properties between the two module types.

4.3.3.2 Encapsulant

To access the EVA directly for chemical analysis, two characteristic modules were broken (shattered) to expose the material for characterization. EVA samples represented the Case A ("best") and several of the Case B ("visibly discolored") modules. This section is based upon more detailed chemical analysis, identifications, and discussion of the changes in the chemistry of the EVA for these two specific cases reported by de Oliveira [20].

The evaluation also used FTIR-ATR for chemical/compositional analysis. The EVA samples are designated EVA 1 (present in module 1 [Case A]) and EVA 2 (present in module 2 [Case B]) encapsulants. A commercial EVA sample was used as reference material from a major PV module manufacturer. The EVA photodegradation was studied by the spectral analysis estimating the carbonyl index (CI) by comparing the intensity ratio of the ester carbonyl absorption band with the reference band at 2850 cm^{-1} [22, 24]. This reference band is characteristic of the deformation of the rocking vibrational mode in the plane to methylene (CH_2–), which is taken as an internal standard in relation to the other bands that compensate the differences of signals. This band is used to normalize the difference in signals that occurs due to surface factors such as the contact angle with the light beam incident on the ATR-FTIR and the variations in the thickness of the samples [22].

The band observed in the region 1735 cm^{-1} was deconvoluted to separate bands that overlapped. The procedure was performed using the OMNIC™ software and employing the Lorentzian mathematical function that better fitted the curve.

The ratio CI (carbonyl index) given by Eq. (4.3) was determined from the ratios between the absorbance value (A) of the stretching vibrational mode of the ester carbonyl group at 1735 cm^{-1} and the symmetrical vibrational mode of stretch of methylene (–CH_2–) in 2850 cm^{-1} [22, 24]. That is,

$$\mathrm{CI} = \frac{\mathrm{A1735}(\text{ester carbonyl})}{\mathrm{A2850}(\text{methylene})} \tag{4.3}$$

The CI was determined to obtain a quantitative comparison of the functional groups of the EVA: commercial copolymer, EVA 1 (module 4), and EVA 2 (module 2). Photovoltaic module 2 showed yellowing and browning of the encapsulant (see Table 4.2). Module 4 was not significantly degraded. Table 4.2 shows the carbonyl index (CI) for the EVA samples extracted from each of the PV modules [22].

Table 4.2 Carbonyl index (CI) value for EVA samples from this study

Encapsulant	CI: A1735/A2850
Commercial EVA (typical)	1.1
EVA 1 (Case A)	1.2
EVA 2 (Case B)	1.7

The highest carbonyl index (CI) of module 2 agrees with the degraded aspect presented by the encapsulant of this photovoltaic module. The yellowing and browning of EVA are the visibly physical aspect due to its degradation [21, 22] (affecting primarily the I_{sc}). EVA 1 (Case 1: module 1) presented a lower CI value than EVA 2 (Case 2: module 2) and close to the commercial EVA. This can be attributed to the presence of additives, such as accelerators, adhesion promoters, initiators, antioxidants, and stabilizers or UV absorbers [10]. These additives are organic compounds that have some functional groups identical to those presented by EVA. Because they are at levels typically less than 1–2% in most EVA, it is difficult to identify these species in ATR-FTIR spectra or by XPS [3, 4].

The exposure to the UV component of solar radiation and high temperature can cause browning of the encapsulant formed by ethylene-vinyl acetate (EVA) copolymer and its delamination. The chemical analysis studies of the encapsulant degradation process of the modules 1, 2, and 3 were made. They showed that with exposure to solar radiation, EVA undergoes deacetylation of the vinyl acetate pendant group with the formation of predominantly acetic acid, acetaldehyde, and polyenes, which are the discoloring chromophores, or cross-links with adjacent chains. Mainly acetic acid and acetaldehyde can catalyze the degradation of EVA beyond they are corrosive of the metallization in crystalline Si PV modules [7, 8]. The metal oxidation also occurs between the metal contacts, from the moisture that enters the PV module due to sealing problems in edges related to its incorrect application or rupture by external agents [8].

The EVA from **Case A** and **Case B** had distinct chemical and physical differences from these analyses. The remaining difference was in the EVA processing. From the manufacturers, **Case A** had EVA that was "fast cured," with controlled Res Gel contents of about 85–90%—both aspects for minimizing the discoloration mechanisms. **Case B** had used slow curing for the EVA (conventional processing by that manufacturer at that time). No data were available on the gel content for comparison, but it would be expected that the gel content would be lower based upon reports from slow-cured EVA in the literature. This major difference in EVA processing is the likely major reason for the differences in the color and associated degradation mechanisms.

4.4 Summary and Conclusions

This paper examined and compared performance losses and degradation mechanisms for modules from two PV systems installed in the state of Minas Gerais, Brazil, for ~15-year periods. This provided the special opportunity to directly evaluate and contrast the losses, reliability, and durability of modules operating for a long period for the same distributed power applications. The focus of these studies was on modules (a) from two installations, (b) operating under the same climate conditions (tropical zone according to Köppen-Geiger), (c) with modules from two different manufacturers and technologies, (d) manufactured and installed at about the same time, *but* (e) with very different resulting observed changes in their operational and physical characteristics over their long-time field exposures.

"What" the differences were and the "why?" underlying these differences followed the characterization process:

- Visual observations ("spreadsheet methodology")
  Case A (polycrystalline or multicrystalline Si cells) had no encapsulant discoloration and few indications of delamination and corrosion. (Initial discoloration of the EVA can be detected under UVB light. This was observed for a small number of modules in this case—but most had no indications of changes.)
  Case B (monocrystalline Si cells) had significant yellowing and browning, widespread areas of delamination, and fairly extensive interconnect corrosion regions.

- Electrical and thermal characterization
  Case A: Average annual loss in power of 0.3–0.7%; thermal mapping of modules showed uniform temperature characteristics with no significant higher-temperature regions.
  Case B: Average annual power losses in the range 2.0–4.0% (or more) over their installation time. Thermography showed many hot spots corresponding to delamination and corrosion defects.

Encapsulant discoloration and delamination provided the first clues and apparent correlation to the measured differences in module performance. The climate conditions of high ultraviolet (UV) radiation and high ambient temperatures correspond to the environment to enable and to accelerate such encapsulant (EVA) degradation.

Because of the differences in these characteristics, the first question addressed was whether the UV was purposely inhibited (using cerium) from reaching the encapsulant for **Case A** versus **Case B**.

- Glass: Chemical properties
  The glass for neither the **Case A** nor the **Case B** modules had cerium (cerium oxide) detected by ATR-FTIR, EDS, or XPS. Neither glass had this UV inhibitor. And both glasses had approximately the same chemical (expected) constituents. It was concluded that the module glass was not responsible for the differences.
- Encapsulant (EVA) properties
  The value found to CI of **Case B** confirmed the degradation suffered by the EVA encapsulant and the presence of carboxylic acid hydroxyl (higher CI), confirming the formation/presence of acetic acid. It was found that the EVA with the highest thermal stability was that of **Case A** (lowest CI conforming to source material before exposure). These results correspond to previous reports in the literature for the discoloration of the EVA under enhanced UV and high-temperature conditions—as well as for the delamination. The photothermal degradation of the EVA encapsulant under high ultraviolet (UV) and high-temperature exposure results in color changes over the areas of the solar cells that progress from yellowing to browning. This discoloration primarily reduces the transmission of light to the solar cells, which directly reduces the short-circuit current (I_{sc})

decreasing the associated module efficiency. This EVA degradation process leads to the production of acetic acid and other volatile gases that get trapped within the module at different interfaces, causing delamination or bubble formation, reducing module performance and reliability [25–28].

The acetic acid attacks the metallic contacts of solar cells and causes corrosion, in turn increasing the series resistance (R_s) and ultimately reducing the power yield [26]. Therefore, the formation of bubbles, delamination, and corrosion, and in some cases the formation of hot spots, are some indirect effects of EVA discoloration that can contribute to the reduction of PV module efficiency and lifetime.

There were some differences between the samples examined between **Case A** and **Case B**. **Case A** EVA has some indications of the presence of very low-level additives, such as accelerators, adhesion promoters, initiators, antioxidants, and stabilizers or UV absorbers [2, 9]. These additives are organic compounds that have some functional groups identical to those presented by EVA. Because they are at levels typically <1% in most EVA, it is difficult to identify these species in ATR-FTIR spectra [3, 4]. However, the initial XPS results show some subtle differences that might provide the possible presence of these low-level additives. However, the manufacturer could not confirm this.

Thus, we have preliminary conclusions for the distinct differences in the long-time field performances between **Case A** and **Case B**:

- The major cause of the degradation of **Case B** modules compared to **Case A** lies with the EVA encapsulation.
- **Case A** EVA was "fast-cured," resulting in a gel content in the >85% range (ideal for this encapsulant); **Case B** EVA was "slow-cured"—a process more prone to polymer degradation.
- UV: The glass from neither case had cerium as a UV inhibitor. Though this is a common additive for module glass, it was sometimes left out in order to increase the light transmission through the glass. From the chemical analyses, there were indications that the EVA for **Case A** might have UV stabilizers, but this needs additional confirmation.

Acknowledgments The authors would like to thank the Energetic Company of Minas Gerais (CEMIG-D), University of São Paulo (USP), and Pontifical Catholic University of Minas Gerais (PUCMINAS) for their support for the development of this project. Also, our team would like to thank FAPEMIG and CNPq for the financial support and guidance. We also gratefully acknowledge the help and inputs of the entire GREEN technical team.

References

1. TamizhMani G, Kuitche J (2012) A Literature Review and Analysis on: Accelerated Lifetime Testing of Photovoltaic Modules, Solar ABCs report
2. Michele CC de Oliveira (2018) PhD thesis, Universidade Federal de Minas Gerais (UFMG)
3. Jiang S, Wang K, Zhang H, Ding Y, Yu Q (2015) Encapsulation of PV modules using ethylene vinyl acetate copolymer as the encapsulant. Macromol React Eng 9:522–529

4. Charki A, Laronde R, Bigaud D (2013) Accelerated degradation testing of a photovoltaic module. J Photonics Energy 3:1–10
5. Oliveira MCC, Diniz ASAC, Viana MM, Lins VFC (2018) The causes and effects of degradation of encapsulant ethylene vinyl acetate copolymer (EVA) in crystalline silicon photovoltaic modules: a review. Renew Sustain Energy Rev 81:2299–2317
6. Czanderna AW, Pern FJ (1996) Encapsulation of PV modules using ethylene vinyl acetate copolymer as a pottant: a critical review. Sol Energy Mater Sol Cells 43:101–181
7. Peike L, Purschke C, Weiss KA, Köhl M, Kempe M (2013) Towards the origin of photochemical EVA discoloration. In: Proceedings of the 39th IEEE photovoltaic specialists conference (PVSC 39). IEEE-PVSC, Tampa, FL, pp 1579–1584
8. Klemchuk P, Erzin M, Lavigne G, Holley W, Galica J, Agro S (1997) Investigation of the degradation and stabilization of EVA-based encapsulant in field-aged solar energy modules. Polym Degrad Stab 55:347–364
9. Dunn L, Gostein M, Stueve B (2013) Literature review of the effects of UV exposure on PV modules. NREL PV Module Reliability Workshop
10. Kempe M (2010) Ultraviolet light test and evaluation methods for encapsulants of photovoltaic modules. Sol Energy Mater Sol Cells 94:246–253
11. Holley WH, Argo SC, Galica JP, Yorgensen RS (1996) UV stability and module testing of nonbrowning experimental PV encapsulants. In: Proceeding of 25th IEEE PVSC. IEEE, Washington, DC
12. Pern FJ (1997) Ethylene-vinyl acetate (EVA) encapsulants for photovoltaic modules: degradation and discoloration mechanisms and formulation modifications for improved photostability. Angew Makromol Chem 252:195–216
13. Jordan DC, Kurtz SR (2013) Photovoltaic degradation rates—an analytical review. Prog Photovolt 21:12–29
14. Wohlgemuth J (2012) Standards for PV modules and components—recent developments and challenges. NREL Report number: NREL/CP-5200-56531
15. Packard C, Wohlgemuth J, Kurtz S (2012) Development of a visual inspection data collection tool for evaluation of fielded PV module condition. National Renewable Energy Laboratory (Technical Report: NREL/TP-5200-56154)
16. Chattopadhyay S, Dubey R, Kuthanazhi V, Zachariah S, Bhaduri S, Mahapatra C, Rambabu S, Ansari F, Chindarkar A, Sinha A, Singh HK, Shiradkar N, Arora BM, Kottantharayil A, Narasimhan KL, Sabnis S, Vasi J (2016) All-India Survey of photovoltaic module reliability: 2016, National Center for Photovoltaic Research and Education (NCPRE). IIT Bombay and National Institute for Solar Energy (NISE), Mumbai, Gurugram. http://www.ncpre.iitb.ac.in/research/pdf/All_India_Survey_of_Photovoltaic_Module_Reliability_2016.pdf
17. Kottek M, Grieser J, Beck C, Rudolf B, Rubel F (2006) World map of the Köppen-Geiger climate classification updated. Meteorol Z 15(3):259–263
18. ATLAS SOLARIMÉTRICO DE MINAS GERAIS. Belo Horizonte. Companhia Energética de Minas Gerais (CEMIG) (2012). http://www.cemig.com.br/pt-br/A_Cemig_e_o_Futuro/inovacao/Alternativas_Energeticas/Documents/Atlas_Solarimetrico_CEMIG_12_09_menor.pdf
19. Francisco Hering Alves de Freitas Souza, Inspeção e Monitoramento do Desempenho de Sistemas Fotovoltaicos Conectados a Rede Elétrica: Estudo de Caso Real (2014) Master thesis, Pontificia Universidade Catolica de Minas Gerais, Programa de Pós-Graduação em Engenharia Mecânica, Minas Gerais
20. Diniz ASAC, Lauro VBMN, Camara CF, Morais P, Cabral CVT, Filho DO, Ravinetti RF, França ED, Cassini DA, Souza MEM, Santos JH, Amorim M (2011) Review of the photovoltaic energy program in the state of Minas Gerais, Brazil. Renew Sustain Energy Rev 15:2696–2706
21. ASSOSSIAÇÃO BRASILEIRA DE NORMAS TÉCNICAS, ABNT NBR–15572:2013—Ensaios não destrutivos—Termografia Guia para inspeção de equipamentos elétricos e mecânicos (2013)
22. ASSOSSIAÇÃO BRASILEIRA DE NORMAS TÉCNICAS, ABNT NBR–15866:2010. Ensaio não destrutivo—Termografia—Metodologia de avaliação de temperatura de trabalho de equipamentos em sistemas elétricos

23. Ali S, Ali A, Saher S, Saif Agha H, Majeed H b A, Mahmood FI, TalizhMani G (2018) A comprehensive study of 18–19 years field aged modules, for degradation rate determination along with defect detection and analysis using IR, EL, UV. In: Proceeding of 5th International Bhurban Conference on Applied Sciences and Technology (IBCAST). IEEE, New York. Also, Personal Communication with A.S.A.C. Diniz (2017)
24. Jin J, Chen S, Zhang J (2010) UV-aging behaviour of ethylene-vinyl acetate copolymers (EVA) with different vinyl acetate contents. Polym Degrad Stab 95(5):725–732
25. Gagliardi M, Lenarda P, Paggi M (2017) A reaction-diffusion formulation to simulate EVA polymer degradation in environmental and accelerated ageing conditions. Sol Energy Mater Sol Cells 164:93–106
26. Sinha A, Sastry OS, Gupta R (2016) Nondestructive characterization of encapsulant discoloration effects in crystalline-silicon PV modules. Sol Energy Mater Sol Cells 155:234–242
27. Igor Alessandro Silva Carvalho, Preparação estudo da fotodegradação de compósitos de matriz polimérica para encapsulamento de módulo fotovoltaico (2007) Master thesis, Pós-Graduação em Engenharia de Materiais da REDEMAT, Minas Gerais
28. Kojima T, Yanagisawa T (2005) Ultraviolet-ray irradiation and degradation evaluation of the sealing agent EVA film for solar cells under high temperature and humidity. Sol Energy Mater Sol Cells 85:63–72
29. Jin J, Chen S, Zhang J (2010) Investigation of UV aging influences on the crystallization of ethylene-vinyl acetate copolymer via successive self-nucleation and annealing treatment. J Poly Res 17:827–836

Chapter 5
The Bioclimatic Approach in Developing Smart Urban Isles for Sustainable Cities

D. K. Serghides, S. Dimitriou, I. Kyprianou, and C. Papanicolas

5.1 Introduction

The intense urbanisation observed from 1950 (30% urban dwellers) to today (55% urban dwellers), in conjunction with increasingly energy-intensive lifestyles, has led to voracious patterns of global energy consumption, especially in heavily urbanised areas [1–3]. Indicatively, cities are estimated to consume 75% of the natural resources and emit between 60 and 80% of the global greenhouse gas (GHG) emissions, while occupying 3% of the Earth's land surface [4]. These patterns indicate not only a steep increase in the urban population globally, but also drainage of natural resources and simultaneous increase of pollution caused by urban centres.

The approach of bioclimatic architecture is well established. Bioclimatic design considers the building totally from the stage of its inception as a place of energy exchange between the indoor and the external environment, natural and climatic, as a dynamic structure which utilises the beneficial climatic parameters (solar radiation for winter, sea breezes for summer, etc.), while avoiding the adverse climatic effects (cold winds for winter, solar radiation for summer) [5]. Bioclimatic buildings aim to address three main challenges encountered in urban centres: energy conservation, health and wellbeing of inhabitants and sustainability [6]. In addition to the structural design of the building, determining is the human factor. It was shown that the ability of users to manually operate elements of the building (e.g. windows) improves thermal comfort and energy conservation [7]. The bioclimatic design and functionality of buildings can and should be extended to the urban area surrounding them. The need for such approaches is critical in urban settings, due to

D. K. Serghides (✉) · S. Dimitriou · I. Kyprianou · C. Papanicolas
The Cyprus Institute/EEWRC, Nicosia, Cyprus
e-mail: d.serghides@cyi.ac.cy; s.dimitriou@cyi.ac.cy; i.kyprianou@cyi.ac.cy; president@cyi.ac.cy

A. Sayigh (ed.), *Renewable Energy and Sustainable Buildings*, Innovative Renewable Energy, https://doi.org/10.1007/978-3-030-18488-9_5

the urban heat island (UHI) effect. UHI is a phenomenon characterised by increased urban temperatures relative to the surrounding rural and sub-urban spaces. Some of the impacts of the UHI effect include high energy consumption for cooling, low thermal comfort levels and increased mortality and morbidity rates [8].

The bioclimatic approach has the capacity to improve indoor and outdoor thermal conditions. Components of bioclimatic design with regard to outdoor open spaces include urban canopy and geometry, surface materials utilised as well as vegetation and water surfaces of the locale [9, 10]. Of course, bioclimatic design of outdoor areas has an impact on the energy needs of adjacent buildings. For example, a case study performed in low-income dwellings in Cyprus showed that the orientation of buildings affects indoor temperatures, illustrating that urban geometry can influence the building's energy consumption and indoor thermal comfort [11]. The existing literature on bioclimatic design primarily considers case studies of certain regional characteristics, with the results being applicable mainly to areas with similar attributes. For example, in a refurbishment study, it was estimated that adding movable shading devices to an existing single family house of the Mediterranean region was the optimal solution for energy conservation and economic viability [12]. A similar study in a Cyprus dwelling also suggested external bioclimatic interventions (horizontal overhangs) within a range of measures that could be applied to refurbish an existing dwelling into a nearly zero energy building (nZEB), with a payback period of less than 1 year [13].

The present study will explore bioclimatic applications on an existing urban isle located in the coastal city of Limassol, in Cyprus. The urban isle is defined as an area around public buildings, which makes use of synergies amongst different building functions and scale advantages for energy solutions. The case study for this urban isle is part of the project "Smart bioclimatic low-carbon urban areas as innovative energy isles in the sustainable city" (SUI). This project adopts a threefold approach: Smart Grids (decentralised renewable energy generation, energy storage and distribution), Management Platform (manage and provide optimal energy flow) and Bioclimatic Design and urban planning [14]. Essentially, bioclimatic units of the urban fabric (urban isles) are examined in relation to their inter-connectability and management through an intelligent platform.

5.2 Description of Case Study

The aim of this investigation was to determine the human comfort conditions in the case study of a (public) building and area and calculate their energy efficiency. This section describes the case study building and area, as well as the key performance indicators related to energy consumption and emissions in the area.

5.2.1 The SUI Public Building

The public building under study is a four-floor (ground floor, mezzanine, first floor and second floor) University Administration building constructed in 1979 and renovated in 2010 (Fig. 5.1, left). It is equipped with various services, including approximately 60 workplaces, a lecture room for 30 people and two meeting rooms. It is located in the old town neighbourhood of Limassol, near the seafront, a commercial area with up to seven-storey buildings surrounding it. Although recently refurbished, the building is not insulated. Details on the thermal performance of the envelope elements can be found in Table 5.1.

The case study building's load-bearing structure is reinforced concrete. The main wall structure is reinforced concrete with brickwork, cladded partially outside with travertine marble, whereas the main roof structure is reinforced concrete, polystyrene, cement screed and waterproof membrane. The floor is comprised of reinforced concrete, screed concrete, raised floor with final layer of parquet in the ground floor or fitted carpet. For the simulation with Design Builder, the heating and cooling temperatures were set at 22 °C and 26 °C, respectively, based on the onsite observation and the thermal comfort survey.

Fig. 5.1 (1) Limassol building (left); (2) Limassol urban isle area (inside red circle) (right)

Table 5.1 Limassol building envelope thermal transmittance values

Envelope element	U-value (W/m²K)
Main concrete wall	3.33
Secondary brick	1.18
Roof	1.36
Floor	0.87
Windows—double glazed	2.00
Windows—single glazed	6.00

The building was modelled with both Design Builder and iSBEM software. The latter is the official tool for the energy certification of buildings in Cyprus. The results indicate that the simulation generated by Design Builder gives smaller deviation from the real consumption than the one generated by iSBEM (Fig. 5.2), since Design Builder is a dynamic simulation program.

The breakdown of energy consumption per energy service for the case study is illustrated in Fig. 5.3, as simulated with Design Builder. Currently, the most energy-consuming service is cooling, followed closely by lighting and equipment. Surprisingly, heating is only responsible for about 12% of the total consumption of the building, suggesting that the proposed measures should focus on the more draining energy services of the building.

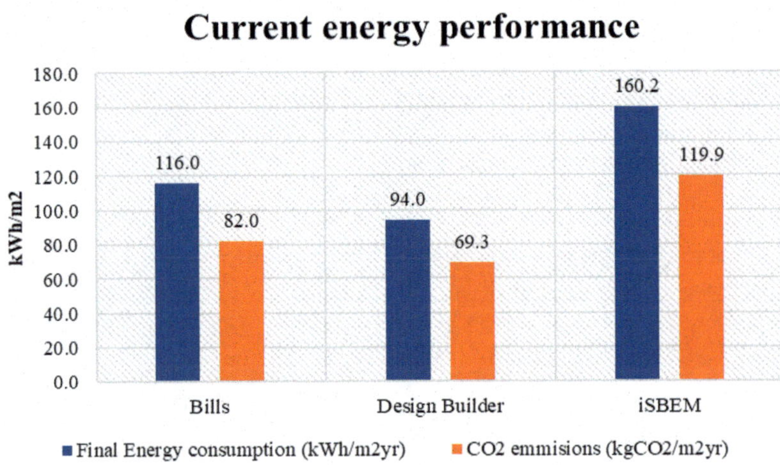

Fig. 5.2 Limassol building current energy performance

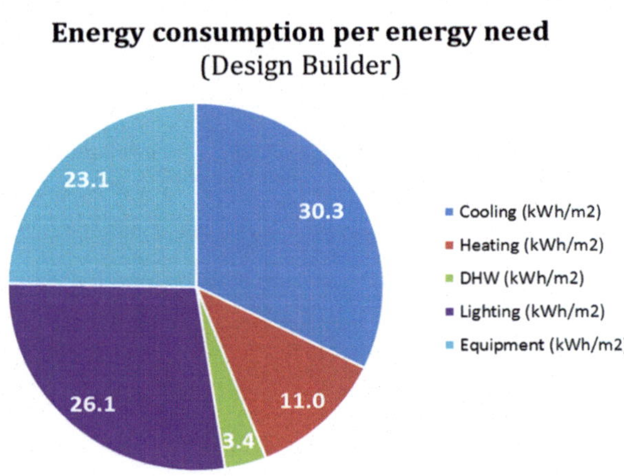

Fig. 5.3 Limassol building energy consumption breakdown (Design Builder)

5.2.2 The SUI Area Buildings

The urban isle is located in the built-up area of the town of Limassol, adjacent to the main commercial street (Anexartisias street) and the Headquarters of the Cyprus University of Technology (Fig. 5.1, right). There are various open spaces near-by, such as squares and parking lots. There are old stone buildings of one and two storeys and newer concrete structures of more than three floors. The area is occupied by retail stores, university buildings, cafeterias and bars.

Real consumption data from a report provided by the Ministry of Commerce, Industry and Tourism [15] were used to determine the energy profile of the 20 non-public buildings of the isle (Table 5.2). From this information, the energy consumption per energy need was also calculated for the entire urban isle (Fig. 5.4). The largest share of energy consumed in the investigated urban isle is attributed to lighting and equipment (jointly 56%), followed by cooling (33%) and finally heating (11%), in line with the energy demands of the specific case study building previously described.

5.2.3 The Key Performance Indicators

As expected, the CO_2 emissions and energy consumption accounted for the case study building and the buildings of the surrounding area do not meet the in-force regulation (Directive 432/2013) regarding the minimum energy performance of

Table 5.2 Energy profiles of the isle buildings

Emissions	Number of properties	20
	CO_2 emissions (ton/year)	451.6
	Average CO_2 emissions per building (kg/year)	22,942
Electricity	Total electricity consumption (MWh/year)	710.9
	Average electricity consumption per building (kWh/year)	35,544

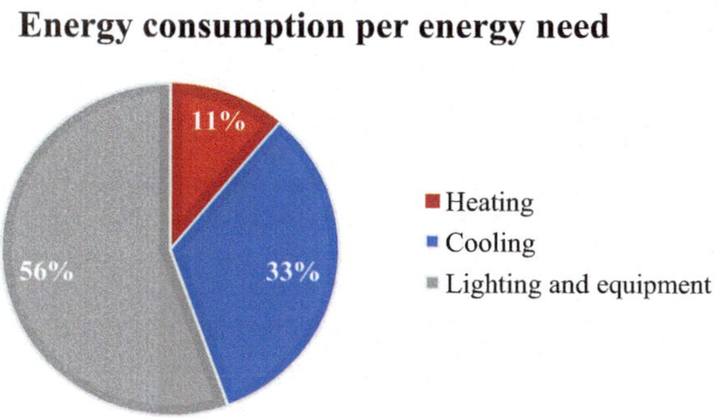

Fig. 5.4 Limassol building energy consumption breakdown (Design Builder)

Table 5.3 Limassol energy profile—CO_2 emissions (left); Limassol energy profile—energy consumption (right)

SUI Energy Profile_ Limassol CO2 emissions (kgCO2/m2)
Building
79
Area
142

SUI Energy Profile_ Limassol final energy (kWh/m2)
Building
115
Area
220

buildings. For the key performance indicators (Table 5.3), the actual energy consumption data were used. The case study building currently accounts for more than half the CO_2 emissions and final energy consumption of the total area.

5.3 Building Energy Retrofitting Scenarios and Results

Cyprus has been a member of the EU since 2004 and is therefore bound to comply with the EU legislation. From 2006 onwards, the Republic of Cyprus has been publishing Laws/Directives/Regulations/Rules and National Action Plans concerning the minimum energy requirements to be fulfilled by all new buildings and those falling under the category of "major refurbishment". The legislation follows the Directive 2002/91/EC of the European Commission and after 2010 the Directive 2010/31/EU of the European Parliament on the energy performance of buildings.

The effectiveness of individual measures was explored, and a Smart SUI scenario was developed for the case study building. For the energy improvement scenarios and simulations of the Limassol case study building, iSBEM was used to calculate the energy performance of the building. Moreover, Design Builder was used to evaluate the energy consumption and performance of the case study building before and after the energy refurbishment measures.

In addition, measures towards the improvement of outdoor thermal comfort were investigated. The improvement of outdoor thermal comfort was promoted though these measures, which lead to decrease in the energy cooling needs. ENVI-met was used with Biomet, as a post-processing tool to calculate human thermal comfort indices for the simulation data of ENVI-met, and the results were visualised with Leonardo.

5.3.1 *Energy Retrofitting Scenarios*

The minimum energy performance requirements towards an nZEB, as specified by the National Legislation, reduce effectively the heating need (≈90%), but fail to address as effectively the cooling need (≈50% reduction), while lighting and equipment are even less affected. The same shortcomings regarding cooling reduction apply for the Label B Scenario refurbishment. The energy retrofitting scenarios were developed taking into consideration the high energy consumption due to cooling and lighting, which are not effectively addressed by the typical nZEB and Label B refurbishments. The energy and cost-effectiveness of the measures below aiming to combat the aforementioned were investigated on an individual basis.

5.3.1.1 The Optimal Building Retrofitting Scenario

The energy refurbishment measures below, related to different envelope elements, are used as variables, initially to be examined individually in simulations using the Design Builder. Consequently, the most effective measures are selected to build an optimised scenario, termed "the SUI scenario".

In order to determine which measures to include in the SUI scenario, the measures must also be examined in terms of cost-effectiveness. The most cost-effective measures with a payback period less than 15 years, in the order of most to least effective, are:

- Add window film shade in the single glazed windows (5 years): A window film is added on the external surface of the single glazed windows to intercept the solar heat from entering the building. According to simulations, the total solar transmission value decreases sevenfold when the solar film is applied. The result is 1.6% reduction in energy needs and 2.6% reduction in CO_2 emissions.
- Substitution of the existing bulbs with LED (10 years): When substituting the existing lighting with LED technology, approximately 76% lower normalised power density (irradiance) is simulated. This results in 4.4% reduction in energy demand and 5.3% reduction in CO_2 emissions.
- Control the indoor temperature (12 years): The comfort temperature range is increased from 22–26 °C to 19–27 °C, decreasing the heating set point by 3° and increasing the cooling set point by one [7]. The temperatures are controlled using a programmable Thermostat, and this measure results in 4.7% reduction in energy needs and 2.6% reduction in CO_2 emissions.
- Cool paint for the roof (15 years): Painting the roof with a cool paint prevents the building from overheating and results in almost 1% reduction in energy consumption and CO_2 emissions.

The above measures were combined in order to build the SUI scenario. The results of the simulations of this scenario are presented in Fig. 5.5.

The energy consumption of the case study building is reduced by approximately 12% according to the SUI scenario, similar to the cutbacks in CO_2 emissions. As for the individual energy services of the building (Fig. 5.6), the highest percentage of energy savings occurs from cooling (19%), followed by lighting (16%) and then heating (12%). There is no RES contribution in this scenario, and the payback period is estimated to be 10 years.

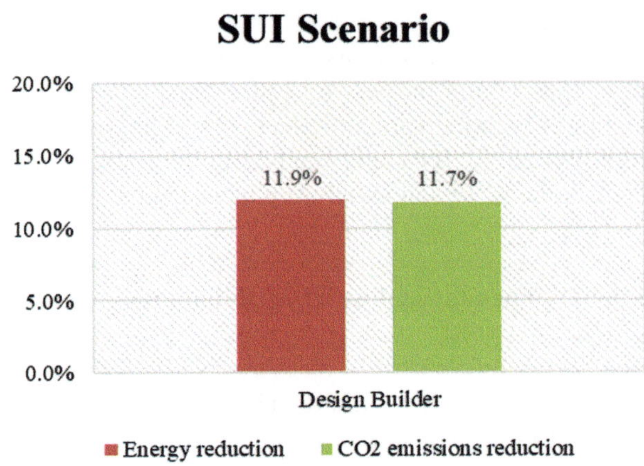

Fig. 5.5 Limassol building SUI scenario—percentage energy and CO_2 reduction (DesignBuilder simulation)

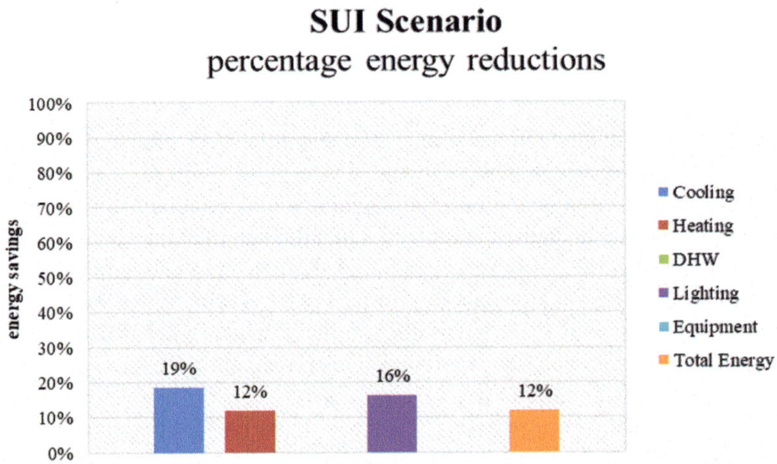

Fig. 5.6 Limassol building SUI Scenario—percentage energy savings (DesignBuilder simulation)

5.3.1.2 Proposed Surrounding Area Retrofitting Scenario

During a typical summer day, the PMV (predicted mean vote—methodology that defines thermal comfort standards) does not fall below +2, ranging from +2.1 to +5.7, being over +4 (classified as extremely hot) at 2 pm. The most compromising time to be outside is between 1 and 2 pm.

A scenario for the enhancement of the outdoor comfort was developed, using replacement of asphalt and pavement with granite single stones and doubling the number of trees in the urban isle (Fig. 5.7). The results indicate 0.8 K reduction in air temperature on average and expected decrease in cooling demand by 1.5%–2.0% [16].

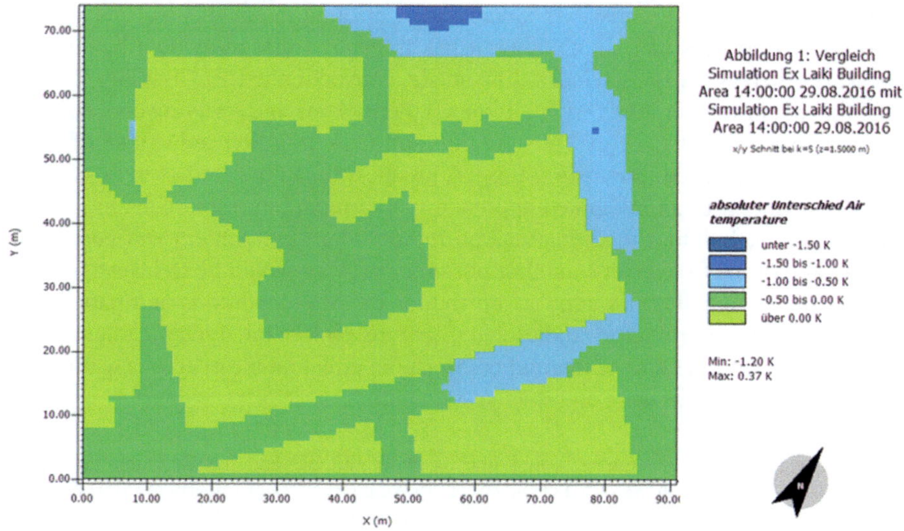

Fig. 5.7 Absolute difference in temperature in winter at 2 pm—comparison before and after the measures

5.3.2 Bioclimatic Scenarios Combined Impact

The SUI refurbishment of the public building can lead to approximately 28 MWh yearly reductions in its energy consumption. With the outdoor area retrofitted, a reduction of 2% is expected on the cooling demand of all the SUI buildings. This will further decrease the energy consumption of the isle buildings by 5830 kWh per year. The subsequent total reduction in CO_2 emissions will be up to almost 24 tonsCO_2/year, which corresponds to an overall percentage decrease of 4%.

5.4 Conclusion

The current case study of the urban isle with the buildings has indicated that major energy consumption patterns are observed in the cooling, lighting and equipment services of the constructions in Cyprus. Therefore, energy refurbishments in Cyprus must focus primarily on minimising the cooling needs of the buildings. The results of the study show that addressing these energy draining issues by following the directives for the nZEB and Label B certification as guidelines is not cost-effective. Furthermore, the results demonstrate that the buildings can reduce their energy consumption by means of easily applicable, bioclimatic, cost-effective measures. Improving the energy efficiency of the buildings in combination with the outdoor materiality induces higher, more generalised and permanent reductions in energy consumption and CO_2 emissions. Specifically, through the analysis of the various developed scenarios and simulations, it has transpired that individual bioclimatic measures such as applying window film shade, conversion to LED lighting, controlling the indoor temperatures with automated thermostats and applying cool paint on the roof can result in energy consumption and CO_2 emission reductions of more than 10%, in this Mediterranean setting. A prospect for future studies in this field is the development of an integrated approach with a model that simulates the effects of bioclimatic building modifications on the outdoor environment and pedestrian thermal comfort. Conversely, another possible pathway would be to investigate the impact of outdoor thermal comfort on indoor thermal conditions and patterns of energy consumption. Future studies can therefore explore the dynamic relationship between indoor and outdoor thermal comfort, the effect on the energy consumption and the consequent bearing on urban sustainability.

Acknowledgements The authors would like to thank the EU and the Research Promotion Foundation for funding, in the frame of the collaborative international Consortium (SUI) financed under the ERA-NET Cofund Smart Cities and Communities Cofunded Call. This ERA-NET is an integral part of the 2014 Joint Activities developed by the Joint Programming Initiative Urban Europe (JPI Urban Europe).

References

1. UN (2018) World Urbanization Prospects: The 2018 Revision
2. Grubler A, Bai X, Buettner T, Dhakal S, Fisk DJ, Ichinose T et al (2012) Urban energy system. Cambridge University Press, Cambridge
3. Friedman K, Cooke A (2011) City versus national energy use: implications for urban energy policy and strategies. Procedia Eng 21:464–472. https://doi.org/10.1016/j.proeng.2011.11.2039
4. UNEP (2012) Global initiative for resource efficient cities: engine to sustainability. https://europa.eu/capacity4dev/unep/document/global-initiative-resource-efficient-cities-brief
5. Serghides D (2010) The open access. Open Construct Build Technol J 4:29–38
6. Metallinou VA (2006) Ecological propriety and architecture. In: Eco-architecture. Harmonisation between architecture and nature, WIT Transactions on The Built Environment. p 15–22. doi:https://doi.org/10.2495/ARC060021
7. Serghides D, Dimitriou S, Kyprianou I, Papanicolas C (2017) The adaptive comfort factor in evaluating the energy performance of office buildings in the mediterranean coastal cities. Energy Procedia 134:683–691. https://doi.org/10.1016/j.egypro.2017.09.588
8. Santamouris M (2016) Innovating to zero the building sector in Europe: minimising the energy consumption, eradication of the energy poverty and mitigating the local climate change. Sol Energy 128:61–94. https://doi.org/10.1016/j.solener.2016.01.021
9. Chatzidimitriou A, Yannas S (2016) Microclimate design for open spaces: ranking urban design effects on pedestrian thermal comfort in summer. Sustain Cities Soc 26:27–47. https://doi.org/10.1016/j.scs.2016.05.004
10. Karakounos I, Dimoudi A, Zoras S (2018) The influence of bioclimatic urban redevelopment on outdoor thermal comfort. Energ Buildings 158:1266–1274. https://doi.org/10.1016/j.enbuild.2017.11.035
11. Chatzinikola C, Serghides DK, Santamouris M (2015) Effect of the orientation on the winter indoor temperatures and on the energy consumption of low-income dwellings in the Mediterranean region. In: International Conference with Exhibition S.ARCH—Environment and Architecture, Montenegro p 1–9

12. Serghides DK, Dimitriou S, Katafygiotou MC, Michaelidou M (2015) Energy efficient refurbishment towards nearly zero energy houses, for the Mediterranean region. Energy Procedia 83:533–543. https://doi.org/10.1016/j.egypro.2015.12.173
13. Serghides DK, Michaelidou M, Christofi M, Dimitriou S, Katafygiotou M (2017) Energy refurbishment towards nearly zero energy multi-family houses, for Cyprus. Procedia Environ Sci 38:11–19. https://doi.org/10.1016/j.proenv.2017.03.068
14. Urban Europe, Smart Urban Isle. Accessed 23 May 2018. https://jpi-urbaneurope.eu/project/smart-urban-isle/
15. MECIT (2017) 4th National Action Plan for Energy Efficiency of Cyprus (in Greek), Nicosia. http://www.mcit.gov.cy/mcit/EnergySe.nsf/All/EF97759A9580E25EC22581C500345706/$file/4oΕΣΔΕΑ_19_09_17.pdf
16. Akbari H, Pomerantz M, Taha H (2001) Cool surfaces and shade trees to reduce energy use and improve air quality in urban areas. Sol Energy 70:295–310

Chapter 6
Plug&Play: Self-Sufficient Technological Devices for Outdoor Spaces to Mitigate the UHI Effect

Alessandra Battisti, Flavia Laureti, and Giulia Volpicelli

6.1 Introduction

This contribution regards the technological innovation and the transformation of outdoor spaces within the consolidated city where the theme of the research focuses on environmental recovery practices and adaptation to climate change in the open urban public space in the Mediterranean area.

In particular, certain innovative technologies are investigated, verifying their effectiveness and impact on environmental sustainability within the urban tissue [1], with the ultimate aim of ensuring the user's thermo-hygrometric comfort. in line with directions of "strategies for the adaptation to and mitigation of climate change" defined by the Europe-ADAPT Platform (2012), URBACT III (2014–2020), and the UHI Project (2011–2014), according to the directions of Target 20-20-20 and of the Paris Agreement (COP21, 2015), ratification national law 204/2016 and UN framework.

The research presented focuses on different tools for monitoring, mapping, and modeling (Ecotect and Envimet software), probing the relationship between morphology, urban microclimate, and well-being, and analyzing the relationship between urban design and urban climatology at the micro, meso, and macro scales, especially as regards the adoption of strategies to adapt to the climate and mitigate the phenomenon of urban heat island (UHI) [2], particularly felt in Mediterranean cities characterized by consolidated and dense historical urban tissues [3].

Furthermore, attention is focused on social aspects of urban quality, which, especially in the Mediterranean area, are deeply linked to outdoor thermal comfort [4], an area where spaces with good thermal-hygrometric well-being encourage community living, promoting, where well-designed, urban identity and sociality among the inhabitants.

A. Battisti (✉) · F. Laureti · G. Volpicelli
Università Degli Studi di Roma Sapienza, Rome, Italy
e-mail: alessandra.battisti@uniroma1.it; flavia.laureti@uniroma1.it

A. Sayigh (ed.), *Renewable Energy and Sustainable Buildings*, Innovative Renewable Energy, https://doi.org/10.1007/978-3-030-18488-9_6

A further technical direction of ongoing transdisciplinary research [5] is on materials, innovative technologies, and paving and covering components that concern the relationship between morphological configuration of spaces, characteristics of materials, comfort, and energy savings associated with them.

The importance of integrating the microclimatic analysis into a design phase for a better understanding of the formal choices, configuration of the components of urban design spaces, and technological choices is emphasized [6].

The aim is to bring greater awareness in urban redevelopment practices of open spaces, in light of the current environmental and social crisis, and, through the in-depth study of certain technologies, to test their effectiveness and open a broader reflection on the energy flows of the outdoor space, on the behavior of materials, and on a probable relationship (pattern) that binds material and surface in the in-between space [7].

By means of simulations in Envimet software, a comparative analysis between two scenario solutions was done, taking into account mean radiant temperature, relative humidity, wind velocity, and hydrological variables, in particular, in light of the concept of energy redevelopment and the Italian scenario of the consolidated and historical tissues. Simulations of a typical summer day in Rome La Sapienza (15 July) were assessed to determine UHI reductions for two different scenarios with roof and pavements realized in green or cool materials and with the use of blue-green infrastructures, in order to evaluate pedestrians' comfort levels and the effectiveness of selected mitigation techniques for the investigated case.

The numerical simulation results confirm that an appropriate selection of pavements and roof materials and green infrastructures contribute to the reduction of the negative effects of UHI, but they cannot reduce the perceived temperature by more than 4°. Therefore, to try to offer spaces of comfort and well-being, Plug&Play systems with radiant panel floors and roofs have been designed, a sort of "well-being urban islands" that can offer the inhabitants thermal comfort areas.

6.2 Case Study: La Sapienza

The case study chosen is the University City of La Sapienza of Rome (coordinates 41° 54′10.32″ N 12° 30′51.72″ E, 45m altitude), with a Mediterranean climate, belonging to the dry-subtropical climate [8] with intermediate seasons, spring and autumn, characterized by mild and warm temperatures and concentrated rainfall during the months between November and April. The winter season is mild, while the summer season is warm and humid, characterized by low rainfall.

The University City of La Sapienza of Rome (about 44 ha) is especially important within the city of Rome for its cultural, historical, social, and architectural significance [9]. Furthermore, it is located in the central district of San Lorenzo, near the historic city center and close to the Roma Termini railway station and the Verano Monumental Cemetery, which with its 83 ha means a coherent green area for the microclimate of the district. Finally, the consolidated fabric of La Sapienza represents a masterpiece of modern Italian architecture whose energy efficiency must be improved as foreseen in the 2016–2021 University Strategic Plan [10].

Over the period since the foundation (1938) of the University "La Sapienza" of Rome to date, progressive expansion of the urbanized areas to its surroundings has led to a strong acceleration of the soil sealing process, implemented with artificial materials such as asphalt or concrete used for the construction of buildings and settlement roads. The climate surveys show that around the city of studies in the last 50 years, the climate has become warmer due to the less vegetal transpiration and evaporation and the wider waterproofed surfaces. In fact, despite the parts with vegetal covering and their evapotranspiration, the heat produced by the air conditioning in synergy with the city traffic and with the absorption of solar energy by dark surfaces in asphalt and concrete contributes to the local climate changes of La Sapienza, causing in summer the effect of "urban heat island" (UHI). The research project took as its starting point the current condition of the building heritage of the university campus by defining a framework of outdoor interventions aimed at transforming it into a resilient university city, able to protect its inhabitants—students and professors in charge of the work—the soil, and internal infrastructure from the risks of the UHI [11]. The study focuses on two areas of the university: the central avenue, Viale Piero Gobetti in the south-east area, which goes from the main entrance of Piazzale Aldo Moro to Piazzale della Minerva (Fig. 6.1), where the Aula Magna and the Rectorate are located. In the first instance, the materials that make up the intermediate spaces of the built environment are analyzed, and then two different scenarios (Figs. 6.2 and 6.3) for the mitigation of the effect of the urban heat island are developed through the design renovation of floors and building roofs with cool materials [12], PCM, and thermal reflective materials in combination with vegetation use, green infrastructure, and blue strategies, verifying through the

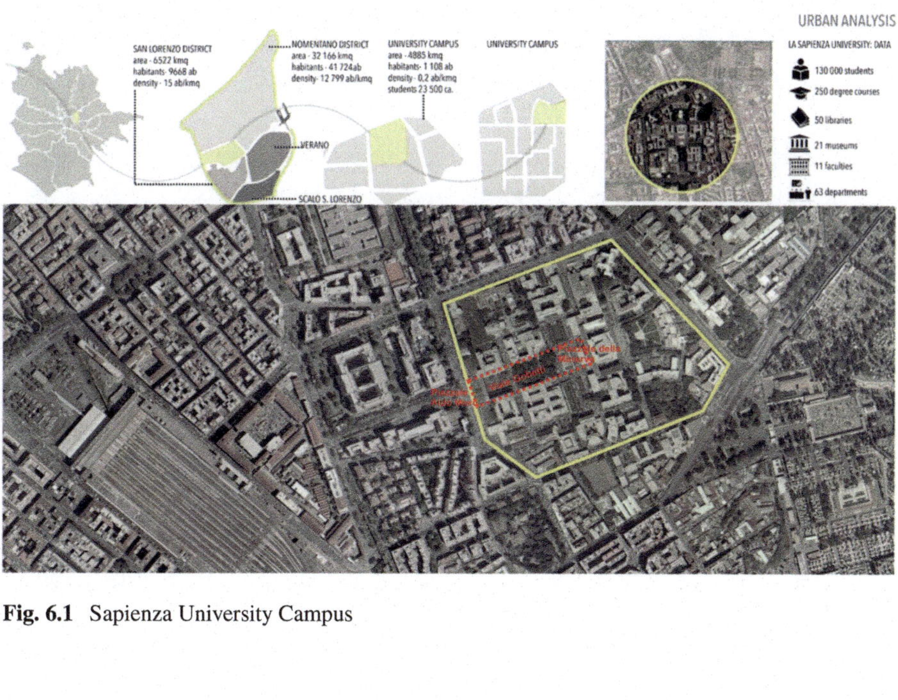

Fig. 6.1 Sapienza University Campus

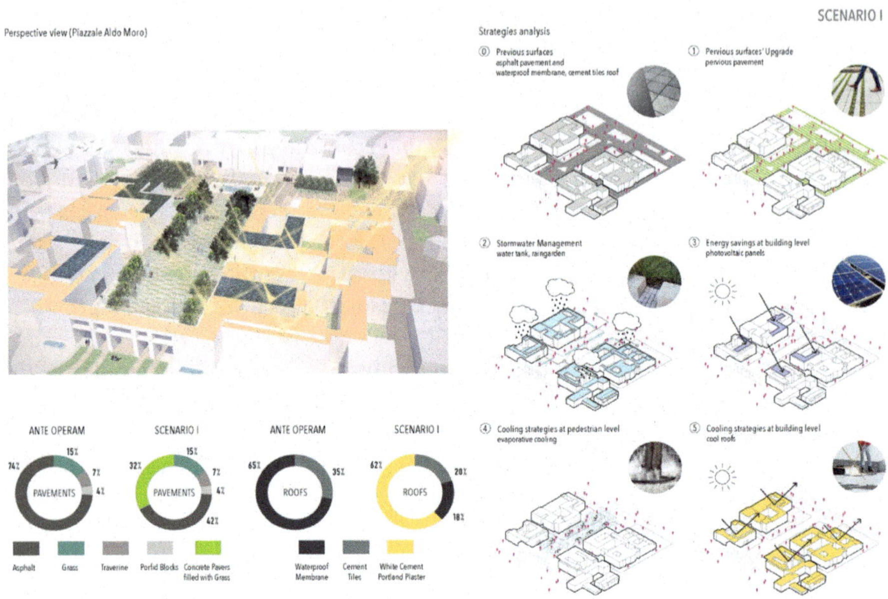

Fig. 6.2 First scenario proposed

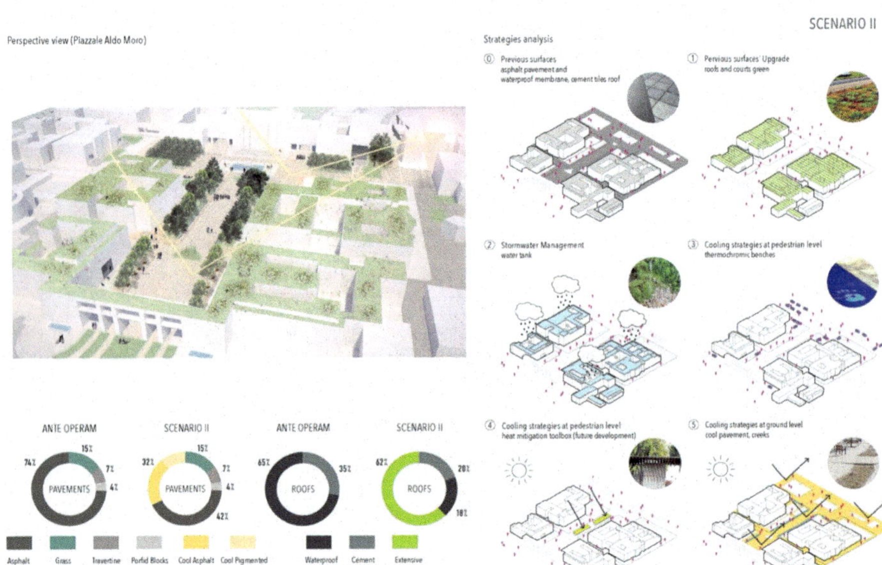

Fig. 6.3 Second scenario proposed

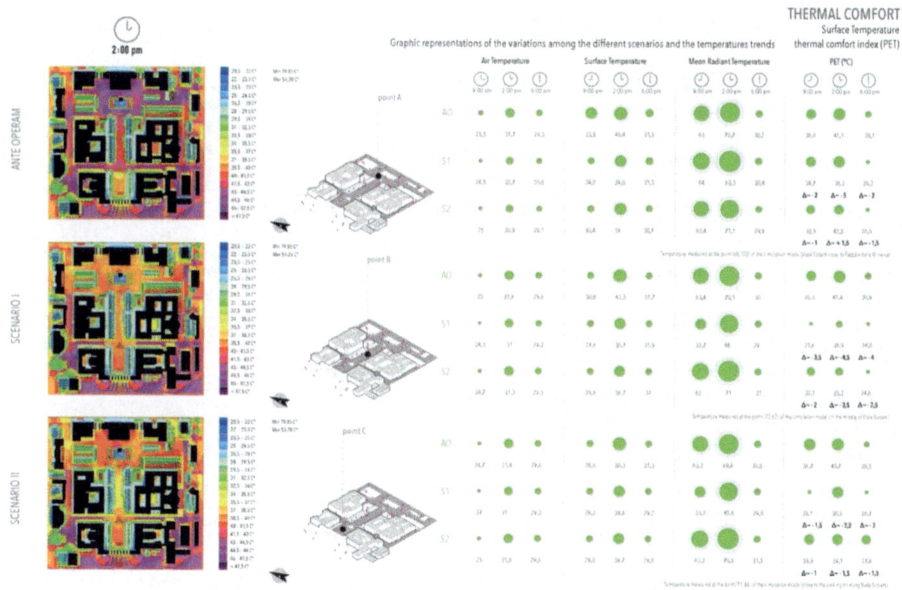

Fig. 6.4 Comparative analysis between two scenario solutions

Envimet software the different microclimatic conditions and the external thermal comfort of pedestrians in the *ante operam* scenario in comparison with the results of two different scenarios proposed (Fig. 6.4).

6.3 Mitigation and Adaptation Measures

The solar radiation analysis of a typical summer day in Rome La Sapienza (15 July) developed by Ecotect software shows many urban open spaces within the University Campus exposed to direct solar radiation, especially at 2:00 p.m. It is important to underline how the presence of a diffuse system of trees and vegetated area can improve the shadow effect on pedestrian spaces, especially early in the morning (8:00 a.m.) and during the evening (7:00 p.m.). Nevertheless, the effect of trees and greenery is local, and their shadow patterns are not so relevant for what concerns Piazzale della Minerva, whereas for the main central boulevard, Viale Gobetti, where the presence of different tree typologies along the sideways must be considered a good solution for the *ante operam* condition, the initial configuration results in a varied distribution of the overlapping shadows (Fig. 6.5).

In the *ante operam* scenario in Piazzale della Minerva, the Tair value distribution shows a relevant presence of higher values (+31 °C) in those areas with higher SVF and characterized by impervious and low albedo materials such as asphalt and porfid blocks in Viale Gobetti.

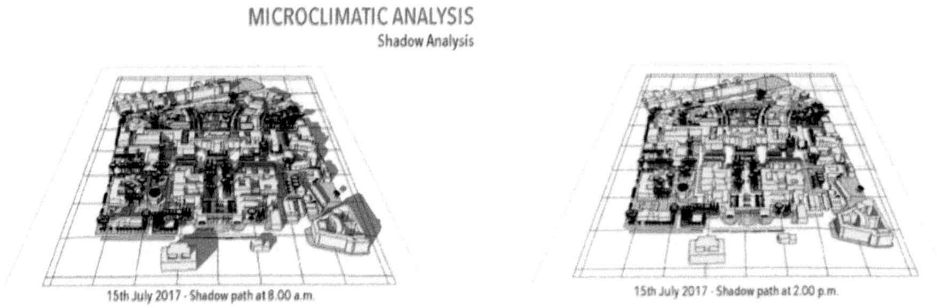

Fig. 6.5 Shadow analysis

The entrance to the central boulevard shows higher Tair values (+31 °C). The presence of rows of trees has most likely a local effect for the Tair, whereas a better performance can be observed for those areas shadowed by a building (Fig. 6.4).

Based on the existing knowledge thus, several questions concerning mitigation and adaptation strategies for counteracting UHI and climate change risks have been raised from this special urban design perspective [13]. In particular, in light of the concept of energy redevelopment and the Italian scenario of the consolidated and historical tissues, the historical heritage view, both as single building and as of district network, seems to be an intriguing field of experimentation [14].

The present study aims to investigate among the numerous strategies, a meta-design framework, in particular the two proposed scenarios that follow two criteria: the minimal intervention and the combination of green-blue and smart material technologies, in order to make urban open spaces more resilient to climate change and UHI and increase thermal outdoor comfort for the creation of mixed and vibrant public spaces [15].

The outdoor spaces were designed with the aim to promote the vitality and sociality among students, and to this end, external spaces and rest areas have morphological properties that favor psychophysical well-being.

In the proposed synergistic vision, the redevelopment intervention of the public open space represents an activity that is incorporated by a very dense context renewing itself through continuous contextual micro-decisions in spatially open areas, according to a process of "urban acupuncture" [16]. The two different scenarios are configured as a meticulous work of recombination and manipulation of paving, green and water that allow to identify specific and flexible solutions oriented to well-being.

In order to mitigate climate change phenomena, the two proposed design scenarios among the geo-engineering-based solutions used high reflective materials, also referred to as "cool materials" for their capacity to present low surface temperatures, retro-reflective materials, and phase change materials (PCM) [17].

As second type of mitigation and adaptation measures, the two scenarios (designed with different solutions and ratio in the two) present green infrastructures that offer different benefits related to clean air, natural ventilation, and the psychological effects. They comprise the increase of vegetated surfaces, such as trees, green roofs, and green spaces.

In addition, they include "urban blue spaces" based on the evaporative cooling. Considering the minimal intervention measures for this historical context, there were used spray fountains and water mist cooling systems. In particular, direct cooling with mist is expected to relieve thermal stress through two main routes: reducing the air temperature and light wetting of the skin with evaporation, both effects promoted by the forced convection from the fan.

6.4 Plug&Play Systems

The numerical simulation results confirm that an appropriate selection of pavements and roof materials and green infrastructures contribute to the reduction of the negative effects of UHI, but they cannot reduce the perceived temperature by more than 4° (Fig. 6.4).

In addition to the definition of a meta-design framework, the research proposes a micro-architecture, self-sufficient from the energy point of view, able to contribute to the mitigation of the UHI phenomenon, thanks to the creation of a functional space serving the university community whose internal microclimate acts on Universal Thermal Comfort Index (UTCI) values. This Thermal Active Plug&Play system offers spaces of comfort and well-being in summer and winter with radiant panel floors and ceilings, a sort of "well-being urban islands" that can offer the inhabitants thermal comfort areas.

The architectural device is characterized by a smart floor or roof through which the production and storage of electricity required for the operation of the air handling system for heating and cooling of indoor spaces are ensured.

The integrated technological system, called Plug&Play islands, combines passive mitigation strategies such as shadowing and natural ventilation, with an electric heating/cooling system powered entirely by a renewable energy source such as solar radiation (Fig. 6.6).

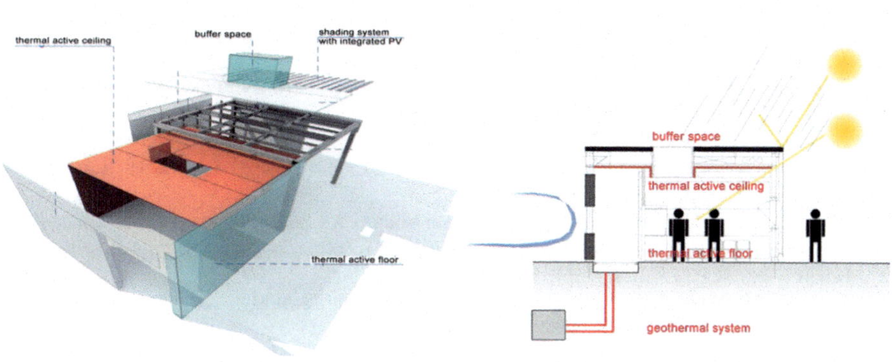

Fig. 6.6 Plug&Play self-sufficient island

6.5 Conclusions

The integrated design of the technological system allows to locate the mechanical components in the outer layer of the roof and on the floor, ensuring ease of installation, repair, and monitoring. In addition, the Thermal Active Plug&Play system aims to expand the traditional concept of the roof and floor, from a simple barrier of protection against atmospheric agents to an active device able to produce energy and control indoor comfort. The micro-architecture proposed, thanks to the prefabrication and integration of the components, can be applied in different urban scenarios to ensure simple functionality (e.g., outdoor study spaces, info point, small shops) and to provide thermal comfort. The benefits of Thermal Active Systems are not just about energy saving and well-being but concern the concept of innovative design of small buildings which are cost-effective and psychologically convenient and represent a new flexible and durable model for a sustainable architecture.

References

1. Santamouris M (2007) Heat island research in Europe: the state of the art. Adv Build Energy Res 1(1):123–150
2. Nikolopoulou M-H, Baker N, Steemers K (2001) Thermal comfort in outdoor urban spaces: understanding the human parameter. Sol Energy 70:227–235
3. Martinelli L, Lin T-P, Matzarakis A (2015) Assessment of the influence of daily shadings pattern on human thermal comfort and attendance in Rome during summer period. Build Environ 92:30–38
4. Emmanuel MR (2005) An urban approach to climate-sensitive design: strategies for the tropics, 1st edn. Spon Press, London and New York
5. Shahidan MF, Jones PJ, Gwilliam J, Salleh E (2012) An evaluation of outdoor and building environment cooling achieved through combination modification of trees with ground materials. Build Environ 58:245–257
6. Aljawabra F, Nikolopoulou M (2010) Influence of Hot arid climate on the use of outdoor urban spaces and thermal comfort: Do cultural and social backgrounds matter? Intellig Build Int 2:198–217
7. Erell E, Pearlmutter D, Williamson TJ (2011) Urban microclimate: designing the spaces between buildings. Earthscan, London and Washington, DC
8. Koeppen W, Geiger R (1936) Hundbuch der Klimatologie, Verlag von Gebrueder Borntraeger, Berlin
9. https://www.uniroma1.it/en/pagina/strategic-plan
10. Piano strategicolSapienza Università degli Studi di Roma. Accessed 4 Jun 2018. https://www.uniroma1.it/it/pagina/piano-strategico
11. Salata F, Golasi I, Treiani N, Plos R, de Lieto Vollaro A (2018) On the outdoor thermal perception and comfort of a Mediterranean subject across other Koppen-Geiger's climate zones. Environ Res 167:115–128
12. Zinzi M (2010) Cool materials and cool roofs: potentialities in Mediterranean buildings. Adv Build Energy Res 4(1):201–266
13. Yang J, Mohan Kumar DI, Pyrgou A, Chong A, Santamouris M, Kolokotsa D, Lee SE (2018) Green and cool roofs' urban heat island mitigation potential in tropical climate. Sol Energy 173:597–609

14. de Lieto Vollaro R, Evangelisti L, Carnielo E, Battista G, Gori P, Guattari C, Fanchiotti A (2014) An integrated approach for an historical buildings energy analysis in a smart cities perspective. Energy Procedia 45:372–378
15. Gehl J (2010) Cities for people. Island Press, Washington, DC
16. Lerner J (2003) Acupuntura urbana. Editora Record, Rio de Janeiro
17. Pisello AL (2017) State of the art on the development of cool coatings for buildings and cities. Sol Energy 144:660–680

Chapter 7
Smart Homes and Regions as Building Blocks for 100% Renewable Energy Globally

Winfried Hoffmann

7.1 Introduction

Only a few years ago it was difficult to imagine a world powered 100% by renewables only—not only electricity, but also all secondary energy needs, including transportation, heating, and cooling, as well as industry needs. The major reason for this was that the production cost for electricity from wind and solar was much too high and electricity storage was not seen as an option at all. Only with large centralized solar concentrating systems together with hydrogen production (hydrogen economy) or a worldwide distribution system (using high voltage DC current) a renewably powered world was seen potentially possible in the long run.

Until today the energy world was top-down: few big and centralized power stations produce the base load power which is transported with a high-voltage grid network. Peak power stations provide the electricity needs for all consumers at any time. This situation is shown in the left graph of Fig. 7.1. With the rapidly declining cost per produced kWh from PV and wind, together with the push from the automotive sector for Li-ion batteries which also results in a quick reduction of electricity storage cost, there is now—including IoT (Internet of Things)—a new horizon opening how to serve mankinds' energy needs rather quickly. This enables a complete new situation: the consumers become also producers of energy, and most of the needs can be served locally; centralized systems—e.g., large off-shore wind and/or big PV farms—are only needed for the energy intensive industries. This is summarized in the right graph of Fig. 7.1.

W. Hoffmann (✉)
Hanau, Germany
e-mail: winfried@hoffmann-ase.de

A. Sayigh (ed.), *Renewable Energy and Sustainable Buildings*, Innovative Renewable Energy, https://doi.org/10.1007/978-3-030-18488-9_7

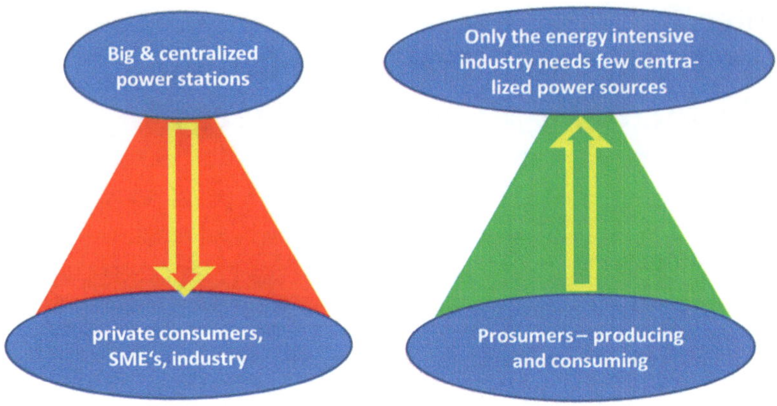

Fig. 7.1 Old top-down (left) and new bottom-up (right) energy world

7.2 Three Building Blocks

7.2.1 First Building Block - Smart Home

Individual houses, offices, and stores are seen as **the first building block**, where on-site as much as possible renewable energy is produced by PV and together with batteries (k-/MWh) self-consumption is as high as possible. With proper insulation and heat pumps (also including thermal modules where appropriate), the heating and cooling energy needs are minimized. This situation is summarized in Fig. 7.2. The integration of a small battery increases significantly the self-consumption in individual homes from ~20% without up to 80% (if an electric car is integrated) on an annual basis. It is interesting to note that the battery not only shifts PV electricity from the afternoon to the evening hours, but also can shift cheap night electricity - as long as it is available - to the more expensive morning peaks as seen in Fig. 7.3.

Next step is the integration of adjacent buildings for exchanging energy. As the number of communicating buildings increases, the self-consumption of the quartier is also increasing and always higher than the sum of the individual homes. This concept is already in place in Italy and Switzerland, and many others have still bureaucratic hurdles for not allowing this useful and simple procedure.

7.2.2 Second Building Block - Smart Energy Region

The second building block is the municipal community which transforms their territory into a smart energy region (Fig. 7.4). They will run in the nearby region wind mills to leverage the seasonal variations from wind and solar. Also they

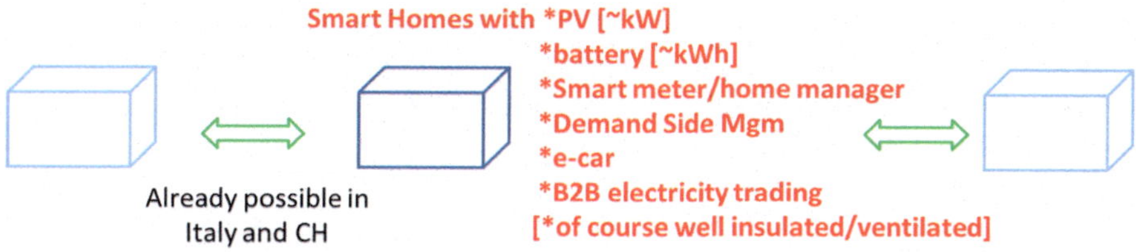

Fig. 7.2 First building block consists of individual and adjacent smart homes (family houses, offices, Small and Medium Enterprises (SMEs))

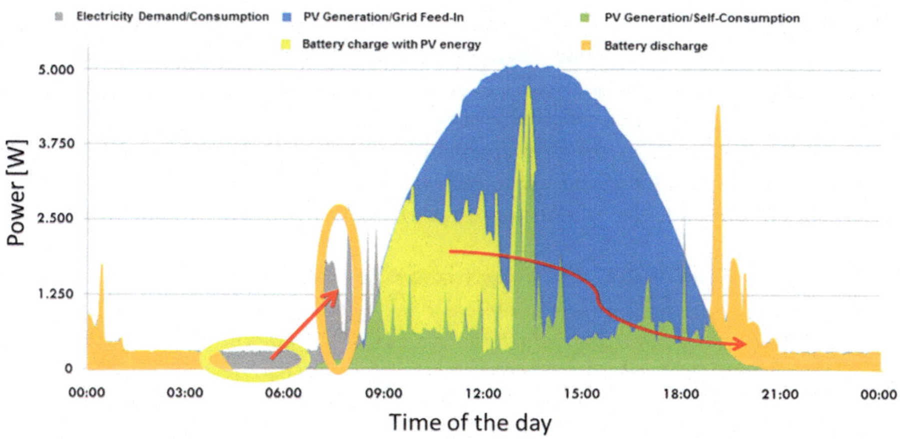

Fig. 7.3 Load profile for a German home and PV—profile (blue) [1]; shift of PV electricity from morning/afternoon hours to evening peaks and night electricity to morning peaks by battery storage is indicated (red arrows)

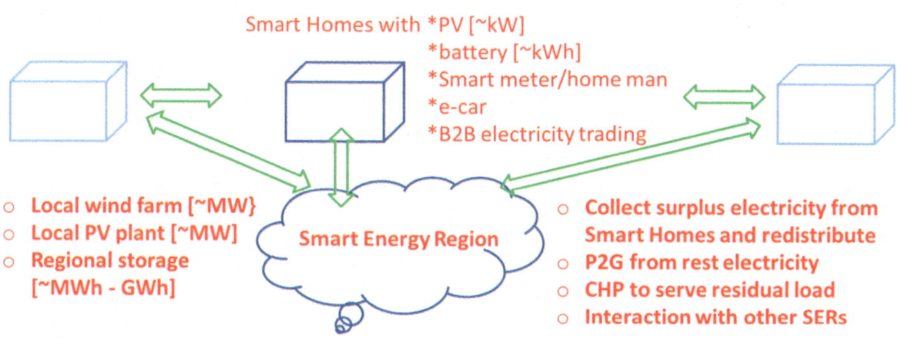

Fig. 7.4 Second building block consists of smart energy regions

collect the surplus of not self-consumed electricity to store in larger battery systems (M-/GWh) and redistribute in their area. A small rest of energy which can still not be satisfied cost efficiently by the described steps can be provided by P2G (power to gas), run by surplus of wind and PV electricity. The example of Germany, where

150 energy regions have already firm plans to cover their energy needs by renewables, only shows that beside the technological components, there are also political will and support from many citizens that the described path is not an idea only but already taken in reality.

7.2.3 Third Building Block - Smart Country

The third building block is the communication with adjacent municipalities/smart energy regions first and on a national level the integration of powering the needs for the big industries by renewables only. The result of this will be a smart country. As shown for Germany, there are well-documented studies (e.g., from the Fraunhofer ISE) which conclude that a renewable scenario for heat and electricity is (1) possible by 2050, (2) does not cost more than the business as usual case but offers much more internal value-added, and—last, but not least—(3) helps the planet to survive without CO_2 emissions from transportation, electricity production, and heating and cooling needs.

The three building blocks are summarized schematically in Fig. 7.5.

The three building blocks are complemented by GW-scale off-shore wind and/or PV farms for energy-intensive industries (e.g., *steel, aluminum, silicon) resulting in a smart country.*

In order to minimize the needed storage capacity for a region (or country), it is important to have a locally dependent ratio of PV and wind systems installed. The monthly variation for a unit power of PV and wind for Germany is shown in Fig. 7.6. While in winter months there is high output for wind and low for PV, the situation reverses in summer months. If we add the monthly power of wind and PV, there is

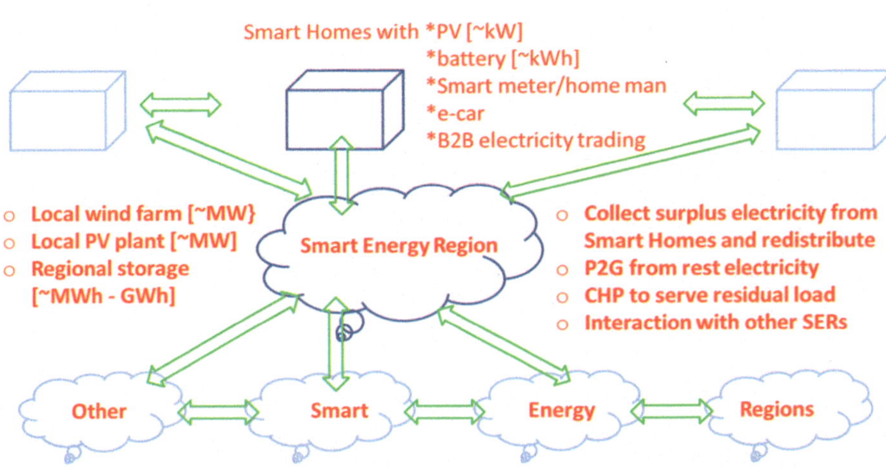

Fig. 7.5 Third building block is characterized by communication between adjacent smart energy regions

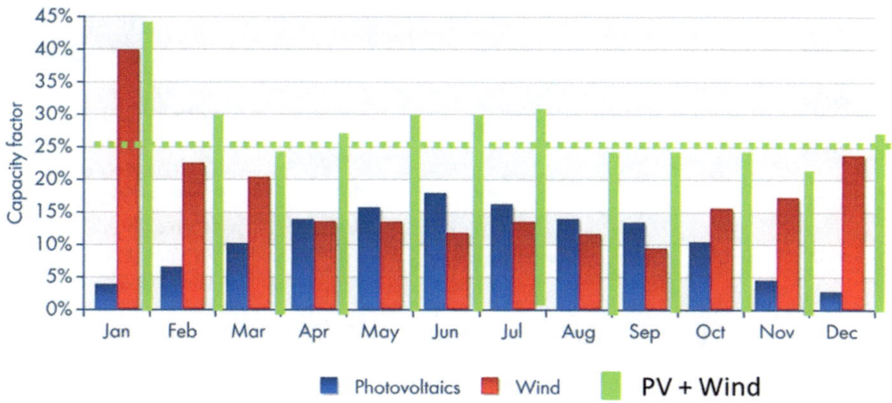

Fig. 7.6 Average monthly grid feed-in of equal PV and wind power in Germany [1]

a balanced monthly power output throughout the year. The consequence is that we do not need a seasonal storage but only a balance within a month. With one unit of power (W), we produce one unit of energy (kWh) for PV and two units for wind. Hence, the ~600 TWh electricity used today in Germany could be produced with 250 GW PV (=250 TWh) and 200 GW wind (=400 TWh). The surplus of 50 TWh is foreseen to balance the losses for storage and P2G. In the future, when mobility is also fuelled by electricity (or fuel made by electricity), these numbers may increase by 50%, maximally double. The sun belt region worldwide, where most people will live in the coming decades, will favor even more PV relative to wind, due to the much less monthly variation and ~doubling of energy output for PV (~2 kWh/W_{PV}).

7.3 Economical Considerations

If the renewable world were only good for the environment but would be more expensive compared to the traditional business as usual, it is sad to say that we then would have to wait for quite some time—until the externalities like sea level rise, heavier storms, etc., would dictate the change, unfortunately too late by then. Fortunately, the cost development for the two major renewable technologies, wind and PV, has demonstrated in recent years a dramatic decrease. It is often argued that competitiveness of PV on-site generated electricity compared with grid electricity is only reached with support measures like FiT (Feed-in Tariff), especially in countries with minor sunshine like Germany (~1 kWh/W_{PV}). In the following, it is shown (see Fig. 7.7) that in a situation as of today, the annual cost for electricity is not only cheaper with a PV system with FiT compared with grid electricity but even more when a battery is included. The benefit for the battery system increases further if an annual increase for the grid electricity is assumed as indicated.

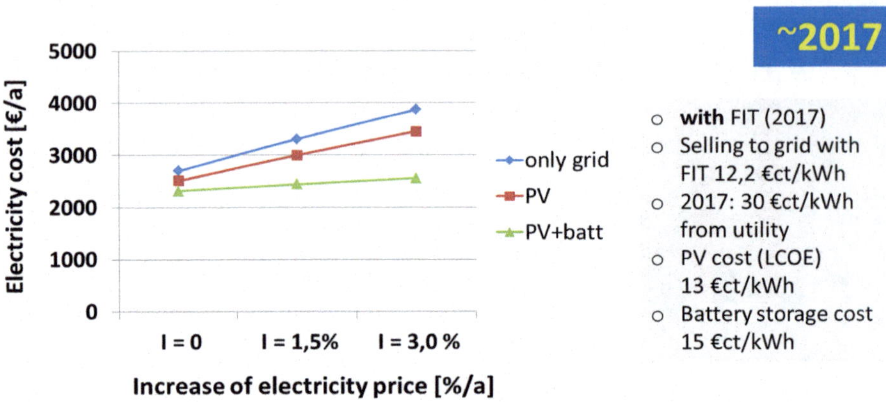

Fig. 7.7 Annual cost for 9000 kWh household (**a**) only utility, (**b**) +PV (20% self-consumption), and (**c**) +PV + batt (80% self-consumption); cases (**b**) and (**c**) with FIT

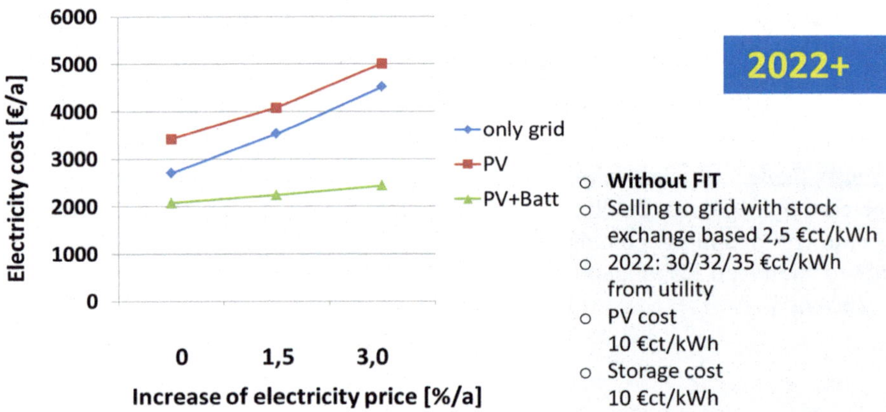

Fig. 7.8 Annual cost for 9000 kWh household (**a**) only utility, (**b**) +PV (20% self-consumption), and (**c**) +PV + batt (80% self-consumption); cases (**b**) and (**c**) with FIT

This situation changes significantly if we assume a situation without FiT which in Fig. 7.8 is taken arbitrarily in the year 2022. Here the PV system alone is more expensive than grid electricity due to the assumed small 20% self-consumption. The addition of a battery, however, demonstrates the clear advantage compared to grid electricity: the annual cost is bisected if a realistic annual grid price increase is considered.

Mass production driven by a phenomenal market growth at a global level has by now proven that even without the inclusion of externalities, the levelized cost of electricity (LCOE) for MW-sized PV systems is already today lower than new nuclear and clean fossil. However, even for experts in the field, it is quite difficult to calculate a market-based LCOE in order to be able to compare realistically different

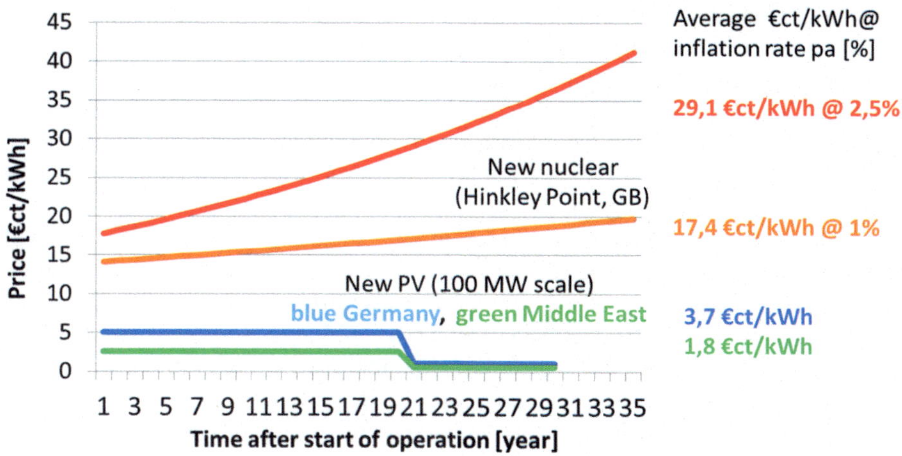

Fig. 7.9 Average LCOEs for PV and new nuclear including inflation

Table 7.1 Boundary conditions for new nuclear and PV power plants

	Base year	Earliest delivery	Duration (years)	Initial price (€ct/kWh)	Yearly inflation
2 × 1600 MW nuclear (Hinkley Point in GB)	2012	2029	35	12 [2]	Yes
~100 MW PV green-field in Germany (~1 kWh/W_{PV})	2018	2019	20	~5 [3]	No
~100 MW PV green-field in Middle East (~2 kWh/W_{PV})	2018	2019	20	~2.5 [4]	Depending on contract

technologies. In recent years, a much easier cost number has evolved, the price for a kWh electricity based on a power purchase agreement (PPA). In this case, the investor for a new power plant (PV, wind, nuclear, or fossil) signs a contract with the electricity receiving party (utility, SME, industrial company) with well-defined boundary conditions: date of the contract, start of delivery, duration time, price for a kWh electricity, and yearly adjustment for inflation to name the most prominent ones. As the negotiation is done with experts from the respective field, the easy-to-understand kWh price can be used effectively to compare different technologies. This is done for PV and new nuclear as shown in Fig. 7.9 based on boundary conditions summarized in Table 7.1. In the case of PV, it has been further assumed that after the typical agreement for new PPAs of 20 years, there is at least another 10 years of "Golden Age." This is based on warranty time for quality modules which today is up to 30 years. Hence, if we take very conservatively a maintenance and repair cost, including an additional inverter to meet increasing grid services, of 1 €ct/W in Germany (0.5 €ct/kWh in Middle East) for the third decade of producing electricity, we arrive over the 30 year lifetime of a PV system at an average cost of 3.7 and 1.8 €ct/kWh for countries with German (~1 kWh/W_{PV}) or Near-/Middle

East (~2 kWh/W_{PV}) insolation conditions, respectively. If these numbers are compared with the average price for new nuclear electricity, which is between 17 and 29 €ct/kWh for an annual inflation rate of 1 and 2.5%, respectively, the true competitiveness of solar is clearly seen: the addition of storage, which in the future will be below 10 €ct/kWh, brings solar (and wind) away from the old obstacle of being only an intermittent electricity source and is still well below the price for new nuclear (where the price for high-voltage grid lines has also to be added).

The comparison with new clean fossil plants shows also the superior competitiveness of new solar: if we consider CCS (carbon capture and storage), there is widespread belief that the price for a kWh from such plants is way above 10 €ct. Alternatively, if we assume a carbon price of 50 €/t CO_2, the production cost has to be increased by 6 €ct/kWh for lignite plants, 4.5 €ct/kWh for hard coal, and 2 €ct/kWh for gas power stations. Many experts strongly advocate a carbon price of more than 100 €/t in order to combat global warming.

7.4 Outlook

The countries which follow quickly the renewable path will demonstrate to those continuing the traditional way using fossil and nuclear for electricity production and oil for transportation that a replication makes most sense (1) economically, (2) socially, and (3) from an energy security point of view. This will trigger the transformation toward a 100% renewably powered world quicker as many expect also on a global level.

References

1. Courtesy of SMA Solar Technology AG (2017)
2. BBC News. March 19, 2013
3. https://www.bundesnetzagentur.de average prices for PV PPA's developed from 5.66 to 4.33 €ct/kWh for the bidding process in June 2017 and February 2018, respectively
4. ENEL Green Power Peru 180 MW project, Dubai Electricity and Water Authority 800 MW project

Chapter 8
Climate Change: An Overview of Potential Health Impacts Associated with Climate Change Environmental Driving Forces

Jazla Fadda

8.1 Introduction

The climate is changing in a very quick manner, which needs effective responsive actions. The global temperature is expected to rise above 2 °C, and that will lead to serious environmental consequences causing serious health problems. Climate change is a critical public health problem. It could make existing diseases becoming worse, or else introducing new pests and pathogens into the environment. For example, if planet warms, the sea level rises, and that leads to floods and drought. Heat waves, rising sea-levels, changes in precipitation resulting in flooding and droughts, intense hurricanes, and degraded air quality, affect directly and indirectly the physical, social, and psychological health of humans. Moreover, climate change can be a driver of disease migration, as well as releasing of toxic air pollutants resulting in many health problems. Climate change health outcomes will basically affect specific vulnerable groups: elders, children, and people with respiratory problems. That would in return stress out the healthcare infrastructure and healthcare delivery systems, as people will seek more healthcare services. Strategies to mitigate and adapt to climate change can prevent morbidity and mortality and protect the environment and health of future generations. Such actions are basically related to mitigate emissions of green houses and enhance the sinks that trap or remove carbon from the atmosphere. Adaption with climate changes refers to lessening the impact on health and the environment that cannot be prevented through mitigation, such as changes in our public health and healthcare infrastructure.

Climate change is expected to cause approximately 250,000 additional deaths per year between 2030 and 2050, approximately 87 million people were displaced due to extreme weather events between 2008 and 2011. The World Health Organization (WHO) reports that climate change is responsible for at least 150,000

J. Fadda (✉)
American University in Dubai, Dubai, United Arab Emirates
e-mail: jazla.fadda@vmprofessionals.ae

A. Sayigh (ed.), *Renewable Energy and Sustainable Buildings*, Innovative Renewable Energy, https://doi.org/10.1007/978-3-030-18488-9_8

deaths per year, a number that is expected to double by 2030. The effects of global warming will cause dire health consequences.

> "The complex nature of climate change and its environmental and social manifestations result in diverse risks to human health"

[1]. Climate change has many recent adverse health impacts and further projected effects are anticipated in the upcoming future [2].

Human actions transformed one-third and one-half of the land surface, and since the beginning of the Industrial Revolution, we have 30% increase of the carbon dioxide concentration in the atmosphere. On the other hand, more than half of all accessible surface fresh water is used by humanity, and about one-quarter of the bird species on Earth have been driven to extinction. Therefore, we do believe that we live on a human-dominated planet" [3].

The scale of environmental health problems has expanded from:

- Household (e.g., Indoor air pollution) to
- Neighborhood (domestic refuse) to
- Community (urban air pollution) to
- Regional (acid rain) to
- Global level (climate change).

8.2 The Review Procedures

8.2.1 Objectives

The main objective of the review is providing an overview of variety of evidence related to the impact of climate change on physical, mental, and social health, through the different environmental outcomes of climate change. Specifically, the main sub-objectives of this paper are:

(a) To review the direct and indirect impact of climate change on health
(b) To provide evidence of climate change and related infectious diseases
(c) To highlight on the serious diseases correlated with climate extremes
(d) To provide an evidence for estimating disease burden and risk factors correlated with climate change
(e) To review climate change health impacts on developing countries with more focus on children
(f) To review climate change health impacts on Middle East and UAE

A literature review was used to generate a conceptual framework of the direct and indirect effects of climate change, and how it can impact health directly and indirectly depending on some vulnerability factors. The review was held to answer the following questions:

- What are the direct and indirect effects of climate change on health?
- What are the climate change-related infectious diseases?

- What are the serious diseases correlated with climate extremes?
- What is the global and regional estimated disease burden of climate change?
- What are the climate change health impacts on developing countries with more focus on children?
- What are the climate change health impacts on Middle East and UAE?

A search has been conducted to generate variety of sources, using computerized resources (Medline, Scopus, and pro-quest); different key words were used to synthesize the inputs, such as climate change, impacts, direct effects, indirect effects, infectious disease, flooding and health, drought and health, burden of disease, and natural disasters.

About 249 search results were generated from the previous search engine, the relevant studies were reviewed to match this paper questions to find the association between climate change driving forces and the exact health impacts for each.

8.2.2 Inclusion and Exclusion Criteria

The paper reviews the studies and journal articles that address the climate change-related mortality and morbidity, considering the local and regional and global statistics.

This paper excludes the economic loss and financial expenditure correlated with climate change direct and indirect effects and associated health impacts.

8.3 Literature Review and Key Findings

Global warming can result in many serious alterations to the environment, eventually impacting human health [4].

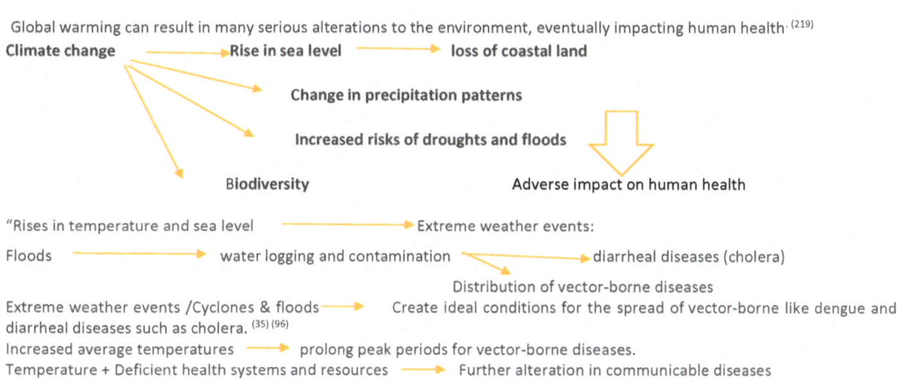

According to WHO, climate change causes around 150,000 deaths worldwide every year in low-income countries. Main causes of death are:

- Crop failure
- Malnutrition, floods
- Diarrheal diseases such as cholera
- Malaria

South-East Asia Region has 26% of the world's population and 30% of the world's poor. Fourteen million deaths occur each year, and 40% are attributable to communicable diseases, leading to higher burden of communicable diseases [6].

Main health impacts of global warming:

- *Infectious diseases*
  Increase in temperature leads to increase in **mosquito populations** means higher risk of malaria, dengue, and other insect-borne infections. **Legionnaires' diseases and bacterial lung infection attacked UK** in 2006. WHO states, *"global warming will also cause a major increase in insect-borne diseases in Europe. Countries like Azerbaijan, Tajikistan, and Turkey might already be in the danger zone for mosquito-borne malaria*" [4]. Use of more technology to contain temperature rise, such as p air conditioners and minimizing heat retention-based constructions, besides the strong public health systems and infrastructure, will prevent such outbreaks to happen.
- High temperatures lead to heat waves that strongly affect health. One study conducted in UK at "Hadley Center for Climate Prediction and Research" showed that the most common health of heat waves which will be fatal if not treated is **hyperthermia or heatstroke** [4].
- Global warming strongly affects people with heart problems due to **stressing their cardiovascular** system to keep their bodies cool. Furthermore, high temperatures increase the ozone concentration that damage people's lung tissue, leading to complications for patients with asthma and other lung diseases.
- **Food insecurity**, in turn, can lead to resource conflicts due to the loss of basic needs. "Climate change is an act of aggression by the rich against the poor," stated by Ugandan President Yoweri Museveni. According to IPCC report, by 2020, rising temperature will affect crop productivity and lead to food shortage, impacting almost 75 million to 250 million people in Africa, and 130 million people in Asia [4].

Figure 8.1 shows the Interrelationships between major types of global environmental change, including climate change. Note that all impinge on human health and—though not shown here explicitly—there are various interactive effects between jointly acting environmental stresses. The diverse pathways by which climate change affects health are the subjects of much of the remainder of this volume [7]

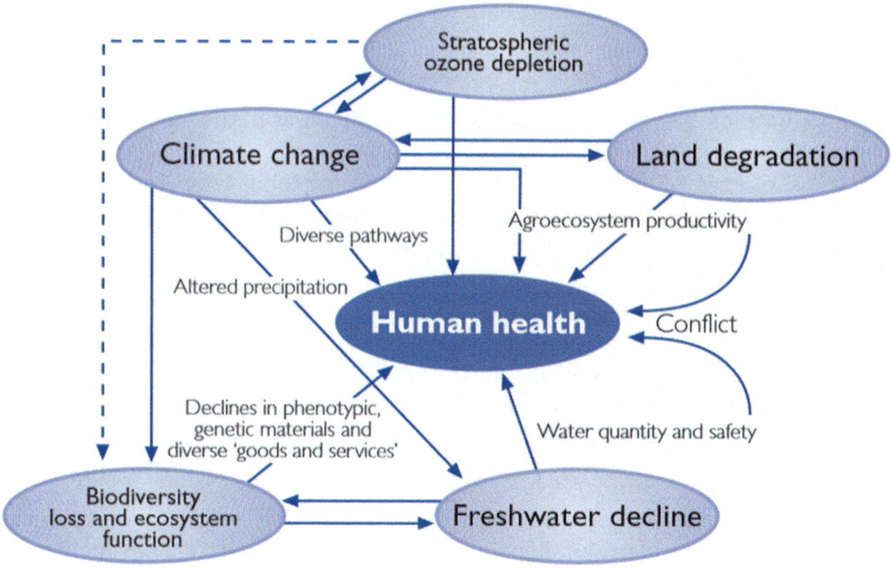

Fig. 8.1 Illustrates processes and pathways through which climate change influences human health [1]

Climate change main diseases:

- Asthma, respiratory allergies, and airway diseases
- Cancer and CVD
- Vector-borne and zoonotic diseases
- Foodborne diseases and nutrition
- Mental health and stress-related disorders
- Neurological diseases and disorders
- Waterborne diseases
- Weather-related morbidity and mortality

8.3.1 Direct Impacts of Climate Changes on Health

8.3.1.1 Weather Extreme: (Heat Waves and Winter Cold)

Burning fossil fuels releases carbon dioxide (CO_2) in higher concentration and leads to rise in Earth's temperature as a blanket trap in heat, which in turn distributes the balance of the environment, such as changes in wind, moisture, and heat circulation patterns [5].

Health Impacts of Heat Waves

Weather extremes leading to dangerous health outcomes might be even fatal.

- More hospital admissions for health-related illness, cardiovascular and respiratory disorders.
- Heat stroke: when the body becomes unable to control its temperature.
- Body temperature rises rapidly, the sweating mechanism fails, and the body cannot cool down, causing death or permanent disability if emergency treatment is not given.
- Infants, the elderly, people with chronic diseases, low-income populations, and outdoor workers have higher risk for heat-related illness.
- Higher temperatures and respiratory problems are also linked. One reason is because higher temperatures contribute to the buildup of harmful air pollutants [5].
- In countries with a high level of excess winter mortality, such as the United Kingdom, the beneficial impact may outweigh the detrimental [8, 9] (Fig. 8.2).

Increase in frequency and intensity of heat waves uplifts the risk of death and serious illness, particularly in the older age groups, those with preexisting cardio-respiratory diseases, and the urban poor. The effects of an increase in heat waves often would be exacerbated by increased humidity and urban air pollution [10].

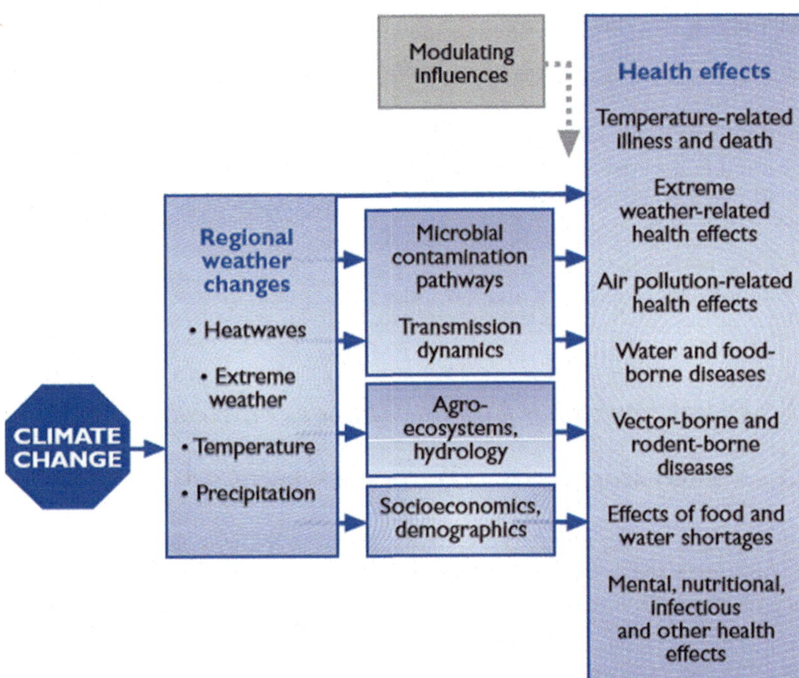

Fig. 8.2

Moreover, climate change has impact on thermal stress-related mortality in cities in developing countries. Warmer winters and fewer cold spells will decrease cold-related mortality in many temperate countries; sometimes, in the temperate zones, reduced winter deaths probably would outnumber increased summer deaths [11].

The impact of heat wave events on mortality heat waves can kill. In July 1995, a heat wave in Chicago, USA, caused 514 heat-related deaths (12 per 100,000 population) and 3300 excess emergency admissions [12]. During heat waves, excess mortality is greatest in the elderly and those with preexisting illness [13]. Much of this excess mortality is due to cardiovascular, cerebrovascular, and respiratory disease. A heat wave in India in June 1998 was estimated to have caused 2600 deaths over 10 weeks of high temperatures [14].

In the United States in 1994, 221 heat-related deaths were recorded, but only 101 (46%) were due to ambient weather conditions. Heat-related mortality due to weather conditions was 2.7–3.7 per million population in the four highest reporting states (Arizona, Arkansas, Kansas, and Missouri). Most of these deaths occurred in the over-55 age group [15].

Rooney et al. estimated the excess mortality associated with the 1995 heat wave in the United Kingdom [9]. Excess mortality occurred in England and Wales, and in Greater London, during the 1995. An estimated 619 extra deaths (8.9% increase) were observed relative to the expected number of deaths, based on the 31-day moving average for that period. Individual risk factors for dying in the heat wave were identified: chronic illness, confined to bed, unable to care for themselves, isolated, and without air conditioning. A comparison of mortality rates in three Illinois heat waves (1966) found that women and white people were at more risk [16].

Winter Mortality

Death rates during winter are 10–25% higher than those in summer. The major causes of winter death are cardiovascular, cerebrovascular, circulatory, and respiratory diseases [17, 18].

Annual outbreaks of winter diseases such as influenza, which have a large effect on winter mortality rates, are not strongly associated with monthly winter temperatures [11].

Sensitivity to cold weather (measured as the percentage increase in mortality per 1 °C change in temperature) is greater in warmer regions. Mortality increases to a given fall in temperature in regions with warmer winters, in populations with less home heating and where people wear lighter clothes [19].

The elderly (aged 75 and over) are particularly vulnerable to winter death, having a winter excess of around 30%.

Excess winter mortality is an important problem in the United Kingdom where there has been much debate about the role of poor housing, fuel poverty, and other socioeconomic issues for the elderly population [20]. A small-area study found that inadequate home heating and socioeconomic deprivation were the strongest independent predictors of ward-level variation in excess winter death in England and Wales [21].

8.3.1.2 Extreme Events "Disasters"

Flooding

Floods can cause contamination of public water supplies with both ***bacteria and parasites*** as surface discharge flows into rivers and reservoirs. Flood events and windstorms where the actual deaths and injuries are directly caused by the event are small compared to the problems that arise as a result, including deaths from communicable diseases and the economic losses sustained [22–24].

Indirect effects are largely death and injuries from drowning and being swept against hard objects. Local infrastructure damage to buildings and equipment includes materials and supplies, flood damage to roads and transport, problems with drainage and sewerage, and damage to water supply systems.

Diarrhea was the most common illness, followed by respiratory infection. Watery diarrhea was the most common cause of death for all age groups under [24]. In developed countries, both physical and disease risks from flooding are greatly reduced by a well-maintained flood control and sanitation infrastructure and public health measures, such as monitoring and surveillance activities to detect and control outbreaks of infectious disease [24].

Ecuador and northern Peru experienced severe flooding and mudslides along the coastal regions which severely damaged the local infrastructure. In Peru, 9.5% of health facilities were damaged, including 2% of hospitals and 10% of other health centers [24].

Mechanism of how above average rainfall can affect health

Description	Potential health impacts	
Hydrological	River/steam over tops its banks	– Changes in mosquito abundance – Contamination of water surface
Social	Property or crops damage	– Changes in mosquito abundance – Contamination of water with fecal matter and rat urine – (Leptospirosis)
Catastrophic flood disaster	Floods cause >10 killed or >200 affected people Government called for external assistance	– Contamination of water with fecal matter and rat urine – Increased risk of respiratory and diarrheal diseases – Death due to drowning – Injuries health effects – Population displacement – Loss of food supply – Psychological impacts

Drought

Droughts can increase the concentration of pathogens in the limited water supplies. A reduction in the availability of clean water increases the risk of drinking contaminated supplies and reduces the amount of water available for personal hygiene, thus leading to ***skin infections*** [25].

Floods and droughts are each associated with an increased risk of diarrheal diseases. Bacterial infections such as cholera and typhoid, bacteria such as cryptosporidium, *E. coli*, giardia, and shigella, and viruses such as hepatitis A cause diarrhea. Outbreaks of cryptosporidiosis, giardia [25, 26].

The health impacts of drought:

- Lack of food production leads to famine.
- Malnutrition and diseases resulting from malnutrition.
- Malnutrition increases susceptibility to infection.
- Adverse environmental conditions and political, environmental, or economic crises can trigger a collapse in the food marketing system—famine and conflicts.
- Lack of hygiene increases the risk of diarrheal diseases (due to fecal contamination).
- Water washed diseases (trachoma, scabies).

Outbreaks of malaria can occur due to changes in vector breeding sites [27, 28].

Mechanism of how below-average rainfall can affect health

Type of effect	Description	Potential health impacts
Meteorological	– Evaporation exceed water absorption – Soil moisture decreases – Several indices developed ex: palmer	– Changes in vector abundance – Vector breeds in dried up river beds
Agriculture	– Drought severity index – Drier conditions lead to less crop production	– Depends of other sources of food – Food shortage illness – Malnutrition – Increase risk of infections – Increased risk of lack of water and hygiene.
Social	Reduction of food supply Reduction of water supply Poor water quality	– Contamination of water with fecal matter and rat urine – Increased risk of respiratory and diarrheal diseases – Death due to drowning – Injuries health effects – Population displacement – Loss of food supply – Psychological impacts
Food shortage Famine/Drought Disaster	Death due to food shortage >10 killed or >200 affected Government call for external assistance	– Deaths caused by starvation – Malnutrition – Increased risk of infections – Health impacts associated with displacement

Ref: Geneva, Switzerland, World Health Organization 1998 [29]

The Health Impacts of Natural Disasters

Besides direct impacts of natural disasters such as death, injury, and morbidity, climate natural disasters can affect health indirectly through:

- Loss of shelter; population displacement and overcrowding; and increases in respiratory and diarrheal diseases due to crowding of survivors, often with limited shelter and access to potable water. Many studies have shown the association between dampness correlated with flooding and a variety of respiratory symptoms. A Canadian study found that flooding was linked significantly with childhood experience of cough, wheeze, asthma, bronchitis, chest illness, upper respiratory symptoms, eye irritation, and non-respiratory symptoms [30].
- Contamination of water supplies; increased risk of water-related and infectious diseases due to disruption of water supply or sewage systems.
- Loss of food production (leading to hunger and malnutrition); decreases in nutritional status, especially in children.
- Increased risk of infectious disease epidemics (including diarrheal and respiratory diseases).
- Damage to infrastructure for provision of health services.
- Ecological change leads to vector-borne diseases.
- Impairment of the public health infrastructure, psychological and social effects, and reduced access to healthcare services, Release and dissemination of dangerous chemicals from storage sites and waste disposal sites into flood waters [31].
- Floods also cause psychological morbidity. Following flooding in Bristol, UK, primary care attendance rose by 53% and referrals and admissions to hospitals more than doubled [32], and also psychological effects were found following floods in Brisbane, Australia, in 1974 [33]. Moreover, further psychological symptoms and post-traumatic stress disorder, including 50 flood-linked suicides, were reported in the 2 months following the major floods in Poland in 1997 [34].

8.3.1.3 Forest Fires

- The direct effects of fires on human health are burns and smoke inhalation.
- Loss of vegetation on slopes may lead to soil erosion and increased risk of landslides, especially when an urban population expands into surrounding hilly and wooded areas.
- Air pollution is linked to increased mortality and morbidity in susceptible persons.
- Risk of hospital and emergency admissions.
- Assessments are being undertaken of the short-term impacts on mortality and morbidity associated with the 1997 El Niño episode.
- Forest fires lead to short- and long-term effects due to exposure to air pollution, acute respiratory infections, death from chronic diseases. Long-term effects are related to carcinogenic components such as polycyclic aromatic hydrocarbon, whose effect appears after few years [35, 36].

8.3.1.4 Sea Level Rise

Rising seas means rising health risks, flooding and storm are associated with sea level rise projecting direct and indirect risks. The direct health risks are those associated with drowning, injury, and displacement. The indirect risks are associated with the intrusion of saltwater into fresh water of groundwater basins and well water. That entails the reduction of crop yields and safe drinking water and increases the risk of vector-borne and diarrheal disease. Coastal flooding and storms are also associated with indoor mold due to dampness, which in turn will be a risk factor of related respiratory disease.

Moreover, rising in sea levels affect coastal tourism and then the economic status. In California, losses in coastal tourism could cost California beach communities many thousands of jobs and hundreds of millions of dollars. 260,000 people and $50 billion are now at risk for a 100-year flood [37].

If population, development, and climate change stay on track, by 2100 those numbers will nearly double, with 480,000 people and $100 billion in property at risk [37].

8.3.2 Indirect Impacts of Climate Changes on Health

8.3.2.1 Climate Change and Air Quality

Air pollution influences the transportation of air-borne pollutants, pollen production, and levels of fossil fuel pollutants resulting from household heating and energy demands. *Aeroallergens Experimental research has shown that doubling CO_2 levels from about 300 to 600 ppm induces a fourfold increase in the production of ragweed pollen* [38, 39].

How climate change impacts air quality:

- Higher temperatures lead to an increase in allergens and harmful air pollutants.
- Longer warm seasons can mean longer pollen seasons,
- Increased allergic sensitizations and asthma episodes, and diminished productive work and school days.
- Higher temperatures associated with climate change can also lead to an increase in ozone, a harmful air pollutant, which increases the concentration of ground level ozone as well [40]. The potential health consequences of stratospheric ozone depletion are increase in incidence of skin cancer in fair-skinned populations and eye lesions such as cataracts [41].
- Climate change increases the risks of forest fires and associated smoke hazards, associated with increases in respiratory and eye symptoms, acute (or sudden onset) respiratory illness, respiratory and cardiovascular hospitalizations, and lung illnesses [42].

8.3.2.2 Climate Change and Water Quality

Higher sea surface temperature which occurs due to high temperature leads to more hospitable environment for some harmful algae and other microbes to grow. Warmer waters and flood conditions introduce some of the public health concerns.

- Group of diseases caused by bacteria such as ***cholera and typhoid*** as well as ***parasitic diseases such as amoebiasis, giardiasis, and cryptosporidium***.
- **Vibrio parahaemolyticus**: diarrheal illnesses linked with consuming raw or undercooked oysters.
- **Vibrio vulnificus** causes vomiting, diarrhea, and abdominal pain associated with seafood-related deaths in healthy adults.
- **Naegleria fowleri** (sometimes referred to as a "brain-eating ameba") is a microbe that can be present in soil and warm freshwater. It usually infects people when contaminated water enters the body through the nose. Changes in surface water quality and quantity are likely to affect the incidence of diarrheal diseases [43].

Climate can increase **directly the amount of pathogen** in the water through increasing the biotic reservoir of the infectious agent (cholera) or **by decreasing the amount of water** in a river or a pond and thus raising concentration of the bacteria (typhoid). Infections with cholera and typhoid bacteria are dependent on the concentration of the pathogens in water or food. Currently the World Health Organization *(WHO) estimates more than one billion people worldwide to be without access to safe drinking water, and that every year approximately 1.7 million die prematurely because they do not have access to safe drinking water and sanitation [44].*

Effects of social and economic disruptions: In some settings, the impacts of climate change may cause severe social disruptions; local economic decline and population displacement lead to both physical and mental health problems [45] (Table 8.1).

Rainfall and diseases spread by insect vectors which breed in water, depending on surface water availability. Main species of interest are mosquitoes, which spread malaria and viral diseases such as dengue and yellow fever.

There are complex relationships between human health and water quality, water quantity, sanitation, and hygiene. Increases in water stress are projected under climate change. Heavy rainfall events can transport terrestrial microbiological agents into drinking-water sources resulting in outbreaks of ***cryptosporidiosis, giardiasis, amoebiasis, typhoid***, and other infections [26, 46–48].

Recent evidence indicates that copepod zooplankton provide a marine reservoir for the cholera pathogen and thereby facilitate its long-term persistence and disseminated spread to human consumers via the marine food web. Cholera epidemics are also associated with positive surface temperature anomalies in coastal and inland lake waters [49]. An association between drinking water pollution and gastrointestinal illness has been reported [50]. Disease outbreaks in the United States has shown that about half were significantly associated with extreme rainfall [51].

Table 8.1 Water and food-borne agents: connection to climate

Pathogen groups	Pathogenic agent	Food-borne agents	Water-borne agents	Indirect weather effect	Direct weather effect
Viruses	Enteric viruses (e.g. hepatitis A virus, Coxsackie B virus)	Shellfish	Groundwater	Storms can increase transport from faecal and waste water sources	Survival increases at reduced temperatures and sunlight (ultraviolet)[a]
Bacteria Cyanobacteria Dinoflagellates	Vibrio (e.g. *V. vulnificus, V. Parahaemolyticus, V. cholerae* non-01; *Anabaena* spp., *Gymnodinium, Pseydibutzschia* spp.)	Shellfish	Recreational, wound infections	Enhanced zooplankton blooms	Salinity and temperature associated with growth in marine environment
Protozoa	Enteric protozoa (e.g. Cyclospora, Crytosporidium)	Fruit and vegetables	Recreational and drinking water	Storms can increase transport from faecal and waste water sources.	Temperature associated with maturation and infectivity of Cyclospora

Source: Reproduced from Ref. [20]
[a]Also applies to bacteria and protozoa

8.3.2.3 Climate Change and Ecological Change

Climate change affects physical factors such as temperature, precipitation, humidity, surface water, and wind and biotic factors such as vegetation, host species, predators, competitors, parasites, and human interventions.

Climate Change and Vector-Borne Diseases

Higher temperatures in combination with conducive patterns of rainfall and surface water, climate variability, and poor public health infrastructure would alter the geographical range and seasonality of distribution of vector organisms (e.g., malarial mosquitoes) [52]. Alternatively, a change in ecology of the natural predators may affect mosquito vector dynamics; mosquito populations recover more quickly than their predator populations following a dry year. Temperature-related changes in the life-cycle dynamics of both the vector species and the pathogenic organisms (flukes, protozoa, bacteria, and viruses) would increase the potential transmission of many vector-borne diseases such as malaria (mosquito), dengue fever (mosquito), and leishmaniasis (sand-fly)—although schistosomiasis (water-snail) may undergo a net decrease in response to climate change [53, 54].

How temperature affects Vector population:

- Temperature reduces the time taken for vector populations to breed.
- Increases in temperature also decrease the incubation period of the pathogen (e.g., malaria parasite, dengue, or yellow fever virus), meaning that vectors become infectious more quickly [55].
- However, hot, dry conditions can reduce the lifetime of mosquitoes.
- Temperature also may affect the behavior of the vector and human populations, affecting the probability of transmission.
- Warmer temperatures tend to increase biting behavior of the vector and produce smaller adults which may require multiple blood meals to reproduce.

Climate change and transmission of vector-borne diseases include:

- Survival and reproduction rate of the vector
- Time of year and level of vector activity, specifically the biting rate
- Rate of development and reproduction of the pathogen within the vector

The most influential climatic factors for vector-borne diseases include:

- Temperature and precipitation.
- Sea level elevation.
- Wind and daylight duration are additional important considerations.
- Extreme temperatures often are lethal to the survival of disease-causing pathogens but incremental changes in temperature may exert varying effects.
- A vector lives in an environment of low mean temperature; a small increase in temperature may result in increased development, incubation, and replication of the pathogen [56, 57].
- Temperature may modify the growth of disease-carrying vectors by altering their biting rates, as well as affect vector population dynamics and alter the rate at which they contact humans. Finally, a shift in temperature regime can alter the length of the transmission season [58].
- Disease-carrying vectors may adapt to changes in temperature by changing geographical distribution. An emergence of malaria in the cooler climates of the African highlands may be a result of the mosquito vector shifting habitats to cope with increased ambient air temperatures [59].
- Another possibility is that vectors undergo an evolutionary response to adapt to increasing temperatures [60].

Types of Vector-borne diseases influenced by climate change:

- Ticks, their animal hosts (deer), tick-borne diseases ex; encephalitis.
- Lyme disease caused by bacteria "borrelia type" spread by ticks.
- West Nile virus has no medications or vaccines to prevent, recovery takes weeks or months.
- Dengue.
- Leishmaniasis [5].

"Climate change scenarios over the coming century would cause a small net increase in the proportion of the world population living in regions of potential transmission of malaria and dengue" [61–63].

Malaria Status in the World

- Currently 40% of the world population lives in areas in which endemic malaria occurs.
- Over 2.5 billion people are at risk, and there are estimated to be 0.5 billion cases and more than 1 million deaths from malaria per year [29].
- Malaria incidence is influenced by the effectiveness of public health infrastructure, insecticide and drug resistance, human population growth, immunity, travel, land-use change, and climate factors.
- Very high temperatures are lethal to the mosquito and the parasite and malaria transmission would therefore decrease.
- At low temperatures a small increase in temperature can greatly increase the risk of malaria transmission [56].

  Malaria's sensitivity to climate is illustrated in desert and highland fringe areas where rainfall and temperature, respectively, are critical parameters for disease transmission [64].
- Also, drought may reduce malaria transmission resulting in a reduction in herd immunity in the human population. Therefore, in the subsequent year the size of the vulnerable population is increased [65].
- Famine conditions may have contributed to excess mortality during historical epidemics of malaria.
- Drought-breaking rains increased vector abundance, exacerbated by population movement and the concentration of people in feeding camps [66].

Drought in the previous years has been identified as a factor contributing to increased malaria mortality. There are several possible reasons for this relationship. Drought-related malnutrition may increase an individual's susceptibility to infection [67].

Dengue

- Dengue is the most important arboviral disease of humans, occurring in tropical and subtropical regions worldwide, and an increasing urban health problem in tropical countries. The main vectors of dengue are the domesticated mosquito *Aedes aegypti* and *Aedes albopictus*. They breed in urban environments in artificial containers that hold water. The latter can tolerate colder temperatures and spread because of ineffective vector and disease surveillance; inadequate public health infrastructure; population growth; unplanned and uncontrolled urbanization; and increased travel [68]. Dengue is seasonal and usually associated with warmer, more humid weather.

  There is evidence that increased rainfall in many locations can affect the vector density and transmission potential. ENSO may act indirectly by causing changes in water storage practices brought about by disruption of regular supplies. A study of dengue in Viet Nam found that the number of cases increased in El Niño years [69]. Many countries in Asia experienced an unusually high level of dengue and dengue hemorrhagic fever in 1998, some of which may be attributable to El Niño-related weather [55].

Climate Change and Zoonotic Diseases

Hantavirus is naturally maintained in rodent reservoirs and can be transmitted to humans at times of increased local abundance of the reservoir [70]. Rabies is another directly transmitted zoonosis that naturally infects small mammals, although with very little opportunity for widespread transmission, being highly pathogenic to its vertebrate host [71]. TB and HIV originally emerged from animals as well.

Climate Change and Rodent-Borne Diseases

- *Rodent-borne diseases* associated with flooding include leptospirosis, tularemia, and viral hemorrhagic diseases.
- Other diseases associated with rodents and ticks include plague, Lyme disease, tick-borne encephalitis (TBE), and hantavirus pulmonary syndrome (HPS).
- Rodent populations have been shown to increase in temperate regions following mild wet winters [72].
- One study found that human plague cases in New Mexico occurred more frequently following winter-spring periods with above-average precipitation [70].
- These conditions may increase food sources for rodents and promote breeding of flea populations.
- Infection by hantaviruses mainly occurs from inhalation of airborne particles from rodent excreta.
- The emergence of the disease hantavirus pulmonary syndrome in the early 1990s in the southern United States has been linked to changes in local rodent density [73, 74].

  Drought conditions had reduced populations of the rodents' natural predators; subsequent high rainfall increased food availability in the form of insects and nuts.
- Above-average precipitation during the winter and spring of 1992–1993 may have increased rodent populations and thereby increased contact between rodents and humans and viral transmission [75].

8.3.2.4 Climate Change and Social Factor

Climate change will have mixed effects on **food production** globally. Most of the research to date has focused on cereal grain production—an important indicator of total food production, since it accounts for around 70% of global food energy. The probability of reduced food yields is, in general, greatest in developing countries where it is estimated that approximately 790 million people currently are undernourished. Populations in isolated areas with poor access to markets will be particularly vulnerable to local decreases or disruptions of food supply [76].

There is a web of interactions between ecosystems, climate, and human societies, which influences the occurrence of infections health infrastructure, poor water and sanitation, overcrowding, and lack of shelter. Communicable diseases are related to social factors like and limiting social development as well [77]. Social factors that increase possibilities of communicable diseases are population growth, urbanization, changes in land use and agricultural practices, deforestation, international travel, and breakdown in public health infrastructure [78].

8.3.3 Climate Change-Related Infectious Diseases (Emerging and Remerging of Infectious Diseases)

Climate is one of several important factors influencing the incidence of infectious diseases, like the non-climatic factors such as

- Sociodemographic influences such as human migration and transportation
- Drug resistance and nutrition
- Environmental influences such as deforestation, agricultural development, water projects, and urbanization

Classification of infectious diseases:

- **By mode of transmission**:
  Spread directly from person to person through direct contact or droplet exposure, indirectly through an intervening vector organism—*biological* (mosquito or tick) or *non-biological* physical vehicle (soil or water).
- By **Infectious agent**: (protozoa, bacteria, viruses, etc.).
- **By reservoir**.
- **Anthroponoses** (Human reservoir) is a class of diseases defined by pathogen transmission between two human hosts by either a physical vehicle (soil) or a biological vector (tick) (Fig. 8.3).

  These diseases require *three components* for a complete transmission cycle: the pathogen, the physical vehicle or biological vector, and the human host.

  Most vectors require a blood meal from the vertebrate host to sustain life and reproduce and are transmitted between human hosts by vector organism (mosquitoes, ticks, sandflies, etc.) vectors (vector-borne disease).
- **Zoonoses (animal reservoir)**.
  Question related to infectious diseases
  - How climatic variations or trends affect the occurrence of infectious diseases?
  - Is there any evidence that infectious diseases have changed their prevalence in ways that are reasonably attributable to climate change?
  - How future scenarios of different climatic conditions will affect the transmissibility of infectious diseases?

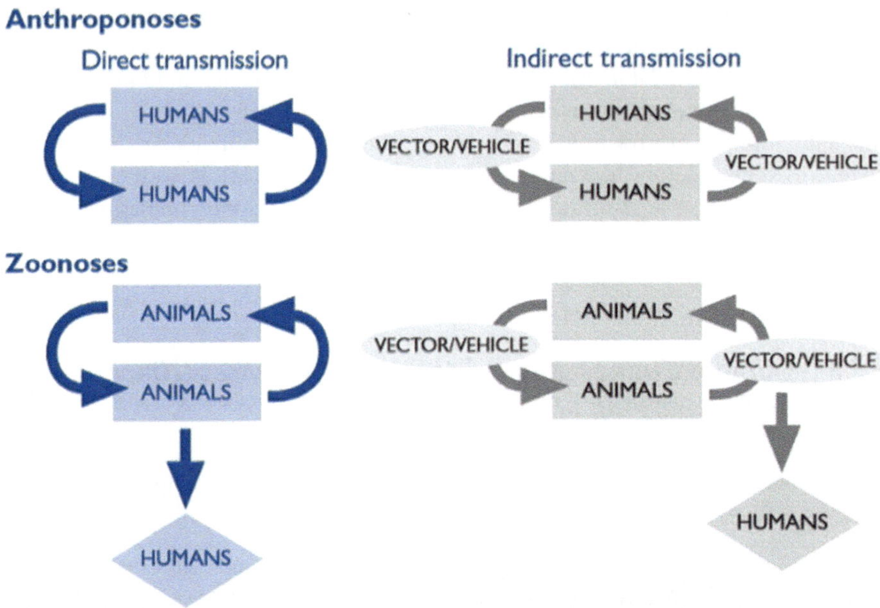

Fig. 8.3 Four main types of transmission cycle for infectious diseases (Source: reproduced from Ref. [3]; Wilson, M.L. (2001). Ecology and infectious disease. In: Ecosystem change and public health: a global perspective)

Clinicians are concerned with clinical manifestation to treatment of infected patients.

Microbiologists are concerned with characteristics of the microorganisms, such as viral or bacterial.

Epidemiologists are concerned with mode of transmission of the pathogen and its natural reservoir, to control spread and outbreaks [79].

How climate change affects reservoir:

Agents are spread naturally among animal reservoirs and the infection of humans is a result of an accidental human encounter. The persistence of these pathogens in nature is largely dependent on the interaction of the animal reservoir and external environment which can impact the rate of transmission, host immunity, rate of reproduction, and species death, rendering these diseases more susceptible to effects of climate variability.

Climate affects transmission cycles that require a vector or non-human host and such diseases are more susceptible to external environmental influences than those diseases which include only the pathogen and human.

8.3.3.1 Climate Sensitivity of Infectious Disease and Seasonality of Infectious Disease

Winter mortality and infectious disease using the example of cyclic influenza outbreaks occurring in the late fall, winter and early spring in North America. This disease pattern may result from increased likelihood of transmission due to indirect social or behavioral adaptations to the cold weather such as crowding indoors. Another possibility is that it may be attributed directly to pathogen sensitivities to climatic factors such as humidity. Epidemics of other infections (e.g., meningococcal meningitis) tend to erupt during the hot and dry season and subside soon after the beginning of the rainy season in sub-Saharan Africa [80].

Seasonal fluctuations of infectious disease occurrence imply an association with climatic factors. However, to prove a causal link to climate, non-climatic factors must be considered. Furthermore, to assess long-term climate influences on disease trends, data must span numerous seasons and utilize proper statistics to account for seasonal fluctuations.

Temperature and fluid levels of those agent and organisms are affected by local climate. Incubation time of a vector-borne infective agent within its vector organism is typically very sensitive to changes in temperature, level of precipitation, sea level elevation, wind, and duration of sunlight.

8.3.4 Climate Change Estimated Burden of Disease

The burden of disease refers to the total amount of premature death and morbidity within the population [1]. DALY can be considered one lost year of "healthy" life. The sum of DALYs in a population, or the burden of disease, can be considered a measure of the gap between current health status and an ideal health situation in which the entire population lives to an advanced age, free of disease and disability [81]. DALY's = years of life lost due to premature death (YLL) "using standard units of health deficit" + years of life lived with disability (YLD) "Considering the severity/disability and duration of the health deficit" [1] (Tables 8.2 and 8.3).

Table 8.2 Estimates for the impact of climate change in 2000 in thousands of DALYs, given by applying the relative risk estimates for 2000 to the DALY burdens for specific diseases quoted in the World Health Report (2002) [2]

	Malnutrition	Diarrhoea	Malaria	Floods	Total	Total DALYs/ million population
African region	616	414	860	4	**1894**	**3071.5**
Eastern Mediterranean region	313	291	112	52	**768**	**1586.5**
Latin American and Caribbean region	0	17	3	72	**92**	**188.5**
South-East Asian region	1918	640	0	14	**2572**	**1703.5**
Western Pacific region[a]	0	89	43	37	**169**	**111.4**
Developed countries[b]	0	0	0	8	**8**	**8.9**
World	2847	1460	1018	192	**5517**	**920.3**

[a] Without developed countries
[b] Cuba

Table 8.3 Main vector-borne diseases: populations at risk and burden of diseases

Disease	Vector	Population at risk	Number currently infected or new cases per year	Disability adjusted life years lost[a]	Present distribution
Malaria	Mosquito	2400 million (40% world population)	272,925,000	42,280,000	Tropics/subtropics
Schistosomiasis	Water snail	500–600 million	120 million	1,760,000	Tropics/subtropics
Lymphatic filariasis	Mosquito	1000 million	120 million	5,644,000	Tropics/subtropics
African trypanosomiasis (sleeping sickness)	Tsetse fly	55 million	300,000–500,000	1,598,000	Tropical Africa
Leishmaniasis	Sand Fly	350 million	1.5–2 million	2,357,000	Asia, Africa, Southern Europe, Americas
Onchocerciasis river blindness	Black fly	120 million	18 million	987,000	Africa, Latin America, Yemen
American trypanosomiasis (Chagas' disease)	Triatomine bug	100 million	16–18 million	649,000	Central and South America
Dengue	Mosquito	3000 million	Tens of millions	653,000[b]	All tropical countries
Yellow fever	Mosquito	468 million in Africa	200,000	Not available	Tropical South America and Africa
Japanese encephalitis	Mosquito	300 million	50,000	767,000	Asia

Based on data from Ref. [1], with updated DALY estimates from Ref. [16]

Gubler, D.J. & Meltzer, M. The impact of dengue/dengue hemorrhagic fever in the developing world [82]

[a] The Disability–Adjusted Life Year (DALY) is a measure of population health deficit that combines chronic illness or disability and premature death [17]. Numbers are rounded up to nearest 100,000

[b] Other analyses suggest this value could be as high as 1,800,000 [18]

The three main categories, with sub-conditions, are:

- Health impacts of extreme weather events, including heat-waves, cold-waves, floods, and windstorms
- Respiratory diseases, related to ground-level ozone, particulate matter and allergens
- Infectious diseases, including vector-borne, waterborne, and foodborne diseases

Disease is then described by incidence or prevalence:

- Disease incidence = new cases per year, usually acute health conditions
- Disease prevalence = cases that tend to last longer or are underlying conditions that are never or rarely treated (e.g., malnutrition)

"Current disease burdens may be lower in the future not because of climate adaptation or response measures but due to a strengthened health system and greater resilience to the impacts of climate change" [81].

Estimating disease burden and risk factors correlated with climate change (climate *change-related diseases are illustrated in Appendices 1 and 2)*
World Health Organization quantified the amount of premature morbidity and mortality due to a range of risk factors, including climate change. In the year 2000, climate change is estimated to have caused the loss of over 150,000 lives and 5,500,000 DALYs (0.3% of deaths and 0.4% of DALYs, respectively) [7, 83]. The World Health Organization (WHO) estimates that every year about 150,000 deaths occur worldwide in low-income countries due to adverse effects of climate change, malnutrition, floods, diarrheal diseases, and malaria. This is due to the limited financial and human capacity to respond to the impacts of climate change. FAO estimates that by 2030, up to 122 million more people could be forced into extreme poverty because of climate change—many of them women [81].

8.3.5 Climate Change Health Impacts in Developing Countries

The characteristics of developing countries:

- Limited social safety nets
- Widespread poverty
- Fragile healthcare systems
- Weaker governmental institutions
- Weaker institutional structures
- Weaker infrastructure
- Higher birth rates
- High ratio of children to adults
- More children are at risk there than in the developed world
- Less-adequate health systems may make it harder to mitigate the effects of temperature

- Have warmer climates than richer ones do
- Temperature changes may affect health and agriculture more severely
- Developing countries are likely to face a disproportional share of extreme changes in weather
- Energy infrastructure lags
- Weaker labor and credit markets, which may make it hard for families to adapt to losses caused by climate change not have enough funds to sustain the family and invest in agricultural inputs, such as seeds and fertilizer, for the following year

Response to climate change: developing countries vs. developed countries:
Developing countries still depend primarily on agriculture as a source of income; excessive heat or rain will reduce crop production and then directly threaten the livelihoods of developing-country families and their ability to feed their children. In developed countries, electrification and greater access to healthcare have greatly reduced mortality from heat [84, 85]. Moreover, developed countries can get loans, for healthcare, compared with developing countries who have limited access to banks and formal financing; loans may be unavailable. In developing countries families might not only lose income from the crop loss and further sustained losses of income as they can't invest in future production. Government safety net programs may offer emergency help, but again, such programs are more limited in the poorest countries [86].

8.3.5.1 Climate Change Impacts on Children in Developing Countries

In developing countries, children who are physically weaker and less tolerable to heat are likely to face the following:

- More severe threats from both air and water pollution
- Vulnerability to infectious and parasitic diseases carried by insects or contaminated water; triggered by climate change by environmental factors, entails displacement, migration, and violence, entails

Environmental factors affecting through pregnancy and early childhood have negative outcomes in adulthood. "Healthier children grow into healthier, wealthier, and more educated adults" [87].

How Climate Change Affects Children's Health in Poor Countries:
Climate change is likely to exacerbate conditions that bring forth infections and parasites, threatening health of children in poor nations. Children can be exposed to vector-borne illnesses such as malaria that are transmitted by insects and other pests. New health threats emerge from disturbed ecosystems such as skin cancer [86]. Infant mortality rate is a leading indicator of child health and access to healthcare. In 2013, the infant mortality rate was ten times higher in low-income countries than it was in wealthy ones: 53 vs. 5.3 deaths per 1000 births, respectively. In poor countries [86] (Table 8.4).

Table 8.4 Top causes of death among children under five (2012)

WHO Africa Region	WHO Southeast Asia Region	United States
Acute lower respiratory infections (15.9%)	Prematurity (25.2%)	Congenital anomalies (19.4%)
Malaria (14.7%)	Acute lower respiratory infections (14.1%)	Prematurity (15.0%)
Prematurity (12.3%)	Birth asphayxia and birth trauma (11.4%)	Accidents/unintentional injuries (9.0%)
Birth asphyxia and birth trauma (10.9%)	Diarrheal diseases (9.8%)	Sudden infant death syndrome (6.0%)
Diarrheal diseases (10.3%)	Other communicable perinatal, and nutritional conditions (8.5%)	Maternal pregnancy complications (5.4%)

Note: WHO is the World Health Organization
World Health Organization, "Global Health Observatory Data Repository" [88]

Direct Effects of Climate Change on Children in Developing Countries

Effect of Hot Days

Developing countries will see a disproportional rise in warm days (defined as days above 30 °C, or 86 °F). Hot days and heat waves can directly affect children's health through increases in rashes, heat exhaustion, temporary loss of consciousness (syncope), and heat stroke [86].

The additional warm day affects children mortality seven times greater in rural areas of developing countries than it is in the United States. This would be attributed to technologies that are unavailable to most people in the developing world.

Effect of Water Shortage

Climate change will lead to an increase in water shortages, which means less and lower-quality drinking water, and spending more time and money to collect water from sources farther from their homes [89], and causes further cases of diarrhea [90].

Climate Change and Infections

Healthcare, sanitation, and pest control can reduce human vulnerability to infections and vector-borne diseases. However, they are still among the primary killers of children in developing countries.

- Infections Short Run Effects:

  Heavy rainfall, flooding, and rising water temperature cause an increase in heavy precipitation leading to cholera outbreaks [91, 92]. According to UNICEF large population of children are vulnerable (about 1 in 10 experienced cholera in 2012) [93].
- Infections Long-Run Effects:

  Parasitic diseases have long-term effects on health, schooling, then future income. 25% reduction in moderate to heavy infections of intestinal worms (such as hookworm, roundworm, whipworm, and the worms that cause schistosomiasis) reduced school absenteeism by one-fourth and had long-run impacts on adult outcomes [94, 95].

Vector-Borne Diseases

The optimal incubation temperature for the malaria parasite is likely to be exceeded, as many of the insects transmit malaria to human's nurture in warm night and winter temperatures [96]. Malaria, dengue, and other vector-borne illnesses will expand to previously unaffected areas that will become suitable habitats for malaria transmitting insects [97, 98] and will affect particularly people whose immune systems have never been exposed to such parasites.

Impacts of Flooding and Natural Disaster

Children in developing countries are more impacted with flooding and natural disasters for these reasons:

- First, climate change will make extreme precipitation near the centers of tropical cyclones on poor regions.
- Second, mortality rates due to disasters, such as floods and windstorms, are much higher in low-income countries compared with high-income countries.
- Third, lack of complete credit and insurance markets. Few people have private insurance that can't cover to rebuild or resettle after a disaster.
- Fourth, poorly functioning credit markets make it difficult to borrow funds. International aid can help, but again the limited institutional capacity and infrastructure hurdle the response to human needs [99].

Research shows that income growth can initially increase the death toll from environmental disasters. However, even with the increase of per capita incomes to $5000 in the next few decades, we'll witness a peak of vulnerability to environmental disasters at a time when such disasters become more common [100].

Impact of Air Pollution

Increased drought leads to uncontrollable wildfires and dust storms, which in turn causes higher concentrations of several air pollutants, including ozone and particulate matter. This could be found especially in some regions such as Asia and Eastern and Southern Africa [101]. Moreover, precipitation patterns may increase infant mortality and reduce general health, especially for fetus in wombs [102]. Droughts can also increase the amount of dust over large areas [103]. Dust storms and wildfires produce inhalable, coarse particles and a smaller share of fine particles, lead to increased mortality from heart disease and respiratory disease, especially among infants and the elderly. On the other hand, air pollution affects both adults and children, by reducing their performance of everyday activities, such as carrying heavy objects [104, 105].

Further Health Problems

Emerging and reemerging infections:
Rapid changes in temperature and precipitation patterns can lead to emerging diseases due to ecological changes such as destabilization of animals. Hantavirus, which causes dangerous pulmonary disease in humans, was a serious problem in the US Southwest after an El Niño event in the early 1990s [106].

However, Drew Harvell et al. in their report "Climate Warming and Disease Risks for Terrestrial and Marine Biota" stated that predictions about emerging diseases are uncertain, and sound empirical evidence is scarce [107]. Developing countries face difficulties in coping with these new infectious diseases due to the weak healthcare systems, like what happened in West Africa in 2014, when faced the spread of Ebola outbreak.

Skin cancer "Melanoma":
High temperatures are threatening health in developing countries by also having more cases of skin cancers "melanoma". Skin cancer could become 21% more common if global temperatures rise by 2 °C, and 46% more common with a temperature increase of 4 °C. Even with underreported cases, melanoma is becoming more common in Africa and Southeast Asia [86].

Indirect Impacts of Climate Change on Developing Countries

Income Loss, Food Insecurity, and Malnutrition

According to the World Bank, 75% of the world's poor who depend on farming to live are in rural areas. Those people are supported by the government in developed countries but have insufficient credit markets and social safety net program in developing countries [108]. Poor countries are more susceptible to climate shocks and low urban wages and purchasing power, which means that children in this urban area will also have less to eat [109]. Moreover, chronic degenerative diseases can be attributed to mom's nutrition during pregnancy. Seema Jayachandran and Rohini Pande in their report "Why Are Indian Children So Short" pointed out that malnutrition in the early stages of life can contribute to stunting, which is an indicator of adult health; it can also increase vulnerability to other health shocks. For example, Indian children born second tend to be much shorter than firstborns; children born third are even shorter, and so forth [110]. Another example of nutrition-related problem is a cleft palate which is easily treated in rich countries [111].

Extreme weather such as typhoons, hurricanes, and droughts can be disastrous, wiping out entire crops, housing, health centers, and schools, and lead to famine affecting people in developing countries [112].

Education

Climate change may affect education through:

- Shocks that affect income and wages, lowering school enrollment, and force children to work.
- Poor nutrition because of income losses.
- Direct effects of climate changes like pollution or heat.
- Children are forced to work to help provide for the family. Evaluations of programs that gave additional income to families in developing countries or directly helped them with school fees show that income changes have large positive effects on school attendance [113].
- In contrary, excessive rain can improve school attendance by reducing wages, thus making jobs scarce for women and children [114].

Climate change affects education depending on income losses to families and changes to children's wages. Thus, climate shocks, which affect both the labor market and wages at the same time, may have different effects than does loss of income alone.

Poor health because of low income can affect fetus in the womb, and infants; this later affects children's school attendance. Taking birth weight as an indicator of mothers' nutrition and health, according to Janet and Rosemary study (1999), US children whose birth weight is low (defined as less than 2500 g, or about 5.5 pounds) are much less likely later in life to pass high-school-equivalency exams and to be employed [115]. Another study done by Erica, Maximo and Omar (2009) indicates that mothers' caloric intake and their overall nutritional status during pregnancy is important; for instance, in Tanzania, when intensive iodine supplementation program was introduced for mothers, children achieved 0.36–0.56 additional years of schooling [116]. Another factor that affects schooling was related to the weather. More rain, means more crops, better nutrition, better school attendance; nevertheless, no big difference has been shown in boys, as families in developing countries are investing more on boys rather than on girls [117, 118].

Extreme events causing massive impacts on human capital accumulation will also affect education. For example, the famine that took place in China in 1959–1961 affected pregnant mothers; men born were 9% more likely to be illiterate, and girls were 6% more likely to be illiterate. Less illiterate means higher rates of unemployment and changes in marital patterns [119]. Similarly, Greece famine that took place in 1941–1942 influenced education with more illiteracy and fewer years of education [120]. On the other hand, extreme weather damaged the infrastructures [121].

Extreme events mean more children's parents die that will affect schooling. "Children whose mothers die tend to attain fewer years of schooling and have less money for their education" [122]. Heat can also affect the economic status in adults in adults and then affect children's schooling [123].

Displacement and Migration

Extreme events force families to be displaced putting them under poor conditions such as refugee camps, with low sanitation, poor access to food, healthcare, and education [124]. Permanent migration could reduce the adverse bad consequences of natural disasters [125]. In many cases, when people moved to cities, the latter will be overburdened with more migrants. Such migration might affect the crops and the families [126]. Specifically, children's health and schooling will be affected if the increase in urban populations is not going along with more social services [127]. Sometimes the migration might have positive impact on income loss, if the weather helps, a high proportion of jobs might be integrated in the industries, such as tourism [128]. Some studies have found that children who were left behind had better schooling outcomes [129].

Psychological Impacts

Climate change has tremendous effects on children's psychological health. Mental health affects schooling, education, and employment in the future. Nilamadhab Kar et al. (2007) indicate that children exposed to more natural disasters are more likely to suffer posttraumatic stress [130]. Studies also showed that children exposure to famine and war can cause antisocial behaviors and schizophrenia in adulthood [131].

Climate change will affect adults' mental health as well. Poor economic status causes more distress and less ability to decision-making among adults [132].

Violence

Climate change increases violence as the following:

Greater heats generate more aggression and increase household violence.

- Climate shocks lead to economic stress which will increase violence.
- Job loss can lead to child abuse, child transactional sex for money, service, or goods [133].
- Climate change raises crime rates and increases the probability of conflict and war; both low rains and high temperature increase the rates of property crimes [134].

For instance, job loss among US men was correlated with child abuse [135] and hurricanes have been linked to rise the cases of traumatic brain injury among young children [136]. Moreover, mothers stressed from violence during pregnancy could lead to lower birth weight [137]. Some studies also correlate the rainfalls shock with increase in the prevalence of HIV due to risk sexual behavior [138].

Natural Disasters and Long-Lasting Effects on Survivor's Health

Natural disasters kill people directly, 2.47 deaths and 2.7 injuries occurred from hurricanes took place from 1970 till 2002. Hurricanes estimated that from 1970 to 2002, they killed 2.47 million people and injured another 2.7 million. However, the effects of natural disaster are also long-run ones; poor sanitation, destroyed infrastructures, weak healthcare services will also cause further deaths [139–141]. Poor health in the adulthood will also be attributed to bad nutrition both in wombs and early childhood occurred due to natural disasters. Famine that happened in China in 1959–1961 generated shorter children, 1.2 in. shorter [142].

Finally, "*Climate change will likely impair child nutrition and health, offsetting the gains that children in developing countries have made in recent decades*" [86].

Policies to Mitigate the Climate Change Implications on Children Might Include

- Expanding electrification and developing new technology, such as air conditioning which mitigate the harm happens due to hot days inventing more weather resistance crops
- Reducing our carbon output
- Improving access to clean water
- Increasing foreign aid during disasters
- Expanding safety net programs to support families
- Introducing more-weather resistant crops
- Support families in adjusting to the new situations and challenges arising from climate change

Climate Change Health Impacts on Middle East:
Middle East going to experience climate change threatening basic life.

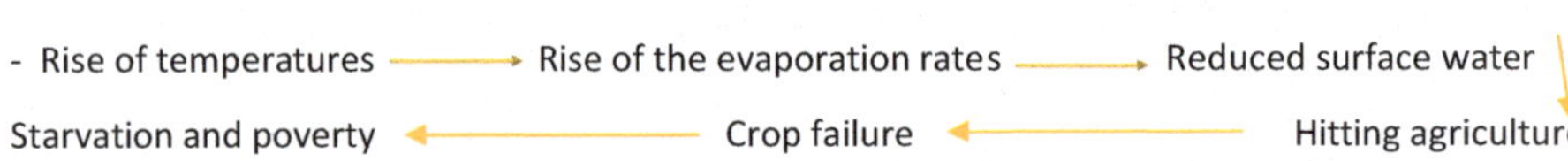

How Climate Change Will Affect People's Health in The Middle East:

- Rise of temperature leads to lack of drinking water, which will in turn increase cholera and other waterborne diseases
- The government could respond by importing more water at a financial cost but also an environmental one
- Importing water will require transport, causing further release of CO_2, of putting the region under climate change challenges again
- Food prices will rocket due to import, as seen in Iraq 1999/00 and 2000/01 drought caused non-irrigated wheat production to decrease by 90% [143].

- Lack of enough drinking water will lead to more cholera outbreaks and other waterborne diseases [144].

8.3.6 Climate Change in UAE

UAE has one of the highest levels of GHG emissions per capita, consistently ranking second or third in the world over the past two decades.

Health impacts of climate change in the UAE: climate change is likely to have only limited impacts on infectious and diarrheal diseases in the UAE due to relatively low baseline levels of these climate-sensitive diseases. However, climate change is more likely to increase heat stress and possibly increased water- and vector-borne diseases, which are more controllable in UAE.

There are 410 additional healthcare facility visits and three additional deaths from cardiovascular disease in the UAE in 2008 due to the added risks of climate change [144].

8.3.7 Actions to Be Done

- Promoting active modes of transportation to reduce CO_2 emission released from traffic-related air pollution, and at the same time promoting better health associated with more physical activity
- Create supportive environment through enhancing sanitation and hygiene to reduce vector-borne diseases
- Encourage people in the risky region vulnerable to insects' bites to cover their exposed skin to avoid insects' bites
- Enhance intersectoral cooperation at different governmental level to expand more mosquito-controlled programs to prevent several infectious outbreaks
- Inspect the compliance of all country levels to take the appropriate measures protecting public health

8.4 Conclusion

Climate change affects the social and environmental determinants of health—clean air, safe drinking water, enough food, and secure shelter. Climate change has enormous and diverse effects on human health. Rises in temperature and sea level and extreme weather events such as floods cause water logging and contamination, which in turn exacerbate diarrheal diseases. Vector-borne diseases and other communicable disease will be the result of climate change. Poor and poorer nations will be more impacted due to fragile health systems and resources.

Appendix 1: Extreme Weather Events and Related Health Problems

	Primary environmental effects	Social effects	Health impacts		Vulnerable groups
Heat-waves	Reduced crop productivity	Increased food prices	Increased total mortality	Diabetes	Elders
	Animals death		Cause-specific mortality	Mental health	Children
	Fires		Respiratory diseases	CVD	People with pre-existing chronic diseases
			Rental diseases		
Floods	Increase bacteria	Infrastructure damage	Mortality due to drowning	Injuries	Elders
	Parasites	Mobility	CVD	Intoxication	People live in flood plains
		Relocation	Injuries	Mental health	Repair workers
		Interruption of healthcare provision	Respiratory infections		
Droughts	Reduced crop productivity	Relocation	Protein-energy malnutrition	Skin diseases	Children
	Inadequate water supply	Inadequate sanitation services	Micronutrient deficiency		Women
	Lack of water	Lack of sanitation	Skin diseases		
	Animals deaths		Infectious diseases		
			Acute respiratory infections		
	Fires		Measles		
Fires	Toxic and pollutants in the air	Infrastructure damage	Deaths		Young children
			Burning injuries		Pregnant women
			Asthma	Eyes irritations	Elders
			Chronic diseases		CVD patients
			CVD		Respiratory patients
			Mental health		Fire fighter

Appendix 2: Increase of Temperature and Changes in Precipitation and Related Health Problems

	Primary environmental effects	Social effects	Health impacts		Vulnerable groups
Glacier melting and Sea level rise	Floods risk	Relocation	Injuries	Morbidity	Everyone
	Breaking down of coastal settlement		Morbidity	Mortality	
	Water quality		Waterborne diseases	Diarrheal diseases Cholera/***typhoid***	
	Higher sea surface temperatures increase the occurrence of algal		Poisoning	Algal blooms	
			Foodborne diseases "Salmonella"	Parasitic diseases such as amoebiasis, giardiasis and cryptosporidium	
Reduced food yields		Changes in food availability	Malnutrition	Child development	Children
		Change in food price	Micronutrient deficiencies		Pregnant women
Ecosystem changes	Microbial ecology change		Vector- borne diseases	Leishmaniasis	Everyone
	Host animals		Rodent borne diseases	Dengue	
	Vectors			Tickborne diseases	
	Pathogens multiplications				
Heavy Storms	Air pollution		Morbidity	CVD	People with chronic diseases
			Mortality	Respiratory diseases	
		Displacements	Mental health		Children
		Loss jobs	Other related risks to lack of accessing health services		Children, Elders
		Loss of livelihood			
Heavy rainfall	Transfer terrestrial microbiological agents into drinking-water sources		***Cryptosporidiosis, giardiasis, amoebiasis, typhoid***		Everyone

References

1. McMichael AJ (2013) Globalization, climate change, and human health. N Engl J Med 368:1335–1343
2. Confalonieri U et al (2007) Human health. Climate change 2007: impacts, adaptation and vulnerability. Contribution of Working Group II to the fourth assessment report of the Intergovernmental Panel on Climate Change. In: Parry ML et al (eds) Fourth assessment report of the Intergovernmental Panel on Climate Change. Cambridge University Press, Cambridge, pp 391–431
3. McMichael AJ et al (2000) The sustainability transition: a new challenge (Editorial). Bull World Health Organ 78:1067
4. UNchronicle (2007) The health effects of global warming: developing countries are the most vulnerable. The magazine of United Nations, vol XLIV(2)
5. CDC (2016) Climate effects on health. Available on: https://www.cdc.gov/climateandhealth/effects/default.htm
6. Kumaresan J, Sathiakumar N (2010) Climate change and its potential impact on health: a call for integrated action. World Health Organization Center for Health Development, Kobe, Japan. Bull World Health Organ 88:163. https://doi.org/10.2471/BLT.10.076034. Available at: http://www.who.int/bulletin/volumes/88/3/10-076034/en/#R1
7. McMichael AJ et al. Climate change and human health: risks and responses. WHO Library Cataloguing-in-Publication Data. ISBN: 92-4-156248-X
8. Langford IH, Bentham G (1995) The potential effects of climate change on winter mortality in England and Wales. Int J Biometeorol 38:141–147
9. Rooney C et al (1998) Excess mortality in England and Wales during the 1995 heatwave. J Epidemiol Commun Health 52:482–486
10. Kalkstein LS, Greene JS (1997) An evaluation of climate/mortality relationships in large U.S. cities and the possible impacts of a climate change. Environ Health Perspect 105:84–93
11. Donaldson GC et al (2001) Heat and cold related mortality and morbidity and climate change. In: Health effects of climate change in the UK. Department of Health, London
12. Whitman S et al (1997) Mortality in Chicago attributed to the July 1995 heatwave. Am J Public Health 87(9):1515–1518
13. Kilbourne EM (1989) Heat waves. In: Gregg MB (ed) The public health consequences of disasters. US Department of Health and Human Services, Centers for Disease Control, Atlanta, GA, pp 51–61
14. Kumar S (1998) India's heatwave and rains result in massive death toll. Lancet 351: 1869–1869
15. Centers for Disease Control and Prevention (CDC) (1997) Heat-related deaths—Dallas, Wichita, and Cooke Counties, Texas, and United States, 1996. J Am Med Assoc 278:462–463
16. Bridger CA, Helfand LA (1968) Mortality from heat during July 1966 in Illinois. Int J Biometeorol 12:51–70
17. Donaldson GC et al (1998) Winter mortality and cold stress in Yekaterinberg, Russia: interview survey. Br Med J 316:514–518
18. West RR, Lowe CR (1976) Mortality from ischaemic heart disease: inter-town variation and its association with climate in England and Wales. Int J Epidemiol 5(2):195–201
19. Donaldson GC et al (2001) Outdoor clothing and its relationship to geography, climate, behaviour and cold related mortality. Int J Biometeorol 45:45–51
20. Mitchell R (2001) Short days—shorter lives: studying winter mortality to get solutions. Int J Epidemiol 30:1116–1118
21. Wilkinson P et al (1999) Case-control study of hospital admission with asthma in children aged 5–14 years: relation with road traffic in north-west London. Thorax 54(12): 1070–1074

22. Glantz M, Jamison DT (2000) Societal response to Hurricane Mitch and intra-versus inter-generational equity issues: whose norms should apply? Risk Anal 20(6):869–882
23. Munich re group topics: natural catastrophes 2000, 2001
24. Pan American Health Organization (PAHO) (1999) Disasters and health in 1998: a report of the Pan American Health Organization's Emergency Preparedness and Disaster Relief Coordination Program. PAHO, Washington, DC
25. Banse M, Van Meijl H, Tabeau A, Woltjer G (2008) Will EU biofuel policies affect global agricultural markets? Eur Rev Agric Econ 35:117–141
26. Curriero F et al (2001) The association between extreme precipitation and waterborne disease outbreaks in the United States, 1948–1994. Am J Public Health 91(8):1194–1199
27. Bouma MJ, van der Kaay HJ (1996) The El Niño Southern Oscillation and the historic malaria epidemics on the Indian subcontinent and Sri Lanka: an early warning system. Trop Med Int Health 1(1):86–96
28. McMichael AJ et al (1996) Climate change and human health: an assessment prepared by a task group on behalf of the World Health Organization, the World Meteorological Organization and the United Nations Environment Programme. World Health Organization, Geneva
29. World Health Organization (WHO) (1998) El Niño and its health impacts. Wkly Epidemiol Record 20:148–152
30. Dales RE et al (1991) Respiratory health effects of home dampness and molds among Canadian children. Am J Epidemiol 134:196–203
31. Greenough G et al (2001) The potential impacts of climate variability and change on health impacts of extreme weather events in the United States. Environ Health Perspect 109(Suppl 2):191–198
32. Bennet G (1970) Bristol floods 1968: controlled survey of effects on health of local community disaster. Br Med J 3:454–458
33. Abrahams MJ et al (1976) The Brisbane floods, January 1974: their impact on health. Med J Aust 2:936–939
34. International Federation of Red Cross and Red Crescent Societies (IFRC) (1999) World disasters report. International Federation of Red Cross and Red Crescent Societies, Geneva
35. Sastry N (2002) Forest fires, air pollution and mortality in south-east Asia. Demography 39(1):1–23
36. Schwela DH et al (1999) Health guidelines for vegetation fire events. Institute of Environmental Epidemiology (IEE), Singapore
37. Public Health Institute, Central Climate Change and Health (2016) Sea level rise
38. Ziska LH, Caulfield FA (2000) Pollen production of common ragweed (Ambrosia artemisiifolia) a known allergy–inducing species: implications for public health. Aust J Plant Physiol 27:893–898
39. Ziska LH, Caulfield FA (2000) The potential influence of raising atmospheric carbon dioxide (CO2) on public health: pollen production of common ragweed as a test case. World Resour Rev 12:449–457
40. World Health Organization (WHO) (1999) Health guidelines for episodic vegetation fire events. World Health Organization, Geneva. (WHO/EHG/99.7)
41. Shindell DT et al (1998) Increased polar stratospheric ozone losses and delayed eventual recovery owing to increasing greenhouse gas concentrations. Nature 392:589–592
42. Patz JA et al (2000) The potential health impacts of climate variability and change for the United States. Executive summary of the report of the health sector of the U.S. National Assessment. J Environ Health 64:20–28
43. Lipp EK, Rose JB (1997) The role of seafood in foodborne diseases in the United States of America. Revue Scientific et Technicale. Office Internationale des Epizootics 16:620–640
44. World Health Organization (WHO) (2002) The world health report 2002. World Health Organization, Geneva

45. Prothero RM (1994) Forced movements of populations and health hazards in tropical Africa. Int J Epidemiol 23:657–663
46. Atherholt TB et al (1998) Effects of rainfall on giardia and cryptosporidium. J Am Water Works Assoc 90:66–80
47. Lisle JT, Rose JB (1995) Cryptosporidium contamination of water in the US and UK: a mini-review. Aqua 44:103–117
48. Rose JBS et al (2000) Climate and waterborne outbreaks in the US: a preliminary descriptive analysis. J Am Water Works Assoc 92:77–86
49. Colwell RR (1996) Global warming and infectious diseases. Science 274:2025–2031
50. Schwartz J, Levin R (1999) Drinking water turbidity and health. Epidemiology 10:86–89
51. Clayton S, Manning CM, Hodge C (2014) Beyond storms & droughts: the psychological impacts of climate change. American Psychological Association and ecoAmerica, Washington, DC, 51 pp
52. World Health Organization (WHO) (1998) Malaria—WHO fact sheet no. 94. World Health Organization, Geneva
53. Martens WJM (1998) Health and climate change: modelling the impacts of global warming and ozone depletion. Earthscan, London
54. Patz JA et al (1996) Global climate change and emerging infectious diseases. J Am Med Assoc 275:217–223
55. MacDonald G (1957) The epidemiology and control of malaria. Oxford University Press, Oxford
56. Bradley DJ (1993) Human tropical diseases in a changing environment. Environmental change and human health. In: Ciba Foundation Symposium 175. Press, Oxford, pp 146–162
57. Lindsay SW, Birley MH (1996) Climate change and malaria transmission. Ann Trop Med Parasitol 90(6):573–588
58. Gubler DJ et al (2001) Climate variability and change in the United States: potential impacts on vector- and rodent-borne diseases. Environ Health Perspect 109(Suppl 2):223–233
59. Bradshaw WE, Holzapfel CM (2001) Genetic shift in photoperiodic response correlated with global warming. Proc Natl Acad Sci USA 98(25):14509–14511
60. Cox J et al (1999) Mapping malaria risk in the highlands of Africa. MARA/HIMAL technical report, p 96
61. Jetten TH, Focks DA (1997) Potential changes in the distribution of dengue transmission under climate warming. Am J Trop Med Hyg 57:285–287
62. Martens WJM et al (1995) Climate change and vector-borne diseases: a global modeling perspective. Global Environ Change 5:195–209
63. Martens WJM et al (1995) Potential impact of global climate change on malaria risk. Environ Health Perspect 103:458–464
64. Bouma MJ, van der Kaay HJ (1995) Epidemic malaria in India's Thar Desert. Lancet 373:132–133
65. Bouma MJ, Dye C (1997) Cycles of malaria associated with El Niño in Venezuela. J Am Med Assoc 278:1772–1774
66. Diaz HF et al (2000) Climate and human health linkages on multiple timescales. In: Jones PD et al (eds) Climate and climatic impacts through the last 1000 years. Cambridge University Press, Cambridge
67. Gill CA (1920) The relationship of malaria and rainfall. Indian J Med Res 7(3):618–632
68. Gubler DJ (1997) Dengue and dengue hemorrhagic fever: its history and resurgence as a global public health problem. In: Gubler DJ, Kuno G (eds) Dengue and dengue hemorrhagic fever. CAB International, New York, pp 1–22
69. Lien TV, Ninh NH (1996) In: Glantz MH (ed) Currents of change: El Niño's impact on climate and society. Cambridge University Press, Cambridge
70. Parmenter RR et al (1999) Incidence of plague associated with increased winter-spring precipitation in New Mexico. Am J Trop Med Hyg 61:814–821

71. Wilson ML. Ecology and infectious disease(2001) In: Aron JL, Patz JA (eds) Ecosystem change and public health: a global perspective. John Hopkins University Press, Baltimore, MD, pp 283–324
72. Mills JN et al (1999) Long-term studies of hantavirus reservoir populations in the southwestern United States: rationale, potential and methods. Emerg Infect Dis 5:95–101
73. Hjelle B, Glass GE (2000) Outbreak of hantavirus infection in the Four Corners region of the United States in the wake of the 1997–1998 El Niño-Southern Oscillation. J Infect Dis 181(5):1569–1573
74. Wenzel RP (1994) A new hantavirus infection in North America. N Engl J Med 330: 1004–1005
75. Engelthaler DM et al (1999) Climatic and environmental patterns associated with hantavirus pulmonary syndrome, Four Corners region, United States. Emerg Infect Dis 5:87–94
76. Food and Agricultural Organization of the United Nations (FAO) (1999) The state of food insecurity in the world 1999. FAO, Rome, p 32
77. Sachs J, Malaney P (2002) The economic and social burden of malaria. Nature 415:680–686
78. Intergovernmental Panel on Climate Change (IPCC) (2001) Climate change 2001: IPCC third assessment report. Intergovernmental Panel on Climate Change, Geneva
79. Nelson KE (2000) Early history of infectious disease: epidemiology and control of infectious diseases. In: Nelson KE et al (eds) Infectious disease epidemiology. Aspen Publishers Inc., Gaithersburg, MD, pp 3–16
80. Moore PS (1992) Meningococcal meningitis in sub-Saharan Africa: a model for the epidemic process. Clin Infect Dis 14(2):515–525
81. WHO (2013) Climate change and health: a tool to estimate health and adaptation costs
82. Gubler DJ, Meltzer M (2000) The impact of dengue/dengue haemorrhagic fever in the developing world. In: Maramorosch K et al (eds) Advances in virus research, vol 53. Academic Press, San Diego, CA, pp 35–70
83. Campbell-Lendrum D et al (2003)
84. Barreca A et al (2013) Adapting to climate change: the remarkable decline in the U.S. temperature—mortality relationship over the 20th century, Working paper no. 18692. National Bureau of Economic Research, Cambridge, MA
85. Almond D, Chay KY, Greenstone M (2006) Civil rights, the war on poverty, and black-white convergence in infant mortality in the rural South and Mississippi, Working paper no. 07-04. Department of Economics, Massachusetts Institute of Technology, Cambridge, MA
86. Hanna R, Oliva P (2016) Implications of climate change for children in developing countries. Future for children.org 26(1)
87. Almond D, Currie J (2011) Killing me softly: the fetal origins hypothesis. J Econ Perspect 25(3):153–172. https://doi.org/10.1257/jep.25.3.153
88. World Health Organization. Global Health Observatory data repository. http://apps.who.int/gho/data/node.main.CM300REG6
89. Hijioka Y et al (2014) Asia. In: Barros VR et al (eds) Climate change 2014: impacts, adaptation, and vulnerability. Part B: regional aspects. Cambridge University Press, Cambridge, pp 1327–1370
90. Gamper-Rabindran S, Khan S, Timmins C (2010) The impact of piped water provision on infant mortality in Brazil: a quantile panel data approach. J Develop Econ 92:188–200. https://doi.org/10.1016/j.jdeveco.2009.02.006
91. Orlowsky and Seneviratne. Global changes
92. Rodó X et al (2002) ENSO and cholera: a nonstationary link related to climate change? Proc Natl Acad Sci 99:12901–12906. https://doi.org/10.1073/pnas.182203999
93. UNICEF. Bangladesh—statistics. http://www.unicef.org/infobycountry/bangladesh_bangladesh_statistics.html. Accessed 26 Jan 2015
94. Miguel E, Kremer M (2004) Worms: identifying impacts on education and health in the presence of treatment externalities. Econometrica 72:159–217. https://doi.org/10.1111/j.1468-0262.2004.00481.x

95. Baird S et al (2015) Worms at work: long-run impacts of a child health investment, Working paper no. 21428. National Bureau of Economic Research, Cambridge, MA
96. Epstein PR (2001) Climate change and emerging infectious diseases. Microb Infect 3:747–754. https://doi.org/10.1016/S1286-4579(01)01429-0
97. Rogers DJ, Randolph SE (2000) The global spread of Malaria in a future, warmer world. Science 289:1763–1766. https://doi.org/10.1126/science.289.5485.1763
98. Martens WJM, Jetten TH, Focks DA (1997) Sensitivity of malaria, schistosomiasis and dengue to global warming. Clim Change 35:145–156. https://doi.org/10.1023/A:1005365413932
99. Strömberg D (2007) Natural disasters, economic development, and humanitarian aid. J Econ Perspect 21(3):199–222. https://doi.org/10.1257/jep.21.3.199
100. Kellenberg DK, Mobarak AM (2008) Does rising income increase or decrease damage risk from natural disasters? J Urban Econ 63:788–802. https://doi.org/10.1016/j.jue.2007.05.003
101. Flannigan MD, Stocks BJ, Mike Wotton B (2000) Climate change and forest fires. Sci Total Environ 262:221–229. https://doi.org/10.1016/S0048-9697(00)00524-6
102. Jayachandran S (2009) Air quality and early-life mortality: evidence from Indonesia's wildfires. J Human Resour 44:916–954. https://doi.org/10.3368/jhr.44.4.916
103. Prospero JM, Lamb PJ (2003) African droughts and dust transport to the Caribbean: climate change implications. Science 302:1024–1027. https://doi.org/10.1126/science.1089915
104. A. Pope III (2000) Epidemiology of fine particulate air pollution and human health: biologic mechanisms and who's at risk? Environ Health Perspect 108:713–723. https://doi.org/10.2307/3454408
105. Chay KY, Greenstone M (2003) The impact of air pollution on infant mortality: evidence from geographic variation in pollution shocks induced by a recession. Quart J Econ 118:1121–1167. https://doi.org/10.1162/00335530360698513
106. Luis AD et al (2010) The effect of seasonality, density and climate on the population dynamics of Montana deer mice, important reservoir hosts for Sin Nombre hantavirus. J Anim Ecol 79:462–440. https://doi.org/10.1111/j.1365-2656.2009.01646.x
107. Harvell D et al (2002) Climate warming and disease risks for terrestrial and marine biota. Science 296:2158–2162. https://doi.org/10.1126/science.1063699
108. Chaudhuri S, Paxson C (2002) Smoothing consumption under income seasonality: buffer stocks vs. credit markets, Discussion paper 0102-54. Department of Economics, Columbia University, New York
109. Currie J, Vogl T (2013) Early-life health and adult circumstance in developing countries. Annu Rev Econ 5:1–36. https://doi.org/10.1146/annurev-economics-081412-103704
110. Jayachandran S, Pande R (2014) Why are Indian children so short? Women and Public Policy Program, John F. Kennedy School of Government, Harvard University, 3 Nov 2014
111. Currie J, Vogl T (2012) Early-life health and adult circumstance in developing countries, Working paper no. 18371. National Bureau of Economic Research, Cambridge, MA
112. Sen A (1981) Ingredients of famine analysis: availability and entitlements. Quart J Econ 96:433–464. https://doi.org/10.2307/1882681
113. Gertler P (2004) Do conditional cash transfers improve child health? Evidence from PROGRESA's control randomized experiment. Am Econ Rev 94(2):336–341. https://doi.org/10.1257/0002828041302109. Demography 43 (2006): 401–20, https://doi.org/10.1353/dem.2006.0022
114. Shah M, Steinberg BM (2013) Drought of opportunities: contemporaneous and long term impacts of rainfall shocks on human capital, Working paper no. 19140. National Bureau of Economic Research, Cambridge, MA
115. Currie J, Hyson R (1999) Is the impact of health shocks cushioned by socioeconomic status? The case of low birthweight. Am Econ Rev 89(2):245–250. https://doi.org/10.1257/aer.89.2.245
116. Field E, Robles O, Torero M (2009) Iodine deficiency and schooling attainment in Tanzania. Am Econ J: Appl Econ 1(4):140–169

117. Glewwe P, Jacoby H, King E (2001) Early childhood nutrition and academic achievement: a longitudinal analysis. J Public Econ 81:345–368. https://doi.org/10.1016/S0047-2727(00)00118-3
118. Maccini S, Yang D (2009) Under the weather: health, schooling, and economic consequences of early-life rainfall. Am Econ Rev 99:1006–1026. https://doi.org/10.1257/aer.99.3.1006
119. Almond D et al (2010) Long-term effects of early-life development: evidence from the 1959 to 1961 China famine. In: Ito T, Rose A (eds) The economic consequences of demographic change in East Asia, NBER-EASE, vol 19. University of Chicago Press, Chicago, pp 321–345
120. Neelsen S, Stratmann T (2011) Effects of prenatal and early life malnutrition: evidence from the Greek famine. J Health Econ 30:479–488. https://doi.org/10.1016/j.jhealeco.2011.03.001
121. Gould ED, Lavy V, Paserman MD (2011) Sixty years after the magic carpet ride: the long-run effect of the early childhood environment on social and economic outcomes. Rev Econ Stud 78:938–973. https://doi.org/10.1093/restud/rdq038
122. Case A, Ardington C. The impact of parental death on school outcomes: longitudinal evidence from South Africa
123. Heal G, Park J (2013) Feeling the heat: temperature, physiology and the wealth of nations, Working paper no. 19725. National Bureau of Economic Research, Cambridge, MA
124. Currie J et al (2009) Does pollution increase school absences? Rev Econ Stat 91:682–694. https://doi.org/10.1162/rest.91.4.682
125. Hornbeck R (2010) The enduring impact of the american dust bowl: short- and long-run adjustments to environmental catastrophe. Am Econ Rev 102:1477–1507. https://doi.org/10.1257/aer.102.4.1477
126. Bohra-Mishra P, Oppenheimer M, Hsiang SM (2016) Nonlinear permanent migration response. Implications of climate change for children in developing countries. 26(1):131
127. Dinkelman T, Schulhofer-Wohl S (2012) Can migration reduce benefits of spatial programs? A model of congestion externalities with evidence from South Africa, Working paper no. 700. Federal Reserve Bank of Minneapolis, Minneapolis, MN
128. Hsiang SM (2010) Temperatures and cyclones strongly associated with economic production in the Caribbean and Central America. Proc Natl Acad Sci 107:15367–15372. https://doi.org/10.1073/pnas.1009510107
129. Mansuri G (2006) Migration, school attainment, and child labor: evidence from rural Pakistan, Working paper no. 3945. World Bank, Washington, DC
130. Kar N et al (2007) Post-traumatic stress disorder in children and adolescents one year after a super-cyclone in Orissa, India: exploring cross-cultural validity and vulnerability factors. BMC Psychiat 7:8. https://doi.org/10.1186/1471-244X-7-8
131. Neugebauer R, Hoek HW, Susser E (1999) Prenatal exposure to wartime famine and development of antisocial personality disorder in early adulthood. J Am Med Assoc 282(5):455–462. https://doi.org/10.1001/jama.282.5.455
132. Mani A et al (2013) Poverty impedes cognitive function. Science 341:976–980. https://doi.org/10.1126/science.1238041
133. Baird SJ et al (2012) Effect of a cash transfer programme for schooling on prevalence of HIV and Herpes Simplex Type 2 in Malawi: a cluster randomised trial. Lancet 379(9823):1320–1329. https://doi.org/10.1016/S0140-6736(11)61709-1
134. Blakeslee DS, Fishman R (2014) Weather shocks, crime, and agriculture: evidence from India, Working paper. Social Science Research Network, Rochester, NY
135. Sekhri S, Storeygard A (2013) Dowry deaths: consumption smoothing in response to climate variability in India, Online paper 407. Department of Economics, University of Virginia, Charlottesville, VA
136. Keenan HT et al (2004) Increased incidence of inflicted traumatic brain injury in children after a natural disaster. Am J Prev Med 26:189–193. https://doi.org/10.1016/j.amepre.2003.10.023
137. Mansour H, Rees DI (2012) Armed conflict and birth weight: evidence from the al-Aqsa Intifada. J Develop Econ 99(1):190–199. https://doi.org/10.1016/j.jdeveco.2011.12.005

138. Burke M, Gong E, Jones KM (2014) Income shocks and HIV in Africa, Paper 55392. Munich Personal RePEc Archive, Munich
139. Yang D (2008) Coping with disaster: the impact of hurricanes on international financial flows, 1970–2002, B.E. J Econ Anal Policy 8:1–45. https://doi.org/10.2202/1935-1682.1903
140. Costello A et al (2009) Managing the health effects of climate change. Lancet 373:1693. https://doi.org/10.1016/S0140-6736(09)60935-1
141. Anttila-Hughes JK, Hsiang SM (2013) Destruction, disinvestment, and death: economic and human losses following environmental disaster, Working paper. Social Science Research Network, Rochester, NY. https://doi.org/10.2139/ssrn.2220501
142. Chen Y, Zhou L-A (2007) The long-term health and economic consequences of the 1959–1961 Famine in China. J Health Econ 26:659–681. https://doi.org/10.1016/j.jhealeco.2006.12.006
143. Ivanic M, Martin W (2008) Implications of higher global food prices for poverty in low-income countries. http://www-wds.worldbank.org/external/default/WDSContentServer/IW3P/IB/2008/04/16/000158349_20080416103709/Rendered/PDF/wps4594.pdf. Accessed 16 Dec 2009
144. Essays, UK (2013) Effects of climate change on the Middle East, Nov 2013. Retrieved from https://www.ukessays.com/essays/environmental-studies/the-effects-of-climate-change-on-the-middle-east-environmental.php?vref=1
145. Birmingham ME et al (1997) Epidemic cholera in Burundi: patterns in the Great Rift Valley Lake region. Lancet 349:981–985
146. Rigau-Perez JG et al (1998) Dengue and dengue haemorrhagic fever. Lancet 352:971–977
147. National Center for Injury Prevention and Control. 10 Leading causes of death by age group, United States—2012. http://www.cdc.gov/injury/wisqars/pdf/leading_causes_of_death_by_age_group_2012
148. Huq A et al (2005) Critical factors influencing the occurrence of Vibrio cholerae in the environment of Bangladesh. Appl Environ Microbiol 71:4645–4654. https://doi.org/10.1128/AEM.71.8.4645-4654.2005
149. Lavy V, Ebenstein A, Roth S (2014) The impact of short term exposure to ambient air pollution on cognitive performance and human capital formation, Working paper no. 20648. National Bureau of Economic Research, Cambridge, MA
150. Al-Midfa AA, Nation R. QATAR, Introduction. http://www.wetlands.org/RSIS/WKBASE/MiddleEastDir/QATAR.htm. Accessed 17 Jan 2010
151. Adhvaryu A, Fenske J, Nyshadham A (2014) Early life circumstance and adult mental health, Working paper no. 2014-03. Centre for the Study of African Economies, Oxford
152. Dinar A et al (1998) Measuring the impact of climate change on Indian agriculture. Technical paper no. 402. World Bank, Washington, DC, p 12
153. Asian Development Bank (2009) The economics of climate change in Southeast Asia: a regional review. ADB, Manila. Available from: http://www.adb.org.Documents/Books/Economics-Climate-Chant3-SEA.default.asp. Accessed 14 May 2009
154. Atherton F et al (1995) An outbreak of waterborne cryptosporidiosis associated with a public water supply in the UK. Epidemiol Infect 115:123–131
155. Bates BC, Kundzewicz ZW, Wu S, Palutikof JP (eds) (2008) Climate change and water. Technical paper of the Intergovernmental Panel on Climate Change. IPCC Secretariat, Geneva, 210 pp., pp 15–16, 20, 29–30, 40. http://www.ipcc.ch/pdf/technical-papers/climate-change-water-en.pdf. Accessed 27 Dec 2009
156. BBC. Ben Thompson. http://news.bbc.co.uk/1/hi/business/8144171.stm. Accessed 13 Jan 2010
157. Jones BF, Olken BA (2010) Climate shocks and exports. Am Econ Rev: Papers Proc 100:454–459. https://doi.org/10.1257/aer.100.2.454
158. Hansen B (2013) In: National Bureau of Economic Research (ed) Economic conditions and child abuse, Working paper no. 18994, Cambridge, MA. Rema Hanna and Paulina Oliva 132 THE FUTURE OF CHILDREN

159. Bentham G, Langford IH (1995) Climate change and the incidence of food poisoning in England and Wales. Int J Biometeorol 39:81–86
160. Bentham G, Langford IH (2001) Environmental temperatures and the incidence of food poisoning in England and Wales. Int J Biometeorol 45(1):22–26
161. Berman I, Wihbey PM. The new water politics of the Middle East. http://www.iasps.org/strategic/water.htm. Accessed 27 Dec 2009
162. Orlowsky B, Seneviratne SI (2012) Global changes in extreme events: regional and seasonal dimension. Clim Change 110:669–696. https://doi.org/10.1007/s10584-011-0122-9
163. Brown ME et al (2015) Climate change, global food security and the U.S. Food System. U.S. Global Change Research Program, 146 pp
164. Gautier C (2008) Oil, water and climate. Cambridge University Press, Cambridge, pp 4–34, 81–137, 168–221
165. CDC (2015) Lyme disease: data and statistics: maps—reported cases of Lyme disease—United States, 2001–2014. Centers for Disease Control and Prevention
166. Centres for Disease Control and Prevention (1998) Community needs assessment and morbidity surveillance following an ice storm—Maine, January 1998. Morb Mort Wkly Rep 47:351–354
167. Centres for Disease Control and Prevention (1995) Heat-related deaths: Chicago July 1995. Morb Mort Wkly Rep 44:577–597
168. Gray CL, Mueller V (2012) Natural disasters and population mobility in Bangladesh. Proc Natl Acad Sci 109:6000–6005. https://doi.org/10.1073/pnas.111594410
169. Climate change and human health in Asia and the Pacific: from evidence to action. Report of the: Regional Workshop, Bali, Indonesia, 10–12 Dec 2007. World Health Organization, Geneva, 2007. Available from: http://www.who.int/globalchange/publications/en/. Accessed 4 May 2009
170. Climate of Saudi Arabia, Dec 1992. http://ancienthistory.about.com/od/aneplacesgeography/qt/ClimSaudiArabia.htm. Accessed 14 Jan 2010
171. St. Clair D et al (2005) Rates of adult schizophrenia following prenatal exposure to the Chinese famine of 1959–1961. J Am Med Assoc 294(5):557–562. https://doi.org/10.1001/jama.294.5.557
172. Robine JM, Cheung SL, Le Roy S, Van Oyen H, Griffiths C, Michel JP et al (2008) Death toll exceeded 70,000 in Europe during the summer of 2003. C R Biol 331(2):171–178
173. Dedet JP et al (1995) Leishmaniasis and human immunodeficiency virus infections. Presse Med 24:1036–1040
174. Bankher KA, Al-Harthi AA. Earth fissuring and land subsidence in Western Saudi Arabia. http://www.springerlink.com/content/k24237u861k4rwh2/fulltext.pdf. Accessed 14 Jan 2010
175. Economics and Development Resource Center (2008) Basic statistics. Asian Development Bank, Manila. Available from: www.adb.org/statistics/pdf/basic-statistics-2008.pdf. Accessed 28 May 2009
176. Fann N, Nolte CG, Dolwick P, Spero TL, Curry Brown A, Phillips S, Anenberg S (2015) The geographic distribution and economic value of climate change-related ozone health impacts in the United States in 2030. J Air Waste Manag Assoc 65:570–580. https://doi.org/10.1080/10962247.2014.996270
177. Frykberg M (2010) Gaza's water supply near collapse. http://www.commondreams.org/headline/2009/09/17-7. Accessed 14 Jan 2010
178. Glass GE et al (2000) Using remotely sensed data to identify areas at risk for hantavirus pulmonary syndrome. Emerg Infect Dis 6:238–247
179. WHO (2008) Global health statistics. World Health Organization, Geneva
180. Hegerl G, Luterbacher J, González-Rouco F, Tett SFB, Crowley T, Xoplaki E (2011) Influence of human and natural forcing on European seasonal temperatures. Nat Geosci 4:99–103
181. WHO (2009) Impact of climate change on communicable diseases. World Health Organization, Geneva. Available from: http://www.searo.who.int/EN/Section10/Section2537_14458.htm. Accessed 28 May 2009

182. Industrial diversification lifts off in Saudi Arabia. http://www.ameinfo.com/65654.html. Accessed 13 Jan 2010
183. IPCC (2014) Summary for policymakers. In: Edenhofer O, Pichs-Madruga R, Sokona Y, Farahani E, Kadner S, Seyboth K, Adler A, Baum I, Brunner S, Eickemeier P, Kriemann B, Savolainen J, Schlömer S, von Stechow C, Zwickel T, Minx JC (eds) Climate change 2014: mitigation of climate change. Contribution of working group III to the fifth assessment report of the Intergovernmental Panel on Climate Change. Cambridge University Press, Cambridge
184. van der Leun JC, de Gruijl FR (2002) Climate change and skin cancer. Photochem Photobiol Sci 1:324–326. https://doi.org/10.1039/b201025a. … skin cancer could become 21 percent
185. Kyselý J, Beranová R (2008) Climate-change effects on extreme precipitation in central Europe: uncertainties of scenarios based on regional climate models, 20 Jan 2008. http://www.springerlink.com/content/g352151n17746722/fulltext.pdf
186. Gibson J, McKenzie D, Stillman S (2011) The impacts of international migration on remaining household members: omnibus results from a migration lottery program. Rev Econ Stat 93:1297–1318. https://doi.org/10.1162/REST_a_00129
187. Angrist JD, Lavy V (1999) Using Maimonides' rule to estimate the effect of class size on scholastic achievement. Quart J Econ 114:533–575. https://doi.org/10.1162/003355399556061
188. Duflo E, Dupas P, Kremer M (2010) Peer effects, teacher incentives, and the impact of tracking: evidence from a randomized evaluation in Kenya. Am Econ Rev 101:1739–1774. https://doi.org/10.1257/aer.101.5.1739
189. Kalkstein L, Smoyer KE (1993) The impact of climate on Canadian mortality: present relationships and future scenarios. Canadian Climate Centre Report 93-7. Atmospheric Environment Service, Downview, ON, p 50
190. Katsouyanni K et al (1993) Evidence for interaction between air pollution and high temperature in the causation of excess mortality. Archit Environ Health 48:235–242
191. Kovats R (1999) El Niño and health. World Health Organization, Geneva
192. Kríz B (1998) Infectious diseases: consequences of the massive 1997 summer flood in the Czech Republic (EHRO 020502/12 Working group paper)
193. Kríz B et al (1998) Monitoring of the epidemiological situation in flooded areas of the Czech Republic in 1997. In: Proceedings of the conference DDD'98 11–12 May 1998, Podebrady, Czech Republic
194. Lake IR et al (2012) Climate change and food security: health impacts in developed countries. Environ Health Perspect 120:1520–1526. https://doi.org/10.1289/ehp.1104424
195. Landefield J, Seskin E (1982) The economic value of life: linking theory to practice. Am J Public Health 72:555–566
196. Lindhjem H, Analyse V, Navrud S (2012) Meta-analysis of stated preference value of a statistical life studies: further model sensitivity and benefit transfer issues. Working Party on National Environmental Policies, Organization for Economic Co-operation and Development, Paris. (ENV/EOPC/WPNEP(2010)10/ Final)
197. Ramos MC, Martínez-Casasnovas JA (2008) Impacts of annual precipitation extremes on soil and nutrient losses in vineyards of NE Spain. http://www3.interscience.wiley.com/cgi-bin/fulltext/121412168/PDFSTART. Accessed 21 Jan 2010
198. MacKenzie WR et al (1994) A massive outbreak in Milwaukee of cryptosporidium infection transmitted through the public water supply. N Engl J Med 331:1529–1530
199. Manitoba Water Commission (1998) An independent review of action taken during the 1997 Red River flood. Minister of Natural Resources, Winnipeg, MB
200. Martens P (1997) Climate change, thermal stress and mortality changes. Soc Sci Med 46:331–334
201. Martens WJM et al (1999) Climate change and future populations at risk of malaria. Global Environ Change 9(Suppl):S89–S107
202. Kahn ME (2005) The death toll from natural disasters: the role of income, geography, and institutions. Rev Econ Stat 87:271–284. https://doi.org/10.1162/0034653053970339

203. McMichael AJ, Kovats S (1998) Assessment of the impact on mortality in England and Wales of the heat wave and associated air pollution episode of 1976. Report to the Department of Health. London School of Tropical Medicine, London
204. McMichael AJ (2002) Population, environment, disease, and survival: past patterns, uncertain futures. Lancet 359:1145–1148
205. Melillo JM, Richmond TC, Yohe GW (eds) (2014) Climate change impacts in the United States: the third National Climate Assessment. U.S. Global Change Research Program, Washington, DC, 841 pp. https://doi.org/10.7930/J0Z31WJ2
206. Middle East Central Asia Economy (2010). http://www.economywatch.com/world_economy/middle-east-central-asian/
207. Middle Eastern Tourism Analysis, 1 May 2007. http://www.tourism-review.com/top_weekly_full.php?id=305. Accessed 2001
208. Miller A et al (2000) What's fair? Consumer and climate change. Redefining Progress, San Francisco, CA, p 67
209. International Federation of Red Cross and Red Crescent Societies (IFRC) (1996) World disaster report 1996. Oxford University Press, Oxford
210. Türkeş M, Tatlı H (2009) Use of the standardized precipitation index (SPI) and a modified SPI for shaping the drought probabilities over Turkey, 8 Jan 2009. http://www3.interscience.wiley.com/cgi-bin/fulltext/121685339/PDFSTART. Accessed 21 Jan 2010
211. Narain JP (2009) Climate change brings natural disasters and disease. Science and Development Network, London. Available from: http://www.scidev.net/en/health/climate-change-insect-borne-disease/opinions/climate-change-brings-natural-disasters-and-disease.html. Accessed 6 Oct 2009
212. Hildebrandt N, McKenzie DJ (2005) The effects of migration on child health in Mexico, Working Paper no. 3573. World Bank, Washington, DC
213. Brown O, Crawford A. Rising temperatures, rising tensions, climate change and the risk of violent conflict in the Middle East. http://www.iisd.org/pdf/2009/rising_temps_middle_east.pdf. Accessed 27 Dec 2009
214. O'Riordan T (2000) Environmental science for environmental management, Essex, Chapter 7: Climate change, pp 178–182
215. Pan American Health Organization (PAHO) (1998) El Niño and its impact on health
216. Patz J et al (2000) The potential health impacts of climate variability and change for the United States: executive summary of the report of the health sector of the US National Assessment. Environ Health Perspect 108:367–376
217. Franco AA et al (1997) Cholera in Lima, Peru, correlates with prior isolation of Vibrio cholerae from the environment. Am J Epidemiol 146:1067–1075
218. Pittock AB (1999) Coral reef and environmental change: adaptation to what? Am Zool 39:10–29
219. Zhou XN, Yang GJ, Yang K, Wang XH, Hong QB, Sun LP et al (2008) Potential impact of climate change on schistosomiasis transmission in China. Am J Trop Med Hyg 78(2):188–194
220. Bohra-Mishra P, Oppenheimer M, Hsiang SM (2014) Nonlinear permanent migration response to climatic variations but minimal response to disasters. Proc Natl Acad Sci 111:9780–9785. https://doi.org/10.1073/pnas.1317166111
221. Durao R, Pereira MJ, Costa AC, Côrte-Real JM, Soares A (2009) Indices of precipitation extremes in Southern Portugal—a geostatistical approach. http://www.nat-hazards-earth-syst-sci.net/9/241/2009/nhess-9-241-2009.pdf. Accessed 17 Jan 2010
222. Samuels R, Rimmer A, Alpert P (2009) Effect of extreme rainfall events on the water resources of the Jordan River. http://www.sciencedirect.com/science?_ob=MImg&_imagekey=B6V6C-4WNRK7SR&_cdi=5811&_user=486651&_orig=search&_coverDate=09%2F15%2F2009&_sk=996249996&view=c&wchp=dGLbVtb-zSkzV&md5=def4e3d16a3559b93c6763ed0353c362&ie=/sdarticle.pdf. Accessed 18 Jan 2010
223. Royal Society (2009) Reaping the benefits: science and the sustainable intensification of global agriculture. Royal Society, London

224. Soyupak S, Mukhallalati L, Yemişen D, Bayar A, Yurter C (1996) Evaluation of eutrophication control strategies for the Keban Dam reservoir. http://www.sciencedirect.com/science?_ob=MImg&_imagekey=B6VBS-3SWJMJ2-M-3&_cdi=5934&_user=486651&_orig=search&_coverDate=04%2F15%2F1997&_sk=999029998&view=c&wchp=dGLbVzz-zSkWb&md5=48bf35a8ccce93e5d25cbc0d67be5c64&ie=/sdarticle.pdf. Accessed 17 Jan 2010
225. Puri S, Klingbeil R. Aquifers in the Middle East Region. http://www.geopotenzial-nordsee.de/nn_459046/EN/Themen/TZ/TechnZusammenarb/Politikberatung__GW/Downloads/pbgw__puri,templateId=raw,property=publicationFile.pdf/pbgw_puri.pdf. Accessed 14 Jan 2010
226. Lindo JM, Schaller J, Hansen B (2013) Economic conditions and child abuse, Working paper no. 18994. National Bureau of Economic Research, Cambridge, MA
227. Smith AB, Katz RW (2013) US billion-dollar weather and climate disasters: data sources, trends, accuracy and biases. Nat Hazards 67:387–410. https://doi.org/10.1007/s11069-013-0566-5
228. Smith JB, Lenhart SS (1996) Climate change adaptation policy options. Clim Res 6(2):193–201
229. Adhvaryu A, Kala N, Nyshadham A (2014) In: University of Michigan (ed) The light and the heat: productivity cobenefits of energy-saving technology, Ann Arbor, MI
230. Stern N (2007) The economics of climate change: the Stern review. Cambridge University Press, Cambridge
231. Dasgupta S, Laplante B, Meisner C, Wheeler D, Yan J (2007) The impact of sea level rise on developing countries: a comparative analysis, Feb 2007. http://www-wds.worldbank.org/external/default/WDSContentServer/IW3P/IB/2007/02/09/000016406_20070209161430/Rendered/PDF/wps4136.pdf. Accessed 18 Jan 2010
232. Kovats S (2000) El Niño and human health. Bull World Health Organ 78(9):1127–1135
233. Dinkelman T (2013) Mitigating long-run health effects of drought: evidence from South Africa, Working paper no. 19756. National Bureau of Economic Research, Cambridge, MA
234. The World Bank, Middle East & North Africa. http://web.worldbank.org/WBSITE/EXTERNAL/TOPICS/EXTWAT/0,,contentMDK:21634952~menuPK:4708988~pagePK:210058~piPK:210062~theSitePK:4602123,00.html. Accessed 16 Dec 2009
235. Turkey's Climate Regions. http://www.turkeytravelplanner.com/WhenToGo/ClimaticRegions.html. Accessed 14 Jan 2010
236. Turner BL et al (2003) A framework for vulnerability analysis in sustainability science. Proc Natl Acad Sci USA 100:8074–8079. https://doi.org/10.1073/pnas.1231335100
237. United Nations Intergovernmental Panel on Climate Change (2007) Climate change 2007. IPCC fourth assessment report. World Meteorological Association, Geneva. Available from: http://en.wikipedia.org/wiki/IPCC_Fourth_Assessment_Report. Accessed 28 Oct 2009
238. United States Department of Agriculture, Foreign Agriculture Service (2008) IRAQ: drought reduces 2008/09 winter grain production, 9 May 2008. http://www.pecad.fas.usda.gov/highlights/2008/05/Iraq_may2008.htm. Accessed 27 Dec 2009
239. Kharin VV, Zwiers FW, Zhang X, Hegerl GC. Changes in temperature and precipitation extremes in the IPCC ensemble of global coupled model simulations. http://ams.allenpress.com/archive/1520-0442/20/8/pdf/i1520-0442-20-8-1419.pdf. Accessed 14 Jan 2010
240. Wallis PM et al (1996) Prevalence of giardia cysts and cryptosporidium oocysts and characterization of giardia spp. Isolated from drinking water in Canada. Appl Environ Microbiol 62:2789–2797
241. Woodward A et al (1995) Tropospheric ozone: respiratory effects and Australian air quality goals. J Epidemiol Commun Health 49:401–407
242. Petroeschevsky A et al (2001) Associations between outdoor air pollution and hospital admissions in Brisbane, Australia. Arch Environ Health 56:37–52
243. World Atlas. http://www.worldatlas.com/webimage/countrys/me.htm. Accessed 14 Dec 2009

244. World Health Organization (WHO) (2000) In: Corvalán C et al (eds) Decision-making in environmental health: from evidence to action. Published On Behalf of WHO by E & FN Spon, London
245. World Health Organization (WHO) (2000) World health report 2000: health systems: improving performance. World Health Organization, Geneva
246. World Health Organization (2007) Climate change and human health in Asia and the Pacific: from evidence to action. Report of the: Regional Workshop, Bali, Indonesia, 10–12 Dec 2007. WHO, Geneva. Available from: http://www.who.int/globalchange/publications/en/
247. World Health Organization (2009) Impact of climate change on communicable diseases. WHO, Geneva. Available from: http://www.searo.who.int/EN/Section10/Section2537_14458.htm. Accessed 28 May 2009

Chapter 9
Simultaneous Clean Water and Power Production from Seawater Using Osmosis: Process Simulation and Techno-economic Analysis

Elena Barbera, Luca Zorzetto, Adel O. Sharif, and Alberto Bertucco

9.1 Introduction

Global energy demand is currently mostly satisfied by oil, coal, and natural gas. Fossil fuels are non-renewable sources and are the main responsible for greenhouse gas (GHG) emissions and air pollutants [1, 2]. Moreover, because of the increasing population and the rapid economic development, the world is facing unprecedented challenges regarding energy supply. Since the Kyoto Protocol in 1997, the need to reduce GHG emissions from fossil fuels has been amplified all over the world, exploiting several alternative renewable energy sources such as solar, wind, geothermal energy, etc. for global sustainable development.

A novel renewable technology for power production based on osmosis principles has emerged between the late 1990s and the beginning of the new millennium with the construction of the first osmotic power testing facility in Norway by Statkraft [3]. The entropy of mixing water at different salt concentrations is used to enhance osmotic power: when a semi-permeable membrane is placed between compartments containing water with different salt concentrations, a net flow of water permeates the membrane from the low concentrated side to the high concentrated one because of osmosis. Considering a fixed chamber volume in the higher concentrated side, the pressure will increase until a theoretical maximum of about 27 bar if seawater and freshwater are used, corresponding to a water head of 270 m [4–7]. The increased hydraulic pressure obtained could be used in several ways, one of which is power production through a hydro-turbine [4, 8, 9]. Considering typical operating

E. Barbera (✉) · L. Zorzetto · A. Bertucco
Department of Industrial Engineering DII, University of Padova, Padova, Italy
e-mail: elena.barbera@unipd.it

A. O. Sharif
Chemical and Process Engineering Department, Centre for Osmosis Research & Applications, University of Surrey, Surrey, UK

A. Sayigh (ed.), *Renewable Energy and Sustainable Buildings*, Innovative Renewable Energy, https://doi.org/10.1007/978-3-030-18488-9_9

conditions and available membranes, a hydrostatic pressure corresponding to 100 m of water head can generate electricity in a hydropower station by osmotic power technology (pressure retarded osmosis, PRO). It has been estimated that a global power production potential of about 2000 TWh per year could be generated [3], with 170 TWh per year only in Europe [10].

This power production potential could be further increased if a suitable draw solution (DS) is used in place of high concentrated seawater, thanks to the high osmotic pressure achievable with low draw solute concentration. Hydro-osmosis power generation (HOP) emerged as a novel closed-cycle process based on osmosis phenomena [4]. Similar to PRO, forward osmosis principles are harnessed to generate high water permeation through a semi-permeable membrane, from a lower osmotic pressure feed side to a higher osmotic pressure draw solution side, in order to pressurize the DS stream as a result of the osmotic pressure difference between the feed and the draw solutions and to use it to produce energy by means of a hydro-turbine.

Low-molecular-weight alcohols are suitable draw agents to be used in the HOP process, thanks to the high osmotic pressure that is obtained at acceptably low concentration levels. However, other volatile water-soluble components can also be considered (e.g., isopropanol and dimethyl ether). In a closed-loop process perspective, the draw solution needs to be regenerated in a suitable separation unit.

Considering that water stress represents one of the most alarming problems for humankind, the HOP process can be modified for simultaneous power and clean water production, by using wastewater [11], brackish water, or even seawater as feed [7]. In fact, according to the World Resources Institute, it was estimated that 37 countries are currently facing extremely high levels of freshwater scarcity, which mostly affects public health as well as national prosperity and economy. Even if water covers about 70% of the world surface, just a small percentage is usable as freshwater to satisfy the water demand. Due to the uneven distribution of freshwater, more than one billion people currently live in water-stress regions, with an expectation for this number to increase about three times by 2050, due to the increasing world population [12–14]. As the demand for freshwater continues to grow, it is clearly necessary to find new alternatives for freshwater supply.

The modified HOP process for power generation and clean water production aims to represent a renewable alternative in the energy and freshwater industry. Energy is produced taking advantage of the clean water permeation through the semi-permeable membrane. On the other hand, the diluted draw solution is regenerated to the inlet conditions by separating the permeated water, and producing clean water, in a suitable regeneration unit. Although a few experimental and theoretical investigations are available in the literature, assessing the performances and the potential of hydro-osmotic power processes [4–7, 10, 11], detailed material and energy balances aimed at overall process optimization need to be carried out to determine the large-scale feasibility and profitability.

The objective of this work is to perform accurate process simulation based on literature data and to investigate whether the simultaneous clean water production and power generation are feasible in a profitable way. The draw solution and the regeneration unit optimization were carried out by means of sensitivity analyses on the main process variables in order to find out the best conditions to optimize the proposed process.

9.2 Process Model Development

9.2.1 Process Flow Sheet

A plant size corresponding to 1 MW power generation was taken as a reference. The process was simulated using Aspen Plus v.9. According to the process flow sheet displayed in Fig. 9.1, seawater (at a concentration of 35 g/L of sodium chloride) is fed to the osmotic unit countercurrent to the draw solution, which is at relatively high hydraulic pressure (15 bars), and has a higher osmotic pressure to guarantee a net water flux through the membrane. In particular, a hollow fiber membrane module is considered, with the seawater feed flowing in the outer shell (with pressure drops equal to 0.1 bar) and the draw solution inside the fibers (with pressure drops equal to 1 bar). Seawater is thus fed at 1.1 bar, while the draw solution is at 16 bar. The models used for osmotic pressures calculation and for the osmotic unit (OU) simulations are detailed in Sects. 2.2 and 2.3, respectively.

Ethanol was chosen as the draw agent, according to the screening criteria presented by Kim et al. [15], in particular its low molecular weight and viscosity, boiling point <100 °C, which provide a greater ease of separation from water in the regeneration unit, and non-toxicity in water at a concentration <1 ppm.

After the membrane unit, the diluted draw solution is split into two streams: one (*DDS-T15*) goes through the hydro-turbine (whose efficiency η_T is assumed equal to 90%) for power generation, while the other (DDS-PX) is sent to a pressure exchanger (PX) device for pressure recovery [16]. The PX device exploits part of the high-pressure diluted draw solution to directly pressurize the regenerated draw solution (DRAW-2). The energy transfer efficiency of this unit is assumed equal to $\eta_{PX} = 95\%$.

The equation governing the pressure exchange is:

$$\dot{V}_{DDS}\left(P_{DDS-PX} - P_{DDS-PX1}\right)\eta_{PX} = \dot{V}_{DRAW}\left(P_{DRAW-X} - P_{DRAW-2}\right) \tag{9.1}$$

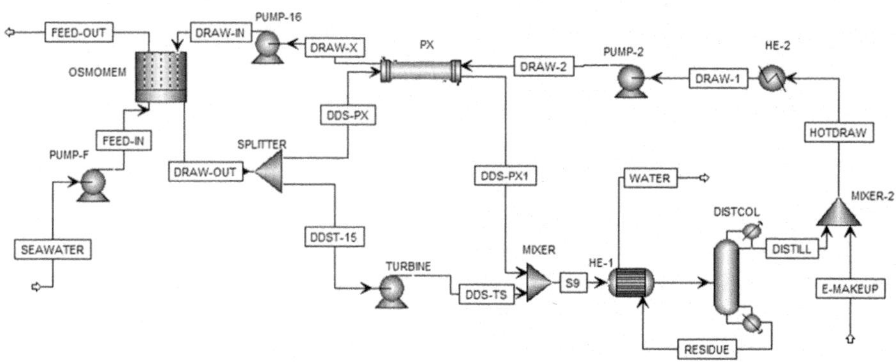

Fig. 9.1 HOP process flow sheet

where $\dot{V}_{DDS}$ and $\dot{V}_{DRAW}$ are the volumetric flow rates of the diluted draw solution entering the PX unit and that of the regenerated draw solution (m^3/h). The diluted draw solution enters the PX device at $P_{DDS\text{-}PX} = 15.0$ bar, while the regenerated draw solution has to be pressurized to a final pressure of $P_{DRAW\text{-}IN} = 16$ bar. A pressure difference of 0.5 bar between the DDS-PX and DRAW-X streams is assumed; hence, $P_{DRAW\text{-}X}$ is set to 14.5 bar, which is then brought up to the final pressure of 16 bars by means of a booster pump (*PUMP-15*). The pressure of the DRAW-2 stream entering the PX device is set equal to 2.0 bar, by means of a supply pump. The PX device is simulated in Aspen Plus by a *User Model* unit. The calculated split ratio for the DRAW-OUT stream resulted equal to 0.48 (i.e., DDS-PX = 0.48·DRAW-OUT).

The two fractions of diluted draw solution, after the hydro-turbine and pressure exchanger units, respectively, are mixed and delivered to the regeneration unit. Different separation techniques, such as a flash unit, distillation, and liquid–liquid extraction, were investigated with the aim of minimizing the energy needed for regeneration. Among these, only distillation resulted in the recovery of a clean water stream of sufficient purity (<1 ppm of ethanol [17]). The distillation unit was simulated at both atmospheric pressure and under vacuum. The latter configuration would in fact allow the use of low-grade waste heating sources. In this case, the power energy required by the vacuum pump was also taken into account, according to [18]:

$$W = \left(\frac{\gamma}{\gamma-1}\right) \times R_g \times T_{in} \times \left[\left(\frac{P_{atm}}{P_v}\right)^{\frac{\gamma-1}{\gamma}} - 1\right] \times \frac{\dot{n}}{1000 \times \eta_{iso} \times \eta_{mec}} \tag{9.2}$$

where $\gamma = c_p/c_v = 1.4$ is the specific heat ratio of air, $R_g = 8.314$ J/mol/K is the ideal gas constant, T_{in} is the temperature of the extracted air, P_v is the desired vacuum pressure, $\dot{n}$ the extracted air molar flow rate, and $\eta_{iso} = \eta_{mec} = 0.8$ are the pump isentropic and mechanical efficiencies.

In both cases, the column was simulated with 13 ideal stages. A specification was set to recover a water flow rate in the residue equal to that extracted in the osmotic unit. The clean water residue stream is used to preheat the diluted draw solution feed, which helps in reducing the thermal duties. Finally, the regenerated draw solution distillate is cooled down to the OU operating temperature (25 °C), brought up to a pressure of 16 bars, and recirculated back to the membrane unit.

9.2.2 Thermodynamic Model Valuation

Osmotic pressure is the key variable governing the performances of the membrane unit. The choice of an appropriate thermodynamic model to evaluate the osmotic pressure of both the seawater feed and the ethanol draw solution is crucial to obtain reliable results.

The osmotic pressure of a solution is thermodynamically defined as:

$$\pi = -\frac{R_g T}{\overline{V}_{pA}} \ln a_A \tag{9.3}$$

where R_g is the ideal gas constant, a_A is the activity of the solvent, and $\overline{V}_{pA}$ is the partial molar volume of pure solvent at temperature T [19]. In the case of non-ideal solutions involving electrolyte solutes, osmotic pressure is determined as:

$$\pi = i_v \Phi \frac{R_g T x_B}{V_{pA}} \tag{9.4}$$

with i_v being the dissociation parameter, i.e., the number of ions per mole of solute produced by its dissolution, and Φ the osmotic coefficient, which takes into account the deviation from ideal behavior [19, 20].

An extensive vapor–liquid equilibrium (VLE) dataset exists for the ethanol–water binary system. Accordingly, the UNIQUAC-HOC activity coefficient model was used to describe the water–ethanol draw solution, while ELECNRTL, which takes into account ions dissociation, was selected for seawater, approximated by a water–NaCl solution. The validity of the two models was verified against experimental data.

9.2.3 Model of Membrane Unit

The osmotic unit is the core of the HOP process. The water flux permeated through the membrane ultimately defines the performances and the profitability of the process, as both power and clean water production depend on its value. Hence, its simulation is fundamental in order to understand the effect of the operating variables to optimize the process conditions and maximize the water flux.

The net water flux J_w ($m^3/h/m^2$) permeating through the membrane and the corresponding flow rate $\dot{V}_w$ (m^3/h) are defined as:

$$J_w = A_w \left(\pi_{DS,m} - \pi_{FS,m} - \Delta P \right) \tag{9.5}$$

$$\dot{V}_w = J_w S_m \tag{9.6}$$

where A_w is the membrane permeability ($L/m^2/h/bar$), $\pi_{DS,m}$ (bar) is the osmotic pressure on the draw side of the membrane active layer, and $\pi_{FS,m}$ (bar) is the osmotic pressure on the feed side of the membrane active layer. ΔP (bar) is the hydraulic pressure difference between the two compartments, and S_m (m^2) is the membrane module surface. Here, a fixed membrane permeability coefficient equal to 0.2 $L/m^2/h/bar$ was considered [4]. We assumed that concentration polarization

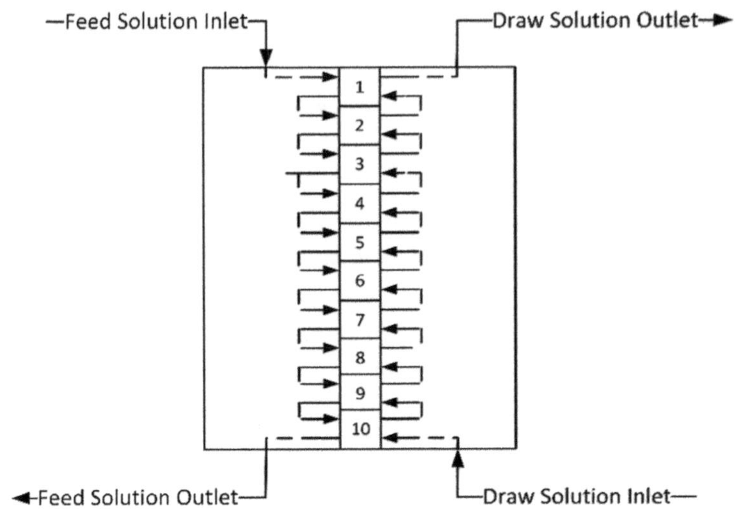

Fig. 9.2 Scheme of the membrane unit subdivision

phenomena are included in this value. Accordingly, the values of osmotic pressures π_{DS} and π_{FS} in Eq. 9.5 are taken as the bulk ones. Moreover, a membrane surface of 1800 m^2 was used for a single membrane module [21].

As described in Sect. 2.1, the seawater feed and ethanol draw solution enter the osmotic unit in a counter-flow configuration. In order to take into account the progressive concentration of the seawater stream and corresponding dilution of the draw solution along the membrane module and the related effect on the osmotic pressure difference between the two streams, the membrane unit was subdivided into ten partitions, each characterized by a surface $S_{m,i} = S_m/10$ (Fig. 9.2). Material balances and net water flux evaluation are then applied to each partition, approximating a plug-flow behavior to allow a better description of the local driving force and the corresponding water flux. This model was implemented in Aspen Plus by means of a *User Defined* Excel unit.

9.3 Process Simulation Results

9.3.1 Osmotic Pressure Validation

Prior to performing process simulations and optimization, the validity of the thermodynamic models chosen in terms of osmotic pressure calculations was verified with experimental data [20, 22]. Figure 9.3 shows the results for the water–NaCl (a) and for water ethanol (b) systems, respectively, at $T = 25$ °C and $P = 1$ bar.

The evaluation of the water–NaCl solution was performed up to a solute molar fraction of 0.1, which corresponds to the salt solubility in water at $T = 25$ °C and

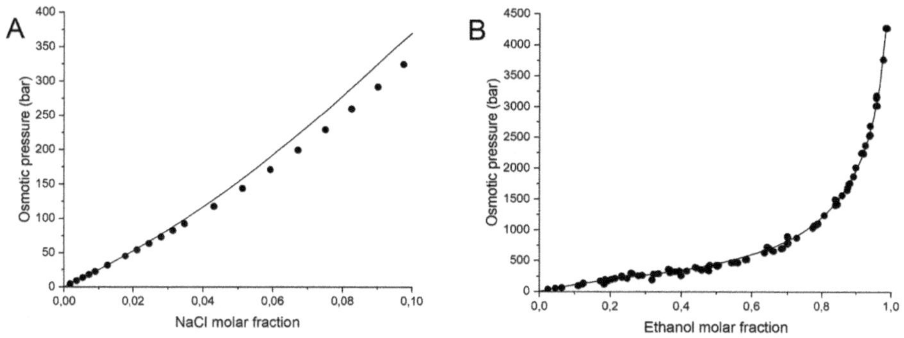

Fig. 9.3 Osmotic pressure of water–NaCl (**a**) and water–ethanol (**b**) solutions as a function of solute concentration. Dots represent experimental values, while continuous lines are the result of Aspen Plus calculations

$P = 1$ bar. The osmotic pressure calculated with the ELECNRTL method fits well the experimental data for low values of NaCl molar fractions, with an increasing deviation (up to 10%) towards higher concentrations. However, the seawater composition corresponds to a salt molar fraction of roughly 0.01 (i.e., 35 g/L), while the maximum molar fraction of NaCl in the concentrated brine outlet obtained from the process simulation is 0.05 (i.e., 160 g/L), with an error of 6.7%, which makes the model acceptable to our scope.

Concerning the water–ethanol binary mixture, the osmotic pressure calculated using the activity coefficient-based model UNIQUAC-HOC matches the experimental values reported by McCormick et al. [22]. It was also verified that this thermodynamic model is able to reproduce experimental VLE data (i.e., the activity coefficients to be used in Eq. 9.3) of the mixture up to hydraulic pressures of 20 bar, which makes it suitable to be used at the operating conditions of the HOP process [23].

9.3.2 *Regeneration Unit*

The draw solution regeneration is the other key operation unit of the HOP process. In order to obtain a profitable process, it is desirable to select a technology with low thermal or power requirements. In addition, since the goal is to recover clean water from the process, the regeneration unit should achieve a complete separation of the extracted water from the draw solution. Among different possibilities investigated, distillation was identified to satisfy both the desired separation and water purity criteria (<1 ppm of ethanol). Considering the boiling point of water at atmospheric pressure (100 °C), low-pressure steam is required to satisfy the thermal duty, with consequently high operating costs. Alternately, performing the distillation operation under vacuum would allow the use of low-grade waste heat sources, which would greatly offset the process operating costs [23]. For this reason, simulations were performed both at atmospheric pressure ($T_{b,wat} = 100$ °C) and at vacuum pressures corresponding to $T_{b,wat} = 50$ °C and 40 °C, which would require waste heat available at $T = 60$ °C and 50 °C, respectively. Considering the vapor pressure of water, the corresponding distillation pressures are 0.124 bar at 50 °C and 0.074 at 40 °C. Table 9.1 summarizes the results obtained in terms of specific thermal duties (kWh/m^3 of extracted water) and power required to achieve the desired vacuum pressure. These regeneration unit simulations, aimed at identifying the most suitable technology, are referred to a non-optimized single module osmotic unit base case (S_m = 1800 m^2), which produces a net power of 2.68 kW and a water flux of 9.52 m^3/h, resulting in an energy density equal to 0.282 kWh/m^3.

Our results show that the specific power consumption required to make the desired vacuum accounts for just 8.5–9% of the net energy produced by the HOP process. The thermal duties slightly vary with respect to atmospheric distillation. If waste heat sources could be used to supply the heat duty at the reboiler, this term would no longer represent a process cost. However, the savings related to the reboiler duties might be offset by the need of using refrigerated cooling water at the condenser. The major drawback of vacuum distillation is related to the column size, which is between two and three times larger with respect to atmospheric distillation. In a plant aimed at the production of 1 MW of power energy, the flow rates involved make this solution technically unfeasible. For these reasons, atmospheric distillation was eventually selected as the draw solution regeneration unit.

9.3.3 Sensitivity Analysis and Optimization

A sensitivity analysis was carried out in order to optimize the process conditions and maximize its performances, which are directly related to the water flux through the osmotic unit. The process variables affecting the water permeation driving force are the draw solution flow rate and concentration and the hydraulic pressure. The optimization was carried out with reference to a single membrane module (S_m = 1800 m^2) and a seawater feed flow rate of 40.4 m^3/h. The optimum condition found was then used to scale up the process for the production of 1 MW net power energy. Figure 9.4 displays the behavior of the permeated water flow rate (a) and of the net power density (W/m^2 of membrane area) (b) as a function of the draw solution flow rate, at three values of inlet ethanol concentration.

The permeated water flow rate increases with the draw solution inlet flow rate and concentration, as a result of the higher osmotic pressure. In particular, using higher flow rates reduces the concentration variation due to dilution of the draw solution along the membrane module, keeping a higher osmotic pressure difference with the seawater feed also at the module outlet. On the other hand, the net power density, i.e., the difference between the power produced by the hydro-turbine and the power consumptions of the pumps, shows a maximum, due to a lower efficiency of the turbine with respect to hydraulic pumps. Higher draw solution flow rates have a negative impact on the performances of the regeneration unit as well, increasing the thermal duties and column size, as can be seen in Table 9.2.

Table 9.1 Specific thermal and power consumption of atmospheric and vacuum distillation units for draw solution regeneration

	$P = 1$ bar		$P = 0.124$ bar		$P = 0.074$ bar	
	Reboiler	Condenser	Reboiler	Condenser	Reboiler	Condenser
Temperature	100 °C	86 °C	50 °C	40 °C	40 °C	31 °C
Heat duty	828 kWh m^{-3}	−670 kWh m^{-3}	747 kWh m^{-3}	−702 kWh m^{-3}	733 kWh m^{-3}	−708 kWh m^{-3}
Power duty	–		0.024 kWh m^{-3}		0.025 kWh m^{-3}	

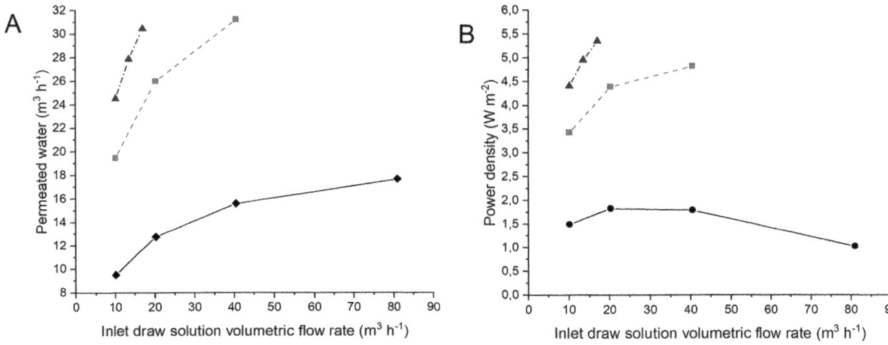

Fig. 9.4 Permeated water flow rate (**a**) and net power density (**b**) as a function of the draw solution inlet flow rate, at different ethanol molar fractions (black diamonds correspond to $x_{\text{ethanol}} = 0.1$, light gray squares to 0.25, and dark gray triangles to 0.4)

Table 9.2 Summary of sensitivity analysis results

Ethanol molar fraction	DS flow rate (m^3/h)	Permeated water (m^3/h)	Net power generated (kW)	Power density (W/m^2)	Energy density (kWh/m^3)	Condenser duty (kW)	Reboiler duty (kW)
0.10	10.1	9.52	2.68	1.49	0.282	−6374	7885
	20.2	12.8	3.29	1.83	0.258	−12,747	15,195
	40.4	15.6	3.23	1.80	0.207	−25,495	29,499
	80.9	17.7	1.85	1.03	0.105	−51,011	57,785
0.25	10.1	19.5	6.16	3.42	0.317	−5018	7315
	20.2	26.0	7.91	4.40	0.304	−10,036	13,467
	40.4	31.2	8.68	4.82	0.278	−20,071	25,058
0.4	10.1	24.5	7.93	4.41	0.323	−4092	6804
	13.5	27.9	8.92	4.96	0.320	−5463	8645
	17	30.4	9.63	5.35	0.316	−6879	10,466

It should be noted that, in order to avoid exceeding the solubility limit of NaCl on the seawater side, the maximum allowable water flow rate across the membrane is equal to about 36 m^3/h. Accordingly, optimum draw solution inlet conditions are identified in correspondence to a volumetric flow rate of 13.5 m^3/h and ethanol molar fraction equal to 0.4. However, it should be specified that the ethanol back flux across the membrane, which was neglected in these simulations, might become relevant at such high concentrations.

Overall, the power density achieved in these conditions is equal to 4.96 W/m^2, which is comparable to that reported in the literature for pressure retarded osmosis processes [4, 9, 21].

Increasing the hydraulic pressure of the draw solution reduces the driving force for water permeation through the membrane. At the same time, the power production by the hydro-turbine increases, thanks to a higher pressure drop [24]. In fact, it is reported

Table 9.3 Stream table of the osmotic unit for a 1 MW power production

	Feed-in	Draw-in	Feed-out	Draw-out
Temperature (°C)	25	25	25	25
Pressure (bar)	1.01	16	1.01	15
Solute molar fraction	0.0109	0.4	0.0347	0.0823
Volumetric flow (m^3/h)	4571	1526	1444	4673
Osmotic pressure (bar)	27.8	355	98.5	96.1

Table 9.4 Overall performances of the 1 MW HOP process

Osmotic membrane area (m^2)	203,400
Total net power generated (kW)	1008
Power density (W/m^2)	4.96
Energy density (kWh/m^3)	0.320
Condenser duty (kW)	−617,330
Reboiler duty (kW)	734,967
Permeated water (m^3/h)	3150
Water purity (ppm of ethanol)	0.923

that the power density is maximized when the hydraulic pressure is approximately equal to half the osmotic pressure difference between the feed and draw solution [4, 5, 10, 21]. In this work, a pressure of 15 bar was chosen, considering the performances of membranes that are currently commercially available [4, 6, 7].

As reported in Table 9.2, a single membrane module operated in these conditions generates 8.92 kW of net power energy. Therefore, in order to obtain 1 MW, 113 modules are required, for a total membrane area of 203,400 m^2. The main results are summarized in Tables 9.3 and 9.4.

The total power generated shows values competitive with those of PRO processes. As well, also clean water production is comparable with most of the existing industrial plants placed all around the world [25, 26]. However, due to the very high flow rates involved (around 4600 m^3/h of draw solution to be regenerated), the thermal duties are consistently high, and more than one column is required to achieve the draw solution regeneration (Sect. 4.2).

9.4 Economic Analysis

9.4.1 Economic Assumptions

The economic analysis carried out to obtain a preliminary evaluation of process profitability, with reference to the process flow diagram in Fig. 9.1, is based on the approach proposed by Turton et al. [27], which provides estimates accurate in the range of +40% to −25%.

The total capital investment (TCI) is calculated as the sum of the fixed capital investment (FCI), working capital (WC), and start-up costs (StC):

$$\mathrm{TCI} = \mathrm{FCI} + \mathrm{WC} + \mathrm{StC} \tag{9.7}$$

Both working capital and start-up costs are estimated based on the fixed capital investment [28], which is represented by direct and indirect costs for building the plant.

The FCI of the pumps and the heat exchangers is calculated based on their bare module cost, i.e., the purchase cost of the equipment at reference conditions (carbon steel material and atmospheric pressure) multiplied by a cost factor, which takes into account the specific construction material and the operating pressure [27]. On the other hand, due to exceeding sizes, the FCI of distillation columns, osmotic unit, pressure exchanger, and hydro-turbine units are estimated using equations proposed by Towler et al. [29]. In addition, the cost of seawater pretreatment is taken into account, and it was assumed equal to 4% of the sum of the total grassroots costs [7]. All the calculated costs are referred to 2017 using the corresponding Chemical Engineering Plant Cost Index (CEPCI).

The cost of manufacturing without depreciation ($\mathrm{COM_d}$) is estimated according to [27]:

$$\mathrm{COM_d} = 0.180\mathrm{FCI} + 2.73C_{\mathrm{OL}} + 1.23\left(C_{\mathrm{UT}} + C_{\mathrm{RM}}\right) \tag{9.8}$$

where C_{OL}, C_{UT}, and C_{RM} are the cost of operating labor, cost of utilities, and cost of raw materials, respectively.

A stream factor of 0.95 is considered for the calculation of annual operating costs and revenues, i.e., the plant operates for 345 days per year.

FCI and $\mathrm{COM_d}$ are strongly affected by the location of the plant. Here, the economic analysis is developed for a plant built in Saudi Arabia, a region with the highest water scarcity level and with a fast growing population [12]. Accordingly, the FCI is multiplied by a location factor $F_{\mathrm{L}} = 0.97$, to take into account the influence of the specific location with respect to the Gulf Coast of the United States, used as a Ref. 30.

The annual revenues are calculated accounting for both the net energy generated and the freshwater incomes. The unit selling price of electricity is taken equal to 0.05 €/kWh [31], while for freshwater, a range from the current value of 1.5 €/m^3 [31] to a realistic future unit selling price of 2 €/m^3 was considered, since in the last years, its value increased about three times because of the higher water demand as well as the growing population.

The economic profitability of the process is evaluated by means of discounted cash flow analysis, according to four indexes: the discounted payback period (DPBP), the net present value (NPV), the internal rate of return (IRR), and the present value ratio (PVR).

Further assumptions are as follows:

- The useful life of the plant is 25 years, the first one of which is used for its construction [4].
- The cost of land is 2% of FCI, and it is purchased at year 0 [28].

- At the end of year 1, the working capital is invested to start the operation of the plant; it is assumed to be equal to 20% of FCI without the cost of the land (FCI_L) [28].
- The whole fixed capital investment is invested in the first year of plant construction.
- An income tax rate of 20% is considered based on the Saudi Arabian legislation [32].
- Depreciation of FCI_L in the first 10 years of plant operations is evaluated by straight line depreciation method, considering a salvage value equal to 3% of the FCI_L [28].
- Cash flow is discounted back to year 0 considering an interest rate equal to 2% for Saudi Arabia [32, 33].

9.4.2 Results

As the first result of the techno-economic analysis, the capital (CAPEX) and operating (OPEX) expenses of the process are estimated (Fig. 9.5). As displayed in Fig. 9.5a, the capital expenses related to the distillation unit for draw solution regeneration, including column body, trays, reboiler, and condenser, represent the largest share (65.3%) of the total costs. In fact, considering the high draw solution flow rate to be processed (around 4600 m^3/h, Sect. 3.3), 5 parallel distillation columns of diameter $D = 5$ m and height $H = 36.7$ m are necessary. The membrane unit, on the other hand, whose unit cost is equal to 2.45 €/m^2 (i.e., 3 \$/m^2 [4]), only accounts for roughly 9% of the CAPEX, which amount to 33,714,648 €. This results in a projected cost of electricity equal to 152 €/MWh.

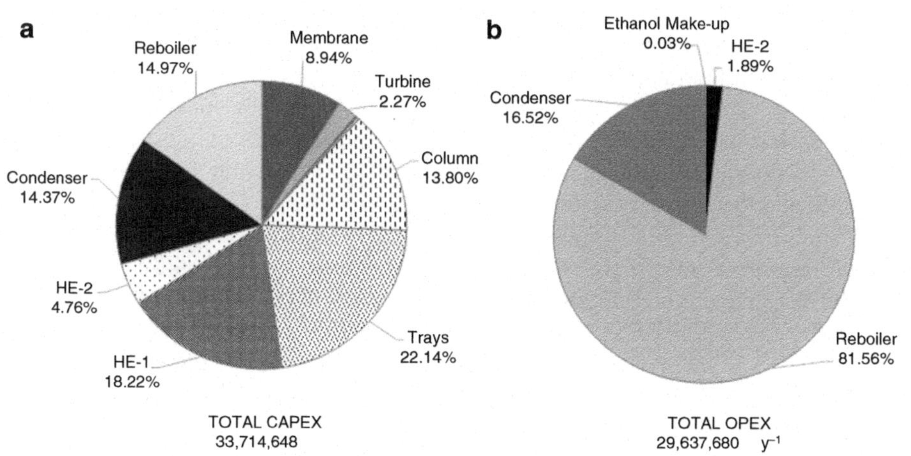

Fig. 9.5 Capital (**a**) and operating (**b**) expenses of the HOP process

Table 9.5 Summary of contributions affecting the cost of manufacturing

	Unit costs	Total cost
Cost of utilities (€/year)		**29,627,985**
HE-2	0.0333 (€/m^3)	558,805
Reboiler	0.00249 (€/kg)	24,171,598
Condenser	0.0333 (€/m^3)	4,897,582
Cost of raw materials (€/year)		**9695**
Ethanol make-up	323 (€/m^3)	9695
Cost of operating labor (€/year)		**475,000**
Operators (19)	25,000 (€/year)	475,000
COM$_d$ (€/year)		**43,819,733**

Table 9.6 Process profitability at different water selling prices

Water selling price (€/m^3)	1.5	1.73	2.0
Energy revenues (€/year)	419,393	419,393	419,393
Water revenues (€/year)	39,315,408	45,343,770	52,420,544
Total revenues (€/year)	39,734,801	45,763,163	52,839,937
DPBP (years)	–	25	4.44
NPV (€)	−91,642,981	0	109,029,608
IRR	–	–	18.8%
PVR	–	–	3.73

Moreover, as shown in Fig. 9.5b, the distillation unit strongly affects the operating costs as well, with 81.56% of the total OPEX represented by the steam required in the reboiler (see Table 9.5) and 16.52% by the cooling water at the condenser, accounting altogether for the 98.08% of the total operating expenses.

The ethanol make-up cost is instead negligible. Thus, distillation strongly affects both CAPEX and OPEX costs, but it appeared as the only way that allows achieving good performances in terms of both draw solution regeneration and freshwater production, with high purity (<1 ppm of ethanol).

The cost of manufacturing (COM$_d$) corresponding to the operating expenses shown in Fig. 9.5 results to be about 43.8 M€ and includes the cost of operating labor (19 operators with an average salary of 25,000 €/year [34]), the cost of utilities, and the cost of raw materials [30] (Table 9.5).

The revenues coming from the sale of electricity amount to only 419,393 €/year, which represents less than 1% of the total revenues, as displayed in Table 9.6. On the other hand, the profitability of the process results to be highly dependent on the freshwater selling price. Considering the average current water price (1.5 €/m^3), the process is in fact not profitable, since the total revenues are lower than the cost of manufacturing, leading to a negative gross profit (Table 9.6). Anyhow, it was calculated that with a selling price of 1.73 €/m^3 (15% higher than the current one), the fixed capital investment would be entirely recovered at the end of the project life (NPV = 0), while with a price of 2 €/m^3 (33% higher), the process would be highly

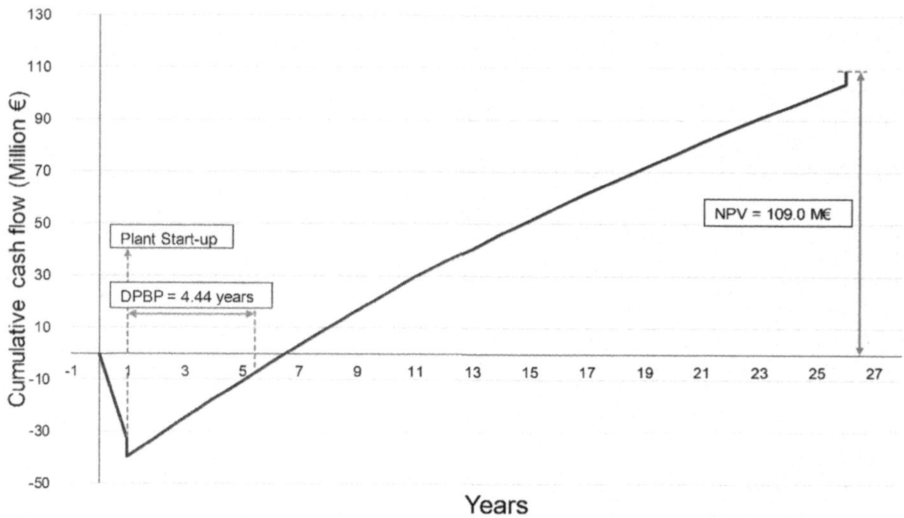

Fig. 9.6 Discounted cash flow of the HOP process with water selling price equal to 2 €/m³

profitable. A drinking water price of 2 €/m³ can be considered a realistic expectation in the years to come, given that this value increased about three times in the last few years and considering the increasing water demand.

The resulting cash flow plotted in Fig. 9.6 shows that with these assumptions, the process is feasible and quite attractive, with a discounted payback period after the plant start-up of less than 5 years and a net present value of about 109 M€, with a present value ratio of 3.74. The internal rate of return is 18.8%, which is higher than the threshold limit of 15% corresponding to the minimum interest for a low-risk profitable project.

9.5 Conclusions

This work aimed at developing process simulation and techno-economic analysis of an osmotic process for simultaneous power generation and clean water production from seawater using an ethanol draw solution, in order to verify its feasibility at large scale. Among the several draw solution regeneration units investigated, distillation resulted the best one in order to obtain a sufficiently high purity of the extracted water, despite having high thermal duties. The key process variables of the osmotic unit have been optimized to maximize the water production and power generation, minimizing the distillation duties, hence the energy consumptions.

An economic analysis referred to the Saudi Arabia region showed that the process profitability is strongly affected by the water selling price, since the revenues coming from power energy account for just 0.8% of the total. In particular, the cost of low-pressure steam to extract sufficiently pure water residue affects for the most the process feasibility. Hence, considering the current water price (1.5 €/m^3), the revenues would not be sufficient to cover the costs. It emerged that water price should be at least 1.73 €/m^3 in order to have the cumulative cash position equal to zero at the end of the process life. Nevertheless, both water and power could be industrially produced in a profitable way (*DPBP* less than 5 years) considering water selling price equal to 2 €/m^3, which is about 33% higher than the current price yet a realistic value for the near future.

Overall, the work can be a starting point for further optimizations, considering the fast progress in the osmotic technologies and the need to find new solutions for freshwater production and power generation.

References

1. IPCC (2014) Climate Change 2014 Synthesis Report
2. IEA, World Energy Outlook (2017). http://www.iea.org/publications/freepublications/publication/WEB_WorldEnergyOutlook2015ExecutiveSummaryEnglishFinal.pdf
3. Aaberg RJ, Statkraft SF (2003) Osmotic power. Refocus:48–50. https://doi.org/10.1016/S1471-0846(04)00045-9
4. Sharif AO, Merdaw AA, Sanduk MI, Al-Aibi SM, Rahal Z (2011) The potential of chemical-osmotic energy for renewable power generation. World Renew Energy Congr 44:2190–2197
5. Achilli A, Cath TY, Childress AE (2009) Power generation with pressure retarded osmosis: an experimental and theoretical investigation. J Membr Sci 343:42–52. https://doi.org/10.1016/j.memsci.2009.07.006
6. Altaee A, Sharif A (2015) Pressure retarded osmosis: advancement in the process applications for power generation and desalination. Desalination 356:31–46. https://doi.org/10.1016/j.desal.2014.09.028
7. Altaee A, Zaragoza G, Sharif A (2014) Pressure retarded osmosis for power generation and seawater desalination: performance analysis. Desalination 344:108–115. https://doi.org/10.1016/j.desal.2014.03.022
8. Skilhagen SE, Dugstad JE, Aaberg RJ (2008) Osmotic power—power production based on the osmotic pressure difference between waters with varying salt gradients. Desalination 220:476–482. https://doi.org/10.1016/j.desal.2007.02.045
9. Gerstandt K, Peinemann KV, Skilhagen SE, Thorsen T, Holt T (2008) Membrane processes in energy supply for an osmotic power plant. Desalination 224:64–70. https://doi.org/10.1016/j.desal.2007.02.080
10. Thorsen T, Holt T (2009) The potential for power production from salinity gradients by pressure retarded osmosis. J Membr Sci 335:103–110. https://doi.org/10.1016/j.memsci.2009.03.003
11. Han G, Zuo J, Wan C, Chung T-S (2015) Hybrid pressure retarded osmosis–membrane distillation (PRO–MD) process for osmotic power and clean water generation. Environ Sci Water Res Technol 1:507–515. https://doi.org/10.1039/C5EW00127G
12. Badran A, Murad S, Baydoun E, Daghir N (2017) Water, energy & food sustainability in the middle east. Springer International Publishing, New York
13. Greenlee LF, Lawler DF, Freeman BD, Marrot B, Moulin P (2009) Reverse osmosis desalination: water sources, technology, and today's challenges. Water Res 43:2317–2348. https://doi.org/10.1016/J.WATRES.2009.03.010
14. Elimelech M, Phillip WA (2011) The future of seawater desalination: energy, technology, and the environment. Science 333:712–717. http://science.sciencemag.org/content/333/6043/712.abstract

15. woo Kim T, Kim Y, Yun C, Jang H, Kim W, Park S (2012) Systematic approach for draw solute selection and optimal system design for forward osmosis desalination. Desalination 284:253–260. https://doi.org/10.1016/j.desal.2011.09.008
16. Geisler P, Hahnenstein FU, Krumm W, Peters T (1999) Pressure exchange system for energy recovery in reverse osmosis plants. Desalination 122:151–156. https://doi.org/10.1016/S0011-9164(99)00036-3
17. World Health Organization (2011) Guidelines for drinking-water quality, 4th edn. WHO, Geneva. https://doi.org/10.1016/S1462-0758(00)00006-6
18. Grande CA, Rodrigues AE (2007) Biogas to fuel by vacuum pressure swing adsorption I. behavior of equilibrium and kinetic adsorbents. Ind Eng Chem Res 46:4595–4605
19. Smith JM, Van Ness HC, Abbott MM (2005) Introduction to chemical engineering thermodynamics. McGraw-Hill Education, New York
20. Hamer WJ, Wu Y (1972) Osmotic coefficients and mean activity coefficients of uni-univalent electrolytes in water at 25°C. J Phys Chem Ref Data Monogr 1:1047–1100. https://doi.org/10.1063/1.3253108
21. Straub AP, Deshmukh A, Elimelech M (2016) Pressure-retarded osmosis for power generation from salinity gradients: is it viable? Energy Environ Sci 9:31–48. https://doi.org/10.1039/C5EE02985F
22. McCormick P, Pellegrino J, Mantovani F, Sarti G (2008) Water, salt, and ethanol diffusion through membranes for water recovery by forward (direct) osmosis processes. J Membr Sci 325:467–478. https://doi.org/10.1016/j.memsci.2008.08.011
23. Sharif AO, Arayfar M (2013) A Thermal regeneration forward osmosis process, UK patent application number GB1321711.2
24. Sabah M, Atwan AF, Mahood HB, Sharif A (2013) Power generation based on pressure retarded osmosis: a design and an optimisation study. Int J Appl Innov Eng Manag 2:68–74
25. Fritzmann C, Löwenberg J, Wintgens T, Melin T (2007) State-of-the-art of reverse osmosis desalination. Desalination 216:1–76. https://doi.org/10.1016/J.DESAL.2006.12.009
26. Dehwah AHA, Missimer TM (2016) Subsurface intake systems: green choice for improving feed water quality at SWRO desalination plants, Jeddah, Saudi Arabia. Water Res 88:216–224. https://doi.org/10.1016/J.WATRES.2015.10.011
27. Turton R, Bailie RC, Whiting WB, Shaeiwitz JA, Bhattacharyya D (2012) Analysis, synthesis, and design of chemical processes. Prentice-Hall, Upper Saddle River, NJ
28. Douglas J (1988) Conceptual design of chemical processes. McGraw-Hill, New York
29. Towler G, Sinnot RK (2012) Chemical engineering design: principles, practice and economics of plant and process design, 2nd edn. Butterworth-Heinemann, Oxford
30. https://www.intratec.us/. Accessed Feb 2018
31. https://www.modon.gov.sa/. Accessed Feb 2018
32. AlYahya S, Irfan MA (2016) The techno-economic potential of Saudi Arabia's solar industry. Renew Sustain Energy Rev 55:697–702. https://doi.org/10.1016/J.RSER.2015.11.017
33. Ramli MAM, Hiendro A, Al-Turki YA (2016) Techno-economic energy analysis of wind/solar hybrid system: case study for western coastal area of Saudi Arabia. Renew Energy 91:374–385. https://doi.org/10.1016/J.RENENE.2016.01.071
34. https://www.payscale.com/. Accessed Feb 2018

Chapter 10
Cooling of PV Panels for Performance Enhancement of Grid-Connected Systems

Kamaruzzaman Sopian, Ali H. A. Alwaeli, Ali Najah Al-Shamani, and Mohd Yusof Hj. Othman

10.1 Introduction

The rapid growth in population and corresponding growth in industrial and service sectors have reached to new heights in 2018. This opted for an increase in the energy demands. Fossil fuels have served during the industrial revolution as the primary source of energy, and they still are. However, due to climate-change treaties and the continuous search for a healthy source of energy that is more stable, renewable energies have emerged. Technologies like geothermal, wind and solar energies have been rising in efficiency and popularity, across many parts of the world (e.g. Europe, North America and east Asia). Solar energy has gone through massive jumps in efficiency rise and cost reduction from the 1970s to current date [1]. Solar energy can be classified based on the part of solar spectrum that is utilized. The two main classifications are the photovoltaic and solar thermal collector. Photovoltaic (PV) panels are devices that utilize the light of the sun to produce electricity. While, solar thermal collectors capitalize on its heat by storing it and extracting that heat using a base fluid. PV panels are negatively affected by the rise of its temperature. This is not the case for the solar thermal collector. This means to maintain the electrical efficiency of PVs it is preferable to provide them with cooling. Another device in solar energy is the photovoltaic thermal (PV/T) which is a hybrid that mixes both photovoltaic and solar thermal collector. This hybrid is particularly designed to provide cooling for PV; the thermal collector is attached to its back and will draw heat

K. Sopian (✉) · A. H. A. Alwaeli · M. Y. Hj. Othman
Solar Energy Research Institute (SERI), The National University of Malaysia (UKM), Bangi, Malaysia
e-mail: ksopian@ukm.edu.my

A. N. Al-Shamani
Al-Furat Al-Awsat Technical University, Kufa, Iraq

A. Sayigh (ed.), *Renewable Energy and Sustainable Buildings*, Innovative Renewable Energy, https://doi.org/10.1007/978-3-030-18488-9_10

from it, therefore providing cooling. This will raise the electrical efficiency during operation and simultaneously raise the thermal efficiency; giving that, it will draw more heat at its current predicament. Photovoltaic systems can be classified by whether they are connected to the grid or independent from it. The latter is called a stand-alone PV system, while the former is a grid-connected or grid-tied PV system. The grid-connected PV systems are usually designed to feed the grid with excess electrical power: feeding the local load as a priority and feeding the grid as second priority [2]. This paper presents novel methods of cooling PV panels that are connected to the grid. This can serve solar industry well, as it is useful to achieve highest performances and maintain low costs through novel methods.

10.2 Cooling of PV and Grid Connection

Different cooling techniques are conceptualized and tested in various research for PV systems. One method is to use phase change material (PCM) to control the temperature of the PV panel, in particular, organic PCM such as paraffin wax. This is because they are available in different melting points and selected based on the temperature of the application and also due to their relatively cheap price and market availability [3]. PCM are capable of absorbing and storing high amounts of heat due to latent heat absorption. Other advantages are associated with PCM cooling technique such as the absence of noise, electricity requirement and maintenance, in addition to relatively high heat transfer rates. Nanofluid-based PV/T collectors have also been implemented to raise the heat transfer of the PV/T system. Nanofluids are chosen because they exhibit better thermophysical properties than water as base fluids [4]. All the abovementioned techniques can serve PV panels immensely. Grid-connected PV systems generally do not require the use of batteries. Direct conversion of the PV's DC current using an inverter is performed to produce AC current that is synchronized with the grid [5]. More research and innovation need to be directed to study the effect of these cooling techniques on selection or optimization of grid-connected PV systems.

10.3 Grid Connected with Nanofluids

A 1.2 kWp roof top grid-connected photovoltaic thermal system (GCPV/T) with SiC nanofluid for cooling has been designed, fabricated and evaluated under the tropical climate conditions. The performance was compared to a conventional photovoltaic grid-connected system (GCPV). The PV/T collectors consist of specially designed rectangular tube absorbers with a height of 15 mm and a width of 25 mm and attached under an array of ten pieces of photovoltaic modules. Figure 10.1 shows a schematic diagram of the proposed system.

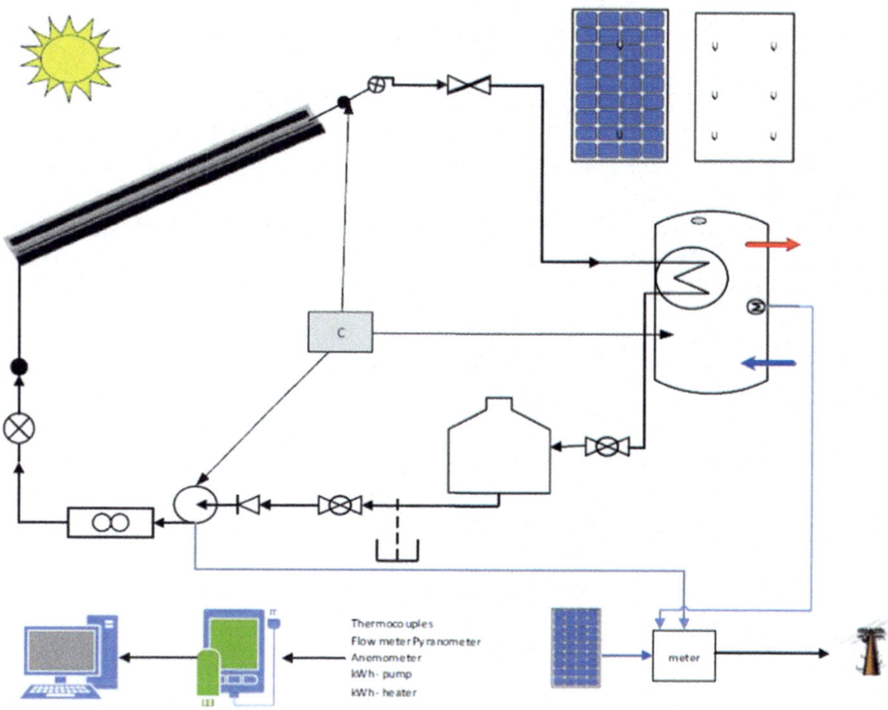

Fig. 10.1 Schematic diagram of GCPV/T with nanofluids

The energy balance analysis requires a number of factors to be calculated; the equations to these factors are included below, which show the combined PV/T efficiency, solar irradiance absorbed by a collector per unit, the electrical efficiency and the energy balance equation for the collector:

$$\eta_{\text{combined}} = \eta_{\text{th}} + \eta_{\text{el}} \tag{10.1}$$

The total efficiency of the collector is the summation of electrical and thermal efficiencies which are η_{el} and η_{th}, respectively.

$$S = (\tau\alpha)_{\text{PV}} I_{(t)} \tag{10.2}$$

S denotes the solar irradiance absorbed by the collector, $(\tau\alpha)_{\text{PV}}$ is the transmittance-absorbance product and $I_{(t)}$ is the solar radiation.

$$\eta_{\text{el}} = \eta_{\text{r}} \left(1 - \beta\left(T_{\text{pm}} - T_{\text{r}}\right)\right) \tag{10.3}$$

η_{r}, β, T_{pm} and T_{r} denote the reference efficiency of PV module, temperature coefficient, temperature of solar cell and reference temperature, respectively.

$$Q_u = A_C F_R \left[S - U_L \left(T_i - T_a \right) \right] \tag{10.4}$$

Q_u, A_c, F_R, U_L, T_i and T_a denote the useful heat, collector area, heat removal factor, heat loss coefficient, fluid inlet temperature and ambient temperature, respectively. After data collection and analysis, the system was compared to the conventional PV system in terms of power production and total yield, which is provided in Fig. 10.2. This will help provide a clearer image to how PV performance is enhanced. The efficiency of the PV module, system and inverter is illustrated in Eqs. 10.5–10.7, respectively.

$$\eta_{PV} = \left(\frac{E_{DC}}{H_t \times A_C} \right) \times 100(\%) \tag{10.5}$$

$$\eta_{SYS} = \left(\frac{E_{AC}}{H_t \times A_C} \right) \times 100(\%) \tag{10.6}$$

$$\eta_{INV} = \left(\frac{E_{AC}}{E_{DC}} \right) \times 100(\%) \tag{10.7}$$

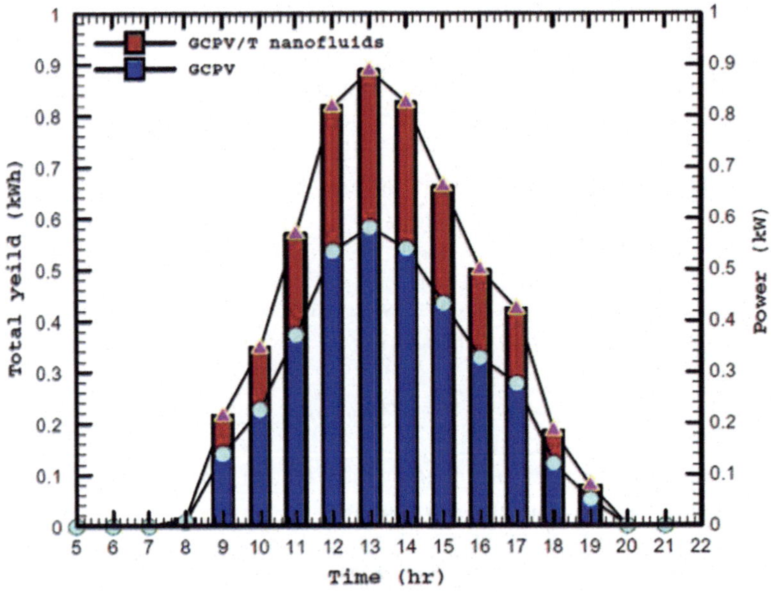

Fig. 10.2 GCPV and GCPVT-SiC nanofluid system total yield and power for 1 day

In the equations above, E_{DC}, E_{AC}, H_t and A_C denote DC energy, AC energy, total in-plane solar radiation and collector area.

From Fig. 10.2, the hourly energy production of the two systems over a sunny day demonstrates the superiority of the nanofluid-cooled PV system over a conventional PV system, due to the enhancement of heat transfer by the SiC nanofluid of the PV/T module. The reason for this was due to the lower module's temperature and thus increasing the electrical efficiency of the PV. The conventional grid-connected PV showed a maximum yield factor, performance ratio, inverter efficiency, PV efficiency and system efficiency of 100.538 kWh/kWp (in March), 77.138% (in March), 95.87% (in January), 8.8% (in April) and 8.41% (in April), respectively. The findings for nanofluid-cooled PV system parameters are illustrated in Fig. 10.3. From Fig. 10.3 it is observed that during March the highest values for monthly array yield and final yield were obtained with around 157.32 and 152.71 kWh/kWp, respectively. On the other hand, during February the lowest values were recorded for the monthly array yield and final yield with around 145.11 and 140.71 kWh/kWp, respectively. This low yield is attributed to lower solar irradiation. As for the performance ratio (PR) of the nanofluid-cooled GCPV system is around the range of 95.49–95.92%, in comparison to the range of 70.65–74.71% for conventional grid-connected panel. Finally, the electrical efficiency of the cooled GCPV system was around 13.5–8.41% of the conventional GCPV.

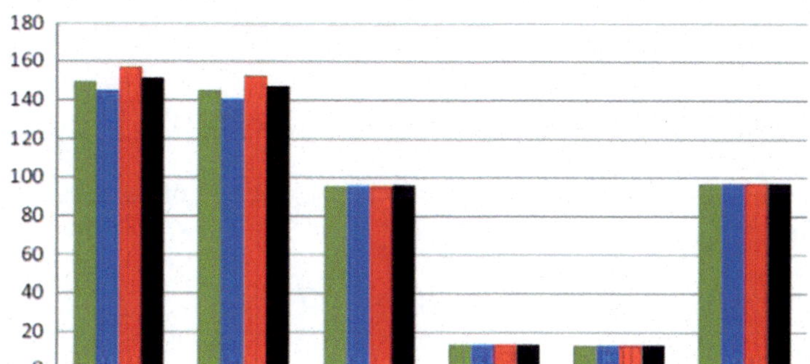

	Ya, kWh/kWp	Yf, kWh/kWp	PR, %	PV eff., %	Sys. Eff., %	Inver. Eff., %
January	149.705	145.015	95.489	13.513	13.094	96.900
February	145.110	140.713	95.713	13.524	13.110	96.970
March	157.326	152.711	95.762	13.526	13.117	96.975
April	151.39	147.145	95.922	13.528	13.125	97.023

Fig. 10.3 The monthly energy yield, system performance ratio (PR) and (PV, system, inverter) efficiency for GCPVT-SiC nanofluid system

10.4 Grid Connected with Nano-PCM and Nanofluid

PCM are effective to control the temperature of PV panels. However, they exhibit low thermal conductivity. This issue could be fixed by employing nanoparticles to enhance their thermal conductivity [6]. This design is proposed to mix two cooling techniques: nano-PCM and nanofluid cooling of PV panels. This is achieved by attaching a nano-PCM container to the back of the panel; within the container is a layer of nano-PCM and collector tubes submerged with nanofluid flowing within them. The used nanofluid is SiC-water nanofluid, and the used PCM is paraffin wax. The cross-section diagram of this collector is shown in Fig. 10.4. The proposed collector is evaluated based on theoretical and experimental results. The system was validated using CFD analysis. This requires a set of governing equations: continuity, momentum and thermal energy which are described in Eqs. 10.8, 10.9 and 10.10, respectively.

Continuity equation:

$$\frac{\partial}{\partial x_i}(\rho u_i) = 0 \tag{10.8}$$

Momentum equation:

$$\frac{\partial(\rho u_i u_j)}{\partial x_i} = -\frac{\partial \rho}{\partial x_i} + \frac{\partial}{\partial x_j}\left[\mu\left(\frac{\partial u_i}{\partial x_j} + \frac{\partial u_j}{\partial x_i}\right)\right] + \frac{\partial}{\partial x_j}\left(-\rho \overline{u_i' u_j'}\right) \tag{10.9}$$

The energy equation:

$$\partial h / \partial t + \partial H / \partial t + \nabla(\vec{V}h) = \nabla(k_{\text{eff}} / (\rho cp)_{\text{nf}} \nabla h) \tag{10.10}$$

The enthalpy of the material is computed as the sum of the sensible enthalpy, h, and the latent heat, ΔH:

$$H = h + \Delta H \tag{10.11}$$

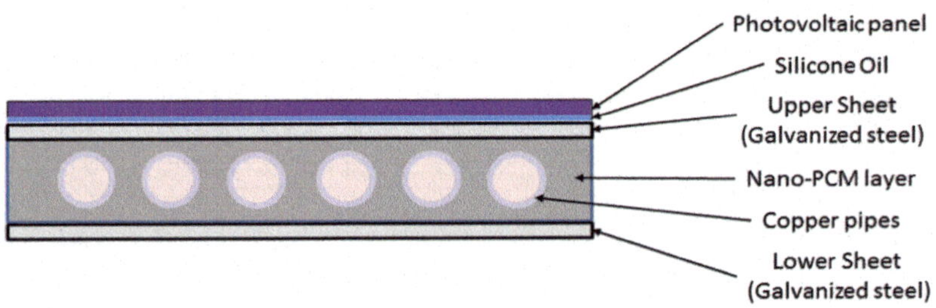

Fig. 10.4 Cross-section drawing of nano-PCM and nanofluid-based PV/T collector

where

$$h = h_{\text{ref}} + \int_{T_{\text{ref}}}^{T} C_p \, dT \tag{10.12}$$

Due to implementing nanofluids, and employing nanoparticles in PCM, the following equations are used:

The effective thermal conductivity of the nanofluid is

$$k_{\text{eff}} = k_{\text{nf0}} + k_{\text{d}} \tag{10.13}$$

The thermal conductivity enhancement term due to thermal dispersion is given by

$$k_{\text{d}} = C\left(\rho c_{\text{p}}\right)_{\text{nf}} \left|\vec{V}\right| \varphi d_{\text{p}} \tag{10.14}$$

The system was compared to a conventional PV system, both tested under Malaysian climatic condition. Figure 10.5 illustrates the obtained efficiency of the proposed collector and a conventional PV across the period of experiment.

From Fig. 10.5, it can be observed that the efficiency gap is quite clear between conventional PV and nano-PCM and nanofluid cooled PV panel. The solar radiation peaks around 710 W/m^2 at 1:30 PM. Figure 10.6 shows a comparison between power production of GCPV panel and nano-PCM and nanofluid PV/T collector.

Figure 10.6 shows the monthly production and yield factor of the GCPVT system, which shows consistent figures. The maximum and minimum energy generation

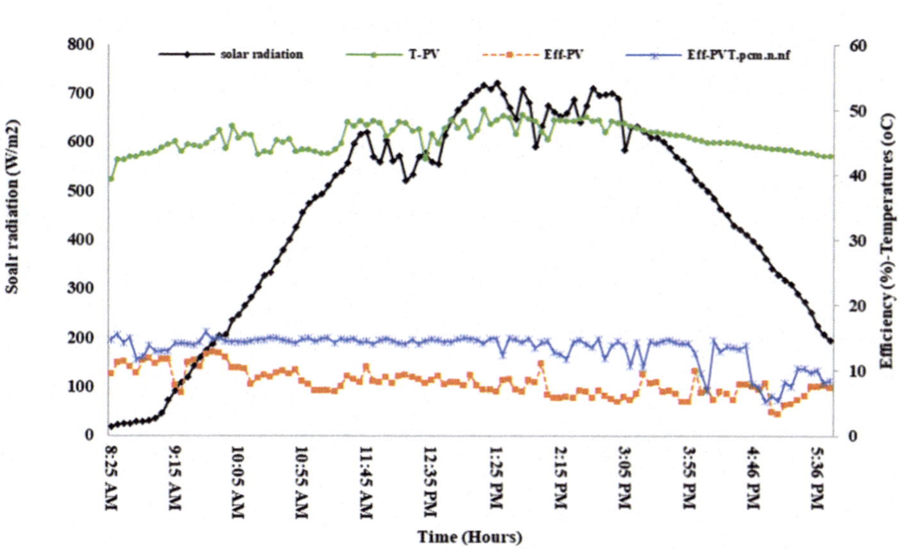

Fig. 10.5 Efficiency and temperature of conventional PV and proposed nano-PCM and nanofluid cooled PV over a sunny day

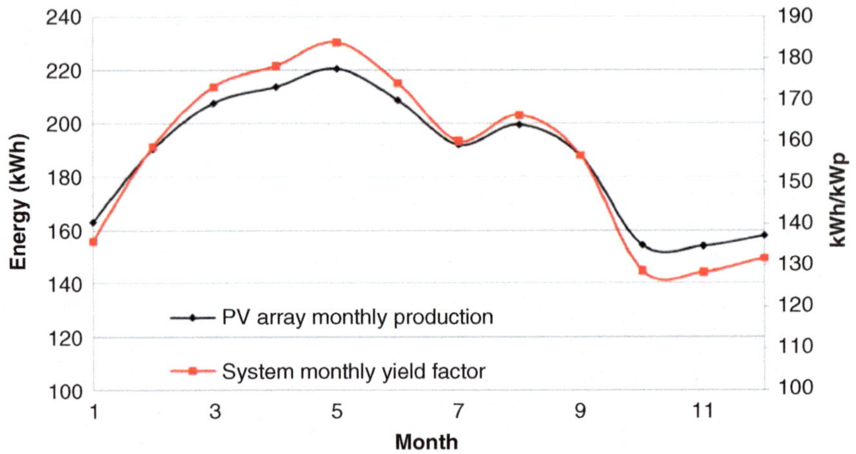

Fig. 10.6 PV power production of GCPV and proposed nanofluid and nano-PCM GCPV/T

occurred in May and October, respectively. However, the curve changed due to the change in solar radiation. It is seen that from March to August the monthly production and yield factor is higher than the period of September to February. The maximum and minimum energy production was found to be 220.51 and 154.02 kWh, respectively. However, the monthly and annual yield factor was found to be (128.34–183.75) kWh/kWp and 1875.12 kWh/kWp, respectively.

10.5 Conclusions

In conclusion, this paper presents novel methods of cooling grid-connected PV panels. The cooling is achieved through different processes of heat transfer. The use of cooling techniques has proven to be highly efficient for PV panel's electricity production. This is achieved through using nanofluid cooling and nano-PCM and nanofluid cooling. The higher thermal conductivity and capacitance of the system, the higher heat transfer is carried out throughout the process. In all cases, cooled PV panels produce higher open-circuit voltage than conventional PV panels. The main consideration for each of the designs is to maintain moderate cost and reduce cost of energy.

References

1. Green MA (2009) The path to 25% silicon solar cell efficiency: history of silicon cell evolution. Prog Photovolt 17(3):183–189
2. Kjaer SB, Pedersen JK, Blaabjerg F (2005) A review of single-phase grid-connected inverters for photovoltaic modules. IEEE Trans Ind Appl 41(5):1292–1306

3. Zalba B, Marın JM, Cabeza LF, Mehling H (2003) Review on thermal energy storage with phase change: materials, heat transfer analysis and applications. Appl Therm Eng 23(3):251–283
4. Jang SP, Choi SUS (2004) Role of Brownian motion in the enhanced thermal conductivity of nanofluids. Appl Phys Lett 84(21):4316–4318
5. Verma D, Midtgård Ole-Morten, Sætre Tor O (2011) Review of photovoltaic status in a European (EU) perspective. In 2011 37th IEEE Photovoltaic Specialists Conference (PVSC). IEEE, Washington, p 003292–003297
6. Karunamurthy K, Murugumohankumar K, Suresh S (2012) Use of CuO nano-material for the improvement of thermal conductivity and performance of low temperature energy storage system of solar pond. Dig J Nanomater Biostruct 7(4):1833–1841

Chapter 11
Plasmonic Coupling Enhanced Absorption and Fluorescence Emission in Thin Film Luminescent Solar Concentrator

S. Chandra and S. J. McCormack

11.1 Introduction

The active optics-based luminescent solar concentrator (LSCs) is employed to concentrate solar radiation from a large area to onto a smaller area of solar cell to enhance the solar cell output [1–3]. This technology uses polymer or glass as a matrix material doped with luminescent molecules to collect and waveguide the light; hence, it can replace the expensive silicon material and as a result reduces cost of solar energy production. In LSCs, incident solar radiation is absorbed by luminescent material, and subsequently, light is re-emitted over all solid angles. The re-emitted light hitting the waveguide at angles larger than the critical angle is guided via total internal reflection (TIR) to the edges where solar cells are attached. The active optics of luminescent material and diffractive optics of polymer in the LSC have extended several advantages: can concentrate both direct and diffuse radiation, not limited solar concentration ratio [4]; narrow emission band of concentrated solar radiation can be matched to the spectral response of solar cells [5]; spectral overlap to solar cell minimizes thermalization losses [5]; solar spectrum can be separated and concentrated in different wavelength band by tuning luminescent material optical properties [6]; most important and significant that their static nature makes them an ideal candidate for building integrated photovoltaic (BIPV) [7–9] to achieve zero-carbon buildings.

The LSC was proposed as a alternative and promising solar concentrator technology; however, the optical efficiency is subverted by several energy losses phenomena: Stokes shift in the re-emitted radiation [10], re-absorption [10–12], escape cone losses [13], scattering [12], narrow absorption range of luminescent material [14, 15], chemical and photo stability of the luminescent material [16].

S. Chandra (✉) · S. J. McCormack
Department of Civil, Structural and Environmental Engineering,
Trinity College Dublin, Dublin 2, Ireland
e-mail: schandra@tcd.ie

A. Sayigh (ed.), *Renewable Energy and Sustainable Buildings*, Innovative Renewable Energy, https://doi.org/10.1007/978-3-030-18488-9_11

To minimize scattering and re-absorption losses process, the concept of thin film LSCs was introduced. In addition, decreasing the thickness of LSCs increased the solar concentration ratio [17]. The thin film LSC consists of a heavily doped luminescent material thin film coated on a highly transparent thicker substrate of the same refractive index to the thin film [15]. Theoretically the quantity of the luminescent material in the thin film LSC is of the same size as in that of homogenous LSC [18, 19]. The principally fluorescence emission was trapped in the substrate and waveguided by total internal reflection to the edge of the plate. Since the substrate is transparent, hence parasitic losses of re-absorption and scattering were reduced [18].

Besides that thin film LSCs have extended more advantages: reduced fabrication cost [20], luminescent material can be protected against UV radiation damage, and easily achieved high doping concentration [8]. However, high doping concentration in thin film LSC introduced non-linearity in optical properties, hence, the optical efficiency of thin film LSC [17]. This research article exploited a plasmonic approach to increase the absorption and consequently fluorescence emission in thin film LSCs without increasing doping concentration of the luminescent material by employing plasmonic coupling.

11.2 Plasmonic Coupling

The fluorescent materials excitation rate is highly a function of local photon mode density (PMD) and electric field (E_{loc}) intensity and, therefore, can be manipulated through modification in the local PMD [21]. The localized surface plasmon resonance (SPR) in metal nanoparticles (MNPs) can be modified and control the local PMD. When MNPs have dimensions smaller than the wavelength of the excitation light, energy can be confined in a small spatial region through a localized surface plasmon resonance (SPR) phenomenon [22]. The SPR phenomenon can dramatically amplify the incident PMD and consequently electric field intensity (E_{loc}) around the MNPs [23]. This enhanced electric field vector decays exponentially away from the surface of MNPs. The interaction between the fluorescent material and localized electric field (E_{loc}) generated by SPR in the MNPs is known as plasmonic coupling [24]. The plasmonic coupling modifies the optical properties of fluorescent material, absorption, and fluorescence emission [25], excitation rate [26], fluorescence lifetime and quantum yield [27], and photostability [24]. Plasmonic interaction between a fluorescent and MNPs can be controlled by spacing between them, the refractive index of the embedding medium, the degree of spectral overlap of the SPR wavelength of the MNPs, and the absorption/emission wavelength of fluorescent material. The spectral overlap between the SPR peak of MNPs and the absorption spectra of fluorescent material leads to enhancement of both the absorption and excitation [26], hence, the total fluorescence emission rate.

This research employed the plasmonic coupling in the thin film LSC. The plasmonic coupling in the plasmonic composite structure is monitored locally and macroscopically to predict the behavior of the device. Then, optimized plasmonic composite is used to fabricate a plasmonic thin film LSC device; subsequently, the electrical characterization is carried out.

11.3 Experimental Material and Method

Perylene-based Lumogen F Red305 dye (BASF, Germany) was used as a fluorescent material in this research (referred as red dye molecules)—typical absorption and fluorescence emission spectra presented in Fig. 11.1.

The layered plasmonic composites for thin film LSC in which polymer spacer layers controlled spacing between red dye molecules and Au NPs film. The configuration was designed and fabricated with varying spacing thickness of 0–60 nm between the red dye molecules and Au NP film, as shown in Fig. 11.2.

The plasmonic composites were fabricated using spin coating (G3 P-8 *Specialty Coating System (SCS)*) on glass substrate of 25 × 25 × 1 mm and refractive index of $n \approx 1.5$. Each component of composite plasmonic structures was separately optimized prior to their assembling into the plasmonic composite structure. The red dye molecules thin film LSC of doping concentration 0.7 wt% was subjected to plasmonic coupling [17].The spacer layer thicknesses of 0.0, 30 ± 5 nm, and 60 ± 5 nm were separately optimized by spin coating of 2 wt% Plexit 55 in toluene. The thickness of the various films were measured using an atomic force microscope (AFM).

The spherical Au NPs were synthesized using colloidal chemistry, which is reported elsewhere [28]. Au NPs synthesized in ethanol were spin coated on a glass substrate of the same size used for thin film LSC fabrication. The deposited films were optically and microscopically characterized by measuring absorption in UV/Vis/NIR and SEM images, respectively, reported in the earlier work [28] in Fig. 11.3. The SPR peak is slightly red-shifted from 556 to 562 nm in the thin film, which may be due to the size re-distribution through agglomeration and a change in surrounding medium from solution to glass substrate.

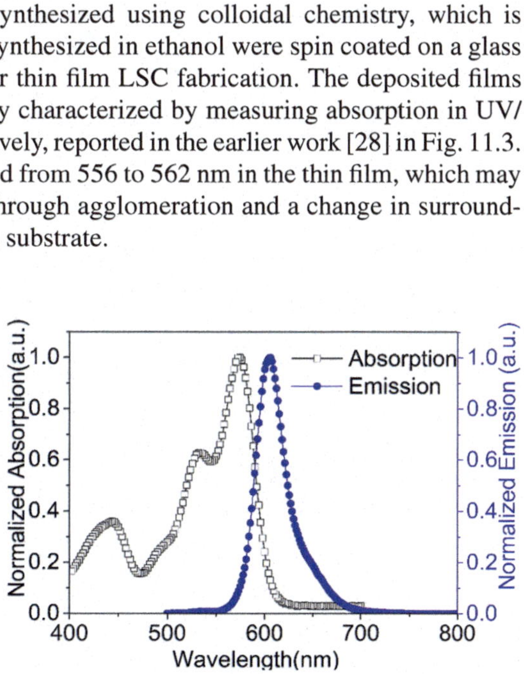

Fig. 11.1 Normalized absorption and fluorescence emission of 0.001 wt% red dye molecules in toluene, the peak absorption and emission are 575 and 606 nm, respectively

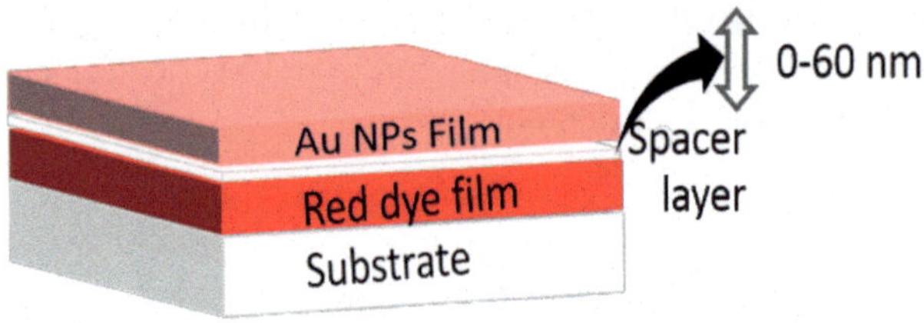

Fig. 11.2 Schematic of plasmonic composite structure with varying spacing (0, 30, and 60 nm) between red dye molecules and Au NPs film

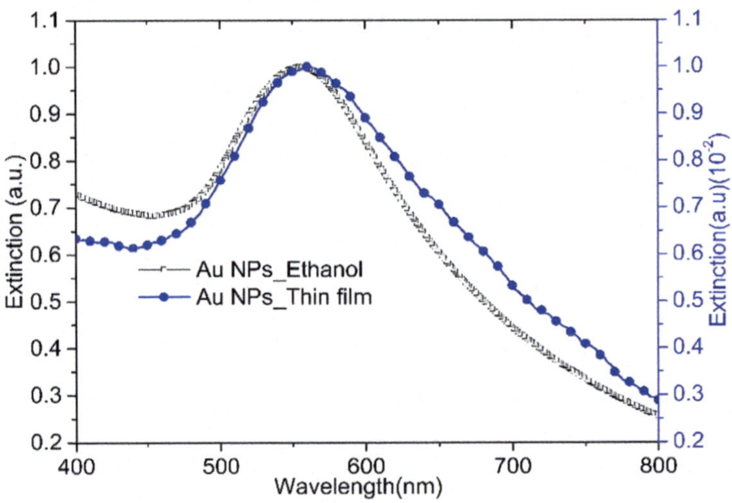

Fig. 11.3 Extinction spectra of Au NPs in ethanol and spin coated on glass substrate

The proposition of increased excitation rate of red dye in the presence of the SPR enhanced localized electric field of Au NPs was studied and verified separately by studying homogenous plasmonic composites, with 0.1 wt% of red dye molecule and varied Au NP concentration from 10 to 60 ppm. The lower red dye molecules concentration was selected for this study since it was observed that for higher doping concentration of red dye molecules in thin film LSCs have adverse optical losses [17].

The absorption and extinction measurement was carried out using a Perkin Elmer Lambda 900 UV/Vis/NIR spectrometer. The AvaSpec-2048 (*Avantes, UK*) fiber optic spectrometer setup was used to record the fluorescence emission at the edge of the red dye molecules film and of their plasmonic composite thin film LSCs; the optical setup detail is reported elsewhere [17]. The Zeiss LSM 510 confocal laser scanning microscope is used to measure the local fluorescence emission of the red dye molecules film and their plasmonic composite structures. More details of measurement are reported in earlier work [17].

11.4 Results and Discussions

The optimized red dye molecules doping concentration of 0.7 wt% thin film LSC was used for the plasmonic composite structure since the efficiency of LSC is dependent on doping concentration, as reported in earlier work [17]. The control LSC 0 is the red dye molecules thin film LSC of 0.7 wt% doping concentration. The plasmonic composite LSC 1, 2, and 3 have a spacing of 0, 30, and 60 nm, respectively. The absorption of the plasmonic composite red dye molecule film LSC significantly increased by 12% and is independent of the spacer layer thickness in Fig. 11.4.

The increased absorption of red dye molecules in the thin film plasmonic composite is contributed by the Au NPs SPR phenomena modified incident solar radiation. Light encounters the Au NPs before reaching the red dye molecules film and excites the SPR, which increases the scattering cross section of Au NPs which exceeded the geometric dimension of the particles [29]. The increased absorption is contributed to several factors. The Au NP is sandwiched between air and PMMA film, and this leads to scattering of light into a higher refractive index material [30], and hence more light is coupled to the red dye molecules film. The overall red dye molecules film absorption depends on the optical-pathlength for a constant doping concentration. The SPR enhanced scattering in Au NPs increases the optical-path length of incident light in the red dye molecules film. The scattering cross section of Au NPs is most significantly pronounced at the SPR wavelength, and the red dye molecules absorption band substantially overlaps to the SPR wavelength of Au NPs. These three effects work together to increase the local electric field

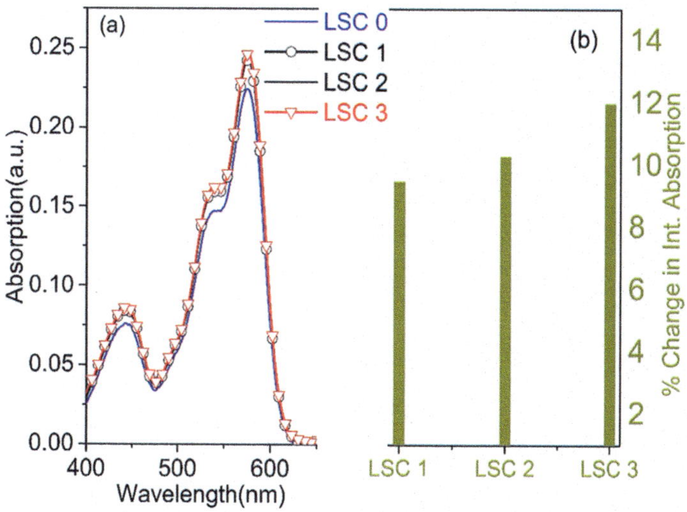

Fig. 11.4 (**a**) Optical absorption of plasmonic composite structure, and (**b**) % change in integrated absorption. (Note: % change was calculated with respect to control sample red dye molecules thin film of 0.7 wt%). (LSC 0; control sample, LSC 1, 2, and 3 have a spacing between red dye film and Au NPs of 0, 30, and 60 nm, respectively)

intensity (E_{loc}) through modified local photon mode density of incident light in the red dye molecules film, hence, increase the exciting rate and consequently absorption. The spacer layer thickness independence of absorption enhancement can be explained; the PMMA is transparent and the refractive index closely matched the glass substrate, and therefore there are no reflection losses at the interface by adding the spacing layer.

The fluorescence emission measured at the edge of LSC enhanced by 10–25% as presented in Fig. 11.5. This changed fluorescence emission is due to both increased absorption which is independent of spacer layer thickness (Fig. 11.4) and the increased excitation rate (multiple excitation) of red dye molecules by experiencing enhanced localized electric field (E_{loc}) of Au NPs. The enhanced localized electric field (Eloc) at red dye molecules led to increased exciation rate of red dye molecules.

The SPR modified E_{loc} intensity decays away from Au NPs surface, and therefore the presence of the spacer layer can be used to control the intensity local E_{loc} in addition to the volume of red dye molecules experiencing plasmonic coupling. In the absence of a spacer layer, the certain fraction of red dye molecules comes in contact with the Au NPs and the possibility of non-radiative relaxation in red dye [31]. This is a short range effect and leads to quenching of fluorescence emission by non-radiative energy transfer from red dye molecules to the Au NPs through the near-field component and compensates the effect of the enhanced emission rate. The spacer layer when introduced this non-radiative effect is minimized, and at the same time it decreases the volume of red dye molecules experiencing enhanced E_{loc}; however, still it is beneficial since the overall competition between these two effects

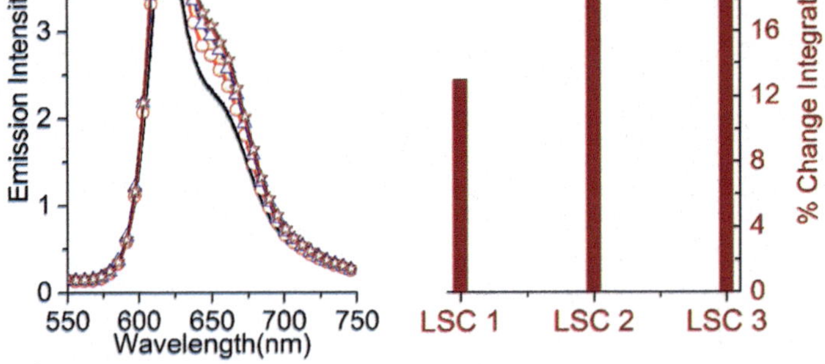

Fig. 11.5 (**a**) Fluorescence emission spectra at the edge of plasmonic composite LSC, and (**b**) % change in integrated emission. (Note: the % change was calculated with respect to control LSC 0 which is red dye thin film LSC)

gives rise to increased fluorescence emission for the cases of the 30 and 60-nm spacer layer, respectively.

The spectroscopic characterization presents the overall modified optical properties of plasmonic composite structure. The enhanced fluorescence emission in the plasmonic composite structure led to higher fluorescence coupling to the glass substrate for solar radiation concentration. However, the only portion of fluorescence emission in the TIR range is coupled to the glass substrate. Therefore the actual fluorescence emission enhancement could be significantly higher.

It is proposed that enhanced emission is contributed by the increased excitation rate of red dye molecules and the overall enhancement is much higher than the recorded at the edge of LSCs. This is tested by monitoring localized plasmonic coupling. The proposition of the increased red dye molecules excitation rate in the presence of the SPR enhanced localized electric field of Au NPs was studied and verified separately for thin film plasmonic composites. The doping concentration of the red dye molecules was kept low to prevent re-absorption and confirm that change in fluorescence emission is solely caused by plasmonic interaction increased excitation rate. The simultaneously recorded images are presented in Fig. 11.6 and their corresponding florescence spectra in Fig. 11.7. The brightness of images varied with Au NP concentration, and 40 ppm is the brightest.

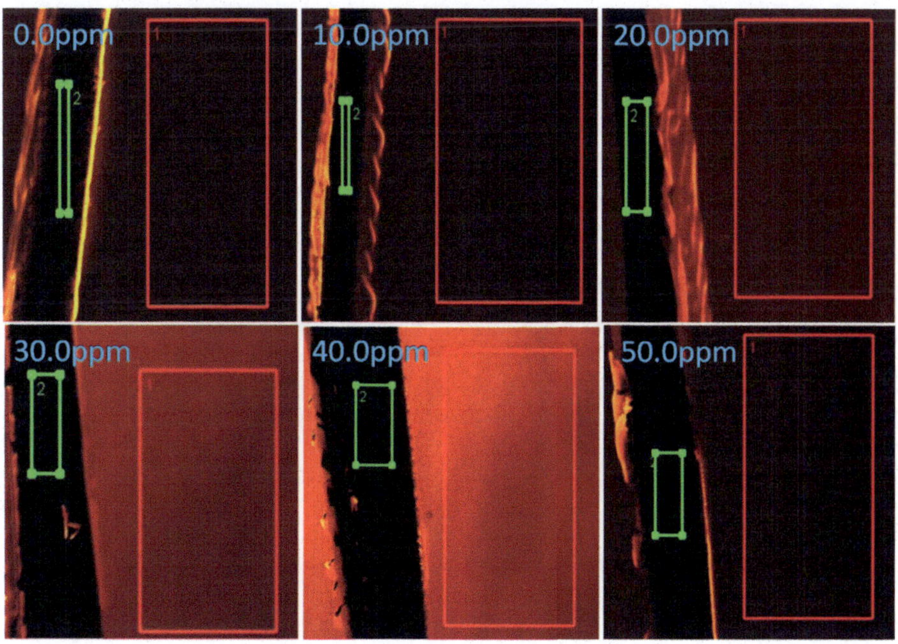

Fig. 11.6 Confocal images recorded from the top surface of plasmonic composite thin film of 0.1 wt% red dye molecules and Au NPs (varied from 10 to 50 ppm). The black thick line is the background glass substrate and focal plane lies at the interface of red dye thin film and glass substrate. The red and green rectangles represent the area selected for fluorescence emission measurement

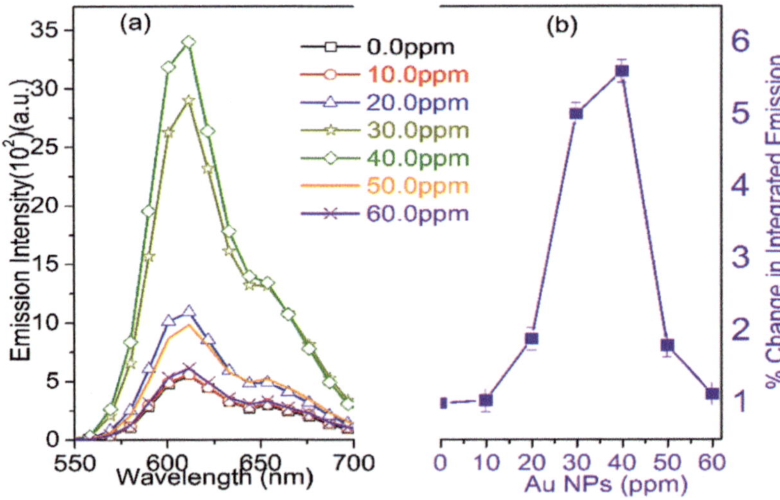

Fig. 11.7 Fluorescence emission measured in confocal from the top surface of 0.1 wt% red dye molecules and its plasmonic composite with Au NPs (varied from 10 to 50 ppm), (**a**) fluorescence emission spectra profile (**b**) relative integrated fluorescence emission (Note: relative change was calculated with respect to 0.0 ppm plate, which is red dye molecules film only)

Fluorescence emission of the red dye molecules in the plasmonic composite film follows the images trend, with the highest intensity seen for 40 ppm in Fig. 11.7. The change in brightness and fluorescence emission intensity as a function of Au NP concentration can be explained. Since the red dye molecules has a very high FQY ~95%, therefore the change in fluorescence emission intensity is solely contributed by an increased excitation rate of the red dye molecule due to the presence of the surface plasmon enhanced local field (E_{loc}) of the Au NPs [32]. The enhanced excitation rate leads to higher emission and is compensated by increased non-radiative relaxation in the red dye molecules at higher concentration of Au NPs. Since the higher concentration of Au NPs reduced the spacing between Au NPs and red dye molecules, it induced a non-radiative phenomenon which is a short-range effect and dominates when spacing between red dye molecule and Au NPs is less than a few nanometers.

11.4.1 Electrical Characterization of Plasmonic Thin Film LSCs

The solar cell was attached to one of the edges of plasmonic thin film LSC. The short circuit current (I_{sc}) gain of the plasmonic composite thin film LSC is presented in Fig. 11.8. The current gain closely correlates with the fluorescence emission

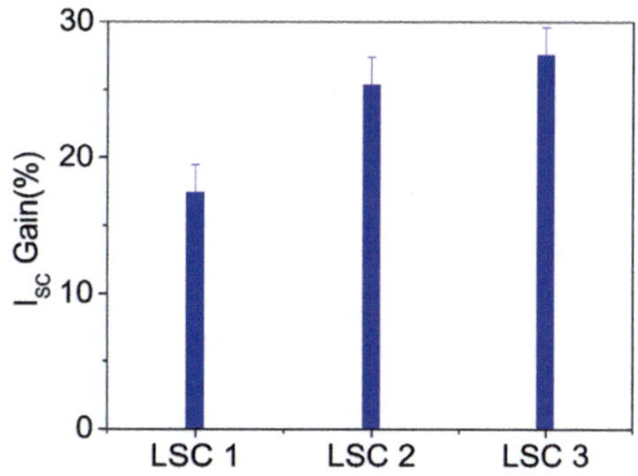

Fig. 11.8 Current gain for plasmonic composite thin film LSCs plate and % gain calculated with respect to control LSC 0, which is only red dye molecules thin film LSCs

trend with a small deviation. This deviation can be associated with optical coupling efficiency between LSCs edge and solar cell. Since the area solar cell matches to the edge area of LSCs, hence, the coupling efficiency is better than the detector used in optical measurements.

11.5 Conclusions

The spin coating technique was successfully employed to fabricate plasmonic composite thin film LSCs. The optimum red dye molecules (0.7 wt%) doping concentration thin film was subjected to plasmonic coupling with surface plasmon enhanced local electric E_{loc} of Au NPs. The plasmonic coupling was manipulated by incorporating a spacer layer between the red dye molecules and the Au NPs with varied spacer layer thickness; 0.0, 30 ± 5, and 60 ± 5 nm.

The SPR phenomena of Au NPs led to higher forward scattering, increase scattering cross section, and longer optical path-length of excitation light in the composite plasmonic thin film. The optical absorption of the red dye molecules film is increased by ~12% and independent of spacer layer thickness. Increased absorption and excitation rate of the red dye molecules enhanced the fluorescence emission by 13, 20, and 25% at the edge of plasmonic composite thin film LSCs for spacer layers of thickness 0.0, 30 ± 5, and 60 ± 5 nm, respectively. The local increased excitation rate of red dye molecules by SPR enhanced local electric field of Au NPs was independently examined. The local enhancement in the fluorescence of red dye molecules is much higher than overall enhancement recorded at the edge of thin film LSCs that indicate only a portion of enhancement fluorescence emission is collected and concentrated.

In summary, the plasmonic coupling can be manipulated in plasmonic composite thin film LSC, and correlation between spectroscopy and microscopy characterization can predict plasmonic composite optical properties. This could be a vital step in improving the optical efficiency plasmonic coupling-based LSC device.

Acknowledgements This research work is funded by the European Research Council under the project Plasmonic Enhancement and Directionality of Emission for Advanced Luminescent Solar Devices (PEDAL).

References

1. Weber WH, Lambe J (1976) Luminescent greenhouse collector for solar radiation. Appl Opt 15:3–4
2. Goetzberger A, Greube W (1977) Solar energy conversion with fluorescent collectors. Appl Phys 14:123–139
3. Rapp CF, Boling NL (1978) Luminescent solar concentrator. In: Proceeding of the 13th IEEE PVSC. IEEE, Washington, DC, pp 690–693
4. Smestad G, Ries H, Winston R, Yablonovitch E (1990) The thermodynamic limits of light concentrators. Sol Energy Mater 21:99–111
5. Batchelder JS, Zewai AH, Cole T (1979) Luminescent solar concentrators 1: theory of operation and techniques for performance evaluation. Appl Opt 18:3090–3110
6. Earp AA, Smith GB, Franklin J, Swift P (2004) Optimisation of a three-colour luminescent solar concentrator daylighting system. Sol Energy Mater Sol Cells 84:411–426
7. Debije MG, Verbunt PPC (2012) Thirty years of luminescent solar concentrator research: solar energy for the built environment. Adv Energy Mater 2:12–35
8. Wiegman J, van der Kolk E (2012) Building integrated thin film luminescent solar concentrators: detailed efficiency characterization and light transport modelling. Sol Energy Mater Sol Cells 103:41–47
9. Norton B, Eames PC, Mallick TK, Huang MJ, Mccormack SJ, Mondol JD, Yohanis YG (2011) Enhancing the performance of building integrated photovoltaics. Sol Energy 85:1629–1664
10. Olson RW, Loring RF, Fayer MD (1981) Luminescent solar concentrators and the reabsorption problem. Appl Opt 20:2934–2940
11. Wilson LR, Rowan BC, Robertson N, Moudam O, Jones AC, Richards BS (2010) Characterization and reduction of reabsorption losses in luminescent solar concentrators. Appl Opt 49:1651–1661
12. Chandra S, McCormack SJ, Kennedy M, Doran J (2015) Quantum dot solar concentrator: optical transportation and doping concentration optimization. Sol Energy 115:552–561
13. Debije MG, Verbunt PPC, Rowan BC, Richards BS, Hoeks TL (2008) Measured surface loss from luminescent solar concentrator waveguides. Appl Opt 47:6763–6768
14. Leow SW, Corrado C, Osborn M, Isaacson M, Alers G, Carter SA (2013) Analyzing luminescent solar concentrators with front-facing photovoltaic cells using weighted Monte Carlo ray tracing. J Appl Phys 113:214510–243502
15. Reisfeld R, Shamrakov D, Jorgensen C (1994) Photostable solar concentrators based on fluorescent glass films. Sol Energy Mater Sol Cells 33:417–427
16. Griffini G, Brambilla L, Levi M, Del Zoppo M, Turri S (2013) Photo-degradation of a perylene-based organic luminescent solar concentrator: molecular aspects and device implications. Sol Energy Mater Sol Cells 111:41–48
17. Chandra S, Rafiee M, Doran J, Mc Cormack SJ (2018) Absorption coefficient dependent non-linear properties of thin film luminescent solar concentrators. Sol Energy Mater Sol Cells 182:331–338
18. Dienel T, Bauer C, Dolamic I, Brü D (2010) Spectral-based analysis of thin film luminescent solar concentrators. Sol Energy 84:1366–1369
19. Griffini G, Levi M, Turri S (2014) Novel high-durability luminescent solar concentrators based on fluoropolymer coatings. Prog Org Coatings 77:528–536
20. Griffini G, Levi M, Turri S (2015) Thin-film luminescent solar concentrators: a device study towards rational design. Renew Energy 78:288–294

21. Barnes WL (1998) Fluorescence near interfaces: the role of photonic mode density. J Mod Opt 45:661–699
22. Barnes WL, Dereux A, Ebbesen TW (2003) Surface plasmon subwavelength optics. Nature 424:824–830
23. Calander N, Willander M (2002) Theory of surface-plasmon resonance optical-field enhancement at prolate spheroids. J Appl Phys 92:4878–4884
24. Muskens OL, Giannini V, Sánchez-Gil JA, Gómez Rivas J (2007) Strong enhancement of the radiative decay rate of emitters by single plasmonic nanoantennas. Nano Lett 7:2871–2875
25. Tam F, Goodrich GP, Johnson BR, Halas NJ (2007) Plasmonic enhancement of molecular fluorescence. Nano Lett 7:496–501
26. Chen Y, Munechika K, Jen-La Plante I, Munro AM, Skrabalak SE, Xia Y, Ginger DS (2008) Excitation enhancement of CdSe quantum dots by single metal nanoparticles. Appl Phys Lett 93:1–4
27. Thomas M, Greffet J, Carminati R (2004) Single-molecule spontaneous emission close to absorbing nanostructures. Appl Phys Lett 85:3863–3865
28. Chandra AJCS, McCormack SJ, Doran J, Kennedy M (2010) New concept for luminescent solar concentrator, in: Proceeding of the 25th European Photovoltaic Solar Energy Conference. WIP Renewable Energies, München, p 759–762
29. Bohren DRHF (1983) Absorption and scattering of light by small particles, 1st edn. Wiley, Hoboken
30. Catchpole KR, Polman A (2008) Design principles for particle plasmon enhanced solar cells. Appl Phys Lett 93:23–25
31. Kulakovich O, Strekal N, Yaroshevich A, Maskevich S, Gaponenko S, Nabiev I, Woggon U, Artemyev M, Kulakovich O, Strekal N, Yaroshevich A, Maskevich S, Gaponenko S, Nabiev I, Woggon U, Artemyev M (2002) Enhanced luminescence of CdSe quantum dots on gold colloids. Nano Lett 2(12):1449–1452
32. Gryczynski I, Malicka J, Shen Y, Gryczynski Z, Lakowicz JR (2002) Multiphoton excitation of fluorescence near metallic particles: enhanced and localized excitation. J Phys Chem B 106:2191–2195

Chapter 12
Addressing the Energy Challenges: Needs and Perspectives

Osman Benchikh

12.1 Meeting the Basic Energy Needs: A Priority

Our ways of producing and using energy have directly influenced the structure and development of all societies. For three centuries the human activities have been consuming ever more energy, even as governments adopt increasing vigorous policies for its conservation. Thus, the wide use and application of alternative and sustainable energy forms are necessary in order to fulfil the world's ever-growing needs.

Sustainable development underlines the need to manage energy resources judiciously; on the other hand, no development can occur without access to basic energy services. Access to energy is required in all economic and social sectors, and inequality in capacity to access energy resources and to utilise them for development purposes, results in further inequality in wealth distribution, be it in terms of social welfare or economic competitiveness. The international community has thus institutional, scientific, financial and ethical responsibilities to respond to basic energy needs in an inclusive and sustainable way. This is particularly important for those with poor access to energy resources.

Recognizing that access to energy services in developing countries is essential for the achievement of the Sustainable Development Goals, which would help to reduce poverty, the United Nations launched the *'Sustainable Energy for All'* Decade. Furthermore, the revived interest in alternative and renewable energy resources, combined with the substantial advances in the related science and technology in recent decades, justifies the priority given to addressing energy sustainability and access also reflected in the new set of SDGs that includes energy (Goal 7).

O. Benchikh (✉)
Head Energy Programme, EX UNESCO, Paris, France

ISEDC, Moscow, Russia

Energy Watch Group, Berlin, Germany

A. Sayigh (ed.), *Renewable Energy and Sustainable Buildings*, Innovative Renewable Energy, https://doi.org/10.1007/978-3-030-18488-9_12

Energy is seen as an essential component in any economic and social development processes, and the capacity to meet the basic energy needs of all citizens should be a key objective. Besides, it constitutes a prerequisite to reaching the Sustainable Development Goals, energy being one of them, and constitutes a major multiplier of the other Goals as few, if any of them, can be achieved without it.

In most of the developing countries, public electric utility grids and improved infrastructure for energy supply exist mainly in urban and peri-urban areas. Globally over 1.3 billion people still remain without electricity access, and more than 95% of them are either in Africa or Asia. Some 85% of those people leave in peripheral urban and rural areas of developing countries. In Sub-Saharan Africa, approximately 70% of the population does not have access to electricity, and this figure rises up to 80% for the rural areas (Fig. 12.1). Households may spend as much as 30% of their disposable income on fuel-based lighting, and women devote at least a quarter of total household labour to wood collection, while those consumers receive poor services in return.

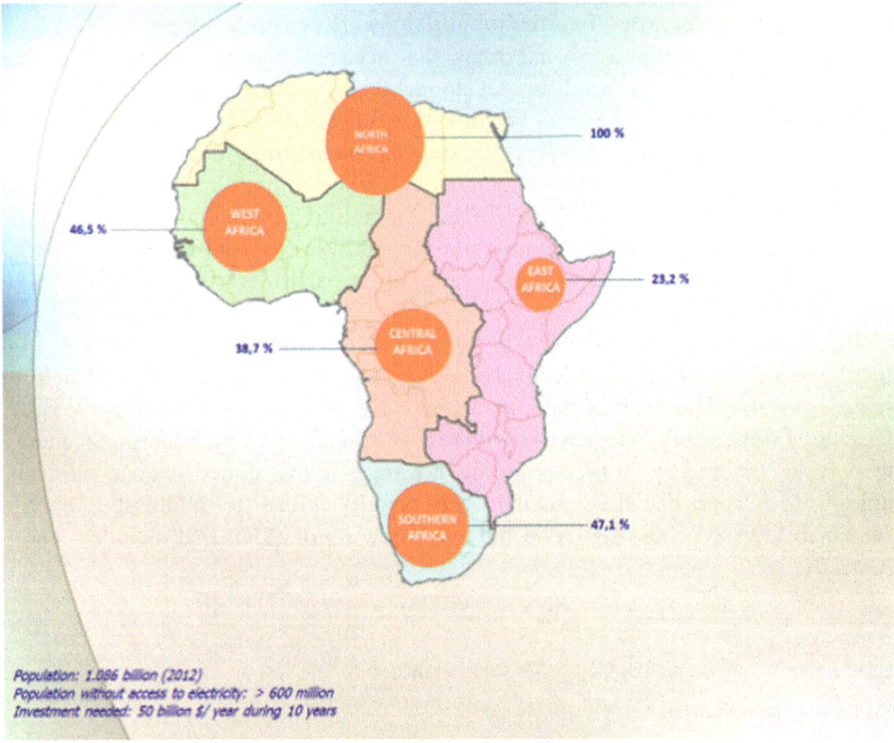

Fig. 12.1 Needs in term of energy access in Africa

12.2 Energy: Human Development–Climate Nexus

The quest for long-term sustainable energy supplies and the imperative for environmentally responsible energy use are the two main interrelated aspects of the vital energy component that underpins today's global agenda. Energy considerations are of paramount importance to key global concerns such as climate change, sustainable development, poverty reduction and economic growth. In this regard, climate change has, over the last decades, evolved into an issue of global concern. In this regard, there is increasing consensus in both the scientific and political communities that significant reductions in emissions are necessary in order to limit climate changes to manageable levels.

Concurrently, influence of some selected energy indicators on Human Development Index (HDI) components highlights the strong impact of energy on Sustainable Development. More specifically, the role of access to electricity services and the electricity consumption per capita constitutes the key factors that have a stronger positive influence on the achievement of the Sustainable Development and related targets. The figures below (Figs. 12.2, 12.3 and 12.4) show for example the evident correlation that has the energy indicators with the life expectancy as one of the HD. Similarly, the electricity consumption and access to electricity services have a strong emphasis on the literacy rate as well as on the leaving conditions and the social development including the gender development particularly in the developing countries.

While progress will occur and the number of people with access to modern energy services will increase, population growth will erase most of these gains, meaning that it is unlikely that the target of the SDGs can be met by 2030, if no appropriate measures are taken.

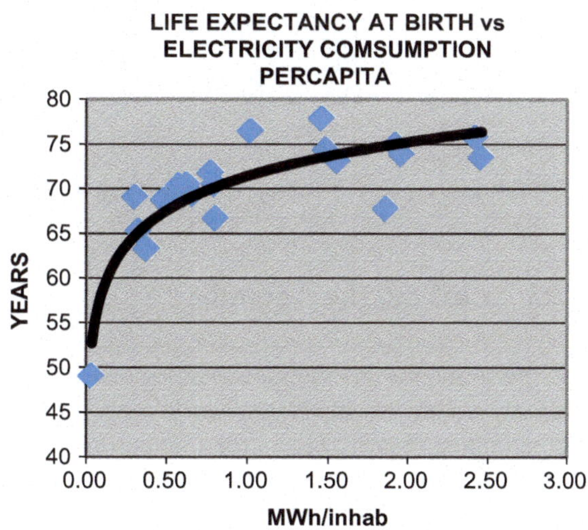

Fig. 12.2 Life expectancy at birth vs. electricity consumption percapita

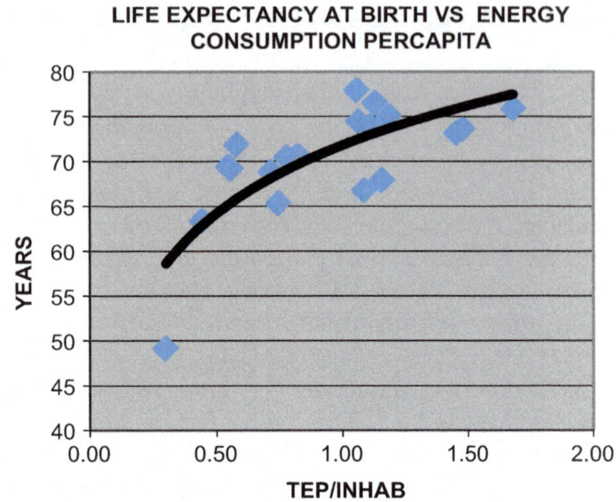

Fig. 12.3 Life expectancy at birth vs. energy consumption percapita

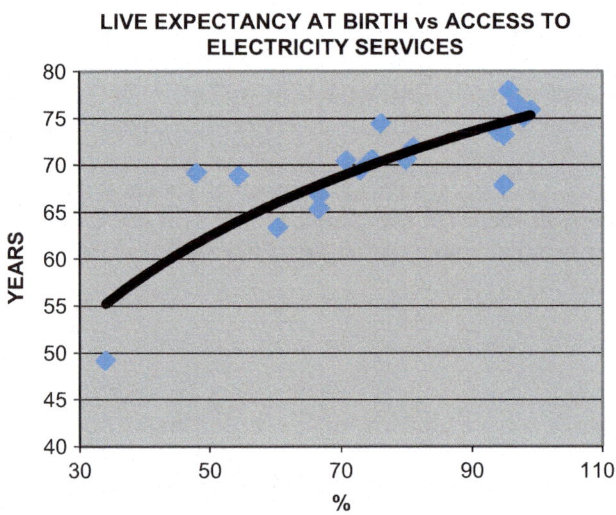

Fig. 12.4 Life expectancy at birth vs. access to electricity services

12.3 Linking the Agendas

Energy strategies must be governed by a wide array of intertwined issues. It is no longer possible to satisfy the world's energy demands by the exploitation of a limited range of energy resources. To ensure a viable socioeconomic future for all, the international community needs to expend more effort on the development and wide use of renewable and environmentally sound sources of energy as a viable complementary alternative to the conventional methods of energy generation and promote efficient use of the latest. For low density rural populations, decentralised energy technologies based on renewable energy sources offer a viable alternative to grid extensions; as such, they constitute a valuable component in the global efforts aiming at addressing the global sustainability.

Responding to the energy challenges requires the diversification of energy sources. It also calls for building a knowledge base, disseminating relevant technical and scientific knowledge and promoting appropriate energy policies and choices as a foundation for increased use and application of the various environmentally sound energy technologies. Furthermore, joint international efforts, effective dialogue and cooperation are needed in order to find solutions to complex energy issues.

The process of moving from the concept of energy development focusing on economic growth to a new one centred on global sustainable development including environment protection will thus be made possible only if the current energy development paradigm is changed.

We stand at the crossroads where we believe that it is vital to promote a comprehensive, holistic approach to affordable and sustainable energy access, sustainable development and climate change.

Chapter 13
The Impact of the Rise of Using Solar Energy in GCC Countries

W. E. Alnaser and N. W. Alnaser

13.1 Introduction

GCC countries had the passion in using renewable energies since 1970s. This is no surprise for GCC countries as it is blessed with an abundance of solar energy as the annual average solar radiation within the GCC countries is relatively equal to 1.1 barrel of oil equivalent per m^2. The radiation is the highest in Kuwait, in June–July (8200 kWh/m^2), and the lowest in Oman (6400 kWh/m^2). The radiation is low in January–December (4200 kWh/m^2 in the UAE and 3200 kWh/m^2 in Bahrain) [1]. The first large solar energy project in the region was in Kuwait (by Kuwait Institute for Scientific Research, KISR), followed by Saudi Arabia (by King Abdulaziz City for Science and Technology, KACST) in the 1980s, United Arab Emirates in the 1990s (by Masdar), and Bahrain (by Bahrain Petroleum Company and now by the Sustainable Energy Unit SEU), Oman and Qatar in twenty-first centuries. Among the old, but advanced, PV projects in the GCC countries the full solar air-conditioned school (The British School) in Kuwait using PV and solar thermal applications in 1970s, full PV operated fuel filling station in Bahrain at Al Hora area in Manama, Bahrain, in 1980, The Solar Village PV system 350 kW (2155 MWh) to provide AC/DC electricity for remote areas from 1981 to 1987) [3].

Now all GCC countries had conducted a relatively large project in Solar and Wind Energy, especially Kuwait (currently about 70 MW among a plan of 2000 MW by 2030), UAE (currently about 300 MW among a plan of 2500 MW by 2030) and

W. E. Alnaser (✉)
Department of Physics, College of Science, University of Bahrain,
Zallaq, Kingdom of Bahrain
e-mail: walnaser@uob.edu.bh

N. W. Alnaser
Department of Architecture and Interior Design, College of Engineering,
University of Bahrain, Zallaq, Kingdom of Bahrain

A. Sayigh (ed.), *Renewable Energy and Sustainable Buildings*, Innovative Renewable Energy, https://doi.org/10.1007/978-3-030-18488-9_13

Table 13.1 Summary of solar PV development in the GCC countries [2]

Country	Solar energy target share in total installed capacity	Primary energy sources	Maximum solar insolation (kWh/m²/year)	Current operational solar projects	Future projects
Bahrain	5% of all renewable including solar (2020)	Oil, gas	2180	0.5 MW solar plant at the Bahrain University 5 MWp Bapco pilot plant	5 MW solar project at Al Dur power and water plant
Kuwait	1% in 2015, 10% by 2020 and 15% by 2030	Oil, gas	2200	See Table 13.1	50 MW solar project across the country
UAE	7% of all renewables by 2020, 25% 2030 and 75% by 2050	Oil, gas	2285	See Table 13.2	1000 MW Mohammed bin Rashid Al Maktoum Solar Park
Qatar	20% of all renewables with solar incorporated by 2030 and 15% solar by 2018	Oil, gas	2113	Establishment of solar PV manufacturing plant in Ras Lafin	Stadiums with solar technology cooling for the Football World Cup 2022; 1000 MW solar PV plant in Doha
Oman	10% by 2020	Oil, gas	2500	See Sect. 3.5	300 kW solar project at Al Mazyonah-Dhofar; 1000 kW solar project in Haima, 28 kW PV and storage system at Al Mathfa-Dhofar; 100 kW PV in Hijji
Kingdom of Saudi Arabia	54 GW from renewables including solar technology	Oil, gas	2200	See Table 13.3	1000 MW solar plant in Makkah; 10 MW Al-Khafji plant; 1.8 MW King Abdullah Petroleum Studies and Research Centre

Saudi Arabia (with an ambitious renewable energy target of 3450 GW by 2020 with a further 6000 GW envisioned by 2023 and to 200,000 MW by 2030). Table 13.1 sheds light on current large projects [2]. Figure 13.1 illustrates the current and future planned solar projects in the pipeline in the GCC countries [2]. However, it is thought that more major solar projects—at GW scale—will be announced officially, i.e. an installation of the world largest PV project (GW size) during the visit of the Chinese president to UAE in middle of July 2018. Currently, Saudi Arabia was in

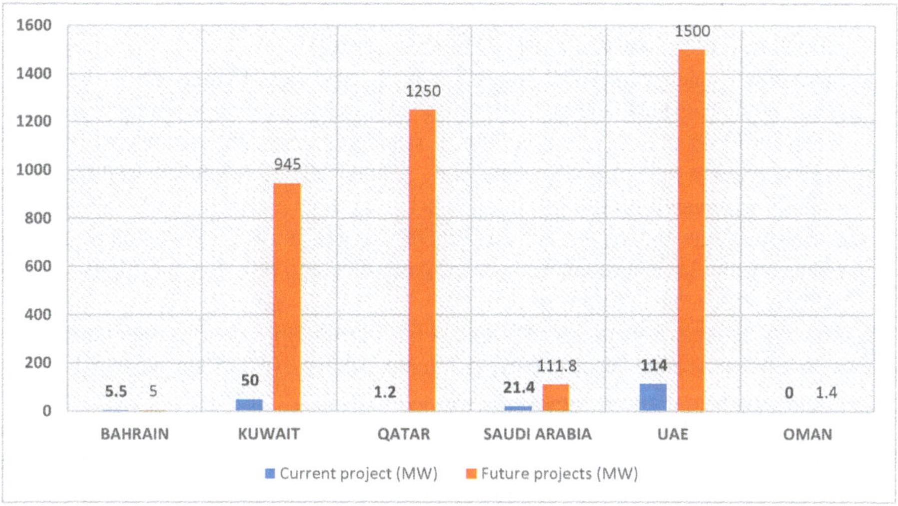

Fig. 13.1 Current and future planned solar projects in the pipeline in the GCC countries [2]

talks to develop the world's largest solar power plant in entering an agreement to develop a $200 billion solar power plant that's anticipated to be up and fully running by 2030 (200 GW) of solar power, which is approximately double the amount of solar energy produced globally last year and more than 100 times as large as the new next biggest solar plant project which would generate more than 100,000 jobs [4].

What has been noted is there is a great development of solar energy in GCC countries that have been done as current projects or future projects in accelerated steps without underpinning it with strategic policy development plans, which will surely have an impact. This paper is attempted to explore these impacts.

13.2 Technical and Strategical Issue on Solar Technology Utility

Before involving in discussing the impact of the rise of using solar energy in GCC countries it sounds better to highlight some technical and strategical issues on solar technology utility in these countries. These are the followings:

1. Real investments and plans to use other alternative energy sources such as solar, wind, hydrogen and geothermal in GCC countries gain a momentum, but recent developments in the technology used for extracting gas and oil from shale rock formations place a big question mark on the GCC countries' energy plans including those relating to alternative and renewable sources of energy [5].
2. Electricity demand in the region has been increasing at an annual rate of around 8%—faster than the growth recorded in any other region of the world. It is estimated that over the next decade the GCC countries will need to add 100 GW

of additional power to support their economies which is running at twice the growth rate of major advanced word economies [5].

3. Sustainable energy in GCC states has been considered since 2008. Resource efficiency programs, clean technology research, alternative energy projects, green building codes, green economy strategies and public transport systems have become a part of mainstream news. The GCC states have started to have a stake in the transition to sustainable energy and have already undertaken some policy, projects and sector-wide efforts which will boost the investment, and its effect on the environment. However, many of these policies and plans are not actually enforced or imposed [6].
4. The region is appropriate for PV and solar thermal applications on a large scale with little presence of clouds do not exceed only 20% of the year [6] which highly qualify these countries to use solar heating and cooling applications, Concentrated Solar Power (CSP) and PV applications since the global radiation is relatively high (from 1920 to 2450 kWh/m^2/year). The summer demand in the region is about 10.8 GW, and this demand is supplied from combined-cycle turbines fueled with cheap legacy gas and expensive imported LNG. Using solar power plants can introduce about 3.5 GW of nominal PV capacity and in part can be stored to overcome evening peak (10:00 to 11:00 pm) where thermal storage, energy storage (e.g. pumped storage where available), production of portable water at off-peak times and electricity trading with neighbors at irregular peak times can be utilized. These solutions have become economically attractive as they have less capital cost compared to gas turbines [7]. For this reason, GCC countries have launched a sense of the importance of converting part of its electrical energy produced by fossil fuels to alternative energies.
5. The six GCC countries are located in an arid and solar zone. They occupy Latitude 17° N to 32° N and Longitude 35° E to 60° E. The following information imposed itself for inclusion to allow comparison of technical, economic and policy issues between these countries:

   (a) The mean yearly solar radiation in Bahrain is, approximately, 2180 kWh/m^2 [1]. The share of RE in Bahrain is planned to reach between 5% and 7% from electricity power generating capacity by 2030 [8].
   (b) Saudi Arabia has a yearly average solar radiation of 2200 kWh/m^2 [9, 10]. Research and development activities show that solar energy has diverse practical applications in the KSA. The share of Renewable Energy is 30% Electricity generation.
   (c) UAE has huge solar potentials with the mean annual solar radiation of 2285 kWh/m^2 and average sunshine of 10 h per day [11, 12]. However, high humidity and dust particles in the atmosphere reduce the solar intensity [13]. However the government of the UAE had set a target of 7% energy mix by the year 2020, 25% by 2030 and 75% by 2050 [14].
   (d) Kuwait's annual solar radiation is predicted to be between 2100 and 2200 kW/m^2 [15]. The average daily sunshine hours/year for Kuwait range from 7 to 12 [16]. Kuwait plans to expand the share of RE in the country's energy mix by 1% in the year 2020 and up to 15% by 2030 [17].

(e) Oman has an annual solar radiation in Oman and varies from 2200 to 2500 kWh/m^2 [18]. The highest solar radiation in Oman is in Marmul, followed by Fahud, Sohar and Qairoon Hairiti [19].
(f) Qatar is a country with abundant solar resources, reported to be up to 2113 kWh/m^2/year for the ground measured yearly average, calculated considering the 4-year period of coincident data [20]. Studies show that the presence of high-ambient temperature in Qatar coupled with dusty conditions affects the performance and reliability of PV panels [21]. A recent study shows that Qatar solar potentials are equivalent to 1.5 million barrels of crude oil annually with an average sunshine hours in Qatar are 9.5 across the year [22].

6. It had been reported that the drop in the output of the PV electricity (PVD), as a function of dust accumulation—measured as mass per unit area (d_s)—which is nearly applicable to all GCC countries, can be calculated from the following relation [23]:

$$\mathrm{PVD}(\mathrm{in}\%) = \left(\mathrm{PV_d} / \mathrm{PV_{nd}}\right)_{\mathrm{drop}} = 100 - 99.66\exp\left(-0.038 d_s\right), \qquad (13.1)$$

where PV_d is the output of the dusty and PV_{nd} is the output of non-dusty (clean) PV.

7. The actual solar electricity gained from each kW installation of PV in GCC countries can be estimated from the following relation which take into consideration the dust accumulation and sky turbidity [24].

$$\begin{aligned}&\text{Actual solar electricity produced in kWh}\\&= \text{Installed Power}(\text{in kW}) \times 3.6 \times 365,\end{aligned} \qquad (13.2)$$

while the maximum estimated solar electricity yield—which assumes a regularly cleaned PV and clear sky with no turbidity—can be estimated from the following relation:

$$\left.\begin{array}{l}\text{The maximum estimated}\\ \text{solar electricity produced in kWh}\end{array}\right\} = \text{Installed Power}(\text{in kW}) \times 4.6 \times 365 \qquad (13.3)$$

Figures 2.8 and 4.6 are related to the system performance which is related to PV efficiency and its performance under warm–hot environment like GCC countries. Table 13.2 shows the PV installation in GCC countries with cost per solar Watt. It clearly shows that each installed kW is expected to produce, annually, about 1650 kWh of solar electricity, i.e. daily 4.5 kWh [25].

8. Nearly all of the GCC countries have set relatively low prices of electrical energy unit (kWh) not to encourage more energy consumption and waste but rather to assist the citizens in overcoming the harsh–hot and warm weather that nearly prevail for 7 months. Nearly 70% of household electricity is used for air-conditioning. The temperature in GCC countries reaches up to 60 °C (Kuwait)

Table 13.2 PV installation in GCC countries with cost per solar Watt [25]

	Project name	Location	Type of installations	Capacity (max peak DC)	Year built	Cost	Annual generation	Normalized annual generation
1	Masdar	Abu Dhabi, UAE	Ground Mount	10 MWp	2009	\$5/W	17.5 GWh	1.75 MWh/MW
2	KAUST Solar Park	KAUST Saudi Arabia	Rooftop	2 MWp	2010	\$7/W	3.3 GWh	1.65 MWh/MW
3	Saudi Aramco Solar Car Park	Dhahran Saudi Arabia	Mixed (car park, rooftop, trees, pole mount)	10.5 MWp	2012	\$10/W	17.5 GWh	1.67 MWh/MW
4	KAPSARC Project-1	Riyadh Saudi Arabia	Ground Mount	3.5 MWp	2013	\$4.5/W	5.8 GWh	1.66 MWh/MW
5	KAPSARC Project-2	Riyadh Saudi Arabia	Ground Mount	1.8 MWp	2014	\$3.5/W	2.9 GWh	1.61 MWh/MW
6	Kuwait Oil company	Umm Gudair Kuwait	Ground Mount	5 MWp	2014	\$5.6/W	8 MWh	1.68 MWh/MW
7	BAPCO Pilot Project	BAPCO, Bahrain	Mixed (14 Locations)	5 MWp	2014	\$5/W	8.3 GWh	1.66 MWh/MW
8	TATWEER	Bahrain	Ground Mount	1 MW	2016	\$1.5–1.8	1.9 GWh	1.9 MWh/MW

in mid-summer and relative humidity of about 75% like Bahrain and UAE. It's only few months where people in GCC countries do not use air condition.

In 2011, energy subsidies accounted for 15%, 32%, 28% and 15% of all government expenditures respectively in Kuwait, Saudi Arabia, the UAE and Qatar. Since that time, however, the GCC region has been accelerating a trend to develop more open energy markets with less subsidies in order to ease the burden on government budgets already strained by the prolonged global oil price slump. The prices of electricity had increased from 2011 to now about four to six times.

The average cost of kWh in GCC countries is now 6 c\$, while in 2011 it was only 2 c\$ for residential while for commercial it was 6 c\$ in 2011 and jumped now to 12 c\$ [26]; this is still much less than the tariff in USA (13 c\$/kWh). The relatively low tariff for kWh in GCC countries is not an encouraging factor to install PV systems on rooftop to produce solar electricity. This price makes the pay-back period (PBP) too large (about 100 year for a 7 kW with cost of USD 3/W). PBP can be easily estimated by using the following relation:

$$\mathrm{PBP} = \left[\text{cost of total PV system}/\left(\text{solar electricity produced in kWh} \times 3.0 \times 365\right)\right] \quad (13.4)$$

For example, in Bahrain, for a full 7 kW photovoltaic system the customer will pay USD 9100 (PV panels plus invertors and charge regulators). If the government purchase each kWh at a cost of 1 c$ (3 fils) then the PBP will be 118 years, while if the tariff is purchased at a cost of 26 c$ (100 fils)—if Feed-in-Tariff is initiated—the PBP is 4.6 years. More accurate calculation of PBP can be made by referring to ref. 27.

9. All GCC countries had signed the Kyoto Protocol and there acceleration in using solar and other renewable energy is driven mainly by their respect and obligations toward the climate change. The previous delay in using renewable energy at a large scale is attributed to lack of sufficient awareness, high capital cost, fear of moving from secure and conventional energy resources to uncertain and irregular clean renewable energy sources, lack of clear regulations and incentives, lack of industrial motivation, lack of expertise and know-how and lack of adequate of information about these new technologies [6, 28].
10. The average electricity consumption per capita in GCC countries exceeds 10.000 kWh; this is mainly due to harsh environment, lack of natural vegetation, lack of natural water, dusty nature of region and high solar radiation. All these factors impose the GCC citizen to become heavily dependent to electricity utility especially to air conditioning. Here comes the importance of investing and supporting research in solar cooling to make the cost of purchase of such devices affordable, reliable and low tariff; such product must be subsided initially, in a similar manner of subsidizing the LED lights in China during its initial launching.

The electricity consumption per capita in Bahrain is 19,592 kWh (CO_2/capita = 23.4 tons), in Saudi Arabia is 9444 kWh (CO_2/capita = 19.5 tons), Oman 6554 kWh (CO_2/capita = 15.4 tons), Qatar 15,309 kWh (CO_2/capita 45.4 tons), UAE 11,264 kWh (CO_2/capita = 23.3 tons), Kuwait 15,213 kWh (CO_2/capita 25.2 tons) and average world average is 3125 kWh (CO_2/capita 5.0 tons) [29]. It is very noticeable that the per capita consumption in Bahrain is very high which is attributed to warm and humid weather—nearly throughout the year.

Figure 13.2 shows the sectoral energy consumption in 2015, which is assumed none has been changed much [30]. The GCC produced 1.17 billion tons of CO_2—equivalent in 2012 [29]. The bulk of CO_2-equivalent emissions in GCC are related to energy production and consumption with 95% of emissions are energy related, with agriculture, waste and land-use changes making up the remainder [31] and of this amount, the power and water sectors emitted 438 million tons of CO_2-equivalent.

All GCC counties have a mix of power and water desalination capacity (Fig. 13.3) with Saudi Arabia more than 80.5 GW of power generation and UAE with the second-largest (almost 29 GW) while other countries have smaller capacities (Kuwait 18.3 GW, Qatar 8.6 GW, Oman 7.8 GW and Bahrain 2.8 GW) [30].

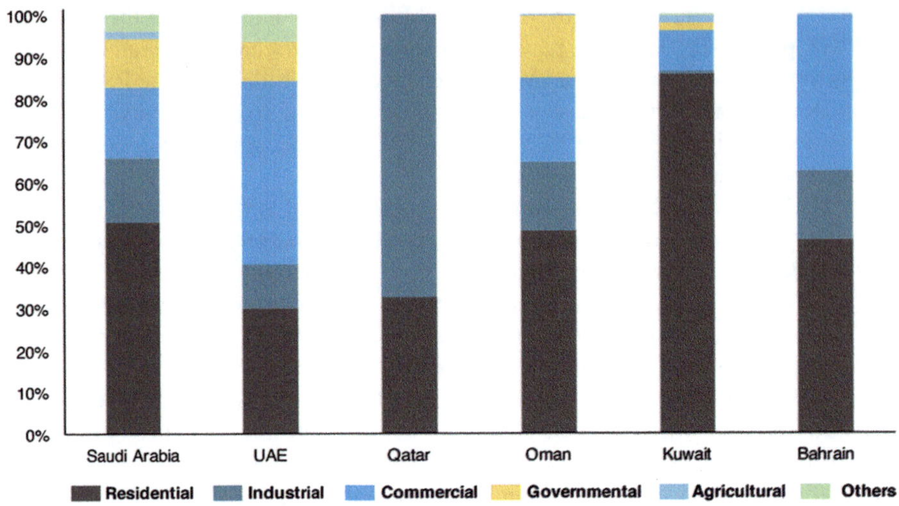

Fig. 13.2 Sectoral energy consumption in 2015 [30]

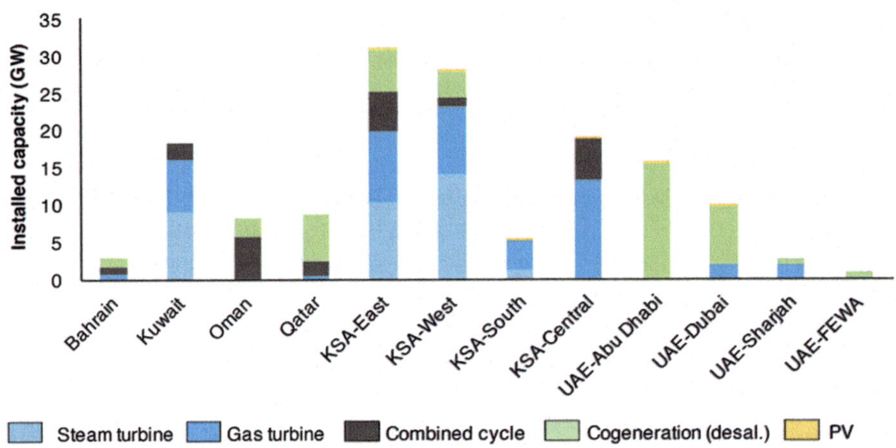

Fig. 13.3 Power and water capacity by technology type [31]

GCC countries use a large amount of electricity to produce desalinated water (about 16 million m^3 per day) [32]. Notably, Bahrain, Oman and the three coastal regions of Saudi Arabia utilize RO technology. Kuwait and Qatar do not have any RO plants. A small number of thermal water-only plants, utilizing MED or MSF processes, exist in Kuwait and in the southern region of Saudi Arabia.

The magnitude of electrical production, consumption and peak loads is shown in Table 13.3 [30]. Saudi Arabia produces over half of all electricity in the GCC.

Table 13.3 Electricity production in terawatt-hours (TWh), consumption (TWh) and peak load (GW) [32]

Country	Production	Consumption	Peak load
Bahrain	14.1	12.6	2.9
Kuwait	68.3	60.5	12.8
Oman	31.3	31.3	6.1
Qatar	38.7	36.1	6.7
Saudi Arabia	304.2	274.5	56.6
UAE—Abu Dhabi	70.9	52.8	9.0
UAE—Dubai	39.6	38.4	7.2
UAE—Sharjah	5.7	10.2	2.2
UAE—FEWA	0.40	10.3	2.2
Total GCC	573.2	526.7	

13.3 The Impact of the Rise of Using Solar Energy in GCC Countries

It has to be understood that GCC countries are of need to use renewable energy in general and in particular solar energy for many reasons. These are: (a) Vulnerable to the threat of rising sea levels. (b) Industrial areas are expanding. (c) High annual energy demand growth rate which ranges from 7 to 10%. (d) Need of natural gas to meet its future energy demand. (e) Have very high solar radiation levels. (f) Set a Renewable Energy target in their vision 2030. (g) Need to open a green job. (f) Allow exchange of knowledge, technology and mutual cooperation. Such an accelerated use of solar energy after long slow maneuver to use this energy may have the following impacts:

1. Major reduction in the solar electricity prices in the region

   The Renewable Energy Project Development Office's (REPDO) 300 MW Sakaka project in Saudi Arabia reported the lowest PV LCOE (2.34 US$ cents per kWh) [33]. Also, Masdar and its French partner EDF submitted the lowest offer for the project at 1.79 US cents per kilowatt hour (kWh)—as illustrated in Fig. 13.4 [34]. The prices of PV panels had dropped from US$ 76 to only US$ 0.3 per Watt [35].

   This drop in the prices is due to the availability of 100's of PV producers, offering more than 30 years warranty on PV panels and drops in solar system components. Table 13.4 lists the PV pipeline projects while Table 13.5 for CSP pipeline projects.
2. More investors in renewable energy business

   GCC countries could reap multiple benefits from scaling up renewable energy use, including reducing water withdrawal by 11 trillion liters (or 16%) and saving 400 billion barrels of oil in the power sector, creating 2000 direct jobs and reducing the region's per capita carbon footprint 8% by 2030 [36]. This will attract investors to further use solar energy as they are witnessing this acceleration in

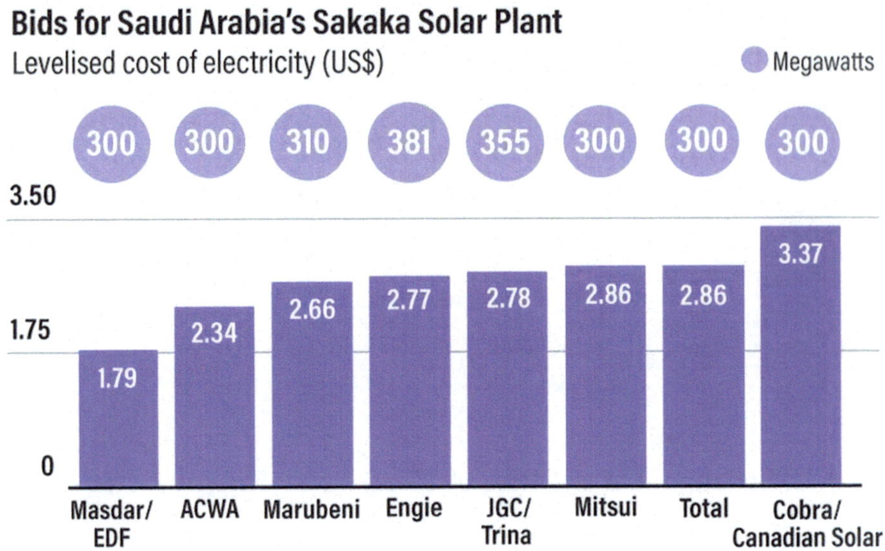

Fig. 13.4 Bid for Saudi Arabia Sakaka Solar Plant. It's a 300 MW solar photovoltaic plant. The tarrif for each kWh solar is little as 6.7 halalas or 6.7 fils (or 1.79 cents) [34]

Table 13.4 PV pipeline projects in Arab Region [33]

Country	Country	Capacity	Status	Client
Bahrain PV	Bahrain	200	Announced	EWA
Solar IPP (West Nile)	Egypt	600	Prequalification	NREA
Solar IPP (Kom Ombo)	Egypt	200	Bid stage	EETC
Round 3	Jordan	200	Bid stage	MEMR
RISHAPV	Jordan	50	Financial close	NEPCO
Water Authority Jordan	Jordan	30	Prequalification	Water Authority Jordan
KNPC	Kuwait	1000	Bid stage	KNPC
Noor Midelt	Morocco	300	Bid stage	MASEN
IBM	Oman	SOO	Prequalification	OPWP
PDO-100MW	Oman	100	Prequalification	PDO
Qatar PV	Qatar	200	Announced	Kahramaa
Multiple Sites	Saudi Arabia	6.4QU	Announced	HLPUO
Tunisia PV	Tunisia	70	Bid stage	STEG
Tozeur PV	Tunisia	10	Bid stage	STEG
Sweihan II	UAE	1200	Announced	ADWEA
DEWA Phase V	UAE	300	Announced	DEWA
Total		11,860		

Table 13.5 PV pipeline projects in Arab Region [33]

Country	Country	Capacity	Status	Client
TAQA CSP	Egypt	250	Announced	TAQA Arabia
West Nile CSP	Egypt	100	Announced	
Lebanon CSP	Lebanon	50	Announced	
Noor Midelt	Morocco	800	Bid stage	MASEN
Total		1200		

executing major successful solar project with relatively low cost. This might encourage other developer in renewable energy to introduce themselves to GCC market and offer attractive prices and facilities.

Currently, China accounted for a record 45% of the global investment total, up from 35% in 2016 followed by Europe (15%), United States (14%), Asia Oceania, excluding China and India, (11%), Americas, excluding Brazil and the United States (5%), India (4%), Middle East and Africa (4%) and Brazil (2%) [37]. In 2017, Solar PV and wind power had dominated the investment, accounting for roughly 57% and 38%, respectively. Solar power was the only technology to witness an increase in 2017, with new investment up 18% relative to 2016, to USD 161 billion. The investment in Middle East and Africa had increased by 11% in 2017, to USD 10.1 billion, with substantial increases in Egypt (USD 2.6 billion), and the United Arab Emirates (USD 2.2 billion) [37].

3. Rise of innovative design of houses to utilize its structure in installation and integration of renewable energy devices

   The installation of PV systems on a rooftop—making the building integrated with photovoltaic—for power up to 7 kW for a relatively low cost (BD 3500 or US$ 9100)—such as in Bahrain—is attractive especially if (a) the customer receives financial support or loan with minimal interest and (b) the government buys the generated solar electricity from the rooftop with a generous cost (Feed-In-Tariff). This will encourage house developers to make the roofs more fit to allow large utilization of the roof to account for PV installation with no obstacles. Currently, in GCC countries, the available roof area for PV panels is about 18% only for flats and 25% for villas in Khobar, Saudi Arabia [38, 39]. Researchers in the GCC countries started to pay more attention for the houses design to account for possible integration with PV [40]. It might also be taken into consideration in making use of integration with other promising and brand new technologies and renewable energy technology such as wind, geothermal, biofuels, biomass and hydro [41].

   It has to be noted that it has been already reported the advantage and disadvantage of using the rooftop. It was the 9.6 PV installed on the rooftop, in a house in Greece, reduces the heat in the building in summer by 17.8% but requires more heating load in winter by 6.7%! [42].

4. Rise of many service and maintenance of solar technology companies

   Unfortunately, not all governments of the GCC countries had Net Metering facility; it's only made in Bahrain. It is believed that with this accelerated use of solar energy utility all GCC countries will introduce such a facility. Net Metering is a metering and billing arrangement designed to compensate distributed energy generation system owners for any generation that is exported to the utility grid. It allows residential and commercial customers who generate their own electricity from solar power to feed electricity they do not use back into the grid [43]. A more attractive facility is to have a Feed-In-Tariff that enables households to have a payback not more than 5 years. Unfortunately, no GCC country had announced the Feed-In-Tariff facility; only UAE had offered what is called "Grants and subsidies" which is legislated from the Renewable Energy Deployment Strategy (2009, 2011 and last updated in 2015) [44]. Among Arab countries who have Feed-In-Tariff are Egypt and Tunisia [44].

This means any interruption in the PV system will lead to money loss and waste. This will lead to the birth and establishment of many maintenance or Energy Saving Companies (ESCo). This will create more green jobs and green economy. This rapid hop toward use of solar technology will encourage the investor to open business in this solar energy business including PV panels, Charge regulators (MPPT), PV aluminum frames and structure, PV auto cleaning using nanotechnology, AC/DC invertors and promotion and media companies in such field.

The impacts of different renewable energy pathways on ecosystems and biodiversity, and the implications of these impacts for transitioning to a Green Economy have been discussed in detail elsewhere [45].

5. Establishment of new academic programs and courses in solar and other renewable energy technologies in the higher education institute and technical organizations

   The accelerated use of solar technology in GCC countries will attract students to get registered in Science, Technology, Engineering and Mathematics (STEM). Universities and Academia must prepare to such move and should align their programs with such a need. It will be an opportunity to establish long-term partnership between universities in GCC countries with the other international universities in science and innovation collaborations. This will lead to boost in faculty and student exchange program and flying professor program as well as condensed international courses and programs. It will also lead to working strategically with key stakeholders bilaterally across the world. This will foster closer people to people contacts in our shared interest in stability and prosperity.

   It has to be noted that in USA there are only 49 Bachelor's degrees programs in Sustainable Energy and 39 Sustainable Energy Master's degrees. In GCC countries there is no single BSc in renewable energy or a diploma although a set of renewable energy education is proposed before 20 years through ISESCO by the current author [46].

   Kandpal and Broman [47] had made a review of published literature (376 references) on renewable energy education initiatives across the globe, challenges faced and potential approaches toward efficient and effective solutions as in the past 30 years a large number of countries across the globe have initiated academic programs on renewable energy technologies and related aspects.

6. More concern and studies on disturbances to the main grid due to solar connection

   There are many reasons for the fast dissemination of solar energy in the GCC countries after a long lag, among them is renewable generation is made possible to connect to the grid, customers are made to receive incentives and are paid, and appropriate deployment. Using solar electricity can be difficult to steer accurately and can cause surges, i.e. very rapid growth over a few months then nothing for a while) which is difficult for the installers industry.

Network integration is an important issue that must receive great attention in opting for large and fast use of solar electricity. Electricity networks were designed for loads not generators. Therefore, power can flow backward where networks do this naturally and do it very well [48]. Network integration has the advantage of eliminating the need for batteries and the need for local balancing.

Voltage control is important when using solar electricity at a large scale because embedded generators cause voltage rise. In countries where networks are short and well-engineered, then this is not an issue; otherwise, it is significant (Saudi Arabia, Kuwait and UAE).

PV electricity (considered herein as distributed generators, DG) is normally designed to be grid connected and not to Island. The reason it is considered as DG is because the PV system is producing and distributing electrical power back into our utility grid. Therefore, Islanding refers to the condition of a DG generator that continues to feed the circuit with power, even after power from the electric utility grid has been cut off. Islanding can pose a dangerous threat to utility workers, who may not realize that a circuit is still "live" while attempting to work on the line [48] . Designing to allow islanding is expensive and unlikely to be justified with a reliable grid. Anti-islanding protection can lead to "block-tripping" which is bad for the grid. This risk can be reduced by careful choice of inverters and their settings and having fault ride-through capabilities. This is significant when there is a large demand in solar electricity (more than 20% of the main installed conventional electricity of the grid). The conventional system relies on the rotating mass of all its conventional generators to provide inertia and keep the frequency steady at 50 Hz while PV electricity provides no inertia. This becomes a concern at high penetrations as once conventional is significantly reduced and taken offline. In fact operating low-inertia power systems is a current research challenge. An interconnector will help or expand the challenge [49].

The GCC countries should work hard from now on to make their interconnection with solar electricity have the following features [50]:

(a) Prevents the penetration of the harmonics into the power grid
(b) Avoids the requirement of having controllable at the DC side of the power inverter
(c) Controls the power factor level at the point of interconnection with the grid
(d) Relaxes the constraint of synchronizing the inverter output voltage with the power grid voltage
(e) Extracts (if possible) maximum power from PVG panels

7. Major reduction in Carbon footprint per capita in GCC countries.

   According to Arab Petroleum Investments Corporation, the GCC countries represent 47% or 148 GW of the current MENA power-generating capacity [51]. The harsh environment of GCC countries with natural water scarcity and the vision of the leaders of these nations for development and prosperity have led to a rise in the GCC's demand for energy. GCC countries would require US$ 85 billion for the addition of 69 GW of generating capacity and another US$ 52 billion for T&D over the next 5 years. The GCC power capacity needs to expand at an average annual pace of 8% between 2016 and 2020! Now assuming that if by year 2020 the solar electricity represents 5% of the installed electric power—which is 10.85 GW out of 217 GW—then the saved CO_2 emission will be nearly 95 million tons! This means that from year 2020 to year 2030 the saved CO_2 emission will be about 1 billion tons! This is a considerable amount if we know that GCC countries had produced 1.17 billion tons of CO_2—equivalent in year 2012 [29]! This will make the carbon footprint is less than the current. It has to be noted that carbon dioxide (CO_2)-global energy-related CO_2 emissions—grew by 1.4% in 2017, reaching a historic high of 32.5 Gt [52]. The GCC countries emission may have reached about 1 Mtons.

8. Use of more efficient and low consumption household and industrial devices

   Due to the future reliability on solar electricity for use in houses, building and industry in GCC countries, it is expected that consumers will only purchase low consumption and efficient appliances and devices since each kW of solar electricity offers maximum of 1680 kWh (costing about US$ 3000) annually, while for this energy we need only 5732.4 ft^3 of natural gas which costs about US$ 7 only (knowing that each 1000 ft^3 cost about US$4). People will purchase five-star appliances. The air-condition technology will witness great attention in GCC countries so that each thermal ton should consume less than 1 kW of electricity. A current air condition will operate for only 40 min if fed with 1 kWh, 1 h for washing machine (rated 1 kW) and 125 h for an eight LED light [53]! Remarkable effort, business and investment will be on reducing the electricity consumption to make such appliance cope with the relatively expensive solar electricity.

9. Boost in battery industry for solar electricity storage

   With the large dissemination and fast growth of solar electricity, which is sustainable and free (after the subsidized pay-back period), investors and developers will think seriously in standalone solar houses or premises or factories, especially, for rural but tourist attraction location. This will make them to invest solar electricity storage devices, in particular batteries. The storage will be important to overcome the mismatch between the time of generation and demand, although in GCC countries the time difference is mostly just a few hours, while in UK and Ireland it is much larger [49]. Storing cold, either in the building fabric (concrete) or in ice (same water each day which not consuming water), will be also an option for cool air storage at night making use of the excess of solar electricity. This technology is straightforward and proven low cost. It is also of low value when gas powered but will become very valuable.

   In 2016, the first hints of a storage-driven transformation of the electricity business had come in Hawaii where Sunrun company offered their Brightbox product—a combination solar-plus-battery product—with a price of 19 cents per kWh, which is almost 50% cheaper than grid electricity. By 2018, one in five new residential Sunrun solar customers in California were choosing to add storage [54]! The situation will be the same soon in GCC countries.

The prospects for solar plus storage are even more remarkable in the near future which is forecasted to witness steep declines in battery costs—by 50% in the next 5 years and by 70% by 2030. According to reports batteries aren't just getting cheaper but will also reach far outstripping predictions. Bloomberg projected batteries crossing the US$300 per kWh threshold in 2022 and will reach US$ 109/kWh in 2015 and further to as low as US$ 73/kWh.

13.4 Conclusion

The fast and accelerated execution of large PV technology projects in GCC countries will have an impact in the region. Besides the reduction in the carbon footprint and further cost of the solar W (low cost of kWh), many new businesses will flourish such as battery storages for PV electricity, low power consumption air conditions and devices, boost in green job and green economy, but policy makers must pay great attention to disturbance in interconnection to grids.

References

1. Alnaser WE, Alnaser NW (2011) The status of renewable energy in the GCC countries. Renew Sust Energ Rev 15:3074–3098
2. Mas'ud AA, Wirba AV, Alshammari SJ, Sukki FM, Mu'azu Mohammed Abdullahi MM, Albarracín R, ZiaulHoq M (2018) Solar energy potentials and benefits in the gulf cooperation council countries: a review of substantial issues. Energies 11:372. https://doi.org/10.3390/en11020372
3. Hepbasli A, Alsuhaibani Z (2011) A key review on present status and future directions of solar energy studies and applications in Saudi Arabia. Renew Sust Energ Rev 15:5021–5050
4. Stinson L (2018) Saudi Arabia is planning to build the world's largest solar power plant, Green Design News, Apr 3 2018. https://www.curbed.com/2018/4/3/17186292/saudi-arabia-largest-solar-power-plant
5. Sultan N (2013) The challenge of shale to the post-oil dreams of the Arab Gulf. Energy Policy 60:13–20
6. Al-Maamary H. M. S, Kazem, H.A, .Chaichan, M.T. (2017) The impact of oil price fluctuations on common renewable energies in GCC countries, Renew Sust Energ Rev, Vol 75, pp. 989–1007
7. Burger S (2013) The emerging opportunities in Saudi Arabia's solar market. GreenTechMedia. Com, Boston, MA
8. Al-Shalabi A, Cotteret N, Menichetti E (2014) EU-GCC cooperation in energy. In: Colombo S (ed) Bridging the Gulf: EU-GCC relations at a crossroads. Edizioni Nuova Cultura, Roma, p 357. ISBN 8868122847
9. Sunwater Solar. What is solar thermal?. Accessed 16 Nov 2017. http://sunwatersolar.com/solar-thermal/whatis-solar-thermal
10. Holland+You (2016) Opportunities for Dutch Business in the Gulf Region; Holland+You: The Hague, The Netherlands
11. Mokri A, Ali MA, Emziane M (2013) Solar energy in the United Arab Emirates: a review. Renew Sust Energ Rev 28:340–375

12. Assi A, Jama M, Al-Shamisi M (2012) Prediction of global solar radiation in Abu Dhabi. ISRN Renew Energy 2012:1–10
13. Zell E, Gasim S, Wilcox S, Katamoura S, Stoffel T, Shibli H, Engel-Cox J, Subie MA (2015) Assessment of solar radiation resources in Saudi Arabia. Sol Energy 119:422–438
14. Masdar (2017) UAE Projects. Accessed on 21 Nov 2017. http://www.masdar.ae/en/energy/uae-projects
15. Alsharhan AS, Rizk ZA, Nairn AEM, Bakhit DW, Alhajari SA (2001) Hydrogeology of an arid region: the Arabian gulf and adjoining areas, 1st edn. Elsevier Science, Amsterdam. ISBN 9780080534329
16. Al-Hasan AY (1997) Electricity generation cost between proposed photovoltaic station and conventional units in Kuwait. Renew Energy 12:291–301
17. International Energy Agency (IEA). Kuwait Renewable Energy Target. Accessed 21 Nov 2017. https://www.iea.org/policiesandmeasures/pams/kuwait/name-157557-en.php
18. Al-Badi AH, Malik A, Gastli A (2011) Sustainable energy usage in Oman—opportunities and barriers. Renew Sust Energ Rev 15:3780–3788
19. Al-Badi AH, Malik A, Gastli A (2009) Assessment of renewable energy resources potential in Oman and identification of barrier to their significant utilization. Renew Sust Energ Rev 13:2734–2739
20. Bachour D, Perez-Astudillo D (2013) Measurement GHI map for Qatar. Energy Procedia 49:2297–2302
21. Martinez-Plaza D, Abdallah A, Figgis BW, Mirza T (2015) Performance improvement techniques for photovoltaic systems in Qatar: results of first year of outdoor exposure. Energy Procedia 77:386–396
22. Energy Research Centre of the Netherlands (ECN) (2015) Private sector engagement—Qatar case study. In: NDE Workshop Yerevan. ECN, Petten, pp 1–14
23. Alnaser NW, Dakhel AA, Al Othman MJ, Batarseh I, Lee JK, Najmaii S, Alnaser WE (2015) Dust accumulation study on the Bapco 0.5 MW_p PV project at University of Bahrain. Int J Power Renew Energy Syst 2(1):39–54
24. Alnaser NW (2017) First smart 8.64 kW BIPV in a Building in Awali Town at Kingdom of Bahrain. Renew Sustain Energy Rev 82:205–214
25. Alnaser WE, Alnaser NW, Batarseh I (2014) Bahrain's BAPCO 5MWp PV grid–connected solar project. Int J Power Renew Energy Syst 1(2):72–84
26. http://cebcmena.com/wp-content/uploads/2018/03/CEBC-Energy-Efficiency-in-the-GCC-January-2018-REV2.pdf
27. Haji S, Durazi A, Al-Alawi Y (2018) Feed-in tariff structure development for photovoltaic electricity and the associated benefits for the Kingdom of Bahrain. Int J Sust Energy 37(5):479–497. https://doi.org/10.1080/14786451.2017.1304940
28. EPIA (European Photovoltaic Industry Association) (2012) Global market outlook for photovoltaics until 2016. http://www.pv-magazine.com/fileadmin/uploads/PDFs/Global_Market_Outlook_2016.pdf
29. The World Bank, World Development Indicators: Power and communications. http://wdi.worldbank.org/table/5.11
30. Wogan D, Pradhan S, Albardi S (2017) GCC energy system overview—2017, October 2017/KS-2017--MP04, King Abdullah Petroleum Studies and Research Center (KAPSARC). https://www.kapsarc.org/wp-content/uploads/2017/11/KS-2017-MP04-GCC-Energy-Overview-2017.pdf
31. WRI (World Resources Institute) (2015) CAIT Climate Data Explorer. http://www.wri.org/resources/data-visualizations/cait-climate-data-explorer
32. Kingdom of Bahrain Electricity & Water Authority, Kuwait Ministry of Electricity & Water, Oman Power and Water Procurement Company, Qatar Electricity & Water Corporation, Saudi Electricity Company, Electricity & Cogeneration Regulatory Authority, Abu Dhabi Water and Electricity Company, Dubai Electricity and Water Authority, Sharjah Electricity and Water Authority, Federal Energy and Water Authority, KAPSARC. https://www.kapsarc.org/wp-content/uploads/2017/11/KS-2017-MP04-GCC-Energy-Overview-2017.pdf

33. Prominent Mesia Member Companies (2018) Solar Outlook Report 2018, Middle East Solar Industry Association. https://www.mesia.com/wp-content/uploads/2018/03/MESIA-OUTLOOK-2018-Report-7March2018.pdf
34. https://www.thenational.ae/business/energy/world-s-cheapest-prices-submitted-for-saudi-arabia-s-first-solar-project-1.663842
35. https://commons.wikimedia.org/wiki/File:Price_history_of_silicon_PV_cells_since_1977.svg
36. Ferroukhi R, Arslan Khalid A, Hawila D, Divyam Nagpal D, El-Katiri L, Fthenakis V, Al-Fara A (2016) Renewable energy market analysis: The GCC region, International Renewable Energy Agency (IRENA) Publication, UAE. http://irena.org/publications/2016/Jan/Renewable-Energy-Market-Analysis-The-GCC-Region
37. Renewables (2018) Global Status Report (2018), Renewable Energy Policy Network for the 21st Century. http://www.ren21.net/wp-content/uploads/2018/06/17-8652_GSR2018_FullReport_web_final_.pdf
38. Asif M (2018) Prospects of building rooftop Application of solar pv in gcc countries, Solar Utility Network Conference, 7th Feb 2018, Bahrain International Exhibition Centre, Kingdom of Bahrain
39. Khan MMA, Asif M, Stach E (2017) Rooftop PV potential in the residential sector of the kingdom of Saudi Arabia. Buildings 7(2):46–53. https://doi.org/10.3390/buildings7020046
40. Al-Ghamdi SA, Alshaibani KA (2017) The potential of solar energy in Saudi Arabia: the residential sector. J Eng Arch 5(2):32–53
41. Arif M, Hussain A, Aslam M (2017) Emerging renewable and sustainable energy technologies: state of the art. Renew Sust Energ Rev 71:12–28
42. Kapsalis,V and Karamanis , D.(2015) On the effect of roof added photovoltaics on building's energy demand, Energ Buildings, Vol 108, pp 195–204
43. Al-Samahiji A (2018) Net metering policy and wind power opportunity SEU Solar Energy Engineer, Water and Energy Sustainable Innovation and Industry –WESII 2018, 22–23 April 2018, University of Bahrain, Kingdom of Bahrain. http://wesii.uob.edu.bh/images/Presentations/Netmeteringpolicyandwindpoweropportunity.pdf
44. Atalay Y, Kalfagianni A, Pattberg P (2017) Renewable energy support mechanisms in the Gulf Cooperation Council states: analyzing the feasibility of feed-in tariffs and auction mechanisms. Renew Sust Energ Rev 72:723–733
45. Gasparatos A, Doll NH, Esteban M, Ahmed A, Olang TA (2017) Renewable energy and biodiversity: implications for transitioning to a Green Economy. Renew Sust Energ Rev 70:161–184
46. Alnaser WE, Al-Kalak A, Al-Azraq MAT (1991) The efforts of the Arab League for Education, Culture and Scientific Organization (ALECSO) in the field of Renewable Energy. Renew Energy 6(5–6):649–657
47. Kandpal TC, Broman L (2014) Renewable energy education: a global status review. Renew Sust Energ Rev 34:300–324
48. https://www.wholesalesolar.com/solar-information/anti-islanding
49. Thomson M (2018) Integration of renewable energy in Bahrain, water and energy sustainable innovation and industry conference. University of Bahrain, Kingdom of Bahrain
50. Taleb M, Al Salam F, Saleh F, Mohamed N (2018) A grid connected photovoltaic system based on current conditioning, water and energy sustainable innovation and industry conference. University of Bahrain, Kingdom of Bahrain
51. Ventures Onsite For Middle East Electricity (2017) Gcc Power Market Report 2017, VENTURES Onsite Publication, Middle East and North Africa Project Tracker. https://www.middleeastelectricity.com/content/dam/Informa/Middle-East-Electricity/en/pdf/GCC%20Power%20Market%20Report%20-%20Oct17.pdf
52. https://www.iea.org/publications/freepublications/publication/GECO2017.pdf
53. http://www.ecgonline.info/index.php/customer-care/info-center/energy-saving-tips.html
54. https://ilsr.org/solar-plus-storage/0

Chapter 14
Architectural Education for Sustainability

Derya Oktay

14.1 Introduction

A prevailing problem with the current architectural practice worldwide is the design approach which takes the visual aesthetics as the main and only objective, denying the environment and environmental conditions. In this context, the majority of architects unfortunately continue to see architectural design as the design of an 'object', although it is an undeniable reality that building design cannot be isolated from its environment, and an architect, while designing a building, affects the existing environment in a positive or negative way.

As the changes on the natural and built environment and the effects of global warming are becoming more clear and the ecological concerns and the concept of sustainability come to the forefront of the architectural and design agenda, it is obvious that professionals who are environmentally responsible and efficient are needed more than ever. In this context, in order to understand the reasons why architects cannot develop appropriate responses in the context of a socially and ecologically sustainable environment, the educational systems and approaches need to be interrogated, in addition to other factors. Although this issue is partially related with the preuniversity education, we, as the architectural educators, are bound to spend efforts to fulfill the requirements for the education of environmentally responsible architects. As an architectural education facilitates the development of critical thinking abilities, which can be applied to solving problems and addressing situations beyond design, our responsibility is not limited to the needs related to the built environment or environmental issues alone. The critical thinking abilities of future architects can also be valuable in designing an organisation or setting strategic goals and implementation plans.

The demand for change in the context of architectural education for sustainability is not limited to the expectations of the institutions and/or educators. A comprehensive

D. Oktay (✉)
Faculty of Architecture, Ondokuz Mayis University, Samsun, Turkey
e-mail: derya.oktay@omu.edu.tr

A. Sayigh (ed.), *Renewable Energy and Sustainable Buildings*, Innovative Renewable Energy, https://doi.org/10.1007/978-3-030-18488-9_14

survey research presented by the American Institute of Architecture Students at the 'Architectural Education After 50 Years' Congress held in Oxford, UK, revealed that the first three items in the list of nine expectations of the future architects from architectural education and architecture were 'ecological diversity and sustainability', 'social responsibility' and 'global change'. These results could be considered a radical change for Western societies, which were deeply influenced by modernism for many decades. On the other hand, architecture programmes of the universities in the West allocate a significant amount of resources towards study trips to countries like Turkey, India, Mexico and the far-eastern countries, which are rich in terms of traditional environment and historic heritage, together with research units, which are founded for the study of regional architecture in order to create milieus for teaching design with environmental responsiveness.

Today, in most architectural practices, the buildings are designed based on a shape-oriented design approach as 'objects' and their use and users are often ignored (Fig. 14.1). In fact, as Hagemann and Peets [1] highlight, the fundamental unit in architecture is not the building itself, but the city; and as stated by Churchill (1874–1965), 'we first shape our buildings, and then they shape us'; a good urban form can be a force to create a cohesive community as well.

Professional associations such as *The Architects' Directive 85/384/EEC*, *AIA* (USA) and *RIBA* (UK) highlight the importance of the issues of urban and social dimensions in both undergraduate and graduate programmes in architecture. The Association of Collegiate Schools of Architecture (ACSA), the American Institute of Architects (AIA), the National Council of Architectural Accrediting Board (NAAB), the Architects' Council of Europe (ACE) and the European Association for Architectural Education all agree that the philosophy of sustainability and sustainable design must be imbedded in the architectural design programme. The International Union of Architects (UIA) and the United Nations Educational Scientific and Cultural Organisation (UNESCO) have also published a document claiming sustainability as the heart of architectural educa-

Fig. 14.1 (**a, b**) An example of shape-oriented building designed with the intention of creating a focal point through its form: The Guggenheim Museum Bilbao, by Frank Gehry (D. Oktay Archive)

tion (UIA and UNESCO 1996). This document highlights the importance of environmental, social, cultural, economic and aesthetic needs and the development of built environments that have sensitivity to ecology and sustainability.

In many architectural schools in the world, a rapid progress is observed in the area of computer sciences and electronic media [5–7]. However, it cannot be said that there is an advancement in the area of teaching related to the ecological sensitivity and design in a local context.

In line with this, it is not surprising to see that the majority of the graduates are interested in individual success and self-promotion, instead of long-term contributions to the society. The problem here can be explained by the fact that the concepts of sustainability and ecology, hence the concept of holistic design, have not been fully understood yet, because most of the architectural products presented as 'ecological architecture' are limited to those 'green-washed buildings' decorated with green plants all over, and it is very rare to see those examples which safeguard regional architecture, consider climatic design and are resources of renewable energy (Figs. 14.2 and 14.3). In fact, sustainability is a more comprehensive issue

Fig. 14.2 An example of 'green-washed' building in Vietnam

Fig. 14.3 An example of 'green-washed' buildings in Switzerland

when compared to ecological design and is not limited to physical and natural sustainability; it includes and requires dimensions of social-cultural sustainability as well [2–4]. Therefore, as highlighted by Wright (2003) [9], architectural schools need to recognize sustainability as a core issue of architecture, and to invest more effort in order to ease their graduates into adopting a philosophy of sustainable design.

14.2 Architectural Schools and Environmental Issues: Exemplary Cases

It is certain that the vision, orientations and applications at the institutional level would have an encouraging role on the efficiency and effectiveness of an education programme regarding environmental issues. An exemplary case is the Taubman School of Architecture and Urban Planning, where I was a visiting scholar in 2005. The University of Michigan in Ann Arbor, USA, has also been leading in the field of sustainability. The reflections of this perspective can be observed both on the campuses and in the city. The presence of the School of Natural Sources, the use of natural materials in the refurbishment of the historic college buildings, the design of some of the new buildings and the floor covering and decoration and the use of the recycled toilet system are the indicators of the environment-friendly and energy-conscious approach of the university, together with the 'Energy Fest' organised by the university students, which prove the students' sensitivity to environmental issues and energy conservation [8] (Figs. 14.4, 14.5, 14.6 and 14.7).

The Michigan Solar House (MiSo), the completely solar-powered, renewable energy-dependent 459 square foot innovation, designed and built by the students for the 2005 US Department of Energy's Office of Energy Efficiency and Renewable Energy's Solar Decathlon, was an exciting competition project that challenged collegiate teams to design and build energy-efficient houses powered by the sun over the course of 2 years. More than 150 students and faculties from various university colleges, including the Taubman College of Architecture and Urban Planning (the main group) and 3 others, designed the house over the course of 4 years. Renewable technologies being implemented in the solar house include a solar chimney which captures heated air from the sun into glass spaces and vacuum tubes that transfer solar heat into a closed-circuit water loop via heat pipes. This brings hot air inside the house in the winter and expels it in the summer, to maintain desirable temperatures. The project was said to have provided students with skills, mindsets and perspectives different from those they would encounter in a classroom setting. It enabled them to work together in a collaborative environment, which is a necessity in the workforce today (Figs. 14.8 and 14.9).

Fig. 14.4 The Energy Fest—University of Michigan (D. Oktay Archive)

Fig. 14.5 The Energy Fest—University of Michigan (D. Oktay Archive)

The concept of sustainability has been integrated into our teaching endeavours too. We have a theory course devoted to ecological issues in architecture, a theory course devoted to urban design and, in connection with this, a studio course dealing with architectural design in urban context. In this particular studio I have coordinated in two schools of architecture, Eastern Mediterranean University in North Cyprus and Ondokuz Mayis University in Turkey, we aimed to provide students with a comprehensive overview in designing for user and community, while enabling them experimenting to overcome the conventional design issues of an architectural/urban studio project including physical and ecological aspects of indoor, outdoor and in-between space configurations; functions, overall form, structure, construction techniques and site and city context; and the integration of psychological, sociocultural factors to the design process and development (Figs. 14.10 and 14.11). This

Fig. 14.6 The Energy Fest—University of Michigan (D. Oktay Archive)

Fig. 14.7 The Energy Fest—University of Michigan (D. Oktay Archive)

Fig. 14.8 The Michigan Solar House (MiSo) during its construction (D. Oktay Archive)

Fig. 14.9 The Michigan Solar House (MiSo) during its construction (D. Oktay Archive)

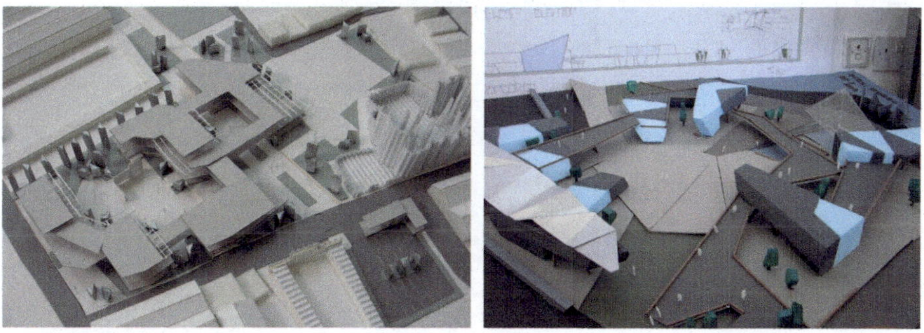

Fig. 14.10 (**a**, **b**) Exemplary student projects of ARCH 302 'Architectural Design in urban context' (Saya Kaynama and Azadeh Amirmashuri. EMU)

approach provides an opportunity to designing for the people and society that we are living with and searching various types of architectural design themes fitting to this local context; hence, it supports the vision of the school regarding architectural education for sustainability.

14.3 Measuring the Environmental Efficiency of a School of Architecture

Understanding the level of a school's environmental sensitivity naturally requires information as to what extent the programme connects with environmentally responsible design and sustainable design philosophy. In line with the observations and analysis in various institutions, establishment of a measurement method, through which various aspects can be measured at various levels, emerges as a new requirement.

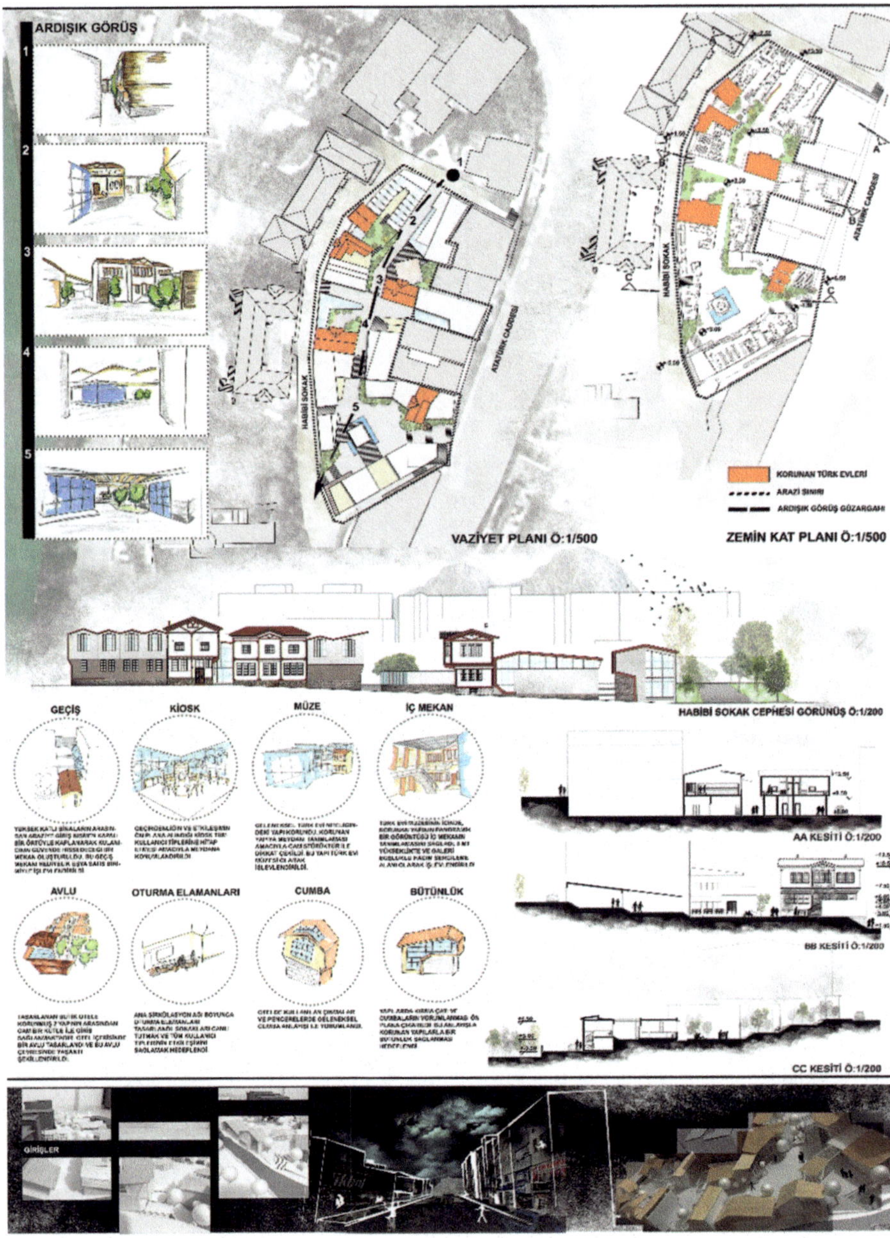

Fig. 14.11 An exemplary student project of ARCH 302 'Architectural Design in urban context' (Oguzhan Bayrak, OMU)

In this context, the following criteria are proposed to frame the issues (Table 14.1). These criteria attempt to sample a programme across several dimensions, including faculty, facility, students, curriculum and institutional setting.

Table 14.1 A guidance model for measuring the efficiency of a school of architecture on sustainability

Faculty members	Students	Institution	Curriculum	Facilities
The number of academic staff who are members of certain associations specialised in environmental design (LEED-AP, LEED-ND, so forth)	The number of students participating in conferences focusing on energy-conscious design and sustainability	The number of memberships in environment, ecology and design associations	The existence of a design studio course focusing on ecology and sustainable design	The existence of campus buildings with 'green' certificates such as LEED-ND, LEED-EB and so forth
The number of journal articles and conference papers of faculty members on the themes environment and sustainability	The number of student memberships in environment, ecology and design associations and other societies	The existence of a sustainability strategy/plan at the official level	The existence of a design studio elective course focusing on ecology and sustainable design	Respect to energy density index
The number of faculty members participating in conferences focusing on the themes environment and sustainability	The number of student societies linked to associations with expertise in the field of environmental and energy-efficient design	The existence of university target towards more environment-friendly architectural environment both for the old and the new buildings	The existence of a theory course focusing on ecology and sustainable design	Respect to water density index
The number of faculty members participating in professional societies and activities focusing on the themes environment and sustainability	The number of students who carried out funded research projects focusing on energy-conscious design and sustainability	The existence of seminar and workshop activities focusing on energy-conscious design and sustainability both in general and at the level of departments	The existence of an elective theory course focusing on ecology and sustainable design	Respect to waste density index
The number of funded projects focusing on the themes environment and sustainability		Existence of a website allocating a page to activities and opportunities related to sustainability and energy-efficient design	Existence of courses focusing on environmental evaluation systems, i.e. LEED	Existence of an active recycling programme

14.4 Conclusion

In the era of globalisation, as big changes in cities cause environmental and social problems, it is inevitable for architectural institutions to pioneer for consciousness and orientations towards better relationships between architecture, city, society and nature together with a practice of ecological design at all levels.

As we move forward, building curriculums to bring the lessons of sustainability to all students, we need to bring the perspectives, contents and understanding of all disciplines to bear on the problem. While architecture is generally seen as a professional, not academic, field, the architect's experience and expertise relate directly to many of the issues we face. Their experience with buildings, energy consumption, materials and durability is critical to sustainability. That expertise needs to be included, along with the expertise of other disciplines, in assisting students to better understand the changing world.

A number of architectural educators are concerned about architectural education for sustainability. The currently available rankings of sustainable architecture programmes do not dispel such concerns. Because of the substantial impact of the built environment on resource consumption and carbon emissions, students about to embark on a career in architecture should be able to access consistently applied and meaningful evaluations of an architecture programme's engagement with green/ sustainable design. We believe that the sustainable programme rating system proposed herein is a step in the right direction.

As presented in the proposed guidance model, the credibility of a school of architecture in terms of sustainability has various dimensions. These dimensions include qualities and performances at the levels of faculty members, students, institution, curriculum and facilities/opportunities. The university's commitment to sustainability would help support the culture of sustainability at all levels. The respect to integrating the sustainability agenda into the architectural curriculum and ethos of the campus is a must. Design-build studios; community-based projects, in local context; urban and architectural conservation courses; quality of life and healthcare; social equity; and sustainability are some of the aspects that should be available to students.

In this context, what is understood from a more sustainable architectural education is important; its meaning should be made clear and a set of criteria needs to be defined within the limits of existing conditions. On the other hand, students should be well taught about the significance of research for ecologically and socially sustainable architecture, and a certain period in the course of design studio should be allocated for research. Further, the importance of interdisciplinary collaborations should be highlighted, and an institutional structure should be created in order to support interdisciplinary research and development projects.

Without doubt, the application of this holistic approach, without compromising creative and innovative attitudes, would not be easy, but be very valuable.

References

1. Hagemann & Peets (1993) The American Vitruvius. Princeton Architectural Press, New York
2. Oktay D (2011) Sustainable Urbanism revisited: a holistic framework based on tradition and contemporary orientations. In: Ercoskun OY (ed) Green and ecological technologies for urban planning: creating smart cities. IGI-Global, Pennsylvania
3. Oktay D (2004) Urban design for sustainability: a study on the Turkish City. Int J Sust Dev World Eco 11:24–35
4. Oktay D (2001) Planning housing environments for sustainability: assessments in Cypriot settlements. YEM, Istanbul
5. Dunham-Jones E (2004) In: Bothwell S et al (eds) The modernist model, the Windsor forum on design education. New Urban Press, Washington, DC
6. Glasser DE (2000) Reflections on architectural education. J Archit Educ 53(4):250–252
7. Oktay D (2008) Architectural education and social perspectives: establishing the human context of everyday urban life through design studios, the Oxford conference 2008: 50 years on: re-setting the agenda for architectural education. University of Oxford, Oxford. (poster)
8. Oktay D (2007) Üniversite—Kent İlişkisi ve Değişen Boyutlar (University—City Relationship and Changing Dimensions). Ocak 302:42–47
9. Wright J (2003) Introducing sustainability into the architecture curriculum in the United States. Int J Sustain High Educ 4(2):100–105

Chapter 15
Dutch Efforts Towards Sustainable Schools

Wim Zeiler

15.1 Introduction

In Europe, more than 64 million students and almost 4.5 million teachers work inside a school in pre-primary, primary and secondary schools (EU 2015). School buildings represent a significant part of the building stock (17% of the non-residential sector in Europe) and also noteworthy part of total energy use (12%) [1]. In existing school buildings, there are very often nonoptimized systems in terms of energy consumption [2]. As the results of the energy use of the built environment, which is around 40% of the total energy consumption, become more clear (depletion of fossil fuel and global warming), there is a demand for energy reduction. Sustainability is a crucial issue for our future and architecture has an important role to direct sustainable development. The ultimate goal is to create an energy-neutral built environment. To reduce the high energy demand and pollution of greenhouse gasses, the performance of the buildings has to be improved. The future policy timeline to reach for all new buildings to be nearly zero-energy buildings is ambitious: in 2021 all new buildings should be nZEB, and in 2030 30% of all buildings renovated (after 2015) have to be nZEB. The definition of a 'nearly zero-energy building' is described within the EPBD recast of the EU which specified that by 31st of December 2020, all new buildings shall be 'nearly zero-energy buildings'. Governmental buildings occupied and owned by public authorities will have to be 'nearly zero-energy buildings' by 31st of December 2018 according to the EPBD recast. The actual definition of nZEB is given in Article 9 of the EPBD [3]: 'Nearly Zero Energy Building (nZEB): Technical and reasonably achievable national energy use of >0 kWh/(m^{2}y) but no more than a national limit value of non-renewable primary energy, achieved with a combination of best practice energy efficiency measures and renewable energy technologies which may or may not be cost optimal'.

W. Zeiler (✉)
Department of the Built Environment, TU Eindhoven, Eindhoven, Netherlands
e-mail: w.zeiler@bwk.tue.nl

A. Sayigh (ed.), *Renewable Energy and Sustainable Buildings*, Innovative Renewable Energy, https://doi.org/10.1007/978-3-030-18488-9_15

This clearly shows that especially at schools there is a need for a strong reduction of the energy use. This led to a development trying to reduce the energy demand of schools as much as possible and apply renewable energy. The first Dutch net zero-energy school was already built in 2000 [4–6]. The school's electricity consumption is covered by 145 m^2 PV panels on the roof, while its heating needs around 16.000 kWh were covered by participation in a wind turbine park. After this the development of low-energy schools came temporarily to a stop due to lack of funding by the government. This article will describe the latest developments and results of renewed initiatives by a new stimulating funding program.

15.2 Methodology

Traditionally, the potential for energy reduction in the Netherlands was mainly determined by the applications of building energy reduction measures according to the Trias Energetica method; *see* Fig. 15.1. In the first design step, the energy demand is reduced as much as possible by avoiding waste of energy to the environment by applying high levels of insulation and implementing energy-saving measures such as lowering the temperature in winter or increasing the temperature in summer slightly. In the second step, all possibilities to apply renewable energy sources are investigated. In the third step, the remaining energy demand is covered by using fossil energy solution as efficiently as possible.

An adapted version of the Trias Energetica method, the five-step method, has been developed and could be used in the future adding the integration of user behaviour as well as energy exchange and storage systems (buffer tanks, aquifer thermal energy storage). In this way the energy demand becomes more flexible by its process

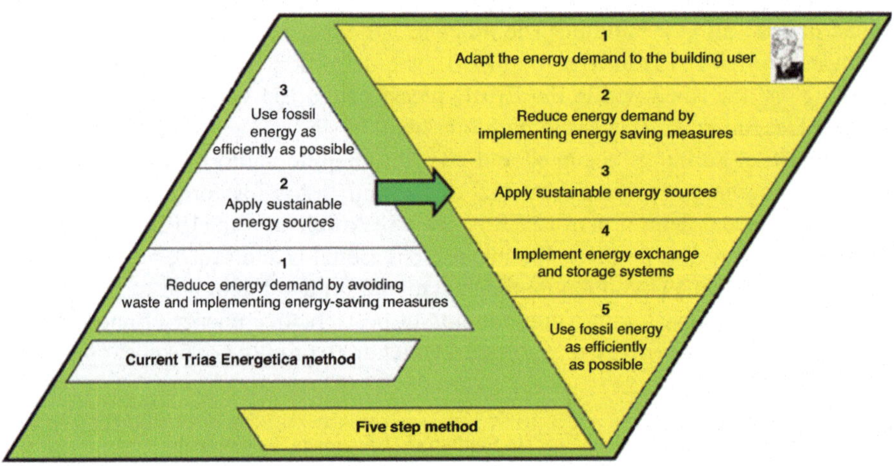

Fig. 15.1 The Trias Energetica method and the five-step method

control that is increasingly bottom-up rather than top-down. As a result the influence of the building's user interactions is included in efforts towards unitary performance metrics that capture both demand side (users) and power network side (grid).

In 2009 the Dutch government started their so-called UKP NESK program to stimulate innovation for energy-neutral buildings. UKP means unique chances projects, and NESK means 'Towards energy neutral schools and offices' (Naar Energie neutrale Scholen en Kantoren). In the NESK from 2009 program, there were eight new school buildings additionally funded of which six new schools were really built and one school renovated. Unfortunately it took for some project quite a long time to get finished. For example, the school 1 was officially opened December 2015. Recently the Rijksdienst voor Ondernemend Nederland (RVO) published their results on the top 15 most energy-conscious schools in the Netherlands for 2014 and 2016 [7, 8]; *see* Figs. 15.2 and 15.3. A special case is the school 16; it was the only renovation project within the NESK program. The other NESK schools are 1, 3, 7, 9, 12 and 15.

The total energy consumption is determined to calculate the energy use for heating, hot water, cooling, ventilation and lighting. A more detailed study especially concerning the indoor air quality of the 6 new NESK schools and other from the top 15 can be found in [7, 8]; here we will focus only on the energy aspects of the Dutch most sustainable and energy-conscious schools of the Netherlands up to 2016.

Quite remarkable is that the most energy-conscious school one of the original NESK projects is school 1. As can be seen from Fig. 15.4, there is quite a difference between the schools according to their energy use, ranging from 60 kWh/m²y

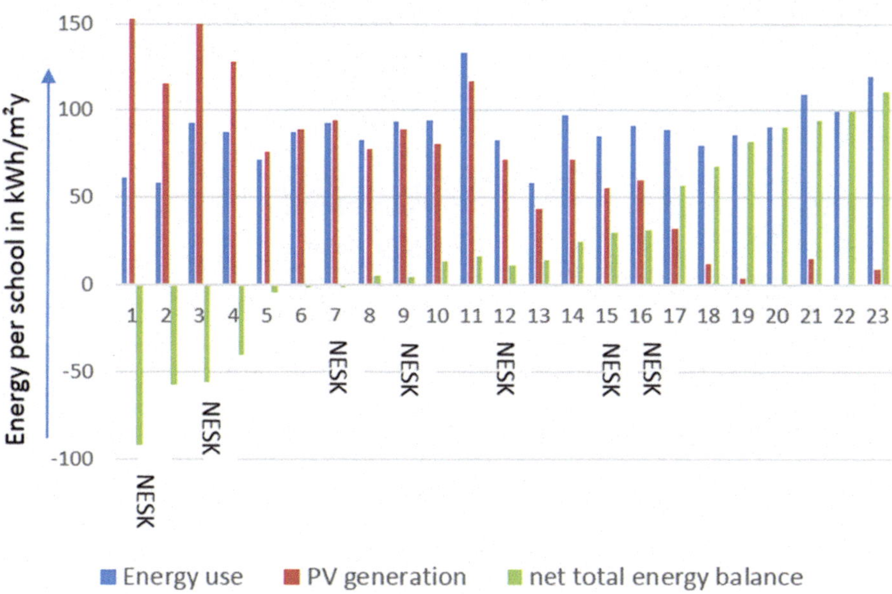

Fig. 15.2 Overview energy balance of the seven NESK schools within the top 15 energy-efficient schools of 2014 and 2016 [7, 8]

1		Energy conservation and efficiency are the main focuses of the project. It resulted in a significant reduction in energy consumption and CO_2 emissions, as well as a healthy indoor climate. It is equipped with ventilation system Dc mechanical supply and drain – centralized and it is CO_2 controlled. It uses HP & ATES systems for cooling/heating EPC: -0.745 Size: 1575 m²
2		This school utilises mixed ventilation system whereas supply is mechanical and exhaust is done naturally in each classroom. On the roof are more than 1,500 net square meters of PV panels. In addition, the building has excellent thermal insulation and heat pumps with soil thermal bank. This school utilizes HP & ATES to provide cooling/heating EPC: -0.585 Size: 4059 m²
3		This school uses CO_2 controlled ventilation system that is centralized, four main air supplies are installed with the functionality of steering air flow. The centre is designed according to the BREEAM methodology and several durable techniques have been applied to achieve average RC value of 6 sq.m.K / W. Cooling and heating are provided using HP and boiler. EPC: -0.2 Size: 1524 m²
4		Equipped with fully mechanical and centralized ventilation system plus there are 268 solar panels on the roof of the school which makes the building positive energy with 180 Kwh extra power production. It is heated with two heat pumps connected to a closed source system of 15 vertical, 150 meter deep pipes. During winter, cold is accumulated and stored in the soil and used in the summer for cooling. EPC: - 0.327 Size: 1276 m²
5		Equipped with fully mechanical ventilation system and mostly passive techniques that are used for cooling and district heating and electric water boiler are used for heating. In heating seasons due to in depth design process of rooms there is no need for mechanical cooling EPC: -0.027 Size: 1062 m²
6		The building had virtually no mechanical ventilation, poorly insulated and had relatively much outside surface and high energy and maintenance costs. After the renovation the building greatly improved with added insulations and measures to increase energy conservation. Ventilation system is fully mechanical and centralized. Heating and cooling will be done by creating thermal bank in soil. The required electrical power for this installation is largely generated by solar panels on site. EPC: 0.010 Size: 2096 m²
7		Centralized mechanical ventilation is installed with good heat recovery system. The school is energy-neutrally built according to the Trias Energetica. The building houses two schools with outdoor space and gymnasium. A combination of heat pumps does the heating that can also cool in warm weather. There are approximately 780 m² of solar panels installed. A solar water heater will heat the water for the showers in the gym. EPC: 0.008 Size: 2521 m²
8		The building is completely energy-neutral, ventilation system is fully mechanical and centralized. The requirement for the energy consumption for heating is set at up to 15 kWh / m². Energy consumption will be nearly 40% lower than the passive home concept. This building uses HP & ATES for heating, the cooling is done with passive solutions. EPC: 0.064 Size: 2738 m²
9		Centralized mechanical ventilation is used with high heat recovery and for insulation three-layer glass, high air density, compact building, balance ventilation with WTW (CO2- steering per room), underground thermal energy storage with heat pump, underfloor heating, wood pellet boiler for hot tap water and as a heating system for heating, use of energy efficient lighting with daylight-dependent control and presence detection and PV panels to compensate for the building-based energy consumption of the school. EPC: 0.15 Size: 4803 m²
10		Ventilation is set up fully mechanical and well balanced with centralized layout. The heating is achieved with the combination of HP & underground thermal storage. This school highly insulated with RC value of 6.5 for walls, roofs and façade. Heat recovery for ventilation and heat pumps are designed for improved energy efficiency. EPC: 0.122 Size: 1112 m²
11		The building is thermally well insulated with triple layer glazing windows. In addition, the school uses a balanced ventilation system with CO2 control. The building is equipped with sensors and display screen to fully monitor the air quality of outdoors. EPC: 0.07 Size: 1577 m²
12		It is the first school in the Netherlands designed and built according to the "passive building" principle and has received the approval mark of the Passive Building Foundation. Heating is provided with HP and cooling is done via opening the windows. Ventilation is hybrid between natural and mechanical, it is equipped with heat recovery utilities to minimize heat loss. EPC: 0.135 Size: 2137 m²
13		District heating with concrete core activation is used for heating and as for cooling concrete core activation and using the river nearby. Good insulated to counter energy loss with average RC value of 5.0 m².K / W. Ventilation system is balanced, centralized and mechanical which is also equipped with heat recovery system. EPC: 0 Size: 6500 m²
14		The school building is connected to the heat supply in the area and concrete core activation has been applied. In addition, the complex is insulated with high quality triple glass and the required electricity is generated on the roof by PV cells. It is equipped with fully mechanical ventilation system that is decentralized. EPC: 0.315 Size: 1112 m²
15		PV panels are installed without additional investment through co-operation with an energy company. The school building is connected to district heating and has a low-pressure ventilation system with heat recovery (WTW), energy consumption. The building's outside surface is well insulated with a good air seal. Thanks to the thermal open ceilings, the concrete structure acts as a dampening of temperature fluctuations. EPC: 0.25 Size: 2868 m2
16		This primary school was built in 1950 and renovated to a sustainable school. Heating and cooling is done by a floor system, which is based on a closed thermal energy storage system with a heating pump. The ventilation system is a mechanical system, in each room windows can be opened and the ventilation system is CO2 controlled. EPC: 0.16 Size: 1088 m2
17		The school building is connected to the heat supply in the area and concrete core activation has been applied. In addition, the complex is insulated with high quality triple glass and the required electricity is generated on the roof by PV cells. It is equipped with fully mechanical ventilation system that is decentralized. EPC: 0.37 Size: 3738 m²
18		The school is designed according to the Cradle-to-Cradle, C2C-principle. Which means to use as much recyclable material as possible and to think about of upgrading of used material. The heating and cooling is achieved with the combination of Heat Pump & underground thermal storage. This school has RC value of 5 and walls, roofs and façade are good insulated. Heat recovery for ventilation and heat pumps are designed for improved energy efficiency EPC: 0.45 Size: 5713 m²
19		The school has classrooms with a greenery, as such it is a building within a building which makes it quite unique.  EPC: 0.54 Size: 6116 m²
20		The Energy School is the first real-called Passive School of the Northern Netherlands. The school is one of the first with a special two-channel ventilation system BaOpt. Which functioning unfortunately is quite problematic in practice. EPC: 0.47 Size: 1557 m²
21		The school is equipped with an energy roof, which means that sun sensitive layers of the roof can provide heat-energy and electric energy. This energy is directly used by the building, any surplus of energy is used by the adjacent dwellings and a sports hall. The indoor climate is regulated with a floor heating / cooling system and a balanced ventilation system. EPC: 0.61 Size: 6851 m²
22		It is the first passive school in the Netherlands: thermal insulation with a thickness of 30 – 40cm and an average Rc-value of 10m2K/W, combined with a high air tightness placed at the inside of the façades and roofs. The balanced ventilation system keeps the temperature inside the classrooms balanced and is able to cool down the classrooms during summer. EPC: 0.48 Size: 1342 m²
23		School with special air inlet system through climate floor, airConomy-system.  Heat pump with closed ground loop as source EPC: 0.61 Size: 2010 m²

Fig. 15.3 The combined list of most energy-conscious schools in the Netherlands for 2014 and 2016 [7, 8]

Fig. 15.4 Overview energy generation by PV panels of school 1 [9]

(school 2) towards 130 MJ/m^{2}y (school 11). The number one school in the top 15 of the combined 2014 and 2016 list [7, 8] is a primary school located in a multifunctional building (MFA) including a childcare, gym, school of music, a cafe and meeting rooms. Constructional and installation technical measures contribute to the degree of energy efficiency of the building. Sustainability has played a major role in determining the materials. In addition, a high thermal conductivity is obtained in the structural elements to prevent large heat losses. The open façade parts consist of frames with three-layer glass. Solar screens are provided to prevent overheating. Not only the prevention of energy loss is involved, but also the way of generation, transport and delivery is important. The school uses a 150 kW biomass boiler fed with wood pellets. Besides this the school building generates a lot of energy due to the large amount of solar panels on the roof, in total 135 kWp: 41% of the generated energy is used by the school, and the rest is delivered to the grid; *see* Fig. 15.4.

A first step to minimize the energy demand is to look at the actual behaviour and occupancy in relation to the ventilation need, which is directly related to the number of pupils in a specific classroom and their CO_2 production. So this demand-driven ventilation is a good step to reduce the energy demand by the ventilators. A similar measure in this step is the occupancy detection for the control of the lighting and daylight control. This is to minimize the use of artificial lighting, thus reducing the electricity demand.

An important second step to reach nZEB is the application of the passive house strategy. This leads to energy savings on heating of 80% compared to conventional standards of new buildings. The general definition of a passive house is that the energy consumption is limited to around maximum 15 kWh/m^{2}y for space heating and around maximum total of 120 kWh primary energy/m^2 for heating, domestic hot water and electrical consumption by electrical equipment and lighting. To meet these criteria, the passive house concept focuses first and foremost on reducing the energy demand of the building. Analysis of passive house solutions shows high priority with regard to the performance of the thermal envelope: high insulation of walls, roofs, floors and windows/doors, thermal bridge-free construction and air tightness.

When we look at the different energy needed for heating, there is also some quite remarkable difference, for example, school 12 with 7 kWh/m^{2}y compared to 33–34 kWh/m^{2}y for schools 4 and 23. This is mainly due to the fact that some schools were built according to the passive house standard (schools 12, 14, 17, 20 and 22) which means high level of insulation (R_c value up to 10) accordingly low energy losses and low heating demand.

In Table 15.1 an overview is provided of the different heating concepts, as well as focus on the concept of the energy system. Quite remarkable, that besides the four schools which are connected to a district heating system, all other schools are using a heat pump besides school 1 which uses only a biomass pellet boiler.

Table 15.1 Overview of different techniques used for heating/cooling and principle ATES

School	HP	ATES	Pellet boiler	Borehole HE	District heating	HR boiler	Ventilation system	CO_2 control
1			X				MS/E + HR	+
2	X	X					MS/E + HR	+
3	X					X	MS/E + HR	+
4	X						MS/E + HR	+
5					X		MS/E + HR	+
6	X	X	X				MS/E + HR	+
7	X	X					MS/E + HR	+
8	X	X					MS/E + HR	+
9	X	X					MS/E + HR	Partly
10	X			X			MS/E + HR	+
11	X		X				MS/E + HR	Partly
12	X						MS/E + HR	Partly
13					X		MS/E + HR	+
14					X		MS/E + HR	+
15					X		MS/E + HR	–
16	X	X					MS/E + HR	+
17					X		MS/E + HR	Partly
18	X	X					MS/E + HR	Partly
19	X	X					MS/E + HR	–
20							MS/E + HR	–
21	X	X					MS/E + HR	–
22						X	MS/E + HR	+
23	X	X					MS/E + HR	–
Total	15	10	3	1	5	2	23	

HP heat pump, *ATES* aquifer thermal energy storage, *HE* heat exchanger, *HR* heat recovery, *MS/E* mechanical supply/exhaust

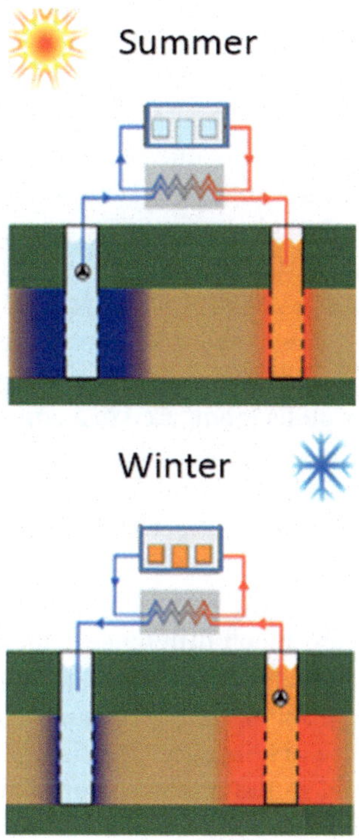

Schools have typically have a high internal heat load (heat generated by people, lighting and buildings), as a result, the required amount of cooling is significantly higher than the average building. Nowadays modern school buildings have reached the insulation quality at which the amount of cooling required during the summer roughly equals the amount of heating needed during the winter. Hypothetically this means that if all heat could be stored within the building, no external heat source would be needed throughout the year. The storage of such large amounts of (low-quality/low-temperature) heat within the building would require vast amounts of high heat capacity materials like thick stone walls (as in churches, castles) or phase change materials (PCMs). A more feasible option is to store the energy outside the building. An increasingly popular solution is energy storage in the groundwater below the building. This groundwater is stored in porous sand layers, called aquifers. Therefore, this method is called aquifer thermal energy storage (ATES). It is a different way of generating cold where water from a well in the ground is being cooled in winter, stored and used in summer. Often this system works in combination with a heat pump in order to ensure the required temperature difference for cooling is reached. The principle of an ATES system is based on transferring groundwater between two separated storage wells. During summertime water is extracted from the coldest well and used to cool the building. During cooling, the water temperature increases from approximately 8 to 16 °C. The heated water is injected in the warmer well and stored until winter season. During winter the extraction/injection flow is reversed, and the heated water (which still has a temperature of approx. 14 °C) is pumped back to the building. Using a heat pump, the heat is extracted and converted to higher temperatures to heat the building. The water is cooled to approx. 6 °C and is injected in the cold well. A heat exchanger between the groundwater and the building system water is used to avoid contamination of the water. An optimal performing ATES system can deliver very efficient cooling. The case-study system, for example, uses a 1 kW well pump that delivers 10 m^3/h of cooling water with a ΔT of 8 K between extraction and injection. This equals roughly 100 kW of cooling power and a coefficient of performance (COP) of 100. For comparison, a regular (compression-based) cooling system reaches a COP between 4 and 6. The energy gains (compared to a conventional system) for heating are not that significant, because the stored low-temperature heat is not directly applicable in the building. Because of these favourable conditions, the use of ATES systems in the Netherlands has become increasingly popular since the first installations in 1990.

When looking at the average yearly energy consumption for lighting of the different schools, there is not too much difference with exception of school 11; *see* Fig. 15.6. The amount of installed lighting power is reduced in the last 5 years from 10 W/m^2 to around 4 W/m^2 through the use of more efficient LED lighting armatures. Lighting controls reduce or eliminate electric lighting in response to daylighting to encourage natural light as the primary source of illumination in spaces and learning streets; *see* Fig. 15.5.

For ventilation, *see* Fig. 15.7 quite large difference occurs: schools 1 and 13 use only 8–10 kWh/m^2y, whereas school 10 uses not less than 42 kWh/m^2y, so more than four times more. This is rather strange as the program of demand for ventilation of classrooms is rather fixed, and therefore you would not expect such large differences. So clearly the difference has to come of the ventilation effectiveness within the classroom as well as the optimizing of the pressure losses within the ducts as well as in the air handling unit (Fig. 15.7).

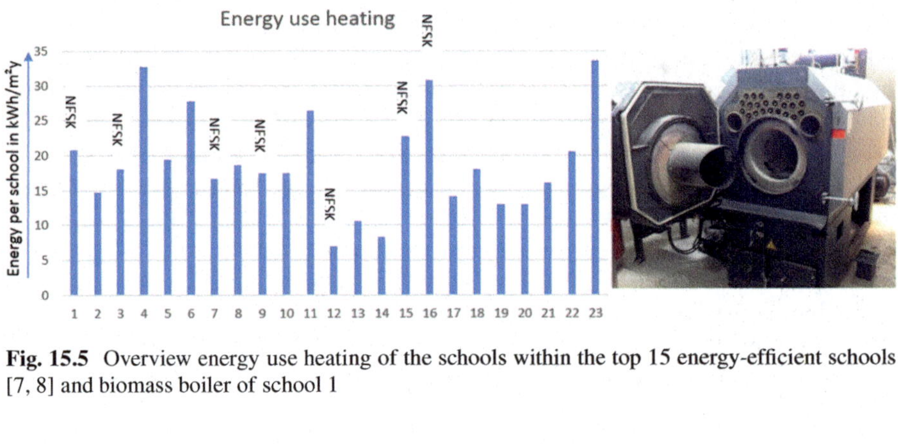

Fig. 15.5 Overview energy use heating of the schools within the top 15 energy-efficient schools [7, 8] and biomass boiler of school 1

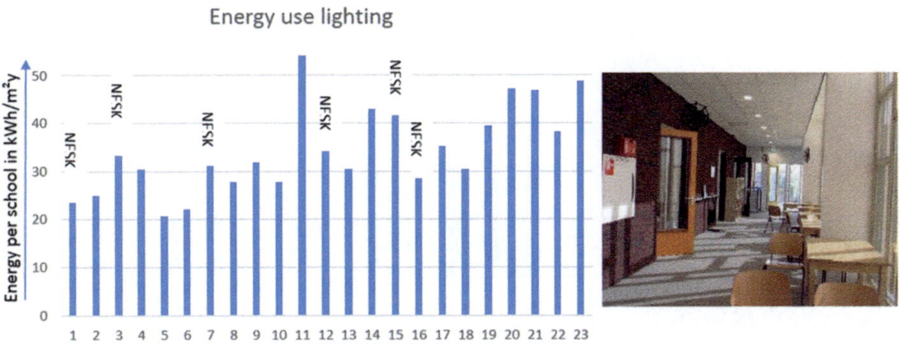

Fig. 15.6 Overview energy use lighting of the seven NESK schools within the top 15 energy-efficient schools of 2014 and 2016 [7, 8] and daylight illuminated 'learning streets'

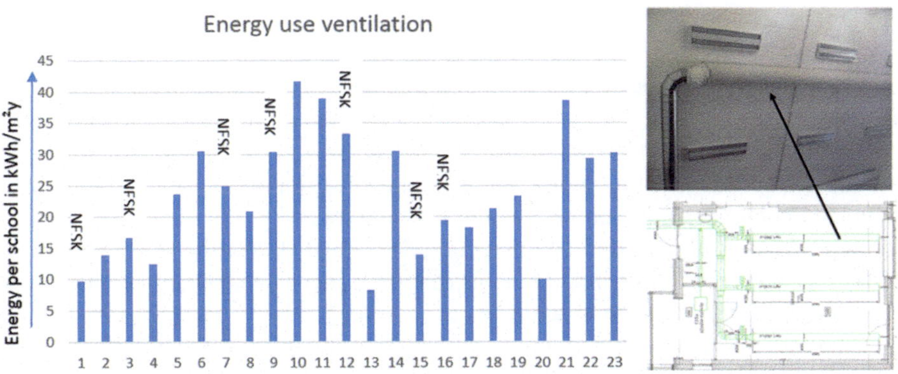

Fig. 15.7 Overview energy use for ventilation of the seven NESK schools within the top 15 energy-efficient schools of 2014 and 2016 [7, 8] and the textile supply ducts of school 1

15.3 Discussion and Conclusions

School buildings on account of their public educational purpose have a major social responsibility. Therefore their energy performance is of great importance [10]. All energy comparisons with the new nZE Dutch schools are based on data provided by RVO from 2014 and 2016 [7, 8] as an overall estimation of how much energy each school is expected to use per square meter; these were compared to results from other studies [11–16]; *see* Fig. 15.8.

Within the Interreg SUDOE ClimACT project (www.climACT.net) [12], pilot schools were selected in Spain, Portugal and France to evaluating their initial energy performance. Spanish pilot schools, located in a warmer climate, present the lowest energy consumption 56 kWh/m^{2}y, Portuguese schools consume an average 72 kWh/m^{2}y, and French schools 208 kWh/m^{2}y. Clearly, energy use varies widely

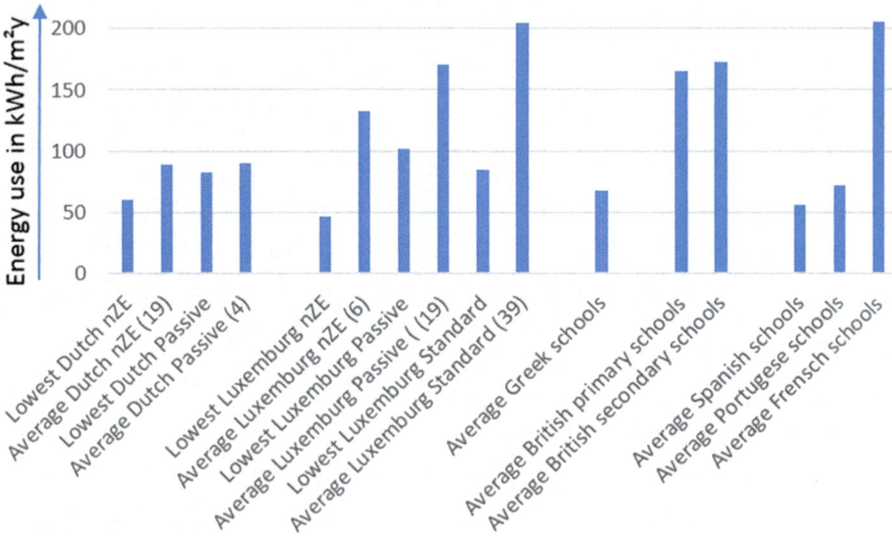

Fig. 15.8 Comparison energy consumption of different school types and different countries [7, 8, 10–16]

with the specific school typology, age and climate region. For example, in hot dry climates, cooling can account for a major part of total energy consumption, whereas schools in a cool climate don't have cooling demand at all. For the heating demand, things are the other way round. Schools in cloudy areas consume more energy for lighting than those situated in sunny areas.

The energy consumption in the real situation might diverge due to different use, operation, maintenance and weather conditions. As it was impossible to get reliable real energy consumption data, the data from RVO was used. All the NESK schools were intended as energy-neutral concepts, but did not succeed always. However this first series of nZEB schools led to an intensive development in energy-conscious schools, and now there are even net positive energy schools. The Rijksdienst voor Ondernemend Nederland (RVO) has made a white paper for the ministry of Internal Affairs and Kingdom relations about the experiences of four of these nZEB schools. This proved that the schools were actually performing rather well, but still it is very important to commission them well and to have kind of continuous commissioning as well to keep everything working.

References

1. VV.AA Europe (2011) S Buildings under the Microscope. A country-by-country review of the energy performance of buildings. In Buildings performance institute Europe (BPIE)
2. Thewes A, Maas S, Scholzen F, Waldmann D, Zürbes A (2014) Field study on the energy consumption of school buildings in Luxembourg. Energ Buildings 68:460–470

3. European Parliament (2012) Directive 2012/27/EU of the European Parliament and of the Council of 25 October 2012 on energy efficiency. Off J Eur Union Dir:1–56
4. van Weenen JC, Dettmers WJM, Overtoom MSJ, Poldermans HGJM (2000) Development of the First Sustainable Elementary School of The Netherlands. In: Proceedings International Conference Sustainable Building 2000, 22–25 October. Æneas Technical, Maastricht, pp 651–653
5. Agentschap NL, 2011, Eerste duurzame school Nederland nog steeds fris, Publicatienummer: 2EGOU1103
6. Duurzaam gebouwd 2018, Duurzame keuzes bewijzen zich bij De Sokkerwei, https://www.duurzaamgebouwd.nl/project/20180704-duurzame-keuzes-bewijzen-zich-bij-de-sokkerwei
7. RVO (2015b) Top 15 energiezuinige scholen onderzoeksrapport 2014, Publicatienummer RVO-042-1501-/FD-DUZA. Rijksdienst Voor Ondernemend Nederland, Utrecht
8. RVO (2016) Top 15 energiezuinige scholen onderzoeksrapport 2016. In: Rijksdienst voor Ondernemend Nederland. Rijksdienst Voor Ondernemend Nederland, Utrecht
9. SCX Solar, 2014, https://www.scx-solar.eu/projecten/zonnestroominstallatie-op-mfaschool-hart-oijen/
10. Pereira LD, Raimondo D, Corgnati SP, de SMG (2014) Energy consumption in schools—A review paper. Renew Sust Energy Rev 40:911–922
11. Advanced Energy Retrofit guide (2013) Practical ways to improve energy performance. Ke12 schools. U.S. Department of energy. Energy Efficiency & Renewable Energy
12. SUDOE ClimACT project (www.climACT.net)
13. Lizana J, Serrano-Jimenez A, Ortiz C, Becerra JA, Chacartegui R (2018) Energy assessment method towards low-carbon energy schools. Energy 159:310–326. https://doi.org/10.1016/j.energy.2018.06.147
14. Ouf MM, Issa MH (2017) Energy consumption analysis of school buildings in Manitoba, Canada. Int J Sust Build Environ 6:359–371
15. Vourdoubas J (2016) Creation of Zero CO_2 emissions school building due to energy use in crete-greece. Open J Energy Effic 5:12–18
16. Hong S-M, Paterson G, Mumovic D, Steadman P (2013) Improved benchmarking comparability for energy consumption in schools. Build Res Info 42:47–61

Chapter 16
Demand-Side Energy Flexibility Management of Office Buildings

Wim Zeiler

16.1 Introduction

People need buildings to protect them against the environmental conditions to be able to work and live. Building services make it possible to provide comfort and an acceptable indoor air quality for building occupants. However, with 40% of the energy use within the developed world and 36% of the CO_2 emissions, the built environment is one of the most important areas for sustainable development [1]. The increase of power grid connected renewable energy sources (RES) has resulted to change in power generation characteristics and grid operation from 'passive' to 'active' [2] due to the stochastic nature of sources like the sun and wind. Buildings with integrated RES become energy producers and possibly supply energy towards the grid, creating multidirectional energy flows on lower grid levels (Fig. 16.1a) [3–5]. The 'active' role of the energy consumers is crucial in the success of smart grids [1, 6], and all actors should participate. Balancing the energy mismatch between supply as well as from 'bottom-up' demand-side; the building, *see* Fig. 16.1b, is essential. Subsequently, improved 'flexible' energy management between the power grid and connected buildings is mandatory. Load flexibility, due to controllable load systems and battery energy storage systems (BESS), can help in handling the energy supply and demand mismatch by switching off or on for certain periods of time, when congestions may occur on grid level. With these interventions the grid reliability increases; however the load shifting operations come with indoor comfort requirements and rebound effects as will be discussed along this paper. To validate this strategy, field experiments were conducted in an office building to evaluate alternative methods; more details can be found in [7, 8].

W. Zeiler (✉)
Department of the Built Environment, TU Eindhoven, Eindhoven, Netherlands
e-mail: w.zeiler@bwk.tue.nl

A. Sayigh (ed.), *Renewable Energy and Sustainable Buildings*, Innovative Renewable Energy, https://doi.org/10.1007/978-3-030-18488-9_16

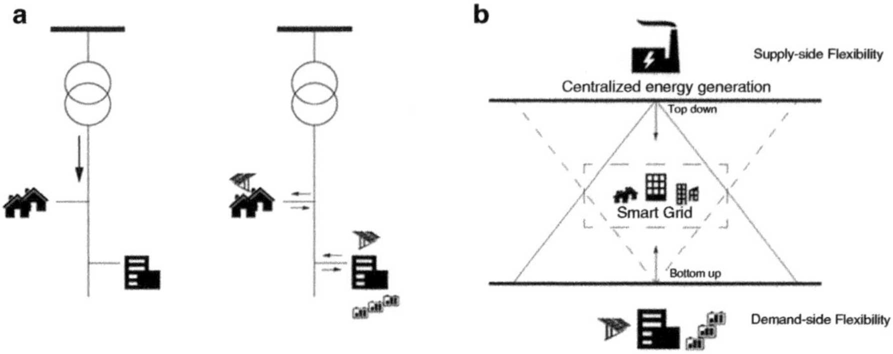

Fig. 16.1 (**a**) Multidirectional energy flow. (**b**) The smart grid approach [5]

16.2 Methodology

Traditionally, the potential for nZEBs in the Netherlands was mainly determined by the possible applications of building energy reduction measures according to the Trias Energetica method; *see* Fig. 16.2. An adapted version of the Trias Energetica method could be used in the future adding the integration of user behaviour as well as energy exchange and storage systems (smart grids). Especially these possibilities become crucially important for nZEB because of the intermittent characteristics of most renewable energy sources. Energy exchange has great potential for reducing energy demand, especially when buildings with a specific heat or cold demand are combined (e.g. nursing homes, ICT data centres, swimming pools or other sports facilities like ice rinks, offices and houses). Coping with complex and unpredictable factors related to decentralized renewable energy source (DRES) and the grid requires a more flexible approach to process control that is increasingly bottom-up rather than top-down. As a result the influence of the building's design and its users' interactions becomes more important. It is widely recognized that increasing energy flexibility is key for the reliable operation of future power systems with very high penetration levels of RES [9]. In general two kinds of flexibilities can be distinguished in energy infrastructures [10]:

1. Architectural, enabling modifications of system configurations to future uncertainty
2. Operational, allowing energy modification by operating strategies without changes of the systems themselves

Performance results for operation of thermal comfort systems in demand-side flexibility modes indicate that improvements are needed for both models of flexibility towards unitary performance metrics that capture both demand side (buildings) and power network side (grid). Clearly the energy demand characteristics of buildings, available from building energy management systems (BEMS), are very valuable information on flexibility for grid optimization. Smart control of energy

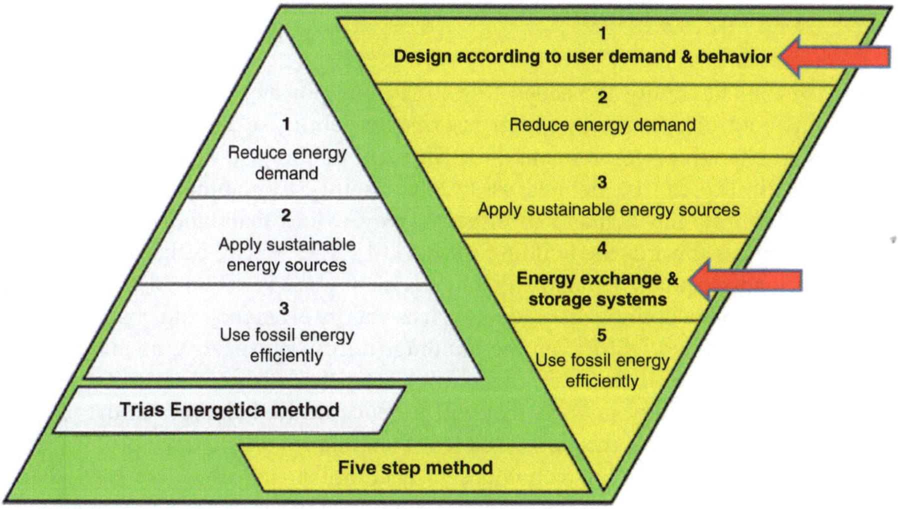

Fig. 16.2 The Trias Energetica method and the five-step method

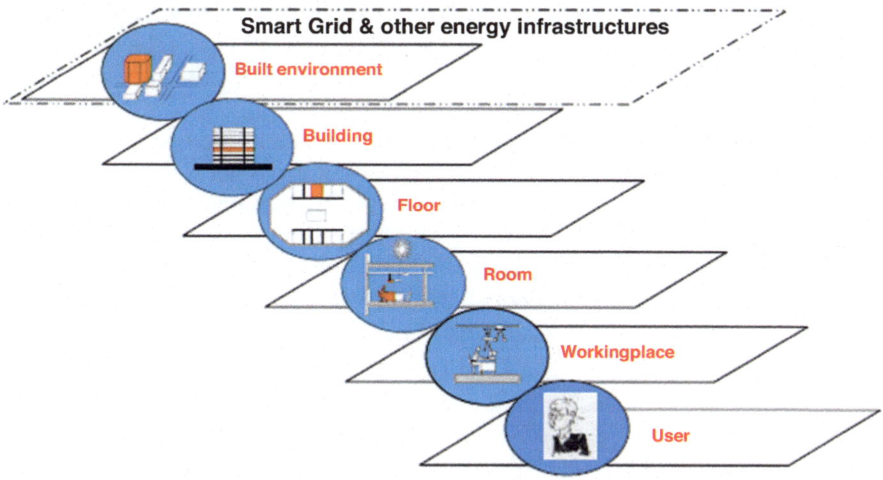

Fig. 16.3 IEA Annex 67 Energy Flexible Buildings proposed functional layered approach

consumption and generation inside (nanogrid) and around buildings (microgrid) can provide major flexibility contributions to address the imminent energy problems within the total energy infrastructure, the smart grid (SG), within the boundaries of acceptable comfort levels.

The IEA Annex 67 defines building energy flexibility [11]: its ability to manage energy demand and generation according to local climatic conditions, occupant needs and energy grid requirements. To cope with the total complexity, a functional layered approach is proposed; *see* Fig. 16.3.

16.3 Energy Flexibility

Power system flexibility is needed to mitigate the power grid uncertainties [12]. Flexibility control is a strategy which describes the ability of a system to handle and respond to changing requirements [13]. In view of the grid environment, power system flexibility can be defined as system's ability to continuality balance the electricity supply and demand in effective way, while maintaining the required comfort services towards the building's users [14]. The 'active' office building with 'flexible' loads requires demand-side management (DSM) to control building load systems and realize optimization goals such as energy efficiency and cost reduction. Gellings [15] defined DSM from the building (utility) perspective as the planning and implementation of those electric utility activities designed to influence customer uses of electricity in ways that will produce desired changes in the utility's load shape. As a result of the performed DSM techniques, new energy profiles arise. DSM includes a variety of techniques to accomplish the objective load shape; examples are peak shaving and valley filling. In [16] six DSM techniques are described; this article focuses on two of them: load shifting and flexible load shape; *see* Fig. 16.4. Load shifting describes the technique in which loads are rescheduled from peak to off-peak periods. The DSM technique flexible load shape presents controllable load systems during critical periods. This technique relies on a set of actions that respond to the need of the grid; the actions are related to power system reliability of the moment.

For the DSM load shifting techniques, building systems are operational during shifted periods of time, preferably off-peak. However, this is not always possible due to the 'rebound' of building load demand units [17]. For example, the demand load chiller operates on highly reduced capacity for half an hour; within this half hour, the indoor comfort is maintained. After the half hour, the chiller has to operate on full capacity to compensate the downtime and keep comfort parameters between the required boundaries, creating a rebound effect as shown in Fig. 16.5. These new

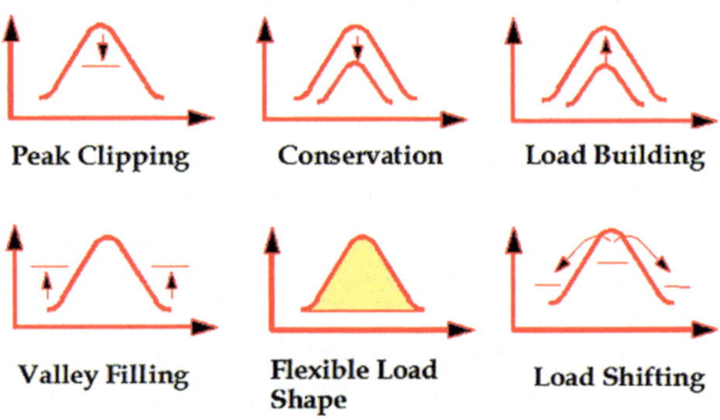

Fig. 16.4 Six DSM techniques [14]

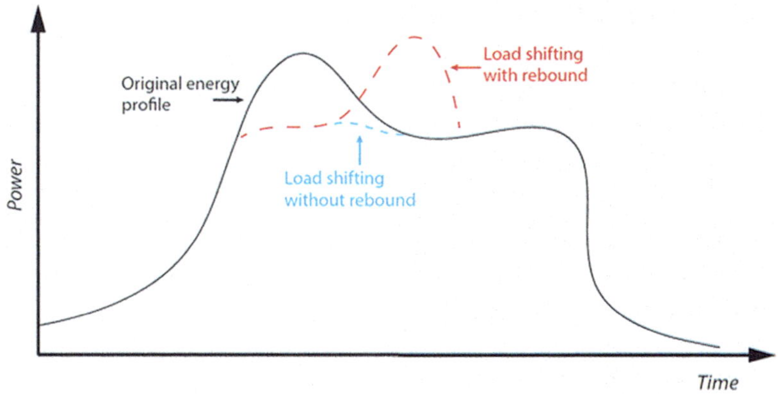

Fig. 16.5 Rebound effect to compensate downtime

peaks may disturb the power grid balance and have to be taken into account for DSM technique development. The consequence of rebound demand may be significant at grid level as a result of cascaded effect of multiple buildings [18].

16.4 Multi-agent System (MAS)

New integral approaches are needed to increase the buildings' energy flexibility towards the smart grid. Traditionally the energy approach towards the built environment is top-down (centralized energy generation/distribution through the smart grid). Instead, we want to use a middle-out approach (control on building level by the BEMS) as well as a bottom-up approach (demand driven by the human needs for energy/comfort); *see* Fig. 16.2. Energy infrastructure's functionalities boil down to energy management making use of the flexibilities of all grid-connected systems which will lead to a better balanced and controlled network at all levels. The energy demand characteristics of buildings available in BEMS represent crucial information for grid optimization [19] to activate participation of buildings in the grid. Intelligent agent technology offers intriguing possibilities to implement advanced DSM control strategies and management of the smart grid [20, 21]. In artificial intelligence, agents are physical or virtual entities that intelligently interact in an environment by both perceiving and affecting it. Consequently, an agent can be described as a computational system with a high degree of autonomy performing actions based on the information received from the environment. Within a MAS, agents interact to achieve cooperative (e.g. distributed problem-solving) or competitive (e.g. coalition formation, auction) group behaviour. Agents with various functionalities achieve this by sharing a minimum amount of information between modules and asynchronous operation implemented via message exchanges [22]. The agents are designed using an open-source web-based and fully decentralized agent design platform called EVE [23] and couples to the BEMS.

16.5 Test-Bed Office Building

The test-bed building is a three-floor office building covering a space of approximately 1450 m^2. A constant air volume (CAV) ventilation system, which comprises an air-handling unit (AHU), heat recovery wheel and an electrical steam humidifier, provides space ventilation and air-conditioning. The building is composed of three thermal zones having both open-plan flexi workspaces and cell offices with building occupancy ranging from between 20 and 35 occupants. The test-bed building had an averaged peak load of 36 kW in 2014 and a peak load of 42 kW composed of 12 kW, 5 kW, 15 kW and 10 kW for lighting, ventilation, humidification and appliance load, respectively. Besides 65 photovoltaic panels, the building has a quite unique battery electric storage system (BESS). This BESS is based on a modular bipolar nickel metal hydride (NiMH) battery. It is 'completely' (99%) recyclable and suitable for a large number of cycles. System nominal voltages are possible in steps of 12 VDC sizing to any required voltage. One module consists of ten cells, of 1.2 VDC; together this is one module. One pack can be made of 10 modules in series for larger capacity and higher voltage to 120 VDC. When these packs are placed in a string, the nominal system voltage can be increased. Since the stack is flat, each end plate is an electrode pole (bipolar). Details are given in Table 16.1 and Fig. 16.6. The bipolar technique has uniquely low internal resistance between the internal celss due to the use of the whole plate surface as electrode pole for a uniform current flow. High charge and discharge rates are possible up to 3C (three times the battery capacity).

Table 16.1 Main components of on-site EES [7]

Component	Function/comments
Schneider Electric 32 A control box	Control system; connects to the building's electrical installation infrastructure/metre box. Also equipped with MCB and energy metre
25 kVA transformer	Transformer; steps down the voltage from 500–800 VDC to 300–400 VDC
KEB F5 23 kVA bidirectional inverter	AC/DC bidirectional inverter; converts DC power from the battery to AC power output and vice versa
Eaton PLC Battery Management Unit	Battery management unit; manages the power quality, charge and discharge operations Relies on Modbus RS-485 connection for communication; this is converted to TCP/IP port to communicate with the existing building management system (Insiteview BlvIS) Also fitted with sensors for monitoring battery voltage, pressure, temperature and current
Battery storage units	Nickel metal hydride (NiMH) battery type; 500–800 VDC, 42 kWh

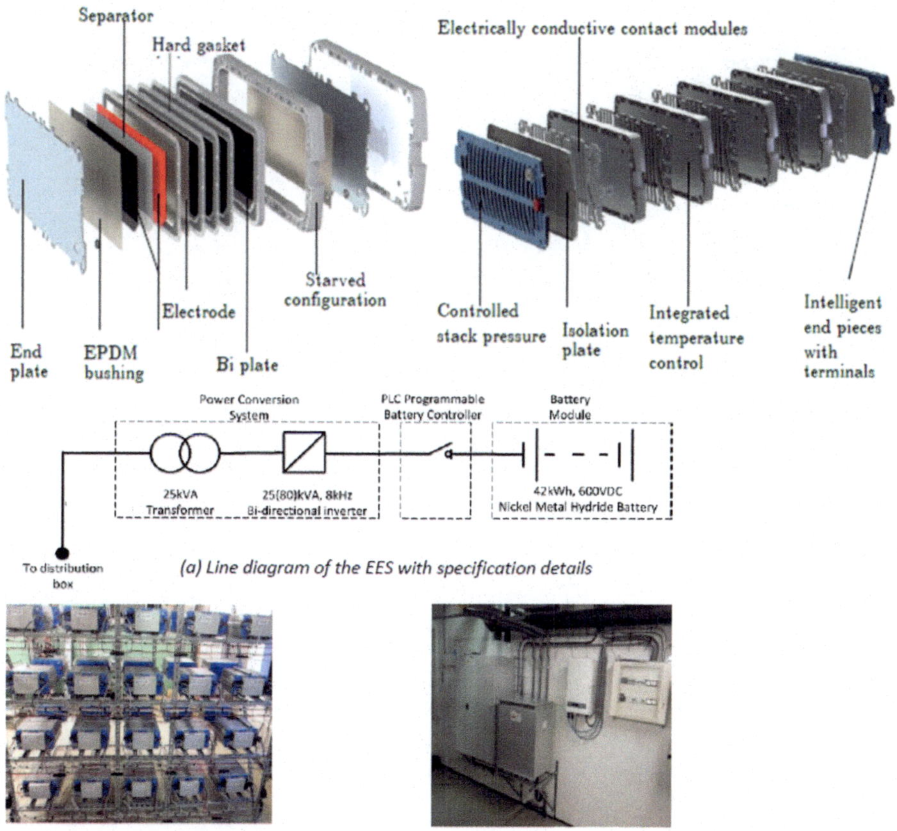

Fig. 16.6 Nilar modular battery technology [24] and diagram and photograph of installed EES

16.5.1 Chiller Load Shifting Experiment

During the summer period, different tests were conducted to obtain knowledge about the flexibility of the chiller by switching the load system chiller off for certain periods of time. Thereafter the effects of the load shifting strategies on the indoor temperature were analysed. The following three strategies were applied: (1) two load shifting periods of half an hour, (2) a load shifting period of 1 h and (3) a load shifting period of 1 h and two shifting periods of 15 min. The measured results of the strategy are shown in Fig. 16.7. In this figure three additional temperatures are shown, which are the room temperature, the outdoor temperature and the supply air temperature. Figure 16.7 shows that the room temperature as well as the supply temperature is increasing when the chiller is switched off. The measured room temperatures can be found in the 80% lower tier and 90% acceptance tier of the ATG thermal comfort limits (Fig. 16.8). These results are shown in Table 16.1; for all three strategies, the highest percentage of the temperature is addressed to class A (thermal comfort can be considered good).

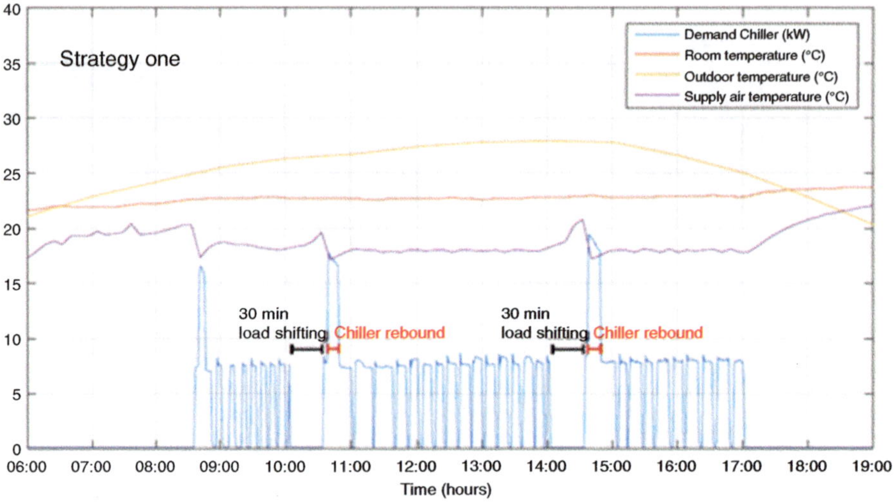

Fig. 16.7 The measured results of strategy 1 [5]

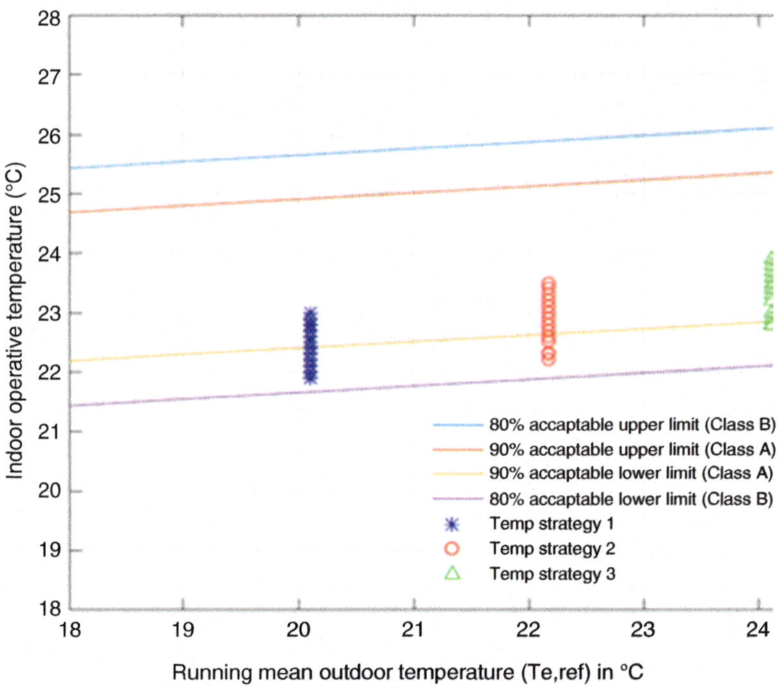

Fig. 16.8 ATG temperature limits with the temperature results

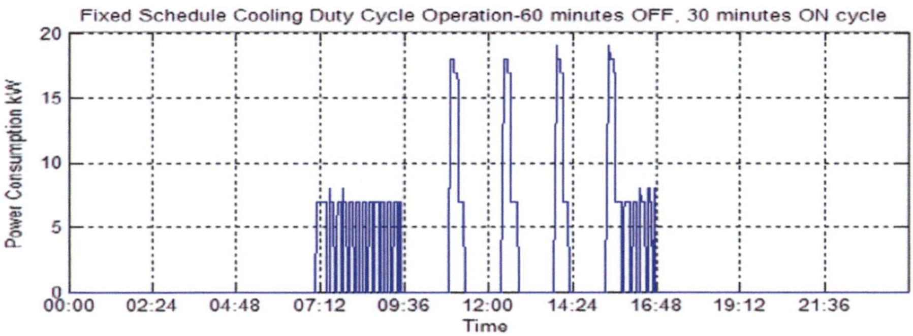

Fig. 16.9 Power characteristics FSCD—60 min OFF and 30 min ON duty cycle [7]

The rebound effects are found in Table 16.1. The rebound power output ranges from 17.1 to 21.3 kW; the capacity fluctuates from 2.9 kWh up to 9.1 kWh. What can be seen is the power and capacity of the rebound are depending on the moment during the day and the load shifting time interval. During Fixes Schedule Cooling Duty (FSCD) tests, rebound power demand was experienced of more than double the cooling demand at thermostatic setting (Fig. 16.9). The resulting rebound power demand is attributed to fulfilment of delayed cooling requirement built up during period of 'off' cycle. Avoidance of rebound power demand during FSCD is only possible with designing duty cycling periods in a manner that ensures the thermal storage is balanced with overall heat gains in the building. Specific time characteristics during the FSCD tests are presented in Table 16.2. On period = comfort recovery and OFF period = available period. Combining FSC duty cycling (FSCD) with EES operation mode I for the actual installation improves flexibility (demand reduction) by a margin of 2–10% depending on the profile and hour of the day; flexibility (demand reduction) upwards by a margin of 0–4% above that for FSCD only the flexibility offer by the building improves as a result of integrated uniform discharge of storage, and FSC duty cycle improves the time characteristics of the building for emulated power flexibility activity.

Subsequently, relatively stable and continuous time characteristics result; availability period is extended by 160 min and response time improved to under 1 min. However, increase in system size beyond the actual installation (i.e. 42 kWh) does not improve power flexibility potential for the case of uniform discharge at 5 kW for 3 h when combined with cooling. Details are illustrated in Fig. 16.10. Evidently, battery operation modes significantly reduce or compensate for the instantaneous rebound cooling demand that results after 60 min of the cooling system being switched 'off'.

It is also shown in Fig. 16.10 that use of EES systems lengthens the period of grid support activity service. This is especially important because FSCD can only be available for a period of 20–40 min just like services such as load following, transmission and distribution infrastructure deferral, demand shifting and peak reduction

Table 16.2 Thermal comfort and rebound effects of the three strategies

Shifting interval	Running outdoor temp.	Class B/Class A	Power/capacity
Strategy 1 1 × 30 min 1 × 30 min	20.11 °C	12%/88%	17.1 kW/2.9 kWh 18.7 kW/3.1 kWh
Strategy 2 1 × 60 min	22.18 °C	21%/79%	19.2 kW/6.1 kWh
Strategy 3 1 × 60 min 1 × 15 min 1 × 15 min	24.12 °C	9%/91%	18.7 kW/4.4 kWh 21.3 kW/7.4 kWh 21.0 kW/9.1 kWh

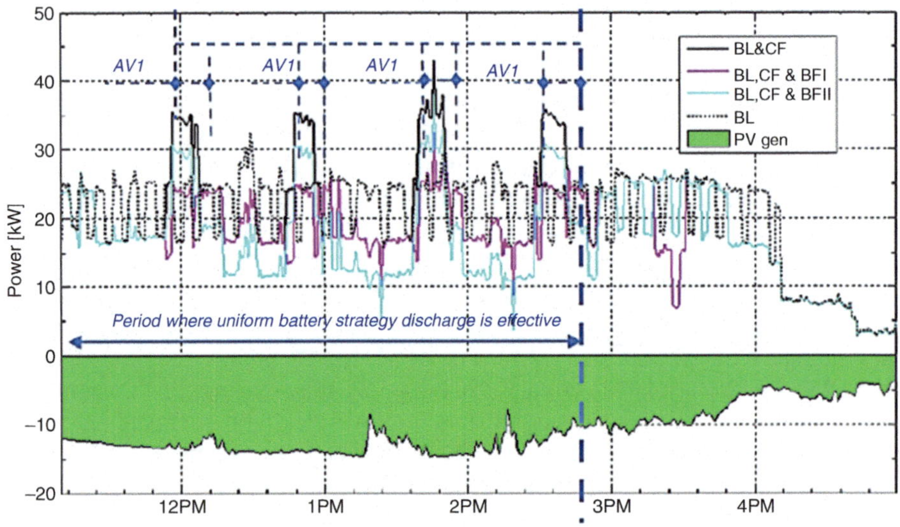

Note:
1. BL & CF: Gross building load with FSCD taken into account
2. BL, CF & BFII: Gross building load with FSCD & battery flexibility mode I taken into account
3. BL, CF & BFII: Gross building load with FSCD & battery flexibility mode II taken into account
4. BL: Gross building load
5. PV gen: PV electricity generation
6. AV1 & AV2 are the availability period for FSCD and battery flexibility II respectively

Fig. 16.10 Rooftop PV generation, building load profile and power flexibility using cooling duty cycling and electrical storage discharge for summer/spring time [7]

and variable supply source integration. Spinning reserve and non-spinning reserve require continued presence of over 1 h. In the process, self-consumption of generated power is intensified. By increasing self-consumption in the building, rather than power allowing flow back to the grid, chances of peak shaving of on-site PV generation are increased by at least 10% of the installation load along with costs deduction in form of network charges.

16.6 Conclusion

Demand-side management strategies such as demand-driven control and demand response have the potential to contribute worthwhile improvement to a building's energy performance as well as in the management and control of the smart grid. While advancement in process control has improved the quality of information obtainable from buildings promoting improved energy efficiency through demand-driven control, on the other hand, buildings participation in demand response events is still vastly limited.

In this article we proposed a bottom-up push strategy based on the actual components and comfort demands aimed at improving buildings participation and responsiveness to DR events by DSM. For participation in DR, there is often a minimum amount of load required by the utility companies for participation. As observed from the test-bed office building, the available service used in populating was below 20 kW. Although additional services from lighting loads could be achieved, being a medium-sized building, its contribution might not be sufficient as around 100 kW is often required for participation.

Due to the limitations posed by the centralized nature of the HVAC system in the building, further experimental studies are being conducted in office buildings having a more decentralized system.

Acknowledgements This research would not have been possible without the financial support of the Province Noord-Brabant for the research development of a microgrid strategy for process control on room level within the context of the project Smart Energy Regions-Brabant, SER-B.

References

1. Abdulaal A, Asfour S (2016) A linear optimization based controller method for real-time load shifting in industrial and commercial buildings. Energy Build 110:269–283
2. Manditereza PT, Bansal R (2016) Renewable distributed generation: the hidden challenges—a review from the protection perspective. Renew Sust Energ Rev 58:1457–1465
3. Aduda KO, Zeiler W, Boxem G, de Bont K (2014) On the use of electrical humidifiers in once buildings as a demand side resource. Proc Comput Sci 32:723–730
4. Veldman E, Gibescu M, Slootweg HJ, Kling WL (2013) Scenario-based modelling of future residential electricity demands and assessing their impact on distribution grids. Energy Policy 56:233–247
5. Voerman M, Bont K, de Zeiler W (2018) Active participation of buildings as a source of energy flexibility, Proceedings ASHRAE conference, Houston, US
6. Shomali A, Pinkse J (2016) The consequences of smart grids for the business model of electricity firms. J Clean Prod 112:3830–3841
7. Aduda K (2018) Smart grid-building energy interactions, demand side power flexibility in office buildings. PhD thesis, TU Eindhoven, Eindhoven, Netherlands
8. Labeodan T (2017) A multi-agents and occupancy based strategy for energy management and process control on the room-level. PhD thesis, TU Eindhoven, Eindhoven, Netherlands
9. Papaefthymiou G, Grave K, Dragoon K (2014) Flexibility options in electricity systems, Project number: POWDE14426, Ecofys 2014 by order of: European Copper Institute, Woluwe-Saint-Pierre

10. Melese YG, Heijnen PW, Stikkelman RM (2014) Designing networked energy infrastructure with architectural flexibility. Proc Comput Sci 28:179–186
11. IEA, (2015) Annex 67 Factsheet. http://www.ieaebc.org/fileadmin/user_upload/docs/Facts/EBC_Annex_67_Factsheet.pdf
12. Ulbig A, Andersson G (2015) Analyzing operational flexibility of electric power systems. Int J Elect Power Energy Sys 72:155–164
13. Saleh JH, Mark G, Jordan NC (2009) Flexibility: a multi-disciplinary literature review and a research agenda for designing flexible engineering systems. J Eng Des 20(3):307–323
14. Lund PD, Lindgren J, Mikkola J, Salpakari J (2015) Review of energy system flexibility measures to enable high levels of variable renewable electricity. Renew Sust Energ Rev 45:785–807
15. Gellings C., 1985 The concept of demand-side management for electric utilities, Proc IEEE 73.10: 1468–1470
16. Macedo MNQ, Galo JJM, de Almeida LAL, de Lima AC (2015) Demand side management using artificial neural networks in a smart grid environment. Renew Sust Energ Rev 41:128–133
17. Palensky P, Dietrich D (2011) Demand side management: demand response, intelligent energy systems, and smart loads. IEEE Trans Ind Info 7(3):381–388
18. Motegi N, Piette MA, Watson DS, Kiliccote S, Xu P (2007) Introduction to commercial building control strategies and techniques for demand response, vol 500. Ernest Orlando Lawrence Berkeley National Laboratory, Berkeley, pp 1–107
19. Wang S (2013) Intelligent building electricity demand management and interactions with smart grid, Proceedings of Clima 2013, Prague
20. Basso G, Gaud N, Gechter F, Hilaire V, Lauri F (2013) A Framework for qualifying and evaluating smart grids approaches: focus on multi-agent technologies. Smart Grid Renew Energy 4(4):333–347
21. Jarvis D, Jarvis J, Rönnquist R, Jain L (2013) Multiagent systems and applications. Intell Syst Ref Libr 46:1–12
22. Kofler MJ, Reinisch C, Kastner W (2012) A semantic representation of energy-related information in future smart homes. Energ Buildings 47:169–179
23. de Jong J, Stellingwerff L, Pazienza GE (2013) Eve: A novel open-source web-based agent platform. In: Proceeding of the 2013 IEEE Int. Conf. Syst. Man, Cybern. SMC. IEEE, Hoboken, p 2013
24. NILAR (2017) Nilar modular battery and Schematic overview of bi-polar technology. http://www.hannovermesse.de/product/nilar-nimh-bi-polar-energy-battery/2276335/V548074

Chapter 17
Toward Green Building and Eco-cities in the UAE

Riadh H. AL-Dabbagh

17.1 Three Climate Change Priorities

By looking at end use categories for energy, we see that we must focus on cleaner building and vehicle technologies. In both cases, it is of course a choice between systems using carbon fuels and systems using clean electricity [3]. Very simply, we need clean tech buildings, clean tech vehicles, and clean electricity generation [4]. Promoting these three should be seen as most important when we ask "What should we do?"

The most proven non-carbon ways to generate electricity are solar energy, hydro generation, and wind turbines. Equally important is conservation which is using less energy, while still heating and cooling buildings and keeping the lights on. As world population grows, it may seem impossible to conserve energy, but we can, if we adopt green building technologies [12].

17.2 Buildings Are the Key

Buildings are responsible for more emissions than any other contributors, represent about half of our energy use, and also have the potential to provide a dominant everyday reminder of the importance of change and our ability to triumph over the difficulties facing our planet. Superior approaches to human shelter will contribute more than any other initiatives to the psychosocial momentum now needed to collectively alter our course toward a better tomorrow (Fig. 17.1).

R. H. AL-Dabbagh (✉)
Ajman University, Ajman, UAE

A. Sayigh (ed.), *Renewable Energy and Sustainable Buildings*, Innovative Renewable Energy, https://doi.org/10.1007/978-3-030-18488-9_17

Fig. 17.1 Development of buildings in the UAE

17.3 The UAE Vision 2021

The National Agenda of the UAE focuses on:

1. Improving the air quality
2. Preserving water resources
3. Increasing the contribution of clean energy
4. Implementing green growth plans

1. *Improving the Quality of Air*
   The establishment of a national air emissions inventory is expected to provide useful, up-to-date, and comprehensive data on pollutant emissions, that maintaining an accurate, reliable, and comprehensive inventory is crucial to inform policies on air quality management going forward. Improving air quality is one of the goals of the UAE Vision 2021, where the National Agenda aims at raising the rate of air quality in the UAE to 90 percent by 2021.

Fig. 17.2 Improving air quality

In collaboration with the Global Green Growth Institute, GGGI, the project will be implemented in two phases. The first phase will involve developing a methodology for an air emissions inventory for the UAE based on best practice and stakeholder consensus. The second phase of the project will aim to deliver a full-scale national inventory that covers all major point, line, and area source (Fig. 17.2).

2. *Preserving Water Resources*
   An indicator that measures water overuse by monitoring freshwater usage (including surface water, renewable water, and fossil water) as a percentage of overall renewable water in the UAE. The result is weighted to take into account desalination and waste water treatment (Fig. 17.3).
3. *Increasing the Contribution of Clean Energy*
   The UAE government wants to ensure sustainable development while preserving the environment and to achieve a perfect balance between economic and social development (Fig. 17.4).
4. *Implementing Green Growth Plans*
   The UAE aims to showcase the country's major achievements and approach in transforming into a green economy [6], diversifying economy and energy sources, and integrating environmental aspects within all development plan (Figs. 17.5, 17.6 and 17.7).

Strategic objectives for UAE regulations:

1. Reduce UAE Ecological Footprint
   The UAE ecological footprint dropped to 7.75 ha per person last year, down from 11.68 ha in 2006; the decrease was because of improved environmental sustainability, through the country's adoption of the Ecological Footprint initiative. The UAE is the third country in the world to do so after Switzerland and Japan. According to the 2014 Living Planet Report, five countries—China, the United States, India, Brazil, and Russia—account for 47.2% of the world's ecological footprint. Despite UAE urban growth, which brings more carbon emissions and a greater ecological footprint, efforts to promote a greener economy and environmental policies have improved the situation. The UAE has set

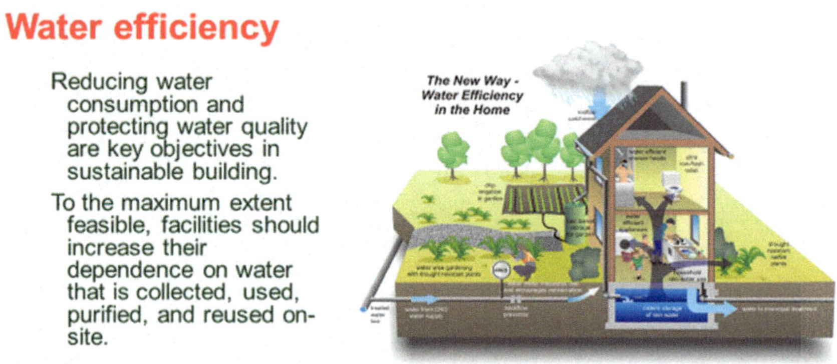

Fig. 17.3 Water efficiency

Fig. 17.4 Moving toward depending on renewable energy

standards to ensure lighting product imports are energy-efficient, especially for the housing sector [7, 8], which represents 57% of the country's ecological footprint. The government also aims to develop standards for fuel and cars, to reduce carbon emissions (Fig. 17.8).

2. Reduce Electrical Consumption by 20%
   The UAE aims to cut energy consumption by 20% by 2020, as it enforces an energy-efficient building project, "Buildings use 70% of the energy that the country produce," and "To get the country green, that needs to target the biggest consumer—the buildings, whether residential or commercial." The efficiency of the program, called the green building project, was started as an initiative in 2008 by the Supreme Council of Energy [10], Electricity and Water Authority [3]. In 2011, it became mandatory for all government buildings to adhere to the

Fig. 17.5 Major achievements and the approach in transforming into a green economy

Designing "Green" doesn't mean "Green"

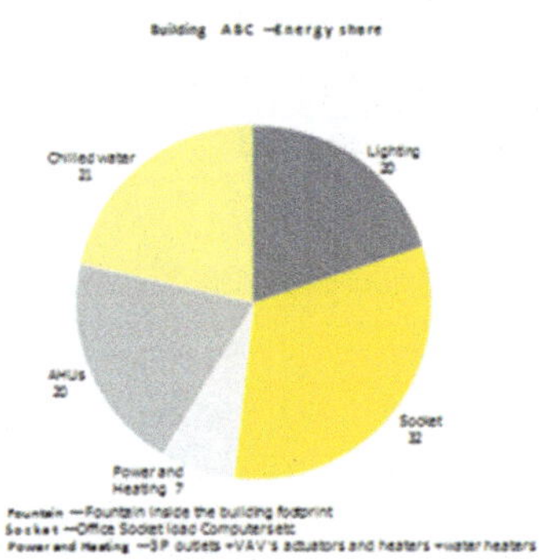

Building	Actual EUI		Design EUI		%Increase from design
	[illegible]	[illegible]	[illegible]	[illegible]	
Lighting	85.32	27.05	34.05	10.79	150.8
Socket	137.00	43.43	166.70	52.84	(17.8)
Power and heating	29.44	9.33	5.22	1.65	464.1
MCC-AHUs	85.16	27.00	49.70	15.76	71.3
Fountain	-	-	-	-	0
Chilled water	92.06	29.18	83.63	26.51	10.1
Total	428.97	135.98	339.30	107.56	26.4%

*based on an occupancy of full building capacity

- External benchmark (EnergyStar):
  - Median property-158.4 kStu/sq.ft.
  - LEED score - 16

Fig. 17.6 Green building design based on an occupancy of full building capacity

program. However, this year it was extended, requiring all new buildings to be energy-efficient (Fig. 17.9).

3. Reduce Water Consumption by 18%

   The Federal Electricity and Water Authority (FEWA) has managed to save 53.44 million gallons of water last year [3]. The 18% reduction in consumption, as compared to that of 2014, followed a nationwide conservation award that saw a stiff competition from housewives, schools, and mosques. Schools set a big example by saving up to 35.34 million gallons of water that represented 25% cut

Fig. 17.7 Dubai will become the capital of green economy

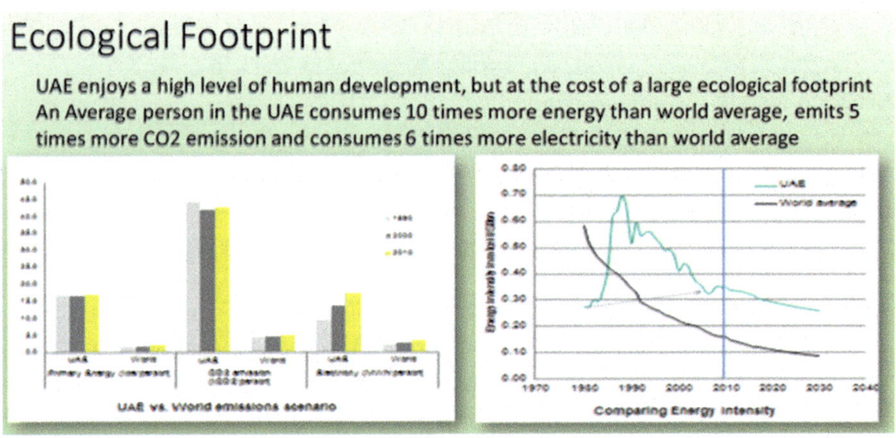

Fig. 17.8 UAE ecological footprint

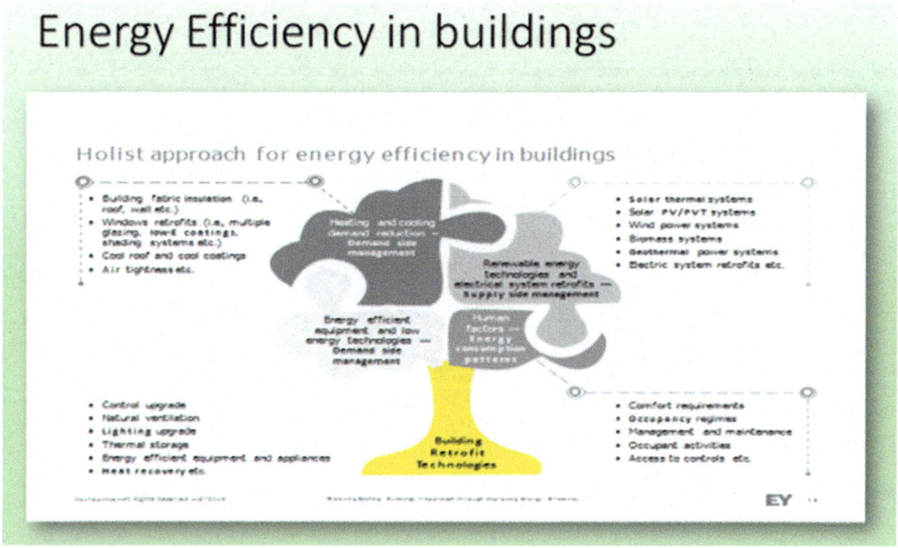

Fig. 17.9 Holist approach for energy efficiency in buildings

as compared to 2014's consumption. Those were followed by the mosques category, which saved 10.04 million gallons of water or 18% drop in comparison compared to 2014's water consumption. "Housewives came third after saving 8.1 million gallons of water that represented nine per cent reduction in consumption compared to 2014's water consumption, The Eastern region (B) led the country in rationing with 13.42 million gallons of water saved or 30 per cent drop as compared to 2014's consumption." However, the Western region (A) managed to save 16.71 million gallons of water, but that represented 22% drop in comparison to 2014's water consumption. The Ministry of Justice, represented by the courts in the emirates of Fujairah, Umm Al Quwain, and Diba Al Hisn, in Sharjah saved 584,186 gallons of water, whereas the Ministry of Interior in Fujairah saved 188,380 gallons of water. The Emirates Conservation Award is meant to promote conservative consumption of water and electricity and instill the values of rationing in young generations. Participation is open to all nationalities. The list of winners included Emiratis, Arabs, and Asians. "The targeted conservation was 12%, however, some families managed to reduce consumption by up to 60 and even 80%. Fewa, in collaboration with strategic partners, has earlier installed 25,000 water-saving gadgets at 64 mosques and 60 schools. The award, The first edition of the Emirates Conservation Award saw 2810 residential participants, including 506 Emiratis and 2304 expatriates, but only 1124 were qualified. The list of top winners included 18 homemakers—three in each of the six regions involved in the competition (Figs. 17.10 and 17.11).

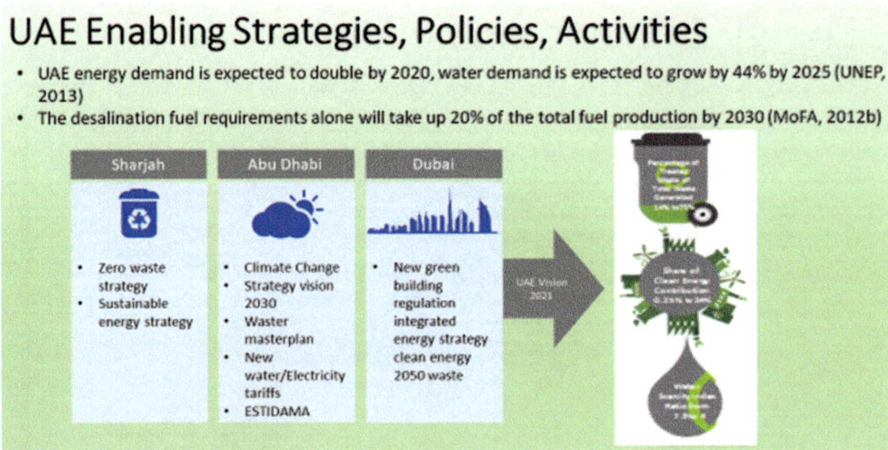

Fig. 17.10 Water demand growth

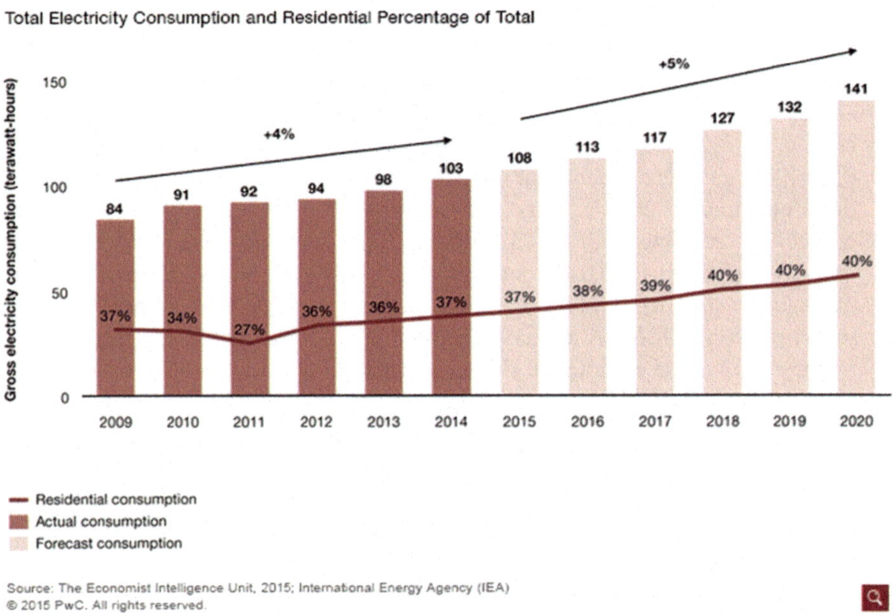

Fig. 17.11 Total electricity consumption, with residential percentage of total

17.4 Smart Cities

The UAE is one of the most urbanized countries in the world, with 85 percent of its population living in urban areas in 2014. The urban population is forecast to continue growing and should reach 91 percent of the UAE population by 2050, according to the UN. The UAE therefore needs to have integrated infrastructure planning as a prerequisite to any urban master plan [7].

Smart cities allow city planners to improve efficiency at the intersection of different infrastructure sectors.

Perhaps the best way of doing this, which takes advantage of digitization and the big data that urban populations produce, is through the so-called smart city approach. Although the term smart city does not have a formal definition, it typically entails integrated infrastructure planning that applies digital technologies to deliver better services and, critically for the UAE, reduce energy consumption [1]. The shift to smart cities is still in its early stages, and the overall relationship between improved urban design and the total sustainability effect of those initiatives has yet to be measured. The transition to smart cities entails overcoming the inertia that stems from ingrained habits, whether in terms of urban planning or citizens' behaviors. It also requires the development and implementation of coherent government policies that will deal with the complex changes that smart cities engender. Nonetheless, there is a broad consensus that smart cities will improve energy efficiency. By adopting a more comprehensive approach, smart cities allow city planners and managers to improve efficiency at the intersection of different infrastructure sectors, such as electricity, water, transport, telecommunications, cooling, and waste [2]. The number of smart cities is projected to triple worldwide by 2025 because of these advantages.

The number of smart cities will more than treble over the next decade (forecast totals) (Fig. 17.12).

One advantage of smarter urban electrical grids is that they monitor usage over time and so can encourage the more efficient consumption habits that lead to significant decreases in usage. For example, electricity demand in Abu Dhabi has been growing by more than 10 percent a year, which exceeds the emirate's remarkable annual population growth rate of 9 percent. In response, the Abu Dhabi government

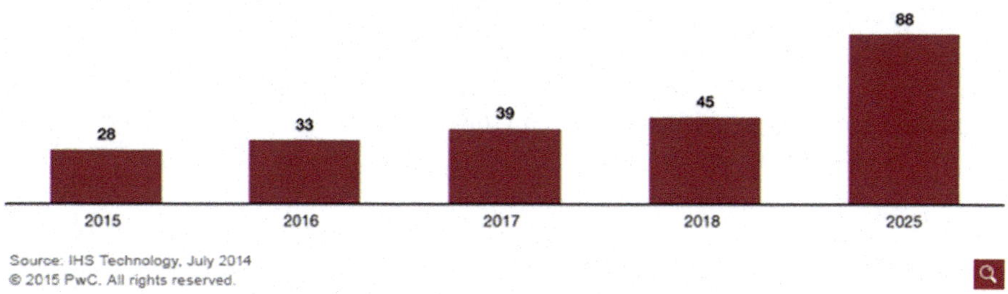

Fig. 17.12 The future growth of smart cities

has deployed an advanced metering system that offers a wide range of functionalities and rates to encourage customers to reduce consumption or shift it to off-peak hours. Similarly, Dubai now has several interrelated programs in place to substantially reduce electricity consumption, including smart metering, demand-side management to encourage energy users to consume more efficiently, and distributed generation (which involves generating energy on-site to reduce transmission losses). From 2009 through 2014, these measures have saved more than 1100 GW h of electricity and 5.4 billion gallons of water and reduced carbon dioxide emissions by over 536,000 tons [2].

Another important aspect of smart cities is transportation, an area in which technology has advanced dramatically. The UAE is encouraging the use of electric cars and building the necessary infrastructure to support them. The country currently has in place 16 power charging stations for electric vehicles, and it expects to have more than 100 by the end of 2015. However, electric vehicles are only a small part of the solution. To meet the transportation needs of an increasingly urban population, the UAE still must address broader challenges, such as improving the fuel efficiency and emissions of imported cars, persuading residents that they may not need to own a vehicle, and reducing the dependence on cheap gasoline.

To that end, awareness campaigns should be part of a smart city initiative and represent a softer means of influencing citizens' behaviors. The UAE has already begun allocating budgets for such campaigns. For example, the Dubai Supreme Council of Energy has launched the Emirates Energy Award, which offers monetary prizes for innovative work in projects of all sizes across both the public and private sectors [9]. The Dubai Electricity and Water Authority (DEWA) has also launched customer awareness campaigns that have, among other things, persuaded residents and businesses to set their air conditioners to warmer settings. In all, DEWA's initiative has saved more than 600 million AED (US$163 million) since 2009.

17.5 Building Efficiency

Building efficiency—commonly referred to as "green" or "sustainable" construction—refers to structures that are designed and built with improved energy efficiency as a key design constraint. The concept aims to reduce the environmental impact of buildings and to improve the well-being of their occupants [12].

The UAE already has a range of building efficiency measures in place. For example, in Abu Dhabi, a program called Estidama (the Arabic word for sustainability) regulates the design, construction, and operation of buildings through phased approvals. Estidama also uses an assessment scale called the "Pearl Rating System," which measures the sustainability performance of villas, buildings, and communities [13]. In Dubai, the government has issued a set of green building regulations and specifications that cover planning, the use of resources, materials, and waste. Notably, the regulations are intended to improve the sustainability performance of buildings throughout their entire life cycles, from design through construction, operation, and ultimate teardown (Fig. 17.13).

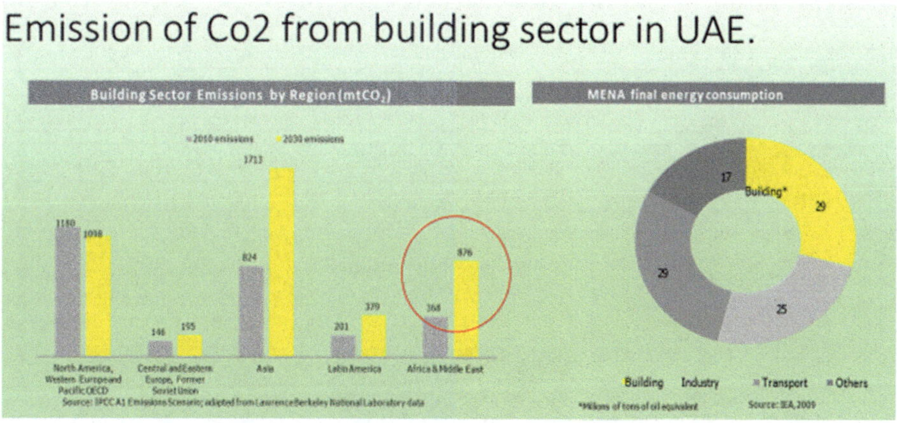

Fig. 17.13 CO_2 emission from building sector in the UAE

Government authorities throughout the UAE will still need to develop more detailed regulations and frameworks that dictate energy efficiency in buildings.

Government authorities throughout the UAE will still need to develop more detailed regulations and frameworks that dictate energy efficiency in buildings, particularly during construction. These regulations can use a range of levers: licensing contractors based on energy performance; mandating procurement procedures that factor in total life-cycle cost, rather than just initial construction; and requiring advance modeling and simulations to predict the energy consumption footprint prior to construction, as part of the procurement process (a tool known as building information modeling).

In addition, district cooling holds significant potential to increase the energy efficiency of UAE urban areas [4]. Currently, the cooling of buildings represents about 70 percent of the peak electricity load in the GCC [5]. District cooling is a network-based system that uses centralized infrastructure to provide air conditioning to multiple buildings at once. Such a system is far more efficient than the current approach to air conditioning because it pools the demand for cooler air. Increasing the use of district cooling would reduce the burden of the cooling load on overall energy consumption. In the UAE, various regulatory authorities are exploring ways to mandate and regulate district cooling, and the technology should be considered for any new urban planning concept (Fig. 17.14).

What are also needed are enhanced communication and information campaigns that persuade real estate companies to improve the energy efficiency of their projects without regulation. For example, such campaigns can highlight the efficiency benefits of new technologies, designs, and materials, as well as compare building performance among similar buildings and award energy performance certificates that include incentives for builders.

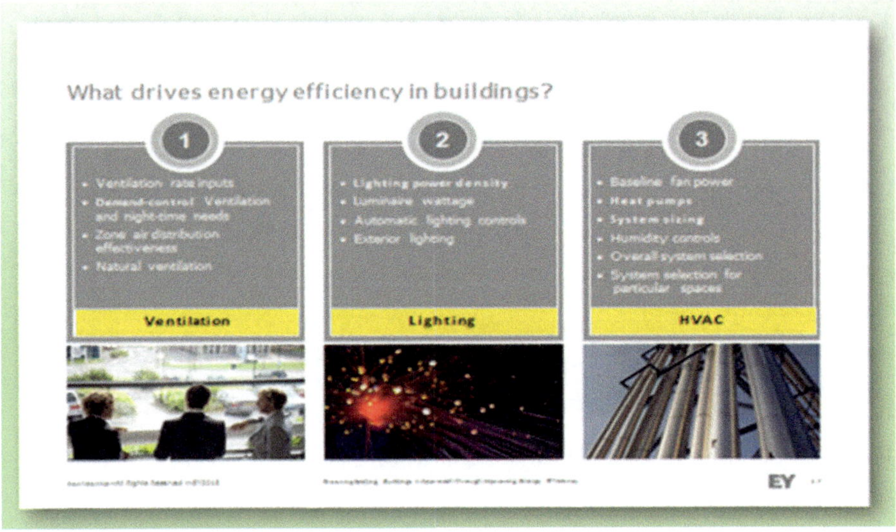

Fig. 17.14 Energy efficiency in buildings

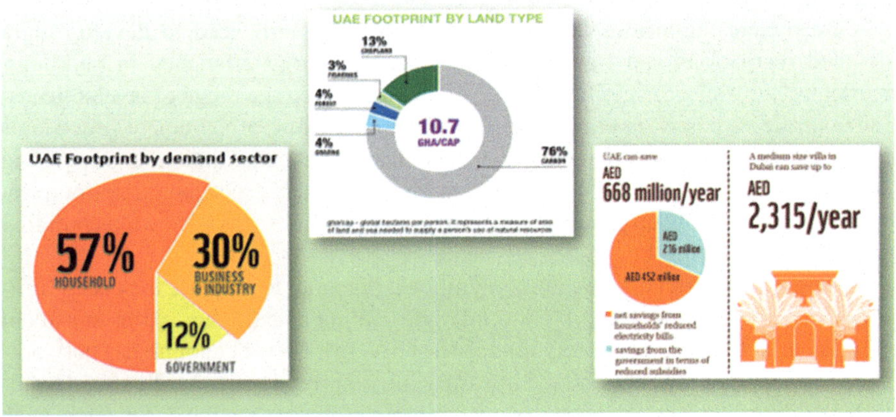

Fig. 17.15 UAE footprint by demand sector

Finally, the government should encourage the use of innovative building technologies, materials, and systems. Advanced concrete mixes, wall insulation, glass glazing, and coatings that are adapted to the region's requirements would substantially improve the efficiency of buildings. These materials are already used in other markets as they deliver benefits in building efficiency as soon as they are in place. The UAE government should ensure that these materials are considered–or required—during the design and construction of new projects. Once construction is complete, projects can still generate gains due to increased energy efficiency, through tools such as advanced building automation, integrated applications, and monitoring and reporting, which increase energy awareness among occupants and facility managers and drastically reduce consumption (Fig. 17.15).

17.6 Conclusion

Green buildings and the deployment of renewable energy systems are some of the solutions that could insure a considerable amount of carbon emission. Energy efficiency assessment for buildings is an important tool that controls the right implication of the proposed practices. It has been concluded that the UAE is developing a good program to reduce the footprint.

For decades, the abundance of easily accessible hydrocarbon resources meant that energy efficiency was not a pressing topic in the UAE. However, steady population and economic growth have changed attitudes. Today, sustainability is a critical issue that is becoming more urgent each year. The time is ripe for the government of the UAE to intervene and create a more sustainable future for the country. By leveraging other countries' experience and focusing on the priorities discussed in this paper, the UAE can ensure that it meets the needs of its citizens, while also safeguarding the environment for upcoming generations.

References

1. Building next-generation sustainable cities: A road map for economic, social, and environmental sustainability, Strategy 2014
2. DEWA saves Dh752m through energy efficiency measures, Emirates 24/7, April 8, 2015
3. DEWA extends its 24 °C campaign to the commercial sector, DEWA website, August 13, 2014
4. Unlocking the potential of district cooling: The need for GCC governments to take action, Strategy, 2012
5. Achieving a sustainable water sector in the GCC: Managing supply and demand, building institutions, Strategy, 2014
6. Al-Dabbagh R, AL-Dabbagh RH (2012) Potential role of UAE in promoting environmental technology. Eur Aff Soc Sci 2:2012
7. United Nations Environmental Program (UNEP) (2013) Annual report
8. Ministry of Foreign Affairs (MOFA) (2012) act amendment
9. Intergovernmental Panel on Climate Change (IPCC) (2007)
10. Super Council of Energy Report (2008)
11. United Nations Environmental Program (UNEP) (2011) Path towards sustainable development
12. NAHB, Icc (2007), National green building standards
13. LEED (1998) USA green building council leadership in Energy & Environmental design. Internet Resources
14. www.Siemens.com/energyefficiency
15. www.diag.org.Uk/documents.jsp
16. www.carbontrust.co.Uk/publications
17. www.greenbuilding.com
18. www.total.com/energy/Efficiency
19. www.dubaicityguide.com/. Dubai City Guide

Chapter 18
Renewable Energy in Africa: Changing Support Systems

Terence Cook and David Elliott

18.1 Introduction

Given that there are 54 very different countries in the continent of Africa, ranging from the very poor to the relatively affluent, with varying levels of political stability and economic development, it is difficult to make generalisations. However, one thing is clear: Africa has a very large renewable energy potential, which, if properly developed, can help mitigate some of the major social, environmental and economic problems it faces, including rising energy demand and climate change, poor energy access and environmental pollution, while creating sustainable employment.

The International Renewable Energy Agency (IRENA) says that Africa has the potential and the ability to utilise its renewable resources to fuel the majority of its future growth with renewable energy. It adds 'doing so would be economically competitive with other solutions, would unlock economies of scale, and would offer substantial benefits in terms of equitable development, local value creation, energy security, and environmental sustainability' [1].

This paper looks at the various approaches that have been adopted to making that a reality, set within the aim of the UN-backed Sustainable Energy for All programme to double energy efficiency and double the use of renewable energy globally by 2030.

In the context of Africa, the Clean Development Mechanism (CDM) carbon market system has only played a small role, but there are more direct renewable energy-related support programmes and financial initiatives that have been more significant. They include the various UN, World Bank, IMF and EU initiatives, the most recent of the latter being set within the EU's wider External Investment Plan, with an initial €4.1 billion allocation.

T. Cook · D. Elliott (✉)
Department of Engineering and Innovation, The Open University, Milton Keynes, UK
e-mail: d.a.elliott@open.ac.uk

A. Sayigh (ed.), *Renewable Energy and Sustainable Buildings*, Innovative Renewable Energy, https://doi.org/10.1007/978-3-030-18488-9_18

The starting point for many of the support schemes in the past has been direct financial aid, usually coupled with technical support, but of late the emphasis has begun to shift from 'aid' to 'trade', with the aim being to create new markets and stimulate the investment of private finance. That has already been the approach adopted by China, via its expanding energy-related investments of around $13 billion so far, about half of this being for renewable energy projects, mostly hydro [2]. The US-led 'Power Africa' programme, with $7 billion committed, has focused more on the new renewables. It is still being maintained under the Trump administration, possibly since much of the funding comes from the private sector [3].

There have been concerns expressed about the dominance of market-based private investment approaches [4, 5], but with the scale, direction and effectiveness of aid programmes sometimes being challenged, views on what should be done in the future seem to be changing. The European Commission has claimed that 'development cooperation needs to evolve. Traditional assistance in the form of grants remains essential—but must be complemented with other tools and sources of finance in order to reach the ambitious targets set by the Sustainable Development Goals. The international community agreed in 2015 on an innovative agenda on financing for development, which calls for new partnerships, notably to mobilise private resources and to apply innovative financing models. The External Investment Plan is part of the EU's contribution to these commitments' [6].

18.2 The New Approach

The new EU approach to support for Africa has been championed by the German Federal Ministry of Economic Cooperation and Development's proposals, 'Africa and Europe—A new partnership for development, peace and a better future', in a so-called Marshall Plan for Africa. It says that 'we need to move away from the donor-recipient mentality that has predominated for many decades and shift towards an economic partnership based on initiative and ownership' and that 'it's not the governments that will create all the long-term employment opportunities that are needed, it's the private sector. So it's not subsidies that Africa needs so much as more private investment.'

However, it is not seen as just a matter of commercial investment: 'The Marshall Plan is powered by a new kind of economic policy—one focused on economic diversification, the establishment of production chains, targeted support for agriculture and small and medium-sized businesses, enhanced status for trades and crafts and thus the creation of a new SME sector.' In particular, it is stressed that 'it is vital that Africa's young people can see a future for themselves in Africa. The average age in Africa is 18. Soon Africa's population will top 2 billion. That means that 20 million new jobs will be needed each year, in both urban and rural settings. Developing the necessary economic structures and creating new employment and training opportunities will be the central challenge. Africa's young people also need contact and interaction with Europe. Europe must develop a strategy that allows for legal migration whilst combating irregular migration and people smuggling' [7].

The emphasis on job creation in Africa has been picked up as a key new focus by the EU, faced as it is with a migration crisis. The new External Investment Plan aims to 'tackle some of the root causes of irregular migration'. It is noted that 'Instability and conflicts in Africa and the EU Neighbourhood have been aggravated by the global economic crisis, reducing access to finance for much needed investment. Instability and conflict have also exacerbated the ongoing migration crisis with more people than ever on the move in Africa and in the Neighbourhood' [8].

Making the links between development support, climate issues and migration concerns clear, German Development Minister Gerd Müller said: 'Germany and Europe have an interest to save people's lives, to limit the effects of climate change and avoid "climate refugees," to prevent mass migration and to help create a future for Africa's youth' [9].

Some of the proposals that emerged under the Marshall Plan for Africa emphasised what could be done to this end at the local level. For example, in parallel with, and as part of, the drive on youth employment, local community energy initiatives are seen as a key, with Germany's energy cooperatives set to serve as a model. According to DW News, 'The German Development Ministry is keen to use this expertise to help create similar groups in Africa. Over the next five years, the plan is to set up 100 partnerships in which German cooperatives can share their knowledge and experience with African community groups—in eight countries across the continent' [10].

Clearly then, the new approach includes a social policy element, with, for example, there also being an increased focus on gender equality in EU external support, aiming to improve job prospects for women in the new energy sector. However, the main focus, seen as essential to make the whole process work, is to stimulate, and remove obstacles to, private sector investment in renewable energy in Africa, so as to create new markets and jobs.

The main driver is thus competition, with contract auctions/project tendering being favoured, over Feed-In Tariffs [11]. This is despite the fact that FiTs were initially seen as the way ahead for Africa, much as they had been in the EU [12]. While FiTs worked well in the EU, where there was an affluent consumer base, leading to large numbers of projects going forward (e.g. 100 GW of wind and PV solar in Germany, including many small PV projects), it was felt that they might not work well in Africa, where there were fewer well-off consumers.

However, auction/tendering systems also have problems. They tend to work best for large projects/companies that can afford the relatively high transaction costs, so smaller projects would be disadvantaged. There is also the risk that companies would submit bids that were lower than was actually viable, so that, although they might win contracts, in the event they might find it hard to actually deliver the capacity. That had happened in the case of the NFFO price/capacity auction system at one time used in the UK, and, although auctions have recently led to some very low price contracts around the world, it is not yet clear if they will all be able to deliver. Penalties for nondelivery have been proposed to limit this risk, but it is early days as yet for the project auction approach in Africa. Until recently, most projects in Africa, for example, hydro schemes, were funded via a negotiation process between vendors and state utilities and/or the government. But with new smaller wind and PV solar technologies now entering the market and independent power producers playing more of a role, new project funding patterns are emerging, including auctions.

18.3 The Technological Focus and National Shares

Hydro has been the dominant renewable in Africa in the past. It has been quite extensively developed, with, by 2016, 32.6 GW installed in all, supplying 20% of its power and nearly all the electricity in some countries. However, there may be limits to its use and expansion. Droughts have reduced its reliability in some countries, and climate change could have an increasingly significant impact on this source [13, 14]. There can also be local social and environmental issues with large hydro projects, including, in some locations, methane production from trapped biomass [15]. There are proposals for more schemes, some of them very large, such as the 42 GW Grand Inga scheme on the Congo River, but some argue that, in the absence of a fully developed grid system, large centralised projects like this are unlikely to offer many benefits to rural off-grid area, although smaller projects can be useful [16].

At present solar and wind generation in Africa are still mostly at a low level, with a peak output of only around 3.6% of the total current electricity plant output [17]. However, unlike large hydro, these sources can supply power direct to local users, solar PV especially, and, with costs falling, they are likely to be widely adopted both for off-grid and on-grid delivery. Wind is in the lead. By 2016, there was 3.7 GW of wind capacity in place in Africa, with South Africa leading at 1.5 GW, followed by Egypt and Morocco, each at just over 700 MW [18]. Some large projects are underway, including the 300 MW Lake Turkana wind farm in Kenya.

Solar PV is catching up at 2.9 GW total in 2016, with many new projects underway, some of them quite large, following on from the 175 MW plant in South Africa and the 155 MW plant under development in Ghana. Smaller off-grid PV solar projects are also proliferating, with a key debate being whether that should be the main emphasis in the drive to improve local energy access. PV has been relatively expensive in the past, although still not as expensive as building grids to link to, and cover, remote low population density areas. However, now PV costs are falling, as are battery costs, so the off-grid option is more attractive, especially if linked up in local mini-grids, which can provide shared storage backup and balancing services, making solar energy more reliable [19, 20]. At the utility-linked scale, concentrated solar-thermal focusing plants (CSP) are also being developed, in North Africa especially, offering output 24/7, since solar heat can be stored for power generation overnight.

In addition to these solar electricity options, solar water heating is also an obvious option for much of Africa and ought to be preferable to the widespread but inefficient and unhealthy use of traditional biomass for heating and cooking. Geothermal heat and power is another option available in some locations, notably in the Great Rift Valley. Modern biomass and waste-fired plants can also offer significant potentials for heat and power production, if care is taken to source the biomass sustainably. That is even more important for biofuel production for vehicle use: there are major land-use and biodiversity issues.

As can be seen, there are many renewable supply options, with some well established. Taken together, by 2016, there was around 38 GW of renewable capacity in Africa, with Ethiopia in the lead, just ahead of South Africa and Egypt. See Table 18.1.

Looking further ahead, IRENA has estimated that, assuming energy efficiency is improved, renewables could supply 22% of Africa's *energy* (heat and power) needs by 2030, with the share of modern renewables in the *power* sector reaching 50% by 2030 and maybe up to 60% of total generation [22, 23]. Looking longer term, a recent study claimed that the sub-Saharan Africa renewable electricity generation share could reach over 92% by 2050 [24].

At present, significant funding is still being directed to hydro projects, as in the case of China's investment programme (it has been responsible for 58% of all new hydro projects recently), but wind and PV solar are catching up, with off-grid and mini-grid PV projects attracting more attention. The EU funding system is seeking to reduce the financial risks faced by private sector investors in developing smaller schemes. The EU-backed Electrifi funding system offers recyclable loans, while the GETFIT system run in several African countries also offers initial start-up grants to help smaller projects. Independent power producers (IPPs) often prefer dealing with

Table 18.1 Renewable generating capacity in megawatts: top 20 countries in Africa in 2016, IRENA [21]

	MW
Ethiopia	4188
South Africa	4064
Egypt	3666
Congo DR	2579
Zambia	2436
Morocco	2309
Mozambique	2200
Nigeria	2062
Kenya	2057
Sudan	1793
Ghana	1612
Angola	936
Zimbabwe	898
Uganda	796
Cameroon	730
Tanzania	659
Cote d'Ivoire	604
Reunion	378
Malawi	373
Guinea	370

conventional large grid companies and utilities rather than with local community projects and mini-grids, so as to limit power off-taker risks, and the EU is trying to deal with that by offering special off-taker guarantees.

18.4 Analysis

The EU does seem to be making a major effort to entice private capital to invest in new energy technologies in Africa, and some might ask whether it is right to provide public support to entice private IPPs to adopt better approaches. More generally, privatisation and a shift to mainly private sector-led projects, with market power shaping developments and maybe operating via deficit financing mechanisms, open up fears of economic exploitation and even the risk of creating a new version of 'third world' debt!

The idea of moving 'from aid to trade' is a classic neo-liberal recommendation, which not all subscribe to. Some worry that corporate private sector involvement will co-opt aid and distort or undermine the social development process. Certainly there have been concerns about China's pragmatic, 'no strings' commercial investment approach, in terms of proper attention being given to local accountability, environmental impacts and also international trade rules.

For example, the IEA notes that China is not a member of the OECD Development Assistance Committee (DAC), which has grouped the world's main aid and development support donors, defining and monitoring global standards in key areas of development. The IEA says that 'while DAC members typically link their assistance to certain criteria such as effective governance, transparency or anti-corruption, Chinese institutions offer support to national governments with fewer conditions' [25].

This may of course offer them some economic advantages over DAC members, who also have to engage with the various social and environmental impact assessment processes adopted by the Western aid agencies and programmes. The fear is that the Chinese approach will not only cut corners but will also corner markets and undermine Western development efforts and the West's (rival) pursuit of markets.

However, a recent study claims that the Chinese programme has not so far led to significant problems for the Western developers. It 'did not impair the effectiveness of grants and loans from Western donors and lenders', and there was 'no evidence that Chinese aid is inferior to aid from established donors on economic growth grounds' [26]. That does not mean that all the projects were necessarily environmental and socially appropriate (one of its main focuses has been on coal projects, as well as large hydro), but certainly China has delivered renewable capacity, and it has made some efforts to engage with community-level projects.

Moreover, the Chinese approach arguably does not differ very much in some of its practices from that adopted by the West. For example, of the £6.1bn of UK energy spending in developing countries between 2010 and 2014, 46% was on oil-, coal- and gas-fired schemes, compared with 22% for renewable energy projects. Even in the case of overseas aid by the UK Department for International Development, although 32% of its support for energy projects went to renewables, 22% still went to support fossil fuel schemes [27].

While some of the differences in practice may thus not be that great, some say the Western aid-based approach is inefficient and bureaucratic and are unsure whether aid is fundamentally the right way to help countries to develop, given that, they claim, aid can sap self-reliance, undermine local initiatives and lead to a dependency culture.

The aid versus trade debate, like the wider state versus market debate which underlies it, will no doubt continue and shape views as to which type and scale of technology to develop and how to support it. In practice both approaches may be needed. *Properly targeted* aid can help countries and communities develop their own capacities, with education and training being a key need, while *properly managed* markets could create opportunities for local-level developments in the expansion of decentralised renewables, aided by the advent of new cheaper energy supply technologies and improved energy efficiency to cut demand.

However, there may be resistance to the implicit changes in technological approach. For example, at the national level, countries in Africa that are at present heavily reliant on coal and other fossil resources will want to continue to use them. Indeed, in some cases, the exploitation of new oil and gas resources and the export revenue received are seen as vital to their economies. That may be a false vision, or at least one fraught with political, social and economic problems and risks. Even ignoring the negative environmental impacts and costs of this trade, these resources will not last forever, and plans are needed for the future. With that in mind, one interim option, meanwhile, might be to tap off some of this fossil fuel export income to develop renewables—building up the infrastructure of the future [28].

As that future emerges, and as the benefits of renewables and downsides of fossil fuel become clearer, the pace of change may accelerate. While global markets for fossil fuel trade remain attractive, that may not continue for much longer, with political pressures mounting worldwide to reduce reliance on polluting and climate-impacting energy options and to cut investment in unabated coal use in particular [29]. Nuclear power meanwhile still remains in the equation, with nuclear vendors lobbying hard in Africa, as one of the few markets left to them. However, it seems unlikely to prosper, given its high costs and the rapidly falling costs of renewables [30].

18.5 Conclusions

Until recently, Africa's carbon emissions had been relatively low, given its relatively low level of energy-using consumer activity [31]. However, with energy use in sub-Saharan Africa rising by 45% since 2000, emissions are rising, adding to Africa's many problems. Not all of them are of its own making. There has been a long history of exploitation and consequently of skewed investment patterns, which still continues. As the International Energy Agency notes, 'Sub-Saharan Africa is rich in energy resources, but very poor in energy supply' with one reason being that 'two out of every three dollars put into the sub-Saharan energy sector since 2000 have been committed to the development of resources for export' [32].

However, as we have seen, changes are underway, although arguably not fast enough. Coal is still the main source of power, supplying 90% of electricity in South Africa. The majority of the population in Africa is still off-grid, with poor energy access, especially in rural areas. Climate change is likely to hit Africa hard, with temperature rises and droughts, but also floods, being more common, undermining agricultural activities and much else [33, 34]. Already stressed in many ways, Africa may therefore face even more problems. But renewables can help improve energy access and, along with improved energy efficiency, cut the use of fossil fuels and provide the employment needed for its growing population. It will not be easy, especially if population continues to boom: it could double by 2050. With climate change issues also sometimes being seen as marginal and less pressing than more immediate and urgent economic and social concerns in Africa, making the necessary social and technological changes will be a struggle, but there seems to be no sensible sustainable alternative [35].

In terms of support systems, Africa as a whole has hardly benefitted at all from the Kyoto CDM system, attracting just 3% of global cumulative investment under the CDM by the end of 2015 [36]. As we have seen, the various more direct aid/grant programmes are now being adjusted to focus more on trade and private sector-led market creation, with competitive auctions being favoured, but with the outcome being as yet uncertain. There are concerns about continued conflicts between private and public sector motivations and aims, with some also regretting the demise of Feed-In Tariffs [37].

However, with renewable costs now falling rapidly, it could be that aid and subsidies will be less needed longer term, and new more decentralised economic and social patterns may emerge. But there is a way to go. Certainly, given the major social, environmental, economic and political challenges Africa faces, aid and other forms of support will be needed for some while yet, while adjustments will have to be made to market-based systems to ensure that they are compatible with wider social and environmental sustainability objectives [38].

References

1. IRENA (2013) Africa's renewable future: the path to sustainable growth. International Renewable Energy Agency, Abu Dhabi. http://www.irena.org/publications/2013/Jan/Africas-Renewable-Future-the-Path-to-Sustainable-Growth
2. IEA (2016) Boosting the power sector in sub-Saharan Africa: China's involvement. International Energy Agency, Paris. https://www.iea.org/publications/freepublications/publication/Partner_Country_SeriesChinaBoosting_the_Power_Sector_in_SubSaharan_Africa_Chinas_Involvement.pdf
3. Onwuemenyi O (2017) Donald Trump set to continue Obama's 'Power Africa' program, Sweet Crude Reports, Nigeria. http://sweetcrudereports.com/2017/09/20/donald-trump-set-to-continue-obamas-power-africa-program-official/. Accessed 21 Sept
4. Newell P, Bulkeley H (2017) Landscape for change? International climate policy and energy transitions: evidence from sub-Saharan Africa. Clim Pol 17(5):650–663. http://www.tandfonline.com/doi/full/10.1080/14693062.2016.1173003

5. Ockwell D, Byrne R (2017) Sustainable energy for all: innovation, technology and pro-poor green transformations. Routledge, London
6. Europa (2017) EU external investment plan, Europa EC news service fact sheet. https://eeas.europa.eu/sites/eeas/files/factsheet_us_format_eu_external_investment_plan_en.pdf
7. BMZ (2017) Africa and Europe—a new partnership for development, peace and a better future. Federal Ministry for Economic Cooperation and Development. https://www.bmz.de/en/publications/type_of_publication/information_flyer/information_brochures/Materialie270_africa_marshallplan.pdf
8. Europa op cit ref 6
9. Pelz D (2017) Germany's 'Marshall Plan' for Africa unveiled. DW News, Deutsche Welle, Bonn. http://www.dw.com/en/germanys-marshall-plan-for-africa-unveiled/a-37178506. Accessed 18 Jan
10. Wright L (2017) Germany to boost green energy in Africa—but will it work? DW News, Deutsche Welle, Bonn. http://www.dw.com/en/germany-to-boost-green-energy-in-africa-but-will-it-work/a-39485293. Accessed 30 Jun
11. IRENA (2017) Rethink energy 2017. International Renewable Energy Agency, Abu Dhabi. http://www.irena.org/publications/2017/Jan/REthinking-Energy-2017-Accelerating-the-global-energy-transformation
12. FoE (2012) Powering Africa through feed in tariffs, friends of the earth/WFC/HPF. https://www.foe.co.uk/sites/default/files/downloads/powering_africa_summary.pdf
13. Yamba FD, Walimwipi H, Jain S, Zhou P, Cuamba B, Mzezewa C (2011) Climate change/variability implications on hydroelectricity generation in the Zambezi River Basin. Mitig Adapt Strateg Glob Chang 16:617–628
14. AFP (2017) Malawi suffers blackouts as drought exposes 98% reliance on hydro power. The Guardian/Agence France-Presse, Paris. https://www.theguardian.com/world/2017/dec/08/malawi-blackouts-drought-hydro-power. Accessed 8 Dec
15. Brown P. (2014) Tropical dams an underestimated methane source, Climate Central. http://www.climatecentral.org/news/tropical-dams-methane-18019. Accessed 14 Sept
16. International Rivers (2014) Grand Inga Dam, DR Congo, International Rivers group overview. https://www.internationalrivers.org/campaigns/grand-inga-dam-dr-congo
17. Gies E (2016) Can wind and solar fuel Africa's future? Nature 539(7627):20–22. https://www.nature.com/news/can-wind-and-solar-fuel-africa-s-future-1.20907-/power
18. IRENA (2017) Renewable capacity statistics 2017. International Renewable Energy Agency, Abu Dhabi. http://www.irena.org/publications/2017/Mar/Renewable-Capacity-Statistics-2017
19. Castellano A, Kendall A, Nikomarov M and Swemmer T (2015) Brighter Africa: the growth potential of the sub-Saharan electricity sector, McKinsey report. http://www.mckinsey.com/industries/electric- power-and-natural-gas/our-insights/powering-africa
20. Avila N., Carvallo J, Shaw B, and Kammen D (2017) The energy challenge in sub-Saharan Africa: a guide for advocates and policy makers Part 1: generating energy for sustainable and equitable development, Oxfam America backgrounder. https://www.oxfamamerica.org/static/media/files/oxfam-RAEL-energySSA-pt1.pdf
21. IRENA op cit ref 18
22. IRENA (2015) Africa 2030: roadmap for a renewable energy future. International Renewable Energy Agency, Abu Dhabi. http://www.irena.org/publications/2015/Oct/Africa-2030-Roadmap-for-a-Renewable-Energy-Future
23. IRENA (2015) Africa power sector: planning and prospects for renewable energy: synthesis report. International Renewable Energy Agency, Abu Dhabi. http://www.irena.org/publications/2015/Mar/Africa-Power-Sector-Planning-and-Prospects-for-Renewable-Energy-synthesis-report
24. LUT/EWG (2017) Global energy system based on 100% renewable energy. Lappeenranta University of Technology and Energy Watch Group, Berlin, presentation on sub Saharan Africa. https://www.researchgate.net/publication/320758165_Global_100_RE_System_Sub-Saharan_Africa
25. IEA op cit ref 2

26. Dreher A, Fuchs A, Parks B, Strange A, and Tierney, M (2017) Aid, China, and growth: evidence from a new global development finance dataset. AidData Working Paper No.46, Williamsburg, VA. http://docs.aiddata.org/ad4/pdfs/WPS46_Aid_China_and_Growth.pdf
27. Cafod (2017) UK support for energy in developing countries. Cafod. https://cafod.org.uk/About-us/Policy-and-research/Climate-change-and-energy/Sustainable-energy/Analysis-UK-support-for-energy
28. Collier P and Laroche C (2015) Harnessing natural resources for inclusive growth. International Growth Centre, London School of Economics. http://www.theigc.org/reader/harnessing-natural-resources-for-inclusive-growth/. Accessed 24 Mar
29. ETC (2017) Better energy, greater prosperity. Energy transitions commission. http://energy-transitions.org/sites/default/files/BetterEnergy_Executive Summary_DIGITAL.PDF
30. Elliott D (2017) Nuclear power: past, present and future. Institute of Physics Publications, Bristol
31. Boden TA, Marland G, Andres RJ (2011) Global, regional, and national fossil-fuel CO2 emissions. Carbon Dioxide Information Analysis Center, Oak Ridge National Laboratory, U.S. Department of Energy, Oak Ridge. http://cdiac.ornl.gov/trends/emis/tre_afr.html
32. IEA (2014) Africa energy outlook. International Energy Agency, Paris. https://www.iea.org/publications/freepublications/publication/WEO2014_AfricaEnergyOutlook.pdf
33. IPCC (2014) Fifth assessment report, intergovernmental panel on climate change, Geneva. https://www.ipcc.ch/pdf/assessment-report/ar5/wg2/WGIIAR5-Chap22_FINAL.pdf—page=4
34. Serdeczny O, Adams S, Baarsch F, Coumou D, Robinson A, Hare W, Schaeffer M, Perrette M, Reinhardt J (2017) Climate change impacts in Sub-Saharan Africa: from physical changes to their social repercussions. Reg Environ Chang 17(6):1585–1600. https://link.springer.com/article/10.1007/s10113-015-0910-2
35. AIDC (2016) One million climate jobs—moving South Africa forward on a low-carbon, wage-led and sustainable path. Alternative Information & Development Centre, Cape Town. http://aidc.org.za/download/climate-change/OMCJ-booklet-AIDC-electronic-version.pdf
36. Ockwell, and Byrne op cit ref 5
37. Fell H-J (2017) The shift from feed-in-tariffs to tenders is hindering the transformation of the global energy supply to renewable energies. Energy watch group policy paper. http://energywatchgroup.org/wp-content/uploads/2017/09/FIT-Tender_Fell_PolicyPaper_EN_final.pdf
38. Cook T, Elliott D (2018) Renewable energy: from Europe to Africa. Palgrave Pivot, Basingstoke

Chapter 19
Measure the Embodied Energy in Building Materials: An Eco-Sustainable Approach for Construction

Francesca Scalisi and Cesare Sposito

19.1 Introduction

According to the definition of the Bruntland Commission, the sustainable development consists in satisfying the current needs without compromising the possibility of future generations of satisfying theirs [1]. The term sustainability, initially used to define fairest economic development dynamics, has quickly become of common use and was, then, used to identify not only the development methods but also each action or situation in which this model is applied and put into effect, starting from sustainable production and sustainable market to the sustainable building: a building that does not use polluting materials and prefers natural organic materials, recycled materials and components, that limits the use of fossil fuels and that reduces the production of waste and energetic consumption during operation.

Therefore, energy and environment are the two subjects which architecture and building need to take into account as the building sector is responsible for 40% of global energy consumption and 30% of greenhouse gas emissions [2]. The total energy used in the life cycle of a building is determined by the sum of the embodied energy and the operational energy [3–5]. The definition of operational energy is clear, it defines the quantity of energy requested during the operation of buildings in heating, cooling or ventilation, the production of domestic hot water and illumination [6]. The embodied energy can be generically defined as the energy used in the production stage of the material [7], even if an accurate definition should consider the whole life cycle of the material, with an analysis cradle-to-grave that includes

F. Scalisi (✉)
DEMETRA Ce.Ri.Med., Euro-Mediterranean Documentation and Research Center, Palermo, Italy

C. Sposito
Department of Architecture, University of Palermo, Palermo, Italy
e-mail: cesare.sposito@unipa.it

A. Sayigh (ed.), *Renewable Energy and Sustainable Buildings*, Innovative Renewable Energy, https://doi.org/10.1007/978-3-030-18488-9_19

the energy needed for the extraction of raw materials, processing and transport, periodic energy for maintenance and final energy for disposal [8, 9]. The analysis cradle-to-grave is very complex, mostly because finding the data related to periodic energy for maintenance and final energy for disposal is difficult. Most of the studies on energy efficiency of buildings focuses on the reduction of operational energy, as it affects the most the total energy consumption of a building [10].

Opting for high-performance solutions and using high-performance materials definitely allows to create buildings classified as "Nearly Zero Energy Buildings". This might activate an increase of consumption, creating the "Jevons Paradox" [11]. The economist William Jevons observed that technological improvements that increase a resource's efficiency can cause an increase of this resource's consumption instead of its decrease, because higher efficiency results into a decrease of costs and, therefore, a growth in consumption.

And this paradox is followed by another, according to which the necessity of using high-performance materials to get energy efficiency during operation causes the rise in the necessary embodied energy to create those materials.

The acknowledgment of the idea of a higher embodied energy could be justified only by a significant decrease of energy during the operation of the building, thanks to which a positive balance between embodied energy and operational energy is maintained [12–14].

Many studies question this balance, arguing that the extreme growth of embodied energy is not always adequately compensated by the decrease of operational energy, since embodied energy might represent almost half of the total energy used in a building life cycle and, sometimes, it exceeds operational energy [15–18]. In this sense, the research for efficiency in the operational phase can be counterproductive for the total energy consumption because it doesn't supply enough advantages from a sustainable environmental point of view [11].

19.2 Methods to Measure Embodied Energy

It is necessary to consider the data on embodied energy since the start of the decision-making process, fundamental to facilitate optimal project choices for the environment. But the calculation for embodied energy is complex, especially for two reasons: the lack of data and the conflicting measurement methods.

Langston and Langston [19] state that measuring operational energy is less complicated than establishing the embodied energy: a complex operation requiring a longer time.

The calculation of embodied energy has two fundamental steps: the choice of the system limit and the method according to which the data in the system was elaborated.

The first one represents one of the most important and controversial points, because establishing the limit of the system determines the amount of information provided: it is quite clear that the limit should always be cradle-to-grave, in order to

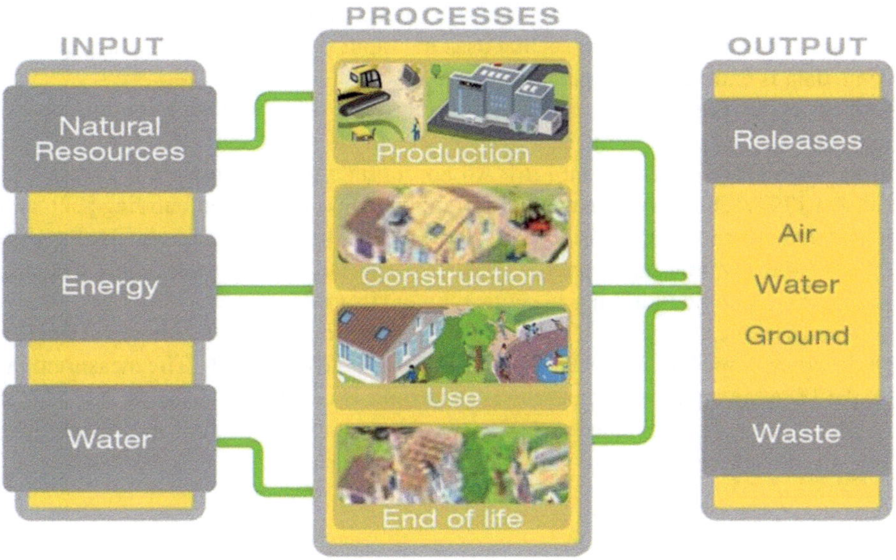

Fig. 19.1 Flow diagram of the life cycle (Source: EPD® Isover Saint-Gobain)

understand the whole life cycle, but basically the most used system limit is cradle-to-gate. This idea will be extensively analysed in the following paragraphs, while now, we will focus on the measurement methods of embodied energy (Fig. 19.1).

The measurement of embodied energy is complex and until now a generally accepted method has not been defined to distinctively measure it [19–22]. The main methods used to measure embodied energy are process analysis, input-output analysis and the hybrid method.

Process analysis allows to identify and quantify direct and indirect energy required by the work processes. Even if it is highly demanding, the process analysis is really widespread because it allows to get accurate and specific results. It consists in the decomposition of the production process of a good in individual activities and quantification of energy fluxes necessary to its realization. The final value of used energy is known as process energy requirements (PER) [9].

The input-output analysis is a macroeconomic technique created in the 1970s by the Nobel Prize Leontief. Through the cross-sector matrix the energy directly and indirectly used in the economic system is measured. The limit of this approach is to get data grouped in a sector in which different goods and/or services are produced. The analysis carried out with this technique was born from the need of putting into the final balance also the items usually ignored in the process analysis, that is the effect of the services indirectly involved in the production and all the works preceding the point to which the process analysis might decide to stop. The final quantity of energy calculated with the input-output analysis is known as gross energy requirements (GER) [9, 23, 24].

The hybrid method has the main characteristics of the two aforementioned methods, since it is a process analysis that uses the input-output analysis when acquiring some data is extremely complex [25].

The use of more sustainable building materials and building techniques is an important contribution to the eco-efficiency in the building industry, therefore, to a more sustainable development [26]. An appropriate building materials choice can cause a 17% reduction of the energy used in the construction of a building [27], and it can reduce the CO_2 emissions by 30% [28]. Having reliable databases with updated data on embodied energy is an efficient help to whoever wants to make aware project choices, aiming to protect the environment.

One of the best known and used databases is the Inventory of Carbon and Energy (ICE) created by Hammond and Jones from Bath University [29]. The measurement method used is input/output and the system limit is cradle-to-gate, even if for some materials the consumed energy during the delivery of the products to the building site was considered. In Table 19.1 some of the materials of the ICE are listed.

19.3 TC 350 Standards

In 2011–2012 have been published the standards developed by CENT TC 350, a Technical Committee developing the standards that can define a harmonized methodology to assess the environmental performance and the costs of the life cycle of buildings. It is made up by three Working Groups: WG1—Environmental Performance of Buildings, WG2—Building Life Cycle Description, WG3—Product Level. The purpose of the latter is to define the Product Category Rules (PCR) to elaborate the Environmental Product Declaration (EPD) for building materials. The Environmental Product Declaration (EPD) is a voluntary-based technical document verified by a certification body that accompanies the marketing of a product. The TC 350 Standards define the impact from cradle-to-grave of buildings and products using as measurement method the process analysis.

The TC 350 Standards for building products, in compliance with EN 15804: 2012 are divided in four stages:

Product stage, construction stage, use stage, end-of-life stage, and an optional module reuse-recovery (D).

In the table of the Fig. 19.2 the mandatory and optional stages are listed, according to the considered system limit.

Product, construction, use and end-of-life stages are strictly linked to the building. As a matter of facts, the type of building in which a product must be installed will establish the scenarios of the building's life cycles, and these will determine the scenarios that must be evaluated for the installation, the usage pattern and the end of life, but the final disposal will depend on the product.

In the analysis cradle-to-gate only the product stage (A1–A3) is considered and, therefore, is mandatory. In the analysis cradle-to-gate with options the product stage (A1–A3) is mandatory while all the other stages are optional. In the analysis cradle-to-grave all the stages are mandatory but D is optional.

Table 19.1 Inventory of Carbon and Energy (ICE) of Bath University [29]

Materials	EE—MJ/kg	EC—$kgCO_2/kg$	Materials	EE—MJ/kg	EC—$kgCO_2/kg$
Aggregate	0.10	0.005	General insulation	45.00	1.86
General sand	0.10	0.005	Lead virgin	49.00	2.61
Rammed soil	0.45	0.023	Stainless	56.70	6.15
General concrete	0.95	0.130	Natural latex rubber	67.60	1.63
General stone	1.00	0.056	General paint	68.00	3.56
General plaster (gypsum)	1.80	0.12	Polyurethane	72.10	3.00
Marble	2.00	0.112	General Carpet	74.40	3.89
Common Brick	3.00	0.22	PVC general	77.20	2.41
Typical cement	4.6	0.83	Expanded Polystyrene	88.60	2.50
Plasterboard	6.75	0.38	PVC Injection Moulding	95.10	2.20
General timber	8.50	0.46	ABS	95.30	3.10
General steel Recycled	9.50	0.43	Polycarbonate	112.90	6.00
General ceramics	10.00	0.65	Polypropylene, Injection Moulding	115.10	3.90
Lead Recycled	10.00	0.53	Synthetic rubber	120.00	4.02
General glass	15	0.85	Nylon 6	120.50	5.50
Paperboard	24.80	1.32	Nylon 6.6	138.60	6.50
Linoleum	25.00	1.21	Epoxide Resin	139.30	5.91
Fibreglass	28.00	1.53	Aluminium Virgin	218	11.46
Aluminium Recycled	28.8	1.69	Titanium virgin	361 to 745	–
General steel virgin	35.30	2.75	Titanium recycled	258.00	–

Miscellaneous

	Embodied Energy—MJ	Embodied Carbon—kg CO_2
PV Modules	MJ/sqm	kg CO_2/sqm
Monocrystalline	4750 (2590 to 8640)	242 (132 to 440)
Polycrystalline	4070 (1945 to 5660)	208 (99 to 289)
Thinfilm	1305 (775 to 1805)	67 (40 to 92)

The product stage includes the supply of all material and energy, the disposal of final waste during the production stage. In detail, it includes:

A1—extraction and processing of raw materials, including processing of secondary materials

A2—transport of raw materials and secondary material to the producer

A3—manufacture of the construction products, and all upstream processes from cradle to gate

The construction stage also includes the waste handling and the disposal of final waste. The transport and installation of the building process depend from the context of the building in which the building product is used. Specifically, it includes (Fig. 19.3):

Type of EPD	A1	A2	A3	A4	A5	B1	B2	B3	B4	B5	B6	B7	C1	C2	C3	C4	D
	BUILDING ASSESSMENT INFORMATION																
	Building life cycle information																Supplementary information beyond the building life cycle
	A1-3			A4-5		B1-7							C1-4				D
	Product stage			Construction process stage		Use Stage							End of life stage				Benefit and loads beyond the system boundary
	Raw material supply	Transport	Manufacturing	Transport	Construction	Use	Mintenance	Repair	Replacement	Refurbishement	Operational energy use	Operational water use	Demolition	Transport	Waste processing	Disposal	Reuse/Recovey/ Recycling potential
				Scenarius		Scenarius							Scenarius				
Cradle to gate[1]	M	M	M														
Cradle to gate with options[2, 4]	M	M	M	O	O	O	O	O	O	O	O	O	O	O	O	O	O
Cradle to grave[3, 4]	M	M	M	M	M	M	M	M	M	M	M	M	M	M	M	M	O

M = Mandatory O = Inclusion Optional
[1]for a declared unit, [2]for a declared unit or functional unit, [3]for a functional unit
[4]Reference Service Life to be included only if all scenarios are included

Fig. 19.2 Information modules for construction products, adapted from EN 15804:2012

Fig. 19.3 Flow diagram of the life cycle (Source: EPD® Gyprom Saint-Gobain)

A4—transport of products from the manufacturer to the construction site
A5—the installation/construction of the building

The use stage can be divided into two groups: use stage related to building fabric (B1–B5) and use stage related to the operation of the building (B6–B7).

The use stage related to building fabric includes the following steps:

B1—use of the product, service or appliance installed
B2—product maintenance
B3—product repair
B4—product replacement
B5—product renovation

The module B1 refers to the releases into the environment. At product level, the maintenance (B2) includes all the operations for maintenance of the product installed in a building during the service life of the product. The impacts of this stage are directly linked to the context of the building and must be evaluated with well-defined settings. This is true also for the repair (B3), replacement (B4) and renovation (B5). The evaluation should include production, transport, energy and water use and any associated wastage and end-of-life processes.

The use stage related to the operation of the building must include the use of energy and water during the operation of the product, including the production and transport of any waste in the location produced by energy/water use. It includes, specifically:

B6—use of operational energy
B7—operational water use

As listed in the EN 15804: 2012, the end-of-life stage of a construction product begins when it is replaced, dismantled and does not have any function. Specifically, it includes:

C1—demolition of the building/construction product
C2—transport of demolition waste including the end-of-life construction product in the waste treatment plant
C3—waste treatment operations for reuse, recovery or recycling
C4—disposal and linked processes

During the end-of-life step, everything coming out from the system (that is the building) is considered waste until it gets the status of end-of-waste. The end-of-waste status is reached if one of these materials or products meet one of these requirements: it is commonly used for specific purposes; there is a market or a demand for it; it meets technical requirements for specific purposes; its use will not lead to negative effects.

The module D refers to the possible reuse/restoration/recycling.

19.4 EDP Analysis

The implementation of EPD according to these standards can be a valuable contribution for building professionals in search of reliable data on embodied energy of materials. Even if there are several doubts [30, 31]: first, the voluntary nature of this tool and the steps envisaged according to the system limit, which leaves a wide margin of discretion on what must be the steps considered (apart from the mandatory requirement in each step, if the limit of the chosen system is cradle-to-grave).

The existing research shows the analysis of the EPDs made by the producers on building materials, to quantify, first of all, what type of system limit was chosen. In this way, it can be understood which steps are excluded the most and if this can determine a low estimate of the embodied energy of a product.

The data in this paper refer to the existing EPDs on the International EPD® system website [32].

The geographical area of reference is Europe; on the website there are 395 EPD files divided in 24 European countries as shown in Table 19.2.

The analysis of the 395 EPD files shows that in most cases the system limit is cradle-to-gate with options, precisely in 46% of the cases; in 29% of the cases the system limit is cradle-to-gate and only in 25% of the cases the system limit is cradle-to-grave (Table 19.3).

In the case of EPDs with a system limit cradle-to-gate, the steps A1–A3 have been considered in all the files. In the files declaring that their system limit is cradle-to-gate

Table 19.2 395 EPD files divided in 24 European countries

Country	Construction products	Country	Construction products
Italy	75	Germany	5
United Kingdom	61	Hungary	5
Spain	53	Lithuania	5
Turkey	48	Finland	4
France	32	Ireland	3
Sweden	32	Luxembourg	3
Belgium	18	Czech Republic	2
Norway	14	Bulgaria	1
Denmark	9	Croatia	1
Romania	9	Netherlands	1
Switzerland	7	Poland	1
Russia	6	Portugal	1

Table 19.3 System boundaries of EPD

System boundaries	Quantity	%
Cradle-to-gate	115	29
Cradle-to-gate with options	182	46
Cradle-to-grave	98	25

with options, it was examined which steps were considered, among the optional ones. Of course, the steps A1–A3 were considered in each file. The following steps have revealed that:

- Most of the cases have data only for the optional step A4
- A small percentage have data only for the optional step A5
- Few cases have data for steps B1–B7 and C1–C4
- The same happens for step D that is rarely considered, even in the cradle-to-grave analysis

Therefore, when in the EPD is declared that an analysis cradle-to-gate with options was made, the optional steps considered are A4 and A5, rarely the others.

In the cradle-to-grave system limit the phases are all mandatory, except for D, but it should be noted that on most of the files the values of the phases B1–B7 and C1–C4 are equal to zero or not relevant.

19.5 Conclusions

The paper has highlighted the complexity of the embodied energy measurement even if this comes with the understanding that it increasingly is a fundamental aspect of studies on energy saving in buildings. The TC 350 standards have been published, defining a harmonized methodology to assess the environmental performance and the costs of the life cycle of buildings. It represents an important tool, even if in practice it has several limits.

The analysis of the Environmental Product Declaration (EPD), made with the TC 350 Standards, shows how in practice few steps are included in their life cycle, mainly the product stage (A1–A3) and partially the construction stage (A4–A5). The impacts coming from the use stage, both the use stage related to building fabric (B1–B5) and use stage related to the operation of the building (B6–B7), are mostly omitted. It happens the same for the end-of-life settings. Rarely, they go after the end-of-life settings, with reuse, recovery or recycling (D). These results lead to some considerations that are also important challenges to be carried on in research:

- First, we should reconsider the voluntary nature of this tool that is definitely a limitation.
- If the goal is energy saving, in the interest of environmental sustainability, the data should be thoroughly as possible.
- The current lack of data prevents us from knowing the real impact of the use and end-of-life steps in energy consumption and consequently prevents us from assessing whether in the overall energy balance it is negligible.
- Many steps, especially in the cradle-to-gate system with options, are optional: it is therefore necessary to elaborate on whether this is a counterproductive choice for this tool.

- Involve more the academic world, in order to deepen the subject of research and foster debate, the building industry and end users to raise mutual awareness by offering and requesting materials and building components of certain environmental sustainability, and finally the building professions so that they can work since the creation stage with tools and data that allow to know exactly the overall energy balance of the building.

This research aims to collect data and their critical analysis, useful for evaluating future development actions and policies, but it is still in its initial stage; it will continue, first, by examining the files on the International EPD system website for non-European geographical areas. After the initial analysis on the phases of the considered life cycle, we will continue analysing more in detail the data provided for the individual stages and the type of materials in the EPDs.

The contribution, resulting from a common reflection, is to be assigned in equal part to both Authors.

References

1. WCED (1987) Our Common Future
2. UNEP SBCI (2009) Buildings and Climate Change. Summary for Decision-Makers, United Nations Environment Program Sustainable Buildings and Climate Initiative
3. Nassen J, Holmberg J, Wadeskog A, Nyman M (2007) Direct and indirect energy use and carbon emissions in the production phase of buildings: an input output analysis. Energy 32(9):1593–1602
4. Gonzalez MJ, Navarro JG (2006) Assessment of the decrease of CO_2 emissions in the construction field through the selection of materials: practical case studies of three houses of low environmental impact. Build Environ 41(7):902–909
5. Treloar GJ, Love PED, Holt GD (2001) Using national input output data for embodied energy analysis of individual residential buildings. Constr Manag Econ 19(1):49–61
6. Barucco MA, Verde F, Scalisi F (2016) Innovazione tecnologica di sistemi, componenti e materiali/Technological innovation of systems, components and materials. In: Lucarelli MT, Mussinelli E, Trombetta C (eds) Cluster in progress. La Tecnologia dell'architettura in rete per l'innovazione/The Architectural technology network for innovation. Maggioli, Santarcangelo di Romagna (RM), Italy, pp 103–108
7. Koskela L (1992) Application of the new production philosophy to construction, CIFE technical report 72. Stanford University, California
8. Hammond GP, Jones CI (2008) Embodied energy and carbon in construction materials. In: Proceedings of the institution of civil engineers, energy, vol 161. ICE Publishing, London, pp 87–98
9. Dixit MK, Fernández-Solís JL, Lavy S, Culp CH (2010) Identification of parameters for embodied energy measurement: A literature review. Energ Buildings 42:1238–1247
10. Verbeeck G, Hens H (2010) Life cycle inventory of buildings: a contribution analysis. Build Environ 45:964–967
11. Copiello S (2017) Building energy efficiency: a research branch made of paradoxes. Renew Sust Energ Rev 69:1064–1076
12. Hammond GP (2007) Industrial energy analysis, thermodynamics and sustainability. Appl Energy 84(7–8):675–700
13. Sartori I, Hestnes AG (2007) Energy use in the life cycle of conventional and low-energy buildings: a review article. Energ Buildings 39:249–257

14. Ibn-Mohammed T, Greenough R, Taylor S, Ozawa-Meida L, Acquaye A (2013) Operational vs. embodied emissions in buildings—a review of current trends. Energ Buildings 66:232–245
15. Chastas P, Theodosiou T, Bikas D (2016) Embodied energy in residential buildings- towards the nearly zero energy building: a literature review. Build Environ 105:267–282
16. Blengini GA, Di Carlo T (2010) The changing role of life cycle phases, subsystems and materials in the LCA of low energy buildings. Energ Buildings 42(6):869–880
17. M. Paleari, M. Lavagna, A. Campioli, Life Cycle Assessment and Zero Energy Residential Buildings, Proceedings of the PLEA2013-29th Conference, Sustainable Architecture for a Renewable Future. http://mediatum.ub.tum.de/doc/1169326/file.pdf. Accessed Feb 2 2018
18. Crawford RH, Bartak EL, Stephan A, Jensen CA (2016) Evaluating the life cycle energy benefits of energy efficiency regulations for buildings. Renew Sust Energ Rev 63:435–451
19. Langston YL, Langston CA (2008) Reliability of building embodied energy modelling: an analysis of 30 Melbourne case studies. Constr Manag Econ 26:147–160
20. Miller AJ (2001) Embodied energy a life cycle of transportation energy embodied in construction materials. In: COBRA Proceedings of the RICS Foundation Construction and Building Research Conference. Metropolitan University, Leeds
21. P. Crowther (1999) Design for disassembly to recover embodied energy, in: Proceedings of the 16th annual conference on passive and low energy architecture, Melbourne/Brisbane/Cairns
22. Lenzen M (2006) Errors in conventional and input output base life cycle inventories. J Ind Ecol 4:128–148
23. Crawford RH, Treloar GJ (2003) Validation of the use of Australian input output data for building embodied energy simulation. In: Eighth International IBPSA Conference. Eindhoven, Netherlands
24. Hammond GP, Jones CI (2008) Embodied energy and carbon in construction materials. In: Proceedings of the Institution of Civil Engineers, Energy, vol 161. ICE Publishing, London, pp 87–98
25. Alcorn JA, Baird G (1996) Use of a Hybrid Energy Analysis Method for Evaluating the Embodied Energy of Building Materials. Center for Building Performance and Research, Victoria University of Wellington, New Zealand, Wellington
26. Pacheco-Torgal F, Jalali S (2012) Earth construction: Lessons from the past for future eco-efficient construction. Constr Build Mater 29:512–519
27. Thormark C (2006) The effect of material choice on the total energy need and recycling potential of a building. Build Environ 41:1019–1026
28. Gonzalez M, Navarro J (2006) Assessment of the decrease of CO_2 emissions in the construction field through the selection of materials. Build Environ 41:902–909
29. Hammond G, Jones C (2010) Inventory of Carbon and Energy (ICE), Version 1.6a. Sustainable Energy Research Team (SERT), Department of Mechanical Engineering, Bath, UK: University of Bath, Bath
30. De Wolf C, Pomponi F, Moncaster A (2017) Measuring embodied carbon dioxide equivalent of buildings: A review and critique of current industry practice. Energ Buildings 140:68–80
31. Moncaster AM, Symons KE (2013) A method and tool for 'cradle to grave' embodied carbon and energy impacts of UK buildings in compliance with the new TC350 standards. Energ Buildings 66:514–523
32. The International EPD system®. https://www.environdec.com. Accessed on July 21 2018

Chapter 20
Optimizing Building Form for Integration of Solar Photovoltaic in the Design of a Textile Industry in Katsina, Nigeria

Amina Batagarawa, Yusuf Ahmed Abdulkarim, and Musa Lawal Sagada

20.1 Introduction

Energy plays the most vital role in the economic growth, progress and development of any nation; uninterrupted energy supply is a vital issue for all countries today [1]. Energy as the backbone of world economic growth and development has resulted to depleting fossil fuel reserves, heavy CO2 emissions, climate change and a continually growing energy demand, a large part of which is consumed by the industrial sector. Mitigating these problems and provision of energy security calls for adopting clean and renewable energy effectively and extensively. According to the U.S. Energy Information Administration [2], the industrial sector consumed 52% of global delivered energy in 2010, and its energy consumption grows by an average of 1.4% per year. In Nigeria this sector's growth suffers due to insufficient power generation and supply [3] which in turn increases running cost and hampers growth of industries.

Textile industries consumes a large portion of delivered energy use in the sector; over 25% of the delivered electricity use of non-energy intensive manufacturing industries is consumed by textile factories [4]. Indigenous textile factories are experiencing high operational cost due to high cost of electricity with factories spending over 150 million naira in energy bills from the grid and another 60 million on diesel. These factories extensive envelopes remain unexploited and latent even as various available technologies and strategies exist to produce cost-effective energy solutions.

Nigeria lies within a high sunshine belt and thus has enormous solar energy potentials. Integrating solar PV into the building envelope of Nigeria's textile

A. Batagarawa (✉) · Y. A. Abdulkarim · M. L. Sagada
Faculty of Environmental Design, Department of Architecture, Ahmadu Bello University, Zaria, Kaduna, Nigeria

A. Sayigh (ed.), *Renewable Energy and Sustainable Buildings*, Innovative Renewable Energy, https://doi.org/10.1007/978-3-030-18488-9_20

industrial buildings to harness this solar energy potential would provide valuable clean energy and diversify the energy supply mix.

Compared to large scale power plants located far away from cities, BIPV systems would not require additional land or built structure and is more valuable as it eliminates transmission loss. It has been demonstrated in studies that one unit of on-site renewable energy is around three times the value on energy produced from remote sources [5]. Also, renewable energy integration could have social consequences, encouraging businesses to conserve energy, since its production and performance can be gauged and understood on a micro scale [6].

Architectural integration of solar photovoltaic is the result of a controlled and coherent integration of the solar PV simultaneously from all points of view, functional, constructive and formal (aesthetic) [7].

Building-Integrated Photovoltaics (BIPV) are dual-purpose PV systems, they serve as both the outer layer of a structure and generate electricity for on-site use or export to the grid. BIPV systems can provide savings in materials and electricity costs, reduce pollution, and add to the architectural appeal of a building. The greatest value for BIPV systems is realized by including them in the initial building design [8].

20.1.1 Aim

To provide a design framework for optimizing building form for architectural integration of Building-Integrated Photovoltaic (BIPV) in the design of a textile factory.

20.1.2 Objectives

1. To assess the integration potential of BIPV in the building

   Envelope of existing textile factories in northern Nigeria.
2. To optimize the energy yield from BIPV through optimized building

   Form in textile factories.

20.2 Methodology

The research is an evaluative research, due to its aim, objectives and the nature of data acquisition and analysis. The research uses the case study approach. The cases selected for this study are:

1. Funtua Textiles Limited (FTL), Funtua, Katsina.
2. United Nigerian Textile PLC (UNTL), Kaduna.
3. Kaduna Textiles Limited (KTL), Kaduna.

Tools for data collection and analysis include visual survey, physical measurement, interview and simulation The variables considered in this research are design dependent variables which are required in the integration of BIPV in the building envelope of selected cases and design considerations required for optimum performance of PV systems. They include:

Integration	Variable		Data analysis
Potential of BIPV variables	Roof type		Descriptive analysis
	Façade type		
	External building features		
	Orientation of Buildings		
Optimizing BIPV performance variables	Building Form		Simulation
	Surface tilt		
	Local climate	Modelling information	
	Building size		
	Self-shading		
	Site features		

20.3 Results and Discussion

This section assesses the renewable energy potential of BIPV in Katsina state. It attempts to draw out building forms of textile industries in northern Nigeria and the integration potential of BIPV through examining existing cases and simulating against varying design parameters. It highlights ways to achieve optimum energy yield from integrated photovoltaic systems.

Three cases were visited and studied, which are Funtua Textile Limited (FTL), Funtua, Katsina, United Nigerian Textiles Plc, Kaduna and Kaduna Textile Limited (KTL), Kaduna. These cases were selected based on their functions and climatic zones.

The results obtained are presented below under the following headings based on the objectives of the study:

1. Integration potential of BIPV in the building envelope.
2. Optimizing BIPV performance.

20.4 Integration Potential of BIPV in the Building Envelope

The results and assessment of the variables under this objective for each case studied are (Plate 20.1):

In United Nigeria Textile, the roof types found are gable roof and shed roof. Long span corrugated aluminium roofing sheets are used in all cases; illustrations would be used to represent the roof as camera wasn't allowed into the complex (Plate 20.2).

In Kaduna Textiles Limited, the roof types found are gable roof and saw-tooth roof. The saw-tooth roof covers the spinning and weaving department while the gable roof covers the administrative, engineering and finishing department (Plate 20.3).

Plate 20.1 A gable roof, monitor roof and shed roof at Funtua textile limited (Source: Author's field work)

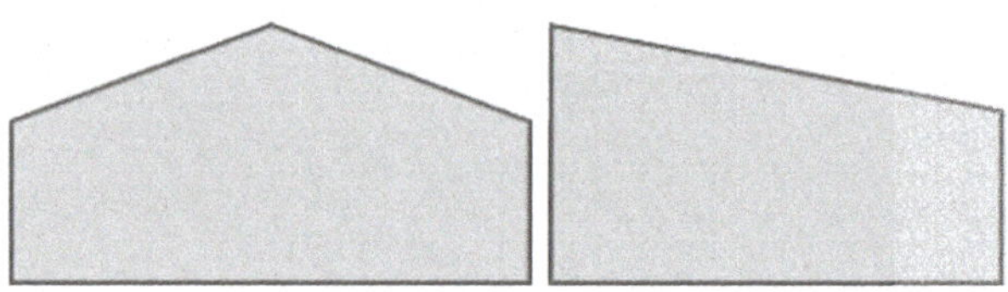

Plate 20.2 Gable style roof and Lean-to roof used at UNTL, Kaduna (Source: Author's field work)

Plate 20.3 Saw-tooth roof and gable roof covering buildings at Kaduna Textiles Limited (Source: Author's field work)

Plate 20.4 Façade at Funtua textile limited (Source: Author's field work)

20.4.1 Façade Type

Façade type of a building could vary widely; this could depend on the building type, location, architect's preference or need based on other factors. The façade type could determine the type or even the possibility of BIPV integration.

In Funtua Textiles Limited, the walls are constructed with hollow sandcrete blocks with the upper portion cladded with long span corrugated aluminium sheets. This façade type can be integrated with BIPV systems (Plate 20.4).

In United Nigeria Textile, the factory building's walls here are majorly constructed with hollow sandcrete block and a few are cladded with long span corrugated aluminium sheets making it possible to be integrated with BIPV systems. Images are not available as camera was disallowed into the premise.

In Kaduna Textiles Limited, the walls of the buildings at the Kaduna Textile Limited are constructed with hollow sandcrete block, solid stone and the upper portion of some buildings cladded with long span corrugated aluminium sheets. The sandcrete area and area cladded with aluminium sheets can be integrated with BIPV. The stone façade used widely for the material aesthetic appeal would lose this appeal but still serve its thermal insulation purpose if integrated with BIPV.

- Roof type.
- Façade type.
- External building features.

20.4.2 Roof Type

The roof surface is among the most important features of a building for the integration of solar BIPV. The type of roof to a large extent determines the potential for BIPV integration in a building.

In Funtua Textiles Limited, the roof types are gable roof, gable with monitor roof and lean-to/shed roof. The lean-to roof covers the engineering department while the monitor roof covers the ginnery and dyeing and finishing department with the remaining building covered with gable roof. Long span aluminium roofing sheets were used in covering these roofs (Plate 20.5).

20.4.3 External Building Elements

This could include shading devices, balconies, parking structures or any external structure supporting the main structure. They provide additional surfaces for integration of BIPV in industrial facilities.

Plate 20.5 Façade at Kaduna textile limited (Source: Author's field work)

In Funtua Textiles Limited, the Factory buildings has little external devices, consists majorly of plain facades. External device found on site is a covered vehicular parking, an external building element with integration potential (Plate 20.6).

In United Nigeria Textile, the complex has no external device present, presenting no external surface for BIPV integration.

In Kaduna Textiles Limited, the factory buildings have external devices, consists majorly of plain facades. External device found on site is a covered vehicular parking which is an external element with integration potential (Plate 20.7).

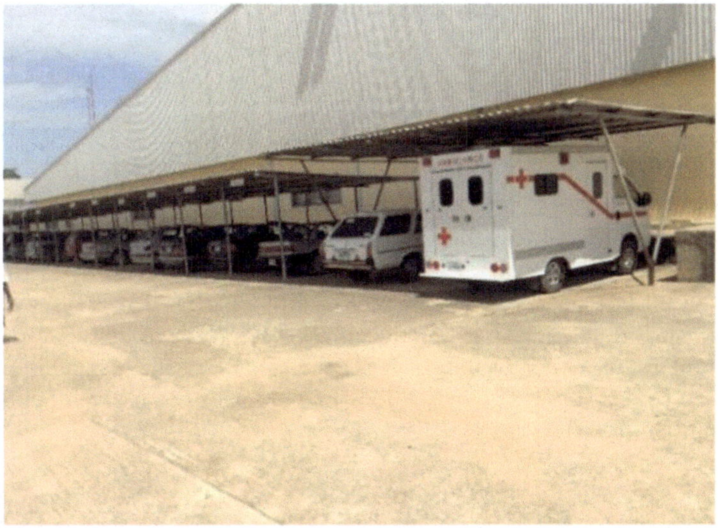

Plate 20.6 External parking shed (Source: Author's field work)

Plate 20.7 Covered vehicular parking (Source: Author's field work)

20.4.4 Integration Potential of BIPV

The results obtained from the case studies shows that five building envelope components of the six building envelope components considered for this study are present within the case studies with the possibility of BIPV integration.

Building envelope component	Funtua Textiles Limited (FTL)	United Nigeria Textiles (UNTL)	Kaduna Textiles Limited (KTL)
Tilted roof	+	+	+
Flat roof	–	–	–
Skylight	+	–	+
Façade walls	+	+	+
Façade glazing	+	+	+
External device	+	–	+

The absence of flat roof for integration is observed throughout the cases studied. With the possibility to integrate BIPV in the Façade walls and tilted roof of the factories in all instances. Skylight is present in both FTL and KTL, with that of KTL north facing making it an inefficient choice for integration. Façade glazing adequate for BIPV is found solely at KTL, the lack of façade glazing is as a result of the need to control internal conditions of the factory to a given humidity and temperature level.

20.5 Optimizing BIPV Performance

The results and assessment of the variables and modelling characteristics studied for each case are:

Local climate
Building size
Site features
Building form
Building orientation
Surface tilt

20.5.1 Local Climate

The irradiation values for the local climates were gotten from PVGIS-CMSAF solar radiation database. The rainfall duration and cloud cover values would be determined from existing literature.

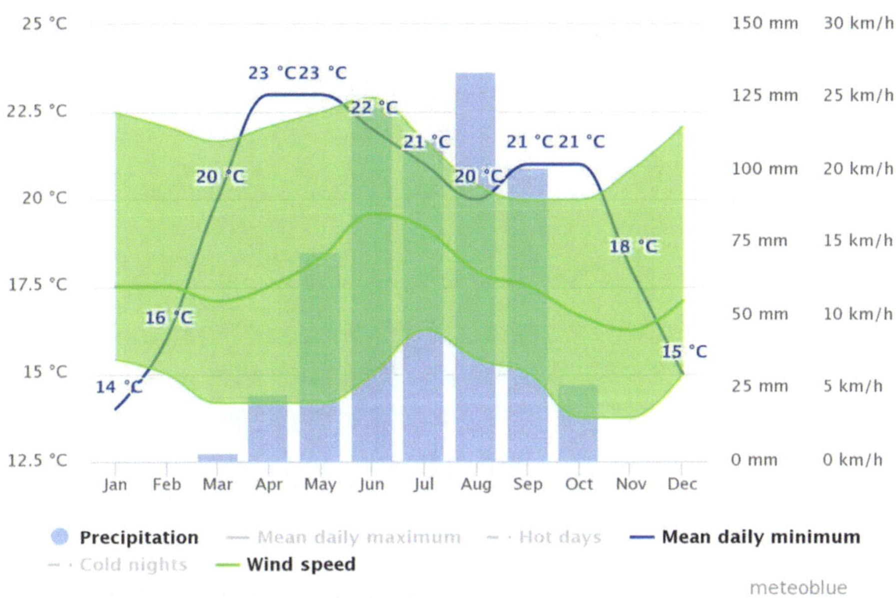

Fig. 20.1 Precipitation, mean temperature and wind speed data for Funtua. Source: (Meteoblue, 2017)

In Funtua Textiles Limited, there is more rainfall during the summers than the winter. It receives an average of 5970 Wh/m^2 per day of irradiation on the horizontal plane. The average annual temperature in Funtua is 24.8 °C and about 1024 mm of precipitation falls annually.

In United Nigeria Textile, the local climate is tropical dry having more rainfall during the summers than the winter. This also goes for Kaduna Textiles Limited as they fall within the same local climate. They receive an average of 5770 Wh/m^2 per day of irradiation on the horizontal plane (Figs. 20.1, 20.2, 20.3 and 20.4).

Case studies	Solar irradiation	Annual mean temperature	Precipitation
Funtua Textiles Limited	5970 Wh/m^2	24.8 °C	1024 mm
United Nigeria Textile	5770 Wh/m^2	25.2 °C	1211 mm
Kaduna Textile Limited	5770 Wh/m^2	25.2 °C	1211 mm

20.5.2 *Site Features*

These include soft and hard landscape and could include natural features of the site. They could be of positive impact or negative impact to the solar power generation potential depending on how they relate to the building on site. Their presence could

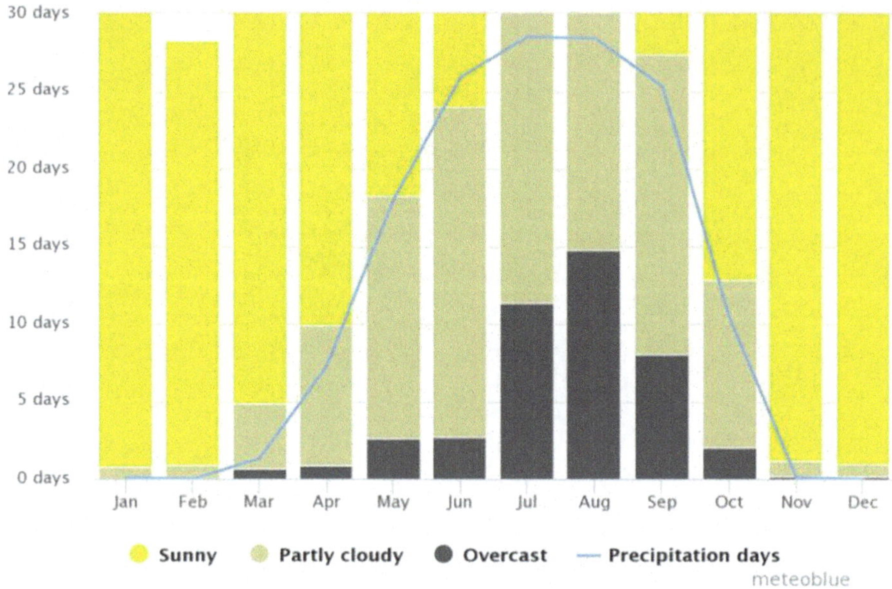

Fig. 20.2 Cloud cover data for Funtua, Katsina state. Source: (Meteoblue, 2017)

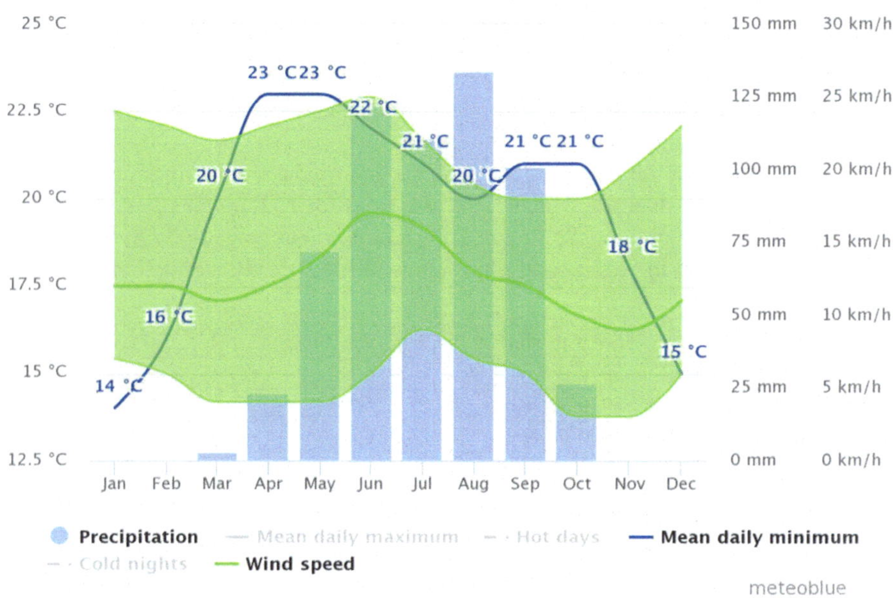

Fig. 20.3 Precipitation, mean temperature and wind speed data for Kaduna. Source: (Meteoblue
Fig. 20.4: Cloud cover data for Kaduna. Source: (Meteoblue, 2017)

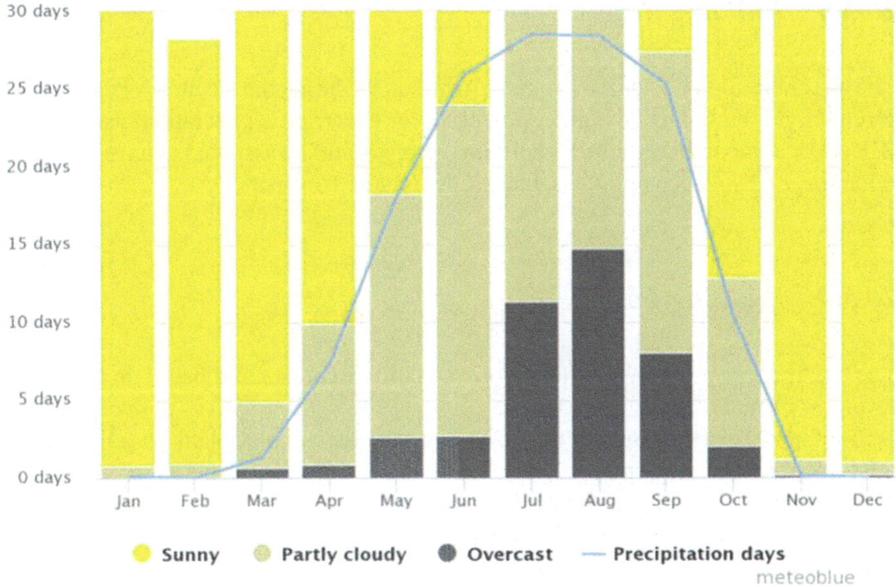

Fig. 20.4 Cloud cover data for Kaduna. Source: (Meteoblue, 2017)

be detrimental when they shade surfaces integrated with BIPV. The site features as observed in the cases studied are discussed.

In Funtua Textiles Limited, the site is laid in an L-shaped pattern with buildings along this path with the surrounding covered in hard and soft landscape. The drive-way and loading bays are finished in concrete with flowers, shrubs and short trees along the sides. Large trees are planted along the east end of the site and in green pockets between the weaving and dyeing department. The trees are positioned at a distance from the building which ensures no shading on building surfaces.

In United Nigeria Textile, buildings are arranged around a central concrete paved yard serving as parking and loading bay with trees heavily scattered around the site. This shades the façade area extensively during the morning and evening period.

In Kaduna Textiles Limited, the site rekindles a trapezium with buildings placed adjacent one another. The access route and loading bays are finished in concrete with few trees within the site and scattered flowers and shrubs around. The administrative area has closely planted trees shading a portion of it during the day, with the trees around the production facility at a distance its shadow falls away from the building.

20.5.3 Building Size

The larger the size of solar PV exposed to the sun the larger the electricity production, given larger buildings a higher potential for electricity generation using solar BIPV. For comparable results all building forms would be assumed to have the same floor area of 1 m^2 each and the same wall height of 1 m each.

20.5.4 Building Form

The building forms of the textile factories would be categorized based on their plan layout, roof type.

In Funtua Textiles Limited, the buildings have rectangular plans with individual total floor area ranging from 13500 m^2 to 860 m^2 in size and a height of about 4 m for the shortest point and 11 m for highest point. The buildings have tilted roofs, the roof types consist of gable roof, monitor roof and lean-to roof.

In United Nigeria Textile, the buildings have rectangular plans with heights reaching 10 m at highest point for the weaving and spinning department while the finishing department reaches 6 m (highest point). The buildings are roofed with gable roof and lean-to/shed roof. Illustrations are used to represent these as taking of images within the premises was disallowed (Plate 20.8).

Plate 20.8 Factory buildings with rectangular plans and tilted roofs (Source: Author's field work)

In United Nigeria Textile, the buildings have rectangular plans with heights reaching 10 m at highest point for the weaving and spinning department while the finishing department reaches 6 m (highest point). The buildings are roofed with gable roof and lean-to/shed roof. Illustrations are used to represent these as taking of images within the premises was disallowed (Fig. 20.5).

In Kaduna Textiles Limited, the factory buildings have rectangular plans with the spinning department roofed with a saw-tooth roof while the weaving and finishing departments are roofed with gable roof (Plates 20.9 and 20.10).

Other plan layouts common to industrial buildings would be included in the study and analysed to determine self-shading factor and size available for integration. The plan layouts include:

Linear layout
L-shape layout
U-shape layout

Fig. 20.5 Building forms at UNTL (Source: Author's Fieldwork)

Plate 20.9 Factory building with rectangular plan and saw-toothed roof (Source: Author's field work)

Plate 20.10 Factory building with rectangular plan and gable roof (Source: Author's field work)

20.5.5 Summary

This section summarizes the building forms established in the cases studied categorized based on their floor plans and roof type. Four categories are established:

1. Rectangular plan with gable roof

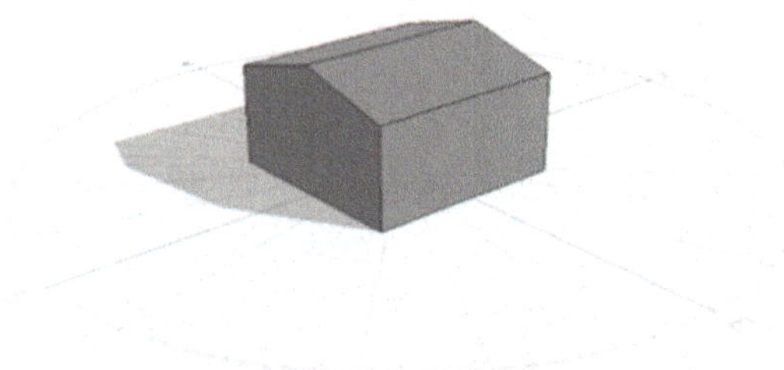

2. Rectangular plan with lean-to roof

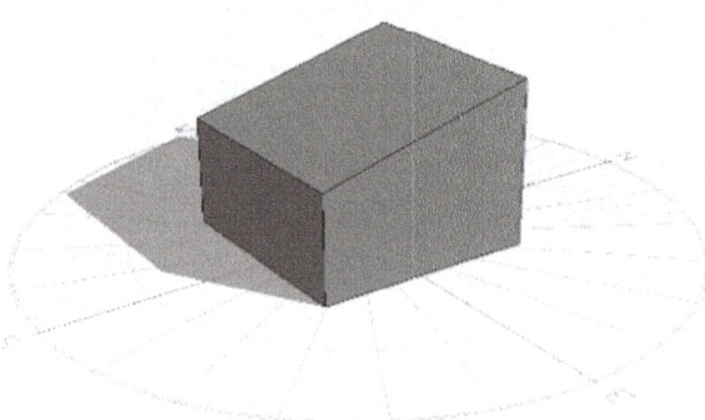

3. Rectangular plan with monitor roof

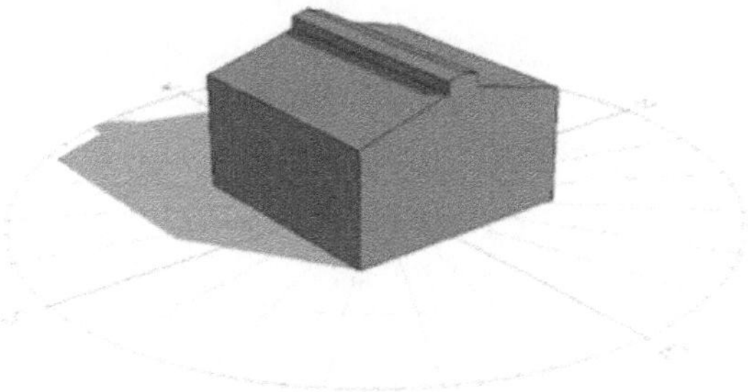

4. Rectangular plan with saw-tooth roof

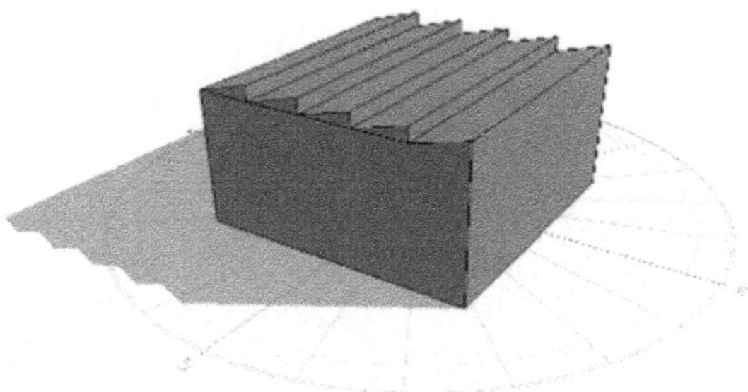

20.5.6 Self-Shading Factor

Even Partial shading can cause significant efficiency losses in BIPV module output. Considering this, a shadow range analysis is carried out on the different roof types of textile industries using Autodesk® Ecotect® Analysis software to determine areas of self-shading.

20.5.6.1 Gable Roof

From the above figure the entire northern side of a 'rectangular plan and gable roofed' building is observed to be covered in shadows, with no area of the building envelope being self-shaded (Fig. 20.6).

20.5.6.2 Lean-to/Shed Roof

Result from shadow range analysis for a 'rectangular plan with lean-to roof' building shows the entire northern side of the building covered in shadows, with no area of the building envelope self-shaded (Fig. 20.7).

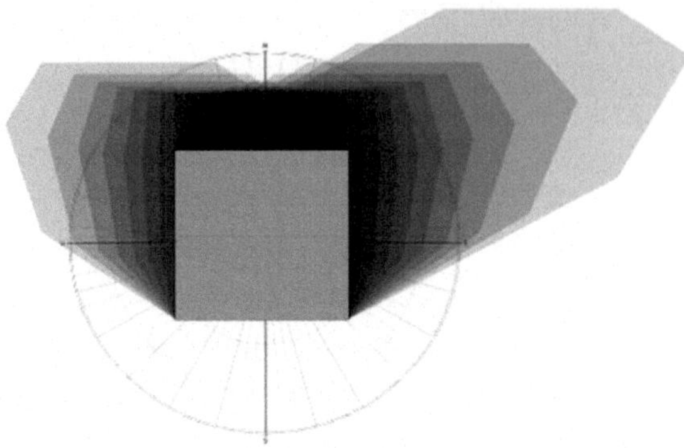

Fig. 20.6 Shadow range analysis from Jan 1st to Dec 31st (Source: Autodesk Ecotect output)

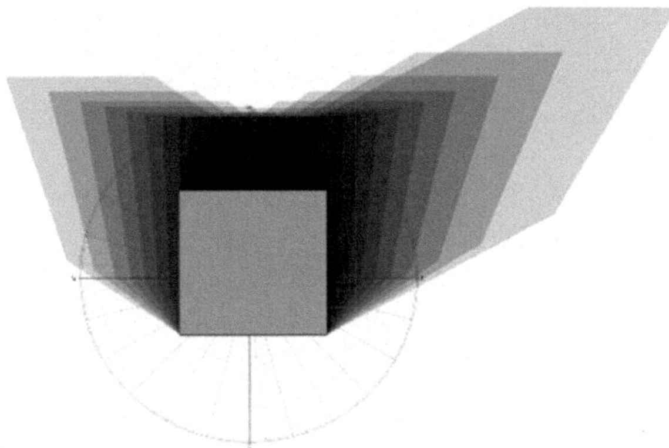

Fig. 20.7 Shadow range analysis from Jan 1st to Dec 31st (Source: Autodesk Ecotect output

20.5.6.3 Monitor Roof

From the above figure the entire northern side of a 'rectangular plan and monitor roofed' building is observed to be covered in shadows, with about 19% of the northern roof self-shaded during the months of November through December, January to February and over 11% self-shaded in the months of March and October. While up to 6% of the southern roof gets shaded in the months of June and July with no self-shading during the months of April and September (Fig. 20.8).

20.5.6.4 Saw-Tooth Roof

From the above figure the entire northern side of a 'rectangular plan and monitor roofed' building is observed to be covered in shadows, with about 19% of the northern roof self-shaded during the months of November through December, January to February and over 11% self-shaded in the months of March and October. While up to 6% of the southern roof gets shaded in the months of June and July with no self-shading during the months of April and September (Fig. 20.9).

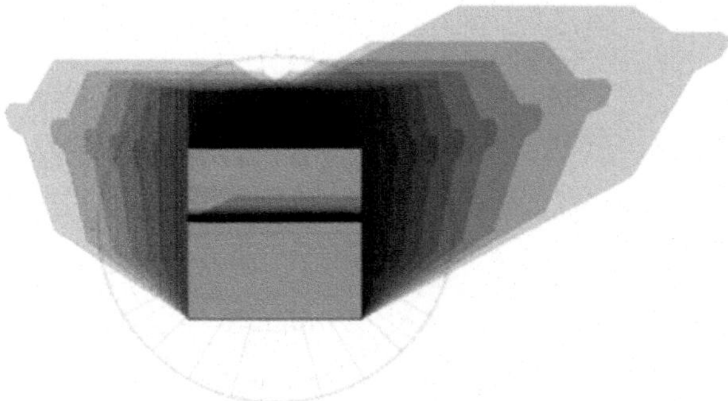

Fig. 20.8 Shadow range analysis from Jan 1st to Dec 31st (Source: Autodesk Ecotect output)

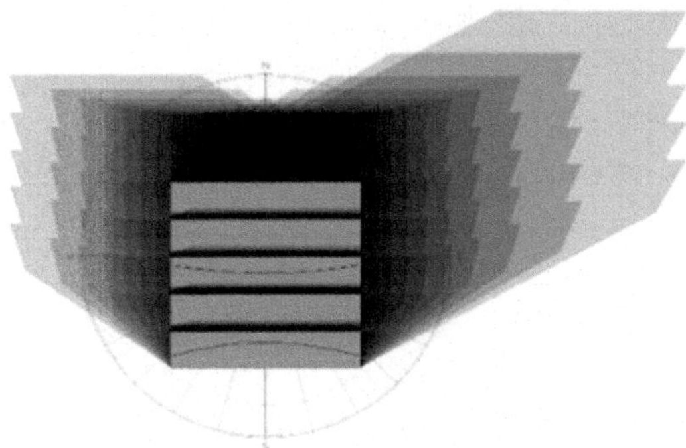

Fig. 20.9 Shadow range analysis from Jan 1st to Dec 31st (Source: Autodesk Ecotect output)

From the above figure the entire northern side of a 'rectangular plan and saw-toothed roofed' building is observed to be covered in shadows, with about 25% of the northern roof self-shaded during the months of November through to February, over 13% is self-shaded in the months of March, October and no shading during the months of April, June, July and September.

20.5.7 Building Form Performance

Based on the results obtained from the shadow analysis carried out on the established building roof types which are gable roof, lean-to roof, monitor roof and saw-tooth roof, lean-to and gable roofs perform better with no self-shading, while monitor roof performs better than the scissor-tooth roof having about 19%, 11% and 6% shaded during the months of November to February, March/October and June/July, respectively. While scissor-tooth roof has about 25%, 13% and 8% of its surface shaded during the months of November to February, March/October and June/July, respectively.

Building forms	Orientation	Self-shading	Generation potential
	100% of roof surface facing south	No Self-shading	47.40 kWh
	Only 50% of roof surface faces south 50% faces north	No Self-shading	45.15 kWh
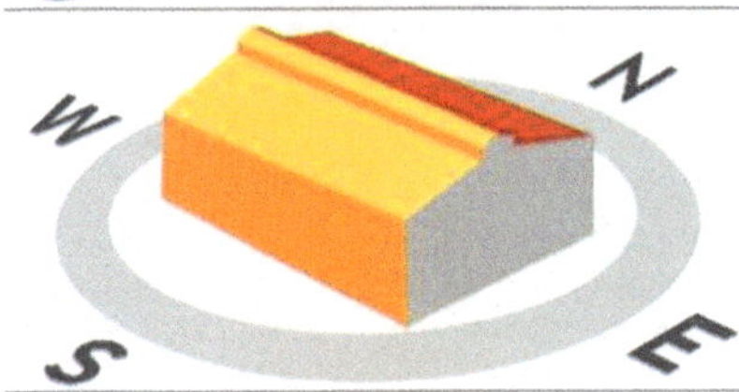	Only 50% of roof surface faces south 50% faces north	• 16% shaded November • 9% shaded March • 5% shaded July	39.80 kWh
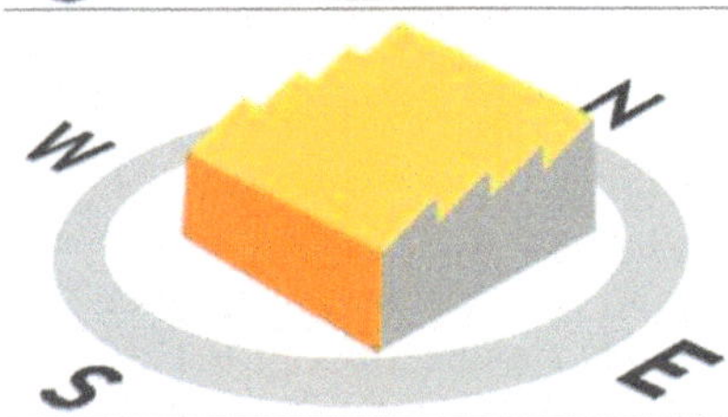	100% of roof surface facing south	• 25% shaded November • 15% shaded March • 9% shaded July	37.92 kWh

20.5.8 Orientation

Orienting the building so the longer side or side with the largest area faces the orientation with highest energy potential is key. A simulation analysis is carried out to determine this orientation. The simulations test for the solar energy potential for surfaces facing North, South, East and West, respectively, and for the exact azimuth that gives the highest potential.

20.5.8.1 Simulation Results

The energy potential of the BIPV systems is tested at four different orientations (East, West, South and North). A crystalline silicon cell is used with an assumed efficiency of 15% (average commercial efficiency). The simulation analysis was conducted using PVGis® tool by Joint Research Center (JRC).

Assumptions while carrying out tests are:

Nominal power of BIPV system: 1.0 kW
Efficiency of BIPV cell: 15%
Losses (cables, inverters): 14%
Surface area of BIPV system: 1m^2
Tilt angle of BIPV system: 15°

The results/performance are presented in graphical charts below:

The figure above represents crystalline silicon cell performance when placed facing four different orientations (East, South, West and North). It performs best when faced due south producing an average of 18.95 kWh/m^2 of electricity each month (Fig. 20.10).

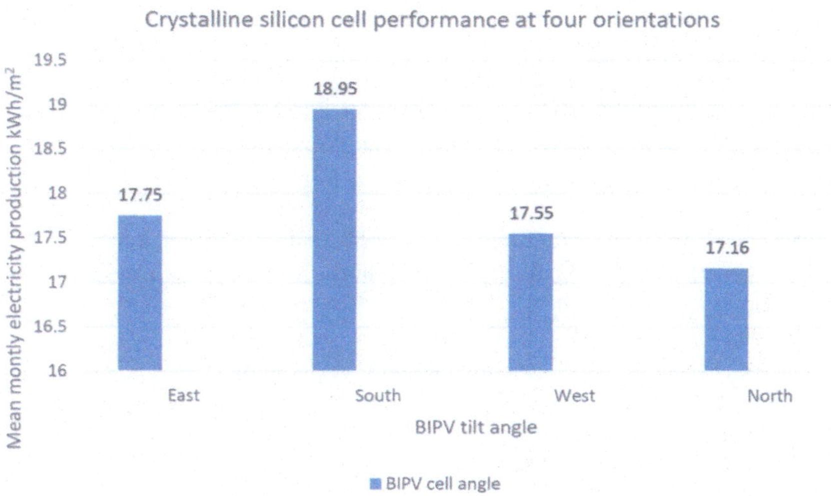

Fig. 20.10 Crystalline silicon cell performance at the various orientations

20.5.9 Tilt Angle

BIPV performance for different tilt angles at different orientations would be studied. This would be used to define the angles at which the building surfaces would be tilted. Documentation would be made on this and the simulation analysis result presented for each respective tilt angle to show performances.

20.5.9.1 Simulation Results

The energy potential of the BIPV systems are tested at varying tilt angles for each of the four orientations (East, West, South and North). A crystalline silicon cell is used with an assumed efficiency of 15% (average commercial efficiency). The simulation analysis is conducted using PVGis® tool by Joint Research Center (JRC).

Assumptions while carrying out tests are:

Nominal power of BIPV system: 1.0 kW
Efficiency of BIPV cell: 15%
Losses (cables, inverters): 14%
Surface area of BIPV: 1m^2

The results/performance are presented below:

The figure above compares the mean monthly electricity production of commercial crystalline silicon cells at various tilt angles when placed facing due south, west, north and east, respectively. It could be observed that the cells, when placed due south, perform best at all tilt angles. The peak electricity production of 18.95 kWh/m^2 occurs when the cells are placed due south and tilted at 16°.

The results shows that a square metre of BIPV cells when faced south would produce up to 18.95 kWh of electricity when tilted at 16° which is optimal for this orientation, this value varies with varying tilt angle as shown in Fig. 20.11. At 90° it produces 9.5 kWh of electricity.

A west facing crystalline silicon BIPV cell as shown in Fig. 20.11 produces up to 17.55 kWh/m^2 of electricity tilted at 15° and 8.5 kWh/m^2 tilted at 90°.

From Fig. 20.11 it can be observed that a north facing crystalline silicon BIPV cell produces 17.16 kWh/m^2 of electricity when tilted at 15° and 4.24 kWh/m^2 tilted at 90°.

A crystalline silicon BIPV cell oriented due east as shown in Fig. 20.11 produces 17.75 kWh/m^2 of electricity at a 15° tilt and 8.91 kWh/m^2 when tilted at 90°. The production when placed facing south is 1.2 kWh/m^2 better than when oriented due east, 1.4 kWh/m^2 better than when oriented due east and 1.79 kWh/m^2 better than when oriented due north.

The results show that optimum electricity for crystalline silicon BIPV cells is generated at a module tilt angle of 16° and a module orientation of −4° having the potential to generate an average of 18.95 kWh/m^2 of electricity monthly.

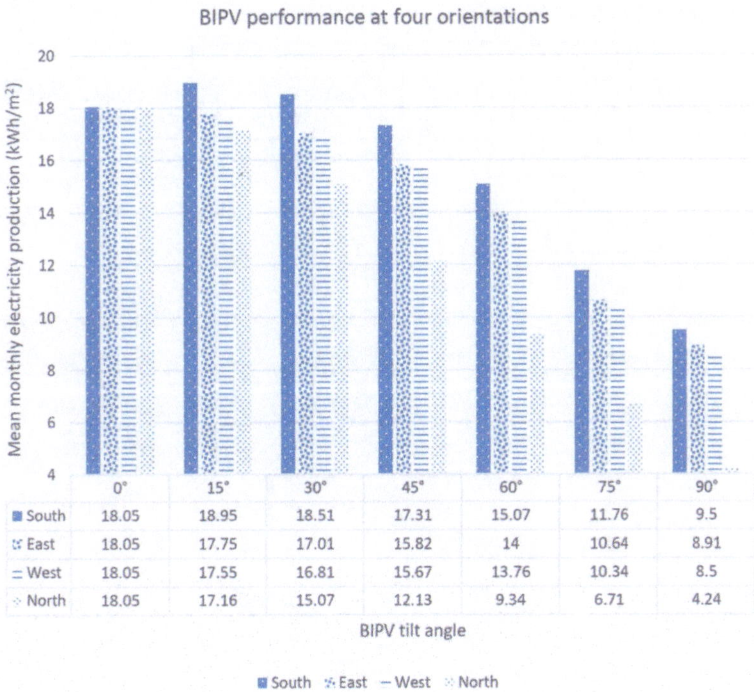

Fig. 20.11 BIPV cells performance at four different orientations

20.6 Summary of Findings

- The results show that solar photovoltaic systems could be integrated into the tilted roof, skylight, façade walls, façade glazing and external device of the factory's building envelope.
- Rectangular plans with gable, monitor, lean-to and scissor-tooth roofs are the building forms predominantly used for textile factories in tropical dry climates.
- Crystalline silicon cell has the potential to generate an average monthly of 18.95 kWh/m^2 of electricity.
- Optimum electricity is generated when modules are placed on surfaces with a tilt angle of 16° and an orientation of −4° (South −0°).
- Lean-to roof with rectangular plan performs best with no self-shading and provides the largest surface for BIPV integration (Fig. 20.12).

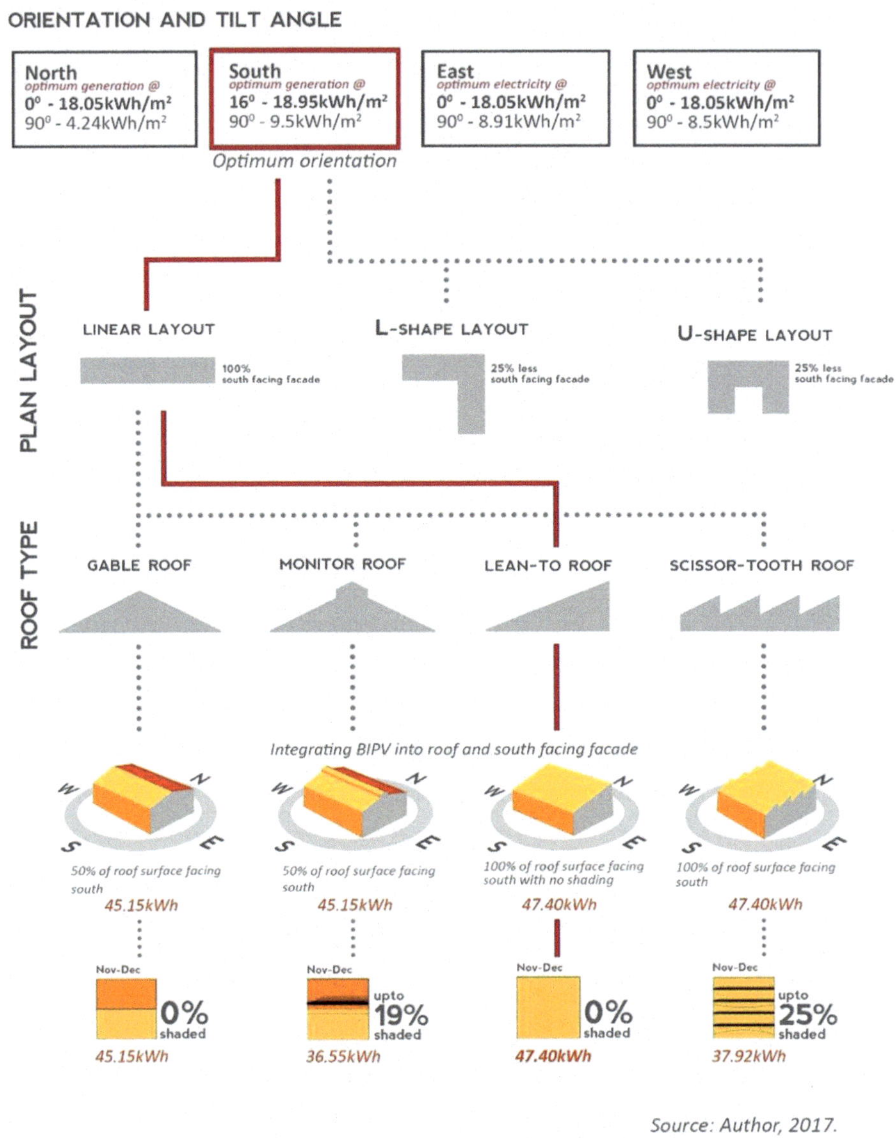

Fig. 20.12 Design framework for optimizing building form (Source: Author, 2017)

References

1. Oyedepo SO (2012) Energy and sustainable development in Nigeria: the way forward. Energy Sustain Soc 2:15
2. U.S. Energy Information Administration (2013) International energy outlook 2013. U.S. Energy Information Administration, Washington, DC
3. Ibikunle TS (2014) Problems crippling manufacturing in Nigeria. In: Ibikunle TS IPWA Plc MD, Vanguard
4. Energy Information Administration (EIA) (2013) Electricity use by machine drives varies significantly by manufacturing industry. U.S Energy Information Administration, Washington, DC
5. Allen SR, Hammond GP (2008) Energy analysis and environmental life cycle assessment of a micro-wind turbine. J Power Energy 222(A7):669–684
6. Lazarus N (2009) A practical guide to producing affordable carbon neutral developments. BioRegional Development Group
7. International Energy Agency (2013) Report T.41.A.3/2: IEA SHC task 41 designing photovoltaic systems for architectural integration. SHC Programme
8. European Photovoltaic Technology Platform (2007) Building integrated photovoltaics: a new design opportunity for architects. European SUNRISE Project
9. Aaditya G, Mani M (2013) Climate-responsive integrability of building-integrated photovoltaics. Int J Low Carbon Technol 8(4):271–281
10. International Renewable Energy Agency (IRENA) (2012) Solar photovoltaics. Renewable energy technologies: cost analysis series, Volume 1: Power Sector (4/5)
11. Uzoma CC, Nnaji CE, Ibeto CN, Okpara CG, Nwoke OO, Obi IO et al (2011) Renewable energy penetration in Nigeria: a study of the south-east zone. Continental J Environ Sci 5(1):1–5
12. Waseef A (2014) Tilt and orientation: a preference factor among photovoltaic roof systems. Int J Innov Res Sci Eng Technol 3(3):10750–10758

Chapter 21
Storage for Community Electricity: A Comparison Between Batteries and Mini Pumped Hydro

Gülce Onbaşılı, Arthur Williams, and Sandeep Dhundhara

21.1 Introduction

This investigation is focussed on the study of microgrids operating in island mode for remote communities that are beyond the reach of national transmission systems. We assume that the electricity is supplied from renewable energy sources, which could be solar or wind or a hybrid system, where both resources complement each other to provide higher reliability levels. Wind and solar are intermittent resources; therefore they need additional systems in order to be able to supply the loads and to control the stability of the system. In some cases diesel or petrol generators are used as a backup for the peak load and/or to work as a backup when the renewable generation is scarce. However, in these remote locations, they are not recommended due to the uncertainty in the supply and high cost of fuels in those isolated regions [1]. Maintenance costs for backup generators are also high in remote locations, so battery storage is the most common method to supply reliable electricity for these situations.

For national electricity grids, pumped hydro is the main source of storage, as it the most cost-effective option, but at much smaller scale, this technology is not competitive with batteries. There are two reasons for this: firstly, cost per unit of hydro storage increases significantly for smaller systems; secondly, it is difficult to maintain system efficiency above that of battery storage when scaling down.

G. Onbaşılı
School of Engineering, University of Lincoln, Nottingham, UK

A. Williams (✉)
Department of Electrical and Electronic Engineering, University of Nottingham, Nottingham, UK
e-mail: Arthur.Williams@nottingham.ac.uk

S. Dhundhara
Department of Electrical and Electronics Engineering, University Institute of Engineering and Technology, Panjab University, Chandigarh, India

A. Sayigh (ed.), *Renewable Energy and Sustainable Buildings*, Innovative Renewable Energy, https://doi.org/10.1007/978-3-030-18488-9_21

The aim of this paper is to identify the size at which mini pumped hydro becomes competitive with battery storage.

The paper employs case studies where mini pumped storage would be an option, and compares the cost-effectiveness relative to batteries. Two communities were studied, one in Nepal and the other in Indonesia, where solar PV could provide the main electricity demand, boosted by some wind power. Daily energy storage was designed based on pumping water up from a river to a dedicated storage tank situated at the top of each village. Detailed costs and efficiencies were calculated for pumped hydro installations of different capacities, and compared with battery costs (for both lead-acid and lithium-ion technologies) using HOMER software.

21.1.1 Battery Characteristics for Energy Storage

In comparing batteries, it is important to take into account parameters that determine the effective capacity and economic use of the battery. Battery manufacturers specify the coulometric capacity (ampere-hours), which is defined by the discharge from a fully charged state until the terminal voltage drops to its cut-off voltage. The ampere-hours capacity is variable as a function of the discharging current rate and the depth of discharge (DoD). DoD is the conjugate of the state of charge (SoC) or the battery, where SoC is defined as the ratio of the remaining capacity to the fully charged capacity. Thus, when a battery is fully charged, the SoC is 100% and DoD is 0% and vice versa. For lead-acid batteries, the economic DoD is around 50%, otherwise the lifecycle of the battery is significantly reduced. This is the limit that has been applied in the case studies.

Lead-acid batteries are the most commonly used for village microgrids. They have been the lowest cost battery technology and are well established in the market. Lithium-ion technology is a new contender for this application. It has high energy densities in the range of 80–150 Wh/kg, making it the technology of choice for electronic equipment such as laptops, cameras, cell phones, etc. and also in electrical vehicles. It is a developing technology with high potential for new advances. The advantages of this technology over lead-acid are the higher energy density, higher lifecycle, higher DoD (typically 90%) and efficiency (up to 95%). Normally the use of Li-ion batteries requires a control strategy and a security system including overload, excessive discharge, overcurrent, short circuit, high temperatures and overvoltage. The main disadvantage currently is the greater initial cost per unit of storage, impeding its greater penetration into off-grid systems. Figure 21.1 presents a comparison between these technologies and others available in the market.

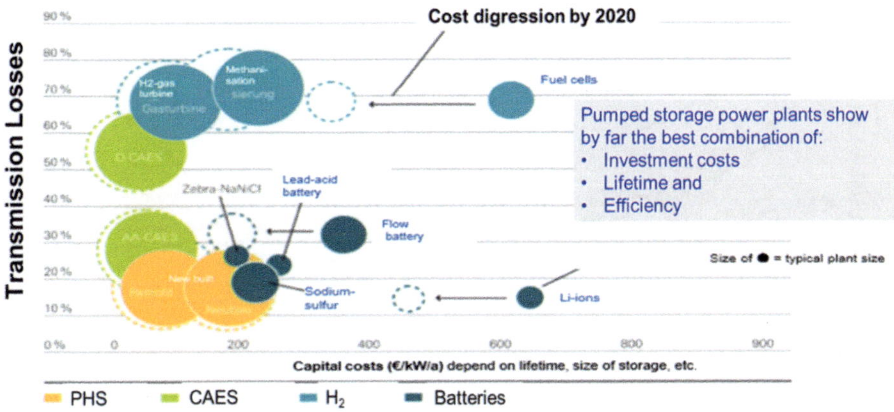

Fig. 21.1 Comparison between different storage technologies (adapted from [2])

21.1.2 Pumped Hydro Storage

Pumped hydro storage (PHS) allows excess electricity generated at off-peak times to pump water from a river or lower reservoir up to a higher reservoir for using when it is needed. In large-scale electricity grids, they are widely accepted as the main method of storage due to high efficiency and relatively low cost. Fast response times of the order of a few minutes mean that they can work well to balance intermittent renewable sources, solar and especially wind. Pumped hydro storage plants are generally more expensive than conventional large hydropower schemes with storage, and the costs are site specific.

The technology has significant economies of scale, which is why the minimum size for grid-connected PHS plants is around 200 MW. Small hydro projects have generally higher investment costs per kW of capacity, increasing if the plant has lower head, as shown in Fig. 21.2. Plant efficiency decreases with decreasing size, so it is expected that for very small-scale energy storage, pumped hydro will not be competitive with batteries.

Once commissioned, hydropower plants require little maintenance, and operation costs will be low. Annual O&M costs are often quoted as a percentage of the investment cost per kW per year. Typical values range from 2.2% for large hydropower and 2.2% to 3% for smaller projects [3]. Some studies indicate fixed O&M costs represent 4% of the total investment cost which may be more suitable number for small-scale production. Lifetime is relatively long, particularly for large-scale systems, with 30 years or more for the electromechanical equipment and 50 years or more for the refurbishment of penstocks and tail races.

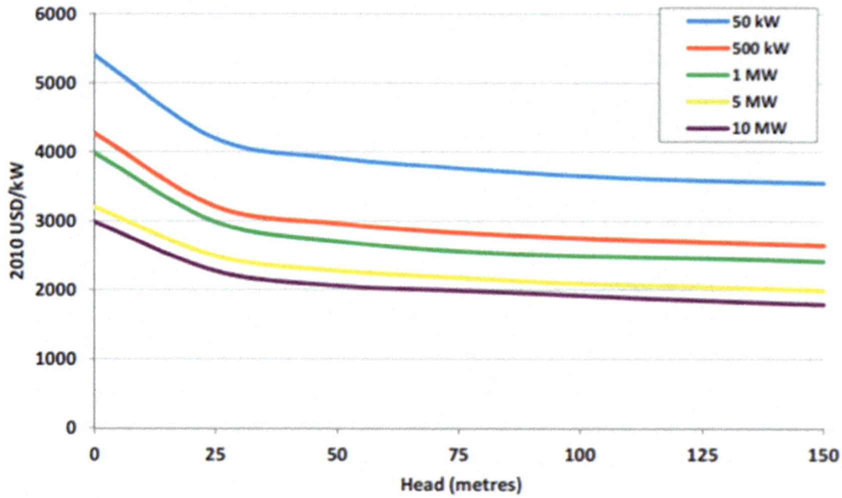

Fig. 21.2 Investment costs for small hydro as a function of installed capacity and turbine head [3]

21.2 Methodology

HOMER, the Hybrid Optimization of Multiple Energy Resources, is the most widely used software for simulation and optimization of renewable energy based mini-grids [4]. This software is suitable for the current investigation as it is aimed at economic optimization. It uses hour-by-hour modelling of available resources and loads, which is a reasonable time step for consideration of daily storage of energy. Two separate case studies, based on actual locations, were chosen to carry out the study, one in Nepal and one in Indonesia. Different sizes of storage were then compared using PHS and batteries to assess the capacity at which the costs overlap.

At the village Hanku (Nepal), the main sources of energy are kerosene and fuel wood for lighting and cooking. Radios are powered with batteries. Hanku's population consists of 2200 inhabitants and 440 households, with the houses located close by each other. It is in the district of Jumla, which is one of the areas with lowest electrification rate in Nepal. The site is suitable for micro hydro pumped storage as it has a lake and, as with many villages in Nepal, a high head available between the lake and the lowest part of the village.

The village chosen in Indonesia is in East Papua, which has the lowest electrification rate in the whole country, at 30% [5]. The specific village chosen is Tambat in Tanah Miring district, due to availability of water sources: a lake and sufficient head for PHS application. The village is larger than Hanku with nearly 4500 inhabitants in 920–950 households. Nevertheless, in the costings, materials and labour for constructing an upper storage tank are included in each case, as few villages will have a natural water store at the high level.

For the HOMER simulations, the solar radiation and wind speed have been uploaded for the exact geographic coordinates of the villages' locations. Solar outputs were based on a recommended tilt angle of 30°, which gives good output and enables some self-cleaning of the modules. Wind data was adjusted for the actual turbine hub heights (10 or 15 m). Some assumptions were made about the loads to be supplied. The programme was run with the load sensitivity selected from different ranges between 0.11 kWh per household per day and 0.57 kWh per household per day for both villages. Average load was decided as 0.37 kWh per household per day, which is fairly generous for village electrification, allowing for additional loads over and above lighting since one LED of 10 W used for 5 h is only 0.05 kWh/day. Other case studies in the literature note that demand tends to grow once electricity is installed. For Nepal, with Hanku having a colder climate than Indonesia, additional load is based on small mat heaters or electric blankets that are 70 W. For Tambat, the hot climate all year round requires fans of a similar rating. Other types of loads with low consumption, such as radios and mobile phone chargers, could be available in every house.

Time series data of the loads have to be indicated to HOMER. For this study, a synthetic load from a typical "community" profile was chosen. The load is projected to have a cyclic annual variation taking into account the month expected to have the highest energy consumption, which is January for Hanku to fit with ambient temperature profiles in Nepal. Figure 21.3 presents the load profile of the community with the mentioned seasonal profile variation; the average consumption per day of this profile is 165 kWh with average load of 6.88 kW and peak of 24.57 kW.

Random variability of loads is accounted for by using the 'day-to-day' option, which causes the size of the load profile to vary randomly from one day to the next, but keeping the same shape. The 'time step' disturbs the shape of the load profile without affecting its size. Values of 10% and 20%, respectively, are considered in this work. Lastly, HOMER permits scaling the baseline profile to perform sensitivity analyses. For this, scaled daily average values of 50 kWh, 100 kWh, 165 kWh and 250 kWh were set, which takes into account uncertainty in the consumption profiles.

HOMER uses costs and performance based on actual batteries, solar modules and wind turbines. For the PHS, detailed costings were carried out based on the size of the storage tank and length and diameter of penstock pipe, using local labour and materials costs in each case.

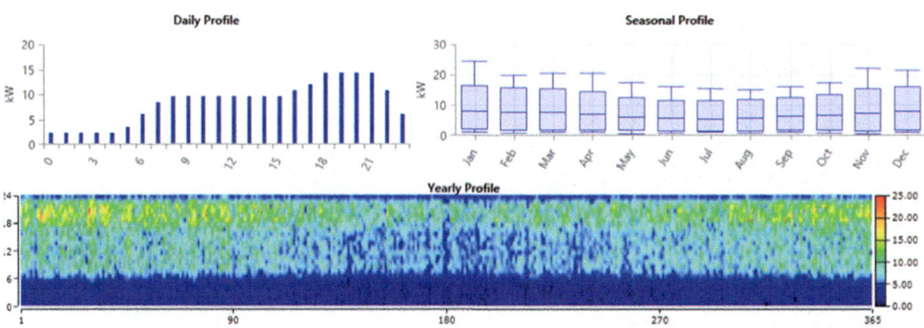

Fig. 21.3 Community load profile

21.2.1 Pumped Hydro Configuration

The main components of a pumped storage hydro scheme are shown in Fig. 21.4. For most modern systems, the turbine and pump are a single reversible unit, which is directly connected to the generator/motor. Some schemes use separate units for pump and turbine, which can be connected to the generator/motor alternately. This system is more complex but may have a higher efficiency, since unit units can be hydraulically optimized. The layout with a reversible pump/turbine reduces the construction cost up to 30% but may have lower efficiency (around 2% for large systems) [6].

In large-scale pumped storage systems, the reversible pump/turbines are designed for the specific site and have variable guide vanes to control the flow. Some systems have also incorporated one unit with variable speed in order to improve efficiency at part flow, the range of head and the dynamic response time of the plant [7]. Using a pump as turbine (PaT) is a cost-effective solution for electrification for isolated communities and is therefore an obvious choice for use as a pump-turbine in a mini pumped storage project. A previous case study compared crossflow turbines against a PaT taking into account energy generation equipment, civil works and energy distribution system. This showed that a 2 kW PaT system offers a 53% cost reduction for energy generation equipment compared to crossflow turbine [8]. The efficiency of pump in turbine mode is usually lower than that of conventional hydro turbines; however, the advantages of installation simplicity (a reversible pump-turbine) outweigh this limitation. Particularly up to 100 kW capacity power plants, the use of PaT may be justifiable because, even though efficiency of PaT is lower, its use will lead to significant reduction in the capital cost of the plant. In this range, the investment cost for conventional hydro turbines is relatively high, and the payback period can be as high as 15 years, which can be reduced to 3 years using PaT [9].

For the application in mini pumped hydro, a centrifugal pump was selected that can operate in turbine mode without any modification of mechanical components. A standard centrifugal pump works in turbine mode economically in the range from 13 to 75 m of head. Above this range the efficiency reduces, but lower heads mean higher flows, which result in a large and expensive pump. Actual pumps were selected for this application from the KSB range.

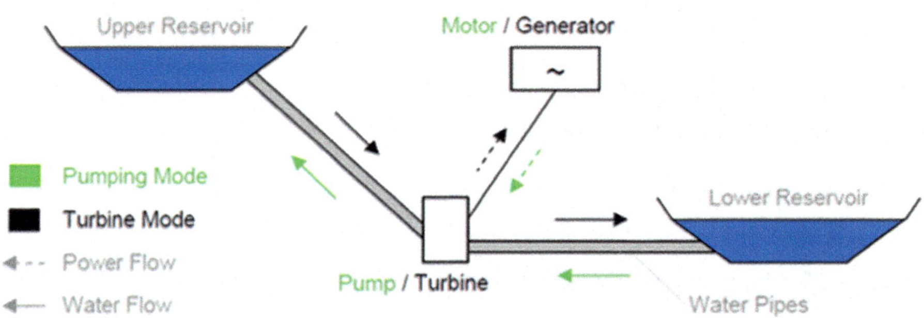

Fig. 21.4 Main components of a pumped hydro storage (PHS) system [6]

Each mode of operation of the reversible PaT has a different operating characteristic. As a turbine, the head-flow characteristic of the PaT follows a rising curve, which is fixed for each rotational speed, since there are no adjustable guide vanes. Below a certain flow, there is no output, because the turbulence losses are greater than the fluid input power.

An example set of performance curves is shown in Fig. 21.5. In turbine mode, a standard pump operates with higher head *(H)* and discharge (flow rate, *Q*) for the same rotational speed *(N)* relative to its pump performance. The efficiency of the turbine is typically 3% less than the pump's efficiency [11].

A PaT system generally uses the pump's induction motor as an AC generator. For stand-alone installations, capacitors are required to provide reactive power that allows the pump's motor to generate AC electricity. Turbine performance of the selected pumps was predicted using Sharma's method [11].

Sharma's prediction was expanded to take into account the changing speed using the pump affinity laws, i.e.

$$Q_t = N_t / N_p \times Q_{\text{bep}} / \eta_{\max}{}^{0.8}$$

$$H_t = \left(N_t / N_p\right)^2 \times H_{\text{bep}} / \eta_{\max}{}^{1.2}$$

where bep refers to pump best efficiency.

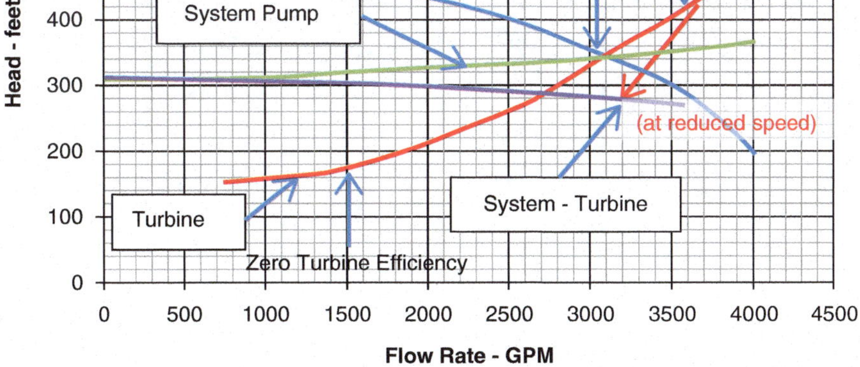

Fig. 21.5 Example pump and PaT curves (head vs. flow rate) [10]. Note: *bep* best efficiency point

21.3 Results and Discussion

The simulations of the microgrids to supply electricity for the basic needs of the villagers in Nepal and Indonesia are presented in this section. For Hanku in Nepal, the average daily load has been set to 165 kWh/day. Three different simulations were set up to compare lead-acid against lithium batteries, PHS against lead-acid and PHS against Li-ion batteries. Two of the architectures are shown in Fig. 21.6, with arrows showing the power flow directions. Using HOMER, the most economic choice of generator capacities was selected in each case, by changing the number of PV modules, batteries, etc.

After running around 800,000 different simulations over 6 h, the results for the battery comparison were obtained. Figure 21.7 shows the levelized cost of energy (LCOE) based on economically optimized system components. The LCOE is similar for the two battery types across a range of loads, despite the lithium technology being approximately 2.6 times more expensive than the lead one. This is due to the higher efficiency and greater DoD for the Li-ion battery, so less batteries and less generating capacity are needed for the same load. Another advantage of Li-ion batteries is their longer life. A recent study looking at four other case studies confirms the economic benefit of Li-ion batteries for microgrids [12].

For the comparison of lead-acid batteries with PHS, around 450,000 different simulations were run, using the architecture shown in Fig. 21.6b. The PHS has been placed in the DC exclusively because HOMER does not allow setting it up as an AC component. However, since the important results from this simulation are the energy balances, the impact of this point over the sizing of the microgrid is only the sizing of the converter, making the overall PHS system slightly more expensive than it would be in reality.

The same process was applied to obtain the results shown in Fig. 21.8. Here there is a difference in the LCOE between the storage technologies: from a daily load of

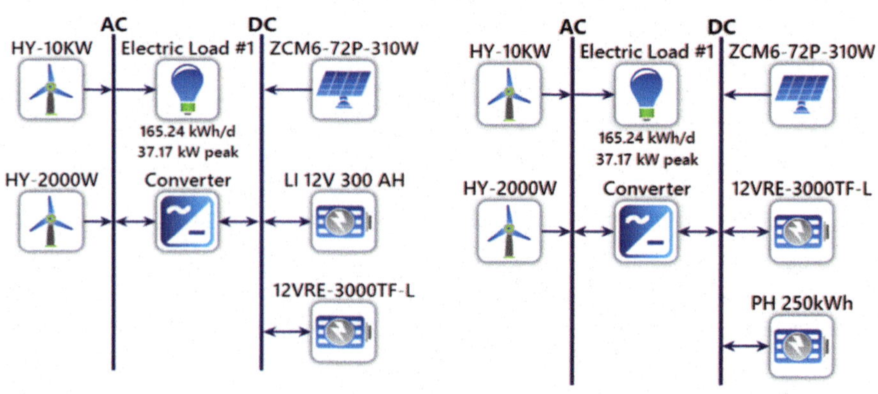

Fig. 21.6 Nepal hybrid system architecture for (**a**) lead vs. lithium batteries and (**b**) battery vs. PHS simulations

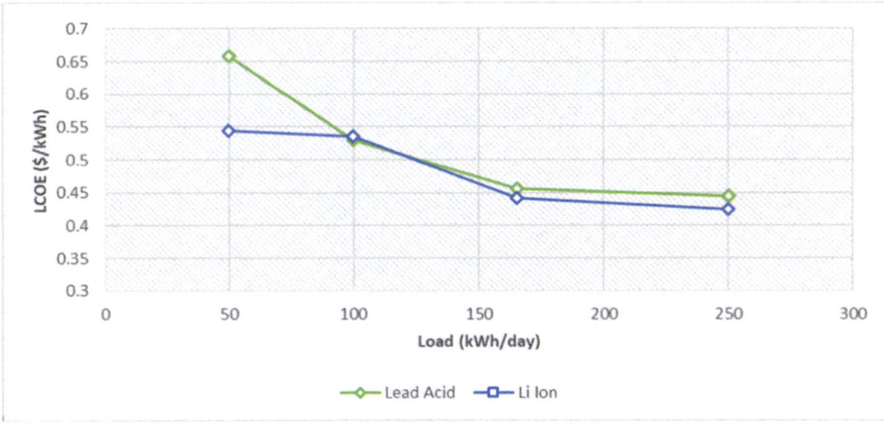

Fig. 21.7 Comparison of LCOE between lead-acid and lithium batteries using Hanku simulations

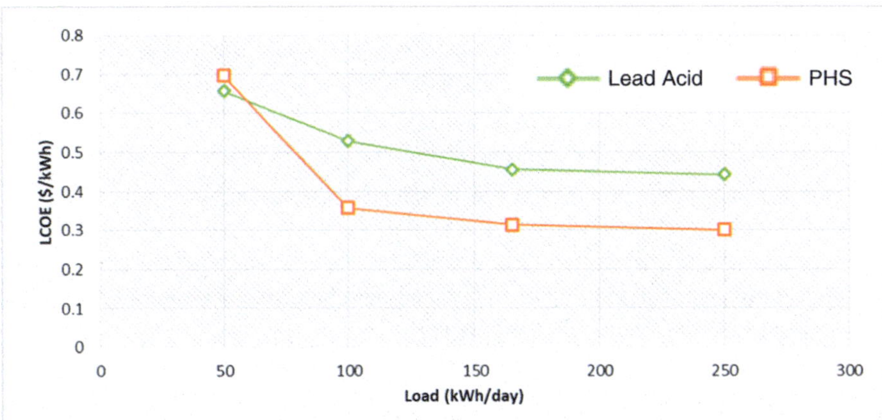

Fig. 21.8 Nepal LCOE for PHS 250 kWh vs. lead battery simulations

around 100 kWh/day, a PHS 250 kWh is cheaper than using lead-acid batteries, for example, for a load of 165 kWh/day, the unit electricity cost with PHS is around 10% lower. The net present cost of the PHS scheme is US $278,000 (USD) (see Table 21.1) rather than $408,000 required for a system with batteries.

As efficiencies for 165 kWh/day are comparable between PHS and lead-acid batteries, both use the same size of PV (93 kW), but the PHS has four 2 kW wind turbines, instead of five. A comparison between Li-ion batteries and PHS showed

Table 21.1 Sensitivity analysis for different loads

Sensitivity	Architecture							Cost			
Electric Load #1 Scaled Average (kWh/d)	ZCM6-72P-310W (kW)	HY-10KW	HY-2000W	12VRE-3000TF-L	PH 250kWh	Converter (kW)	Dispatch	COE ($)	NPC ($)	Operating cost ($)	Initial capital ($)
1.00	1.55			1		1.00	CC	$8.27	$44,837	$134.22	$42,842
50.0	23.3				1	10.0	CC	$0.614	$166,589	$350.46	$161,380
100	46.5				1	20.0	CC	$0.352	$190,691	$600.92	$181,760
165	93.0		5		1	30.0	CC	$0.310	$277,896	$1,551	$254,838
250	170	3	1		1	50.0	CC	$0.300	$406,042	$2,872	$363,353

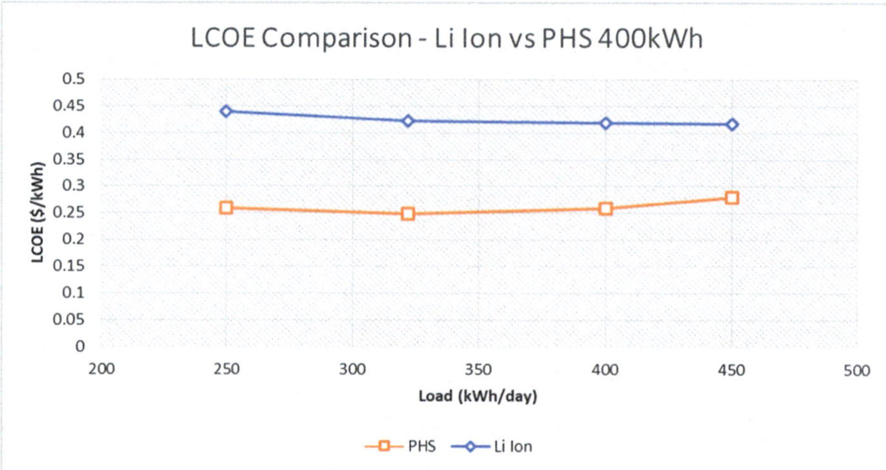

Fig. 21.9 Indonesia LCOE for PHS 400 kWh vs. lithium battery simulations

similar net present cost and LCOE to the lead-acid batteries, with the LCOE crossover against PHS at around 100 kWh/day.

A similar set of simulations were carried out for Tambat in Indonesia. Since it is a larger village, the average daily load is 322 kWh/day, based on scaling the profile in Fig. 21.3 and considering a peak load in July due to increased fan use at higher temperatures. The difference between the two battery types was negligible. Comparing PHS and battery, the LCOE values across a range of daily loads are shown in Fig. 21.9. Note that the LCOE for the PHS is similar to Hanku in Nepal. The increasing cost at 450 kWh/day load is due to the limit of 200 kW for the PHS power.

It is interesting to compare the breakdown of costs for the different sizes of PHS, which are shown in Fig. 21.10.

As the PHS size decreases, the cost of the penstock becomes a significant part of the total cost. The penstock pipe diameter was chosen in order to keep the head loss down to less than 1.5% of total head. For the 200–250 kWh systems, a 75 kW pump was chosen, which has an efficiency of 84% and motor efficiency of 95%, resulting in round-trip efficiency for the PHS of 59%. The round-trip efficiency falls to 44% for the smallest size considered in this study. Full results of efficiency and capital cost comparison are shown in Fig. 21.11.

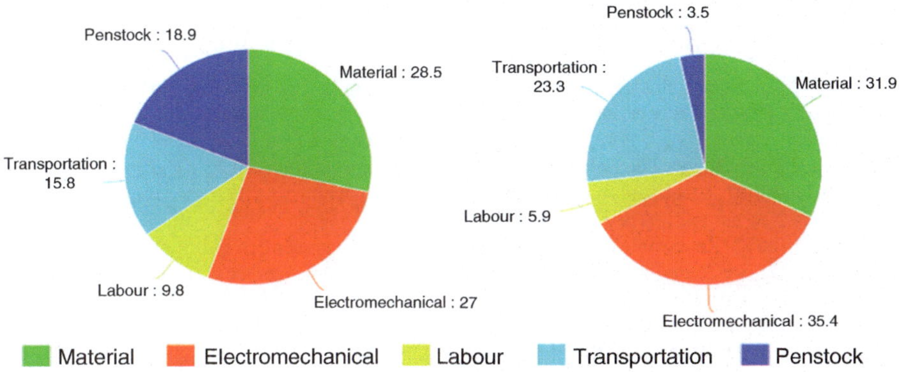

Fig. 21.10 Cost breakdown for PHS (as % of total), with installed capacities of 50 kWh and 400 kWh, respectively

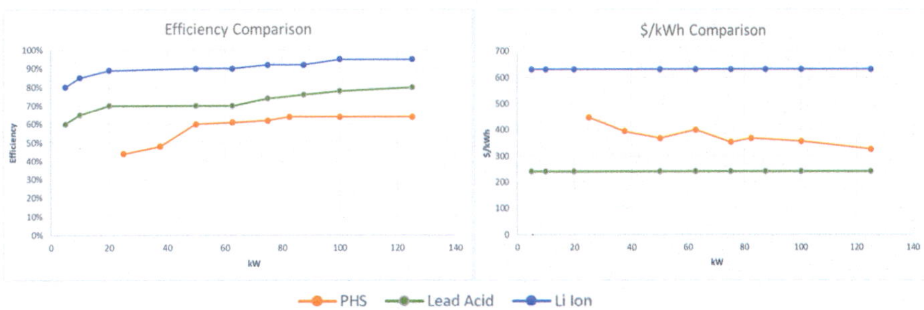

Fig. 21.11 Efficiency and capital cost vs. rated power for the three storage technology options

21.4 Conclusions

This study on small-scale pumped hydro schemes shows that they can be cost-effective in comparison to batteries. Based on the assumptions made, which have been designed to reduce head losses, PHS can be feasible storage systems for rural off-grid electrification, instead of batteries. This applied for places with daily average load of 100 kWh and above. Three constraints have been identified that can affect this conclusion. Firstly, PHS is highly site dependent and may not be feasible for every location. The communities selected for the study are representative of remote locations where grid power is unlikely to be available, but a reasonable head difference is required. Secondly, the pump as turbine with pole-changing machine may be a cheap and robust option, but it cannot be directly compared with batteries in terms of dynamic response. Variable speed technology, and maybe a separate pump and turbine, would increase the flexibility of the system but also increase cost and complexity. This technology is used for larger schemes to help stabilize the grid frequency, but is not available off the shelf for smaller systems. Finally, the costs of batteries, especially Li-ion, are decreasing, which means that they will be competitive for larger and larger storage systems.

References

1. Nepal Solar Energy Society (2014) Application of clean energy in Nepal: prospects and problems. SAR/StANCE Meeting, November 2014
2. Dietmar Reiner, Hydro Pumped Storage/ Lessons Learned from large-scale Alpine Pumped Storage Power Plants The Future Role of Energy Storage in South Eastern Europe,21.08.2014, Available:https://setis.ec.europa.eu/system/files/Hydropumped%20storage%20-%20lessons%20learned%20-%20%20REINER.pdf
3. IRENA (2012) Renewable energy technologies: cost analysis series. Hydropower 1(3/5). http://www.irena.org/documentdownloads/publications/re_technologies_cost_analysis-hydropower.pdf
4. Sunanda S, Shyam C (2014) Review of software tools for hybrid renewable energy systems. Renew Sust Energ Rev 32:192–205
5. Sinaga E Development of freight transport in Indonesia: towards sustainability. http://www.uncrd.or.jp/content/documents/7EST-P5-2.pdf
6. StoRe D2.1 Report summarizing the current status, role and costs of energy storage technologies. Dec. 2011. http://www.store-project.eu/
7. Chang X, Han M, Zheng CH (2015) Power control analysis for variable speed pumped storage with full-size converter. IECON2015-Yokohama, November 9–12, 2015, Paper 001327 (IEEE)
8. Arriaga M (2010) Pump as turbine—a pico-hydro alternative in Lao People's democratic republic. Renew Energ 33(5):1109–1115
9. Motwani KH, Jain SV, Patel RN (2013) Cost analysis of pump as turbine for pico hydropower plants—a case study. Procedia Eng 51:721–726
10. Budris AR (2010) Optimizing efficiency when pumps are operated as both pumps and power recovery turbines. WaterWorld. https://www.waterworld.com/articles/2010/10/optimizing-efficiency-when-pumps-are-operated-as-both-pumps-and-power-recovery-turbines.html
11. Williams A (1997) Pumps as turbines: a user's guide. Practical Action Publications, Rugby
12. Dhundhara S, Verma YP, Williams A (2018) Techno-economic analysis of the lithium-ion and lead-acid battery in microgrid systems at the final stages of review for publication in Energy Conversion and Management

Chapter 22
Urban Microclimate and Thermal Comfort in the Social Housing Districts of Rome: The Combined Effect of Built Form and Urban Materials

Michele Morganti and Federica Rosso

22.1 Introduction

Nowadays, cities are undergoing many challenges, given the unprecedented urban growth [1]. Urban heat island (UHI) effect has been studied by researchers in the past decades and has been demonstrated capable to increase urban temperatures up to around 10 °C [2]. As a consequence, UHI mitigation is of fundamental importance in order to ameliorate citizen's well-being and outdoor thermal comfort, by also increasing the time they spend outside, which on its turn could benefit social life. Moreover, this action has a positive effect on reducing energy consumptions, one of the primary objectives of the EU. In particular, in the Mediterranean area, outdoors are a direct continuation of citizens' houses, in the way that a large portion of time is daily spent outside, due to favourable climate conditions. Thus, the quality and improvement of comfort in the outdoor environment are not only a social but also a cultural factor that requires attention and appropriate design solutions focusing on UHI mitigation in the Mediterranean cities.

Previous studies investigated the effect of morphology and materials on energy consumptions and comfort, demonstrating that these characteristics of the built environment are able to influence both [3–5]. In this research, built form and

M. Morganti (✉)
Department of Archtecture and Urban Studies, Politecnico di Milano, Milano, Italy

Department of Architectural Technology, Architecture and Energy School of Architecture, UPC, Barcelona, Spain
e-mail: michele.morganti@uniroma1.it

F. Rosso
Department of Civil, Construction and Environmental Engineering, Sapienza University of Rome, Rome, Italy

A. Sayigh (ed.), *Renewable Energy and Sustainable Buildings*, Innovative Renewable Energy, https://doi.org/10.1007/978-3-030-18488-9_22

materials are taken into account with respect to their effect on urban microclimate and thermal comfort for pedestrians in the twentieth-century social housing districts of Rome, selected as representative of many others of Central-Southern Europe. Recently, given the unprecedented growth of cities and land consumption, urban areas and the existing building stock underwent a renovation process. Unfortunately, such renovation mainly focused on energy retrofit and building component and it was completely unaware of its environmental performance, missing the opportunity to apply existing knowledge and innovative design strategies in order to ameliorate urban microclimate conditions and to reduce building energy demand.

Considering that many inhabitants live in the above-mentioned social housing districts, which account for a large part of the urban footprint (from 9% to 15% [6]), studying effective strategies to renovate these urban areas would be beneficial, not only for environmental but also for social and economic reasons. Moreover, most of these districts were designed by celebrated architects and have unitary and recognizable characteristics in terms of urban form and construction features. Studies on these districts were carried out with respect to architectural and technological features, as well as to design their sustainable transformation and regeneration [6, 7].

Our work takes into account social housing districts of Rome: three case studies (*Pineto*, *Vigne Nuove* and *Torrevecchia*), characterized by similar features in terms of construction period and technology, urban fabric and climate conditions are analysed, with the aim of mitigating UHI through site-appropriate design renovation strategies for the outdoor areas between the buildings, which are a fundamental portion of the neighbourhood life. It is important to highlight that the microclimatic interaction occurring in the built environment requires to not disjoint these strategies from the building retrofit themselves: interventions on buildings' envelope will have an impact on the outdoors [3], while interventions on the outdoors will have an impact on buildings' energy demand [8, 9]. Simple interventions on the building envelope are hypothesized, while their effect is not investigated with respect to energy consumption, but only to outdoors thermal conditions.

In greater detail, based on previous studies, two improved scenarios were designed to assess the effectiveness of determined strategies in mitigating UHI and improving outdoor conditions for pedestrians. These strategies are (i) the creation of irrigated green areas, introducing additional trees and water ponds [10], and (ii) the modification of the outer layer of buildings' envelope from traditional material to cool paint. Our design proposal allows to exploit evapotranspiration (green and water), shading from direct solar radiation (trees) and cooler surface temperatures (cool paintings) and are considered with respect to the Mediterranean climate. Therefore, the aim of this preliminary research is to promote climate-sensitive urban renovation design strategies, taking into account the distinctive features of the social housing districts, and to improve their environmental, social and economic sustainability.

22.2 Case Study Districts

The districts selected as case studies—*Pineto*, *Vigne Nuove* and *Torrevecchia*—are located in Rome; thus they are subjected to the same climate conditions, typical of the Mediterranean climate. Moreover, these districts were developed by means of public investments as social housing, complying with Italian regulations: L.167/1962 about urban planning and L.373/1976, concerning building energy consumptions. Indeed, the construction period is that of the 1970s–1980s. More precisely, *Pineto* was built in 1975–1978, *Vigne Nuove* in 1972–1977 and *Torrevecchia* in 1979–1986. The common construction period and regulations allowed having a homogeneous urban layout, as well as peculiar construction systems and building technologies (Fig. 22.1). The districts consist of residential buildings, while only few buildings on the main axes have commercial uses on the ground floor. The construction system is based on reinforced concrete structure; prefabricated concrete slab as finishing façade layer, with internal insulation layer; and insulated flat roof. Public space between buildings is characterized by untended green areas, which are arid during the warm season, asphalt roads and sidewalks and generally asphalt parking areas. Trees are distributed in these green areas, usually close to the buildings and along the main street axes, while missing on the secondary roads, e.g. in *Torrevecchia*, and in smaller public areas. Actually, the outdoor areas of the districts are not completely exploited by citizens living there, while we base our study on the hypothesis that by improving the outdoor, public spaces conditions, we could increase the usability of these areas and thus benefit the social, economic and environmental sustainability of the neighbourhoods.

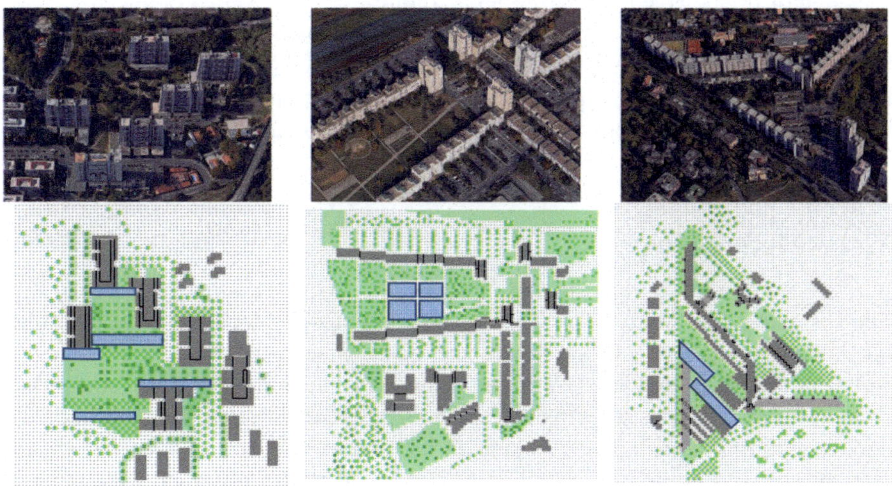

Fig. 22.1 Aerial views of the social housing districts selected as case studies and model plans of the proposed scenarios with urban design strategies for UHI mitigations: left, Pineto; middle, Torrevecchia; right Vigne Nuove

22.3 Urban Design Strategies for Improving Outdoor Urban Microclimate

Research has been conducted on the possible strategies to improve outdoor urban microclimate conditions, towards pedestrians' comfort conditions. Indeed, as described in the introduction section, the UHI is posing serious issues in cities, worsening citizens' comfort and threatening their health, as well as increasing energy consumption for cooling in buildings. The main physical factors influencing outdoor human comfort are air and mean radiant temperature, relative humidity, global solar radiation and wind velocity. Indeed, the built environment is able to strongly influence all these variables, depending on urban morphology and materials composing urban areas [11].

Among the most common solutions to improve outdoor thermal comfort, there are the use of green, under the form of grass, urban parks or trees; the use of water, under the form of fountains and ponds both in parks and squares/urban canyons; and the use of cool materials, under the form of external finishing layer of buildings' envelope or urban paving. In greater detail, greenery allows both shadings from solar radiation and cooling by means of evapotranspiration. Studies evaluated that while grassland is more effective to reduce mean radiant temperatures during the night-time (−1.5 K), trees and bushes are more effective during the daytime, lowering mean radiant temperature up to 6 K [12, 13]. Water body effectiveness in reducing heat stress in persons staying outdoor is due to their thermal capacity and to the cooling of urban temperatures by means of evapotranspiration [14]. Particularly, the positive effect of water ponds depends on their dimension: larger ponds' effect extends on wider areas, even if also 4-square-meter-large pond has been demonstrated able to lower air temperature.

Finally, cool materials on both vertical and horizontal urban surfaces reflect a large part of the solar radiation and thus maintain lower surface temperatures [15]. Depending on the urban form, especially in narrow urban canyons, they can cause thermal stress in pedestrians, when multiple reflections happen [3].

22.4 Methodology

The first part of the study consisted of characterizing key factors that affect urban microclimate for case studies. Then, the districts were modelled on *ENVI-met* software to carry out the assessment taking into account built form and urban materials. For each case study three scenarios, based on the same design criteria but in consideration of each district's peculiarities, were analysed: (1) *Scenario 0*, the reference case, i.e. the current situation, which has been observed by means of in-field surveys, maps and 3D tools; (2) *Scenario 1*, acting on the public space through the conversion of untended green areas into irrigated grassland, the addition of new trees and water ponds in the public areas; and (3) *Scenario 2*, applying cool paints on buildings' envelope and roofs.

More in details, the following design criteria have been applied to *Scenario 1*: (a) when missing, to add trees along sidewalks and parking areas; (b) to add large, but proportioned to the available space, water ponds in the abandoned parks, as close as possible to the surrounding buildings, in order to favour energy demand reduction as a side benefit; (c) to substitute untended green areas—arid during the hot season—with irrigated grassland areas; and (d) to add trees inside the park, close to the main paths. *Scenario 2* dealt with buildings' external envelope, by adding cool materials as finishing layer on all the vertical and horizontal buildings' surfaces. This choice was led by the observation that a large variety of cool materials exists; thus by applying a generically high value for solar reflectance on the simulation software, we did not apply any constraint to the built environment with respect to architectural variety. Indeed, even if cool materials were most often associated with light-coloured materials, recently many cool-coloured materials were developed.

The simulations werc implemented by means of the numerical three-dimensional microclimate model ENVI-met (v. 4.3.2). Details about models, simulation input data, building properties and simulation outputs are shown in Table 22.1. Results are presented considering the key factors that determine microclimate and outdoor thermal comfort: air temperature (*Ta*), surface temperature (*Ts*), relative humidity (*RH*), sensible heat flux (*SHF*) and PMV index [16].

22.5 Results and Discussion

The output of the simulations showed consistent results among the three districts. Specifically, considering *Pineto*, it has been demonstrated that all the above-mentioned factors were modified by these interventions, which were effective in improving thermal conditions for pedestrians during the 21st of June as a representative summer day, simulated at 15:00 in the afternoon (Fig. 22.2). In particular, the difference in *Ta* is equal to −1 °C in the park, and *Ts* were decreased up to 20 °C, while the difference in *RH* is about 6 points. The most effective measure in lowering temperatures resulted to be the presence of shading against solar radiation, whereas the addition of trees was particularly effective on the sidewalks.

In *Torrevecchia*, air temperatures were lower than in the other considered districts: we hypothesize this effect is due to the wind, which is allowed into the neighbourhood due to the open morphology. However, the above-described improvements permitted to lower significantly both air and surface temperatures. In terms of *Ta*, −2 °C is observed in the park, close to the ponds; −1 °C on the sidewalks close to the buildings; and −3 °C on the sidewalks close to the driveways. With respect to *Ts*, improvements are even more significant: up to −15 °C in the park and − 3 °C close to the buildings and on the sidewalks. The main positive effect can be noticed in the square enclosed by the buildings, where grassland and trees allow to reach −10 °C: this is a very good result, considering that this square is frequented, due to the position. The difference in *RH* is above 4 points.

Finally, in *Vigne Nuove*, the same results were assessed, with a decrease in air temperature up to 1 °C due to the addition of trees. Surface temperatures in the park

Table 22.1 Main settings of the microclimate simulation for Scenarios 0, 1 and 2.

Simulation model size	500 × 500 × 90 m			Simulation date June 21			Duration 06:00–16:00		
Grid cell size	Horizontally 5 m; vertically 3 m; telescoping factor $z = 0.20$						**Location** Rome (41.54 N, 12.30 E)		
Climate conditions	Wind speed (10 m height): 3.59 m/s Wind direction: 243°						**Climate data** NCEP		
	Scenario 0			**Scenario 1**			**Scenario 2**		
Soil properties									
Soil wetness (upper-middle-lower layer)	*10-10-30%*			*60-70-80%*			*10-10-30%*		
Vegetation and water ponds *(Pineto-Torrevecchia-Vigne Nuove)*	*P.*	*T.*	*V.*	*P.*	*T.*	*V.*	*P.*	*T.*	*V.*
Irrigated grassland (m^2)	*0*	*0*	*0*	*16,140*	*48,670*	*26,900*	*0*	*0*	*0*
Nonirrigated grassland (m^2)	*16,140*	*48,670*	*26,900*	*0*	*0*	*0*	*16,140*	*48,670*	*26,900*
Trees[a](number)	*183*	*420*	*353*	*457*	*816*	*676*	*183*	*420*	*353*
Water ponds (m^2)	*0*	*0*	*0*	*4040*	*5040*	*4780*	*0*	*0*	*0*
Building materials									
Absorption	*0.70*			*0.70*			*0.15*		
Reflection	*0.30*			*0.30*			*0.85*		
Output simulation data June 21 at 15:00	**Air temperature (Ta) Surface temperature (Ts)**			**Wind flow Sensible heat flux (SHF)**			**Relative humidity PMV index (post-proc.)**		

[a]*Populus alba*, *Pinus pinea*, pine and cypress for Scenarios 0 and 2; *Populus alba*, *Pinus pinea*, pine, cypress, privet, *Tilia cordata* and *Populus alba* for Scenario 1

were lowered up to −15 °C by inserting irrigated grassland, trees and water ponds. A noticeable improvement in terms of surface temperatures can be observed due to the addition of trees, lowering street temperatures from over 47 °C down to around 34 °C. The difference in *RH* is up to 6 points. Indeed, the most critical part of the urban layouts were those spaces enclosed by buildings, e.g. the square in *Torrevecchia* and the urban canyons both in *Torrevecchia* and *Vigne Nuove*. Here, the positive effects of the improvement strategies acting on public spaces were demonstrated.

The design solutions of Scenario 2 (cool materials) had no significant effect on outdoor conditions, due to the open morphology of the considered districts, which permits wind to flow by while avoiding solar trapping.

The Scenario 1 introduces the most effective UHI mitigation strategies for all the case studies. The results of the key factors and their spatial distribution proved a significant positive effect of the urban design solutions on the local microclimate (Fig. 22.2 and 22.3). Shading, wind direction and speed are crucial in determining the most favourable microclimate condition at different hours.

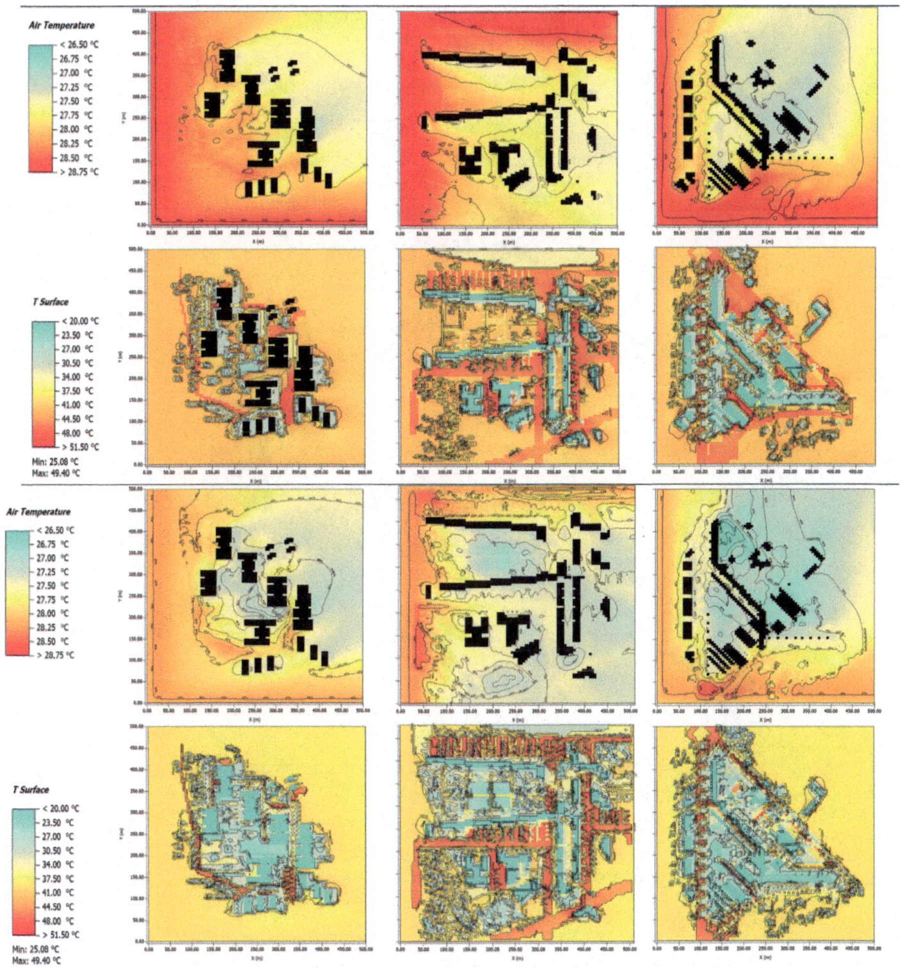

Fig. 22.2 Maps of Ta and Ts + SHF for Scenarios 0 (top) and Scenarios 1 (bottom) at 1.5 m height: Pineto (left), Torrevecchia (middle) and Vigne Nuove (right)

22.6 Conclusions

This study allows opening fruitful discussion about the possible strategies to adopt for improving the livability of social housing districts built in the 1970s–1980s, which constitute a large portion of Italian suburban building stock. The effectiveness of different urban design strategies has been investigated, in order to foster outdoor comfort conditions and to mitigate UHI. The microclimate simulations of three case study districts, performed with ENVI-met software, demonstrated a significant variation of relative humidity and air and surface temperatures due to the implementation of site-appropriate urban design strategies, such as green and blue

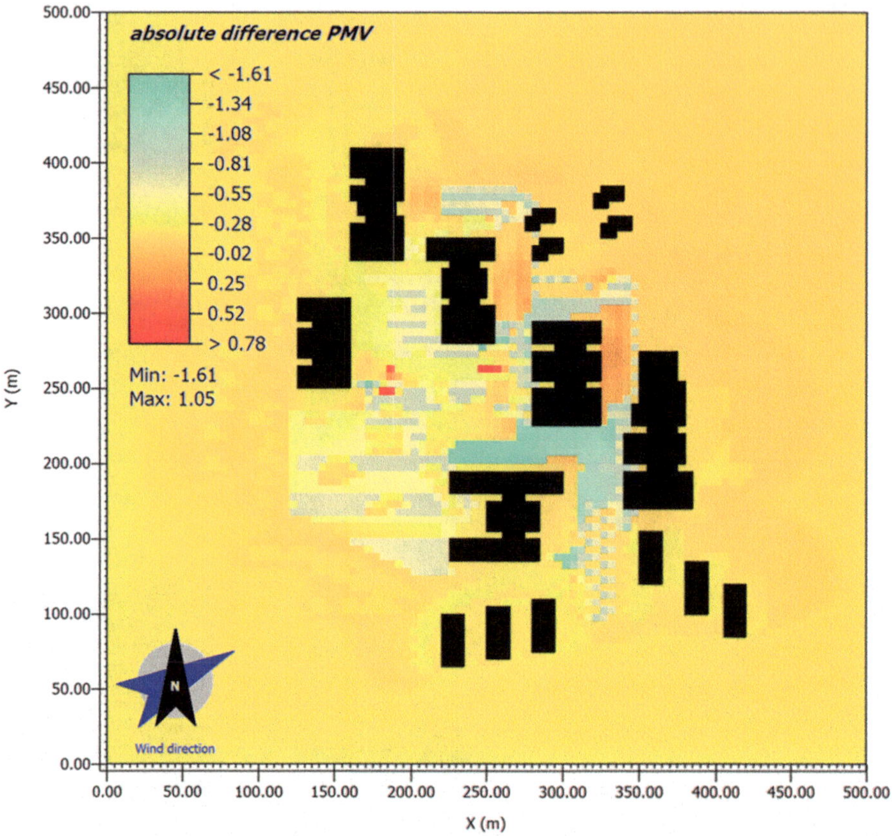

Fig. 22.3 Absolute difference of PMV index between Scenarios 0 and 1 in the case of Pineto district

surfaces. The improvement of outdoor thermal comfort key factors (*Ta*, *Ts*, *RH*, *SHF*, *PMV*) cannot be considered negligible in this case.

Based on the results, a microclimate-conscious urban renovation process of such social housing districts should take into account the following actions, namely, (1) the addition of trees inside the park, close to the main paths and recreational areas; (2) the addition of trees along sidewalks and parking areas (when missing); (3) the irrigation of public grassland areas; and (4) the addition of large water ponds in the abandoned gardens. Moreover, the availability of unbuilt space between the buildings, which nowadays in most of the cases is not maintained, opens up the possibility to renovate such public areas for the social, economic and environmental life of the inhabitants. The tailored interventions allowed improving the usability of the parks, by providing shelter from heat by means of trees, grassland and ponds, and also that one of the sidewalks and squares.

Most of the UHI mitigation measures are designed at city or regional scale rather than at the district scale, often neglecting the peculiar behaviour of each urban fabric and its contribution to the urban system behaviour as a whole. When applied at the district scale, as in the case of these social housing districts, the benefits of suitable design solutions for UHI mitigation were magnified.

Due to this peculiar open morphology, the addition of cool materials on the external envelope of buildings, while beneficial to reduce energy consumptions, allows not to ruin outdoor thermal comfort. Considering that cool materials were demonstrated as able to lower the energy consumption of buildings and that a certain part of the literature found that cool materials were deleterious for outdoor comfort in dense environments due to multiple reflections, we can conclude that even if not beneficial for the outdoors, such materials are hypothesized to lower energy consumptions without negatively impacting outdoor thermal comfort in these neighbourhoods.

Findings will contribute to promoting a climate-sensitive design perspective towards urban renewal strategies, taking into account the Mediterranean climate and the distinctive features of the above-mentioned districts. With this work, we highlight the need to consider the peculiarity of each existing area of intervention, in order to tailor the different strategies and materials to employ with respect to the specific morphology and in consideration of the identity of the place. Specifically, we aim to start a fruitful and much-needed discussion on a specific, disadvantaged part of our cities, in consideration of ongoing and future phenomena challenging our urban areas.

Acknowledgements Authors would like to thank Sapienza University of Rome for supporting this work with fund no. AR21715C82226BCC.

References

1. Angel S, Parent J, Civco DL, Blei A, Potere D (2011) The dimensions of global urban expansion: estimates and projections for all countries, 2000–2050. Prog Plann 75:53–107. https://doi.org/10.1016/j.progress.2011.04.001
2. Oke TR (1982) The energetic basis of the urban heat island. Q J R Meteorol Soc 108:1–24. https://doi.org/10.1002/qj.49710845502
3. Rosso F, Golasi I, Castaldo VL, Piselli C, Pisello AL, Salata F et al (2018) On the impact of innovative materials on outdoor thermal comfort of pedestrians in historical urban canyons. Renew Energy 118:825–839. https://doi.org/10.1016/j.renene.2017.11.074
4. Morganti M, Isalgue A, Cecere C, Coch H (2015) Urban Morphology and energy performance in the 20th century public city: the case of Rome. Plea 2015 Bol. Archit. (R)evolution. Ass. Building Green Futures, Bologna
5. Morganti M, Salvati A, Coch H, Cecere C. Urban morphology indicators for solar energy analysis. 9th Int. Conf. Sustain. Energy Build. SEB-17, 5-7 July 2017, Chania, Crete, Greece, vol. 134, Elsevier Ltd; 2017, p. 1–8. doi:https://doi.org/10.1016/j.egypro.2017.09.533
6. Di Biagi P, Marchigiani E (2008) Città pubbliche: linee guida per la riqualificazione urbana. Bruno Mondadori, Milano
7. Santangelo A, Yan D, Feng X, Tondelli S (2018) Renovation strategies for the Italian public housing stock: applying building energy simulation and occupant behaviour modelling to

support decision-making process. Energ Buildings 167:269–280. https://doi.org/10.1016/j.enbuild.2018.02.028
8. Zinzi M, Carnielo E, Mattoni B (2018) On the relation between urban climate and energy performance of buildings. A three-years experience in Rome, Italy. Appl Energy 221:148–160. https://doi.org/10.1016/j.apenergy.2018.03.192
9. Toparlar Y, Blocken B, Maiheu B, van Heijst GJF (2018) Impact of urban microclimate on summertime building cooling demand: a parametric analysis for Antwerp, Belgium. Appl Energy 228:852–872. https://doi.org/10.1016/j.apenergy.2018.06.110
10. Imam Syafii N, Ichinose M, Kumakura E, Jusuf SK, Chigusa K, Wong NH (2017) Thermal environment assessment around bodies of water in urban canyons: a scale model study. Sustain Cities Soc 34:79–89. https://doi.org/10.1016/j.scs.2017.06.012
11. Nikolopoulou M (2011) Outdoor thermal comfort. Front Biosci S3:1552. https://doi.org/10.2741/245
12. Lee H, Mayer H, Chen L (2016) Contribution of trees and grasslands to the mitigation of human heat stress in a residential district of Freiburg, Southwest Germany. Landsc Urban Plan 148:37–50. https://doi.org/10.1016/j.landurbplan.2015.12.004
13. Salata F, Golasi I, De Lieto Vollaro E, Bisegna F, Nardecchia F, Coppi M et al (2015) Evaluation of different urban microclimate mitigation strategies through a PMV analysis. Sustain 7:9012–9030. https://doi.org/10.3390/su7079012
14. Ishii A, Iwamoto S, Katayama T, Hayashi T, Shiotsuki Y, Kitayama H et al (1991) A comparison of field surveys on the thermal environment in urban areas surroundings a large pond: when filled and when drained. Energ Buildings 16:965–971. https://doi.org/10.1016/0378-7788(91)90091-G
15. Rosso F, Pisello AL, Castaldo VL, Fabiani C, Cotana F, Ferrero M et al (2017) New cool concrete for building envelopes and urban paving: optics-energy and thermal assessment in dynamic conditions. Energ Buildings 151:381–392. https://doi.org/10.1016/j.enbuild.2017.06.051
16. Golasi I, Salata F, De Lieto Vollaro E, Coppi M, de Lieto Vollaro A (2016) Thermal perception in the mediterranean area: comparing the mediterranean outdoor comfort index (moci) to other outdoor thermal comfort indices. Energies 9. https://doi.org/10.3390/en9070550

Chapter 23
A Technical Evaluation of Performance Characteristics for Pump as Turbine Application

Ombeni J. Mdee, Cuthbert Z. M. Kimambo, Torbjorn K. Nielsen, and Joseph Kihedu

23.1 Introduction

Centrifugal pumps are widely available in developing countries to pump water for domestic and industrial use in irrigation, livestock, and sewage systems [1, 2]. The same pump works as a turbine to generate mechanical rotational energy when water flows in the reverse direction. The concept of reversing the pump has been developed since the mid- and late 1970s, especially for micro and mini hydropower development [3]. The pump as turbine (PAT) system has advantages for stand-alone micro hydropower compared with conventional water turbines that include availability in large numbers with a range of capacities and heads, low cost, and easy availability of spare parts such as bearings and seals [4–6]. However, PAT performance in terms of head, flow rate, power, and efficiency is not yet well defined related to off-the-shelf pumps [7–11].

Experimental evidence shows that the pumps run at different performance characteristics in turbine mode at the same rotational speed [12]. But, PATs are very sensitive to changes of head, flow rate, power, and efficiency within fixed geometric conditions, which shift the operating point, leading to undesired outputs or failure of the scheme [13, 14]. Therefore, the inability to easily acquire pump performance characteristics in reverse mode limits as the turbine options. This review chapter presents the evaluation of the operational parameter characteristic, determination of the performance characteristics by modified pump components examination of the internal characteristics, as well as assessment of the conversion methods used for off-the-shelf pump selection.

O. J. Mdee (✉) · C. Z. M. Kimambo · J. Kihedu
Department of Mechanical and Industrial Engineering, University of Dar es Salaam, Dar es Salaam, Tanzania

T. K. Nielsen
Department of Energy and Process Engineering, Norwegian University of Science and Technology, Trondheim, Norway

A. Sayigh (ed.), *Renewable Energy and Sustainable Buildings*, Innovative Renewable Energy, https://doi.org/10.1007/978-3-030-18488-9_23

23.2 Operation Parameter Characteristics

The common basic operational parameter characteristics were described below:

23.2.1 *Head, Flow Rate, Power, and Efficiency Characteristics*

The tests were performed on 32 PATs used to describe the main characteristics including head, power, and efficiency versus the flow rate coefficients that were plotted in Fig. 23.1 [15]. The PAT characteristics were normalized to get the fit lines for head ranges from 1.99 m to 99.52 m, the rotating speed ranges from 750 rpm to 2445 rpm, specific speed ranges from 0.17 to 2.39, and flow rate ranges from 0.008 to 0.222 m³/s. The head coefficient (ϕ), flow rate coefficient (φ), and power coefficient (π) were taken as the ratio to its maximum value. The results indicated the head and power coefficients increased by adding the flow rate coefficient and the maximum point determined by the efficiency curve.

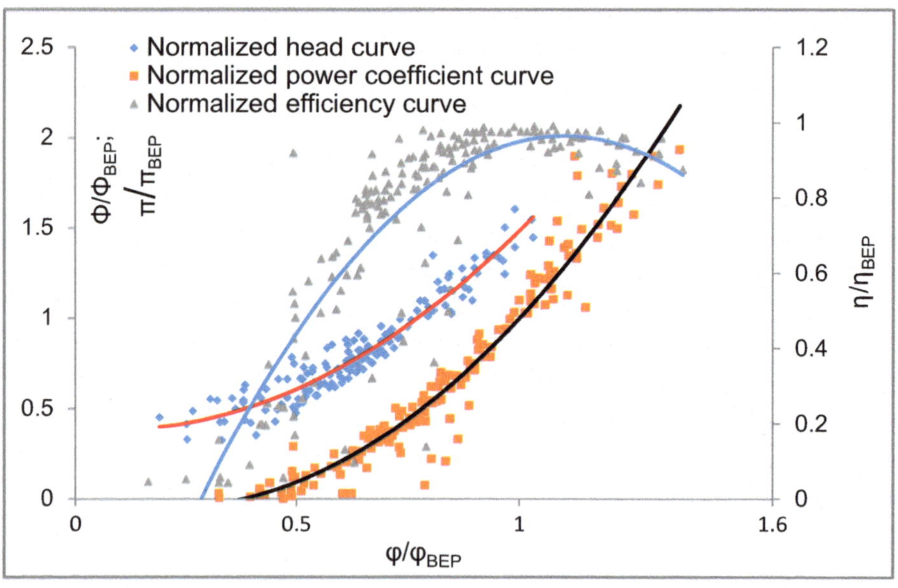

Fig. 23.1 PAT performance curves for different head, power, and efficiency versus flow rate coefficients [15]

23.2.2 Comparison of PAT and Pump Characteristics

The comparison of PAT and pump was conducted by considering the head ratio, flow rate ratio, efficiency ratio, and specific speed ratio at the best efficiency point (BEP). The ratio is obtained by taking the PAT characteristic divided by the pump characteristic. Twenty-eight (28) pump characteristic curves in turbine and pump modes were collected from [1, 9, 13, 16–20]. Figure 23.2 presents the variation of head ratio and efficiency ratio versus the flow rate ratio of specific speed range from 9.08 to 79.21 rpm. The head ratio, efficiency ratio, and flow ratio were obtained either directly or by interpolating from curves at the BEP.

Also, Fig. 23.3 presents the specific speed ratio versus the pump specific speed. Each pump in the market has different values of specific speed defining at the BEP [21]. The specific speed of PAT also can be deduced from the conversion equations from [16, 22, 23]. The n_s presents the specific speed, and subscripts t and p present the turbine mode and pump mode.

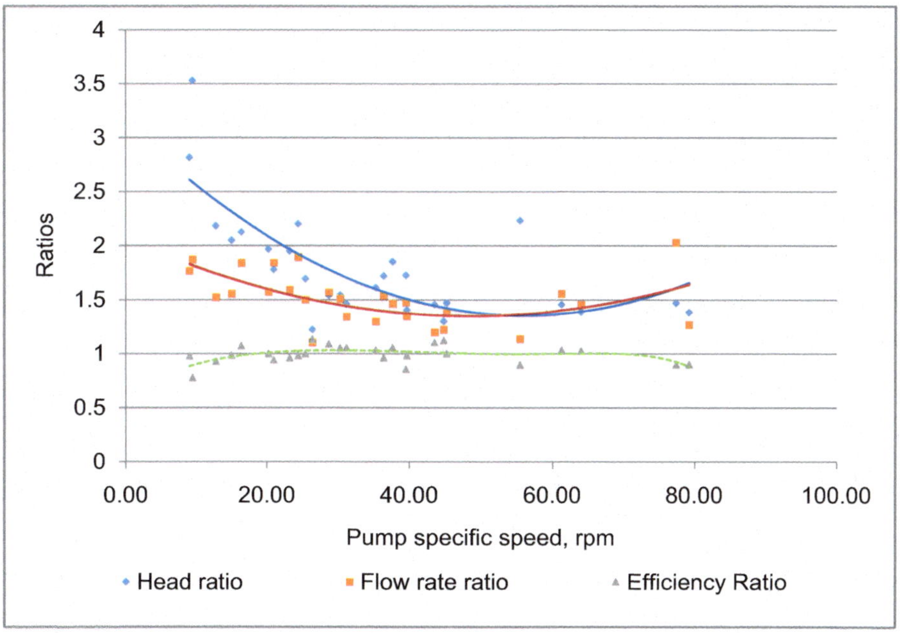

Fig. 23.2 Variation of head ratio, flow rate ratio, and efficiency ratio against pump specific speed

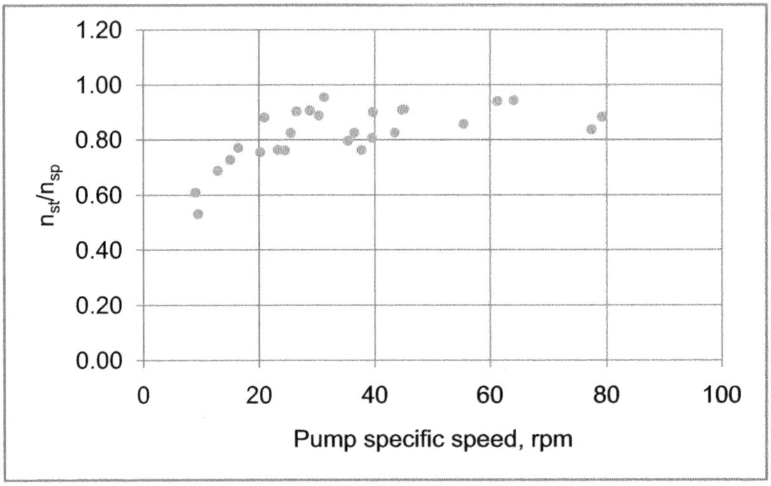

Fig. 23.3 Variation of specific speeds

23.3 Modification of Pump Components

Many experimental works have been conducted to test the performance characteristics of PATs as described in [16–18, 24, 25]. Computational fluid dynamics (CFD) is quite helpful rather than experimental work in case of investigating the internal hydraulic losses, i.e., losses in the impeller, draft tube, volute casing, and flow pattern [26, 27]. Also, changing the blade inlet angle between 70° and 135° for forward-curved blade improved the efficiency [28]. Table 23.1 presents the pump components with performance characteristics, methods, and tools as collected in [13, 29–33].

23.4 Internal Characteristics

The PAT has the following internal characteristics as described below from different research works:

1. Single-stage pump with seven twisted backward blades [34] and multistage pumps with eight twisted blades [35] indicated to have swirl flow and flow separation characteristics between impeller blades.
2. Discharge part indicated more swirl flow by using a single pump of four blades [36].
3. The occurrence of flow separation and flow recirculation can damage the pump components during the part load condition [22, 37, 38].

Table 23.1 Performance characteristic summary of modifying the pump components

Pump component modification	Methods and tools	Performance characteristics
Changing the profile of impeller with blades [28]	The original impeller with backward blades was modified to be forward blades and tested both in laboratory test rig and numerical simulation	1. Efficiency increased by 8.26% when applied CFD simulation 2. Efficiency increased by 8.11% with experimental data
Rounding blades, inner and outer shroud plate [29]	Mechanical cutting edges and tested in laboratory test rig	1. To reduce the losses by 5–10% 2. To increase the efficiency up to 2.5%
Guide vanes [30]	Attached the guide vane to the volute casing and tested in laboratory test rig	PAT has less efficiency than pump runs in pump mode, but guide vanes attached to the casing increased the efficiency
Sizing of impeller diameter [31]	Three impellers tested in laboratory test rig with single-stage centrifugal pump	Increased the efficiency up to 10.26% when impeller diameter increased from 215 to 255 mm. Total hydraulic loss indicated to drop as impeller size increased
Redesign of blades, rounding leading edges of blades, shroud and hub plates [32]	Impeller with blades designed by gradient algorithm coupled with 3D Navier-Stokes flow solver and then manufactured and tested in the laboratory test rig	It increased the head coefficient up to 5.5%, flow coefficient up to 19.8%, power coefficient up to 36.1%, and efficiency up to 5.5%
Grinding the impeller tips [33]	Percentage grinding the inlet edges and tested in laboratory test rig	It increased the efficiency up to 3.2%
Rounding blades, inner and outer shroud plate, modified inlet casing rings and eye enlargement [13]	Mechanical percentage cut edges and tested in laboratory test rig	All three cases indicated the efficiency rise from 1% to 2.5%

4. The total time dependent on flow circulation from a complete cycle of first and last gap around the radial gap between the impeller size and volute casing is 0.8 times the rotor rotation period [39].
5. Radial forces show minimum averaged amplitude of the unsteady radial force at the BEP [34]. PAT radial force varies with changing the relative position of impeller blades [20]. Unbalanced impeller forces acting in the axial direction are resolved by the bearings [40].
6. The hydraulic losses and pump geometrical shape lower the PAT efficiency. The conversion equations of hydraulic losses in the discharge and inlet parts, impeller, volute casing, and vaneless diffuser were described and experimentally tested in [1, 41].
7. The critical effect of cavitation is obtained by reducing head to 2–3% for pumps and turbines, respectively, at the BEP [13, 42]. Small PATs should have the same

Thoma number as the pump, and the rest can lie between pump and turbine at the BEP [40, 43].

23.5 Conversion Methods for PAT Characteristic Prediction

The conversion methods are used to predict the PAT characteristics. The minimum ratios of 1 for head and flow rate are the designed point of pump to run in turbine mode at the BEP. But, PAT runs above or away from the pump BEPs. The total of 23 conversion methods available in [13, 19, 42, 44–47] used 2 common input variables or constant values. Two input variables are (1) efficiency (most of the conversion method in terms of pump efficiency converged to minimum ratios of 1 for head and flow rate as the efficiency increased to 100%) and (2) specific speed of pumps with the range between 9 and 94.4 rpm. It indicated as the specific speed increased the ratios of head and flow rate decreased to minimum ratio point of 1. The conversion method used the constant values of head ratio and flow rate ratio from 1.25 to 1.5 [42, 47]. Pallabazzer and Sebbit [48] used the rule-of-thumb technique by reducing the hydropower plant head and flow rate by 25% to select a pump to run as turbine.

Chapallaz et al. [40] developed curves for head and flow rate ratios versus pump specific speed by using the performance pump datasheet. The pump specific speed is calculated by taking PAT specific speed divided by constant factor of 0.89. The low specific speed of 10–20 rpm has the head ratio of greater than 1.65 and low efficiency from 70% to 55%, while flow rate ratio is greater than 1.4. The high specific speed of greater than 30 rpm has low ratios of head and flow rate of less than 1.65 and 1.4, respectively, with efficiency greater than 70%. But, the minimum ratios of head and flow rate versus pump specific speed are greater than 1.25 and 1.2, respectively.

Li [49] indicated the effects of converting the pump head, flow rate, and efficiency to turbine mode parameters by using Reynolds number, i.e., viscosity, with impeller Reynolds number reference (NR_{ef}) of 8,295,768 used to develop the conversion ratios for head, flow rate, and efficiency with errors of 1.67%, 2.57%, and 6.76%, respectively. Barbarelli et al. [1] developed the geometric model for determining the PAT characteristics. The model is related to head shutoff, flow rate coefficient, volute velocity coefficient, and pump diameter ratio variation. Huang et al. [50] developed the rotor-volute matching principle based on pump Euler equation and velocity relations to predict the PAT characteristics. The rotor-volute matching principle worked without using performance datasheet from pump manufacturer. Also, Nielsen [51] derived the equation that governed the pump in turbine mode of operation by means of Euler equation and pump geometry parameters.

Furthermore, Singh and Nestmann [23] developed the model to predict the PAT characteristics at full- and no-load conditions. It is based on the dimensional analysis to derive the head coefficient, flow rate coefficient, specific speed, and specific diameter equations. The PAT head and flow rate will depend on the flow rate coefficient, tangential blade velocity, impeller outlet diameter, and turbine mode specific speed.

23.6 Conclusions

This review chapter presented four research areas for studying the performance characteristics of PAT application. The research areas include operational parameter characteristic, pump component modification, internal flow characteristics, and conversion equations. The performance characteristics differed under full- and part-load conditions when pump is operated in reverse turbine and pump modes. The main operational characteristics of PAT include head, power, and efficiency versus the flow rate that indicated to have similar characteristics with the conventional turbine. Also, PAT and pump were compared by using head, flow rate, efficiency, and specific speed. The PAT indicated to have higher head and flow rate than pump head and flow rate. But, the PAT has less efficiency and specific speed when compared with pump characteristics. Also, the modification of pump components increased the PAT efficiency up to 10.6%. Although, the internal flow characteristics indicated to lower the lifespan of pumps and mostly caused by flow separation and swirl between impeller blades, hydraulic losses, and cavitation effects. Twenty-three conversion equations of head and flow rate reported from different research works. The conversion equations are used to predict the PAT head and flow rate for different pumps.

Acknowledgments This reviewed chapter is part of ongoing PhD study and supported by NORAD under Energy and Petroleum (EnPe) Program at the University of Dar es Salaam, Tanzania, and Norwegian University of Science and Technology, Trondheim, Norway.

References

1. Barbarelli S, Amelio M, Florio G (2016) Predictive model estimating the performances of centrifugal pumps used as turbines. Energy 107:103–121
2. Muttalli RS, Agrawal S, Warudkar H (2014) CFD simulation of centrifugal pump impeller using ANSYS-CFX. Int J Innov Res Sci Eng Technol 3(8):15553–15561
3. Patel JB, Mevada RN, Sardana D, Rajput VP (2015) Experimental and numerical investigation of centrifugal pump performance in reverse mode. Int J Adv Technol Eng Sci 3(1):1066–1072
4. Baburaj E, Sivaprakasam R, Manikandan C, Sunda K (2013) CFD analysis of pump as turbine for micro-hydro schemes. Int J Innov Res Sci Eng Technol 2(3):590–594
5. KSB (2013) Advantages of using pumps as turbines. World Pumps 10:18–20
6. Nautiyal H, Varun V, Kumar A, Yadav S (2011) Experimental investigation of centrifugal pump working as turbine for small hydropower systems. Energy Sci Technol 1(1):79–86
7. Barbarelli S, Amelio M, Florio G (2017) Experimental activity at test rig validating correlations to select pumps running as turbines in microhydro plants. Energy Convers Manag 149(1):781–797

8. Hossain IM, Ferdous SM, Salehin S, Saleque AM and Jamal T (2014) Pump-as-turbine (PAT) for small scale power generation: a comparative analysis. Developments in renewable energy technology (ICDRET), 2014 3rd international conference on the 2014 May 29, pp. 1–5
9. Derakhshan S, Nourbakhsh A (2008) Theoretical, numerical and experimental investigation of centrifugal pumps in reverse operation. Exp Thermal Fluid Sci 32:1620–1627
10. Williams AA (1996) Pumps as turbines for low cost micro hydro power. Renew Energy 9(1–4):1227–1234
11. Carravetta A, Giudice G, Fecarotta O, Ramos HM (2013) Pump as Turbine (PAT) design in water distribution network by system effectiveness. Water 5:1211–1225
12. Popescu D, Duinea A and Rusinaru D (2013) Study of centrifugal pump operating as turbine in small hydropower plants. In: Recent researches in electric power and energy systems, pp 285–288
13. Singh P (2005) Optimization of internal hydraulics and of system design for pumps as turbines with field implementation and evaluation. PhD Dissertation, University of Karlsruhe, Karlsruhe
14. Ramos H, Borga A (1999) Pumps as turbines: an unconventional solution to energy production. Urban Water 1(3):261–263
15. Rossi M, Renzi M (2017) Analytical prediction models for evaluating pumps-as-turbines (PaTs) performance. In: 2nd international conference on advances on clean energy research, ICACER, Berlin, Germany, 7–9 April 2017
16. Tan X, Engeda A (2016) Performance of centrifugal pumps running in reverse as turbine: part ii - systematic specific speed and specific diameter based performance prediction. Renew Energy 99:188–197
17. Qian Z, Wang F, Guo Z, Lu J (2016) Performance evaluation of an axial-flow pump with adjustable guide vanes in turbine mode. Renew Energy 99:1146–1152
18. Couzinet A, Gros L, Pierrat D (2013) Characteristics of centrifugal pumps working in direct or reverse mode: focus on the unsteady radial thrust. Int J Rotating Mach:1–11
19. Derakhshan S, Nourbakhsh A (2008) Experimental study of characteristic curves of centrifugal pumps working as turbines in different specific speeds. Exp Thermal Fluid Sci 32:800–807
20. Fernández J, Blanco E, Parrondo J, Stickland MT and Scanlon TJ (2004) Performance of a centrifugal pump running in inverse mode. In: Proceedings of the institution of mechanical engineers
21. Srinivasan KM (2008) Rotodynamic pumps (centrifugal and axial). New Age International (P) Limited Publishers, New Delhi
22. Yang SS, Derakhshan S, Kong FY (2012) Theoretical, numerical and experimental prediction of pump as turbine performance. Renew Energy 48:507–513
23. Singh P, Nestmann F (2011) A consolidated model for the turbine operation of centrifugal pumps. J Eng Gas Turbines Power 133:1–9
24. Derakhshan S, Kasaeian N (2014) Optimization, numerical, and experimental study of a propeller pump as turbine. J Energy Resour Technol 136:1–7
25. Krivichenko G (1994) Hydraulic machines: turbines and pumps. Lewis, Boca Raton, FLA
26. Nigussie T, Dribssa E (2015) Design and CFD analysis of centrifugal pump. Int J Eng Res General Sci 3(3):668–677
27. Agarwal T (2012) Review of pump as turbine (PAT) for micro-hydropower. Int J Emerg Technol Adv Eng 2(11):163–169
28. Wang T, Kong F, Xia B, Bai Y, Wang C (2017) The method for determining blade inlet angle of special impeller using in turbine mode of centrifugal pump as turbine. Renew Energy 109:518–528
29. Doshi A, Channiwala S, Singh P (2017) Inlet impeller rounding in pumps as turbines: an experimental study to investigate the relative effects of blade and shroud rounding. Exp Thermal Fluid Sci 82:333–348
30. Singh R, Gupta R, Singh A, Subbarao PM (2013) Guided pump-as-turbine: design and development for pico-level energy generation. Int J Appl Res Mech Eng 3(1):71–76

31. Yang SS, Liu HL, Kong FY, Dai C, Dong L (2013) Experimental, numerical, and theoretical research on impeller diameter influencing centrifugal pump-as-turbine. ASME J Energ Eng 139(4):299–307
32. Derakhshan S, Mohammadi B, Nourbakhsh A (2009) Efficiency improvement of centrifugal reverse pumps. ASME J Fluids Eng 131:1–9
33. Suarda M, Suamadwipa N and Adnyana WB (2006) Experimental work on modification of impeller tips of a centrifugal pump as a turbine. In: The 2nd joint international conference on sustainable energy and environment, Bangkok, Thailand
34. Barrio R, Fernandez J, Blanco E, Parrondo J, Marcos A (2011) Performance characteristics and internal flow patterns in a reverse-running pump–turbine. Proc Mech Eng Part C J Mech Eng Sci 226:695–708
35. Sedlar M, Soukal J, Komarek M (2009) CFD Analysis of middle stage of multistage pump operating in turbine regime. Eng Mech 16(6):413–421
36. Rawal S, Kshirsagar JT (2007) Numerical simulation on a pump operating in a turbine mode. In: Proceedings of the 23rd international pump users symposium
37. Guelich JF, Bolleter U (2008) Pressure pulsations in centrifugal pumps. J Vib Acoust 114(2):272–279
38. Wnek TF, Engineer NT and Ine WP (2007) Pressure pulsations generated by centrifugal pumps. Warren Pumps, Inc, Warren, MA. www.warrenpumps.com. Accessed 30 Nov 2017
39. Su X, Huang S, Zhang X, Yang S (2016) Numerical research on unsteady flow rate characteristics of pump as turbine. Renew Energy 94:488–495
40. Chapallaz JM, Eichenberger P and Fischer G (1992) Manual on pumps used as turbines. Vieweg, Braunschweig: Printed in the Federal Republic of Germany by Lengericher Handelsdruckerei, Lengerich
41. Amelio M, Barbarelli S (2004) A one-dimensional numerical model for calculating the efficiency of pumps as turbines for implementation in micro-hydro power plants. In: ASME 7th Biennial conference on engineering systems design and analysis, Manchester, England, 19–22 July
42. Alatorre-Frenk C (1994) Cost minimisation in micro hydro systems using pumps-as-turbines. PhD Thesis, University of Warwick, Faculty of Sciences, Department of Engineering, Coventry, United Kingdom
43. Turton RK (1984) Principles of turbomachinery, 2nd edn. Chapman and Hall, London
44. Bogdanovic-Jovanovic JB, Milenkovic DR, Svrkota DM, Bogdanovic B, Spasic ZT (2014) Pumps used as turbines: power recovery, energy efficiency, CFD analysis. Therm Sci 18(3):1029–1040
45. Nautiyal H, Kumar VA (2010) Reverse running pumps analytical, experimental and computational study: a review. Renew Sust Energ Rev 14:2059–2067
46. Teuteberg BH (2010) Design of a pump-as-turbine micro hydro system for an Abalone Farm, Final Report for mechanical project 878, Department of Mechanical and Mechatronic Engineering, Stellenbosch University
47. Williams AA (1994) The turbine performance of centrifugal pumps: a comparison of prediction methods. Proc Inst Mech Eng Part A J Power Energy 208:59–66
48. Pallabazzer R, Sebbit A (2001) A micro-hydro pilot plant for mechanical pumping. In: Hydropower in the New Millennium. Bergen, Norway, Swets & Zeitlinger
49. Li W (2017) Optimising prediction model of centrifugal pump as turbine with viscosity effects. Appl Math Model 41:375–398
50. Huang S, Qiu G, Su X, Chen J, Zou W (2017) Performance prediction of a centrifugal pump as turbine using rotor-volute matching principle. Renew Energy 108:64–71
51. Nielsen TK (2015) Simulation model for francis and reversible pump turbines. Int J Fluid Mach Syst 8(3):169–182

Chapter 24
Performance Improvement of Contra-Rotating Small Hydroturbine

Toru Shigemitsu, Tomofumi Ikebuchi, Ding Nan, and Takuji Hosotani

24.1 Introduction

Pico hydropower, which generates about 100 W–1 kW, is called new energy in Japan, and there are a lot of places suitable for the pico-hydropower in agricultural water and a small stream. It is better to design the pico-hydroturbine as small as possible to reduce the installation cost and to make the environmentally friendly hydroturbine. Therefore, Darrieus- and gyro-type turbines, suitable for design specifications of a low head in agricultural water and a small river, are investigated, and the performance characteristics and the optimum design parameter are discussed [1, 2]. There are also some extra-low head natural flow streams where the Darrieus-type hydroturbine can be utilized, and in order to avoid the obstacles flowing into the hydroturbine, a waste screening system is designed and tested [3]. In addition to this, an inlet nozzle which is designed for improving the performance of Darrieus-type hydroturbine is investigated by experimental and numerical analysis [4]. Internal flow of a spiral water turbine with wide flow passage, which has small environmental impact, is investigated [5]. On the other hand, a small crossflow turbine used for a small stream as an environmentally friendly pico-hydroturbine and a Savonius turbine with low cost are suggested, and the performance improvement of these turbines by installation and selecting the optimum position of shield plate are reported [6–8]. However, compared with the large hydroturbine, the small one performs lower efficiency, and it can be used in wide flow rates range. Therefore, we adopted contra-rotating rotors, which can be expected to achieve high performance and make the small design possible. In some other references, some

T. Shigemitsu (✉)
Institute of Science and Technology, Tokushima University, Tokushima City, Japan
e-mail: t-shige@tokushima-u.ac.jp

T. Ikebuchi · D. Nan · T. Hosotani
Graduate School of Advanced Technology and Science, Tokushima University, Tokushima City, Japan

A. Sayigh (ed.), *Renewable Energy and Sustainable Buildings*, Innovative Renewable Energy, https://doi.org/10.1007/978-3-030-18488-9_24

basic issues of using the counter-rotating rotors have been discussed, including the characteristic performance of using counter-rotating rotors, potential interference and the optimum value of blade solidity [9–11]. However, it is hard to construct the counter-rotating generator by this small hydroturbine of which diameter is only 60 mm; this problem has to be considered. Therefore, two independent generators in this contra-rotating small hydroturbine may be more suitable. In this study, a significant compact hydroturbine is named as contra-rotating small hydroturbine. Final goal on this study is development of a small hydroturbine like electric appliances, which has high portability and makes an effective use of the unused small hydropower energy resources.

In the previous research, we selected some places in Tokushima Prefecture in Japan, where a small hydropower can be generated, and conducted field tests of the head, flow rate, water quality and capacity utilization [12]. Then we investigated performance of the test model by the numerical analysis, and a prospect to apply this small hydroturbine for pico-hydropower was confirmed [13]. After that, we designed and manufactured the original test turbine, and an experimental apparatus was also designed and assembled to verify the performance of the test turbine. We also investigated the internal flow condition by the numerical analysis results [14, 15]. Because of the large loss between the front and rear rotor, we changed the geometry of the spoke from cuboid to cone. The performance and internal flow condition of the spoke changing model was highly improved [16]. The blade thickness and the shape of front hub were changed from the original test rotor to improve the performance and internal flow condition, and the differences of the internal flow conditions between the original model and new model were clarified [17].

In this paper, the tip clearance and solidity are varied from the base model using the numerical analysis model to improve the performance of small hydroturbine, and the numerical analysis is conducted with steady condition to reduce the computational cost. The solidity of each front and rear rotor is optimized, and the numerical analysis is conducted with the unsteady condition for the best solidity combination model. The pressure distribution of each type of the model is investigated by the numerical analysis results.

24.2 Experimental and Numerical Analysis Condition

24.2.1 Rotor Design Method and Design Parameter

A test turbine diameter of the base model is small as $D = 58$ mm to set up in a pipe of agricultural water with a diameter of about 2 inch and the small-scale water-supply system. Design flow rate and head are $Q_d = 4.825$ l/s and $H_d = 2.6$ m, respectively, based on power ($P = 10$–100 W), head ($H = 1$–4 m), flow rate ($Q = 2$–10 l/s) assumed in the pipe of agricultural water and the small-scale water-supply system. Design head of each front and rear rotor is the same as $H_{\mathrm{df}} = H_{\mathrm{dr}} = 1.3$ m for the base test

Fig. 24.1 Test hydroturbine rotors (base model)

Table 24.1 Primary dimensions of base turbine rotors

		Hub	Mid	Tip
Front rotor	Diameter [mm]	29	43.5	58
	Blade number	4		
	Blade profile	NACA6512		
	Solidity	1.40	1.07	0.85
	Setting angle [°]	25.5	20.0	15.8
Rear rotor	Blade number	3		
	Blade profile	NACA6512		
	Solidity	0.86	0.71	0.59
	Setting angle [°]	44.6	29.1	18.9

rotor, and a rotational speed of each front and rear rotor of the test turbine are $N_f = N_r = 2300$ min^{-1} considering a characteristic of a small generator which can produce about 10–100 W. The rotor and primary dimensions of the base rotor are shown in Fig. 24.1 and Table 24.1, respectively. A mouse is also beside in Fig. 24.1 to compare the size of the test rotor to it. Hub tip ratio of front and rear rotors is $D_{hf}/D_{tf} = D_{hr}/D_{tr}$ = 29 mm/58 mm = 0.50. Each design parameter is determined by the power, head, flow rate and rotational speed. In this study, blade number of each front and rear rotor should be a coprime: the front rotor $Z_f = 4$ and rear rotor $Z_r = 3$ in order to suppress the blade rows interaction of the contra-rotating rotor. The blade type NACA65 serious is chosen as the profile of the blade. A guide vane isn't set at the inlet of the front rotor because the test turbine is designed as compact as possible.

The tip clearance is controlled only by the numerical analysis, and it is conducted by adding or eliminating the rotor outer region in the numerical model, so the rotor region increases with the decrease of the tip clearance and vice versa. The tip clearance of each type of rotor is c = 0.5, 1.0, 1.5, 2.0, 2.5 and 3.0 mm. In the experiment, only one rotor having the tip clearance c = 1.0 mm (base model) is prepared. On the other hand, the solidity of the base model is changed from $\sigma/\sigma_0 = 0.5$ to $\sigma/\sigma_0 = 2.0$ in the numerical analysis to investigate the influence of the solidity on the performance. σ_0 is the solidity of the base model. The blade thickness of each type of the solidity model is kept constant by adjusting the low two digits of the blade profile. The primary dimensions of each solidity model are shown in Table 24.2.

Table 24.2 Primary dimensions of different solidity model

σ/σ_0		0.5		1.0 (base model)		1.5		2.0	
		NACA6524		NACA6512		NACA6508		NACA6506	
Blade profile		Front	Rear	Front	Rear	Front	Rear	Front	Rear
Solidity	Tip	0.395	0.296	0.790	0.592	1.19	0.888	1.58	1.18
	Mid	0.454	0.341	0.907	0.681	1.36	1.02	1.81	1.36
	Hub	0.570	0.428	1.14	0.856	1.71	1.28	2.28	1.71

There is a spacer between the front rotor and rear rotor, and four spokes are distributed around the spacer. It is planed that the generators are put inside of front and rear rotor's hub in a product of this kind of small hydroturbine, and the rotors and generators are supported only by the spokes. Therefore, the spokes are essential for the actual product. The diameter of the spacer is 29 mm, which is the same with the diameter of rotor hub, and the axial length of spacer is 33 mm.

24.2.2 Experimental Apparatus and Method

An assembly of the test rotors and a schematic diagram of the experimental apparatus are shown in Figs. 24.2 and 24.3. A diameter of a casing is $D = 60$ mm and the rotor diameter of the base model is $D_r = 58$ mm because the tip clearance is $c = 1$ mm. A test section is made of a transparent acrylic resin for visualization experiments. The fluid media is water; the static head differences on the casing wall between 2D upstream of the front rotor and 2D downstream of the rear rotor are measured for head evaluation. Each rotor is connected to a respective shaft and driven by a respective motor set upstream and downstream of the test section. The rotational speed of each front and rear rotor is constant as $N_f = N_r = 2300$ min^{-1} in the experiment. The length of the transparent test section is 500 mm, and straight pipes over 4D are kept upstream and downstream of the test rotors to suppress the influence of the swirl flow from the 90° bend near the test section. The flow rate Q is obtained by a magnetic flow meter KEYENCE FD-UH50H installed far upstream of the hydroturbine, and the measurement accuracy of it is ±0.5%. Torque of each front and rear rotor is measured by a respective torque meter SOKKI SS050, and its measurement accuracy is ±0.5%. Then, a shaft power is calculated by the torque and rotational speed measured by a rotational speed sensor ONO SOKKI MP-981. The shaft power is evaluated by the torque eliminating mechanical loss in this performance test. Hydraulic efficiency of the hydroturbine η is calculated as the ratio of the shaft power to the water power. The pressure distribution in axial direction is measured by the pressure sensors installed on the interior of the casing as shown in Fig. 24.4. The pressure sensor holes are set nine points for the front rotor region and eight points for the rear rotor region. The axial positions of the static pressure sensors for the front rotor are 3 mm upstream of the leading edge (LE) of the front rotor hub, seven points with 2 mm interval from LE of the front rotor hub and 3 mm downstream of the trailing edge (TE) of the front rotor hub. The axial positions for the rear rotor are eight points with 3 mm interval from LE of the rear rotor hub.

Fig. 24.2 Assembly of test rotors (base model)

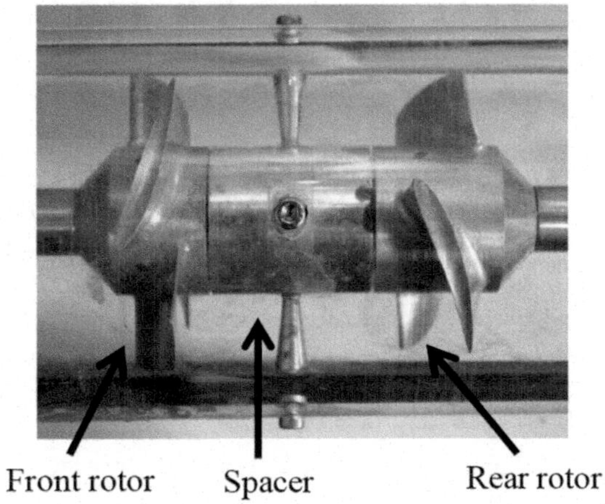

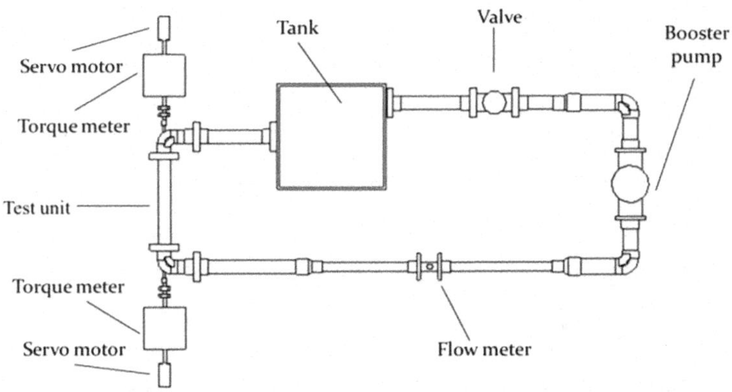

Fig. 24.3 Schematic diagram of experimental apparatus

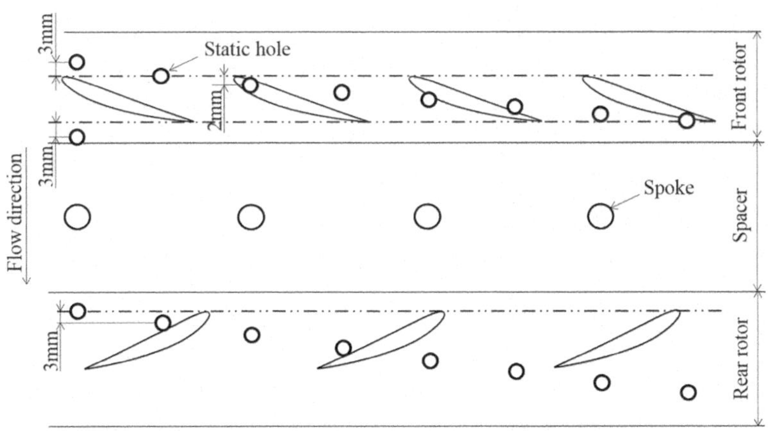

Fig. 24.4 Pressure measurement positions

24.2.3 *Numerical Analysis Conditions*

The commercial software ANSYS-CFX is used in the numerical analysis. Fluid is assumed to be incompressible and isothermal water, and the equation of the mass flow conservation and Reynolds averaged Navier-Stokes equation are solved by the finite volume method. The standard wall function is utilized near the wall, and the standard k-ε model is used as the turbulence model. A numerical model is the same with the test section of the experimental apparatus and includes the inlet and outlet bend of the experimental apparatus. Straight pipes over 4D are set for the inlet and outlet pipes, which are the same with the experimental apparatus, in order to suppress the influence of the bend at the inlet and outlet. Each front and rear rotor's shaft is modelled in the numerical analysis and rotated in the front and rear rotor's rotational direction, respectively. The constant velocity perpendicular to the inlet boundary and the constant pressure having only axial velocity are given as the boundary condition at the inlet and outlet, respectively. The coupling among the front rotor, spokes and rear rotor regions is accomplished by the transient rotor/stator in the unsteady simulation. The time step number per one rotor rotation is 120 and the time step is $t = 2.174 \times 10^{-4}$ s. The data of one rotor rotation is obtained after five rotor rotations in unsteady numerical analysis. The numerical grids of the whole numerical domains used for the numerical analysis are shown in Fig. 24.5. The numerical domains are comprised of the inlet pipe, rotors and outlet pipe regions. The numerical grid points of the base model are 495,241 for the inlet pipe region, 6,891,946 for the rotor region and 316,811 for the outlet pipe region, respectively. The numerical analysis is performed at 11 flow rate points from $0.9Q_d$ to $1.4Q_d$.

24.3 Results and Discussions

24.3.1 *Validity of Numerical Result*

Figure 24.6 shows performance curves of the base model with the tip clearance $c = 1.0$ mm obtained by the experiment. The unsteady numerical analysis results are also shown to compare the numerical results to experimental results. A horizontal axis is the flow rate. First vertical axes are the turbine head and efficiency. Second vertical axis is the shaft power. The rotational speed of each front and rear rotor is $N_f = N_r = 2300$ min^{-1}. The turbine head and shaft power increase according to the increase of the flow rate, and the maximum static pressure efficiency $\eta_{max} = 64.2\%$ is obtained around $1.2Q_d$ for both experimental and numerical analysis results. Figure 24.7 shows the pressure distribution in axial direction obtained by the experiment at the design flow rate 1.0Qd. The unsteady numerical analysis result is also shown in Fig. 24.7. The pressure gradient at each front and rear rotor of the numerical analysis result is similar to the experimental result; however, the static pressure at the outlet of the front rotor is lower than the experimental result. The pressure

Fig. 24.5 Numerical grids

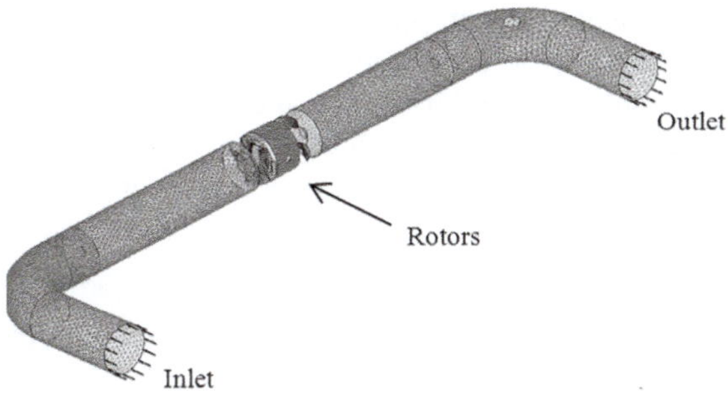

Fig. 24.6 Performance curves

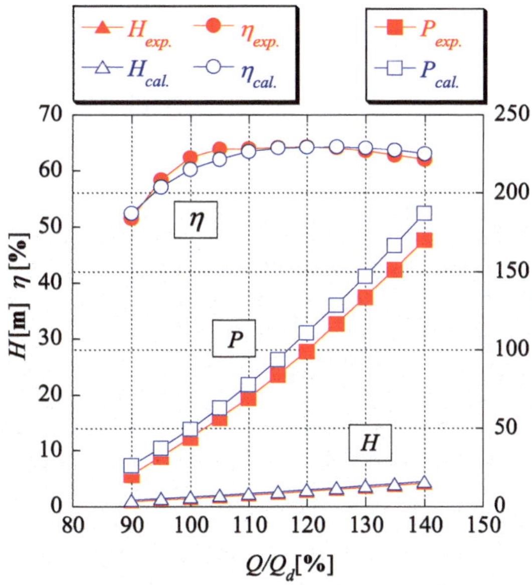

Fig. 24.7 Pressure distribution in axial direction at $1.0Q_d$

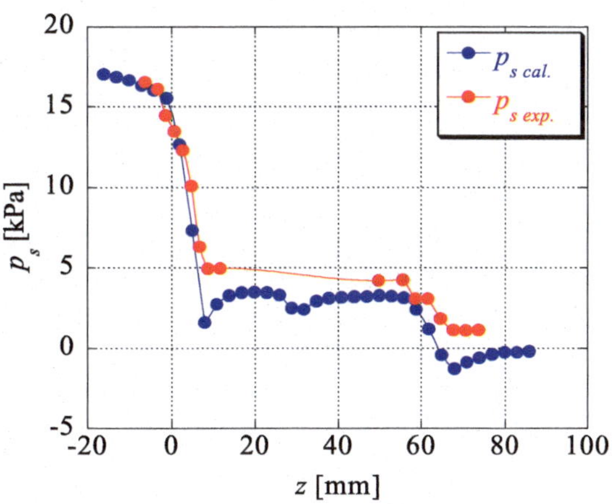

gradient of the front rotor is steeper than that of the rear rotor, and the head decreases considerably at the front rotor. It is found from Figs. 24.6 and 24.7 that the numerical result of the performance and axial pressure distribution represents qualitative tendency of the experimental result. Therefore, the influence of the tip clearance size and solidity on the performance was investigated by the numerical analysis.

24.3.2 Performance Under the Condition of Different Tip Clearance

Figure 24.8 shows performance of each tip clearance at the design flow rate 1.0Qd obtained by the numerical analysis. A horizontal axis is the tip clearance size and vertical axes are the head, power and total pressure efficiency. The head, power and efficiency increase significantly according to the decrease of the tip clearance from $c = 3.0$ mm to $c = 0.5$ mm. The power $P = 73.4$ W is obtained for $c = 0.5$ mm; however, the power of $c = 3.0$ mm is small as $P = 11.0$ W. The total pressure efficiency increases from $\eta = 37.4\%$ to $\eta = 71.3\%$ with the decrease of the tip clearance. It is found that the tip clearance size difference within $c = 0.5$–3.0 mm influences the performance of the small hydroturbine greatly. It is difficult to keep the tip clearance less than 1.0 mm because of the machining cost and assembly accuracy, so $c = 1.0$ mm might be the practical choice for small hydroturbine. Next, the efficiency of each front and rear rotor is investigated. Figure 24.9 shows total pressure efficiency of each front and rear rotor obtained by the steady numerical analysis. The loss of the spokes is not considered in each front and rear rotor total pressure efficiency. The total pressure efficiency of the front rotor increases from $\eta = 49.0\%$ ($c = 3.0$ mm) to $\eta = 73.3\%$ ($c = 0.5$ mm). On the other hand, the total pressure efficiency of the rear rotor increases significantly compared to the front rotor from $\eta = 4.10\%$ to $\eta = 76.6\%$. The influence of the tip clearance becomes large for the rear rotor because the rear rotor gets influenced from the tip leakage flow from the front rotor in addition to the leakage flow of the rear rotor itself.

24.3.3 Performance Under the Condition of Different Solidity

Figure 24.10 shows performance of each solidity model obtained by the steady numerical analysis at 1.25Qd, where the maximum efficiency is obtained. A horizontal axis is the solidity ratio σ/σ_0 and vertical axes are the head, power and efficiency. The head and power increase with the increase of the solidity ratio and the power increases gently at $\sigma/\sigma_0 = 2.0$. The efficiency shows the maximum for the base model ($\sigma/\sigma_0 = 1.0$), so the total pressure efficiency of each front and rear rotor is investigated to clarify the suitable solidity of each rotor. The total pressure

Fig. 24.8 Performance curves of each tip clearance

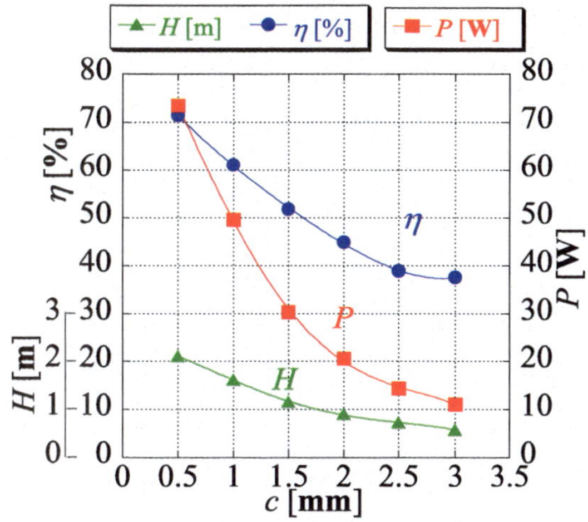

Fig. 24.9 Performance curves of each front and rear rotors

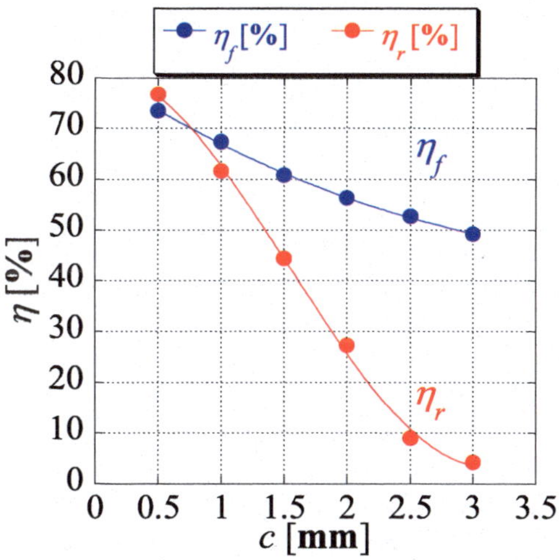

Fig. 24.10 Performance curves of each solidity model

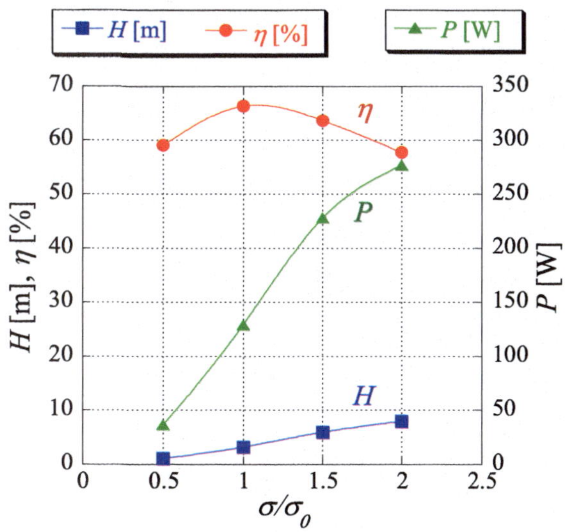

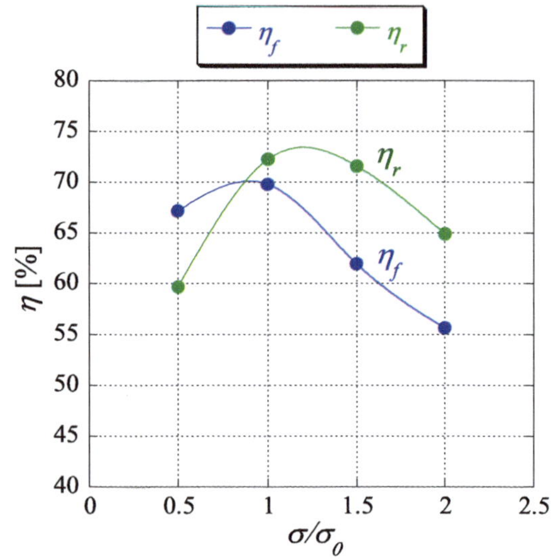

Fig. 24.11 Efficiency curves of front and rear rotors

efficiency of each front and rear rotor at $1.25Q_d$ is shown in Fig. 24.11. The loss of the spokes is not considered in Fig. 24.11. The maximum efficiency is confirmed for the base model ($\sigma/\sigma_0 = 1.0$); the front and rear rotor efficiency shows the maximum at $\sigma/\sigma_0 = 1.0$ for both rotors. However, the maximum efficiency of the front rotor seems to exist between $\sigma/\sigma_0 = 0.5$ and $\sigma/\sigma_0 = 1.0$ and between $\sigma/\sigma_0 = 1.0$ and $\sigma/\sigma_0 = 1.5$ for the rear rotor. Therefore, $\sigma/\sigma_0 = 0.8$ and $\sigma/\sigma_0 = 1.2$ is selected for the suitable solidity of each rotor, respectively, and the steady numerical analysis is conducted. The total pressure efficiency of $\sigma/\sigma_0 = 0.8$(front rotor) and $\sigma/\sigma_0 = 1.2$(rear rotor) is added to the Fig. 24.11 and shown in Fig. 24.12. It can be found that the total pressure efficiency shows the maximum $\eta = 73.4\%$ at $\sigma/\sigma_0 = 0.8$ for the front rotor and $\eta = 73.9\%$ at $\sigma/\sigma_0 = 1.2$ for the rear rotor. Figure 24.13 shows the static pressure distribution in axial direction of each type of the solidity models at 1.25Qd obtained by the steady numerical analysis. The pressure gradient at each rotor becomes small with the decrease of the solidity. It is considered that the pressure gradient of the front and rear rotors matches the best performance at $\sigma/\sigma_0 = 0.8$ (front rotor) and $\sigma/\sigma_0 = 1.2$ (rear rotor), so the unsteady numerical analysis is carried out for $\sigma/\sigma_0 = 0.8$ (front rotor) and $\sigma/\sigma_0 = 1.2$ (rear rotor) to check the accurate performance. The maximum total pressure efficiency of the $\sigma/\sigma_0 = 1.0$ is $\eta = 64.9\%$, and it's improved as $\eta = 68.2\%$ for the suitable solidity model (front rotor $\sigma/\sigma_0 = 0.8$, rear rotor $\sigma/\sigma_0 = 1.2$). Therefore, the performance test using the best model ($c = 0.5$ mm, front rotor $\sigma/\sigma_0 = 0.8$, rear rotor $\sigma/\sigma_0 = 1.2$) will be conducted to validate the numerical analysis results in near future.

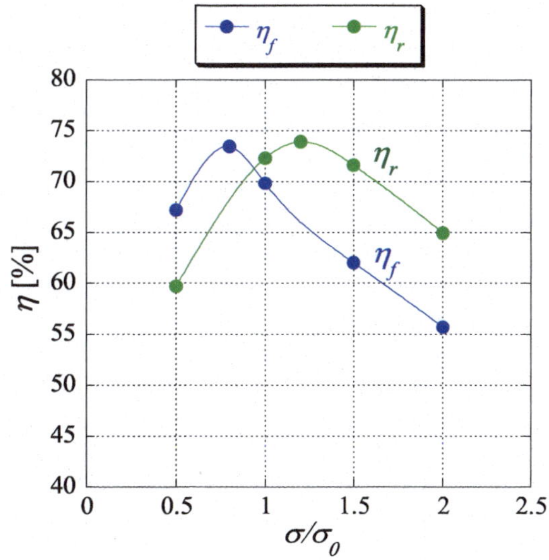

Fig. 24.12 Efficiency curves of front and rear rotors including suitable solidity model

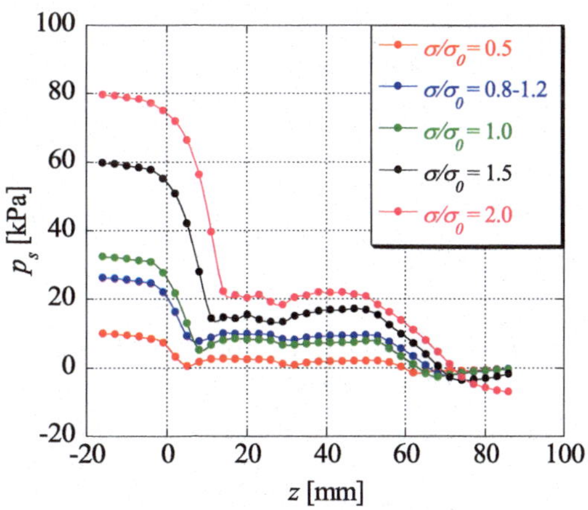

Fig. 24.13 Pressure distribution of each solidity model in axial direction at $1.25Q_d$

24.4 Concluding Remarks

The tip clearance and solidity are varied from the base model using the numerical analysis model to improve the performance of small hydroturbine, and the solidity of each front and rear rotor is optimized, and the numerical analysis is conducted with the unsteady condition for the best solidity combination model. As a result, the following conclusions were obtained.

1. The pressure gradient of the front rotor is steeper than that of the rear rotor, and the head decreases considerably at the front rotor in the case of the base model.

2. The total pressure efficiency increases from $\eta = 37.4\%$ to $\eta = 71.3\%$ with the decrease of the tip clearance. It is found that the tip clearance size difference within $c = 0.5$–3.0 mm influences the performance of the small hydroturbine greatly. The total pressure efficiency of the front rotor increases from $\eta = 49.0\%$ ($c = 3.0$ mm) to $\eta = 73.3\%$ ($c = 0.5$ mm). On the other hand, the total pressure efficiency of the rear rotor increases significantly compared to the front rotor from $\eta = 4.10\%$ to $\eta = 76.6\%$. The influence of the tip clearance becomes large for the rear rotor because the rear rotor gets influenced from the tip leakage flow from the front rotor in addition to the leakage flow of the rear rotor itself.
3. The total pressure efficiency shows the maximum $\eta = 73.4\%$ at $\sigma/\sigma_0 = 0.8$ for the front rotor and $\eta = 73.9\%$ at $\sigma/\sigma_0 = 1.2$ for the rear rotor. The pressure gradient at each rotor becomes small with the decrease of the solidity. It is considered that the pressure gradient of the front and rear rotors matches the best performance at $\sigma/\sigma_0 = 0.8$(front rotor) and $\sigma/\sigma_0 = 1.2$(rear rotor). The maximum total pressure efficiency of the $\sigma/\sigma_0 = 1.0$ is $\eta = 64.9\%$, and it's improved as $\eta = 68.2\%$ for the suitable solidity model (front rotor $\sigma/\sigma_0 = 0.8$, rear rotor $\sigma/\sigma_0 = 1.2$).

Acknowledgements This research was supported by Adaptable and Seamless Technology Transfer Program through Target-driven R&D from JST, JKA and its promotion funds from KEIRIN RACE, the Awa Bank Science and Culture Foundation, the Grant-in-Aid for Young Scientists (B) from Japan Society for the Promotion of Science No. 24760138 and No. 16 K18017, TEPCO Memorial Foundation and Hatakeyama Culture Foundation. We would like to show our special thanks for the foundations.

References

1. Furukawa A, Okuma K, Tagaki A (1998) Basic study of low head water power utilization by using darrieus-type turbine. Trans JSME 64(624):2534–2540
2. Kanemoto T, Inagaki A, Misumi H, Kinoshita H (2004) Development of Gyro-type hydraulic turbine suitable for shallow stream (1st Report, rotor works and hydroelectric power generation). Trans JSME 70(690):413–418
3. Tanaka K, Hirowatari K, Shimokawa K (2013) A study on darrieus-type hydro-turbine toward utilization of extra-low head natural flow streams. Int J Fluid Mach Syst 6(3):152–159
4. Matsushita D, Tanaka K, Watanabe S (2014) Experimental and numerical investigations on performances of darrieus-type hydro-turbine with inlet nozzle. Int J Fluid Mach Syst 7(4):151–159
5. Matsui J (2010) Internal flow and performance of the spiral water turbine. Turbomachinery 38(6):358–364
6. Ikeda T, Iio S, Tatsuno K (2010) Performance of nano-hydraulic turbine utilizing waterfalls. Renew Energy 35:293–300
7. Nakajima M, Iio S, Ikeda T (2008) Performance of savonius rotor for environmentally friendly hydraulic turbine. J Fluid Sci Technol 3(3):420–429
8. Iio S, Uchiyama F, Sonoda C, Ikeda T (2009) Performance improvement of savonius hydraulic turbine by using a shield plate. Turbomachinery 37(12):743–748
9. Kanemoto T, Kaneko M, Tanaka D, Yagi T (2000) Development of counter-rotating type machine for water power generation (1st Report, counter-rotating type generator and axial-flow runners). Trans JSME 66(644):1140–1146

10. Kanemoto T, Tominaga K, Tanaka D, Sato T, Kashiwabara T, Uno M (2002) Development of counter-rotating type machine for hydroelectric power generation (2nd Report, hydraulic performance and potential interference of counter-rotating runners). Trans JSME 68(676):3416–3423
11. Tanaka D, Kanemoto T (2006) Development of counter-rotating type machine for hydroelectric power generation (3rd Report, design materials for solidity of axial flow runners). Trans JSME 72(715):686–692
12. Shigemitsu T, Fukutomi J, Tanaka C (2014) Challenge to use small hydro power by contra-rotating small hydro-turbine. In: Proceedings of World Renewable Energy Congress 2014, London, UK
13. Sonohata R, Fukutomi J, Shigemitsu T (2012) Study on the contra-rotating small-sized axial flow hydro turbine. Open J Fluid Dynam 2(4A):318–323
14. Shigemitsu T, Fukutomi J, Tanaka C (2013) Performance and internal flow of contra-rotating small hydro turbine. In: Proceedings of the ASME fluids engineering division summer meeting, Incline village, Nevada, USA, pp. FEDSM2013–16274
15. Shigemitsu T, Fukutomi J, Tanaka C (2013) Unsteady flow condition of contra-rotating small-sized hydro turbine. In: Proceedings of the 12th Asian international conference on fluid machinery, Yogyakarta, Indonesia, pp. AICFM12–0019
16. Shigemitsu T, Takeshima Y, Tanaka C, Fukutomi J (2015) Influence of spoke geometry on performance and internal flow of contra-rotating small-sized hydroturbine. In: Proceedings of the ASME-JSME-KSME joint fluids engineering conference 2015, Seoul, Korea, pp. AJK2015–22116
17. Nan D, Shigemitsu T, Shengdun Z, Takeshima Y (2018) Study on performance of contra-rotating small hydro-turbine with thinner blade and longer front hub. Renew Energy 117:184–192

Chapter 25
Low-Cost Solar Selective Absorbers by Electrodepositing Technique

Kamil M. Yousif and Sayran A. Abdulgafar

25.1 Introduction

Increasing interest in the exploitation of renewable energy sources led to more interest in study, production, and characterization of high-thermal-stability selective surfaces that have high solar absorptance (α) and low thermal emittance (ε). Selective surfaces have radiative characteristics which vary with wavelength. A spectrally selective surface combines high (α) through the solar wavelength region (0.3–2.5 μm) with low (ε) in the long wavelength region (2.5–50 μm) [1]. Several techniques can be used to produce these coatings, generally divided into chemical, paints, and vacuum deposition techniques. Absorbing paint coatings can be produced cheaply, but the thermal emittance of these paints is high [2]. Absorber coatings obtained by vacuum deposition techniques are more expensive than coatings produced by other methods [3]. In general, the chemical methods have been the most widely used techniques for low temperature solar absorber applications and more economical than the others. Some of the well-known commercial absorbers developed by chemical techniques are black chrome, nickel pigmented aluminum oxide, etc. The black nickel (BN) surface was developed by Tabor group since 1964 [4], coating was prepared by electroplating method on galvanized iron sheet, and values of $\alpha = 0.81$ and $\varepsilon = 0.17$ were obtained. Patel et al. [5] investigated BN coating on copper (Cu) substrate and investigated the optimum optical properties through changing the coating parameters. BN selective absorber coating has been deposited by spray pyrolysis method on commercially available aluminum and galvanized iron substrates [6]. Yousif [7] modified method of electrodeposited (ED) BN. He found that the optical properties of BN coating were improved by using Ni-Zn coating. Wang et al. [8] demonstrate a

K. M. Yousif (✉)
Faculty of Science, Department of Environment Science, Zakho University, Zakho, Iraq

S. A. Abdulgafar
Faculty of Science, Department of Physics, Dohuk University, Dohuk, Iraq

A. Sayigh (ed.), *Renewable Energy and Sustainable Buildings*, Innovative Renewable Energy, https://doi.org/10.1007/978-3-030-18488-9_25

solution-processed plasmonic Ni nanochain-Al_2O_3 selective solar thermal absorber with a high α (>90%) and a low ε (<10%). Zhenxiang et al. [9] reported the preparation of Ni-Al_2O_3 absorbing coatings by a facile aqueous solution-chemical method. During ED of BN, multiple reduction and oxidative processes occurred and produced a complex film, containing α-$Ni(OH)_2$, NiOOH, Ni_2O_3, NiO, and metallic Ni, as indicated by X-ray photoelectron spectroscopy study [10]. Estrella-Gutiérrez et al. [11] show that a metallic Ni interlayer is required in order to achieve thermal stability of BN coatings on copper. Nady et al. [12] reported that a nanoparticles layer of bright Ni base was deposited on copper substrates using ED technique before spraying the paint and will improve the stability of the absorber paint. Nanoparticles of Ni-oxide (nano-NiO_x) have been synthesized using electrochemical and sol-gel routes [13].

In this research, low-cost solar selective absorbers have been obtained by ED. And we tried to get rid of some structures or materials which are harm to the environment, i.e., car batteries, by using lead sheets [from recycled car battery] as an anode during ED of black nickel (BN) coating.

25.2 Experimental Technique

Each sample analyzed was a thin film coating deposited on two types of metallic substrates (copper, and brass). The experimental technique and the results using brass substrates will be published soon.

25.2.1 Sample Preparation

It is well known that metal cleaning and activation play an important role in metal finishing. The smoothness, uniformity, brightness, and adherence of the substrate depend on a clean and active metal surface. The presence of any film (e.g., oxide) between the substrate and coating reduces the adhesion. For this reason, the preparatory treatment for the substrate surface should be done to reduce the presence of such films. Copper (Cu) sheets were polished mechanically with emery paper (320, 400, and 1200) grade to remove the oxide layer and washed with distilled water. They were then chemically polished, cleaned, and dried. This was followed by immersing them in a plating tank for ED of BN coating.

25.2.2 Electrodeposition of Black Nickel Coatings

A schematic representation of the plating equipment is shown in Fig. 25.1. It consists of a low-voltage variable power supply type of Metcix. Plating tank is surrounded by hot water bath type of Clifton with temperature control unit to provide the desired solution temperature. An electronic thermometer type of

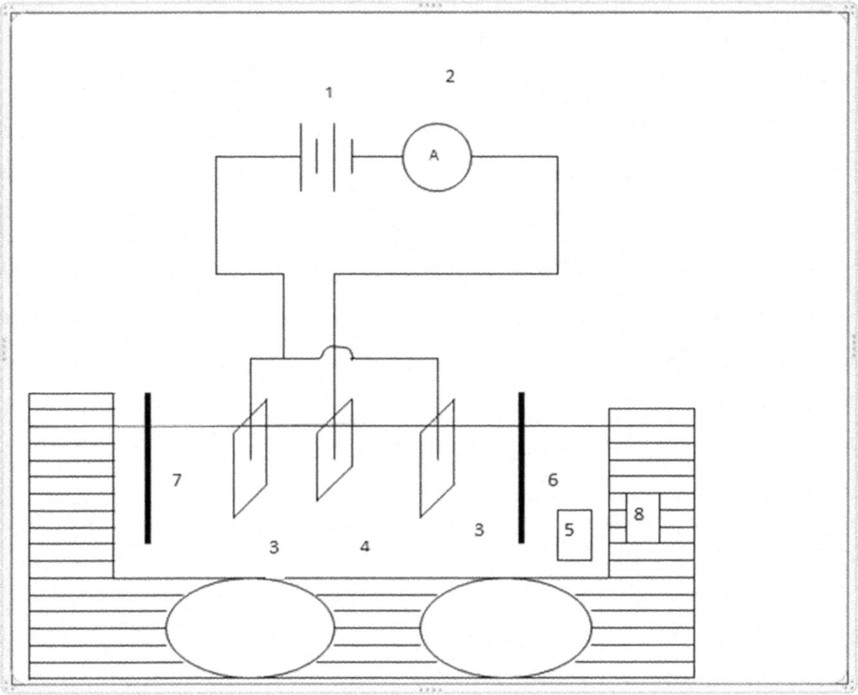

Fig. 25.1 Schematic diagram of electrodeposition apparatus. (1) Low-voltage supply, (2) ammeter, (3) anodes, (4) substrate (cathode), (5) solution tank, (6) thermometer, (7) pH meter, (8) hot water bath

Table 25.1 The plating bath conditions

Parameter	Value	Parameter	Value
Temperature	30 – 40 C°	pH	4 – 5
Current density	1 A/dm^2	time	50 sec

HANNA with stainless steel probe is also used to measure the temperature of the electrolyte solution. Lead sheet [from recycled car battery] were used as anodes. The plating bath conditions are given in Table 25.1. For measuring the pH of the solution, a pH meter type of HANNA is used. Commercially available Cu substrates (5.6 × 5.5 cm^2) were cleaned and then coated with BN film from a bath containing Ni chloride with additions of Zn chloride, ammonium chloride, and sodium thiocyanate. Ni and thiocyanate ions are reduced preferentially at the cathode resulting in the formation of a BN deposit.

25.2.3 Optical and Non-Optical Characterizations

The samples were examined by optical methods mainly to determine the sample efficiency. Non-optical methods were also used and focused on structural, morphology, and surface texture measuring of the films.

25.2.3.1 Measurements of Total Reflectance

For determining absorptance-α or emittance-ε of coatings, it is usually most convenient to measure reflectance (R) to calculate α and ε from it. The measurements of total or diffuse R of samples were performed with instruments equipped with integrating spheres. The Perkin-Elmer (model Lambda 950) double beam spectrophotometer with an integrating sphere coated with barium sulfate [14] used to measure the R in the λ ranges 0.2–0.9 μm. The total R in IR ranges were calculated from the measurements of specular and diffuse reflectance [diffuse (R_d) + specular (R_s)]. Perkin-Elmer spectrum GX (FTIR) system was used for specular R measurements, and Shimadzu Corporation IR-470 was used for R measurements in the IR wavelength range, from 2 to 50 μm.

25.2.3.2 The Topography of the Coatings

The photographs of the films were taken by optical microscopy, as well as scanning electron microscope (SEM).

25.2.3.3 Evaluation of Surface Roughness

The measurement of roughness in this study is taken as the roughness average value, R_a, and is expressed in micrometers (μm). This parameter is also known as the arithmetic average value (AA) or centerline average (CLA) [15]. The roughness of the coatings was measured by a stylus-profilometer type of Tyler Hobson.

25.2.3.4 X-Ray Diffraction (XRD)

The X-ray diffraction for thin films from BN on Cu substrates as plated and after heat treatment (250 °C) were studied by using X-ray diffractometer type of Pw-1840 X-ray diffractometer (Philips) with the following parameters: The target used in x-ray tube is pure Cu, the ray source CuKα with wavelength ($\lambda = 1.54$ Å) using a Ni filter. With voltage (40 KV) and current (20 mA), the scanning speed is (3 deg/min). The range of scanning is ($2\theta = 10^\circ - 90^\circ$). Bragg's law [$n\,\lambda = 2d \sin\theta$] describes the situation when constructive interference occurs. Where n is an integer, λ the wavelength, d the distance between the atoms in parallel consecutive planes, and θ the angle between, these planes and the incident beam. 2θ is the angle between the incident and the diffracted beam.

25.2.3.5 Measuring the Thickness of the Films

Weight method was used to measure the thickness of a coating.

25.2.4 Durability Tests

Samples of BN coatings were subjected to accelerated ageing tests. After measuring the initial radiative properties of the absorbers, i.e., R, α, ε, and the initial characterization of the structure, the samples were subjected to ageing tests, and then they were recharacterized by measuring R and determining the values of α and ε.

25.2.4.1 Constant Temperature Exposure

The samples were placed in a furnace which was thermostatically controlled. The test was carried out in the temperatures 100, 150, 200, 250, and 300 °C during 48 h.

25.2.4.2 Temperature Cycling Test

The durability of the samples was tested by subjecting them to temperature cycling between ambient temperature and 100 °C. The duration of each cycle was about 35 min. An electric oven [medical electronics] type was used. After each 10 cycles, the samples were removed and the values of (α) and (ε) were determined.

25.2.4.3 Humidity Exposure

Samples were placed in a humid environment inside a suitable glass box. The humidity testing (HMT) at (90 ± 5) % using a chamber of (LEYBOLD DIDACTIC GMBH), and the duration of exposure was 24 h.

25.3 Results and Discussion

It was clear from the surface topography of the film that it is a uniform and homogenous film (from microscope photograph). The uniformity of the coating thickness is dependent on the deposition conditions (DC). It has been found that DC play an important role in determining the optical and thermal properties of thin films. As the thickness of this coating increases, the surface topography appears smoother. The average roughness value R_a of the film or centerline average (CLA) obtained by the stylus-profilometer instrument indicates that these coatings are rather smooth and their value is 0.17 μm. The durability tests were carried out on BN/Cu sample in the form of accelerated ageing tests which included constant temperature exposure up to 300 °C, temperature cycling, and HMT. We observed a change in the coatings, color to light black after heat treatment (HT) at 250 °C. There was some voids or pores on the surface. It appears as a rather rough surface whose average roughness value R_a (CLA) was 0.41 μm. The effect of the HMT on the coating was revealed by visible signs that the color of the coating was lightened after this test. This change is linked with changes in the *R* spectra. HMT produces a surface with a speckled appearance as the result of localized condensation and such condensation stains are difficult to be removed. The average roughness value R_a (CLA) of BN /Cu coating after the humidity test was 0.26 µm, which indicates that it is rougher than as prepared film. The total *R* spectrum for BN/Cu before and after the durability tests in the range between 0.2 µm and 0.85 µm is shown in Fig. 25.2. It can be seen from this figure that the *R* for as prepared sample is very small (less than 6%) and almost constant with the wavelength (λ). These films have high absorbing in this region of the spectrum. The *R* for both the samples (after HT at 250 °C and HMT) increased gradually with the λ, and its value is greater than as prepared sample. The total *R* versus a λ in IR region, which was calculated from specular and diffuse spectra, is shown in Fig. 25.3. It can be seen from figure that there is a gradual increase from low to high *R* with the increasing λ. Both (α) and (ε) values and selectivity factor (α/ε) for BN/Cu before and after the durability tests are given in Table 25.2.

The value of α is rather in good agreement with the earlier reports. The increasing in α may be due to diffuse of impurity of some atoms such as Zn from the substrate (ST) as a result of exposure to high temperatures. The value of α increases with increasing the Zn concentration [5], whereas the high temperatures increase the diffusion of atoms and molecules from the ST. We can say that an improvement for the optical properties happened for the coating after the temperature ageing (a slight increasing in selectivity factor α/ε). However, we can consider a BN coating was degraded in humid environments. High humidity may create hydrolytic reactions and is the main source of electrochemical corrosion, which causes an enhanced degree of oxidation of metals. The degradation (on appearance) of black nickel

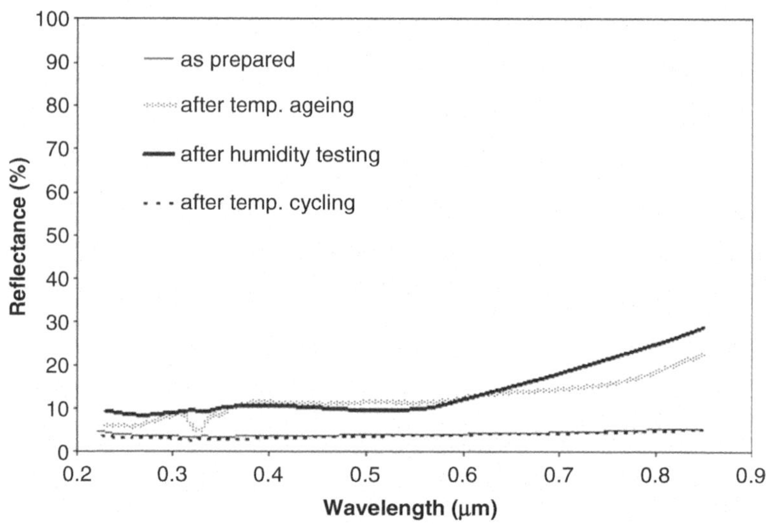

Fig. 25.2 The total reflectance spectrum for BN /Cu before and after the durability tests

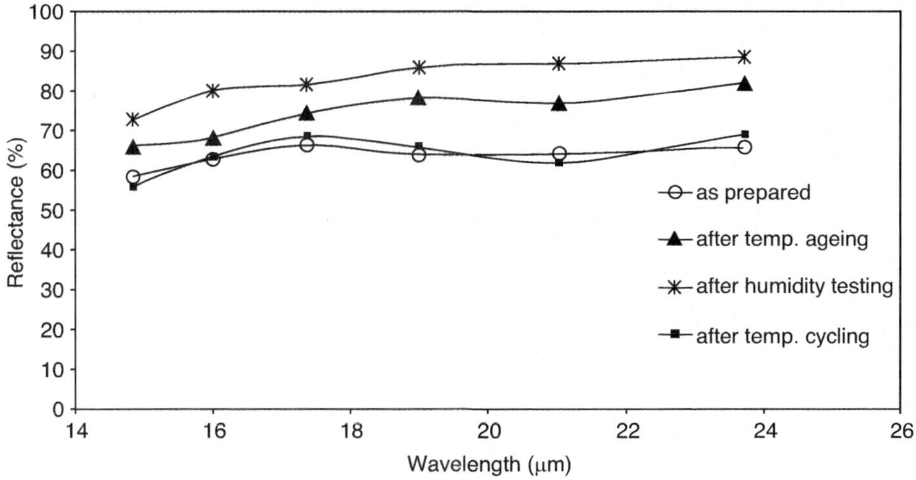

Fig. 25.3 The total reflectance versus a wavelength in IR region

Table 25.2 The values of α and ε (together with selectivity factor α/ε) for BN coating on Cu before and after the durability tests

Black nickel/Cu	α	ε	α/ε
As prepared	0.95	0.36	2.63
After heat treatment	0.84	0.25	3.36
After humidity test	0.81	0.17	5.4
After temp. cycling	0.94	0.35	2.68

(BN) coating, including the presence of some nodules and pitting in the optical microscope photograph. The XRD pattern for electroplated BN coating on Cu ST (see Fig. 25.4a) illustrates that existence of peaks related to the phase of Ni-Zn, such as, Ni_3Zn_{22} phase (JCPDS 10-209) at $2\theta = 42.9°$. The peaks at $2\theta = 73.7°$ and 89.8° are due to mixing for the phases resulting from the ST and Ni_3Zn_{22} component.

XRD pattern for BN coating on Cu ST after thermal exposure at 250 °C (Fig. 25.4b) reveals the presence of a very weak peak at $2\theta = 36°$. It is revealed that it is related to Cu_2O phase (JCPDS 5-667). This result is in a good agreement with the XRD results obtained by Patel et al. [5]. They also observed presence of some Cu oxides such as CuO and Cu_2O after HT, and the coatings increase in the oxide content with the increasing HT. Formation of this oxide layer is associated with a decrease in the α of the coating. The thickness of BN films on Cu ST was ~0.83 ± 0.2 μm.

The efficiency (η) for the coatings under study was calculated using the following equation:

$$\eta = \{I\,\alpha - \varepsilon\,\sigma\,(T_p^4 - T_w^4)\}/I$$

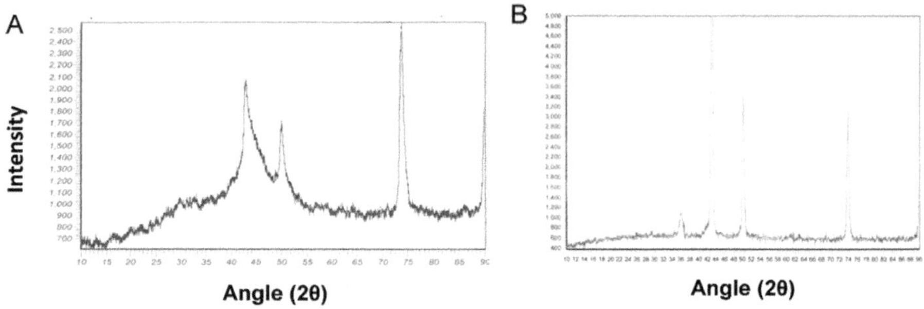

Fig. 25.4 (**a**) XRD pattern for electrodeposited of black nickel (BN) on Cu substrate. (**b**) XRD Pattern for BN on Cu substrate after thermal exposure at 250 °C

where $I\alpha$ is the absorber flux, α solar absorptance; I is the solar insolation; ε is the thermal emittance; T_p and T_W are the temperatures of the plate absorber and the enclosure wall, respectively (in K); and σ is the Stefan-Boltzmann constant (5.67×10^{-8} W/m^2 K^4). All η values were calculated using $I = 800$ W/m^2 (the value of solar radiation). A reduction occurred in η after durability tests, for BN on Cu ST, η for As-prepared BN = 69 %, while η for BN after thermal ageing (250 °C) = 66 %.

25.4 Conclusions

1. For black nickel (BN) coating on Cu substrate, the values of α and ε were 0.95 and 0.36, respectively, when $\alpha/\varepsilon = 2.63$ and $\eta = 69$ %. The coating shows a degradation in radiative properties after the thermal exposing at 250 °C due to decreasing in α which was 0.86. The efficiency of the coating decreased to η = 66%. The degradation of the coating after the thermal exposing may be related to the following: oxidation of the Cu substrate, thermochemical or topographical changes, diffusion effects, and phase transformations.
2. The coating also shows a little degradation in radiative properties after humidity testing (relative humidity, RH = 90%) due to decreasing in α (α = 0.81). This may be related to electrochemical corrosion on the surface of the coating.
3. Temperature cycling test (at 100 °C for 10 cycles) does not show a noticeable influence on the optical properties of the coating.
4. The X-ray analysis data for as prepared BN coating indicated that it consisted essentially from the phase opulent with nickel-zinc (Ni_3Zn_{22}). XRD pattern for BN coating on copper substrate after thermal ageing at 250 °C reveals the presence of Cu_2O phase.

References

1. Seraphin BO, Hahn RE (1978) Physics of thin films, vol vol. 10. Academic, New York, pp 1–69
2. Orel ZC (1999) Sol Energy Mater Sol Cells 57:291
3. Wäckelgård E, Hultmark G (1998) Sol Mater Sol Cells 54:165
4. Tabor H (1978) Status report on selective surfaces. In: Sun-mankind's future source of energy, vol vol. 2. Pergamon Press, Elmsford, NY, pp 829–836
5. Patel SN, Inal OT, Singh AJ, Scherer A (1985) Optimization and thermal degradable study of black Ni solar coll. coatings. Sol Energy Mater 11:381
6. Madhusudana M, Sehgal HK (1982) Appl Energy 10:65
7. Yousif KM (2007) Modification of electrodeposited black nickel selective coating for solar photo thermal conversion. JAAUBAS 4:228–236
8. Wang X, Li H, Yu X, Shi X, Liu J (2007) High-performance solution-processed plasmonic Ni nanochain-Al2O3 selective solar thermal absorbers. Applied Phys Lett 101:20
9. Zhenxiang Li, Zhao J, Ren L (2012) Aqueous solution-chemical derived Ni–Al_2O_3 solar selector absorption coatings. Sol Energy Mater Sol Cells 105:90
10. Lizama-Tzec FI, Macías JD, Estrella-Gutiérrez MA, Cahue-López AC, Arés O, Coss R de, Alvarado-Gil JJ, Oskam G (2015) Electrodeposition and characterization of nanostructured black nickel selective absorber coatings for solar–thermal energy conversion. J Mater Sci Mater Electron 26(8):5553–5561
11. Estrella-Gutiérrez MA, Lizama-Tzec FI, Arés O, Oskam G (2016) Influence of solar absorber coating based on BN electrode onto Cu. Electro Acta 213(20):460
12. El- Nady J, Kashyout AB, Ebrahim S, Soliman MB (2016) Nanoparticles Ni electroplating and black paint for solar collector applications. Alex Eng J 55(2):723
13. Danial AS, Awad MI, Al-Odail FA (2017) Effect of different synthesis routes on the electrocatal. properties of NiOX nanoparticles. J Mol Liq 225:919
14. Labsphere (2017) Technical guide, integrating sphere theory & applications. Labsphere Inc, North Sutton, NH. www.labsphere.com
15. Whitehouse D (2012) Surfaces and their measurement. Butterworth-Heinemann, Boston. ISBN 978-0080972015

Chapter 26
Effects of Mirror Geometry on the Optical Efficiency of a Linear Fresnel Reflector (LFR)

M. P. G. Sirimanna and J. D. Nixon

26.1 Introduction

The power generation industry has been influenced by a significant growth of renewable on-site power plants in recent years. Solar thermal power plants, frequently known as concentrating solar thermal power (CSP) plants, will play an important role [1]. CSP is an electricity generation technology that converts solar irradiation concentrated onto a small area to heat energy. This technology is particularly promising in regions where a high amount of direct normal solar irradiance is received. According to the International Energy Agency Concentric Solar Power roadmap, CSP can be a competitive source of bulk power in peak and intermediate loads in the sunniest regions by 2020 and of base load power by 2025–2030. With appropriate support, by 2050, CSP could provide 9.6% of global electricity from solar power alone [2].

In CSP plants, sunlight is reflected to a receiver, where generated heat is collected by a carrier fluid, and this heat energy is subsequently used to power a turbine and generate electricity. A CSP plant consists of concentrators and a receiver, where different mirror numbers, widths, spacing and types can be used. All these parameters impact thermal performance and optical efficiency. Designing CSP systems is difficult as they are sophisticated optical and heat exchanger systems, where complex interplay of all heat transfer modes is present. As far as the operation of a CSP system is concerned, they undergo lengthy start-up and shut down operations due to the variation of solar radiation during the day. Due to this discontinuity, analysis of plant operation has to consider not only the design of the components but also the analysis of the performance of the system under different operating conditions. The linear Fresnel reflector (LFR), also known as linear Fresnel collector (LFC), is a CSP technology highlighted due to its low capital cost, simplicity and robustness amongst the other

M. P. G. Sirimanna · J. D. Nixon (✉)
Faculty of Engineering, Environment and Computing, Centre for Fluid and Complex Systems, Coventry University, Coventry, UK
e-mail: jonathan.nixon@coventry.ac.uk

A. Sayigh (ed.), *Renewable Energy and Sustainable Buildings*, Innovative Renewable Energy, https://doi.org/10.1007/978-3-030-18488-9_26

common three technologies: parabolic trough collector (PTC), power tower (PT) and parabolic dish. The main drawback of the LFR is a low concentration ratio, which limits the performance of these systems significantly [3, 4] . Therefore, improving the optical performance of these systems is an important step towards promoting and commercialising this technology. Furthermore, the first step to start the analysis of the entire system, as mentioned above, is to have a comprehensive optical modelling of the plant.

Different researchers have addressed optical modelling of LFR to different levels of detail. Nixon and Davies used Optica, a software package developed within Mathematica, to obtain optical efficiencies and an incident angle modifier (IAM) for different spacing arrangement of mirrors [5]. A detailed optical analysis has been carried out by Montes et al. [3], where an in-house code was developed for ray tracing. Simulations on different geometries based on the FRESDEMO prototype located at the Plataforma Solar de Almería, Spain, have been carried out in their work. However the main focus on their research was to compare two LFR configurations, namely, central linear Fresnel reflectors and compact linear Fresnel reflectors. Another optical simulation on the FRESDEMO prototype has been carried out by Moghimi et al. [6] as a reference to their study on optimisation of a trapezoidal cavity absorber for a LFR plant. The purpose of their optical simulation was to obtain the heat flux distribution on the absorber and thereby to find optimum geometrical parameters for the receiver design. Qiu et al. [7] conducted a detailed optical and thermal analysis study for a novel design by developing an optical model using Monte Carlo ray tracing (MCRT). They analysed the solar flux distribution around the absorber tube, and the optical efficiency was calculated. Furthermore, the effects of tracking and slope errors were studied in detail. Abbas and Martínez-Val took a different approach by conducting an analytical study for optical analysis [8]. They specially focused on mirror width and shifts taking the FRESDEMO plant as an example. Further studies in connection to optical analysis are found in the literature such as [9–11]. According to the current literature, there is a need for a comprehensive optical analysis of LFRs considering all the critical geometrical parameters of the mirror field and optical parameters. In this paper, an optical model is developed, and the FRESDEMO prototype is selected as the reference to identify possible optical efficiency improvements. Furthermore, the effects of geometrical and optical parameters on optical efficiency and optical concentration ratio are established.

26.2 Methodology

Optical analysis is started with defining main optical parameters of an LFR system. Starting with the FRESDEMO prototype as a reference, simulations are then carried out by changing mirror number, receiver width and height and mirror focal length.

26.2.1 Optical Modelling

The following definitions and equations are used in this paper:

Optical efficiency, $\eta_{opt}(\theta)$, for a CSP plant is the ratio of the absorbed energy by the receiver to the total solar energy incident on the aperture area of mirrors [12, 13], as indicated in Eq. (26.1):

$$\eta_{opt}(\theta) = \frac{Q_{abs}}{Q_s} \tag{26.1}$$

Geometrical concentration ratio (GCR) is given by Eq. (26.2):

$$\mathrm{GCR} = \frac{\sum_{i=1}^{n} A_{a,i}}{A_r} \tag{26.2}$$

where $A_{a,i}$ and A_r refer to the aperture area of *ith* mirror and the receiver area.

Optical concentration ratio (OCR) of a parabolic trough collector (PTC) is typically expressed as in Eq. (26.3):

$$OCR = \frac{\frac{1}{A_r}\int I_r dA_r}{I_a} \tag{26.3}$$

where I_r and I_a are flux on the receiver and flux on the aperture, respectively. However this definition has to be adapted accordingly for an LFR where there may be a large number of mirrors instead of one parabolic mirror. Assuming an LFR with *n* number of mirrors and having uniform flux on a flat receiver, Eq. (26.4) is obtained:

$$OCR = \frac{I_r}{\left(\sum_{i=1}^{n} I_{a,i} A_{a,i}\right) / \sum_{i=1}^{n} A_{a,i}} \tag{26.4}$$

Optical error (σ_{optical}) is considered to be consisted of slope error and specularity error as given by Eq. (26.5) [14, 15]:

$$\sigma_{\text{optical}} = \sqrt{4\sigma_{\text{slope}}^2 + \sigma_{\text{specularity}}^2} \tag{26.5}$$

Fig. 26.1 View of the liner Fresnel reflector (LFR) plant in Plataforma Solar de Almería(PSA), Spain

26.2.2 Simulation Tool

There are several ray tracing tools used at a commercial and research level such as SolTrace, Optica and Tonatiuh [16]. Some researchers have developed in-house ray tracing codes mainly using MCRT methods [3, 11]. SolTrace is a ray tracing tool which has been widely used and proven to provide accurate optical analyses for a variety of CSP applications [6, 7, 15]. SolTrace is a ray tracing tool developed by the National Renewable Energy Laboratory (NREL) and openly available [14].

26.2.3 FRESDEMO

To overcome the lack of information on LFR plants, a prototype was erected at the Plataforma Solar de Almería (PSA) in Spain, nearly a decade ago as seen in Fig. 26.1. This prototype was known as FRESDEMO plant which is 100 m long and covers a land area of 2100 m^2. 1200 mirrors were installed in 25 parallel rows, and it has a primary mirror surface area of 1433 m^2. The plant generates direct steam and has been designed for a maximum temperature of 450 °C and a pressure of 100 bar [17].

26.2.4 Geometrical and Optical Parameters Selected

Schematic diagram of a LFR plant is indicated in Fig. 26.2. Receiver height from the mirror field, receiver width and mirror width are given by h, w_r and w_m, respectively. Table 26.1 provides a summary of geometrical and optical parameters of the FRESDEMO plant, which have been used in SolTrace simulations.

The receiver was assumed to be a rectangular plane with corresponding dimensions as in [3]. Solar absorptance on the receiver typically decreases with the

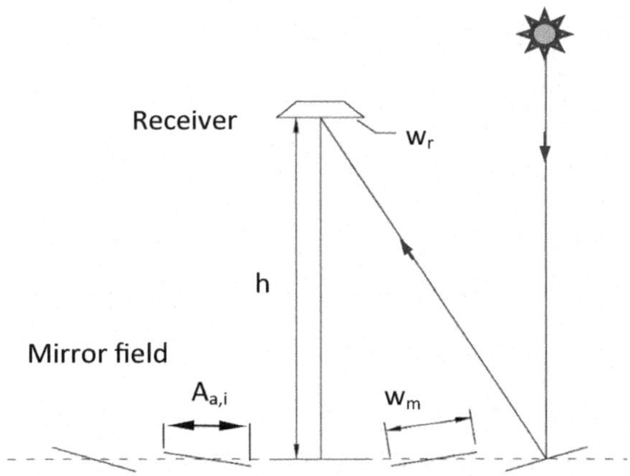

Fig. 26.2 Schematic diagram of a LFR plant

Table 26.1 FRESDEMO plant configuration and optical properties

Mirror width	0.6 m
Number of rows of mirrors	25
Receiver width	0.5 m
Receiver height from the mirrors	8 m
Total plant width	21 m
Plant length	100 m
Plant orientation	N-S mirror axis
Solar reflectivity of mirrors	0.93

increase of angle of incidence, and this change is significant for high incident angles, as seen in Fig. 26.3. Incorporating the indent angle dependency of solar absorptivity to the ray tracing codes is therefore important, particularly for large mirror fields where incident angle can take higher values. This variation has been included in the SolTrace code based on the work of Tesfamichael and Wackelgard [18]. Optical error has been taken as 5 mrad [3, 15, 19] unless mentioned otherwise.

26.3 Results and Discussion

26.3.1 Validation of the Code

Montes et al. modelled the FRESDEMO prototype using SolTrace for validating a ray tracing programme they developed [3]. They obtained 60.7% for optical efficiency for the plant, where this simulation provides 61.0% for the optical efficiency. It indicates an excellent agreement between two simulations. Furthermore, Qiu et al. simulated a LFR plant taking the same parameters from FRESDEMO. The designed LFR could achieve the instantaneous optical efficiency of 65.0% [7].

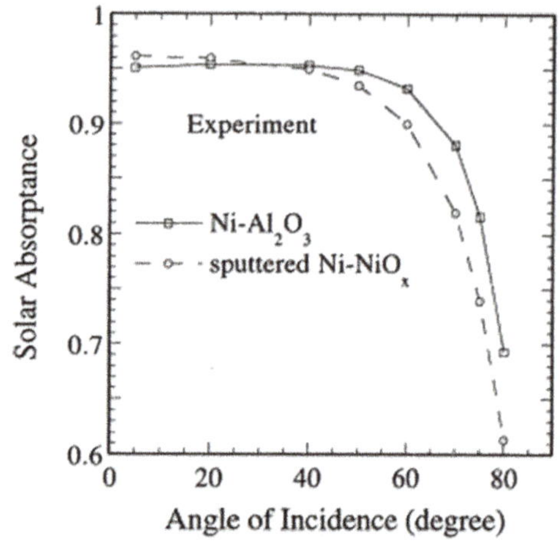

Fig. 26.3 Solar absorptance vs. incident angle for receiver [18]

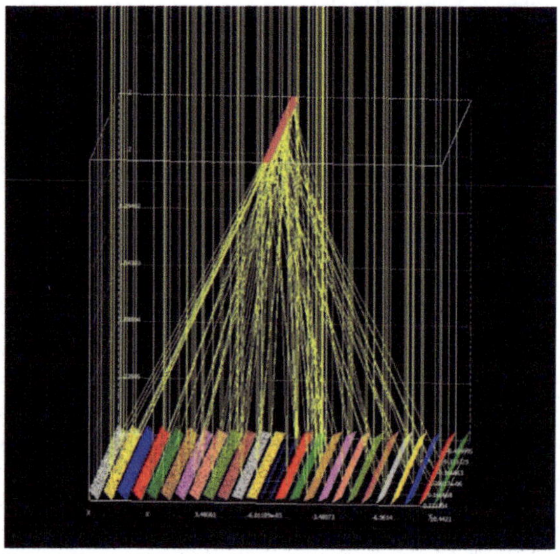

Fig. 26.4 Ray tracing of FRESDEMO prototype using SolTrace

However they used slightly different mirrors in their simulation. All these cases were simulated for solar noon as seen in Fig. 26.4.

26.3.2 Changing Mirror Number

Mirror number of FRESDEMO plant was changed during simulations from 3 to 41. Spacing between two mirror rows and other geometrical parameters were kept constant throughout. Figure 26.5 indicates the effects changing the mirror number of the plant. Initially the plant was simulated without considering incident angle-dependent

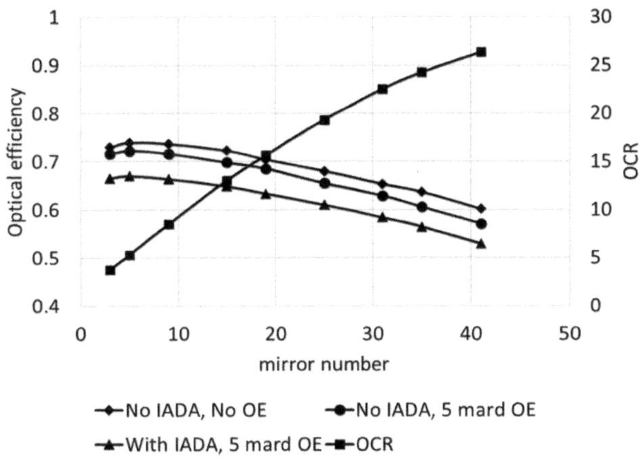

Fig. 26.5 Variation of optical efficiency and OCR against mirror number

absorptivity (IADA), as shown in Fig. 26.3, and optical error (OE). Optical efficiency was 68.0% for this case, and it dropped to 65.5% as a result of adding a 5 mrad OE to the simulation. Optical efficiency was further dropped to 61% by adding IADA to the simulation. When average drop of efficiencies is considered, 2.3–4.8% optical efficiency drops are observed for integrating 5 mrad OE and IADA to the simulation, respectively. This indicates that IADA is more important compared to a standard OE for calculating optical efficiencies. An average 7.0% efficiency drop is observed when both IADA and OE are integrated to the simulation code highlighting their significance to achieve an accurate optical efficiency. Change of optical concentration ratio (OCR) is also indicated in Fig. 26.5. It is noted that OCR for FRESDEMO plant is 19.3.

26.3.3 Changing Receiver Width

Figure 26.6 shows the effects of changing receiver width on optical efficiency and OCR. Optical efficiency is rapidly increasing with the receiver width from the current configuration of the plant and reaching nearly a constant level around 80%. According to the results, a significant 20% of efficiency improvement can be attained by increasing the receiver width by 0.3 m. However this has been attained by an OCR drop of three. Maximum possible OCR for the current plant configuration is 20. It is noted that the FRESDEMO plant has got nearly the best possible receiver width in terms of OCR under this configuration. Figure 26.7 shows OCR and geometrical concentration ratio (GCR) in one diagram. It is seen that both decrease with increase of receiver width. However OCR displays a rapid decrease after 0.6 m, which is equal to the width of the mirrors of FRESDEMO plant.

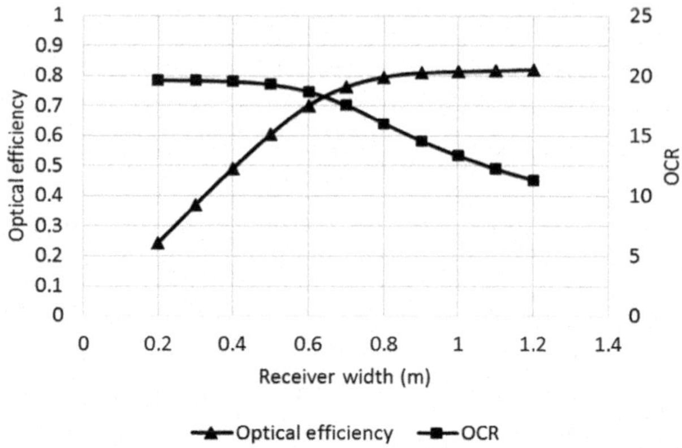

Fig. 26.6 Variation of optical efficiency and OCR against receiver width

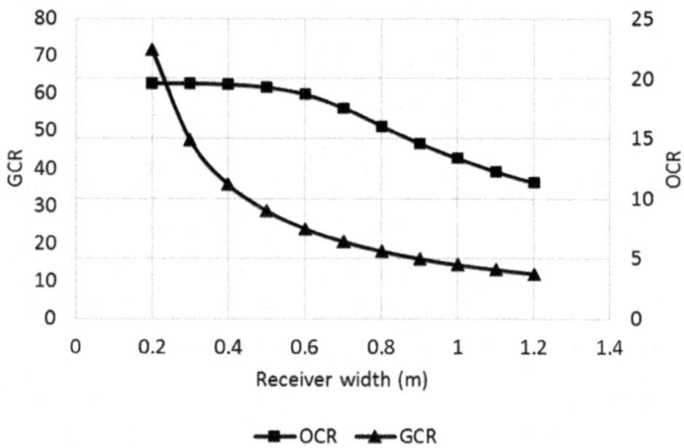

Fig. 26.7 Variation of GCR and OCR against receiver width

26.3.4 Changing Receiver Height

It is interesting to observe the effects of changing receiver height as seen in Fig. 26.8. When optical errors are not included in the simulation, optical efficiency is purely increasing reaching a constant level. However, when the optical error is incorporated as in the case of actual plants, a maximum value of 61.6% is observed for optical efficiency for 12 m receiver height. More importantly, optical efficiency is starting to drop if the receiver height is changed from this value. The same trend is seen for OCR where 20.1 is observed for 12 m receiver height.

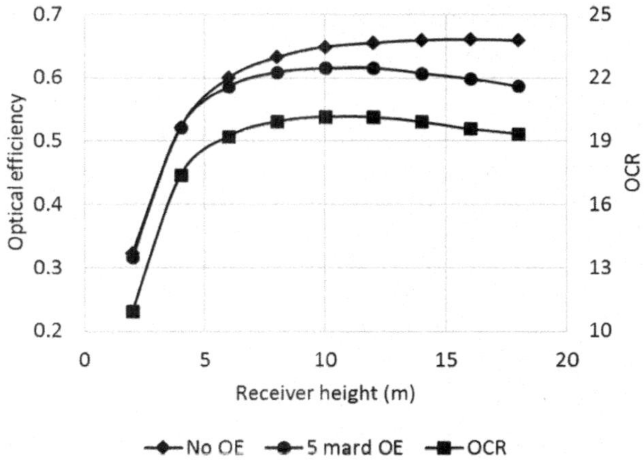

Fig. 26.8 Variation of optical efficiency and OCR against receiver height

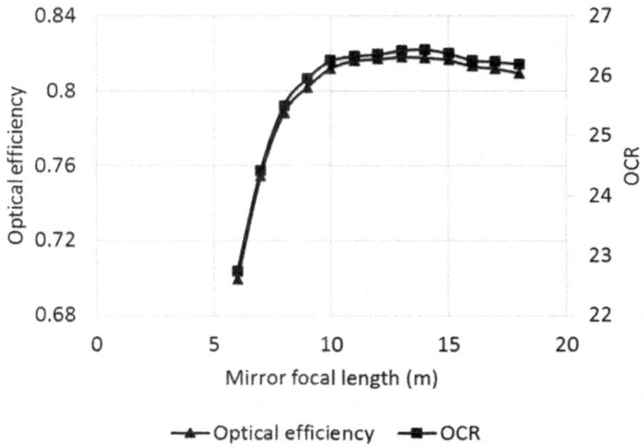

Fig. 26.9 Variation of optical efficiency and OCR against mirror focal length

26.3.5 Changing Mirror Focal Length

Flat mirrors were changed with slightly curved parabolic shaped mirrors during the simulation. The same focal length was used for all mirrors at a time. The range of the focal lengths was selected based on the closest and furthest mirrors to the receiver. Similar focal length values are found in a study of Qiu et al. [7]. It is observed that both optical efficiency and OCR are increased when curved mirrors are used. Maximum optical efficiency and OCR observed are 81.8% and 26.4, respectively, for 14 m focal length. It is also noted that these values show less sensitivity to focal length change between 10 m and 16 m (Fig. 26.9).

26.4 Conclusions and Future Work

Optical modelling has been carried out using SolTrace to identify possible improvement of optical efficiency and optical concentration ratio (OCR) of LFR systems. FRESDEMO prototype has been used as the base case for the simulations, where mirror number, receiver width, receiver height and mirror curvatures were changed. It is noted that the optical efficiency is decreasing with increasing of mirror number indicating that a low number of mirrors is preferred in an LFR plant. Furthermore, results indicate that there is a maximum level of performance that can be obtained by reducing the number of mirrors. However it is noted that OCR is dropped by reducing the mirror number causing low temperatures in the absorber tube penalising the overall plant efficiency. Therefore, deciding the number of mirrors should be followed by both optical and thermal analyses. As far as the receiver width is concerned, it is noted that the FRESDEMO plant has got nearly the best possible receiver width in terms of OCR under the current configuration. It is noted that optical efficiency and OCR follow the same trend with changing of receiver height. Furthermore, they both reach a maximum optical efficiency value indicating that the optimum receiver height for FRESDEMO is reached at 12 m. Using slightly curved parabolic mirrors can improve the optical efficiency significantly. For the configuration considered in the study, this optical efficiency improvement is near 33% from the current value. Since the trend of changing both optical efficiency and OCR with focal length is almost the same, an OCR improvement can be reached at the same time by 37%. This is an interesting scenario where optical efficiency increase is not penalised by OCR improvement. The same behaviour is observed for a change in receiver height, although the performance improvement is relatively low. This work also highlighted the importance of incorporating appropriate optical errors and incident angle-dependent absorptivity to simulations as neglecting those may leads to considerable errors and improper conclusions.

As a future work, the modelling of the receiver can be improved by integrating secondary concentrators and a glass cover. Furthermore, simulation can be extended to include different times and different locations.

References

1. Sørensen K et al (2016) Modelling of a cross flow evaporator for CSP application: analysis of the use of different two phase heat transfer and pressure drop correlations. Int J Therm Sci 107:66–76. http://www.sciencedirect.com/science/article/pii/S1290072916303738. https://doi.org/10.1016/j.ijthermalsci.2016.03.025
2. International Energy Agency (2010) Technology roadmap, concentrating solar power. www.iea.org
3. Montes MJ et al (2014) A comparative analysis of configurations of linear Fresnel collectors for concentrating solar power. Energy 73:192–203. http://www.sciencedirect.com/science/article/pii/S0360544214007087. https://doi.org/10.1016/j.energy.2014.06.010

4. Liu M et al (2016) Review on concentrating solar power plants and new developments in high temperature thermal energy storage technologies. Renew Sust Energ Rev 53:1411–1432. http://www.sciencedirect.com/science/article/pii/S136403211500996X. https://doi.org/10.1016/j.rser.2015.09.026
5. Nixon JD, Davies PA (2012) Cost-exergy optimisation of linear Fresnel reflectors. Sol Energy 86(1):147–156. http://www.sciencedirect.com/science/article/pii/S0038092X11003501. https://doi.org/10.1016/j.solener.2011.09.024
6. Moghimi MA, Craig KJ, Meyer JP (2015) Optimization of a trapezoidal cavity absorber for the Linear Fresnel Reflector. Sol Energy 119:343–361. http://www.sciencedirect.com/science/article/pii/S0038092X15003771. https://doi.org/10.1016/j.solener.2015.07.009
7. Qiu Y et al (2015) Study on optical and thermal performance of a linear Fresnel solar reflector using molten salt as HTF with MCRT and FVM methods. Appl Energy 146:162–173. http://www.sciencedirect.com/science/article/pii/S0306261915001762. https://doi.org/10.1016/j.apenergy.2015.01.135
8. Abbas R, Martínez-Val JM (2015) Analytic optical design of linear Fresnel collectors with variable widths and shifts of mirrors. Renew Energy 75:81–92. http://www.sciencedirect.com/science/article/pii/S0960148114005941. https://doi.org/10.1016/j.renene.2014.09.029
9. Qiu Y et al (2017) Aiming strategy optimization for uniform flux distribution in the receiver of a linear Fresnel solar reflector using a multi-objective genetic algorithm. Appl Energy 205:1394–1407. http://www.sciencedirect.com/science/article/pii/S0306261917313740. https://doi.org/10.1016/j.apenergy.2017.09.092
10. Barbón A et al (2016) Optimization of the length and position of the absorber tube in small-scale Linear Fresnel Concentrators. Renew Energy 99:986–995. http://www.sciencedirect.com/science/article/pii/S0960148116306851. https://doi.org/10.1016/j.renene.2016.07.070
11. Qiu Y et al (2016) A comprehensive model for optical and thermal characterization of a linear Fresnel solar reflector with a trapezoidal cavity receiver. Renew Energy 97:129–144. http://www.sciencedirect.com/science/article/pii/S096014811630475X. https://doi.org/10.1016/j.renene.2016.05.065
12. Wang Y et al (2015) Performance analysis of a novel sun-tracking CPC heat pipe evacuated tubular collector. Appl Therm Eng 87:381–388. http://www.sciencedirect.com/science/article/pii/S1359431115003889. https://doi.org/10.1016/j.applthermaleng.2015.04.045
13. Bellos E et al (2016) Design, simulation and optimization of a compound parabolic collector. Sustain Energy Technol Assess 16:53–63. http://www.sciencedirect.com/science/article/pii/S2213138816300169. https://doi.org/10.1016/j.seta.2016.04.005
14. Wendelin T, Dobos A (2013) SolTrace: a ray-tracing code for complex solar optical systems. National Renewable Energy Laboratory, Golden, CO
15. Zhu G (2013) Development of an analytical optical method for linear Fresnel collectors. Sol Energy 94:240–252. http://www.sciencedirect.com/science/article/pii/S0038092X13001862. https://doi.org/10.1016/j.solener.2013.05.003
16. Cruz NC et al (2017) Review of software for optical analyzing and optimizing heliostat fields. Renew Sust Energ Rev 72:1001–1018. http://www.sciencedirect.com/science/article/pii/S136403211730028X. https://doi.org/10.1016/j.rser.2017.01.032
17. Bernhard R et al (2008) Linear fresnel collector demonstration on the PSA, part I—design, construction and quality control, in March 1, 2008
18. Tesfamichael T, Wäckelgård E (2000) Angular solar absorptance and incident angle modifier of selective absorbers for solar thermal collectors. Solar Energy 68(4):335–341. http://www.sciencedirect.com/science/article/pii/S0038092X00000293. https://doi.org/10.1016/S0038-092X(00)00029-3
19. Chemisana D et al (2013) Optical performance of solar reflective concentrators: a simple method for optical assessment. Renew Energy 57:120–129. http://www.sciencedirect.com/science/article/pii/S096014811300075X. https://doi.org/10.1016/j.renene.2013.01.037

Chapter 27
The Energy Transition in Morocco

A. Laaroussi and A. Bouayad

27.1 Introduction

Renewable energies are an answer particularly adapted to the considerable energy needs of the countries of the world: they make it possible to valorize their natural resources, hydraulics, sunshine, wind, and biomass, to bring the production sites closer to the centers of consumption, and to reduce the dependence of countries with fossil energies. Morocco, like the rest of the world, has realized the importance of renewable energy (RE), especially since it has great potential in the field, mainly in solar energy and wind energy. It is for this reason that a great enthusiasm for the promotion of RE has been felt for some time. The country's interest in the development of green energy stems from its desire to reduce its dependence on fossil energy resources and contribute to the fight against climate change. Indeed, Morocco is a net importer of energy in the form of petroleum products and coal. The country is in energy deficit with an external dependence exceeding 95% [1]: in 2012, the energy bill amounts to the sum of 103 billion dirhams, capturing nearly 26% of total imports. Consumption is constantly increasing, fluctuating between 8% and 11% per year, and local production is insufficient: we import from Spain about 18% of our electricity. The oil bill becomes very burdensome, especially with the increase in demand for electricity and the constant growth in the price of petroleum products. This situation has repercussions on the national economy through a worsening of the trade deficit and a rise in consumer prices [2], thus threatening the financial balance of the state, the competitiveness of national companies, and the power to purchase of citizens. In addition, the acceleration of Morocco's economic and social development has led to a significant increase in

A. Laaroussi (✉) · A. Bouayad
Faculté des Sciences Juridiques, Economiques et Sociales, Université Moulay Ismaïl, Meknès, Morocco

A. Sayigh (ed.), *Renewable Energy and Sustainable Buildings*, Innovative Renewable Energy, https://doi.org/10.1007/978-3-030-18488-9_27

energy demand. To meet its growing energy needs, Morocco has defined a new energy strategy aimed at securing energy supply and optimizing access while rationalizing consumption and protecting the environment.

27.2 Morocco: From Fossil Fuels to Renewable Energy

As a net importer of oil, Morocco is subject to the global oil market. The oil bill becomes very heavy, which affects the national economy through a worsening of the trade deficit and a rise in consumer prices. In recent years, in order to lower public debt and in order to secure and improve its energy supply, Morocco has begun to develop a two-way strategy. On the one hand, the country encourages the local production of fossil fuels through oil exploration to deepwater deposits on its territory; and on the other hand, it is moving toward increasing RE in its energy mix by encouraging investment.

27.2.1 Moroccan Energy Context

Morocco is weakened by its strong energy dependence on the outside, which accounts for 95% of its energy consumption (Morocco produces only 5% of the energy it consumes). Commodity price volatility, combined with growing energy needs per capita, is putting significant pressure on the trade balance and on public finances. The energy bill has recorded an annual growth rate of 18.8% on average since 2002, reaching 107 billion dirhams in 2012, driven by an average annual increase of 5.7% in energy consumption to 17%, eight million tonnes of oil equivalent (Mtoe). Morocco's energy bill has risen sharply in the first 7 months of 2014. In addition, the Moroccan energy bill was 53.392 billion dirhams at the end of July 2014, an increase of 4.3% compared to the same period in 2013 when Morocco had spent only 51.207 billion dirhams. A rather disappointing achievement after the few signs of thinning that had known the net energy bill in January and February. (Fig. 27.1).

In the first 7 months of 2014, energy imports have cost the Morocco 58,250 billion dirhams against timid exports amounting to 4858 billion dirhams. Over the period from January to July 2014, imports have increased by 4.1%, compared with the same period in 2013, and represent 24.8% of the total imports of the kingdom. In addition, the net energy required (energy represents all the quantities of energy made available to domestic energy consumption) increased at the end of July 2014. This consumption amounted to 19,288,5 gigawatt hours (GWh) at the end of July 2014, compared to 18,329 GWh a year earlier, an improvement of 5.3%. This energy production is dominated by thermal energy with 13,733.2 Gwh at the end of July 2014. (Fig. 27.2).

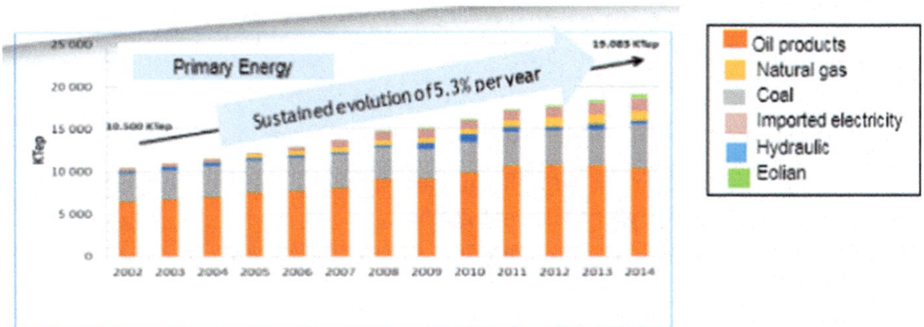

Fig. 27.1 Evolution of energy consumption in Morocco [3]

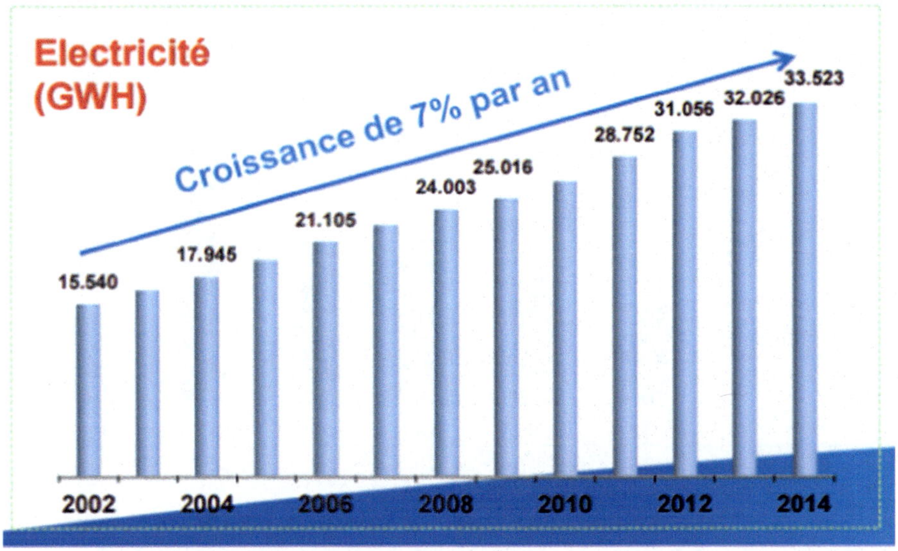

Fig. 27.2 Evolution of electricity demand in Morocco [3]

Most of electricity produced in Morocco comes from thermal power plants, which are largely based on the combustion of imported coal and the rest on imported oil. Moreover, due to the industrial and economic development of Morocco, the demand for electricity has increased on average by 7% per year over the last 10 years, which has resulted in a doubling of demand in 10 years from 17,945 GWh in 2004 to 33,523 GWh in 2014. We therefore note that energy needs are still growing:

- Primary energy: 5.3%
- Electricity: 7%
- Petroleum products: 5%

27.2.2 Challenges Between 2014 and 2025 (Figs. 27.3 and 27.4)

The main challenges to be met between 2014 and 2025 are:

- Satisfaction of energy demand following a sustained growth rate (demand will be increased by 70% between 2013 and 2025)
- Need to accelerate the rate of completion of ongoing projects and launch of new projects
- Need to accelerate the pace of reforms and accompanying measures to enable public and private operators to ensure the supply-demand adequacy of electricity (demand will be doubled between 2013 and 2025)

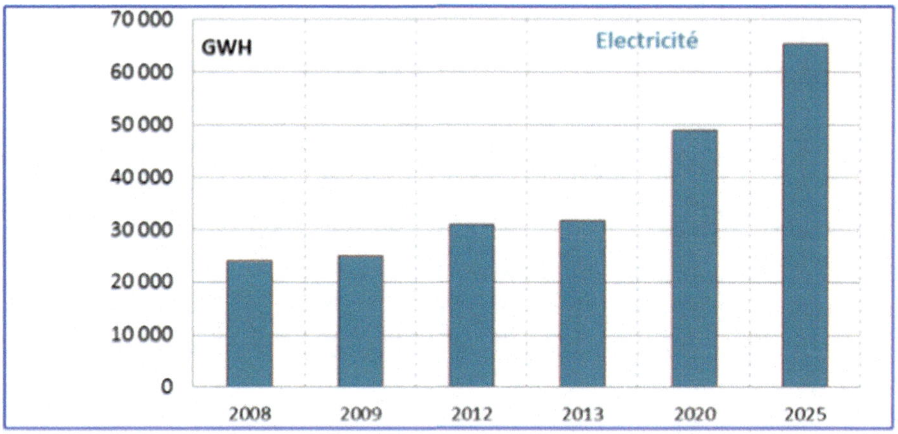

Fig. 27.3 Average change in electricity demand in Morocco [4]

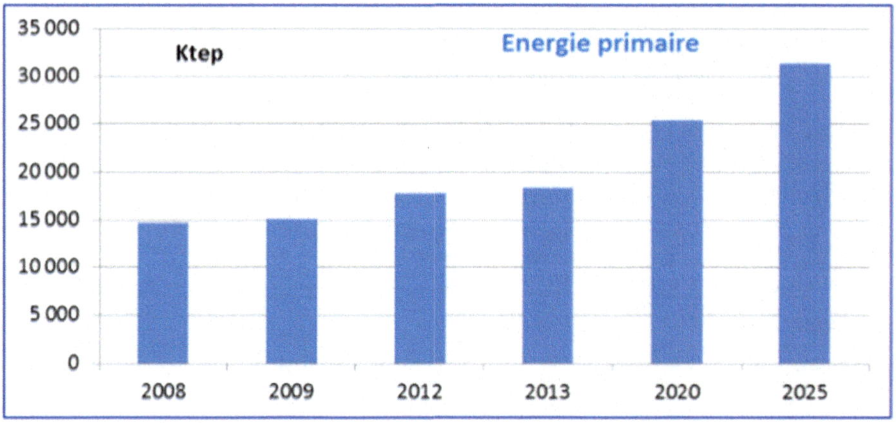

Fig. 27.4 Average change in demand for primary energy in Morocco [4]

- Controlled management of the energy transition in Morocco taking into consideration the need to preserve the balance of the sector
- Morocco's rational interaction with the profound changes in the energy sector at the regional and international levels

27.3 The Energy Cost and Energy Plan in Morocco

Morocco's energy policy since independence has generated a supply structure dominated by conventional energy, which is 96% dependent on imports on the world market. This reality has impacted all the components of the Moroccan economic fabric. National economic growth remains dependent on the international situation. Indeed, the price of energy is highly sensitive to the context of global supply and demand, coupled with the high volatility of the dollar (the global energy market is denominated in dollars). In addition, the move to the floating exchange rate could make the situation even more vulnerable. A complete overhaul of this energy structure is needed to allow the Moroccan economy to free itself from these bottlenecks. The purpose of this presentation is to visualize in a first point the effect of this energy structure on the trade balance and its environmental impact. The second *point* will be devoted to the presentation of the new energy plan as an alternative.

27.3.1 The Energy Cost in Morocco and Trade Deficit

Energy dependence on the outside world has worsened steadily, leading to a growing deterioration of the energy gap, from 70% in 1970 to 83% in 1980 to reach 96% today. This deficit is due to the average annual growth of electricity consumption of 7% (a doubling every ten years), due to the dynamic of economic and social development of the country, due to the rapid expansion of cities, and due to the rural electrification. The energy bill, of which petroleum products are dominant, depends essentially on crude prices, of which Morocco imports in full. Until the early 1970s, the weighted average price over the period 1963–1970 was 3$/barrel, which accounted for between 25% and 33% of export earnings of phosphates. From the first oil shock in 1973, it went up to 12$ per barrel which led to a fourfold increase in the oil bill, and this coverage rate climbed to 81.5% in 1978. The second oil crisis of 1979–1980 hit Morocco hard when the price of crude oscillated in 35$ and 40$ per barrel, equivalent to 112% of the same coverage rate and 36.3% of total exports, which represented almost 9% of gross domestic product (GDP). From 1986, following the discovery of North Sea oil, crude oil prices dropped to 10$ per barrel, climbing exceptionally to 35$ per barrel in 1991 following Iraq's invasion of Kuwait to stabilize between and 22$ per barrel until mid-1999. This drop in price significantly eased the energy bill, or 4% of GDP, which is equivalent to 20% of exports. From 2004 onward, prices have risen to over 111$ per barrel in 2011 and 2012 to reach 98 billion dirhams. According to official sources, the energy bill alone explains 70% of the rise in total exports of Morocco in the first 8 months of 2017, or 45 billion dirhams, which represents an increase of 30% over 2016 over the same period.

27.3.2 Energy Structure and Environmental Impact

In the world, greenhouse gas emissions result from the exploration and exploitation of primary energy sources, conversion of primary energy sources as secondary energy in refineries and power plants, transmission and distribution of fuels, and final fuel consumption in stationary and mobile applications. Emissions related to energy use include CO2, CH4, N2O, nitrogen oxides (NOx), carbon monoxide (CO), and non-methane volatile organic compound (NMVOC) emissions. They also include sulfur dioxide (SO2) emissions. The energy sector emitted 47,890.5 Gr CO2e in 2010 and 55,248.6 Gr CO2e in 2014, an increase of 15% between the 2 years. Total emissions from the energy sector remain dominated by CO2 (>97%) followed by CH4 and N2O (around 1% each). Considering all the emitting sources of the energy sector, including fugitive ones, it is the energy industries which dominate very largely the balance of the emissions with 35–38% of the emissions expressed in CO2e. The road transport sector comes second, but far behind at around 29%. Combustion in industry accounts for 13–15% with 20.8% of emissions from the energy sector, followed by the residential sector (11%) and agriculture/forestry/fishing with almost 5%. All remaining sectors account for approximately 5% of total energy-related emissions.

27.4 Moroccan Energy Strategy: A Transition to Renewable Energy

Renewable energy sources are becoming an increasingly important part of the global energy mix. Morocco is no exception to this trend and has the advantage of enjoying a particularly rich potential compared to other countries, especially since it has a large pool of renewable energy resources. The potential of these resources (solar, wind, biomass, geothermal, hydro, ocean energy, etc.) is more than enough to meet the entire energy needs of our country. The intelligent exploitation of this potential could allow Morocco to completely do without imported fossil fuels, contribute to the protection of the environment, and create new wealth. However, some of the technologies needed to achieve this goal still require further development and cost reduction. A huge research, development, and demonstration effort is therefore needed in both the private and public sectors.

27.4.1 *Resources of Renewable Energy in Morocco*

27.4.1.1 Solar and Wind Resources

The Morocco has a huge potential in RE especially wind power whose global potential is estimated at 25000 MW with 6000 achievable MW in wind speeds exceeding 9 m/s to 40 m of height in several regions and in solar with irradiation of 5 kWh/m^2/year and 3000 h/year of sunshine, an average equivalent to the south of Europe, has a theoretical capacity of more than 20,000 GW (Fig. 27.5).

In the wind, Morocco enjoys a very important potential more particularly along its coasts with wind speeds greater than 6.5 m/s and up to 10 m/s. The technical potential certainly exceeds 10,000 MW of installed capacity.

Solar is certainly the most important source of renewable energy in Morocco. The average incident solar radiation varies from 4.7 to 5.6 kWh per day and per m^2 which represents a sunshine of between 2800 h per year for the less favored regions and more than 3400 h per year for those who are the best exposed. With more than 3000 h/year of sunshine, an irradiation of ~5 kWh/m^2/day, Morocco enjoys a considerable solar field. This source of energy is a particularly important potential, especially in areas with poor electricity production capacity.

27.4.1.2 Hydraulic and Biomass Resources

By 2020, the installed hydroelectric capacity will be increased from 1730 MW to 2700 MW by the construction of new dams and pumping stations (Fig. 27.6).

The Moroccan hydroelectric production is highly variable depending on the annual rainfall. The contribution of hydropower, with a capacity of 1745 MW at the end of 2013, is expected to increase in the coming years to 2700 MW by 2020 through the construction of new dams and pumped energy transfer stations.

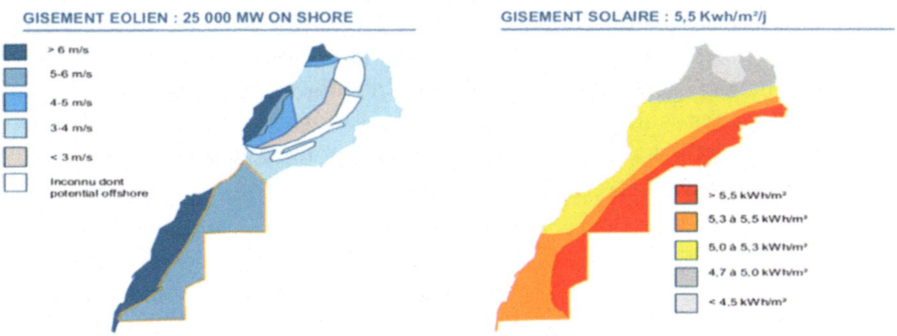

Fig. 27.5 Solar and wind farm in Morocco [5]

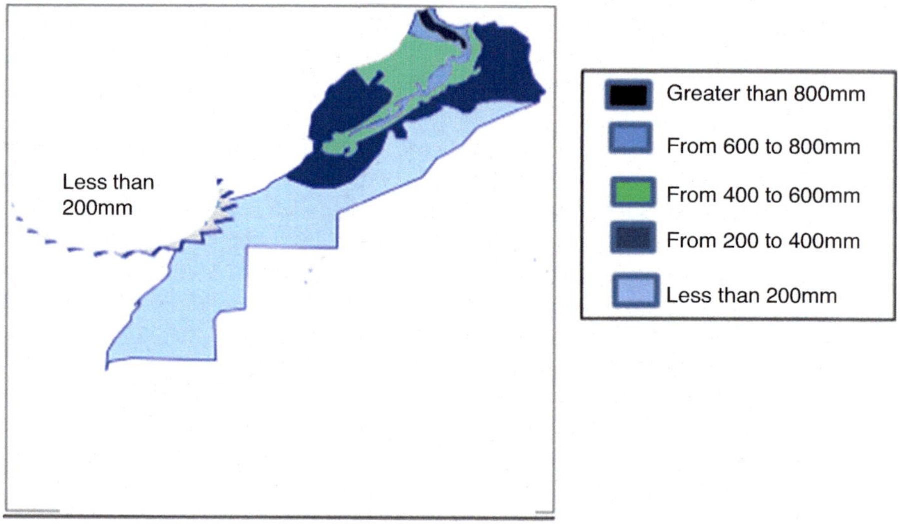

Fig. 27.6 Hydraulic power potential in Morocco [6]

27.4.1.3 Biomass Potential in Morocco

Morocco has a large biomass deposit to exploit. It is mainly forest undergrowth, wood chips from the logging and sawmilling industry, timber from arboriculture and viticulture, stems and leaves from number of crops (for example:cereals, pulses, sunflower, sugar beet, sugar cane, olive trees, vegetable crops), the by-products of agribusiness such as olive cake and vegetable water from the 16,000 traditional mills and of 14 modern oil mills (40 000 tonnes of olive oil are extracted on average annually, 160 000 tonnes for the year 2001), the by-products of the sugar factories (13 sugar factories), the hulls of fines, the seeds of argan tree and seaweed taking advantage of the 3500 km of coastline of the Kingdom [6]. This potential is currently very little developed in Morocco and retains the full attention of national actors. As for the forest, it fulfills an economic role that manifests itself in the production of 600,000 m^3/year of timber and industry, which is 30% of the country's needs. It is also responsible for 10,000,000 m^3/year of fuelwood, accounting for 30% of the global energy balance. Thus, the Moroccan forest contributes nearly 10% to the national agricultural GDP. This contribution considers both the values of direct consumption achieved in the context of market-integrated commercial sectors and income directly derived by local populations in the form of firewood, rangelands, and various menu products.

27.4.2 Renewable Energy and Energy Efficiency Projects in Morocco: Assessment and Prospects

- Noor central photovoltaic (PV) I:

  Power 170 MW: Noor Ouarzazate IV 70 MW; Noor Laayoune and Noor Boujdour, with a total capacity of 100 MW
  Technology: Photovoltaic
  State of progress: Launch of an expression of interest (March 2015)
  Prequalification of 37 bidders (December 2015)

- Noor central Midelt:
- Launching a call for expression of interest in December 2015
- Noor central Tafilalet (central PV in the end of the line Erfoud, Zagora, and Missouri): 75/100 MW

  Date of commissioning: 2017
  Hydroelectric projects: Parks in service

	Pi (MW)
Bine el Ouidane	135
Afourer	94
Tanafnit – El Borj	40
Al wahda	240
Allal el Fassi	240
Al Massira	128
Ahmed el Hansali	92
STEP Afourer	464
Diverses usines	337
Total	1770

The hydraulic is at the origin of the electrical production in Morocco, what is translated by:

- Strengthening of the hydroelectric park in the framework of the policy of large dams launched since the 1960s
- The introduction of the PETS (pumped energy transfer stations) to the regulation of the operation of the national park of production, in particular based on RE
- The renewed interest of the hydraulic with the new energy strategy, including the microphone hydraulic power stations

Noor central Ouarzazate I, concentrating solar power plant (CSP):

Power: 160 MW with 3 h of storage
Technology: Cylindrical parabolic
Developer: Acwa-Sener consortium
Beginning of work: May 2013

Date of commissioning: 2016

Noor central Ouarzazate II (CSP):

Power: 200 MW with storage
Technology: cylindrical parabolic
Developer: Acwa-Sener consortium
Beginning of work: 2015
Expected date of commissioning: 2018

Noor central Ouarzazate III (CSP):

Power: 150 MW with storage
Technology: Tower
Developer: Acwa-Sener consortium
Beginning of work: 2015
Expected date of commissioning: 2018

Development of programs of large- and medium-sized solar photovoltaic power plants:

- PV program NOWE (National Office of Water and Electricity): 400 MW.
- PV program MASEN (Moroccan Agency for Solar Energy): 400 MW.
- PV program private producers very high voltage (VHV), high voltage (HV)
  - Development of projects of solar power plants photovoltaic whose production is intended for consumers connected in the medium voltage.
  - Development of the large-scale use of PV in the residential and tertiary connected in low voltage. The potential is assessed to 4.5 GW.
  - Implementation of programs of accompaniment in matters of industrial integration and R&D dedicated to PV.

National program for promoting solar pumping in irrigation water projects
Energy efficiency, an important pillar of the Moroccan energy strategy

Project Objective

- Promotion of the use of solar pumping systems in irrigation water-saving projects
- Reduction of the energy bill of the farms, as well as the water consumption for the irrigations
- Target: Farms in size to less than 5 ha

Programs and Actions in Progress

- Measures of EE (energetic efficiency)
- EE in the public administrations
- Promotion of EE in the lighting
- Incentive pricing system
- Communication, education, and awareness
- Sustainable urban development of cities, both new and old, integrating the factor EE
- Targets: Saving of 12% to the Horizon 2020

Saving of 15% to the Horizon 2030

- Priority programs targeting consumer sectors: industry, transportation, and building

27.5 Conclusion

Morocco has adopted a transversal, integrated and participatory approach to facilitate transition to green economic development and participate in the global challenge concerning renewable energies and climate change. The country remains convinced that the mobilization of the state, the attractiveness of the energy sector, the general awareness, and the involvement of the whole society are the four engines to make a transition toward a green development and guarantee the resilience of Morocco facing the challenges of climate change and energy dependency. Internally, optimism dominates the RE sector and contributes to its dynamism. Thanks to its geographical position and its potential as natural sources of energy, Morocco is on the right track to succeed in its energy strategy by carrying out major projects on its territory and by putting laws which aim to promote the production of energy from renewable sources, its marketing, and its export by public or private entities. However, the RE sector is fragile in terms of youth and unskilled labor and remains dependent on substantial funding, subsidies, and support. Indeed, most of the costs of the most widespread RE applications are higher than those of conventional energy technologies, which discourages Moroccan companies from specializing in the production of energetic materials. So, a framework of financial and fiscal subsidies for renewable energy generation and encouragement of research and development in RE are needed to improve the sector.

References

1. Report of Ministry of Energy, Mines, Water and Environment (MEMWE) (2014)
2. Ministry of Economy and Finance (MEF) (2010). www.finances.gov.ma
3. Report of the Ministry of Energy, Mines, Water and Environment (2015)
4. MEMWE (2015) The new Moroccan energy model: real investment opportunities. Lisbon
5. Energy Investment Company (EIC) (2015). www.siem.ma
6. National Agency for the development of renewable energies and the energy efficiency (ADREEE) 2014

Chapter 28
Energetic Performance Optimization of a H_2O-LiBr Absorption Chiller Powered by Evacuated Tube Solar Collector

Nasiru I. Ibrahim, Fahad A. Al-Sulaiman, and Farid Nasir Ani

Nomenclature

a_1	First-order heat loss coefficient W/m^2-K
a_2	Second-order heat loss coefficient W/m^2-K^2
A	Area (m^2)
COP	Coefficient of performance
Cp	Specific heat capacity (J/kg-K)
h	Enthalpy (J/kg)
I_G	Incident solar flux (W/m^2)
$\dot{m}$	Mass flow rate (kg/s)
P	Pressure (kPa)
$\dot{Q}$	Heat transfer rate (W)
T	Temperature (°C or K)
UA	Overall heat transfer coefficient, (W/K or kW/K)
$\dot{W}_p$	Pump work (W)
X	Mass fraction of LiBr in solution
η	Efficiency

N. I. Ibrahim (✉)
School of Mechanical Engineering, Faculty of Engineering, Universiti Teknologi Malaysia, UTM, Skudai, Johor, Malaysia

Center of Research Excellence in Renewable Energy (CoRERE), Research Institute, King Fahd University of Petroleum & Minerals (KFUPM), Dhahran, Saudi Arabia

F. A. Al-Sulaiman
Center of Research Excellence in Renewable Energy (CoRERE), Research Institute, King Fahd University of Petroleum & Minerals (KFUPM), Dhahran, Saudi Arabia
e-mail: fahadas@kfupm.edu.sa

F. N. Ani
School of Mechanical Engineering, Faculty of Engineering, Universiti Teknologi Malaysia, UTM, Skudai, Johor, Malaysia
e-mail: farid@mail.fkm.utm.my

A. Sayigh (ed.), *Renewable Energy and Sustainable Buildings*, Innovative Renewable Energy, https://doi.org/10.1007/978-3-030-18488-9_28

Subscripts

a	Absorber, air
c	Collector, condenser
e	Evaporator
g	Generator
p	Pump
sys	System
u	Useful
w	Water

28.1 Introduction

Renewable energy sources are expected to be viable alternative solutions to cover the ever-increasing cooling demand, thereby addressing the issue of high-energy consumption by the conventional cooling systems and negative environmental impacts due to the use of chlorinated fluorocarbon compounds (CFCs) [1]. Solar energy in particular is one of the promising renewable energy sources due to its abundance in many parts of the world and greenness [2, 3]. The role of solar thermal collectors is to convert the solar energy into heat, which is then used to run a thermally driven cooling system, generating cooling effect in terms of chilled water or conditioned air for buildings environment. Closed and open sorption technologies (absorption, adsorption, and desiccant) appear to be more attractive options to replace the conventional vapor compression cooling systems [4, 5]. Absorption technology is leading in terms of worldwide installation for comfort [6]. However, solar-driven absorption chillers are still not widely commercially available, mainly because of system complexity and relatively high initial investment of the solar field, with about 35% of the total cost [6, 7]. Hence, there is a need for proper sizing of the collector to be integrated with cooling systems. Sizing the solar collector field must not be oversized nor undersized for a given chiller capacity. Oversized collector increases the overall cost of the cooling system as well as potential risk of chiller failure, while undersized collector may lead to low cooling output of the system. Deployment of solar thermal collectors as the primary energy input necessitates proper configuration strategies and optimization. System optimization is the usual approach used for sizing the collector field of solar-powered cooling systems [8].

Assilzadeh et al. [9] reported an optimization study of a 3.5 kW H_2O-LiBr absorption system powered by evacuated tube solar collectors for Malaysia and similar tropical regions. Simulation results carried out using TRNSYS program indicate an optimum area of 35 m^2 (10 m^2/kW cooling) evacuated tube solar collector. A similar study revealed that about 3.3 m^2/kW minimum specific collector area based on parabolic trough is needed for a 17.5 kW single-effect lithium bromide-water absorption cooling system in Ahwaz, Iran [10]. Saleh and Mosa [11] optimized the performance of a single-effect H_2O-LiBr absorption system driven by flat plate solar collector at different level of solar radiation and heat source temperature. However, there is no information regarding the corresponding size of the solar collector field. In a related study, Pandya et al. attempted to find optimum generator temperature of a single-effect H_2O-LiBr absorption system at different condenser and evaporator temperature that gives maximum COP or minimum exergy destruction. Kumar et al. [12] presented comparison of performance optimization of single-stage H_2O-LiCl and H_2O-LiBr absorption cooling systems driven by different solar collectors. Optimization parameters considered include heat source temperature, system COP, exergetic efficiency, collector area, and cost to determine the best working pair and collector type. The results from performance indicators showed that the H_2O-LiCl absorption cooling system is better to be powered by evacuated tube solar collector. A similar conclusion was also made in another study by the same group [13].

Literature survey indicates that most of the optimization studies on solar thermal-driven absorption systems focused on external operating parameters such as cooling, heating, and chilled water flow rates and temperature, respectively. The internal operating parameters of the chiller such as temperatures and mass fraction of LiBr or solution concentration need to be considered for better system reliability. This is to minimize failure and ensure reliability when it comes to practical system integration as information such as solution crystallization are very important. Crystallization is associated with the solution concentration (LiBr mass fraction) and temperature. Hence, the main objective of this contribution is to optimize the performance of a solar absorption cooling system for better sizing of the collector field taking into account the internal operating parameters of the absorption chiller.

28.2 System Description

The solar-driven single-effect absorption cooling system consists primarily of a solar thermal collector, an absorption chiller, and other auxiliary components such as pumps and valves, usually accompanied with thermal storage. Figure 28.1 shows a single-effect H_2O-LiBr absorption chiller in its simplest configuration consisting of four main components: evaporator, absorber, generator, and condenser. The heat source can be solar thermal energy, waste heat, geothermal, biomass energy, or electric boilers. For solar-driven absorption system, the chiller receives input energy (hot water) from the solar collector to the generator and converts it to cooling energy in the form of chilled water in the evaporator.

In operation, the generator receives heat input from the solar collector, state (17), which vaporized refrigerant (water) out of the solution, subsequently forming a solution of higher concentration. The vaporized refrigerant exits and enters the condenser (7) where it is condensed. The liquid refrigerant passes through an expansion valve and enters the evaporator at low pressure (8–9), while the regenerated sorbent returns to the absorber through a solution heat exchanger SHX process (4–6). Both pressure and temperature of the refrigerant and solution are decreased after passing through the expansion valves. The refrigerant evaporates in the evaporator as it extracts heat from the cooling demand side (load), producing cooling effect, and exits the evaporator as saturated vapor (10). The concentrated solution in the

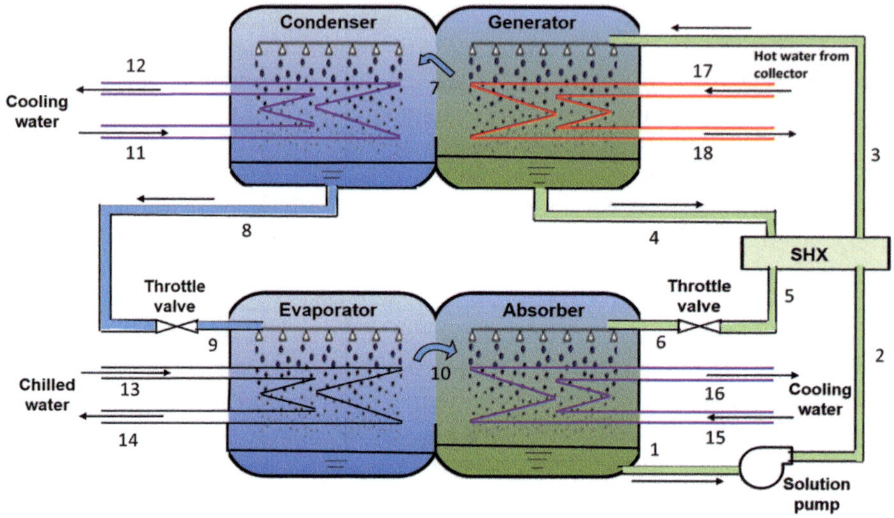

Fig. 28.1 Schematic of a single-effect water-lithium bromide absorption chiller

absorber absorbs the refrigerant vapor coming from the evaporator. Heat must be rejected from condenser and absorber, and this is usually accomplished by cooling tower (not shown in the figure) or ambient cooling, the processes indicated by (11–12, 15–16). The diluted or weak solution in the absorber is pumped to a high pressure back to the generator (1–3), and the cycle continues.

28.3 System Modeling and Method

The system model is developed based on energy and mass conservation applied to each component considering the following assumptions [14]:

- Steady-state operation.
- Saturated liquid refrigerant at condenser exit.
- Refrigerant leaving the evaporator is saturated.
- Negligible pressure drops in heat exchangers and the piping systems.
- Heat losses from the components of the system to the ambient are neglected.
- Kinetic and potential energy terms are neglected in the energy balance equations.

The solar collector is analyzed assuming a steady-state operation. The model of the solar collector is used to determine the useful heat energy required to drive the absorption chiller. The energy supplied by the collector to drive the chiller is given by:

$$\dot{Q}_u = A_c I_G \eta_c \tag{28.1}$$

where η_c is the conversion efficiency of the solar collector given as [15]:

$$\eta_c = \eta_0 - a_1 \frac{T_m - T_f}{I_G} - a_2 \frac{\left(T_m - T_a\right)^2}{I_G} \tag{28.2}$$

and T_m is the mean inlet and outlet temperature of the collector fluid. The values of η_0, a_1, and a_2 for the collector type SEIDO 1–16 are 0.73, 1.5, and 0.0054 W/m^{2}K^2, respectively [16, 17]. Thermodynamic model of the absorption chiller is developed considering four sets of thermodynamic equations, mass balances, energy balances, heat transfer relations and equations describing thermophysical properties of the working fluids [18, 19], and is presented in Table 28.1. The parameters $\dot{m}$, $\dot{Q}$, X, and h represent mass flow rate, rate of heat transfer, mass fraction of LiBr in the solution, and enthalpy, respectively, at various state points (Fig. 28.1).

The model equations are coded and solved in engineering equation solver (EES) [20] software, which contained thermophysical properties of the working fluids. The temperature, pressure, enthalpy, and concentration at each of the state points of the chiller cycle are interrelated as:

$$h = f\left(T,P,X\right) \tag{28.3}$$

Hence, several state property equations are required and can be accessed through H_2O-LiBr property library in EES. The property functions of the H_2O-LiBr solution built in EES are based on the correlations provided by Patek and Klomfar [21].

28.4 Results and Discussion

The analysis is carried out on a WFC10 Yazaki (35.2 kW) absorption chiller driven by evacuated tube solar collector. The input parameters for the simulation are shown in Table 28.2 and are based on the following references [8, 18, 22–24]. The aim of the optimization is to find the optimum collector area at maximum COP. The system optimization problem is a single-objective (COP), multivariable (A_c, I_G, T_a) type; hence genetic method is used. The genetic method is a multivariable optimization technique based on genetic algorithm built in EES [24]. It involves selection of individual population at random from a given range of independent optimization variables, number of generation, and mutation rate.

28.4.1 Optimization Results

Figures 28.2 and 28.3 show the optimization results where Fig. 28.2 represents the objective function (COP) and Fig 28.3 the optimization variables. The ranges of the other optimization variables, that is, solar radiation and temperature, are set to 800–1000 W/m^2 and 30–45 °C, respectively. It is noted from the figures that maximum

Table 28.1 Model of absorption chiller

Chiller parts	Mass balance	Material balance	Energy balance
Generator	$\dot{m}_3 = \dot{m}_4 + \dot{m}_7$	$\dot{m}_3 X_3 = \dot{m}_4 X_4$	$\dot{Q}_g = \dot{m}_4 h_4 + \dot{m}_7 h_7 - \dot{m}_3 h_3$ (4)
Absorber	$\dot{m}_1 = \dot{m}_6 + \dot{m}_{10}$	$\dot{m}_1 X_1 = \dot{m}_6 X_6$	$\dot{Q}_a = \dot{m}_6 h_6 + \dot{m}_{10} h_{10} - \dot{m}_1 h_1$ (5)
Evaporator	$\dot{m}_9 = \dot{m}_{10}$	–	$\dot{Q}_e = \dot{m}_9 \left(h_{10} - h_9\right)$ (6)
Condenser	$\dot{m}_7 = \dot{m}_8$	–	$\dot{Q}_c = \dot{m}_7 \left(h_7 - h_8\right)$ (7)
Pump	$\dot{m}_1 = \dot{m}_2$	$X_1 = X_2$	$\dot{W}_p = \dot{m}_1 \left(h_2 - h_1\right)$ (8)
Solution heat exchanger	$\dot{m}_2 = \dot{m}_3$ $\dot{m}_4 = \dot{m}_5$	$X_2 = X_3$ $X_4 = X_5$	$\dot{m}_2 \left(h_3 - h_2\right) = \dot{m}_4 \left(h_4 - h_5\right)$ (9)
REV	$\dot{m}_8 = \dot{m}_9$	–	$h_8 = h_9$ (10)
SEV	$\dot{m}_5 = \dot{m}_6$	$X_5 = X_6$	$h_5 = h_6$ (11)
Heat transfer equations			
Generator	$\dot{Q}_g = \dot{Q}_u = \dot{m}_{17} C_p \left(T_{17} - T_{18}\right) = UA_g \dfrac{\left(T_{17} - T_4\right) - \left(T_{18} - T_7\right)}{\ln\left(\dfrac{T_{17} - T_4}{T_{18} - T_7}\right)}$ (12)		
Absorber	$\dot{Q}_a = \dot{m}_{15} C_p \left(T_{16} - T_{15}\right) = UA_a \dfrac{\left(T_6 - T_{16}\right) - \left(T_1 - T_{15}\right)}{\ln\left(\dfrac{T_6 - T_{16}}{T_1 - T_{15}}\right)}$ (13)		
Evaporator	$\dot{Q}_e = \dot{m}_{13} C_p \left(T_{13} - T_{14}\right) = UA_e \dfrac{\left(T_{13} - T_{10}\right) - \left(T_{14} - T_9\right)}{\ln\left(\dfrac{T_{13} - T_{10}}{T_{14} - T_9}\right)}$ (14)		
Condenser	$\dot{Q}_c = \dot{m}_{11} C_p \left(T_{12} - T_{11}\right) = UA_c \dfrac{\left(T_8 - T_{11}\right) - \left(T_8 - T_{12}\right)}{\ln\left(\dfrac{T_8 - T_{11}}{T_{18} - T_{12}}\right)}$ (15)		
Solution heat exchanger	$\varepsilon_{SXH} = \dfrac{\left(T_4 - T_5\right)}{\left(T_4 - T_2\right)}$ (16)		
Chiller COP	$\mathrm{COP} = \dot{Q}_e / \dot{Q}_g$ (17)		
System COP	$\mathrm{COP}_{\mathrm{sys}} = \dfrac{\dot{Q}_e}{A_c I_G}$ (18)		

COP (0.754) occurred at optimum solar collector area of about 77.5 m^2, solar radiation of about 902 W/m^2, and ambient temperature of 40 °C. At absorber inlet, there is risk of solution crystallization due to low temperature; hence the solution temperature and concentration (LiBr mass fraction) at the optimized condition are shown in Fig. 28.4. The respective values of the two parameters in Fig. 28.4 indicate

Table 28.2 Input parameters

Parameter	Value
Solar collector area, A_c	60–128 m^2
Collector fluid/heating water mass flow rate, $\dot{m}_{17}$	2.4 kg/s
Heating water inlet temperature, T_{17}	88 °C
Weak solution mass flow rate, $\dot{m}_1$	0.197 kg/s
Effectiveness of solution heat exchanger, ε_{SXH}	0.64
Cooling water inlet temperature, T_{15}	31 °C
Mass flow rate of cooling water, $\dot{m}_{15}$	5.1 kg/s
Mass flow rate chilled water, $\dot{m}_{13}$	1.52 kg/s
Chilled water temperature at evaporator inlet, T_{13}	12.5 °C
Overall heat transfer coefficient of absorber, UA_a	8.251 kW/°C
Overall heat transfer coefficient of condenser, UA_c	7.98 kW/°C
Overall heat transfer coefficient of evaporator, UA_e	14.096 kW/°C
Overall heat transfer coefficient of generator, UA_g	6.34 kW/°C
Number of population	20
Number of generation	80
Mutation rate	0.2

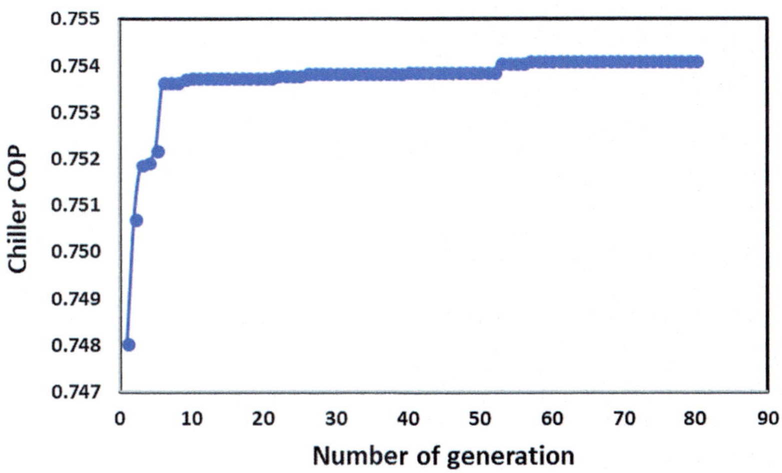

Fig. 28.2 Maximization of chiller COP

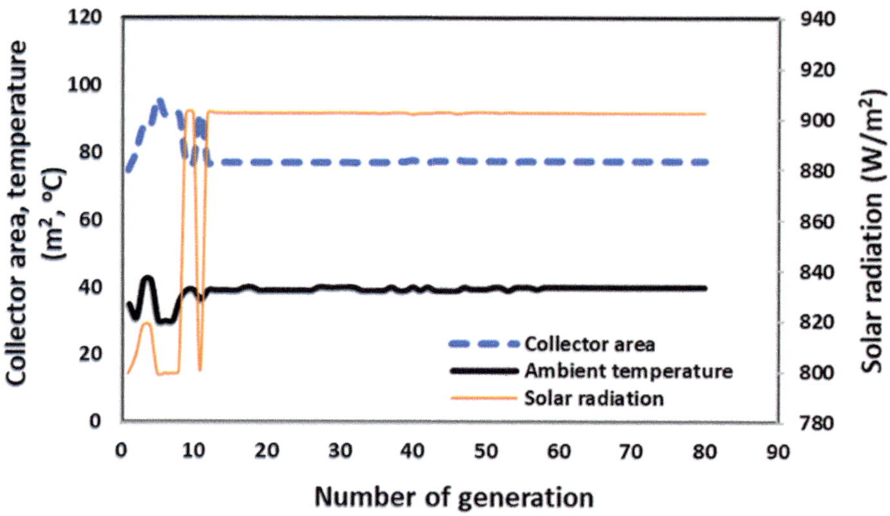

Fig. 28.3 Optimization parameters

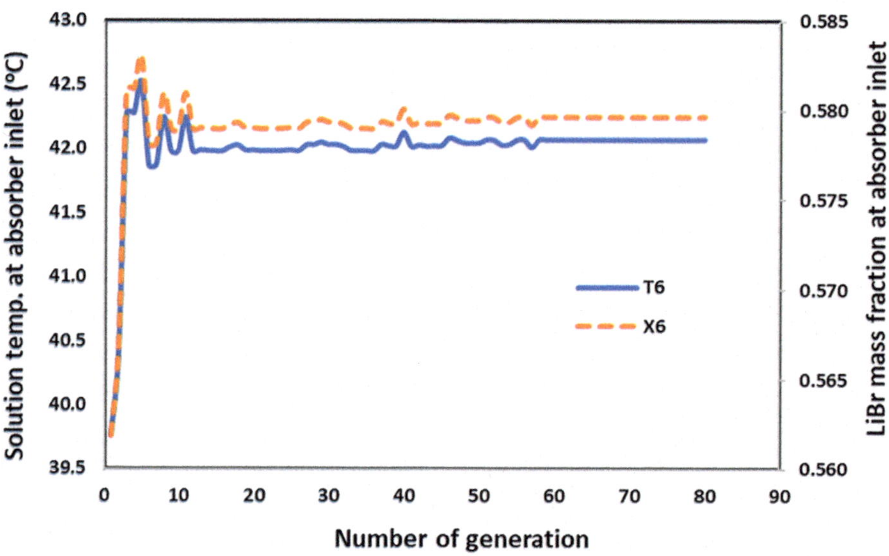

Fig. 28.4 Solution temperature and LiBr mass fraction at absorber inlet

that the optimized condition is obtained away from the crystallization limit. The crystallization limit of the solution at temperature between 40 °C and 60 °C and pressure below 1 kPa is 65% mass fraction of LiBr according to the Duhring chart of H_2O-LiBr solution [25].

28.4.2 Sensitivity Analysis

The purpose of the sensitivity analysis is to study the relative effect of the optimization variables to the chiller performance and operation. This involves varying one optimization parameter and keeping others constant at a time. Figure 28.5 shows variation of COP with area of solar collector for constant ambient temperature and different levels of solar radiation. The figure indicates the same value of maximum COP at three different levels of solar radiation. For solar radiation of 800 W/m^2, the required collector area ranges from 87.5 to 92.5 m^2. For solar radiation 902 W/m^2, the required collector area is 77.5–80 m^2, and for solar radiation 1000 W/m^2, the required collector area is 67.5–70 m^2. Solution temperature and mass fraction of LiBr at the inlet of the absorber are strongly affected by both collector area and solar radiation (Figs. 28.6 and 28.7). These results indicate that the crystallization limit is reached for 1000 W/m^2 solar radiation at solution temperature of 51 °C (Fig. 28.6), corresponding to mass fraction of LiBr of about 0.65 (Fig. 28.7) and collector area 117.5 m^2. This shows that sizing the collector above 117 m^2 in places with high solar radiation up to 1000 W/m^2 may lead to the undesirable phenomena of solution crystallization. The ambient temperature shows insignificant effect on the chiller COP and LiBr mass fraction, respectively, as depicted in Figs. 28.8 and 28.9.

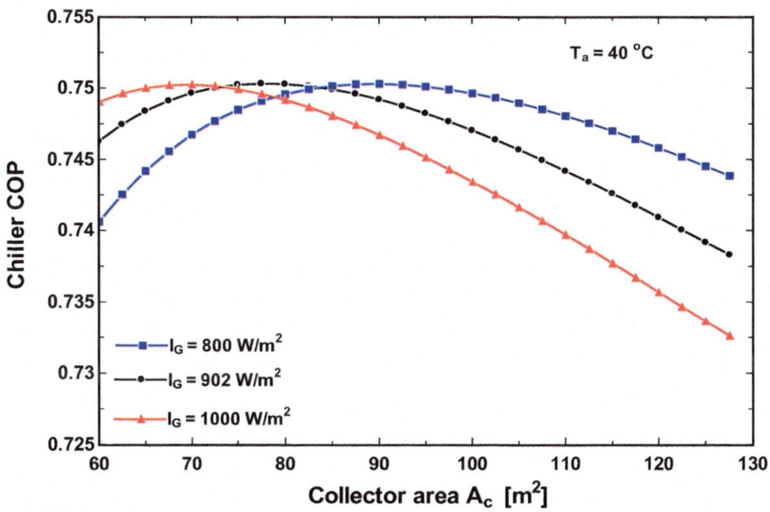

Fig. 28.5 Variation of chiller COP with solar collector area

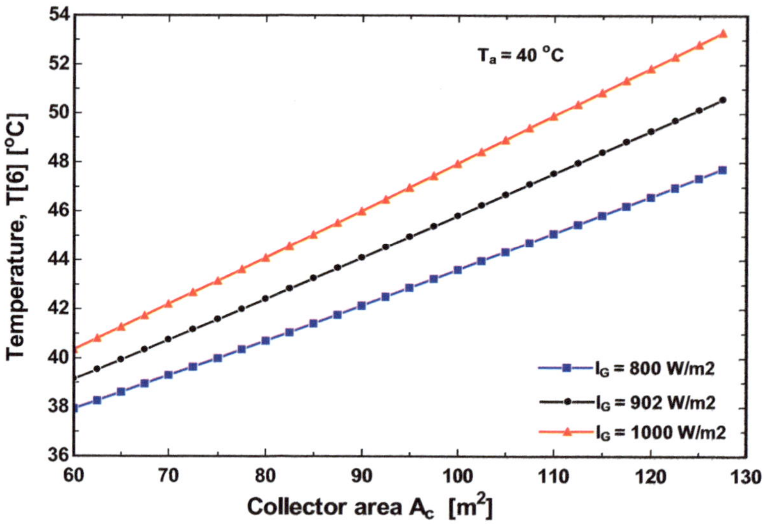

Fig. 28.6 Variation of solution temperature with solar collector area

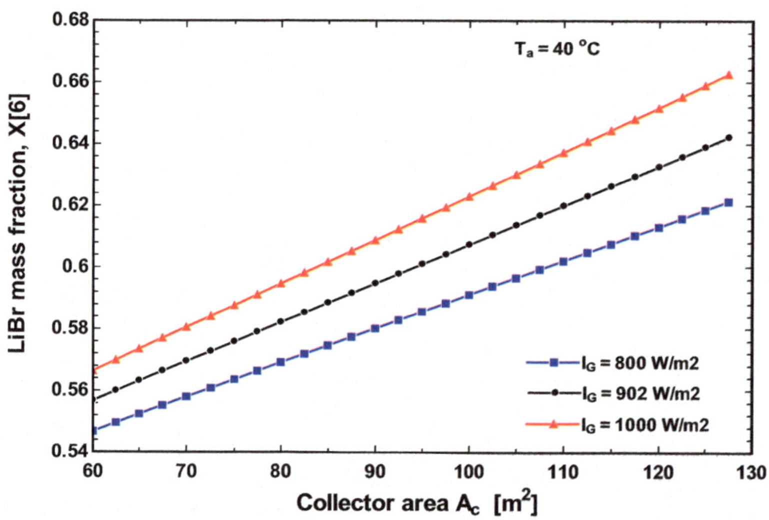

Fig. 28.7 Variation of LiBr mass fraction with solar collector area

28.4.3 Actual Weather Case Study

The system is simulated at optimum collector area using hourly weather data of Subang, Klang Valley, Selangor, Malaysia (longitude 101° 33′, latitude 3° 7′) (Fig. 28.10). The results indicate that there is no significant effect of the weather

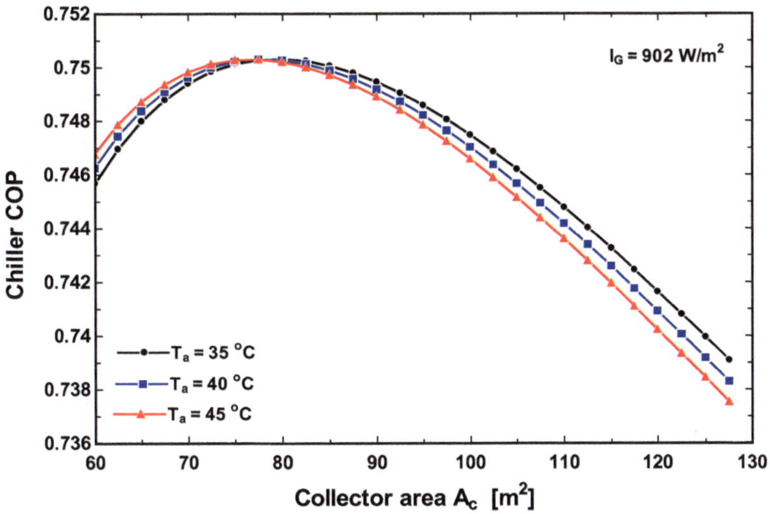

Fig. 28.8 Effect of ambient temperature on chiller COP

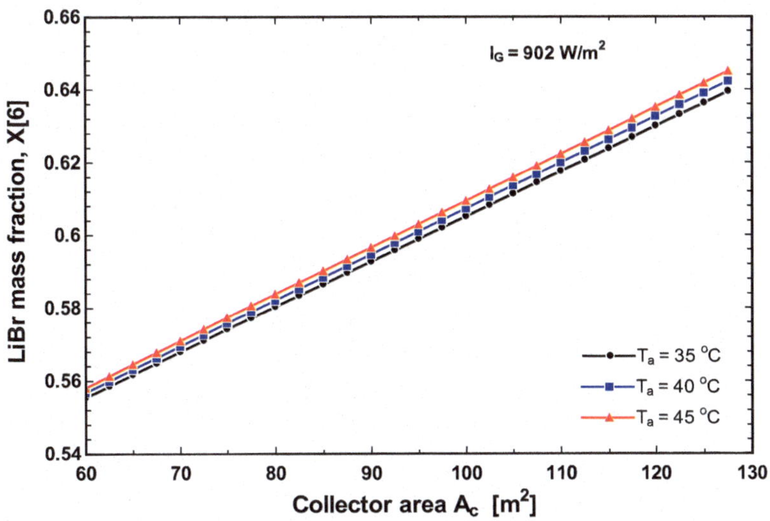

Fig. 28.9 Effect of ambient temperature on LiBr mass fraction

variables on COP which ranges from 0.72 to 0.75 as shown in Fig. 28.11. This is expected because the COP is a function of the cooling effect and the chiller heat input, both of which vary in proportion [19]. As for the COP_{system}, the variation pattern is much similar to that of the solar radiation, reaching peak at the maximum radiation. The rates of heat transfer in the respective heat exchangers are shown in Fig. 28.12. It is clear that the rates of heat transfer are much more affected by the

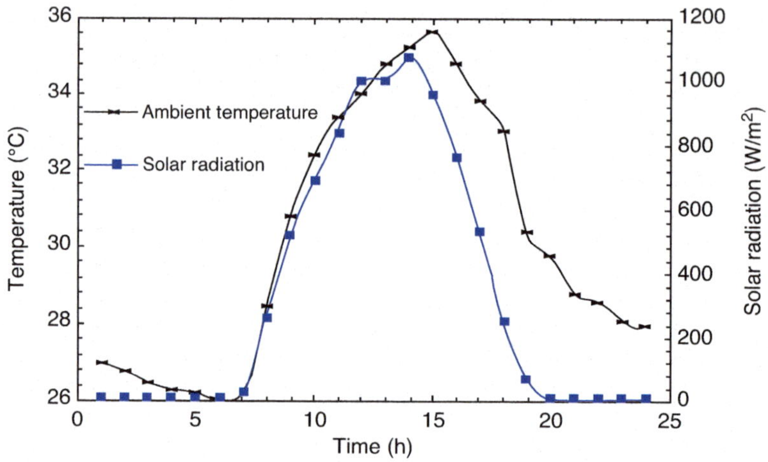

Fig. 28.10 Hourly weather data of Subang

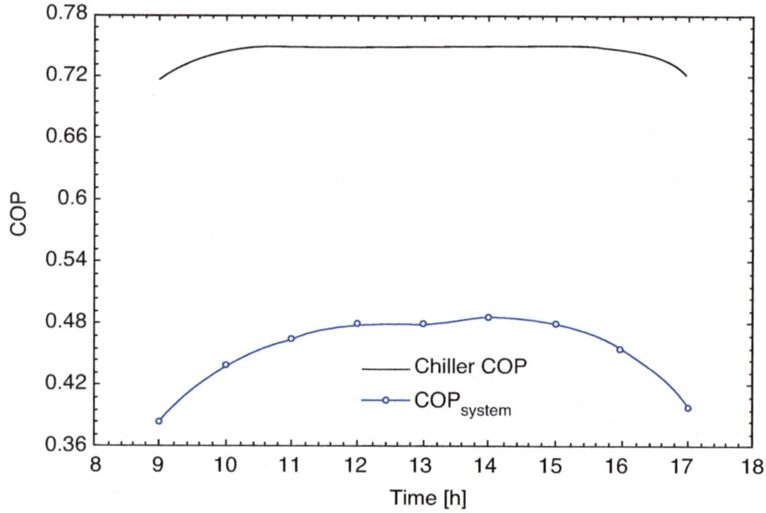

Fig. 28.11 Variation of COP and COP_{system}

solar radiation as their variation pattern follows similar. In order to check for the potential risk of solution crystallization in the absorption chiller, variations of the solution temperature and LiBr mass fraction during the day are plotted in Fig. 28.13. It is observed that there is no risk of crystal formation in the solution based on the conditions under consideration because the maximum mass fraction of LiBr is about 0.59 which is below the crystallization limit.

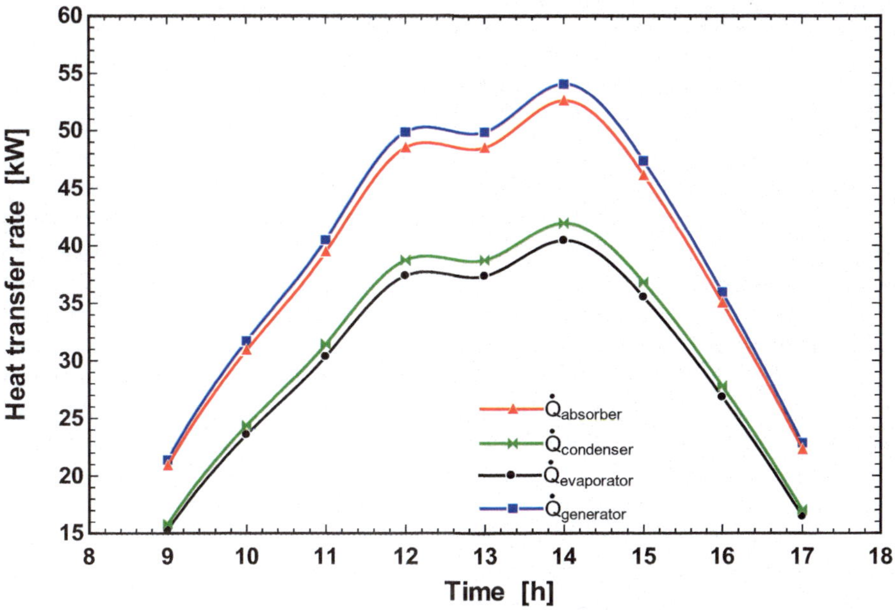

Fig. 28.12 Variation of heat transfer rates

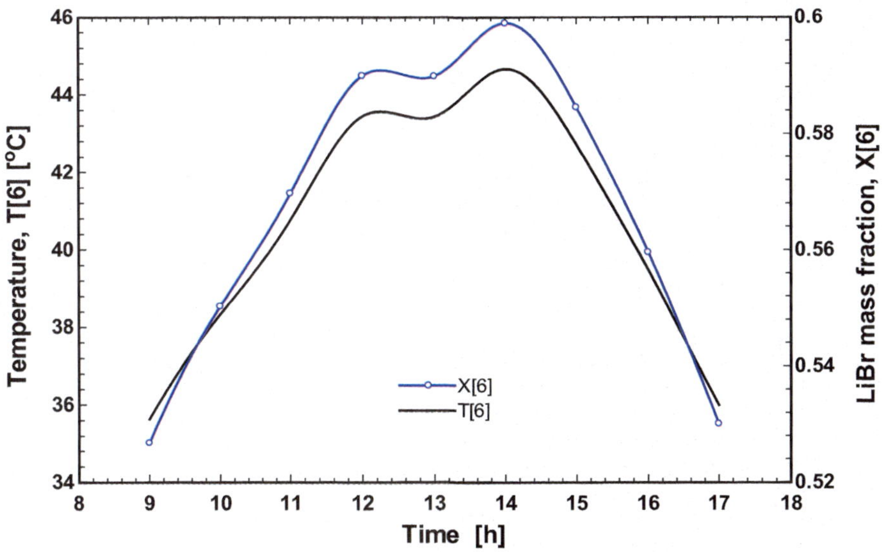

Fig. 28.13 Variation of solution temperature and LiBr mass fraction

28.5 Conclusion

Energetic optimization study of single-effect absorption chiller driven by solar collector is reported in this paper. The aim of the optimization is to find the optimum size of the evacuated tube solar collector according to the chiller nominal capacity at maximum COP. For WFC10 Yazaki (35.2 kW) absorption chiller, the optimization results indicate optimum specific collector area of about 2.2 m^2/kW of cooling at $I_G = 902$ W/m^2 and maximum COP of 0.754. Furthermore, sensitivity analysis revealed that the required collector area ranges from 87.5 to 92.5 m^2 for solar radiation of 800 W/m^2, 77.5 to 80 m^2 for solar radiation of 902 m^2, and 67.5 to 70 m^2 for solar radiation of 1000 m^2. In addition, the results showed that there is risk of solution crystallization by integrating solar collector field larger than 117 m^2 in places where solar radiation is up to 1000 W/m^2. Although the study is not a thorough economic analysis, the outcomes are expected to serve as reference and guide to those deploying solar thermal energy for absorption systems in sizing the solar collector field, which constitute the major share of initial investment.

Acknowledgments The authors are grateful to Universiti Teknologi Malaysia (UTM) for the award of International Doctoral Fellowship (IDF). We also knowledge the support of the Center of Research Excellence in Renewable Energy at King Fahd University of Petroleum and Minerals, Dhahran, Saudi Arabia.

References

1. Prasartkaew B, Kumar S (2014) Design of a renewable energy based air-conditioning system. Energ Buildings 68:156–164
2. Bellos E, Tzivanidis C (2018) Performance analysis and optimization of an absorption chiller driven by nano fluid based solar flat plate collector. J Clean Prod 174:256–272
3. Hepbasli A, Alsuhaibani Z (2011) A key review on present status and future directions of solar energy studies and applications in Saudi Arabia. Renew Sust Energ Rev 15:5021–5050
4. González-Gil A, Izquierdo M, Marcos JD, Palacios E (2011) Experimental evaluation of a direct air-cooled lithium bromide–water absorption prototype for solar air conditioning. Appl Therm Eng 31:3358–3368
5. Ibrahim NI, Al-Sulaiman FA, Ani FN (2018) Solar absorption systems with integrated absorption energy storage-A review. Renew Sust Energ Rev 82:1602–1610
6. Montagnino FM (2017) Solar cooling technologies. Design, application and performance of existing projects. Sol Energy 154:144–157
7. Shirazi A, Taylor RA, Morrison GL, White SD (2018) Solar-powered absorption chillers: a comprehensive and critical review. Energy Convers Manag 171:59–81
8. Eicker U, Pietruschka D (2009) Optimization and economics of solar cooling systems. Adv Build Energy Res 3:45–81 7
9. Assilzadeh F, Kalogirou SA, Ali Y, Sopian K (2005) Simulation and optimization of a LiBr solar absorption cooling system with evacuated tube collectors. Renew Energy 30:1143–1159
10. Mazloumi M, Naghashzadegan M, Javaherdeh K (2008) Simulation of solar lithium bromide-water absorption cooling system with parabolic trough collector. Energy Convers Manag 49:2820–2832

11. Saleh A, Mosa M (2014) Optimization study of a single-effect water–lithium bromide absorption refrigeration system powered by flat-plate collector in hot regions. Energy Convers Manag 87:29–36
12. Kumar V, Pandya B, Patel J, Matawala V (2018) Vapor absorption system powered by different solar collectors types: cooling performance, optimization, and economic comparison. Sci Technol Built Environ 4731:1–14
13. Pandya B, Kumar V, Patel J, Matawala VK (2018) Optimum heat source temperature and performance comparison of LiCl–H_2O and LiBr–H_2O type solar cooling system. J Energy Resour Technol 140:051204
14. Ibrahim NI, Al-sulaiman FA, Saidur R (2016) Performance assessment of water production from solar cooling system in humid climate. Energy Convers Manag 127:647–655
15. Duffie JA, Beckman WA (2013) Solar engineering of thermal processes, 4th edn. John Wiley & Sons, Hoboken, NJ
16. Mohan G, Uday Kumar NT, Pokhrel MK, Martin A (2016) Experimental investigation of a novel solar thermal polygeneration plant in United Arab Emirates. Renew Energy 91:361–373
17. Mohan G, Kumar U, Pokhrel MK, Martin A (2016) A novel solar thermal polygeneration system for sustainable production of cooling, clean water and domestic hot water in United Arab Emirates: dynamic simulation and economic evaluation. Appl Energy 167:173–188
18. ASHRAE (2009) ASHRAE handbook: fundamentals. American Society of Heating, Refrigerating and Air-Conditioning Engineers, Atlanta, GA
19. Ibrahim NI, Al-sulaiman FA, Ani FN (2017) Performance characteristics of a solar driven lithium bromide-water absorption chiller integrated with absorption energy storage. Energy Convers Manag 150:188–200
20. Klein SA, Alvarado FL (2013) Engineering equation solver
21. Pátek J, Klomfar J (2006) A computationally effective formulation of the thermodynamic properties of LiBr-H2O solutions from 273 to 500 K over full composition range. Int J Refrig 29:566–578
22. Yazaki Energy System (2003), Specifications Chillers and Chiller-Heater WFC-SC (H) 10, 20, 30 (n.d.)
23. Martínez JC, Martinez PJ, Bujedo LA (2016) Development and experimental validation of a simulation model to reproduce the performance of a 17.6 kW LiBr–water absorption chiller. Renew Energy 86:473–482
24. Klein S, Nellis G (2012) Mastering EES. F—Cahrt Software, Madison, WI
25. Herold EK, Radermacher R, Klein AS (2016) Absorption chillers and heat pumps, 2nd edn. CRC Press, Boca Raton, FL

Chapter 29
Design and Construction of a Small Stand-Alone Wind Turbine Using Scrap Materials

Ali Hamzeh, Sadeq Hamed, and Zakaria Al-Omari

29.1 Introduction

Wind generation of electrical energy is a promising energy resource for its sustainable nature and its very low net carbon impact. Wind power is cheap and a source of clean energy, so it is of interest to many countries. The new trend is now to produce small turbines that are easy to install and use to convert wind power into electricity. A small wind turbine is a wind turbine used for microgeneration, as opposed to large commercial wind turbines, such as those found in wind farms, with greater individual power output. The Canadian Wind Energy Association (CanWEA) defines "small wind" as ranging from less than 1000 W (1 kW) turbines up to 300 kW turbines. The smaller turbines may be as small as a 50 W auxiliary power generator for a boat, caravan, or miniature refrigeration unit. The advantage in these turbines is that it can be installed anywhere on boats, lighting poles, residential buildings, schools, or industrial facilities. Small wind power systems are very attractive to support the energy demand either for local residential and small businesses or for developing countries where the electrical grid infrastructure is limited [1].

Large wind power systems, with outputs in megawatts, have clear advantages over small wind power systems because of the economy of scale. The cost of electric power generated by small wind turbines is about 50% more than that of large open-space units, but compared to other power sources, it is very cheap, ranging from 15 to 30 cents per kWh. China is a country that produces and benefits greatly from small wind turbines. This method has contributed to the provision of electricity to nearly 1.75 million people who had no electricity in the past. Many countries, such as India and South Korea, are interested in generating wind power through these small turbines, but the United States is progressing significantly in the future industry as US companies produce and export these units as well [2].

A. Hamzeh (✉) · S. Hamed · Z. Al-Omari
Department of Electrical Engineering, Al-Ahliyya Amman University, Amman, Jordan

A. Sayigh (ed.), *Renewable Energy and Sustainable Buildings*, Innovative Renewable Energy, https://doi.org/10.1007/978-3-030-18488-9_29

According to data from the American Wind Energy Association, the rates of energy produced from small wind turbines have doubled in the United States over the last 3 years. The Association expects the rate to triple by 2015. Perhaps the most important reason for interest in this industry in the United States is the desire of many Americans to reduce dependence on nonrenewable energy sources, in addition to the rise in energy prices in general [3].

For a small wind turbine to be effective, it must produce energy across a wide range of wind speeds. It must be able to generate energy from winds that are switching directions and gusting. It must also be very quiet, so that it will not disturb people living nearby, and it certainly helps if it is pleasing to the eye as well [4].

29.2 Wind Energy in Jordan

There are a number of regions in Jordan with acceptable wind speed to generate electricity, where the great potential areas are the northern and southern parts. The country is classified into three wind regions according to prevailing wind speed: less than 4 m/s, between 4 and 6 m/s, and more than 6 m/s for low, medium, and high regions, respectively. But high wind regime is limited to certain districts: most attractive sites are Hofa, in the northwestern corner and Fjeij, near Showbak, and Wadi Araba in the south [5]. Figure 29.1 shows wind speeds of selected areas.

A wind atlas has been prepared, based on an assessment of the available resource, which demonstrates the existence of a potential for several hundred megawatts of

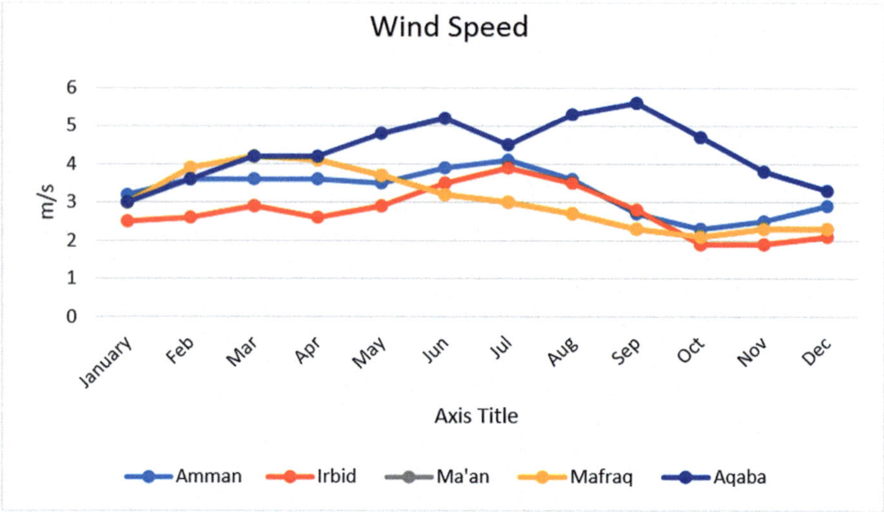

Fig. 29.1 Wind speed of selected cities [6]

wind power installations. Currently there are three operational wind farms in Jordan: Al-Ibrahimiya, with a capacity of 320 kW (4 × 80 kW), established in 1988 in cooperation with a Danish firm and considered as a pilot project; the other, in Hofa, which has a capacity of 1125 kW (5 × 225 kW), established in 1996 in cooperation with the German Government under a program called Eldorado; and the third farm in Altafila which generates 117 MW. Both wind farms are operated and maintained by the Central Electricity Generating Company (CEGCo). During 2008, generation from the two plants totaled 2.92 GWh [7].

In [8], it was found that the wind potential of the Azraq South, Northeast Badia region, could be adequate for stand-alone electrical and mechanical applications, such as wind generation, battery charging, and water pumping as well as agricultural applications.

29.3 Wind Turbine Design

A small wind turbine has five things in common though: a generator, blades, a mounting that keeps it turned into the wind, a tower to get it up into the wind, and batteries and an electronic control system. Most designers use surplus permanent magnet DC motors as generators in their projects. There are probably lots of other brands and models of permanent magnet DC motors available that will work well as generators. Permanent magnet DC motors work as generators, but they weren't designed to be generators. So they aren't great generators. Some types of motor are a lot worse than others. When used as generators, motors generally have to be driven far faster than their rated speed to produce anything near their rated voltage. So what you are looking for is a motor that is rated for high DC voltage, low rpms, and high current. We want a motor that will put out over 12 V at a fairly low rpm and a useful level of current. So a motor rated for say 325 rpm at 30 V when used as a generator could be expected to produce 12+ volts at some reasonably low rpm. On the other hand, a motor rated at 7200 rpm at 24 V probably won't produce 12+ volts as a generator until it is spinning many thousands of rpm, which is way too fast for a wind turbine [9, 10].

As such motors are not available in our market, we decided to use a 600 W, 12 V automobile DC generator. The wind turbine to be built has to deliver 161 kWh in a 30-day month. The design approach comprises the following steps:

1. Calculation of the capacity factor of the WT (CF)
2. Calculation of the rotor diameter (D)
3. Calculation of gear ratio
4. Calculation of the output power and the power curve

29.3.1 Capacity Factor

$$\text{Capacity factor}(\text{CF}) = \frac{\text{Actual energy delivered per month}}{\text{Genrator rated power} \times 24 \times 30} = \frac{161\,\text{kWh}}{0.6 \times 720} = 0.372$$

29.3.2 Rotor Diameter

The average wind speed is assumed to be 6 m/s. The rotor diameter is calculated using the following approximate relationship [11]:

$$\text{CF} = 0.087\left(\text{V}_{\text{avg}}\right) - \frac{\text{P}_{\text{r}}}{\text{D}^2}$$

$$\text{D} = \sqrt{\frac{\text{P}_{\text{r}}}{0.087 \times \text{V}_{\text{avg}} - \text{CF}}} = \sqrt{\frac{0.6\,\text{kW}}{0.087 \times (6\,\text{m/s}) - 0.372}} = \mathbf{2\,m}$$

where:

V_{avg} is the average wind speed m/s, P_r is the rated power of generator (kW), and D is the rotor diameter (m).

29.3.3 Gearbox Ratio

$$\text{Gearbox ratio} = \frac{\textit{Number of teeth}(\textit{large bobbin})}{\textit{Number of teeth}(\textit{small bobbin})} = \frac{\mathbf{76}}{\mathbf{30}} = 2.5$$

This means that the wind turbine shaft speed will be increased 2.5 times.

29.3.4 The Output Power and Power Curve

The power in the wind is calculated as follows:

$$\text{P}_{\text{w}} = \frac{1}{2} \times \rho \times \text{A} \times \text{v}^3$$

$$\text{A} = \frac{\pi}{4} \times \text{D}^2 = \frac{\pi}{4} \times 2^2 = 3.1415\,\text{m}^2$$

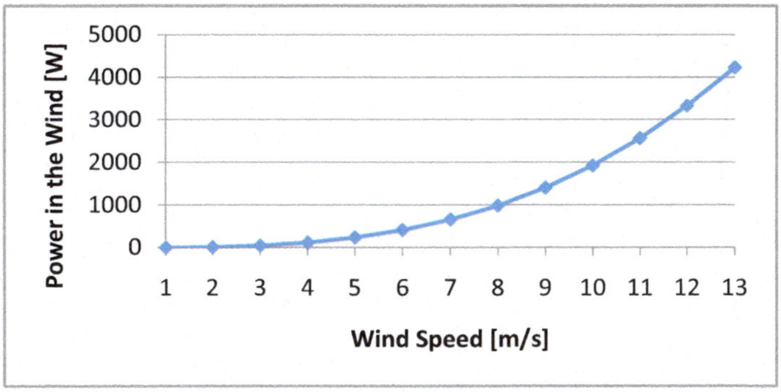

Fig. 29.2 Power in the wind

We find P_w for different wind speeds and we get the curve shown in Fig. 29.2. For example, for a wind speed of 7 m/s:

$$P_w = \frac{1}{2} \times \rho \times A \times v^3$$

$$P_w = \left(\frac{1}{2}\right) \times 1.225 \times 3.1415 \times 7^3 = 659.9\,\text{W}$$

where ρ is 1.225 kg/m^3 at 30 °C and 1 atm.

The output power is:

$$P_{out} = \eta \times P_w = \frac{1}{2} \times \rho \times A \times v^3 \times \eta$$

where η is the overall efficiency of the system (wind turbine, gearbox, generator, and inverter). For a wind speed of 7 m/s and efficiency of 27%, we get:

$$P_{out} = P_w \times \eta = \frac{1}{2} \times \rho \times A \times v^3 \times \eta = 659.99 \times 0.27 = 178.2\,\text{W}.$$

The calculated power curve of the designed WT is shown in Fig. 29.3.

29.4 Mechanical Components

The *tower* is made from a cylindrical iron conduit with the dimensions: length = 254 cm, diameter = 7 cm, and thickness = 5 mm. The tail is made with the dimensions as shown in Fig. 29.4. The main purpose of the tail is to control the direction of the WT rotor due to the wind.

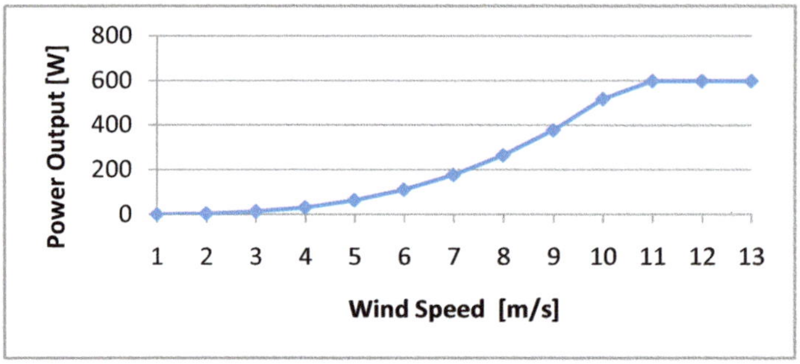

Fig. 29.3 Power curve of the designed WT

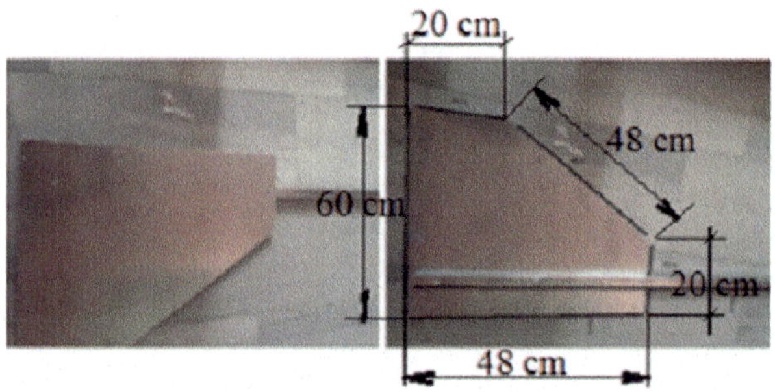

Fig. 29.4 The tail

29.4.1 Blades

We decided to use five blades for the wind turbine. The decision is based on advantages of five-blade small wind turbines over three-blade small wind turbines. These are [12]:

- Five-blade wind turbines will greatly improve annual energy production in low wind conditions. For areas with average wind speeds of 5 m/s, if you compare annual energy output to conventional three-blade wind turbine, there is an increase of annual energy output of more than 60%.
- Five-blade wind turbines will dramatically improve the reliability and safety of wind turbine. The blade rotation speed of a five-blade turbine is 60% of the rotational speed for a three-blade wind turbine. Five-blade wind turbines will greatly reduce chance of overspeed control malfunction. This will ensure operational reliability from a long-term perspective.
- The lower blade rotation speed of five-blade wind turbine will lower wind turbine **noise** and make five-blade wind turbines more community friendly than three-blade wind turbines.

The five blades have been made by cutting sections out of PVC pipes of 4 inches diameter and 1 m long and shaping them into airfoils (Fig. 29.5). We started by placing marks on the perimeter of the pipe equally and cutting it lengthwise into three pieces. Then we cut out one blade and used it as a template for cutting out the others. We then did a little extra smoothing and shaping using a belt sander and palm sander on the cut edges. The dimensions of a blade are shown in Fig. 29.6.

The next step is building a hub to bolt the blades to and attach to the generator. We found a scrap disk of aluminum 26 cm in diameter and 1/4 inch thick that I could bolt the blades onto, but it wouldn't attach to the motor shaft. The simple solution of course was to bolt these two pieces together to make the hub (Fig. 29.7). The hub is attached to a gearbox with a ratio of 1:2.5.

A gearbox is typically used in a wind turbine to increase rotational speed from a low-speed WT rotor to a higher-speed electrical generator, to increase the output voltage of generator. The gearbox contains tow bobbins; a large-diameter bobbin is attached to the WT rotor, and the other small-diameter one is attached to the generator. We use a belt to link the bobbins with each other (Fig. 29.8).

Fig. 29.5 Building steps of the blades out of PVC pipes

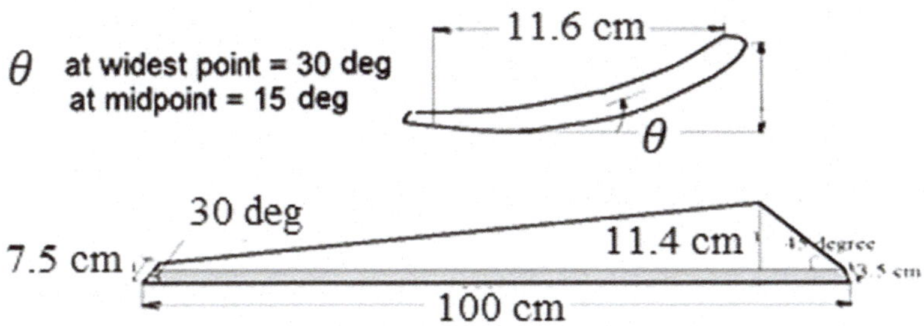

Fig. 29.6 Blade dimensions

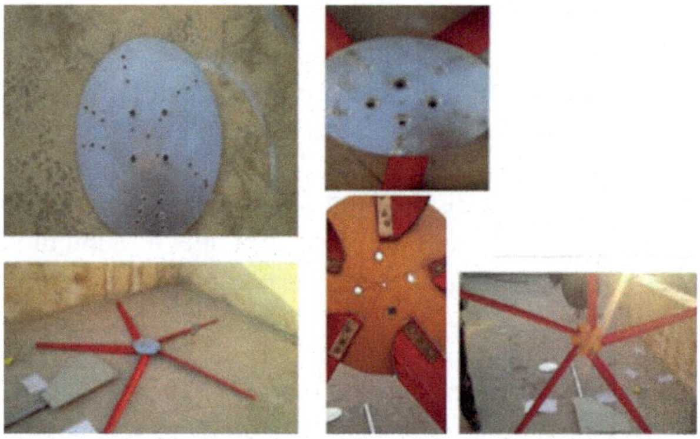

Fig. 29.7 The hub with blades attached

Fig. 29.8 Bobbins and belt

29.5 Electrical Components

29.5.1 Generator

We used an automobile DC generator with voltage rating of 12 V and rated current of 50 A (Fig. 29.9). We gave it a real test by chucking it up in a drill press and connecting it to a dummy load. It works great giving output power of some hundred watts with this setup. We knew then that if we could make a decent set of blades to drive it, it would produce plenty of power.

29.5.2 Charge Controller (Cutout)

The general principle behind the charge controller is that it monitors the voltage of the battery and either sends power from the generator into batteries to recharge them or dumps the power from the generator into a secondary load if the

Fig. 29.9 Used automobile DC generator

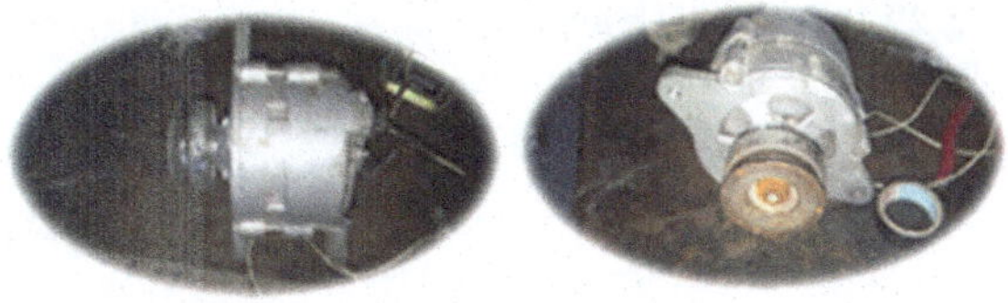

batteries are fully charged (to prevent overcharging and destroying the batteries). We used the controller model New-Era AVR-811, 12 V. All loads are taken directly from the battery. If the battery voltage drops below 11.9 V, the controller switches the turbine power to charging the battery. If the battery voltage rises to 14 V, the controller switches to dumping the turbine power into the dummy load.

29.5.3 Inverter

We selected a 500 W inverter of 12 V DC to 210–230 V AC.

29.6 Wind Turbine Tests

We have performed the following tests on the built wind turbine (Fig. 29.10):

Test 1: The wind turbine is operated without the gearbox at low speed. Output voltage is around 3 V.

Test 2: The wind turbine is operated with gearbox; we notice that the output voltage has increased to 11–13 V.

Test 3: The wind turbine is operated without cutout, and we have an output voltage of 18 V (Fig. 29.11); in this test we used a drill with 2800 rpm to show that the voltage is increasing when the rotational speed of the rotor is increased.

Test 4: This test aims to examine the performance of charge controller (cutout). It shows that the output voltage to the battery is between 12.5 V and 13.5 V and the voltage at inverter input is between 12.5 V and 13 V. The cutout output voltage has been set to protect the battery from a voltage that is higher than 13 V (Fig. 29.12).

Test 5: The WT has been operated by with three blades and with five blades, and we found that the five-bladed WT gave 30% more output power for the same wind conditions, and it worked with less noise.

Fig. 29.10 The built wind turbine

Fig. 29.11 Output voltage reaches 18 V

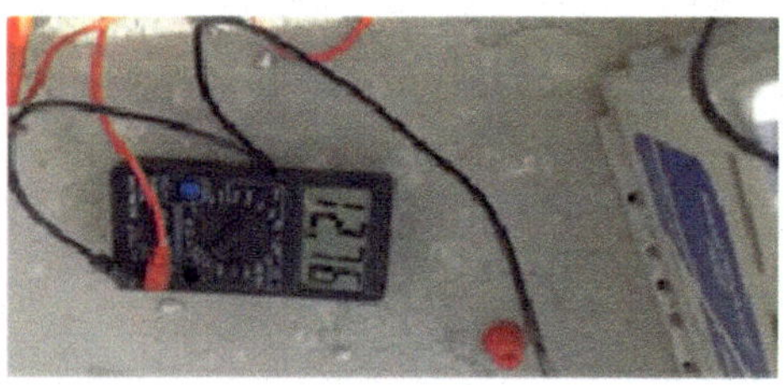

Fig. 29.12 Voltage limitation after cutout connection

29.7 Economic Analysis

The capital cost of the project is shown in Table 29.1.

The levelized cost of electricity (LCOE) [USD/kWh] delivered by the wind turbine is given by [4, 13]:

$$\text{LCOE} = \frac{(\text{Capital cost} \times \text{equity annual return}) + (\text{O}\,\&\,M\ \text{cost})}{\text{Actual energy delivered per year}}$$

$$\text{LCOE} = \frac{389 \times (20\%) + 389 \times (3\%)}{1955\,\text{KWh / year}} = 0.046\frac{\text{USD}}{\text{kWh}} = 4.6\,\text{cents / kWh}$$

The estimated payback period is:

$$\frac{389}{0.046 \times 1955} = 4.33\,\text{years}$$

29.8 Future Work

Future modifications and enhancements we would like to make to the system include:

- Mount the electronics in a weather-proof enclosure.
- Add meters to monitor battery voltage and charge/discharge current.
- Add a tachometer so we know how fast it is spinning.
- Add more batteries to increase reserve storage capacity.

Table 29.1 Total cost of the project

Equipment	Cost (USD)
DC generator	41
Battery	54
Tower	30
Inverter	35
Charge controller	21
Iron conduits	14
PVC pipes for blades	13
Wiring	11
Gearbox	22
Bearing	13
Misc. pipe fittings	35
Labor	100
Total cost	389

- Add a second wind turbine or solar panels to increase power production.
- Get a higher-capacity inverter.
- Some control to automatically brake the unit in high winds.
- A taller tower with steel stakes and steel guy wires.

29.9 Conclusion

The outcome of the performed work is designing and building a home-sized stand-alone wind turbine (WT) using second-hand materials. The built WT is designed to deliver 161 kWh monthly. It is found that a five-bladed WT outperforms a three-bladed one in two features: less noise and around 30% more output power. The developed five-bladed wind turbine is working well, and it could be used in any place as a stand-alone system for cheap electricity supply. The performed economic analysis of the system shows that the levelized cost of energy (LCOE) is about 4.6 USD cents/AC kWh, and the payback period is about 4 years.

The constructed wind turbine from used materials would be an attractive demonstration project of renewable energy, which will encourage the use of stand-alone wind generation in unserved areas in developing countries.

References

1. https://en.wikipedia.org/wiki/Small_wind_turbine
2. Nilivojevic N (2010) Power and energy analysis of commercial small wind turbine systems. IEEE International conference on Industrial Technology, Vina del Mar, Chile
3. https://www.awea.org/market-reports
4. Hamzeh A Lecture of the course "Renewable Energy Systems", 2016–2017 Semester 1
5. Badran OO (2000) Wind energy research and development in Jordan. In: Proceedings of the world renewable energy Congress VI, Brighton, 1–7 July 2000, 2360–2363
6. Baniyounes AM (2017) Renewable energy potential in Jordan. Int J Appl Eng Res 12(19):8323–8331
7. https://www.worldenergy.org/data/resources/country/jordan/wind/
8. Al-Nhoud O, Al Smairan M (2015) Assessment of wind energy potential as a power generation source in the Azraq South, Northeast Badia, Jordan. Modern Mech Eng 5:87–96
9. http://www.mdpub.com/Wind_Turbine/
10. Bumby JR, Maitin R (2005) Axial-flux permanent magnet air-cored generator for small scale wind turbines. IEE Proc Electric Power Appl 152(5):1065–1075
11. Masters GM (2004) Renewable and efficient electric power systems. Stanford University, Wiley, Stanford, CA
12. http://forum.solar-electric.com/discussion/8415/3-blades-vs-5-blades
13. Ackermann T (2005) Wind power in power systems. John Wiley & Sons, Chichester

Chapter 30
RenovaBio Opportunities and Biofuels Outlook in Brazil

Fernando Henriques Salina, Isabela Aroeira de Almeida, and Felipe Ribeiro Bittencourt

30.1 Introduction

In 2015 Brazil committed to the Paris Agreement and defined the specific goals in its Nationally Determined Contribution (NDC). Brazil committed to reduce its greenhouse gas emission (GHG) by 43% below 2005 levels in 2030 [1]. In the transportation sector, Brazil has defined to increase to 18% the share of biofuels on energy matrix. With this goal in mind, the Brazilian government has created the RenovaBio.

RenovaBio's program has already been approved; however, at this moment it is under technical validation. It promotes biofuels by their energetic environmental performance, measured by a life-cycle approach. Therefore, each biofuel type receives a different grade for its sustainability. The Biofuels Decarbonization Credits (CBIO) is generated by the difference between fossil fuel CO2e emission (baseline) and its biofuel substitute. The higher this difference, the more CBIO could be issued and commercialized. With the CBIO, producers can reduce the production costs and develop new technologies to improve the process and the environmental performance. However, the real impact of RenovaBio on the new and current technologies is still uncertain. By these means, the goal of this work is to evaluate the possible impact of RenovaBio on biofuel production cost and to conclude on the viability of current and future biofuels technologies.

F. H. Salina (✉) · I. A. de Almeida · F. R. Bittencourt
WayCarbon, Belo Horizonte, Minas Gerais, Brazil
e-mail: fernando.salina@waycarbon.com; iaroeira@waycarbon.com; fbittencourt@waycarbon.com

A. Sayigh (ed.), *Renewable Energy and Sustainable Buildings*, Innovative Renewable Energy, https://doi.org/10.1007/978-3-030-18488-9_30

30.2 Context

30.2.1 Biofuels Production

Brazil is the second biggest producers of ethanol in the word, seconded only to the USA [2]. The Brazilian ethanol production technology was boosted by Pro-Alcool program in the 1970s. In Brazil, fuel ethanol has two main destinations: for gasoline blend and direct for gas stations. Currently, Brazil has a blend mandate of 27% volume of anhydrous ethanol in the gasoline sold in every gas station. The hydrated ethanol is sold without blends and is mainly used by flex-fuel cars (cars that accept both gasoline and ethanol or any mixture of these). As showed in Fig. 30.1 from 2007 to 2010 the hydrated ethanol production and consumption presented growth. However, after these years the consumption of hydrated ethanol started to decrease. On the other hand, the anhydrous ethanol production and consumption have continued to grow during the last decade.

The profile presented by Fig. 30.1 indicates that gasoline consumption was growing too, as the anhydrous ethanol is only used blended into gasoline. This also indicates that gasoline was a more profitable fuel than ethanol after 2009. The main reason for this was the low oil price delivered by the global economic crisis. By 2009 oil price reached US$ 30.28 per barrel, which was quite low compared to the current US$ 80 per barrel [4].

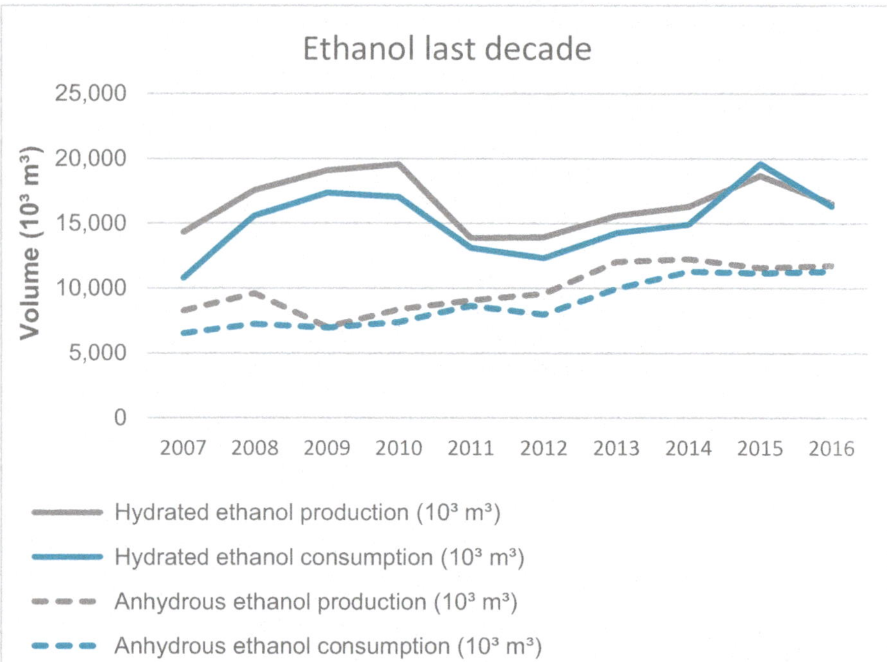

Fig. 30.1 Ethanol production and consumption in the last decade [3]

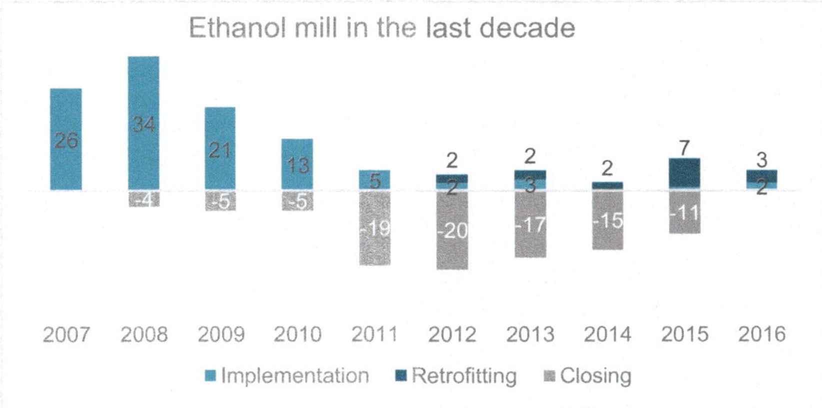

Fig. 30.2 Implementation, retrofitting and closing ethanol mills in the last decade [5]

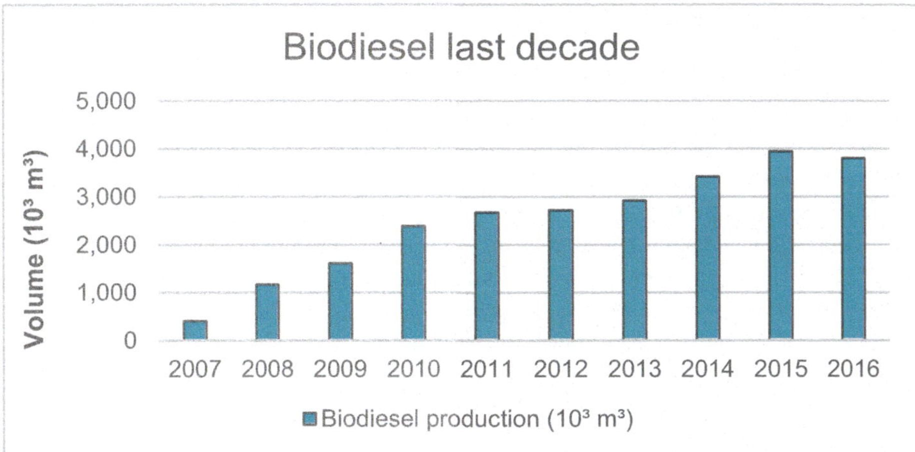

Fig. 30.3 Brazilian biodiesel production in the last decade [3]

As result of the oil price in 2008/2009, many ethanol mills started to close during the 2008/2015, as presented in Fig. 30.2. This shows the actual ethanol scenario in Brazil, a production stagnation. However, some main ethanol producers, such as GranBio and Raizen, are leading developments on sugarcane production and the second generation. Such advances could lead to a real reduction on price and also reduce environmental impact during ethanol life cycle.

In the last decade biodiesel production has increased drastically Fig. 30.3. The majority of biodiesel produced in Brazil is used to compound the diesel mandatory blend [3]. Currently, regular diesel blend uses 8% of biodiesel in its composition.

It is expected that the percentage will rise to 10% in 2019, indicating a real production increase [6].

Brazil is not self-sufficient to produce diesel. In 2016, Brazil imported around 5.9 billion litres of Diesel [3]. Such dependency and volatile oil prices present risk for Brazil. However, it is also potential opportunity to increase biodiesel percentage on Diesel and gives Brazil a local market security.

30.2.2 RenovaBio Description

In Brazil, fuels are produced in petroleum refineries, ethanol mills and other fuel plants. After production, biofuels are sold to fuel distributors and then resold to gas stations. RenovaBio's mechanism proposes that each fuel distributor must have decarbonization goals based on their market share. The individual goals are part of Brazilian national's goal to increase the participation of renewable fuels in transportation sector (Brazil has committed to this specific goal on the Paris Agreement). Each fuel distributor will have to prove a specific quantity of Biofuels Decarbonization Credits (CBIO) to prove that they are meeting decarbonization goals.

CBIO are a financial asset, negotiated in the stock market, emitted by a biofuel producer [7]. The CBIO is generated by the difference between fossil fuel CO2e emission (baseline) and its biofuel substitute. The higher this difference, the more CBIO could be issued and commercialized. Figure 30.4 shows a general mechanism for RenovaBio and CBIO emission and acquisition flow.

RenovaBio intends to improve the environmental performance of biofuels by giving different energetic environmental grade for each producer. As the producers have different efficiency through the life cycle, the one who has a better performance will have the best energetic environmental grade. The higher the note, the more CBIO producer could generate. The energetic environmental grade is measured by a life-cycle assessment (LCA) framework, named the RenovaCalc.

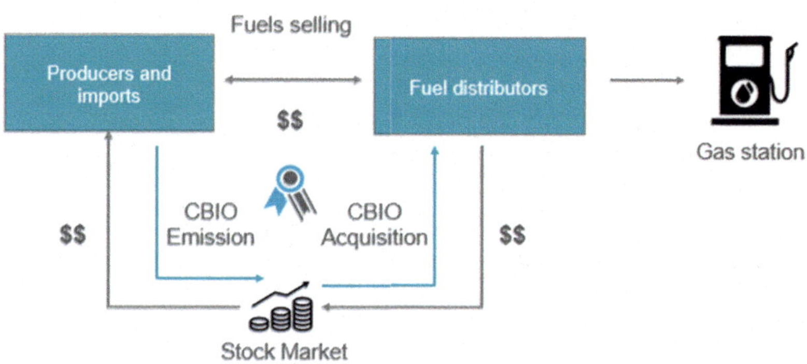

Fig. 30.4 RenovaBio simplified mechanism

The RenovaCalc is on public technical validation and considers only few biofuels (such as sugarcane ethanol and soybean biodiesel). The energetic environmental grade will be certificated by three parts to assure the authenticity [7].

CBIO selling increases revenue, so the government expects to increase the biofuel participation on transportation matrix. At a first moment, the biofuel producers could use increased revenue from CBIO to increase the efficiency of their process to improve carbon emission and reduce losses, strengthening the potential to generate more CBIO. In the ethanol sector, the RenovaBio program could lead a development on second-generation technology. For biodiesel, RenovaBio could attract more investor by showing the need to increase the biodiesel production in the next few years (with the percentage raise on diesel blend) and the potential to generate an extra financial asset.

30.3 Methodology

30.3.1 Financial Projections

To conclude on the financial impact on biofuels, two macro variables were considered: the CBIO price (CP) and the biofuel production cost (BPC). From these two variables, it is possible to imply which type of biofuel will be more cost-effective. This study considers, if the producers generate CBIO, there will be a reduction on the biofuel production cost. By these means, the biofuel apparent cost (BAC) can be calculated by Eq. 30.1.

$$BAC = BPC - CP \tag{30.1}$$

The present study considers 11 different cases for the CBIO price (CP), as showed in Table 30.1. The partial economic model (economic model that considers a constant price of all products to estimate the price of a specific one) was used to calculate CBIO prices. The model considers aspects of the fuel structure in Brazil (such as price, tributes, emissions) and three variables: the oil price, efficiency gain in biofuel production and the level of decarbonization in transportation sector. More details about the premises and the mathematical model to estimate the CBIO price can be found at RenovaBio technical note [7].

The biofuels production cost (BPC) considers all expenses from farming to industrial phase. In this study, five different biofuels were considered as showed by Table 30.2. Table 30.2 also shows the production cost for these biofuels.

The price of corn ethanol was considered the same as sugarcane ethanol, and fuel taxes were not considered in this study. The taxes applied to biofuels and fossil fuel are different in Brazil, so, in order to avoid misinterpretation, none were considered.

Table 30.1 Cases and premises adopted to estimate the CBIO price [8]

Case	Oil price (US$/barrel)	Productive efficiency gain	CBIO price (US$/ ton of CO_2e)
I	20	0%	53
II	20	10%	42
III	20	20%	32
IV	20	30%	20
V	40	0%	36
VI	40	10%	24
VII	40	20%	14
VIII	40	30%	3
IX	60	0%	19
X	60	10%	7
XI	60	20%	0

Table 30.2 Production cost for biofuels in Brazil [9, 10]

Fuel	Feedstock	Production cost (R$/L)	Reference
Hydrated ethanol	Sugarcane	1.159	[9]
Hydrated ethanol	Corn	1.159	[9]
Ethanol second generation	Sugarcane bagasse	1.528	[9]
Biodiesel	Tallow	1.247	[10]
Biodiesel	Soybeans	1.247	[10]

30.3.2 Carbon Footprint

The less biofuel carbon emitted during its life cycle, the more CBIO the producers can generate. By these means, biofuels with better environmental performance will have more advantages according to RenovaBio's premises. To measure the quantity of CBIO that each biofuel can generate, it is necessary to know the energetic environmental grade. It is calculated by the difference between fossil fuel CO2e emission (baseline) and its biofuel substitute.

This study evaluates the CBIO cost impact on five different types of hydrated ethanol and two types of biodiesel, as showed in Table 30.3. Table 30.3 also presents the carbon footprint of each one of the biofuels and fossil fuels. The energetic environmental grade was calculated based on the GHG emission difference between fossil fuels and biofuel. For example, gasoline GHG emission through the life cycle is 87.4 g CO2e/MJ and for hydrated ethanol I is 21.3 g CO2e/MJ; the difference between the emission of these fuels is 66.1 g CO2e/MJ, which is equivalent to the energetic environmental grade.

To convert the energetic environmental grade into fuel litres, the density and lower heating value (LHV) were used. The value for density and LHV used for ethanol and biodiesel are presented in Table 30.4.

Table 30.3 Biofuels carbon footprint and energetic environmental note [7, 11–14]

Fuel	Feedstock	GHG emissions (gCO2e/MJ)	Energetic environmental grade (gCO2e/MJ)	Reference
Hydrated ethanol I	Sugarcane	21.30	66.10	[11]
Hydrated ethanol II	Sugarcane	14.87	72.53	[12]
Hydrated ethanol	Corn	53.80	33.60	[12]
Ethanol second generation	Sugarcane bagasse	7.55	79.85	[13]
Biodiesel	Tallow	14.14	72.36	[14]
Biodiesel	Soybeans	30.73	55.77	[14]
Gasoline	–	87.40	–	[7]
Diesel	–	86.50	–	[7]

Table 30.4 Density and LHV for ethanol and biodiesel [3]

Fuel	Density (kg/m^3)	LHV (kcal/kg)
Ethanol	809	6300
Biodiesel (B100)	880	9000

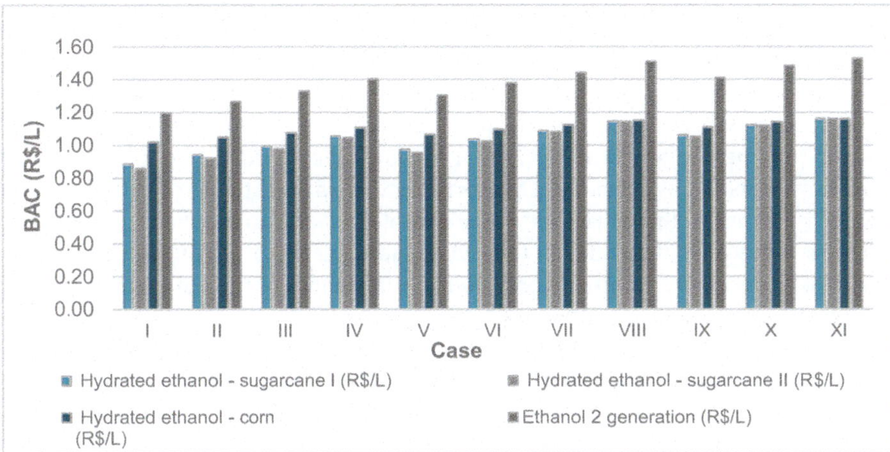

Fig. 30.5 Ethanol apparent cost for different CBIO price

30.4 Results

30.4.1 Impact Projections on Brazilian Biofuel Scenario

Figure 30.5 shows the apparent cost to produce ethanol considering different types of ethanol and CBIO prices. Case I considers the maximum CBIO price, US$ 53.00, and Case XI considers null, the lowest in this case (Table 30.1).

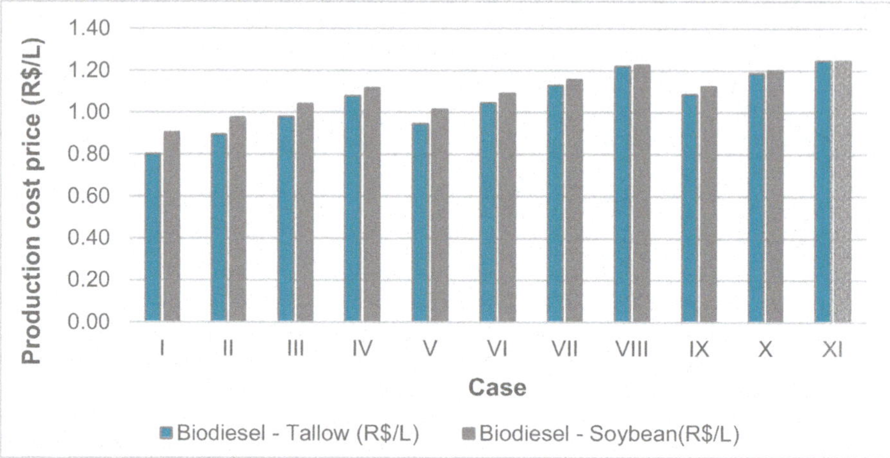

Fig. 30.6 Biodiesel apparent cost for different CBIO price

In Case I, it is possible to see that the sugarcane ethanol II has the lowest price between the four types presented (R$ 0.86 per litre). The sugarcane ethanol (I and II) has low carbon emission during its life cycle. By these means, the cost reduction was more significant when compared to corn hydrated ethanol. Currently, the price to produce second-generation ethanol is R$1.53 per litre. In Case I, second-generation ethanol presents the higher reduction on production cost (R$0.33 per litre), as this fuel has the best environmental performance. However, the production cost of second-generation ethanol is higher than hydrated ethanol in baseline. It means that for the second-generation ethanol to become more attractive, a reduction on production will be necessary; in other words, RenovaBio cannot be the only reason to scale up the second-generation production.

Figure 30.6 shows the apparent cost to produce tallow and soybean biodiesel considering different CBIO prices. Tallow biodiesel has lower GHG emissions as during its life cycle it presents the biggest reduction. The meat industry produces tallow as subproduct, and it has more than one destination (cosmetics and laundry applications). RenovaBio could potentialize the viability to produce tallow biodiesel, diversifying the products portfolio. On the other hand, the soybean biodiesel still has space to increase its production. As tallow is a subproduct from meat industry, its production has limit to grow, different from soybean.

30.5 Conclusion

It is safe to say that RenovaBio will benefit Brazilian biofuel producers. It also important that producers notice which better environmental performance will have more competitive advantages. Second-generation apparent cost production could be reduced by almost 0.33 R$/L in Case I. Although second-generation ethanol has the best potential to generate CBIO, the current production price is still high. These means, new technologies and advances in this route will be necessary to reduce second-generation ethanol production cost. As biodiesel (independent of the feedstock) has a low carbon footprint, which gives this fuel a potential production expansion in the next decades. By the means, RenovaBio will have a positive impact on biofuel sector; however, efforts to improve biofuel technologies are still necessary.

References

1. Federative Republic of Brazil (2015) Intended nationally determined contribution—towards achieving the objective of the united nations framework convention on climate change
2. Irena (2016) Boosting biofuels. Sustainable paths to greater energy security
3. EPE, Brazilian Energy Balance (2017) Ministry of mining and energy
4. West Texas Intermediate (2018) WTI crude oil prices—10 year daily chart. macrotrends. https://www.macrotrends.net/2516/wti-crude-oil-prices-10-year-daily-chart. Accessed 22 Jul 2018
5. EPE (2018) Cenários de oferta de etanol e demanda do Ciclo Otto 2018–2030. p. 14
6. ANP (2017) Biodiesel. http://www.anp.gov.br/biocombustiveis/biodiesel. Accessed Jan 2018
7. MME A (2017) Nota Explicativa sobre a Proposta de Criação da Política Nacional de Biocombustíveis. Minister of Mining and Energy
8. MME B (2017) RENOVABIO—Cenários e Simulação de Impacto. Ministry of Mines and Energy
9. Milanez AY et al (2015) De promessa a realidade: como o etanol celulósico pode revolucionar a indústria da cana-de-açúcar—uma avaliação do potencial competitivo e sugestões de política pública. BNDES Setorial—Biocombustíveis 41:237–294
10. Barros GSC et al (2006) Custos de produção de Biodiesel no Brasil. Politica Agricola 3:36–50
11. Seabra JEA et al (2011) Life cycle assessment of Brazilian sugarcane products: GHG emissions and energy use. Biofuels Bioproducts and Biorefining—Modeling and Analysis 5(5). https://doi.org/10.1002/bbb.289
12. Donke ACG (2016) Environmental and energy assessment of sugarcane, corn and sorghum ethanol production in an integrated plant, according to the life cycle approach. USP University, São Paulo
13. GRANBIO (2014) GranBio inicia produção de etanol de segunda geração
14. CGEE (2012) Sustainability of sugarcane bioenergy. Center for Strategic Studies and Management, Brasilia, p 226

Chapter 31
Evolution of Photovoltaic-Thermal Hybrid Solar Technology for the Tropics: A Case Study of Malaysia

Mohd Yusof Hj. Othman, Kamaruzzaman Sopian, Mohd Hafidz Ruslan, Sohif Mat, and Suhaila Abdul Hamid

31.1 Introduction

Major concerns on energy security, impact on environment and sustainability of high world oil prices support have allow human race to expand the use of nuclear power and renewable energy sources (RES) over the projection. Nevertheless, due to nuclear disasters at Fukushima Daiichi power plants on 11 March 2011, public became aware of the benefits of RES towards the environment [1]. It has attracted both academic and industrial communities' attention. The importance of promoting this alternative energy also was increasing ever since.

The strategic location of Malaysia at equator line allows the implementation of solar energy widely. In the Solar Energy Research Institute (SERI), a variety of research and developments have been undertaken experimentally and theoretically. The main research purpose is to increase the electrical gain by reducing the temperature of solar cells. This is due to evidence reported that an increase of 1 °C on solar cells dropped the electrical efficiency by 0.4–0.5% [2]. Figure 31.1 shows the flowchart of PV/T systems that has been built in SERI, UKM, since the early 1980s.

The objective of this paper is to present an overview of PV/T systems which have been under research activities by SERI, Universiti Kebangsaan Malaysia (UKM). This paper also describes our experience in introducing the system for cottage industry in Malaysia.

M. Y. Hj. Othman (✉) · K. Sopian · M. H. Ruslan · S. Mat · S. Abdul Hamid
Solar Energy Research Institute (SERI), Bangi, Selangor, Malaysia
e-mail: myho@ukm.edu.my

A. Sayigh (ed.), *Renewable Energy and Sustainable Buildings*, Innovative Renewable Energy, https://doi.org/10.1007/978-3-030-18488-9_31

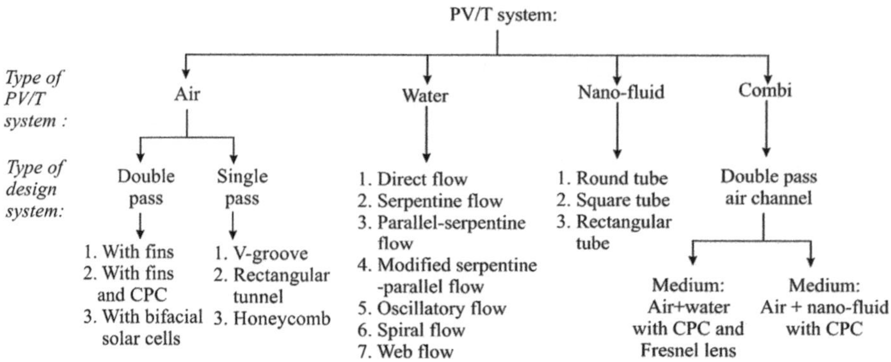

Fig. 31.1 Flowchart of PV/T systems built in SERI, UKM

31.2 Experiments Conducted

31.2.1 Double-Pass PV/T Air-Based System

The construction of the double-pass PV/T system comprised of three essential components: a glazing on the top, a series plate of photovoltaic (PV) cells and an absorber plate. Three distinguished double-pass PV/T systems have been set up in SERI: (a) fin collector, (b) fin-CPC collector and (c) bifacial collector.

31.2.1.1 Double-Pass PV/T Collector with Fins

The ideal setup of fin collector is shown in Fig. 31.2. In order to enhance heat absorption on PV cells, the vertical fins which were made from aluminium sheet were attached underneath the cells and act as heat extractor. The inlet heated air flows from upper channel to the lower channel. The effort of the idea increases the efficiency of the collector.

31.2.1.2 Double-Pass PV/T Collector with Fins and CPC

As pictured in Fig. 31.3, the concept of the collector (fin-CPC collector) was much similar to fin collector with addition of compound parabolic concentrator (CPC). It functions on focusing the sunlight received on its surface before the rays hit solar cells. The multiple reflections occur on CPC that allows the absorption of photon energy on PV cells which was converted to electrical energy. Thus, it yielded an increment of electrical production, and the efficiency of the overall system was improved [3].

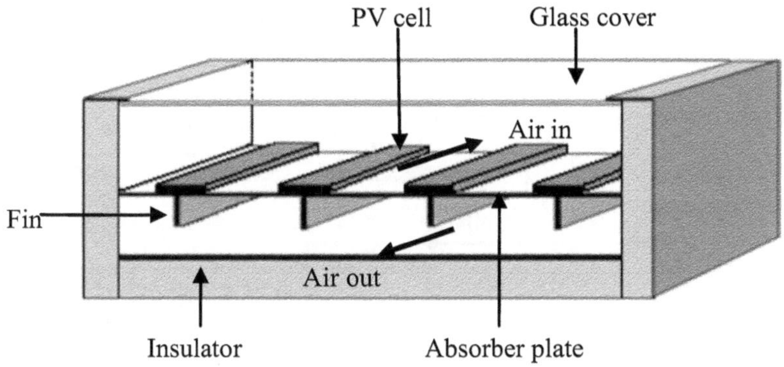

Fig. 31.2 The schematic diagram of double-pass PV/T collector with fins [11]

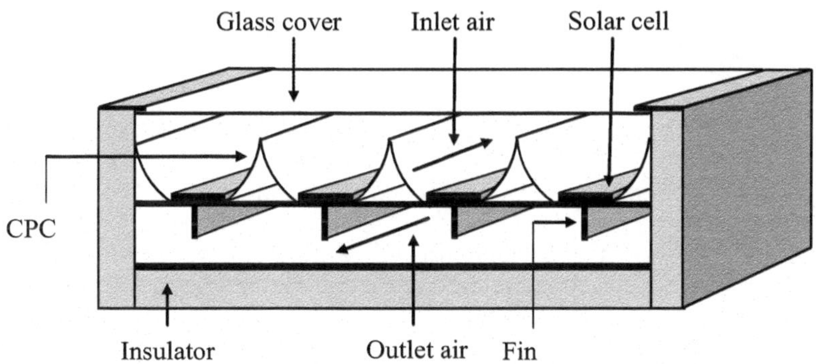

Fig. 31.3 The schematic diagram of PV/T collector with fins and CPC [3]

31.2.1.3 Double-Pass PV/T Collector Using Bifacial Solar Cells

The configuration of this collector depicted in Fig. 31.4 is similar to the previous two designs mentioned earlier. By using bifacial solar cells, the collector (bifacial collector) was capable to absorb photon energy from the top as well as the bottom surface. This invention generates up to 90% more electric energy compared to mono-facial module [4]. Furthermore, due to wide space at the rear surface, the bifacial collector was able to absorb less infrared solar radiation. Eventually, the system operates at lower temperature, and higher electrical efficiency was achieved. A packing factor of 0.7 has been used so that the incident radiation can travelled through the panel reflector by the back reflector. The thermal efficiency was between 50% and 70% for a flow rate of 0.03–0.14 kg/s and solar intensity of 500–900 W/m^2.

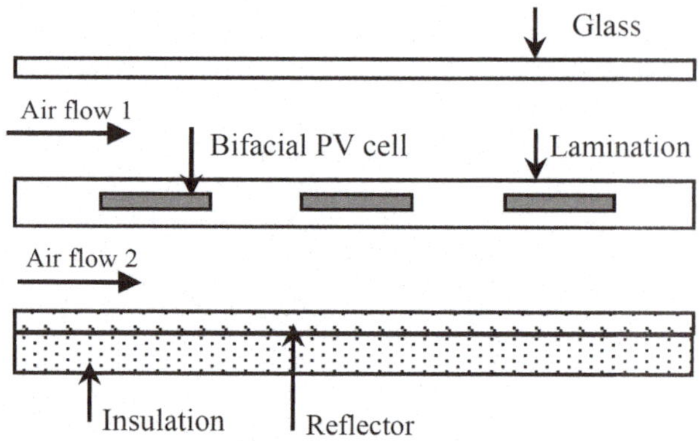

Fig. 31.4 The side view of PV/T collector using bifacial solar cells [4]

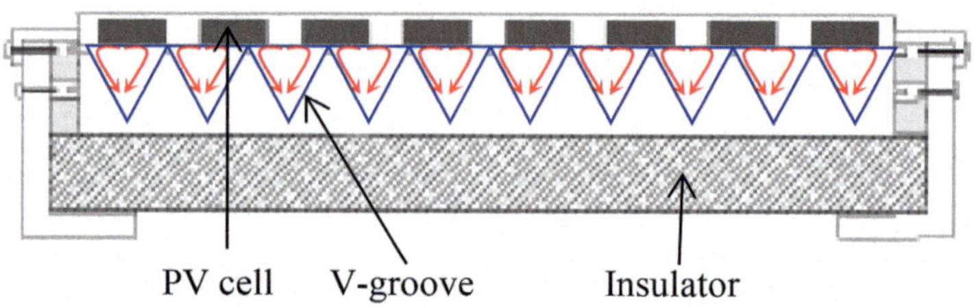

Fig. 31.5 The schematic diagram of single-pass V-groove collector [5]

31.2.2 Single-Pass PV/T Air-Based System

The early study of the single-pass collector was conducted as an alternative use of solar energy for drying processes. It was also to defeat the setback drawn by glazing double-pass collectors which caused deficiency of electricity production by more than 50% [5]. The configuration of the collector allows the air to flow in one direction through the absorber. The single-pass PV/T systems that have been tested in SERI are (a) V-groove absorber, (b) rectangular absorber and (c) honeycomb absorber.

31.2.2.1 Single-Pass PV/T Collector with V-Groove Absorber

The aluminium sheet was folded to form a V shape at an angle of 50.8°. It then was combined together upside down and attached on top of the flat aluminium plate. This construction allows heat to be transferred uniformly throughout the absorber. The PV module was exposed to sunlight with no glazing resulting in an optimum electrical energy conversion. The conceptual design of the collector is presented in Fig. 31.5.

31.2.2.2 Single-Pass PV/T Collector with Rectangular Tunnel Absorber

The function of this collector (rectangular collector) serves the same purpose and operation as V-groove collectors. The innovation made was replacing the V-groove with an aluminium-made rectangular tunnel absorber to flow out the absorbed heat in the tunnel effectively. The diagram is referred at Fig. 31.6.

31.2.2.3 Single-Pass PV/T Collector with Honeycomb Absorber

The absorber of this PV/T collector (honeycomb collector) was using a honeycomb shape with regular hexagonal geometry. The heat exchanger was made by local aluminium sheet and shaped into corrugated sheet. The set of joined sheet was then placed at the back side of PV panel as shown in Fig. 31.7. The operational system of the collector was also met the same as V-groove collector.

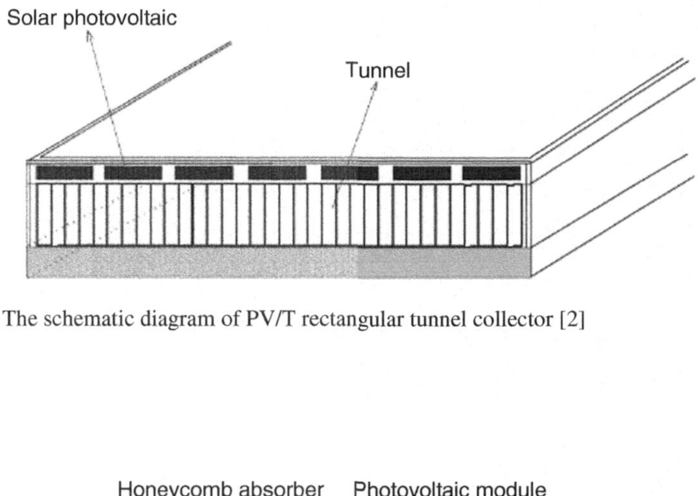

Fig. 31.6 The schematic diagram of PV/T rectangular tunnel collector [2]

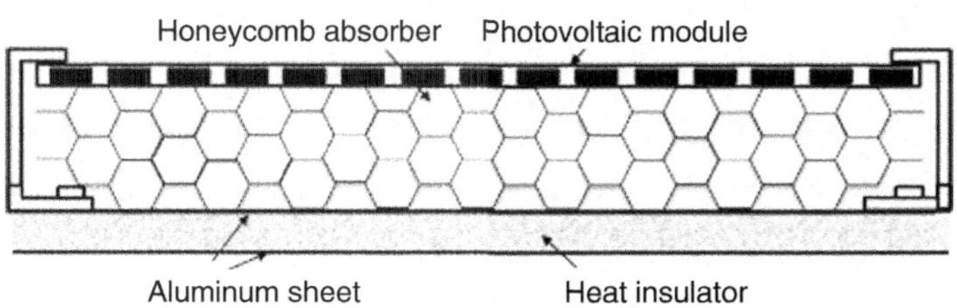

Fig. 31.7 The schematic diagram of PV/T honeycomb collector [6]

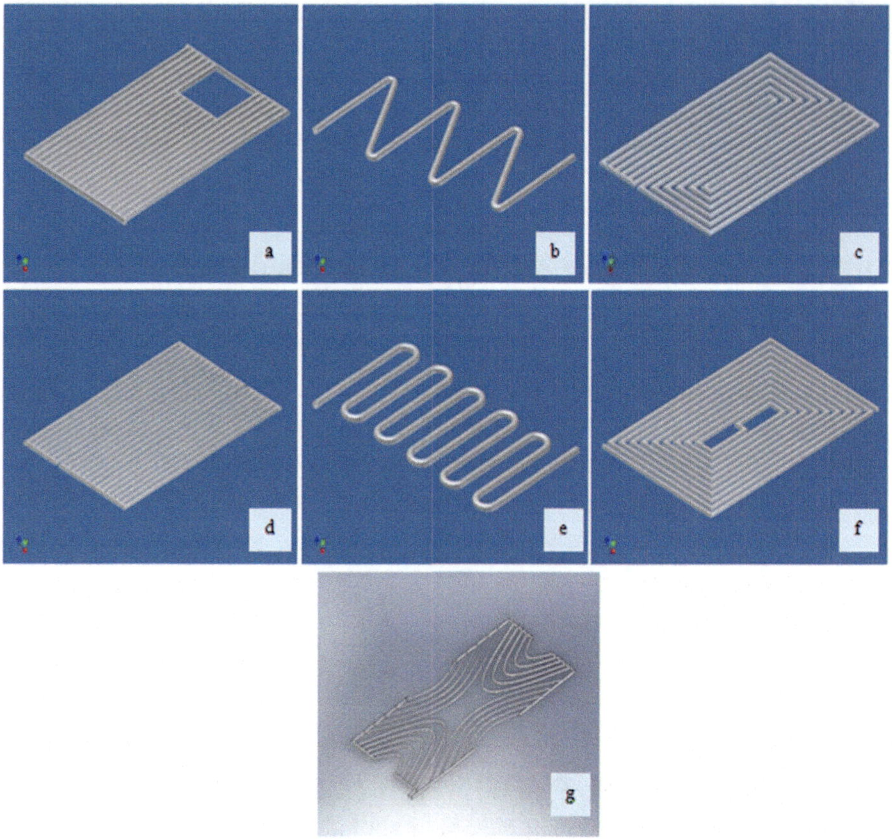

Fig. 31.8 The schematic diagram of PV/T water-based system: (**a**) direct flow, (**b**) serpentine flow design, (**c**) parallel-serpentine flow design, (**d**) modified serpentine-parallel flow design, (**e**) oscillatory flow design, (**f**) spiral flow design and (**g**) web flow design [7]

31.2.3 PV/T Water-Based Collector

The principle of PV/T water-based collector is identical with single-pass PV/T air-based collector. Without glass cover, a flat plate PV module was exposed to sunlight [7]. Heat was extracted by water which flows underneath the module through various designs of tube channel. Figure 31.8 shows the design of seven water-based absorbers studied.

31.2.4 PV/T Nano-Fluid-Based

A PV/T nano-fluid-based system run in SERI was conducted under three different absorbers, round tube, square tube and rectangular tube, with three different types of nano-fluids: SiO_2, TiO_2 and SiC. The operation of the system still follows the

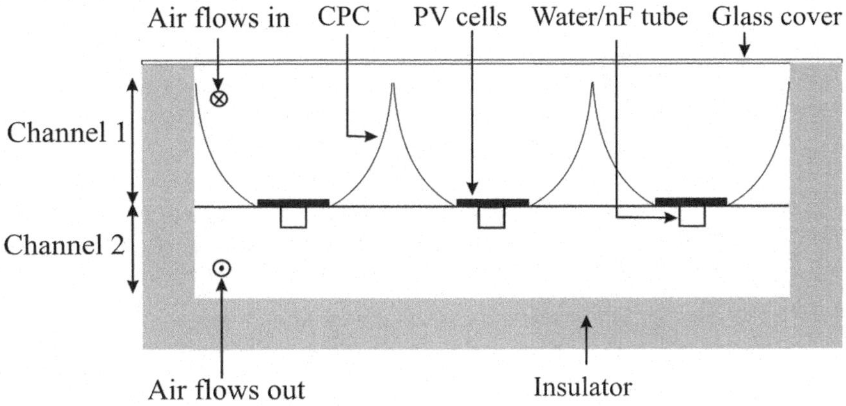

Fig. 31.9 Schematic diagram of PV/T Combi [10]

same rules as PV/T water-based collector [8]. In general, nano-fluids possessed better and higher thermal properties compared to water and air. By using nano-fluids as the medium of heat exchanger, the overall performance of the system was improved especially on the electrical efficiency.

31.2.5 PV/T Combi System

PV/T Combi system is a system combining two fluids operating in one unit. Two designs of double-pass PV/T Combi system were being studied at SERI: (a) PV/T Combi, air and water using CPC and Fresnel lens, and (b) PV/T Combi, air and nano-fluids using CPC. The system can be operated by single mode or dual mode depending on the application needs [9]. Figure 31.9 depicted the schematic diagram of the system.

31.3 Results

All PV/T systems at SERI were tested experimentally and theoretically. The experiment was done either in indoor under controlled parameter environment or outdoor with real-condition environment. The parameters under studied include the mass flow rate of the fluid, solar radiation, wind speed, length of collector and concentration of nano-fluids. The prototype of PV/T system has also been applied for cottage industry especially for drying processes. Table 31.1 summarizes the electrical and thermal performance of each PV/T system discussed in the earlier section at the variation of mass flow rate and solar intensity.

Table 31.1 Performance of PV/T system understudy in SERI

Bil.	PV/T system	Mass flow rate of fluid (kg/s)	Solarintensity (W/m²)	η_{el} (%)	η_{th} (%)
1.	Double-pass PV/T air-based with fins [11]	Air: 0.0316	400	10.50–12.09	17.00–26.43
2.	Double-pass PV/T air-based with fins and CPC [3]	Air: 0.0082–0.1149	700	1.27–1.53	37.27–62.45
3.	Double-pass PV/T air-based with bifacial solar cell [4]	Air: 0.03–0.14	952	11.00–14.00	50.00–70.00
4.	Single-pass PV/T air-based: (a) V-groove absorber [5] (b) Rectangular absorber [2] (c) Honeycomb absorber [6]	Air: 0.02–0.15	828	(a) 7.05–7.04 (b) 6.94–6.87 (c) 6.95–7.25	(a) 29.97–77.87 (b) 27.00–78.00 (c) 30.22–89.12
5.	PV/T water-based [7] (a) Direct flow (b) Serpentine flow (c) Parallel-serpentine flow (d) Modified serpentine-parallel flow (e) Oscillatory flow (f) Spiral flow (g) Web flow	Water: 0.011–0.041	800	(a) 12.69–12.19 (b) 11.68–12.10 (c) 12.80–12.22 (d) 13.25–12.63 (e) 11.77–12.21 (f) 13.81–13.01 (g) 12.37–11.91	(a) 46.43–54.13 (b) 29.24–34.39 (c) 44.23–52.12 (d) 43.32–51.98 (e) 37.17–44.12 (f) 45.00–54.61 (g) 41.11–48.07
6.	PV/T nano-fluid-based [8]: (a) SiO_2 1 wt% (b) TiO_2 1 wt% (c) SiC 1 wt%	Nano-fluids: 0.17–0.68	1000	(a) 10.30–10.09 (b) 10.97–10.58 (c) 13.52–13.07	(a) 65.93–60.54 (b) 66.93–63.67 (c) 68.21–65.88
7.	PV/T Combi with CPC and Fresnel lens [9]: (a) Air fix, water flow (b) Water fix, air flow	Air: 0.0064–0.03 Water: 0.00884–0.0252	650	(a) 12.92–13.12 (b) 12.90–13.20	(a) 48.07–40.79 (b) 45.05–35.38
8.	PV/T Combi with CPC [10]: (a) Air-TiO_2 0.5 wt% • Air fix, TiO_2 flow • TiO_2 fix, air flow (b) Air-SiC 0.5 wt% • Air fix, SiC flow • SiC fix, air flow	Air: 0.0104–0.1248 Nano-fluids: 0.0078–0.25	750	(a) • 11.89–12.55 • 10.88–11.03 (b) • 11.89–12.44 • 11.91–13.14	(a) • 37.91–55.31 • 37.59–47.18 (b) • 24.11–57.71 • 24.11–55.87

31.4 Conclusions

Numerous designs of PV/T collector done in SERI had been described briefly in this paper. The working principles of the collectors were different based on its own specific way of harvesting solar energy to be converted to electrical and heat energy. However, the objective of the study meets the same goal which is to increase electricity production and to improve the overall system performance. The application of each collector also varies to heating and drying purposes. Research and development done at SERI have open many opportunities in terms of knowledge transfer, innovation ideas as well as mature PV/T technology.

Acknowledgement The authors would like to express their gratitude to the University Kebangsaan Malaysia and the Ministry of Higher Education Malaysia for funding this research project through the research grants 03-01-02-SF1145 and AP-2015-003.

References

1. Oskin B (2017) Japan earthquake & tsunami: facts and information. https://www.livescience.com/39110-japan-2011-earthquake-tsunami-facts. Accessed 28 June 2018
2. Othman MY, Ibrahim A, Jin GL, Ruslan MH, Sopian K (2013) Photovoltaic-thermal (PV/T) technology-the future energy technology. Renew Energy 49:171–174
3. Othman MY, Yatim B, Sopian K, Abu Bakar MN (2006) Double-pass photovoltaic-thermal solar air collector with compound parabolic concentrator and fins. J Energy Eng 132(3):116–120
4. Ooshaksaraei P, Sopian K, Zaidi SH, Zulkifli R (2017) Performance of four air-based photovoltaic thermal collector's configurations with bifacial solar cells. Renew Energy 102:279–293
5. Othman MY, Yatim B, Ruslan MH (1996) Preliminary results of a V-groove back-pass solar collector. World Renew Energy Congress 9:622–625
6. Othman MY, Hussain F, Sopian K, Yatim B, Ruslan H (2013) Performance study of air-based photovoltaic-thermal (PV/T) collector with different designs of heat exchanger. Sains Malaysiana 42(9):1319–1325
7. Ibrahim A, Othman MY, Ruslan MH, Alghoul MA (2009) Performance of photovoltaic thermal collector (PVT) with different absorbers design. WSEAS Trans Environ Dev 5(3):321–330
8. Kadhim AN (2017) Performance enhancement of grid connected photovoltaic thermal systems by nanofluids. Institut Penyelidikan Tenaga Suria. Universiti Kebangsaan Malaysia
9. Baljit SSS, Chan H-Y, Audwinto VA, Hamid SA, Fudholi A, Zaidi SH, Othman MY, Sopian K (2017) Mathematical modelling of a dual-fluid concentrating photovoltaic-thermal (PV-T) solar collector. Renew Energy 114:1258–1271
10. Sopian K, Hamid SA, Baljit SSS, Othman MY, Chan H-Y, Zaidi SH (2017) Performance of double pass PV/T Combi solar collector with CPC: Theoretical approach. IEEE international conference in energy and sustainability in small developing economies (ES2DE 2017) on 10th–12th Jul 2017. doi: https://doi.org/10.1109/ES2DE.2017.8015345
11. El Gefi EMA, Sopian K, Othman MY, Yatim B (2008) Experimental investigation of single-pass, double duct photovoltaic thermal (PVT) air collector with CPC and fins. Am J Appl Sci 5(7):866–871

Chapter 32
Feasibility Study into Design, Development and Testing of an Indirect Multi-Rack Solar Dryer for Agricultural Products

Hossein Mirzaii and Juan Pablo Nojek Barbieri

Nomenclature

A_c	Collector area, m^2
C_p	Specific heat at constant pressure, kJ/kg °K
d	Mass of the dry matter in the sample, kg
g	Acceleration due to gravity, 9.81 m/s^2
h	Height of hot air column, m
I	Insolation incident on the absorber's surface, W/m^2
I_T	Total solar radiation incident on the top surface,W/m^2
L	Latent heat of vaporization of water, kJ/kg$_{H2O}$
m	Mass, kg
m_a	Mass of drying air, kg
$\dot{m}_a$	Mass of air leaving the dryer per unit time, kg/s
m_i	Initial mass of the product to be dried, kg
M_i	Initial moisture content wet basis, %
M_f	Final moisture content wet basis, %
m_v	Mass of water removed from product, kg
MC	Moisture content
OS	Open sun
P	Pressure, Pa
Q_u	Useful energy gained by collector, W
Q_{useful}	Useful energy gained by collector, W
Q_{Cond}	Conduction heat losses from the absorber, W
Q_{Conv}	Convective heat losses from the absorber, W
Q_L	Heat losses of the absorber, W
Q_g	Heat gained by the air, W
Q_{Rad}	Radiation heat losses from the absorber, W
Q_{Ref}	Reflection losses from the absorber, W
R	Universal gas constant, 8.3 kJ/k mole K

H. Mirzaii (✉) · J. P. N. Barbieri
Faculty of Science, Engineering and Computing, Kingston University, London, UK
e-mail: H.Mirzaii@kingston.ac.uk

A. Sayigh (ed.), *Renewable Energy and Sustainable Buildings*, Innovative Renewable Energy, https://doi.org/10.1007/978-3-030-18488-9_32

T	Temperature, K
T_a	Ambient air temperature, K
T_c	Temperature of the collector's absorber, K
T_i	Initial temperature of drying air, K
T_f	Final temperature of drying air, K
U_L	Overall heat transfer coefficient of the absorber, W/m^{2}K
V	Volume, m^3
w	Mass of the wet product, kg
α	Solar absorptance of the absorber
ρ_1	Density of air outside the dryer respectively, kg/m^3
ρ_2	Density of air inside the dryer, kg/m^3
η_C	Efficiency of the collector, %
η_d	Drying efficiency, %
$\mathcal{R}$	Reflection coefficient of the absorber
τ	Transmittance of the glazing cover, W/m^2
τ_d	Drying time, s

32.1 Introduction

These days, small farmers use sun-drying techniques to dry harvested crops in order to preserve the excess of production. This procedure is used especially in small to medium scale for domestic or commercial use. It is estimated that, in developing countries, more than 80% of the food produced by small farmers is dried with this technique [1]. The method consists of a large surface outdoors where the product is spread, normally on the ground or on trays, and left there to dry until the product reaches the desired content of moisture. Regularly, during the day, the product has to be turned over to improve the efficiency and homogeneity of the drying. Even though this method requires a low capital investment, the drawbacks are many, considering that the drying times may take from 10 up to 30 days. During this period the food is exposed to dust, rain, insects and other animals [2]. These are the reasons why the rate of loss food is very high.

32.1.1 Drying Principles

The objective of the drying process of a product is to reduce its content of moisture to a certain level that would prevent its deterioration. During this process there are actually two processes taking place. One that involves a heat transfer from the heating source to the product, and the other one that involves a mass transfer of moisture, first from the interior of the product to the surface and then from the surface to the air.

At this point it is important to introduce the two drying rate regimes that take place in agricultural products when they are heated [2, 3], as shown in Fig. 32.1:

- Phase I, the moisture of the surface is evaporated. This phase occurs in a constant rate until there is enough water (saturated moisture) at the surface.
- Phase II, the internal moisture is transferred to the surface, which is not vapour-saturated, and subsequently evaporated. The rate in this period falls and the drying stops when the partial pressure of the vapour in the solid equals the one in the surrounding air. Usually, this second phase is divided into two stages. The first period that involves the drying of the unsaturated surface, and the second one that involves the moisture diffusion to the surface.

While phase I depends on the ambient conditions, such as pressure, temperature, air flow and humidity and exposed surface's area, phase II depends on the product's physical characteristics, its diffusion rate and content of moisture.

Finally, the drying process stops when the rate of moisture desorption from the product to the environment and absorption from the environment are in equilibrium. At this condition the product contains the equilibrium moisture level [4]. At this point the vapour pressure of the moisture in the product equals the vapour pressure of the drying air. If the two vapour pressures are not equal, the product will lose or gain moisture [5, 6].

32.1.2 Aim of the Project

This project is going to be focused on the advantages of using a renewable source of energy in the process of drying agricultural products. Solar dryers give people the possibility to dry their own food or replace traditional drying methods (sun drying) that are used in small villages and communities.

The aim of the project is to design, develop and test an indirect, passive (natural circulation) solar dryer, specifically thought to dry fruits, vegetables or grains. Additionally, the intention of this project is to promote an affordable and easy-to-build solar energy dryer.

Fig. 32.1 Drying rate curve

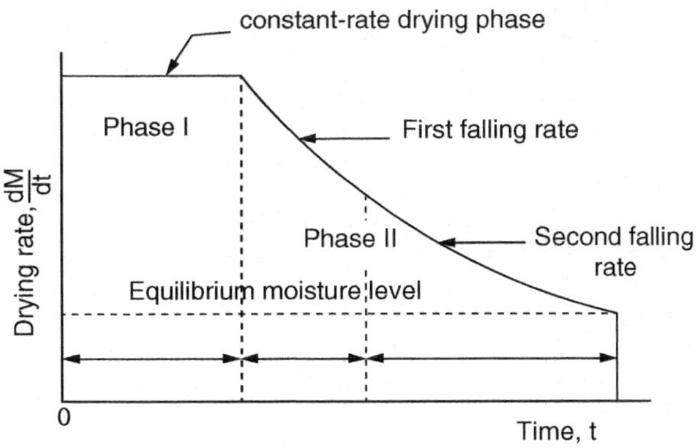

This kind of technology will bring benefits to people who don't have access to more advanced technologies due to economic or energy source limitations. Their benefits will be in terms of sanitation, duration of the drying process, access to an affordable new technology and the possibility of preserving food.

32.1.3 Objectives

The main objectives of this work are:

- To elaborate the design of a natural-convection distributed-type solar energy dryer, to be used with different agricultural products in a small scale.
- To conduct the construction of the designed solar dryer.
- To conduct the drying experiment with the sample product of bananas and analyse the outcome results.
- To compare the experimental results with a simulation model carried out in other project by the use of Computational Fluid Dynamics (CFD) analysis based in the same design.
- To compare the results with traditional ways of drying food and detect the benefits of using this model.

32.1.4 Scope

For the design of the solar dryer, the project was based on other designs reported previously [7] and some modifications were done to achieve a comprehensive and more affordable design.

For the construction and testing of the solar dryer the facilities of the Kingston University Laboratories were used and the authors of this project conducted the required activities.

The results of the experimental tests were compared to a computerized simulation using CFD analysis. This simulation was part of a separate project that went together at the same time, conducted by Ali El Hammoud [8], which was based on the same design of solar dryer.

32.2 Materials and Methods

32.2.1 Approaches to the Design

The energy balance is obtained by equating the total heat gained by the absorber and the total heat that it releases. This can be defined by [9]:

$$I\,A_c = Q_{\text{useful}} + Q_{\text{Cond}} + Q_{\text{Conv}} + Q_{\text{Rad}} + Q_{\text{Ref}} \tag{32.1}$$

The insolation received by the absorber can be expressed in terms of the total solar radiation (I_T) that reaches the top surface and is transmitted (τ) through the glazing cover. Additionally, the heat loss terms (conduction, convection and radiation) are frequently combined into one single term (Q_L), giving the following expression:

$$I\,A_c = \tau\, I_T\, A_c = Q_u + Q_L + Q_{\text{Ref}} \tag{32.2}$$

If reflective losses are:

$$Q_{\text{Ref}} = \mathcal{R}\,\tau\, I_T\, A_c$$

Then, if $\alpha = (1 - \mathcal{R})$, Eq. 32.2 can be reformulated as:

$$Q_u = (\alpha\,\tau)\, I_T\, A_c - Q_L \tag{32.3}$$

It is usual to express $\boldsymbol{Q_L}$ as:

$$Q_L = U_L\, A_c (T_c - T_a) \tag{32.4}$$

Giving the following expression for the useful energy (Q_u) gained by the collector:

$$\boldsymbol{Q_u = (\alpha\,\tau)\, I_T\, A_c - U_L\, A_c (T_c - T_a)} \tag{32.5}$$

32.2.1.1 Moisture Content (MC)

The wet basis moisture content of the product is given by:

$$\text{MC} = \frac{w - d}{w} \times 100 \tag{32.6}$$

where the term $w - d$ is the moisture loss during the process.

32.2.1.2 Energy Balance of the Drying Process

The drying process involves the migration of water from the interior of the product to its surface, followed by evaporation of the water from the surface to the air in the chamber. The energy needed to complete this process is equivalent to the latent heat of evaporation of water (L) and it comes from the air, which was heated throughout the collector. Thus, the energy to evaporate the water equals the heat lost by the air, and this is expressed as:

$$m_v\, L = m_a\, C_p\left(T_i - T_f\right) \tag{32.7}$$

The amount of moisture to be removed (m_v) from the product is calculated by:

$$\boldsymbol{m_v = m_i \frac{\left(M_i - M_f\right)}{\left(100 - M_f\right)}} \tag{32.8}$$

32.2.1.3 Average Drying Rate (d_r)

The drying rate is the amount of moisture removed from the food in a given time, and is calculated as:

$$d_r = \frac{m_v}{\tau_d} \tag{32.9}$$

32.2.1.4 Efficiency of the Collector (η_C)

The thermal efficiency of the collector is defined as $\eta_C = \frac{\textbf{Heat out}}{\textbf{Heat In}}$, therefore [10, 11]:

$$\eta_C = \frac{Q_g}{I_T\, A_C} \tag{32.10}$$

where $Q_g = \dot{m}_a\, C_p\left(T_c - T_a\right)$

32.2.1.5 Dryer Efficiency (η_d)

The dryer efficiency is expressed as [10, 11]:

$$\eta_d = \frac{m_i\, L}{I_T\, A_C\, \tau_d} \tag{32.11}$$

32.2.1.6 Air Pressure Difference (ΔP)

The difference in pressure across the product will be due to the difference between the air density inside the dryer and the ambient air density.

$$\Delta P = \left(\rho_1 - \rho_2 \right) g\, h \tag{32.12}$$

Considering the gas law equation: $P\ V = m\ R\ T$

$$\rho = \frac{P}{R\,T} \tag{32.13}$$

32.2.2 *Construction of the Distributed-Type Solar Dryer*

For each part of the solar dryer a material have been selected, taking mainly two considerations. The first one was its availability at remote locations. As stated, this project aims to give rural communities the opportunity of building their own devices, so its design and materials tries to maintain as simple and easy-to-find as possible.

The other consideration was the cost. As these types of dryers are used both with home or commercial purposes, and especially in rural communities, materials were selected trying to keep the lowest cost. Even though, without putting in compromise the efficiency or increasing the maintenance costs due to deterioration.

Figure 32.2 shows the design that was selected for the distributed-type solar dryer, and the materials for each component are shown in the exploded view of the solar dryer in Fig. 32.4.

Casing: The alternatives that were evaluated to use for the main structure (sides, walls, door and chimney) were Plywood and MDF. Finally, it was constructed using 18 mm thick non-structural hardwood plywood, mainly because:

- Wood is a poor conductor and will reduce losses from sides ($k_{plywood} = 0.1154\ W/mK$).
- Has better response to screw holding in time, reducing maintenance works.
- Has better performance than MDF under moisture conditions.
- Has similar density than MDF ($\rho_{plywood} = 700\ kg/m3$).
- Doesn't contain urea formaldehyde, as MDF does. This makes it less risky when cutting, as this chemical may be released from the material.

Glazing Cover: This component was selected in between glass, acrylic, and polycarbonate. In this case, a 6 mm thick transparent polycarbonate sheet of 0.85 m^2 area was chosen mainly because:

- Less likely to chip and high impact resistance.
- High transmissivity, better than many glass ($t_{polycarbonate} = 89\ \%$).

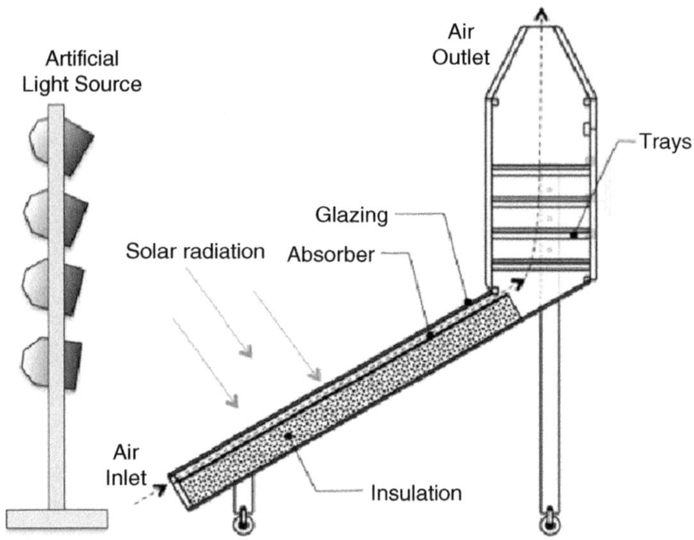

Fig. 32.2 Sectional view of the solar dryer and its components

- Lighter than glass and acrylic ($\rho_{polycarbonate} = 1200\ kg/m3$).
- Good resistance to heat.
- Easier to work with (drilling, cutting, less likely to break).

Absorber Plate: The heat absorber of the collector is a 550 mm wide × 1.55 m long and 1.5 mm thick copper plate; it was sprayed with black enamel paint to increase the absorptance. Table 32.1 shows some of the physical properties of different materials. Also, aluminium and steel have been evaluated as an alternative to be used as absorber plates. Neither of them reached the performance and characteristics that copper offered (Table 32.2).

Other Materials: Aluminium Angles to support the glazing. This helped to seal the joints between the glazing and top of the collector.

Glass Mineral Wool Insulation beneath the absorber plate. A 10 cm loft insulation layer was placed to reduce heat losses from the base ($k = 0.044$ W/mK).

Hinges and Handle for the dryer's door.

Aluminium Mesh. The four rectangular trays were made with pine strip woods, glued together and a 1 mm thick aluminium mesh was stapled to the frame to support the bananas. The top of the chimney was protected with the mesh to avoid insects or animals entering the chamber.

Table 32.1 Physical properties of other materials

Material	Thermal Conductivity at 25 °C (*k*) (W/m K)	Specific Heat Capacity (*Cp*) (kJ/kg K)	Thermal Diffusivity (*a*) (mm²/s)
Silver	420	0.23	165.6
Copper	401	0.39	111
Gold	317	0.13	127
Aluminium	237	0.91	97
Steel	43	0.51	11.7
Stainless Steel	16	0.5	4.2

Table 32.2 Design specifications

Parameter	Value	Unit
Solar Collector Area	0.85	m^2
Air in Vent Area	0.0263	m^2
Air out Vent Area	0.0550	m^2
Height of Air Column	1.89	m
Height of Air Inlet Above Ground Level	0.15	m
Angle of Inclination of the Collector	30	o
Air Duct between Glazing and Absorber	50	mm
Number of Trays	4	u
Vertical Distance Between Trays	140	mm
Thickness of the Sliced Bananas	4	mm

32.2.3 Costs

The total cost of the materials that were used for the construction of the solar dryer was £471. This cost doesn't include any tool or extra cost that the manufacturing could present. The most expensive parts were: Copper sheet, £200, Polycarbonate sheet, £65, and Plywood board, £46.5. In order to reduce these costs alternatively aluminium (£70) instead of copper and 12 mm thick (£27) instead of 18 mm for the plywood could have been chosen. Nevertheless, this would have worsened the performance of the dryer.

32.3 Experiment

32.3.1 Indoor Test

The distributed-type solar dryer was tested during five sessions (days) under two conditions, with and without bananas on the drying shelves. The first two days, the test was conducted to evaluate the performance of the solar dryer without products. Each day, the solar dryer was exposed to 5 h of electromagnetic radiation through an artificial light source. Then, loaded with bananas in the drying chamber, the drying process was carried out within the following three days during 14 h. Additionally, a board with bananas was placed outside the dryer, simulating an open sun (OS) drying.

The setup of the experiment is shown in Fig. 32.2. Figure 32.3 shows the place where the parameters were recorded. These were: temperature, light intensity (solar irradiation), air velocity and weight of the bananas. The light source system of 8 kW was placed 40 cm apart from the solar dryer, giving an average irradiance of 760 W/ m^2.

The first day, the light source was turned on and the parameters recorded after 15, 45 and 75 min to evaluate the warm up of the solar dryer. After this time, the dryer reached a steady state. Once it reached this condition, the trays were introduced to the drying chamber and every hour the parameters were recorded. For weighing the bananas, each tray was weighed outside the dryer and reintroduced to its position after taking the reading. When the session was over, the light source was turned off and the trays left inside until the following day. To start the process again, the lightning system was turned on and the trays were taken outside the solar dryer until the drying chamber has reached the new steady-state condition. From that point on the

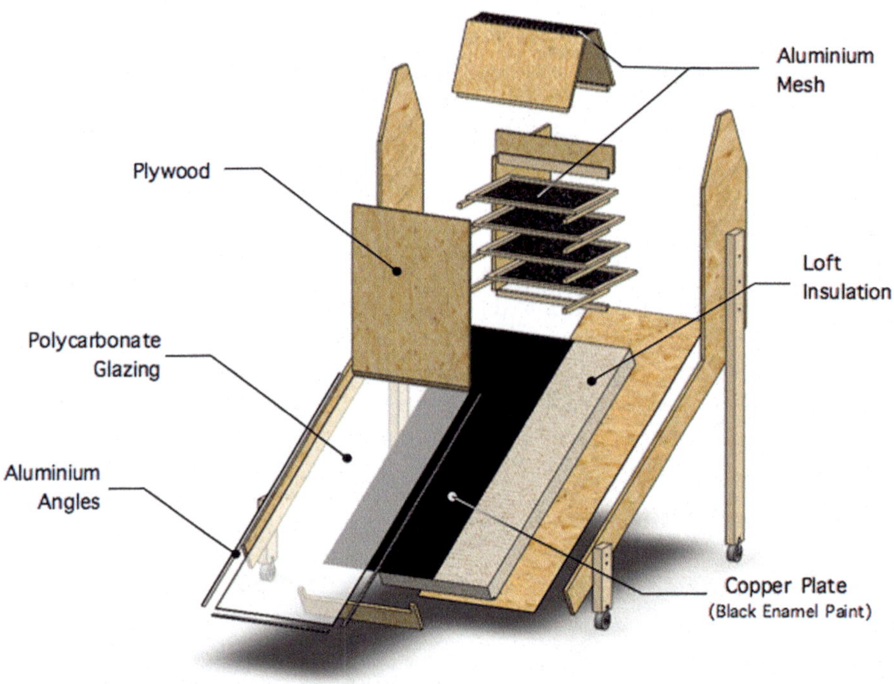

Fig. 32.3 Parameters and place where recorded

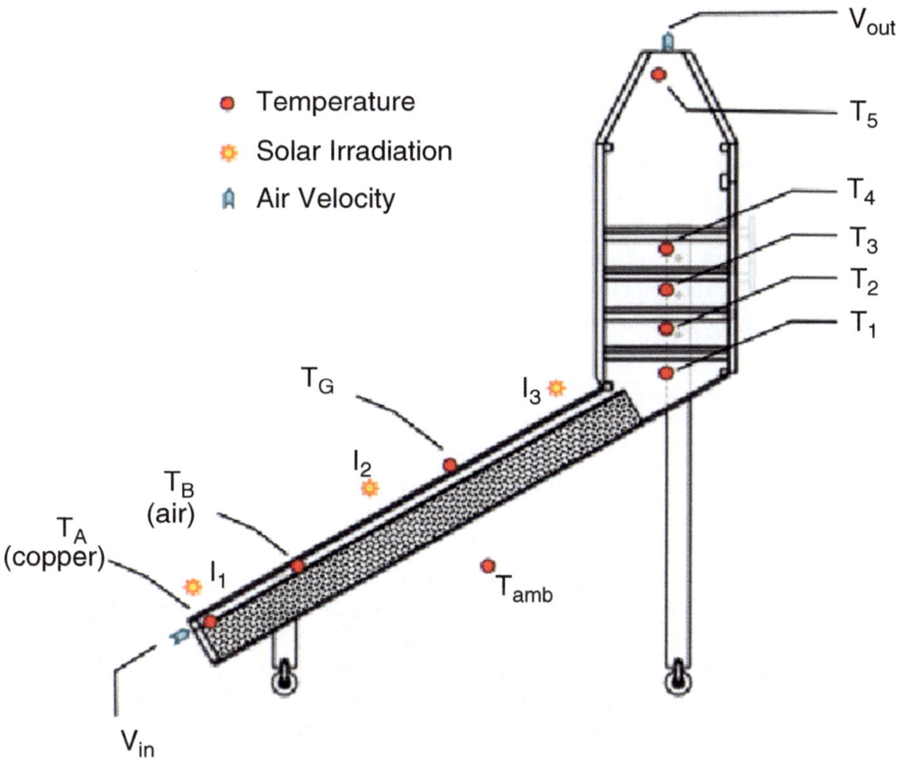

Fig. 32.4 Exploded view of the solar dryer

trays were introduced and every hour new readings were recorded. This same procedure was repeated for the third day until the end of the experiment.

32.3.2 Exterior Test

This part of the experiment was conducted during two sessions (summer days). The solar dryer was only tested with bananas on the shelves and it has not been conducted a no produce test during this part. The whole test was conducted throughout 8 h, during 3 h the first day and 5 the second one. The first day was partly cloudy; therefore, the irradiance changed thus the temperature of the absorber. The second day the sky was clear and this gave more stable readings.

The solar dryer was placed on a south-facing position where no shades from other objects could reach the absorber. The procedure was similar to the one followed for the indoor test, where the readings were recorded on an hourly bases. At the beginning of the session, for both day one and two, the solar dryer was preheated under sun exposure during 1 h, until it reached a steady state temperature. After this, the trays containing the bananas were introduced to the drying chamber.

32.4 Results and Discussions

32.4.1 Indoor Results

Figure 32.5 shows the variation of the temperatures throughout the experiment at different trays (1–4), chimney and ambient conditions. The horizontal axis shows the warm up period (0:00–1:00) and the drying period (0:00–14:00).

From this chart it is observed that the ambient temperature remains almost constant throughout the experiment, while inside the drying chamber the variation is bigger. During the first day, it is observed a drop in the temperature. This shows the moment when the wet bananas were introduced to the chamber, they started absorbing heat and the drying process began.

As time went on, and while the bananas got drier, the temperature of the chamber got more homogeneous. After the first hour of drying, the difference was 16.5 °C in between tray 1 and 4. While at the end of the experiment the difference of temperature in between those trays was 1.5 °C. During the last part of the experiment the temperature inside the drying chamber was over 20 °C above the ambient temperature.

Figure 32.6 shows the amount of moisture removed hourly during the drying process. As expected, and explained in sect. 1.1, the first period of the process is where the external moisture is evaporated, and thus more easily released. Then the internal moisture is removed from the banana, so that is why the amount of moisture removed drops. At the end of the process the amount of moisture that is being evaporated tends to zero. That is when the equilibrium moisture content has been reached.

Figure 32.7 shows the moisture content of the product at the different trays and the OS drying. The x-axis represents the time throughout the drying process, and the

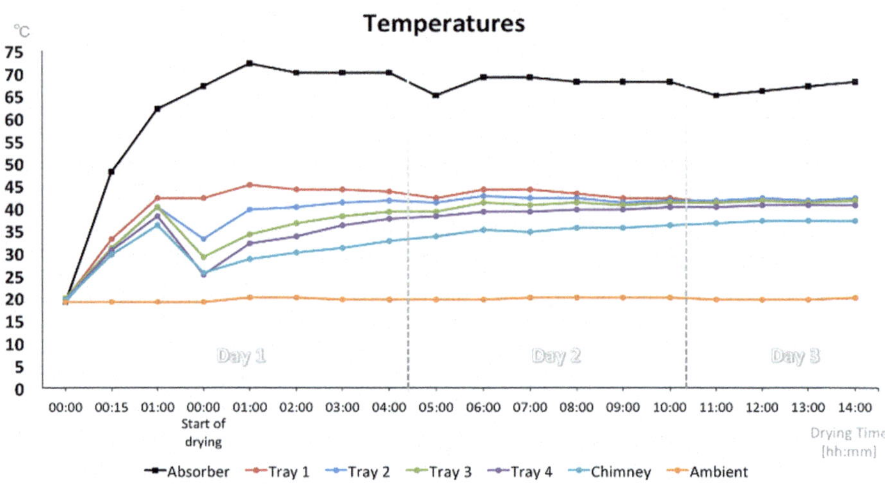

Fig. 32.5 Temperature variation throughout indoors test

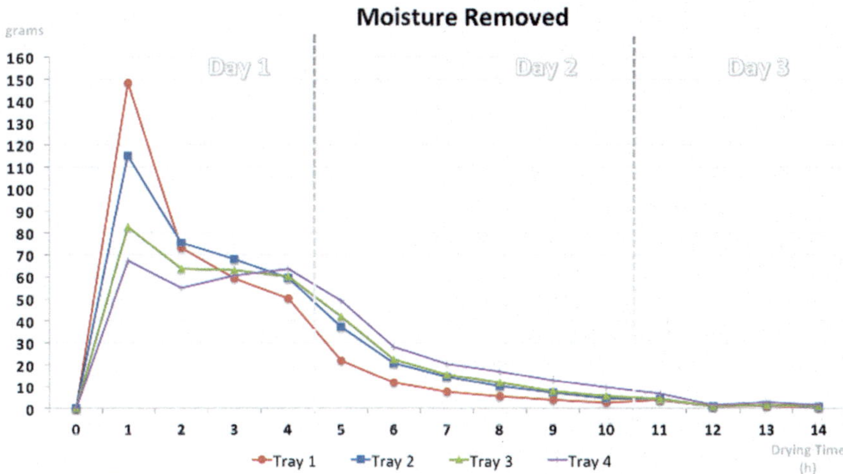

Fig. 32.6 Amount of water removed (grams) hourly during indoor test

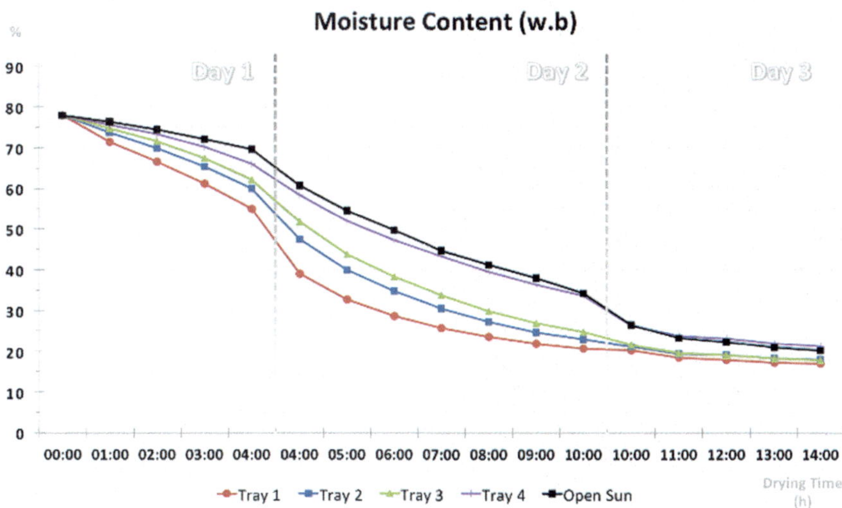

Fig. 32.7 Moisture content throughout indoors test

repeated values (4:00 and 10:00) are shown to give the moisture content reduction that occurred during the night. It is observed that the entire product starts with the same MC (78%), but each tray and the one that is dried outside the dryer have a different variation in time. Even at the end of the experiment, the MC of each sample was different, 16.9%, 17.8%, 17.7%, 21.2% and 20.2% for trays 1, 2, 3, 4 and OS, respectively.

From Fig. 32.7 it can also be observed the time difference, in between trays, that it took to each of them to reach certain MC. For example, tray 1 took 8 h to reach 23%, while tray 4 needed 11 h to reach the same MC; the same time was needed by OS drying to reach that MC.

Both efficiencies, collector and dryer, were calculated using eqs. 32.10 and 32.11 throughout the experiment and they are represented in Fig. 32.8. It is observed from the graph that the collector efficiency remained fairly constant in time, having a mean value of 37%. This can be attributed to the fact that during this test the isolation was generated by the artificial source, which was constant throughout the experiment. Therefore, the difference in temperature between the collector and ambient also remained with small changes.

The dryer efficiency, on the other hand, changed throughout the experiment and shows how the drying efficiency decreased as the product got dryer. This parameter is affected by the properties of the product being dried, i.e. size, physical properties, moisture content and ambient conditions. At the end of the experiment the dryer efficiency was 14%.

32.4.2 Exterior Results

From the exterior test results it is observed that the solar dryer behaved with similar tendencies to the indoor test. This is shown in Fig. 32.9 where the evolution of the temperatures throughout the experiment. Due to weather conditions, especially during the first day, the parameters recorded showed a bigger variation. This is expected as the indoor results came from artificial controlled conditions. From this figure it is also observed that when the bananas were introduced to the drying chamber there was a drop in the temperature and after that point, the difference in temperature in between trays got smaller as the bananas got dryer. At the beginning of the drying process the temperature difference between tray 1 and 4 was 12 °C, while at the end of the experiment that difference was 3 °C.

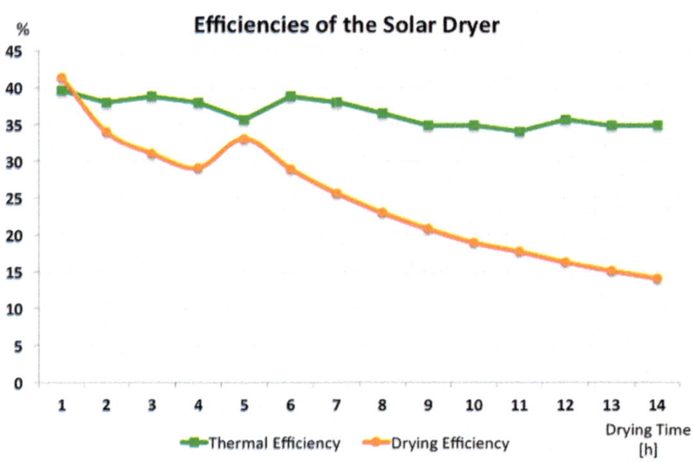

Fig. 32.8 Collector (thermal) and Dryer efficiencies throughout indoors test

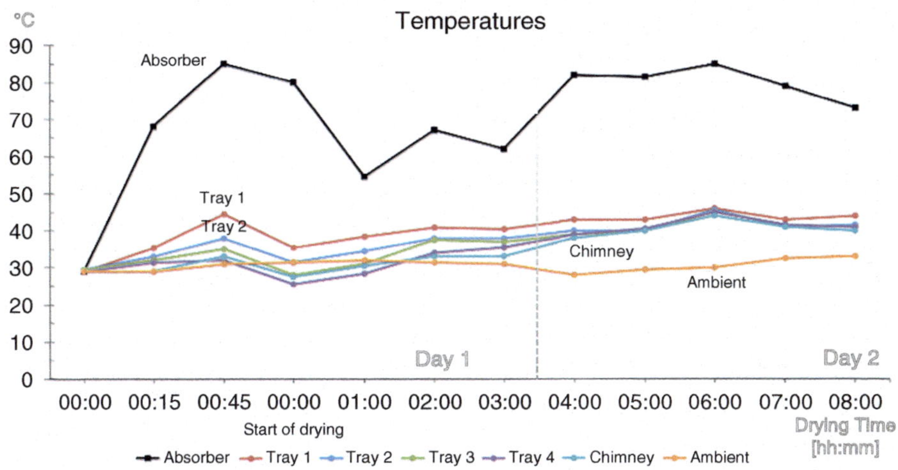

Fig. 32.9 Temperature variation throughout exterior test

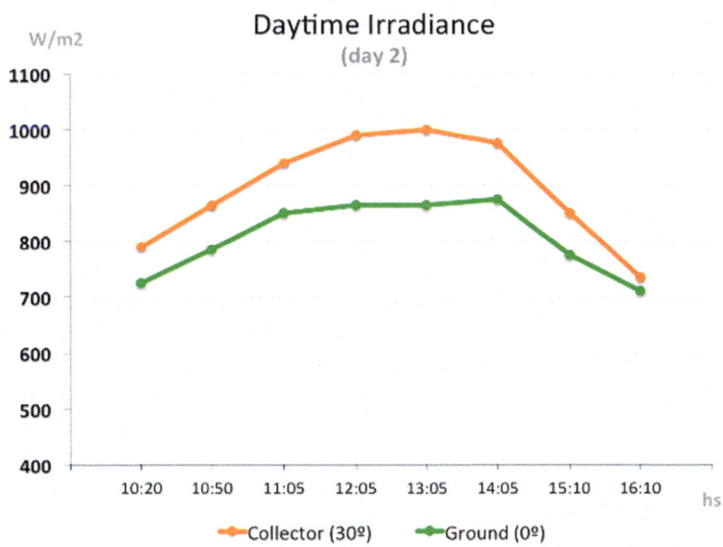

Fig. 32.10 Irradiance on 30° and 0° planes

Also it is observed from Fig. 32.9 that, once the wet bananas were introduced to the chamber, the temperature recorded on trays 3 and 4 was lower than ambient temperature.

Figure 32.10 shows how the irradiance changed throughout the second day of the exterior test both for the collector plane (30° inclination) and on the ground plane (0° inclination).

In Fig. 32.11 below the moisture removed from the product during the exterior test is represented. Following the behaviour of the indoor test, the bananas lose the biggest amount of moisture during the first hours of drying. As they dried, the

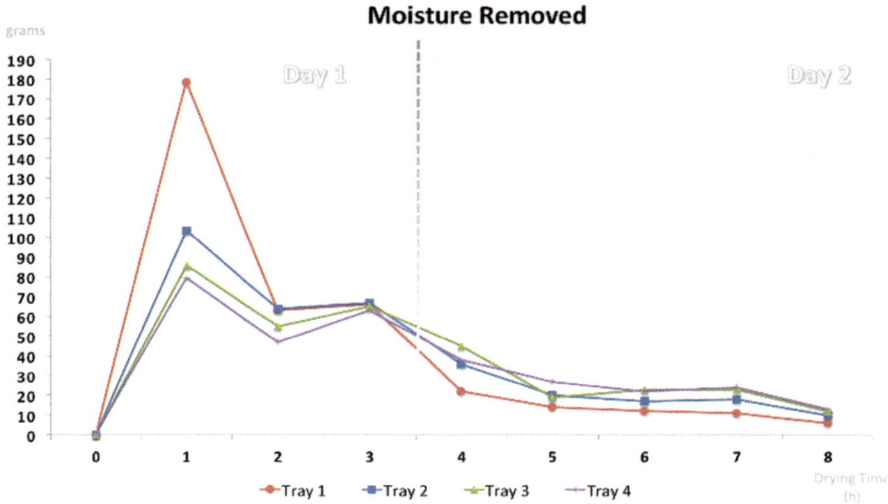

Fig. 32.11 Amount of water removed (grams) hourly during exterior test

amount of water being removed reduced. Also it is observed, from the last values at the end of the test, that the drying process had not finished by the time the experiment was stopped and some more water could have been removed from the product.

Figure 32.12 represents the moisture content of the bananas throughout the exterior experiment. The x-axis represents the time throughout the drying process, and the repeated value (3:00) is shown to give the moisture content reduction that occurred during the night. As in the other test, the amount of water in the product, at different trays, progressed differently. From the figure it is observed that tray 1 is the one that achieved a faster reduction and by the time the test was ended the open sun drying achieved lower moisture content than the product on trays 3 and 4.

32.4.3 CFD Simulation

The results brought by the CFD analysis [8] are represented in the following figures, where the temperatures at the absorber plate and airflow along the solar dryer are shown on a coloured gradient. From Fig. 32.13 it can be observed that the simulation predicted an absorber's temperature in the range between 57 °C and 61.4 °C. This shows that the average temperature of 67 °C achieved during the indoor test exceeded such simulation.

The other area where the temperature was analysed during the CFD simulation was on a vertical plane, covering the space where the air flows from the inlet at the front of the collector to the chimney at the top of the dryer. The air temperature within that area is represented in Fig. 32.14 below.

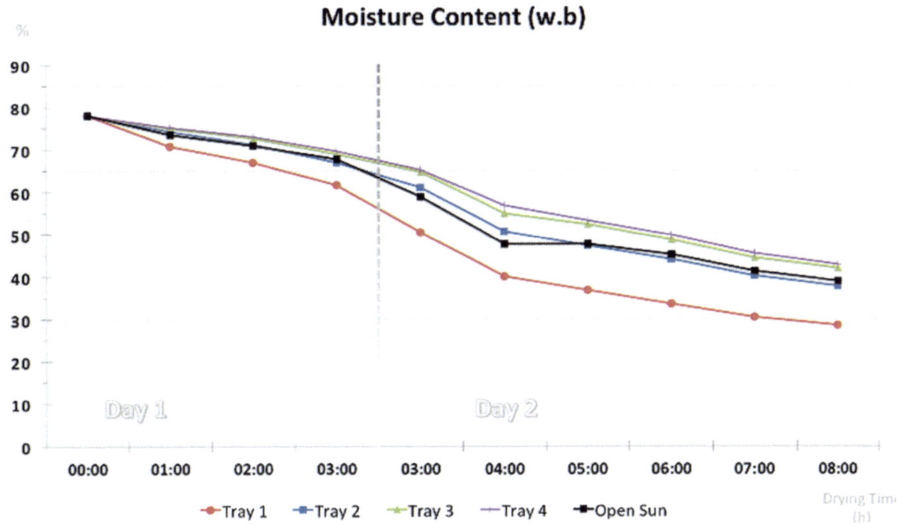

Fig. 32.12 Moisture content throughout exterior test

Fig. 32.13 Temperature at the absorber plate [8]

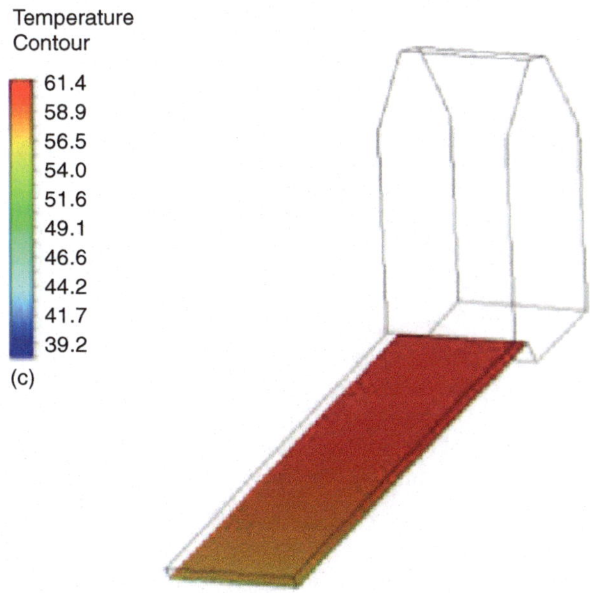

It is observed that it enters the absorber at ambient temperature and increases at the same time as it flows through the absorber. Once in the drying chamber, the air reaches a temperature of 39.2 °C, almost constant, until it leaves through the chimney. This temperature is very close to the one achieved during the indoor test, which was 43 °C.

These differences could be attributed to the fact that some conditions differed when evaluating the simulation. The CFD design did not consider the trays and

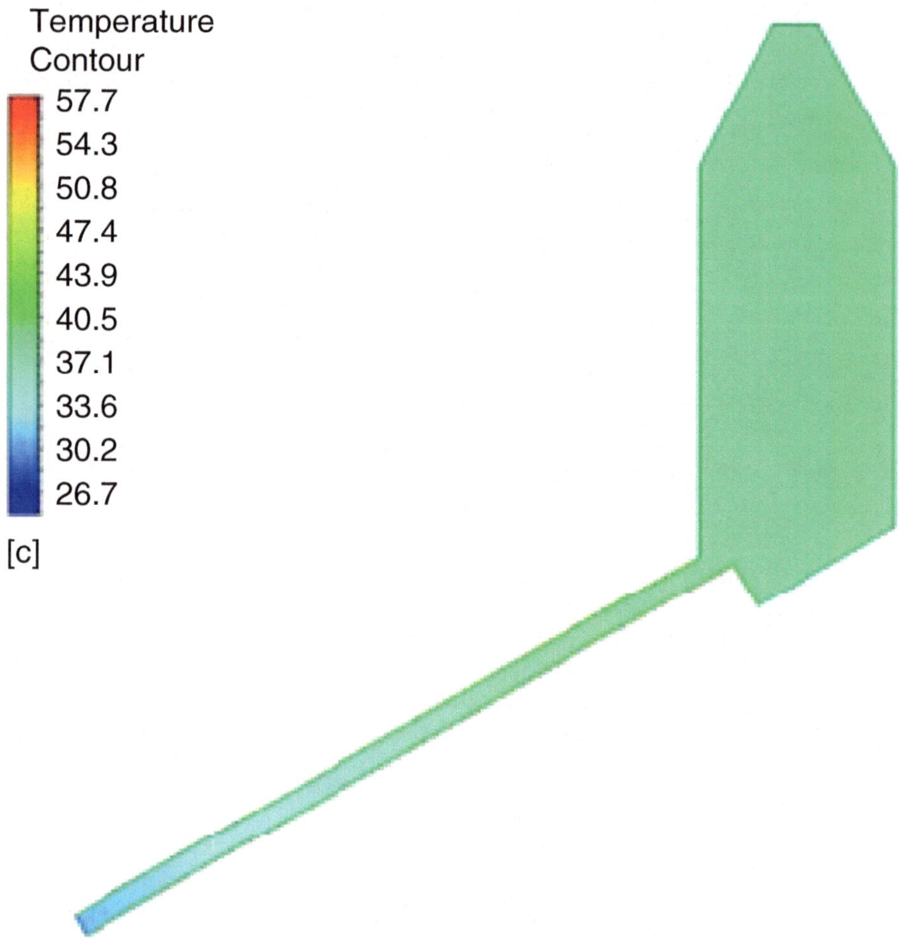

Fig. 32.14 Airflow temperature through the solar dryer [8]

produce inside the drying chamber. Also, the wooden casing was considered to be an insulated boundary. Whereas in the reality, the casing receives part of the radiation that warms it up.

32.5 Conclusions

A simple and inexpensive distributed-type solar dryer was designed, constructed and tested. The materials that were used can easily be found locally.

The indoor experiment was successfully conducted for 14 h and showed that solar drying is a better alternative method than the traditional open sun drying technique. The drying chamber remained with much higher temperatures than the exterior, being over 20 °C the difference in between those by the last part of the experiment. The absorber reached a steady temperature of 67 °C.

During this test, the moisture content of the bananas was reduced from 78% to 16.9%, 17.8%, 17.7%, 21.2% and 20.2% for trays 1, 2, 3, 4 and open sun, respectively. This shows that the performance of the upper tray (tray 4) is worse than the one of OS drying. The contribution here would only be in terms of securing the product from the exterior. While for the rest of the trays the performance was much better than the OS drying. Reaching a 23% of MC is 33%, 22% and 10% slower with OS drying than the performed on trays 1, 2 and 3, respectively.

In terms of efficiency, the dryer efficiency was found to be 14%, while the thermal efficiency was determined to be 37%.

The exterior test results achieved similar tendencies than the ones achieved indoors. This corroborates that the design has a good performance under different conditions. Also, that the manufacturing was conducted successfully.

The performance achieved during the experimental tests and analytical simulation using CFD brought similar results when compared. A difference of 10% was observed in between the experimental and the analytical results. The temperature at the absorber plate was 67 °C and 61 °C, respectively. Similarly, the temperature of the airflow in the drying chamber differed in 10% when comparing the experimental and analytical results, reaching 39 °C and 42 °C, respectively.

References

1. Murthy R (2009) A review of new technologies, modes and experimental investigations of solar dryers. Renew Sust Energ Rev 13(4):835–844
2. Belessiotis V, Delyannis E (2011) Solar drying. Sol Energy 85(8):1665–1691
3. Brennan JG (2006) Food processing handbook. Wiley-VCH Verlag GmbH & Co. KGaA, Weinheim, Germany
4. Ekechukwu OV (1999) Review of solar-energy drying systems: an overview of drying principles and theory. Energy Convers Manag 40:593–613
5. Nahar NM, Garg HP (1981) Selective coatings on flat-plate solar collectors. Renew Energy Rev J 3(1):37–51
6. Ekechukwu OV, Norton B (1999) Review of solar-energy drying systems II: an overview of solar drying technology. Energy Convers Manag 40:615–655
7. Scanlin D (1997) The design, construction and use of an indirect, through-pass, solar food dryer. Homebrew 57:62–72
8. El Hammoud Ali (2018) Design and testing of an indirect solar dryer using computational fluid dynamic. MSc Dissertation. Kingston University
9. Bolaji BO, Olalusi AP (2008) Performance evaluation of a mixed-mode solar dryer. Department of Mechanical Engineering, University of Agriculture Abeokuta, Ogun State, Nigeria; Vol. 11, N 4, pp 225–231
10. Dhanushkodi S, Wilson VH, Sudhakar K (2014) Thermal performance evaluation of indirect forced cabinet solar dryer for cashew drying. Am Eur J Agric Environ Sci 14(11):1248–1254
11. Lingayat A, Chandramohan VP, Raju VRK (2017) Design, development and performance of indirect type solar dryer for banana drying. Energy Procedia Elsevier 109:409–416

Chapter 33
Production of Hydrogen Using Solar-Powered Electrolysis

D. A. Udousoro and Cliff Dansoh

33.1 Background

The use of hydrogen as a fuel rather than conventional gasoline displaces 764,000 kg of CO_2 per year from the environment and production of hydrogen without emissions of carbon dioxide displaces 1.67gallons of gasoline per kg of hydrogen produced [1]. It also has the added advantage of not producing emissions at the point of use, which in London is considered to be a significant part of the cause of up to 40,000 premature deaths per year [2]. An energy system based on hydrogen gas and deployed on a large scale will improve energy security, air quality and reduce greenhouse gas emissions [3].

Currently hydrogen is being produced in the UK via steam methane reforming of natural gas [4]. Although hydrogen fuel does not emit greenhouse gas at the point of use, the process of producing hydrogen gas emits 11,888 kg of CO_2 per kg of hydrogen gas produced and 58,400 kg of hydrogen gas is required to run the existing eight bus fleet in London annually, hence a total of 694,252 kg of greenhouse gas will be emitted into the environment on yearly basis via steam methane reforming of natural gas [5].

There is a need to adopt a new method for hydrogen production that does not emit greenhouse gas into the atmosphere in order to reduce the total emissions in the United Kingdom (UK). The use of solar-powered electrolysis is potentially an effective means of transitioning towards net zero carbon emissions in the public mass transportation sector.

D. A. Udousoro · C. Dansoh (✉)
Faculty of Science, Engineering, and Computing, School of Mechanical and Aerospace Engineering, London, UK
e-mail: c.dansoh@kingston.ac.uk

A. Sayigh (ed.), *Renewable Energy and Sustainable Buildings*, Innovative Renewable Energy, https://doi.org/10.1007/978-3-030-18488-9_33

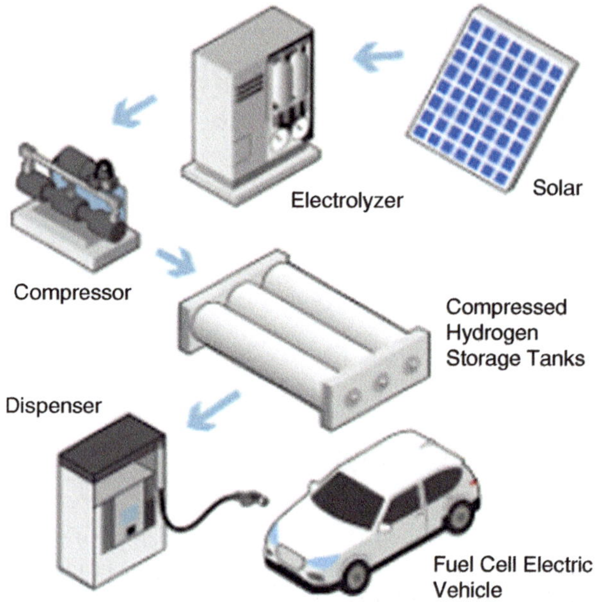

Fig. 33.1 Overview of PV electrolyser system for FC vehicle hydrogen production

33.2 Introduction

Solar-powered electrolysis consists of photovoltaic panels and an electrolysis system to produce compressed hydrogen gas that will be used to run fuel cell buses. Solar energy is used to generate electricity that runs the electrolysis unit to produce compressed hydrogen gas. Electrolysis is the dissociation of water into its elemental components (hydrogen and oxygen) by passing current in the presence of an electrolyte [6] (Fig. 33.1).

The application of electric current causes hydrogen ions to migrate to the cathode and become reduced to hydrogen gas and oxygen molecules migrate to the anode as shown below [7] (Fig. 33.2).

The electrolysis unit consists of an electrolyser, a catalytic recombiner, a hydrogen dryer and a compressor [8]. For this project a solar-powered electrolysis system of the type shown below will be modelled to meet the hydrogen demand of a fuel cell bus using the hydrogen bus fleet at Lea interchange bus garage in London as a case study (Fig. 33.3).

33.3 Modelling of Solar-Powered Electrolysis Unit

In order to model the solar-powered electrolysis unit the energy balance of the electrolysis unit was determined and the photovoltaic system to generate electricity for the electrolysis unit was modelled using system advisor model (SAM). An experiment was conducted using the junior basic solar hydrogen unit. The experiment conducted

Fig. 33.2 PEM Electrolysis Cell

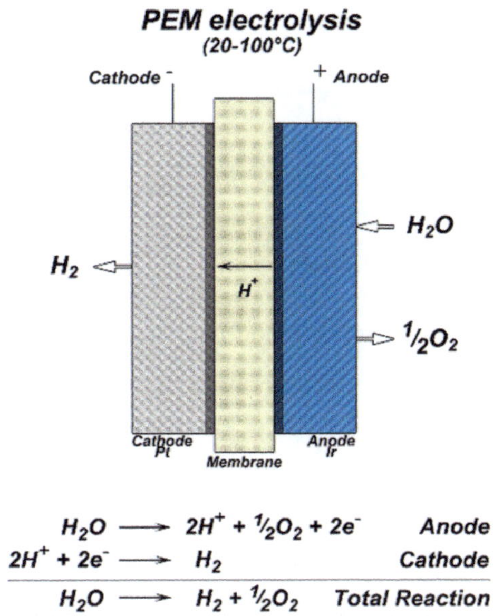

Fig. 33.3 Overview of electrolyser system at Lea Interchange Bus Garage

determined the efficiency of the electrolysis process and the photovoltaic module. The energy balance of the electrolysis unit was calculated in order to size and model the photovoltaic system.

33.3.1 Experiment Conducted

Experiments were conducted to determine the I–V curve of the photovoltaics and electrolysis unit (Fig. 33.4). For the electrolysis unit, the faraday and energy efficiency was calculated at 23 °C and the efficiency of the PV system was also calculated below [9].

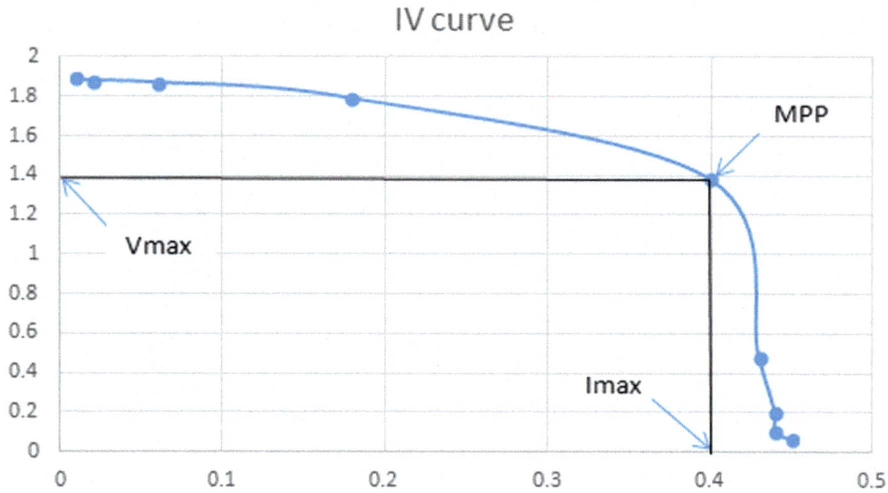

Fig. 33.4 I—V curve for solar cell

$$\eta\text{faraday} = \frac{\text{VH}_2\text{experimental}}{\text{VH}_2\text{theoretical}} = \frac{\text{VH}_2\text{experimental}}{\frac{RITt}{FPZ}}$$

where:

R = Universal gas constant = 8314 J/(mol k),
T = Ambient temperature in kelvin,
I = Current in ampere, t = time in seconds,
P = Ambient pressure,
F = faradays constant = 96,485 C/mol,
Z = number of excess electron = 2,
V = volume of gas produced.

At 23 °C, the theoretical volume of gas produced and the faraday efficiency is as follows:

$$\text{VH}_2\text{theoretical} = \frac{8.314 \times 298 \times 0.17 \times 807}{1013 \times 10^5 \times 96485 \times 2} = 17.4\ \text{cm}^3$$

$$\eta\text{faraday} = \frac{16\ \text{cm}^3}{17.4\,\text{cm}^3} = 0.91 = 91\%$$

The energy efficiency is determined from the ratio of heat energy stored in produced hydrogen and the electric energy used in producing it (Fig. 33.4).

$$\eta\text{energy} = \frac{\text{usable energy}}{\text{energy expended}}$$

$$= \frac{\text{VH}_2 \times \text{H}_o}{\text{voltage} \times \text{current} \times \text{time}}$$

$$= \frac{16 \times 10^{-6} \times 12745 \times 10^3}{1.73 \times 0.17 \times 807} = 0.85 = 85\%$$

$$\text{Efficiency of solar cell} = \frac{I\text{max} \times V\text{max}}{\text{Irradiance} \times \text{Area}}$$

where:

Irradiance was based on tests using natural sunlight at 511 W/m^2.
Area of solar cell employed—0.089m^2

$$= \frac{0.40A \times 1.38V}{511\ \text{W/m}^2 \times 0.089\ \text{m}^2} \times 100 = 12\%$$

$$\text{fill factor} = \frac{I\text{max} \times V\text{max}}{I_{sc} \times V_{oc}}$$

$$\frac{0.552}{0.35A * 2\text{V}} = \frac{0.552}{0.7} = 0.78$$

Using the I–V curve of the solar cell above, an efficiency of 12% and fill factor of 0.78 was obtained for the PV module as shown above. In the graph above, MPP stands for maximum power point, which is the point on the I–V curve which corresponds to the maximum power output (Fig. 33.4).

In order to model the electrolysis unit an energy balance was conducted to determine the energy required to run each of the main component of the electrolysis unit.

33.3.2 Proton Exchange Membrane Electrolyser

PEM cells are electrolytic cells that do not contain liquid electrolyte but possess a solid thin proton conducting membrane which serves as the electrolyte. The DC current splits water into gaseous oxygen and protons at the anode and these protons migrate to the cathode via electro osmosis drag to be desolvated and reduced to molecular hydrogen. Deionized water is mostly used in PEM electrolysis due to sensitivity of materials used in constructing cell components [7].

33.3.3 Heat of Compression Dryer

The dryer proposed for this project is a molecular sieve dryer and the silica gel adsorbent will be regenerated using heat from the compression process that follows [10]. The molecular sieve dryer consists of two beds that operate continuously, one bed acts as an adsorber and the other acts as a regenerator. Each bed will contain a molecular sieve adsorbent to adsorb water from hydrogen gas carried over from the electrolysis stage and then undergo pressurization and adsorption steps followed by a countercurrent blow down and purge process [11].

33.3.4 Compressor

The compressor chosen for this project is a diaphragm compressor based on their ability to function with the intermittent supply of power from solar panels and deliver gases with high pressure and high purity efficiently. The diaphragm compressor is a two-stage compressor delivering hydrogen fuel at 102 bar and 350 bar. The hydrogen gas with pressure of 102 bar is to regenerate the silica gel adsorbent in the dryer. The hydrogen gas with pressure of 350 bar will then be used for the bus services [12, 13].

33.4 Energy Balance

33.4.1 Proton Exchange Membrane Electrolyser

$$\frac{\text{hydrogen required per day} \times \dfrac{\text{kwh}}{\text{kg}} \text{of } H_2}{\text{hours per day} \times \text{efficiency of electrolyser}} = \text{KW of electrolyser}$$

$$\frac{\text{kwh}}{\text{kg}} = \frac{39\text{kwh/kg}}{0.85} = 45.8\text{kwh/kg}$$

$$\frac{160\text{ kg} \times 45.8\text{ kwh/kg}}{24\text{ h} \times 0.85} = 359.2\text{ KW}$$

33.4.2 Heat of Compression Dryer

During electrolysis of water, 1 kg of produced hydrogen gas has 2.18 kg of water. This high moisture content could create problems when used in fuel cell engines [14, 15].

$$\text{specific humidity}\,(\text{w}) = \frac{\text{mass of water vapor in a given volume}}{\text{mass of hydrogen gas in a given volume}}$$

$$w = \frac{2.18\ \text{kg}}{1\ \text{kg}} = 2.18$$

$$\text{absolute humidity}\,(x) = 0.622 \times \frac{P\text{vap}}{P - P\text{vap}}$$

where Pvap = vapour pressure and P = atmospheric pressure (760 mmHg). The moist hydrogen gas obtained from electrolysis is considered to be fully saturated (relative humidity = 100). Vapour pressure was obtained using the pyranometric chart at a temperature of 25 °C and relative humidity of 100%.

$$x = 0.622 \times \frac{24\ \text{mmHg}}{760\ \text{mmHg}} = 0.02$$

The enthalpy (H) of moist hydrogen gas was used as the energy demand for drying hydrogen gas using the following formula [16].

$$H = \left(1.01 + 1.97(x)\right)t + 2493(x)$$

$$H = \left(1.01 + 1.97(0.02)\right)25 + 2493(0.02) = 76\frac{\text{KJ}}{\text{Kg}} = 0.02\frac{\text{kwh}}{\text{kg}}$$

Therefore, the energy used in drying 1 kg of hydrogen gas is 0.02 kwh.

33.4.3 Diaphragm Compressor

$$H_a = Z_{avg} RT1 \frac{K}{K-1}\left(r_P^{\frac{k-1}{k}} - 1\right)$$

where H_a is the enthalpy of adiabatic process, T1 is inlet temperature, R is specific gas constant and Z_{avg} is assumed to be 1. Hydrogen gas constant is 772.5 ft-lb/lb°R. Where T1 is inlet temperature, T2 is outlet temperature, rp is ratio of discharge pressure to inlet pressure and K is specific heat ratio. T1 is atmospheric temperature = 25 °C = 77 °F. Due to heat gain during production process of hydrogen an estimate of 80 F was used as the inlet temperature. T2 is discharge temperature = 150 °C = 302 °F and rp is the ratio of discharge temperature to inlet temperature [12].

$$T2 = 302^{\circ}F + 460 = 762^{\circ}R$$

$$T1 = 80^{\circ}F + 460 = 540^{\circ}R$$

$$r_P = \sqrt[0.286]{\frac{762}{540}} = 3.33$$

$$\frac{K}{K-1} = \frac{1.4}{1.4-1} = 3.5$$

$$H_a = 1.0 \times 772.5 \times 540 \times 3.5\left(3.33^{0.286} - 1\right) = 599558.2\,ft - lb/lb$$

The enthalpy of the process was converted to power with an efficiency of 79% and a weight flow rate of 0.243 lb./min

$$\text{weight flow} = \frac{160\text{ kg/day}}{24\text{ h} \times 60} = 0.11\frac{\text{kg}}{\text{min}} = 0.243\text{ lb/min}$$

$$\text{power} = \frac{0.243 \times 599558.2}{33000 \times 0.79} = 5.59\text{ hp}$$

The power needed for the first stage compression of hydrogen gas from 30 bar to 102 bar is 5.59 hp. For the second stage compression, the inlet pressure is 1495psia (102 bar) and the outlet pressure is 5091psia (350 bar). Hence,

$$r_P = \frac{5091}{1495} = 3.4$$

$$H_a = 1.0 \times 772.5 \times 762 \times 3.5\left(3.4^{0.286} - 1\right) = 863386.4\text{ft} - \text{lb/lb}$$

$$\text{power} = \frac{0.243 \times 863386.4}{33000 \times 0.79} = 8.04\text{ hp}$$

The total power used in the compression process is

$$\text{total hp} = 5.59 + 8.04 = 13.63\text{ hp} = 10163.8\text{ W} = 10.2\text{ KW}$$

$$= \frac{10.2\text{ KW} \times 24\text{ h}}{160\text{ kg}} = 1.56 \approx 2\text{kwh/kg}$$

From the analysis above the total energy required to run the electrolysis unit is 47.82 kwh/k (Table 33.1).

Table 33.1 energy demand for components within hydrogen production system

Electrolysis system components	Kwh/kg
PEM electrolyser	45.8
Heat of compression dryer	0.02
Diaphragm compressor	2
Total	47.82

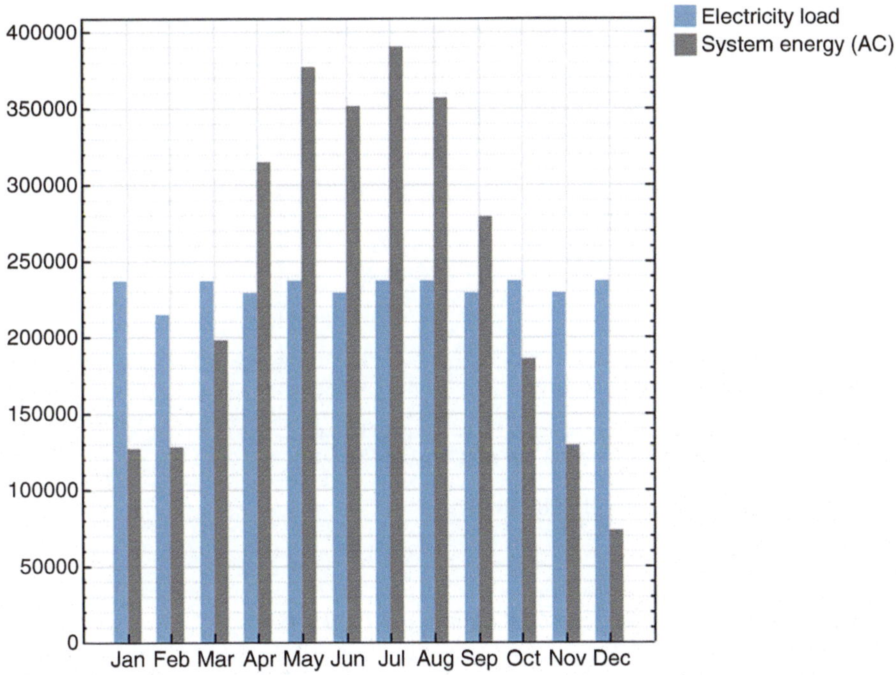

Fig. 33.5 Energy demand of electrolysis unit (blue line) and energy supplied by PV system (grey line)

33.4.4 Modelling Photovoltaic System

The PV module designed for this project was modelled using System Advisor Model software (SAM) to generate enough energy to produce 160 kg of hydrogen per day. Using this software a 2.98 MW of monocrystalline photovoltaic module was modelled to generate 2,910,445 kWh on a yearly basis. The Kwh/kg of hydrogen produced by the various electrolysis components is shown below in Table 33.1.

$$47.82\ \text{kwh} \times 160\ \text{kg} \times 365\ \text{days} = 2792688\ \text{kwh/yr}$$

During the one-year operation there will be a 2-week maintenance period for the solar electrolysis system; hence the system was modelled to have stored hydrogen gas during this period to keep the bus fleet in operation. A period of two weeks will require stored hydrogen of 2462 kg (Fig. 33.5).

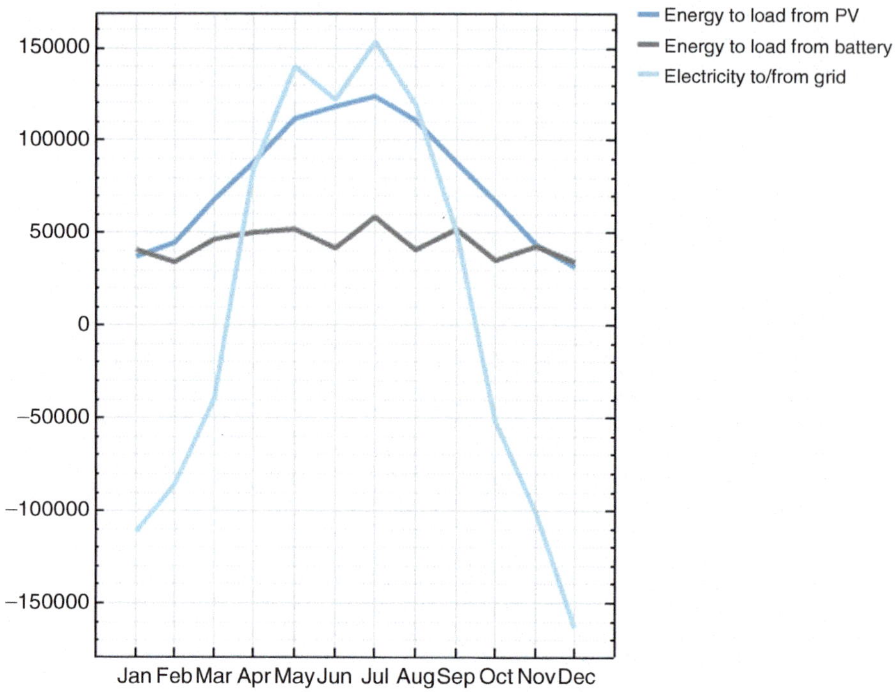

Fig. 33.6 Sources of power for electrolysis system

$$2462\,\text{kg} \times 47.82\ \text{kwh/kg} = 117757\ \text{kwh}$$

$$\text{Total electricity required annually} = 2792688\text{kwh} + 117757\text{kwh} = 2910445\text{kwh/yr}$$

The PV module array to produce this amount of power will require 4.76 ha (11.6 acres) of land. Figure 33.5 shows the energy required by the system and the energy delivered to the system, in Kwh, over the course of one year. Due to the inconsistent supply of solar energy in the UK, a battery system and the grid had to be included in the PV system (Fig. 33.6).

Figure 33.6 shows represents the energy, in KWh, generation from the photovoltaic panels, batteries and electric grid. The light blue line represents electricity from the grid, during the winter months electricity generated by the electric grid is used to power the electrolysis system as indicated by the negative levels of power. During the summer months, from the middle of March to September, more power is being supplied to the electric grid as indicated by the positive levels of power. The grey line represents the power generated from the batteries to the electrolysis system; electricity provided by the battery is fairly constant throughout the year. The dark blue line represents the electricity generated from the photovoltaic panels to the electrolysis system.

Table 33.2 Technical specifications of high pressure gas cylinder [17]

Parameter	
Tensile strength	2000 MPa
Volumetric density	36 kg/m^3
Total volume of cylinder	320m^3
Cylinder material	Aluminium
Gravimetric density	13 mass%
Storage temperature	Room temperature at 25 °C

33.4.5 Solar-Powered Electrolysis (SPE) System

During the winter months, October to March, the electrolysis system is not able to produce enough hydrogen to meet the monthly demand of the fuel cell buses. During those months a total of 11,540 kg of hydrogen gas is needed to meet the monthly demands, for this reason a storage means is required so that excess hydrogen from the summer period can be stored and used during the winter period. During the summer months ranging from April to September the system is able to produce more than required for the fuel cell buses in those months. An excess of 14002.5 kg of hydrogen is produced in these months and this amount is able to offset the amount of hydrogen required during the winter months and still have 2462.5 kg of hydrogen extra. This 2462.5 kg of extra hydrogen gas will be used during the two-week shutdown that will occur during the course of the year for annual maintenance of the electrolysis system. The storage medium considered for this project is a high pressure light weight composite gas cylinder with a total volume of 320 m^3. The specifications for the high pressure gas cylinder is provided below (Table 33.2).

33.4.6 Cost Analysis

The cost analysis was conducted to compare the levelised cost of energy (LCoE) using the photovoltaics with the recent price of electricity from the national grid. The levelised cost of energy for steam methane reforming was compared to that of the solar-powered electrolysis system (SPE) [18].

$$LCoE = \frac{I + \sum_{n=0}^{N} \frac{AO}{(1+d)^n}}{\sum_{n=1}^{N} \frac{Qn}{(1+d)^{n.}}}$$

where LCOE is the levelised cost of energy cost, I = capital investment, AO = Annual operations and maintenance expenditures in the year n (fixed and variable), Qn = Energy generation in the year n, d = discount rate, n = life of the system.

The capital investment was calculated using the system advisor model (SAM) which gave a value of $90,651,528 and yearly output of 2910445 kwh. The operation and maintenance cost was taken to be $16/KW with a 10% discount rate.

$$AO = 2980\ \text{KW} \times \$16 = \$47680$$

$$\sum_{n=0}^{N} \frac{\$47680}{(1+0.1)^{25}} = \$432793\,(\text{total AO during 25yrs period})$$

$$\sum_{n=1}^{N} \frac{2910445kwh}{(1+0.1)^{25}} = 26418226\ \text{kwh}\,(\text{total Qn during 25yrs period})$$

$$\text{LCoE of PV} = \frac{\$90651528 + \$432793}{26418226\ \text{kwh}} = \$3.4/\text{kwh} \cong £2.62/\text{kwh}$$

From the analysis above it is clearly cheaper to obtain electricity from the grid as the levelised energy cost of the PV system is £2.62/kwh and the price of electricity from the grid is 12.6p/kwh [19].

In order to compare the levelised energy cost between steam methane reforming of natural gas and solar-powered electrolysis an annual load of 58,400 kg of hydrogen gas was used in the calculations below. The levelised energy cost was calculated using a period of 20 years and discount rate of 10% (Table 33.3 and 33.4).

$$\sum_{n=1}^{N} \frac{58400kg}{(1+0.1)^{20}} = 49719kg$$

$$\sum_{n=0}^{N} \frac{\$5389}{(1+0.1)^{20}} = \$45879\,(\text{total AO for SMR})$$

Table 33.3 Financial parameters of steam methane reforming [20]

Parameter	Details of cost	Cost
Capital cost	$569/kg/day	$91,040
Operation and maintenance cost	5.92% of capital cost	$5389
Feed stock cost	$5.15/1000 scf	$127,311
Other variables	8% of capital cost	$7283.2
Transport cost	$0.96/kg	

Table 33.4 Financial parameters of solar-powered electrolysis (SPE) system [21]

Parameter	Details of cost	Cost
Capital cost	$940/KW	$337,648
Operation and maintenance cost	$42/KW	$15086.4
Electricity cost	$3.4/kwh	$26,014

$$\sum_{n=0}^{N} \frac{\$15086.4}{(1+0.1)^{20}} = \$128439(\text{total AO for SPE})$$

$$\text{LEC SMR} = \frac{\$91040 + \$45879}{49719kg} = \$0.55 + \$0.96 = \$1.51/\text{kg} \cong £1.14/\text{kg}$$

$$\text{LEC SPE} = \frac{\$128439 + \$337648}{49719\text{kg}} = \$20.8/\text{kg} \cong £15.76/\text{kg}$$

From the analysis it clear that the levelised energy cost of solar-powered electrolysis system is around 15 times the levelised energy cost of steam methane reforming of natural gas. The levelised energy cost (LEC) for steam methane reforming was calculated to be £1.14 while the LEC for solar-powered electrolysis was calculated to be £15.76.

33.5 Countries that Could Benefit from Solar Electrolysis System

It was discovered while undergoing this project that the installation of solar electrolysis plant in London will not be profitable due to the following reasons:

- Established grid network
- Limited land space
- Established natural gas network
- High energy demand

Using the mass and energy balance calculation it was calculated that 47.82 kwh of electricity is required to produce each kg of hydrogen via electrolysis and this will require a significant amount of solar radiation. It was also apparent that the average solar radiation in the UK (3.66 kwh/m^2/day) was insufficient to generate the hydrogen required for the buses within a reasonable amount of space for a densely populated city with high land prices such as London. Hence it is concluded that large-scale solar electrolysis is best located at places with high solar radiation.

Due to the high energy demand a large number of PV cells were required for the project. The total land area required was 11.6 acres of land which is not available within London city and availability of such land capacity cannot be found within close range of the bus station; they can only be found at areas that are remote from the city and it will cost extra money to build transmission lines that will transmit the electricity to the electrolysis system within the garage. This project will be more financially attractive in countries that have densely populated cities and also have enough land to set up photovoltaic systems close to bus stations.

Given below are countries that would be significantly more attractive for installation of solar-powered electrolysis for hydrogen production due to their high solar radiation, availability of land space and lack of established grid network.

33.5.1 South Africa

In the rural sub Saharan part of South Africa there is said to be a lack of transport services within the area. Rural transport is the movement of persons or goods by walking, head loading or on various infrastructures such as roads, tracks or footpaths [22]. More than 50% of people that live within 5 km of a tarred road are limited to walking as they do not have access to transport infrastructures. A proposed solution to this problem is the use of buses within these regions and it would be advantageous if the transport fuels used by these buses were generated within the region in order to create more jobs and cut the cost of importing fuel from other countries. Rural sub Saharan South Africa has large land mass and an average solar radiation of 220 W/m^2 compared to the UK that has an average radiation of 101.2 W/m^2, less PV panels and land space will be required to run the solar electrolysis system in this region [23].

The use of fuel cell buses that utilize hydrogen from solar electrolysis could be economically stimulating to such an area and also reduce the amount of greenhouse gases in South Africa. The total amount of greenhouse gas emissions in South Africa is 391.77MtCO_2e and 12% of which are from the transport sector [24].

33.5.2 Argentina

The national route 40 that crosses the Andes region linking Argentina to Chile is another potential location for fuel cell buses and solar electrolysis system. Argentina does not have a stable grid connection as the demand outstrips supply but it has a high solar energy resource.

The solar radiation in Argentina is 2.2 MWh/m^2/yr and the average radiation in the UK is 1.2 MWh/m^2/yr, hence less PV and land space will be used to implement this project in Argentina. Due to its high solar resource it also reduces the levelised energy cost in solar-powered electrolysis. Argentina has more potential in using a solar-powered electrolysis system than the UK because of its large solar energy resource, inconsistent power supply and large land space [25].

The greenhouse gas emissions in Argentina are 405MtCO_2e, which the use of hydrogen energy could help to reduce. The national route 40 begins from Rio Gallegos in Santa Cruz province and ends in La Quaica in Jujuy province and it is 5000 km long [26]. 5000 km will require approximately 500 kg (10 kg/100 km) of hydrogen gas to run the bus hence saving 6817 kg of CO_2 from being emitted into the atmosphere. There are vast amounts of land around the national route making it worthy of further consideration for this project.

33.5.3 Australia

Australia's total greenhouse gas emissions is 527Mt CO_2e but has experienced 12% decrease in its greenhouse gas emissions since 2005. Most of its electric power is generated from coal. Its total transport emission is estimated to be 79MtCO_2e [27]. The average solar radiation received in Australia is 1.6MWh/m^2/yr. which is relatively high; hence its high potential for installation of a solar-powered electrolysis system [28]. The Simpson Desert is considered as a worthy case study for the installation of solar electrolysis system in Australia.

The land mass of the Simpson Desert in Australia is 176500km^2 and its neighbouring towns at the eastern side in Queensland are Birdsville, Bedourie and Diamantina. These towns could adopt fuel cell buses in their transport sector and generate hydrogen gas from solar-powered electrolysis installed in the Simpson Desert due to its large land space. Installation of a solar-powered electrolysis system in the Simpson Desert would be more financially attractive than in the UK because it will require less PV panels and land space and yield cheaper levelised energy cost due to its high solar irradiation. Also the use of fuel cell buses and solar electrolysis system will reduce the emissions in Australia.

33.6 Conclusion

The use of solar-powered electrolysis does not emit any green house gases or harmful emissions as by-products, hence it is an efficient means of potentially achieving zero carbon emissions in the UK transport sector. Solar-powered electrolysis is also a relatively cost-intensive method of producing hydrogen because it requires significant amounts of electricity. Also the hours of sunlight in the UK, on average is 1400 h, requires that a significant number of PV panels are installed to generate the required electricity for the electrolysis process. So, although it was considered that the system proposed was not financially attractive for deployment to supply hydrogen for the LEA depot in London, a system of this type could be much more financially attractive in other countries such as Argentina, South Africa and Australia. In order to improve the efficiency of solar-powered electrolysis, recent devices such as multi-junction solar cells and photo electrolytic cells can also be used instead of the photovoltaic cells and electrolyser. It is considered that there is benefit in further research into reducing the electricity requirements of the electrolysis process so as to reduce the cost of the electricity used for the system.

It is also advised that further research is conducted in order to minimise permeation of moisture or water vapour into the generated hydrogen and oxygen gas, to eliminate the need for dryers within the system.

References

1. Hydrogen Economy (2003) Weekly compilation of presidential documents. Vol. 39, No. 5. Government Printing Office, Washington, D.C., p 111
2. Royal College of Physics (2016) Every breath we take: the lifelong impact of air pollution. https://www.rcplondon.ac.uk/projects/outputs/every-breath-we-take-lifelong-impact-air-pollution. Accessed 6 Aug 2018
3. Thomas C (2012) Hydrogen untapped energy. Institution of Gas Engineers and Managers, Derby
4. Yorke D (2017) Bus characteristics. https://outlook.live.com/mail/#/conversation/AQQkADAwATYwMAItYmIANWQtYzM1Ni0wMAItMDAKABAAUASt6i3r0Embg7cSgjrUOQ%3D%3D. Accessed 4 Jun 2017
5. Spath P, Mann K (2001) Life cycle assessment of hydrogen production via natural gas steam reforming. National Renewable Energy Laboratory, Golden, CO
6. Frano B (2005) PEM electrolysis for production of hydrogen from renewable energy sources. Sol Energy 78(5):661–669
7. Godula Jopek A, Stolten D (2015) Hydrogen production: by electrolysis. John Wiley & Sons, Hoboken, NJ
8. Bessarabov D, Haijiang W, Hui L, Nana Z (2016) PEM electrolysis for hydrogen production: principles and applications
9. H-TEC Hydrogen Energy System (2017) Model for classroom instruction PEM Power1-Eco
10. Pyle W (1998). http://www.drivehq.com/file/df.aspx/isGallerytrue/shareID452352/fileID9377728?1=1-. Accessed 16 Aug 2017
11. Compressed air and gas drying (2012) Available at: http://www.cagi.org/pdfs/cagiairdryingselectionguide.pdf (Accessed 2 Aug 2017)
12. Brown R (2005) Compressor selection and sizing, 3rd edn. Gulf Professional Publishing, Houston, TX
13. Wanjiku G, Khan M, Barendsea S, Sebiotosi A (2011) Analytical sizing of an electrolyser for a small scale wind electrolysis plant
14. Ivy J (2004) Summary of electrolytic hydrogen production. National renewable energy laboratory
15. Mizraii H (2017) AE7203: solar power engineering. https://canvas.kingston.ac.uk/courses/377/files?preview=16794. Accessed 28 Jul 2017
16. Tamas A, Balan M (2013) Air drying by adsorption on zeolite type dessicants. Chem Bull "POLITEHNICA" Univ (Timisoara) 58(72):1
17. Zuttel A (2004) Hydrogen storage methods. Available at: https://www.academia.edu/4826927/Hydrogen_storage_methods (Accessed 3 Apr 2017)
18. Dansoh C (2017) AE7203: solar power engineering. https://canvas.kingston.ac.uk/courses/377/files? Accessed 19 Aug 2017
19. UKPower (2017) Gas & Electricity tariff prices per kWh. Available at: https://www.ukpower.co.uk/home_energy/tariffs-per-unit-kwh (Accessed 14 Sept 2017)
20. Eichman J, Townsend A, Melaina M (2016) Economic assessment of hydrogen technologies participating in California electricity markets. National renewable energy laboratory
21. Saur G, Ramsden T (2014) Techno-economic analysis of PEM electrolysis for hydrogen production. Available at: https://www.energy.gov/sites/prod/files/2014/08/f18/fcto_2014_electrolytic_h2_wkshp_colella1.pdf (Accessed 21 Feb 2017)
22. Bryceson D, Howe J (1992) Rural household transport in Africa: reducing the burden on women?. Available at: https://www.researchgate.net/publication/222265780_Rural_Household_Transport_in_Africa_Reducing_the_Burden_on_Women (Accessed 24 Mar 2017)
23. Burnett D, Barbour E, Harrison G (2014) The UK solar energy resource and the impact of climate change. Renew Energy 71:333–343. https://doi.org/10.1016/j.renene.2014.05.034
24. Energy (2014) Department of energy republic of South Africa. Available at: http://www.energy.gov.za/files/esources/renewables/r_solar.html (Accessed 2 Sept 2017)

25. Jimeno M, Knaack J, Grundner C, Mayer J, Brückmann R (2015) Enabling PV in Argentina a framework analysis for the conditions of use of solar energy
26. Arias M (2009) Terminar la ruta 40 ahora cuesta el doble. Available at: http://www.lanacion.com.ar/1122178-terminar-la-ruta-40-ahora-cuesta-el-doble (Accessed 29 Aug 2017)
27. Climate action tracker (2017) Available at: http://climateactiontracker.org/countries/australia.html (Accessed 9 Sept 2017)
28. Australian energy resource assessment (AERA) (2015) Solar energy. Available at: https://arena.gov.au/assets/2018/08/australian-energy-resource-assessment.pdf (Accessed 13 Mar 2017)

Chapter 34
City of Sydney Decentralised Renewable Energy Master Plan

Allan Jones

34.1 Decentralised Energy Master Planning

Decentralised energy master planning is an important process for cities who want to implement decisive steps towards a 100% renewable energy future which can be achieved by a combination of energy efficiency and decentralised renewable energy.

Many countries, cities, regions/states and even some utilities have adopted high renewable energy targets. So far, 59 countries, 72 cities, 63 regions/states, 9 utilities and 21 non-profit/educational/public institutions, covering more than 1.8 billion people have shifted or have committed to shifting to 100% renewable energy in at least one sector.[1]

Although these cities, states and countries have adopted 100% renewable energy targets they have no idea what mixture of renewable energy resources, technologies and storage they will actually need to deliver their targets. They also have no idea what renewable energy resources, technologies, regulatory regime or business models they will need to benefit both consumers and prosumers. This is particularly the case for cities.

The laws of physics dictate that electricity will always flow to the nearest demand and the smaller scale of technologies means that most, if not all, of electricity generation in the future will be decentralised and it is important to ascertain the maximum amount of decentralised energy that can be delivered inside the city so that it can be determined what larger scale renewable energy generation needs to be built outside but in proximity to the city. Otherwise, a headlong rush to build large-scale renewables remote from the city without any plan will lead to future stranded renewable energy assets and unnecessary grid infrastructure investment.

[1] Go 100% renewable energy.

A. Jones (✉)
Devon House, London, UK
e-mail: AllanJones@lda.gov.uk

A. Sayigh (ed.), *Renewable Energy and Sustainable Buildings*, Innovative Renewable Energy, https://doi.org/10.1007/978-3-030-18488-9_34

To overcome these issues and to deliver its Sustainable Sydney 2030 energy and climate change targets the City of Sydney, Australia developed and adopted a Decentralised Energy Master Plan—Renewable Energy which established that 100% of the City's local government area electricity, heating and cooling demands could be met by renewable energy resources and reduce 2006 greenhouse gas emissions by 75% by 2030.

This City of Sydney case study not only covers the Renewable Energy Master Plan but also what impacts the City has had on catalysing not only renewable energy in Sydney but also in Australia as a whole, despite opposition from vested interests.

34.2 Background to City of Sydney

Sydney was the first European settlement in Australia with the arrival of the First Fleet in 1788. The City of Sydney was established in 1842 and today, is the local government area (LGA) covering the Sydney central business district and surrounding inner city suburbs of the greater metropolitan area of Sydney. The Lord Mayor is directly elected while the nine other Councillors are elected by the proportional voting method. The current Lord Mayor has been in office since 2004.

Sydney is the state capital of New South Wales, Australia's global city and international gateway with world-renowned tourist attractions and sustained investment in cultural infrastructure and facilities. The City of Sydney is also the nation's economic powerhouse representing around 25% of New South Wales gross domestic product (GDP) and around 8% of Australia's GDP. However, the City is also recognised as the nation's leading environmental performer and one of the world leaders in tackling climate change.

34.3 Sustainable Sydney 2030

Sustainable Sydney 2030[2] is the vision and strategic plan for the City of Sydney to make Sydney a green, global and connected city by 2030. The full spectrum of interested individuals and groups were consulted on Sustainable Sydney 2030 over a period of 18 months making it the most extensive engagement process in the City's history. Of key significance was that 90% of respondents wanted urgent action on climate change. Sustainable Sydney 2030 was adopted by Council in 2008 and provided the mandate for the Lord Mayor and Council to deliver the 10 targets to make Sydney more sustainable by 2030 (Fig. 34.1).

1. The City will reduce greenhouse gas emissions by 70% below 2006 levels by 2030.

[2] City of Sydney—Sustainable Sydney 2030: The Vision 2008.

1. The City will reduce greenhouse gas emissions by 70% below 2006 levels by 2030.
2. The City will meet 100% of electricty demand by local generation by 2030

Source: City of Sydney (2008)

Fig. 34.1 Key energy and climate change targets in sustainable Sydney 2030. Source: City of Sydney (2008)

2. The City will meet 100% of electricity demand by local generation by 2030.

In 2008, 80% of Sydney's greenhouse gas (GHG) emissions come from coal-fired power plants and the 70% reduction in GHG emissions could not be delivered without replacing coal-fired centralised energy generation with low or zero carbon decentralised energy generation. Therefore, the 100% local electricity demand would need to be met principally by decentralised energy—70% from trigeneration and 30% from renewable electricity generation by 2030.

34.4 Green Infrastructure Plan

A key objective in Sustainable Sydney 2030 was to prepare a Green Infrastructure, comprising five Master Plans as follows:

Decentralised Energy Master Plan
Trigeneration
Decentralised Energy Master Plan
Renewable Energy Decentralised Energy Master Plan
Advanced Waste Treatment Decentralised Water Master Plan, and Energy Efficiency Master Plan

In addition, a Climate Change Adaptation Strategy would support and take into account the Green Infrastructure Plan as climate change mitigation and adaptation should be integrated holistically since the green infrastructure that a city needs to function must also be resilient and adaptable to climate change.

34.5 Trigeneration Master Plan

Based on the energy efficiency target in Sustainable Sydney 2030 the Trigeneration Master Plan was the first decentralised energy master plan to be adopted by the City in 2013. The reason for this is that city energy demands are very high in relation to their geographical area and for a city like Sydney those energy demands have very high air-conditioning or cooling demands, particularly in summer. The Trigeneration

Master Plan forecast what the energy demands would be by 2030 on a 'business as usual' basis to ensure that the Master Plan adopted catered for the 2030 energy demand.

The Trigeneration Master Plan broke the city down into energy demand layers and geographical areas to determine the heating and hot water demands and how much of the electric cooling demands could be converted to heat-fired absorption cooling demand. This would have the double benefit of significantly reducing electricity consumption and peak power by switching from electric cooling to thermal cooling. This in turn enabled more local electricity generation from the need for additional waste heat to supply both heating and cooling demands. The Master Plan was then developed into Low Carbon Zones for energy intense inner-city areas, energy intense hot spots outside the inner-city areas and the remainder of the city which were mainly low-rise suburban areas where domestic decentralised energy systems would be more appropriate.

The Trigeneration Master Plan showed that 70% of the City's electricity demands and 100% of the City's heating and cooling demands could be met by trigeneration and reduce the City's greenhouse gas emissions by 31.9%.

Although the initial fuel for the trigeneration network would be natural gas to enable the economic development of the heating and cooling network infrastructure, the City resolved in 2012 that by 2030 renewable gases from waste and other renewable energy resources would replace fossil fuel natural gas in the trigeneration systems. The renewable gas resources necessary to deliver this outcome would be included in the Renewable Energy Master Plan.

34.6 Renewable Energy Master Plan

The Renewable Energy Master Plan was the second decentralised energy master plan to be developed and adopted by the City in 2013. The Renewable Energy Master Plan established that no more than 18.2% of the city's electricity demand could be met by renewable electricity generation, primarily solar PV. The reason for this is that cities have very high energy demands and tall buildings whose roofs are small in comparison to the number of energy consuming floors, small geographical area in relation to the city's energy demands and over-shadowing is more of an issue in cities.

Therefore, the City needed to make up the balance of the 30% renewable electricity generation required from outside the city. However, the City did not want to include renewable electricity generation from Queensland or Victoria or even outback New South Wales whose electricity could never reach Sydney, so the City developed a proximity principle that would only include renewable electricity generation within 250 km of the city. In practice, enough renewable electricity generation could be sourced within 100–150 km of the city to more than make up the 30% renewable electricity generation target. This would also avoid or minimise the very high cost of grid upgrades or reinforcements and future renewable energy stranded assets.

The second stage of the Renewable Energy Master Plan was to identify renewable gas resources derived from waste both inside and within 250 km of Sydney. The renewable gas resources identified comprised virtually all forms of waste that are not otherwise recycled, such as residential, commercial and industrial waste, sewage and landfill. Beyond the city, renewable gases can also be sourced from livestock manure, agricultural stubble and husks from crops or non-native forestry off-cut waste. Energy crops and native woodlands were specifically excluded from the Master Plan to avoid any potential land use conflicts with food crops, water and destruction of native woodlands.

Producing renewable gas from bioenergy, either by anaerobic digestion or gasification, converting into a substitute natural gas for injection into the gas grid and pipelining into the city enables typically 80% of the primary renewable energy resource to be recovered compared with typically 20% for electricity only generation connected to the electricity grid. The renewable gas resources in the Master Plan are all within economic proximity to the gas grid (Fig. 34.2).

The Renewable Energy Master Plan identified that the total residual municipal solid waste (MSW) and commercial and industrial (C&I) waste resource available in New South Wales within 250 km of the City's local government area (LGA) but excluding the City's LGA was around 3.7 million tonnes a year, forecast to grow to 4.6 million tonnes a year by 2030. This was more than enough renewable gas resource required by the City for both trigeneration and other gas uses. Advantages for other local authorities in utilising local advanced waste treatment and renewable

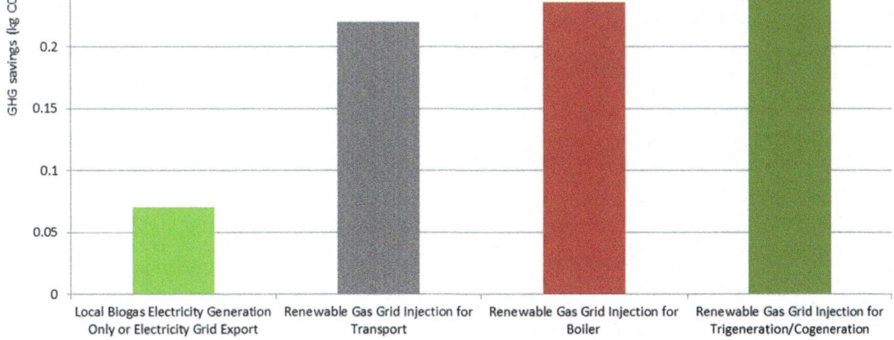

Source: City of Sydney

Fig. 34.2 Greenhouse gas emission savings from renewable gas grid injection compared to electricity generation only and other gas uses. Source: City of Sydney gas grid injection plants to meet the City of Sydney's renewable gas demand would be the virtual elimination of non-recyclable waste going to landfill and the avoidance of the landfill levy which would save local authorities outside of the City of Sydney $177 million a year and businesses $252 million a year.

34.7 Advanced Waste Treatment Master Plan

The Advanced Waste Treatment (AWT) Master Plan was the third decentralised energy master plan to be developed and adopted by the City in 2014. The AWT Master Plan was a subset of the Renewable Energy Master Plan for the renewable gas resources available from the MSW collected by the City and from the C&I waste collected by city business waste contractors. The AWT Master Plan also provided the environmental and financial data to build an advanced waste treatment facility for the City's own MSW and C&I waste.

The AWT Master Plan demonstrated that the diversion of MSW and C&I waste from landfill would increase from 52% in 2012 to 94% by 2030. This would reduce GHG emissions across the City's LGA by 7% below 2006 levels by 2030. The City and the City's LGA businesses would also save in the region of $3.9 million and $18.7 million a year, respectively, in the landfill levy. In addition, using advanced gasification as part of the advanced waste treatment would produce more than enough renewable gas to supply the City of Sydney's own trigeneration and other gas uses.

34.8 Delivering a 100% Renewable Energy Sydney

Taken together, the Trigeneration, Renewable Energy and Advanced Waste Treatment Master Plans would enable 100% of the City's electricity, heating and cooling demands to be met by renewable energy resources and reduce 2006 GHG emissions by 74.6% by 2030 (Fig. 34.3).

Delivering a 100% Renewable Energy Sydney by 2030				
Decentralised Energy Master Plan	Electricity	Heating & Cooling	Gas	Reduction in GHG Emissions
Trigeneration	70%	100%	-	31.9%
Renewable Energy	30%	-	100%	37.5%
Advanced Waste Treatment	Inc. above	Inc. above	Inc. above	5,2%
Total Renewable Energy	100%	100%	100%	74.6%

Source: City of Sydney (2014)

Fig. 34.3 Delivering a 100% renewable energy Sydney by 2030. Source: City of Sydney (2014)

34.9 City of Sydney Projects

The City is implementing its Green Infrastructure Plan at two levels—for the City's own buildings and operations and for its local government area. 'Show by doing' is an important principle since the City cannot expect others to do what it is not prepared to do itself on its own buildings and operations. In support of this principle the City initially implemented four major carbon reducing projects—Building Energy Efficiency Retrofits to 45 of the City's major buildings, LED Street Lighting to replace 6448 street lights with LED's, Solar PV on 31 buildings and the Town Hall Precinct trigeneration and Aquatic Centres cogeneration projects (Fig. 34.4).

These projects were installed from 2011 to 2016 and reduced energy consumption by 16% and GHG emissions by 27.4% below 2006 levels. Other trigeneration, solar PV, energy storage and community renewable energy projects are under way to reduce GHG emissions by 44% by 2021 and 70% by 2030 below 2006 levels.

34.10 City of Sydney Local Government Area Projects

The City of Sydney LGA has reduced energy consumption by 9% and GHG emissions by 19% below 2006 levels by 2016. Some of the projects implemented in the LGA by the City include:

34.11 CitySwitch Green Office Program

The CitySwitch program enables the City to educate tenants and proselytise energy efficiency ratings and emissions reductions as the impact that tenants could have in incentivising landlords to improve the energy efficiency of both the landlord's and

City of Sydney Major Projects on its Own Buildings and Operations			
Project	Energy Savings	Water Savings	Reduction in GHG Emissions
Building Energy & Water Efficiency Retrofits	15.5%	21.8%	15.0%
LED Street Lighting	5.5%	-	5.3%
Solar PV	3.0%	-	3.3%
Trigeneration	3.1%	-	3.0%
Total	27.1%	21.8%	26.6%

Source: City of Sydney (2016)

Fig. 34.4 City of Sydney major projects on its own buildings and operations. Source: City of Sydney (2016)

tenanted parts of their buildings, particularly at new or renewal of or break clause lease contract points, is significant.

The Sydney CitySwitch Green Office currently comprising 144 tenancies occupying 21% of the city's office space have improved their average NABERS rating from 3.8 to 4.6 from 2006 to 2017 and reduced GHG emissions by 65,520 tonnes a year.

34.12 Better Buildings Partnership

In 2011, the City established the Sydney Better Buildings Partnership (BBP) with 13 major landlords who own 50% of the city's commercial floor space. The BBP now has 15 members and 5 associate members who collectively reduced GHG emissions in their property portfolio by 47% and reduced energy bills by $32 million a year from 2006 to 2016.

Similar to the BBP in London, the City identified the number of properties owned by landlords started to taper off into smaller number of properties owned by many smaller landlords. For Sydney, this was 13 major landlords that the City needed to convince to join the BBP and adopt the same energy and climate change targets as the City using Energy Disclosure and NABERS energy ratings as the driver to improve the energy performance of their buildings.

34.13 Environmental Upgrade Agreements

In 2011, the City introduced Environmental Upgrade Agreements (EUA's) in its LGA taking advantage of amendments to the New South Wales (NSW) Local Government Act 1993 to overcome the barriers to implementing energy efficiency and environmental upgrade works in commercial and multi-residential buildings.

The City worked with NSW Government to introduce EUA legislation[3] and the City was the first local authority to take advantage of this with a $26.5 million trigeneration micro-grid scheme serving phase 1 of the new Central Park development. To date, 4 EUA's with a total value of $30.4 million have been signed and a further 7 EUA's are currently being implemented or negotiated for energy efficiency works of the order of $2–$3 million each.

[3] New South Wales Government Environmental Upgrade Agreements.

34.14 Renewable Energy and Climate Change Action in Australia

Although Australia signed the Kyoto Protocol in 1998, Australian Government did not ratify the Kyoto Protocol until 2007 following the election of the Australian Labor Party (ALP) differentiating itself with its climate policies from the outgoing Liberal/National Party (LNP) Coalition government. The election set a new optimism and enthusiasm amongst Australians to tackle climate change and the City of Sydney was one of the first local governments to take action with its Sustainable Sydney 2030 in 2008.

The ALP government introduced a revised Renewable Energy Target (RET) in 2009 which increased the renewable energy target from 9500 GWh by 2010 to 45,000 GWh (20%) by 2020 and a carbon pricing scheme or "carbon tax" in 2011.

The LNP government was re-elected in 2013 and abolished the carbon tax. They also sought to review the RET. The threat of the review led to 15 months of lost investment confidence which forced the industry to agree to a reduced RET of 33,000 GWh by 2020 in 2015.

Meanwhile, the City of Sydney spearheaded a campaign to oppose the proposed increased grid charges to support the 'business as usual' coal-fired centralised energy model. The City commissioned a report[4] from the University of Technology which established that electricity network businesses in Australia were planning to spend over $46 billion from 2010 to 2015. Whilst the campaign did not stop the increases in grid charges it did attract national publicity as consumers energy bills increased by 48.5% from 2007 to 2017 with most of the increase due to increased grid charges of +47.7%. In NSW, Grid charges now represent 50% of the average residential consumer's electricity bill.

34.15 City of Sydney Impact on the Australian Energy Landscape

The increased cost of grid electricity, renewable energy publicity and government's refusal to remove the regulatory barriers to decentralised energy has led to an explosion of 'behind-the-meter' decentralised renewable energy from just under 99,500 installations in 2008 to nearly 2.9 million installations in 2018. Rooftop solar alone generated 7206 GWh in 2016/17.

Australia now has the highest penetration of rooftop solar in the world.

Despite the setback of the reduced RET, large-scale renewables are also enjoying a boom, with 46 projects under construction representing 2590 MW of capacity in 2016/17; 1259 MW of that being large-scale solar projects.

---

[4] Institute for Sustainable Futures, University of Technology Sydney 'Close to Home: Potential Benefits of Decentralised Energy for NSW Electricity Consumers 2010'.

In 2016/17, renewable energy in Australia generated enough electricity to power 7.1 million homes, representing 70% of all households and reducing GHG emissions by 25.4 million tonnes. Australia had a record year in 2017 with $11.5 billion in renewable energy investment, up 150% on 2016 and breaking the previous record of $7.9 billion set in 2011. As a result, the Australian Clean Energy Regulator confirmed that Australia will now meet its 2020 Renewable Energy target.

Quite apart from proselytising decentralised renewable energy throughout Australia, the City's Renewable Energy Master Plan introduced new energy concepts to Australia such as renewable gas derived from waste injected into the gas grid, 'power to gas' technologies and renewable energy mining and exports.

Australia's renewable energy resources are many times greater than Australia's annual energy needs. For example, the annual solar radiation falling on Australia is approximately 58 million petajoules (PJ), about 10,000 times Australia's annual energy consumption. Utilising emerging 'power to gas' and liquefied renewable gas (LRG) technologies would enable Australia to access its vast renewable energy resources in the remotest parts of Australia for itself and for export. The liquefied natural gas (LNG) infrastructure already exists in Australia so LRG can take advantage of this (Fig. 34.5).

Following the City's Renewable Energy Master Plan in 2013, Energy Networks Australia published a report in 2017 on decarbonising Australia's gas distribution

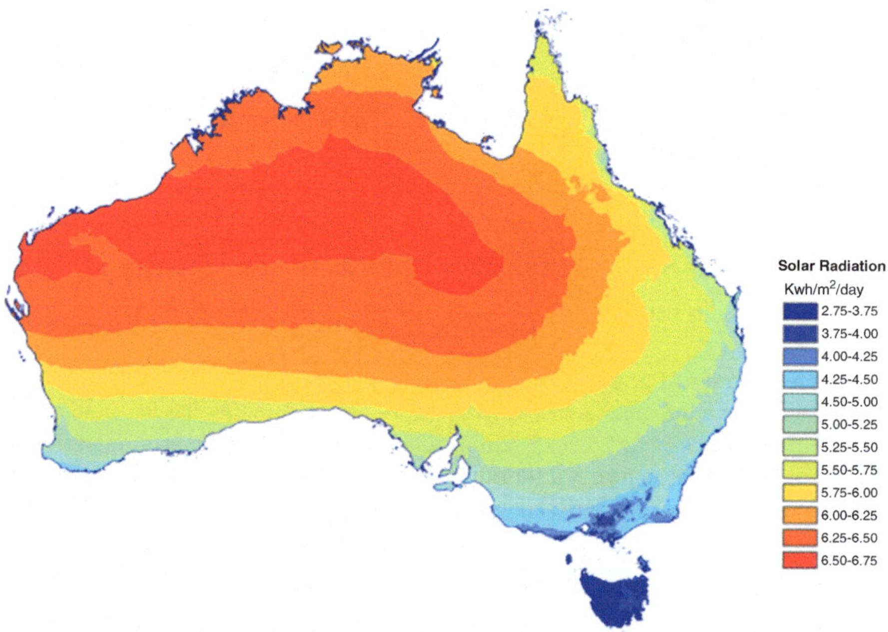

Australian mean annual solar radiation levels (data and map supplied by CRES, Anu)

Source: Renew Economy

Fig. 34.5 Australia solar intensity map. Source: Renew Economy networks which included renewable gas and hydrogen injected into the gas grid. The Australian Renewable Energy Agency is also funding $20 million worth of projects for 'power to gas' and renewable hydrogen exports. These projects demonstrate the impact that the City's Renewable Energy Master Plan has had on the Australian energy landscape in the last 4 years.

Chapter 35
Workshop "Smart Energy Cities and EU GCC Cooperation Opportunities" Within the WREC 2018

Haris Doukas

Date:	Tuesday, 31 July 2018
Venue:	University of Kingston, London, United Kingdom
Time:	14:00–16:00

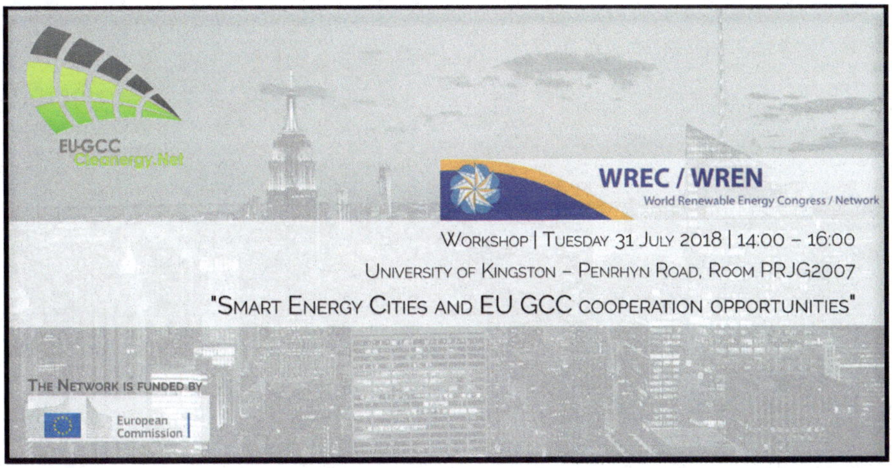

The workshop "Smart Energy Cities and EU GCC Cooperation Opportunities" took place on Tuesday, 31 July 2018 within the framework of World Renewable Energy Congress 2018 (WREC 2018) and was chaired by Dr. Haris Doukas—Network's Research Collaboration Specialist. The workshop dealt with Smart

H. Doukas (✉)
Decision Support Systems Laboratory, School of Electrical and Computer Engineering, National Technical University of Athens, Athens, Greece
e-mail: h_doukas@epu.ntua.gr

A. Sayigh (ed.), *Renewable Energy and Sustainable Buildings*, Innovative Renewable Energy, https://doi.org/10.1007/978-3-030-18488-9_35

Picture 35.1 Dr. Haris Doukas

Picture 35.2 Dr. Hanan Al Buflasa

Energy Cities (SEC) which are expected to play a key role in the efforts towards low-carbon economies in both regions, EU and GCC.

The workshop analysed how ICT can be mostly used by energy/facility managers, energy service companies (ESCOs) and specialists, who take decisions based on the information they get. Moreover, it was investigated how buildings' occupants (energy end-users) can be included in a more efficient way in this process.

Presentations were given by:

- **Dr. Haris Doukas**, Research Collaboration Specialist, EU GCC Clean Energy Technology Network on "Smart Energy Cities and EU GCC Cooperation Opportunities: Setting the scene"
- **Dr. Hanan Al Buflasa**, Assistant Professor, Department of Physics, College of Science, University of Bahrain on "Smart Energy Management Systems for Households in the GCC Countries: Current overview, challenges and opportunities"

Picture 35.3 Tommaso Morbiato and Enrico Fusto

Picture 35.4 Andrew Parker

- **Tommaso Morbiato**, R&D Head and Founder, CEO Windcity srl. Variable geometry for variable flows, and **Enrico Fusto**, Founder and Lead of Strategy and Technology at local flow on "From a power-based energy economy to peer-to-peer energy exchange in the local economies network thanks to the new IoT mini-turbines in the blockchain 2.0"
- **Andrew Parker**, Energy Efficiency and Demand Side Management Convener, EU GCC Clean Energy Technology Network on "Future directions for cooling, EU and GCC innovations".

More specifically **Dr. Haris Doukas** first presented the aim and objectives of the Network and then he presented the workshop's objectives, which were, namely, (a) to analyse how IoT can be mostly used by energy/facility managers, energy service companies (ESCOs) and specialists, who take decisions based on the information they get; (b) to investigate how citizens (energy end-users) can be included in a more efficient way in this process; and (c) to analyse the role that the EU GCC Clean Energy Technology Network could have as a facilitator for successful synergies. Furthermore, he presented the role of cities and how they are moving to becoming smart energy cities, while at the same time, he highlighted the similarities and differences in the approach the EU and GCC countries have on this topic.

Dr. Hanan Al Buflasa presented a short overview of the GCC electricity production and the smart energy solutions for GCC and she also focused on the research opportunities that are available on this topic. Dr. Buflasa also presented the main drives for smart energy management system in the GCC region, which among others are (a) to finite resources and limited access to cheaper hydrocarbons, increased and expensive natural gas import; (b) to better management of the demand-supply gap by compensating for indigenous shortage of oil or gas or both; (c) to align the energy infrastructure for a foreseeable future without oil and gas; (d) to become potential RE exporters with breakthrough technological advancement Economy; and (e) to limit domestic fuel consumption to recover revenues through increasing the share in export.

Tommaso Morbiato and Enrico Fusto presented the developments of their start-ups which are now working together. Their presentation focuses on the peer-to-peer renewable energy exchange in the local economy networks, highlighting that there is an efficiency gap between the renewable power plant CAPEX and the energy records. Moreover, they presented the value proposition for a well-functioning data and energy marketplace for cities. The value proposition is focused on four pillars, namely, (1) infrastructure where sellers offer data and energy in exchange for something of value from buyers; (2) data evaluation and valuation mechanisms; (3) incentives for all market participants to act honestly, including an enforcement system to avoid dishonest behaviour; and (4) incentives for all players to ensure data quality, including an enforcement mechanism to guard against low-quality data.

Andrew Parker finally presented whose presentation for used on the innovation and applied examples for cooling technologies and how these technologies can play a key role in the smart cities era. Moreover, A. Parker presented a project funded by European Commission (EC), the **REMOURBAN (REgeneration MOdel for accelerating the smart URBAN transformation)** which focused on energy efficiency in buildings as well as in system integration, DH, CHP, PV, waste heat, waste to energy, ICT and mobility and transport. The technologies will be applied in pilot cities, which are Nottingham (UK), Valladolid (Spain) and Tepebasi (Turkey).

Chapter 36
A Pricing Method for the Electricity from Renewables to Be Used After Feed-In Tariffs

Yoshihiro Yamamoto

36.1 Introduction

Feed-in tariffs (FITs) have been widely used to encourage investment in renewables. When a targeted amount of investment is achieved, FITs will be decommissioned. A new method of pricing should be introduced in place of FITs because surplus electricity, the electricity generated from renewables but not self-consumed by a facility owner, may be efficiently used by others.

When designing a new pricing method for the surplus electricity generated from some sorts of renewables, we should take account of their distinct features. Consider wind-power or photovoltaic (PV) generation. First, we are unable to control the amount of electricity to be generated; it depends on natural conditions such as wind and solar irradiation. Second, the marginal cost of such power generation is very small; marginal cost pricing will not be appropriate. As a result, demand for electricity may play a critical role because supplies are worth more when demands are larger and demands are worth more when supplies are larger.

The purpose of the study is to propose an effective method of pricing for the surplus electricity from wind-power or PV generation on a market where some participants just consume electricity while others engage in such power generation as well as consuming it.

The study [1] is relevant to our purpose. The problem that those authors addressed is how to distribute a joint cost among a group of producers of electricity from renewables. They introduced seven axioms to characterize just and reasonable allocation rules.

On the other hand, our study proposes a method for pricing the surplus electricity, focusing on PV generation in the residential sector. It considers a period of time. Our model considers some group of households; some households just consume

Y. Yamamoto (✉)
Department of Economics, Takasaki City University of Economics, Gunma, Japan
e-mail: ysyama@tcue.ac.jp

A. Sayigh (ed.), *Renewable Energy and Sustainable Buildings*, Innovative Renewable Energy, https://doi.org/10.1007/978-3-030-18488-9_36

electricity and others generate and consume electricity. This means that the surplus electricity may be priced locally in the group as distributed generation. Note that we do not address how to determine this group in this study. The model is based on a coalitional game. The solution in the game distributes the worth, which all the participants achieve together, among them so that no coalition (subset of the group) forms to deviate from this arrangement of distribution.

The remaining sections are organized as follows. In Sect. 36.2, the model is presented as a coalitional game. Following the model, Sect. 36.3 examines the pricing by the core, one of the solution concepts in a coalitional game. In Sect. 36.4, another solution concept, the Shapley value, is examined. Section 36.5 concludes the study.

36.2 The Model

Let us define a coalitional game $\langle N, v\rangle$ of our problem, basically following the book by Osborne and Rubinstein [2]. Let N be a group of households. The households in N may or may not own PV systems. The group N may be located in a neighborhood because the electricity generated from residential rooftop PV panels should be used locally in terms of efficiency as distributed generation.

It is supposed that a system operator must balance supply and demand every period on the electricity market. We consider this series of time periods.

Let us look at one of the periods. Then, N is divided into two subsets, E and D. The households in E were exporters of surplus electricity and those in D were importers of electricity. Exporters own PV systems, with which they produced electricity and exported the surplus electricity to the grid during that period. On the other hand, importers may or may not own PV systems. If they own, they produced electricity, all of which they consumed themselves. They imported electricity from the grid because the demand was not met by the electricity they produced with their PV systems. If they do not own PV systems, they just imported electricity from the grid during that period.

For any $i \in E$, let e_i be the amount of surplus electricity that i exported. For any $j \in D$, let d_j be the amount of electricity that j imported. We assume $\sum_{i \in E} e_i \le \sum_{j \in D} d_j$. The shortage, $\sum_{j \in D} d_j - \sum_{i \in E} e_i$, is supplemented by backup power sources. The same argument as in the following holds in the case where $\sum_{i \in E} e_i \ge \sum_{j \in D} d_j$ with assuming that the surplus is stored in a battery system.

Let the reservation prices of every $i \in E$ and every $j \in D$ be 0 and 1, respectively. This means that an exporter i is willing to export as long as the price is not lower than 0 and that an importer j is willing to import as long as the price is not higher than 1. We may consider the price of 1 as the electricity retail rate and the price of 0 as the highest price among other opportunities of selling the surplus electricity, e.g., storage.

Let S be any coalition of N (Fig. 36.1). For any nonempty $S \subseteq N$, define a worth $v(S) \in \mathbf{R}_{+}$, which can be divided among all members of S.

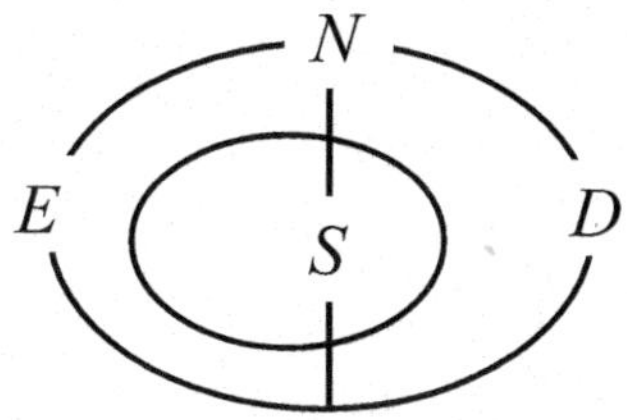

Fig. 36.1 The sets of households

We may define

$$v(S) = \min\left\{ \sum_{i \in S \cap E} e_i, \sum_{j \in S \cap D} d_j \right\}.$$

This means that the worth $v(S)$ is associated with how much of the electricity imported was actually met by the surplus electricity exported within S or the worth is associated with how much of the surplus electricity exported was actually imported within S.

We now have a proposition that coalitional game $\langle N, v \rangle$ is cohesive ([2]): if $N = \cup_{i=1}^{k} S_i$, where $S_i \cap S_j = \varnothing$ for any $i \neq j$, then $v(N) \geq \sum_{i=1}^{k} v(S_i)$. This can be straightforwardly verified. The proposition means that it is optimal that the coalition N of all households forms. Hence, we hereafter confine ourselves to examining how to distribute $v(N)$ among all the households so that no coalition of households forms to deviate from this distribution.

The distribution is represented by a profile $(x_i)_{i \in N}$. It is an S-feasible payoff profile if $\sum_{i \in S} x_i = v(S)$ ([2]). In particular, an N-feasible payoff profile is called a feasible payoff profile, for which we have $\sum_{i \in N} x_i = v(N)$. Accordingly, our goal is to propose a feasible payoff profile x so that no coalition of households forms to deviate from x.

36.3 Pricing by the Core

We examine two solution concepts of coalitional game $\langle N, v \rangle$ in the study. The first solution concept is the core ([2]). A feasible payoff profile x is in the core if no coalition can obtain a payoff that exceeds the sum of its members' payoffs guaranteed by x: for any coalition S, $\sum_{i \in S} x_i \geq v(S)$. This means that there are no incentives for any households to deviate from x by forming a coalition in order to obtain larger payoffs.

A feasible payoff profile x, where $x_i = e_i$ for $i \in E$ and $x_j = 0$ for $j \in D$, is in the core. This can be straightforwardly verified. Each of the exporters obtains the payoff of 1 per unit of surplus electricity it exported, and each of the importers obtains the payoff of 0 per unit of surplus electricity it imported. This means that the surplus electricity is priced at 1. If we assume $\sum_{i \in E} e_i \geq \sum_{j \in D} d_j$, the result is reversed.

Let us illustrate the result by a small example. Let $N = \{1, 2, 3, 4\}$, $E = \{1, 2\}$, and $D = \{3, 4\}$. For the exporters 1 and 2, $e_1 = 1$ and $e_2 = 3$. For the importers 3 and 4, $d_3 = 2$ and $d_4 = 4$. Note that $e_1 + e_2 = 4 < d_3 + d_4 = 6$. Then the feasible payoff profile $x = (1, 3, 0, 0)$ is obviously in the core. Hence, we have the feasible payoff profile per unit of surplus electricity traded $y = (1, 1, 0, 0)$.

Let us discuss the pricing by the core. The total amount of surplus electricity exported and the total amount of electricity imported together determine the price. Whichever group, E or D, has a smaller total amount, every member of the group is more likely to obtain a maximal payoff. Hence, each group has an incentive: on the one hand, if exporters expect that the total amount of export will be larger than that of import, they try to reduce the export by self-consumption. On the other hand, if importers expect that the total amount of import will be larger than that of export, they try to reduce the import by decreasing consumption. In other words, every household may control its consumption anticipating the balance of electricity.

However, contributions by each exporter and each importer to the result are not taken into account. As a result, every exporter equally obtains a maximal payoff and every importer equally obtains nothing, or vice versa.

36.4 Pricing by the Shapley Value

We next propose a pricing method that uses the Shapley value ([2]). Each player's marginal contribution to the coalition is considered in this method. For any coalition S, player i's marginal contribution to S is defined as

$$\Delta_i(S) = v(S \cup \{i\}) - v(S),$$

where $i \notin S$. Then the Shapley value, $\phi_i(N, v)$, of player i is defined as the expected marginal contribution over all orders of i to the set of players who precede i:

$$\phi_i(N,v) = \frac{1}{|N|!} \sum_{R \in \Re} \Delta_i(S_i(R)),$$

where $\Re$ is the set of all orderings of the member in N and $S_i(R)$ is the set of players preceding i in ordering R.

Let us illustrate the result of the Shapley value by the same simple example as in Sect. 36.3. The Shapley values are calculated as $\phi = (0.58, 1.75, 0.42, 1.25)$. Then, the Shapley values divided by the amount of surplus electricity traded are $w = (0.58, 0.58, 0.21, 0.31)$. These are moderate compared to the result y by the core.

Let us discuss the pricing by the Shapley value. The payoff profile is "fair" because a marginal contribution is considered. However, the calculation may be a little complicated because there are $|N|!$ orderings of $|N|$ households. More detailed examination is needed about the Shapley value application in our model.

36.5 Conclusion

We proposed a new method of how the surplus electricity from PV generation is priced for a group of households. The worth of a coalition of households was defined as the minimum between the export of surplus electricity and the import of electricity. The core and the Shapley value were calculated. More detailed investigation is needed for practical use of these methods.

References

1. Chakraborty P et al (2018) Cost causation based allocations of costs for market integration of renewable energy. IEEE Trans Power Syst 33(1):70–83
2. Osborne MJ, Rubinstein A (1994) A course in game theory. MIT Press, Cambridge, MA

Chapter 37
Disclosing the Immaterial Resilience of Sustainable Architecture for a New Renovation Processes of the Inland Mediterranean Areas

Antonella Trombadore

37.1 The Immaterial Resilience as New Driver for Sustainable Renovation

Ecosystems and Mediterranean environmental assets represent a heritage on which to develop knowledge about living by stimulating the cooperation capacity on themes of a sustainable development for regional identity evolution. It is about the distinctive architectural, climatic and cultural elements to which research centres, academia, companies and public administrations must aim to boost competitiveness on an international scale. The inland areas of Mediterranean regions are characterized by high levels of biodiversity and often are placed in fragile contexts. At the same time, they are objects of great economic and political interest, and therefore they are generally identified as the projects difficult to manage, in terms of planning. The small towns of the hinterland are still today chests full of stories, memories, symbols, which, in recent years, have undergone a process of transformation, often uncontrolled, which has altered the profile and urban structure.

Through the years, a lot of small inner small towns, especially in Italy and South Mediterranean countries, had suffered heavy phenomena of exodus in order to supply the lack of development opportunities and possibilities. This had caused a deep demographic impoverishment, with serious negative consequences on the public building heritage and economic-cultural activities of historical settlements. This depletion, together with the inadequate of skills and services management, contributes to reduce further the living "attractiveness" of these areas that instead have great potentialities thanks to their "uniqueness", belonging to the extremely beautiful surrounding landscape and the artistic and cultural value of their small villages.

A. Trombadore (✉)
Architectural Department, University of Florence, Florence, Italy
e-mail: antonella.trombadore@unifi.it

A. Sayigh (ed.), *Renewable Energy and Sustainable Buildings*, Innovative Renewable Energy, https://doi.org/10.1007/978-3-030-18488-9_37

The sustainable development approach has generally oriented the choices of architectural design as well as the dynamics of urban transformation, mainly focusing on the technological solutions, technics and material. After the last disasters' experience in the European countries (seismic hazard in Italy, floods in central Europe, fires in Portugal, etc.), the public and scientific debate stressed the need to enrich the meaning of *resilience.*

If we analyse the problems of the *sensible contexts*, characterized by the large cultural heritage, by the environmental fragility and by the strong anthropic pressures or depopulation (touristic and archaeological areas/the inner villages), beside the material capacity of the cities and buildings to react to the disasters, it has emerged the same priority to implement the meaning of the *resilience* including in a new holistic approach also oriented to the *immaterial* issue.

In order to promote the balanced quality of the cities, to ensure both environmental and social sustainability of built environments and to implement the regeneration of small towns and buildings, the enhancement of *immaterial resilience* has to become the new driver of the holistic approach, not only for the rehabilitation after disasters but to avoid immaterial damages.

In this framework, we need to stimulate the international scientific debate to analyse and implement the new integrated approach based on the main matter of *immaterial resilience*, fostering not only the communities' capacity to take care and regenerate their habitat from an architectural point of view but also their aptitude to preserve the socio-economic context and the specific cultural identity (Fig. 37.1).

Fig. 37.1 General view of Seggiano, in Tuscany, typical small village in inner Mediterranean area

37.2 The Identity of Local Architecture as Resilient, Adaptive, Sensitive to Change

Together with the implementation of local activities and the encouragement of new way for the revitalization of small realities, there is the necessity of individuate and develop new scenarios of innovation to support communities in inner settlements and to promote the competitiveness of small and medium companies; the vocation on territorial innovation raises through material and immaterial interconnectivity, linking together socio-economic development implication and local attractiveness.

There is also the necessity of aim to the accessibility and safety, using environmental and ecological aspects as leverage, to reach a complete integration between places and infrastructures, in a developing model in which the man (communities) can be less "costumer" and more "user". This model has to be the more interactive and integrated as possible, in order to individuate new viewpoints of operability in a well-rounded perspective.

In this sense the importance of local architecture (as vernacular architecture) is well seen; in a world of hyper-technological solutions for every aspect of life, there's the need of coming back to simplicity. As well-known at all levels, the architecture of the past in every single part of the earth teaches us optimal solution for the local problem-solving. In every specific context at every requirement corresponds a real approach that leads to a particular solution. Also, in the name itself "vernacular" (from the Latin *vernaculus* which means domestic, familiar), term coined for the first time in 1964 by Bernard Rudofsky in his "Architecture without architects" [1] is clearly evident the message of this type of method: all the problems regarding the primary necessity of most communities (first of all the need of repairing from weathering, feeding, sleeping, living together) are strictly related to the place where it grows. It can be seen as an answer to all the social and environmental requests. In this sense, we need to be aware of a simple but not trivial assumption: the vernacular architecture, just because of its adaptive character, deeply linked to the construction (in its most imminent meaning), possesses a strictly fragile character but not in the weak sense (as the term might suggest) but rather resilient, adaptive, sensitive to change. It corresponds to a vision of the world (proper to every human group), declined according to particular needs.

These characteristics suggest an idea of architecture (understood as technique, constructive wisdom), absolutely strong and capable of giving a strong message of identity. In contrast to the standardizing tendency of the modern vision of the world (the great metropolises are an example), we try to give new life to an idea of individuality based on the enhancement of uniqueness; uniqueness of the small realities of which the Mediterranean area is full, which, however, are likely to remain crushed forever. In this sense, this idea can be extended to all fields of knowledge, in order to create a model (continuously evolving precisely because of its empirical character) economically and socially sustainable in order to revitalize the most depressed areas of the hinterland. This model can trigger virtuous processes that could be repeated in subsequent experiences, so as to create a real network between the various realities, focusing attention on the theme of tourist enjoyment (Figs. 37.2 and 37.3).

Fig 37.2 Traditional approach of tourism development with street market of handcraft

Fig 37.3 Example of Albergo Diffuso as new approach of tourism development

37.3 New Hypothesis of Renovation Processes for the Inland Areas of Mediterranean Countries: The Vivimed Project Experience

Recently it has been assisted to a general tendency to the growing of touristic request in the inner Mediterranean areas, as well as a crescent research of high-quality services (especially for the quality of life and wellness during the travel). These aspects have led to a major awareness of what the traveler wants to find in the place that has chosen, making the vacation a real "experience" based on particularity. Nowadays people (thanks the opportunity given by the Internet and media) can be aware also of the importance of ecosystems preservation, natural and cultural resources and of the possibility of having a real 360° involvement. The key action for the renovation of the small realities in the inner villages of Mediterranean areas, as a consequence the revitalization of small village buildings countries, has to be found in the solutions given by the vernacular architecture, that is, the real way to find sustainable and responsible answers to the abandonment and death of these places [3].

Among the positive aspects that involve thinking about the territory as an environmental system capable of resilience, one of the main is the reference to an organic vision that allows integration, enhance and direct towards common objectives, solutions and interventions that, alone, risk to create discontinuity. In a dialogue with the different stakeholders, we share the idea of a "eco-sustainable territorial development model",

Fig. 37.4 The *logo* and web page of Vivimed project and the announcement of living lab program

aiming to combine architectural features solutions to new productive, tourist and commercial functions [4]. It will be important to identify the possible drivers of revitalization and imagine the profile of communities capable of triggering the proliferation of a network of social relations (stimulating new economic connection with the local productive fabric) and at the end to redevelop its existing building heritage; these are to be considered the first steps to start a regeneration of the anthropic fabric and a sustainable use of the territory (Figs. 37.4 and 37.5).

The Vivimed is transnational cooperation project financed by Interreg Programme It-Fr. The activities focus on the enhancement of common, tangible and intangible assets, both in this regard to their natural potential (sun, wind, sea, water, land, habitats, plants, animals, ecosystems and landscapes), both in terms of their cultural heritage, which over the centuries has strengthened their uniqueness and the territorial identity. This is the basis for the dissemination of a competitive model, for the development of a network of eco-green infrastructure and sustainable use. A model that can be exported to other contexts present in the Italian and Mediterranean hinterland, where naturalistic potentialities and resources linked to the authenticity of the places are dominant, but not adequately valued in the logic of a green-circular economy (and/or conscious tourism) [5].

The research on the method can be focused on a strongly transversal sector: the tourism/fruition sector. The mapping of the tourism SMEs active on the territories is elaborated to analyse their profiles and services, as well as the typology of the receptive structures (farmhouses, B&Bs, widespread hotels in the historical villages).

Fig. 37.5 Architecture renovation for the Albergo Diffuso *Sotto le Cummerse*, Locorotondo Puglia—Italy

To elaborate the scenarios of innovation, the sectors potentially complementary to the tourism sector, often not considered, as agricultural activities and related historical-building heritage (crushers, mills, chestnut dryers, etc.), floriculture activities and forestry, fish farming in the vicinity of rivers, industries or their historical branches related to the territory (educational path *via della carta* [the paper road] developed with the paper industry in Lucca), craft enterprises (carpentry, glassworks, etc.) or activities completely extinct but of elevated archaeological-cultural interest (system of the icehouses, obsolete after the invention of the refrigerator) will also be included.

Innovative scenarios will be developed for the integrated and multi-theme tourist offer, according to the specificity of the hinterland territories, also in collaboration with regional trade associations and agro-food and social cooperatives and their consortia, through:

- The enhancement of the territorial resources for the development of an integrated and multi-theme tourism offer, flexible in the themes and methods of development by the type of user (e.g. single, families, old people)
- The configurability of the type of involvement with the desired territory: experiential, social, naturalistic/sporting, cultural, historical, artistic tourism
- The promotion of innovative forms of hospitality such as the widespread hotel, network of bike hotels, hospitality in parks (e.g. zero-impact accommodation facilities) and pathways of ecological brands
- The enhancement of educational aspects as promoters of the environmental theme
- The rethinking of the seasonality of tourist packages to ensure affluence throughout the year (e.g. promotion of events related to agricultural activities, e.g. collection of grapes, olives and chestnuts by combining landscape/cultural routes)

- The activation of partnerships with marinas and terminals of sea ports, cruise activities and tourist villages along the coast through rural tourism packages/ excursions
- The promotion of a model of infrastructure recovery (road network, paths, integrated village/territory/landscape systems, historic buildings) and construction of green structures to support tourism

Through a process of cross-fertilization, it could be easier to reach a convergence of territorial development policies that increases the advantages of the alignment of the respective socio-economic systems. In this way, a virtuous process which will lead to the revitalization of the territory could be triggered on the other favours an improvement in the competitiveness levels of local SMEs operating in the green-circular economy chain, focusing above all on technological innovation, multicultural inclusion, the pursuit of quality of life and the application of the concept of "Design for All" (people with reduced mobility, children, old people) as the common thread of the new vision of territorial and business development. To implement an efficient and operational governance mode where all the stakeholders of the territories in question are actively represented and involved, the project envisages a process of public-private participation articulated within living labs that play the role of territorial catalysts. The living labs will involve the four main actors: public administration, local businesses, scientific researchers and the inhabitants of the territory.

The Mediterranean hinterland has as common thread the low competitiveness of the commercial sector and the lack of interconnections (the strong connotation of territorial "islands" predominates with respect to the potential of being an archipelago). The territories of the hinterland need to:

- Increase the awareness of the territorial potential and skills of SMEs in the sector
- Diversify into the types of services and products to offer
- Specialize towards innovative, attractive and tailor-made activities and infrastructures for end users, both in terms of services and eco-friendly reception facilities
- Enhance in terms of the green-circular economy of the resources present in the territory, through the enhancement of the authenticity of the historical cultural heritage, the protection, management and sustainable accessibility to environmental resources

37.4 Intangible Cultural Heritage and Tourism Development

One of the current challenges facing the tourism sector is to contribute to the identification, protection and safeguarding of intangible cultural heritage through tourism development.

Fostering the responsible use of this living heritage for tourism purposes can provide new employment opportunities, help alleviate poverty, curb rural flight migration among the young and marginally employed and nurture a sense of pride among community members. Tourism also offers a powerful incentive for preserving and enhancing intangible cultural heritage, since the revenue it generates can be channelled back into initiatives to aid its long-term survival. Extremely fragile, intangible cultural heritage must be thoughtfully managed if it is to survive in an increasingly globalized world. True partnerships between communities and the tourism and heritage sectors can only occur if all sides develop a genuine appreciation for each other's aspirations and values.

Tourism interests need to acquire an awareness of cultural heritage management concepts, ideals and practices, while heritage managers must endeavour to comprehend the complex phenomenon of tourism and its modus operandi. Through mutual understanding, both can build on their shared interest in intangible cultural heritage, in close consultation with local communities, the ultimate bearers of mankind's intangible cultural legacy. Intangible cultural heritage is embodied in those practices, expressions, knowledge and skills, as well as in associated objects and cultural spaces, that communities and individuals recognize as part of their cultural heritage. Transmitted through generations and constantly recreated, it provides humanity with a sense of identity and continuity [2] (Fig. 37.6).

Fig. 37.6 Example of Vivimed collection and data base of the most important examples of *Albergo Diffuso,* as sustainable accommodation and responsible tourism solution

37.5 The Criteria of Environmental, Architectural and Social Quality as Material and Immaterial Resources Present in the Territory

As part of the Vivimed project, aimed at the definition of innovative services for the development of the tourism sector in the Mediterranean area, it aims at the environmental qualification of the Albergo Diffuso as a tool of great potential to trigger a valorization process in the territory. Both the settlements and the communities, will combine the principles of territorial innovation and social innovation, with a view to a short supply chain and a green-circular economy. Actually, the regulation for the *Albergo Diffuso* is being approved by the region of Tuscany where, in Article 21, the definition of *Albergo Diffuso* [diffused hotel or pervasive hotel] is given and the minimum requirements for the equipment, systems and equipment provided by the diffused hotels and the minimum requirements for the services offered by the diffused hotels are defined. In 2004 ENEA published the report for the application of the Ecolabel mark to tourism services: objectives, principles and main experiences in progress. In particular, we analyse the most successful international experiences in order to demonstrate that careful and aware environmental management does not necessarily involve burdensome commitments but rather a greater attention to quality and the choice of measures and actions that may prove interesting in order to save money, local economic and return image.

The national association of *Legambiente Turismo Bellezza Natura* has already published in 2015 the Decalogo for tourist accommodation facilities in rural areas. Large operators and networks are launching awareness campaigns towards environmental "qualification", as EcoLeader of TripAdvisor who wants to promote the industry's leading facilities in ecological initiatives, launching an excellent system to show travellers the commitment to implement environmental initiatives and to attract potential customers.

Thanks to the results of Vivimed project, we try to enrich and integrate minimum requirements in the *Albergo Diffuso* regulations with the criteria of environmental, architectural and social quality, stimulating the networking of material and immaterial resources present in the territory: the criteria selected by Vivimed project are declined at different scales:

- *Territorial scale*, enhancing the natural, historical and cultural heritage and the potential for networking of accommodation capacities.
- *Urban scale* (historic villages), focusing on the persistence and homogeneity of the urban microsystem, its morphological characteristics and architectural quality, environmental and social sustainability, management of accessibility for all, of soft mobility, the conscious management of natural resources (green water–soil–renewables and green energy) and handmade resources.
- *Building scale*: enhancing the persistence of homogeneity of typologies, the morphological characteristics of historic buildings, construction technologies, use of local materials and vernacular architecture solutions for energy efficiency.

Fig. 37.7 Santa Fiora Village in Tuscany. View of rehabilitation actions of courtyards, towers, walls and house buildings, according to the valorization of environmental performance

- *Horizontal scale for cross-fertilization of intangible resources*, crossing and combining services and networks, giving added value to an integrated system of services and experiences already active in the territory and focusing on the protection and promotion of the local brands, typical food, cultural events and handcrafts or networks of promotion, communication and marketing grid (Figs. 37.7, 37.8 and 37.9).

37.6 Conclusions

The complexity of urban transformation processes, therefore, presents us today with an ethical challenge: building in the built environment through soft and light interventions that respect the local traditions and the identity of places, heave preserving the historical legacy for benefits of the nowadays communities. This also represents a precious resource and a stimulus for designers, who have nothing to invent, but who can and must tap into history, through what we can call a principle of *conscious imitation* (the past that builds the future).

Fig. 37.8 (**a–c**) Different logo to evaluate and define the architectural and environmental quality of historical villa

Fig. 37.9 Examples of material and immaterial resources present in the territory implemented in the touristic network

Moreover, this enables the transitions to a sustainable approach that has a strategic importance especially as regards the need to revitalize the territory. In this context, a main problem to be considered in order to cope with population growth and the consequent increase in the soil and energy consumption is to identify appropriate solutions able to satisfy all the new social needs and based on the sustainable use of natural and human resources that reflect local traditions and cultural heritage. Specially for the sustainable tourism actions, we need to implement a new integrated approach based on the main matter of *immaterial resilience*, fostering not only the communities' capacity to take care and regenerate their habitat from an architectural point of view but also their aptitude to preserve the socio-economic context and the specific cultural identity.

References

1. Rudofsky B (1964) Architecture without architects: a short introduction to non-pedigreed architecture is a book based on the MoMA exhibition of the same name by Bernard Rudofsky originally published in 1964
2. UNESCO Convention for the Safeguarding of Intangible Cultural Heritage (2003). https://ich.unesco.org/en/convention
3. Trombadore A, Marco Sala M (2017) Sustainable eco-architecture for sustainable eco-tourism: the strategic plan and pilot project of Asinara Island. In: PLEA International Conference Design to Thrive, Edinburgh, 2th–5th July 2017, NCEUB 2017, vol. III, p 4357–4364. ISBN: 978-0-9928957-5-4
4. Trombadore A (2018) Green design for a smart island: green infrastructure and architectural solutions for eco-tourism in Mediterranean areas. In: Sayigh A (ed) Seaside building design: principles and practice—buildings in maritime zones. Springer, New York, pp 163–194. ISBN: 978-3-319-67948-8
5. Trombadore A (2016) Mediterranean smart cities. Innovazione tecnologica ed ecoefficienza nella gestione dei processi di trasformazione urbana Altralinea—Firenze

Chapter 38
Application of DC–DC Boost Converter to Photovoltaic Pumping System

Sarah Abdourraziq and Mohamed Amine Abdourraziq

38.1 Introduction

Stand-alone PV water pumping systems photovoltaic pumping systems present a cost-effective solution of solar energy. In the first uses, a DC motor type was a standard. Actually, the use of AC induction motor drive systems is possible to use a more robust and less expensive motor for photovoltaic pumping application [1–4]. Also, several kinds of solar pump are existing in the literature, depending on the application and the water sources (wells, drilling, pumping river, etc.). Therefore, the most PV pumps used are the centrifugal and the volumetric pumps.

To improve the efficiency of the system, a MPPT algorithm is used to reach the MPP of PV panel, periodically with the change of solar irradiation and temperature. A variety of maximum power point tracking (MPPT) algorithms have been proposed including fractional open-circuit voltage [5, 6], fractional short-circuit current, perturb and observe (P&O) [7, 8], incremental conductance (INC) and artificial-intelligence-based algorithms. These algorithms vary in their complexity, efficiency, cost and potential applications. In high-power applications, the cost of MPPT control is dwarfed by the cost of the photovoltaic (PV) array and power converters.

The general diagram of the studied system is presented in Fig. 38.1. It consists of solar panel, power converter, motor-pump set and storage tank. This paper will present a detailed definition of each component, with the presentation of the principle of the MPPT technique (Fig. 38.1).

In the context of the energy sector, EIAs have been widely applied to large-scale energy infrastructure projects such as those associated with oil and gas exploration and development, coal mining and power generation and large-scale hydropower. EIAs have been conducted on onshore wind power generation projects and are

S. Abdourraziq (✉) · M. A. Abdourraziq
LESSI Lab FSDM, REEPER Group, EST, Sidi Mohammed Ben Abdellah University, Fez, Morocco
e-mail: Sarah.abdourraziq@usmba.ac.ma

A. Sayigh (ed.), *Renewable Energy and Sustainable Buildings*, Innovative Renewable Energy, https://doi.org/10.1007/978-3-030-18488-9_38

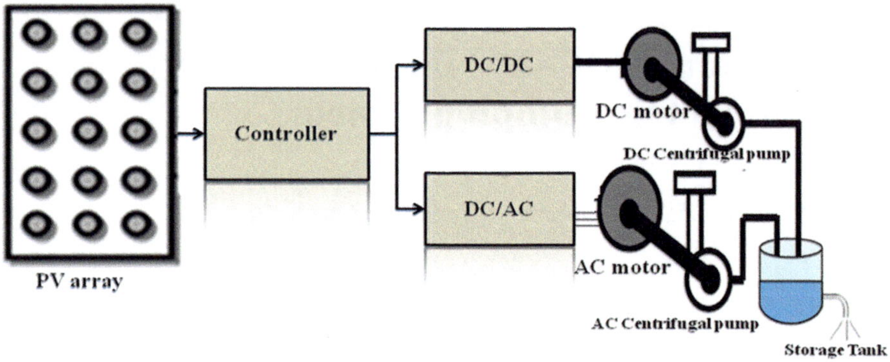

Fig. 38.1 General diagram of PV water pumping system

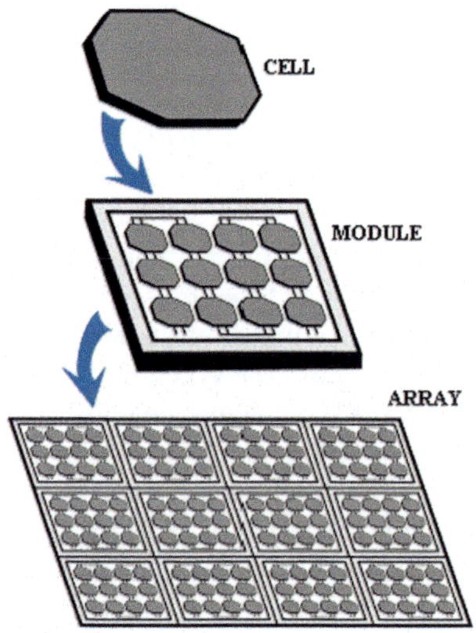

Fig. 38.2 Components of PV panel

increasingly being applied to other renewable energy projects including offshore wind, large-scale solar, biofuels and geothermal. Experience in applying impact assessments to newer renewable technologies such as wave and tidal power in the marine environment and solar thermal arrays is more sporadic. Strategic assessments are increasingly important as a means of addressing the potential cumulative effects of renewable projects as well as a higher-order approach for identifying alternative technologies and systems designs.

38.2 PV Panel Definition

The photovoltaic generator is constituted by modules. Each module is formed by PV cells. The solar cell represents the elementary power conversion unit (Fig. 38.2).

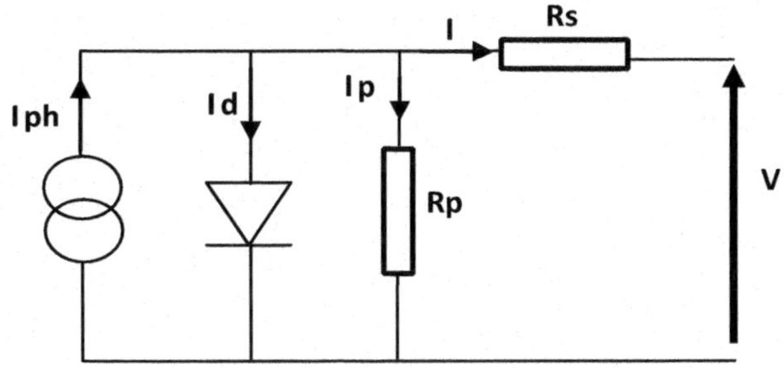

Fig. 38.3 Equivalent circuit of PV cell

38.2.1 PV Cell Based Single-Diode Model

The PV cell consists of a p–n junction based on semiconductor. The equivalent PV cell circuit of single-diode model is presented in Fig. 38.3. It consists of a diode in parallel with a current source and two resistors. The current is proportional to the radiation.

The behaviour of the PV array may be described by equations follows:

$$I = I_{ph} - I_d - I_p$$

with

$$I_d = I_o\left(\exp\left(\frac{V_j \cdot q}{K_o \cdot T}\right) - 1\right)$$

and

$$I_p = \frac{V + R_s \cdot I}{R_p}$$

$$I = I_{ph} - I_o\left(\exp\left(\frac{V_j q}{K_o T}\right) - 1\right) - \frac{V + R_s \cdot I}{R_p}$$

where:

I_{pv} is the cell output current.
V_{pv} is the cell output voltage.
I_o is the saturation current.
R
K is the Boltzmann's constant (1.3806503.1023 J/K).
T is the junction temperature.

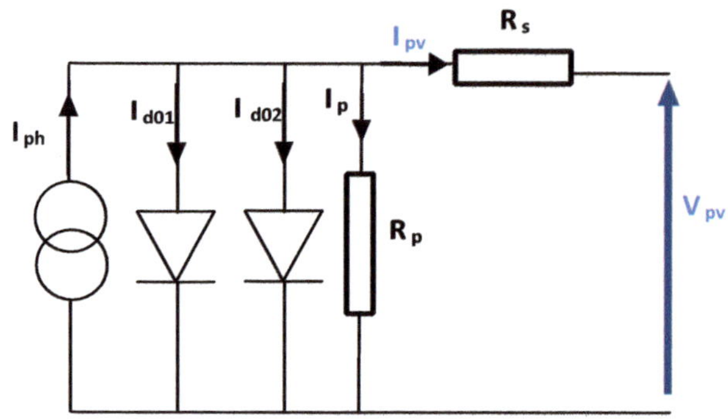

Fig. 38.4 Electrical circuit of PV model

Table 38.1 Output characteristics of studied PV panel

Maximum power (P_{mpp})	310 W
Voltage at MPP (V_{mpp})	36.9 V
Current at MPP (I_{mpp})	8.4 A
Open-circuit voltage (V_{oc})	45.8 V
Short-circuit current (I_{sc})	8.83 A

38.2.2 PV Cell Based Two-Diodes Model

The second type of PV cell model is based in two diodes. This PV cell model consists of a two p–n junction based on semiconductor. This model presents a good compromise between precision and computational complexity of operations. Figure 38.4 shows the electrical PV cell based two diodes.

38.2.3 Atmospheric Condition Effect on PV Cell Characteristics

The output characteristics of current and voltage and power and voltage depend directly on atmospheric conditions (solar radiation and temperature). The electrical characteristics of our studied PV panel (called Bluesun) at irradiation = 1000 W/m^2 and temperature = 25 °C are resumed in Table 38.1.

The simulation results of the output characteristics of PV panel are presented in Fig. 38.5. The simulation is realized at different value of radiations (G_1 = 1000 W/m^2, G_2 = 700 W/m^2, G_3 = 500 W/m^2) and for (T = 25 °C). The results show clearly the influence of the solar irradiation in the output characteristics (P–V) and (I–V). We can notice that the output power (P) and current (I) of the PV panel change simultaneously with the variation of solar irradiance.

Figure 38.6 shows the simulation results in the second case with the variation of the temperature at (T_1 = 25 °C, T_2 = 50 °C and T_3 = 75 °C) for constant solar irradiation (G = 1000 W/m^2). The output characteristics show that the output power and voltage of the PV panel are influenced by the variation of the temperature, whereas the value of the output current is almost constant. From Figs. 38.3 and 38.4, we can deduce

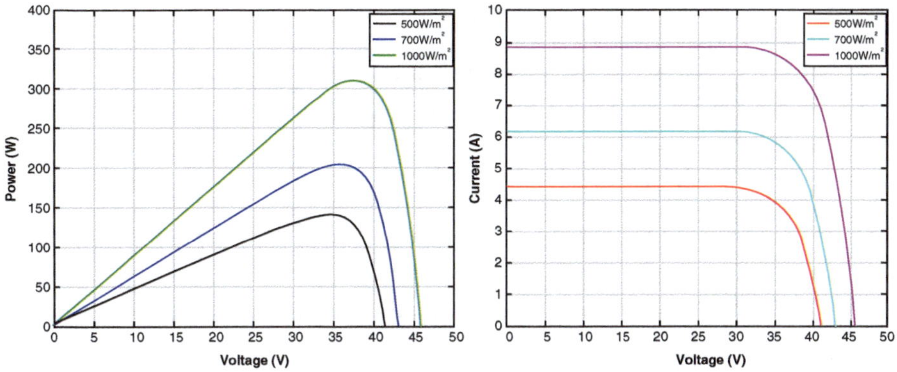

Fig. 38.5 (**a**) *P–V* and (**b**) *I–V* curves for various irradiation ($S = 500$, 700 and 1000 W/m², $T = 25$ °C)

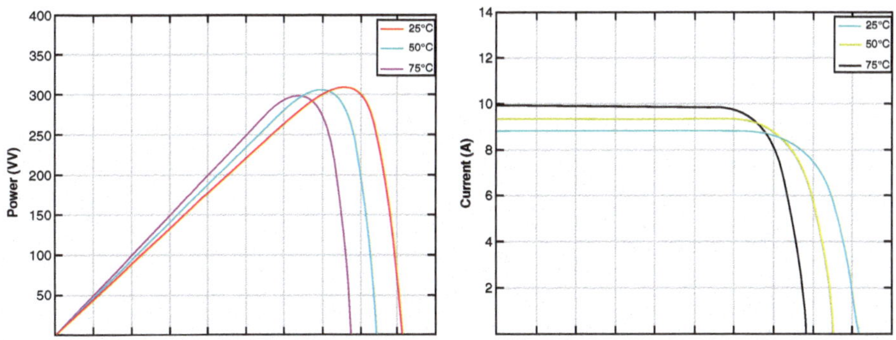

Fig. 38.6 (**a**) *P–V* and (**b**) *I–V* curves for various temperature ($T = 25$, 50 and 70 °C; $S = 1000$ W/m²)

that the variation of radiation and temperature have a direct influence in the output characteristics of the PV panel. This variation can be explained by the dependence of the short-circuit current I_{sc}, with the maximum current I_{mpp}, open-circuit voltage U_{oc} and maximum voltage U_{mpp} on the solar irradiation and cell temperature. The relationships between the different parameters of the PV panel with solar irradiation and temperature is given by:

38.3 DC/DC Converter

The device responsible for maintaining the photovoltaic generator on the maximum power point, MPP, is the DC/DC converter, Fig. 38.7. This component is used as an adapter between PV generator and motor-pump set and characterized by that the output power is higher than the input. The equation characteristic of the DC–DC converter is (Fig. 38.8):

$$V_s = \frac{V_e}{(1-\alpha)}$$

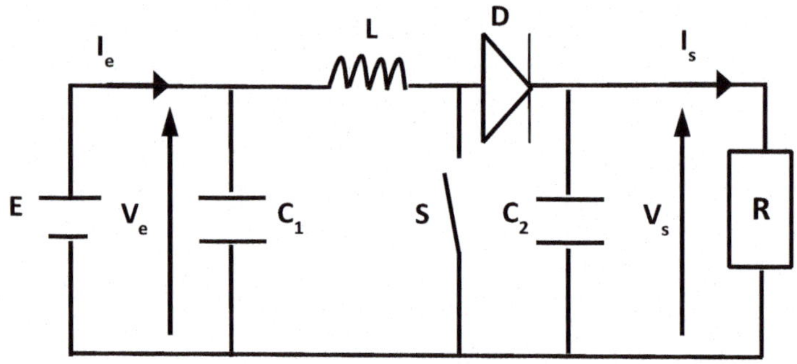

Fig. 38.7 Electrical circuit of DC–DC boost converter

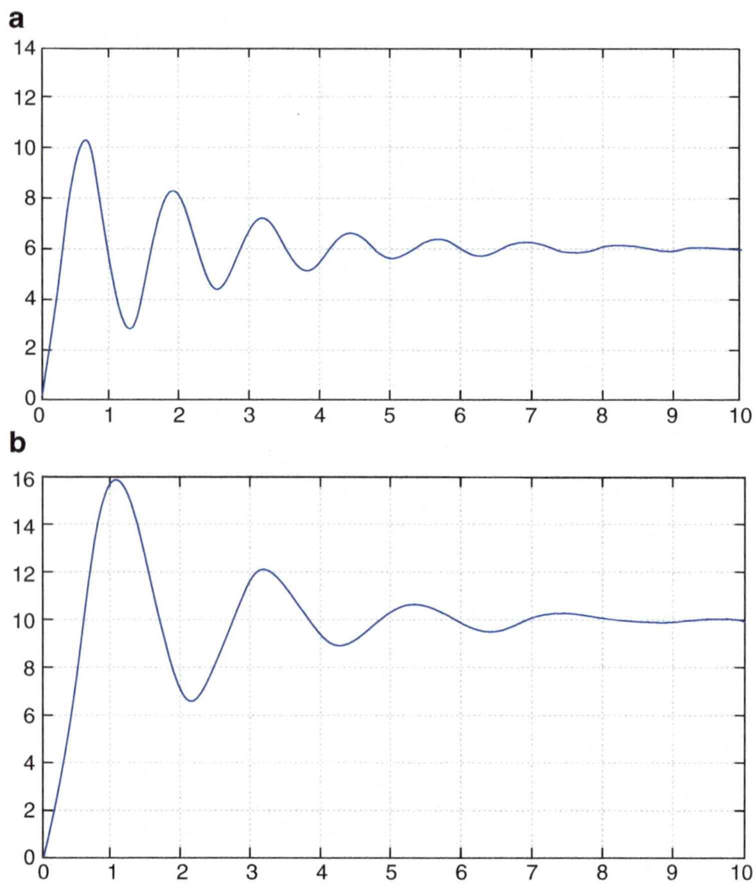

Fig. 38.8 (**a**) Output voltage of the DC–DC boost converter for $V_{pv} = 3$ V and $\alpha = 0.5$. (**b**) Output voltage of the DC–DC boost converter for $V_{pv} = 3$ V and $\alpha = 0.7$

where ton corresponds to the total time interval where both switches are closed (ton = ΔT). Accordingly, with D and the high frequency transformer turn ratio n (n), it is possible to adjust the output voltage (E) as follow.

38.4 Maximum Power Point Technical Research

MPPT algorithm consists of some techniques used to track the MPP of PV panel, and try reach it with the periodically change of solar irradiation and temperature. An important improvement in energy utilization can be achieved by using a simple DC–DC converter controlled by an MPPT algorithm, reducing the required size of the PV array for a given power output. Several researchers have given attention to those techniques. Some techniques are based on one sensor (open-circuit voltage method, short-circuit current method…), and other techniques are based on two sensors (perturb and observe (P&O) technique, incremental conductance (IncCond), P&O with fuzzy logic…).

The simplest and cheapest MPPT control operates the PV array at a constant voltage [3] equal to its standard test conditions (STC) MPP voltage provided by the manufacturer. This approach ignores the effects of solar radiation and temperature changes on the MPP voltage. The utilization efficiency can be further improved, at a small increase in implementation cost, by employing a hill-climbing MPPT technique. These are simple algorithms that do not require previous knowledge of the PV generator characteristics or the measurement of solar intensity and cell temperature.

The most commonly applied hill-climbing MPPT technique is the P&O algorithm. The P&O algorithm perturbs the operating point of the PV generator by increasing or decreasing a control parameter by a small amount and measuring the PV array output power before and after the perturbation. If the power increases, the algorithm continues to perturb the system in the same direction; otherwise, the system is perturbed in the opposite direction. The direction search of the MPP and the algorithm of this method are presented in Fig. 38.9a, b.

38.5 Model of PMDC Motor

Several types of AC and DC motors are available for PV pumping systems. The motor choice depends on many factors: requirements of the size, efficiency, price, reliability and availability (Fig. 38.10).

Among the different kinds of DC existing motors, the permanent magnet motors (PMDC) are the most commonly used in PV pumping systems. They provide a high torque at the start. Brushed motors are rarely used, because their brushes should be changed periodically, and pump must then be removed of the drilling for changing the brushes. Mathematical relationships that describe the model of a DC motor are expressed as follows [7]:

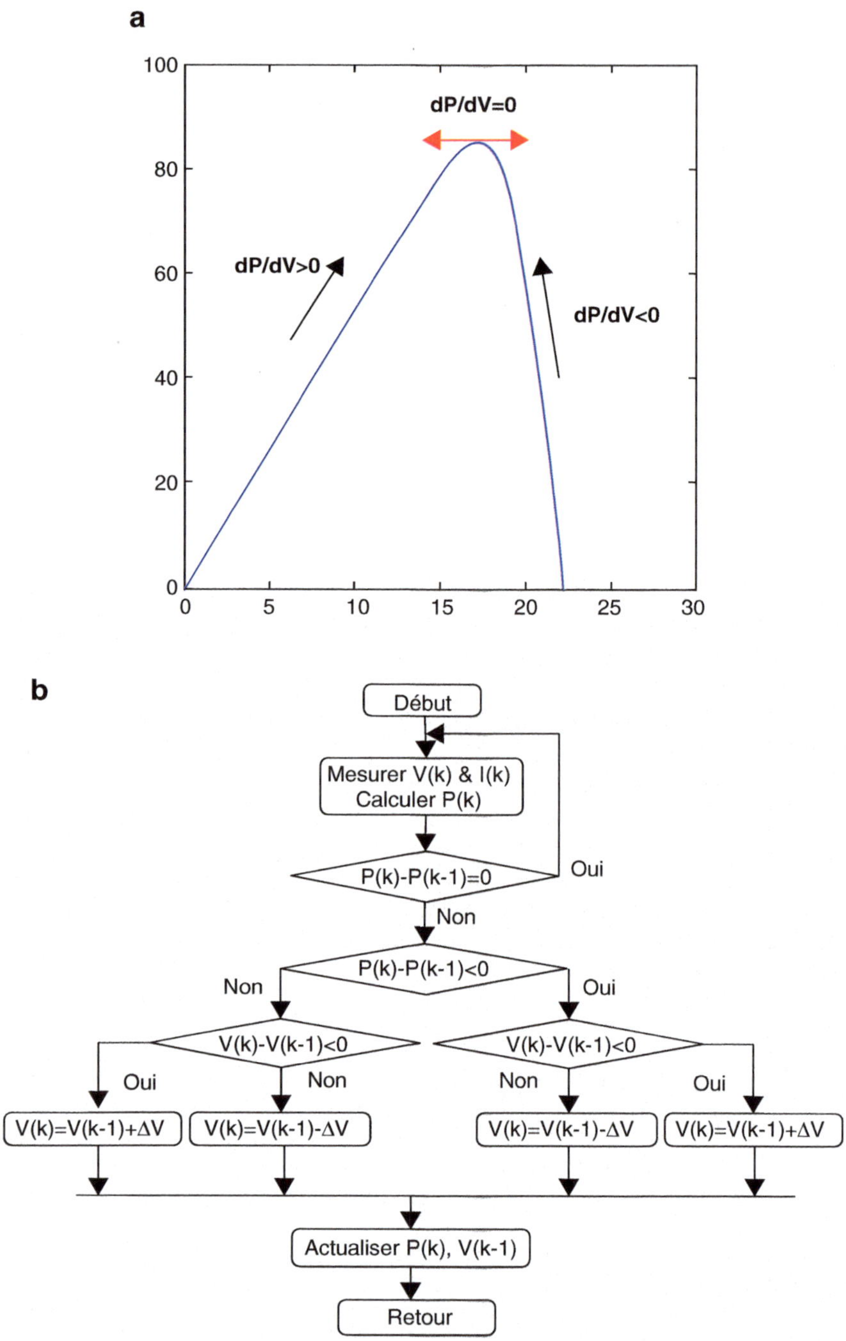

Fig. 38.9 (**a**) Direction search of the MPP. (**b**) Organizational chart of P&O method

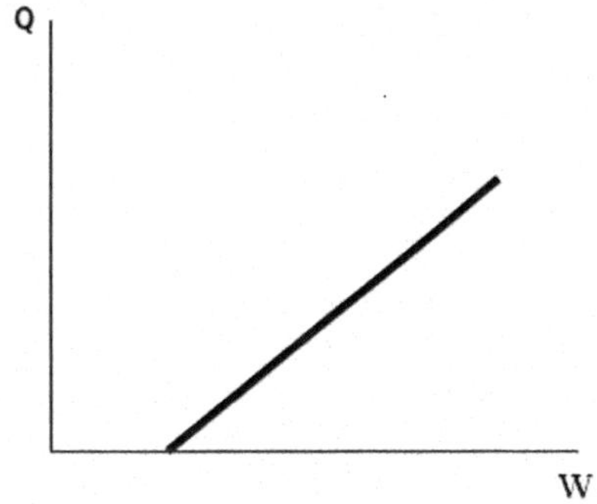

Fig. 38.10 Variation of water flow depending on rotation speed

$$U = \mathrm{RI} + E : \text{terminal voltage of the armature}$$

$$E = K \times \omega : \text{electromotive force}$$

$$Te = K' \times I : \text{electromagnetic torque}$$

where: K' is the constant torque, Φ is the magnetic flux through the turns, K is the constant of electromotive force, R armature resistance and ω is the angular speed of the rotor. The flow rate of the pump is proportional to the rotational speed of the motor:

38.6 Solar Pump

The choice of solar pump depends on the application and different water sources (wells, drilling, pumping river, etc.). In photovoltaic pump, the centrifugal and the volumetric pumps are the most used [8]. The centrifugal pump considered in this study applies a torque proportional to the square of the rotational speed of the motor:

$$Tr = Kc * \times \omega^2$$

where Kc is the proportionality constant [(Nm/rad $s^{-1})^2$] and ω is the rotational speed of the motor (rad s^{-1}).

Any pump is characterized by its output power, which is given by $P = Pu \times \eta$ with:

$$Pu = \rho \times g \times H \times Q$$

where η is the overall performance, ρ is the fluid density (kg/m^3), g is the gravity acceleration (m^2/s), H is the height of rise (m) and Q is the water flow (m^3/s).

The operating point of the motor pump is located at the intersection between the mechanical characteristics of useful torque of motor $Tu = f(\omega)$ and resistive torque of pump $Tr = f(\omega)$.

The flow rate of the pump is proportional to the rotational speed of the motor; however, at start we must have a minimum speed to obtain a flow [6].

38.7 Storage Tank

The methods and materials used to construct the tanks vary according to the local qualifications and materials available. The main materials used in the construction of storage tanks are:

- Ferro-cement
- Masonry or concrete blocks
- Concrete reinforced with bamboo
- Plastic
- Iron or galvanized sheet

Ferro-cement tanks are the most used. Their main advantage is that they consist of commonly available materials (cement, sand, water, wire). Their construction requires manual effort, so that users can help by using basic tools. They are usually cheaper than other types of tanks and their durability has been proven. Reports indicate frequently reservoirs of more than 25 years of use. Construction techniques require few materials, compared to traditional concrete tanks. The minimum thicknesses are 2 cm for a height of 1 m and 4 cm for a tank height of 2 m.

Finally, new tanks made entirely of plastic can be used for small applications. For example, a Ghanaian firm now produces a tank called "Polytank" with dimensions up to 5 m^3. In developed countries, larger plastic tanks are also used for the storage of water.

38.7.1 Distribution Networks

The construction of a distribution network that must bring the water from the reservoir directly to the users requires the services of a professional. Care must be taken to ensure adequate water pressure and circulation throughout the system. However, there are software packages that design of such systems. Networks connected to solar pumping systems are usually simple and small. Since 1985, the World Bank has developed software to assist in the design of simple distribution networks.

38.7.2 Irrigation Canals

For small irrigation systems, the irrigation canals are simply dug on the ground and covered with clay soil, on the ground. For larger irrigation systems, with main and secondary channels, the channels are often made of concrete to counter erosion and reduce losses due to infiltration.

38.8 Conclusion

The use of photovoltaic pumping has increased in the recent year, for irrigation and for personal uses. Therefore, it exists always the problem of efficiency and the beneficial structure. This paper shows a definition of the components of the PV pumping system and presents the MPPT techniques used to improve the efficiency of the system.

References

1. Muljadi E (1997) Pv water pumping with a peak-power tracker using a simple six-step square-wave inverter. IEEE Trans Ind Appl 33:714–721
2. Heng G, Zheng X, You-Chun L, Hui W (2005) A novel maximum power point tracking strategy for stand-alone solar pumping systems. In: Proceeding of the IEEE/PES conference and exhibition in transmission and distribution. IEEE, Piscataway
3. Vongmanee V (2002) The vector control inverter for a PV motor drive system implemented by a single chip DSP controller ADMC331. In: Proceeding of the IEEE. IEEE, Piscataway, pp 447–451
4. Chandrasekaran N, Thyagarajah K (2011) Modeling and MATLAB simulation of pumping system using PMDC motor powered by solar system. Eur J Sci Res 59(1):6–13
5. Learreta AB (2006) Réalisation de Commandes MPPT Numériques. LAAS-CNRS (France), Toulouse
6. El Ouariachi M, Mrabti T, Yaden MF (2011) Design and realization of a photovoltaic system equipped with the analogical and digital MPPT command for better exploitation of solar energy Colloque International Journées des Géosciences de l'Environnement. J Mater Environ Sci 2(S1):538–543
7. Femia N, Petrone G, Spagnuolo G, Vitelli M (2005) Optimization of perturb and observe maximum power point tracking method. IEEE Trans Power Electron 20(4):963–973
8. Elgendy MA, Zahawi B, Atkinson DJ (2012) Assessment of perturb and observe MPPT algorithm implementation techniques for PV pumping applications. IEEE Trans Sustain Energy 3(1):21–33

Chapter 39
The Impact of Dust's Physical Properties on Photovoltaic Modules Outcomes

Hussein A. Kazem, Miqdam T. Chaichan, and Ali H. A. Alwaeli

39.1 Introduction

Photovoltaic systems are considered lately as a solution to reduce the air pollution and the greenhouse effect [1]. Besides, using these systems can reduce the dependence on fossil fuels especially oil, which will protect the oil-importing countries from its prices fluctuations [2]. Although the PV systems have low electrical efficiency less 20%, it can generate electricity to high levels as megawatts [3]. Many researchers have studied the possibility of using PV cells to operate clinics in remote areas [4], pumping systems [5, 6], communication towers [7], and street lighting [8]. PV systems suffer from low efficiency due to many variables that affect the PV technology, such as geographic location and its topography, cell orientation, cell technology, and environment conditions [9]. These factors include temperature [10], humidity [11, 12], wind speed and direction [13], intensity of solar radiation [14], accumulation of dust and dirt on the cell [11]. Several approaches have been used to investigate dust in the last decades [15]. The question of physical properties (such as size, shape, morphology, and composition) [16, 17], chemical (e.g. dust type and its constituent elements) has been discussed for dust in many studies [18]. Many researchers have investigated all the variables that affect PV cells and their relationship to the dust accumulation and their mutual influence on the productivity of the system [19–21].

H. A. Kazem (✉)
Sohar University, Sohar, Oman

Solar Energy Research Institute, National University of Malaysia, Bangi, Malaysia

M. T. Chaichan
Energy and Renewable Energies Technology Center, University of Technology, Baghdad, Iraq

A. H. A. Alwaeli
Solar Energy Research Institute, National University of Malaysia, Bangi, Malaysia

A. Sayigh (ed.), *Renewable Energy and Sustainable Buildings*, Innovative Renewable Energy, https://doi.org/10.1007/978-3-030-18488-9_39

Studies of Refs. [22–24] are considered pioneers in the field of testing and testing the effect of dust accumulation on the performance of solar cells. Sayegh investigated the effect of accumulated dust on solar cells in Kuwait and studied the effect of the tilt angle on the accumulation of this dust. Shubakshi has done a numerical simulation of accumulation of artificial dust such as cement, carbon particles, and limestone with diameters of 10, 50–80 μm, respectively.

Reference [25] experimentally studied the low efficiency of a photovoltaic system due to the accumulation of dust on the surface of the panels. Experiments were performed using different types of PV techniques, using solar simulations within the laboratory. The results obtained confirmed that the accumulation of dust affected the solar cell output voltage. The very high or very low density of the solar radiation intensity caused a significant reduction in the efficiency of the PV unit. The change in cell technology causes dust deposition effects to be different. The study observed a faster decrease in the performance of the polycrystalline silicon coated with epoxy PV module by dust accumulation than the other studied techniques.

Reference [26] studied the impact of human activities on the desertification dilemma in Iraq that led to increase the dust storms in this country. The study evaluated the specifications, causes, and types of accumulated dust. The analyses of the impact of physical and chemical aspects for the Iraqi dust on the PV systems were measured. The study conclusion revealed that PV systems outcomes are affected negatively by the dust deposition.

Reference [27] studied the outdoor conditions and their influence on air pollution and dust deposition on PV module. The study results clarified that the PV module outcomes deteriorated because of dust and traffic's air pollutants suspended in air. The exposure of the PV panels to outdoor conditions caused a reduction of 12% in the generated power, when the panels were exposed to outdoor conditions for 2 months without cleaning. At the same time, the panels naturally cleaned by rain and wind lost about 8% of its performance compared to the clean panel.

Reference [28] examined the reasons for the deterioration of the PV system's performance due to the accumulation of dust and the differences that result from changing the type of the solar collector, geographical location, local climate, and exposure time to external conditions with no manual cleaning. The researchers also studied the advantages of cleaning the PV systems using manual, natural, passive, and automatic methods.

The study aims to investigate the effect of the physical properties of the dust particles accumulated on the solar cell that were collected from variable locations in northern Oman. The study will examine the effect of the size and engineering form of the accumulated dust on the performance of photovoltaic systems. The effect of dust accumulation on experimental photovoltaic output will also be assessed inside the laboratory.

39.2 Studied Region

Oman is in south-west Asia and in the south-eastern corner of the Arabian Peninsula. Oman has a long coastline of 3165 km from Ras Darbat Ali near the southern Yemeni border up to the Strait of Hormuz in the north (Ministry of Tourism). Figure 39.1 shows a map of the Sultanate of Oman [29]. The population

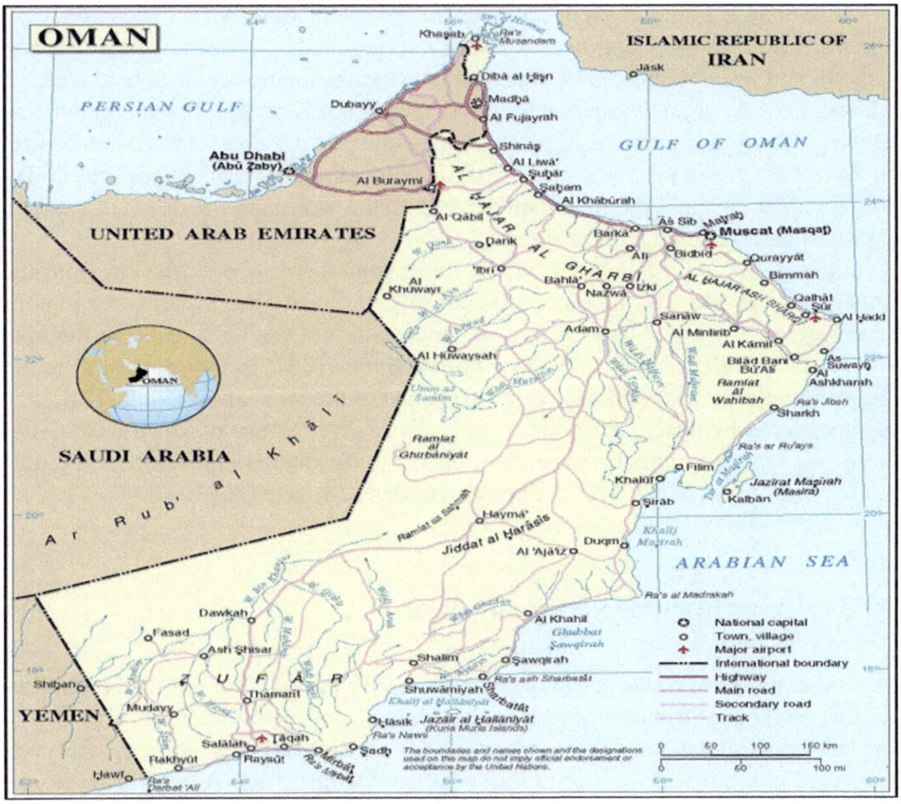

Fig. 39.1 Map of Sultanate of Oman [29]

of Oman, in 2010, was around 3,174,000, including 1,156,000 expatriates. As for housing units in this country, it reaches a total number of approximately 540,770. Individual house and apartments in buildings are included in this number (statistical yearbook). Oman's area is 309,500 km^2, which made its population density at 2010 about 9.3 inhabitants per km^2. This country population annual growth rate is approximately 4.4% [30]. The Omani government did its best to able most of the Omani population to have access to electricity. However, till today, in some parts (rural and remote areas) of the country, the people there have difficulties to access to the electric grid [31].

Oman is in the Sunbelt area making it perfectly suited to this type of solar application. Its geographic location allows it to receive higher solar energy than many regions of the world. Oman has a clear sky (about 342 days/year) [32]. In this country, there are many rural and remote areas characterized by being desert areas where sand and dust storms occur continuously most of the year. The deposition of dust on PV panels results in a deterioration in electricity production [33].

Al Batinah is home to about 55% of Oman's population [34] and consists of 12 administrative districts and two of Oman's largest industrial zones. In this study, volatile and accumulating dust was used for the following areas: Sohar, Wadi Al Mawal, Loy, Awabi, Rustaq, Nakhal, Barka, Suwaiq, Khaboura, Mina, Siham, and Shinas. The studied area is a coastal area extending from a coastal maid in the north to Ras Al-Hamra in the south. The Al Batinah location makes it between the Oman Sea from the east and the mountainous area to the west, and the distance between the coast and the mountains is about 25 km.

Sohar is the main industrial hub in the region and is about 230 km from the Omani capital, Muscat, the second largest city in Oman after Muscat. For regions receiving the highest solar radiation in Oman, Marmol can be considered the first, followed by Fahd, Sohar, and then Giron Hareti [35]. Heavy dust storms are frequent in the interior because of the protection of the mountains, although the area is surrounded by dusty desert areas. The physical properties of volatile dust vary from one region to another depending on the geological composition of the dust source in addition to the local climate conditions of course [36].

39.3 Experimental Setup and Results

Several experiments have been conducted to verify the effect of physical properties of dust on the performance of photovoltaic units. Air conditioning was provided in the laboratory to maintain a temperature of 25 °C, while relative humidity (RH) was measured at around 45%. Four tungsten halogen lamps, 1600 W, were used for the solar simulator to produce a maximum light intensity of around 850 W/m^2. Experiments were conducted in three main axes: The first part concentrated on collecting dust from variable locations in Oman; these locations included Sohar, Muscat, and Al-Buraimi. The samples collected from accumulations of 1 m^2 horizontal glass sheets and for a period of 3 months. The collection was done on weekly basis, using a small plastic container. After dust was collected, it was tested and evaluated for its impact on PV performance. The tests conducted on the samples aimed to study their physical properties, as described below.

39.3.1 Particles Specific Gravity

The standard BS 1377-2:1990:9.3 was used to test the physical properties of the dust [37]. Equation 39.1 evaluates the particle density, ρ_s, which is the mass ratio of a given volume of material to the mass of an equal volume of water:

$$\rho_s = \frac{m_2 - m_1}{(m_4 - m_1) - (m_3 - m_2)} \tag{39.1}$$

where m_1, m_2, m_3, and m_4 are the masses (in grams) of the gas jar and round plate; round plate and soil; round plate, soil, and water; and round plate and water, respectively.

A gas jar (1 L with a rubber cap) was used, a round glass plate used to cover the gas jar, a balance, a thermometer, and a drying oven capable of stabilizing the temperature between 105 and 110 °C in experiments to find the specific density of the dust particles.

39.3.2 Surface Properties and Particle Shape

The geometry of the dust particles was determined using the Mohs hardness test. In this test, the scratch resistance of the dust sample is compared with ten reference metals of previously determined hardness based on the Mohs hardness scale.

39.3.3 Moisture Content, Plastic Limit, and Liquid Limit

Dust is classified based on its water content to four basic classifications: solid, semi-solid, plastic, and liquid. The geometric characteristics, the consistency of the dust, and its behaviour vary depending on the classification within which it falls. The limits of these classifications are defined by changes in the dust behaviour. These classifications can determine the quality of the dust and its impact on the area surrounding the accumulation area. The dust expands when wet with water and keeps a small amount of water. The ability of the dust particle to absorb water and expand the size depends on the type of atoms forming it.

The liquid limit represents the highest moisture content followed by the plastic dust and then the semi-solid dust. Finally, the steel is known as the difference between the liquid limit and the plastic limit as the plasticity index. This test is used for all samples. The tools used in checking liquid limit are the porcelain dish, flat slitting tool with scale, glass plate, balance, spoon, bottle filled with distilled water, and drying oven set at 105 °C. Moisture content is expressed by the mass of water to the mass of solids in the sample.

$$w = \frac{M_W}{M_S} \times 100\% \tag{39.2}$$

where w, M_W, and M_S are the moisture content, mass of water, and mass of solids in a soil sample, respectively.

39.3.4 The Analysis of the Grain Size

The grain size of soil is checked using sieve and hydrometer scale tests. Table 39.1 shows sieve measurements used for grain size analysis. A sieve vibrator was used to distribute the grain regularly. Also, the following devices were used: a balance, a set of sieves, and brush cleaning. The hydrometer test was used to check grains sizes less than 63 μm, as most of the dust particles in air in the desert and arid regions are less than 63 μm. The criterion was followed using 1999: 13.3–1990: 9.3 standard for grain size analysis with sieves and hydrometer scale [37].

In the second part of the practical experiments, the dust was distributed homogeneously on the surface of the photoelectric panel to estimate the effect of accumulating different types of total dust present on the performance of the unit. In these experiments, one photovoltaic panel was used.

The tests showed that the dust accumulates and aggregates more when moisture content is high. The greatest challenge in practical experimentation was the redispersion of dust to produce the same natural distribution. Ensure that the accumulated dust is not accumulated due to moisture content by using a vibrator for the dust contained in the plastic container. The vibrator used in the experiments was type Jamie Orbit Shaker Oven −80 °C with a vibration speed of 10–400 rpm. In the second part of experiments, the natural conditions are re-represented by redistributing the dust in a manner like the natural accumulation on the PV panels by placing a vibrator under the back of a vibrating plate for several minutes to ensure a regular distribution of the dust. The PV panel was placed horizontally, and the dust particles were deposited over it entirely. Table 39.2 shows the specifications of the PV module used in the tests. The outdoor experiments were conducted in the roof of solar cells and photovoltaic laboratory at Sohar University, where the PV module was mounted and connected by a cable to monitor its electrical performance.

The PV module tilt angle was set to 0 degree. The accumulated dust that was collected through a period of 3 months was tested using X-ray diffraction (XRD) (Bruker D8 Advance) and X-ray fluorescence (XRF) (Horiba XGT-7200) to analyse the dust contents materials. Figure 39.2 displays the XRF results. Furthermore,

Table 39.1 The used sieves sizes

Serial no.	Sieve size (holes diameter) (μm)	Sieve no.
1	3350	6
2	2360	8
3	2000	10
4	600	30
5	425	40
6	300	50
7	212	70
8	150	100
9	75	200
10	63	230

Table 39.2 The specification of PV module

Maximum power (P_{max})	125 W
Open circuit voltage (V_{oc})	21.8 V
Short circuit current (I_{sc})	7.45 V
Max. power voltage (V_{mp})	17.2 V
Max. power current (I_{mp})	7.27 A
Cell temperature (T_c)	25 °C
Irradiation (G_a)	1000 W/m^2
PV length	1.43 m
PV width	0.63 m
PV area	0.9 m^2

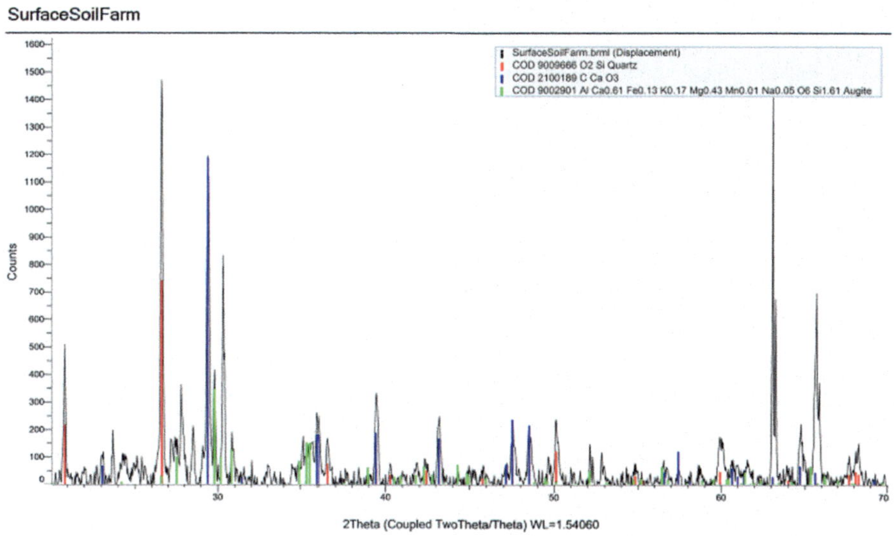

Fig. 39.2 XRD patterns) of collected dust in Sohar City

Fig. 39.3 shows a 50× zoom microscope was employed to investigate the microscopic characteristics of Sohar dust. This is to determine the particle size of the dust.

39.4 Analysis and Discussion

39.4.1 Physical Properties

The XRD patterns of the examined dust, as shown in Fig. 39.2, contain the following components: Al_2O_3, SiO_3, SiO_2, MgO, K_2O, CaO, Cr_2O_3, Fe_2O_3, TiO_2, MnO_2, NiO, SrO, P_2O_3, and Cl. The main contributors are quartz silicates (SiO_2) and calcium oxide (CaO), with total dust proportions of 55.79% and 30%, respectively. Therefore, the physical properties of these two pollutants are ought to be

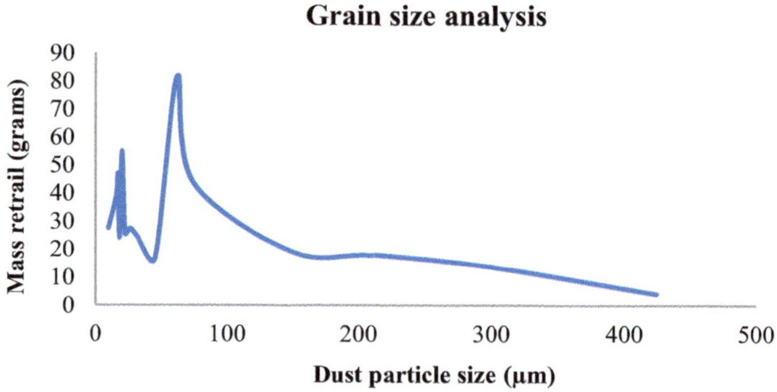

Fig. 39.3 The values of dust grain size analysis for Sohar City

Table 39.3 Values of specific gravity for dust in the studied sites

No.	No. of gas jar	m_1	m_4	m_2	m_3	$\rho_s = (m_2 - m_1)/$ $(m_4 - m_1) - (m_3 - m_2)$
Sohar						
1	545	31.220	80.896	56.021	94.041	2.13
2	315	17.969	42.672	28.720	49.041	2.45
3	48	16.534	41.059	26.607	46.242	2.06
Al-Buraimi						
1	304	16.978	42.190	28.910	47.054	1.68
2	48	16.534	41.059	26.395	46.077	2.03
3	45	16.025	40.751	26.485	44.751	1.62
Muscat						
1	304	16.978	42.190	19.633	42.531	1.14
2	48	16.534	41.059	18.083	41.269	1.15
3	45	16.025	40.751	17.437	41.270	1.58

investigated, such as their specific gravity, moisture content, geometrical shape, and grain size. The complete cleaning of the PV panel used in dust testing is essential in ensuring proper and right measurements of the PV cell outcomes.

Water was not used in the cleaning process to avoid any cell efficiency losses based on the conclusions in Ref. [27]. However, sodium surfactant and alcohol were used as cleaning materials.

The density of a substance divided by the density of water is referred to as the specific gravity of a substance. The specific gravity of the metals that form the dust particles mainly affects the specific gravity of the dust particle, as it refers to particle porosity. It is useful to know the constituent metal of dust to determine its specific gravity. The accumulation of dust on PV panels has a significant impact on unit performance because it reduces the incoming solar radiation and thus reduces the electricity produced. Table 39.3 shows the specific gravity of the dust samples studied. The specific gravity values of the dust collected in Sohar, Muscat, and Berymi were 2.20, 1.29, and 1.70, respectively. The collected dust from Sohar has the highest

Table 39.4 A summary of the results of plastic and liquid limit tests for the collected dust samples

City	Moisture content (%)	Plasticity index (PI)	Type of plasticity
Sohar	45	6.25	High
Al-Buraimi	19	2.623	Low
Muscat	35	5.25	Moderate

specific gravity compared to the dust of the rest regions, which has a relatively low weight. Dust from these cities is easier to be removed by wind or with cleaning, and generally has less influence on PV performance as opposed to dust from Sohar.

The results of plastic and liquid limit tests for the studied locations are shown in Table 39.4. The range of water content, over which the soil is plastic, is described by the moisture content, plastic limit and liquid limit, and plasticity index tests. The range of the moisture content above which dust has entered the plastic state is indicated by the plastic limit. Therefore, the plasticity index shows the plastic state of dust.

The moisture content in Muscat was 35.0% and for Al-Buraimi 19.0%, which can be considered low to moderate. As for the dust collected from Sohar City, the moisture content was high up to 45.0%, which makes the cleaning of the PV panels relatively difficult. The results of the plasticity index (PI) in Eq. (39.3) show that the Sohar City PI index is high (6.25) compared to the low PI index in Al-Buraimi City (2.623).

$$\text{Plasticity index} = \text{liquid limit} - \text{plastic limit} = \text{LL} - \text{PL} \tag{39.3}$$

In engineering, hardness is commonly defined as the resistance of a material to penetration. The resistance of an aggregate is simply the hardness of individual grains that are cemented or interlocked. Therefore, low degree of hardness is associated with soft aggregate particles. The weaker the particles, the poorer they cement. Due to low dust accumulation issues, weaker dust particles impose lesser influence on large-scale PV power plants. Based on the results of the hardness and grain analysis tests, the hardness of all dust measured, regardless of their type, is less than one; hence, the dust is poorly graded. This means that collected dust can be characterized as smooth, very fine, and not hard. In addition, due to low effect of dust accumulation, it cannot cause harmful cracks when accumulated on PV modules.

The overall physical properties of the tested dust samples are shown in Table 39.5. The obtained results show large moisture content in dust from Sohar, consequently having a large negative impact on performance of PV and cleaning methods.

The high moisture content with high plasticity index makes dust bond on the surface of the panels, making the cleaning process difficult, and requires greater amounts of water. The results of the tests show that the Al-Buraimi City is the most suitable area for the use of PV panels due to the low impact of its dust as it has low moisture content, plasticity index, and gravity. The effect of the PV panels in the Al-Buraimi was less than that of Muscat. The impact of the accumulation of dust in these three cities was low when compared to the impact of the dust of Sohar. Dust collected from Muscat can be considered having characteristics located midway between the cities of Sohar and Buraimi.

Table 39.5 Summary of the properties of all types of dust

City	Moisture content (%)	Plasticity index (PI)	Specific gravity (Gs)	Type of plasticity
Sohar	45	6.25	1.1	High
Al-Buraimi	19	2.623	1.7	Low
Muscat	35	5.25	1.29	Moderate

Table 39.6 The impact of adding 100 g and 200 g from the collected dust on its resulted performance

City	Voltage (V)		Ampere (A)		Power (W)	
Dust weight	100 g	200 g	100 g	200 g	100 g	200 g
No dust	20	20	5.3	5.3	106	106
Sohar	15.9	15.1	4.35	4.15	69.165	62.66
Masqat	17.45	16.85	4.35	4.2	75.9	70.77
Al-Buraimi	18	17.65	4.9	4.7	88.2	77.66

39.4.2 Electrical Properties

From the second set of tests, it can be observed that voltage, current, and output power has reduced due to dust accumulation. As the dust particles were deposited on the surface of PV module at a controlled surface mass density, measurements of voltage, current, and power output were taken. The quantities of distributed dust were 50, 100, 150, and 200 g. The impact of distributing dust on performance of PV module with quantities of 100 g and 200 g is shown in Table 39.6. The power of the PV module is at 106 W, when the module is clean. This value decreased, when 200 g of dust was accumulated on the surface, to 77.6 W.

This power reduction is significant as it approximately reached 26.79%. The results of the test agree with the findings of Ref. [25] where dust accumulation highly influenced I_{sc}. However, the second result does not coincide with results of this study, which states that impact of dust on V_{oc} is insignificant. This difference in conclusion can be attributed to differences in the size of the used PV module; Ref. [25] implemented a module with an area of 0.015625 m^2, whereas the area of the module utilized in the current study is 0.9 m^2. Moreover, there are differences in the physical and chemical properties and weight of the dust used in these studies.

The performance of PV modules is greatly affected by the increase of accumulated dust. In all cases, the increase of dust accumulation led to continued decrease in the output of the PV module. The largest difference in output between a clean PV module "no dust" and a dirty one "with dust" was recorded for Sohar, while the minimum difference was recorded for dust in Al-Buraimi. All of the other dust affected the PV performance to varying degrees.

39.5 Conclusions

Dust accumulation causes a decrease in the PV modules' outcomes. An experimental investigation was carried out to study the impact of variable physical properties of dust on the PV panel. These properties included specific gravity, moisture content, hardness, and particle grain size. An accumulated dust from three Omani cities were collected and analysed.

The results of the study showed that the accumulated dust on the photovoltaic cell physical properties differs from one region to another. The dust of Sohar has a high moisture content compared to the other cities. So the dust of Sohar reduces the outcome of photovoltaic units. The dust from Al-Buraimi is characterized as having low moisture content, plasticity index, and specific gravity.

References

1. Al-Maamary HMS, Kazem HA, Chaichan MT (2017) Climate change: the game changer in the GCC region. Renew Sust Energ Rev 76:555–576. https://doi.org/10.1016/j.rser.2017.03.048
2. Al-Maamary HMS, Kazem HA, Chaichan MT (2017) The impact of the oil price fluctuations on common renewable energies in GCC countries. Renew Sust Energ Rev 75:989–1007
3. Al-Waeli AHA, Kazem HA, Sopian K, Chaichan MT (2017) Techno-economical assessment of grid connected PV/T using nanoparticles and water as base-fluid systems in Malaysia. Int J Sust Energy 37:558–575. https://doi.org/10.1080/14786451.2017.1323900
4. Kazem HA, Ali SQ, Alwaeli AHA, Mani K, Chaichan MT (2013) Life-cycle cost analysis and optimization of health clinic PV system for a rural area in Oman. Proceedings of the World Congress on Engineering 2013, vol. II, WCE 2013, London, UK, July 3–5
5. Al-Waeli AHA, Al-Mamari ASA, Al-Kabi AHK, Chaichan MT, Kazem HA (2016) Evaluation of the economic and environmental aspects of using photovoltaic water pumping system. In: 9th International Conference on Robotic, Vision, Signal Processing & Power Applications. Springer, New York
6. Kazem HA, Al-Waeli AHA, Chaichan MT, Al-Mamari AS, Al-Kabi AH (2017) Design, measurement and evaluation of photovoltaic pumping system for rural areas in Oman. Environ Dev Sustain 19:1041–1053. https://doi.org/10.1007/s10668-016-9773-z
7. Chaichan MT, Kazem HA, Mahdy AMJ, Al-Waeely A (2016) Optimization of hybrid solar PV/diesel system for powering telecommunication tower. Int J Eng Sci Emerg Technol 8(6):1–10
8. Chaichan MT, Kazem HA, Mahdy AMJ, Al-Waeely A (2016) Optimal sizing of a hybrid system of renewable energy for lighting street in Salalah-Oman using Homer software. Int J Sci Eng Appl Sci 2(5):157–164
9. Kazem HA, Chaichan MT (2016) The impact of using solar colored filters to cover the PV panel on its outcomes. Bull J 2(7):464–469. https://doi.org/10.21276/sb.2016.2.7.5
10. Al-Waeli AHA, Sopian K, Kazem HA, Chaichan MT (2017) PV/T (photovoltaic/thermal): status and future prospects. Renew Sust Energ Rev 77:109–130
11. Kazem HA, Chaichan MT (2015) Effect of humidity on photovoltaic performance based on experimental study. Int J Appl Eng Res 10(23):43572–43577
12. Kazem HA, Chaichan MT, Al-Shezawi IM, Al-Saidi HS, Al-Rubkhi HS, Al-Sinani JK, Al-Waeli AHA (2012) Effect of humidity on the PV performance in Oman. Asian Trans Eng 2(4):29–32
13. Kazem HA, Chaichan MT (2017) Wind resource assessment for nine locations in Oman. Int J Comput Appl Sci 3(1):185–191

14. Chaichan MT, Kazem HA (2016) Experimental analysis of solar intensity on photovoltaic in hot and humid weather conditions. Int J Sci Eng Res 7(3):91–96
15. Al-Hasan AY (1998) A new correlation for direct beam solar radiation received by photovoltaic panel with sand dust accumulated on its surface. Sol Energy 63(5):323–333
16. Kazem HA, Chaichan MT, Saif SA, Dawood A, Salim SA, Rashid A, Alwaeli A (2015) Experimental investigation of dust type effect on photovoltaic systems in north region, Oman. Int J Sci Eng Res 6(7):293–298
17. Kazem HA, Chaichan MT (2016) Experimental effect of dust physical properties on photovoltaic module in northern Oman. Sol Energy 139:68–80. https://doi.org/10.1016/j.solener.2016.09.019
18. Sueto T, Ota Y, Nishioka K (2013) Suppression of dust adhesion on a concentrator photovoltaic module using an anti-soiling photo catalytic coating. Sol Energy 97:414–417
19. Asl-Soleimani E, Farhangi S, Zabihi MS (2001) The effect of tilt angle, air pollution on performance of photovoltaic systems in Tehran. Renew Energy 24(3):459–468
20. Darwish ZA, Kazem HA, Sopian K, Alghoul MA, Chaichan MT (2013) Impact of some environmental variables with dust on solar photovoltaic (PV) performance: review and research status. Int J Energy Environ 7(4):152–159
21. Khatib T, Kazem H, Sopian K, Buttinger F, Elmenreich W, Albusaidi AS (2013) Effect of dust deposition on the performance of multi-crystalline photovoltaic modules based on experimental measurements. Int J Rene Energy Res 3(4):850–853
22. Sayigh A, Al-Jandal S, Ahmed H (1985) Dust effect on solar flat surfaces devices in Kuwait. In: Proceedings of the workshop on the physics of non-conventional energy sources and materials science for energy, pp. 353–367. ICTP Triest, Itally
23. El-Shobokshy MS, Hussein FM (1993) Degradation of photovoltaic cell performance due to dust deposition on to its surface. Renew Energy 3(6–7):585–590
24. El-Shobokshy MS, Hussein FM (1993) Effect of dust with different physical properties on the performance of photovoltaic cells. Sol Energy 51(6):505–511
25. Jiang H, Lu L, Sun K (2011) Experimental investigation of the impact of airborne dust deposition on the performance of solar photovoltaic (PV) modules. Atmos Environ 45(25):4299–4304
26. Kazem A, Chaichan MT, Kazem HA (2014) Dust effect on photovoltaic utilization in Iraq. Renew Sust Energ Rev 37:734–749
27. Chaichan MT, Mohammed BA, Kazem HA (2015) Effect of pollution and cleaning on photovoltaic performance based on experimental study. Int J Sci Eng Res 6(4):594–601
28. Chaichan MT, Kazem HA (2017) Effect of sand, ash and soil on photovoltaic performance: an experimental study. Int J Sci Eng Sci 1(2):27–32
29. UN Cartographic Section (2004), Map of Oman. http://www.unhcr.org/refworld/docid/460a393b2.html. Accessed 10 Oct 2010
30. Ministry of National Economy (2010) Statistical Yearbook
31. Oman Power and Water Procurement Company (2009) 7-year statement (2010–2016). Oman Power and Water Procurement Company, Muscat
32. Kazem HA, Chaichan MT (2016) Design and analysis of standalone solar cells in the desert of Oman. J Sci Eng Res 3(4):62–72
33. Al-Waeli AHA, Kazem HA, Chaichan MT (2016) Review and design of a standalone PV system performance. Int J Comput Appl Sci 1(1):1–6
34. Ministry of National Economy. http://www.moneoman.gov.om. Accessed 9 Oct 2017
35. Al-Badi AH, Albadi MH, Al-Lawati AM, Malik AS (2011) Economic perspective of PV electricity in Oman. Energy 36(1):226–232
36. Ministry of Tourism. http://www.omantourism.gov.om. Accessed 12 Nov 2017
37. Biryukov S (1998) An experimental study of the dry deposition mechanism for airborne dust. J Aerosol Sci 29(1):129–139

Chapter 40
Performance Simulations of Different Energy Flexibility Sources in a Building with the Electrical Grid

Reino Ruusu, Sunliang Cao, and Ala Hasan

40.1 Introduction

Energy use in buildings represents a major share of consumed energy globally [1]. On the other hand, renewable energy technologies are emerging fast and spreading on the level of single buildings through on-site generation from wind and sun [2, 3]. Examples are photovoltaic panels (PV), solar thermal collectors and small wind turbines. The energy price of electricity is normally fluctuating between different hours of the day. With the use of electricity-based thermal generation (e.g. heat pumps and electric heating elements), together with on-site facilities of small-scale electrical and thermal energy storage, energy use in buildings can be shifted to times when it is favourable to import energy from the grid at a lower price and export at a higher price. This can be best accomplished by using an energy management system (EMS), which uses optimization methods and energy forecasting to find optimum interaction times and quantities of energy between the building's system and the electrical grid [4].

In this study, we present the results of the performance of a developed EMS for the maximization of the economic benefit of the interaction of a building, which is equipped with different means of renewable energy generations and storages, with the electrical grid. The focus of this study is to find the effect of different sizes of on-site energy storages in a battery and a hot-water storage tank on the building's energy cost.

R. Ruusu · A. Hasan (✉)
VTT Technical Research Centre of Finland Ltd., Espoo, Finland
e-mail: ala.hasan@vtt.fi

S. Cao
Department of Building Services Engineering, Faculty of Construction and Environment, The Hong Kong Polytechnic University, Kowloon, Hong Kong

A. Sayigh (ed.), *Renewable Energy and Sustainable Buildings*, Innovative Renewable Energy, https://doi.org/10.1007/978-3-030-18488-9_40

40.2 Case Description

A detailed simulation model is used for a single-family house including multiple on-site renewable energy (RE) generations and storages (Fig. 40.1). The building is a one-storey house (floor area 150 m^2) assumed located in Helsinki-Finland, constructed according to the Finnish Building Regulation D3-2012 [5] and used by four occupants. The RE generation is from PV panels (nominal 3.27 kW), small wind turbine (nominal 4 kW) and solar thermal collectors (8.6 m^2).

In the base case, the nominal capacity of the electric battery is 20 kWh with 60% cycle depth, i.e. 12 kWh of effective capacity, and the capacity of the hot-water storage tank (HWST) is 0.5 m^3. The tank is composed of two connected comportments, where a ground-source heat pump (4.5 kW, COP = 3) supplies heat to the lower compartment of the HWST while an electric heating element (6 kW, efficiency = 0.95) is located in the upper compartment. The lower comportment supplies water at a minimum temperature of 40 °C for space heating load, and the upper comportment supplies domestic hot water (DHW) at a minimum temperature of 60 °C for hygienic purposes (Fig. 40.2).

A developed energy management system (EMS) using an optimization-based model predictive control (MPC) method is utilized to optimize the building's interaction with the electric grid. The aim is to maximize the overall economic income from the energy operation. The optimization method uses successive linear programming (SLP) for continuous approximation of a non-linear discrete control problem. The building is modelled using TRNSYS [6] and a detailed simulation

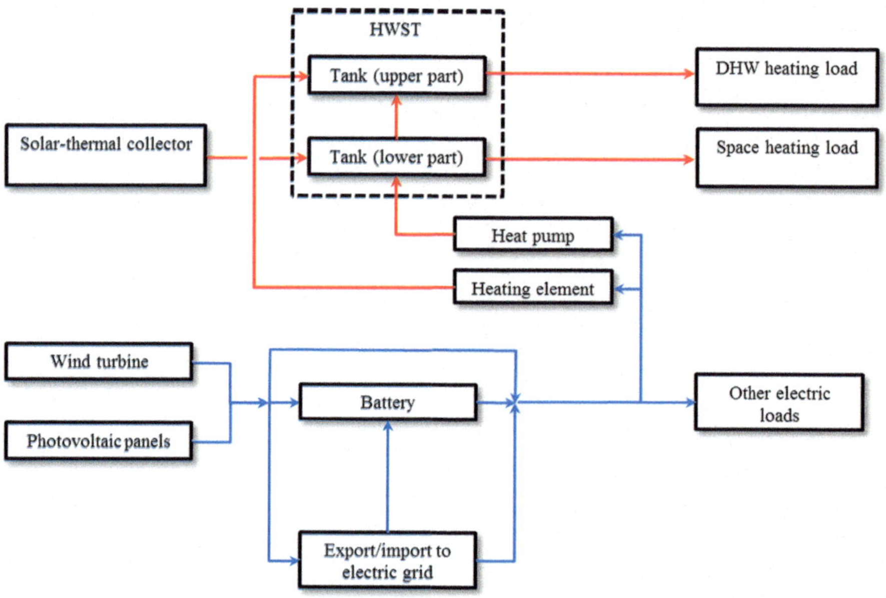

Fig. 40.1 Energy flow in the energy system

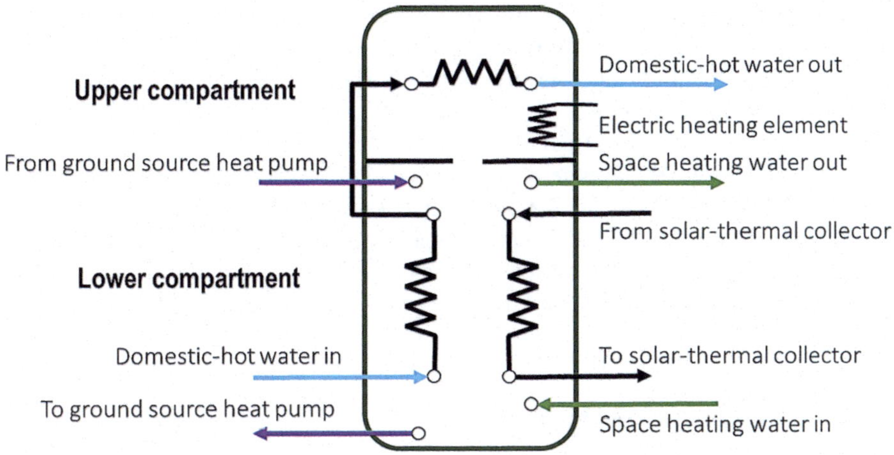

Fig. 40.2 Configuration and connections of the hot-water storage tank (HWST)

model written in MATLAB is used for the energy system components operation. The EMS makes full-year simulation evaluations based on 0.1 h time step. The simulation studies are performed using the weather for year 2015 Helsinki-Finland. The system utilizes prior information of electricity price from the Nord Pool energy market for that year.

Energy generation and demand forecast are assumed by two extreme methods: a perfect forecasting, which uses the real weather data for that year, and an assumption of repetition of the previous 24-h data, which is an example of a low-quality forecasting. Any other assumed forecasting method can be considered located somewhere in between these two assumptions.

The EMS uses the electricity price and the forecast of energy generation and demand at the next 24-h time window. Accordingly, it produces a plan for the optimum decisions for the energy system operation in those 24 h using the optimization method. The decisions for the current time step are implemented. The procedure is repeated in the next 0.1 h, which produces new updated decisions at the next shifted 24-h time window.

The method was used in detailed evaluation of the energy system performance. First results were published [7], and detailed results and a comprehensive analysis were then published [8], where connection to the district heating network was also included. For a lower tariff of the energy export price compared with the import price, the results indicate that for the past 24 h forecasting method, the imported and exported energy cost were 271.8 € and 115.2 €, respectively, and for the perfect forecasting, they were 216.2 € and 103.1 €. The used lower tariff for the energy export refers to only considering the spot market price, while the higher import price includes the transmission cost as well. This is a realistic situation in Finland since no feed-in tariff is implemented, which, for the studied sizes of the RE generation in the house, will result in a negative yearly income (i.e. cost of energy import is higher than the energy export).

The energy flexibility sources are represented by the operation of the heat pump and the electric heating element with respect to the existing loads and the available capacities in the battery and HWST, which will aim at load shifting and higher operational energy efficiency.

In the current study, the effect of changing the capacity of the battery and HWST on improving the yearly income is investigated. The comparison includes six nominal battery sizes (0, 10, 20, 30, 40, 50 kWh) and four HWST sizes (0.3, 0.5, 0.7 and 1 m^3) with the two forecasting methods—perfect forecasting and assumption of repetition of past 24 h. This makes 48 cases of full-year simulations of the building's energy system operation and interaction with the electricity grid.

40.3 Results and Discussion

In Figs. 40.3, 40.4 and 40.5, we can see that the electric battery size has an expected beneficial effect on the total yearly cost of electricity. This effect rapidly saturates after a capacity of about 30 kWh. Looking at the electricity import and export in Fig. 40.3, we can notice that the effect of the battery size is strongly related to improved matching of the on-site produced electric energy with the building's demand, as both imports and exports are reduced. It has a better effect on the total yearly income since a larger battery size reduces the total payment as shown in Fig. 40.4. The accuracy of forecasting (perfect forecasting of energy generation and demand versus forecasting based on the previous 24 h) has an effect on both imports and exports, but its impact is stronger on the import side.

One interesting feature is that when the battery capacity is sufficient, improvement of forecasting quality results in elimination of the effects of the HWST size on electricity imports, while electricity export effects are mostly unaffected (Fig. 40.3). This suggests that increasing HWST capacity provides some advantage in mitigating unintended electricity imports caused by forecasting errors, while electricity export dependency on HWST size is mostly unaffected by the forecasting quality. When looking at the electricity import and export figures individually (Fig. 40.3), we can see that the HWST size has a small effect in terms of energy matching, but looking at the total economic outcome (Fig. 40.4), we can notice that the effect is inverted (except when there is no battery at all), which is an unexpected result. This means that larger tank sizes result in a lower economic outcome compared to smaller tanks. This effect is strongest at the performance bound of perfect forecasting but still present when the low-quality forecasting is applied. The perfect forecasting shows better total yearly income than the other low-quality forecasting (Fig. 40.4). The negative values of the total yearly income in this figure refer to payments due to lower income from energy export compared to import payment.

In the studied hardware configuration, the HWST has a very limited role as a source of flexibility, which is caused by the fact that the domestic hot water (DHW) utilizes heat from both parts of the HWST and the limit of the heat pump supply temperature (<60 °C) does not overlap with the hygienic requirements of DHW

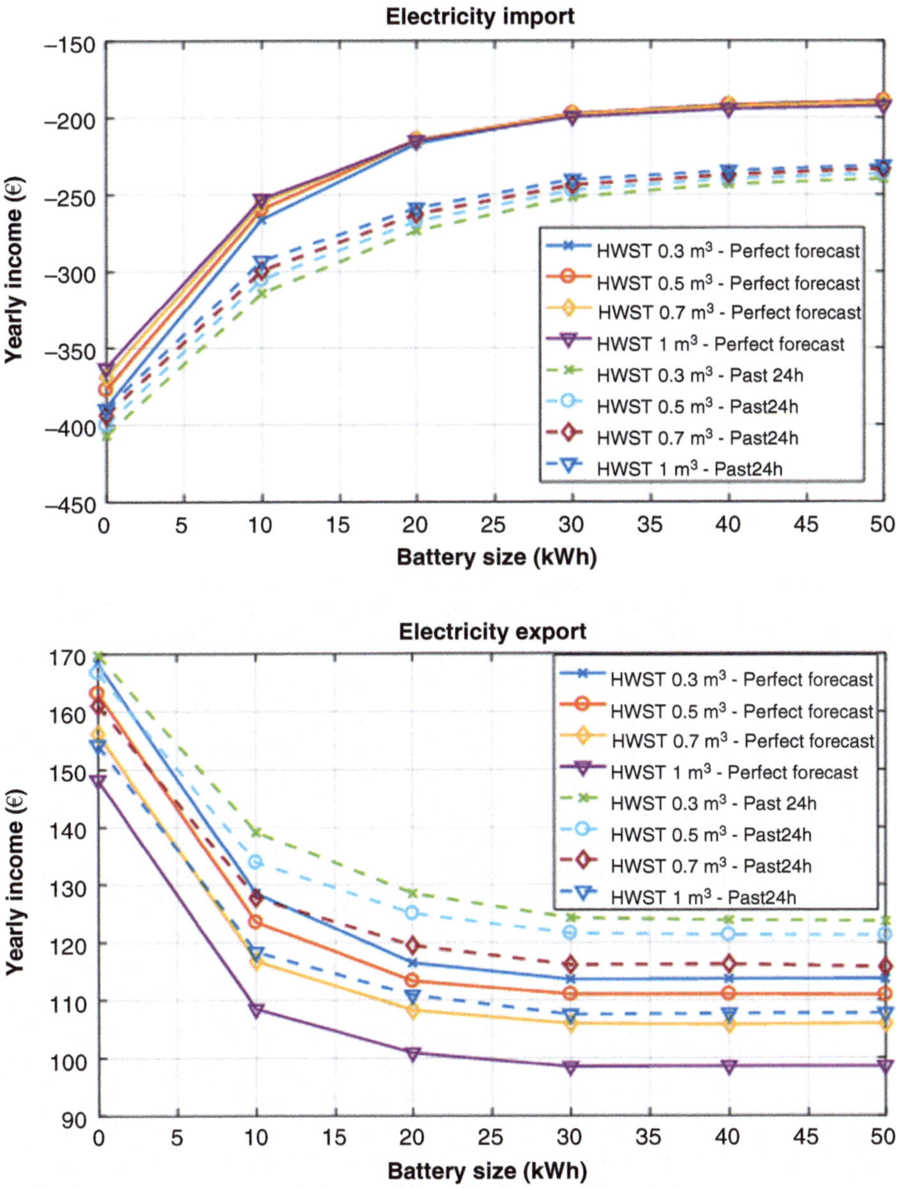

Fig. 40.3 Effect of storage size on electricity import and export costs

(≥60 °C). This means that in order to maximize the overall efficiency of the heating system, the lower part of the tank must be kept as warm as possible (i.e. more use of the heat pump) and the top of the tank be kept at the lower limit of the required temperature for DHW (i.e. lower use of the heating element). On the other hand, the solar heat collector is only connected to the lower part of the HWST, while the

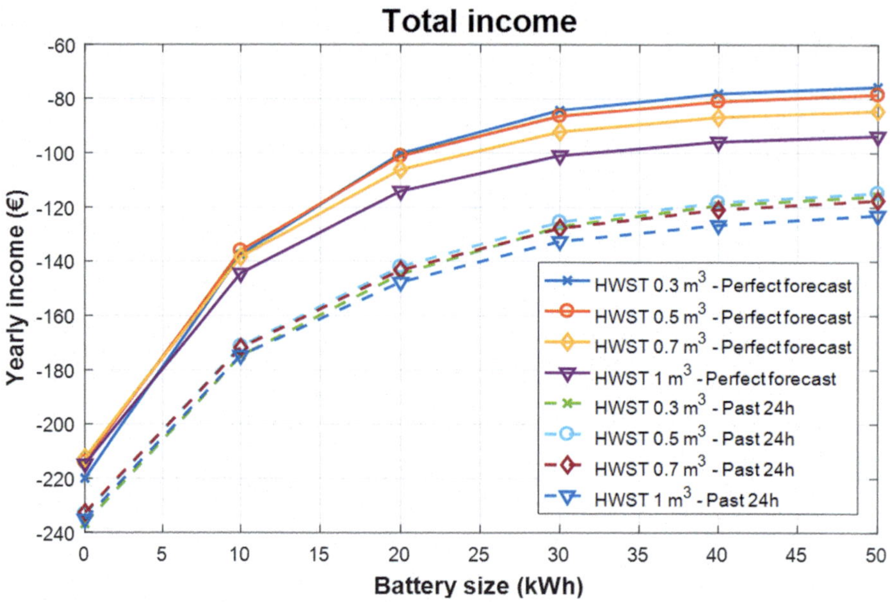

Fig. 40.4 Effect of storage size on the total yearly income

hygienic temperature limit is applied to the top part of the tank. This means that using the electric heating element is necessary even when output from the solar thermal collector is available, until the bottom part of the tank reaches the lower limit of 60 °C. However, in a smaller HWST, the heat from the solar thermal collector can propagate more quickly to the top of the tank, where it is most needed.

Figures 40.5 and 40.6 present the individual effect of the storage on the heating element and heat pump utilization and the average COP of the heating subsystem for the perfect forecasting method. The tank size has a strong adverse effect on the average electric efficiency of the heating system. The larger tank has a higher use of the electric heating element and a lower use of the heat pump. An additional effect arises from the fact that a larger heat tank will have a larger surface area, which results in higher passive heat losses, but this effect is not significant in comparison with the effect of the average COP.

Despite of its strong effect on the average COP of the heating subsystem, the HWST size seems to have a minor effect on the financial outcome when compared to the flexibility and matching potentials provided by the electric battery. This is explained by the fact that the electric battery can be used for responding to price signals by effectively shifting the overall energy demand of the system to times with more favourable prices. We can thus conclude that, in terms of the economic outcome, energy shifting has a more important effect compared to changes in the overall system efficiency.

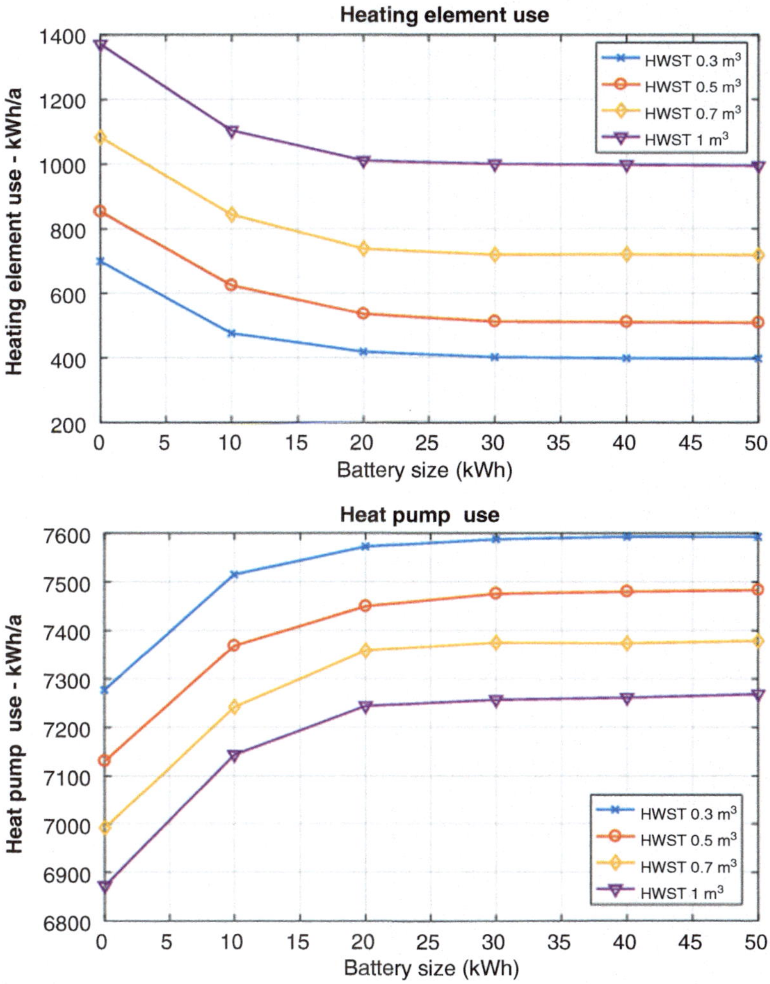

Fig. 40.5 Effect of storage size on the heating element and heat pump use with perfect forecasting

On a closer inspection, faster propagation of solar thermal collector heat to the top of the tank is not the only dynamic that explains the observed effects. Figure 40.7 presents an extreme example of a period from early October using perfect forecasting, when daytime production of the solar thermal collector is still strong on sunny days, but night-time space heating demand is still significant. A smaller tank (0.3 m^3) lets the bottom compartment to cool down more deeply during night, which allows the heat pump (highlighted in the lower figure) to be used for the heating demand. With a larger tank (1 m^3), the bottom compartment only rarely reaches the temperature limit for operating the heat pump, which only leaves the heating element as the means for utilizing the surplus electricity from the PV generation (highlighted in the

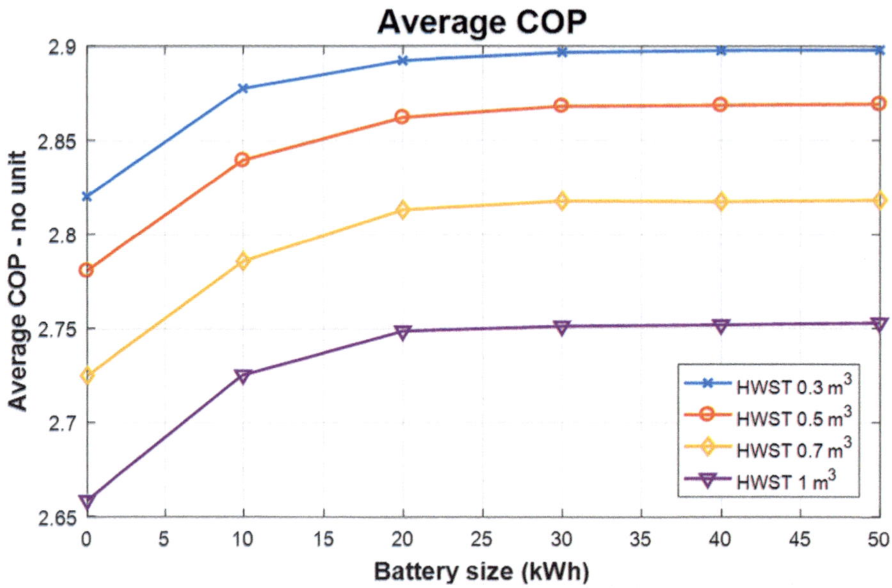

Fig. 40.6 Effect of storage size on the average COP of heating with perfect forecasting

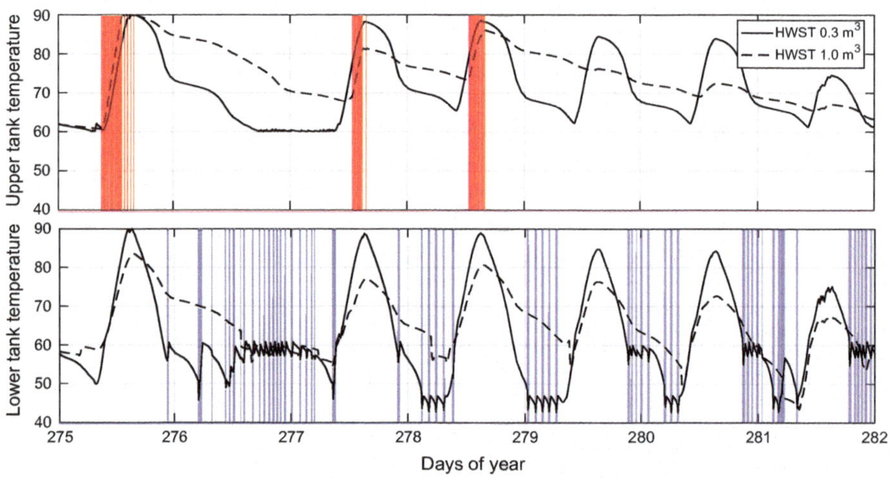

Fig. 40.7 Solar heat collector behaviour in two tanks with perfect forecasting (red highlight, electric heating element operation times for 1 m^3 tank; purple highlight, heat pump operation times for 0.3 m^3 tank)

upper figure). With a smaller tank, the solar thermal collector heats the entire tank to a higher temperature, and use of the electric heating element is entirely avoided. Even though the heat pump operates during night, the battery allows its indirect utilization of the daytime surplus electricity form the PV generation.

40.4 Conclusions

The developed EMS, which is an optimization-based model predictive controller, is shown to be effective in managing the flow of energy between the building's energy system and the electric grid. This is done by optimum utilization of the energy system storages for shifting the electricity and heat demand and best interaction with the grid.

In an energy system that exhibits strong interactions between different components, the effects of storage size on the system flexibility are hard to assess analytically, especially in the presence of a model predictive energy management system. In this case, price-sensitive shifting of energy consumption is more important than improvements in the overall efficiency of the system.

The complex interactions of the combinations of heat consumption and generation modes together with a model predictive energy management system produce an emergent behaviour, which becomes clear after some more detailed analysis of the system behaviour. This outcome is dependent on the exclusive temperature domains of the heat pump operation (<60 °C) and the hygienic requirements of DHW (≥60 °C) and the dual-purpose tank configuration for both space heating and DHW use, which result in limiting its usability as a source of flexibility.

In the studied system, it appears that the HWST flexibility is smaller than the electric battery flexibility and that the effect of the battery size saturates at a nominal capacity of 30 kWh, which is equivalent to 18 kWh effective capacity. In addition, it is found that increasing the HWST size has an adverse effect on the overall economic outcome. Larger tanks result in increased use of the electric heating element and reduced use of the heat pump, thus producing lower average COP for the whole system. In a smaller HWST, heat from the solar thermal collector can flow faster to the top of the tank, avoiding excessive use of the low COP heating element for DHW use. In addition, a larger tank results in increased heat losses due to the larger surface area. All of the above give preference to a smaller HWST.

Acknowledgements This work is a product of the project "Advanced Local Energy Matching in Future Smart Hybrid Networks 2014–2018", which is mainly funded by the Academy of Finland.

References

1. European Commission (2017) Energy efficiency—buildings, European Commission—Policies, Information and Services. https://ec.europa.eu/energy/en/topics/energy-efficiency/buildings . Accessed Feb 14 2018
2. Bajpai P, Dash V (2012) Hybrid renewable energy systems for power generation in stand-alone applications: a review. Renew Sust Energ Rev 16(5):2926–2939
3. Wirth H (2018) Recent facts about photovoltaics in Germany. Fraunhofer ISE, Freiburg, p 92. https://www.ise.fraunhofer.de/content/dam/ise/en/documents/publications/studies/recent-facts-about-photovoltaics-in-germany.pdf

4. Shivarama Krishna K, Sathish Kumar K (2015) A review on hybrid renewable energy systems. Renew Sust Energ Rev 52:907–916
5. Ministry of Environment, Department of the Built Environment. National Building Code of Finland, Part D3, Energy Efficiency in Buildings Regulations and Guidelines 2012. Finland
6. Duffy MJ (2017) A Transient Systems Simulation Program. TRNSYS simulation platform, Solar Energy Laboratory, University of Wisconsin-Madison, Madison. http://sel.me.wisc.edu/trnsys/index.html
7. Ruusu R, Cao S, Hasan A, Karhela T (2016) Developing an energy management system for optimizing the interaction of a residential building with the electrical and thermal grids. In: 12th REHVA World Congress CLIMA 2016. Aalborg University, Department of Civil Engineering, Aalborg
8. Ruusu R, Cao S, Delgado BM, Hasan A (2019) Direct quantification of multiple-source energy flexibility in a residential building using a new model predictive high-level controller. Energy Convers Manag 180:1109–1128

Chapter 41
Mathematical Modeling of Temperature Effect on Algal Growth for Biodiesel Application

S. M. Zakir Hossain, Nader Al-Bastaki, Abdulla Mohamed A. Alnoaimi, Husny Ezuber, Shaikh A. Razzak, and Mohammad M. Hossain

41.1 Introduction

The trend in recent years is replacing fossil fuels by liquid biofuels. First generation biofuels have achieved the required production rate economically. The energy source for this generation was food and oil crops including oil, sugar, maize, and animal fats. It is obvious that the generation and consumption of liquid biofuels will remain the same and might grow in the near future. In general, the utilization of first generation biofuels has formed several arguments, mainly because of their effect on international food markets and on food security, particularly with respects to the most vulnerable regions of the global economy. Relevant questions were appeared about their effectiveness and sustainability of their production. Second-generation biofuels produced fuels from the whole plant matter of dedicated energy crops or agricultural deposits, jungle gathering remains, or lumber processing waste, rather than from food crops. However, the production technologies for the second-generation biofuel in the most part has not met the scales for commercial use and not attracts with the market demand which has so far prevented any substantial exploitation [1]. For a technically and economically feasible fuel supply, some conditions must be satisfied: relatively cheap; should need low or no additional land usage; environmentally friendly and no contribution to air pollution; and should use minimal amount of water. Microalgae exploitation could achieve these requirements and therefore make an important influence to meeting the main energy

S. M. Z. Hossain · A. M. A. Alnoaimi · H. Ezuber
Department of Chemical Engineering, University of Bahrain, Zallaq, Kingdom of Bahrain

N. Al-Bastaki (✉)
Academic Affairs and Scientific Research, Kingdom University, Riffa, Kingdom of Bahrain

S. A. Razzak · M. M. Hossain
Department of Chemical Engineering, King Fahd University of Petroleum & Minerals, Dhahran, Saudi Arabia

A. Sayigh (ed.), *Renewable Energy and Sustainable Buildings*, Innovative Renewable Energy, https://doi.org/10.1007/978-3-030-18488-9_41

demand and, in the meanwhile, providing environmental assistances [2]. Microalgae are all unicellular and simple multicellular microbes, counting both prokaryotic microalgae, i.e., cyanobacteria (Chloroxybacteria), and eukaryotic microalgae, e.g., green algae (Chlorophyta), red algae (Rhodophyta), and diatoms (Bacillariophyta) [2–4]. Algae culture is of increasing value given that: (i) algae can be cultivated on non-agricultural land using wastewater, (ii) algae can provide a high yield on a per unit of light irradiated area, (iii) algae growth requires CO_2 and nutrients that can be obtained from wastewater and fossil fuel combustion, and (iv) algae contains high oil and starch making possible the production of high-quality biodiesel. Thus, algae culture can contribute to CO_2 fixation, wastewater treatment and can be a source of bioenergy [5–12].

Generally, temperature, nutrient (nitrogen and phosphorus) concentration, light intensity, initial CO_2 concentration, and initial inoculum size are the parameters that affect the microalgae growth rate. Among these, temperature and nitrogen concentration are considered the main parameters. It is well known that models of cell culture are useful tools to predict growth kinetics as well as better information in aquatic resource management [13–15]. To our knowledge there is no mathematical model for *Chlorella kessleri* (one of the green microalgae sp.) available in literature. Therefore, this paper develops a mathematical model of *Chlorella kessleri* growth at different temperature and nitrogen concentration. For this, both Monod model and Arrhenius equation were applied. The development of this model could open the opportunity to propose models of auto-mated bioprocess control and optimize productivity at large scale, which in turn is helpful for biodiesel production.

41.2 Materials and Methods

41.2.1 Chemical and Reagents

All the solvents and chemicals used were purchased from Sigma-Aldrich, unless specified otherwise.

41.2.2 Algae Species and Growth Conditions

The microalgae species (*Chlorella kessleri*) were obtained from the University of Texas (UTEX) culture collection, USA. The media used for the algae growth is known as Bold's Basal Medium (BBM). The stock solutions for each component of BBM were prepared according to the recipe listed in the Table 41.4 of Appendix 41.1. A total of 64 mL of concentrated stock solution was mixed with 936 mL of deionized water to make 1 L BBM. The initial pH of the medium was adjusted to 7 by 0.1 M NaOH. Algae species were cultured in batch photobioreactors with

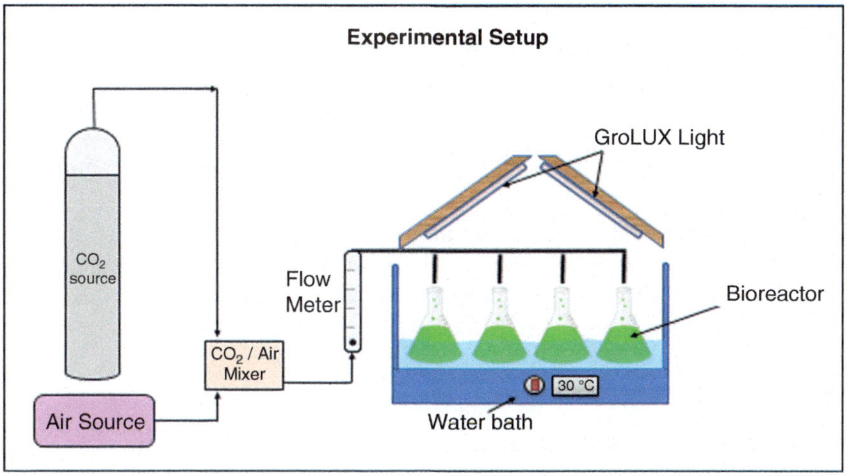

Fig. 41.1 Schematic diagram of the experimental setup used to cultivate microalgae

PYREX 1L Erlenmeyer flasks (Fig. 41.1). The working volume was 500 mL and the initial algal concentration was approximately the same for all the cultivation conditions: 2.2×10^7 cells mL^{-1} (obtained using a hemocytometer). The reactors were placed on a bench where four Grolux fluorescent light bulbs (that provide an average light intensity of 65 μmol m^{-2} s^{-1} measured by a LI-250 Light Meter) were arranged above the water bath. An air mixture of 96% atmosphere air and 4% carbon dioxide was supplied to the cultures.

41.2.3 Quantification of Biomass

For quantification and growth studies, the product samples (1 mL each) were collected every day. Optical densities were measured after appropriate dilution at 690 nm using a UV-spectrophotometer (UNICAM). The biomass (g/L) was determined based on a biomass-optical densities standard curve (Fig. 41.7, Appendix 41.1). Biomass values (g/L) were obtained after vacuum filtration of an appropriate amount of sample (10 mL) at every 3 days, with subsequent drying at 55 °C for 24 h.

41.2.4 Determination of Growth Kinetic Parameters

Algal growth kinetic data, such as specific growth rate (μ or SGR, day^{-1}) and biomass productivity (P, g_{dw} L^{-1} day^{-1}), were calculated from experimental data by using the following Eqs. (41.1) and (41.2), respectively [15–20]:

$$\text{Specific growth rate, } \mu = \frac{\ln(N_2 / N_1)}{t_2 - t_1} \tag{41.1}$$

$$\text{Biomass productivity, } P = \frac{X_t - X_0}{t_x - t_0} \tag{41.2}$$

where N_1 and N_2 are the biomass concentrations (g/L) at the beginning t_1 and at the end t_2 of the exponential growth phase, respectively; X_t is the biomass concentration (g/L) at the end of the cultivation period t_x and X_0 the initial biomass concentration (g/L) at t_0 (day).

41.2.5 Determination of Nitrogen Removal Efficiency

The removal of nitrogen efficiency was estimated by quantifying the level of removal (obtained by subtracting from initial to final values) from the culturing medium. The filtrate (obtained while performing biomass dry weight) was subjected to nitrate test using Chromo tropic acid method with acid persulfate digestion respectively according to procedure given in Hach Company, 2015 [21].

41.2.6 Development of Mathematical Models and Validation

Mathematical model was developed to determine the quantitative linkage between environmental parameters (temperature, nitrate concentration in the medium) and cell growth kinetics. For simplification, prior to formulate model, the following assumptions were made: (i) light intensity was kept constant, (ii) initial inoculum size was kept constant, and (iii) death kinetics were ignored. First, the effect of limiting nutrient (nitrate) on the growth of microalgae was described by the Monod model as shown in Eq. (41.3) [22].

$$\mu = f(S) = \mu_{\mathrm{m}} \left[\frac{S}{K_{\mathrm{S}} + S} \right] \tag{41.3}$$

where μ = specific growth rate, day^{-1}; μ_{m} = maximum specific growth rate, day^{-1}; S = limiting nutrient concentration, mg/L; K_{S} = half-saturation coefficient, mg/L.

The maximum specific growth rate (μ_{m}) which appears in the Monod model is not a function of limiting nutrient concentration, rather, it is a function of temperature and light intensity. Since the light intensity was assumed constant, the maximum specific growth rate can be a function of temperature only and it may be described fairly by the Arrhenius Eq. (41.4) [22].

$$\mu_m = Ae^{-E/RT} \tag{41.4}$$

where A = constant, day^{-1}; E = Activation energy, cal/mol; R = Universal gas constant, cal/K/mol; T = Temperature, K.

Upon combination of Eqs. (41.3) and (41.4), Eq. (41.5) was formulated, which can describe the growth of microalgae as a function of temperature and limiting nutrient concentration.

$$\mu = Ae^{-E/RT}\left[\frac{S}{K_S(T)+S}\right] \tag{41.5}$$

The half-saturation coefficient is an empirical value which can be found by the graphical method, illustrated in Fig. 41.8 of Appendix 41.1. Moreover, the half-saturation coefficient depends on environmental conditions such as the temperature.

For model validation, theoretical specific growth rates (computed using the model equation) vs. nutrient concentrations (S) were plotted for different temperature. These figures were compared with those obtained in experiment.

41.3 Results and Discussion

41.3.1 Growth Profile of Microalgae sp. Chlorella kessleri

To characterize the time-resolved growth phase, it is important to know the growth pattern of microalgae species *Chlorella kessleri* in BBM at different culturing temperature. The population of *Chlorella kessleri* grown at 25 °C in a batch photobioreactor containing BBM is shown in Fig. 41.2. One can see from this figure that

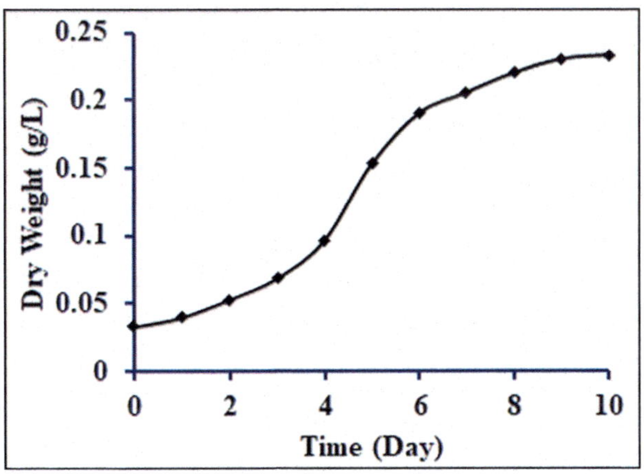

Fig. 41.2 Growth curve for microalgae sp. *Chlorella kessleri* in BBM at 25 °C in a batch bioreactor

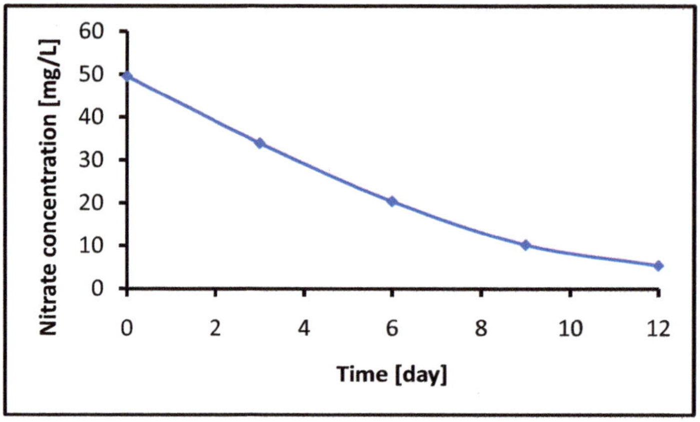

Fig. 41.3 Consumption of nitrate with time at 25 °C

Table 41.1 Data used to fit the Arrhenius model

T (°C)	$\hat{\mu}$[day^{-1}]	Ln ($\hat{\mu}$)	1/T (K^{-1})
25	0.55	−0.605	0.0034
35	0.52	−0.653	0.0033
40	0.44	−0.834	0.0032

for 3 h incubation, the growth of *Chlorella kessleri* was slow. The absorbance then continued to increase until day 9 before reaching a plateau or stationary phase. The logarithmic growth phase was observed within 3–9 days and thereby maximum specific growth rates at different culturing temperature were determined in this region. Limiting nutrients concentration in the form of nitrate (NO_3^-) was monitored by plotting the nitrate concentration [mg/L] vs. time [day]. The decreasing pattern of consumption of limiting nutrient, nitrate by *Chlorella kessleri* is shown in Fig. 41.3. It is important to note that different growth and nitrate consumption patterns were observed at different culturing temperatures. As such, it is important to develop a model that describes the temperature effect on microalgal growth.

41.3.2 Development of Mathematical Model for Algal Growth

In order to obtain the parameters of pre-exponential factor (A) and activation energy (E) in the *Arrhenius* model, maximum specific growth rate was calculated for each temperature and the data were shown in Table 41.1. Figure 41.4 shows a plot of ln($\hat{\alpha}$) vs. 1/T. It was a straight line with the R^2 value of 0.7621, indicating good fit. Since in biological system, the R^2 value should be more than 0.75 to validate the empirical

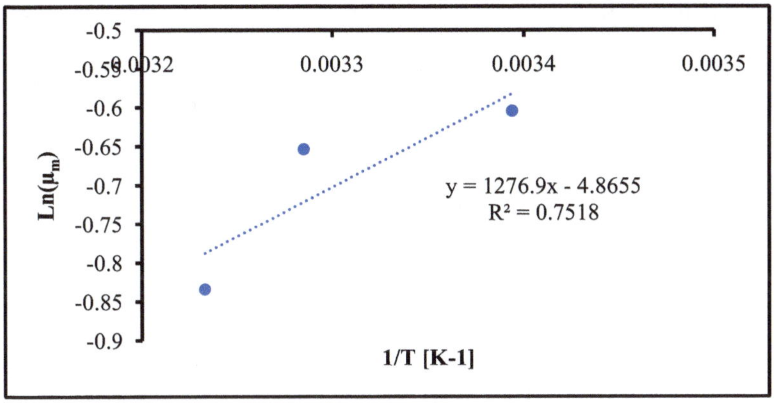

Fig. 41.4 A plot of $\ln(\mu_m)$ vs. $1/T$

Table 41.2 A comparison between experimental and predicted maximum specific growth rates

T (°C)	$\hat{\propto}_{exp}$ (day^{-1})	$\hat{\propto}_{Arr}$ (day^{-1})	%error$_{exp/Arr}$
25	0.55	0.56	2.2%
35	0.52	0.49	6.6%
40	0.44	0.45	4.7%

model [23]. The A and E were calculated to be 0.0077 day^{-1} and −2537 cal/mol. Thus, Arrhenius model was rewritten with these values as:

$$\mu_m = 0.0077\exp\left(\frac{1277}{T[\mathrm{K}]}\right) \quad (41.6)$$

It is important to note that negative activation energy was observed. The probable reason behind this is that increasing the temperature causes increase in the growth of algae to a certain level then decrease in the growth. The model (Eq. 41.6) was used to predict the maximum specific growth rates. A comparison between experimental and predicted values of the maximum specific growth rates in terms of % Errors is shown in Table 41.2. The maximum error between predicted and experimental specific growth rate was less than 7%. Thus, the developed model was reliable.

The next step was to incorporate both temperature and nutrient concentration into one model. The nutrient limitation was modeled by finding the empirical coefficient of the Monod equation, which is the half-saturation coefficient (K_s). The half-saturation coefficient was found graphically from plots of specific growth rate (μ) vs. nutrient concentration (S), as illustrated in Fig. 41.8 of Appendix 41.1. Since the half-saturation coefficient (K_s) is a function of temperature, its values were determined for each temperature as shown in Table 41.3. A relationship between the

Table 41.3 Half-saturation coefficient at different temperatures

T (°C)	K_s (mg/L)
25	48
35	51
40	37

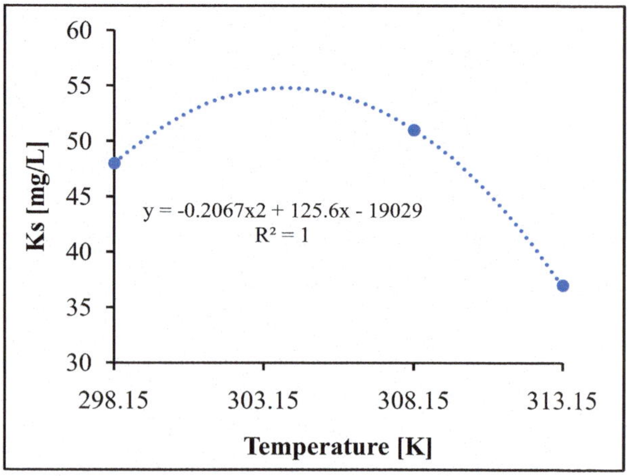

Fig. 41.5 The effect of temperature on the half-saturation coefficient

half-saturation coefficient and temperature is obtained by plotting K_s [mg/L] vs. temperature [K] as depicted in Fig. 41.5. Apparently, a second-order polynomial relationship (as seen in Eq. 41.7) was obtained for describing the effect of temperature on the half-saturation coefficient. It is clear from the figure that at moderate temperature the value of K_s was relatively higher than that of both low and high temperature. Then, Eq. (41.7) was incorporated into the Monod model (Eq. 41.3), and a resultant model was obtained (Eq. 41.8) that describes the specific growth rate as a function of limiting nutrient concentration.

$$K_s = K_s(T) = -0.2067T^2 + 125.6T - 19{,}029 \tag{41.7}$$

$$\mu = \mu_m \left[\frac{S}{\left(-0.2067T^2 + 125.6T - 19{,}029\right) + S} \right] \tag{41.8}$$

Finally, an integrated model was made (as shown in Eq. 41.9) by combining Eqs. (41.8) and (41.6) and it describes the algal growth as a function of temperature and limiting nutrient concentration. This model was used to predict the specific growth rate for any temperature and nutrient concentration. The optimum temperature and nutrient concentration that will maximize the specific growth rate can also be measured.

$$\mu = 0.00771 \exp\left(\frac{1276.9}{T[\mathrm{K}]}\right) \left[\frac{S}{\left(-0.2067T^2 + 125.6T - 19{,}029\right) + S} \right] \tag{41.9}$$

Table 41.4 Composition of Bold's Basal Medium (BBM)

Component	Stock solution (g/L)	Quantity used (mL)
Macronutrients		
$NaNO_3$	25	10
$CaCl_2 \cdot 2H_2O$	2.5	10
$MgSO_4 \cdot 7H_2O$	7.5	10
K_2HPO_4	7.5	10
KH_2PO_4	17.5	10
NaCI	2.5	10
Alkaline EDTA solution		1
EDTA (with 31 g KOH)	50	
Acidified iron solution		1
$FeSO_4 \cdot 7H_2O$ (with 1 mL H_2SO_4)	4.98	
Boron solution		1
H_3BO_3	11.42	
Trace elements solution		1
$ZnSO_4 \cdot 7H_2O$	8.82	
$MnC1_2 \cdot 4H_2O$	1.44	
$CuSO_4 \cdot 5H_2O$	1.57	
$Co(NO_3)_2 \cdot 6H_2O$	0.49	
	Total	**64**

41.3.3 *Model Validation*

The maximum specific growth rate was calculated by model Eq. 41.6 that shows satisfactory agreement with experimental data for all temperature. For integrated model (Eq. 41.9) validation, predicted specific growth rates vs. nutrient concentrations (S) were plotted for different temperatures, separately as shown in Fig. 41.6a–c. These figures were compared with those obtained in experiment in the same figures. The predicted data are relatively deviated from the experimental values indicating fairly predictions in specific growth rate because these models are quadratic models and are perhaps unable to describe the algal growth perfectly in all kinetic regimes. Besides this model is solely a function of temperature and limiting nutrient concentration. Other important factors (e.g., light intensity, initial inoculum size, decay rate) that could possibly affect the specific growth rate of microalgae were ignored during model development.

41.4 Conclusions

In the present study, a mathematical model was formulated by combining Monod model and Arrhenius equation based on some assumptions. It describes the relationship of algal growth rate with culturing temperature and limiting nutrient

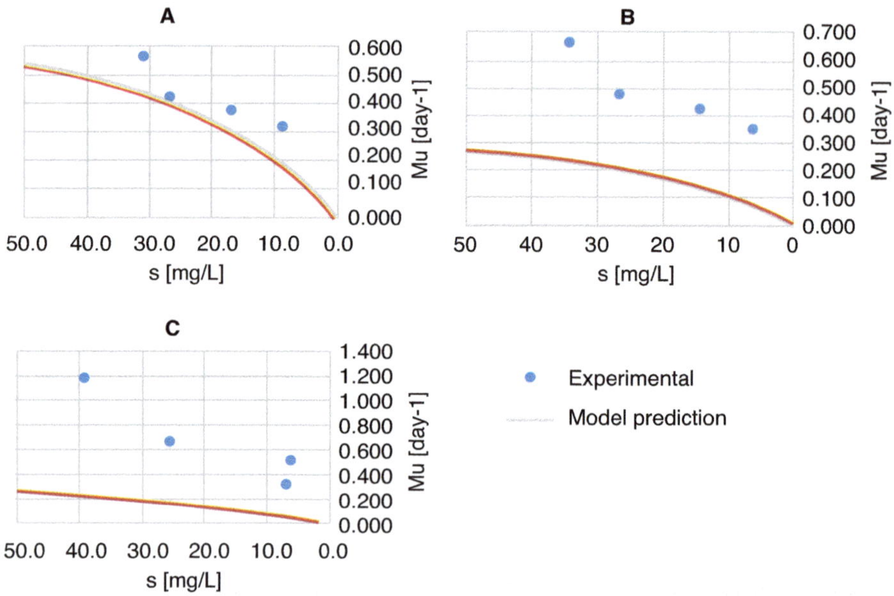

Fig. 41.6 A comparison between experimental and model prediction of specific growth rate at different temperatures: (**a**) 25 °C, (**b**) 35 °C, and (**c**) 40 °C

Fig. 41.7 Calibration curve for microalgae sp. *Chlorella kessleri* BBM at 25 °C in a batch photo bioreactor

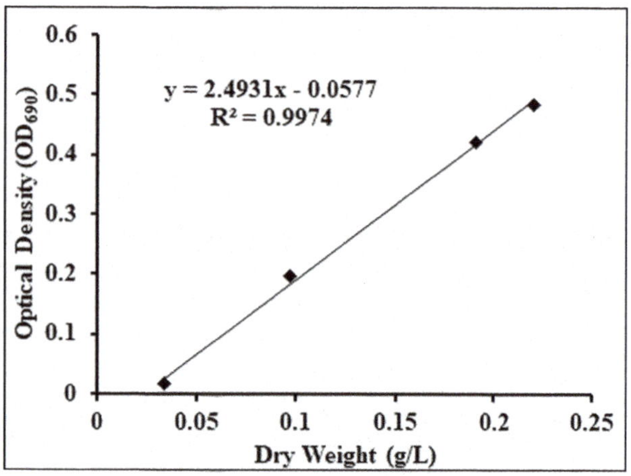

concentration. This model was used to predict microalgal (*Chlorella kessleri*) specific growth rate at different temperatures. The apparent activation energy and pre-exponential factors were calculated to be 2537 cal/mol and 0.0077 day^{-1}, respectively. The developed model agreed fairly with experimental data and could be useful to visualize the effective impacts of temperature on outdoor algae culture. However, to get a perfect prediction of growth rate by the model, it is important to

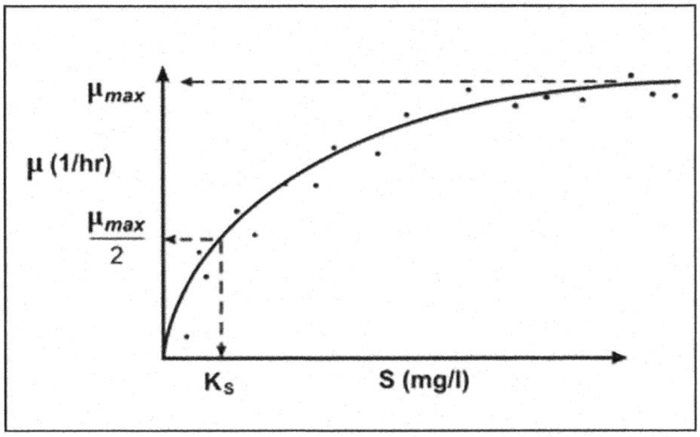

Fig. 41.8 A graphical method to determine the half-saturation coefficient

develop a comprehensive model considering all possible factors such as culture temperature, nutrient (nitrate and phosphate), light intensity, initial inoculum size, CO_2 uptake rate, and death kinetics. Our research is underway in this direction.

Acknowledgments The author would like to gratefully acknowledge the support provided by King Abdulaziz City for Science and Technology (KACST) through the Science & Technology Unit at King Fahd University of Petroleum & Minerals (KFUPM) for funding this work through project No. NSTIP # 13-WAT096-04 as part of the National Science, Technology and Innovation Plan.

41.1 Appendix

References

1. Liam B, Owende P (2010) Biofuels from microalgae—a review of technologies for production, processing, and extractions of biofuels and co-products. Renew Sustain Energy Rev 14:557
2. Wang B, Li Y, Wu N, Lan CQ (2008) CO_2 bio-mitigation using microalgae. Appl Microbiol Biotechnol 79:707
3. Cheng J, Huang Y, Feng J, Sun J, Zhou J, Cen K (2013) Improving CO_2 fixation efficiency by optimizing *Chlorella* PY-ZU1 culture conditions in sequential bioreactors. Bioresour Technol 144:321
4. de Morais MG, Costa JAV (2007) Carbon dioxide fixation by Chlorella kessleri, C. vulgaris, Scenedesmus obliquus and Spirulina sp. cultivated in flasks and vertical tubular photobioreactors. Biotechnol Lett 29:1349
5. Hossain SMZ, Hossain MM, Razzak SA (2018) Optimization of CO_2 biofixation by Chlorella vulgaris using a tubular photobioreactor. Chem Eng Technol 41(7):1313
6. Hossain SMZ, Alnoaimi A, Razzak SA, Ezuber H, Al-Bastaki N, Safdar M, Alkaabi S, Hossain MM (2018) Multiobjective optimization of microalgae (Chlorella sp.) growth in a photobioreactor using Box-Behnken design approach. Can J Chem Eng 96:1903
7. Kazeem MA, Hossain SMZ, Razzak SA, Hossain MM (2018) Application of central composite design to optimize culture conditions of Chlorella vulgaris in a batch photobioreactor. Chem Prod Process Model 13. https://doi.org/10.1515/cp

8. Schenk PM, Thomas-Hall SR, Stephens E, Marx UC, Mussgnug JH, Posten C, Kruse ORJ, Hankamer BD (2008) Second generation biofuels: high-efficiency microalgae for biodiesel production. Bioenergy Res 1:20
9. Dismukes GC, Carrieri D, Bennette N, Ananyev GM, Posewitz MC (2008) Aquatic phototrophs: efficient alternatives to land-based crops for biofuels. Curr Opin Biotechnol 19:235
10. Cantrell KB, Ducey T, Ro KS, Hunt PG (2008) Livestock waste-to-bioenergy generation opportunities. Bioresour Technol 99:7941
11. Chisti Y (2007) Biodiesel from microalgae. Biotechnol Adv 25:294
12. Razzak SA, Hossain MM, Lucky RA, Bassi AS, De Lasa H (2013) Integrated CO_2 capture, wastewater treatment and biofuel production by microalgae culturing—a review. Renew Sustain Energy Rev 27:622
13. Zhang D, Dechatiwongse P, del Rio-Chanona EA, Maitland GC, Hellgardt K, Vassiliadis VS (2015) Modelling of light and temperature influences on cyanobacterial growth and biohydrogen production. Algal Res 9:263
14. Bissinger JE, Montagnes DJS (2008) Predicting marine phytoplankton maximum growth rates from temperature: improving on the Eppley curve using quantile regression. Limnol Oceanogr 53(2):487
15. Ocampo-López C, Colorado-Arias S, Ramírez-Carmona M (2015) Modeling of microbial growth and ammonia consumption at different temperatures in the production of a polyhydroxyalkanoate (PHA) biopolymer. J Appl Res Technol 13:498
16. Mohsenpour SF, Richards B, Willoughby N (2012) Spectral conversion of light for enhanced microalgae growth rates and photosynthetic pigment production. Bioresour Technol 125:75
17. Das P, Lei W, Aziz SS, Obbard JP (2011) Enhanced algae growth in both phototrophic and mixotrophic culture under blue light. Bioresour Technol 102:3883
18. De-Morais MG, Costa JAV (2007) Biofixation of carbon dioxide by *Spirulina* sp. and *Scenedesmus obliquus* cultivated in a three-stage serial tubular photobioreactor. J Biotechnol 129:439
19. Tang D, Han W, Li P, Miao X, Zhong J (2011) CO_2 biofixation and fatty acid composition of Scenedesmus obliquus and Chlorella pyrenoidosa in response to different CO_2 levels. Bioresour Technol 102:3071
20. Yun YS, Lee SB, Park JM, Lee CI, Yang JW (1997) Carbon dioxide fixation by algal cultivation using wastewater nutrients. J Chem Technol Biotechnol 69:451
21. Hach Company (2015) Chromotropic acid method. Test "N Tube Vials 10(3):1–6.
22. Goldman JC, Carpenter EJ (1974) A kinetic approach to the effect of temperature on algal growth. Limnol Oceanogr 19(5):756
23. Haaland PD (1989) Experimental design in biotechnology. CRC Press, Taylor & Francis Group, London

Chapter 42
Waste-to-Energy Solutions in Uppsala, Sweden

Mofoluwake Ishola and Cecilia Tilli

42.1 Introduction

Sweden has a long history of system solutions and policymaking within sustainability, which begins with the first ban on emissions of untreated sewage in 1956 and continues with the first Environmental Protection Act in 1969. Sweden is currently one of the leading countries in the world in regard to sustainability and efficient waste management systems. Recently, it was reported that Sweden will reach its 2030 Renewable Energy Target in 2018 [1]. Currently in Sweden, waste is regarded as an important feedstock from which useful products can be produced. Incineration facilities have even made a business out of importing waste for energy production from neighbouring countries. Over the past 6 years, see Fig. 42.1, more than 50% of the household waste has been recovered as energy, 30% have been recycled as materials, 16% has been used for biogas and compost while only 0.7% goes to landfill [2]. Public policy and investments, research and development, innovations and implementation of the European Union waste hierarchy—reduce, reuse and recovery of energy from waste streams—have aided the realization of clean and sustainable environments in Sweden. This paper presents an overview of strategies used to achieve functioning sustainable systems in Sweden with a focus on Uppsala solutions.

M. Ishola · C. Tilli (✉)
Lindhea, Uppsala, Sweden
e-mail: cecilia.tilli@lindhea.se

A. Sayigh (ed.), *Renewable Energy and Sustainable Buildings*, Innovative Renewable Energy, https://doi.org/10.1007/978-3-030-18488-9_42

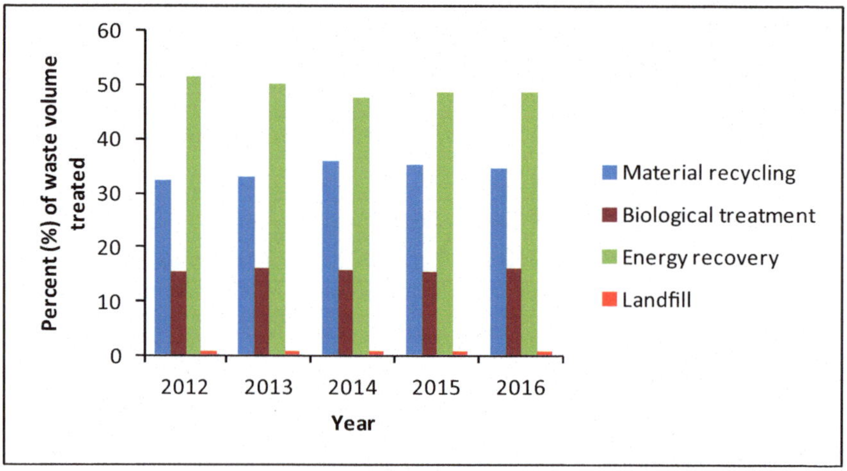

Fig. 42.1 Percentage of Swedish treated waste volumes in 2012–2016 (Data Source: [2])

42.2 Applied Strategies for Sustainable Waste Management in Sweden

The Swedish political structure, policies, legislations and the role of the municipalities have played major parts in the development of the current system. Between 1952 and 1974, there was a process of decentralization of power from the central government to the 290 municipalities. The municipalities received the responsibilities for city planning, waste management, water supply, housing, public transportation, local business development, schools, kindergartens and care of elderly. This decentralization of power has enabled the municipalities to work progressively with sustainable system solutions. This does not only involve solving one particular challenge at a time but also creating solutions that become integrated in many parts of the society, with a likelihood of long-term success. The system solutions require a combination of public and private efforts [3].

42.3 Uppsala: Climate Capital of Sweden

In 2018 Uppsala, a city of 220,000 inhabitants just north of Stockholm, was awarded as the climate capital of Sweden by WWF for the second time [4]. The international jury appraised the approach of the municipality with system solutions and efforts regarding sustainable mobility and investments in renewable energy and energy efficiency.

An example of how Uppsala has approached sustainability issues is the local climate protocol, a project initiated in 2010 using the UN Climate Protocol as a model. It is a voluntary collaboration between private businesses, public organizations and

NGO's with the purpose of fulfilling the local goals for reduction of greenhouse gas emissions. The municipality of Uppsala's climate target is that by 2020; emissions of greenhouse gas per capita in Uppsala will have been halved compared with 1990 levels, and by 2050, emissions should be about half a ton per inhabitant. That means energy use and transport in the geographical area of the municipality must be almost emission-free by then. Milestone goals are set for each three-year period, and so far the milestones have been reached with good margin [5].

42.3.1 Uppsala's Sustainable Waste Management System

Uppsala has an efficient source separation system, whereby wastes are sorted into different fractions close to the households. Source separation is an important part of waste management; this act makes it easy to understand the compositional properties of the different waste streams and also to find the most suitable applications for each waste stream. In Uppsala, wastes are sorted into the following fractions [6]:

1. Organic waste such as food waste, forest and agricultural residues
2. Recyclables items such as plastic packaging, paper and metals
3. Bulky items such as furniture, white goods and tyres
4. Residual wastes such as diapers and tissues
5. Hazardous waste such chemicals, electronics and batteries

An example of Swedish household waste collection can be seen in Fig. 42.2. The organic wastes including food wastes are transferred to the biological processing

Fig. 42.2 Photo of waste collection system in Uppsala with source separation for the households. Photo: C. Tilli

plants for biogas and fertilizer co-production, the recyclables are sent to the material recycling industries outside Uppsala and bulky wastes are sent to appropriate industries for treatment. The residuals wastes are treated in combustion plant for energy recovery, while hazardous wastes are sent to the respective industries for treatment.

42.3.2 Bioprocessing of Organic Waste to Biogas

In Uppsala, food waste, slaughterhouse waste and sewage sludge are sent to the biogas production facility, which is run by Uppsala Vatten och Avfall, a company owned by the municipality of Uppsala. In this facility, the materials are used as raw materials for co-production of biogas and fertilizer in a wet digestion process. The biogas serves as the fuel for city buses, while the remaining digestate from the gas production is used as bio-fertilizer.

Biogas production has a long history in Uppsala, with the first production done from sewage sludge in the 1940s. In 1996 the first gas refinery facility and fuelling station were built and the first biogas fuelled city buses started running. One year later the new biogas production facility was built with manure and slaughterhouse waste as primary feedstocks.

Obligatory sorting of organic household waste was introduced in 2000. During the first 6 years, this fraction was composted, until investments for reception of household food waste to the biogas facility was made in 2006. Further investments in increased capacity has been done since, and the production capacity of Uppsala biogas facility has expanded from 9 GW/yr. in 2008 to 30GW/yr. in 2017 with the two digesters installed. The biogas is distributed through underground pipes to the 63 stations which supply biogas to city buses and some regional buses. Private gas-fuelled cars can also fuel up at these stations. The biogas facility is still expanding as a new digester and a bigger waste storage was built recently [7].

42.3.3 Energy Recovery from Combustible Wastes

The combustibles and the residual waste fraction are sent to the incineration facility in Uppsala, run by Vattenfall AB, a company owned by the Swedish state. In this plant, mixed wastes, combustibles and residuals are incinerated in boilers for energy recovery. The recovered energy is used for district heating, district cooling and electricity production. Of all the housing properties in Uppsala, 95% are being supplied with district heating from the facility. The main fuels at this facility are basically combustible household and industrial wastes. The flue gases are cleaned of hazardous emissions before being released to the atmosphere. Research has been done about the recovery of valuable metals from the ash residue [8].

The incineration facility for district heating was built in the 1970s but was initially run on oil until waste combustion was introduced a decade later. Until 2000 the facility was owned and run by Upplands Energi, a company owned by the municipality. In 2000 Upplands Energi was bought by Vattenfall. Both the capacity and the district cooling network have since been expanded, and the electricity production began in 2010. The plant is currently investing in a completely new incineration facility which will contribute to making the process completely carbon neutral by 2030 [9, 10].

42.4 Uppsala: Centre of Innovation

Uppsala is a centre of innovation with two leading universities, Uppsala University and the Swedish University of Agricultural Sciences. Many innovations from the university's research have been implemented commercially, often through the establishment of private spin-out companies supported by the innovation system.

In Sweden, about 8% of the state funds allocated for research targets early-stage commercialization of research results [11]. This development from academic publications to commercial implementation is crucial to accomplish large-scale impact. It also generates taxable revenue that can be seen as return on the governmental investment in research.

The intellectual property (IP) regulations for Swedish researchers further encourage private commercialization. For most employees in most parts of the world with a job description that includes research and innovation their results, the IP that their work generates is owned by their employer. Swedish researchers have what is called the *teacher exception*, meaning they themselves own all results as private individuals. If they choose to commercialize their results, the university has no claim on the IP.

The two universities in Uppsala specifically encourage and facilitate efforts to create spin-out businesses based on research. At Uppsala University, there is a specific division called UU Innovation that is dedicated to advising and supporting such ventures, and both Uppsala University and the Swedish University of Agricultural Sciences have holding companies that invest in early-stage university spin-outs that might be too high risk for private investors. There are also several other organizations offering free consultations, mentorship programs and or financial support. There is Drivhuset that specializes in students' business ventures, Nyföretagarcentrum that is accessible to all kinds of startups and Almi that is focussed on early-stage financing by grants, loans and investments. Uppsala also has an incubator, Uppsala Innovation Centre, which has been ranked number 4 in the world. There are also a number of co-working spaces for startup companies in Uppsala. An organization called STUNS (Stiftelsen för samverkan mellan universiteten i Uppsala, näringsliv och samhälle) functions as a platform for collaboration to support the development of companies in new fields. STUNS consists of representatives of the universities in Uppsala, Chamber of Commerce Uppsala County, Uppsala County Administrative Board, Region Uppsala and the Municipality of Uppsala.

Since 2010, Sweden has adopted what is called "public innovation procurement", meaning that municipalities and other public entities may request new solutions in public procurement processes. Uppsala is in the forefront of innovation procurement, making Uppsala a "test-bed city". This creates an important possibility for many clean-tech startups that would otherwise have difficulties finding a first customer when they don't have any previous references.

Through the support systems in place, Uppsala has become a city with a growing number of clean-teach startup companies.

42.5 Conclusion

Sustainability strategies in Sweden has evolved over the last 50–60 years, and there are now many examples of system solutions that involve many stakeholders and that address several issues at once. It is primarily at municipal level that sustainable strategies are designed and implemented. While technical solutions are of course necessary, public policy and willingness to invest in new technology and infrastructure are also crucial success factors. Education and engagement of the public is another central aspect.

References

1. Mccarthy J (2018) Sweden will reach its 2030 renewable energy target this year, Global Citizen. https://www.globalcitizen.org/en/content/sweden-reach-renewable-energy-goal-this-year/. Accessed Jul 2018
2. Avfall Sverige (2017) Swedish waste management. https://www.avfallsverige.se/fileadmin/user_upload/Publikationer/Avfallshantering_2017_eng_low.pdf. Accessed Jul 2018
3. Enviro Sweden (2018) Swedish eco-history. http://www.symbiocity.se/en/approach/. Accessed 19 Jul 2018
4. WWF Global (2018) Finalist of the one planet city challenge. http://wwf.panda.org/our_work/projects/one_planet_cities/one_planet_city_challenge/. Accessed Jul 2018
5. Uppsala Klimatprotokoll. https://klimatprotokollet.uppsala.se. Accessed 19 Jul 2018
6. Uppsala Vatten Publications: Ditt avfall är inget skräp. https://www.uppsalavatten.se/hushall/avfall-och-atervinning/kallsortering/. Accessed Jul 2018
7. Uppsala Vatten Publications: Biogasanläggningen vid Kungsängens gård. https://www.uppsalavatten.se/Global/Uppsala_vatten/Dokument/Trycksaker/biogasanlaggning_kungsangens_gard.pdf. Accessed 26 Jul 2018
8. Lassesson H et al (2014) Leaching for recovery of copper from municipal solid waste incineration fly ash: influence of ash properties and metal speciation, 2014, Waste management and research

9. Wolters S (2005) Dyrare fjärrvärme efter fem år med Vattenfall, 2005-02-20, Uppsala Nya Tidning. http://www.unt.se/nyheter/uppsala/dyrrare-fjarrvarme-efter-fem-ar-med-vattenfall-661421.aspx. Accessed 26 Jul 2018
10. Vattenfall publication: Vi värmer Uppsala https://www.vattenfall.se/fjarrvarme/orter/uppsala/. Accessed 23 Jul 2018
11. Statistics Sweden: Statliga anslag till forskning och utveckling 2018. http://www.scb.se/hitta-statistik/statistik-efter-amne/utbildning-och-forskning/forskning/statliga-anslag-till-forskning-och-utveckling/. Accessed 23 Jul 2018

Chapter 43
Geo-Climatic Early-Design Tools and Indicators

Giacomo Chiesa

43.1 Introduction

Several studies have confirmed that the energy consumption for space cooling and ventilation is an increasingly important element in national energy balances [1–4]. Even if the increase in the efficiency of mechanical cooling systems can reduce this consumption, alternative solutions are needed to invert the trend. Natural solutions to passive cooling may be used to prevent, mitigate and dissipate heat gains in buildings [5–7]. In particular, passive-heat-gain dissipation is based on the use of heat sinks, such as air, water, ground and night sky. Unfortunately, passive cooling solutions are climate specific and need an attentive design approach so that they can be effectively applied to the local climate and well integrated into building design.

This paper aims to define a simple method to include, from early-design phases, passive cooling dissipative systems (PCDS) in the design process. This method is conceived to help designers consider PCDS as a potential set of bioclimatic technologies which are able to reduce energy needs without compromising the comfort of users. The development of simple methods is crucial since the applicability of PCDS is very local specific and needs to be considered as early as possible in the design development process to optimise the integration of these systems. The potential of PCDS relates to technological and geo-climatic (environmental) issues, though this paper focuses on the latter. However, a preliminary analysis of technological issues was reported in [8]. Section 2 of the paper describes the proposed simple methodology that is based on a performance-driven approach. In Sect. 2.1, the performance-driven approach is briefly defined in order to include the proposed method in a consolidated background which is devoted to environmental and technological design. Section 2.2 defines a series of potential environmental indicators related to the local climate potential of PCDS. Section 3 reports a sample application of the geo-climatic analysis

G. Chiesa (✉)
Department of Architecture and Design, Politecnico di Torino, Turin, Italy
e-mail: giacomo.chiesa@polito.it

A. Sayigh (ed.), *Renewable Energy and Sustainable Buildings*, Innovative Renewable Energy, https://doi.org/10.1007/978-3-030-18488-9_43

of a sample set of locations in consideration of the indicators described in the previous section. Finally, in Sect. 3.1, the method is applied to analyse the resilience of the PCDS geo-climatic applicability under the effect of climate change. This last analysis is used to define another environmental indicator, namely, the resilience of a specific PCDS in maintaining its environmental potential. Section 4 is devoted to conclusions and limitations.

43.2 Simple Methodology to Consider PCDS in Early-Design Phases

43.2.1 Performance-Driven Approach to PCDS: Requirements for Their Application

The paper focuses on the performance-driven approach to early-building design where a detailed analysis of activities, needs and related requirements is conducted in response to user needs without directly referring to any specific building shape or dimension. In this very early-design phase, called building programming [9, 10], space unit layout and technological strategies are defined to optimise a compliance check based on a list of requirements and virtual performance. Thanks to this analysis—and to the related feedback cycles—, it is possible to further proceed with the definition of a schematic building design in compliance with users' needs including environmental and technological requirements. The performance-driven approach to early-building design has been recognised in the Architectural Technology research field since the 70's—see [7]—and several Italian standards are devoted to regulating this design process: UNI 7867-4:1979—now UNI 10838:1999—, which refers to the terminology; UNI 8290-1:1981; 8290-2:1983; 8290-3:1987; UNI 8289:1981, which refers to the principal-needs classification; and UNI 11277:2008.

This approach is based on a simple process: users → activities → needs → requirements ← performances, in which the last logical passage is constituted by a feedback cycle based on specific indicators of the performance. According to this process, requirements can be defined as the technical transposition of needs connected with the building process—see UNI 8290-2:1983—while needs are what is needed to carry out the general execution of an activity. In this sense, PCDS mainly refer to the classes of needs (for the complete list of recognised classes of needs see UNI 8289), well-being (especially in relation to thermal-comfort need) and environmental safeguards (rational use of climatic and energy resources). Passive cooling systems, however, also need to be correctly integrated and operated. The main benefit of PCDS is therefore the possibility to guarantee thermal comfort conditions without consuming fossil fuels or by reducing this consumption to a minimum—e.g. by only using fan air-movement activation (low energy solutions, hybrid). A specific class of requirements which refer to these technologies may be introduced and named "use of passive/low energy cooling dissipative systems". In a previous work [8]—see also [11]—, a compatible class of requirements named "use of thermal sinks", together with a list of specific requirements related to technological units and technical elements for both the environmental and the technological system were defined. The "use of passive/low energy cooling dissipative systems" class, however, also includes hybrid systems such as ceiling fans for comfort ventilation. The following table reports the previously mentioned outcomes adapted for the proposed class of requirements (Table 43.1).

In this paper, the performance-driven approach is integrated with technological and environmental aspects related to the usage of passive cooling dissipative systems (PCDS). Unlike the analysis of this topic that was elaborated in the previously mentioned studies, the methodology reported here is expanded with a more precise definition of environmental and especially climatic issues to check PCDS potential in very early-design phases.

43.2.2 Environmental Indicators to Assess Geo-Climatic Potential of PCDS

PCDS principally refers to four natural heat sinks in order to perform heat dissipation from a flow (e.g. air, water) and passively cool it. The four heat sinks are air, water, ground and night sky. Nevertheless, for the purpose of this paper, only the first three sinks are considered connecting them with ventilative cooling strategies—see also [12]. The considered flow is therefore the air in all described cases. A detailed analysis of passive cooling systems, including technologies to prevent and modulate heat gains, is reported, among others in [5, 6, 13]. The paper considers, from a larger set of PCDS, the following technologies: controlled natural ventilation (CNV), direct evaporative cooling (DEC) and earth-to-air heat exchangers (EAHX).

The CNV, considered as ventilative microclimatic cooling [14], is a class of technologies that includes three main categories: comfort ventilation, in which the activation of an air-movement near the users' skin generates a perceived reduction in temperature; environmental ventilation, in which external air at a low temperature is used to substitute indoor air—air exchange; and structural ventilation, in which the low external-air temperature is used to cool structural building masses and dissipate the heat absorbed by them during the day when released later (this is also called night cooling) [15].

Each PCDS under consideration acts on the perceived or real air temperature based on the thermal sink used. In order to define the potential impact of these solutions under local climate conditions, a series of indicators have been developed to simulate the virtual effect of the specific PCDS on the external air. Thermal conditions of a resulting, virtually treated airflow are then compared with comfort conditions in order to study the potentiality of PCDS in reducing the original discomfort. For the purpose of this paper, local discomfort was defined by defining the number

Table 43.1 General requirements for the class of requirements "use of passive/low energy cooling dissipative systems"—building programming phase for both environmental and technological systems—modified from [8]

Class of needs: rational use of climatic and energy resources"	Class of requirements	System	Requirements for technological units	Requirements for technical elements
	Passive/low energy cooling dissipative systems	Environmental	The spatial distribution of environmental units must take into consideration microclimatic and local contexts in order to maximise the potential use of local PCDS	Spatial elements have to be distributed and connected in order to maximise the potential use of local PCDS and hybrid solutions
			The spatial distribution of environmental units must be organised to allow for a correct distribution of airflow by ceiling fans for comfort ventilation	
		Technological	Massive internal surfaces (partitions including walls, slabs and ceilings) have to be considered in order to activate thermal masses so that they facilitate structural ventilation. This means that the correct attenuation factor and the time lag of thermal stored release must be defined	Devoted technical elements for passive cooling systems (e.g. openings, vents, massive walls with exposed surfaces) and relevant materials, must allow for controlling in-out and internal airflows by correctly designing their characteristics including shape, dimensions, orientation, positioning and thermal-physical aspects
			Openings devoted to ventilative cooling must be foreseen in order to allow for the control of both natural inlet and outlet air flows in consideration of air temperature (coupled or not with other parameters, e.g. RH%)	
			Internal building partitions have to ensure the passage of natural/hybrid air flows	
			Structural elements to support ceiling fans at a correct distance from people have to be considered (ceilings that are too high; thin false ceilings)	Dimensions and performance of mechanical systems have to guarantee the possible integration of passive techniques
			When needed, integration between mechanical and natural systems has to be guaranteed	
			Passive/low-energy cooling systems may be chosen according to the local climatic potential of each heat sink in order to maximise the possibility to perform thermal comfort via PCDS	

of discomfort hours, calculated by counting the number of hours in which the environmental air is above a comfort threshold temperature. The set comfort temperature can be assumed as 26 °C in accordance with other early-design studies [16], even if different base temperatures may be considered in further design stages in order to include the specific effect of solar and internal gains on a building—i.e. 22 °C [17], 18.3 °C [18] and 15 °C [19]. This calculation will be further implemented with regard to a "virtual" airflow temperature after a PCDS treatment in order to define the number of residual discomfort hours. Consequently, an indicator of PCDS potential is defined as follows:

$$\text{PCDS potential}_{\text{fr}} = \frac{\left(\text{No. discomfort hours}_{\text{e}} - \text{No. discomfort hours}_{\text{PCDS}}\right)}{\text{No. discomfort hours}_{\text{e}}} \quad (43.1)$$

This indicator is principally devoted to free-running buildings, such as residences and small offices where no or only personal cooling systems are used. Nevertheless, different indicators may be used to analyse the effect of PCDS as pre-cooling technologies, to be further connected with mechanical systems (e.g. AHU), or when used in alternation with mechanical solutions, as was described in previous researches by defining the residual Cooling Degree Hours indexes [15, 20, 21].

43.2.2.1 Controlled Natural Ventilation Strategies

The three strategies considered for CNV technology need to be treated separately by defining different approaches to characterise the PCDS potential indicator. Considering the purpose of this paper, this analysis will focus on comfort and environmental ventilation, while structural ventilation is not included. Nevertheless, the PCDS potential of structural ventilation, which refers to the activation of internal thermal masses that store heat during the day and release it in attenuated form during the night in order to dissipate it via nocturnal environmental ventilation when the external air is expected to be lower in temperature, can be estimated by using a CDH daily balance, as was described in the previous mentioned researches.

In particular, comfort ventilation refers to the expected reduction in cooling energy needs due to the decrease in the perceived temperature by users when they are directly influenced by an airflow whose velocity is suggested to be lower than 1 m/s in order to prevent other discomforts, e.g. paper movement [15]. Airflow movement can be naturally activated—i.e. wind-driven and/or stack-driven ventilation—or generated by personal or ceiling fans. According to ASHRAE Handbook of Fundamentals 1989, it is possible to predict the expected reduction in the perceived temperature as a function of the air velocity. This relation can be translated into a specific expression which can be implemented in order to calculate the air temperature after comfort ventilation treatment [15, 22]:

$$\vartheta\text{PCDS,cv} = \vartheta_{\text{e}} - 2.319 \cdot v + 0.4816 \quad (43.2)$$

where v is airflow velocity. Considering the early-design purpose of this indicator, a fixed air velocity of 1 m/s can be assumed, considering that this threshold value may be reached via natural or fan-inducted air movements. Different approaches were described in previous researches by taking into consideration the hourly averaged wind velocity reduced by a discharge coefficient. Furthermore, if personal fans are considered, higher velocities may be assumed because of the possibility to orient the flow and manually vary the velocity.

Environmental ventilation techniques refer to the possibility to exchange indoor air for external air at low temperatures. By controlling the amount of air exchanges per hour, by means of net opening areas and operation times, it is possible to cover different levels of cooling requirements. However, the cooling performance is a function of the difference in temperature between the internal and external airs—see, for example, [7]. To represent simply the potential of environmental ventilation, it is possible to compute the difference between a reference number of discomfort hours, calculated with a base temperature including the negative effect of heat gains—e.g. 22 °C—, and a higher temperature threshold, directly based on the comfort threshold representing the inlet air. The PCDS potential indicator for environmental ventilation is therefore based on the ratio between the difference (No. discomfort hours_{22} − No. discomfort hours_{26}) and the original cooling demand including heat gains (No. discomfort hours_{22}). Of course in advanced design stages, it is necessary to assume more complete calculation modes, including a definition of the internal temperature while considering heat gains, such as the one mentioned in [10] for preliminary design, or advanced balances through dynamic energy simulations using, for example, software based on monozone models, such as Dial+ and NatVent, bi-zonal, such as WindChill [7], or multi-zonal such as CoolVent, Comis and the devoted modules of EnergyPlus and Trnsys—see the description in [12].

43.2.2.2 Direct Evaporative Cooling

The effect of direct evaporative cooling systems (DEC), e.g. downdraught evaporative towers, is based on the conversion of sensible to latent heat with a fixed wet bulb temperature [23]. Hence, the minimum theoretical temperature at which the inlet airflow can be cooled in a DEC system is the wet bulb temperature of the same inlet airflow. The maximum potential of DEC is characterised by the local wet bulb depression (WBD), calculated as the difference between the inlet dry bulb temperature and the inlet wet bulb temperature. Of course higher DEC applicability is expected in hot and dry climates. A detailed description of DEC systems is in [24], while a process to integrate DEC systems into environmental building design was described in [25]. Simplified approaches to DEC applicability were analysed in literature—see, for example, [26, 27]—and are generally based on an average seasonal analysis. For the purpose of this paper, a method detailed in a previous work was assumed [20].

It is possible to estimate the hourly "virtually" treated air temperature by reducing the inlet dry-bulb temperature by a ratio of the local WBD—see expression

Eq. (43.3). This ratio is defined as the effectiveness factor of the DEC system—see the Scott, Parson and Koehler's formula [13].

$$\vartheta \mathrm{PCDS,dec} = \vartheta_{\mathrm{in}} - \varepsilon \cdot \left(\vartheta_{\mathrm{in}} - \mathrm{WBT}_{\mathrm{in}}\right) \tag{43.3}$$

where $\vartheta_{\mathrm{PCDS,dec}}$ is the "virtually" treated air temperature after the DEC system, ϑ_{in} is the air dry bulb temperature at the inlet condition (external air), ε is the DEC effectiveness, $\mathrm{WBT}_{\mathrm{in}}$ is the wet bulb temperature at the inlet condition (external air).

This equation was demonstrated to be effective also in describing the functioning of a DEC tower in comparison with other simplified expressions [28]. It is therefore possible to substitute the hourly values obtained by Eq. (43.3) to calculate the number of residual discomfort hours and furthermore define the PCDS potential indicator for DEC. For a preliminary analysis, an effectiveness value of 0.5 is suggested, in a precautionary approach, which is able to avoid too high RH%.

43.2.2.3 Earth-to-Air Heat Exchangers: EAHX

In order to define the early-design potential of buried pipes under the effect of local climate, it is possible to refer to an approach which is similar to the one presented for DEC systems—see Eq. (43.3). In this case, an effectiveness value of 0.5 can be assumed considering a sample system with a tube of 50 m, a diameter of 25 cm, an airflow rate of 500 m^3/h, a depth of 2.5 m, and a dry clay soil [29]. Nevertheless, for this technology, the theoretical limit in temperature that can be reached by cooling an airflow by using an EAHX system is constituted by the soil temperature at the given depth. This value is assumed to be constant along the entire length of the pipe, and for the purpose of this simplified approach the influence of the EAHX operation on the soil is neglected. The soil temperature at the given depth may be estimated for each hour of the year by assuming the Labs [30] or the Hadving [7] expressions. One can be derived from the other by assuming that the moments of the year in which maximum and minimum soil surface temperatures are registered are specular—distance of 180 days or π. The following expression may be used to define the "virtually" treated airflow temperature by a EAHX system [31] and consequently calculate the PCDS potential indicator for EAHX:

$$\vartheta \mathrm{PCDS,EAHx} = \vartheta_{\mathrm{in}} - \varepsilon \cdot \left(\vartheta_{\mathrm{in}} - \left(\begin{array}{c} \vartheta_{\mathrm{av.soil_{surf.}}} - \Delta\vartheta_{\mathrm{s.year}} \exp\left(-h\sqrt{\pi / \alpha t_0}\right) \\ \cos\left(2\pi / t_0 \left(t - t_\varphi\right) - \frac{h}{2}\sqrt{t_0 / \pi\alpha} \right) \end{array} \right) \right) \tag{43.4}$$

where $\vartheta_{\mathrm{PCDS,EAHx}}$ is the "virtually" treated air temperature; ϑ_{in} is the air temperature at the inlet condition (external air); ε is the EAHX effectiveness; $\vartheta_{\mathrm{av,soil\ surf.}}$ is the average annual temperature of the soil surface; $\Delta\vartheta_{\mathrm{s,year}}$ is the amplitude of surface variation of temperatures; t_0 is the yearly duration in seconds; t is the instant of calculation (from the beginning of the year) in seconds; t_ϕ is the phase constant (moment when the minimum soil surface temperature is expected—from the beginning of the year) [s]; α is the thermal diffusivity of the soil [m^2/s]; and h is the depth of the EAHX.

The average soil temperature at the pipe depth, the amplitude and the phase shift can be calculated using the software SoilSurfTemp, which is also used by EnergyPlus. Nevertheless, for EAHX, it is important to include a "virtual" control effect on the calculation of the residual discomfort hours by choosing the best conditions from the environmental or treated airs since, especially in midseason, this PCDS may be not effective.

43.3 Environmental Sample Application: Mapping the Geo-Climatic Potential of PCDS and Their Resilience

The methodology described in Sect. 2 was further applied to a sample set of 10 locations selected in the area of the Mediterranean Basin: Rome, Bari and Palermo in Italy; Valencia and Madrid in Spain; Athens in Greece; Larnaca in Cyprus; El Cairo in Egypt; Tripoli in Libya; and Algiers in Algeria. For all locations the Typical Meteorological Year—TMY—was produced by using Meteonorm v.7.1—period of reference 2000–2009 (temperatures). The WBT was calculated from DBT and RH% of the air by using the Stull expression [32]. The soil conditions to run the SoilSurfTemp software were assumed to be "heavy and damp" at the pipe depth and "barren and arid" at level ground surface for all locations in consideration of average summer conditions.

Figure 43.1 illustrates the geo-positioning of the chosen locations and also reports a classification of local CDH_{26}—(a). Furthermore, the number of summer discomfort hours was also reported in Fig. 43.1b.

The "virtual" PCDS potential was further calculated for the considered heat sinks. In particular, Fig. 43.2 reports the "virtual" PCDS applicability index for all locations considering the four thermal sinks. Controlled natural ventilation strategies have

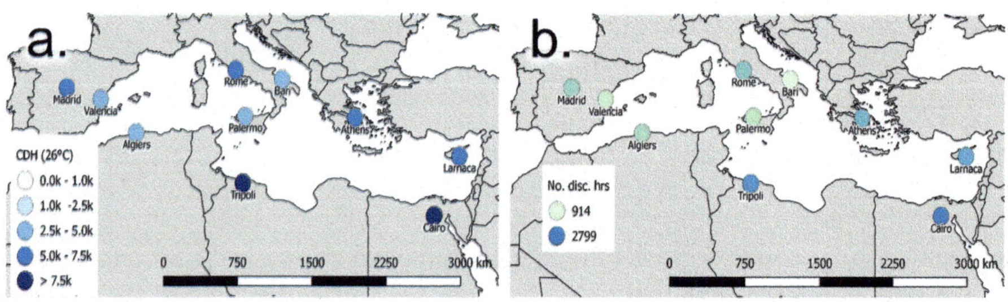

Fig. 43.1 (**a**) Classification of the local CDH26; (**b**) percentage of local summer discomfort hours on the total number of hours (8760)

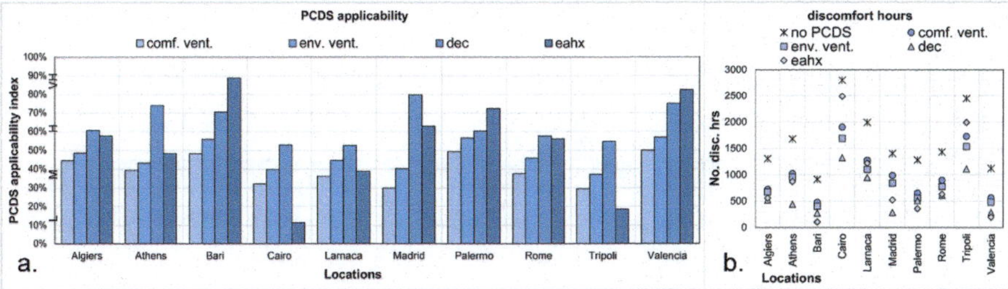

Fig. 43.2 (**a**) PCDS applicability index for the 10 locations considering the four thermal sinks. The percentage refers to the amount of climatic discomfort hours turned to comfort by the PCDS under consideration; (**b**) absolute values of number of discomfort hours of the 10 considered location with and without the PCDS under consideration

good potentiality but in the hottest sites (e.g. Cairo, Larnaca and Tripoli) they are less effective. In these locations EAHX is less effective than in the others, because of the higher soil surface temperature. Also in Madrid CNV strategies are less effective due to higher discomfort intensity, while the average soil surface temperature is lower thus allowing for a high potentiality for EAHX. DEC is effective in the majority of locations according to the local hourly distribution of the WBD.

By comparing the results of different PCDS techniques it is possible to suggest which set of systems may be more adaptable to a specific location with regard to local climate issues. Nevertheless, other aspects reported in Sect. 2.1, related to technological and operational issues also need to be carefully considered from early-design stages. For example, if in a location EAHX are found to have the highest cooling potential but no space is available in the plot area, this technique is not economically viable. Furthermore, especially for DEC and EAHX systems, an economic analysis is suggested to promote the adoption of low-energy, low-cost and effective strategies for space cooling and ventilation.

43.3.1 Resilience Analyses

This section is devoted to analysing the expected effect of climate changes on the PCDS potential. It is important to consider this aspect because the proposed methodology is based on a geo-climatic analysis, and given that PCDS are very local specific changes in local climate conditions may significantly affect their potential in midterm scenarios. A resilience analysis is suggested which implements the method here described to define the PCDS potential while considering a series of morphed TMYs—see also [33, 34]. The IPCC mid SRES scenario A1B, very rapid economic growth and a balance across all energy sources [35], was assumed for this analysis. Two series of morphed TMY were generated for the chosen locations by using the well-known software Meteonorm v7.1.11—see also the software guidelines [36, 37]—for year 2030 and 2050.

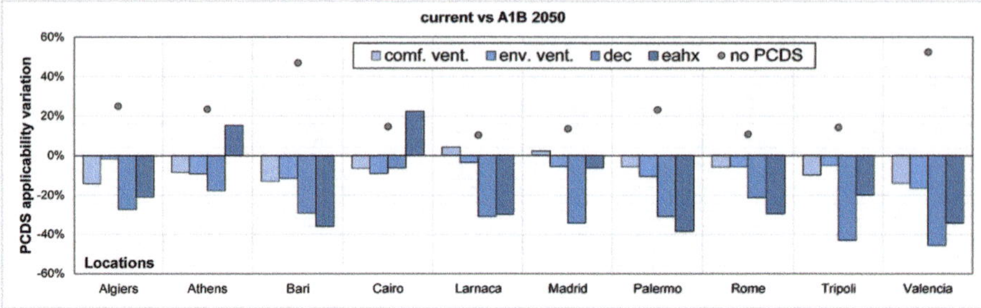

Fig. 43.3 Variation of the PCDS applicability between current and A1B 2050 TM

Table 43.2 compares the number of discomfort hours for different PCDS for current and future conditions. Finally, Fig. 43.3 shows the behaviour of PCDS potential indicators under climate change conditions (A1B 2050). Hence, it is possible to define a further indicator which describes the percentage of variation of the PCDS potential under the effect of climate changes. As is illustrated in this graph, the cooling needs are expected to increase in future years. Comfort and environmental ventilation show good resilience to expected climate change, while DEC shows a reduction in its potential. EAHX shows local behaviour with a reduction in its potential in the majority of locations, although in other sites a change in the soil surface temperature may increase its potential.

43.4 Conclusions and Limitations

The paper illustrates a method to easily define the local geo-climatic potential of different passive cooling dissipative systems in the absence of any building definition from very early-design phases. According to local environmental conditions it is possible to define which PCDS may be considered and designed in detail in further design steps, thus allowing for a better integration and diffusion of these techniques, which are very important in reducing energy needs and related GHG emissions. The described method is based on a performance-driven approach that considers technological and environmental requirements for natural/hybrid space cooling. In addition, the paper defines a series of environmentally related indicators which can analyse

- The impact of local climate conditions on the applicability of a specific PCDS
- The definition of a PCDS potential indicator based on the variation of the climate number of discomfort hours
- The impacts of climate change (A1B SRES scenario) on the geo-climatic applicability

Of course, this method is limited if compared to detailed simulation models for each PCDS since it is based on average design and site conditions, while specific

Table 43.2 The calculated number of discomfort hours under current and future TMYs

	Current					A1B 2030					A1B2050				
Location	Ref	c.v.	e.v.	dec	EAHX	Ref	c.v.	e.v.	dec	EAHX	Ref	c.v.	e.v.	dec	EAHX
Algiers	1304	723	670	515	554	1373	806	685	661	602	1629	1006	851	913	890
Tripoli	2445	1726	953	1109	1989	2551	1834	1005	1664	2067	2792	2053	1258	1923	2375
Rome	1432	896	404	609	630	1343	844	518	680	638	1585	1026	680	868	960
Palermo	1280	650	1689	509	357	1277	634	1751	676	513	1575	844	2051	920	877
Madrid	1402	985	1104	286	522	1318	882	1028	518	1085	1591	1107	1252	761	655
Larnaca	1990	1274	838	946	1218	1908	1166	761	1099	1285	2194	1371	987	1401	1598
Cairo	2799	1900	557	1320	2486	2880	1954	554	1275	2543	3205	2242	780	1619	2767
Bari	914	474	776	272	103	1109	596	705	447	283	1343	781	900	675	580
Athens	1678	1020	1534	439	869	1779	1075	1616	595	673	2070	1327	1804	814	919
Valencia	1120	562	483	281	196	1351	743	627	677	504	1707	976	895	1011	781

aspects may generate an alteration of results. With this in mind, a precautionary approach was followed to define the suggested effectiveness values and the chosen boundary conditions. Furthermore, as regards DEC and EAHX, the application of the residual CDH potential index is also suggested because they are generally used as pre-cooling technologies coupled with AHU. In this case, their potential applicability is considerably higher, because they do not reach comfort conditions directly, but they are able to reduce the airflow temperature. For example, for Cairo the EAHX applicability is low for free-running mode (−11% of discomfort hours), but medium-high for pre-cooling (−54% of CDH). Furthermore, the analysis of the effect of climate change has demonstrated that an attentive analysis of expected impacts is essential from early-design phases, because this perturbation phenomenon may affect the potential of PCDSs during their lifespan. The method proposed here may be used to give a general idea of the expected resilience of PCDS techniques in future years in order to suggest potential changes in the advanced design phases.

References

1. Santamouris M (2016) Cooling the buildings—past, present and future. Energ Buildings 128:617–638
2. Harvey LDD et al (2014) Construction of a global disaggregated dataset of building energy use and floor area in 2010. Energ Buildings 76:488–496
3. Logue JM, Sherman MH, Walker IS, Singer BC (2013) Energy impacts of envelope tightening and mechanical ventilation for the U.S. residential sector. Energ Buildings 65:281–291
4. Isaac M, van Vuuren DP (2009) Modeling global residential sector energy demand for heating and air conditioning in the context of climate change. Energy Policy 37:507–521
5. Santamouris M (ed) (2007) Advances in passive cooling. Heartscan, London
6. Cook J (ed) (1989) Passive cooling. MIT Press, Cambridge
7. Cavaglià G, Ceragioli G, Foti M, Maggi PN, Matteoli L, Ossola F (1975) Industrializzazione per programmi. Strumenti e procedure per la definizione dei sistemi di edilizia abitativa. RDB, Piacenza
8. Chiesa G, Grosso M (2018) Meta-design approach to environmental building programming for passive cooling of buildings. In: Sayigh A (ed) Sustainable building for a cleaner environment. Springer, Cham, pp 285–296
9. Chiesa G, Grosso M (2017) An environmental technological approach to architectural programming for school facilities. In: Sayigh A (ed) Mediterranean green buildings & renewable energy. Springer, Cham, pp 701–716
10. Chiesa G, Grosso M (2017) Environmental and technological design: a didactical experience towards a sustainable design approach. In: Gambardella C (ed) World heritage and disaster. Knowledge, culture and representation. La Scuola di Pitagora editrice, Napoli, pp 944–953
11. Grosso M, Scudo G, Piardi S, Peretti G (2005) Progettazione ecocompatibile dell'architettura. Esselibri, Napoli
12. Kolokotroni M, Heiselberg P (eds) (2015) Ventilative cooling state-of-the-art review. IEA EBC programme Annex, vol 62. Aalborg University, Aalborg
13. Givoni B (1994) Passive and low energy cooling of buildings. Van Nostrand Reinhold, New York
14. Grosso M (2017) Il raffrescamento passivo degli edifici in zone a clima temperato, IV edn. Maggioli, Sant'Arcangelo di Romagna

15. Chiesa G, Grosso M (2015) Geo-climatic applicability of natural ventilative cooling in the Mediterranean area. Energ Buildings 107:376–391
16. Heiselberg P (ed) (2017) Ventilative Cooling Design Guide, IEA EBC programme Annex 62. Aalborg University, Aalborg
17. Büyükalaca O, Bulut H, Yilmaz T (2001) Analysis of variable-base heating and cooling degree-days for Turkey. Appl Energy 69:269–283
18. Lee K, Baek H-J, Cho CH (2014) The estimation of base temperature for heating and cooling degree-days for South Korea. J Appl Meteorol Climatol 53:300–309
19. Day T et al (2006) Degree-days: theory and application TM41: 2006. CIBSE, London
20. Chiesa G (2016) Geo-climatic applicability of evaporative and ventilative cooling in China. Int J Vent 15(3-4):205–219
21. Chiesa G (2019) Calculating the geo-climatic cooling potential of different heat sinks. Build Simul 12(2):157–168
22. Chiesa G, Grosso M (2017) Cooling potential of natural ventilation in representative climates of Central and Southern Europe. Int J Vent 16(2):81–83
23. Erell E (2007) Evaporative cooling. In: Santamouris M (ed) Advances in passive cooling. Earthscan, London, pp 228–261
24. Ford B, Schiano-Phan R, Francis E (eds) (2010) The architecture & engineering of downdraught cooling, a design sourcebook. PHDC Press, London
25. Chiesa G, Grosso M, Bogni A, Garavaglia G (2017) Passive downdraught evaporative cooling system integration in existing residential building typologies: a case study. Energy Procedia 111:599–608
26. Aparicio-Ruiz P, Schiano-Phan R, Salmeron JM (2018) Climatic applicability of downdraught evaporative cooling in the United States of America. Build Environ 136:162–176
27. Chiesa G, Huberman N, Pearlmutter D, Grosso M (2017) Summer discomfort reduction by direct evaporative cooling in Southern Mediterranean areas. Energy Procedia 111:588–598
28. Chiesa G, Grosso M (2015) Direct evaporative cooling of buildings. A comparison amid simplified simulation models based on experimental data. Build Environ 94:236–272
29. Chiesa G (2018) EAHX—earth-to-air heat exchanger: simplified method and KPI for early building design phases. Build Environ 144:142–158
30. Labs K (1981) Regional analysis of ground and above-ground climate, ORNL/Sub-81/40451/1. Oak Ridge National Laboratory, NTIS, Springfield, VA
31. Chiesa G (2017) Climate-potential of earth-to-air heat exchangers. Energy Procedia 122:517–522
32. Stull R (2011) Wet-Bulb temperature from relative humidity and air temperature. J Appl Meteorol Climatol 50:2267–2269
33. Chiesa G (2017) Evaluating early design choices under the potential effects of climate change. SMC 6:24–28
34. Chiesa G, Palme M (2018) Assessing climate change and urban heat island vulnerabilities in a built environment. TECHNE 15:237–245
35. IPCC (2007) Climate Change 2007: Synthesis Report. Contribution of Working Groups I, II and III to the Fourth Assessment Report of the Intergovernmental Panel on Climate Change. IPCC, Geneva
36. Remund J et al (2016) Meteonorm global meteorological database V.7—handbook part I: software. Meteotest, Bern
37. AA.VV (2016) Meteonorm global meteorological database V7—handbook part II: theory. Meteotest, Bern

Chapter 44
New Horizons for Renewable Energies in Morocco and Africa

Hassan Nfaoui and Ali Sayigh

44.1 Introduction

Morocco has a strategic location in the North West of Africa and is only 14 km from Europe, across the Strait of Gibraltar. In addition, it is characterized by a varied hilly relief (the Atlas and Rif Mountains, the Sahara, etc.) and the length of its coasts (the Atlantic Ocean and the Mediterranean Sea constitute 3500 km of coastline). In the Ouarzazate region in central Morocco, for example, the annual average daily global irradiation on a horizontal surface varies from 3.86 to 7.85 kWh/m^2, which means an annual average of 7.9–11.2 h of sunshine per day. Concerning the wind potential, the annual average wind speed, at 10 m, for the Tangier zone in the north is 10 m/s, but for the Dakhla zone in the Laâyoune-Sakia El Hamra region, it varies between 7 and 8.5 m/s.

Morocco imports more than 96% of its energy needs and its electricity consumption increases by about 6% per year. For these reason, in 2009, a large-scale solar energy use strategy was launched by the creation of the "Moroccan Agency for Solar Energy" (MASEN) to develop renewable energies. This politic permits Morocco to promote regional integration between sub-Saharan African and European countries, and strengthen the links with the main players in this sector on a worldwide basis. Thanks to its model of sustainable development, Morocco will be able to identify the latest innovations, energy efficiency and clean technologies.

The new strategy adopted is based on increasing the contribution of renewable energies to the national electricity installed at 52% by 2030, and on the development of a clean energy economy. This will enable Morocco to reduce its energy bill and

H. Nfaoui (✉)
Solar Energy & Environment Laboratory, Sciences Faculty, Mohammed V University in Rabat, Rabat, Morocco

A. Sayigh
WREC, WREN, Brighton, UK

A. Sayigh (ed.), *Renewable Energy and Sustainable Buildings*, Innovative Renewable Energy, https://doi.org/10.1007/978-3-030-18488-9_44

its dependence on energy from abroad. In addition, It will limit its greenhouse gas (GHG) emissions and fight against Climate Change (CC), consequently to contribute to the preservation of the environment.

44.2 Energy Sector

44.2.1 Energy Consumption

The Moroccan energy balance remains characterized by the predominance of petroleum products which represents, in 2016, 54% of the national energy consumption, evaluated at 19.58 Mtoe. Coal ranks second with 26% (Fig. 44.1).

44.2.2 Natural Gas

Natural gas remains one of the cleanest commercial fuels. Figure 44.2 shows that the consumption of natural gas remained almost negligible until the year 2004. It increased significantly when the Maghreb-Europe gas pipeline was putting into service, transporting Algerian gas to Europe (Fig. 44.3), as part of the Moroccan security of energy supply by diversifying its energy sources.

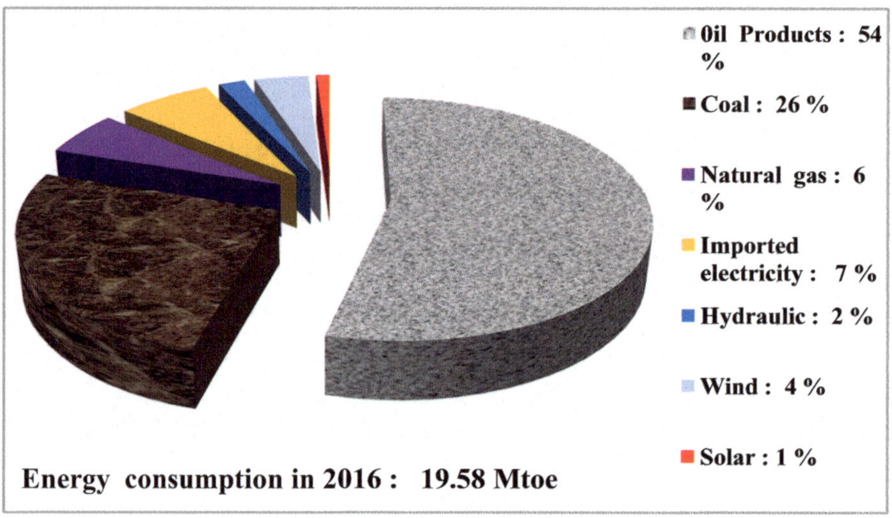

Fig. 44.1 Distribution of national energy consumption in 2016 [1–3]

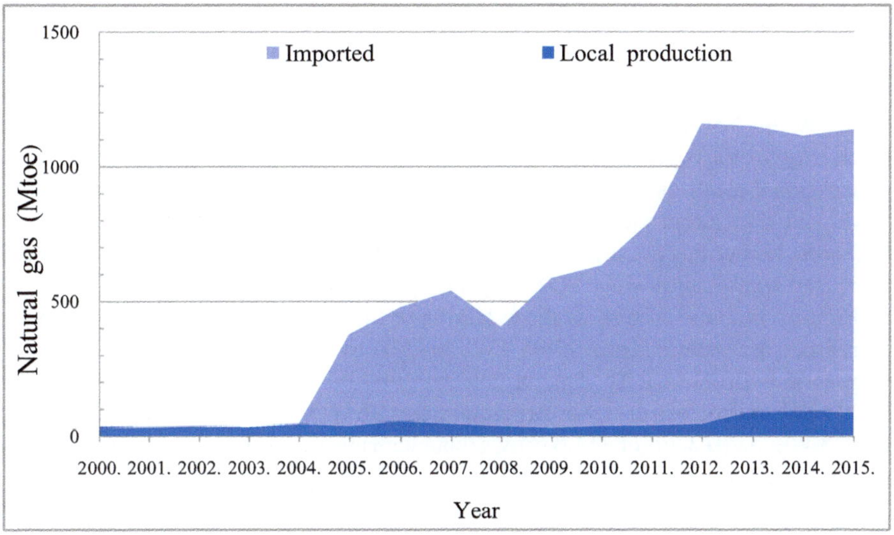

Fig. 44.2 Evolution of national consumption of natural gas [1–3]

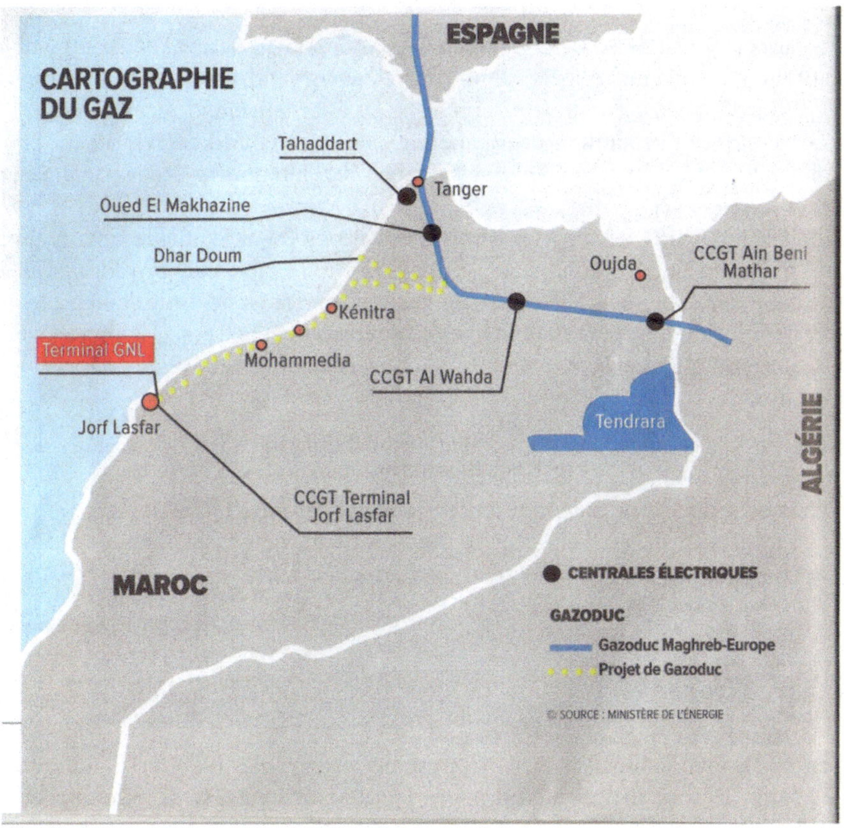

Fig. 44.3 Gazoduc Maghreb—Europe

44.2.3 Electricity

Until 1996, of thermal and hydraulic origin, almost all electricity production is provided by the National Electricity Office (ONE). The gradual liberalization of the electricity sector began in 1997, either by deprivation of refining companies or by concessional electricity production (Fig. 44.4). The electrical power installed by the ONE and the autoproducers is 8155 MW in 2015. Figure 44.4 shows that the hydraulic power has not changed much since 2005. Thermal and wind power have increased significantly since 2008, especially for thermal power.

Figures 44.5 and 44.6 show the evolution of the net electrical energy demand has increased significantly, from 17.95 TWh in 2004 to 34.41 TWh in 2015. The contribution of hydroelectricity depends on rainfall. For the same installed capacity, 1800 MW, hydroelectric power generation was 3631 TWh in 2010 and it dropped to 1816 TWh in 2012. Figure 44.6 shows that electricity consumption varies from year to year. It Is around 6% on Average.

44.3 Renewable Energies

In its speech Royal of 30/07/2008, the King Mohammed 6, recalled that the objective of the solar plan is to increase Morocco's energy independence. The sovereign said: "Morocco must continue efforts to make alternative and renewable energies the cornerstone of its national energy policy". He added: "Morocco made the choice of an energy transition towards sustainable, by integrating Renewable Energy sources into its energy mix".

In 2015, MASEN became the Moroccan Agency for Sustainable Energy. It will be tasked to do the overall strategic planning of renewable energy projects. It will take charge of all projects for electricity generation from renewable sources, including feasibility studies, planning, financing, development and construction as well as operation and maintenance. Among the concerns and objectives of MASEN, we quote also:

- The implementation and the evaluation of the national energy strategy, in particular that related to renewable energies
- The cost-effectiveness appropriation for energy storage technologies
- The effects of the reduction of storage costs on the expectations of electricity demand and the supply strategy of the national market

44.3.1 Hydraulic

Morocco faces imminent danger of shrinking water tables (aquifers) and drainage of its dams. Despite the Moroccan policy to rationalize the use of water, the consequences of global warming will be greater than those of other countries, like for

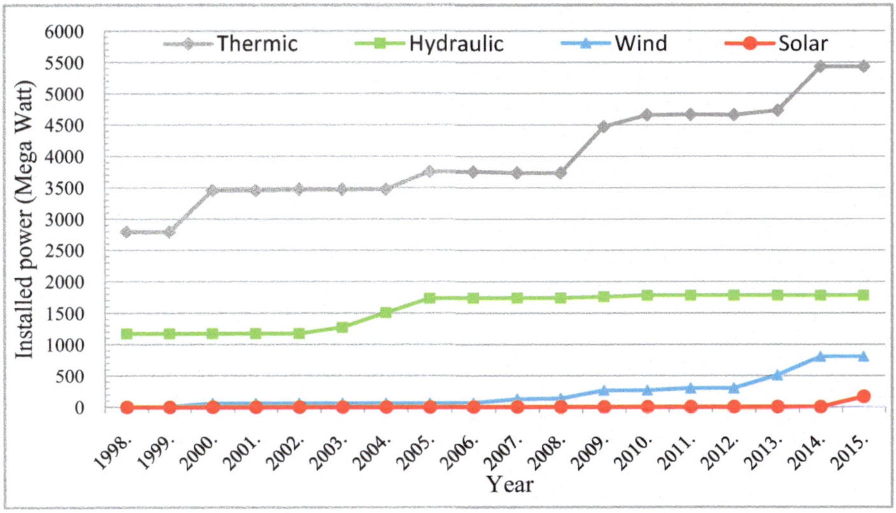

Fig. 44.4 Evolution of the national electric power [1–3]

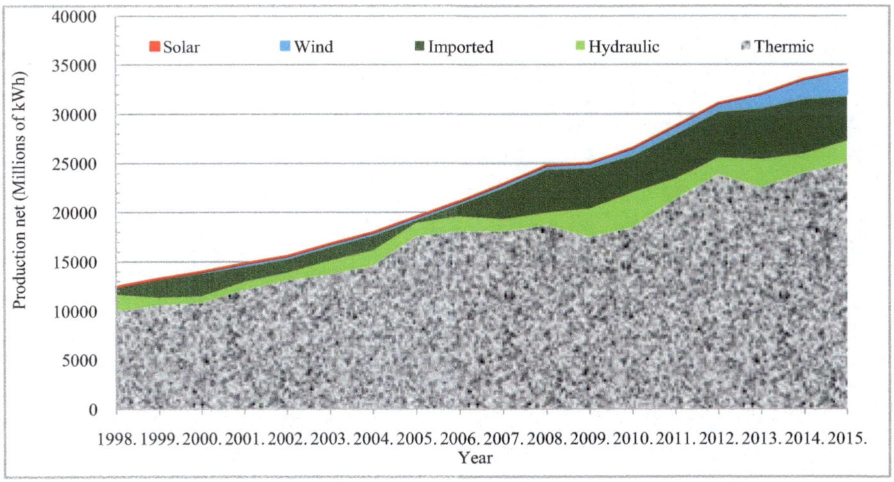

Fig. 44.5 Evolution of the production of national net electricity [1–3]

example Spain and India, due to its location in a geographical area that has suffered for years from the decline of the rains and the acceleration of the desert climate [4]. This has pushed Morocco to rethink its water policy and take into account the climatic specificities of Morocco, by developing a great know-how in the field of feasibility studies of dam constructions, recognized worldwide. Figure 44.9 shows that the surface water resources are in average year at 18 billion m^3. This unequal distribution of water resources over time makes it necessary to store supplies from wet years for use during periods of drought (Fig. 44.7).

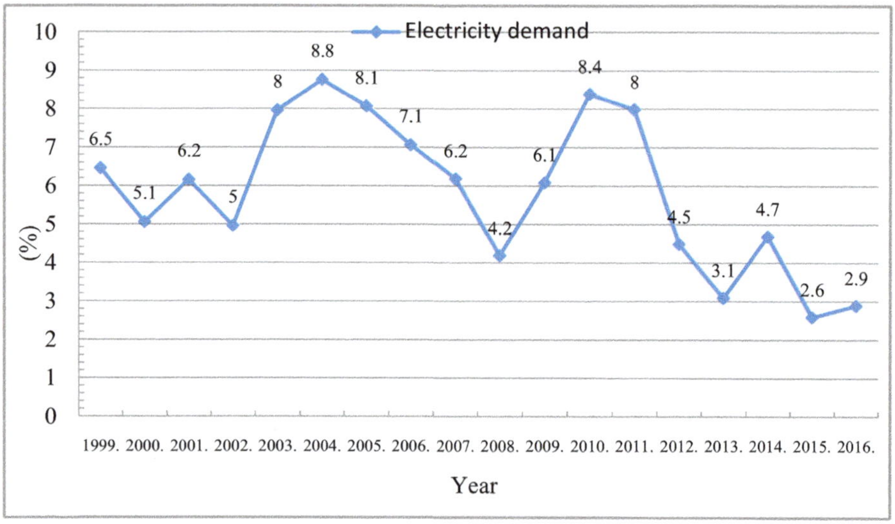

Fig. 44.6 Evolution of electricity demand growth [1–3]

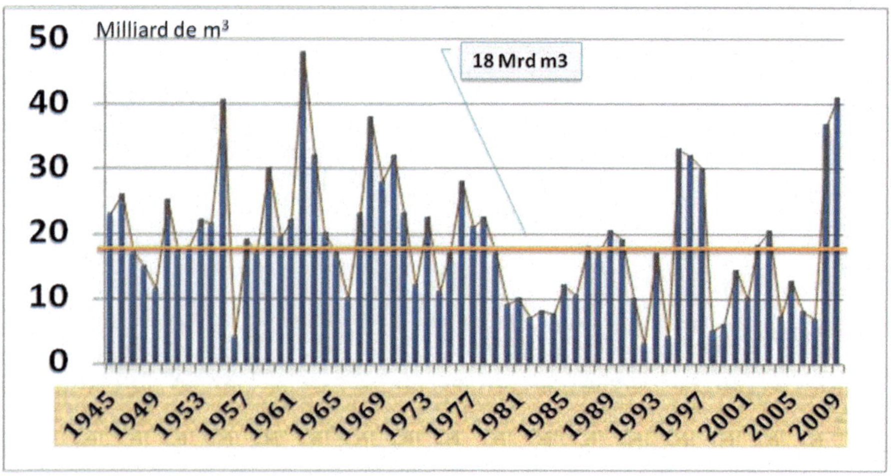

Fig. 44.7 Unequal distribution of water in time [5]

About that, the construction of dams has been the primary orientation of water management in Morocco. In 2015, Morocco managed to build 139 large dams instead of 9 in 1950 (Fig. 44.8).

For example, the Morocco's second large dam, AL Massira shrunk by 60% in 3 years due to recurring drought (Fig. 44.9).

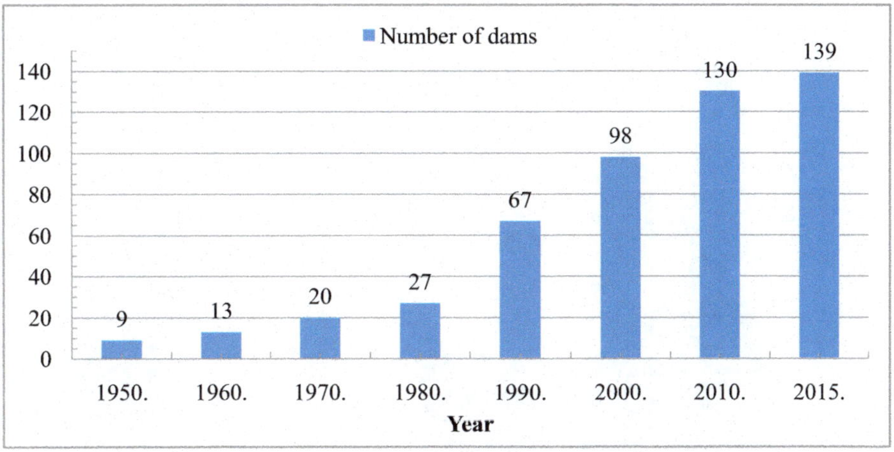

Fig. 44.8 Evolution of dam construction [5]

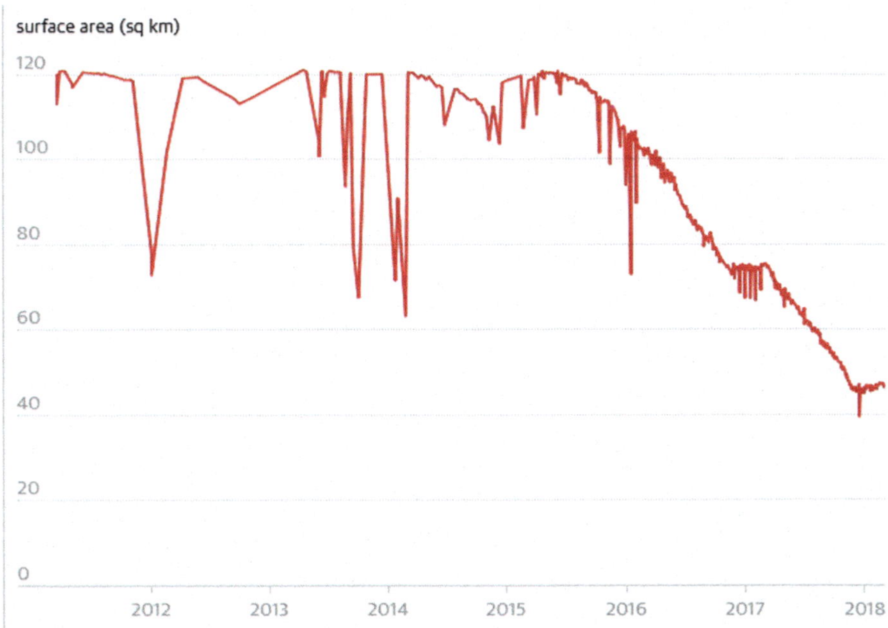

Fig. 44.9 Starkest decline of AL Massira dam [4]

In addition of the development of the skills, research consultancy and Moroccan companies, Morocco succeeded to dispose of a national competence centre that is currently exported to foreign countries wishing to carry out studies or building dams (Fig. 44.10).

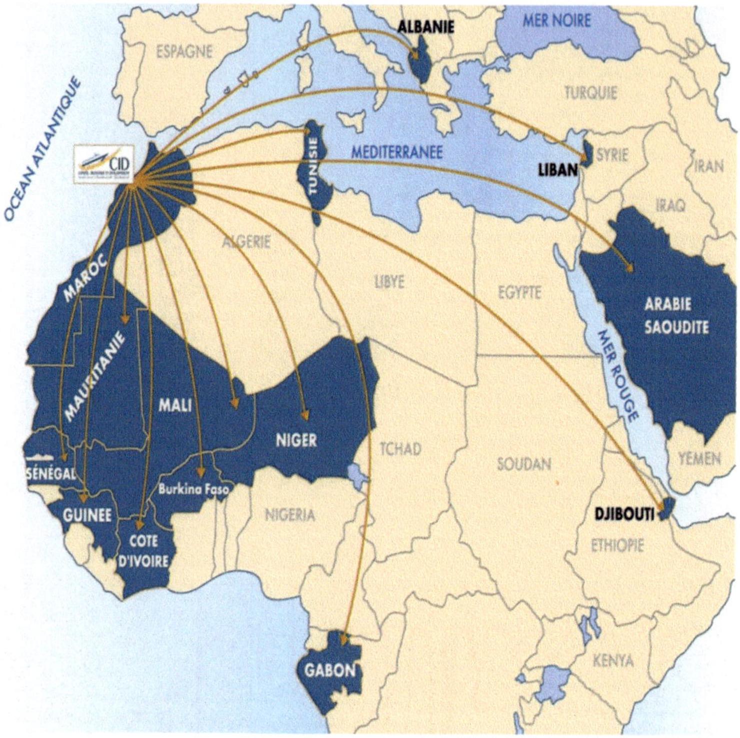

Fig. 44.10 Moroccan cooperation in the hydraulic sector [5]

The hydroelectric sector has also been particularly important since 1960. The contribution of installed hydroelectric power (1800 MW), which depends on rainfall, is between 5.1% and 13.7% of the net electricity consumption (Fig. 44.11).

44.3.2 Solar

The first hybrid power plant (470 MW), incorporating 20 MW of solar power, was built in 2010. It is a Solar-Natural Gas Hybrid Plant. The second giant solar power station "Noor-Ouarzazate" (582 MW), one of the largest solar energy complexes in the world, was installed between 2016 and 2018. It uses concentrated solar technology (CSP) (360 MW), CSP technology with a solar thermal tower and heliostats (150 MW) and photovoltaic technology (72 MW). The electricity storage capacity is 7 h, which will be used during peak hours. It consists of four multi-technology solar sub-plants, which will be developed in full compliance with international standards, both technologically and environmentally, including a platform

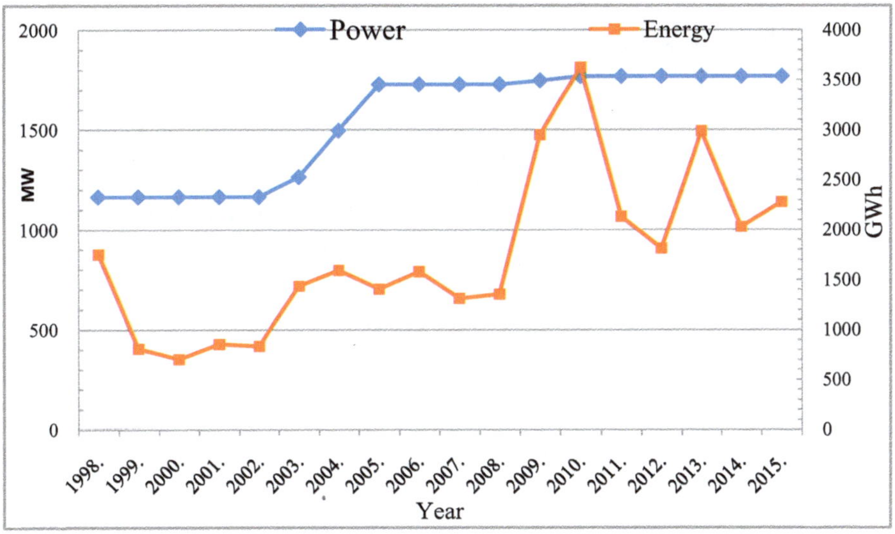

Fig. 44.11 Evolution of hydraulic power and electricity [1–3]

for scientific research, tests and training. For the solar plant "Noor-Ouarzazate", the price of kWh reached $ 0.19, comparable to that of the solar power plant Redstone Arsenal (USA), a competitive price with traditional sources of energy (Fig. 44.12).

44.3.3 Wind

The first wind farm, installed in Morocco in 2000, is Koudia Al Baida (Tetouan), with a capacity of 50 MW. However, the largest wind farm is Tarfaya in southern Morocco (300.1 MW), putting into service in 2014. It is the largest in Africa, outclassing that of "Gulf of EL-Zay" (200 MW) in Egypt. The most recent park, putting into service in 2018, is "Jbel Khalladi" (Tangier), with a capacity of 120 MW.

In 2016, wind power installed in Morocco reached 898.416 MW, or 10.87% of its national power, generating 2858.257 GWh, or 8% of its national electricity consumption (Figs. 44.13 and 44.14). The wind power installed in Morocco exceeded 1.0155 GW in 2018, classified second in Africa, after South Africa (2.2947 GW).

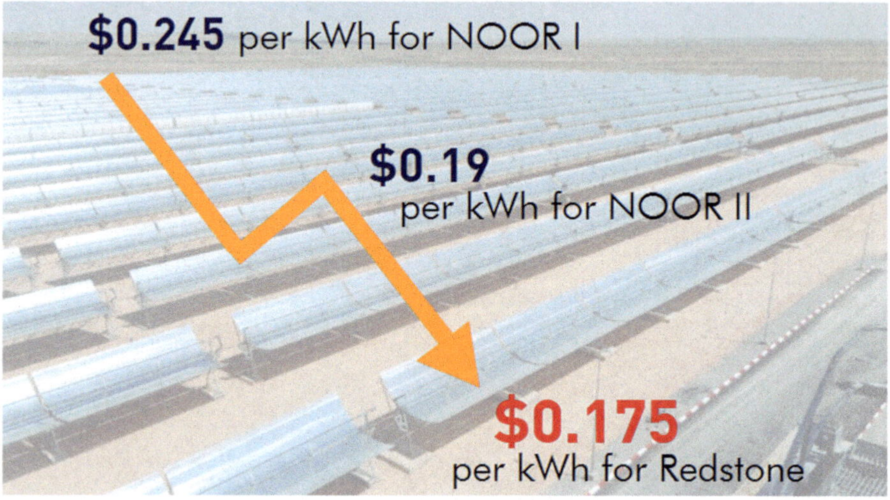

Fig. 44.12 Falling cost of concentrated solar power [2]

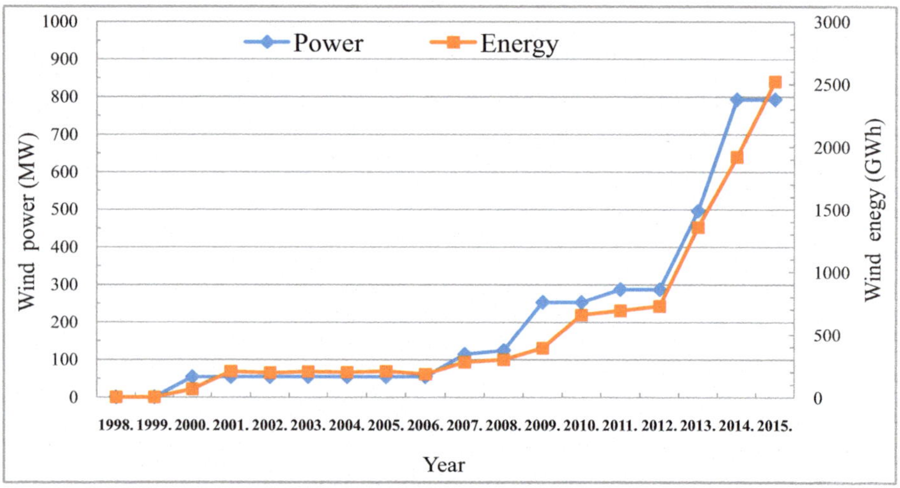

Fig. 44.13 Evolution of wind power and electricity [1–3]

44.4 Morocco: Renewable Energy, an Exemplary Model in Africa

Since the beginning of the twenty-first century, the contribution of renewable energies to the national electricity consumption has become significant and in constant growth, reflected by the completion in 2018 of the installation of "Noor-Ouarzazate"

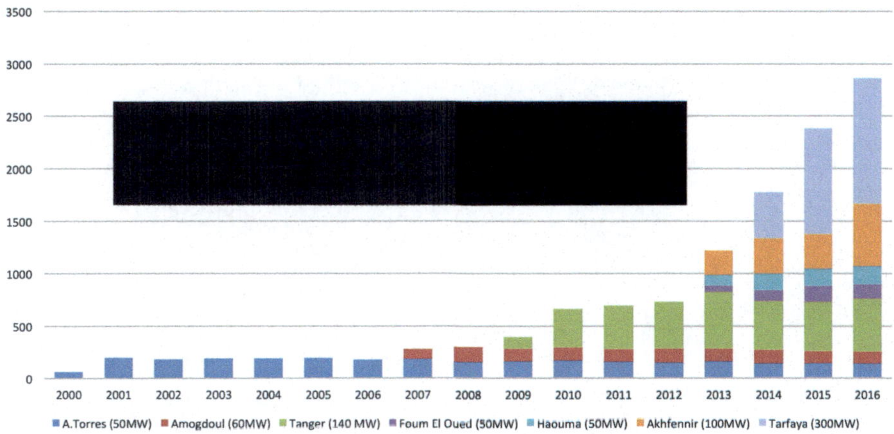

Fig. 44.14 Wind power generation (GWh) [6]

solar power station (582 MW) and the installation of the wind farm "Jbel Khalladi" in Tangier (120 MW).

The share of solar and wind power will be 40% in 2030 compared to 13% in 2016 and only 2% in 2009 (Fig. 44.15). Morocco managed to break through the 1000 MW in wind power installed in 2017, overtaking Egypt and approaching the level of South Africa. The acceleration of the deployment of the national strategy in this sector will consolidate its leading position on African scale.

Morocco, with this energy policy, opens up new horizons for the future of clean energy development in Africa. Its strategy includes, in addition to the development of solar thermal, photovoltaic and wind energy, the introduction of a series of legislative, regulatory and institutional reforms, while improving tax benefits to encourage investment in this sector. Moreover, starting from 2015, for a rapid and successful energy transition, it will have increased the share of renewable energies to 52% of its installed electricity capacity by the year 2030.

It also considers itself to be one of the most active African countries for investment in this sector. This makes it one of the pioneering African countries and it is the first in Africa in the solar energy sector. Morocco has made renewable energies a real lever for South-South cooperation and a vector of development for sub-Saharan African countries, having proven renewable potential. It also plays a leading role in African development through training and experience provided for a range of African countries.

Moroccan renewable energy projects are arousing an increasing interest from international and local investors, allowing Morocco to attract more investment in this sector, such as the installation, in Morocco, of an industrial unit of the group German Siemens, with an investment of nearly 100 million Euros for the manufacture of wind turbine blades.

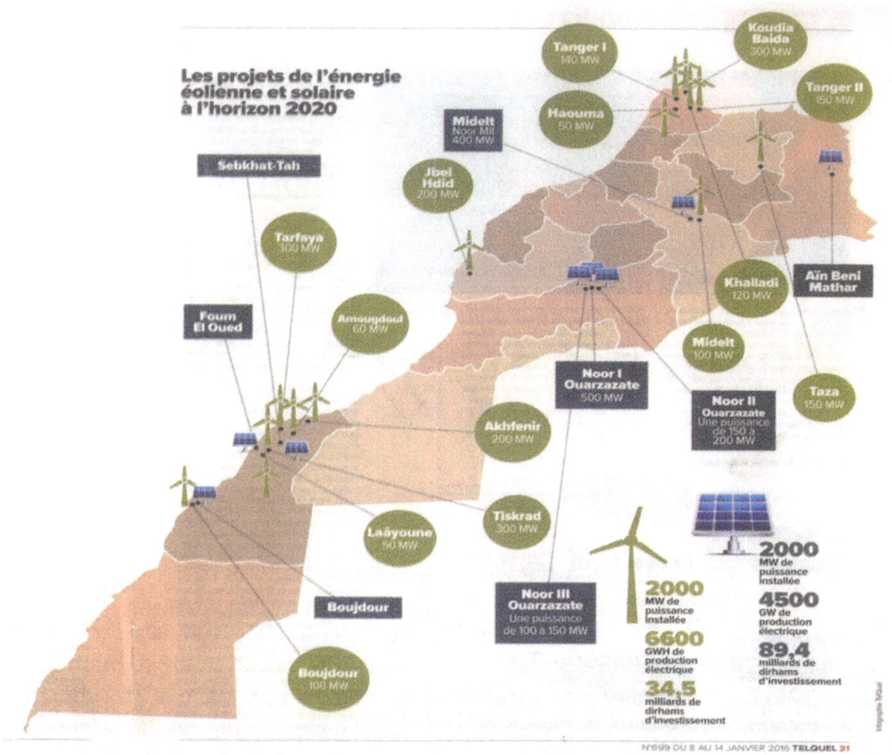

Fig. 44.15 Projects of solar power plants and wind farms in 2030 [1–3]

44.5 Morocco, Driving Force of South-South Cooperation in Africa

44.5.1 New Foreign Policy in Africa

Regaining its membership in the African Unity (AU) in 2017, Morocco set up a new foreign policy in Africa by mobilizing internal cooperation between African countries rather than remain at the mercy of foreign powers. It based on two priorities: sustainable development and economic integration. As examples, the Moroccan Monarch, Mohammed VI, has extended its visits not only to West Africa, or to the Francophone African countries, but even the Anglosaxon ones. Moreover, he participated in person at the Summit of Heads of State at the Climate Committee and "Elazrak" Funds for the Congo Basin, held in April 2018, with different agendas:

1. Supporting the African countries, in order to face the negative impacts of climate change on the African agricultural sector by taking advantage of Moroccan experience in this sector
2. Activating the "Elazrak Funds" for the Congo Basin, in accordance with the recommendations adopted at the African Summit held on the sidelines of the

conference of the of the United Nations (COP22) on climate change in Marrakech, in 2016, as well as the two regional commissions, namely Sahel Committee and Island States Committee [7]

44.5.2 *Transfer of Its Know-How to African Countries*

The Kingdom's national renewable energy model is consolidating its leadership position at African level. It has made this sector a real lever for South-South cooperation. In addition to its know-how and recognized expertise in the field of assistance services, Morocco takes advantage of its Science and Technology Center in Casablanca: It is chosen in 2013 as a centre of excellence by the Association of Electricity Enterprises of Africa. Here are examples of cooperation carried out by Morocco with African countries:

- Improving the energy services provided to African populations by facilitating access to training for technicians and strengthening their capacities
- Management and supervision of energy production, transmission and distribution infrastructure projects
- Completion of specific training courses for African national electricity companies

In addition to cooperation in the water sector through the construction of dams, there is cooperation in the electrical energy sector (organization of training, exchange of experiences, technical assistance by the study feasibility of rural electrification projects, incentives for private investors to carry out electricity generation projects using renewable energies).

44.5.3 *Electricity*

The European electricity grid is the largest interconnected network in the world in terms of power transmitted. Since 1997, the Moroccan power grid is the only African network connected to the Europe one. Morocco's planning is to be a regional centre for the export of competitive clean electricity to its neighbours, Spain and Algeria (Fig. 44.16).

44.5.4 *Natural Gas*

The construction of more than 5000 km of gas pipeline connecting Nigeria, Africa's third largest gas producer, and Europe through Morocco is essential. Gas is less of a pollutant than coal and oil. The realization of this important economic and social project will strengthen Morocco's relations in Africa and in Europe. This strategy gives credibility to its economic policies and creates confidence in its North African diplomacy (Fig. 44.17).

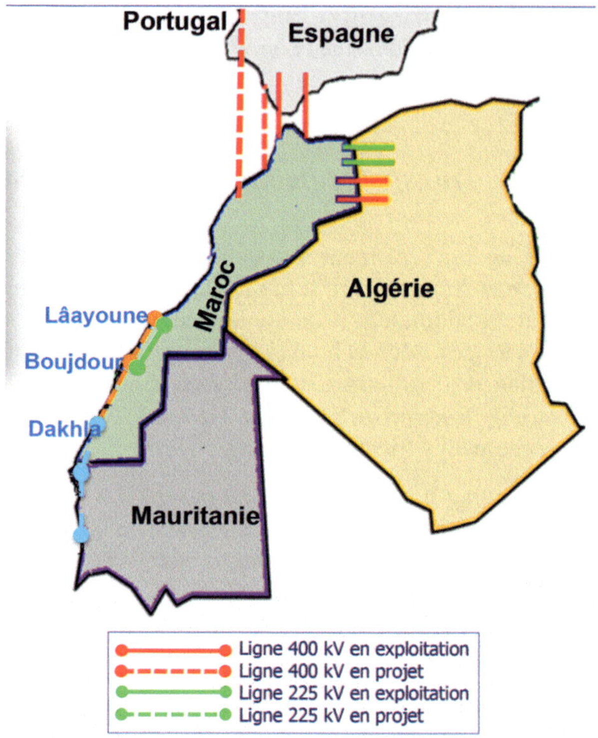

Fig. 44.16 Electrical interconnection [6]

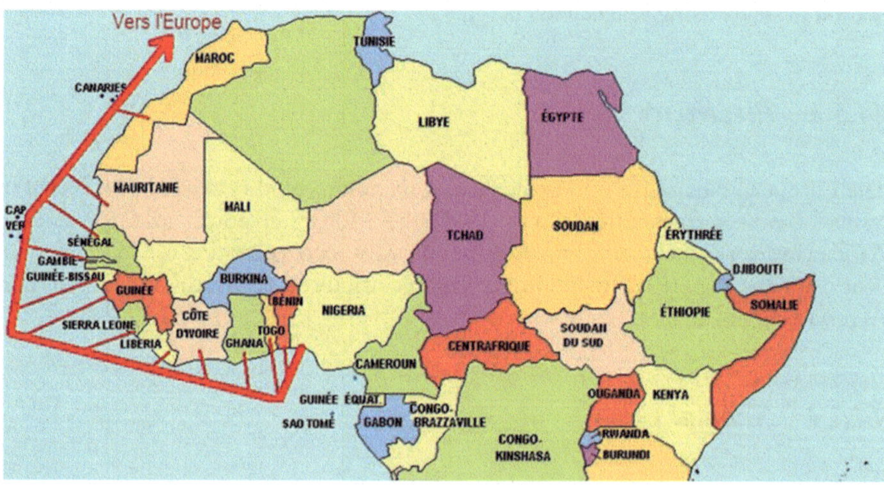

Fig. 44.17 Nigeria-Morocco gas pipeline officially launched

44.6 Conclusion

Morocco has available a significant potential in solar and wind energy. It is characterized by unequal distribution of water resources over times. On top of building more than 139 large dams, Morocco developed a great expertise in the field of studies and construction of dams, recognized worldwide.

By capitalizing on the national expertise, MASEN has developed an exemplary model of renewable energy in Africa. Its strategic geographic position, strengthened by the interconnected electricity transmission with Spain and Algeria, makes the Kingdom a hub for trade with Africa and Europe. Morocco became one of the pioneering African countries investment in renewable energies and first for solar energy use in Africa. For "Noor-Ouarzazate", one of the largest solar plant in the world, the price of KWh reached $ 0.19, a competitive price with traditional sources of energy. Concerning wind energy, the largest wind farm is the Tarfaya one (300 MW), installed in 2004. It is the largest in Africa. The Wind power installed exceeded 1.015 GW in 2018, classified second in Africa, after South Africa (2.295 MW). It generated 2858 GWh in 2018 (8% of Moroccan annual electricity consumption).

Morocco's Renewable Energy model consolidated its leadership position at African level and opened up new horizons for the future of clean energy development in Africa. In 2017, Moroccan adopted a new foreign policy in Africa, based on two priorities: Sustainable development and Economic integration.

Since 1997, the Moroccan power grid is the only African network connected to the Europe one. Morocco's planning is to be a regional centre for the export of competitive clean electricity to its neighbours, Spain and Algeria. The construction of more than 5000 km of gas pipeline connecting Nigeria and Europe strengthens Morocco's relations in Africa and Europe.

Morocco's reintegration of the AU puts at the disposal of Africa its knowledge, skills and experience in dams construction, agri-food industry, renewable energy model, etc. It offers endogenous development and strong economic integration in Africa comparable to that of the European Union.

References

1. Annuaires statistiques du Maroc (1988–2017), Direction de la statistique
2. www.masen.gov.ma, www.one.gov.ma, www.mem.gov.ma
3. Morocco (2014) Energy Policies. Beyond IEA countries. https://www.iea.org/publications/freepublications/publication/Morocco2014.pdf
4. https://www.theguardian.com/environment/2018/apr/11/day-zero-water-crises-spain-morocco-india-and-irak-at-risk-as-dams-shrink
5. https://docplayer.fr/51193702-La-politique-et-les-bienfaits-des-barrages-au-maroc-presentee-par-mr-khalid-el-ghomari-directeur-des-amenagements-hydrauliques.html
6. https://au.int/sites/default/files/documents/32251-doc-lome150317vfhaj_maroc.pdf
7. http://www.renewableenergyworld.com/articles/2017/04/what-will-remain-of-the-climate-challenge-conference-the-fate-of-the-planet-between-paris-and-marrakech.html

Chapter 45
Developing Energy Control and Optimisation Methodology for Built Environment of the Future

Monday Ikhide, Alex Egaji, and Abdullahi Ahmed

45.1 Introduction

A built environment is formed by the buildings and other structures constructed by humans including water and drainage systems, power systems, communication systems and transportation systems, whereas building itself accounts for 30–40% of total global energy consumption [1–3]. Studies have shown that controlling and optimising the amount of energy consumed or utilised in buildings is the first step in realising an energy efficient built environment [4, 5]. This will not only save cost by minimising the level of re-enforcement on the supply system or reduction in the bills incurred by the occupant but ultimately reduce the amount of CO_2 emission. This will result in huge benefit to the building occupants and owners, the energy provider and the government. Generally, non-residential buildings present the highest energy consumption, hence accounting for 19% of the total global energy consumption [1, 2]. In all, non-residential buildings contribute towards 30% of the total CO_2 emissions [2] whereas CO_2 emission is also largely dependent on the energy demand. Therefore, developing measures and strategies to control and optimise the total energy consumption in non-domestic buildings is a major step in reducing CO_2 emission, which ultimately reduces the total energy demand. Studies have also shown that through appropriate control measures and strategies, a carbon reduction of over 70% could be achieved in non-residential buildings at no net cost [4]. A separate study in [5] also revealed that fitting a full set of controls to an older heating system which previously had none could

M. Ikhide (✉) · A. Ahmed
Faculty Engineering, Computing and Environment, School of Energy, Construction and Environment, Coventry University, Coventry, UK
e-mail: monday.ikhide@coventry.ac.uk; abdullahi.ahmed@coventry.ac.uk

A. Egaji
Faculty of Computing, Engineering and Science, Centre of Excellence in Mobile and Emerging Technologies (CEMET), University of South Wales, Wales, UK
e-mail: alexander.egaji@southwales.ac.uk

A. Sayigh (ed.), *Renewable Energy and Sustainable Buildings*, Innovative Renewable Energy, https://doi.org/10.1007/978-3-030-18488-9_45

help save more than 15% of the total energy bills. In the light of this, energy efficiency and saving strategies have become a top priority objective for energy policies in most countries, hence making many countries to introduce a range of legislation which differs from country to country [2, 6]. For example, the EU has embarked on legislative campaigns to promote the improvement of energy performance of buildings. These legislative directives include the 2010 Energy Performance of Buildings Directive [7] and the 2012 Energy Efficiency Directive [8], among others. There have also been increasing demands by building inhabitants for improved indoor environmental comfort while reducing the energy consumption and reducing the CO_2 emissions. Hence, there is a need for an energy and comfort management system (ECMS) which comprises intelligent controls for building using a computer, microprocessor, storage devices and communication links [9]. The ECMS can aid in fulfilling the occupant expected comfort index while reducing the energy consumption with regard to energy price variations. The ECMS often requires some indoor parameters to meet the occupant's expected comfort level, and some of the most significant parameters are thermal, humidity, indoor air quality and illumination levels.

The basic components of a typical building control are shown in Fig. 45.1 [5]. A very key element of building control is the Building Energy Management Systems (BMES). BMES is a centralised system that monitor, control, and record all aspects of building services in buildings to increase energy efficiency, thus reducing environmental impact and increasing savings on the part of building occupants. BEMS ensure that these systems are operated based on occupancy and energy demand. It also monitors the operation of the systems and checks for faults, sends an indication to initiate an action if otherwise. Generally, these devices would also communicate with each other as per Internet of Things (IoT). The IoT is a network of interconnected objects or devices embedded with sensors and mobile devices which can generate data, communicate and share that data with one another [10]. It is envisaged that the proliferation of IoT related technologies such as low-cost sensors and high-speed networking would further foster the adoption rate of "smart built environment" solutions over the next few years [10]. In all, understanding and predicting the energy usage pattern plays a key role in making management and investment decisions. This process is referred to as Load Forecasting (LF), which can either be short term, medium term or long term.

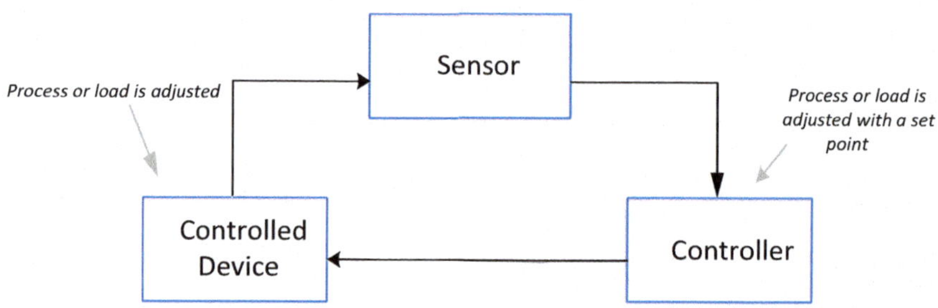

Fig. 45.1 Basic elements of building controls [5]

45.2 The Smart Built Environment

A smart built environment (SBE) deals with "*driving better usage and utilisation of existing and new assets through data and analysis to deliver services as sustainably as possible for the benefit of citizens*" [11]. The SBE will therefore employ modern and complex technology to ensure that future built environment is energy efficient, productive and comfortable place to live, work and travel to or from. Generally, future buildings will be able to integrate with the local supply network and modern transport systems for them to be considered as smart or intelligent. Furthermore, these building will be potential sources of renewable energy while also participating in demand side management (DSM) [11]. The primary drivers for smart built environment are operational efficiency, cost reduction and environmental sustainability. The drive towards a smart built environment will lead to proliferation of intelligent buildings. Intelligent buildings use wireless sensor technology (WST) and data mining methods to increase their energy efficiency over time by anticipating and meeting their occupants' needs [10]. The sensors consist of algorithms that can learn from previous historical data set to predict future energy demand while optimising the total energy usage and ensuring occupants comfort and wellbeing. Generally, an intelligent building will have the ability to coordinate the entire building energy systems as well as improving the overall building efficiency. Considering the increased penetration of renewable energy sources (RES) such as photo-voltaic generation (PV) generation in the built environment, energy storage system (ESS) will also play a key role in intelligent buildings in the efficient utilisation of the energy generated from these RES. Therefore, RES and ESS are key enablers for the realisation of the SBE. Generally, energy storage technologies include mechanical storage (such as pumped hydro storage, compressed air energy storage and flywheels) and electromechanical storage (such as batteries, chemical energy storage (such as hydrogen and synthetic natural gas), elevated temperature thermal storage and electromagnetic storage) [12], with each having their own relative advantages and disadvantages. Generally, enabling the integration of more renewables such as solar and wind will decrease the need to invest in new conventional generation capacity. This ultimately will result in financial savings and reduced CO_2 emissions. Furthermore, it will also improve energy security by optimising the supply and demand, providing system stability during electricity outages and storing energy when prices are low and used on site when they are high, resulting in huge savings and loss reduction [12]. However, to stored energy, the actual energy actual energy consumption must be accurately predicted in the first instance. Accurate LF would help utility supplies to predict the required load and therefore supply the load required. LF also find useful advantages in DSM. DSM would also help in planning, implementation, and monitoring the energy consumption during peak hours [13].

It is also envisaged that future transmission and distribution network will be interconnected to form the so-called Smart Grid (SG). As indicated in [14], a SG is an advanced electricity transmission and distribution network which uses information, communication and control technologies to improve economy, efficiency, reliability and the security of the grid. Key aspect of SG is the ability to predict or forecast future load demand based on existing data; in this way only the actual demand is provided, thus minimising loss and wastage while improving energy efficiency and ensuring the satisfaction of end users. Other key aspects of future SGs include demand side response (DSR), integration and deployment of low carbon technologies to the grid such as solar and wind, energy storage systems, and power quality and power factor improvement, among others. According to [13], the advantages of SG are to make power systems more efficient, reduce peak-to-average load, and minimise the cost of production as well as pave the way for integrating renewable energy resources (RES). Another major aspect of SGs is the smart metering. Generally, the proliferation of smart metres in buildings is a catalyst towards the drive in reducing energy consumption which ultimately will reduce greenhouse emissions. It will also enable better management of energy consumption within the building as well as save cost [2].

Data collection and processing is also an integral aspect in the decision-making and engineering management processes within the built environment. This can either be within the building, within a small community or the city at large [10]. Furthermore, the need for efficient data management will also escalate as the information technology becomes more integrated into the built environment. These include deployment of sensors, advanced communication networks, etc., which will be able to collect the data from the physical environment. Furthermore, and as indicated in [15], these diverse types of sensor technologies coming into the built environment have the capacity to feed almost unlimited quantities of data, so there is need to be selective in what data is collected and processed. Generally, the issue of *big data* and *data analytics* will also come into play; where the former refers to the generic term for an enormous collection of datasets from multiple sources and the latter refers to a set of processing mechanisms, techniques and methods used to analyse and interpret the datasets; for extracting information and meaningful value from them, and in the visualisation of data [15]. Generally, managing these huge amounts of data involves three stages: data collection, data analytics and data use or reuse (Fig. 45.2). The effective analysis and utilisation of big data is a key factor for success in the energy efficiency and optimisation in SGs and the SBE [16].

45.3 Energy Efficiency and Optimisation in a Smart Built Environment

As indicated in [17], one of the greatest challenges facing the EU is how best to design and adapt cities into smart, intelligent and sustainable environments. Generally, a smart built environment will have the same attributes of smart cities, such as [17] "*offering the maximal quality of living to its inhabitants with minimal consumption of resources by intelligently joining of infrastructure (energy, mobility, transport communication, etc.) on different hierarchical levels (building, district, and city)*".

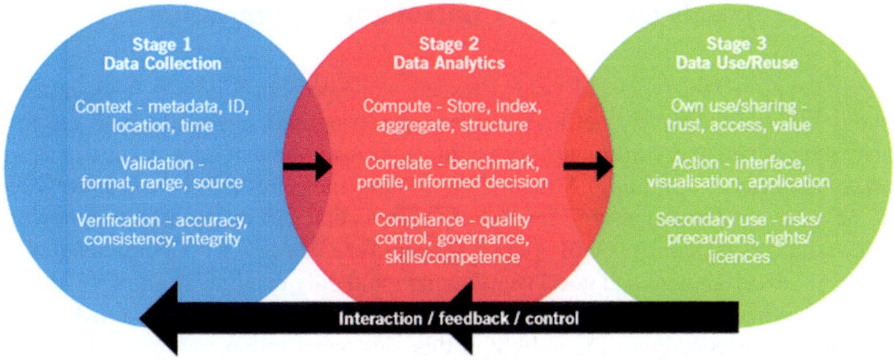

Fig. 45.2 Data lifecycle: collection, analytics, use/reuse and feedback/control [15]

However, a fundamental aspect of the smart built environment is how to efficiently and effectively utilise the energy. In general, the methods of energy efficiency in buildings can be grouped into two broad categories: the passive and active energy efficiency measures. The passive energy efficiency measure focuses on the thermal insulation issues of the building fabrics such as thermal insulation issues (thermal conductivity-values, etc.) insulation, glazing, etc. Active energy efficiency can be achieved when not only are energy saving devices and equipment installed but also that they are controlled and monitored to minimise wastage and loss. This aspect of energy efficiency strategy is very vital since it helps to optimise the use of energy so that only the required energy is used. The three main types of active energy control strategies in buildings are *controlling by time*, *controlling occupancy* and *controlling the condition*. Traditional time control utilises timer switches to turn on/off building appliances (lighting, HVAC systems, etc.) based on the occupant's usage and the activities taken place within the buildings. Types of time switches includes 24-h timers (turning on/off at various set times during the day), 7 day timers (turning on/off devices for specific days in the week). Occupancy control is mainly used in lighting control and in extract fans in bathroom. However, with the advances in technology, they can be used to control the HVAC systems. Notable types of condition control are temperature, daylight levels, humidity and carbon dioxide (CO_2) control. Modern technologies also incorporate more than one type of sensors for better response and improved efficiency, resulting in hybrid sensors. Good passive design ensures thermal comfort, low heating and cooling bills, and ultimately ensures reduced greenhouse gas emissions for the entire lifespan buildings. Passive design takes advantage of natural sources of heating and cooling (such as sunlight and cooling breezes), which is achieved by appropriately orientating the building as well as carefully designing the building envelope (roof, wall, windows and floors) [18]. Passive design strategies can either be incorporated during the design stage or into an existing building as per retrofitting. Aspects of the passive design include [18] design for climate, orientation, shading, passive cooling, passive solar heating, airtightness, thermal insulation, glazing, thermal mass and skylights. However, for optimum results, the building occupants must also understand how the building and associated devices work.

45.4 Proposals for Energy Control and Optimisation

Having explored the types of Energy Efficiency measures in buildings, a generalised and hybrid building energy efficiency strategy is proposed. Generally, both the passive and active energy efficiency strategies must complement each other to ensure an energy efficient building. Although the passive measures will be fixed, however complex control strategies will involve algorithms that will dictate how and in what circumstances the controls will operate. Generally, a building with a high insulation level, good airtightness and good ventilation would only result in an energy efficient and sustainable home if only the occupants are aware and properly trained on the building services gadgets. In the light of this, this paper proposes a generalised and integrated hybrid energy optimisation and energy efficiency strategy that encompasses all three elements which include *passive, active and user awareness* (Fig. 45.3). As indicated in [19], employee or occupant engagement in energy efficiency and carbon reduction can help improve their behaviour in the workplace, thereby reducing unnecessary energy consumption and ultimately cutting organisation's carbon emissions. Generally, this can be achieved through communication and engagement and by adopting any one or a combination of the following measures [20]: information and promotion, training, personal advice and one-to-one engagement, demonstrations, benchmarking, commitment, goal-setting, labelling, prompts, modelling, and feedback. Others include economic incentives and disincentives (such as subsidies, levies, surcharges, taxes, bonuses, tax differentiation, tax refunds, financial instruments such as interest-free loans, rewards and penalties); use of posters and stickers, and regulatory issues [20].

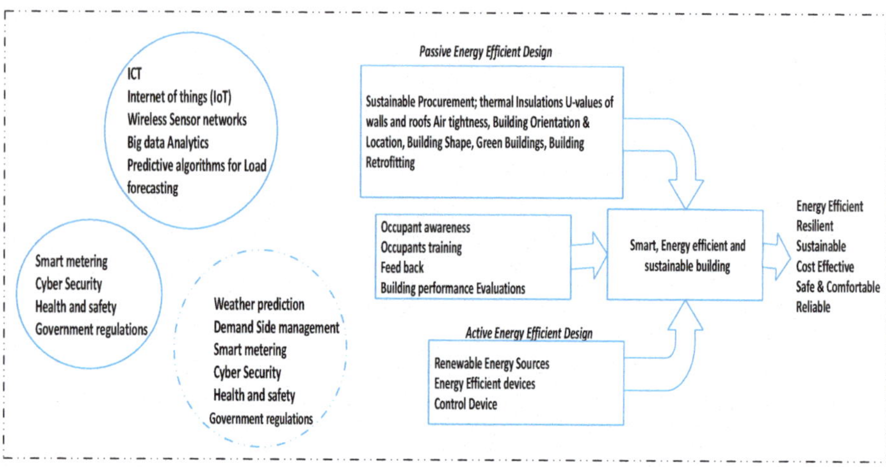

Fig. 45.3 Generalised hybrid model for optimising energy usage in the smart built environment

45.5 Case Study

The case study building used for this study is the new engineering and computing building at Coventry University (Fig. 45.4). The building is equipped with the state-of-the-art facilities which helps to facilitate collaborative learning. It is a multi-purpose building with activities ranging from teaching and learning, research and other administrative activities. The building services are controlled remotely by a Building Management System (BMS); however, the occupants still have some level of control over the facilities such as lighting, window opening, etc. The choice of this building is predicated on the fact that it has been designed and built to be energy efficient with modern BMS. This building has a main supply intake (or the incomer) from the secondary distribution transformer, and with a smart metre installed to monitor the daily electricity consumption. The daily electricity consumption is logged by the BMS systems on a half-hourly basis, which were retrieved for this study.

The methodology adopted in this research involves obtaining the daily electricity data logged by the building BMS system on a half-hourly basis from the selected case study building over a 1-year period (January–December). A three-point moving average prediction technique, where the past observation is weighed equally to predict the future electricity, was used to predict the half-hourly electricity consumption. The actual and predicted half-hourly electricity consumption calculated over a month are given in Fig. 45.5. The plots shown correspond to winter month (January), spring (April), summer (July) and autumn (October) month, respectively. The calculated coefficient of determination (R-Squared) for the four plots are 0.907, 0.941, 0.919 and 0.917, respectively, which implies that the three-point moving average algorithm can reliably predict the half-hourly load demand for the case-study building.

The block diagram for a typical control strategy for optimising energy usage is shown in Fig. 45.6. As shown, during normal building operation and assuming with no renewable source, the building relies only on the grid. However, when the control-

Fig. 45.4 The new Engineering and Computing Building at Coventry University, UK

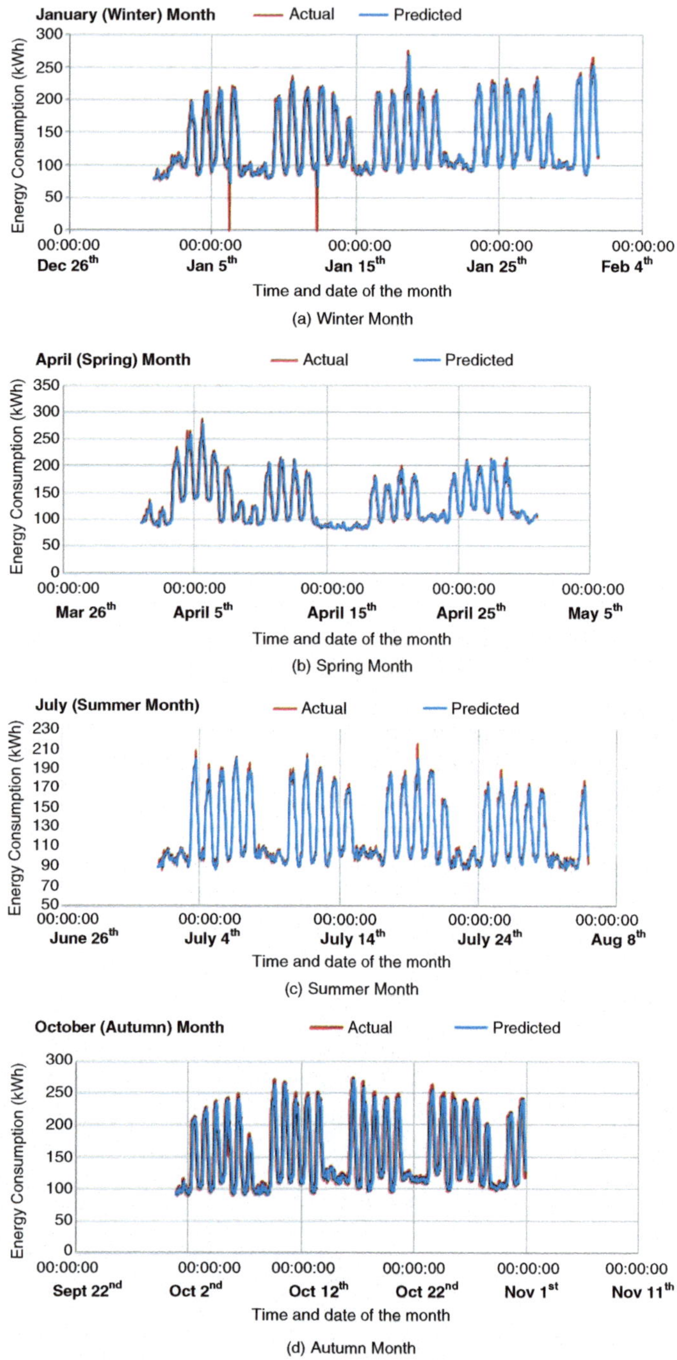

Fig. 45.5 Actual and predicted half-hourly electricity consumption for the case study building

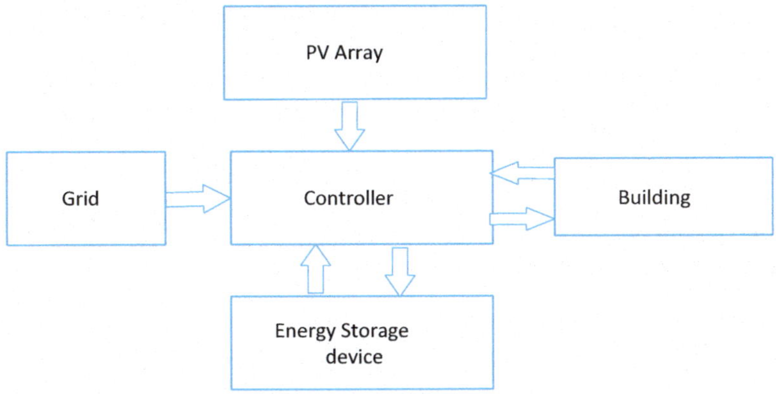

Fig. 45.6 Actual and predicted half-hourly electricity consumption for the case study building

ler senses that the PV has sufficient energy to power the load, it automatically switches to the PV, while the excess energy generated by the PV is stored in the energy storage device. However, when the energy on the PV drops below the required level by the building, the energy on the storage devices will supply the building loads. The entire process is controlled and coordinated automatically by the controller.

45.6 Conclusions

The study explores the various strategies in optimising energy usage in buildings and propose general hybrid strategies encompasses passive, active and the user awareness. By adopting this hybrid integrated approach, modern buildings can be made more energy efficient, sustainable while also ensuring the occupant's well-being and comfort. Existing building could also be retrofitted to meet this demand. The reliability of the generalised model depends largely on the following: the accuracy of the predictive algorithm to forecast the load requirements in the building in real time, the accuracy of the modelling tool to accurately model the operation and control of the entire system, the efficiency and capability of the PV arrays and ESD, the reliability of the sensors to accurately capture and manage the data, the losses (such as switching and harmonic issues) associated with the use of the power electronic devices, the airtightness of the building, the thermal insulations and building fabrics, occupant's awareness, feedback, training, and building performance evaluation, among others. All stakeholders (such as government/policy makers, energy providers, researchers and building owners/occupants) also have a major role to play. The authors hope that this paper does not only contribute to the discussions aimed towards the transition to a carbon-free built environment but would serve as a platform for further research in the area of building energy efficiency and optimisation. Possible research directions include but not limited to research in: energy storage devices and strategies in buildings, predictive algorithms to accurately forecast the building demand, intelligent occupancy sensing techniques in buildings, demand side management, integration of PV cells to minimise losses and ensure optimal efficiency, airtightness in the buildings and overall building thermal insulations, renewable and embedded generation in distribution systems, energy efficient district heating, consumer preference in energy efficiency measures, and building performance evaluations among others.

Acknowledgement Special thanks are due to the staff of Estate department of Coventry University for providing the data on the case study building.

References

1. Gul MS, Patidar S (2015) Understanding the energy consumption and occupancy of multipurpose academic building. Energ Buildings 87:155–165
2. Ahmad MW, Mourshed M, Mundow D, Sisinni M, Rezgui Y (2016) Building energy metering and environmental monitoring—a start of the art review and direction for future research. Energ Buildings 120:85–102
3. Shaikh PH, Nor NBM, Nallagownden P, Elamvazuthi I, Ibrahim T (2014) A review on optimized control systems for building energy and comfort management of smart sustainable buildings. Renew Sust Energ Rev 34(C):409–429
4. Green Building Council (2011) Carbon reductions in existing non-domestic buildings. UK Green Building Council, London
5. https://www.carbontrust.com/media/671511/energy-technology-list.pdf. Accessed 6 May 2018
6. Rocha P, Siddiqui A, Stadler M (2015) Improving energy efficiency via smart building energy management systems: a comparison with policy measures. Energ Buildings 88:203–213
7. EUR-Lex—32010L0031—EN—EUR-Lex. https://eur-lex.europa.eu/legal-content/EN/ALL/?uri=CELEX%3A32010L0031. Accessed 9 Jun 2018
8. Directive 2012/27/EU of the European Parliament and of the Council of 25 October 2012 on energy efficiency, amending Directives 2009/125/EC and 2010/30/EU and repealing Directives 2004/8/EC and 2006/32/EC Text with EEA relevance, vol. OJ L. 2012. Accessed 8 May 2018
9. Levermore GJ (2000) Building energy management systems : applications to low-energy HVAC and natural ventilation control, 2nd edn. E & FN Spon, London
10. http://www.smartbuildingsmagazine.com/features/smart-cities-to-boost-energy-efficiency. Accessed 8 Jun 2018
11. http://www.osborneclarke.com/media/filer_public/5e/5e/5e5e086a-49ae-4b2a-a89a-b27b6dc720ed/oc_future_of_built_environment.pdf. Accessed 2 Jun 2018]
12. https://www.r-e-a.net/upload/rea_uk_energy_storage_report_november_2015_-_final.pdf. Accessed 2 Jul 2018
13. Hossain MR, Too AM Smart Grid. https://www.springer.com/cda/content/document/cda_downloaddocument/9781447152095-c2.pdf?SGWID=0-0-45-1406312-p175138227
14. Khan AR, Mahmood A, Safdar A, Khan ZA, Khan NA (2016) Load forecasting, dynamic pricing and DSM in smart grid: a review. Renew Sust Energ Rev 54:1311–1322
15. https://www.theiet.org/sectors/built-environment/files/digital-technology-pdf.cfm. Accessed 2 Jun 2018
16. Nuaimi EA, Neyadi HA, Mohamed N, Al-Jaroodi J (2015) Applications of big data to smart cities. J Internet Serv Appl 6(1):25

17. European Commission: Energy research challenge for smart cities. https://setis.ec.europa.eu/energy-research/sites/default/files/library/ERKC_PB_Smart_Cities.pdf
18. http://www.yourhome.gov.au/sites/prod.yourhome.gov.au/files/pdf/YOURHOME-PassiveDesign.pdf. Accessed 21 May 2018
19. https://www.carbontrust.com/resources/guides/energy-efficiency/employee-awareness-and-office-energy-efficiency/. Accessed 11 Jul 2018
20. https://www.eea.europa.eu/publications/achieving-energy-efficiency-through-behaviour/file. Accessed 26 May 2018

Chapter 46
Performance Optimization of Concentrated Photovoltaic-Thermal (CPV-T) System Employing Phase Change Material (PCM) in Hot Climate

Shaimaa Abdel Baqi, Ahmed Hassan, and Ali Hassan Shah

Nomenclature

A	Area (m^2)
ACPPVC	Asymmetric compound parabolic photovoltaic concentrators
BIPV	Building integrated photovoltaics
c	Heat capacity
CPV	Solar concentrator photovoltaics
CPV-PCM	Solar concentrator photovoltaics integrated with phase chance material
FF	Fill factor
G	Solar radiation intensity (W/m^2)
I_{sc}	Short circuit current
k	Thermal conductivity (W/mK)
PCM	Phase change material
PV	Photovoltaics
PV-PCM	Photovoltaic integrated with phase change material system
T_{amb}	Ambient temperature (°C)
THM	Temperature history method
T_l	Liquids temperature (°C)
TPCM	PCM temperature (°C)
T_s	Solids temperature (°C)
V_{oc}	Open circuit voltage (V)
η	Energy efficiency
ρ	Density (kg/m^3)

S. A. Baqi (✉) · A. Hassan · A. H. Shah
Department of Architectural Engineering, College of Engineering, United Arab Emirates University, Al Ain, United Arab Emirates
e-mail: 200734406@uaeu.ac.ae

A. Sayigh (ed.), *Renewable Energy and Sustainable Buildings*, Innovative Renewable Energy, https://doi.org/10.1007/978-3-030-18488-9_46

46.1 Introduction

The performance of a photovoltaics solar cell is determined by the parameters named short circuit current (I_{sc}), open circuit voltage (V_{oc}), fill factor (FF), and efficiency (η) [1]. Increased radiation flux helps to improve the power output in concentrated photovoltaic (CPV) systems; however, the concomitant increased temperature affects these parameters and consequently renders additional power losses as well as long-term cell degradation [2], hence lower performance of solar cells. In order to mitigate temperature-based failures in PV systems, incorporation of phase change material (PCM) with a suitable phase transition temperature has been extensively investigated. Passive cooling of PV with solid–liquid PCM has been evaluated using a PCM-paraffin wax and a rectangular aluminum container with a selectively coated front surface to mimic the PV cell [3]. Further, evaluation of two PCMs for thermal management of manufactured aluminum PV has been tested and the thermal performances of different internal fin arrangements for improving bulk PCM thermal conductivity have been presented. The output power has been predicted numerically [4]. Depending on this, the PV-PCM cooling concept was tested using four different fabricated cell sizes in the indoor environment. The results showed that the measured output power values were encouraging, however further temperature reduction was necessary to make the PCM financially viable [5]. Some criteria have been reported to be hugely influencing the performance of PV-PCM system such as PCM containment materials, PCM combinations, and metallic conductivity enhancing fins [6]. Selection of the optimum PCM therefore requires an in-depth understanding of PCM factors affecting phase change process [7]. The PCM cooling system has been evaluated outdoors with two PCMs on a PV panel for few days in two different climates of Dublin, Ireland and Vehari, Pakistan. Employing $CaC_{12}{\cdot}6H_2O$ achieved up to 3–4 °C higher temperature drops and about 3% more power savings than capric–palmitic acid at both sites. However, the system was more effective in hot and stable climatic conditions by achieving higher PV temperature drop in Vehari than in Dublin, reaching up to 21 °C and resulting associated power saving of 13% in Vehari [8]. Subsequently, yearly energy performance of a PV-PCM system was evaluated in hot climate (UAE climate). The PV-PCM system dropped PV temperature at peak by 10.5 °C and increased PV electrical yield by 5.9% on yearly basis [9].

Recent emerging integration of PV-PCMs system for temperature control offers an opportunity for extending its usage to building integrated photovoltaics (BIPV) systems. PCM with melting point 26 °C integrated in BIPV was numerically tested by attaching microencapsulated PCM at PV back to cool to enhance performance of BIPV. The simulation results showed very low temperature drop of only 2 °C and 5 °C in summer and winter condition, respectively, indicating that microencapsulated PCM was least effectiveness in BIPV temperature drop due to lower thermal conductivity of encapsulation materials and the lower mass ratio of PCM contained in microencapsulation [10]. Also, a simulation study using simplified heat balance model to calculate the extra energy gain on an annual basis has been conducted and a moderate increase of up to 3% of the total energy output was expected. However, the system was not economically viable with additional PCM material cost having an acceptable payback period of 10–20 years to be reached [11]. Experimentally, the BIPV-PCM has been analyzed for PV cooling and its impact on building energy efficiency in wintertime. The system reached a maximum electrical and thermal efficiencies of 10% and 12%, respectively [12]. Therefore, thermal energy stored in the PCM behind the PV was utilized for domestic water heating applications in the winter conditions of UAE to deliver heat during water heating demand periods. The proposed system achieved about 1.3% increase in PV electrical conversion efficiency, along with the recovery of ~41% of the thermal energy compared to the incident solar radiation [13].

From the previous studies, although employing PCMs passively in PV panels improves PV efficiency and restricts temperature below safety limits [14], unfortunately, the PV-PCM system as a solar energy resource to produce electricity has some drawbacks of requiring large areas and high capital costs per kWh in the widespread use of PCM integrated in solar photovoltaics [15]. Also, the PCM-based PV cooling approach has been substantially reviewed concluding that the proposed technology has constraints of recovering and utilizing stored thermal energy effectively [16–18] as well as melting re-solidification of lower melting point PCM in hot climates [13] depending on weather conditions, PCM properties, containment properties, the heat extraction mechanism [16], and solar radiation intensities [5]. Therefore, recent studies have been directed to concentrated photovoltaics (CPV) that offer lower capital cost per kWh by using curved mirrors or Fresnel lenses [19] and enhance electrical power output per unit area of the PV panel and total energy efficiency of the system. Further, the PCM-based cooling technique is recently extended to concentrator photovoltaics (CPV) systems in an attempt to benefit from thermal energy at higher operating temperature to avoid higher likeliness of power failure, delamination, and rapid degradation [20]. Initially, the concept was further applied in V-trough concentrator system using Paraffin metal–wax composite PCM with melting range of 56–58 °C and thermally enhanced by embedded aluminum lathe turnings in India. A reduction up to 16 °C in the PV temperature and an improvement of 1.55 times in output power were attained at solar concentrated intensity of 1982 W/m^2 during the day time while the PCM re-solidification process took a place in the evening time [15]. An asymmetric compound parabolic photovoltaic concentrators (ACPPVC) system with a concentration ratio of 2 has been designed, constructed, integrated with PCM RT27, and tested at incident solar radiation of (672 and 280) W/m^2 and incident angle of 0°. The results showed a drop in solar cells temperature up to 18 °C and 7 °C in ACPPVC-PCM system for radiation intensities of (672 and 280) W/m^2, respectively, as compared to system without PCM [21]. A comprehensive transient melting-solidification-thermo-fluid model for CPV-PCM system has been tested at different inclination angles. The results concluded that angle of 45° achieved the best performance with the reasonable uniformity minimum average temperature obtained while the angle of −45° leads to unfavorable performance [22]. Using low-concentrator-based BICPV linked to a designed wax-based organic PCM (RT-42) system has been designed and experimentally tested at 1000 W/m^2 achieving an enhancement of electrical efficiency of the system by 7.7% and an average drop of 3.8 °C in PV temperature [23]. The effects of phase change material (PCM) cooling for a tracking-integrated concentrating photovoltaic-thermal (CPV-T) system has been investigated experimentally showing that the average increases of the electrical, thermal, and overall energy efficiencies for the CPV-T system with PCM cooling reach 10%, 5%, and 15%, respectively, more than the CPV-T system with water cooling [24].

Although a good number of researches into CPV-PCM system were performed, the results indicate the need of further consideration. Deep looking in previous studies highlighted some concerns about the intensities, study duration and weather conditions where the studies conducts. Most of the studies were conducted in low-concentrator-based CPV system (<1000 W/m^2) either experimentally [2, 23, 24] or numerically [22] showing the need of further studies for higher solar intensities. Some experiments were conducted for only 1 day; however, an extended period of time is required to understand the discharging and re-charging of PCM, as full-discharge becomes indispensable for its maximum heat storage capacity during the next charging cycle. Most researches were done considering the cold weather conditions indicating that future studies are required to report the performance of CPV-PCM system in hot climate. The present study is evaluating the performance of high-concentrator based CPV system employing PCM in the PV panel for consecutive three summer days in the hot weather conditions of UAE climate.

46.2 Methodology

The methodology of this paper is composed of integrating the PCM into CPV and compares it with non-tracked and tracked & concentrated PV performance. The PCM was investigated and characterized using temperature history method (THM) to find its thermo-physical properties of interest. The PCM is contained in metallic container with internally fitted fins mounted on the dark side of CPV. The PCM cools the CPV during melting and stores the incoming thermal energy to form a CPV-PCM system. The experiment was conducted for three consecutive days in May with the same initial conditions at the start of each day. The cooling effect quantified by temperature drops at the front surface of the CPV and associated improvement in electrical as well as thermal energy yield of the CPV-PCM are determined. The enhanced CPV power is calculated by employing fill factor (FF) on measured open circuit voltage (V_{oc}) and short circuit current (I_{sc}) of the respective PVs. The thermal energy stored in the PCM is determined by changes in temperature and physical state (solid–liquid) of the PCM.

46.2.1 Experimental Setup

Three 29 cm × 36 cm monocrystalline photovoltaic modules with a rated capacity of 10 W and module efficiency of 15% and Fresnel lens concentrator of 1 m^2 aperture area, fabricated from PMMA with the focal length of 800 mm, were used in the experimental part. All PV modules were installed at a 24° slope facing south at the latitude of Al Ain (24.1° N, 55.8° E) and tracked on two-axis. Two PV modules were fixed 360 mm below the Fresnel lens in the same plane and received uniformly distributed 36 cm diameter spots of concentrated radiation illuminating the entire PV surface. An internally finned container fabricated from a 3-mm thick sheet of aluminum alloy (1050A) and thermally insulated with glass wool. It was filled with PCM and fixed to the dark side of the CPV using epoxy resin glue. The thermophysical properties of the materials used in the experiment are listed in Table 46.1.

The PCM was filled as liquid (40 L) in the container and subsequently cooled until it completely solidified. The solidified PCM left a 7-cm empty space on top of container intended to accommodate volume expansion during PCM melting. A 2.5 cm thick glass wool layer with the thermal conductivity of 0.039 $Wm^{-1} K^{-1}$ [27] was used to thermally insulate the PCM container. Multiple K-type thermocouples tested in ice-bath with a measurement error of ±1.5 °C (Class 1 tolerance) were used to measure temperatures of the absorber front and back PV surfaces. For the CPV-PCM system, the PCM's temperatures inside the container in three different positions were also measured. Three self-powered pyranometers with 0.20 mV per W/m^2 sensitivity and ± 5% calibration uncertainty [28] were installed to measure solar radiation intensity (G) incident on non-concentrated PV and CPV surfaces. A weather station with an accuracy of ±1% [29] was installed to measure ambient temperature (T_{amb}). The tracking system, Data acquisition, weather station, Pyranometers, and thermocouples were tested and confirmed working functionally before starting the experiment. All the sensors were connected to an NI-Compact DAQ [30] to log the data with a time step of 5 min as shown in Fig. 46.1.

46.3 Numerical Model

A two-dimensional heat transfer model is developed and applied to simulate the cooling of CPV by the PCM conceptually illustrated in [8, 31]. The developed model is employed to predict the transient temperature distribution within a two-dimensional region in the CPV-PCM system for different G, T_{amb} and convective and radiative heat transfer boundary conditions. A simulation at summer ambient conditions in May have been carried out employing the material properties listed in Table 46.2. The experimental results are compared with the prediction of the simulation model. The comparison of results obtained for the CPV panel only and PV–PCM are discussed through the paper.

Table 46.1 Thermo-physical properties of materials used in the experiments in solid state

Materials	Thickness d (m)	Area A (m^2)	Thermal conductivity k (W/m K)	Specific heat capacity C_p (J/kg K)	Density ρ (kg/m^3)	Melting point T_m (K)	Solidification point T_s (K)	Heat storage capacity (J/k)
PCM Paraffin wax RT-58 [25]	0.5	0.08	0.2	2000	802	328 (55 °C)	334 (61 °C)	64,160
Aluminum [26]	0.003	1.533	202.4	871	2719	–	–	10,891
Insulation Glass Wool [27]	0.025	20	0.039	0.67	16	–	–	5.36

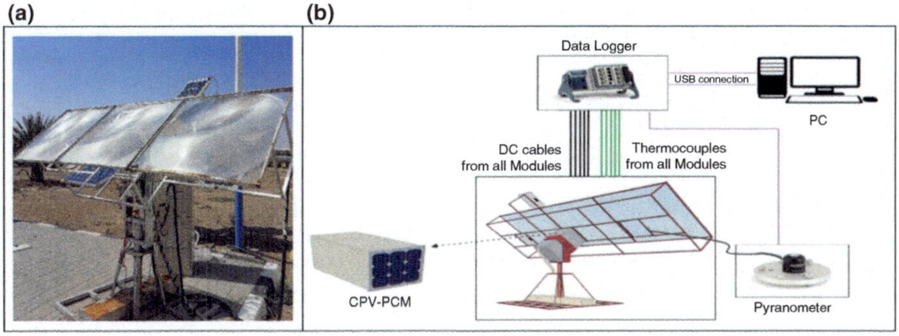

Fig. 46.1 The experimental setup (**a**) consisting of Tracking PV, CPV, and CPV-PCM and (**b**) the schematic diagram for the systems with all connections

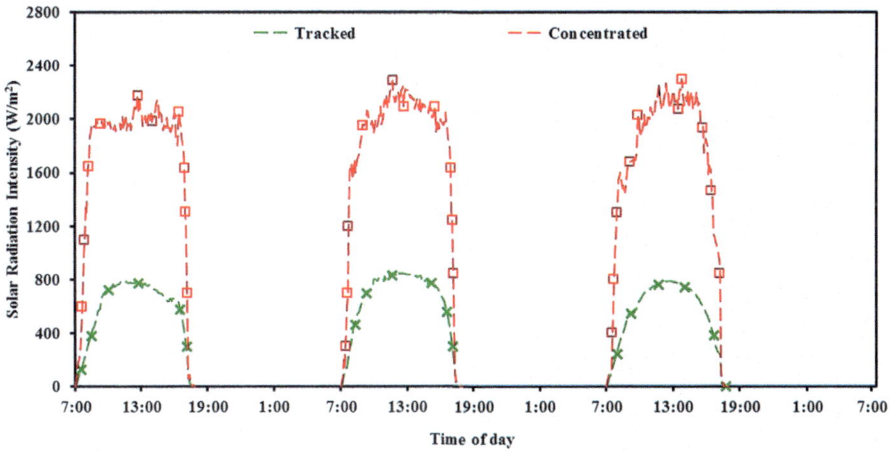

Fig. 46.2 The measured solar irradiance for tracked and concentrated PV panels for consecutive three summer days at the site in Falaj Hazza Campus, UAE University Al Ain—UAE

46.4 Results and Discussion

The total solar radiation received by the PV is evaluated to determine the amount of the PCM required to size the PCM containment. Figure 46.2 presents the on-site measured solar radiation received by tracked and concentrated PV for consecutive three summer days in summer in Al-Ain, UAE.

Figure 46.2 shows that the solar radiation intensity reached the maximum during the noon time for all three systems (tracked, CPV, CPV-PCM) with different magnitudes. It can be seen that the approximately 760 W/m^2 was an average available on tracked PV system from 9:00 to 17:00 while the CPV achieved a solar radiation intensity of 2100 W/m^2 during the same time showing a three times increased radiation on CPV compared to non-concentrated PV of the same size.

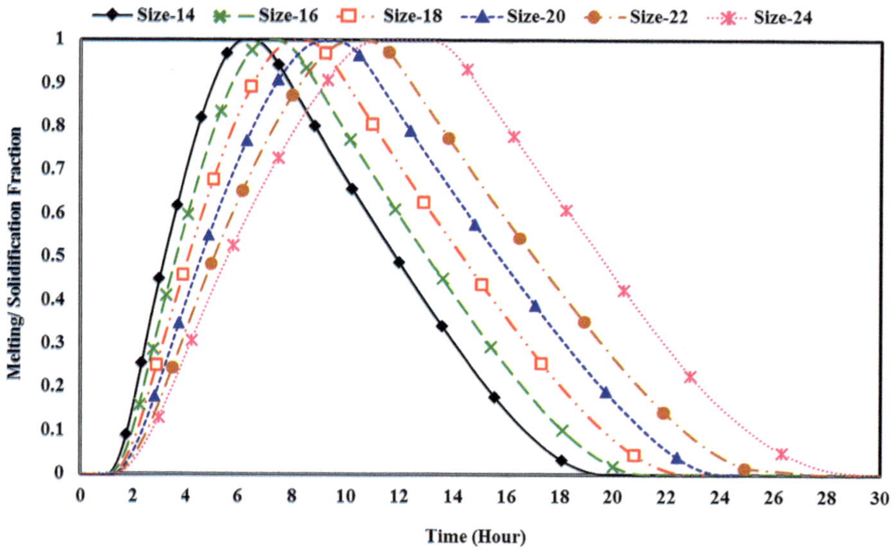

Fig. 46.3 The time needed for the selected PCM to complete melt/freeze cycle for PCM container with depth of 14–24 cm

In order to determine the effect of including PCM on cooling and thermal energy recovery from the CPV, the measured solar radiation data (reported in Fig. 46.2), and paraffin with melting point of 58 °C as a PCM were simulated through the developed numerical model. The initial simulation was aimed at sizing the PCM container at the available intensity. The results are judged based on the criteria of PCM melting within day length (10-h) and a complete melt-freeze cycle within 24 h to determine optimum PCM container size (shown in Fig. 46.3).

Figure 46.3 presents the transient volume averaged melting/solidification fraction of the PCM contained in various enclosure depths ranging from 14 to 24 cm as a representative for the amount of PCM. It can be observed that all the PCM thicknesses completed melting within the average summer day length of 10 h in UAE apart from size 22–24 cm. However, the melting/solidification cycle of the PCM contained in the enclosure depth of 14–18 cm complete in less than 22 h indicating the under-sizing of the PCM container. On the contrary, the containers with PCM thickness of 22 cm and 24 cm could not complete melting within 10 h and consequently the melting/solidification within 24 h representing an oversized container. From Fig. 46.5, the optimum depth for the PCM container was 20 cm that could complete melting within 10 h and the melting/solidification cycle within 24 h.

The measured solar radiation determines the energy incident on the PV and eventually controlling PV panel temperature (T_{PV}). The evolution of temperature on the front surface of the tracked, concentrated, and concentrated-PCM panel for consecutive three summer days in May are presented in Fig. 46.4.

It can be seen that tracked PV system shows the lowest transient T_{PV} temperatures with an average day T_{PV} of 47 °C peaked at 57 °C between 11:00 and 17:00.

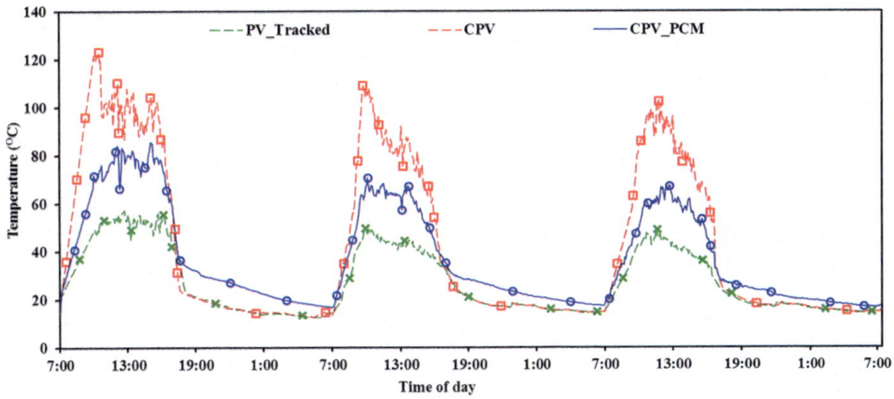

Fig. 46.4 The measured front surface temperatures of tracked, concentrated and concentrated–PCM PV panels for consecutive three summer days in May at the site in Falaj Hazza Campus, UAE University Al Ain—UAE

While the concentrated PV system reached the highest average day T_{PV} of 88 °C and peaked at 123 °C during the same time. Inclusion of the PCM into the system cools the CPV panel and maintains its temperature below the safety limit. The average T_{PV} of concentrated system is higher than the liquids temperature (T_l) of (58 °C) for the selected PCM (Paraffin) indicates that the inclusion of the PCM into the system will complete melting during the daytime by absorbing heat from the CPV and regulate its temperature. As expected, the concentrated PV-PCM rendered a temperature lag behind the reference concentrated PV during the day time and remained below the safety limit of 85 °C with an average T_{PV} of 67 °C yielding a temperature drop up to 38 °C at peak indicating the cooling effect produced by the PCM. The retained heat can be utilized for thermal application.

Figure 46.5 presents the experimental and simulated transient temperature of the contained PCM at different depth positions (front, middle, and back) over 24 h.

It can be observed that during the daytime, the ambient temperature (T_{amb}) remained always lower than PCM liquids temperature (T_l) of 61 °C indicating the importance of solar radiation to trigger the PCM melting. While at nighttime, T_{amb} stayed below PCM solids temperature (T_s) of 58 °C dictates the PCM solid regeneration for the next day at nighttime due to passive heat dissipation to ambient.

From Fig. 46.5, the initial PCM temperature (T_{PCM}) of 37 °C was being substantially lower than T_m pointing out the solid phase of the contained PCM. As the solar radiation received, T_{PCM} sharply increased reaching 58 °C (start of PCM melting) at 8:30, 10:30, and 11:30 in front, middle, and back PCM surfaces, respectively. The gradient of temperature curve decreased during phase transformation due to the latent energy storage reaching 62 °C (end of PCM melting) at 14:00, 15:00, and 16:00 in the front, middle, and back PCM surfaces, respectively, indicating that PCM had completely melted. At nighttime, the T_{pcm} dropped below 58 °C (PCM solidification temperature) at 3:00, 4:00, and 6:00 morning in the front, middle, and back PCM surfaces, respectively, indicating the complete solidification process of the PCM before the start of the next day.

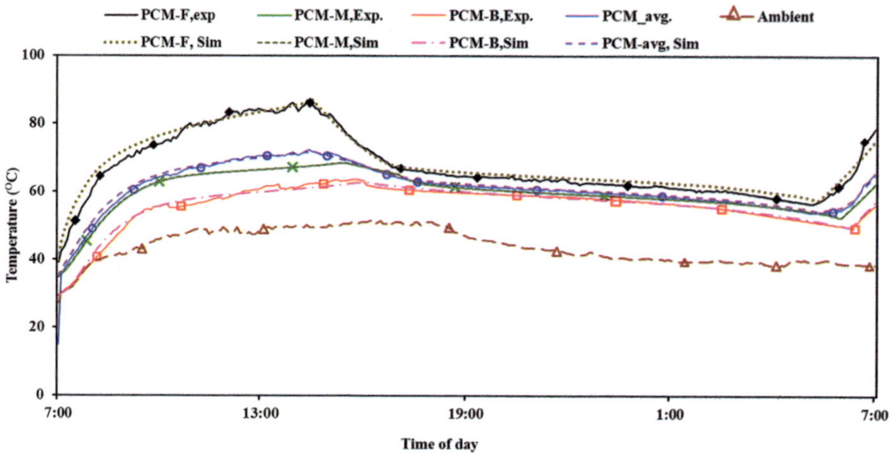

Fig. 46.5 The measured and simulated PCM (Paraffin) temperatures at different depth locations (front, middle, and back surfaces) in summer day in May at the site in Falaj Hazza Campus, UAE University Al Ain—UAE

The results show that the PCM completely melted in 9 h while it took more than 14 h to solidify in nighttime due to high T_{amb} in May. The experimental and simulated temperatures of PCM are in agreement with stander deviation error of 2.5, 1.3, and 1.1% for front, middle, and back surfaces, respectively.

The transient electrical energy (P_e) attained by tracked, concentrated, and concentrated-PCM PV systems over the time for consecutive three summer days in May was calculated using the experimentally measured open circuit voltage (V_{oc}) and short circuit current (I_{sc}) with applying the fill factor (FF) of 0.72 deduced from the PV catalogue data as given by Eq. (46.1).

$$P_e = V_{oc} \times I_{sc} \times \text{FF} \tag{46.1}$$

The cumulative electrical, thermal, and total energy produced over the time by tracked, concentrated, and concentrated-PCM PV panels for 3 days are shown in Fig. 46.6.

Figure 46.6a illustrate the cumulative electrical energy produced by three proposed system showing that the power output of tracked PV was gradually increased reaching up to ~115 Wh, while concentrating the solar radiation was sharply increasing this cumulative electrical energy reaching up to 350 Wh. Further improvement in the cumulative electrical energy was obtained by inclusion of the PCM to the system reaching up to 370 Wh. Figure 46.6b clearly showed that tracked PV system is producing very small commutative thermal energy (maximum of 20 Wh) during noon time (13:00) then decreasing and stayed zero over the night time. Concentrated PV system shows higher commutative thermal energy peaked at ~50 Wh between 10:00 and 17:00 local daytime. While the CPV-PCM yield the highest cumulative thermal energy peaked at 290 Wh in noon time due to temperature regulation and

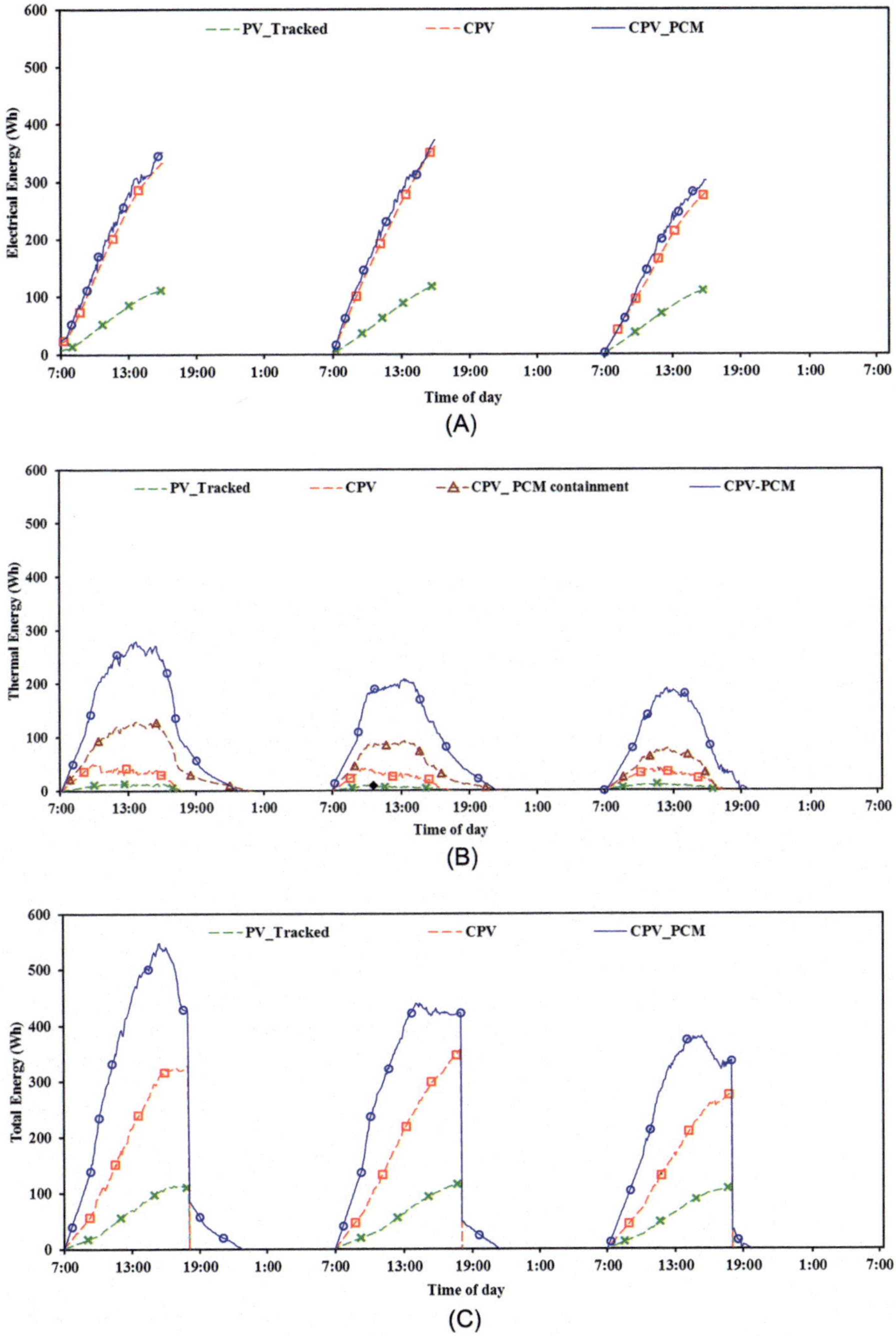

Fig. 46.6 The cumulative electrical energy (**a**), thermal energy (**b**), and total energy (**c**) produced by tracked, concentrated, and concentrated PCM-PV panels over the time for consecutive three summer days in May at the site in Falaj Hazza Campus, UAE University Al Ain—UAE

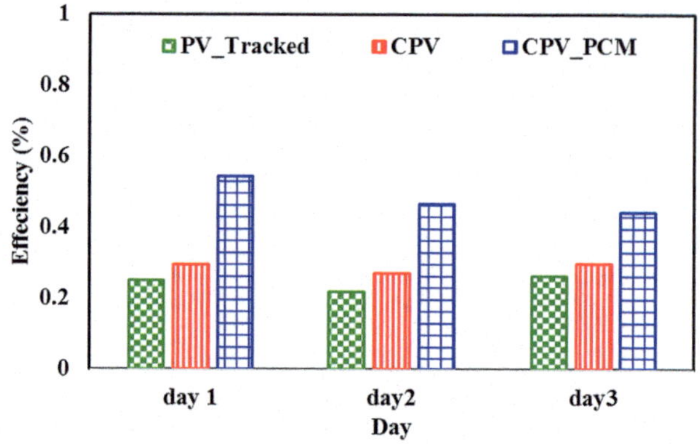

Fig. 46.7 The calculated energy efficiency achieved by tracked, concentrated, and concentrated PCM-PV panels over the time for consecutive three summer days in May at the site in Falaj Hazza Campus, UAE University Al Ain—UAE

latent heat absorption in PCM indicating the optimum time to utilize the energy produced before the heat released to ambient. Thereby, adding commutative electrical and thermal energies resulting in the cumulative total energy as shown in Fig. 46.6c. The highest cumulative total energy was achieved by concentrated-PCM PV system (peaked at 550 Wh) as compared with tracked and reference concentrated PV that produced maximum cumulative total energy of 110 and 330 Wh, respectively. Thus, it can be concluded that the PV-PCM system can increase electrical output as well as thermal energy of CPV system by virtue of maintaining lower temperatures.

The energy efficiency (η) achieved by tracked, concentrated, and concentrated–PCM PV systems for consecutive three summer days are presented in Fig. 46.7. The energy efficiency (η) is calculated by dividing the cumulative Q_{tot} for each system by the solar radiation received by that system produced by reference PV given in Eq. (46.2).

$$\eta = Q_{\text{tot}}\,\text{commutative} / G \tag{46.2}$$

As expected from the previous results, the tracked PV system shows the lowest energy efficiency (maximum of 25%) followed by concentrated PV with small increasing in efficiency reached approximately up to 30%. Cooling CPV panel using PCM double the energy efficiency of the system up to 60% due to virtue of temperature drop.

46.5 Numerical Optimization

The validated numerical model was used to optimize the PCM container size for different intensities varied from 1 to 12k, thereby predict the optimum material amount that can fully melted within the day time (9 h) and completely solidified over the night for next day (Fig. 46.8).

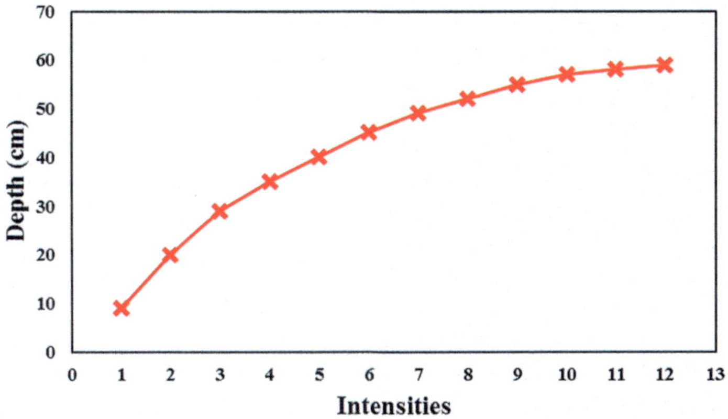

Fig. 46.8 The optimum PCM (paraffin)'s container size (depth) to completely melt/solidify at different intensities ranged from 1 to 12k

Only the depth of the PCM container has been changed and no alteration happens to container's width (25 cm) or length (25 cm) attached to the PV panels during the optimization process. The predicted optimum container depth for intensity of 1k is 9 cm which is almost half of the container's depth needed for 2k (20 cm). As the intensity increased, the container size increased, however, the required supplemental depth decreases and the difference between container depths become unnoticeable at the high intensities (The optimum container depth for intensity of 11k and 12k are 59 cm and 60 cm, respectively).

An assessment for different PCM materials named paraffin, palmitic acid, and SP salt was numerically conducted to optimize the amount of each material (container size) that completely melt and solidified within 24-h (1 day) for optimum performance as shown in Fig. 46.9.

Figure 46.9 presents the time needed for the cyclic melting and solidifying process for different amount of PCMs named paraffin, palmitic, and SP salt. It is generally noticeable that as the amount of PCM (container size) increases, the time required for any of studied materials to completely melt and solidify increases, however, with different gradient magnitude for each PCM. Comparing the masteries, same amount of SP will need triple the time taken by paraffin to completely melt and solidify, while the palmitic will need double the time for the same container size. A container depth of 6 cm SP, 10 cm palmitic, and 18–20 cm paraffin will be required to completely melt and solidify the contained PCM within 24-h (1-day) indicating that SP is the optimum material; however, SP salt have some demerits of lower thermal stability, chemical instability, corrosion to metals, incongruent melting, and sub-cooling below the solidification temperature [13].

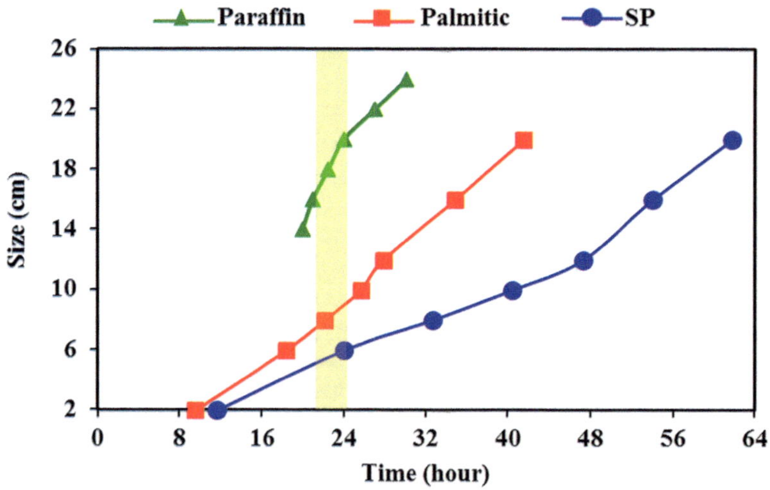

Fig. 46.9 The numerical prediction of time needed for cyclic melting and solidifying process of different materials (paraffin, palmitic acid, and SP salt) contained in different inclusion sizes ranged from 2 to 24 cm

46.6 Conclusion

The CPV-PCM system employing optimum melting point paraffin-based PCM is studied in the summer season in Al Ain, UAE. It is observed that the PCM achieved the peak-temperature drop of 38 °C and average temperature drop of 22 °C. The energy performance of CPV-PCM system is compared to non-concentrated (tracked) and concentrated (CPV) systems. The increased electrical and thermal energies due to cooling produced by PCM are analyzed. The tracked, CPV, and CPV-PCM yielded total energy efficiency of 25%, 30%, and 60%, respectively, concluding that inclusion of PCM double the performance of the system. The CPV-PCM system enhanced the maximum commutative electrical and thermal energies to 370 Wh and 290 Wh, respectively. An experimentally validated model predicted the PCM melting and solidification times and melt fractions for different intensities, and different materials. The optimum PCM melted and solidified within 24-h found to be SP salt contained in enclosure with depth of only 6 cm compared to paraffin contained in enclosure with depth of 20 cm, however, disadvantages of salt SP need to be considered.

References

1. Singh P, Ravindra NM (2012) Temperature dependence of solar cell performance—an analysis. Sol Energy Mater Sol Cells 101(2012):36–45
2. Saly V, Ruzinsky M, Redi P (2001) Indoor study and ageing tests of solar cells and encapsulations of experimental modules. In: 24th International Spring Seminar on Electronics Technology: Concurrent Engineering in Electronic Packaging. IEEE, Calimanesti-Caciulata
3. Huang MJ, Eames PC, Norton B (2004) Thermal regulation of building-integrated photovoltaics using phase change materials Int. J Heat Mass Trans 47(12–13):2715–2733
4. Huang MJ, Eames PC, Hewitt NJ (2006) The application of a validated numerical model to predict the energy conservation potential of using phase change materials in the fabric of a building. Sol Energy Mater Sol Cells 90:1951–1960
5. Hasan A, McCormack SJ, Huang MJ, Norton B (2010) Evaluation of phase change materials for thermal regulation enhancement of building integrated photovoltaics. Sol Energy 84:1601–1612
6. Huang MJ, Eames PC, Norton B, Hewitt NJ (2011) Natural convection in an internally finned phase change material heat sink for the thermal management of photovoltaics. Sol Energy Mater Sol Cells 95:1598–1603
7. Hasan A, McCormack SJ, Huang MJ, Norton B (2014a) Characterization of phase change materials for thermal control of photovoltaics using Differential Scanning Calorimetry and Temperature History Method. Energy Convers Manag 81:322–329
8. Hasan A, McCormack SJ, Huang MJ, Sarwar J, Norton B (2015) Increased photovoltaic performance through temperature regulation by phase change materials: materials comparison in different climates. Sol Energy 115:264–276
9. Hasan A, Sarwar J, Alnoman H, Abdelbaq S (2017) Yearly energy performance of a photovoltaic-phase change material (PV-PCM) system in hot climate. Sol Energy 146:417–429
10. Ho CJ, Tanuwijaya AO, Lai C-m (2012) Thermal and electrical performance of a BIPV integrated with a microencapsulated phase change material layer. Energ Buildings 50:331–338
11. Hendricks JHC, Sark WGJHM (2013) Annual performance enhancement of building integrated photovoltaic modules by applying phase change materials. Prog Photovol 21:620–630
12. Aelenei L, Pereira R, Gonçalves H, Athienitis A (2014) Thermal performance of a hybrid BIPV-PCM: modeling. Des Exp Invest Energy Proc 48:474–483
13. Hasan A, Alnoman H, Shah AH (2016) Energy efficiency enhancement of photovoltaics by phase change materials through thermal energy recovery. Energies 9(10):1–15
14. Bahaidaraha HMS, Baloch AAB, Gandhidasan P (2016) Uniform cooling of photovoltaic panels: a review. Renew Sust Energ Rev 57:1520–1544
15. Maiti S, Banerjee S, Vyas K, Patel P, Ghosh PK (2011) Self regulation of photovoltaic module temperature in V-trough using a metal-wax composite phase change matrix. Sol Energy 85:1805–1816
16. Browne MC, Norton B, McCormack SJ (2015) Phase change materials for photovoltaic thermal management. Renew Sust Energ Rev 47:762–782
17. Ma T, Yang H, Zhang Y, Lu L, Wang X (2015) Using phase change materials in photovoltaic systems for thermal regulation and electrical efficiency improvement: a review and outlook. Renew Sust Energ Rev 43:1273–1284
18. Islam MM, Pandey AK, Hasanuzzaman M, Rahim NA (2016) Recent progresses and achievements in photovoltaic-phase change material technology: a review with special treatment on photovoltaic thermal-phase change material systems. Energy Convers Manag 126:177–204
19. Du B, Hu E, Kolhe M (2012) Performance analysis of water cooled concentrated photovoltaic (CPV) system. Renew Sust Energ Rev 16(9):6732–6736
20. D'Orazio M, Di Perna C, Di Giuseppe E (2013) Performance assessment of different roof integrated photovoltaic modules under Mediterranean Climate. Energy Procedia 42:183–192

21. Wu Y, Eames P (2011) Experimental characterization of an asymmetric compound parabolic PV concentrator complied with a phase change material. In: Presented at 7th photovoltaic science, applications and technology conference. Heriot-Watt University, Edinburgh
22. Emam M, Ookawara S, Ahmed M (2017) Performance study and analysis of an inclined concentrated photovoltaic-phase change material system. Sol Energy 150:229–245
23. Sharma S, Tahir A, Reddy KS, Tapas KM (2016) Performance enhancement of a Building-Integrated Concentrating Photovoltaic system using phase change material. Sol Energy Mater Sol Cells 149:29–39
24. Su Y, Zhang Y, Shu L (2018) Experimental study of using phase change material cooling in a solar tracking concentrated photovoltaic-thermal system. Sol Energy 159:777–785
25. Rubitherm GmbH (2018) PCM Paraffin wax RT-58. https://www.rubitherm.eu/. Accessed 15 May 2018
26. Aalco Metals Ltd (2018) Aluminum. http://www.aalco.co.uk/. Accessed 17 May 2018
27. Osking L (2018) Glass Wool Insulation Price_Langfang Osking Business Co., Ltd. http://osking-insulation.com/Glass_Wool/Glass_Wool_Blanket/13.html
28. Apogee Instruments (2018) SP-110: Self-Powered Pyranometer. http://www.apogeeinstruments.co.uk/sp-110-self-powered-pyranometer/. Accessed 29 April 2018
29. TMART.com (2018) Professional Weather Station. http://www.tmart.com/WS1041-Professional-Weather-Station-with-PC-Link_p238349.html. Accessed 21 April 2018
30. National Instruments (2018) CompactDAQ. http://www.ni.com/data-acquisition/compactdaq/. Accessed 7 April 2018
31. Sarwar J (2012) Experimental and numerical investigation of thermal regulation of photovoltaic and concentrated photovoltaic using phase change materials. Ph. D Thesis. Dublin Institute of Technology

Chapter 47
Assessment of Habitants' Thermal Comfort Through Different Treatments of Flat Rooftops in Residential Buildings

Dana K. Amro and Suheir M. S. Ammar

47.1 Introduction

The components of a building envelope are typically walls, floors, roofs, and fenestrations. These components affect the thermal comfort in addition to other variables related to humans such as age, health situation, clothes, and fatness. It is well known to architects that building's shape, orientation, windows and doors materials and areas, surrounding environment affect the internal temperature. The most common treatment in the area rooftops without any treatment is the lowest efficiency treatment concern the habitants' thermal comfort. Rooftops are a critical part of these components that are highly susceptible to solar radiation and other environmental changes, thereby influencing the indoor comfort conditions for the occupants. In a study aimed to define the proper rooftops constructions in tropical areas such as India, Madhumathi et al. [1] claimed that about 50% of the heat inside a building is from the roof. More specifically for those who live in the highest floors because of the insufficient insulation properties of roof systems that exposed to solar radiation during the daytime. This study aimed to evaluate the efficiency of four suggested treatments of rooftops including a rooftop with shed, a green rooftop, a rooftop with compact insulation, a rooftop with tile to be compared with a rooftop without any treatment in achieving indoor thermal comfort. The results would be important for Amman in Jordan and other similar cities.

D. K. Amro (✉)
Collage of Fine Arts and Design, University of Sharjah "UOS", Sharjah, UAE

Faculty of Architecture and Design, Alahliyya Amman University "AAU", Amman, Jordan
e-mail: d.amro@ammanu.edu.jo; damro@sharjah.ac.ae

S. M. S. Ammar
Faculty of Engineering, Islamic University of Gaza "IUG", Gaza, Palestine

A. Sayigh (ed.), *Renewable Energy and Sustainable Buildings*, Innovative Renewable Energy, https://doi.org/10.1007/978-3-030-18488-9_47

47.2 Treatments for Thermal Comfort in Amman

This study is conducted using Amman climate. It is the capital of Jordan which lies at the heart of the Middle East and the Arab world, extending between latitudes of 29°11′N and 33°22′N. The climate is influenced by the location of Jordan between the subtropical arid regions of the Arabian Desert and the subtropical humid regions of the eastern Mediterranean. January is the coldest month, with temperatures ranging from 5 °C to 10 °C, whereas August is the hottest month, with temperatures ranging from 20 °C to 35 °C. Daily temperatures can be extremely high, particularly during summer. On some days, the temperature can reach 40 °C or higher, particularly when a hot, dry southerly wind blows. These winds can sometimes be extremely powerful and cause sandstorms. In a study about annual heating and cooling loads of residential buildings in Amman, Qawasmeh et al. [2] found that these loads depend directly on the building direction, building properties which include building materials for roofs and walls, window type, and insulation thickness.

The study area, Amman city, displays a distinct contrast between a relatively wet rainy season from November to April and the rest of the year, which is relatively dry. Referring to Jordanian Code for thermal insulation, the overall U-value for the roof should be less than or equal to 0.55W/m^2 K. U-value history within regulations of UK building shows a decrease since 1965 until 2010 from 1.4 W/m^2 K to 0.25 W/m^2 K. Currently, 0.25 W/m^2 K or less is required for all new buildings in the UK. This reduction in the U-value over the years emphasizes the significance of thermal performance of roofs in the effort to increase the overall thermal performance of buildings. There are some practices to reduce U-value of the rooftops. Most studies tried methods and materials that are used or adequate to their context. Locally, sloped roofs are used slightly for decoration purposes only, while flat rooftops are affordable and commonly used. A flat rooftop can be used without any treatments, or can be customized to have a shed, green cover, compact insulation, or tile [3].

47.3 The Different Treatments of Rooftops

The literature of indoor thermal comfort shows different treatments of rooftops to eliminate the indoor temperature. Jayasinghe et al. [4] argued that the use of appropriate materials and insulation are significant in achieving indoor thermal comfort and the use of light colors roof can enhance it, while the roof orientation was not significant. They used north-south and east-west ridge roof and a flat one during December, March, and June. Hui [5] recommended to set up government policies for using green roofs, develop educational programs and guidelines, to do researches to determine the best-required roof, and how to design it, and to stimulate the market. From an environmental and economic point of view, Hui [5] stated that buildings are an area of potential importance to minimize the energy consumption.

47.3.1 Simple Rooftop Without Any Treatment

Simple flat rooftop consists of solid reinforced concrete slabs, or hollow concrete block slabs (Ribbed Slabs). Based on a local survey conducted in Jordan, the most common types of roof built are ribbed slab with 18 cm thick hollow concrete block covered by 6–8 cm thick reinforced concrete layer. The upper face of the roof left without treatment and has the grayish color of concrete does not require any additional costs. Maintenance is very rear when the roof is constructed without any defects. The U-value for the basic solid slab roof is 3.7 W/m^2 K, and for ribbed slabs is 3.1 W/m^2 K. The reduction of U-value in ribbed slabs compared to the solid slabs is due to the cavity that is part of hollow block structure; however it is still excessive value compared with global codes (Jordanian Green Building Guide, 2013).

47.3.2 Rooftop with Shed

In a study aimed to reduce heat gain for low cost residential buildings in Cairo in Egypt, Dabaieh et al. [6] used an experimental simulation for 37 alternative solutions combining shapes and properties of construction materials.

They used five shapes including vault and ventilated vault, flat and double roofs, and dome. They also used four materials such as insulation, albedo coating, an air gap, and a water pond. Using DesignBuilder software, the results showed that the best alternative, which reduced 53% the needed cooling hours, was using vault roof with rim angle vault roof of 70° and coated with high albedo coating. This solution increases the cost of constructing the roof 3% (Fig. 47.1). In addition, they found that using roof shading or albedo is better than using roof material with lower U-value.

They declared that shed rooftop is less investigated in literature.

However, roof shed is used in Amman city; mostly wooden or coated aluminum pergolas are the common design structures for shading the rooftops. They may have cloth or polycarbonate sheet 16 mm to increase the shading from sunrise (Fig. 47.2).

47.3.3 Green Rooftop

Thermal comfort for the inner of the building has many solutions in the literature. Green rooftop is one of them. Many studies covered the experiences of cultivating over roofs. This may be related to its additional advantages, known to architects and others, which cannot be estimated financially. In his study about the benefits of green rooftops, Theodosiou [7] assured the scientific literature agreement on its importance in individual buildings and on an urban scale. This importance is related to their interior design, reducing the impact of climate change, storm water

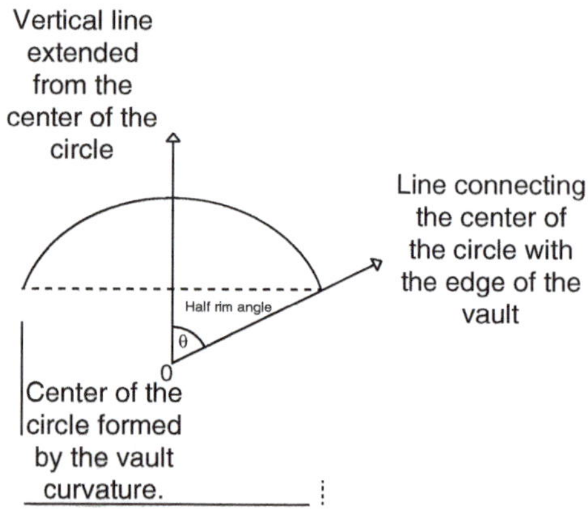

Fig. 47.1 Vault roof and the rim angle vault roof

Fig. 47.2 Shed rooftop

management, minimizing the Urban Heat Island, a source for healthy safe food, social relationships among neighbors, esthetical aspect, providing job opportunities, and reducing the air pollution and noise. France approved a law in March 2015 for new buildings in commercial zones to use partially green roofs or solar panels as cited in ([8]. Niachou et al. [9] found that the effect of the green roof is clear when used with moderate insulation or without insulation.

Ferrante et al. [10] used a DesignBuilder simulation for a building to compare a conventional roof with a rooftop with a green cover. The results show a good rate of energy saving so they concluded that the green roof is one of good choices to improve the energy performance of buildings; however, the amount of contribution depends on the climate of the zone. Karzam [11] stated that using green roofs over a thick insulation layer reduces temperature in the upper floor and minimize energy consumption in Gaza during summer. In a comparative study between two studies in Greek and France, Semaan and Pearce [8] demonstrated that using rooftop garden decreases the indoor air temperature 2 °C. Another study by Hui [5] about the effect of green roof on the attic of a pitched roof showed an actual effect in reducing the indoor heat gain. During hot summer days, the attic temperatures with green roof was around 30 °C and for typical tiled roof, it was 40–60 °C.

Kumar and Mahalle [12] studied the effect of green floor in reducing indoor temperature using both simulation and experiment in mild warm climate of India. They used two models with dimensions of 1.5 m × 1.5 m × 2.2 m and 19 cm concrete slab and 15 cm hollow brick wall for two rooms; one with a plain concrete roof and the other has a green roof. They found that simulation reduced temperature 3.1 °C while the experiments shows a reduction of 4 °C comparing with conventional roof. In addition, they found that by increasing the depth of the soil layer, the lower the indoor temperature.

In a warm humid climate, Dareeju et al. [13] used four small-scale models to study the effect of using green roof on the indoor temperature. The results showed that, the minimum indoor temperatures, during daytime, was for the model with soil cover 75 mm followed by soil covers of 50 mm, 25 mm, and without soil cover. The indoor temperatures of the models were 31.3 °C, 31.5 °C, 32.2 °C, and 34.0 °C, respectively. However, the indoor temperatures during the nighttime were approximately the same. The indoor temperature diminishes with the increase of the soil cover thickness. The indoor temperature of the green roofs is below the other roofing systems. In addition, the indoor temperature of the green roofs was below the other traditional roofing systems such as flat slabs, clay tile roofs, and asbestos roofs. The flat slab was the worst. They claimed that, with time the absorptivity of flat slabs, asbestos sheets and clay tiles increase as a result of change in roof color, while green roof did not show this variation.

Other studies investigated more details about the green roof such as Vera et al. [14] and Theodosiou [15]. Vera et al. [14] investigated the effect of plant and substrate characteristics of a green roof of one-storey supermarket using Design Builder simulation. They studied four parameters, irrigation schedules, plant height, substrate type, and leaf area index, which is the ratio between the leaves' surface and other parts of soil and the plant beneath (LAI). They found that increasing LAI reduces the consumption of energy for cooling between 39.5% and 59.5% and it was the most effective factor. Theodosiou [15] and Dominique et al. [16] assured the effect of leaf area index and added that this effect is related to transpiration and shading influences. Dominique et al. added that the reduction in temperature was between 5 °C and 7 °C. Similar results were achieved by Niachou et al. [9]. They evaluated the effect of a green roof experimentally and mathematically for different times over year using Transient System Simulation Program depending on U-value of materials. They found that using thick dark green type of vegetation decrease the indoor temperature better than that which do not have green roof or that have sparse red vegetation.

Another effective factor of increasing the amount of irrigation system contributes to reduce cooling loads [14]. However, plant height and thermal inertia of the substrate were not effective. Theodosiou [15] shared them about the lack of influence of plant height and added that using insulation minimize the ability of cooling the interior of the green roof.

Gargari et al. [17] and Chenani et al. [18] demonstrated that many researches assured the importance of using green rooftops for indoor thermal comfort, but the attention to its disadvantages is less. Gargari et al. [17] verified that it has higher cost in maintenance operation which includes using of fertilizers, consumption of water, any replacement procedures, and emissions to water and air. Furthermore, the efficiency of green roof depends on the used growing medium and its ability to be recycled. Chenani et al. [18] stated that layers of materials used in green roofs have different effects on environment. They preferred to avoid using several artificial layers that have great environmental impact.

47.3.4 Rooftop with Tile and Compact Insulation

In their study about the thermal behavior of using different covering for a roof of classrooms at a school Curiel-Razo et al. [19] found that decreasing or increasing the inside temperature depends on the rooftop covering, rather than the building material of the roof. They used four colors, red bitumen-based asphalt, white bitumen-based asphalt, black bitumen-based asphalt; each of them covers a 10 cm reinforced concrete slab. The fourth cover was brown clay roof that consists of three layers: waterproof acrylic, waterproofing membrane, and substrate clay thickness 12 cm. The brown roof was used to view a behavior as a thermal insulation material in the region. Additionally, they measured the temperature 0.4 m above each roof and 0.15 m below the ceiling. For the inside temperature, they found that the red color gives the least temperature followed by the brown clay roof and the black one gives the maximum with a difference of 6.5 °C. This study examines just the effect of the color for insulation, but there should be other studies to compare using insulation with other treatments. Another study conducted by Mohamed et al. [20] confirmed the importance of adding white coating over a tiled roof. They stated that the external reinforced concrete roof typically, in their case study, consists of reinforced concrete, bitumen, soil, cement, and roof tile. There was no insulation. From inside there are gypsum plaster followed by smooth plaster and emulsion paint. They used EnergyPlus V8.2 for simulation and simulated eight orientations, the four main ones and the four among them for row houses. They found that the choice in which the main elevation faces the east was most energy-consuming orientation. As such, they used this orientation for four different roof covers as follows: typical construction, adding white coating above, radiant barrier placed between the roof and ceiling, and interior radiation control coating underneath of the roof. The second choice gives the best results of reducing the space cooling loads during the summer by 13.1%, the third and fourth choices give a reduction of 7.7% and 7%, respectively.

47.4 Methodology

Using computer simulation software such as Design Builder software version 2.2.5.004 and Energy Plus 4.0.0.024 were the main tool of acquiring the results. The simulation tools were used for assessing the environmental and energy performance. It provides different types of data related to energy consumption, occupant comfort, and daylighting analysis.

Design Builder software requires specific information about the site and buildings to conduct the simulation process. The site information includes the location specification and the related weather data. The building information includes activity, construction, openings, lightings, HVAC systems, and CFD-Computational fluid dynamics. Amman city in Jordan is chosen as the location of the conducted model in the study area. The climate in Amman is considered hot and dry in summers with cool evenings and cold with barely any rainfall in winter which is between 271 mm (10.7 in.) and 65 mm (2.4 in.) per year [21]; the weather data used in the simulation is in Table 47.1.

DesignBuilder activity templates allow the researchers to choose from a group of functions such as bedroom, office, living room, and dining room spaces. The main attributes that must be set for the assigned template are occupancy density, occupancy schedules, environmental control, heating and cooling set point and set back temperature, lighting target and illuminance level.

The building function assigned as residential building with area 180 m^2. The occupancy rate for the room equal 0.27 (people/m^2), where the number of people in the zone equal 5 persons as the average household size in Jordan is 4.82 persons (census, 2016). The occupancy hours were assigned to be from 6.00 am to 12.00 am. The set points temperature for heating and cooling were fixed with 20 °C heating, 10 °C heating setback, 23 °C cooling, and 26 °C cooling setback temperatures.

Table 47.1 The location's weather data

	Model characteristics	Values of characteristics
Location	Location	Jordan/Amman
	Longitude	35.93
	Latitude	31.95
	ASHREH climate type	3B
Summer	Start month	June
	End month	August
	Extreme hot week, starting	20 July
	Typical hot week, starting	13 July
	Cooling degree days (base 10 °C)	3068
Winter	Start month	Dec.
	End month	Feb.
	Extreme cold week, starting	13 Jan.
	Typical cold week, starting	27 Jan.
	Heating degree days (base 18 °C)	1168

Table 47.2 Building envelop characteristics of the case study

Elements	Materials	Thickness (cm)	Conductivity (W/mK)	Density (kg/m^3)
Exterior wall	Stone	5	1.53	2200
	Concrete	10	1.17	2000
	Insulation	3	0.33	30
	Block	10	0.49	1400
	Finishing	1	0.53	1570
Floor	Ceramic tiles	3	1.200	2000
	Cement mortar	2	0.103	555
	Gravel	10	0.52	2050
	Waterproof asphalt	2	0.17	1100
	Reinforced concrete	15	1.17	200
	Earth	15	1.28	1460
Internal partitions	Block	10	0.49	1400
	Finishing	1	0.18	600

Construction templates also allow the researcher to determine the elements of the model structure. The main attributes that must be set for the assigned template are external wall, internal walls, floor, and roofs with their related layers, thicknesses, and specifications as described in Table 47.2.

Opening templates allow the software user to define specific attributes, including glass type, window-to-wall ratio, window height, and sill height. Window types were assigned as double-glazed window with operation schedule as the windows were opened from 7.00 a.m. to 7.00 p.m. Lightings templates are provided with a set of templates related to the luminaire type and control type. Lighting requirements was with target illuminance of 300 Lux for residential buildings. And normalize power density of 3.0 W/m^2—100 Lux. The luminaire type was chosen to be suspended with lighting control type linear/off. The air ventilation was set equal to 0.35 air change/hour (ach). The HVAC templates include data about the mechanical ventilation in heating and cooling systems, and natural ventilation.

It is clear through reviewing the literature that there are many studies regarding using different treatments for rooftops. Each study follows the circumstances in its area and tried possible solutions there. Actually, different thicknesses, different layers, and different arrangements for the layers were used. Some conditions such as the orientation of the building, controlling ventilation and openings and examining for different times over the year were used. Benefiting from previous studies, this study defines its methodology depending on the context of the construction system in Amman. These treatment styles have different construction and maintenance cost. Table 47.3 describes five sections for five types of flat rooftops treatments used in Amman city. They are rooftop with shed, green rooftop, rooftop with compact insulation, rooftop with tile, to be compared with a simple rooftop without any treatment (the common treatment used in Amman).

Table 47.3 Flat rooftop types used in the simulation

	Rooftops styles	Rooftops sections	Constriction cost JD/m²	Maintenance cost Jd/m²
1	Rooftop with shed (coated aluminum pergola with polycarbonate sheet 16 mm)	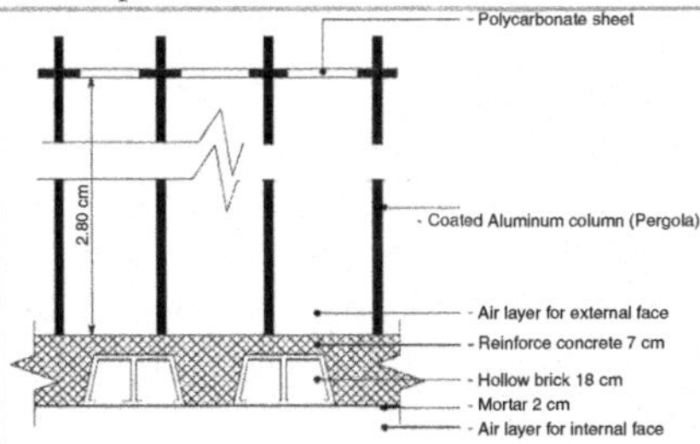	90 JD to 110 JD/m² without concrete roof[a]	Nothing
2	Green rooftop	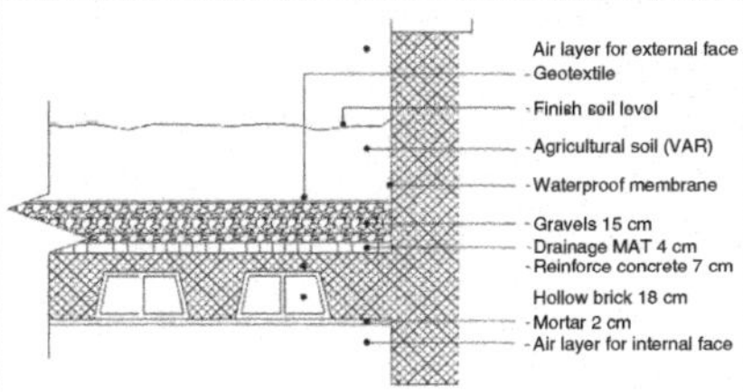	40–47 JD/m² without concrete roof[a]	Depends on the plantation, no maintenance cost for the civil works. It could be 50–5000 JD/month for the project depending on the area, type location of plantation etc.
3	Rooftop with compact insulation	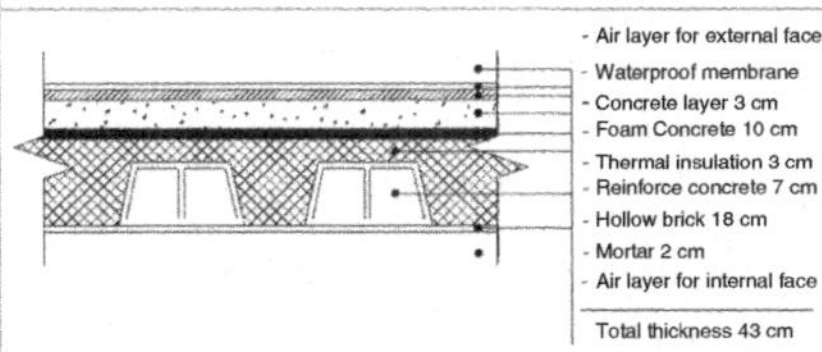	28 JD/m² without concrete roof[a]	Almost nothing, only if the tiles need repair each 5 years
4	Rooftop with tile	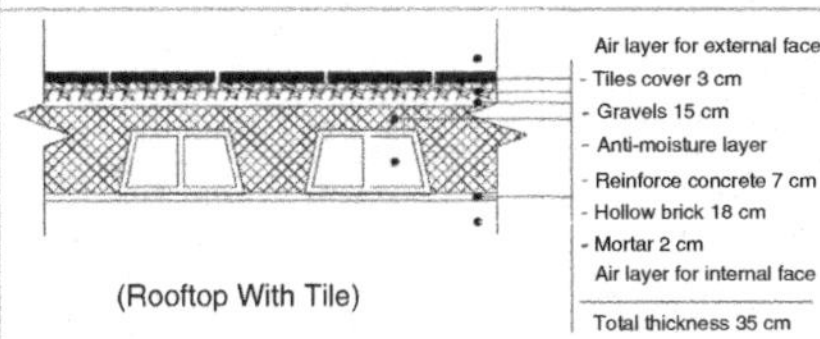	Tiles 13 JD/m² without concrete roof[a]	Almost nothing, only if the tiles need repair each 5 years
5	Simple rooftop without any treatment	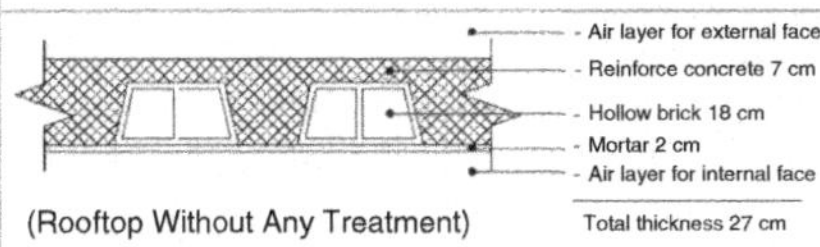	28 JD/m²	Almost nothing

[a]Reinforced Concert slap roof costs 28 JD/m²

47.5 Results and Discussion

The researchers applied these different five flat rooftop treatments on the same model structure simulation that is mentioned previously. The DesignBuilder simulation results for each treatment shows the heating load and the cooling load during the 12 months according to the program simulation data.

The shed is supposed to shade the whole roof. The results of the rooftop with shed thermal performance in Table 47.4 show that the maximum heating load is in January and reaches 3591.57. As for the cooling load, the maximum load is in August and reaches 4348.74.

The results of the green rooftop thermal performance in Table 47.5 show that the maximum heating load in January reaches 2863.31. As for the cooling load, the maximum load is in August and reaches 4491.05.

The results of the rooftop with compact insulation thermal performance in Table 47.6 show that the maximum heating load in January reaches 2977.09. As for the cooling load, the maximum load is in August and reaches 4673.95.

The results of the rooftop with tile thermal performance in Table 47.7 show that the maximum heating load in January reaches 3257.10. As for the cooling load, the maximum load is in August and reaches 4576.04.

The results of the rooftop without any treatment thermal performance in Table 47.8 show that the maximum heating load in January reaches 3557.10. As for the cooling load, the maximum load is in August and reaches 4952.84.

The Designbuilder analysis results show differences in the five rooftops types in heating and cooling load total as shows in Table 47.9. The results showed that the green rooftop is the best treatment that offers a high thermal comfort to the habitants beneath, which shows the lowest total load (26,400.51). However, this treatment is considered the highest maintenance cost on the long run which depends on the plantation and maintenance cost for the civil works. It could be 50 JD upto 5000 JD/month for the project depending on the area, location, type of plantation, etc.), especially with the problems that might arise from water leakage and the construction cost considers high (40–47 JD/m^2).

The best second treatment is rooftop with compact insulation treatment; the total load is (27,460.32). Moreover, rooftop with tile has close results (27,786.27). Nevertheless, the difference between the green rooftop and rooftop with compact insulation in the construction cost is more reasonable (28 JD/m^2) with almost no maintenance cost. On the other side, rooftop with shed results showed the lowest cooling load (17,224.04) that consider high efficiency in the summer only. This treatment considers the highest construction cost between the other treatments (90 JD to 110 JD/m^2) with no maintenance cost. The highest total load is rooftop without any treatment, the most common treatment in the area, (30,218.55), which is considered inconvenient for the indoor thermal comfort level that affect habitant discomfort. However, this treatment considers the cheapest treatment of all (12 JD/m^2).

Table 47.4 Rooftop with shed (coated Aluminum pergola with (polycarbonate sheet 16 mm)

	Jan	Feb	Mar	Apr	May	Jun	Jul	Aug	Sep	Oct	Nov	Dec
Heating load	**3591.57**	2790.80	1297.07	265.84	23.78	0.0	0.0	0.0	0.0	13.91	630.30	2178.99
Cooling load	0.0	0.03	69.69	335.21	1246.61	2646.36	3899.05	**4348.74**	2983.23	1610.28	84.84	0.0
Operative temp.	17.30	17.74	19.47	21.51	23.27	24.72	25.52	25.29	24.50	23.39	20.15	17.87

Table 47.5 Green rooftop

	Jan	Feb	Mar	Apr	May	Jun	Jul	Aug	Sep	Oct	Nov	Dec
Heating load	**2863.31**	2196.90	874.84	137.32	8.34	0.0	0.0	0.0	0.0	4.34	353.86	1584.50
Cooling load	0.0	0.57	107.87	398.10	1373.08	2806.73	4009.75	**4491.05**	3237.33	1845.88	106.72	0.0
Operative temp.	18.42	18.77	30.49	22.42	23.73	24.93	25.55	25.35	24.79	23.94	21.09	19.05

Table 47.6 Rooftop with compact insulation

	Jan	Feb	Mar	Apr	May	Jun	Jul	Aug	Sep	Oct	Nov	Dec
Heating load	**2977.09**	2273.56	879.55	126.82	6.13	0.0	0.0	0.0	0.0	4.45	387.36	1719.18
Cooling load	0.0	0.42	107.30	420.84	1464.14	2969.79	4191.54	**4673.95**	3331.30	1826.61	100.30	0.0
0perative temp.	18.27	18.66	20.54	22.57	23.97	25.18	25.75	25.57	24.94	23.92	20.94	18.77

Table 47.7 Rooftop with tile

	Jan	Feb	Mar	Apr	May	Jun	Jul	Aug	Sep	Oct	Nov	Dec
Heating load	**3257.10**	2498.08	1018.80	167.71	9.50	0.0	0.0	0.0	0.0	7.47	496.18	1953.75
Cooling load	0.0	0.16	89.78	383.52	1367.15	2861.94	4098.50	**4576.04**	3217.73	1698.80	84.06	0.0
Operative temp.	17.77	18.22	20.11	22.19	23.67	25.02	25.66	25.47	24.79	23.60	20.46	18.26

Table 47.8 Rooftop without any treatment

	Jan	Feb	Mar	Apr	May	Jun	Jul	Aug	Sep	Oct	Nov	Dec
Heating load	**3557.10**	2671.96	1071.97	176.66	9.54	0.0	0.0	0.0	0.0	9.12	582.63	2158.19
Cooling load	0.0	0.14	97.54	436.37	1536.29	3168.62	4466.90	**4952.84**	3462.74	1765.94	94.01	0.0
Operative temp.	17.36	18.05	20.26	22.52	24.27	25.75	26.42	26.20	25.35	23.83	20.37	17.92

Table 47.9 The flat rooftops styles total load

	Rooftop styles	Heating load	Cooling load	Total load
1	Rooftop with shed	10,792.25	17,224.04	28,016.29
2	Green rooftop	8023.42	18,377.09	**26,400.51**
3	Rooftop with compact insulation	8374.13	19,086.19	27,460.32
4	Rooftop with tile	9408.57	18,377.70	27,786.27
5	Rooftop without any treatment	10,237.17	19,981.38	30,218.55

47.6 Conclusion

The green rooftop treatment is considered the best efficient solution in cooling and heating load on the habitants' thermal comfort during the year time. However, the construction and maintenance cost are very high which is considered not affordable to the common habitants. The second best efficient treatment to habitant comfort level is the rooftop with compact insulation with fordable cost and rooftop with tile. However, the rooftop with the shed is an efficient treatment only in summer followed by green rooftop and rooftop with tile. The most common treatment in the area rooftops without any treatment is the lowest efficiency treatment concern the habitants' thermal comfort.

References

1. Madhumathi A, Radhakrishnan S, Priya RS (2014) Sustainable roofs for warm humid climates-a case study in residential buildings in Madurai, Tamilnadu, India. World Appl Sci J 32(6):1167–1180
2. Qawasmeh B, Al-Salaymeh A, Ma'en SS, Elian N, Zahran N (2017) Energy rating for residential buildings in Amman. Int J Ther Environ Eng 14(2):109–118
3. Jordan Green Building Guide (2013) Jordan green building guide. The Construction and Sustainable Building Center, Amman
4. Jayasinghe M, Attalage R, Jayawardena A (2003) Roof orientation, roofing materials and roof surface colour: their influence on indoor thermal comfort in warm humid climates. Energy Sustain Dev 7(1):16–27
5. Hui D (2006) Benefits and potential applications of green roof systems in Hong Kong. Evaluation 11:12
6. Dabaieh M, Wanas O, Hegazy MA, Johansson E (2015) Reducing cooling demands in a hot dry climate: a simulation study for non-insulated passive cool roof thermal performance in residential buildings. Energ Buildings 89:142–152
7. Theodosiou T (2009) Green roofs in buildings: thermal and environmental behaviour. Adv Build Energy Res 3(1):271–288
8. Semaan M, Pearce A (2016) Assessment of the gains and benefits of green roofs in different climates. Proc Eng 145:333–339
9. Niachou A, Papakonstantinou K, Santamouris M, Tsangrassoulis A, Mihalakakou G (2001) Analysis of the green roof thermal properties and investigation of its energy performance. Energ Buildings 33(7):719–729
10. Ferrante P, La Gennusa M, Peri G, Scaccianoce G, Sorrentino G (2015) Comparison between conventional and vegetated roof by means of a dynamic simulation. Energy Procedia 78:2917–2922
11. Karzam G (2014) Growing plants on building's roof is an essential component of green building. Prospec Environ Dev 66. Retrieved from http://www.maan-ctr.org/magazine/article/370/
12. Kumar V, Mahalle A (2016) Investigation of the thermal performance of green roof in Indian climate. Int J Renew Energy Res 6(2):487–493
13. Dareeju BSSS, Meegahage JN, Halwatura RU (2010) Indoor thermal performance of green roof in a tropical climate. Int Inst Infra Renew Reconstr, Sri Lanka
14. Vera S, Pinto C, Victorero F, Bustamante W, Bonilla C, Gironás J, Rojas V (2015) Influence of plant and substrate characteristics of vegetated roofs on a supermarket energy performance located in a semiarid climate. Energy Procedia 78:1171–1176

15. Theodosiou TG (2003) Summer period analysis of the performance of a planted roof as a passive cooling technique. Energ Buildings 35(9):909–917
16. Dominique M, Tiana RH, Fanomezana RT, Ludovic AA (2014) Thermal behavior of green roof in Reunion Island: contribution towards a net zero building. Energy Procedia 57:1908–1921. https://doi.org/10.1016/j.egypro.2014.10.055
17. Gargari C, Bibbiani C, Fantozzi F, Campiotti CA (2016) Environmental impact of Green roofing: the contribute of a green roof to the sustainable use of natural resources in a life cycle approach. Agric Sci Proc 8:646–656
18. Chenani SB, Lehvävirta S, Häkkinen T (2015) Life cycle assessment of layers of green roofs. J Clean Prod 90:153–162
19. Curiel-Razo YI, Soto-Zarazúa GM, Rivas-Araiza E, Chavez-Alegria O, Toledano-Ayala M (2011) Thermal behavior in different roof coverings on classrooms. Int J Phys Sci 6(35):7889–7897
20. Mohamed HI, Lee J, Chang JD (2016) The effect of exterior and interior roof thermal radiation on buildings cooling energy. Procedia Eng 145:987–994
21. www.weather-and-climate.com. Accessed 20 Jul 2018

Chapter 48
Integrating Sustainability and Renewable Energy Systems with Architecture Form and Urban Design: Greening the Central District of Irbid, Jordan

Anwar Ibrahim and Hikmat Ali

48.1 Introduction

Sustainable architecture is a trending architectural movement that is concerned with the technical aspects of architecture with an aim of creating a devoted environmental friendly architecture. Mainly, it pursues an architecture with further consciousness to environmental concerns, an optimized physical performance accompanied by a minimum consumption of traditional energy substances and essentially relying on sustainable natural resources. This can be achieved intrinsically by a careful consideration of the building form, the utilized construction materials, along with installing active renewable energy systems, such as photovoltaic panels as energy supplier. Applying sustainable development concepts to design, construction, and operation of buildings can enhance both the economic well-being and environmental health of communities worldwide [1].

It is important for design concepts to invent a publicly appealing sustainable architecture by developing a convincing aesthetically visual presence. However, this matter still forecasts a challenge. Sustainable architecture with the applied renewable energy systems did not materialize itself in a unique architectural formal identity [2]. Unlike earlier technological revolutions like concrete and steel, sustainable architecture has not yet had any significant, self-identifying formal consequences. Instead, the experience of sustainable forms and spaces has to be hyper-mediated [2]. The need to use photovoltaic and solar panels, new building materials, as well as assembly and disassembly techniques and the need to optimize the architectural form in order to maximize the reception of solar radiation are destined to create major implication in the architecture design and form [3]. Enhancing sustainability apart from architecture design process is at stake here. Developing an integrated

A. Ibrahim (✉) · H. Ali
Jordan University of Science and Technology, Ramtha, Jordan
e-mail: afibrahim@just.edu.jo

A. Sayigh (ed.), *Renewable Energy and Sustainable Buildings*, Innovative Renewable Energy, https://doi.org/10.1007/978-3-030-18488-9_48

conceptual framework that harnesses sustainability to architecture is a must and might start from higher education in design studios.

The fast-paced development of renewable energy systems is mainly accomplished by engineers and driven by maximizing their efficiency which makes the aesthetic aspect relatively irrelevant. It seems that the high interest in renewable energy systems in architecture did not motivate a mutual cooperation between the technology manufacturers and the architects. Yet, as an interface material that cannot be concealed, photovoltaic panels have a high potential of integration into buildings as a multifunctional element that may contribute to a modern architecture that acquires its own aesthetic quality and spectacle.

48.2 Sustainable Architecture in Jordan

Despite its small area, Jordan is a pioneer country in the domain of sustainable architecture and renewable energy systems. It has ranked first in the Middle East and North Africa in renewable and clean energy growth and third out of 71 countries worldwide after China and Brazil [4]. Unsurprisingly, Jordan has one of the highest annual daily averages of solar radiance in the world with an estimated 330 days of sunshine per year. In the last 5 years, installing photovoltaic cells became rampant and constitutes a profitable business that helps to support the national economy. However, installing these systems in a fragmented way resulted in creating a visually unpleasant urban spectacle: a stained urban scene altered by aesthetically unpleasing blue panels on inclined metal frames (Images 48.1 and 48.2).

The goal of this paper is to present an academic educational experience inspired by the above status. An urban design project was assigned to architecture students in the fourth-year level of 5-year program at the College of Architecture and Design/ Department of Architecture at Jordan University of Science and Technology which aimed at urban renewal of the central district of Irbid by integrating sustainable principles with the design from its early stages. The assigned term-long project had to be conducted in a group work through three phases before working individually in the fourth one.

48.3 Integration, Sustainability, and Architecture Education

Integrating architecture and urban design with sustainability has to take into consideration the relationship between the architecture product, the environment, and the flow of resources and energy. It requires the holistic integration of all architectural activities into the design process throughout the life cycle of the building [3]. On the other hand, architecture and urban design require the appropriate planning, selection,

Image 48.1 An urban view from the city of Amman. Photo Credit: Zaid Marji

Image 48.2 A mosque with installed photovoltaic panels. Photo Credit: Zaid Marji

configuration, and implementation of architectural systems and technologies. The core idea of design is to establish major architectural goals and objectives of a project and then commands the process of accomplishing them. Therefore, integration is about bringing all of the building components together in a way and emphasizing the synergy of the parts without compromising the integrity of the pieces [5].

The most fundamental goal of integrated building and urban design systems' design is the elimination of redundant resources, usually achieved through strategic combinations of the systems that are deployed in concert with their shared mandates of space, image, or function [3]. It is important that integration measures provide these and other sufficient tangible benefits to justify the effort involved. Integration forms in different levels of architecture can take limitless possibilities [5, 6].

Universities have sometimes demonstrated support towards sustainable design integration in academia. Undeniably, architectural education merits should embrace the concern between disciplines, across the different disciplines, and beyond all disciplines. Over the past decades, several new approaches have emerged for the incorporation of sustainability and renewable energy into architectural education such as educating architects in the design of buildings which are empowered by renewable forms of energy for heating, lighting, cooling, and ventilation [7]. However, utilizing such knowledge to be translated in integrated architecture design courses is still not sufficiently delegated and should be advised.

At Jordanian universities, the scope of sustainability and renewable energy systems education in architecture curricula emphasizes the technical aspects in both graduate and undergraduate levels. However, enhancing issues of sustainability in design courses does not take a systemized official form, nevertheless, operates on an autonomous mode. As design courses are usually thematic and have a predetermined scope that is clearly stated in the respective university curriculum, other themes get the priority over sustainability. Thus, identifying a sustainable dimension in the assigned design projects is minor and sometimes suggested shallowly by some interested students.

Organizing competitions may create a vibe and increase the impulse for investing serious efforts on integrating sustainability with architecture design. In Jordan, two annual national competitions are organized with a particular scope on sustainable architecture: Jo Green and iSustain. While Jo Green proposes architecture projects that should be designed on a sustainable base by a team of university students from different fields with giving interest to technical calculation, iSustain is more concerned about creating projects that take sustainability as an aspect of design.

48.4 Integrating Sustainability with Architecture and Urban Design: Case Study of Irbid

According to the curriculum, design V is a fourth-year level design course that has to tackle the issues of urban design. The main pedagogical aim of the course is introducing students to urban design problems, strategies, and solutions with a focus on developing thinking processes, enhancing research abilities, and providing students with necessary skills to design in urban contexts. The course highlights the role of urban design and planning in proposing highly efficient and effective urban and architectural design solutions for current environmental, sociopolitical, and economic problems that are relevant to specific types of contexts.

In the fall of 2016, the instructors of the course agreed on proposing a project where sustainability and environmental awareness constituted an essential aspect in the design, thus to be integrated in the urban design process from the very early stages. The suggested project title is "Greening the Central District of Irbid." In general, greening the central district in cities (GCD) is important physical, natural, and cultural aspects of cities. Greening process plays a significant role in sustainable urban development and urban ecology by virtue of multiple environmental and socioeconomic benefits. It offers an opportunity to create environmentally efficient buildings and urban settings by using an integrated approach of design so that the negative impact of building on the environment and occupants is minimized (Architectural Institute of Japan, 2005). In the case of Irbid, GCD would result in three types of urban benefits: environmental, economic, and social. Environmental benefits would include improved local economic activity, reduced health care costs as a result of human health and well-being benefits, value of avoided construction and management costs of gray infrastructure, financial value of saved energy and reduced CO_2 emissions, and value of increased use of walking and cycling transportation.

48.4.1 *Project Background*

Irbid city is a border city in the north side of Jordan and located 70 km far from the capital Amman. It accommodates the second largest metropolitan population in Jordan. Due to the rapid growth of the city in the last three decades, a group of commercial hubs evolved in scattered locations in the city. Those diffused hubs forecasted two serious ramifications, a rupture and dilution on the definition of the city center and abandoned undeveloped buffer zones in between. Therefore, the assigned project entailed to initially locate and then urban renewal the designated central district of the city based on sustainability principles. The purpose is to fill the gap between old and new urban districts of the city through developing a well-searched integrated framework of sustainable development processes and strategies.

The project had forethoughts to expand the utilization of renewable energy systems and promote sustainable architecture to be integrated with the urban design. Therefore, as an additional challenge, students were required to conceptually develop possible ways of integrating the renewable energy systems within the architectural form and landscape design of the urban setting. This was not a mandatory requirement, but only for those who wanted to implement those systems in their projects. Measuring the efficiency of the utilized renewable energy systems and by all means green techniques was out of the scope of this project. The limited availability of technology necessary to run such measures and time limitations obligated these restrictions.

48.4.2 Methodology and Design Process

The conceptual framework of the methodology was extracted from a previous research about the integration of architecture and sustainability that built its framework on the Soft Systems Methodology (Fig. 48.1). Soft Systems Methodology (SSM) deals with complex situations of elements that involve lacking of common agreement on what constitutes the problem to be addressed. The aim of SSM is to bring about improvements in a situation perceived as problematic [8]. By itself, it does not seek to solve "the problem" but to facilitate a learning process which allows its users to gradually develop a more comprehensive understanding of the situation under study [3]. The proposed integration of conceptual model worked on embedding the different sustainability dimensions within the different stages of the building life cycle starting from the preliminary stage through the conceptual design, construction to the post-construction and occupation of the building. This model was customized to suit the proposed urban design project and the time framework of the project.

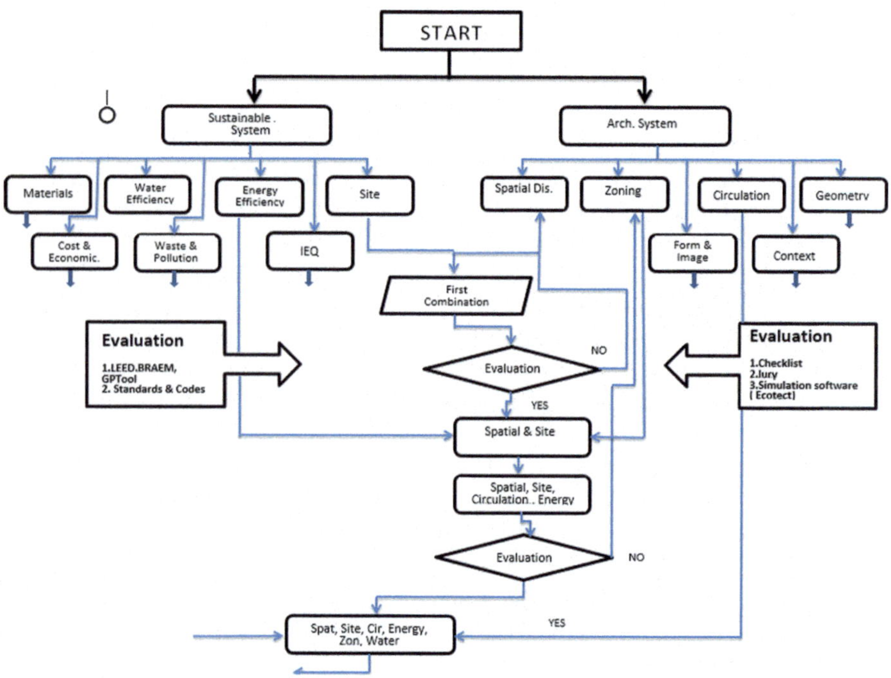

Fig. 48.1 A diagram that represents the proposed integrating methodology. Source ref. [3]

Accordingly, the students were provided with a project brief that described the main goals and objectives of the project and mapped out four phases: preliminary stage, analysis and evaluation, urban design intervention, and finally proposing an architectural design for a new building or adaptive reuse for an existing one.

Phase I: Preliminary Site Survey and Literature Review

Urban design and architecture should start with site survey and urban condition analysis. This requires full diagnosis of the urban character in terms of uses, circulation, urban fabric, urban image, history, urban regulations, and development trends. It also includes reviewing the recent thinking trends and approaches of design and urban design with summary of these trends and experiences. The results of this analysis should be developed in a matrix of urban analysis of the city through selecting representative samples of the city. Spatial, morphological, and socioeconomic analysis for each sample were required. Major synthesis of the entire samples and interpretations of the analysis is a must to build a general framework for the next phases. Under the sustainable design strategy, students had to examine the environmental effects of the materials used and the structure's design, orientation, and impact on the landscape. In addition to given regular lectures, students had to do a thorough research about the meaning of green architecture and renewable energy systems and the ways they could be integrated with architectural form and landscape elements.

Phase II: Analysis and Evaluation

Phase II was dedicated for inducing strategies and applying evaluative criteria. Students had to conduct a SWOT analysis and analyze a minimum of three similar international cases. Selection criteria were based on the previously mentioned strategy as well as the similarity to the site. The analysis had to be based on the surveyed principles of urban design and on the abovementioned diagnosis items of the site's urban character.

Phase III: Urban Design Intervention

In this phase students had to develop conceptual and design proposals for urban design development within different scopes of proposed scenarios based on the results of investigation and concept formation at the predesign phase. Students had to generate an integrated urban design theme of the selected area under the cover of green systems. This phase is characterized as being of a high level of complexity. It has to be elaborated through ideological concepts concerning the local context, concepts concerning needed services and functions, relatively diffused concepts concerning expression and spatial organization, diffused concepts regarding technical systems, and concepts regarding environmental impacts. Students got an intense follow-up to improve their concepts and designs to achieve the stated goals of the project.

Phase IV: Architecture Design

In this final conclusion phase, each student was asked to develop a conceptual architectural design for a single building or adaptive reuse for an existing one. The holistic approach was required to achieve the integration of the various building systems: architectural design, structural design, mechanical design, etc., and landscape into the surrounding environment. The concept and the design should be in harmony with the overall proposed urban design in the third stage. In addition, the student had to study the impact of the proposed design on the new proposed urban fabric.

48.5 Results and Discussion

Based on the conducted surveys and analysis, each group of students delineated what they decided would define the central district of Irbid. The various defined zones ranged from reemphasizing the essentiality of the old city center to nominating other zones to be the central district of Irbid. Accordingly, the proposed designs adopted two contrasting conceptual orientations; a conservative approach minimized the intervention and showed sympathy with the existing urban fabric with a low interest in adding and/or integrating extra elements that might alter the architectural characteristics of the designated district. A conscious avoidance of integrating sustainable technology like the photovoltaic panels and a preference of implementing other sustainable and greening forms like expanding the green areas horizontally and vertically showed clear evidence on these projects. On the other hand, the other orientation was revolutionary by all its means. It proposed a contemporary image of the central district while preserving the old city center intact. Students who adopted this approach proposed a severe intervention and maximum exploitation of the renewable energy resources and sustainable technologies. In contrast to the other students, they considered utilizing sustainable technology a device for fostering a contemporary image of the city. The following two examples illustrate the two orientations.

48.5.1 Project 1: Breathing City

The students who designed this project chose "breathing city" as a main theme for the project, which proposed a synergic city. Social, economic, and environmental aspects were involved together to make the city breathe and live. The proposal did not call for major changes that could potentially meet a rejection from the society; however, they provided solutions to seek advantages from existing elements: social, economic, and environmental. The overall design suggested a variety of possibilities such as creating a dialogue of green areas and water which intervenes with the

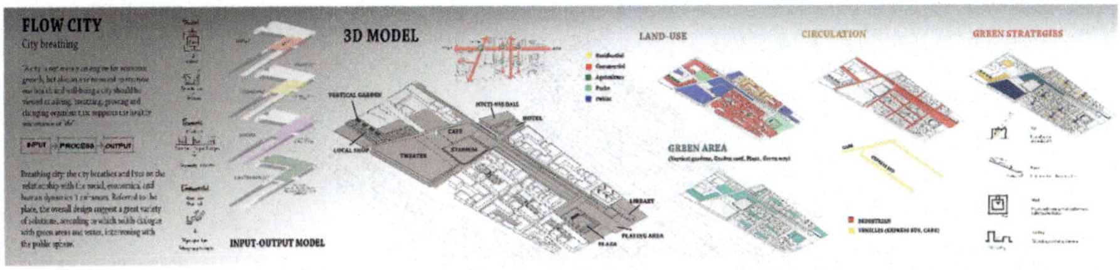

Image 48.3 Breathing city, concept formation

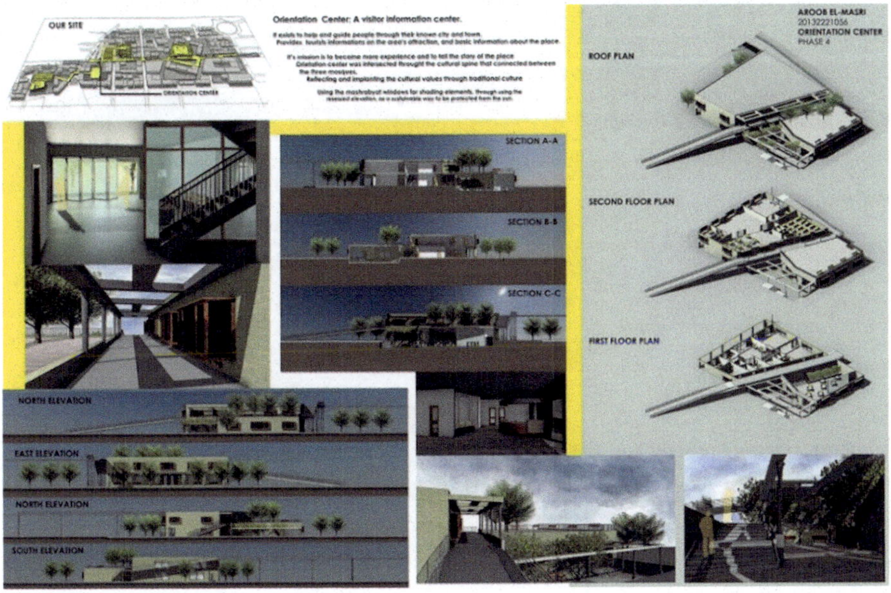

Image 48.4 Breathing city, single building design

social sphere. They adopted what they called input–output model to apply to the breathing city concept (Images 48.3, 48.4, and 48.5). On the environment aspect, green façades cover old and new buildings. Greening roofs is encouraged to achieve the concept of "breathing city."

48.5.2 Project 2: Rejuvenating Irbid

The main idea of the project was to rejuvenate the old city center of Irbid by creating an almost tabula rasa in the areas with insignificant abandoned old structures to be totally demolished and incorporating adjacent empty areas for an opportunity to create open space for the public (Image 48.6). A large-scale multifunctional grid

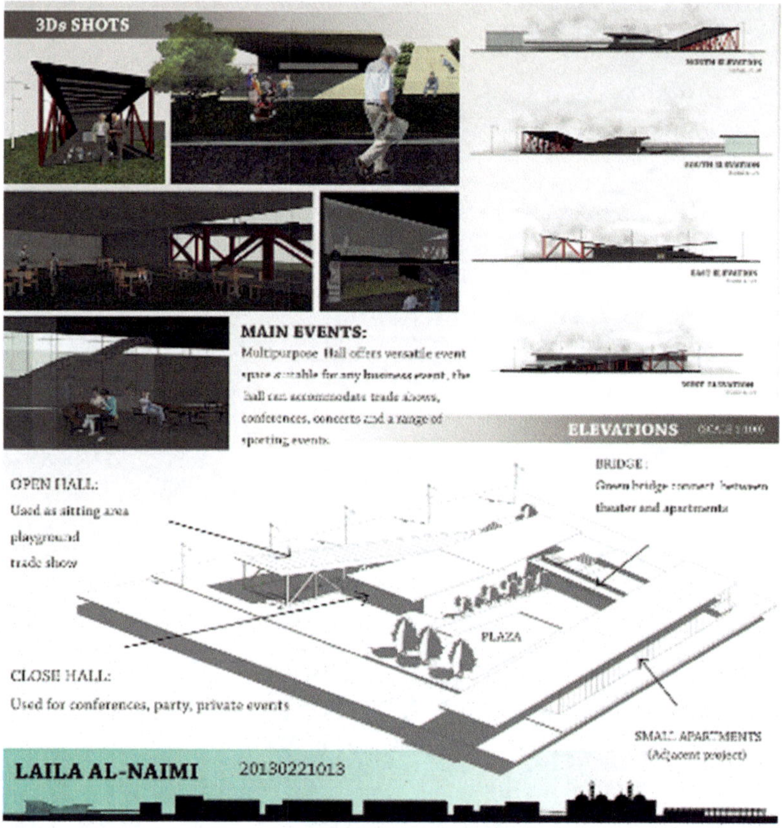

Image 48.5 Breathing city, single building design

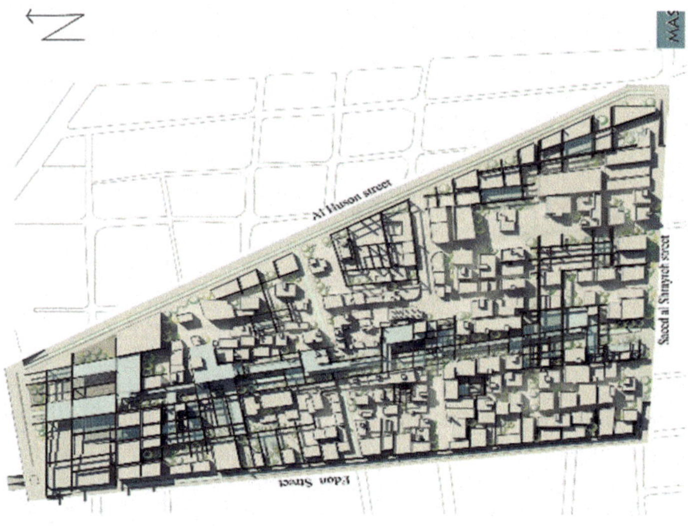

Image 48.6 Rejuvenating Irbid, concept formation

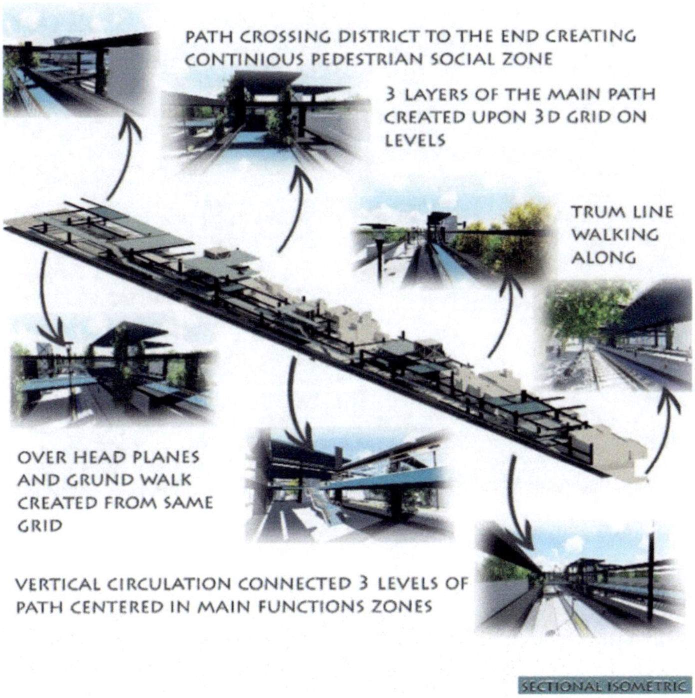

Image 48.7 Rejuvenating Irbid, the central path with the structure holding PV panels

structure of posts and beams laid down over the whole designated area to work as a platform for sustainable solutions (Image 48.7 and 48.8). The structure would visually unify the district: the old, the renovated, and the new structures under one image to be visually and physically connected. In addition to working as rainwater collectors, posts and beams will perform as a platform for hanging walls to expand the very small green area of the city which will help in creating visually pleasant healthy urban setting. But most importantly, the spans between the beams will be used as frames for photovoltaic panels to generate green electrical energy and to provide integrated shaded pathways for pedestrians. This new modern image of the city center will create opportunities to add modern buildings in harmony with the grid in terms of forms and materials.

48.6 Conclusion

In the shade of the current environmental conditions, the exigency of green architecture and cities is elevated. As such, integrating architecture and urban design and planning with the competencies of renewable energy systems development would be highly desirable and urged. As these domains are performing in a segregated

Image 48.8 Rejuvenating Irbid, a public space view overshadowed by PV panels

mode in architecture academic and practical life, they should be merged in a framework that defines efficient and sustainable integrated strategies that are built on a comprehensive approach base where environment, technology and architecture, and urban design and planning are all forged in an essential integrated vision. Institutionalizing this integration in architecture curriculum is an important primary step in this direction. In addition to theory courses, design courses should emphasize at some level the concept of sustainable architecture, thus work on officially stated design course description with a scope on green architecture. This course has to be instructed by architects with high-level training and deep knowledge in sustainability and renewable energy systems. This also might be done by a mutual cooperation between architects and professional specialist engineers. Such a mutual cooperation will introduce both involved parties, architects and engineers, to the other unseen facet. It will enhance the merits and technical requirements of the technology to the sensibilities of architecture principles and concerns, particularly the aesthetic aspect. As these systems' imperatives necessitate their bold manifestation on architecture, the cooperation will be highly fruitful. This interdisciplinary approach will enhance the capacities of the graduated architects and increase their employment opportunities. The outcome is professional architects with theoretical and design qualifications and capacities to deal with sustainable technological solutions.

References

1. Public Technology Inc. and USGBC (1996) Sustainable building technical manual: green building design, construction, and operation. Public Technology Inc., Washington, DC
2. Heymann David (2017) The ugly pit. Places. https://placesjournal.org/article/the-ugly-pet/
3. Ali H (2012) Integrated framework for implementing sustainability into architectural design process: case of Saudi architectural practices. J Architect Plan 24(2):119–134
4. The Jordan Times (2018) Jordan 3rd in renewable energy growth globally, 1st in region. http://www.jordantimes.com/news/local/jordan-3rd-renewable-energy-growth-globally-1st-region%E2%80%99
5. Bachman LR (2003) Integrated buildings: the system basis of architecture. John Wiley & Sons, Inc, Hoboken, NJ
6. Zunde J, Bougdah H (2006) Integrated strategies in architecture. Taylor & Francis, New York
7. Taleghani M et al (2010) Renewable energy education for architects: lessons from developed and developing countries. Int J Sustain Energy 29(2):105–115
8. Checkland P, Poulter J (2010) Soft systems methodology. Springer, London

Chapter 49
Abu Dhabi Schools: Optimization of Building Form for Energy Efficiency

Meriem Rahmani and Khaled A. Al-Sallal

49.1 Introduction

Local and international statistics around the world revealed the adverse impact of humans' immature practices on the environment. The higher reliance on inefficient energy systems for instance causes a parallel increase in greenhouse gas (GHG) emissions, with high risks on the environment and human health. In the United States, 75% of the total consumed electricity is attributed to buildings only [1]. In UK, school buildings were classified as the third most energy consuming after commercial buildings and offices [2]. School buildings in the USA consume 10.8% of the total electricity consumed by buildings, while in UK, they are classified as the third most energy consuming [3]. On the local level, 80% of the total consumed energy in UAE is attributed to buildings' sector, with an average of 220–360 kWh/m^2 per year. Recent statistics have stated that commercial buildings consume 48.2% of the total consumed energy while governmental buildings consume 9.3% of the total consumed energy [4].

Improper building form could lead to higher reliance on mechanical and electrical systems for the provision of cooling, lighting, and ventilation and thus higher energy consumption. Statistics across the world indicate substantial levels of building energy consumption, which also cause significant increase in greenhouse gas (GHG) emissions. Other research showed the risks of unhealthy indoor educational environment on the student's health, performance, and even attitude.

For sustainability targets, Abu Dhabi 2030 vision launched several initiatives that were designed to enhance sustainable approaches. Abu Dhabi Department of Education and Knowledge (ADEK) launched a new initiative, known as the New

---

M. Rahmani (✉) · K. A. Al-Sallal
UAE University, Al-Ain, United Arab Emirates
e-mail: 201570116@uaeu.ac.ae; k.sallal@uaeu.ac.ae

A. Sayigh (ed.), *Renewable Energy and Sustainable Buildings*, Innovative Renewable Energy, https://doi.org/10.1007/978-3-030-18488-9_49

School Models (NSM), with the objective to enhance the sustainability levels and overall quality of Abu Dhabi schools. The NSM is a student-centered learning approach that employs technology and modern teaching facilities in an integrated sustainable environment. The NSM recommends the reliance on passive design strategies and the integration with other energy-efficient environmental systems. The finger-plan typology was adopted, with each finger representing a learning community (LC). The LC is used as an organizational approach to divide of the school into smaller units, with each LC including one or multiple grade levels [5]. This research integrates with the efforts of ADEK with a specific focus on the impact of the building form on energy consumption in the NSM.

49.2 Literature Review

The considerable impact of building form on energy consumption was confirmed by previous research [6–10]. According to Al-Sallal [11], among the most important considerations to reduce heat gain is to optimize the building form with regard to its self-shading and compact design. Compacted forms proved to prevent heat gain in hot dry climates due to reduction of external surfaces exposed to harsh outdoor conditions. In addition, building forms should be designed in a way that can maintain the passage of the cool breezes. In hot humid climates the spread-out forms are the most effective while the introvert forms or the courtyard designs can provide many benefits in hot dry climates [11]. Al Anzi et al. [6] stated that there are three factors related to the impact of the building from on energy consumption, these are the relative compactness (RC), the window-to-wall ratio (WWR), and the solar heat gain coefficient (SHGC). Many researchers have used the RC to express how form configuration affects building energy. According to Mahdavi and Gurtekin, the RC of a form is derived as the ratio of its volume to surface area, an approach that is useful when compared to that of the most compact shape that has the same volume [12]. Previous studies confirmed that an increase of the RC generally leads to a decrease in the energy consumption [8, 13].

49.3 Methodology and Results

49.3.1 School Survey

The conducted survey counted all ADEK schools in Al Ain and Abu Dhabi cities by using an online application (School Finder) developed by ADEK. The total number of schools found in both cities was equal to 411 schools, 244 of which were in Abu Dhabi and the rest (167) were in Al Ain. These figures included the old and newly constructed schools. The private schools were excluded at a later stage. The paper focused on the public schools since they are constructed in large numbers, which make their energy demands and impact on the environment much higher.

Prototype	Form	Prototype	Form
1	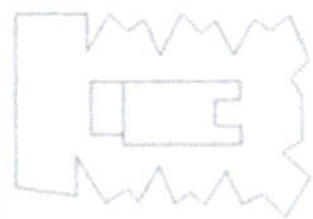	2	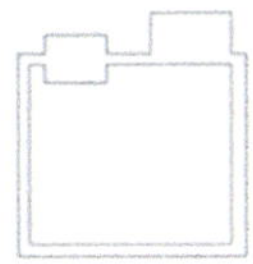
Description: LCs surrounding one small covered courtyard		Description: Classrooms surrounding one big uncovered courtyard.	
3	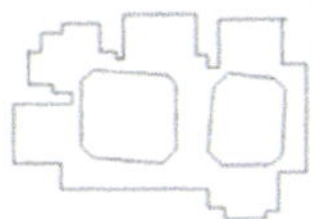	4	
Description: Classrooms surrounding 2 courtyards		Description: LCs adjacent.	
5	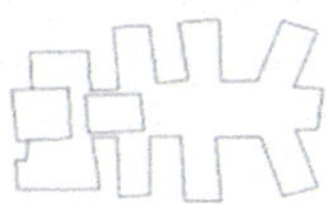	Description: LCs separated by coutyards.	

Fig. 49.1 The most frequent school forms in Abu Dhabi and Al Ain cities

Moreover, public schools generally follow specific design characteristics set by responsible authorities, which results in more common school designs. Therefore, the number of schools considered for the survey decreased to 118 schools in Abu Dhabi and 108 schools in Al Ain. To find the most dominant forms, the research team conducted a second survey using the GPS school code provided by the School Finder application along with Google maps. The results of this stage revealed that 79 schools in Abu Dhabi city and 72 schools in Al Ain city have forms that can be described as most dominant (see Fig. 49.1), compared to other ones that were eliminated from the study because their forms were uncommon. The interview with ADEK staff revealed that old school forms were no longer authorized for sustainability reasons; hence they were also eliminated. Therefore, the research focused only on the public new school models (NSM).

49.3.2 *Case Study Analysis*

ADEK until this date prepared six new school models, which the majority of these models follow the "finger-plan school design." One of these models was designated to KG only, while the other five models were designed for all grades (starting from

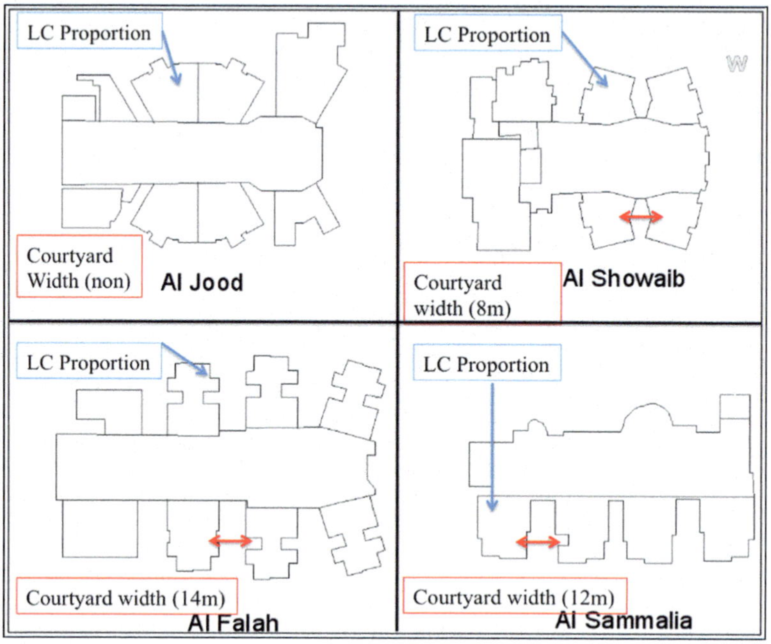

Fig. 49.2 Case study analysis

Table 49.1 Potential form variables with their ranges

Variable	Range
Learning community (LC) width and length proportions	Varied from 1:1.38 to 1:1.5
The space width between the learning communities "courtyard (C)"	Varied from 0 to 14 m
The number of floors	Varied from 2 to 3 floors
The form axis direction	Varied between (N/W–S/E) and (N/E–S/W)
WWR	17–32%

grade 1 to 12). The main difference between the presented models was the capacity of the school to host certain number of students. The sixth model is still under development; therefore, it was not analyzed here. The remaining four models presented by four different case studies, namely Al Jood School, Al Showaib School, Al Falah School, and Al Sammalia School, were analyzed (see Fig. 49.2). The purpose of the case study analysis was to identify which design variables could help optimize the building form; second to define the range of these variables with consideration to maintain the design requirements set by ADEK. The analysis revealed five potential form design variables. These are shown in Table 49.1 and presented graphically in Fig. 49.3.

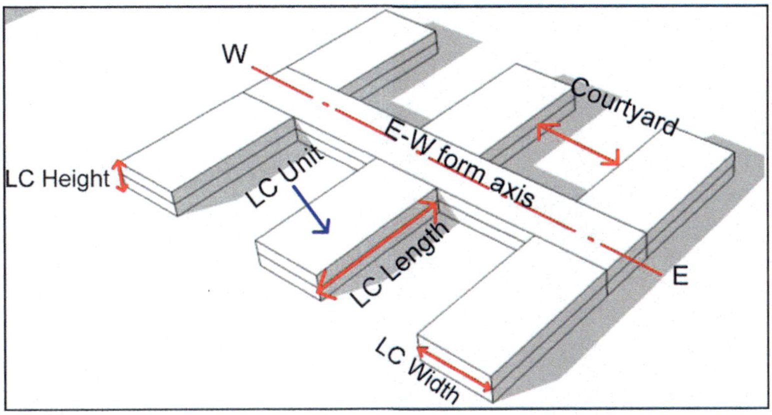

Fig. 49.3 Representation of the potential variables in each case

Table 49.2 Case study variables with their values

Case studies	No. of floors	LC dimensions	C dimensions	WWR	Form axis direction
Al Jood	2	31 × 22.5 m	0 m	40%	E–W
Al Showaib	2	33 × 22 m	8 m	30%	E–W
Al Sammalia	2	30 × 25 m	12 m	30%	E–W
Al Falah	3	38 × 27 m	14 m	20%	E–W

49.3.3 Simulation

To test the impact of the previously mentioned variables and their impact on energy consumption, a pilot experiment was conducted. The experiment covered the four case studies (Al Jood, Al Showaib, Al Sammalia, and Al Falah schools). The cases differ in terms of the number of floors, learning communities (LC) and courtyard (C) dimensions, the window-to-wall ratio (WWR), and the form axis direction. The rest of the design criteria were fixed according to ADEK requirements among all the tested cases. The following table demonstrates the approximate main differences between the cases (Table 49.2).

The energy simulation results proved that Al Falah School performs better compared to the other cases, followed by Al Showaib, then Al Sammalia, and finally Al Jood. The energy reduction between Al Jood (which recorded the highest rate of energy consumption) and Al Showaib equals 3%, while between Al Jood and Al Sammalia equals 4%, yet this reduction reaches up to 37% between Al Jood and Al Falah (which recorded the lowest rate of energy consumption). Figure 49.4 demonstrates the energy simulation results.

For the CO_2 emissions, results showed some consistency with the amount of energy consumption, which as expected revealed a direct relation between energy

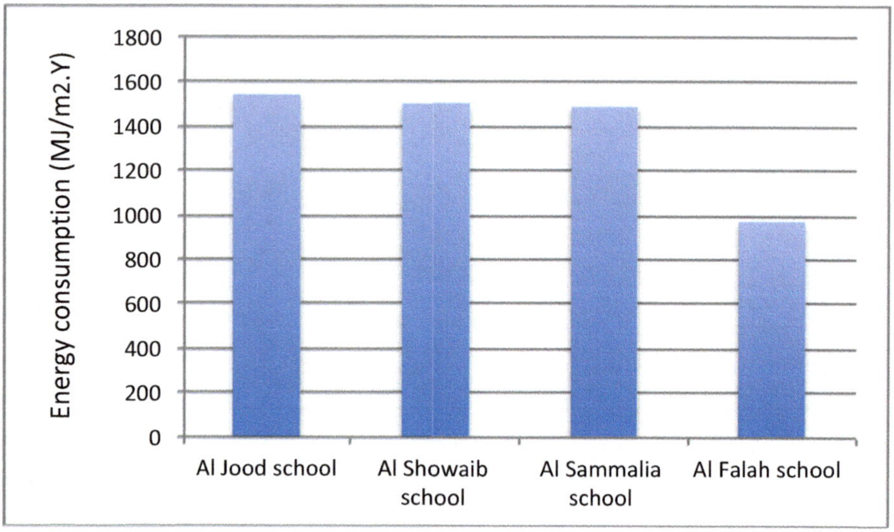

Fig. 49.4 Energy consumption of the four cases

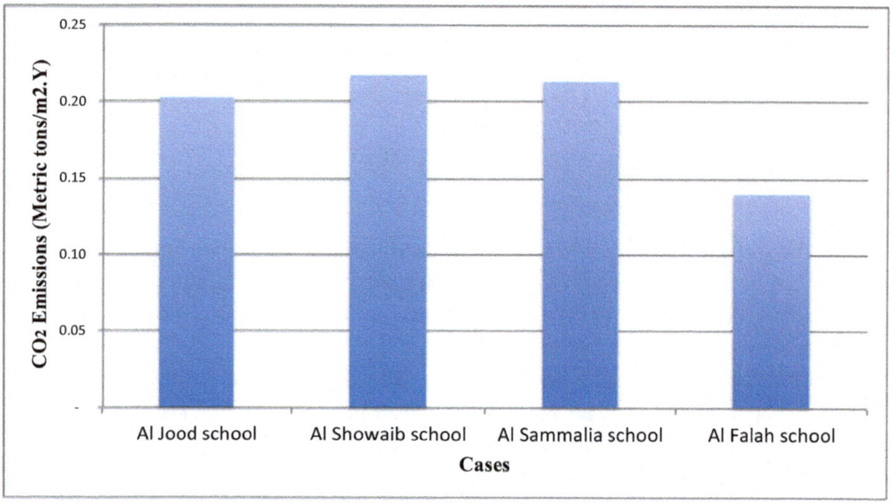

Fig. 49.5 CO_2 emissions of the four cases

consumption and CO_2 emissions. Hence, the amount of energy being saved from one case to another equals in parallel better chances for improving the environment. Figure 49.5 demonstrates the results of CO_2 emissions.

49.4 Discussion

The variation of the results indicates the impact of the selected building form variables on energy consumption. The substantial reduction of energy consumption in Al Falah School might be attributed to the number of floors (height) since it is the main variable that distinguishes Al Falah from the other schools. By calculating the verticality (which is calculated by dividing the vertical surface area over the total surface area) and horizontality (which is calculated by dividing the horizontal surface area over the total surface area) of each case and linking it with the energy consumption as showed in Figs. 49.6 and 49.7, the results indicate a negative relation between energy consumption and verticality. When the form verticality of a building is increased, its energy consumption becomes less. It can also be expressed in an opposite way if we use form horizontality concept; when the form horizontality of a building is increased, its energy consumption becomes more. Vertical forms help to create more compacted forms that have less horizontal surfaces exposed directly to the perpendicular sun radiation during the hot months which results in less energy needed to cool the buildings. Simulation results demonstrate the importance of compactness and self-shading provided by the form itself as stated before by Al-Sallal [11]. The current work explored only the potentials of some design variables to improve energy savings in schools. More detailed research that tests the impact of each variable individually is needed.

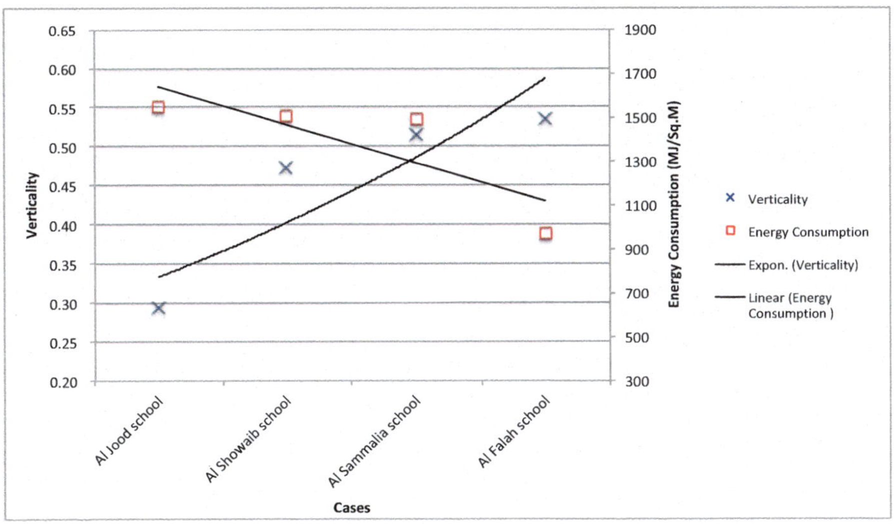

Fig. 49.6 The correlation between verticality and energy consumption

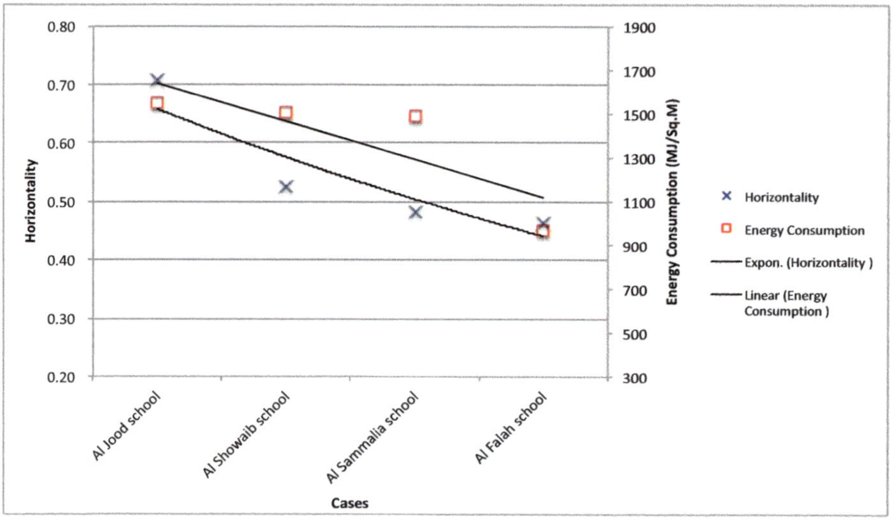

Fig. 49.7 The correlation between horizontality and energy consumption

49.5 Conclusion

The case study analysis helped to address the potential form design variables that could improve energy performance in school buildings. These variables are the learning community, the courtyard, the number of floors (height), the form axis direction, and finally the window-to-wall ratio (WWR). The ranges of the previous variables that were obtained from the case studies helped to maintain the requirements set by ADEK. The outcomes of the analysis will be used for future research paper to design several cases for simulation, each case differ from the other cases in terms of the previously mentioned design variables or their ranges. In each case only one variable will be manipulated while the rest of the variables will be fixed in order to understand the impact of each variable separately. This will help classify the most influential design parameters for future.

Acknowledgments This paper is part of an MSc thesis that was supported by travel fund from the College of Engineering, UAE University.

References

1. Architecture 2030 (2016). http://architecture2030.org/. Accessed 20 Nov 2016
2. Department of Energy and Climate Change (DECC) 2015. https://www.gov.uk/2050-pathways-analysis
3. Pérez-Lombard L, Ortiz J, Pout C (2008) A review on buildings energy consumption information. Energ Buildings 40(3):394–398. https://doi.org/10.1016/j.enbuild.2007.03.007

4. Clarke K (2016) 80% energy consumed by buildings. https://www.khaleejtimes.com/nation/abu-dhabi/80-energy-consumed-by-buildings. Accessed 26 Dec 2017
5. ADEK (2013) Educational Facilities Design Manual (Rep.). Abu Dhabi
6. Al Anzi A, Seo D, Krarti M (2009) Impact of building shape on thermal performance of office buildings in Kuwait. Energy Convers Manag 50(3):822–828. https://doi.org/10.1016/j.enconman.2008.09.033
7. Catalina T, Virgone J, Iordache V (2011) Study on the impact of the building form on the energy consumption. Proceedings of building simulation. http://www.ibpsa.org/proceedings/BS2011/P_1563.pdf
8. Depecker P, Menezo C, Virgone J, Lepers S (2001) Design of buildings shape and energetic consumption. Build Environ 36(5):627–635. https://doi.org/10.1016/S0360-1323(00)00044-5
9. Koranteng C, Abaitey EG (2010) The effects of form and orientation on energy performance of residential buildings in Ghana. J Sci Technol (Ghana) 30(1). https://doi.org/10.4314/just.v30i1.53940
10. Ourghi R, Al-Anzi A, Krarti M (2007) A simplified analysis method to predict the impact of shape on annual energy use for office buildings. Energy Convers Manag 48(1):300–305. https://doi.org/10.1016/j.enconman.2006.04.011
11. Al-Sallal KA (2016) Low energy low carbon architecture: recent advances & future directions. CRC Taylor and Francis Group. Carbon-Architecture-Recent-Advances—Futur Directions/AlSallal/9781138027480. https://www.crcpress.com/Low-Energy-Low
12. Mahdavi A, Gurtekin B (2002) Adventures in the design-performance space. European Meetings on Cybernetics and Systems Research Conference 2002, Vienna, 2–5 April, 2002
13. Pessenlehner W, Mahdavi A (2003) Building morphology, transparency, and energy performance. ResearchGate. transparency_and_energy_performance. https://www.researchgate.net/publication/237581710_Building_morphology_

Chapter 50
Efficiency of Microstructured Sunlighting Systems in Different Climatic Zones

Helmut F. O. Mueller

50.1 Introduction, State of the Art

Consumption of electrical energy for room lighting can be reduced by daylighting systems redirecting direct sunlight to a reflective ceiling and the depth of the room. Artificial light can be dimmed or switched off for large room areas during periods of insolation on windows, while solar heat gains and glare are avoided. No direct glare can occur as the microstructured components for sunlight redirection are fixed in the upper window area above eye level (Fig. 50.1). A solar protective coating on the glass unit reflects the infrared spectrum and transmits only the visible spectrum in a selective manner. Because of the microstructure the light-redirecting glass unit in the upper window area cannot be looked through in contrast to the lower window area. For this part of the window any kind of conventional solar protection can be applied, like shading systems or solar protective glazing.

Prototypes of these microstructured window components have been manufactured, measured, and monitored under natural conditions in test rooms of the Fraunhofer IBP in Stuttgart, Germany [1, 2]. Depending on the task (location, window orientation, and resulting solar altitude) as well as manufacturing technology of microstructures (hot- or UV-embossing), different solutions have been developed. The objective of the research project is to develop one microstructure for a maximum range of solar altitudes that can be applied in many different geographic latitudes and in different climates, from temperate to tropical. Figure 50.2 shows simulated results of the optimization process with high efficiencies for sunlight redirection for solar heights from 15° to 70°. As the light reflection at the glass surface is increasing with the angle of incidence, a tilted sunlighting component will improve the efficiency for solar altitudes > 65° (compare Fig. 50.3). If solar altitudes < 15° can cause glare, e.g., in high-rise buildings, suitable measures for glare avoidance will be applied.

H. F. O. Mueller (✉)
Green Building R&D GmbH, Duesseldorf, Germany
e-mail: helmut.mueller@greenbuildig-rd.com

A. Sayigh (ed.), *Renewable Energy and Sustainable Buildings*, Innovative Renewable Energy, https://doi.org/10.1007/978-3-030-18488-9_50

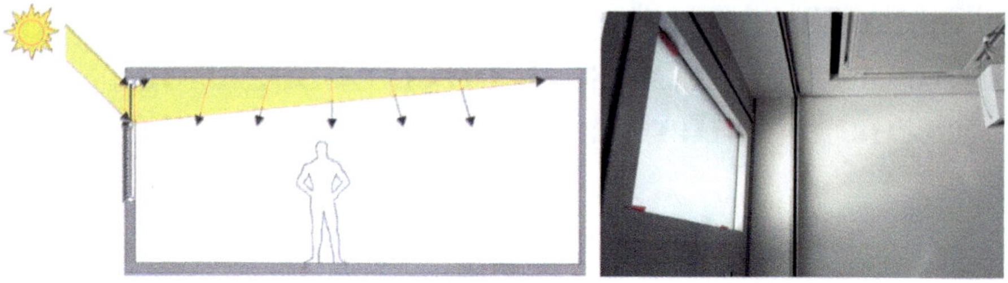

Fig. 50.1 Left: Principle of redirecting solar light from the upper window area to a reflecting ceiling and deep into the room. Solar protection in the lower window area, e.g., shading or glass coating. Right: Prototype in test room of Fraunhofer IBP, Stuttgart

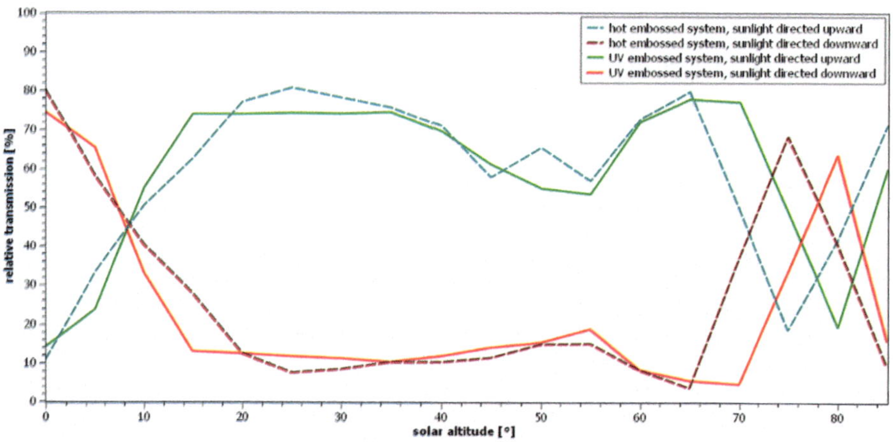

Fig. 50.2 Optimization of sunlight redirection. Relative transmission of sunlight to the ceiling and to the floor over angle incidence, i.e., solar altitude ([2], Jakubowsky, M., Neyer, A., RIF)

The magnitude of energy savings by sunlighting in different geographic latitudes is a crucial requirement for the early building design phases. An estimation can be made by the methodology of sun-path diagrams, offering a visual information about the potential of sunlighting for the months and daytimes of the year. A comparison of different latitudes is possible and optimal window orientations can be recommended.

The actual hours of insolation and light redirection, taking into account the annual time of covered sky, are defined by the local climatic data. The product of annual sunlighting hours and specific wattage of room illumination, e.g., 10 W/m^2, indicates the lighting energy savings. The high quality of sunlighting in comparison to artificial lighting as to illuminance and spectral distribution is an additional advantage.

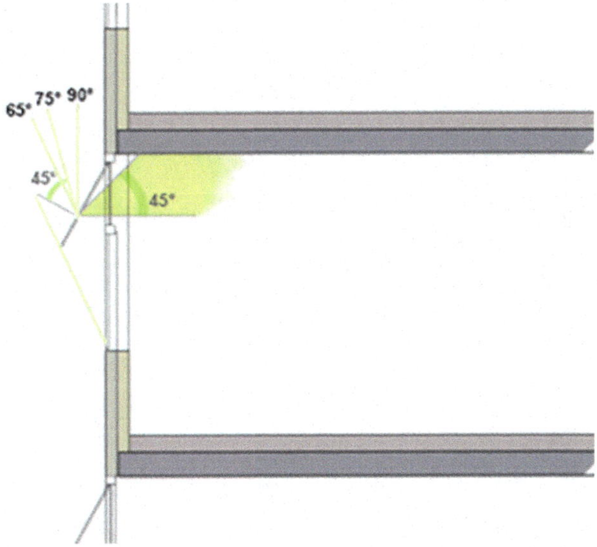

Fig. 50.3 Tilted sunlighting component improves light redirection for solar altitude > 65°

50.2 Methodology and Results

50.2.1 Insolation of Windows According to Sun-Path Diagrams

The annual hours of direct sunlight on windows with specific orientations can be defined by sun-path diagrams. They map the sky dome over a circular chart (polar). The position and path of the sun is projected onto the diagram at different times of the year, shown by 12 date lines for a particular day of each month and by hour lines indicating the day time. Web-based sun-path tools can be used [3].

Examples of sun-path diagrams for the latitudes of 50° N and 0° are given in Figs. 50.4 and 50.5. They show the sunlighting potential for different window orientations. Solar altitudes are considered from > 10°, because of shading effects of buildings in the neighborhood. Vertical and horizontal angles of light incidence > 80° are neglected because of high reflection losses by the window glass.

In most geographical locations only three window orientations allow for sunlighting, but in equatorial locations four orientations can be utilized. A building at the equator with north-/south-orientation of windows, e.g., has a total sunlighting potential of 2 × 1530 h/a = 3060 h/a compared to 2827 h/a at 50° latitude. And a building with east-/west-facing windows has a total sunlighting potential of 2 × 1795 h/a = 3590 h/a at the equator compared to 2 × 1645 h/a = 3290 h/a at 50° latitude.

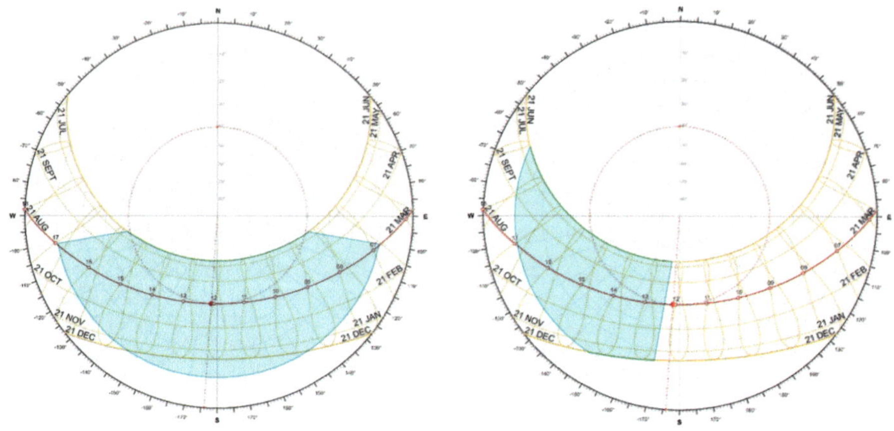

Fig. 50.4 Sun-path diagrams [3] with potential of annual redirection of sunlight for geographical latitude of 50° N. Left: south-facing windows (2827 h/a). Right: west- (east-) facing windows (1645 h/a)

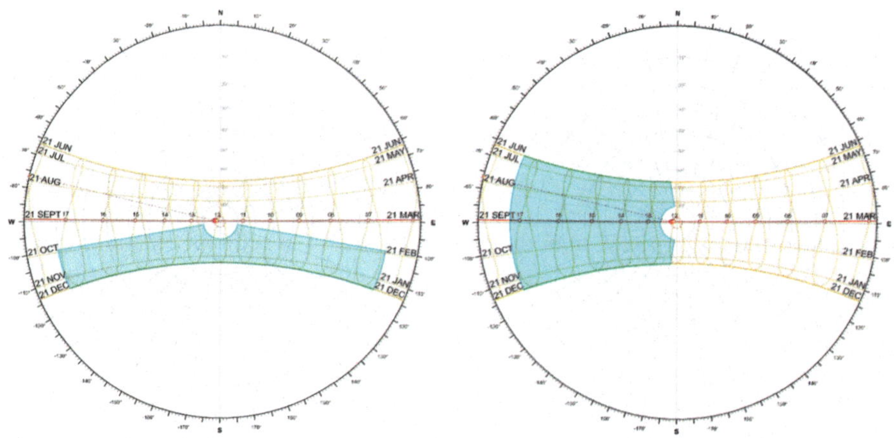

Fig. 50.5 Sun-path diagrams [3] with potential of annual redirection of sunlight for geographical latitude of 0°. Left: south- (north-) facing windows (1530 h/a). Right: west- (east-) facing windows (1795 h/a)

50.2.2 Realistic Insolation Considering Cloudiness of Climatic Regions

A survey of the sunlighting potential of four different latitudes, based on sun-path diagrams, is given in Table 50.1. It shows a realistic estimate of sunlighting hours beside the theoretical one, taking into account the cloudiness for exemplary climatic regions by sunny percentage of annual daylight hours. The overview demonstrates that sunlight redirection can be applied advantageously in geographical latitudes from 0° to 60° and under varying cloudiness. For equatorial regions the sunlighting hours for north- and south-facing windows of a building can be cumulated in contrast to other locations.

Table 50.1 Theoretical and realistic hours of sunlight redirection for latitudes, orientations, and exemplary locations

	Theoretical insolation for window orientations		Realistic insolation for exemplary climates			
Latitude (approximately)	North/south (h/a)	East/west (h/a)	Location, country	Sunny % of daylight hours (%)	North/south windows (h/a)	East/west windows (h/a)
0°	1530	1795	Singapore, SG Kampala, UG	45.4 54.9	695 840	815 985
24°	2427	1684	Hong Kong, HK Alice Springs, AU	44.8 78.9	1087 1915	754 1329
50°	2827	1645	London, UK Berlin, DE	33.3 39.7	941 1122	548 653
60°	2438	1342	Stockholm, SE Anchorage, US	45.0 45.4	1097 1107	604 609

50.2.3 Illuminance and Energy Saving in Sunlit Rooms

The illumination of rooms by redirected sunlight can be calculated for defined solar positions at a specific time of the year and day. Simulation tools like DIALUX or RELUX use the measured data of sunlight redirecting systems. In the TaLED project [2] goniometrical measurements of prototypes of hot embossed microstructures have been carried out und used for the simulation of room illuminance by DIALUX. Table 50.2 shows simulation results for a typical office room with south-facing windows in different latitudes and at different times of the year and day.

The examples demonstrate that the illuminance by sunlighting is sufficient in nearly all situations with insolation on the windows, i.e., latitudes from 0° to 51°, seasons from winter to summer, and daytimes from morning to afternoon. The mean horizontal illuminance mostly is higher than the minimum of 300 lx to 500 lx defined in standards for room lighting. This means that the artificial lighting system is switched off by the automatic control and electricity is saved. Depending on the wattage of the lighting system, e.g., 10 W/m^2 for LED lighting, the energy savings can be calculated according to Table 50.3. The energy savings for lighting are economical (2-year return of investment is realistic), and the improved lighting quality will increase work efficiency of occupants.

Table 50.2 Simulated illuminance distribution in office room with south-facing windows for different vertical and horizontal angles of sunlight incidence

Latitude and day of the year	Daytime and angles of sunlight incidence (vertical, horizontal) on south-facing window		
	8 h	12 h (noon)	16 h
0° Dec. 21st	Vert. 28°, Horiz. 17° 	Vert. 67°, Horiz.	Vert. 28°, Horiz.
20° N Sept. 21st	Vert. 26°, Horiz. 80° 	Vert. 71°, Horiz.	Vert. 28°, Horiz.
51° N Sept. 21st	Vert. 16°, Horiz. 66° 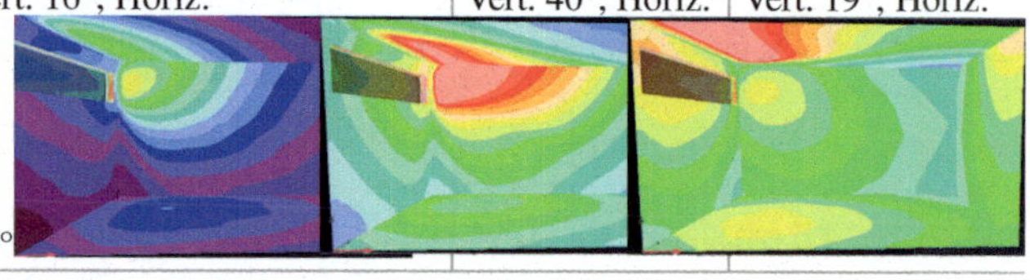	Vert. 40°, Horiz.	Vert. 19°, Horiz.

Illuminance

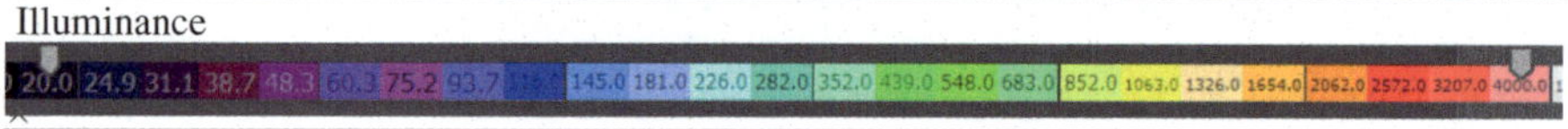

Room dimensions l × w × h = 5 m × 4.8 m × 3 m, window height 0.5 m ([2], Hubschneider, C., Fraunhofer IBP)

Table 50.3 Examples of energy savings and investments for sunlighting systems

Location, country	Annual sunlighting hours (h/a)	Annual lighting energy saving		Max. costs per window area for a 2-year return of investment (€/m^2)
		Energy (kWh/a)	Running costs (€/a)	
London, UK	941	461	92	53
Alice Springs, AU	1915	938	188	107

Office or classroom, 7 m × 7 m, 3 m high, south-facing, redirecting window area 3.5 m^2, installed artificial lighting 10 W/m^2, energy price 0.20 €/W

The calculated savings allow for a rough indication and comparison in the early design stages. In the later design stages simulations should be done more accurately by suitable tools. Studies for different architectural applications have shown that sunlighting systems can illuminate classrooms, shopping centers, and workshops up to a depth of 10 m from the windows, depending on the height of room and windows. Other architectural applications are high and narrow rooms like atria sunlit by skylights [4].

50.3 Conclusions

Micro-optical structures for sunlight redirection can be applied in a wide range of latitudes from the equator (0°) up to 60° N and S. Sun-path diagrams are a simple tool to estimate the annual time of utilization, especially in combination with cloudiness of climatic regions and resulting sunny percentage of daylight hours. Significant energy savings for artificial lighting can be achieved. The quality of room illumination is increased by the daylight spectrum and high illuminances well beyond standard minimum performance. The current research project [2] will provide optimized microstructures to be manufactured in hot- or UV-embossing technology. Large-scale production and wide range of geographic applicability will reduce the unit costs and increase the economy of energy saving and work efficiency of occupants by better lighting.

Acknowledgements The joint research project TaLED, running 2016–2019, is funded by the German Federal Ministry for Economic Affairs and Energy and the partners [2].

References

1. Jakubowsky Michael, Neyer Andreas, Müller Helmut (2018) Microstructured Façade elements for energy efficient office room illumination by sunlight combined with LED light. South African Solar Energy Conference (SASEC2018), Durban, SA. p 25–27
2. Research Reports of Project TaLED (Energy- and cost efficient, façade integrated Day- and LED lighting based on micro-optical components) (2016–2019) Partners: RIF e.V., Karl Jungbecker GmbH, Temicon GmbH, Durlum GmbH, Saint-Gobain Securit GmbH, Green Building R&D GmbH, ai3, SSP AG. Funding: German Federal Ministry for Economic Affairs and Energy. Co-ordination: Fraunhofer Institute Building Physics, Stuttgart, Germany
3. Andrew J (2014) Marsh: sun-path diagrams. http://andrewmarsh.com/apps/releases/sunpath2d.html
4. Mueller HFO (2017) Micro-optical structures for daylighting and LED systems. Renew Energy Environ Sustain 2:29. www.rees-journal.org
5. Klammt S, Müller HFO, Neyer A (2012) Microoptics for efficient redirection of sunlight. Appl Opt 51(12):2051–2056

Chapter 51
Hybrid Method to Solve a Two-Stage Stochastic Biofuel Hub-and-Spoke Network Problem

Mario Aboytes-Ojeda and Krystel K. Castillo-Villar

51.1 Introduction

In recent years, several works have been conducted in the area of production and distribution of biofuels from second-generation biomass (i.e., nonfood sources) to minimize cost, maximize profit, and minimize environmental impact, among others. The random nature of biomass needs to be considered in the design and implementation of supply chains due to their impact in investment and operational decisions. Stochastic models have been employed since they allowed the inclusion of biomass random properties in the modeling stage. Castillo et al. [1] propose a two-stage stochastic programming model for local biofuel production that considers random variation in the content of moisture in order to determine the optimal depot and biorefinery location as well as conversion technology selection.

To address large-scale instances, a two-stage stochastic hub-and-spoke network for the implementation of a supply chain has been introduced by Aboytes et al. [2] with the purpose of minimizing the transportation costs, the moisture content, and the biomass quality-related costs. According to previous numerical experimentation reported in Aboytes et al. [2], the moisture and ash content have an influence in the solution, specifically, in the network topology and the flows within the arcs of the network. The authors solve an instance of Texas with an L-shaped algorithm; however, the L-shaped algorithm exponentially increases the solution time as the size of the problem increases. Hence, solving regional instances poses a challenge. The p-hub problem is proved to be NP-hard [3]; therefore, metaheuristic approaches are suitable solution methods to find good solutions with a reasonable computational burden. In this work, a metaheuristic-based method inspired by Benders decomposition is presented in order to solve large-scale instances (i.e., case study of Texas).

M. Aboytes-Ojeda · K. K. Castillo-Villar (✉)
Texas Sustainable Energy Research Institute/Department of Mechanical Engineering, University of Texas at San Antonio, San Antonio, TX, USA
e-mail: Krystel.Castillo@utsa.edu

A. Sayigh (ed.), *Renewable Energy and Sustainable Buildings*, Innovative Renewable Energy, https://doi.org/10.1007/978-3-030-18488-9_51

51.2 Literature Review

In the field of biofuels, stochastic problems have primarily been solved with variations of the standard L-shaped algorithm [4], and different strategies to reduce the solution space have been implemented. For instance, Castillo et al. [1] solved a two-stage stochastic biofuel production problem with the implementation of various cut strategies in the standard L-shaped such as trust region cuts and multicuts.

Marufuzzaman et al. [5] present a two-stage stochastic model to design and manage biodiesel supply chains; the model is solved utilizing a combination of Lagrangian relaxation and L-shaped solution methods. The authors applied the solution approach to a case study of Mississippi State. Aboytes et al. [2] introduce a reformulation and an L-shaped algorithm based in connectivity constraints to solve a stochastic programming problem that considers quality-related properties. The algorithm shows a good performance for a case study in the state of Texas, but it presented limitations when more than two states are considered in the case study.

A review of closely related papers shows that the solution procedures perform well for problem of moderate size (i.e., state level). Nevertheless, issues with computational memory and time when scaling-up instances urge to develop solution procedures able to solve national and/or interstate biofuel supply chain networks.

51.3 Methodology

In order to introduce the TS-SM algorithm, the formulation of the hub-and-spoke model proposed in Ref. [2] follows:

$$\min : \sum_{i \in D} \xi_i W_i + \sum_{i \in B} \sum_{k \in K} \varrho_{ik} \beta_{ik} + \sum_{i \in D} \sum_{j \in B} \psi_{ij} A_{ij} + \tag{51.1}$$

$$\sum_{o \in \Omega} p(o) \left[\sum_{i \in P} \sum_{j \in D} \sum_{k \in K} c_{ij}^{'T} x_{ijk}(o) + \sum_{i \in D} \sum_{j \in B} \sum_{k \in K} c_{ij}^{R} y_{ijk}(o) + \alpha \Pi(o) \right] \tag{51.2}$$

Subject to:

$$\sum_{j \in D} \sum_{k \in K} x_{ijk}(o) \leq s_i(o) \quad \forall i \in P, \quad o \in \Omega \tag{51.3}$$

$$\sum_{i \in P} (1 - e_i) x_{ijk}(o) = \sum_{i \in B} y_{jik}(o) \quad \forall j \in D, \quad k \in K, \quad o \in \Omega \tag{51.4}$$

$$\sum_{i \in D} \sum_{j \in B} \sum_{k \in K} g_{jk} y_{ijk}(o) + \Pi(o) = d \quad \forall o \in \Omega \tag{51.5}$$

$$\sum_{k \in K} y_{ijk}(o) \leq v_{ij} A_{ij} \quad \forall i \in D, \;\; j \in B, \;\; o \in \Omega \tag{51.6}$$

$$\sum_{i \in P} \sum_{k \in K} x_{ijk}(o) \leq u_j W_j \quad \forall j \in D, \;\; o \in \Omega \tag{51.7}$$

$$\sum_{i \in D} g_{jk} y_{ijk}(o) \leq q_{jk} \beta_{jk} \quad \forall j \in B, \;\; k \in K, \;\; o \in \Omega \tag{51.8}$$

$$x_{ijk}(o) \geq 0 \quad \forall i \in P, \;\; j \in D, \;\; k \in K, \;\; o \in \Omega \tag{51.9}$$

$$y_{ijk}(o) \geq 0 \quad \forall i \in D, \;\; j \in B, \;\; k \in K, \;\; o \in \Omega \tag{51.10}$$

$$\Pi(o) \geq 0 \quad \forall o \in \Omega \tag{51.11}$$

$$W_i \in \{0,1\} \quad \forall i \in D \tag{51.12}$$

$$\beta_{ik} \in \{0,1\} \quad \forall i \in B, \;\; k \in K \tag{51.13}$$

$$A_{ij} \in \{0,1\} \quad \forall i \in D, \;\; j \in B \tag{51.14}$$

The objective of the formulation is to minimize the first- and second-stage cost (Eq. 51.1) and (Eq. 51.2) within the network. The following set of constraints are defined: (Eq. 51.3) limits supply capacity for every county within the network *N*; (Eq. 51.4) represents the mass balance in every depot; (Eq. 51.5) ensures the demand is met, either by the SC network or by a third-party supplier; (Eq. 51.6) ensures the capacity in the arcs is not exceeded; (Eq. 51.7) ensures the capacity at the preprocessing depots is not exceeded; (Eq. 51.8) ensures the production capacity at each biorefinery with a fixed technology is not surpassed; (Eqs. 51.9–51.11) enforce non-negativity; and (Eqs. 51.12–51.14) represent the binary nature of decision variables.

This model was previously solved using an enhanced L-shaped algorithm. In this paper, we utilize a decomposition framework with tabu search (TS) [6] to facilitate faster convergence and improved computational efficiency for tackling regional or, potentially, national instances. Jourdan et al. [7] conducted a literature review of hybrid algorithms combining metaheuristics and exact methods to solve network configurations. Remarkably, 12 out of 18 works utilize TS or local search. Hence, the literature points to TS as a suitable metaheuristic.

51.3.1 Decomposition Framework

The proposed solution procedure is a hybrid between TS and exact optimization methods (e.g., simplex method). The hybrid method uses a decomposition framework to divide the original problem in smaller problems: the master

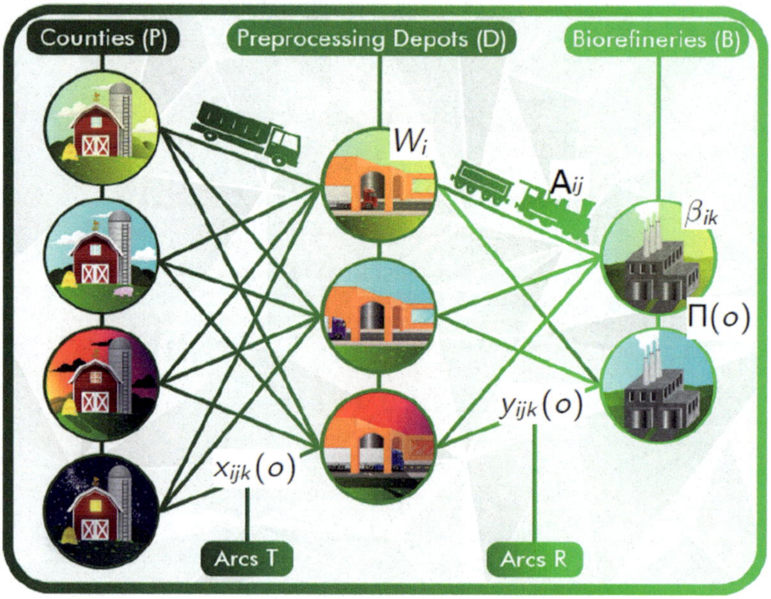

Fig. 51.1 Hub-and-spoke network

problem (MP) and the subproblems (SPs). The MP contains all the depot and biorefinery potential locations as well as the arcs that connect those facilities (modeled as binary variables) and the SPs that include the flow of biomass along arcs and the third-party supply (modeled as continuous variables) as shown in Fig. 51.1.

The model's objective is the minimization of the total cost due to the investment, transportation, and biomass purchase/processing. The proposed hybrid method utilizes the TS approach to explore the topology neighborhood of the logistics network (i.e., variables W_i, β_{ik}, and A_{ij} within the master problem in the first stage as shown in Fig. 51.1). The search procedure generates a solution for the network, and then, it incorporates the aforementioned topology into the model as fixed values. In the second stage, the SPs are solved with the simplex method to find second-stage variables.

51.3.2 *Initial Solution, Solution Structure, and List of Candidates*

The initial cost is calculated with the assumption that all the demand is satisfied by the third-party supplier ($c_0 = \alpha \times d$ and $\Pi(o) = d, \quad \forall\, o \in O$). The best solution achieved with the algorithm c^* at time 0 is set equal to the initial cost c_0. TS-SM

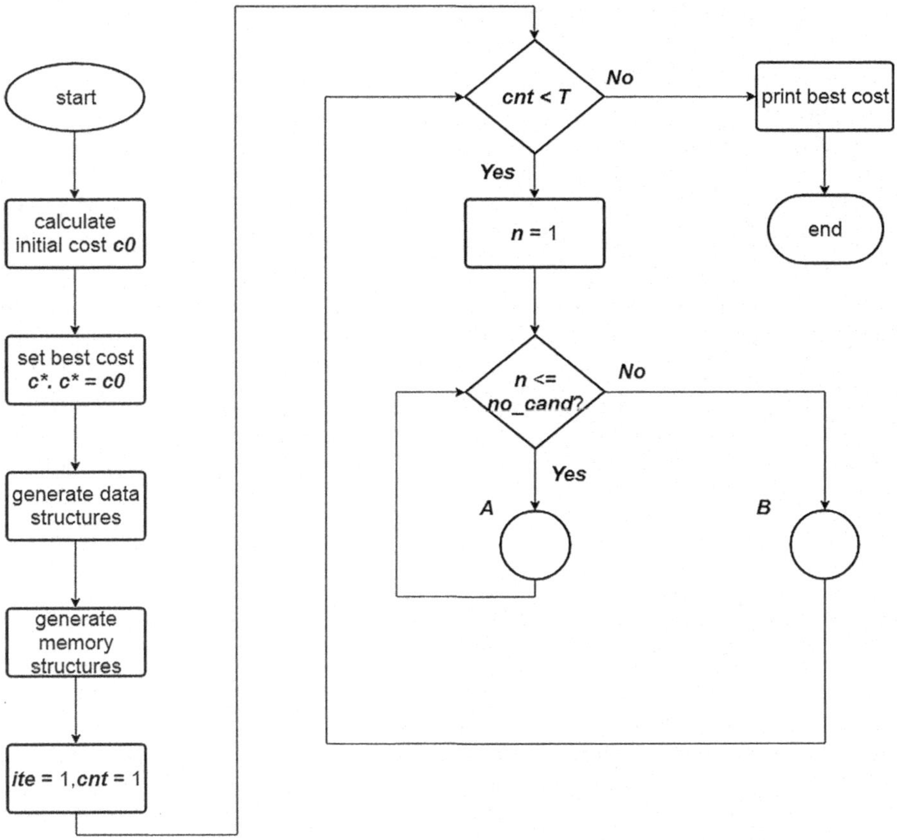

Fig. 51.2 Tabu search flowchart

explores network configurations that decrease the overall cost. The solution structure introduces a list of origin–destination pairs (i.e., arcs from depots to biorefineries) of size n, where n represents the maximum number of arcs required to satisfy the demand with available resources. These arcs are selected from a pool of origin–destination pairs that have at least *minimum-allowed* distance between each one of these pairs. The selection of the index position to perform an operation with an origin–destination pair is random. The list of candidates is created with neighborhood operations randomly selected; every change in the network topology and the corresponding result is stored in the list. When the list is completed, the next step is to select the best candidate according to the criteria. Figure 51.2 shows the general diagram for the TS proposed in this work.

51.3.3 Neighboring Function

The proposed metaheuristic performs local search utilizing the TS-SM approach with the introduction of exploration operations such as (1) *add*, (2) *remove*, (3) *depot switch*, and (4) *biorefinery switch* as illustrated in Fig. 51.3. The *add* operation consists of introducing an arc in the network with their corresponding depot (origin) and biorefinery (destination). The *remove* operation deletes an existing arc from the network. The *switch* operation modifies the network structure by changing the depot or biorefinery location. Figure 51.4 shows the methodology to select the proposed operation; a random number *r1* is generated to select a position within the solution list. If the connection position is empty, then TS-SM will add a connection between a potential depot and biorefinery; otherwise, the algorithm will randomly select between deleting the connection and switching an origin/destination point. If a switching movement has been selected, then, another random selection is made to choose either switching depot or biorefinery. This selection is repeated until the candidate list is completed.

51.3.4 Solution Selection Process

The solution selection process is performed according to the quality, recency, and frequency of the solution. If the solution quality is better than the best current solution, then, the candidate is accepted without checking the recency and frequency memory of previous operations. If the best solution candidate is not better than the best current solution but it does not violate the recency criteria, the candidate is selected to continue the neighborhood search. The solution candidate is selected according to the frequency criteria, if the recency rule is violated. Figures 51.5 and 51.6 will show the process selection of the search strategies as well as their definition.

Var	Description. *=1st stage, **=2nd stage.
W_i	potential depot location $i \in D$ selection*
β_{ik}	potential biorefinery location $i \in B$ selection with tech k*
A_{ij}	Arc $(i,j) \in R$ depot-biorefinery railroad transportation *
$x_{ijk}(o)$	Flow along arc $(i,j) \in T$ under scenario $o \in \Omega$**
$y_{ijk}(o)$	Flow along arc $(i,j) \in R$ under scenario $o \in \Omega$**
$\Pi(o)$	Third party supply under scenario $o \in \Omega$**

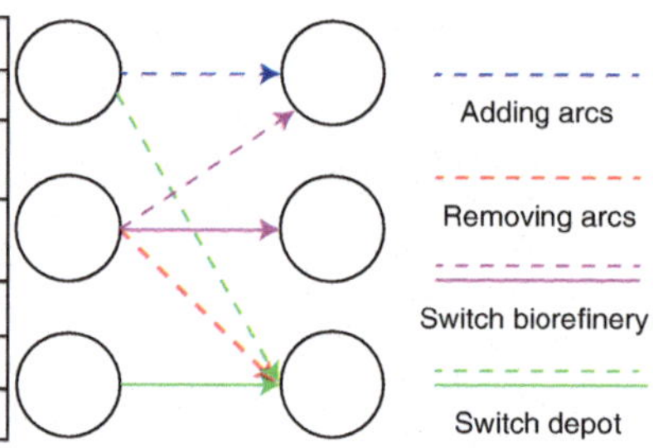

Fig. 51.3 Local search operations

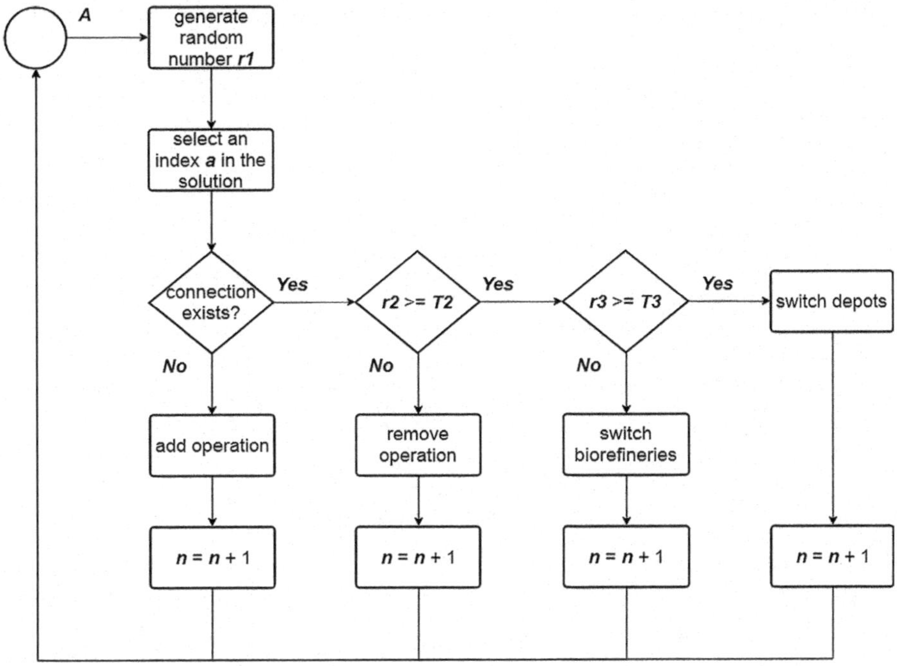

Fig. 51.4 Neighboring operation selection

Fig. 51.5 Strategy selection of tabu search

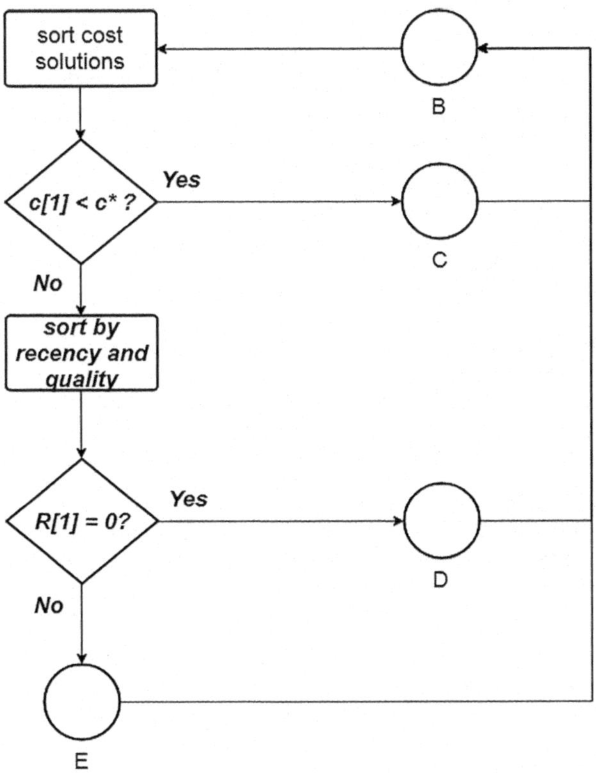

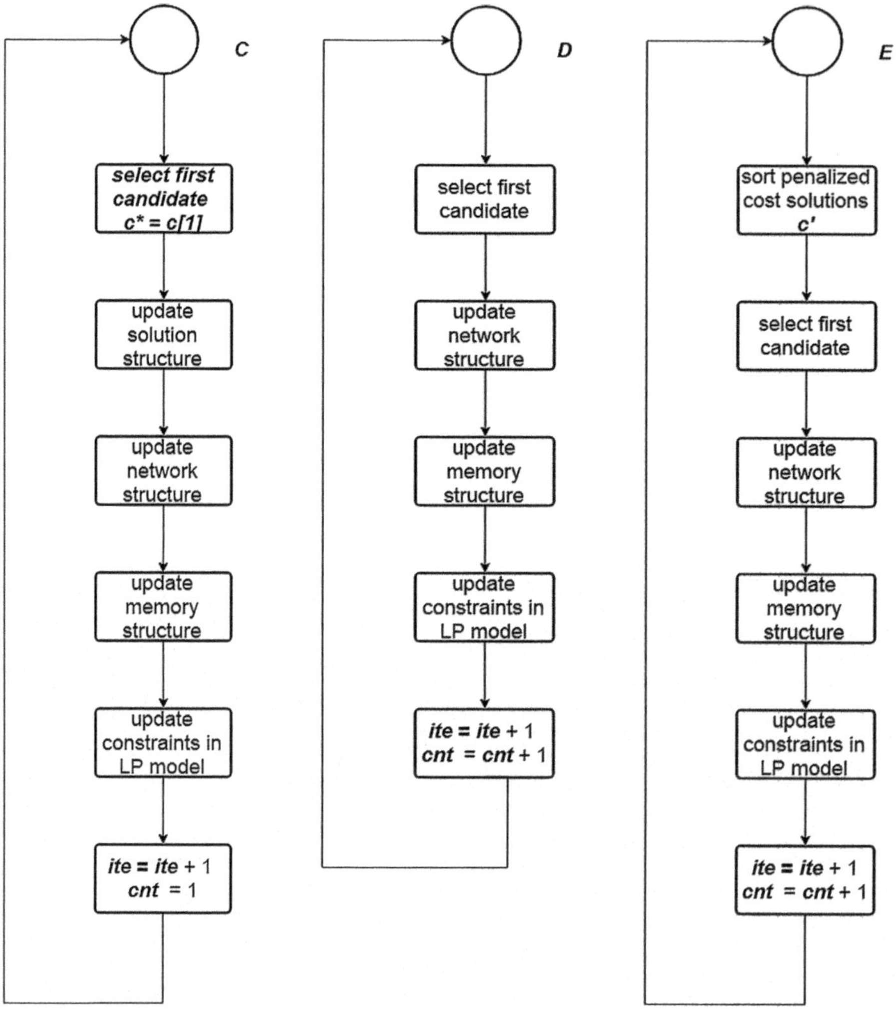

Fig. 51.6 Definitions of search strategies

51.3.5 Stopping Criteria

The first-stage variables are used in the SP as fixed parameters. Then, an exact method (in this case, the simplex algorithm) is called. This process is repeated until the candidate list is completed. The solution obtained in every iteration leads the search of near-optimal solutions. Every time that the algorithm finds a better solution from all performed iterations, the counter is reset. The algorithm stops if no further improvement is achieved after a fix number of iterations. The overall performance of the hybrid method TS-SM is tested by comparing it against LS using an instance of the state of Texas.

51.4 Results

In this paper, six problems for the state of Texas are introduced to test and compare the performance of the proposed solution procedure. TS-SM is compared with LS; quality and solution times are presented in this section. The problems were created with information of moisture and ash content obtained from works of Yu et al. [8] and Aboytes et al. [9] as well as Castillo et al. [1]. Table 51.1 shows the description of the problems that were solved in this work. It provides the information about the content of moisture and ash within the biomass.

The set of problems presented in this work consists of combinations of random moisture content, which follows a triangular distribution, with six different ash concentrations. In Table 51.2, the solutions of TS-SM regarding the total cost for the biofuel supply chain implementation and operation are presented.

The solutions for Problem 0 and 6 open the same number of biorefineries. The moisture content impacts the storage need and the biomass weight; therefore, the investment cost to open the depots and the transportation costs are higher in Problem 6 as we can observe in Table 51.2. Figure 51.7 shows different topologies due to biomass moisture content, and the solution quality and solution times of the TS-SM and LS are presented in Tables 51.3 and 51.4.

Table 51.1 Moisture and ash content for different distributions

Problem	Particle size (cm)	Min (%)	Avg (%)	Max (%)	Ash (%)
0	243.84	0	0	0	0
1	243.84	26	27	29	1.33
2	243.84	26	27	29	4.53
3	243.84	26	27	29	0.71
4	243.84	26	27	29	3.79
5	243.84	26	27	29	0.82
6	243.84	26	27	29	3.49

The moisture content (min, avg, max) follows a triangular distribution

Table 51.2 Breakdown of cost ('000 USD)

Problem	Depot	Biorefinery	Conn	Flow T	Flow R	I&OP
0	$34,762	$523,827	$30,668	$60,405	$36,659	$686,321
1	$31,286	$392,870	$27,601	$96,067	$25,453	$573,277
2	$31,286	$392,870	$27,601	$110,123	$30,636	$592,516
3	$31,286	$392,870	$27,601	$94,670	$27,775	$574,203
4	$31,286	$392,870	$27,601	$98,936	$30,513	$581,207
5	$31,286	$392,870	$27,601	$93,186	$26,810	$571,753
6	$41,715	$523,827	$36,801	$134,253	$42,654	$779,250

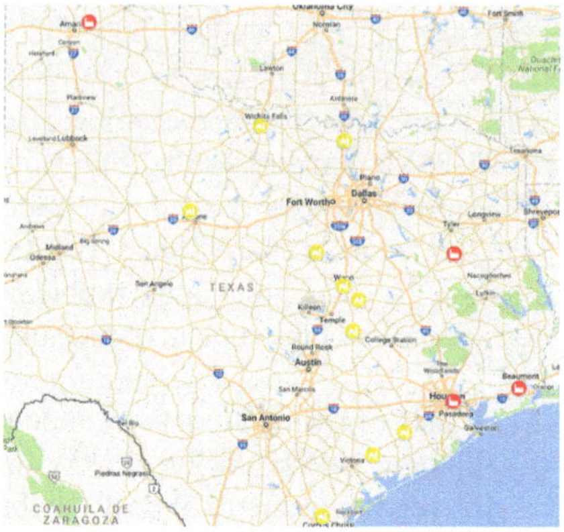

(a) Problem 0

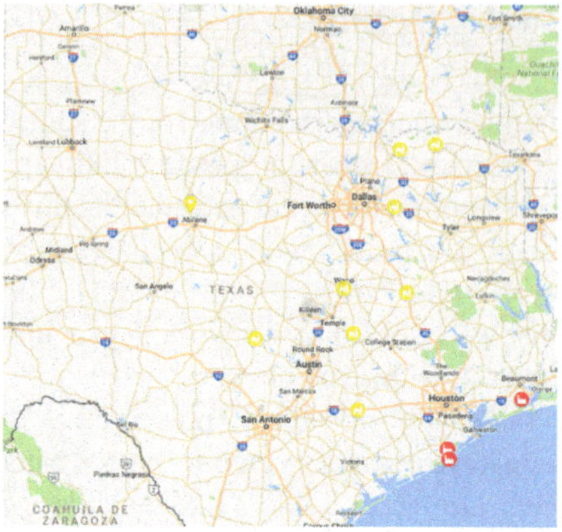

(b) Problem 2

(c) Problem 6

Fig. 51.7 Hub-and-spoke network for Texas

Table 51.3 Solution quality comparison

Problem	TS-SM[a]	LS[a]	Deviation[a]
0	2,070,157	2,089,659	−19,502
1	2,155,220	2,156,233	−1,013
2	2,176,247	2,164,450	11,797
3	2,155,599	2,192,554	−36,955
4	2,160,000	2,168,808	−8,808
5	2,154,810	2,158,636	−3,826
6	2,127,286	2,156,850	−29,564
AVG	2,127,286	2,153,058	−12,553

[a](‘000 USD)

Table 51.4 Solution times comparison

Problem	TS-SM[a]	LS[a]	Deviation[a]
0	2,679	700	1,979
1	7,433	31,049	−23,616
2	4,703	25,753	−21,050
3	3,673	65,348	−61,675
4	2,271	59,277	−57,006
5	8,104	17,584	−3,826
6	3,501	21,827	−18,326
AVG	4,623	31,648	−27,025

[a]Seconds

The difference in the total cost found by TS-SM is on average about $12.5 million dollars less than the cost obtained by LS, in all the problems. The proposed algorithm outperforms LS regarding the solution quality. Regarding solution time, Table 51.4 shows a reduction in the average solution time of approximately 27,025 CPU s or 7.51 CPU h. In summary, the TS-SM presents a better performance in both, solution quality and computational time.

51.5 Conclusions and Future Research

The TS-SM algorithm solved Texas instances in a reasonable amount of time. The quality of the solutions is comparable with the quality of the solutions obtained with benchmark solution procedures. The quality of the solutions obtained with the proposed algorithm outperforms the ones obtained with LS by approximately 0.58% on average using the six preliminary instances. TS-SM is a promising alternative to solve large-scale instances since it has reduced LS average solution times by approximately 85%.

Future lines of research involve finding an alternative stop criteria in order to reduce the computational time (in some instances, TS-SM finds the solution at an early stage of the local search) and implementing a better initial solution strategy for TS-SM such as the relaxation of the original problem to lead the local search to a faster algorithm convergence.

Acknowledgements This project was supported by Agriculture and Food Research Initiative Competitive Grant no. 2014-38502-22598 and by the Hispanic Serving Institutions Education Grants Program no. 2015-38422-24064 from the USDA National Institute of Food and Agriculture.

References

1. Castillo-Villar KK, Eksioglu SD, Taherkhorsandi M (2017) Integrating biomass quality variability in stochastic supply chain modeling and optimization for large-scale biofuel production. J Clean Prod 149:904–918. https://sciencedirect.com/science/article/pii/s0959652617303463. Accessed 4 Mar 2019
2. Aboytes-Ojeda M (2019) Stochastic Programming Models to Integrate Biomass Quality Variability in the Design of Biofuel Logistics Networks (Doctoral dissertation, The University of Texas at San Antonio).
3. Ernst AT, Hamacher H, Jiang H, Krishnamoorthy M, Woeginger G (2009) Uncapacitated single and multiple allocation p-hub center problems. Comput Oper Res 36:2230–2241
4. Van Slyke RM, Wets R (1969) L-shaped linear programs with applications to optimal control and stochastic programming. SIAM J Appl Math 17:638–663
5. Marufuzzaman M, Eksioglu SD, Huang Y (2014) Two-stage stochastic programming supply chain model for biodiesel production via wastewater treatment. Comput Oper Res 49:1–17. https://sciencedirect.com/science/article/pii/s0305054814000653. Accessed 4 Mar 2019
6. Glover F, Laguna M (1998) Tabu search. In: Handbook of combinatorial optimization. Springer, New York, pp 2093–2229
7. Jourdan L, Basseur M, Talbi E-G (2009) Hybridizing exact methods and metaheuristics: a taxonomy. Eur J Oper Res 199:620–629
8. Yu TE, Larson JA, English BC, Boyer CN, Tyler DD, Castillo-Villar KK (2015) Influence of particle size and packaging on storage dry matter losses for switchgrass. Biomass Bioenergy 73:135–144
9. Aboytes-Ojeda M, Castillo-Villar KK, Yu T-h E, Boyer CN, English BC, Larson JA, Labbé N (2016) A principal component analysis in switchgrass chemical composition. Energies 9:913

Chapter 52
The Conflict Between Aesthetics and Sustainability: Empowering Sustainable Architecture with Aesthetics to Enhance People's Lifestyle and Sustainable Behavior

Shaden Abusafieh

52.1 Introduction

Global climate change has the potential to completely disrupt the whole world. It would touch every person on Earth, and as such, it is not just an issue for world leaders, politicians, and business executives. Everyone has a responsibility; this responsibility extends especially to the profession of architecture as it is a major contributor of greenhouse gases. For this moment and despite many intensive efforts, architectural profession still lacks any cohesive approach to the problem of sustainability. We need a new paradigm for architectural design [1].

52.2 Architecture and Global Climate

The Fourth Assessment Report that was issued by the United Nations' Intergovernmental Panel on Climate Change concluded that human activity has been the trigger of global climate change over the past 50 years. Now that there is an understanding of where the problem comes from, it is time to devise solutions and to take an action [2].

Recently, policy makers and leaders are focusing on factories, power plants, and vehicle emissions, while the real problem is the buildings that we live and work in every day. Ed Mazria's study concluded that architecture consumes 48% of all energy in the United States and is responsible for more CO2 emissions than transportation or industry [3].

S. Abusafieh (✉)
Department of Architectural Engineering, Faculty of Architecture and Design, Al-Ahliyya Amman University, Amman, Jordan

A. Sayigh (ed.), *Renewable Energy and Sustainable Buildings*, Innovative Renewable Energy, https://doi.org/10.1007/978-3-030-18488-9_52

Nowadays, society has some buzzwords like sustainability, green, and eco-friendly. For that, architectural profession has been influenced by the US Green Building Council and their "Leadership in Energy and Environmental Design" (LEED) rating system. But, sustainability remains largely on the fringes of architectural practice today. That was because sustainable design as a whole is more of criteria and codes that architects must comply with. So, this model has not produced a unified architectural solution. Implications of sustainability require a new way of practicing architecture.

52.3 A Cohesive Sustainable Philosophy

Throughout architectural history, architectural styles that have been founded with strong theoretical, social, and moral principles were the most powerful and potent styles. The Arts and Crafts and Modern movements and the Post-Modern movement are strong examples of this [4].

In this era, sustainability presents the new step in architecture styles history. It shows a huge responsibility to the moral and social challenges of climate change, trying to prevent catastrophic consequences. But, sustainable architecture has a great challenge to be effective and desirable, to be holistic [5]. It must go beyond checklists, codes, materials, and techniques. To obtain that, architecture needs a cohesive sustainable philosophy that embedded a sustainable aesthetic philosophy.

52.3.1 Sustainable Architectural Aesthetics

Architects should fight the rigid stylistic dogma that sustainability might burden architecture with. There must be no fixed design rules or visual norms. Since architecture is a practical field and has real and usable products, architectural aesthetic must be a practical philosophy related to the practice and process of architecture [6].

Although assessment of beauty and personal experience are necessary aspects of aesthetics, the nature of sustainable architecture requires that it takes unique and various forms based on different circumstances, making it impossible to know exactly what sustainability will look like in any given circumstance. David Douglass suggested a definition of sustainable aesthetics in the field of architecture. He assessed that "three principles of a sustainable aesthetic are established: that sustainability represents a practical philosophy, that sustainability serves as the concept generator in the design process, and that sustainable architecture is universally specific to the constraints of its site" [7].

52.3.2 Revealing the Conflict Between Aesthetics and Sustainability

According to the US Green Building Council, "LEED promotes a whole-building approach to sustainability by recognizing performance in five key areas of human and environmental health: sustainable site development, water savings, energy efficiency, material selection, and indoor environmental quality" [8].

In accordance with the potential usage and methods of sustainability, it faces some limitations and challenges. For example, where and how we have to make compromises using the alternative solutions? Where to sacrifice aesthetics for efficiency, and how much we lose technically in choosing something nicer?

Sustainable design as a whole remains little more than a basic set of criteria, yet it is another code that architects must comply with. This model has not produced a unified architectural solution. Therefore, the potential impact and implications of sustainability require a new way of practicing architecture. To create an ideal architecture, a building should have "soul" that we can discover from its aesthetics values and "body" that we can feel by usage and accommodate. Architecture would be appreciated when it is simultaneously an object of art and a useful tool [9].

Roger Scruton in his research described the necessity of understanding a building's utility in order to truly understand its value [10]. He asserted that classical architectural styles are morally superior to Modern styles that neglect the social and moral basis that Modernism was founded on.

As Langer described the buildings, "That is the image of life which is created in buildings; it is the visible semblance of an 'ethnic domain', the symbol of humanity to be found in the strength and interplay of forms" [11]. This confirms that all art is human creation. But Langer lightens a point that separates architecture from the other arts. Art may symbolize the life of a culture, but architecture is the embodiment of that life.

52.3.3 Case Studies: Evaluating Aesthetics Needs

In order to evaluate the aesthetics needs level, this research examined some sustainable solutions from an aesthetical point of view. This was through questionnaires and interviews within the city of Amman, Jordan.

A number of techniques, methods, and materials involved in green architecture were tested to investigate their potential in architecture design. Three questions were defined during this process: What limitations are there in alternative design usage of sustainable architecture? Where and how we have to make compromises using the alternative solutions? Where to sacrifice aesthetics for efficiency, and vice versa?

52.3.3.1 Place Attachment Through Aesthetic Appreciation

Place attachment is a field of study in environmental psychology that can guide designers and architects to produce places that are meaningful and therefore it will influence their behavior.

Place attachment is the emotional bond between person and place which produces place loyalty via cognitive and affective processes. Empirical studies show that place attachment predicts proenvironment intentions [12] and guides behavior [13]. According to a 3-year study that explored what emotionally attaches people to their community, conducted by Gallup across communities in the United States, three issues mattered most to form place attachment: social offerings, openness-welcomeness, and aesthetics that includes physical beauty and green spaces [14].

During an interview with Mrs. Yasmeen, owner of Kamaliya Residence-Green Home, we learned that since she was not involved in the earlier design stage of the house, she felt many features would have changed. This creates a lack in "place attachment" to her house and so affected her behavior in accepting the house that would be the best in saving energy as it is classified in Amman.

She mentioned that she would have made the living space's current two windows (Fig. 52.1) into one large French-style window that framed a better view to the front garden. The lady's desire towards larger windows ties to the innate human propensity towards connection with nature, with an indoor-outdoor visual linkage. Such aesthetic vision is legitimate as long as it is balanced by higher thermal performance of glazing.

The proclivity for aesthetics, activated by an individual's need to belong and emotionally attach to the place, also appears in the fireplace that was located in the living room's corner. As we were sitting in the living space, we noticed a fireplace with a sign written on top of it saying "No Burning Please." Even though it is discouraged to have a fireplace in a green home, due to its carbon emissions from wood

Fig. 52.1 The living space's current two windows in Kamaliya Residence (source: researcher)

Fig. 52.2 The fireplace with a sign written on top of it saying "No Burning Please" (source: researcher)

burning and its energy inefficiency, Mrs. Yasmeen was adamant about including a fireplace in her dream home. Dr. Abu Dayyeh accommodated his wife's request but ensured that it is never operated as long as the family owns this home. The fireplace stands as a cosmetic aesthetic element that was never operated (Fig. 52.2).

52.3.3.2 Aesthetic Taste Can Hinder Sustainable Behaviors

Aesthetic taste can hinder sustainable behaviors if the market does not afford environmentally responsible choices and if it would not match the culture aesthetics values. Here are some facts that founded during some interviews were held with people, professionals, and researchers in the field of sustainability in Amman:

- HABITAT wanted to build efficient homes for a group of economically strained families in Al-Ghor area. It was proposed to be built from clay, and they were faced with fierce resistance to the choice of material as it expressed a reduced social status. After consulting with the future homeowners, HABITAT built the houses from concrete blocks instead, a material that was approved aesthetically by the people.
- Before yellow LEDs were introduced to the Jordanian market, an initiative was undertaken by a nonprofit organization to exchange existing lighting in nearby suburbs of Amman with energy-saving CFLs. After visiting and getting the

approval of the community leader to showcase the new lamps in the Mukhtar's official headquarters, the lamps were bought and installed. The shape and color of the CFLs that substituted the incandescent lamps of the chandelier were rejected profusely, and the project was dropped since the yellow glow lights reflect nobility and white glow lights would reflect low income in our culture. Same incident happened in Kamaliya Residence.

52.3.3.3 Sacrificing Efficiency for Aesthetics

Can we neglect efficiency, saving energy for example, to keep and protect some of the aesthetic aspects in a built environment? Seventy-three percent of randomly 120 samples answered with yes, in a questionnaire that was designed to test how aesthetics is important to us compared to saving energy. Forty-three percent of them in the second round of the discussion that tried to trigger them towards saving energy answered: yes, we would go with saving energy if the gain will be more than 50% of our total electrical bill. Figures 52.3, 52.6, 52.7 are showing good examples hat have been appreciated by people. While figures 52.4, 52.5, 52.8, 52.9, 52.10, 52.11, 52.12 are showig a bad examples hat have been rejected by people.

The second part of the questionnaire was related to various photos that represent some solar sustainable methods widespread in Amman. Seventy-five percent of the sample strongly rejected photos that have non-aesthetic solar treatments. Forty-five percent of them rejected photos that have solar treatments with no helpful usage or that have wasted usable areas (Figs. 52.3 and 52.12).

Fig. 52.3 A good example of the photovoltaic panels placed on the roof of a passage at the Hashemite University. It was highly accepted by people (source: albosala.com)

Fig. 52.4 Villa in Amman: the PV panels are misplaced; this treatment was rejected by people since it doesn't match the roof and it distorts the house (source: researcher)

Fig. 52.5 PV panels were misplaced next to student's plaza at Applied Science Private University; it was highly rejected by people (source: researcher)

Fig. 52.6 A good example of the photovoltaic panels placed on the roof of a building at Al-Ahliyya Amman University; it was accepted by people (source: research)

Fig. 52.7 A good example of the photovoltaic panels placed on the roof of a shelter and benches at Applied Science Private University; this was a student project that has been appreciated by people (source: research)

Fig. 52.8 PV panels were misplaced on usable land between olive trees at Applied Science Private University; it was rejected by people (source: research)

To elucidate the power of aesthetics and its effect in driving behaviors, T. Heberlein concludes in his book "*Navigating Environmental Attitudes*" that "We often forget that attitudes have two parts, the cognitive and the emotional, and too often research focuses on the cognitive [15]. The real power of attitudes is the emotional dimension."

Fig. 52.9 PV panels are misplaced; this treatment was highly rejected by people since it doesn't match the mosque's dome and it distorts the view (source: research)

Fig. 52.10 PV panels are misplaced; this treatment was highly rejected by people since it distorts the view and ruined the elevation (source: research)

Fig. 52.11 PV panels were misplaced on usable agricultural land; it was rejected by people (source: research)

Fig. 52.12 A bad example of the photovoltaic panels placed on the roof of a building at Al-Ahliyya Amman University and affected the elevation; it was rejected by people (source: research)

52.4 Conclusion

As mentioned by Panagopoulos, a psychophysical preference modelling as a quantitative holistic technique is needed to evaluate sustainable buildings, and if it is used in combination with indirect aesthetic evaluation methods, it might create new standards and protocols for techniques of objectively estimating public perception of aesthetic quality and thus to enhance social sustainability in architecture [16].

The delicate objective of sustainability first and foremost is to delicately satisfy the economic, social, and environmental needs, and in this case to aesthetically gratify the users' "wants" while meeting the economic and environmental "needs."

Many researches present some observations about repositioning the research on enterprise sustainability in a new direction [17]. It suggested focusing on external and internal spaces at the same time; internal spaces should be revealed in the human mind and emotions. This would be a try to evoke a sustainable aesthetics philosophy. This new philosophy in sustainable architecture should create an invitation to look at the less examined but essential emotional and embodied aspects of sustaining humans on Earth. Aesthetics as a repository of emotions can empower the passion to sustainable architecture and a green Earth.

This research does try to aesthetically judge some sustainable treatments and how they influence the occupants' behaviors. It opens an inquiry to courageous venture into the unknown world of aesthetic influences on human behavior. Future studies might explore the importance of involving occupants in designing their spaces and consider their aesthetic judgment to foster an emotional bond to the place, in order to affect their behaviors and to obtain sustainable behaviors.

It may be true that one has to choose between ethics and aesthetics, but whichever one chooses, one will always find the other at the end of the road. Jean-Luc Godard

References

1. David BD (2008) Defining a sustainable aesthetic: a new paradigm for architecture. Dissertation, University of California
2. Intergovernmental Panel on Climate Change Fourth Assessment Report (2007) Climate change 2007: synthesis report. United Nations
3. Hawthorne C (2003) Turning down the global thermostat. Metropolis
4. Corbusier L (1986) Towards a new architecture. Translated from the thirteenth French edition by Frederick Etchells. Dover, New York.
5. Bachman L (2007) Eco-aesthetics: bridging architectural and ecological motivations. A paper presented at the solar 2007 conference of the American Solar Energy Society, Cleveland
6. Forster M, Gjesdal K, Guyer P (2015) Aesthetics. In: Forster MN, Gjesdal K (eds) The Oxford handbook of German philosophy in the nineteenth century. Oxford University Press, Oxford
7. Douglass D (2008) Defining a sustainable aesthetic: A new paradigm for architecture. University of Southern California, ProQuest Dissertations
8. Jodidio P (2009) Green architecture now! Taschen GMBH, Cologne
9. Rader M (1964) A modern book of esthetics. Holt, Rinehart and Winston, New York
10. Scruton R (1979) The aesthetics of architecture. Princeton University Press, Princeton
11. Langer SK (1953) Feeling and form. Charles Scribner, New York
12. Halpenny EA (2010) Pro-environmental behaviours and park visitors: the effect of place attachment. J Environ Psychol 30(4):409–421
13. Scannell L, Gifford R (2010) The relations between natural and civic place attachment and pro-environmental behavior. J Environ Psychol 30:289–297

14. Gallup, Inc. Social offerings, openness key to community attachment. Gallup.com, The Knight Foundation, 15 Nov 2010
15. Heberlein TA (2012) Navigating environmental attitudes. Oxford University Press, Oxford
16. Panagopoulos T (2009) Linking forestry, sustainability and aesthetics. J Ecol Econ 68(10):2485–2489
17. Shrivastava P (2012) Enterprise sustainability 2.0: aesthetics of sustainability. In: The Oxford handbook of business and the natural environment. Oxford University Press, Oxford

Chapter 53
The Influence of Roof Fenestration on Daylight Distribution in an Atrium Space Under Tropical Sky Conditions

J. Yunus, A. Zain-Ahmed, and S. S. Ahmad

53.1 Introduction

In the design of atrium buildings, the crucial elements are the shape of the atrium, roof configuration, building orientation, roof transmittance, internal surface reflectance and glazing. When daylighting is incorporated in the design, the most important consideration is the external or outdoor sky condition [1]. The main aim of this study is to investigate how modern construction techniques especially roof configurations typical to Malaysia attenuate daylight transmittance. This chapter reports on the evaluation of daylight performance in atrium buildings in Malaysia, focusing on the atrium space with four different roof configurations and under a range of sky conditions. The scope of this study is the interaction between selected roof configurations in relation to the daylight conditions. The IES-3178 VE-Pro (IESRadiance) software was utilised to achieve the objective of the study.

53.2 Methodology

Initially a typological study on existing atrium and atrium roof configurations in Malaysia was conducted. It was found that the preferred design is the top-lit and enclosed rectangular shallow atrium with an average height of four storeys and well index of one (WI = 1) and supported by other works [2–6]. A field study was then

J. Yunus · S. S. Ahmad
Faculty of Architecture, Planning & Surveying, Universiti Teknologi MARA, Shah Alam, Malaysia

A. Zain-Ahmed (✉)
INTEC Education College, Shah Alam, Malaysia

A. Sayigh (ed.), *Renewable Energy and Sustainable Buildings*, Innovative Renewable Energy, https://doi.org/10.1007/978-3-030-18488-9_53

conducted using scale models under real sky conditions as reported by Yunus et al. [7]. Four models were developed to investigate the daylight performance of structured flat (Fig. 53.1), pyramidal-gridded (Fig. 53.2), pitched (Fig. 53.3) and saw-tooth roof (Fig. 53.4), respectively. The results of the field study were then applied in the simulation to predict the daylight performance of atrium buildings that used the four types of roof configurations.

Fig. 53.1 Simulated model of atrium with structured flat roof

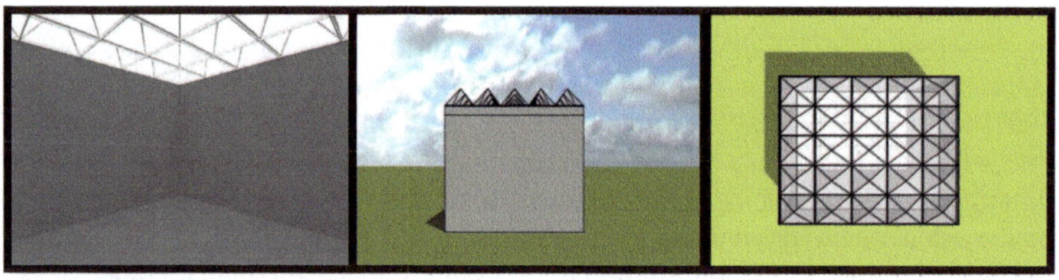

Fig. 53.2 Simulated model of atrium of pyramidal-gridded roof

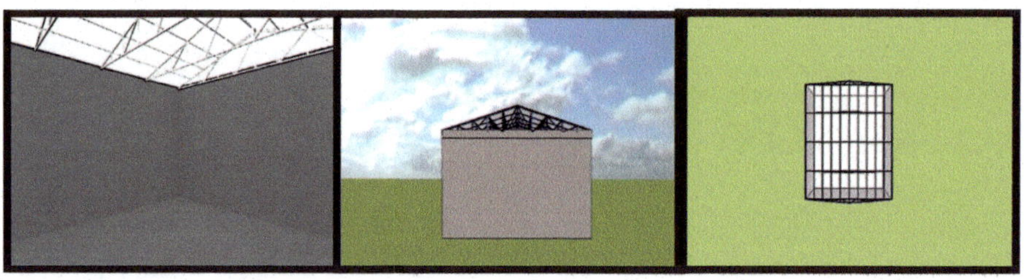

Fig. 53.3 Simulated model of atrium with pitched roof

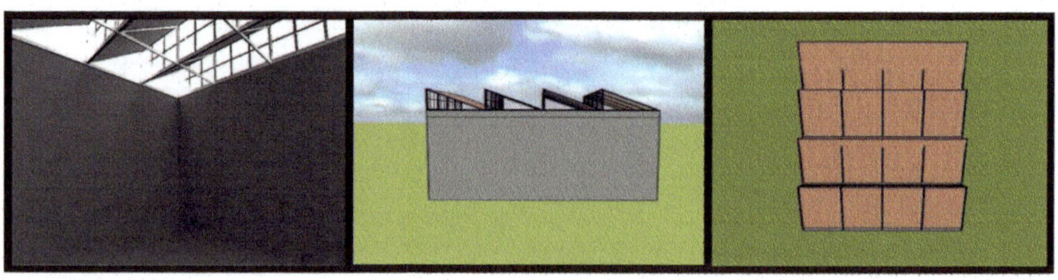

Fig. 53.4 Simulated model of atrium with saw-tooth roof

The sky types covering Malaysia are, namely, the Commission Internationale de L'Eclairage (CIE) standard overcast sky, intermediate with sun and clear sky which are 83% of the total frequency of occurrence. Two out of the three sky types were used in the simulation study of daylight performance in the atrium. The IESRadiance module of the software was used to simulate the illuminance and daylight factors (DF) in the atrium. The IESRadiance is one of the internationally recognised virtual environment simulation tools. The software was validated prior to the study using simulations. The days chosen for the simulation were 21st of March, 21st of June and 22nd of December which represent the typical hottest, average and coolest months of the year [8]. Simulations were done for the specific times at 09:00 h, 12:00 h and 18:00 h which represented morning, noon and evening. Mousavi et al. [9] had also validated the software especially for the daylight factor (DF) and daylight ratio (DR) which they proved to be more accurate than workplace illuminance and external illuminance. The daylight performance of the atrium buildings is measured in terms of illuminance, daylight factor and average daylight factor (ADF).

The simulation study was conducted in three phases. The first phase was a simulation study of a base case model with the exact specification as the scale model used in the field experiment. This base case model was tested with absolute illuminance (no roof) under overcast sky conditions. In this case the illuminance at the centre of the horizontal atrium floor was recorded. In the second phase, simulations were conducted for the four types of roof configurations, and the final phase simulations were done for the four types of roof configurations but with different surface reflectances. For phase two and phase three, results of illuminances were recorded at three positions on the horizontal floor, namely, the centre, central edges and corners. Figure 53.1 shows the positions on the horizontal floor of the atrium where the illuminances were recorded.

53.2.1 Software Validation

In order to validate the IESRadiance tool, a comparative analysis between measured data from the field study and simulated data was carried out. Daylight distribution in the atrium space was simulated for the base case model with and without roofs. The daylight distribution on the horizontal atrium floor was evaluated in terms of DF. The surface reflectance used was 0.02 or 2% (for matte black surfaces). This assumption is used for the base case model and there are minimum reflectances occurring during the simulations. Two types of sky conditions were used in the simulation study which are intermediate with sun and CIE standard overcast sky. From the comparative analysis, it was found that the values of divergence were in the range of 5 and 15%. The divergence values at the corners of the atrium floor were least at 5% while the biggest value of divergence occurred at the centre of the atrium floor of 15%. The average divergence was 8%. These figures concur with the findings and validations done by Calcagni and Paroncini [10], Ghasemi et al. [11] and Thanachareonkit [12, 13] where the divergences achieved were 10% on average and a maximum of 25%.

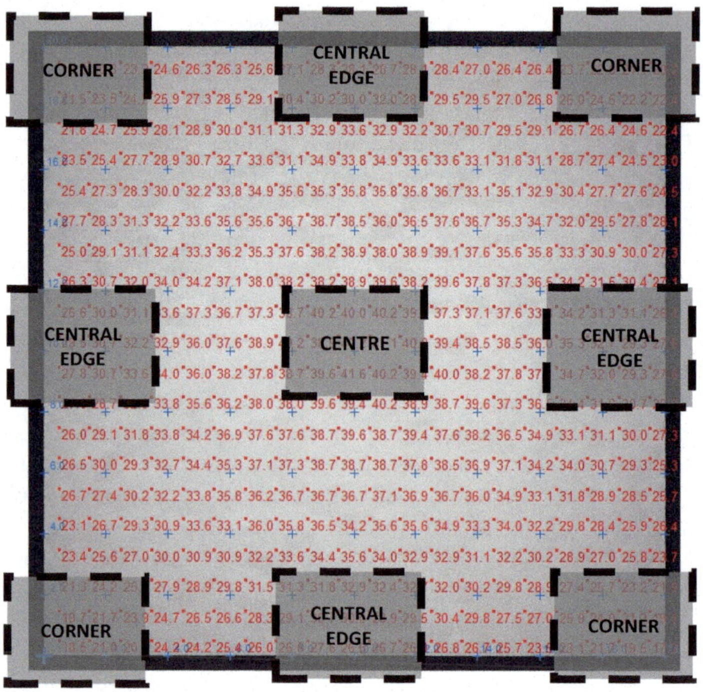

Fig. 53.5 Measured points on atrium horizontal plan

53.2.2 Simulation Study

After validating the software, the daylight performance was simulated for an unobstructed top-lit atrium (no roof) and other models with different roof configurations as already mentioned. The impact of all the types of roof fenestration systems on daylight illuminance distribution was examined at the centre, central edges and corner positions as shown in Fig. 53.5 on the atrium floor at different orientations (north, south, east and west). The DF and ADF were calculated from the illuminance values obtained in the simulations.

53.3 Results

53.3.1 Daylight Performance of Atrium Under Overcast Sky

The results corresponded with the measurements taken under real sky experiments. It was noted that the DF values at the centre floor position were always higher than other positions. For all roof systems, the DF always decreases according to measurement location with the centre having the highest value, followed by the central

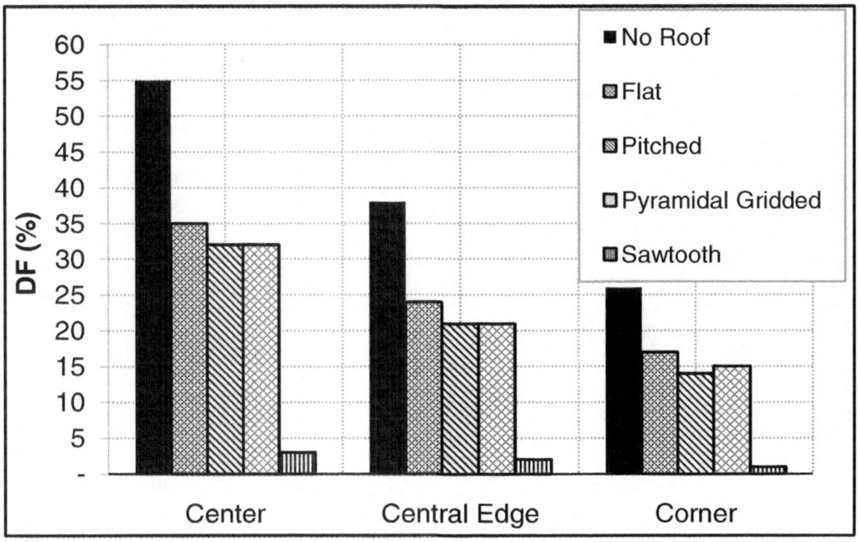

Fig. 53.6 DF distribution on atrium floor for various roof configurations under overcast sky

edge and corner. Although the results are comparable (similar distribution pattern) with the measurements from the experiment conducted under real sky conditions, the levels of light transmitted to the atrium models were not similar. For instance, for all three atrium spaces' positions, the cut-off lights were only one-quarter with roof obstructions under the simulation, while there were more than half with the real condition results. Meanwhile, the corner position was constantly affected by the roof obstructions at all orientations, which brought its DF down to lower levels. Figure 53.6 shows the DF values for all types of roof fenestrations. The flat, pitched and pyramidal-gridded roof performed better than the saw-tooth type. The saw-tooth roof recorded DF values of less than 5% which may be insufficient for certain space functions [14] and the results are reported elsewhere. Overall, the structured roof forms reduce daylight levels in the atrium well by 55%. The transmittance of the atrium roof structure also decreased the illuminance level at the lowest corners of the atrium space by 50%. The pyramidal-gridded atrium roof received more daylight penetration in the morning and later in the afternoon compared with the other two roofs due to the structures obstructing daylight penetration into the atrium.

53.3.2 Daylight Performance of Atrium with Varying Surface Reflectances

The simulations in Sect. 3.1 were repeated except for the saw-tooth roof, using different surface reflectance. The reflectance values used in the simulation were 25%, 50% and 75%. Figure 53.7 shows the effect of atrium interior wall surface and its reflectance on the average daylight factor (ADF) of various atrium floor positions

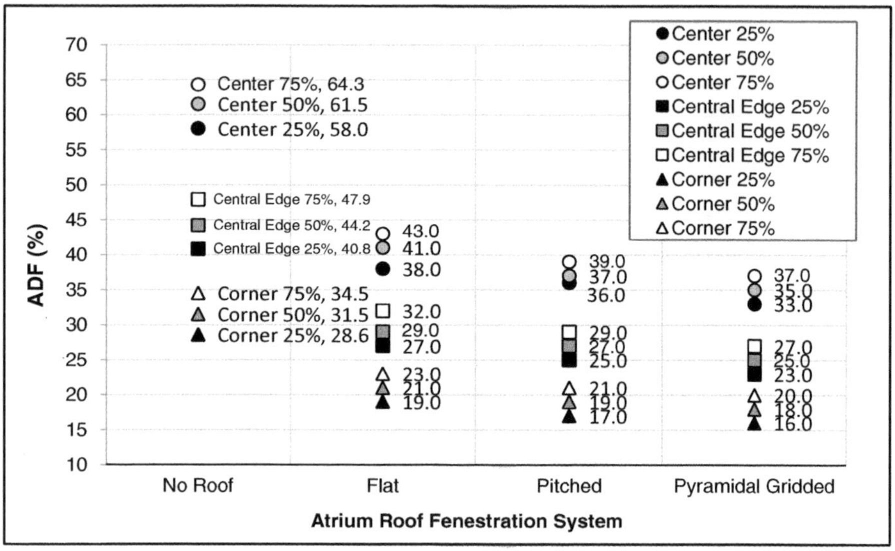

Fig. 53.7 Daylight performance of various atrium roof fenestration systems and reflectance values compared with unobstructed (no roof) atrium

with various roof fenestration systems. It was then compared with the unobstructed atrium base case model. A maximum DF of 43% was observed for the roofed atrium floor, especially for the structured flat roof, and DF of 64% for the unobstructed roof (no roof). The maximum ADF values were achieved from the 75% reflectance for the internal atrium surface. Configurations with roof increase the illuminance in their atrium spaces up to 44% for the three reflectance categories of the atrium horizontal floors. From the results, it was evident that the amount of daylight distribution increases with the increase in the value of surface reflectance for all roofs.

53.3.3 *Daylight Performance of Atrium Under Intermediate Sky (with Sun)*

The daylight performance for three types of roof fenestration systems, i.e. structured flat roof, structured pitched roof and structured pyramidal-gridded roof, was simulated at 09:00 h, 12:00 h and 15:00 h on 21st of March, 21st of June and 22nd of December under intermediate sky (with sun) conditions. The structured flat roof with uniform surface reflectance produced a significant uniform distribution when the external illuminance of at least 100 klx occurred more than 80% of the day during the March equinox. The overall indoor illuminance was as expected much less in December, and the corner positions were most affected, especially at 15:00 h. The distribution patterns for each type of roof are not the same for all.

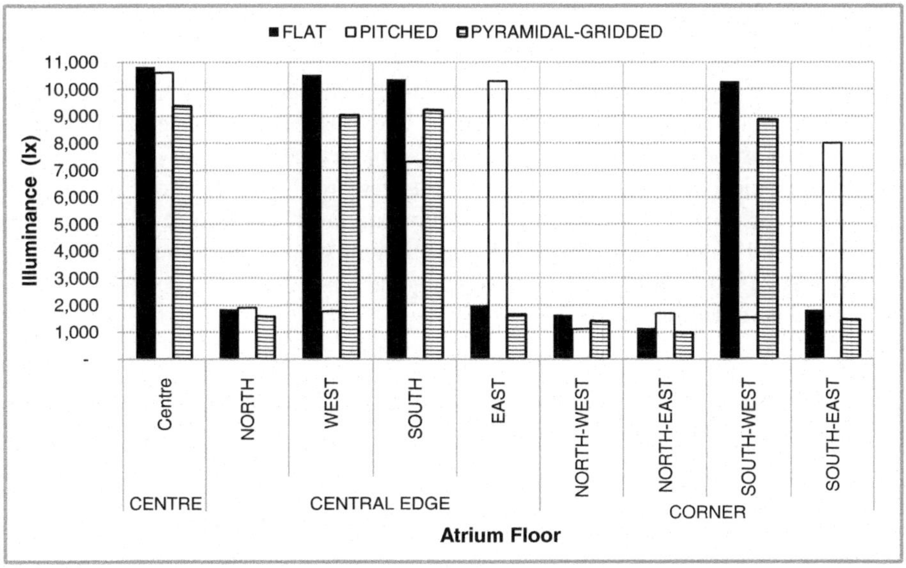

Fig. 53.8 Illuminance against atrium roof obstruction type at various floor positions simulated under intermediate sky (with sun) at 12:00 h on 21st of June

An example of the results of daylight distribution for three types of roofs at 12:00 h on 21st of June is shown in Fig. 53.8. The maximum daylight transmitted on the floor of the atrium buildings at the values equal to or more than 10,000 lx occurs at the centre of the atrium floor at 12:00 h for the flat and pitched roofs. For the pyramidal-gridded roof, a maximum of 10,000 lx is transmitted at the centre of the atrium floor at 15:00 h. The minimum daylight transmitted on the atrium floor is less than 700 lx occurring at 09:00 h for all the three types of roof fenestrations. Nevertheless, in the worst-case scenario, the daylight transmitted for all roof types is still satisfactory as it is well within the accepted visible range.

In summary, the results show that the presence of obstructions on the horizontal aperture of atrium roofs affects daylighting performance positively in terms of illuminance values. Each situation is quite unique and has individual value; hence it is difficult to offer a single solution that is appropriate for each condition.

An obvious condition which is derived from both experiment and simulation programs is during noon time (as shown in Fig. 53.8) which may be used to balance the overall results and prepare for worst conditions. The saw-tooth type of fenestration may not be considered as a suitable roof for tropical skies unless supplemented by artificial lighting. Additionally, when the simulations were compared with field experiment results, illuminance values produced by the simulations were overestimated. However, the simulated data are appropriate enough to estimate average daylight factors and reliable enough to estimate the performance of atrium buildings under tropical skies.

53.4 Conclusions

The results of daylight performance simulation, the shape and structure of the atrium roof, orientation to the sun, sky conditions and surface reflectance are important parameters for the optimum daylighting design of atrium buildings. The simulations done for tropical sky conditions proved that these roof configurations do have an impact on the daylight performance of atrium buildings. The choice of fenestration would depend on the use and function of the atrium space. This in turn will have an effect on the indoor thermal comfort, lighting and air-conditioning needs in the case of hot and humid climates [15–17].

Acknowledgement The authors thank the Faculty of Architecture, Planning and Surveying, Universiti Teknologi MARA, INTEC Education College and ISESCO for the support of this work and the production of this chapter.

References

1. YunusJ, AhmadSS, Zain-AhmedA (2010) Analysis of atrium's architectural aspects in office building under tropical sky conditions. Paper presented at the International Conference In Social and Science Research (CSSR2010), Kuala Lumpur
2. AhmadMH (1996) The influence of roof form and interior cross section on daylighting in the atrium spaces in Malaysia. PhD, University of Manchester
3. Ahmad MH, Mohamad Rasdi MT (2000) Design principles of atrium buildings for the tropics. UniversitiTeknologi Malaysia, Malaysia
4. Kuan SL (2006) Daylighting of atria in Singapore (Master of Arts (Architecture)). National University of Singapore, Singapore
5. Othman AR, Zain-Ahmed A, Ayop SM (2007) Comparative analysis of top-lit atriums in Kuala Lumpur with respect to thermal comfort & illumination level. Abstracts of CSSR 06'07. Conf. on Scientific and Social Research 2007. UPENA.
6. OthmanAR, Zain-AhmedA, AyopSM (2006) Study of sun-exposure hours of three roof form types for BIPV application in Malaysia. Paper presented at the International Symposium & Exhibition on Sustainable Energy & Environment (ISESEE2006), Kuala Lumpur, Malaysia
7. YunusJ, AhmadSS, Zain-AhmedA (2013) Analysing the impact of roof obstructions on daylight levels in atrium buildings: physical scale model under real sky conditions. Paper presented at the 3rd International Building Conference 2013,Universiti Malaya, Malaysia
8. Zain-Ahmed A, Sopian K, Othman MY (2002) Availability of daylight from tropical skies—a case study of Malaysia. Energy Conversion Manage 25(2002):21–30
9. Mousavi M, Hayat Khan T, Lim YW (2016) Empirical validation of radiance-IES daylight simulation for furnished and unfurnished rooms under tropical sky. Int J Sustain Build Technol Urban Develop 7(1):1–9
10. Calcagni B, Paroncini M (2004) Daylight factor prediction in atria building designs. Sol Energy 76(6):669–682
11. Ghasemi M, Kandar MZ, Noroozi M, Yazdipur F, Namaziyan S, Roshan M (2013) Capability of computer simulation software for predicting average daylight factors in a vertical top-light atrium. J Basic ApplSci Res 3(11):96–105
12. Thanachareonkit A (2005) Comparing daylighting performance assessments of buildings in scale models and test modules. Sol Energy 79(2):168–182

13. Thanachareonkit A, Scartezzini JL (2010) Modelling complex fenestration systems using physical and virtual models. Sol Energy 84:563–586
14. Malaysian Standards (2014) Code of practice MS1525:2014. Energy efficiency and renewable energy applications in non-residential buildings. SIRIM, Malaysia
15. Zain-AhmedA (2000) Daylighting and shading for thermal comfort in Malaysian buildings. PhD, University of Hertfordshire
16. Zain-Ahmed A, Sopian K, Othman MY, Sayigh AAM, Surendran PN (2002) Daylighting as a passive solar design strategy in tropical buildings: a case study of Malaysia. Energy Conversion Manage 43(2002):1725–1736
17. Othman AR, Zain-Ahmed A, Ayop SM (2006). Development of top-lit atrium design in Kuala Lumpur with respect to daylighting and thermal comfort. In Zain- Ahmed A, Abdul Rahman S, Shaari S (eds). Energy in Buildings: Sustainable Symbiosis. UPENA

Chapter 54
Techno-Economic Study of a Biogas-Based Polygeneration Plant for Small Dairy Farms in Central Bolivia

J. Villarroel-Schneider, Brijesh Mainali, J. Martí-Herrero, Anders Malmquist, Andrew Martin, and Lucio Alejo

54.1 Introduction

The use of alternative sources of energy in productive activities is an aspect that should be considered when promoting a sustainable development of a country. In the case of rural and semi-urban areas, the provision of energy services also generates benefits in social, education and health aspects [1]. Bolivia, as many other developing countries, is in its early steps to implement energy policies to promote a sustainable use of its energy resources. The new "Constitution of the Plurinational State of Bolivia" supports "new forms of alternative energy production" considering the conservation of the environment. It also emphasized the communitarian labour in rural economic activities through productive organizations [2].

J. Villarroel-Schneider (✉)
Department of Energy Technology, School of Industrial Technology and Management (ITM), KTH Royal Institute of Technology, Stockholm, Sweden

Facultad de Ciencias y Tecnología (FCyT), Universidad Mayor de San Simón (UMSS), Cochabamba, Bolivia
e-mail: jhonnyvs@kth.se

B. Mainali
Department of Built Environment and Energy Technology, Linnaeus University, Växjö, Sweden

J. Martí-Herrero
Building Energy and Environment Group, Centre Internacional de Mètodes Numèrics en Enginyeria (CIMNE), Barcelona, Spain

A. Malmquist · A. Martin
Department of Energy Technology, School of Industrial Technology and Management (ITM), KTH Royal Institute of Technology, Stockholm, Sweden

L. Alejo
Facultad de Ciencias y Tecnología (FCyT), Universidad Mayor de San Simón (UMSS), Cochabamba, Bolivia

A. Sayigh (ed.), *Renewable Energy and Sustainable Buildings*, Innovative Renewable Energy, https://doi.org/10.1007/978-3-030-18488-9_54

Decentralized and combined energy systems can be applied in productive activities when arranged as a multiservice (polygeneration) plant (MP) providing services simultaneously (electricity, cooling, heating, clean water, dry processes, etc.). However, the application of these systems is seen as a complex technology by policy makers and investors. On the other hand, the lack of knowledge and mistrust about the technology is still the main obstacle in its development [3]. The benefits of implementing this type of systems in productive activities while using the local available energy sources are in the line of the Sustainable Development Goals (SDG's). We can mention the 7th goal (Affordable and clean energy), 9th goal (Industry, innovation and infrastructure), 13th goal (Climate action) and 15th goal (Life and land) among others [4]. Using these systems generates a positive impact on the technological, economic, social and environmental aspects when targeted at low-income populations. It is even better if waste resources are used to meet the energy demands while reducing fossil fuel consumption and, consequently, greenhouse gas emissions (GHG).

The application of these concepts can be applied when, for example, farmers' associations are involved. These community associations may also be able to manage small industries by making use of their own resources, generating direct and indirect benefits. The associations of dairy farmers in central Bolivia (Cochabamba) are examples of community work. They have the milk as final product and the cow dung as the main waste. Although the farms are close to each other, which can facilitate the collection of the manure, there is no biogas (BG) production, in spite of the big potential. The manure is just accumulated in an open space, sometimes it is sold to other farmers and/or used in agriculture as fertilizer when growing maize, alfalfa and some vegetable crops [5]. Apart from the bad smell that affects the nearby residents, there is a high risk of diseases for humans and animals [6]. Also the accumulation of animal waste in open space generates more GHG than those resulting from anaerobic digestion while producing biogas [7, 8]. On the other hand, the waste from the anaerobic digestion process allows obtaining a high quality fertilizer [9].

Biogas can be used for cooking but also for producing electricity, refrigeration, heating and fertilizers. It can contribute to reduce or avoid the grid dependence and to reduce GHG using waste resources. That is important considering that most of the electricity produced in Bolivia comes from thermoelectric plants based on natural gas [10], which is subsidized by the government [11, 12] like LPG tanks (used for cooking).

Romero Padilla [5] has presented an economic-environmental and competitiveness analysis of the dairy industry chain in Bolivia. That study describes the existing policies, economic aspects and the environmental impacts of the milk industry chain. It is a general overview of the state of the Bolivian dairy sector. The purpose of that study is to inform that there is no proposed solutions for the problems that the sector is currently facing. One of these problems can be identified as the lack of attention in the exploitation of a waste resource that, if used in a multiservice plant, can provide diverse services while providing additional benefits.

54.1.1 *Objective and Scope*

This paper presents a techno-economic feasibility study of a multiservice plant (MP) aimed to be implemented in a milk storage (refrigeration) centre for dairy farms located in "Alba Rancho" central Bolivia, a traditional area of dairy farmers. Cow dung from farms will be the source for the production of biogas, electricity, cooling and fertilizer using Anaerobic Digesters (AD), an Internal Combustion Engine (ICE), an Absorption Refrigeration System (ARS) and a bio-slurry dryer (BSD). These services are a necessity for the farmers including the fertilizer which is dried to facilitate its handling and transport when it is sold to farmers from remote regions.

The results of the present study will be the levelized costs of the proposed services under the particular conditions of the Bolivian market. These results can be considered as referential for the specific application in the dairy sector where the main energy resource utilized is the farm waste (cow dung). However, the same methodology can be followed when considering the application of polygeneration systems in other productive (or not) sectors.

54.2 Methodology

The methodology follows these steps: data collection about energy situation (fieldwork, interviews with related institutions and literature review); quantification of biogas production potential using cow dung from farms; quantification of energy demands and proposal of a multiservice polygeneration plant; and finally techno-economic evaluation of the proposed system.

The techno-economic study to determine the levelized costs of the proposed services will be done considering data collected from fieldwork, referential equipment prices in the Bolivian market, investment capital, operation and maintenance costs. A sensitivity analysis for the services prices will be presented considering the variation of: the cost of feedstock handling cost (for biogas) and subsidies in the investment capital. These prices are then compared to the current subsidized and non-subsidized prices of similar competitive services in the Bolivian market. The techno-economic study ends by defining sale prices of the services which allow to determine a payback period of the initial investment capital cost for the proposed system.

54.2.1 *Potential of Biogas Production, Cooling and Biogas (for Cooking) Demand*

For this study the electrical and thermal energy demands have been previously determined to propose a polygeneration plant. The thermal demands consider the average of the monthly temperatures of the last 20 years in the zone [13].

The input demands for sizing the plant are the cooling capacity for milk refrigeration and the biogas for cooking required by the 30 families that are part of the association.

The refrigeration capacity was determined for the 5100 L of milk per day (current storage capacity of the centre). The initial temperature of the milk is assumed to be the ambient temperature while the final is set at 4 °C. The number of cows owned by the 30 associated families is 364 heads (255 producing milk). It is assumed a daily production of 35 kg of manure per cow but only 28 kg are used for the production of biogas (assuming losses in the collection). The biogas potential produced was determined using parameters from previous studies in the area whit a yield of 0.335 m^3 of biogas per kilogram of volatile solids (VS) [14] while the energy potential of the biogas is assumed to be 6 kWh/m^3 for a biogas with 60% methane content [15].

It is proposed to meet the refrigeration demand using an absorption refrigeration system (ARS), which uses the recovered heat (from the combustion gases) of a biogas power generator, in this case an internal combustion engine (ICE). The biogas demand for the power generator plus the biogas for cooking demand is the total biogas that should be produced in the plant. From the calculations the production of biogas cannot meet the demand (when considering the number of cows of the association), then it is proposed to increase the number of farmers (not associated) for the contribution with additional manure. The total number of cows is 471.

54.2.2 A Proposed Polygeneration System and Technical Characteristics

The proposed system considers an internal combustion engine (ICE) which can be imported from China [16] to Bolivia. This prime movers is proposed to be configured with an absorption refrigeration system (ARS), Pink Chiller PC19 [17], and heat exchangers in such a way that the plant can provide electricity, cooling and heat services. The polygeneration plant shown in Fig. 54.1 requires a cow dung/water mixture, which produces biogas (BG) and bio-slurry. Biogas is used as gas for cooking and for producing electricity and heat (in the flue gas) using the prime mover (ICE) after being cleaned and pumped. The heat is recovered in water heat exchangers and used for driving the refrigeration system (ARS), for drying the bio-slurry (BSD), and, optionally, to regulate the temperature inside the biodigesters. The final products of the plant are biogas, electricity, refrigeration and dry/liquid fertilizers.

Internal Combustion Engine (ICE): The engine does not need a compressor but a biogas pump and a cleaning stage (to remove solid particles, sulphur and water remains). The nominal power output is 40 kW_{el} while its electrical efficiency is 37%. It is assumed that the nominal data is determined for ISO conditions. The electrical efficiency is diminished by the ambient conditions of the installation site and due to the use of a part of the power produced in the biogas pump. Although some manufacturers of large gas engines say that the altitude does not

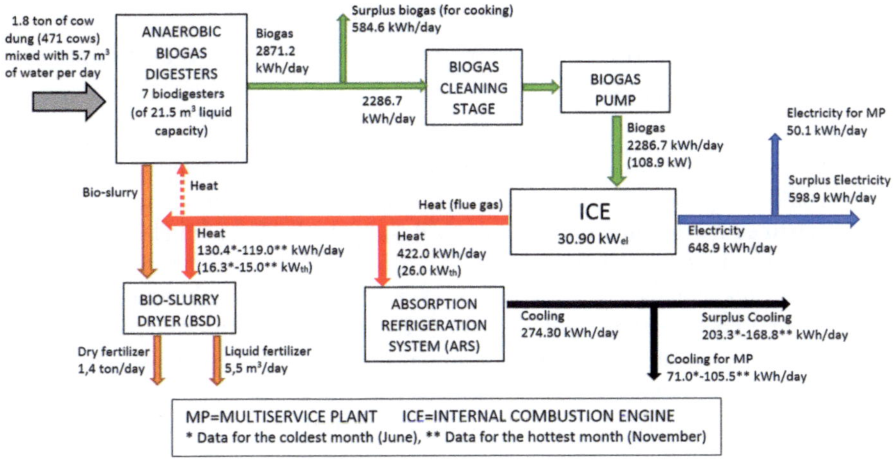

Fig. 54.1 Diagram of the proposed polygeneration plant

affect the power of the engine [18], for this work we will assume a referential derating factor [19]. The actual resulting electrical power is 30.9 kW_{el}, with an electric efficiency of 34%.

Biogas Production System: The incursion of biogas technology in Bolivia has been scarce. Most of the few implementation projects have focused on the production of biogas for cooking in rural areas where the lack of basic services is evident. The technology used for these small systems has been called "low cost tubular digester" because the materials used are relatively cheap when compared to other technologies [20, 21]. Being the only technology that has worked in the region, this study proposes its utilization while considering improvements in the quality of materials [22]. The hydraulic retention time for biogas production is set at 20 days. The calorific energy of the biogas is 21.6 MJ/m^3 [15].

Absorption Refrigeration System: This system (ARS) is marketed in Europe as Pink Chiller PC19 [17]. It requires heat for its operation that is given by a flow of hot water gained in a water heat exchanger. According with data sheet of PC19 the cooling capacity depends of the inlet/outlet temperature of the hot water circuit, they are set at 95/88 °C (flow rate of 3.2 m^3/h). Then the cooling capacity is 17 kW_{th}, the heat required by the ARS's generator (hot water circuit) is 26 kW_{th} and the COP is 0.65. The cooling is given in a cold-water circuit which is proposed to be used in the evaporators of the milk tanks.

Bio-slurry drying system (BSD): In a first stage, the bio-slurry drying system drains the liquid from the bio-slurry keeping a semi-dry product inside a rotating drum. This drum consists of a mixing system while a permanent flow of warm flue gas produces the drying of the product. The final product is a consistent organic fertilizer for use in agriculture. The initial temperature of the bio-slurry is the ambient temperature while the final is set at 95 °C.

54.2.3 Capacity of Production of the MP, Costs of the Production Units (Subsystems) and Investment Cost of the MP

Table 54.1 shows the annual capacity production of the services supplied for the MP when considering the ICE as prime mover. It also shows how much of the production is destined for the functioning of the MP and for sale. The annual rate considers 355 days, which are the operation days fixed for the MP. The rest of the days are assumed for maintenance and equipment revision, also unexpected failures.

Table 54.2 presents the cost data of the MP used to determine the levelized costs of the services, which also includes a raw estimation of the costs of other items involved and the total investment capital cost. The costs of the energy sources considered for the subsystems are cow dung for the production of biogas; biogas for

Table 54.1 Annual capacity production of the multiservice plant when using an ICE

		MP-ICE		
Service	Unit	For MP	For sale	Total
Biogas	kWh	811,763.77	207,527.28	1,019,291.04
Electricity	kWh	17,767.75	212,591.75	230,359.50
Cooling	kWh	31,236.47	65,760.69	96,997.15
Dry fertilizer	ton	–	507.11	507.11
Liquid fertilizer	m^3	–	1954.97	1954.97

Table 54.2 Costs/investment for the polygeneration plant and its production units

Description	Unit	MP-ICE
Biogas production—investment cost	**USD**	**69,638.00**
Cost of the land	**USD**	**50,500.00**
Tubular biodigesters of 7 × 22.3 m^3 (liquid volume)	USD	14,000.00
Pipes, accessories & minor components	USD	5138.00
Feedstock handling cost	USD/ton	10.00
Electricity and heat production—investment cost	**USD**	**29,770.83**
Internal combustion engine	USD	20,167,66
Treatment biogas system (cleaning and pumping)	USD	9603.17
Cooling production—investment cost	**USD**	**65,139.00**
Absorption refrigeration system	USD	38,025.00
Milk tanks, including evaporators	USD	25,500.00
Pipes, accessories and minor components	USD	1614.00
Fertilizer production—investment cost	**USD**	**6360.00**
Bio-slurry drier	USD	5000.00
Pipes, accessories and minor components	USD	1360.00
Total investment cost of the MP	**USD**	**170,907.83**

the production of electricity and heat; and electricity and heat for the production of refrigeration and fertilizer. Therefore, first the price of biogas has to be determined, then the price of electricity and, finally, the price of cooling and fertilizers.

54.2.4 *Economic Analysis, Levelized Cost of the Services and Payback Period*

The method of Levelized Cost of Electricity (LCOE) proposed by Mainali and Silveira [23] will be used (excluding the environmental factor of the equation). That will be done considering the capacity production of the services and the costs presented in the previous section. The result of LCOE calculation is a price that serves to evaluate the proposed technology when compared to alternative technologies, which also generate electricity. Such technologies can be of smaller or larger scale, have different investment costs and/or periods of operation. The LCOE considers the total life cycle cost of the project instead of the simple comparison of the capital costs [15, 23]. The LCOE is defined by Eq. (54.1).

$$\text{LCOE} = \frac{\text{Total life time cost of the project}}{\text{Total life time useful electricity produced}} \tag{54.1}$$

To calculate the annual costs, a project time period, a discount rate and a scaling factor of the prices were defined to be 20 years, 10% and 5% [15, 23], respectively.

This method is also applied to determine the Levelized cost of biogas (LCOB) [15], cooling (LCOC) and fertilizer (LCOF). The units of the services costs are USD/kWh for biogas and refrigeration, and USD/ton and USD/m^3 for dry and liquid fertilizer, respectively.

The payback period is also calculated. That is, the time (in years) in which the sum of the net cash flows (revenues) of the MP equals the initial investment capital.

54.3 Results and Discussion

54.3.1 *Levelized Costs of the Services Supplied by the MP and Costs of the Current Services in the Bolivian Market*

Table 54.3 shows the levelized costs of the services determined for the proposed system, they are shown together with the prices of similar competitive services in the Bolivian market. They correspond to the current subsidized (S) and non-subsidized (NS) prices for electricity [10], refrigeration and LPG [24, 25]. The non-subsidized prices are approximations based on studies conducted in Bolivia that try to predict

Table 54.3 Services costs of the MP and subsidized (S)/non-subsidized (NS) prices in Bolivia

Service			MP-ICE	S-Prices		NS-Prices	
Biogas/LPG	USD/kWh	**LCOB**	**0.018**	S-LPG	0.025	NS-LPG	0.038
Electricity	USD/kWh	**LCOE**	**0.088**	S-EL	0.100	NS-EL	0.160
Cooling	USD/kWh	**LCOC**	**0.102**	S-CO	0.091	NS-CO	0.128
Fertilizer	USD/ton	**LCOF**	**3.13**	–	–	–	–

what the real price of electricity would be [26, 27]. The non-subsidized price of LPG corresponds to the exportation price [28, 29] while the cooling prices were determined using the cost of electricity used in conventional systems, the cost of the refrigeration equipment and the refrigeration demand.

In Bolivia the price of natural gas used in thermoelectric plants is subsidized. Therefore, the price of electricity for the final users, indirectly, has a reduced cost [26, 27]. The refrigeration, considered as a final service, which is obtained with subsidized electricity is consequently subsidized. In the case of LPG used for cooking, the price has been fixed for several years [11, 12], while the exportation price from Bolivia is much higher [28]. There is no evidence of commercialization of dry fertilizers from bio-digestion processes, so prices were not found.

From results shown in Table 54.3 it can be seen that the price of biogas of the proposed system is much lower when compared to the equivalent LPG price subsidized or not. The electricity produced in a MP-ICE is even lower than the S-price. The price of the cooling service produced is slightly higher than the S-price of the cooling service in the market. The only reference with which the price of the dry fertilizer can be compared is with the cow dung price in its natural state, where the selling price was found to be 10 USD/ton for the MP (for biogas production) and for other customers as well (according to consultations). In the present study, the prices determined for the dry fertilizer are much lower, which allows having a greater margin of profit when selling it. This considering that the fertilizer produced in the MP has better quality than the cow dung in its natural state [9], which increases its economic value.

From Table 54.3 it can be seen that the levelized costs of the services of the MP-ICE are attractive and competitive especially when considering a scenario without subsidies.

54.3.2 Sensitivity Analysis for the Cost of Services Supplied by the Proposed MP

In this section the levelized cost of biogas will be identified as LCOB, in similar way for the rest of the services. S-LPG and NS-LPG are the reference prices for the subsidized and non-subsidized prices of LPG, respectively. In similar way for the electricity (EL) and cooling (CO) services.

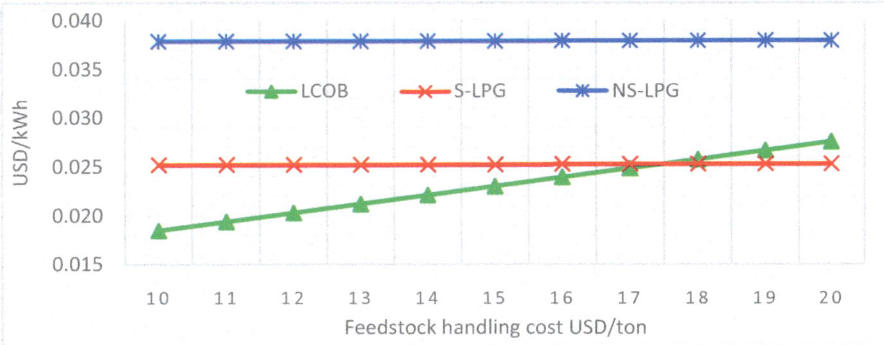

Fig. 54.2 Influence of feedstock handling cost on the LCOB

54.3.2.1 Influence of the Feedstock Handling Cost on the Price of Biogas

The collection and handling of cow dung can be seen as a dirty and unattractive task. This can be counteracted by offering higher payments for the product while improving the economy of the farmers. It is illustrated in Fig. 54.2 when the feedstock handling cost increases from 10 (the current price) to 20 USD/ton. The price of biogas (LCOB) equals the subsidized LPG price (S-LPG) at a feedstock handling cost of around 18 USD/ton. However, it never reaches the non-subsidized price (NS-LPG) even if the feedstock handling cost is 20 USD/ton.

54.3.2.2 Effect of Subsidizing the Initial Investment Capital of the MP on the Services Costs

The prices of most of the services were found lower than the referential prices of the market. However, the cost of cooling was found slightly higher than the subsidized price of cooling in the market, it can be seen in Table 54.3. Although the source of energy for the production of cooling in the MP is the heat whose price is zero, its final price results to be almost similar to the price of cooling produced in conventional systems that use electricity. This is due to the cost of the technology used for the production of cooling, the proposed system uses an absorption system whose cost is quite high compared to the compressor-driven system. Subsidizing the initial investment cost of the MP can be a way to reduce this cost while making the proposed MP more competitive. This subsidy can be also interpreted as a reduction in the technology cost that affects the final levelized costs of the services.

Cooling (refrigeration): When there is no subsidy the determined price (LCOC) is found between the calculated prices of conventional refrigeration system (S-CO and NS-CO). Figure 54.3 shows that this price decreases drastically when the investment capital subsidy increases, this is because it is almost fully dependent

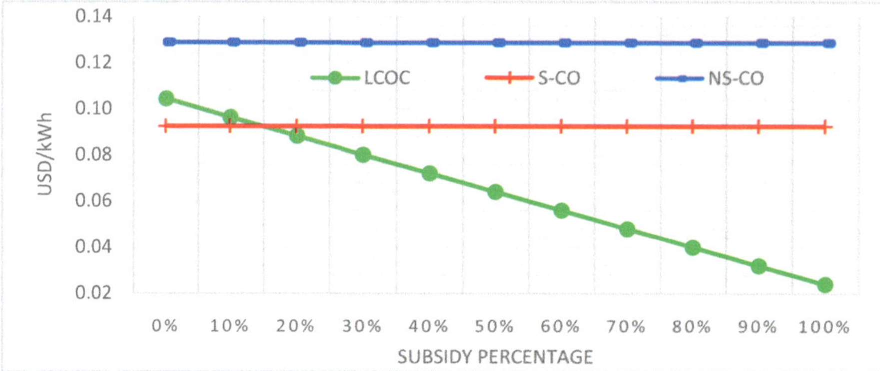

Fig. 54.3 Influence of subsidizing the investment capital cost on the LCOC

on the cost of the technology since the cost of the energy source (heat) is zero. At about 15% of subsidy for the MP, the cooling price (LCOC) equals the subsidized price (S-CO), reducing even more when the subsidy increases.

Applying this subsidy also means a reduction of the levelized costs of the rest of the services.

54.3.3 Sale Prices of Services/Products and Payback Period for the Proposed Systems

Three scenarios are presented for the sale prices of the services/products of the proposed system, they allow to determine a cash flow and the payback period for the investment capital.

In scenario 1 the sale prices are defined to be equal as the determined levelized costs (LCOB, LCOE and LCOC) for the proposed system (MP-IFMT). In scenario 2 and 3 the sale prices are defined to be the same as the subsidized prices (S) and non-subsidized prices (NS) of the conventional services, respectively. All these prices are shown in Table 54.3.

For these scenarios the cost of the dry fertilizer is set at 15 USD/ton. The resulting liquid from the bio-slurry dryer system (BSD) is also considered as (liquid) fertilizer which is proposed to be sold at a price of 10 USD/m^3 in all the cases. The payback period (in years) for the proposed system in the different scenarios is presented in Fig. 54.4.

Scenario 1 is only referential because of the slightly higher cost of the cooling when compared to the subsidized price. It is not feasible to sell this service at a higher cost than the service cost in the market. In scenario 2, the subsidy for the investment capital has to be only 15% which allows to sell the services at the same current subsidized prices of the market services. That is the most promising scenario which can be implemented under the actual conditions of the Bolivian market. In scenario 3,

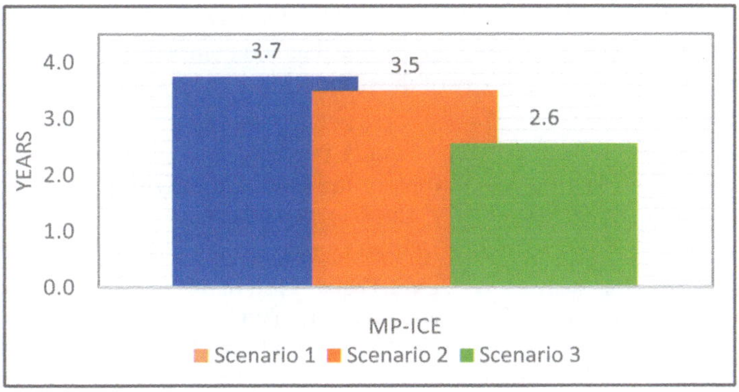

Fig. 54.4 Payback period for the proposed system (for different sale prices)

the supposed case of a non-subsidized market, the MP-ICE does not require any subsidy in the investment capital having a great advantage over the other presented scenarios. That is reflected in the shorter payback periods shown in Fig. 54.4.

54.4 Conclusions and Future Work

A multiservice plant has been proposed for the dairy sector in central Bolivia considering an internal combustion engine as prime mover, an absorption refrigeration system and a bio-slurry dryer. The final services from the plants are biogas, electricity, refrigeration and fertilizers.

The economic analysis has focused on determining the levelized cost of the services. The resulting prices were compared with the current subsidized and non-subsidized prices of similar competitive services existing on the market. The price determined for biogas was found lower when compared to the subsidized LPG price in terms of its energy value. The price of electricity was found lower than the subsidized price. The cost of cooling was found to be slightly higher than the subsidized price in the market. Sensitivity analyses were presented to see the effect on the levelized cost of biogas when increasing the feedstock handling cost and the cost of cooling when a subsidy is applied on the investment capital in order to reach the subsidized referential price. It was shown that the feedstock handling cost can be increased from 10 USD/ton up to 18 USD/ton while maintaining a biogas price lower than the non-subsidized price of LPG. That means better payments for the farmers while improving their economic status. As expected, the prices of the services tend to decrease when a subsidy is assumed in the cost of investment capital. Finally, different scenarios were presented to determine the payback period of capital recovery. From this analysis the plant with a subsidy in the investment capital of 15% and sale prices of the services equal as the subsidized prices of the market services has shown a payback period of 3.5 years being a promising option.

It is also shown that a market without (or with reduced) subsidies for the conventional services makes the proposed system more feasible and competitive. However, the high investment cost and the complexity of the technology can be seen as barriers for a real implementation.

As future work, it is necessary to study possible sources of financing and the compatibility with existing regulations in the Bolivian energy sector to determine the existing gaps.

Acknowledgement This study has been carried out as a part of SIDA's research cooperation with Universidad Mayor de San Simon. The strategic research project STandUP for Energy has contributed to supervision of the work. The association and federation of dairy farmers (APL and FEPROLEC) from Bolivia have contributed with valuable data collected in fieldwork and interviews.

References

1. Cabraal RA, Barnes DF, Agarwal SG (2005) Productive uses of energy for rural development. Annu Rev Environ Resour 30:117–144
2. Constituent Assembly of Bolivia, "Plurinational state of Bolivia' s constitution of 2009 (translated from spanish)," 2009. . https://www.constituteproject.org/constitution/Bolivia_2009.pdf
3. Wu DW, Wang RZ (2006) Combined cooling, heating and power: a review. Prog Energy Combust Sci 32(5–6):459–495
4. United Nations (2015) Transforming our world: the 2030 agenda for sustainable development
5. Romero Padilla C (2007) Competitividad económica-ambiental para la cadena de lácteos de Bolivia. KIPUS, Cochabamba
6. Pell AN Manure and microbes: public and animal health problem? J Dairy Sci 80(10):2673–2681
7. Mainali B, Emran SB, Silveira S (2017) Greenhouse gas mitigation using poultry litter management techniques in Bangladesh. Energy 127:155–166
8. Holly MA, Larson RA, Powell JM, Ruark MD, Aguirre-villegas H (2017) Agriculture, ecosystems and environment greenhouse gas and ammonia emissions from digested and separated dairy manure during storage and after land application. Agric Ecosyst Environ 239:410–419
9. Gomez CDC (2013) Biogas as an energy option: an overview. In: The biogas handbook. http://linkinghub.elsevier.com/retrieve/pii/B9780857094988500014
10. AE-Bolivia (2017) Anuario Estadístico 2016-Autoridad de Fiscalización y Control Social de Electricidad Bolivia. AE-Bolivia, La Paz
11. Coady D, El-Said M, Gillingham R, Kpodar K, Medas P a, Newhouse DL (2006) The magnitude and distribution of fuel subsidies: evidence from Bolivia, Ghana, Jordan, Mali, and Sri Lanka, IMF Work Paper, vol 06. International Monetary Fund, Washington, p 39
12. Arze del Granado FJ, Coady D, Gillingham R (2012) The unequal benefits of fuel subsidies: a review of evidence for developing countries. World Dev 40(11):2234–2248
13. SENAMHI (2018) Servicio Nacional de Meteorología e Hidrología-Bolivia. http://senamhi.gob.bo/index.php/inicio. Accessed 11 Jul 2018
14. Martí-Herrero J, Ceron M, Garcia R, Pracejus L, Alvarez R, Cipriano X (2015) The influence of users' behavior on biogas production from low cost tubular digesters: a technical and sociocultural field analysis. Energy Sustain Dev 27:73–83
15. Khan EU, Mainali B, Martin A, Silveira S (2014) Techno-economic analysis of small scale biogas based polygeneration systems: Bangladesh case study. Sustain Energy Technol Assess 7:68–78
16. Shenzhen Puxin Technology (2018) Industrial biogas plant-Shenzhen Puxin Technology Co. Ltd. http://en.puxintech.com/portablemediuandlarge. Accessed 11 Jul 2018

17. Pink GmbH (2017) Absorption Chiller PC 19 Data Sheet. http://www.pink.co.at/inc.download.php?dlf=325. Accessed 21 Nov 2017
18. General Electric Company (2016) More innovation, power & efficiency. https://www.ge.com/content/dam/gepower-pgdp/global/en_US/documents/product/ReciprocatingEngines/Jenbacher/Type6/GE_J920FleXtra_Brochure_US_2016_rz.pdf
19. Caterpillar (2018) Gas Engines|Caterpillar. https://www.cat.com/en_US/products/new/power-systems/industrial/gas-engines/. Accessed 10 Jul 2018
20. J. Martí-herrero (2015) Transfer of low-cost plastic biodigester technology at household level in Bolivia Experiencia de transferencia tecnológica de biodigestores familiares en Bolivia transfer of low—cost plastic biodigester technology at household level in Bolivia abstract
21. Martí J (2015) Desarrollo, difusión e implementación de tecnologías en el área rural: Biodigestores en Bolivia, vol 1. EnDev-Bolivia GIZ y CIMNE, La Paz
22. E. Group (2016) Digesters in cold climate and high altitude: history, state of the art and challenges, no. 1987
23. Mainali B, Silveira S (2013) Alternative pathways for providing access to electricity in developing countries. Renew Energy 57:299–310
24. Agencia Nacional de Hidrocarburos—Bolivia (2017) Precios Finales al Consumidor Octubre de 2017. http://www.anh.gob.bo/index.php?N=dre. Accessed 11 Oct 2017
25. Troncoso K, Soares da Silva A (2017) LPG fuel subsidies in Latin America and the use of solid fuels to cook. Energy Policy 107:188–196
26. Salinas San Martin L (2011) Determinación del precio de Gas Natural para el Sector Eléctrico Boliviano y su efecto en las tarifas a consumidor final. Investig Desarro 11:136–147
27. Herbas MT (2013) Determinación de un adecuado precio del gas natural para el sector eléctrico boliviano. Scielo 19:99–123
28. Diario Pagina Siete (2017) YPFB venderá GLP a Paraguay a $us 470 la tn durante invierno. http://www.paginasiete.bo/economia/2017/7/4/ypfb-vendera-paraguay-durante-invierno-143325.html. Accessed 28 Jun 2018
29. CAMBIO-Periódico del Estado Plurinacional de Bolivia (2017) Bolivia será el exportador nato de GLP en la región desde 2018. http://www.cambio.bo/?q=node/37046. Accessed Jun 28 2018

Chapter 55
Performance Gap and nZEB Compliance of Monitored Passivhaus in Northern Ireland, the Republic of Ireland and Italy

S. Colclough, V. Costanzo, K. Fabbri, S. Piraccini, P. Griffiths, and Neil J. Hewitt

55.1 Introduction

The legislative framework existing in the EU is driving the development of low-energy building standards including the recast Energy Performance of Buildings Directive (EPBD) [1]. Near Zero Energy Buildings (nZEB) are required for public buildings from the end of 2018 and for all buildings from 2020. The Passive House has been shown to potentially be a key enabler for the nZEB standard, and as such it offers potential, as it represents a well-proven methodology for compliance [2].

Analysis of case studies across a number of regions in the EU, where the deployment of the Passive House standard has successfully taken place, has demonstrated that considerable benefits are seen to have accrued [3]. These include the development of local markets, the creation of jobs and the consequential increase in innovation, enterprise, research and resources at a regional level. A key enabler for the adoption of the Passive House standard was the involvement of forward-thinking municipalities.

Given the adoption in 2016 of the Passive House standard in Dun Laoghaire-Rathdown County Council and the requirement to build ca 30,000 dwellings per annum to the nZEB standard in the Republic of Ireland, considerable potential exists for the architects, builders and equipment suppliers for the Passive House.

S. Colclough (✉) · P. Griffiths · N. J. Hewitt
University of Ulster, Newtownabbey, Co Antrim, Northern Ireland, UK
e-mail: s.colclough@ulster.ac.uk

V. Costanzo
School of the Built Environment, University of Reading, Reading, RG1 5DU UK, UK
e-mail: v.costanzo@reading.ac.uk

K. Fabbri · S. Piraccini
Department of Architecture, University of Bologna, Bologna, Italy
e-mail: kristian.fabbri@unibo.it; steafano.piraccini4@unibo.it

A. Sayigh (ed.), *Renewable Energy and Sustainable Buildings*, Innovative Renewable Energy, https://doi.org/10.1007/978-3-030-18488-9_55

This is particularly the case given that the nZEB standard has been demonstrated to be achievable via the passive house route at no extra cost compared to the current building regulations [4, 5].

This monitoring and analysis exercise has quantified and highlighted some advantages offered by the Passive House standard in achieving nZEB. The Passive House has been demonstrated as providing a superior indoor climate while consuming approximately a third of the energy required for heating compared with contemporaneous dwellings built complying with the prevailing minimum building regulations.

Further, at a societal level, the data already gathered has the potential to facilitate an evidence-based discussion on the future of the housing stock. In particular, this study has shown that the Passive House requires only 38% of the energy required by houses built to the current building regulations. Further, the potential with respect to fuel poverty is self-evident given that space heating costs for a 1000 ft^2 passive house can be approximately £200 per annum.

The implications of Brexit are yet to be fully understood within Northern Ireland and the UK [6]. One key aspect that needs to be considered is the possibility that the EU's Directive will not apply to the building stock in Northern Ireland. On the other hand, Italy will firmly be part of the EU in the years to come, and as highlighted by some authors, the techno-economic feasibility of nZEB can be achieved by energy-efficient solutions at both the building [7, 8] and urban scales [9].

This chapter examines if existing passive houses built in the island of Ireland and in Italy perform as expected in the operational phase and thus provide a "toolkit" for designers not only to achieve the nZEB energy requirements but also to achieve good Indoor Environmental Quality (IEQ) conditions in different climates. A special focus is given to bedroom conditions because the very strict requirements set by low-energy standards in terms of air infiltration and ventilation rates could be detrimental in terms of overheating and high CO_2 concentrations.

55.2 Methodology

55.2.1 Characteristics of the Monitored Houses

Ulster University has carried out monitoring of four passive houses in NI, hereafter called PH 1–4, and a number of passive houses in the Republic of Ireland, with PH 6 reported on in this chapter. An evidence-based analysis has been employed to quantify the performances of the houses, carried out for a 12-month period including both the summer (2016) and the winter (2016/17) seasons.

In order to appreciate climate-induced differences on the performances of a passive house, a house located in the Mediterranean climate of Cesena in Italy (PH 5) has been monitored as well for an entire year (from April 2016 to April 2017) (Table 55.1).

Table 55.1 Characteristics of monitored passive houses

House	Building type	Location	Size (m^2)
PH 1	2 storey, detached,	NI	158
PH 2	Bungalow, detached	NI	220
PH 3	Bungalow, detached	NI	145
PH 4	2 storey, detached	NI	247
PH 5	4 storey, detached	Italy	317
PH 6	Bungalow, detached	RoI	182

All these buildings have been designed using the Passive House Planning Package (PHPP) software to comply with the Passivhaus standard prescriptions [10], namely:

- Space heating demand lower than 15 kWh $m^{-2}y^{-1}$ or space heating load lower than 10 W/m^2
- Total primary energy needs, including space heating, domestic hot water production and electrical appliances, lower than 120 kWh $m^{-2}y^{-1}$
- Air infiltrations lower than 0.6 h^{-1} at 50 Pa pressure difference
- Mechanical Ventilation with Heat Recovery (MVHR) system with efficiency higher than 75%
- Opaque envelope components with U-values lower than 0.18 $Wm^{-2}K^{-1}$ and glazing components with U-values lower than 0.8 $Wm^{-2}K^{-1}$
- High-efficiency lighting and equipment

Over the 26-year period since the passive house standard was first espoused, many examples have been built throughout Europe and the world. After the implementation in the EU of the aforementioned EPBD Directive that prescribes the construction of nZEB from 2020, the opportunity afforded by utilising a well-established standard became apparent, and some authors started exploring its feasibility mainly by means of dynamic building performance simulations [11–13].

The Passive-On project [14] guidelines, as adopted by the Passivhaus Institut, introduced further prescriptions for houses located in warm climates:

- Energy demand for space heating/cooling lower than 15 kWh $m^{-2}y^{-1}$
- Air infiltration lower than 1 h^{-1} at 50 Pa pressure difference, if outdoor air temperature does not fall below 0 °C

Total primary energy needs should now account also for space cooling yet still have to remain below the 120 kWh $m^{-2}y^{-1}$ threshold. The indoor maximum temperature requirements remain unaltered (no more than 10% of the hours in a year with temperatures exceeding 25 °C in living areas).

Table 55.2 Characteristics of the monitoring instruments

Physical variable	Measurement range	Accuracy range
Indoor temperature	0–50 °C	±0.3 °C
Relative humidity	0–100%	±3%
CO_2 meter	0–5000 ppm	±50 ppm

55.2.2 Monitored Parameters

The Indoor Environmental Quality (IEQ) parameters monitored in all the houses are:

- Indoor temperature (°C)
- Relative humidity (%)
- CO_2 concentrations (PPM)

The metrics have been gathered at 5-min intervals, and then averaged over a 1-h time step. The relevant measurement and accuracy ranges for every instrument are summarised in Table 55.2. Threshold values have been set for each metric according to Passivhaus design requirements and national building regulations, and the percentage of time when these thresholds are exceeded is considered as well. The heating energy data has been gathered via real-time monitoring for the houses located in the island of Ireland, whereas these figures were not available for the case study in Italy at the time of writing this chapter.

55.3 Results and Discussion

55.3.1 Indoor Temperatures

Figures 55.1, 55.2 and 55.3 report on the proportion of time-specific temperature ranges that are experienced in the monitored bedrooms together with the relative average temperatures for the summer and winter, respectively. Consideration of the summer temperatures allows investigation into the extent of overheating, while consideration of the winter temperatures allows the "strength" of the good thermal performance of passive house to be examined.

Considering the summer temperatures, PH 1–4 (Northern Ireland) and PH 6 (Republic of Ireland) experience more moderate temperatures than those exhibited by PH 5 (Italy). However, it is evident that—as can be seen in PH 2—overheating can occur, albeit significantly below the PH threshold of 10% of occurrences of temperatures greater than 25 °C in the year.

As for the Italian case (PH 5), temperatures are within the range of 20–24 °C for almost 60% of time, with no occurrences of too cold temperatures (lower than 18 °C). On the other hand, temperatures higher than 25 °C are recorded for 19% of the time, thus exceeding the PH prescriptions about comfort conditions (even though this

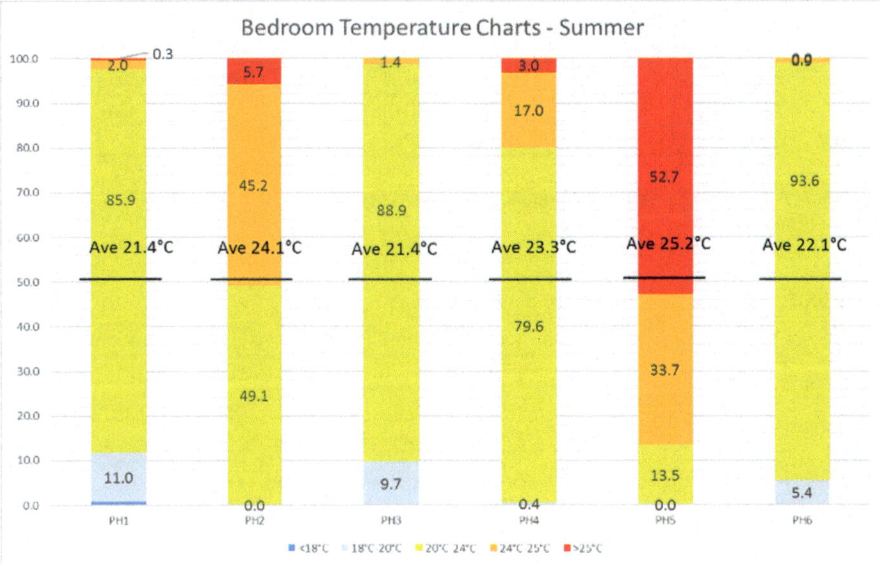

Fig. 55.1 Summer bedroom temperatures for PH 1–6

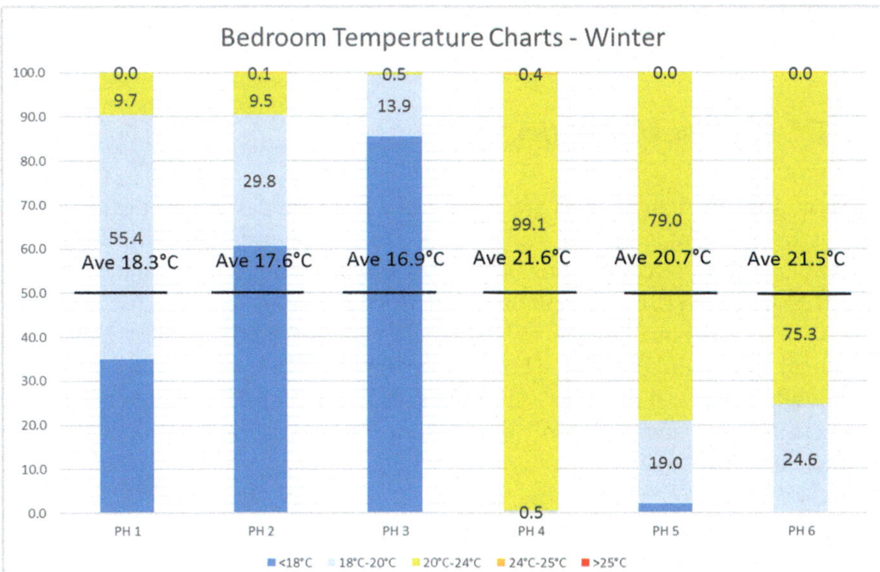

Fig. 55.2 Winter bedroom temperatures for PH 1–6

RH {%}	PH 1	PH 2	PH 3	PH 4	PH 5	PH 6
Average	54.1	48.6	54.1	53.0	47.84	57.0
Max	71.0	59.0	66.0	66.0	76.64	69.0
Min	38.0	36.0	40.0	40.0	19.78	46.0
30-60 {%}	76%	100%	88%	87%	80%	75%

Fig. 55.3 Annual, max, min and frequency of occurrence of between 30% and 60% for bedroom relative humidity

Table 55.3 Annual average values of the IEQ metrics

House	Indoor temperature (°C)	Relative humidity (%)	CO_2 concentrations (ppm)
PH 1	18.4	50.7	721.2
PH 2	17.6	48	596.8
PH 3	16.9	50.9	493.7
PH 4	21.6	48.2	946.9
PH 5	22.6	47.8	660.7

requirement strictly applies to living rooms only) and representing a potential overheating issue. The peak temperature recorded was of 27.8 °C.

Significant variations in temperature are also exhibited in the PH bedrooms during winter. The average temperatures range from 16.9 °C to 21.6 °C. It is noted that the warmest bedroom (PH 4) is located in Northern Ireland and that the bedroom located in Italy experiences a broader range of temperatures than PH 4.

55.3.2 Relative Humidities

While no specific requirements are put in place from the Passivhaus Institute about relative humidity ranges, it is well known that this physical quantity determines thermal comfort in conjunction with temperature. Considering the ASHRAE 55-2013 standard [15] as a reference, comfort conditions (i.e. relative humidity values within the boundaries of 30–60%) are experienced for at least 75% of the time in all the monitored passive houses (see Table 55.3).

Overall, it is found that relative humidity is not an issue for the passive houses under examination due to the MVHR system operation, ensuring a sufficient number of air changes per day.

55.3.3 CO_2 Concentrations

The indoor air quality (IAQ) in bedrooms is of particular interest when considering the overall indoor environment. This is because bedrooms are typically inhabited for an uninterrupted 7 or 8 hours per day during which time there is little adaptation

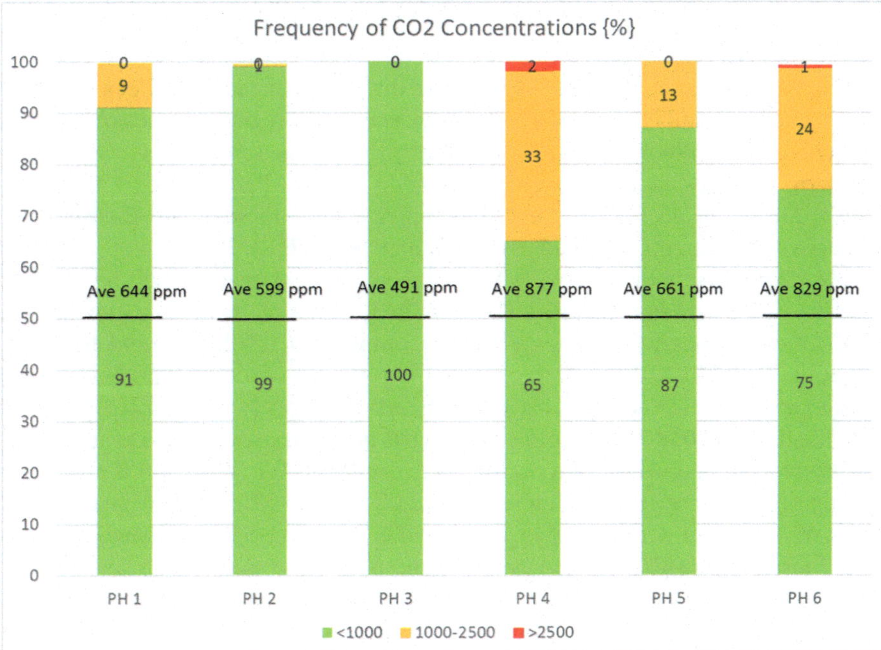

Fig. 55.4 Bedroom CO_2 concentrations for 12-month period

of the indoor environment (e.g. by opening windows or doors). Recent publications have highlighted the link between sleep quality and indoor air quality [16]. In addition, given the reduced air infiltration of nZEB dwellings, there is potentially less ventilation in the buildings of the future especially when compared with the building standards of the past.

The Passivhaus Institute does not specify acceptable levels of CO_2 concentrations. However, carbon dioxide concentrations are the subject of increasing focus, with acceptable levels recognised as typically below 1000 PPM, with levels over 2500 ppm considered as undesirable.

As can be seen from Fig. 55.4, the average carbon dioxide concentrations vary significantly, from the lowest in PH 2 (599 ppm) to the highest in PH 4 (877 ppm). Of particular note is the frequency of occurrence of relatively high carbon dioxide concentrations in PH 4 and PH 6. Given that the bedroom is occupied for only one-third of the 24-h period, the bedrooms clearly exhibit carbon dioxide concentrations in excess of 1000 PPM for the vast majority of the occupied period, with levels exceeding the undesirable 2500 PPM threshold in both PH 4 and PH 6.

Further investigation is needed to determine the root cause, as the Mechanical Ventilation with Heat Recovery systems should be calibrated to provide sufficient ventilation (as is clearly the case in the remaining monitored dwellings).

55.3.4 Energy Demand and nZEB Compliance

While there is uncertainty caused by Brexit, the nZEB standard has been proposed for Great Britain and NI as requiring primary energy for regulated loads of less than or equal to 43.6 kWh $m^{-2}y^{-1}$ (or 44 kWh $m^{-2}y^{-1}$ to the nearest integer). Of the four Northern Ireland passive houses monitored, Energy Performance Certificates (EPCs) are available for the three and indicate the "approximate Energy Use" primary energy unregulated loads as 40 kWh $m^{-2}y^{-1}$ (PH 2), 39 kWh $m^{-2}y^{-1}$ (PH 3) and 44 kWh $m^{-2}y^{-1}$ (PH 4). While further more detailed analysis will be required once the nZEB standard for the UK is finalised, based on the primary energy use metric, the three passive houses are seen to comply with nZEB requirements.

For the Republic of Ireland, there is an indicative regulated load primary energy consumption of below 45 kWh $m^{-2}y^{-1}$ for nZEB purposes (with the nZEB standard to be finalised in 2018). Monitoring of PH 6 has taken place over a 12-month period, and the total purchased electricity was recorded at 4063 kWh. When this is multiplied by the primary energy conversion factor of 2.08 (to give 9575 kWh primary energy demand), the specific primary energy demand is seen to be 52.6 kWh $m^{-2}y^{-1}$ (based on a floor area of 182 m^2). This is below the 61.61 kWh $m^{-2}y^{-1}$ predicted by the National Energy Rating software for regulated loads alone (which comprises heating, ventilation and fixed lighting). The overall consumed energy is approaching the requirements to comply with the forthcoming nZEB specifications with respect to regulated loads.

55.4 Conclusion

Monitoring campaigns carried out during a whole year period (2016/2017) in five passive houses located on the island of Ireland (temperate maritime climate) and one passive house located in Italy (Mediterranean climate) showed that, overall, they perform as per design and represent a viable manner in which to achieve the nZEB target. It has been demonstrated that three of the dwellings already comply with the nZEB primary energy consumption targets, despite not having been designed with the target in mind.

Some potential issues have been recorded in bedrooms, especially for the Italian case, where temperatures are higher than the 25 °C threshold for 19% of year (and 53% of the summer period), but also in the temperate maritime climate of Northern Ireland, where PH 2 recorded temperatures of greater than 25 °C for 6% of the time during the summer. In addition, the CO_2 concentrations can reach values as high as 2371 ppm, with PH 4 and PH 6 recording average CO_2 concentrations of greater than 800 ppm and exceeding 1000 ppm for 35% of the time and 25% of the time, respectively.

While the high summer temperatures in PH 2 are due to a lack of shading (deviating from the plan), the elevated carbon dioxide concentrations may be correlated with a higher than expected occupancy density or the operation of the MVHR system coupled with the intrinsic airtightness of the envelope prescribed by the Passivhaus standard. Future work should thus explore the use of different ventilation techniques, namely, natural and hybrid ventilation, for passive cooling and IEQ purposes.

References

1. EU (2010) Directive 2010/31/EU of the European parliament and of the council of 19 may 2010 on the energy performance of buildings (recast). OJEU 153:13–35
2. Chel A, Janssens A, De Paepe M (2015) Thermal performance of a nearly zero energy passive house integrated with the air-air heat exchanger and the earth-water heat exchanger. Energ Buildings 96:53–63
3. Passive House Institute (2015) PassREg: final report. https://ec.europa.eu/energy/intelligent/projects/en/projects/passreg. Accessed 8 Nov 2017
4. Colclough S, Mernagh J, Sinnot D, Hewitt NJ, Griffiths P (2018) The cost of building to the nearzero energy building standard—a financial case study. Sustainable Building for a Cleaner Environment:71–78
5. ColcloughS, O'LearyT, HewittNJ and GriffithsP (2017) The near zero energy building standard and the Passivhaus standard—a case study.Design to Thrive: Proceedings Volume 1, PLEA 2017 conference, 3rd to 5th July 2017, NCEUB 2017. ISBN 978–0–9928957-5-4, pp. 385–392
6. Pitts A (2017) Passive house and low energy buildings: barriers and opportunities for future development within UK practice. Sustainability 9:272
7. Ferrante A, Cascella MT (2011) Zero energy balance and zero on-site CO2 emission housing development in the Mediterranean climate. Energ Buildings 43:2002–2010
8. Evola G, Margani G, Marletta L (2014) Cost-effective design solutions for low-rise residential Net ZEBs in Mediterranean climate. Energ Buildings 68:7–18
9. Aste N, Adhikari RS, Buzzetti M (2010) Beyond the EPBD: the low energy residential settlement Borgo Solare. Appl Energy 87:629–642
10. Passivhaus Institute http://passivehouse.com/02_informations/02_passive-house-requirements/02_passive-house-requirements.htm. Accessed 11 Nov 2017
11. Consoli A, Costanzo V, Evola G, Marletta L (2017) Refurbishing an existing apartment block in Mediterranean climate: towards the Passivhaus standard. Energy Procedia 111:397–406
12. Bruno R, Arcuri N, Carpino C (2015) The passive house in Mediterranean area: parametric analysis and dynamic simulation of the thermal behaviour of an innovative prototype. Energy Procedia 82:533–539
13. Fokaides PA, Christoforou E, Ilic M, Papadopoulos A (2016) Performance of a passive house under subtropical climatic conditions. Energ Buildings 133:14–31
14. Ford B, Schiano-Phan R, Zhongcheng D, Passive-On Project (2007) The Passivhaus standard in European warm climates: part 3 design guidelines for comfortable low energy homes. School of the Built Environment, University of Nottingham, Nottingham
15. ANSI/ASHRAE Standard 55–2013 (2013) Thermal environmental conditions for human occupacy, Atlanta
16. Piraccini S, Fabbri K (2018) Building a passive house, The Architect's logbook. Springer, Berlin

Chapter 56
How Can Reflected Light Modify Solar Gains in a Compact Urban District?

Benoit Beckers, Jairo Acuña Paz y Miño, and Claire Lawrence

56.1 Introduction

In this paper two different possibilities of urban retrofitting in a compact urban district based on architectural decisions have been explored. A 3D model able to reproduce the desired geometric and semantic level of detail for a full balance thermal simulation was built. The reflectance of materials was measured on-site by photographic techniques in order to nourish the 3D model. Simulations were performed for this scene in an urban context by ray tracing to calculate the irradiance at representative points over a year.

56.2 Construction of an Urban Model for Thermal Studies

On October 19, 2017, a three-dimensional information acquisition campaign was carried out in the Petit Bayonne. For this campaign, the methods of land mobile laser and aerial photogrammetry were combined. The result was two clouds of points. The first contains all the information available from the pedestrian's point of view. The second contains the information corresponding to the covers and the horizontal surfaces visible from a bird's eye. Both point clouds were processed and combined into a single cloud of 200 million (186 550 580) points with RGB color information, georeferenced with an accuracy of 3 cm (Fig. 56.1a).

The complexity of the urban scene, the amount of elements that overlap, and their different material characteristics make automatic reconstruction impossible at

B. Beckers (✉) · J. A. Paz y Miño · C. Lawrence
Urban Physics Joint Laboratory, Université de Pau et des Pays de l'Adour, Anglet, France
e-mail: benoit.beckers@univ-pau.fr

A. Sayigh (ed.), *Renewable Energy and Sustainable Buildings*, Innovative Renewable Energy, https://doi.org/10.1007/978-3-030-18488-9_56

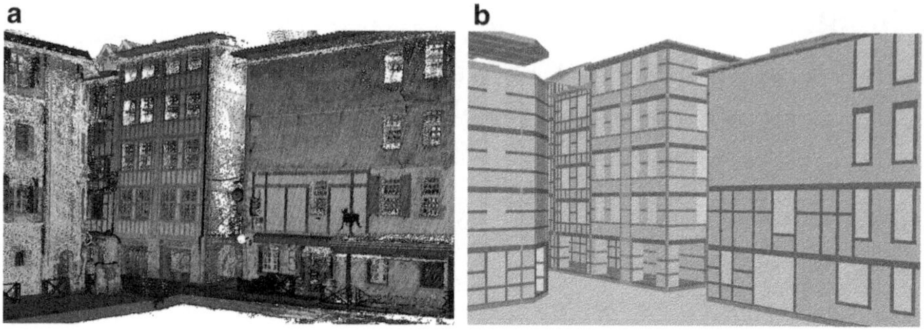

Fig. 56.1 (a) Point cloud. (b) 3D model

Fig. 56.2 Reflectance coefficient map

the moment [1–3]. Taking this information as a reference, a 3D model was built (Fig. 56.1b) with the objective of performing a complete thermal calculation using a finite element method [4].

When solving the complete heat balance equation, two types of phenomena must be distinguished: those that act on the volume and those that act on the surface of each element [5]. The 3D model must be equally flexible.

Sun gain is the shortwave radiation provided by the sun on a unit of surface [6], which affects a building entering through openings such as a window or indirectly through the envelope. Therefore, the appropriate 3D model to simulate these sun gains needs to represent only the surfaces exposed to these contributions. However, the surfaces must accurately represent the reflectance of their materials. To carry out simulations in the urban environment, the common thing is to use a representative reflectance value for all the facades; however, its impact in the calculation of solar contributions is important. In an urban scene, there is a great variety of reflection coefficients (Fig. 56.2). Simplifying them is an incorrect hypothesis [7].

56.3 Reflectance at the Urban Scene

When the radiant energy hits a body, it is absorbed (α), reflected (ρ), and transmitted (τ). In an urban scene, most solids are opaque to radiation. Except for glasses, some plastic, and mineral materials, it is possible to consider that the transmissivity is null.

$$1 = \alpha + \rho + \tau \quad (56.1)$$

$$\tau = 0 \Rightarrow \alpha + \rho = 1 \quad (56.2)$$

For this study, the specular component of the reflection is not taken into account. All surfaces are considered as perfectly diffuse reflectors, except for the windows, where a bidirectional scattering distribution function *BSDF* model was used in order to take into account the angular dependence of the glass.

$$L = \frac{\rho}{\pi} E \quad (56.3)$$

In order to demonstrate the influence of reflectance for this case study, a protocol has been carried out to estimate the reflectance of the surfaces by photographic means (Fig. 56.3).The procedure is divided into three parts. The first is the realization of the captures in RAW format. The second is the post-processing of the captured images and finally the calculation of the reflection coefficients of the surfaces that interest us on the image. Other studies have determined that using a spectrometer [8] may be the most accurate method to find this information; however, doing it at the urban scale is difficult since to make these measurements it is necessary to be centimeters from the surface.

Fig. 56.3 Surfaces and gray card with known reflectance

The reflectance is calculated on a grayscale image using the difference between the luminance observed in the photograph and comparing it with the luminance of a surface of known reflectance, considering that both are under the same lighting conditions. The luminance at a point in the scene was estimated by relating the luminance of this point with the L value in CIELab[1] color space of the pixel that represents it in the image. An error is possible since the color of each surface does not give us any information about the texture of the material, and the specular component is not considered. To minimize the margin of error, the photographs were made under a heavily overcast sky to maintain a homogeneous illumination.

$$E_1 = E_2$$

$$\frac{\pi L_1}{\rho_1} = \frac{\pi L_2}{\rho_2} \tag{56.4}$$

Urban scenes composed of perfectly homogeneous facades do not exist, and simulations with such facades do not represent the influence that each material has on the scene [9]. Facades include a variety of elements with different materials and characteristics that have an impact on the radiosity of the facade. The inclusion of windows in a case study affects the radiance of the façade, decreasing its emittance due to the fact that part of the radiation enters the building.

The reflectivity of the street was also measured. The contribution of the street in the illuminance of the facades is important. Most of the studies consider a reflection factor of 20% for the pavement [9], others have considered 80% [10], while the average value obtained for the Rue des Tonneliers is 41%.

56.4 Simulation

There are two main methods to simulate solar radiation: radiosity and ray-tracing. Both have different philosophies. The first needs to discretize the environment into several surface elements and quantify the exchanges that occur between each element (form factor) and store them in a matrix. The whole problem will then be solved by a simple matrix inversion. The other main method, ray-tracing, consists in emitting and tracking rays through the environment rays in order to link sources (sky and Sun) with the aimed surfaces (work plane). RADIANCE, the software used in this study, belongs to the latter category. One of the advantages of ray-tracing is the ability to take into account a multitude of materials and their appropriate reflection pattern. In this study, different types of materials are present: masonry filling walls, timbers, ground, and roofs are represented by perfectly diffuse materials with their own reflectance. The windows are represented by a BSDF model of a

[1] CIELab: This color space represents human perception in two parameters L = luminous intensity, ab = chrominance

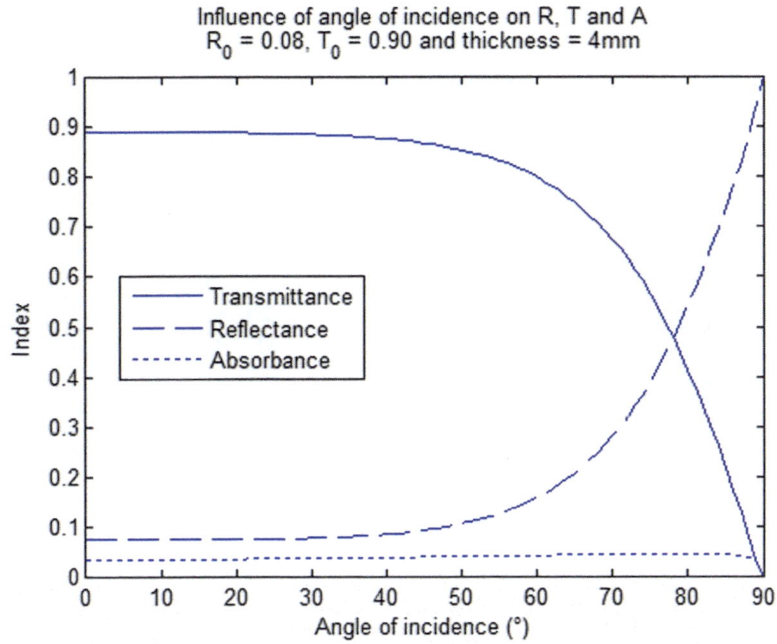

Fig. 56.4 Influence of angle of incidence over glazing surface

simple glazing from the WINDOW software database [11]. This BSDF model allows to simulate both reflectance (BRDF) and transmittance (BTDF) by taking into account a singular property of glazing which is the Fresnel phenomenon. Reflectance and transmittance of glass materials depend on the incidence angle (Fig. 56.4). Given the quantity of glazing in the street, this consideration of the reflected part of the windows is key. With the daylight coefficients method (also called two-phase method), the irradiances during the year were calculated according to the Bordeaux weather data (.epw file). All these considerations are taken into account since the ratio between the energy reflected within the urban scene and the incident energy decreases with the increase in the glazing rate. This decrease varies according to the height of the Sun.

56.5 Results

The following figures were carried out using the RADIANCE software. In Fig. 56.5, the direct irradiance was accumulated over a year. The north facade receives less direct energy, especially in the lower part.

Isolating a building on the north façade (Fig. 56.6) shows that the reflection represents about 50% of the energy received in the less favored part. Therefore, reflection coefficients are important and simplifying them is an incorrect hypothesis.

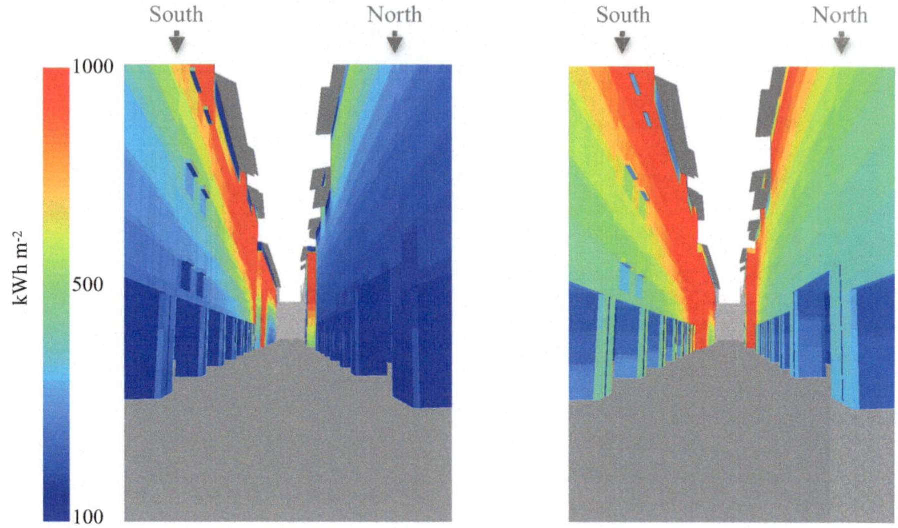

Fig. 56.5 Cumulative annual irradiance at the Rue des Tonneliers (left $\rho = 0$; right $\rho = 0.8$). In both cases the reflectance values are the same for the whole scene

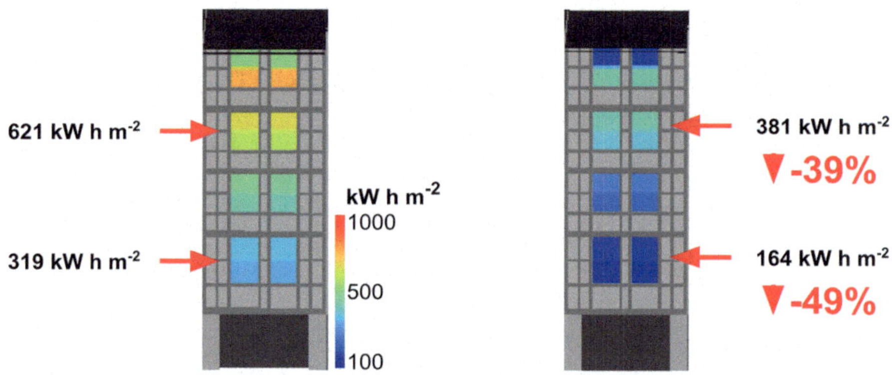

Fig. 56.6 Influence of reflection on the cumulative annual irradiance transmitted through the windows of one building on the north facade

The base case was made taking into account the reflection coefficients measured in situ including that of the street surface. From the base case, two possibilities to modify an urban district were evaluated, first by increasing the reflectance coefficient of the whole scene, and second by doubling the window area of the scene. Painting the white facades increases the reflection coefficient of the surfaces ($\rho = 0.80$). The exchange by interreflections increases; therefore, it is possible to get higher solar gains (Fig. 56.7). As can be seen in the Fig. 56.5, a very high reflection coefficient in a compact urban district can make uniform the distribution of irradiances

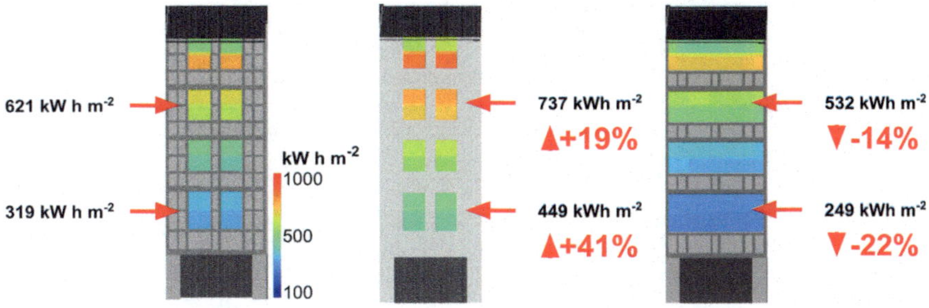

	Actual state	High Reflectivity (ρ= 0.80)	Greater WWR (by doubling the glazing surface)
Higher Story	621 kWh m-2	737 kWh m-2	532 kWh m-2
Lower Story	319 kWh m-2	449 kWh m-2	249 kWh m-2

Fig. 56.7 Comparison of the cumulative annual irradiance transmitted through the windows of one building on the north facade

on the scene. In this case, the idea of the south façade as an advantage is relativized since each of the facades depends on the other. On the other side, enlarging the windows to obtain a higher income from solar gains seems logical; however, in Fig. 56.7, it can be seen that doubling the window area (increasing the WWR) penalizes the whole scene but with greater force the less favored areas, decreasing the contributions received in more than 20%.

Increasing the window area with respect to opaque surfaces for the purpose of increasing solar inputs can be misleading in a compact urban environment. A greater number of openings on the facades mean less reflection surface. The energy that enters the buildings enters more in the upper floors and decreases considerably in the first floors.

56.6 Conclusion

As presented on the results (Fig. 56.7), the reflection represents about 50% of the energy received in the lower part of the facades. Reflection coefficients are important, and simplifying them is an incorrect hypothesis since urban scenes composed of perfectly homogeneous facades do not exist. It does not represent the influence that each material has on the scene. Therefore, results calculated by simplifying reflectance will be misleading, especially on complex urban scenes.

We have seen that a single value is rarely representative of a facade, either because of the differences in the surface between the ground floor and the upper floors or because of the presence of glass (windows, showcases). Multiple reflection factors on all facades should be used for precise calculation.

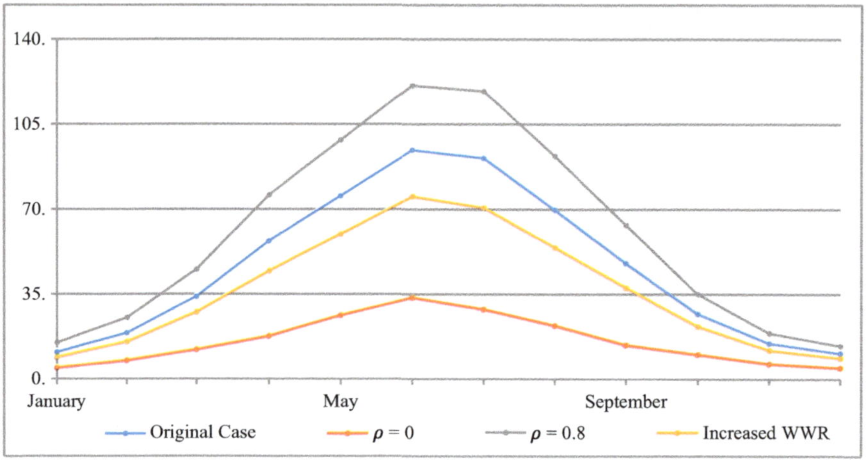

Fig. 56.8 Average annual irradiance

In a compact urban district, the increase of the reflectance coefficient implies also an increase of the irradiance on the facades, due to multiple reflections. A very high reflectance can make uniform the distribution of irradiance over the scene and also help keep lower temperatures [12] during the summer period (Fig. 56.8).

A bigger WWR on a compact district can severely penalize the irradiance distribution over the facades. A greater number of openings on the facades mean less reflection surface. The energy that enters the buildings is more important in the upper floors and decreases considerably in the first floors. This could be misleading, as larger window surfaces mean also bigger losses by the envelope.

References

1. Lafarge F, Mallet C (2012) Creating large-scale city models from 3D-point clouds: a robust approach with hybrid representation. Int J Comput Vis 99(1):69–85
2. Frueh C, Jain S, Zakhor A (2005) Data processing algorithms for generating textured 3D building facade meshes from laser scans and camera images. Int J Comput Vis 61:159–184
3. Musialski P, Wonka P, Aliaga DG, Wimmer M, Van Gool L, Purgathofer W (2013) A survey of urban reconstruction. Comput Graph Forum 32(6):146–177
4. Acuña J, Lefort V, Lawrence C, Beckers B (2018) Maquette Numérique d'une rue du vieux Bayonne pour son étude thermique par éléments finis. In: EduBIM2018: Journées de l'Enseignement de la Maquette Numérique et du BIM en France—4ème édition, 15–16 mai 2018. Clermont-Ferrand, France
5. Beckers B (2012) Worldwide aspects of solar radiation impact. In: Solar energy at urban scale, chapter 5. John Wiley and Sons, Inc, Hoboken, NJ, pp 99–118
6. Beckers B, Beckers P (2014) Chapter 4: Reconciliation of geometry and perception in radiation physics. John Wiley and Sons, Inc, Hoboken, NJ, p 192
7. Beckers B, Aguerre J, Besuievsky G, Fernández E, Garcia-Nevado E, La Borderie C, Nahon R (2018) Visualizing the infrared response of an urban canyon throughout a sunny day. In: Sustainable building for a cleaner environment. Springer, Cham, pp 277–284

8. Patriarche M (2014) Mesure et modélisation de la lumière naturelle dans les canyons urbains. PhD thesis, ENTPE, Vaulx-en-Velin (France), September 9, 2014
9. Brotas L, Wilson M (2006) Daylight in urban canyons: planning in Europe. PLEA, Geneva
10. Bozonnet E, Belarbi R, Allard F (2005) Modelling solar effects on the heat and mass transfer in a street canyon, a simplified approach. Sol Energy 79(1):10–24
11. Klems JH (1994) A new method for predicting the solar heat gain of complex fenestration systems II. Detailed description of the matrix layer calculation. ASHRAE Transactions V.100, Pt. 1
12. Doulos L, Santamouris M, Livada I (2004) Passive cooling of outdoor urban spaces. The role of materials. Sol Energy 77:231–249

Chapter 57
Study of Corrosion Effect of Micronal® Phase Change Materials (PCM) with Different Metal Samples

Rebeca Salgado, Hoda Akbari, Maria C. Brown, Isobelle Reid, and Sarah J. McCormack

57.1 Introduction

Solar energy is one of the most assuring alternatives to the conventional fossil resources. Nevertheless, this resource is intermittent and actually the efficiency of the solar cells is lower; thus was borne the need to incorporate devices, on the solar cells, to improve their efficiencies and to store thermal energy during the sun hours. One of the most useful devices that cover these requirements is the phase change materials (PCMs); these open the opportunity to maintain the solar cell in an efficient temperature and at the same time to store thermal energy [3–5].

For this application is used low-temperature PCM, where the melting point (Tm) range is between 25 °C and 40 °C. For this reason, the most common PCMs are fatty acids and hydrated salts, where the phase change is solid to liquid. This introduces the need to encapsulate the PCM and open the question of what is the most useful material to do the encapsulation. G. Raam et al. mention the container has to be lightweight, of low cost, with high thermal conductivity and non-corrosive in their study of identification and development of new phase change materials for solar thermal energy storage [6].

Actually, there are many studies on the compatibility of inorganic PCMs with different materials as A.J. Farrell et al. studied the corrosion effects of different salt hydrates on aluminium and copper [7]. M. Liu et al. studied the long-term stability of eutectic $NaCl\text{-}Na_2CO_3$ and corrosion behaviour in contact with SS316. They showed that thermos-physical properties do not change in 1000 cycles and the corrosion rate grows with the increase in cycles until stabilized at 50 mg/cm^2 [8].

R. Salgado (✉) · H. Akbari · M. C. Brown · I. Reid · S. J. McCormack
Department of Civil, Structural and Environmental Engineering, Trinity College Dublin, University of Dublin, Dublin 2, Ireland
e-mail: salgador@tcd.ie

A. Sayigh (ed.), *Renewable Energy and Sustainable Buildings*, Innovative Renewable Energy, https://doi.org/10.1007/978-3-030-18488-9_57

Moreover, Cabeza L.F et al. studied the corrosion effect of zinc nitrate hexahydrate, sodium hydrogen phosphate dodecahydrate and calcium chloride hexahydrate on aluminium, copper, brass, stainless steel and steel in short, middle and long term, incorporating the pH measure of the PCM. The process involved immersing the metal sample in melted PCM in a test tube which remained in a water bath for the duration of the experiment to ensure the PCM remained in a liquid state. They saw that the zinc nitrate hexahydrate emerged as a very corrosive material as aluminium and steel were highly corroded. Calcium chloride hexahydrate produced a pitting effect in aluminium and stainless steel [9–11].

In a lesser portion there are studies on the corrosion effect of fatty acids as Sari A. et al.'s study on the compatibility of palmitic, myristic, lauric and stearic acid with stainless steel, steel, aluminium and copper and how they change the latent heat fusion of the PCMs [12]. G. Ferrer et al. presented the study of four different PCMs, two eutectic fatty acids and two salts hydrated, in contact with five different metals, stainless steel SS316 and SS304, carbon steel, copper and aluminium. The study was done for 84 days and every week measures were taken. The study concludes that SS316, SS304 and copper are recommended as a container material for the hydrated salts tested, and in the case of fatty acids non-one metal was considered corroded despite copper showing a little sight of corrosion [13].

The aim of this project is to study the corrosion resistance of aluminium, stainless steel, steel, copper and brass immersed in Micronal PCM in a period of 1073 days.

57.2 Materials Experiment

The most common materials used to encapsulate the PCM in TES systems were tested in this experiment. The materials tested were copper, brass, aluminium, stainless steel and mild steel. The samples were cut from large pieces of materials, and each had dimensions of 50 mm × 14 mm and 2 mm of thickness. The chemical compositions of the materials used are the same as on the study of Maria C. Browne et al. [14].

The PCM used was the Micronal® that is a commercial PCM classified as an organic paraffin, with a phase range from 27.5 °C to 32.7 °C and a heat fusion of 167 kj/kg.

57.3 Procedure

The methodology used in this study was based on the Standard Practice for Preparing, Cleaning, and Evaluating Corrosion Test Specimens (ASTM G1-01) [15] and has been applied to corrosion investigations of PCM and metals.

Table 57.1 Guide for corrosion weight loss used in the industry

mg/cm²yr	mm/yr	Recommendation
>1000	2	Completely destroyed within days
100–999	0.1–1.99	Not recommended for service greater than one month
50–99	0.1–0.19	Not recommended for service greater than one year
10–49	0.02–0.09	Caution recommended, depending on the specific application
0.3–9.9	–	Recommended for long-term service
<0.2	–	Recommended for long-term service; no corrosion, other than as a result of surface cleaning as evidenced

All the metal samples were weighed in an Ohaus Explorer balance (four decimals) before starting the corrosion test. Once the specimens had been weighed, they were immersed in glass test tubes containing PCM to combine each metal specimen with the five different PCM formulations. All tubes were sealed to avoid contact with environmental agents, and as the phase change temperature of all the PCMs was around 20–30 °C, they were kept in a water bath at 40 °C, ensuring all PCMs were always at liquid phase.

The methodology implies that we measure the weight every week for 1073 days (almost 3 years); to do it we need to clean the samples with acetone to remove the excess PCMs on the surface. In addition to this the samples were visually evaluated for bubbles, precipitates, surface changes and pitting.

The mass loss (Δm) [g] and the corrosion rate (CR) [mg/cm²year] were calculated to evaluate the behaviour of the samples in each PCM. To calculate it we follow the ASTM G1-01; for the mass loss we used Eq. 57.1 and for the corrosion rate Eq. 57.2:

$$\Delta m = m_{t_0} - m_t \quad (57.1)$$

$$CR = \frac{\Delta m}{A \cdot (t_0 - t)} \quad (57.2)$$

To fulfil the corrosion test, we carried out an analysis of the corrosion surface of the samples using the secondary electron microscope (SEM), where we determined the morphology of the damage on the samples.

The evaluation and recommendations were made based on the results of the corrosion tests and suggested which materials are most compatible as a containment. The corrosion weight loss guide used in industry practices was used as a reference, as shown in Table 57.1.

57.4 Results and Discussion

On the following section, the obtained results are analysed, which were described in the previous chapter. The results are also discussed to extract the conclusion.

57.4.1 Corrosion Test

The following graphs Figs. 57.1 and 57.2 show the dependence of the mass loss and the corrosion rate with time.

In the mass loss graph, Fig. 57.1, it is detected that the most affected metal sample was the mild steel which reached a final mass loss of 0.5 g. Copper and brass have similar results in front of the time with a final mass loss of 0.08 g. However, the metals that have the lower values were the aluminium and the stainless steel, where the aluminium reached a value of 0.03 g and stainless steel 0.01 g.

On the corrosion rate graph, Fig.57.2, the mild steel has the worst results. In the beginning it reaches a maximum value of 28 mg/cm^{2}year; between 200 and 800 days the value is stabilized at 15 mg/cm^{2}year and after 800 days at 12 mg/cm^{2}year. Copper and brass have similar results with an average corrosion rate of 1.8 mg/cm^{2}year. The values of aluminium and stainless steel were negligible due to low results.

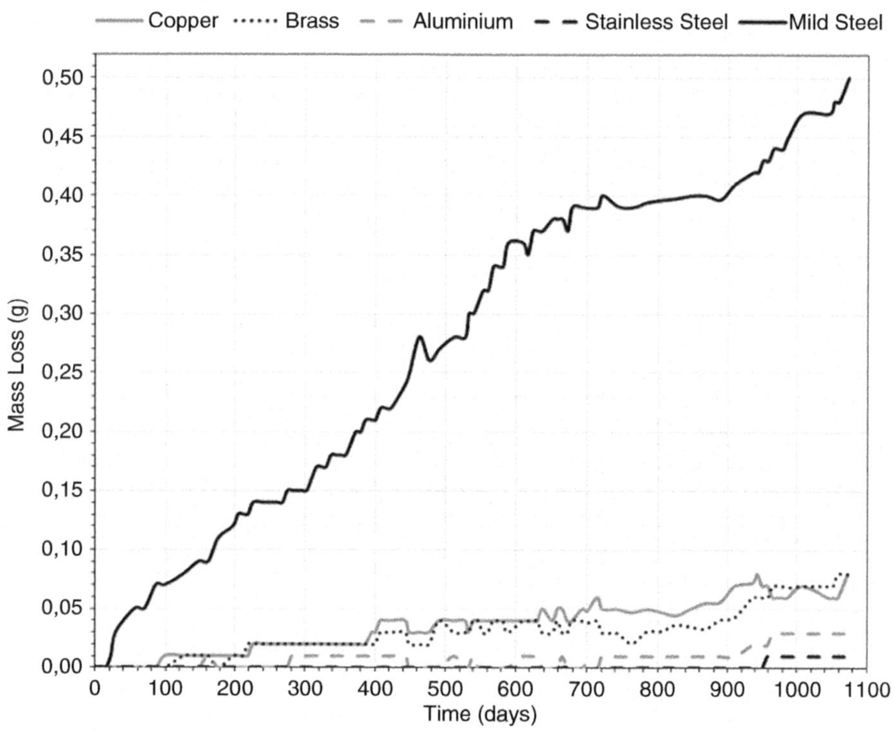

Fig. 57.1 Mass loss due to the immersion in Micronal®

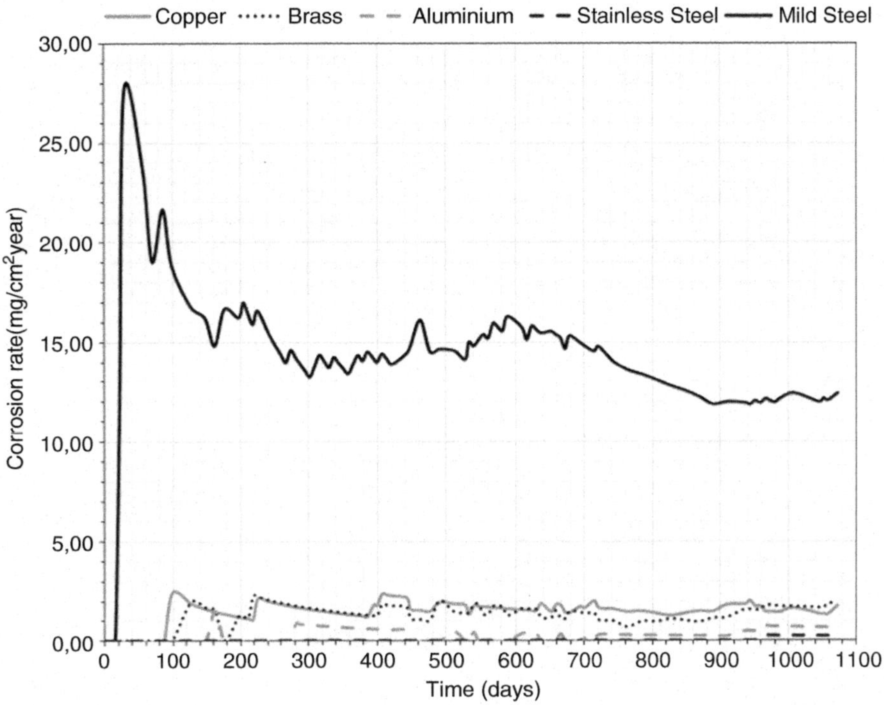

Fig. 57.2 Corrosion rate during immersion tests of material samples with Micronal®

57.4.2 Surface Study

The metal samples were investigated under visual observation and microscope for signs of corrosion on days 1, 540 and 1073. However, stainless steel has not been affected by the Micronal® and as a result there will be no presentation of images of it.

As shown in Fig. 57.3 copper was affected by the Micronal® PCM during the time; with visual observation, the metal samples gradually darken more with time. This means that a corrosion effect is happening in the sample and is related with the microscope observation. Pitting corrosion can be seen with the removal of the vertical indents, as seen on day 0 and day 540, which reflected the mass loss of 0.05 g at this stage of the test considering the initial mass was 10.71 g. The full surface layer on the copper has been corroded and the pits increase their size by day 1073 along with a mass loss of 0.08 g.

Likewise, the brass sample darkens with time as seen in the visual observation (Fig. 57.4), which had an initial mass of 10.25 g and was corroded in the same behaviour as copper. However, at day 540, with a mass loss of 0.04 g, its vertical

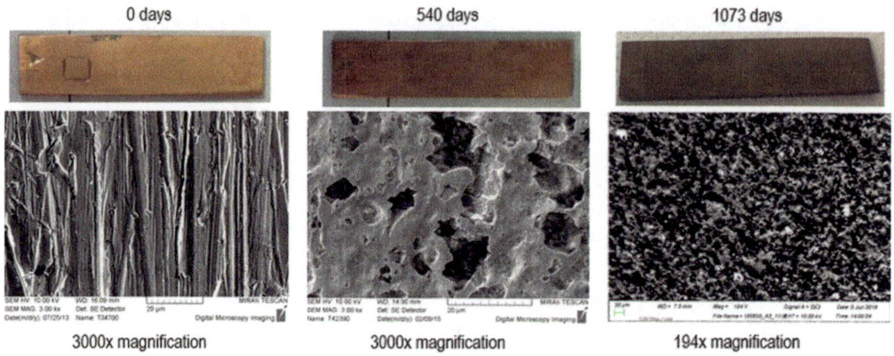

Fig. 57.3 Copper immersed in Micronal®

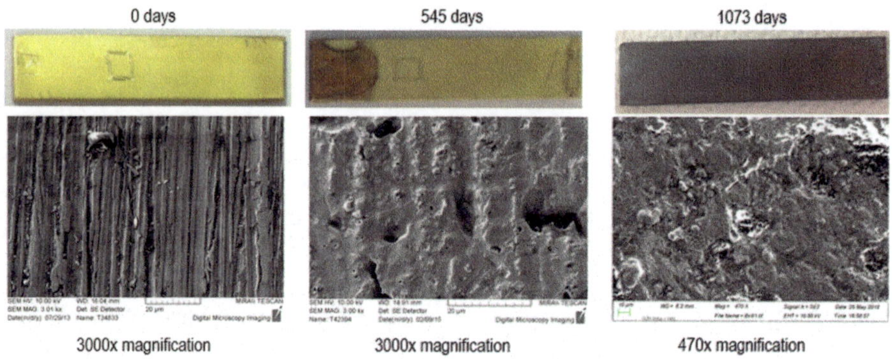

Fig. 57.4 Brass immersed in Micronal®

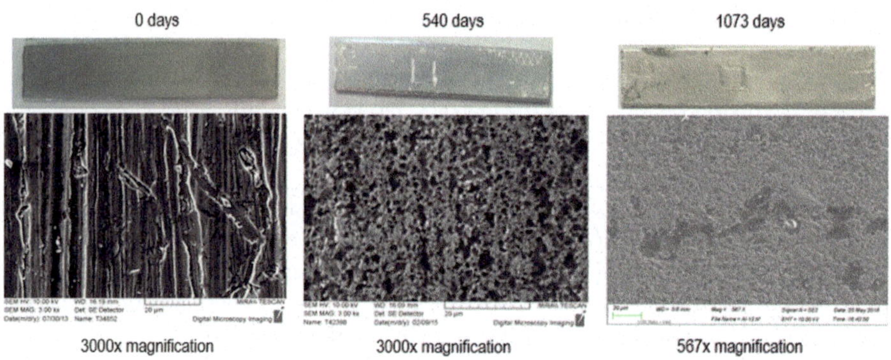

Fig. 57.5 Aluminium immersed in Micronal®

indents were removed (Fig. 57.4), which means that the corrosion occurs at a slower rate, and at day 1073 the full sample was corroded.

Aluminium did not present any change during visual observation (Fig. 57.5). However, pitting was clearly seen in the images taken on day 540, which reflected the

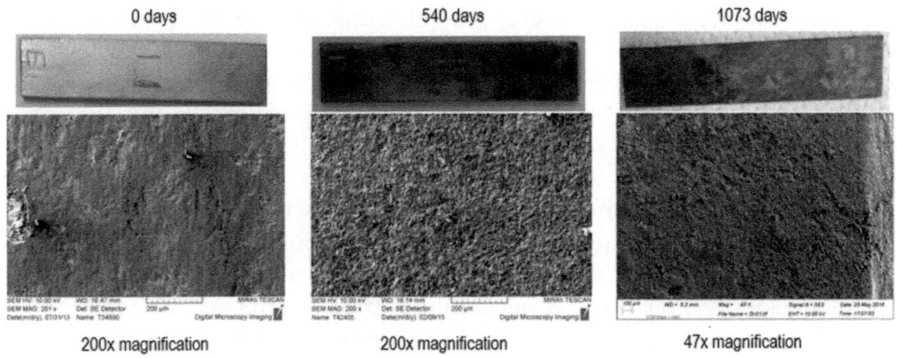

Fig. 57.6 Mild steel immersed in Micronal®

Table 57.2 Corrosion results obtained by gravimetric analysis after 1073 days and recommendation

Material	Mass loss (g)	Corrosion rate (g/cm^{2}year)	Recommendation
Copper	0.08	1.805	Recommended with caution
Brass	0.08	1.844	Recommended with caution
Aluminium	0.03	0.679	Recommended
Stainless steel	0.01	0.226	Recommended
Mild steel	0.50	12.415	Caution

mass loss of 0.01 g at this stage of the test considering the initial mass was 3.23 g had an initial mass loss of 3.23 g, but at day 1073 the corrosion got progressively worse as the mass loss was 0.03 g.

Mild steel was adversely affected by the Micronal® getting the worst result on the gravimetric analysis. The images shown in Fig. 57.6 reflect this. On visual observation, the sample darkened with time, and with microscope observation, we were able to see the corrosion surface at low magnification at day 540 which is related with a mass loss of 0.3 g on this stage of the test considering the initial mass was 9.76 g. Corrosion became more aggressive on day 1073, where at lower magnification we were able to see the corrosive surface and the mass loss was 0.5 g.

57.5 Conclusions

The final results of the gravimetric analysis, mass loss and corrosion rate of each material immersed in Micronal® after 1073 days are summarized in Table 57.2 including an evaluation of the suitability of the tested metal having in count the guide for corrosion weight loss used in the industry (Table 57.1) and the metallographic investigation.

Without any doubt, stainless steel is an ideal candidate for the encapsulation of Micronal® which was shown to have the lowest results of mass loss and corrosion rate, and no corrosion effects were detected with the SEM observation. However, aluminium got the second-best results with a final mass loss of 0.01 and a corrosion rate of 0.679; it is also recommended despite the slight corrosion detected by the SEM.

Copper and brass were shown to have a similar behaviour in the gravimetric analysis as well as in the microscope observation. In relation to the guide for corrosion weight loss used in the industry, these materials are recommended for long-term services, but due to detection of moderate corrosion behaviour on the surface of the metal, it is recommended that copper and brass are to be used with caution.

Finally, mild steel was shown to have the highest result of mass loss and corrosion rate, and an aggressive corrosion behaviour was detected by the SEM. Thus caution is recommended for the encapsulation of Micronal® application.

References

1. Cabeza LF, Castellón C, Nogués M, Medrano M, Leppers R, Zubillaga O (2007) Use of microencapsulated PCM in concrete walls for energy savings. Energ Buildings 39(2):113–119
2. Micronal PCM BASF, Intelligent temperature management for buildings. 2012
3. Browne MC, Quigley D, Hard HR, Gilligan S, Ribeiro NCC, Almeida N, McCormack SJ (2016) Assessing the thermal performance of phase change material in a photovoltaic/thermal system. Energy Procedia 91:113–121
4. Browne MC, Norton B, McCormack SJ (2015) Phase change materials for photovoltaic thermal management. Renew Sust Energ Rev 47:762–782
5. Browne MC, Lawlor K, Kelly A, Norton B, Cormack SJM (2015) Indoor characterisation of a photovoltaic/thermal phase change material system. Energy Procedia 70:163–171
6. Dheep GR, Sreekumar A (2018) Investigation on thermal reliability and corrosion characteristics of glutaric acid as an organic phase change material for solar thermal energy storage applications. Appl Therm Eng 129:1189–1196
7. Farrell AJ, Norton B, Kennedy DM (2006) Corrosive effects of salt hydrate phase change materials used with aluminium and copper. J Mater Process Technol 175(1–3):198–205
8. Liu M, Bell S, Segarra M, Steven Tay NH, Will G, Saman W, Bruno F (Oct. 2017) A eutectic salt high temperature phase change material: thermal stability and corrosion of SS316 with respect to thermal cycling. Sol Energy Mater Sol Cells 170:1–7
9. Cabeza LF, Roca J, Âs MN, Mehling H, Hiebler S (2002) Immersion corrosion tests on metal-salt hydrate pairs used for latent heat storage in the 48 to 58°C temperature range. Mater Corros 53:902–907
10. Cabeza LF, Illa J, Roca J, Badia F, Mehling H, Hiebler S, Ziegler F (2001) Middle term immersion corrosion tests on metal-salt hydrate pairs used for latent heat storage in the 32 to 36 °C temperature range. Mater Corros 52:748–754
11. Cabeza LF, Roca J, Nogueés M, Mehling H, Hiebler S (2005) Long term immersion corrosion tests on metal-PCM pairs used for latent heat storage in the 24 to 29°C temperature range. Mater Corros 56(1):33–38
12. Sari A, Kaygusuz K (2003) Some fatty acid used for latent heat storage: thermal stability and corrosion of metals with respect to thermal cycling. Renew Energy 28:939–948

13. Ferrer G, Sol A, Barreneche C, Martorell I, Cabeza LF (2015) Corrosion of metal containers for use in PCM energy storage. Renew Energy 76:465–469
14. Browne MC, Boyd E, Mccormack SJ (2017) Investigation of the corrosive properties of phase change materials in contact with metals and plastic. Renew Energy 108:555–568
15. ASTM Standard G1-01 (2001) Standard practice for preparing, cleaning, and evaluation corrosion test specimens. ASTM B Stand 1:1–8

Chapter 58
Metal Coordination Complexes: A Bottom-Up Approach Tailored Towards Solar Energy Applications

James Walshe, Pauraic Mc Carron, Hind Ahmed, Sarah Mc Cormack, and John Doran

58.1 Introduction

As photovoltaic (PV) technologies begin to play a more decisive role in the electrical power network, contributing a larger proportion of a country's energy demands, the inherent loss mechanisms within the technology and the need to address them become more prevalent [1]. A major limiting factor, effecting the market penetration and cost-efficiency of all commercial and emerging PV technologies, is the limited bandwidth of frequencies over which the conversion of light directly into electricity can occur [1, 2]. The PV materials characteristics dominate the persistence of sharp cut-off points in the devices spectral responsivity, particularly towards the infrared and ultraviolet regions of the spectrum, irrespective of the type of PV technology upon which the device is based.

Spectral converters, a passive design approach in which a photoluminescent layer is retrofitted directly onto a solar cell, showcase a huge potential to curtail and overcome these spectral losses. The photoluminescent material absorbs photons from a region of the spectrum inefficiently utilized by the underlying cell and re-emits them at wavelengths more favourable to the underyling PV technologies spectral responsivity. Spectral converters are divided into two main categories

J. Walshe (✉)
Dublin Energy Lab, Dublin Institute of Technology, Dublin, Ireland
e-mail: James.walshe@mydit.ie

P. Mc Carron
The Centre for Biomimetic & Therapeutic Research, FOCAS Research Institute, Dublin Institute of Technology, Dublin, Ireland

H. Ahmed · S. Mc Cormack
School of Engineering, Trinity College Dublin, Dublin, Ireland

J. Doran
School of Physics and Clinical & Optometric Sciences, Dublin Institute of Technology, Dublin, Ireland

A. Sayigh (ed.), *Renewable Energy and Sustainable Buildings*, Innovative Renewable Energy, https://doi.org/10.1007/978-3-030-18488-9_58

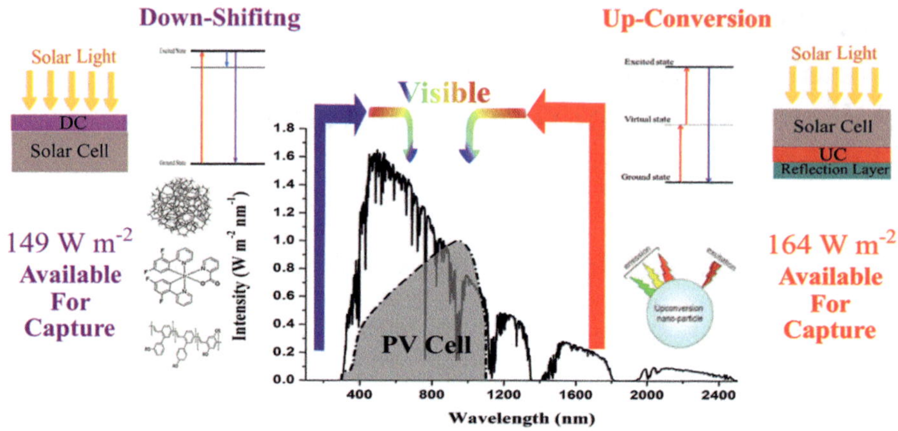

Fig. 58.1 Schematic overview of the spectral conversion approach for photovoltaic technologies. Indicating the solar irradiance available for capture (black line) as well as the two main types of spectral converters available (downshifting—violet, and up converters—red) and their corresponding electronic band structure. The total amount of irradiance available for capture and conversion with each type of converter is also presented. Please note the Wm^{-2} values presented are reflective of the those available for mc-Si technology, and are not reflective of the energy available for capture by alternative PV technologies, whose spectral responsivity profile will be different

(see Fig. 58.1), based on the region of the spectrum in which the photoluminescent material absorbs (1) up converters (UC)—whose absorption is confined within the infrared region—and (2) down converters or luminescent downshifters (LDS), which absorb primarily in the ultraviolet region of the spectrum [2].

A lack of control over the synthesis and optical properties of up-converting materials has led to a more focused approach on improving the library of material candidates for the LDS application [1–3]. Previously limited through the sheer lack of material candidates available, coupled with 'poor' optical properties exhibited upon their inclusion in a transparent host material, the LDS application has undergone a renaissance within the last decade due to advancements in synthetic chemistry and nanomaterials [2, 4]. Incorporating nanomaterials whose optical properties can be tuned through functionalization via one of the many fabrication routes available pushed the enhancements offered by the design approach. The nanomaterial, exposes the embedded luminescent material, to fluctuations in the highly localized electromagentic fields established on the nanostructures surface through the surface plasmon resonance (SPR) phenomena: augmenting the photoluminescent properties of the luminescent species in the process [4, 5]. However, with each new addition to the overall system comes additional potential loss pathways, which need to be overcome for the fitted structure to have a positive impact on the energy performance of the underlying device [1–3].

The primary loss mechanisms introduced due to the addition of an LDS layer directly onto a PV system are the 'poor' optical properties and stability of the host material when exposed to the in situ environmental conditions and the nonideal nature of luminescent materials [1, 2]. This doesn't even begin to consider the lack of control or irreproducibility granted over the optical properties of the nanomaterials utilized, as well as their tendency to aggregate upon exposure to fluctuations in the surrounding environmental conditions (thermal, electrical, etc.).

Hence the focus is once again brought back to the design, characterization and identification of the best performing luminescent species and corresponding host material combinations. Metal coordination complexes functionalized to absorb across a wide range of frequencies and whose emission profile is tailored towards the specific responsivity of the PV technology offer a cost-effective approach to drive down the costs of LDS devices. Paving the way for upscaling of this spectral conversion approach, which has to a large extent been limited in application to small form factor cells [1, 2]. This study explores the evaluation of a selection of 22 fluorescent heteroaromatic ligands, through characterization of their absorption and emission properties, for the formation of metal coordination complexes based around a metal (Eu, Tb, etc.) centre.

58.2 Methods and Materials

A family of 22 different phenanthroline-based heteroaromatic ligands were studied and evaluated for their possible inclusion in the formation of a metal coordination complex specifically designed for Si-based PV technologies. Solutions containing each compound at a concentration of 1.25×10^{-6}M were prepared in DMSO, and their absorption and emission spectra were determined using a Perkin Elmer Lambda 900 UV/VIS/NIR and LS 55 spectrometer. Evaluating each compound under the same concentration and measurement parameters (slit widths, PMT acceleration voltage, etc.) enabled an intercomparison to be performed, between the absorbing and emission capabilities of each compound, with each material evaluated under the following criteria:

- Absorption bandwidth and inclusion of SPR λ of nanomaterial candidates
- Absorption capacity
- Emission bandwidth
- Emission capacity
- Degree of spectral overlap between absorption and emission bands
- Functionalization possibilities
- Novelty

With the absorption and emission capacity, quantified through integrated spectroscopy. From these 22 different ligands, the top 8 candidates, drafted under the criteria established, were further investigated in terms of their luminescent quantum yield (LQY) in solution. The LQY provided a quantitative comparison of the photoluminescent efficiency of the top eight ligands identified during the initial trails conducted. The comparative method [6] was adopted utilizing quinine sulphate (quinine hemisulphate salt monohydrate—bioreagent grade—Sigma-Aldrich) as a fluorescent standard, due to its well-established absorption range and photostability in 0.1 M H_2SO_4 solution [6, 7]. All measurements were repeated independently in triplicate, with the following strict confidence criteria established throughout the measurement:

- Absorption at $\lambda_E < 0.1$ A.U. using 10 mm path length cuvettes
- 95% or greater confidence interval on all fit parameters used
- 10% tolerance level on standard deviation across the trio of measurements

58.3 Results and Analysis

The absorption spectra across the range of ligands explored (Fig. 58.2a) exhibited a confinement to the ultraviolet part of the spectrum, with some of the ligands extending the absorption into the visible region (Fig. 58.2b): specifically, PMCC 166, PMCC 282 and PMCC 183. An overview of the absorption capacity as determined through integration spectroscopy is presented in Fig. 58.2b, showing the top performers in terms of the amount of light they are absorbing. A key criterion for the

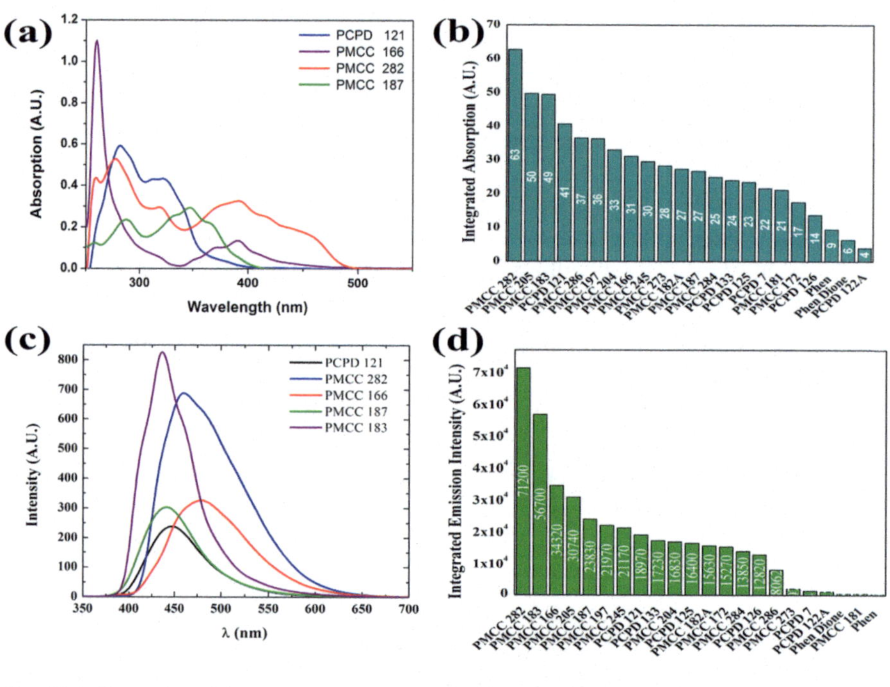

Fig. 58.2 Examples of some of the phenanthroline-based heteroaromatic ligand absorption (**a**) and emission spectra (**c**) at a concentration of 1.25×10^{-6} M. The corresponding intercomparison between each ligand absorption and emission capacity is also presented in (**b**) and (**d**), respectively: the values presented being the areas under the respective spectroscopic curves

potential of a photoluminescent material as an LDS candidate is the molecules' ability to effectively convert the light to a wavelength more favourable to the underlying PV cells' material properties. This is highly dependent upon the technology as newer noncommercial technologies such as dye-synthesized cells have peak responsivities lying in the 600 nm region, while older more developed technologies such as Si-based devices have theirs located at close to 1000 nm. This is further coupled, with a better understanding and mitigation of the loss mechanism in the more developed technologies, at the architecture level, potentially limiting the enhancement offered by conventional LDS approaches for these technologies.

The emission spectra of some of the ligands studied, shown in Fig. 58.2c, indicate that there is very little spectral overlap existing between the absorption and emission bands of the best performing candidates. Another plus from a design perspective, as typically ligands suffer some major drawbacks in terms of their small stokes shift, which can lead to fluorescent quenching, increasing the energetic favourability of non-radiative decay modes in the process. The emission bandwidth spanned by the range of ligands investigated in this study makes them potential candidates for new and emerging technologies such as dye-synthesized cells, whereby an LDS add-on could serve to prolong the lifespan of the device while also improving its efficiency. This, coupled with the potential of such materials to be enhanced through the inclusion of metal nanoparticles, such as silver, whose SPR peak lies within the absorption band of the top three candidates (PMCC 183, PMCC 282 and PMCC 205), could provide even more enhancement to the photostability of these emerging PV technologies.

Examining the relationship between the absorption (Fig. 58.2b) and emission (Fig. 58.2d) capacity of the ligands investigated, it is evident that the molecules which provide the best candidates in terms of absorption do not necessarily correlate with those which yield the highest fluorescent intensity. The origin lies in the relationship between the molecule structure and the probability of non-radiative decay pathways within its band structure. However, in solar thermal applications, whereby thermal energy is either directly or indirectly used to generate electricity, the occurrence of non-radiative decay modes in the micrometre- or nanometre-sized materials typically suspended in modern heat transfer fluids is welcomed. Relating the structural properties of the ligands with the trends observed in the spectroscopic data presented herein, there is a clear trend in terms of their structural activity. The results of this and other investigation will be presented in future publications.

The top eight performing ligands identified through the initial intercomparison in absorbing and emitting performance shortlisted the candidates for the next stage of testing involving the determination of the luminescent quantum yield. Stock solutions at 1×10^{-4} M of both the fluorescent standard (quinine sulphate dissolved in 0.1 M H_2SO_4) and the ligand under consideration (dissolved in DMSO) were diluted using a dilution factor of 2, until the absorption at the selected excitation wavelength (λ_E) of 380 nm reached ~0.1. Careful monitoring of the absorption and emission spectra was carried out at each concentration (Fig. 58.3a quinine sulphate and 58.3b PMCC 205) to facilitate the identification of possible aggregation and minimize its impact on the results obtained. The variation in the integrated fluorescent intensity below this point, in accordance with the absorption at λE, providing a

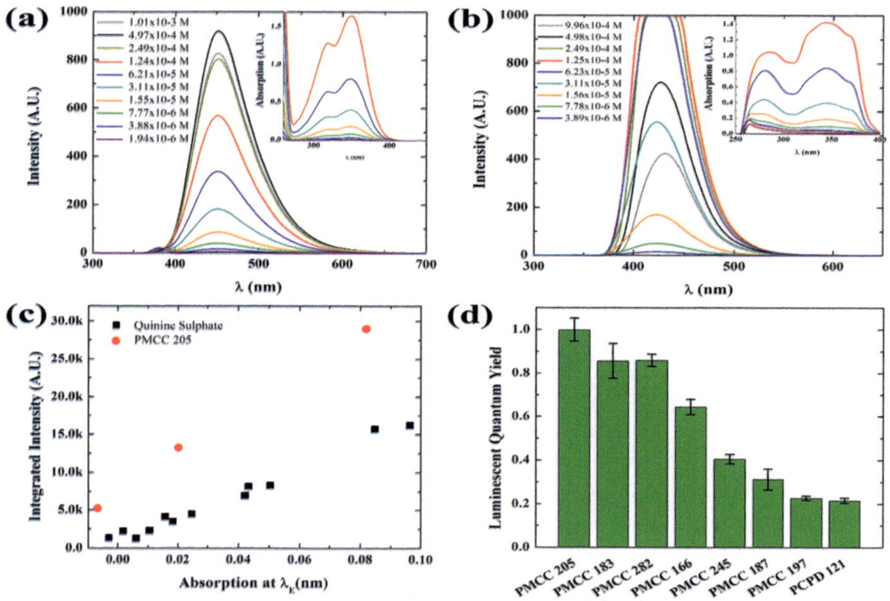

Fig. 58.3 (**a**) Emission spectra of the fluorescent standard (quinine sulphate) used in the determination of the luminescent quantum yield (LQY) of the heteroaromatic ligands via the comparative method: with the absorption spectra shown as an inset. (**b**) Example of the emission of one of the ligands (PMCC 205) as a function of concentration, with the corresponding absorption spectra shown in the inset. (**c**) Example of the fitting process involved in the determination of the LQY, in which the variation in the integrated intensity as a function of the absorption at the excitation wavelength ($\lambda_{E=}$ 380 nm) is monitored. (**d**) The LQY values determined for the top eight performing ligands ± the standard deviation observed across three independent measurements

comparison between the ligands photoluminescence efficiency and that of the standard materials utilized.

Three of the ligands (PMCC 183, PMCC 282 and PMCC 205) exhibited close to unity quantum yield values, making them ideal candidates for absorbing light below 500 nm and transferring it larger wavelengths where a solar cell can more efficiently harness the energy: ideally one of the many emerging technologies. The remaining ligands exhibited rather 'poor' LQY values, making their potential role in solar thermal applications more appealing and rewarding. Although no coordination complexes were synthesized at this point in time, the study conducted herein categorized a family of fluorescent ligands, identifying a number of suitable candidates for the formation of complexes based around a metal centre in the process. The spectroscopic data gathered forming the basis of a solid underpinning of the structural activity of these structures, as well as the relationship each chemical alteration had on the overall photoluminescence properties and performance of the molecule. These discoveries will prove vital in the development of a targeted photoluminescence candidate design, allowing the final product to be well matched to the underlying PV technologies' spectral responsivity profile.

58.4 Conclusion

A series of phenanthroline-based fluorescent ligands were characterized and catalogued for their potential role in solar energy applications. The series of ligands heavily absorbing within the active region (300 nm–500 nm) where most commercial and noncommercial PV technologies suffer low capture and conversion capabilities. Three structures in particular were identified as ideal candidates for LDS applications, in terms of their ability to capture and convert light from this spectral region to larger wavelengths, with a high degree of efficiency as evidence through their resolved quantum yield values (0.80–0.95). The three structures represent the candidates which will act as the sensitizing agents when forming metal-based coordination complexes going forward. Other ligands within the family characterized displayed luminescent quantum yield values much lower in magnitude (0.20–0. 60), making them more suited in solar thermal applications whereby localized heat transfer is a priority.

With these photoluminescent materials rapidly and cost effectively produced using a scalable method, there is precedent to explore their integration into a suitable host candidate and investigate their impact on a range of different PV technologies' short-wavelength response.

References

1. Klampaftis E, Ross D, McIntosh KR, Richards BS (2009) Enhancing the performance of solar cells via luminescent down-shifting of the incident spectrum: a review. Sol Energy Mater Sol Cells 93(8):1182–1194. https://doi.org/10.1016/j.solmat.2009.02.020
2. McKenna B, Evans RC (2017) Towards efficient spectral converters through materials design for luminescent solar devices. Adv Mater 29(28):1606491. https://doi.org/10.1002/adma.201606491
3. de la Mora MB, Amelines-Sarria O, Monroy BM, Hernández-Pérez CD, Lugo JE (2017) Materials for downconversion in solar cells: perspectives and challenges. Sol Energy Mater Sol Cells 165:59–71. https://doi.org/10.1016/j.solmat.2017.02.016
4. Li J-F, Li C-Y, Aroca RF (2017) Plasmon-enhanced fluorescence spectroscopy. Chem Soc Rev 46(13):3962–3979. https://doi.org/10.1039/C7CS00169J
5. Lakowicz JR, Ray K, Chowdhury M et al (2008) Plasmon-controlled fluorescence: a new paradigm in fluorescence spectroscopy. Analyst 133(10):1308–1346. https://doi.org/10.1039/B802918K
6. Würth C, Grabolle M, Pauli J, Spieles M, Resch-Genger U (2013) Relative and absolute determination of fluorescence quantum yields of transparent samples. Nat Protoc 8:1535. https://doi.org/10.1038/nprot.2013.087
7. Velapoldi RA (1972) Considerations on organic compounds in solution and inorganic ions in glasses as fluorescent standard reference materials. J Res Natl Bur Stand Sect A Phys Chem 76A(6):641–654. https://doi.org/10.6028/jres.076A.056

Chapter 59
External Quantum Efficiency Measurements and Outdoor Characterisation for PV Luminescent Downshifting Devices

H. Ahmed, J. Doran, and S. J. McCormack

59.1 Introduction

As part of the European Union energy and climate goals, 20% of the final energy consumption should come from renewable technology, and by 2030 the target has been set to 27% [1]. Harvesting solar energy has the potential to reduce carbon emissions and to provide clean energy contributing to sustainable development. The solar spectrum received at the Earth surface covers a wide range of wavelengths from 290 nm to 3790 nm. In an ideal situation, the absorption spectrum of PV materials should perfectly match the entire solar spectrum in order to convert the maximum photons from solar radiation to electricity. However, there is a large mismatch between the solar emission spectrum and the absorption properties of the present PV materials. At short wavelengths, each photon has a large energy, and therefore the ratio of photons to power is reduced. Any energy in excess of the band gap energy of the solar cell materials is not utilised by the solar cell and instead goes to heating the solar cell and is therefore wasted [2, 3]. Loss mechanisms in photovoltaic represent a practical limit to the solar cell efficiency, and the potential remains to make a better use of the short wavelength radiation. In order for high efficiencies to be achieved in PV technologies, energy loss mechanisms must be reduced. Luminescent downshifting (LDS) layer is a method which aim to convert non-absorbable solar radiation in the UV (290–400 nm) into absorbable incoming

H. Ahmed (✉)
Dublin Energy Lab, Technological University Dublin, Dublin, Ireland

Trinity College Dublin, Dublin, Ireland
e-mail: HAHMED@tcd.ie

J. Doran
Dublin Energy Lab, Technological University Dublin, Dublin, Ireland

S. J. McCormack
Trinity College Dublin, Dublin, Ireland

A. Sayigh (ed.), *Renewable Energy and Sustainable Buildings*, Innovative Renewable Energy, https://doi.org/10.1007/978-3-030-18488-9_59

radiation in the visible (400–700 nm) via florescence phenomena, hence increasing the solar optical response for short wavelength radiation [4–13, 15]. It has been proposed in the late 1970s when Hovel et al. realised that instead of concentrating sunlight high-energy photons could be converted to low energy to overcome the poor solar cell performance in UV, blue light. It involves a luminescent species that is applied in a transparent polymer/glass host material on top of the PV cell [7].

59.2 Experimental

59.2.1 LDS Layer Preparation and Device Fabrication

LDS layers with different concentrations (0.082, 0.44 and 0.170% w/w) of down-shifter compound were fabricated. Naphthalimide-based Lumogen Violet organic dye [BASF] was used for inclusion in LDS layers. The LDS layers were prepared by dissolving the Violet dye in PMMA (Carl Roth GmbH+Co.KG) solutions. The dye was first dissolved in toluene, added to the PMMA and stirred for 45 min resulting in homogeneous solutions. Uniform thicknesses of (100 ± 05 μm) were achieved by drop casting the solutions onto glass plates. Monocrystalline silicon cells (2×2 cm, Sunrydz, Germany) were used to assess the downshifting effect of Lumogen Violet in PMMA. LDS layer was attached to the c-Si cell using a thin layer (<50 μm) of epoxy as adhesive. A total of four devices were fabricated and are presented in Fig. 59.1. The performances of the c-Si cells, before and after the LDS encapsulation, were characterised outdoors, and the external quantum efficiency was measured. Results of both measurements are discussed below in details.

	Reference device	c-Si + Lumogen Violet LDS Devices		
#	C1	C2	C3	C4
Components	c-Si + Blank PMMA	c-Si + LDS (0.082%w/w)	c-Si + LDS (0.044%w/w)	c-Si + LDS (0.170%w/w)
Photo under room light	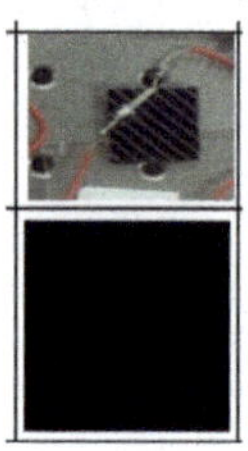	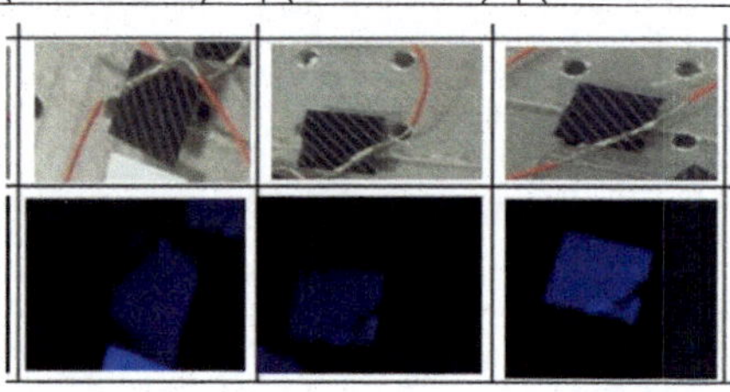		
Photo under UV light				

Fig. 59.1 Monocrystalline silicone cell with C1 blank PMMA C2, C3 and C4 Lumogen Violet/PMMA (concentrations of 0.082, 0.044 and 0.170% w/w) layer under room light and under UV lamp light

59.2.2 Absorption and Emission Measurements of LDS Layers

UV/Vis/NIR absorption spectroscopy was used to measure the absorption characteristics of the downshifting layers. The UV/Vis/NIR absorption spectrometer used was a Perkin Elmer Lambda 900. The emission spectra of the LDS layers were measured by optically pumping samples using a monochromated light source using a luminescence spectrometer Perkin Elmer LS55B.

59.2.3 External Quantum Efficiency Measurements

The EQE is the ratio of the number of charge carriers that are collected by the solar cell to the number of photons of a given wavelength entered into the solar cell [14]. The QEX10 Quantum Efficiency (PV Measurements, Inc., USA) was used to measure the EQE curves of the silicon cell/LDS device. It is a spectral response and incident photon conversion efficiency-based measurement system which uses a xenon arc lamp source, monochromator, filters and reflective optics under bias light. The monochromatic beam area is 1 cm^2 and directed at the center of the solar cells.

59.2.4 Outdoor Measurements of the LDS Devices and Performance Characteristics

Outdoor measurements were carried out on the four devices characterised over the same time period. IV performance characteristics in Table 59.1 were based on eight sets of hourly averages, recorded on January 3 and 4, 2014, on the Technological University Dublin roof setup. It consists of two Kipp and Zonen CM-6b pyranometers aligned side by side to measure the global radiation (Gh) and a EKOMS 710 spectroradiometer to record the solar spectrum from 337 nm to 1100 nm. A data logging system using a DeltaTDL2e included in the setup to permit recording of values on a minute-by-minute basis. Also a horizontal platform adjacent to the

Table 59.1 Average outdoor electrical characteristics

PV/ LDS	8Hrs AVG LDS Devices	Pmax(mW)	Imp(mA)	Vmp(V)	FF	Isc(mA)	Voc(V)	Eff%
C1	cSi+ PMMA blank	7.39	16.64	0.43	0.67	19.44	0.55	13.00
C2	cSi+Lum V (0.082 w/w%)	8.45	17.93	0.47	0.75	19.72	0.56	15.00
C3	cSi+Lum V (0.044 w/w%)	7.21	16.05	0.44	0.70	18.15	0.55	13.00
C4	cSi+Lum V (0.17 w/w%)	6.87	15.34	0.44	0.71	17.18	0.55	12.00

pyranometers hosts the solar cell and a Keithley 2400 source measurement unit was used to perform a voltage sweep on the solar cell. The four-point probe (Kelvin) method was used to determine the corresponding current, yielding the I-V curve for the solar cell. The setup permits comparison of a number of devices within the same time frames useful for comparison of hourly average performance characteristics. Electrical performance parameters such as Isc, Voc and FF are reported. Typically three sweeps per minute are recorded, and hourly averages of the global radiation and IV characteristics are used to calculate the efficiency (η) of the solar cell. The efficiency of the cell was calculated using the equation below:

$$\eta = \left(\frac{P_{\max}}{Gh.A} \right) \times 100,$$

where P_{max} is the maximum power calculated in mW, *Gh* is the global radiation (the total short-wave radiation falling onto a horizontal surface) measured in Wm^{-2} and *A* is the cell area given in m^2. The *Gh* was measured 142.12 Wm^{-2} (1/7th of one sun), and the cell area was 0.0004 m^2.

59.3 Results and Discussions

Absorption and emission measurements of the layers are shown in Fig. 59.2. Absorbance of blank PMMA is also shown. It is clear that Lumogen Violet layers have retained their optical properties in PMMA films since the optical spectra are similar to those in solution.

In Fig. 59.3, the measured EQE of LDS devices with different compositions of Lumogen Violet and their equivalent c-Si cell with blank cover PMMA are presented. An expanded view for EQE in the UV region is shown in Fig. 59.4.

An enhancement has been observed in EQE for Lumogen Violet/PMMA LDS devices (C2 and C3) relative to the blank PMMA device (C1) while C4 has shown a decrease in EQE. The enhancement was noticeable for wavelengths where the violet dye is absorbing at 290–420 nm. Lumogen Violet devices have shown decrease in EQE from 380 nm to 420 nm; this is where the absorption and the emission spectra of Lumogen Violet are overlapping which might cause self-absorption. Also, the blank PMMA layer was found to have slightly absorbing between 300 nm and 370 nm which is penalising the c-Si cell efficiency at these UV wavelengths.

The average of 8-hour outdoor electrical characteristics of all PV/LDS devices are presented in Table 59.1.

The highest gain in overall efficiency η (%) was recorded in Lumogen Violet (C2, 0.082%) device which was 15% relative (2% absolute) higher than the corresponding reference cell C1 (η = 13%). Lumogen Violet (C3, 0.044%) device shows an increase of 0.5% while the Lumogen Violet (C4, 0.170%) device shows a decrease of 0.4% relative to the reference cell.

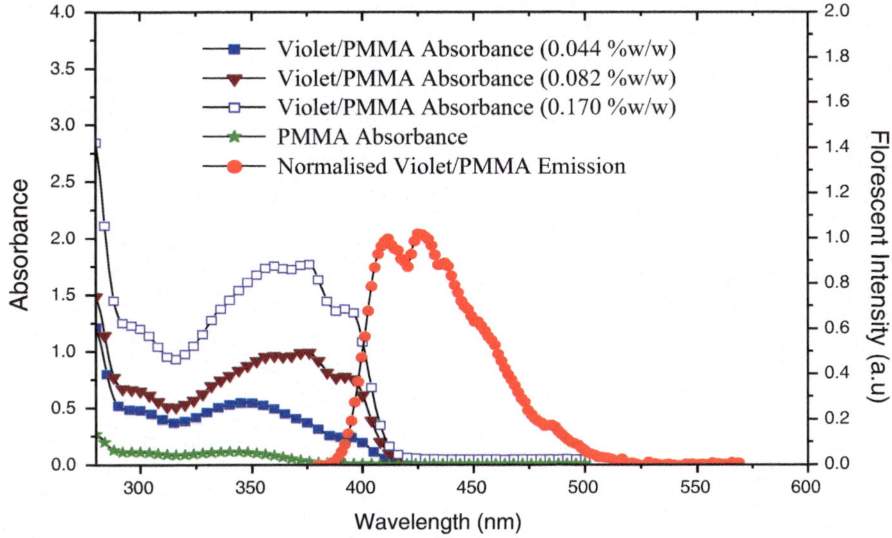

Fig. 59.2 Absorption and emission measurements of varying concentrations of Lumogen Violet in PMMA. Absorbance of blank PMMA is also shown

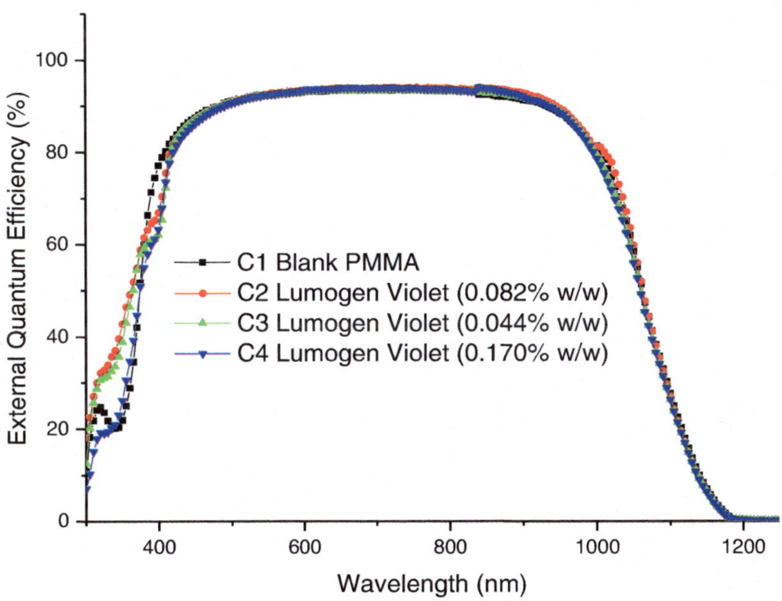

Fig. 59.3 Comparison of EQE curves for c-Si cells encapsulated with LDS layers of Lumogen Violet dye and PMMA. The EQE for the blank PMMA layer is also shown

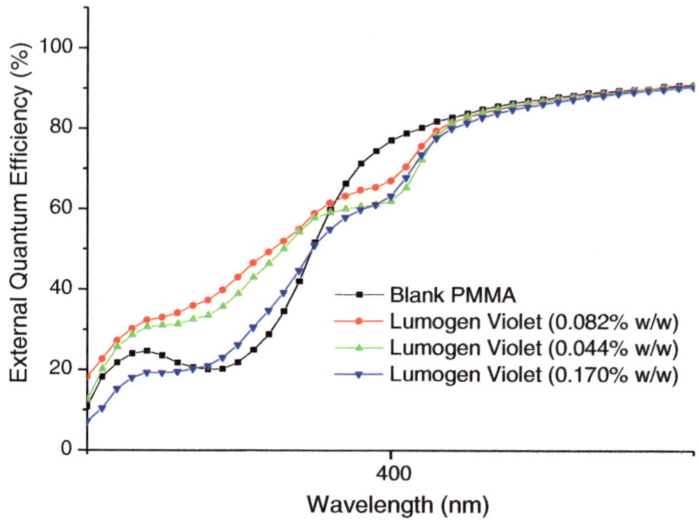

Fig. 59.4 An expanded view for EQE in the UV region comparing EQE curves for c-Si solar cells encapsulated with LDS layers of Lumogen Violet dye and PMMA. The EQE for the blank PMMA layer is also shown

The short-circuit current Isc measured for PV/LDS C2 device was 19% relative higher than the reference cell.

59.4 Conclusion

In this paper, Lumogen Violet downshifting organic dye has been characterised for its inclusion in luminescent downshifting (LDS) layers. Monocrystalline silicone (c-Si) solar cells were attached to Lumogen Violet LDS layer of different concentrations (0.082, 0.044 and 0.17% w/w) in PMMA films. The performance of the c-Si was characterised with and without LDS layers. An increase in EQE was demonstrated for c-Si devices with Lumogen Violet/PMMA layers of lower concentrations (4% for C2 (0.082% w/w) and 0.5% for C3 (0.044% w/w)), relative to the blank PMMA layer, while no enhancement was observed for LDS layer of higher concentration (C4, 0.17% w/w). The highest gain in solar cell efficiency was observed for the Lumogen Violet/PMMA with concentration of 0.082% w/w due to the action of the LDS layer compared to blank PMMA layer where a 15% relative increase has been measured. The short-circuit current, Isc, for this device was 19% relative higher than the reference cell.

Acknowledgements The authors would like to acknowledge the funding from the European Research Council grant (639760) entitled PEDAL: Plasmonic enhancement of advanced luminescent solar devices and funding from Science Foundation Ireland (SFI). Also, this research was partly supported by the European Commission as part of the Framework 7 integrated project EPHOCELL grant (227127) and by the Higher Education Authority of Ireland. Authors would also like to acknowledge Dr. Sergi Riera-Galindo for his help with the EQE measurements and Dr Thomas Confrey for his help with the outdoor measurements.

References

1. Climate and Energy Framework 2030 https://eur-lex.europa.eu/legal-content/EN/TXT/PDF/?uri=CELEX:52014DC0015&from=EN
2. Green MA (2003) Crystalline and thin-film silicon solar cells: state of the art and future potential. Sol Energy 74:181–192
3. Green M (2002) Third generation photovoltaic solar cells for 2020 and beyond. Physica E Low Dimens Syst Nanostruct 14:65–70
4. Ahmed H, Kennedy M, Confrey T, Doran J, McCormack S, Galindo S, Voz C, Puigdollers J (2012) Lumogen violet dye as luminescent down-shifting layer for c-silicon solar cells. EU Photovoltaic Solar Energy Conference and Exhibition (PVSEC), Frankfurt, Germany, pp 24–28
5. Ahmed H, McCormack S, Doran J (2017) Plasmonic luminescent down shifting layers for the enhancement of CdTe mini-modules performance. Sol Energy 141:242–248
6. Alonso-Álvarez D, Ross D, Klampaftis E, McIntosh KR, Jia S, Storiz P, Stolz T, Richards BS (2014) Luminescent down-shifting experiment and modelling with multiple photovoltaic technologies. Prog Photovolt Res Appl 23:479–497
7. Hovel HJ, Hodgson RT, Woodall JM (1979) The effect of fluorescent wavelength shifting on solar cell spectral response. Solar Energy Mater 2:19–29
8. Kennedy M, Ahmed H, Doran J, Norton B, Bosch- Jimenez P, Della Pirriera M, Torralba-Calleja E, Tauste DG, Aubouy L, Daren S, Solomon-Tsvetkov F, Galindo S, Voz C, Puigdollers J (2015) Large stokes shift downshifting Eu(III) films as efficiency enhancing UV blocking layers for dye sensitized solar cells. Phys Status Solidi A 211:203–210
9. Klampaftis E, Ross D, McIntosh KR, Richards B (2009) Enhancing the performance of solar cell via luminescent down-shifting of the incident spectrum: a review. Sol Energy Mater Sol Cells 93:1182–1194
10. McIntosh KR, Lau G, Cotsell JN, Hanton K, Batzner DL (2009) Increase in external quantum efficiency of encapsulated silicon solar cells from a luminescent down-shifting layer. Prog Photovoltaics Res Appl 17:191–197
11. Ross D, Alonso-Álvarez D, Klampaftis E, Fritsche J, Bauer M, Debije MG, Fifield RM, Richards BS (2014) The impact of luminescent down shifting on the performance of CdTe photovoltaics: impact of the module vintage. IEEE J Photovoltaic 4:457–464
12. Rothemund R, Kreuzer S, Umundum T, Meinhardt G, Fromherz T, Jantsch W (2011) External quantum efficiency analysis of Si solar cells with II-VI nanocrystal luminescent down-shifting layers. Energy Procedia 10:83–87
13. Strümpel C, McCann M, Beaucarne G, Arkhipov V, Slaoui A, Švrček V, del Cañizo C, Tobias I (2007) Modifying the solar spectrum to enhance silicon solar cell efficiency- an overview of available materials. Sol Energy Mater Sol Cells 91:238–249
14. Nelson J (2003) The physics of solar cells. Imperial College Press, London
15. Ahmed H, Doran J and McCormack S (2016) Increased short-circuit current density and external quantum efficiency of silicon and dye sensitised solar cells through plasmonic luminescent down-shifting layers. Sol Energy 126:146–155

Chapter 60
Domestic Demand-Side Response: The Challenge for Heat Pumps in a Future UK—Decarbonised Heating Market

Neil J. Hewitt, Nik Shah, Donal Cotter, Chris Wilson, Khoa Le, Raymond Byrne, Paul MacArtain, and Ming Jun Huang

60.1 Introduction

The Clean Growth Strategy of the UK in 2018 illustrated three 2050 pathways which are:

1. The Electricity Pathway—where transport and heating are primarily fuelled by electricity, virtually all of which is from clean sources
2. The Hydrogen Pathway—using hydrogen to heat our homes and buildings, to fuel vehicles and to power UK's industry
3. The Emissions Removal Pathway—where biomass power stations are used in parallel with carbon capture, utilisation and storage (CCUS) [1]

This can be compared with an Irish dimension where in addition to the EirGrid [2] equivalent of the UK's Future Energy Scenarios [3], there has been some thoughts on how an often thought of as a traditional agricultural society will utilise biogas [4].

The European Union (EU) has also set out Energy 2050 pathways that include scenarios as to what the EU will attain in terms of reductions in fossil fuel use by 2050 [5]. PricewaterhouseCoopers (PwC) calculated a low carbon index of nations [6] that revealed that the UK has some of the most progressive policies and is underway in implementing them. Returning to the UK's Clean Growth Strategy, the UK Energy Minister Claire Perry in 2018 stated that the Electricity Pathway (transport and heating are primarily fuelled by electricity, virtually all of which is from clean

N. J. Hewitt (✉) · N. Shah · D. Cotter · C. Wilson · M. J. Huang
Ulster University, Centre for Sustainable Technologies,
Newtownabbey, Co. Antrim, Northern Ireland, UK
e-mail: nj.hewitt@ulster.ac.uk

K. Le · R. Byrne · P. MacArtain
Dundalk Institute of Technology, Dundalk, Co. Louth, Ireland

A. Sayigh (ed.), *Renewable Energy and Sustainable Buildings*, Innovative Renewable Energy, https://doi.org/10.1007/978-3-030-18488-9_60

sources) will have consequences for the low-voltage electricity network managed by Distribution System Operators (DSOs). These will now be discussed.

60.2 Impacts of the Electrification Pathway

Decarbonisation of space heating will see the use of biogas- or biohydrogen-fired sorption units or biogas or biohydrogen micro-combined heat and power or fuel cell units in our homes. Resistive heating, i.e. storage heaters, will have a role to play, but with a coefficient of performance of 1.0, their use may be constrained to areas of strong low-voltage network. Rural areas with weaker distribution networks may utilise bioenergy in combustion systems. However, the electric vapour heat pump when combined with thermal storage does offer a route to the Electricity Pathway. This route can be facilitated by (a) building insulation improvements, (b) integration of domestic renewable energy and (c) superior performing and cost-effective heat pumps.

60.3 Building Insulation Improvements at a Domestic Level

Gillot et al. [7] with fabric, airtightness and mechanical ventilation heat recovery (MVHR) saw phased retrofits over 2 years including numerous "fixes" of airtightness having the costs of £18,500 for a significant saving of 23,196 kWh per year. This took the Nottingham E.ON test house from zero retrofit measures (i.e. single glazed, no insulation, etc.) to an insulated and relatively airtight home. Jones et al. [8] highlighted the costs of domestic retrofit and the undoubted advantages in terms of fuel poverty (through a reduction in energy costs). The costs ranged from £23 k to over £30 k per home, CO_2 emission reductions were estimated to be in the range of 50–75% and cost savings of £402 to £621 per year. These included PV and batteries although their future participation as aggregated demand-side units was unclear. These projects point rightly to a "fabric first" approach, but Malby and Owen [9] highlight the UK's focus on measures delivering the quickest financial returns, e.g. PV panels when supported by a "Feed-In Tariff". UK retrofits also tend to be those that have external and low levels of disruption involved for homeowners/occupiers. This is contrary to European experiences where deep retrofits are carried out and sees contributing factors in the UK caused by high levels of home ownership. The UK "Green Deal" which closed in 2015 was an attempt to facilitate domestic retrofit through initially loans and then grants with an initial focus on building fabric. Gooding and Gul [10] suggested that the failings in the Green Deal were partially attributable to the lack of training to successfully retrofit homes in the UK.

60.4 Integration of Domestic Renewable Energy

In addition to Jones et al. [8] (PV and batteries), Boßmann et al. [11] stated that in countries with high shares of wind power (such as the UK), heating technologies may facilitate the integration of domestic renewable electricity, especially in the absence of alternative flexibility options. This is supported by Vorushylo et al. [12] who reveal the significant potential of heat pump electrification, delivering at least two and three times less carbon emissions, respectively, when compared with conventional options such as gas or oil for 20% of domestic sector.

60.5 Superior Performing and Cost-Effective Heat Pumps

Kelly and Cockcroft [13] illustrated air-source heat pump coefficients of performance of less than 3.0, while Minic et al. [14], Wilson et al. [15] and Hewitt et al. [16] illustrated drops in performance due to heat losses from tanks and loss in performance due to a lack of weather compensation but a very successful peak electricity time avoidance with 600 L of water acting as a thermal store for a high-temperature air-source heat pump (Fig. 60.1). Khoa et al. [17] validated models in TRNSYS of building, heat pump and thermal storage performance.

UK heat pump schemes are well reported and Love et al. [18] is a strong example. The conclusion from this study is that the impact of the heating sector is more predictable as the morning peak of the heat pump load profile is coincident with or begins just before the morning rise in load on the electricity grid, and the evening

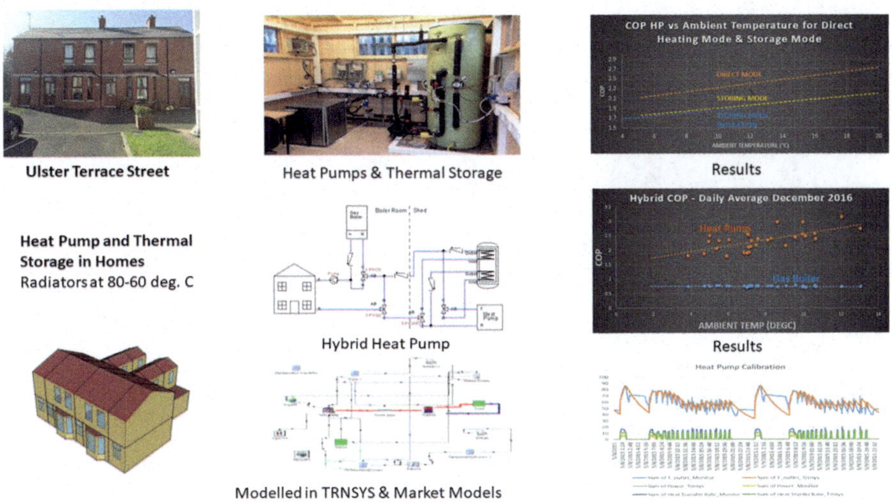

Fig. 60.1 Integration of high-temperature heat pumps and thermal storage at Ulster University Terrace Street

peak of the heat pump load profile is coincident with the evening peak in the electricity grid. In considering how to improve heat pumps, low global warming potential (GWP) working fluids must be considered. A traditional heat pump i.e. that heating an underfloor hydronic system may replace R410a with R32 will see a drop in GWP from 2088 to 675 and a COP that is similar [19] or superior [20]. Retrofit heat pumps that utilise existing high-temperature hydronic radiators (in keeping with the current lack of UK participation in deep retrofits due to high rates of home ownership with its consequential lack of corporate investment) could utilise R410a at the lower temperatures encountered, e.g. 60 °C, and see superior performance with R32 and its superior critical temperature, heat transfer and lower pressure drop [21]. This compares favourably to other proposed replacements as the modelled results indicate (Fig. 60.2).

But cost-effective heat pumps need to be driven by more than higher performance (new working fluids, component optimisation, the rise of low cost power electronic and controls to optimise the use of variable compressors, pumps and fans, etc.). Mass manufacturing requires a policy basis and the "Electrification Pathway" will trigger that to an extent. However, other pathways are also appropriate, and the concluding approach will be a combination of these for overall carbon reductions of "near" 100%.

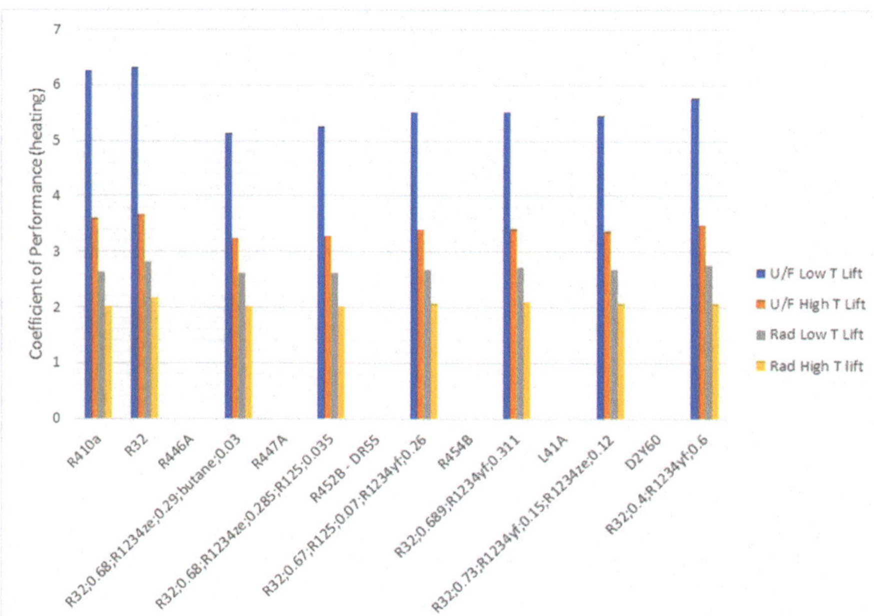

Fig. 60.2 Comparison of R410a alternatives in domestic heat pump applications

60.6 Utilisation of Low-Temperature Heat Networks for Domestic Space Heating

A low-temperature heat network (or fourth-generation heat network) is one that can supply low-temperature district heating for space heating and domestic hot water (DHW) to buildings,

has low grid losses, uses heat from low-temperature sources and integrates renewable heat sources such as solar and wind, be part of a smart energy system and be flexible in the future [22]. Low temperature is defined by the Danish as >50 °C and ultralow networks as <45 °C [23]. The latter would require heat pumps for upgrading to traditional space heating temperatures and for domestic hot water use.

UK low-grade waste represents a significant energy resource. While there are discrepancies on the actual values, industrial heat represents 300 TWh/year in 2016 [24], and 1/6th of the waste heat is available for waste heat recovery [25], i.e. 50 TWh/year. Furthermore, 28 TWh/year has been identified for "potential" district heating networks, and assuming a current household gas price of 3.69 p/kWh, utilisation of 28 TWh/year will displace over £1Bn of gas purchases annually. However, the UK has the equivalent of 13 TWh of existing heat networks [26], and assuming network construction costs are £1500/MWh, new network costs for 15 TWh/year lead to £22Bn investment. As gas is assumed to produce 0.19 kg/kWh of CO_2 and if the UK can utilise 28 TWh of waste heat displacing gas for space heating, 5,320,000 tonnes of CO_2 will be saved. As UK residential buildings currently emit 63.4 million tonnes of CO_2/year, 10% of residential CO_2 emissions can be saved if waste heat is fully utilised.

With respect to heat pump development for waste heat recovery, the EU is cutting the availability of hydrofluorocarbons (HFCs) by 79% between 2015 and 2030 [27] due to high global warming potential, and alternative fluids must be evaluated. The vapour compression heat pump operating in conjunction with lower-grade heat networks, e.g. below 30 °C, can uplift such heat to underfloor heating temperatures (<40 °C) with a coefficient of performance >10 (very low running costs) and provide domestic hot water temperatures avoiding legionella concerns (~60 °C) with a coefficient of performance >5.0 with current refrigerants, e.g. R410a. However, revised heat exchangers must be developed [28]. For thermal networks, alternative fluids such as R245fa have realised coefficients of performance of >7.0 for ~60 °C (Fig. 60.3).

Extending the temperature lift further sees low GWP fluids such as R1233zd [29] being able to upgrade low-grade waste heat to process temperatures. Higher temperature heat pumps have been successful and new fluids such as R1233zd (>150 °C, low GWP) are looking promising. However, compressor lubrication strategies for higher temperatures with existing equipment have not been explored to give commercial confidence (Fig. 60.4).

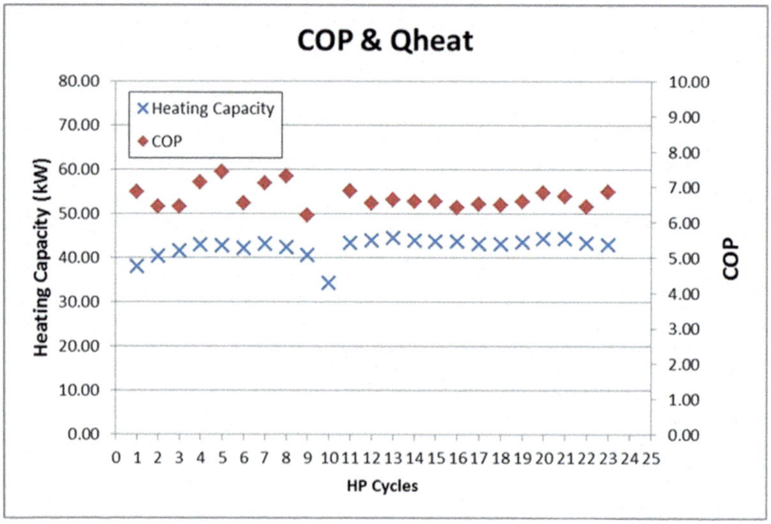

Fig. 60.3 R245fa performance for a high evaporator temperature heat pump providing heat to hydronic radiators at > 60 °C

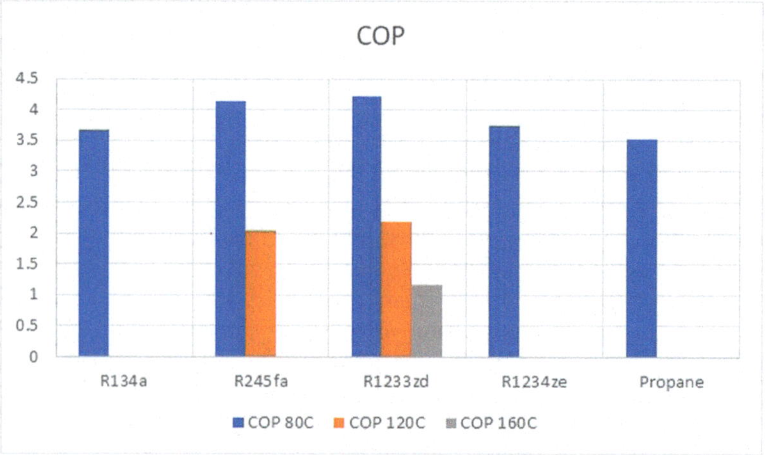

Fig. 60.4 Alternative refrigerant performance at high temperatures

60.7 Hybrid Heat Pumps as a Domestic Retrofit

Hybrid heat pumps initially could be thought of as interim technologies with respect to CO_2 reduction with the use of gas acting as the thermal storage counterpart in that gas would operate at times where it was economic to do so (e.g. electricity costs at peak times) and/or when thermodynamically optimised (e.g. very cold weather for

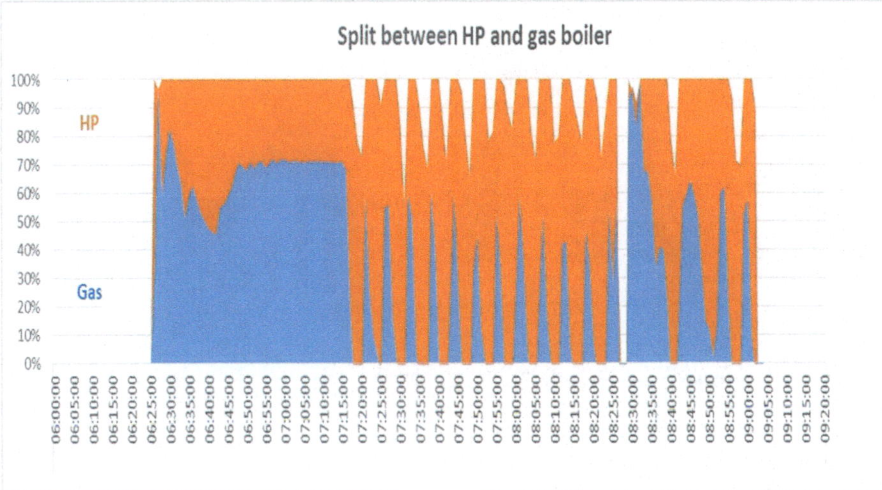

Fig. 60.5 Hybrid heat pump heat allocation example

an air-source heat pump and/or the delivery of hot water over 60 °C for legionella purposes). Thermal storage may also have a role, albeit in a reduced volume to take advantages of any future aggregated distributed energy resource tariff support. Experiments were conducted combining a high-temperature heat pump with the existing gas boiler in Ulster University's "Terrace Street" to explore the broader flexibility agenda (Fig. 60.5).

The gas element contributed approximately 55% of the heat in this example and its performance (and running cost per kWh) is relatively unaffected by ambient conditions. Such a system sees a 6% reduction in running costs over a traditional heat pump only system. However, the main impact might be in the reduction of natural gas use and its impact on the decarbonisation of gas supplies. Gas decarbonisation is of increased interest in areas with strong agriculture industry such as Ireland. It has been reported that approximately one million homes in Ireland could be served by gas in the future, and while the report seems able to acquire the ability to split commercial from household usage, the use of biomethane (from anaerobic digestion, etc.) could deliver 28% of Ireland's natural gas needs [4]. Hybrid heat pumps could form an interesting alternative in terms of the challenge to low-voltage electricity networks that traditional heat pumps can cause.

60.8 Photovoltaics (PV), Batteries and Domestic Applications

A final element of this study is to support electrically driven heat pumps with lower-cost batteries derived from their growing use in the automotive industry. Smart homes become a necessity as which loads should be used by what power source.

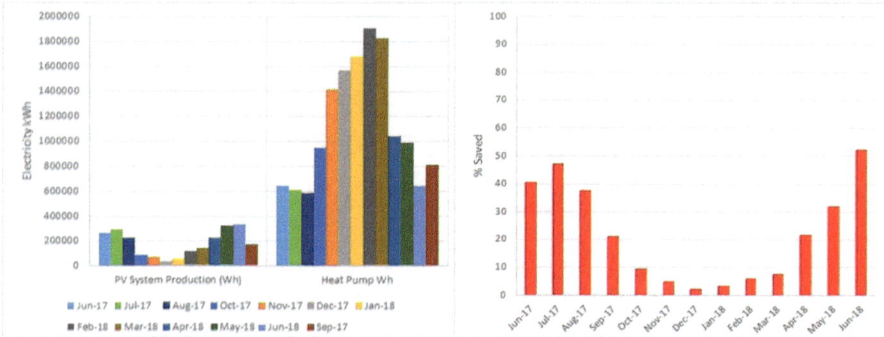

Fig. 60.6 Heat pump/PV interactions and energy saved

PV will charge a battery and/or aid in running a heat pump. Unfortunately, PV in winter conditions is unlikely to meet much of the heat pump demand. However, batteries may become part of the domestic aggregated distributed energy response management system and be charged by the electricity network (with excess wind power for example) to meet peak demands. Figure 60.6 illustrates how smart households may integrate a range of technologies to address decarbonisation and raises an important question regarding the deployment of batteries in homes. Does a battery of a sufficient size negate the need for behavioural change when using PV? The use of dishwashers, washing machines, etc. may require less education with a battery of sufficient size. But the arrival of the electric car may challenge the low-voltage network further, and ultimately would a large static battery charge the smaller battery of the electric vehicle? Would it also assist in running the heat pump? From a PV only perspective, the heat pump received electricity from a roof installed 2.2 kW_{peak} array as shown in Fig. 60.6. The size of PV array was appropriate for the available roof installation. However, if a larger array for this particular time period was available, e.g. 4.2 kW_{peak}, the heat pump would engender net zero electricity export (Fig. 60.7) but would only save 4% of the heat pump running costs in December. The unanswered questions are that of the level of sufficiency and affordability to warrant such domestic demand reductions and their impact on low-voltage electricity network upgrading.

60.9 Conclusions

Electrically driven heat pumps are not a universal panacea when it comes to decarbonisation of heat in household and industrial waste heat recovery/upgrading processes. What they do offer is a link to variable, non-dispatchable renewable energy through distributed energy resource management systems (DERMs) when coupled with thermal storage. All homes and businesses would benefit from energy efficiency programmes that reduce heating demand. New buildings may need to be

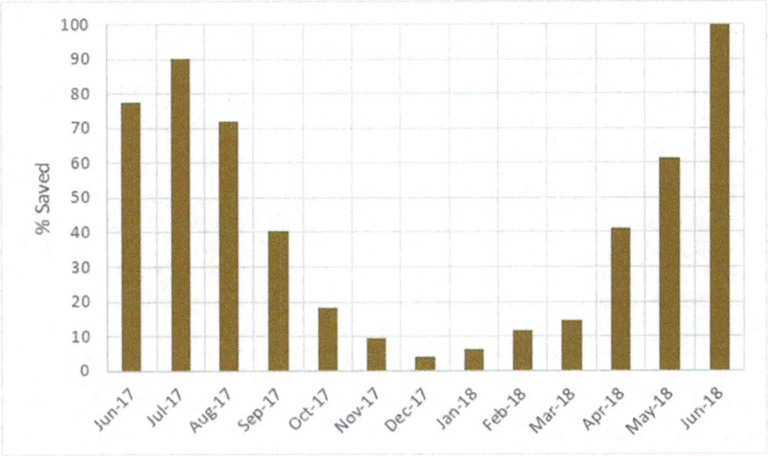

Fig. 60.7 Impact of a larger PV array

underfloor heating "ready" as there is a 50% improvement in heat pump performance by effectively halving the condensing temperature. Hybrid heat pumps offer an alternative flexibility model that is undiscussed in this paper; the role of variable speed compressors in smart capacity management and domestic hot water delivery is feasible. Integration of non-dispatchable renewable energy will see electric heat pumps with compact thermal storage deliver aggregated DERM responses to Distribution System Operator demands but will also need to understand how to accommodate electric vehicle demands. Domestic PV may not be sufficient in winter to support such demands.

Acknowledgements The authors would like to acknowledge the support of EPSRC i-STUTE, EPSRC 4s-DHW, FP7 Einstein, H2020 CHESS-SETUP and INTERREG VA SPIRE 2 in bringing together the concepts into new ideas.

References

1. https://www.gov.uk/government/publications/clean-growth-strategy
2. http://www.eirgridgroup.com/site-files/library/EirGrid/EirGrid-Tomorrows-Energy-Scenarios-Report-2017.pdf
3. http://fes.nationalgrid.com/
4. Decarbonising Domestic Heating in Ireland (2018), KPMG, Eriva, June 2018
5. Communication from The Commission to The European Parliament, The Council, The European Economic and Social Committee and The Committee of The Regions, Energy Roadmap 2050, 2011
6. https://www.pwc.co.uk/services/sustainability-climate-change/insights/low-carbon-economy-index.html

7. Gillott MC, Loveday DL, White J, Wood CJ, Chmutina K, Vadodaria K (2016) Improving the airtightness in an existing UK dwelling: the challenges, the measures and their effectiveness. Build Environ 95:227–239
8. Jones P, Li XJ, Perisoglou E, Patterson J (2017) Five energy retrofit houses in South Wales. Energ Buildings 154:335–342
9. Maby C, Owen A (2015) Installer Power—The key to unlocking low carbon retrofit in private housing. University of Leeds
10. Gooding L, Gul MS (2017) Achieving growth within the UK's domestic energy efficiency retrofitting services sector, practitioner experiences and strategies moving forward. Energy Policy 105:173–182
11. Boßmann T, Elsland R, Klingler A-L, Catenazzi G, Jakob M (2015) Assessing the optimal use of electric heating systems for integrating renewable energy sources. Energy Procedia 83:130–139
12. Vorushylo I, Keatley P, Shah N, Green R, Hewitt N (2018) How heat pumps and thermal energy storage can be used to manage wind power: a study of Ireland. Energy 157:539–549
13. Kelly NJ, Cockroft J (2011) Analysis of retrofit air source heat pump performance: results from detailed simulations and comparison to field trial data. Energ Buildings 43:239–245
14. Minic I, Hewitt N, Huang M, Ramirez M, ICR (2015) Design of phase change thermal energy stores (pcm-tes) for residential heat pump applications, Paper 642. The 24th IIR International Congress of Refrigeration August 16–22, 2015
15. Wilson C, Shah N, Hewitt N, Huang M, ICR (2015) Considerations for the use of heat pumps with combined thermal storage as a DSM tool in a domestic retrofit setting, Paper 279. The 24th IIR International Congress of Refrigeration August 16–22, 2015
16. Hewitt N, Huang M, Ramirez M, ICR (2015) High temperature heat pumps for seasonal thermal energy storage and district heating systems, Paper 259, The 24th IIR International Congress of Refrigeration August 16–22, 2015
17. Le KX, Shah N, Huang MJ, Hewitt NJ, Member IAENG (2017) High temperature air-water heat pump and energy storage: validation of TRNSYS models. In: Proceedings of the World Congress on engineering and computer science 2017 Vol 2, WCECS 2017, October 25–27, 2017, San Francisco, USA
18. Love J, Smith AZP, Watson S, Oikonomou E, Summerfield A, Gleeson C, Biddulph P, Chiu LF, Wingfield J, Martin C, Stone A, Lowe R (2017) The addition of heat pump electricity load profiles to GB electricity demand: evidence from a heat pump field trial. Appl Energy 204:332–342
19. Alabdulkarem A, Eldeeb R, Hwang Y, Aute V, Radermacher R (2015) Testing, simulation and soft-optimization of R410A low-GWP alternatives in heat pump system. Int J Refrig 60:106–117
20. Mota-Babiloni A, Navarro-Esbrí J, Makhnatch P, Molés F (2017) Refrigerant R32 as lower GWP working fluid in residential air conditioning systems in Europe and the USA. Renew Sust Energ Rev 80:1031–1042
21. Botticella F, de Rossi F, Mauro AW, Vanoli GP, Viscito L (2018) Multi-criteria (thermodynamic, economic and environmental) analysis of possible design options for residential heating split systems working with low GWP refrigerants. Int J Refrig 87:131–153
22. Lund H, Werner S, Wiltshire R, Svendsen S, Mathies BV (2014) 4th Generation District Heating (4GDH): integrating smart thermal grids into future sustainable energy systems. Energy 68:1–11
23. Guidelines for Low-Temperature District Heating, A deliverable in the project financially supported by the Danish Energy Agency in the R&D programme EUDP (Energiteknologisk Udviklings-og Demonstration Program): "EUDP 2010-II: Full-Scale Demonstration of Low-Temperature District Heating in Existing Buildings", April 2014
24. https://www.gov.uk/government/statistics/energy-chapter-1-digest-of-united-kingdom-energy-statistics-dukes

25. The potential for recovering and using surplus heat from industry, Final Report for DECC, 05/03/2014, Led by Element Energy Limited
26. https://www.gov.uk/government/publications/summary-evidence-on-district-heating-networks-in-the-uk
27. https://www.gov.uk/guidance/hfc-phase-down-in-the-eu-how-it-works-and-exemptions
28. Lee H et al (2013) Experimental investigation of novel heat exchanger for low temperature lift heat pump. Energy 51:468–474
29. Ju F et al (2017) Performance assessment of heat pump water heaters with R1233zd(E)/HCs binary mixtures. Appl Therm Eng 123:1345–1355

Chapter 61
Microwave-Assisted Hydrothermal Valorisation of Rapeseed Meal for the Co-Production of High Purity Lignin and Saccharide-Rich Aqueous Solutions

Javier Remón, Avtar S. Matharu, and James H. Clark

61.1 Introduction

Rapeseed, the third largest source of vegetable oil in the world, is currently used for the production of both edible oil and biodiesel [1]. During the processing of rapeseed seeds to produce the oil, around 65 wt.% of the feedstock is converted into a lignocellulosic solid residue called rapeseed meal or rapeseed cake [2, 3]. This solid material is mainly composed of cellulose, hemicellulose, lignin and proteins; the precise chemical composition of the residue depends on the type of rapeseed plant and extraction process [2].

An interesting option for the valorisation of this residue that has not been considered before is the simultaneous production of saccharides and pure lignin. However, the extraction of polysaccharide-free lignin from biomass is very challenging because lignin is strongly covalent bonded to cellulose and hemicellulose, which hinders the selective extraction of pure lignin. Therefore, the development of a suitable method for lignin isolation is of paramount importance for the production of pure lignin from biomass. As part of this, the use of microwave heating has recently appeared as a new and promising alternative. Microwave heating is based on the high-frequency rotation of polar molecules, which produces a quicker and higher heating of the species with higher polarity within the biomass structure [4]. As lignin has a higher aromaticity, i.e. lower polarity, than cellulose and hemicellulose, it is less active during microwave heating [5]. This could allow the separation of cellulose and hemicellulose from the biomass without significantly altering the lignin structure, thus allowing a high purity lignin to be produced. In addition, as water is

J. Remón (✉) · A. S. Matharu · J. H. Clark (✉)
Department of Chemistry, Green Chemistry Centre of Excellence,
University of York, York, UK
e-mail: javier.remonnunez@york.ac.uk; jremon@icb.csic.es; james.clark@york.ac.uk

A. Sayigh (ed.), *Renewable Energy and Sustainable Buildings*, Innovative Renewable Energy, https://doi.org/10.1007/978-3-030-18488-9_61

highly effective in microwave energy absorption, the combination of hydrothermal conditions together with microwave-assisted heating might be an interesting new technology for the valorisation of rapeseed meal.

Given this background, this work addresses the valorisation of rapeseed meal by means of a microwave-assisted hydrothermal process catalysed by acetic acid, a much safer and greener alternative to mineral acids, for the simultaneous production of pure lignin and polysaccharide-rich aqueous solutions. In particular, the effects of the temperature (150–210 °C), time (0–1 h) and catalyst (acetic acid) amount (1–4 mol/L) together with all the possible interactions between these variables on rapeseed meal valorisation have been thoroughly analysed.

61.2 Experimental

61.2.1 *Microwave Experiments*

A CEM Discover II microwave facility was used for the experiments. The experiments were conducted in a 30 mL batch reactor using a maximum power of 300 W. For each experiment, 0.5 g of biomass was placed in the reactor along with 15 mL of solvent (CH_3COOH/H_2O). Before placing the reactor inside the microwave unit, the reaction mixture was pre-stirred at room temperature for 2 min. After reaction, the reactor was cooled down from the reaction temperature to 60 °C. Subsequently, the reactor was opened and its content, consisting of a mixture of liquid and solid, was transferred to a centrifuge tube. Centrifugation was used to separate the solid from the liquid. Then, the solid residue obtained after centrifugation was dried overnight at 105 °C and the liquid phase obtained was stored for further characterisation.

61.2.2 *Response Variables, Analytical Methods and Statistical Analysis*

Several response variables were used to analyse the effect of the operating conditions on the process. Table 61.1 summarises the response variables and the analytical methods used for their calculation. The solid fractions (both the original biomass and the solids produced) were characterised by means of ultimate and fibre (cellulose, hemicellulose, lignin and protein) analyses. Proximate and ultimate analyses were performed according to standard methods (ISO-589-1981 for moisture, ISO-1171-1976 for ash and ISO-5623-1974 for volatiles). Elemental analysis was carried out using an Exeter Analytical (Warwick, UK) CE440 Elemental Analyser, calibrated against acetanilide with a S-benzyl-thiouronium chloride internal standard. Fibre characterisation was performed by using the chemical titration method described by Hu et al. [6] to determine the amount of cellulose and hemicellulose,

Table 61.1 Response variables. Definitions and analytical techniques used in their determination

Product	Response variable	Analytical method
Liquid	$\text{Liquid yield}(\%) = \frac{\text{liquid compounds}(g)}{\text{mass of biomass}(g)} 100 = 100 - (\text{Gas yield} + \text{Solid yield})$	Balance
	$\text{Composition}(\text{C}-\text{wt.}\%) = \frac{\sum \text{mass of C of each compound}(g)}{\text{total mass of C in solution}(g)} 100$	GC/MS-FID and HPLC
Solid	$\text{Solid yield}(\%) = \frac{\text{mass of solid}(g)}{\text{mass of biomass}(g)} 100$	Gravimetric
	$\text{Fibre Composition}(\text{wt.}\%) = \frac{\text{mass of structural component}(g)}{\text{mass of solid residue}(g)} 100$	Chemical titration, Tappi T222 Method
	HHV (MJ/kg) = 0.3491 C (wt.%) + 1.1783 H (wt.%) – 0.1034 O (wt.%) – 0.015 N (wt.%) + 0.1005 S (wt.%)	Estimated
	$\text{C,H,O}(\text{wt.}\%) = \frac{\text{mass of C,H,O}(g)}{\text{mass of solid}(g)} 100$	Elemental Analysis
Gas	$\text{Gas yield}(\%) = \frac{\text{mass of gas}(g)}{\text{mass of biomass}(g)} 100$	Gravimetric

wt.% = weight percentage; C-wt.% = percentage in carbon basis

Table 61.2 Feedstock characterisation

Proximate analysis (wt.%)		Fibre analysis (wt.%)		Elemental analysis (wt.%)	
Moisture	7.26	Cellulose	12.41 ± 0.33	C	41.54 ± 0.19
Ash	1.31	Hemicellulose	7.16 ± 0.26	H	6.29 ± 0.17
Volatiles	45.09	Lignin	32.39 ± 2.47	N	6.32 ± 0.19
Fixed carbon	32.04	Protein	39.47 ± 1.17	O[a]	45.86 ± 0.17

[a]Oxygen was calculated by difference

while the lignin content was determined by the standard TAPPI T222 method. High-performance liquid chromatography (HPLC), gas chromatography (GC/MS-FID) and elemental analysis (described above) were used for the characterisation of the liquid phase. The experiments were planned according to a 2-level 3-factor Box–Wilson central composite face-centred (CCF, α: ±1) design. The results were analysed with an analysis of variance (ANOVA) with 95% confidence. In the interaction Figures, the evolution of these variables obtained from the ANOVA analysis of all the experiments performed was represented. In addition, when possible, some experimental points were added.

61.2.3 Rapeseed Meal Characterisation

The rapeseed meal used in this work was provided by Croda International (Widnes, UK). The most important physiochemical properties of the material are listed in Table 61.2.

61.3 Results and Discussion

61.3.1 Yields to Gas, Liquid and Solid

The effects of the operating variables and the most important interactions detected with the ANOVA analysis are plotted in Fig. 61.1. The effect of the temperature on the product distribution (yields to gas, liquid and solid) depends on the reaction time. For a short reaction time (0 min), the temperature does not significantly influence the gas yield and a negligible gas formation takes place regardless of the concentration of acetic acid. Conversely, the reaction temperature exerts a significant effect on the liquid and solid yields. In particular, an initial increase in the temperature from 150 to 190 °C increases the liquid yield and decreases the solid yield, while a further increase in the temperature up to 210 °C does not greatly modify liquid or solid production. These developments in the liquid and solid yields are accounted for by the positive kinetic effect of the reaction temperature on biomass

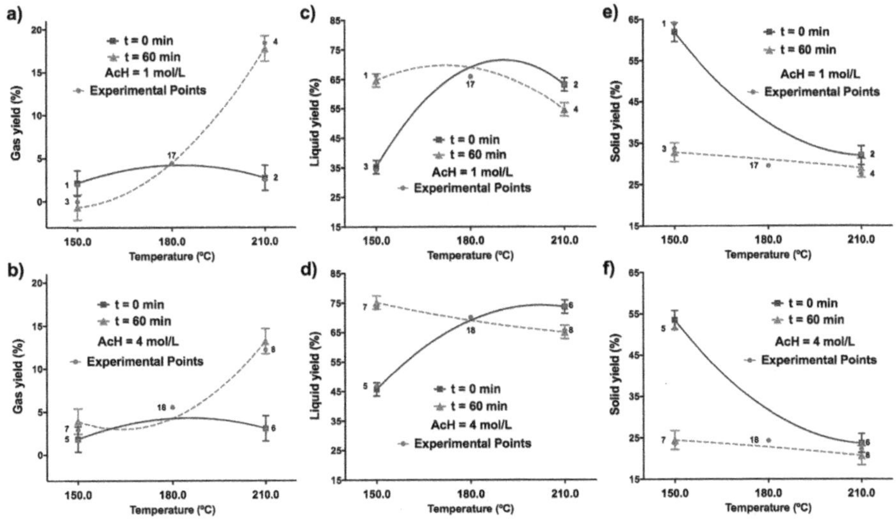

Fig. 61.1 Influence of the operating conditions on the gas (**a** and **b**), liquid (**c** and **d**) and solid (**e** and **f**) yields. Bars are LSD intervals with 95% confidence

solubilisation, thus increasing and decreasing the liquid yield and solid yield, respectively. This leads to the solubilisation of the cellulosic and hemicellulosic matter in the liquid without significantly solubilising the lignin content of the solid, thus allowing a high purity lignin solid fraction to be produced.

An increase in the reaction time modifies the effect of the temperature on the product distribution. Regardless of the concentration of acetic acid, the gas yield is negligible and unaffected by the reaction time (0–60 min) between 150 and 180 °C. However, an increase in the gas yield occurs between 190 and 210 °C and when the reaction time increases from 0 to 60 min. The combination of both high temperatures and long reaction times favours the formation of gases from some of the species produced during biomass hydrolysis secondary reactions such as carboxylic acids, ketones and furans through decarboxylation reactions [2, 3]. In addition, gas formation could also be produced from the thermal decomposition of the proteins present in the solid via deamination [3]. As a result, for a long reaction time (60 min), an increase in the temperature between 180 and 210 °C produces a sharp increase in the gas yield. Conversely, the effect of the reaction time on the yields to liquid and solid is more marked at low temperature (150–190 °C) than at high temperature (190–210 °C). At low temperature, an increase in the reaction time from 0 to 60 min leads to a sharp increase in the liquid yield along with a pronounced decrease in the solid yield. This same increment in the reaction time between 190 and 210 °C slightly decreases the liquid yield; the solid yield remaining unaffected. This development might be accounted for by the long reaction time employed in the experiment, which is high enough to kinetically control the process. In addition, this also shows the high efficiency of microwave heating [5, 7–9].

As a result, when a long reaction time (60 min) is used, the effect of the temperature on the solid yield is negligible. The liquid yield slightly decreases with increasing temperature due to the sharp increase occurring for the gas yield. This might indicate that part of the liquid products is converted into gases if long reaction times and high temperatures are used [3].

The effect of the concentration of acetic acid on the yields to gas, liquid and solid can be studied by comparing Fig. 61.1 a, c and e with b, d and f, respectively. For a short reaction time, an increase in the concentration of acetic acid between 1 and 4 mol/L does not significantly modify the gas yield and a negligible gas formation takes place regardless of the concentration of acetic acid. On the contrary, the concentration of acetic acid has a significant influence on the gas yield when the reaction time increases and different trends for this variable are observed depending on the temperature. When a 60 min reaction time is used, an increase in the acid concentration from 1 to 4 mol/L leads to an increase in the gas yield between 150 and 180 °C, while this same increase reduces gas formation between 180 and 210 °C. Acetic acid exerts a significant catalytic effect on the process by producing a greater spread of decarboxylation reactions, which leads to an increase in gas formation from biomass secondary decomposition products and proteins [2, 3]. At high temperature, gas formation decreases probably because of the formation of humins and char from the furfural and HMF obtained from sugars at high temperature [10–12]. The influence of the concentration of acetic acid on the liquid and solid yields does not depend on the temperature or the reaction time, and similar trends are observed regardless of the temperature and time used in the experiments. In general, an increase in the concentration of acetic acid from 1 to 4 mol/L leads to an increase in the liquid yield and a decrease in the solid yield. The positive catalytic effect of the acid enhances the dissolution of the cellulose and hemicellulose fractions [5, 8, 13], which increases the liquid yield and decreases the solid yield.

61.3.2 Solid Properties

61.3.2.1 Proportions of Cellulose, Hemicellulose and Lignin

The amounts of cellulose, hemicellulose, lignin and proteins in the solid fractions vary as follows: 0–28%, 0–28%, 26–88% and 11–28%, respectively. Figure 61.2 shows the effect of the operating variables and the most important interactions detected with the ANOVA analysis on the fibre analysis of the solid. The effect of the temperature on the fibre analysis of the solid product depends on the reaction time and the concentration of acetic acid. For a short reaction time (0 min) when a diluted (1 mol/L) acid solution is used (Fig. 61.2 a, c, e and g), an increase in the temperature from 150 to 180 °C leads to a sharp decrease in the proportions of cellulose, hemicellulose and proteins together with a pronounced increase in the proportion of lignin of the solid, where a maximum is reached. This allows the production of relatively high purity lignin (85 wt.%) from rapeseed meal, proteins being the only

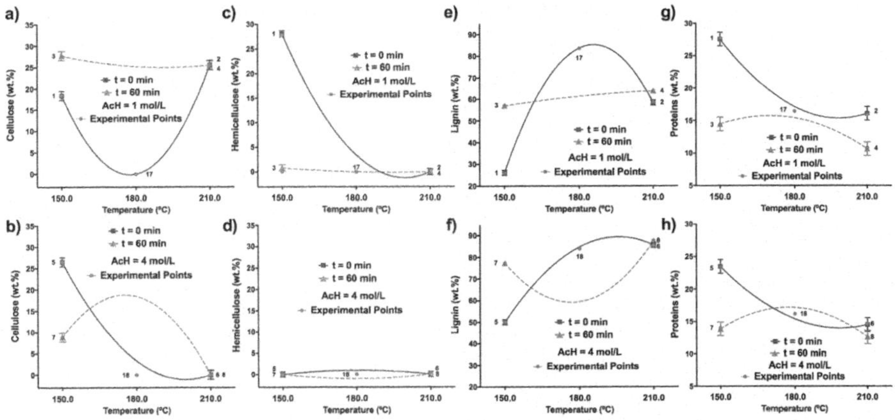

Fig. 61.2 Influence of the operating conditions on the proportions of cellulose (**a** and **b**), hemicellulose (**c** and **d**), lignin (**e** and **f**) and proteins (**g** and **h**). Bars are LSD intervals with 95% confidence

impurity present in the solid. This development accounts for the solubilisation of the cellulosic and hemicellulosic matter without significant lignin solubilisation during microwave hydrothermal treatment [5]. In addition, the protein content of the solid decreases due to the decomposition of the proteins into liquid and gaseous products via decarboxylation and deamination reactions [2, 3]. A further increase in the temperature up to 210 °C increases and decreases the proportion of cellulose and lignin, respectively, without altering the relative amounts of hemicellulose and proteins. An increase in the temperature might promote the formation of humins and char from the sugars produced during the dissolution of cellulose and hemicellulose [10–12]. The presence of humins and char can interfere with the chemical titration method and humins might have been identified as cellulose in the analysis, thus producing and artificial increase in the cellulose content of the solid [6].

The comparison between Fig. 61.2 a, c, e and g with b, d, f and h shows that an increase in the concentration of acetic acid from 1 to 4 mol/L (for 0 min reaction time) decreases the proportion of hemicellulose and proteins and increases the relative amount of lignin regardless of the temperature (150–210 °C) due to the positive catalytic effect of acetic acid in the process. In addition, and very interestingly, in this work acetic acid also exerts an inhibitory effect on humins and char formation, which allows a polysaccharide-free lignin, with relatively high purity (88 wt.%) to be produced between 190 and 210 °C. The reaction time modifies the effects of the temperature and concentration of acetic acid on the fibre composition of the solid product. When a diluted acid solution is used, an increase in the reaction time from 0 to 60 min leads to a decrease in the proportions of hemicellulose and proteins together with a decrease in the relative amount of cellulose of the solid (Fig. 61.2 a, c, e and g). Long reaction times favours the solubilisation of cellulose and hemicellulose even at the lowest temperature used in this work (150 °C) due to the efficiency of microwave heating [5]. However, this also produces the formation of humins and char from some of the species solubilised in the liquid product, which leads to an artificial increase in the relative amount of the cellulose content of the solid. The effect of the reaction time on the proportion of lignin depends on the temperature. While an increase in the reaction time increases the relative amount of lignin between 150 and 165 °C, this same increment decreases the proportion of lignin between 165 and 200 °C. At low temperature, the formation of humins and char takes place to a lesser extent, which result in a higher proportion of lignin in the solid. Conversely, elevated temperatures together with long reaction times increase the production of humins and char [10–12].

An increase in the concentration of acetic acid from 1 to 4 mol/L when a long reaction time is used exerts a significant effect on the proportion of cellulose and lignin, without modifying the proportions of hemicellulose and proteins (Fig. 61.2 a, c, e and g vs. b, d, f and h, respectively). As described earlier, the formation of humins and char is inhibited when a concentrated (4 mol/L) solution of acetic acid is used, and therefore, an increase in the reaction time from 0 to 60 min leads to an increase in the relative amount of lignin together with a decrease in the proportion of cellulose in the solid product. In addition, the effect of the temperature on the proportions of hemicellulose and proteins is negligible. Conversely, the temperature exerts a significant influence on the proportion of cellulose and lignin when a long reaction time and a concentrated (4 mol/L) solution of acetic acid are used. Between 150 and 180 °C, the proportion of cellulose and lignin increases and decreases respectively, while the opposite trend takes place between 180 and 210 °C, i.e. an increase in the relative amount of lignin together with a decrease in the proportion of cellulose due to the lesser humins formation occurring when a concentrated acid solution is used.

61.3.2.2 Elemental Analysis and Calorific Value

The relative amounts (wt.%) of C, H, O and N in the solid fraction shifted between 46 and 63, 5.8 and 6.4, 28 and 42, and 2 and 6, varying the higher heating value (HHV) of the solid between 19 and 26 MJ/kg. Figure 61.3 shows the effect of the operating variables and the most important interactions detected with the ANOVA analysis. The effect of the temperature on the elemental analysis of the solid fraction depends on the reaction time and the concentration of acetic acid. When a short reaction time is used, the concentration of acetic acid does not greatly influence the elemental analysis or the HHV of the solid and similar results are obtained regardless of the acid concentration. In particular, for a short reaction time (0 min), an initial increase in the temperature between 150 and 180 °C increases the proportion of C and the HHV and decreases the relative amounts of H and O. Conversely, further increasing the temperature up to 210 °C has the opposite effect, i.e. the amount of C and the HHV decrease and the proportions of H and O in the solid increase. The temperature does not influence the N content of the solid when short reaction times are used, while a small increase takes place when the temperature increases from 150 to 210 °C and a long reaction time is used. The effect of the reaction time depends on the concentration of acetic acid. When a diluted acid solution (1 mol/L)

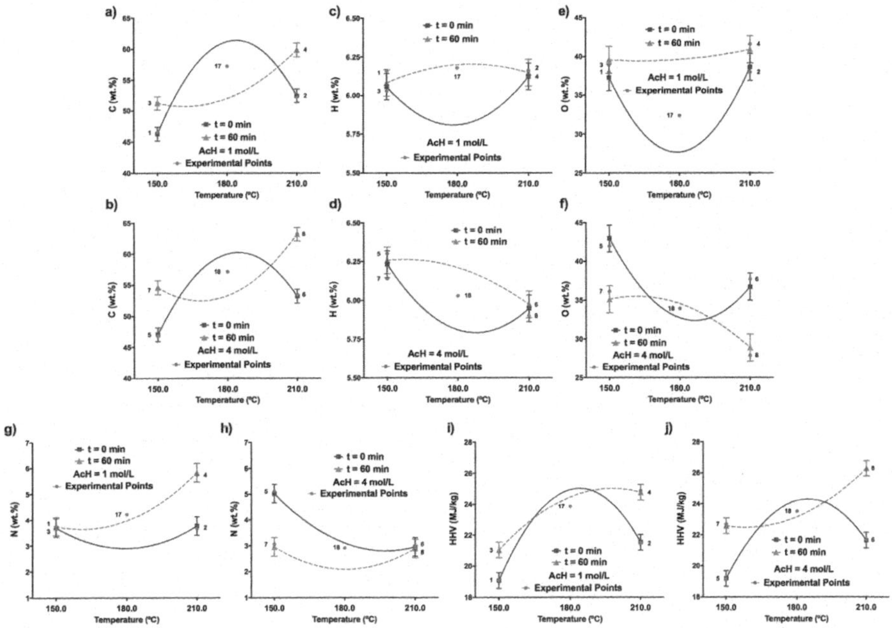

Fig. 61.3 Influence of the operating conditions on the concentrations of C (**a** and **b**), H (**c** and **d**), O (**e** and **f**), N (**g** and **h**) and HHV (**i** and **j**). Bars are LSD intervals with 95% confidence

is used, an increase from 0 to 60 min increases the relative amounts of H, O and N and decreases the proportion of C in the solid when a temperature ranging from 165 to 195 °C is used. In addition, an increase in the reaction time progressively decreases the effect of the temperature on the elemental analysis and the HHV of the solid product due to the positive kinetic effect of the reaction time [5]. As a result, when a reaction time of 60 min is used, the reaction temperature does not affect the proportions of H and O, while a small increase occurs for the relative amounts of C and N when increasing the temperature from 150 to 210 °C. The HHV slightly increases between 150 and 180 °C and remains steady with a further increase up to 210 °C. An increase in the concentration of acetic acid when a long reaction time is used modifies the effect of the temperature on the proportions of H, O and N and the HHV of the solid. In this case, an increase in the temperature from 150 to 210 °C decreases the proportions of H and O. This leads to a decrease in the HHV of the solid.

61.3.3 Liquid Chemical Composition

The liquid product consists of a mixture of oligosaccharides (DP2–6 and DP > 6) and mono-/disaccharides, carboxylic acids, ketones, furans and nitrogen compounds; their relative amount (in carbon basis, C-wt.%) in the liquid product

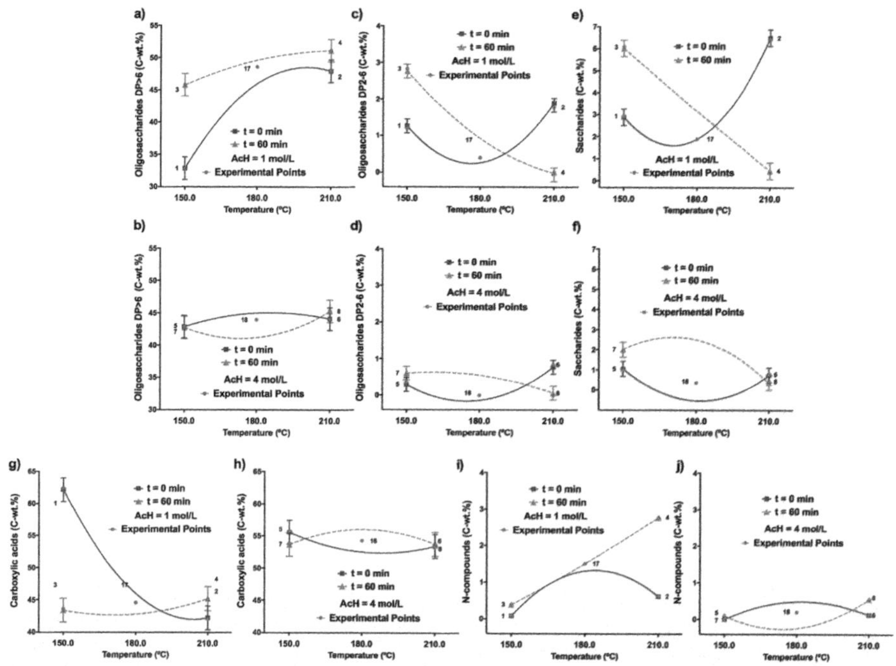

Fig. 61.4 Influence of the operating conditions on the concentrations of oligosaccharides DP > 6 (**a** and **b**), oligosaccharides DP 2-6 (**c** and **d**), saccharides (**e** and **f**), carboxylic acids (**g** and **h**) and N-compounds (**i** and **j**). Bars are LSD intervals with 95% confidence

varying as follows: 33–51%, 0–3%, 0–6%, 40–62%, 0–1%, 0–3% and 0–3%. Saccharides include cellobiose, xylose, glucose, fructose, mannose, arabinose, rhamnose and levoglucosan. Carboxylic acids comprise lactic, formic, levulinic, glucuronic, galacuronic and acetic acids. Acetic acid is the major compound for this family as it was initially loaded and used as a catalyst. Figure 61.4 shows the effect of the operating variables and the most important interactions detected with the ANOVA analysis on the most abundant compounds present in the liquid phase. The effect of the temperature on the chemical composition of the liquid phase depends on the reaction time and the concentration of acetic acid. In this respect, while the reaction time has a very important influence for diluted acid solution, it has a negligible influence when the concentration of acetic acid increases up to 4 mol/L. When a diluted acetic acid concentration (1 mol/L) and short reaction time are used, an increase in the reaction time from 150 to 190 °C sharply increases and decreases the relative amounts of oligosaccharides and carboxylic acids, respectively. In addition, the proportions of DP2–6 oligosaccharides and saccharides slightly decrease, while the relative amount of N-compounds increases. A further increase in the temperature up to 210 °C exerts a negligible effect on the proportions of oligosaccharides and carboxylic acids, while the relative amounts of DP2–6 oligosaccharides and saccharides increase slightly and the N-compounds decreases.

An increase in the reaction time when a diluted acid solution (1 mol/L) is used has two different consequences for the composition of the liquid phase depending on the temperature. On the one hand, at temperatures lower than 190 °C, an increase in the reaction time from 0 to 60 min significantly increases the proportions of oligosaccharides (DP > 6 and DP2–6) and decreases the relative amount of carboxylic acids, without modifying the relative amount of N-compounds. An increase in the reaction time produces a greater spread of the hydrolysis reactions, which experimentally increases the proportion of cellulose- and hemicellulose-derived species in the liquid, thus decreasing the relative amount of carboxylic acid (mostly acetic acid). On the other hand, when a temperature between 190 and 210 °C is used, this same increase in time does not modify the proportions of DP > 6 oligosaccharides or carboxylic acids. In addition, the proportions of DP2–6 oligosaccharides and saccharides decrease and the relative amount of N-compounds increases due to the solubilisation of proteins in the liquid when long reaction times are used [3, 14–16]. This lower influence of the temperature on the composition of the liquid phase produces that an increase in the temperature from 150 to 210 °C does not greatly modify the proportions of oligosaccharides (DP > 6) or carboxylic acids when a 60 min reaction time is used. An increase in the concentration of acid (Fig. 61.4 a, c, e and g vs. b, d, f and h) diminishes the influence of the temperature and reaction time on the composition of the liquid phase as stated earlier. For a 4 mol/L acetic acid concentration, oligosaccharides and carboxylic acids are the most abundant compounds in the liquid phase, their composition being around 43 C-wt.% and 55 C-wt.%, respectively, regardless of the temperature and reaction time. Furthermore, the relative amounts of oligosaccharides (DP2–6) and saccharides are very low (<2%) and the effects of the temperature and reaction time, although being statistically significant, are not important from a practical point of view.

61.3.4 Theoretical Optimisation

Optimum conditions were sought for the selective co-production of high purity lignin and soluble oligosaccharides from rapeseed meal. The optimisation predicts an optimum at 186 °C using a concentration of acetic acid of 1 mol/L for a total reaction time of 2 min, i.e. only the ramping time (2 min ramping with a holding time of 0 min). Under such conditions, it is possible to selectively convert 36% of the original feedstock into relatively high purity (85 wt.%) lignin, the rest (63%) being converted into a mixture of soluble oligosaccharides containing the acetic acid used in the experiment. This indicates that all the lignin and the vast majority of the proteins initially presented in the biomass remained in the solid during the isolation process. In addition, to increase the effectiveness and sustainability of this process, acetic acid can be recovered from the oligosaccharide solution by vacuum distillation (for example) and used again for further experiments. This strategy allows the simultaneous production of sugar-free, relatively high purity lignin (85 wt.%) along a sugar-rich solid fraction comprising oligosaccharides and mono-/disaccharides (92 C-wt.%) with several applications in the chemical and biological industries.

61.4 Conclusions

This work addresses a novel microwave-assisted, acid-catalysed process for the selective production of lignin and oligossacharides from rapeseed meal. The most important conclusions are summarised as follows:

1. The gas, liquid and solid yields are significantly influenced by the operating conditions, their yields varying by 0–18%, 22–64% and 34–74%, respectively. Increasing the temperature or time increased the liquid yield and decreased the solid yield due to the progressive solubilisation of the cellulosic and hemicellulosic contents of the original feedstock, which resulted into the production of a rich lignin solid in some cases.
2. The solid fraction consisted of high purity lignin (26–88 wt.%) together with unreacted cellulose (0–28 wt.%), hemicellulose (0–28 wt.%) and proteins (11–28 wt.%). An increase in the temperature or reaction time decreased the amount of cellulose and hemicellulose and increased the lignin purity of the solid when temperatures lower than 190 °C were used. A further increase up to 210 °C led to a decrease in the lignin content of the solid due to the formation of humins. However, acetic acid displayed an inhibitory effect on humins formation, which allowed high temperatures and reaction times to be used when using concentrated acid solutions.
3. The liquid phase was made up of oligosaccharides (DP2–6 and DP > 6) and mono-/disaccharides, carboxylic acids, ketones, furans and nitrogen compounds. Their relative amount (in carbon basis, C-wt.%) varied by 33–51%, 0–3%, 0–6%, 40–62%, 0–1%, 0–3% and 0–3%. DP > 6 oligosaccharides and carboxylic acids were strongly influenced by the operating conditions, while the variations observed for the other species were less important. An increase in the temperature and reaction time led to an increase in the proportion of oligosaccharides and decreased the relative amount of carboxylic acids in the liquid.
4. An optimum for this process was found at 186 °C using a concentration of acetic acid of 1 mol/L and employing a total reaction time as short as 2 min. These conditions maximise the solubilisation of cellulose and hemicellulose and minimise lignin solubilisation, thus allowing the selective and simultaneous production of a rich (85 wt.%) lignin solid and a oligossacharide-rich water solution. In addition, acetic acid could be recovered from the sugar mixture, which not only can improve the economy and efficiency of the process but also allows the production of high purity saccharides (92 C-wt.%).

Acknowledgements This research has been funded by the Industrial Biotechnology Catalyst (Innovate UK, BBSRC, EPSRC) to support the translation, development and commercialisation of innovative industrial biotechnology processes (EP/N013522/1). EPSRC for research grant number EP/K014773/1.

References

1. Giannakopoulou K, Lukas M, Vasiliev A, Brunner C, Schnitzer H (2010) Conversion of rapeseed cake into bio-fuel in a batch reactor: effect of catalytic vapor upgrading. Microporous Mesoporous Mater 128:126–135
2. Egües I, Alriols MG, Herseczki Z, Marton G, Labidi J (2010) Hemicelluloses obtaining from rapeseed cake residue generated in the biodiesel production process. J Ind Eng Chem 16:293–298
3. Pińkowska H, Wolak P, Oliveros E (2014) Hydrothermolysis of rapeseed cake in subcritical water. Effect of reaction temperature and holding time on product composition. Biomass Bioenergy 64:50–61
4. de Melo EM, Clark JH, Matharu AS (2017) The Hy-MASS concept: hydrothermal microwave assisted selective scissoring of cellulose for in situ production of (meso)porous nanocellulose fibrils and crystals. Green Chem 19:3408–3417
5. Zhou L, Budarin V, Fan J, Sloan R, Macquarrie D (2017) Efficient method of lignin isolation using microwave-assisted acidolysis and characterization of the residual lignin. ACS Sustain Chem Eng 5:3768–3774
6. Hu L, Luo Y, Cai B, Li J, Tong D, Hu C (2014) The degradation of the lignin in *Phyllostachys heterocycla* cv. pubescens in an ethanol solvothermal system. Green Chem 16:3107–3116
7. Briens C, Piskorz J, Berruti F (2008) Biomass valorization for fuel and chemicals production—a review. Int J Chem React Eng 6:51
8. Li T, Remón J, Jiang Z, Budarin VL, Clark JH (2018) Towards the development of a novel "bamboo-refinery" concept: Selective bamboo fractionation by means of a microwave-assisted, acid-catalysed, organosolv process. Energy Convers Manag 155:147–160
9. Li T, Remón J, Shuttleworth PS, Jiang Z, Fan J, Clark JH et al (2017) Controllable production of liquid and solid biofuels by doping-free, microwave-assisted, pressurised pyrolysis of hemicellulose. Energy Convers Manag 144:104–113
10. Remón J, García L, Arauzo J (2016) Cheese whey management by catalytic steam reforming and aqueous phase reforming. Fuel Process Technol 154:66–81
11. Remón J, Laseca M, García L, Arauzo J (2016) Hydrogen production from cheese whey by catalytic steam reforming: preliminary study using lactose as a model compound. Energy Convers Manag 114:122–141
12. Remón J, Ruiz J, Oliva M, García L, Arauzo J (2016) Cheese whey valorisation: production of valuable gaseous and liquid chemicals from lactose by aqueous phase reforming. Energy Convers Manag 124:453–469
13. Briones R, Serrano L, Llano-Ponte R, Labidi J (2011) Polyols obtained from solvolysis liquefaction of biodiesel production solid residues. Chem Eng J 175:169–175
14. Rogalinski T, Herrmann S, Brunner G (2005) Production of amino acids from bovine serum albumin by continuous sub-critical water hydrolysis. J Supercrit Fluids 36:49–58
15. Sato N, Quitain AT, Kang K, Daimon H, Fujie K (2004) Reaction kinetics of amino acid decomposition in high-temperature and high-pressure water. Ind Eng Chem Res 43:3217–3222
16. Yoshida H, Terashima M, Takahashi Y (1999) Production of organic acids and amino acids from fish meat by sub-critical water hydrolysis. Biotechnol Prog 15:1090–1094

Chapter 62
Gas-Fired Heat Pumps as a Replacement for the Condensing Boiler

R. E. Critoph, S. J. Metcalf, and A. Rivero Pacho

62.1 Introduction

62.1.1 Technologies

In the past there have been many approaches to the provision of gas-driven heat pumps that have reached varying stages of development. Most, but not all, have not gone beyond RD&D to commercial launch. Systems can be categorised as engine-driven or sorption.

Engine-driven systems can range from conventional gas-fuelled engines driving conventional vapour compression heat pumps to novel engine types, sometimes combined with novel heat pump cycles. In the past, extensive research went into Stirling-Stirling, Rankine-Rankine and other concepts. Whilst some of the more adventurous ideas may still have some potential, the only ones to survive today as commercial systems are conventional engine types (spark ignition or diesel) driving conventional heat pump cycles via open compressors and recovering engine waste heat as useful output. The vast majority of these systems are sized for larger buildings (30 kW+) where a strict maintenance schedule can be enforced and economies of scale work well. There are four major manufacturers, all Japanese, and European sales are estimated at around 10,000 in total [1]. Sales in Japan are an order of magnitude higher. The focus of this work is on domestic applications perhaps extending to light commercial, a size range less suited to the complexity and maintenance of engine-driven systems, although in Japan there are domestic products championed by Osaka Gas. Within the EU there is consensus that the way forward for domestic units is through sorption systems.

R. E. Critoph (✉) · S. J. Metcalf · A. R. Pacho
School of Engineering, Coventry, UK
e-mail: R.E.Critoph@warwick.ac.uk

A. Sayigh (ed.), *Renewable Energy and Sustainable Buildings*, Innovative Renewable Energy, https://doi.org/10.1007/978-3-030-18488-9_62

Absorption, adsorption cycles: What all sorption systems have in common is that they use the ability of a liquid or solid sorbent to ab/adsorb large amounts of refrigerant to be used in a 'thermal compressor', analogous to the mechanical compressor in an electric heat pump. The refrigerant is ab/adsorbed at a low pressure and temperature, and then the sorbent is heated (in our case by burning gas) and the refrigerant is then driven out (desorbed) at high pressure. The fundamental thermodynamics are the same whether the refrigerant gas is being absorbed into a liquid or adsorbed into a solid. There are engineering advantages and disadvantages in using solids or liquids but neither is inherently 'better', and the choice of one or the other in a particular application will depend on the way they are implemented and the resulting efficiency, cost, size, etc.

Refrigerants: The refrigerants used by either absorption or adsorption cycles need the same characteristics (high latent heat being the major one), and the three main refrigerants considered for both are the same: water, methanol and ammonia. Water has the highest latent heat by far and so for the same complexity and cost of machine gives higher COPs [COP is defined as heat output to load/heat input from gas]. Water also has the benefits of being low cost, non-toxic, environmentally friendly, etc. but has one major disadvantage; its very low operating pressure. From a practical point of view, it is not possible to boil water in the evaporator of a heat pump at less than about 5 °C. In the case of an air source heat pump, this would restrict the ambient temperature at which any heat pumping could be carried out to more than 10 °C which is clearly not sensible.

Ammonia has much lower latent heat but none of the other disadvantages of water. It is the refrigerant of choice in large industrial systems, is not a global warming or ozone depleting gas and can be used at ambient temperatures below −20 °C. Its higher pressure allows smaller pipe sizes, thereby reducing cost. Its disadvantages are that it is chemically incompatible with copper or brass and that it is toxic; however neither of these problems are insurmountable, and existing regulations govern the design of refrigerators or heat pumps that use ammonia refrigerant.

Methanol is (in pressure terms) between ammonia and water, and although methanol-based systems are well below atmospheric pressure, it should be possible to use it to extract heat from ambient air at around 0 °C. It is environmentally friendly but has the disadvantage that it decomposes at temperatures between 120 and 190 °C depending on the chemical environment. High efficiency systems tend to need heat input from the burning fuel at higher than 120 °C.

62.1.2 Existing and Near-Market Products

There are at present two products on the market, both manufactured by Robur [2]. The Robur product is an ammonia-water absorption heat pump (i.e. ammonia refrigerant, water absorbent) which offers air, water and ground source options. It is a development of technology previously used for air-conditioning and thus is

Fig. 62.1 Robur gas heat pump

comparatively mature. The present air source machine can deliver domestic hot water at 65 °C (gross COP 1.24) and will supply 38 kW to radiators (supply temperature 50 °C) with a COP of 1.52 (gross) and 1.38 (net). This represents a saving of about 40% in gas consumption compared to a condensing boiler. The unit is a single module intended to be positioned outside the heated building and 854(w) × 1256(d) × 1281(h). An 18 kW unit, more suited to typical UK dwellings, is also available. The cost of the 40 kW unit is c. £12,000 (Fig. 62.1).

Other products under development: There are at least two other manufacturers developing ammonia-water absorption machines: Stone Mountain in the USA and a European manufacturer. ClimateWell is researching an ammonia-salt adsorption concept.

62.1.3 Research at the University of Warwick

The University of Warwick is currently developing a very compact carbon-ammonia adsorption cycle gas-fired heat pump to be a box-for-box exchange with a conventional domestic gas boiler, with much reduced size and capital cost compared to alternative machines on the market (<4 times). Also for ease of retrofit, it is an air source machine. It must provide at least 10 kW of heat output, which is the requirement for an average semi-detached UK household. The working conditions of the machine are heating water to 50 °C for low-temperature radiators, and the evaporating temperatures vary with the ambient temperature but can range from 0 to 10 °C. The target of internal COP achieved by the machine should be higher than 1.4 so that the gas utilisation efficiency is higher than 1.25. With this predicted delivered COP, the machine will reduce the household gas consumption by 30% compared to a conventional condensing boiler.

The refrigerant used is ammonia since it has the ability to operate below 0 °C evaporating temperature which makes it suitable for air source operation. Also, the high working pressures of the ammonia in the system enables a compact machine, and the achievable power density is almost entirely heat transfer, rather than mass transfer limited. Ammonia is paired with active carbon, as adsorbate, which has been chosen since it is stable at high driving temperatures. It does have lower and more gradual concentration change than adsorbents with S-shaped curves or salts but does not suffer from a sudden 'switch off'.

Figure 62.2 shows the schematics of the heat pump. On the left-hand side, there is the burner that provides heat to the heat transfer fluid (pressurised water). The heat transfer fluid, at a high and at a low temperature, is pumped through two sorption generators, one being cooled, adsorbing ammonia and the other being heated, desorbing ammonia. The ammonia that has been desorbed from Bed 1 first passes through the condenser, releasing heat to the house heating system, then through the expansion valve and then through the evaporator, where it absorbs heat from the ambient air, and finally it is adsorbed in Bed 2. After (typically) 2 or 3 minutes, the beds are switched, Bed 2 being heated and Bed 1 being cooled. There is heat and mass recovery between the beds to improve the COP.

The sorption generator is effectively a thermally driven compressor and is the most critical part of the design. The challenge is to design and build a sorption generator that has excellent heat transfer, not just to reduce the size and cost of the machine but also to allow effective transfer of heat between the beds in a high COP regenerative cycle. However, it must also have low thermal mass since it is repeatedly cycled in temperature and the sensible heat used, whilst partly recoverable is not multiplied by the heat pump COP. In addition, the generator must be capable of reliable mass production at a cost of at most a few tens of Euros. Without this

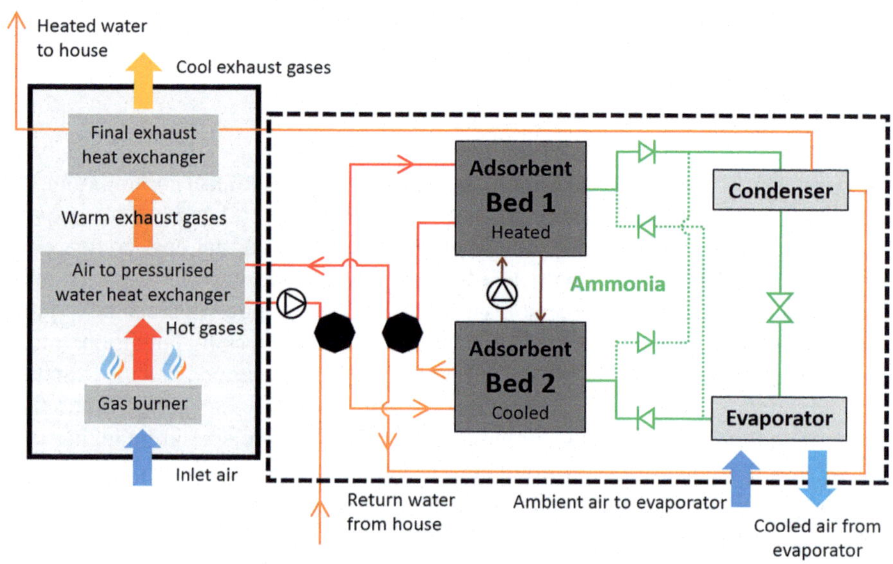

Fig. 62.2 Warwick adsorption heat pump schematic

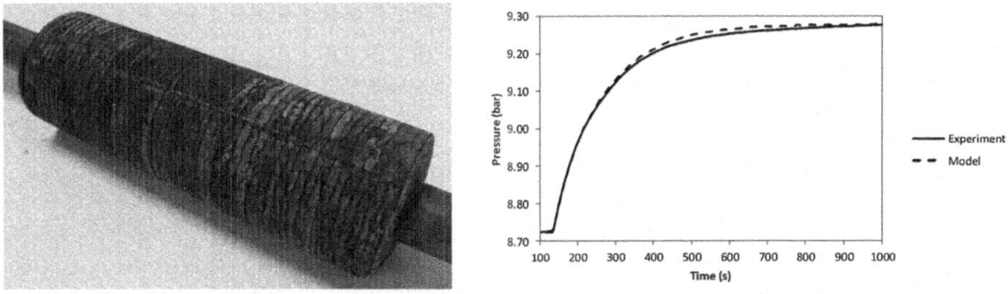

Fig. 62.3 (**a**) Carbon laminate adsorption generator, (**b**) comparison between large temperature jump experiment and computational model

challenge being met, it is unlikely that a domestic adsorption heat pump to replace a gas boiler will ever be viable.

Figure 62.3a shows a sample of the carbon laminate used in the construction of the adsorption generators. The carbon is a monolithic type utilising a binder. The laminate is made by compression in a die before pyrolysis in a furnace to bind the composite into a monolithic layer between the fins. The thermal performance of these laminate heat exchangers has been verified by performing a large temperature jump experiment on a test section and comparing it to a transient model developed in MATLAB. Figure 62.3b shows the comparison and good thermal properties match between the large temperature jump experiment and the model.

A two-dimensional finite difference simulation model in MATLAB has been written to explore how varying the geometry of the generator, dimensions and control parameters affect the coefficient of performance (COP) and power output under specified conditions and sorption materials. In all the simulation runs, in order to compare designs, the cooling phase assumes an inlet water temperature of 50 °C, in the heating phase the water inlet is at 170 °C, the evaporating temperature is 5 °C, and the condensing temperature is 50 °C. The heating/cooling cycle times and recovery times used in the simulations vary between 50 and 200 s and between 0 and 200 s, respectively.

Figure 62.4 show how the performance of the machine is affected by various geometric variables such as the pitch between tubes (where the heat transfer fluid flows), fin thickness and laminar carbon thickness. The performance of the machine also varies depending on the heating, cooling and heat recovery times, and there is a trade-off between COP and heating power. This allows the machine to deliver a higher heating power at lower efficiencies if necessary at a certain time. The performance map of the chosen geometry corresponds to the upper line. By varying control parameters, it should be possible, for example, to have an internal COP of 1.39 delivering 10 kW or 1.34 at 24 kW.

The heat pump prototype has been constructed to test the performance of the generators and to validate the developed computational model of the machine and is currently under test. The machine is being tested with a system of electrical heating and cooling baths that provide up to 170 °C of pressurised water, heat transfer fluid, for high temperature heat input. It also allows 10 kW of heat rejection for the condenser and cooler and provides a glycol flow down to −10 °C for the evaporator (Fig. 62.5).

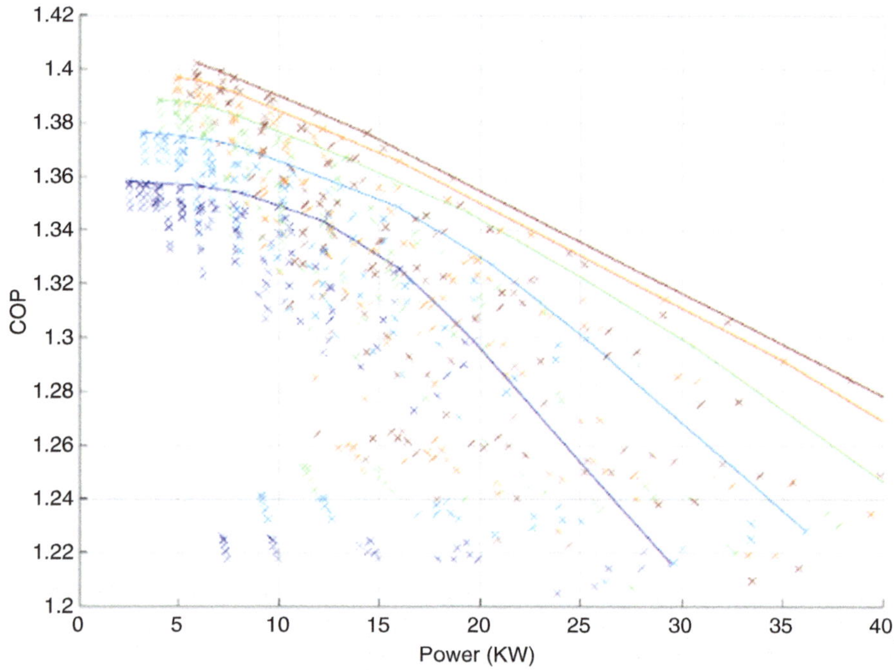

Fig. 62.4 Performance envelopes for different tube pitches

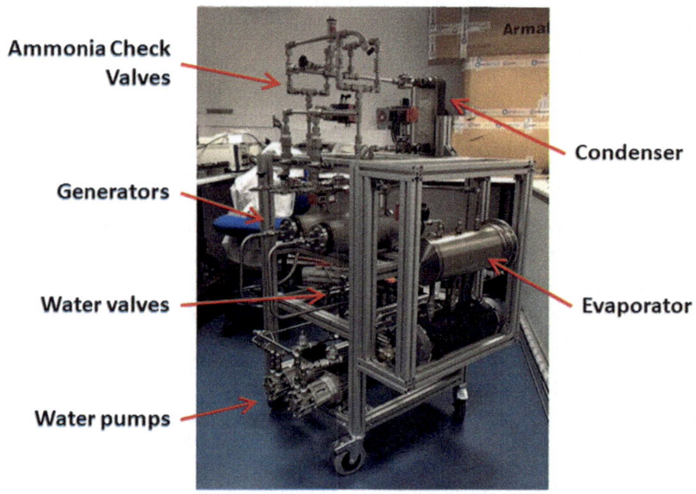

Fig. 62.5 Heat pump assembly in laboratory

62.2 Gas Heat Pump Market Potential in the Short Term

The information presented below is largely adapted from studies [3, 4] carried out for Sorption Energy on the market for its heat pump concept, plus a report by Delta-ee [5]. The premise is that in the short term gas heat pumps must compete with condensing gas boilers rather than electric heat pumps and that any system that has the potential for major impact must be suitable for retrofit and be air source. Whilst ground source heat pumps and new-built homes may be important niche markets, they are dwarfed by the retrofit market size.

The world domestic boiler market is estimated at between 10 and 12 million units per annum. This includes all fuels: gas, oil and electricity. The UK is the biggest market in terms of volume and value, followed by South Korea and Italy. The size of the UK market is around 1.5 million domestic boilers units per annum. UK sales were 1.4 M in 2012 [6]. Historically, regulation has driven technology uptake and witnessed the adoption of Band A condensing boilers influenced by building regulation which saw condensing boiler sales go from zero to about one million units over a 6-year period [7].

The route by which boilers pass from manufacturer to private householder is dominated (79%) by contractor/installers who buy from builder merchants, and this must be the target route if there is to be a widespread uptake of gas heat pump technology. In private replacements, the installer/contractor generally has most influence over the boiler selected as he/she advises the householder. The factors that influence the installer/contractor are:

- Familiarity with boiler (installed them before, received training from manufacturer, accredited installer).
- Easy and quick to install (weight/fittings). Must be a 'one man fit'.
- Good back-up support from manufacturer should there be a problem, e.g. spares available.
- Performance/reliability (if the boiler breaks down, it will be under warranty (2 years generally) from the manufacturer, but the installer will be called back).
- Incentives: manufacturers sometimes give vouchers to installers for each boiler installed.

Social housing replacements will be determined by the provider's purchasing manager, and decisions are based on 'best lifetime value'. Deals will be in place with OEMs. Large social housing groups generally have one or more boiler brands which they have selected as 'preferred' for installation. The small range of boilers means they are not completely reliant on one supplier but also do not have a large variety of boilers installed which would be more expensive and complex to maintain. Decisions are made on 'best lifetime value', i.e. minimising capital cost/installation cost plus lifetime maintenance costs.

To summarise, private new-built and retrofit homes are best target markets. Retrofit is by far the largest market and essential if gas heat pumps are to become mass market.

The payback times depend on the size of house and comfort standards since in the domestic range the capital cost of the heat pump is not strongly dependent on the nominal power output. Assuming economies of scale in production, payback times of 3 years seem feasible for larger houses and perhaps 5 years in smaller ones. In the UK, the RHI (Renewable Heat Incentive) could improve the situation and make gas heat pumps a very attractive replacement for conventional gas boilers, depending on how it might operate. Initially it would be sensible to target larger owner occupied homes (4 bed+, 3 m homes or 12% of the total) where residents tend to have longer investment horizon.

In private new-built homes (150,000 p.a.), gas heat pumps will be most efficient with lower-temperature water output. This requires a well-insulated home with modern radiators or underfloor heating. This means it is well suited to a new build or major refurb.

62.3 Market Potential for Gas Heat Pumps in the Longer Term

It is a commonly stated criticism of the concept of gas-fired heat pumps that in the longer term there will be an all-electric renewable energy system; electric heat pumps will be ubiquitous, and there will be no place for burning fuel. Whether this is a likely outcome depends both on the ultimate technical potential of gas heat pumps and more difficult questions relating to the energy (electricity, gas, heat) supply grid and its development.

Seasonal COPs of present gas-fired systems range from 1.1 to 1.5, and the theoretical limit could exceed 3, but keeping within what might be possible in systems we can envisage now, a seasonal COP in the UK of 2.0 [gross CV, air to water at 50 °C] is certainly not unreasonable. Machines that can achieve these efficiencies will use thermodynamic cycles that are more complex, with cascaded temperature levels, improved heat regeneration, etc. and will probably be larger per unit of heat delivered. At the moment gas heat pumps must compete with gas boilers rather than electric heat pumps, but in the future we can imagine that they will have to compete more with advanced electric heat pumps and a COP of 2.0 might be necessary as well as desirable. The TINA summary report on heat [8] suggested that 1.6 was a lower limit to be of interest in 2050.

Both the National Grid [9] and Delta-ee [10] have looked at scenarios for 2050 and beyond that encompass a role for both gas and electric heat pumps. The National Grid GG (Gone Green) scenario energy demand/duration mix has demand for heat being much 'peakier' than demand for electricity. Demand for electricity excludes heating, which ranges from a maximum of 1110 GWh/day to a minimum of 690 GWh/day (62% of peak daily heat demand). The heating load ranges from a maximum of 3000 GWh/day to a minimum of 500 GWh/day (17% of peak daily heat demand). As a result of the large range in the heat demand, it would be very difficult to electrify the entire heat load.

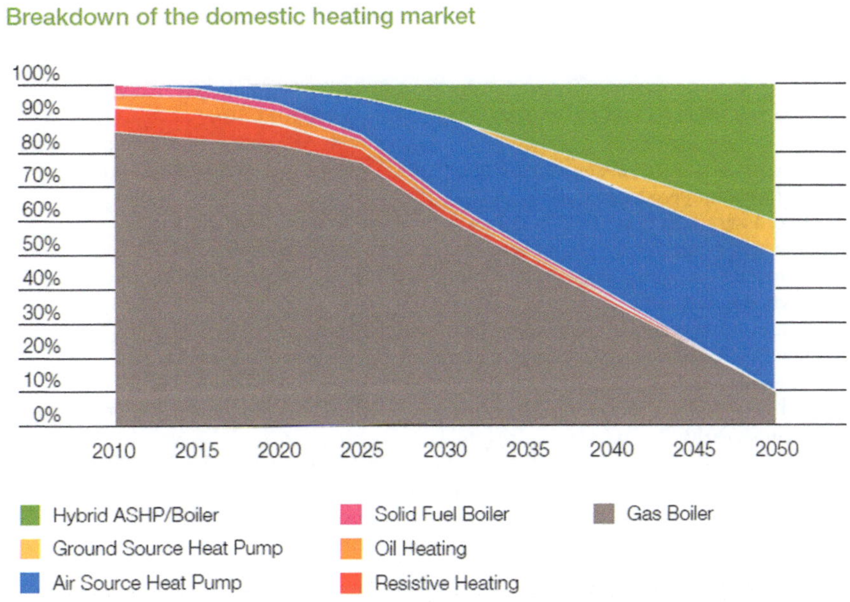

Fig. 62.6 Breakdown of domestic heating market (National Grid, Future Energy Scenarios 2012 [11])

Electricity in heat pumps has higher carbon emissions than gas condensing boilers until now, and gas heat pumps can decarbonise heat quicker than electrification up to 2030. Dual fuel [electric heat pump + gas or gas heat pump] can deliver similar carbon benefits to full electrification. The all-electric solution needs a 100 GWe capacity, whereas a hybrid or gas heat pump mix needs only 13 GWe. It is anticipated that gas heat pumps would be best used in older-, larger-, difficult-to-insulate properties with electric heat pumps in new-built and smaller properties, apartments, etc. The breakdown of the evolving market is shown in Fig. 56 of the FES [11], Fig. 62.6 below. In domestic and commercial premises, there are marginal improvements in energy efficiency between 2030 and 2050 with most of the more cost-effective refurbishments of the UK's buildings being done by 2030. Between 2030 and 2050, most efficiency gains are made by the changing way heat is supplied in buildings with a shift to heat pumps for well-insulated new properties with hybrid heat pump (gas or electric) and boiler systems being used in buildings in which the difference in seasonal heat requirements makes them less suited to be supplied by heat pumps alone. Note that gas boilers still make up some 10% of the mix in 2050 and that gas will also be used in the hybrid heat pump/boiler or gas heat pump systems.

Acknowledgements This work was funded by RCUK project EP/K011847/1 (i-STUTE) and DECC/BEIS contract TRN:478/09/2012 in support of IEA Heat Pump Annex 43.

References

1. Gas heat pumps: product overview, Juliette Promelle, GDF SUEZ, Research division, presented to Marcogaz Gas Heat Pump Workshop, 1,2nd December 2011, Paris
2. Robur – Integrated heating and cooling solutions with Absorption Heat Pumps powered by natural gas and renewable energies, Robur S.p.A. Downloaded 25 July 2013 http://www.robur.com/documenti_prodotto/ROBUR_ABSO_EN13_07_c-20130715090407.pdf
3. Market assessment in the field of domestic heating, Optimat, June, 2009
4. Market assessment for sorption energy, angle technology, December 2010
5. Gas-driven heat pumps: opening opportunities in the UK retrofit sector? Delta-ee Whitepaper, Lindsay Sugden, 2012
6. Private communication, Stephen Marland, National Grid
7. Policy Brief: Improving the energy performance of domestic heating and hot water systems, July 2008, DEFRA
8. The Low Carbon Innovation Coordination Group TINA Summary Report on Heat of September 2012
9. The Future of Heating: Meeting the challenge, DECC, March 2013, URN: 13D/033
10. UK Future Energy Scenarios, UK gas and electricity transmission, September 2012, National Grid. http://www.nationalgrid.com/NR/rdonlyres/2450AADD-FBA3-49C1-8D63-7160A081C1F2/56766/UKFutureEnergyScenarios2014.pdf
11. 2050 Pathways for Domestic Heat Final Report, Delta-ee for the Gas Futures Group of the Energy Networks Association, 16th October 2012

Chapter 63
Cross-Border Education in the Field of Renewable Energies Using a Dynamic Simulation Software

Andreas Wolf, Andreas Witzig, and Daniel Moreno

63.1 Introduction

Globalization is not only affecting trade and commerce, but also education. An increasing number of universities around the world offer Master courses for international students. Due to very different technical, cultural, and social backgrounds many challenges are encountered by students and teachers alike. Advanced didactics and innovative teaching tools are required to master these problems. In the field of renewable energies, dynamic simulation software can offer interesting solutions.

63.2 Increasing Student Mobility Around the World

The fast economic growth of many countries in the world in the last two decades, especially in densely populated Eastern China, has led to higher living standards, more ambitious personal goals, and has strengthened the belief that higher education is one of the most important prerequisites to reach those goals. With English as

A. Wolf (✉)
Vela Solaris AG, Winterthur, Switzerland
e-mail: andreas.wolf@velasolaris.com

A. Witzig
ICP Institute of Computational Physics, School of Engineering, Zurich University of Applied Sciences, Winterthur, Switzerland
e-mail: andreas.witzig@zhaw.ch

D. Moreno
Facultad de Ingeniería, Universidad Autónoma del Estado de México, Cerro de Coatepec S/N, Ciudad Universitaria, Toluca, Mexico
e-mail: dmorenol@uaemex.mx

A. Sayigh (ed.), *Renewable Energy and Sustainable Buildings*, Innovative Renewable Energy, https://doi.org/10.1007/978-3-030-18488-9_63

Top outbound markets and forecast demographic change

	Country	Outbound internationally mobile students	Demographic change (forecast growth in 15-24 year olds 2016-2026)	Top three destinations
1	CHINA	712,200	-12%	US, Australia, Japan
2	INDIA	181,900	2%	US, Australia, UK
3	GERMANY	119,100	-9%	Austria, Netherlands, UK
4	SOUTH KOREA	116,900	-30%	US, Japan, Australia
5	FRANCE	84,000	4%	Belgium, UK, Canada
6	SAUDI ARABIA	73,600	20%	US, UK, Canada
7	USA	60,300	-1%	UK, Canada, Germany
8	MALAYSIA	56,300	-12%	UK, Australia, US
9	VIETNAM	53,500	-9%	US, Australia, France
10	NIGERIA	52,100	35%	UK, Ghana, US

Source: UNESCO, Oxford Economics

Fig. 63.1 Top countries of origin and destinations of internationally mobile students in 2013 [1]

the most important foreign language and the key to commercial success in a globalized economy, the demand for higher education not only in the country's mother tongue but also in English has increased. This has led to a spread of international student mobility, preferably to countries in which English is a national language or at least widely spoken. In consideration of the above mentioned, the most important hosting countries are the English-speaking countries and the most important sending countries are located in Asia (Fig. 63.1):

Of all international students, the Chinese have become the largest group in many nations, especially in the USA (25.4%), Great Britain (15%), Canada (15%), and Australia (30%) [2]. It cannot be neglected that especially Chinese students have become an essential economic factor for many universities. In the academic year 2014/15, the Universities of Liverpool and Manchester both had more than 3000 Chinese students enrolled, the Universities of Nottingham, Birmingham, and Sheffield each more than 2500 Chinese students [3].

Four different types of cross-border education can be found, excluding the special case of international online and distance learning:

1. International joint and dual degree programs (collaborative academic degree programs by two or more partner universities located in different countries; in case of the joint degree only one diploma is awarded by all partner universities, in case of the dual degree (also called "double degree") separate diplomas by the partner universities are issued)
2. International branch campuses (IBC), physical university branches abroad, mostly established by Western universities in Arab and East Asian countries;
3. Internationally funded projects establishing degrees or offering courses in foreign countries (usually know-how transfer from developed countries to developing countries)
4. International Master courses offered by universities in their countries of origin and widely attended by students from abroad (Fig. 63.2)

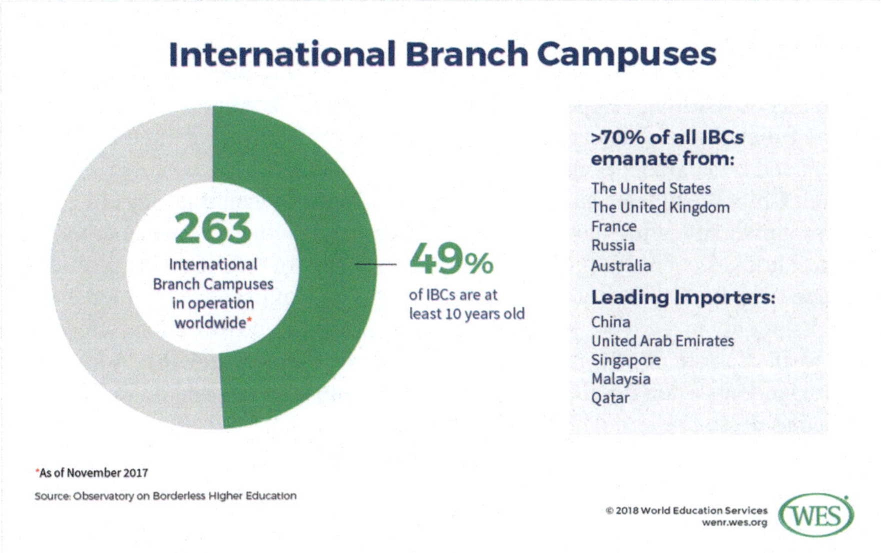

Fig. 63.2 International branch campuses [4]

63.3 Didactic, Cultural, Linguistic, and Social Challenges

The fast-developing cross-border education is a prosperous business, but full of challenges encountered by students and teachers alike. As opposed to national programs, students come from diverse technical backgrounds from the Bachelor courses they attended in their home countries and are not necessarily well prepared for the Master program. They will find following the lectures difficult, additionally they usually face general communication problems, cultural and social challenges. The teacher has to deal with very heterogeneous classes, which will require a lot of didactical talent and experience to ensure that all students benefit from the lectures in the same way.

The United Arab Emirates is one of the countries with the most foreign branch campuses in the world. Although they are affiliated to European or American universities, many students perceive that the branch campuses do not prepare them well enough for universities in Western countries, in case they wish to continue their studies there. One of the IBC stakeholders states that "Branch campuses, they are here to make money. They don't believe in education." [5]. A student previously educated in the UK observes "The main difference in the classroom... between the university here and the university in Britain is the lack of critical thinking or being able to think for yourselves..." [5].

A different way of teaching and thinking is being experienced by Chinese students enrolled in British universities, too. Chinese students are usually reported as being quiet in classroom, not active in group discussion or group work, passive in tutorials, etc. In China education is teacher-centered, textbook-based, and examination-oriented. Students are encouraged to listen to teachers' teaching, acquire knowledge through memorization and repetition, and present their learning outcome through paper-based examinations [6]. Wu [7] defines this type of education "dialectic learning," opposed to "dialogic learning" common in Western countries, in which the students are encouraged to ask questions, challenge the ideas of teachers and other students and express their own ideas.

Often Chinese students suffer from a learning shock when they start to attend a Western university after having studied previously in their home country. This learning shock is even stronger when the students cannot follow the lectures due to language barriers. Especially the speed and accent of English mother-tongue teachers and students make it very difficult and sometimes impossible for Chinese students to participate in the lecture and conversation in general [8]. Additionally Chinese students often do not mingle well with British people and do not even feel the need to do so [7].

63.4 The Situation in the Field of Renewable Energies

Especially in the field of renewable energies, the basic knowledge of international students is extremely diverse: many universities do not offer any undergraduate classes in renewable energy systems, but only teach basic technological principles which do not go beyond the electrical processes within the photovoltaic cell for PV or heat-transfer equations for solar thermal systems. If the student does not even understand the components themselves, there will be a considerable gap between the knowledge of the student and the expectations of a teacher trying to explain the interaction of the components of a system. In many cases theoretical, purely academic teaching and contents prevail instead of a practical approach.

Even if the student has already studied the design of renewable energy systems, especially in solar thermal technology system designs vary greatly between single countries. Something the student has learned in his home country can suddenly be considered wrong during the Master program creating tensions and misunderstandings between student and lecturer.

Some examples: In Mexico solar thermal systems for single-family dwellings are usually low-pressurized thermosiphon systems connected to a cold water tank *(tinaco)*. Automatic domestic hot water mixing valves are not common, usually the hot water is mixed directly at the tap [9]. Also in China thermosiphon systems are very frequently installed, for larger systems unpressurized, open hot water tanks with variable water levels and cold water inlets at the top are used. Heat exchangers are rare, while in Europe the standard is a forced-circulation system, strictly separating the solar loop from the potable water in the tank or a fresh-water unit. Even within Europe there are important differences: in Central Europe the back-up heating loop is connected directly to the solar tank, while in Spain this practice is not allowed by the national *Código Técnico de la Edificación (CTE)* and the *Reglamento de Instalaciones Térmicas en los Edificios (RITE)* in its currently valid versions (Fig. 63.3).

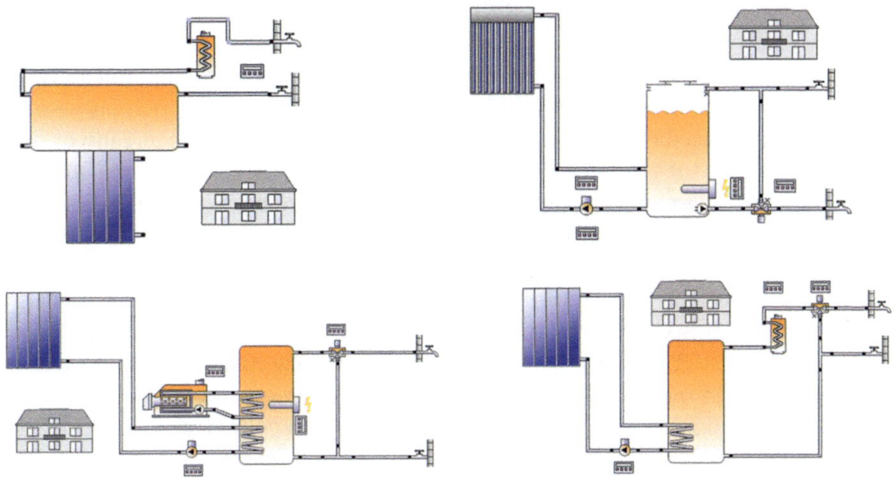

Fig. 63.3 Examples of different solar thermal system designs in Mexico (top left), China (top right), Central Europe (bottom left), and Spain (bottom right) [10]

Furthermore, hardly any student has ever had any practical experience in the field and has had an insight into the market of renewable energies which makes solar energy rather an abstract concept to them.

63.5 The Use of Advanced Simulation Software as a Solution

During the past three decades, hardly any other technology has developed as fast and has changed so deeply our daily life and work as computer technology. Although at the beginning of the digital revolution the capability of software was very modest compared to today, the chances of using it also for teaching were understood early on [11].

At the University of Wisconsin-Madison, Sanford A. Klein's computational research for his Ph.D. thesis lead to the development of TRNSYS which was the first simulation software for solar energy systems and has been commercially available since 1976. William A. Beckman, the second director of the UW-Madison's Solar Energy Lab (SEL) established in 1954, said that "TRNSYS immediately became an invaluable educational tool for UW-Madison's engineering students" [12]. However, in the beginning the usage of TRNSYS in solar education was limited to a small group of people due to the lack of computers [13].

When in the 1990s computers became more accessible, several other simulation tools for renewable energy systems were developed, most of them with reduced functionality compared to TRNSYS. Some of them have successfully been applied in teaching solar energy (e.g., free: RETscren, PVGIS; commercial: T*Sol, Polysun, PVSyst, Insel) [13].

One of the first European universities, which used advanced simulation software for teaching in the field of renewable energies, was Dalarna University in Sweden which has introduced the software Polysun in a solar thermal course already back in 2002 and has used it ever since [14]. Dalarna University is generally considered a forerunner for international courses in renewable energies. Since 1998 its Solar Energy Research Center SERC is hosting the European Solar Engineering School ESES.

Polysun is an innovative tool used to design and optimize decentralized energy systems including solar thermal, photovoltaic, heat pump, and geothermal components. Originally created in 1992 at the Institute for Solar Technology SPF, part of the Technical University of Rapperswil (Switzerland), the software has continuously been setting new and higher industry standards ever since. Since 2007 the company Vela Solaris is developing and marketing the software in close cooperation with universities and manufacturers.

As an example of a special kind of cross-border education mentioned in Chap. 2, the software is being used in international branch campuses of Heriot-Watt University Edinburgh, located in Dubai (United Arab Emirates) and Putrajaya (Malaysia). Polysun is a teaching tool in international Master courses in renewable energies offered by many universities (e.g., Dalarna University in Sweden, Technical University of Denmark, University of Freiburg in Germany, Kingston University London and others). It has been essential for internationally funded projects such as SOLEDA ("Solar Energy System Design using Advanced Learning Aids," a Tempus project in collaboration with Egyptian universities funded by the European Union (2012–2016); [15] and RETAC ("Renewable Energy Training and Certificate Program in China" [16]. RETAC is a Swiss-Chinese know-how transfer project by Vela Solaris AG, the Zurich University of Applied Sciences ZHAW and NERCRE (National Engineering Research Centre for Renewable Energy of China) which has been co-financed by the Swiss government agency REPIC (Renewable Energy & Energy Efficiency, Promotion in International Cooperation) from 2015 to 2017 and is now being funded by the project partners themselves.

Advanced simulation software is an indispensable tool to resolve the problems mentioned in Chap. 3 of this paper. First of all the in-built feature of changing the language of the tool will enable the students to switch to their mother tongue or at least to another, better understood language, if certain English terms in the course are not instantly clear to them.

In the times of social media based on visual communication like Instagram and Facebook, pictures have become increasingly popular and are at the same time internationally understood, too. They help to clarify abstract concepts and give an idea of systems which students might not have seen yet in real life.

True-to-reality modeling and depiction of solar systems enhance the understanding of thermo-technical and electrical processes. Different system configurations can be compared with one another easily and without the necessity to arrange an expensive laboratory. As described above, national standards or customs strongly influence the system design. Thanks to the software physics and thermodynamics will determine which system is more efficient, national standards remain in the background. Students are invited to open their minds and not rely on traditions, but to evaluate systems based on scientific principles.

Comprehensive databases for components, materials and systems ensure a good overview of the products and systems commonly used in the business. The interactive user interfaces of advanced software tools turn even shy students into protagonists of the lecture. This aspect is particularly important, as they resolve problems which have been observed frequently especially with Chinese students which are often considered too passive in class [6]. "Learning by doing" will add considerable value to an otherwise merely academic education.

Today's students belong to the so-called Generation Z, born from the mid-1990s to the middle of the first decade of the twenty-first century. Most of the members of Generation Z use computers on a daily basis, not only for communication and studies, but also in their free time. They are generally comfortable with technology and with interacting on social media. Many are familiar with computer games. Simulation software can be used as a tool for game-based learning [17], applying two different approaches:

- Simulation competitions, rewarding those students who design the most efficient and/or economically most convenient renewable energy systems which can be evaluated quickly and easily through clear simulation results
- The practical exercises can include tasks which let the students perceive the software more like a game, creating this way an attractive didactic tool for them

Often it is observed that students stay in the class after the lecture has finished and continue to play with the software in their free time. This gamification effect increases the engagement and motivation of the students.

References

1. Savills World Research (2016) World student housing 2016/17, London
2. Yin S (2013) The economic impact of Chinese international students in the United States, Yale Economic Review. http://www.yaleeconomicreview.org/archives/294. Accessed 3 Aug 2013
3. Times Higher Education: uk-universities-most-dependant-on-chinese-students-260516-large.jpg (source: British Council, as quoted in the China Daily newspaper, and the Higher Education Statistics Agency)
4. Lane JE (2018). International relations and signals of change in international education. https://wenr.wes.org/2018/02/international-relations-signals-of-change-in-international-education. Accessed 6 Feb 2018
5. Farrugia CA (2012) International branch campus quality in a segmented quality assurance environment, Sheikh Saud Bin Saqr Al Qasimi Foundation for Policy Research, Policy Paper No. Accessed 5 Sept 2012
6. Zhu C, Gao Y (2012) Communication with Chinese International Students: understanding Chinese International Students' learning difficulties and communication barriers, Huddersfield

2012. In: Presented at the British Educational Research Association Annual Conference, University of Manchester, 4–6 September 2012
7. Wu W (2006) East Asian international postgraduate students in one UK university: a cross-cultural perspective, University of Warwick 2006. In: Presented at the 36th Annual SCUTREA Conference, Trinity and All Saints College Leeds, 4–6 July 2006
8. Yang C (2016) Problems Chinese international students face during academic adaptation in english-speaking higher institutions. University of Victoria (Canada)
9. Reyes T (2014) Ernestina (Centro de Evaluación y Capacitación en Energías Renovables (RENOMEX), Deutsche Gesellschaft für Internationale Zusammenarbeit (GIZ)). In: Guía de instalación de sistemas de calentamiento solar de agua para vivienda unifamiliar, México D.F
10. Simulation Software Polysun, Vela Solaris AG, Winterthur 2018
11. Dönhoff HU (1991) Modellbildung und Simulation im Unterricht. In: Gorny P (ed) Informatik und Schule 1991. Informatik-Fachberichte, vol 292. Springer, Berlin, Heidelberg
12. Schmidt S (2014) Solar prophets: a history of UW-Madison's solar energy lab. https://energy.wisc.edu/news/solar-prophets-history-uw-madisons-solar-energy-lab. Accessed 4 Dec 2014
13. Witzig A, Prandini M, Wolf A, Kunath L (2016) Teaching renewable energy systems by use of simulation software: experience at universities of applied sciences, in in-service training, and from international know-how transfer, Winterthur 2016. In: Presented at the 11th International Solar Energy Society (ISES) EuroSun Conference, Palma de Mallorca
14. Bales C, Lorenz K (2016) Application of Polysun in teaching courses in Sweden and in the PhD Program SHINE, Högskolan Dalarna, Falun 2016. In: Presented at the 1st Conference on the simulation of building-integrated energy systems, Winterthur
15. Ghannam R, Khayet M (2016) Solar energy system design using advanced learning aids: an EU Tempus Project, Cairo/Madrid 2016. In: Presented at the 1st conference on the simulation of building-integrated energy systems, Winterthur
16. Wolf A, Zhu J, Prandini M, Witzig A (2016) Renewable energy training and certificate program in China, Winterthur/Beijing 2016. In: Presented at the 1st conference on the simulation of building-integrated energy systems, Winterthur
17. Breuer J (2010) Spielend lernen? Eine Bestandsaufnahme zum (Digital) Game-Based Learning, Landesanstalt für Medien Nordrhein-Westfalen (LfM), Düsseldorf

Chapter 64
A Bi-fluid PV/T Solar Collector and Its Potential Application in Solar Drying

Hasila Jarimi, Mohd Nazari Abu Bakar, and Saffa Riffat

64.1 Introduction

Theoretical and experimental studies of PV/T solar collectors were carried out and documented as early as in the mid-1970s. Wolf [1] (1976) investigated the performance of a combined solar heating and photovoltaic electric power generation system for a single family residence. The study was carried out in Boston, USA. Florschuetz [2] came up with the famous extension of the well-known Hottel-Whillier-Bliss model for thermal analysis of flat-plate collectors to the analysis of combined photovoltaic/thermal collectors. More studies were carried out by S. D. Hendrie [3] to prove the technical feasibility of the collectors. Since then, studies have continued on collector designs and performance analysis, including development of thermal and electrical modelling together with the influence of various parameters on the overall performance of the collectors. This chapter reviews the development of some of the important and related PV/T solar collector which may include four key elements, namely, the design concept, type of modelling, experimental analysis, and the collector overall performance. The PV/T systems being reviewed are mainly segregated into three groups based on the collector's working fluid, namely, air type, water type, and bi-fluid type (dual mode). While the first two types of PV/T solar collectors have been widely researched, the bi-fluid type (with dual mode of operation) was first introduced by Tripanagnostopoulos [4], in the 17th European PV Solar Conference in Munich (Assoa et al. [5]). Tripanagnostopoulos [4] proposed three different basic configurations of PV/T dual

H. Jarimi (✉) · S. Riffat
Buildings, Energy and Environment Research Group, Faculty of Engineering, University of Nottingham, Nottingham, UK
e-mail: Hasila.jarimi@nottingham.ac.uk

M. N. A. Bakar (✉)
Department of Physics, Universiti Teknologi MARA (Perlis), Arau, Perlis, Malaysia
e-mail: mohdnazari@perlis.uitm.edu.my

A. Sayigh (ed.), *Renewable Energy and Sustainable Buildings*, Innovative Renewable Energy, https://doi.org/10.1007/978-3-030-18488-9_64

mode operations which are: the water heat exchanger (WHE) placed directly behind the PV module, placed inside the air channel, and placed on the back plate of air channel. In order to test the designed solar collector, experiments were performed with the solar collector. However, to the best of this author's knowledge, the focus of his research is the independent mode of fluid operation such that the mass flow rate of each of the fluid is fixed at 0.02 kg/s. The experimental results indicate that, when operated independently, the first design with WHE placed directly behind the PV module, the electrical efficiency, and thermal efficiency of the system are considered as satisfactory for the PV component and both water and air, respectively.

Also, Assoa et al. [5] conducted a study on a PV/T solar collector which utilized both air and water as the working fluids. The PV cells act as the thermal absorber component for the air heat exchange component but not for the water heat exchange component. In their study, a 2D steady-state analysis is developed. Indoor experimental studies were performed, and the results were compared with the simulation results and shown to be in good agreement. Several parametric studies were conducted using the validated mathematical model. The significant advantage of this configuration is that high water temperature production is feasible when compared to the conventional hybrid solar collector without causing undesirable effect on the PV cells. In addition, such system can also be integrated with solar cooling devices which are extremely appealing during summertime. Later, some of the current authors [6–8] have introduced a bi-fluid PVT system with both water and air heating component. The researchers introduced the design concept of the collector, and its performance was simulated using 2D finite difference equations solved in MATLAB and validated against both indoors and outdoors testing. Air, water, and the simultaneous air and water mode were considered for the analysis. The primary energy savings of bi-fluid PVT system were 58.10% and 62.31% for air and water mode, respectively. In bi-fluid mode, it was observed that, as the fluid flow of one fluid increased, the thermal efficiency of other fluid tends to decreased. When the flow rate of water was varied and air flow rate was fixed, the total experimental thermal efficiency was increased from 51.88% to 65.70% and primary energy saving was increased from 64.02% to 77.90% in comparison to water mode. However, when flow rate of air was varied keeping constant water flow rate, the total experimental thermal efficiency was increased from 51.88% to 66.12% and primary energy saving increased from 64.02% to 78.98%.

Othman et al. [9] proposed a new design of photovoltaic/thermal solar collector integrating transparent solar panels with double-pass flat-plate solar air heating system and water heating segments made of copper pipe. The performance of the system is calculated based on Hottel-Whillier-Bliss equation. Under controlled indoor environment at radiation level of 800 W/m^2, air flow rate at 0.05 kg/s, and water flow rate at 0.02 kg/s, the average outlet temperature indicated reading of 27.4 °C. Su et al. [10] introduced a photovoltaic/thermal (PV/T) solar collector with dual channels for different fluids. In their research, when the PV/T solar collector was operated under water mode configuration, the performance of the PV/T collector is the highest and the performance of the bi-fluid configuration always dominates. Baljit et al. [11] investigate a PV/T bi-fluid solar collector that integrates a PV panel with a serpentine-flow stainless steel tube as the water-heating component and a double-pass air channel as the air-heating component. They have also used a Fresnel lens as the glazing and primary concentrator. Meanwhile, compound parabolic concentrators (CPCs) are used as the secondary concentrator. The system can simultaneously generate hot air and hot water in addition to electricity, and the total energy generated per unit area is higher than that of a single-fluid system. The simulated results have shown that the total equivalent efficiencies for single fluid condition have ranged from approximately 30 to 60% and increased to a maximum efficiency of near to 90% for the case of the dual fluids. They have concluded that the dual fluids operation mode has reduced the solar cells temperature and hence increased the electrical output.

Bi-fluid type PVT solar collector may be considered as the most recent developed PVT techniques [12]. Researchers are interested to investigate this type of PVT collector since the idea of integrating both type of the common working fluids, air and water, into the same PV/T solar collector is considered appealing due to its space-saving design and potential applications. The use of both fluids (bi-fluid) also creates a greater range of thermal applications and offers options in which three modes of fluid operation, namely, the air mode, the water mode, and the simultaneous mode (air and water), can be produced depending on the energy needs and applications. However, similar to typical PVT technologies, the main limitation to the wide use of PV/T collector is being the environmental conditions especially in cold climates. Although some researchers claimed that the low grade air and water have limited applications, the statement are not very accurate since the thermal output of the collector can be at accepted level depending on the climatic conditions. Based on the current authors' published articles, this paper simulates the performance of a glazed single-pass bi-fluid PV/T solar collector at 2 m^2 aperture area under tropical climate conditions. From the simulation, we then evaluate the potential application of the bi-fluid type PV/T solar collector.

64.2 Design Concept and Mathematical Model

The design concept of the bi-fluid type PV/T solar collector is illustrated in Fig. 64.1. The collector comprises of a single glazing with a single-pass air channel and serpentine-shaped copper tube as the water heating element.

The current authors have discussed the mathematical modelling of the collector design without the glazing in detail in several publications [6, 7]. A set of energy balance equations were developed and solved in MATLAB using finite difference equations. In this paper, in order to take into account the additional single glazing element of the PV/T solar collector, the energy balance equations in the aforementioned publications were modified. Equations (64.1) and (64.2) represent the energy balance for the single glazing and the modified equation for the PV layer of the solar cells components of the PV/T solar collector.

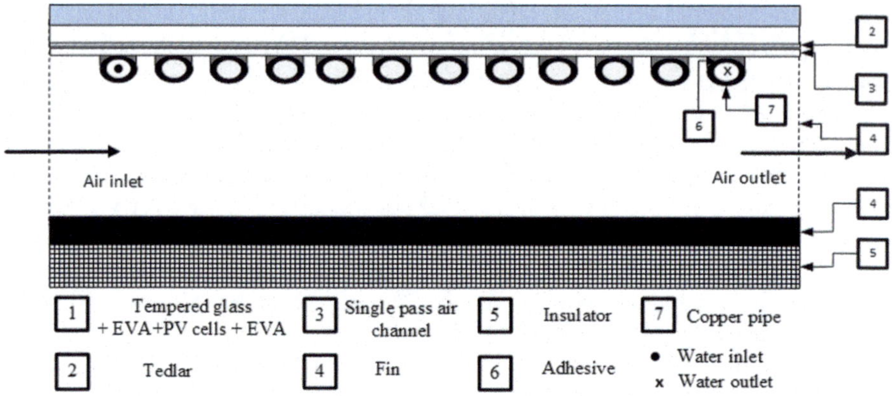

Fig. 64.1 The design concept of the single-pass glazed PV/T bi-fluid solar collector

$$\underbrace{\alpha_g G}_{1} = \underbrace{U_{t,m}\left(T_{gm} - T_a\right)}_{2} + \underbrace{h_{rpg,m}\left(T_{g,m} - T_{PV,m}\right)}_{3} \tag{64.1}$$

where α_g is the absorptance of the glass sheet, G is the incident solar radiation, $U_{t,\,m}$ is the overall top heat loss coefficient for sub-segment, and $T_{g\ m}$, $T_{PV,\,m}$ are the temperature nodes of the glass sheet and the PV cells (at the centre of the cells), respectively.

The heat transfer terms are defined as follows:

(1) The rate of the solar energy absorbed by the single glazing, (2) the rate of the top heat lost to the ambient per unit area, and (3) the rate of radiative heat transfer between the single glazing and the PV panel.

$$\begin{aligned} &\underbrace{\tau_g \alpha_c (\beta_c) G}_{4} - \underbrace{\tau_g \alpha_c \beta_c (\eta_{ele}) G}_{5}{}_{5} + \underbrace{h_{rpg,m}\left(T_{g,m} - T_{PV,m}\right)}_{3} \\ &= \underbrace{h_{pvbs,m}\left(T_{pv,m} - T_{bs,m}\right)}_{6} \\ &\underbrace{-\frac{k_{PV}\delta_{PV}\left(T_{PV,m+1} - T_{PV,m}\right)}{(\Delta x)^2} + \frac{k_{PV}\delta_{PV}\left(T_{PV,m} - T_{PV,m-1}\right)}{(\Delta x)^2}}_{7} \end{aligned} \tag{64.2}$$

where τ_g is the glazing transmittance, α_c is the abosrptance of solar cells, (β_c) is the solar cells packing factor, η_{ele} is the PV cells electrical efficiency, $h_{pvbs,\,m}$ is the conduction heat transfer coefficient from the centre of PV cell to the back surface of Tedlar, k_{PV} is the thermal conductivity of photovoltaic cells (W/m K), δ_{PV} is the thickness of the PV module, Δx is the distance between temperature nodes (Δx= 1 cm), and $T_{bs,\,m}$ is the temperature nodes of the back surface of Tedlar.

The heat transfer terms are defined as follows:

(4) The rate of the solar energy received by solar cells of the PV module after transmission per unit area; (5) the rate of electrical energy available per unit area; (6) the rate of heat transferred to the back surface of Tedlar per unit area; and (7) the rate of heat conducted along the x-direction aside the tube per unit area.

64.3 PVT Bi-fluid Potential Application

64.3.1 Simulation Analysis

The energy balance equations were solved using the inverse matrix solution procedure, with Newton Raphson iteration implemented to compute the unknown temperature nodes using MATLAB. The performance of the PV/T solar collector was simulated for total collector area of 2 m^2 in size, against hourly weather data for Arau, Perlis, Malaysia (see Fig. 64.2), with a latitude of 6.5224816° N and a longitude of 100.2308224°E, between 9:30 a.m. and 5:30 p.m. in a typical day of January 2015. The simulation is important to get an overview on the thermal and electrical performance of the PV/T solar collector to determine its potential application.

Figure 64.3 shows an hourly variation of solar cell temperature and its electrical efficiency. From morning to afternoon, while the mean temperature of the PV cells increases, the electrical efficiency η_{ele}clearly decreases with time, and this condition becomes very obvious during peak solar radiation between 11:00 a.m. and 3:00 p.m. The electrical η_{ele} gives a negative parabolic curve with time of the day. This is due

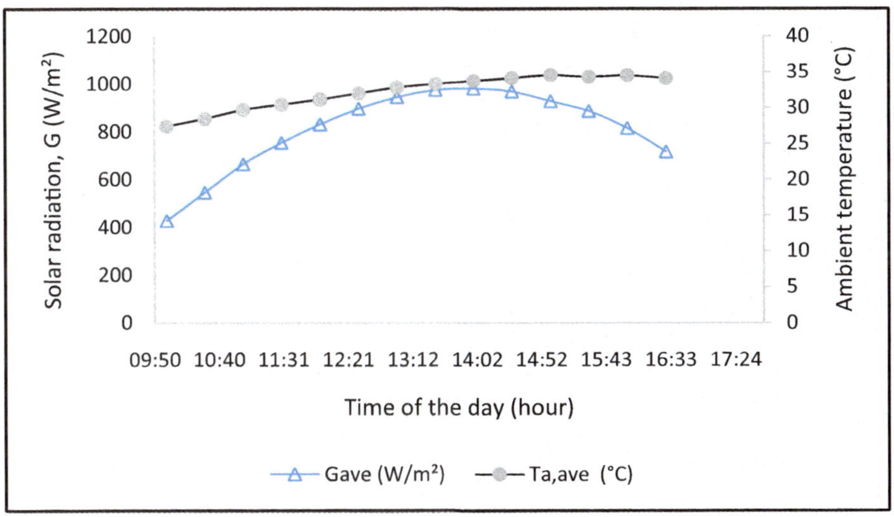

Fig. 64.2 Hourly variation of average solar radiation G_{ave} for the solar collector facing south and average ambient temperature $T_{a,ave}$ on 13th of January 2015

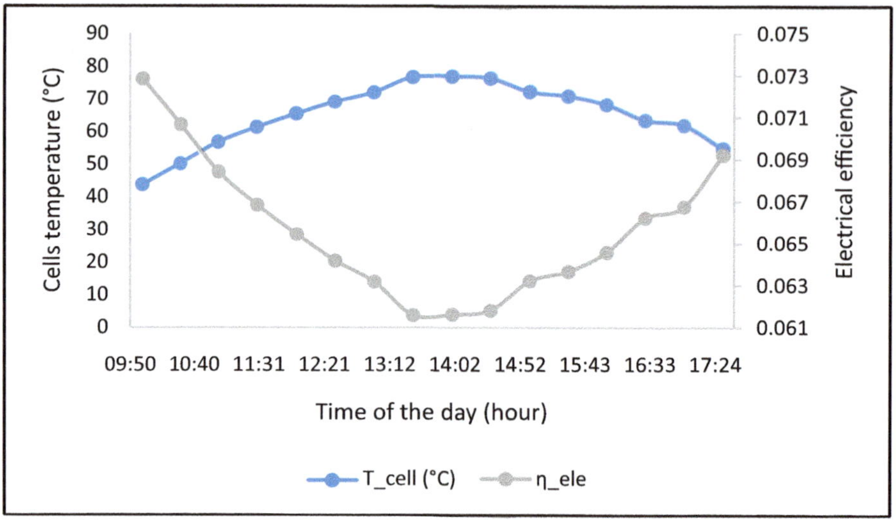

Fig. 64.3 The hourly variation of solar cell temperature T_cell and electrical efficiency with time of the day η_ele

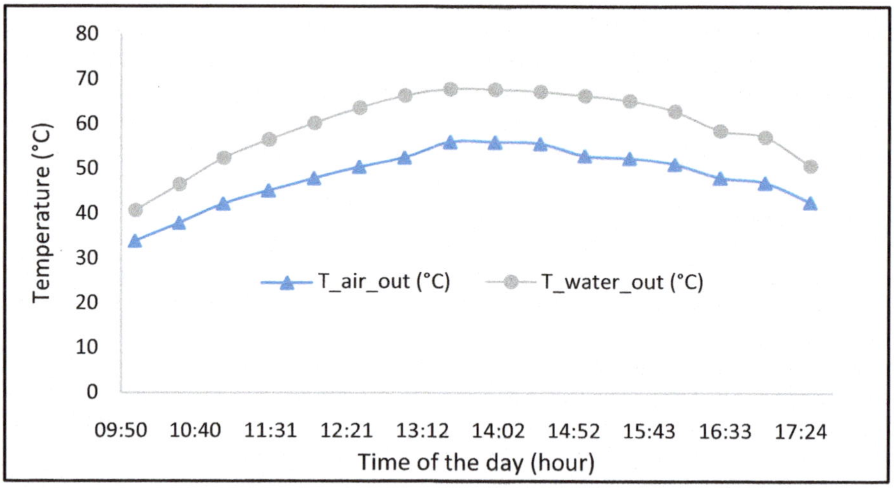

Fig. 64.4 The hourly variation of air T_air and water output T_water temperature

to the temperature dependency of the open circuit voltage of the PV cells with its temperature which in turn influences the amount of power produced by the collector.

Figure 64.4 shows the hourly variation of outlet air and outlet water temperature from the PV/T solar collector at the fluids' optimum flow rates which are 0.0066 kg/s and 0.027 kg/s, respectively. Clearly, the output temperature of air is lower in

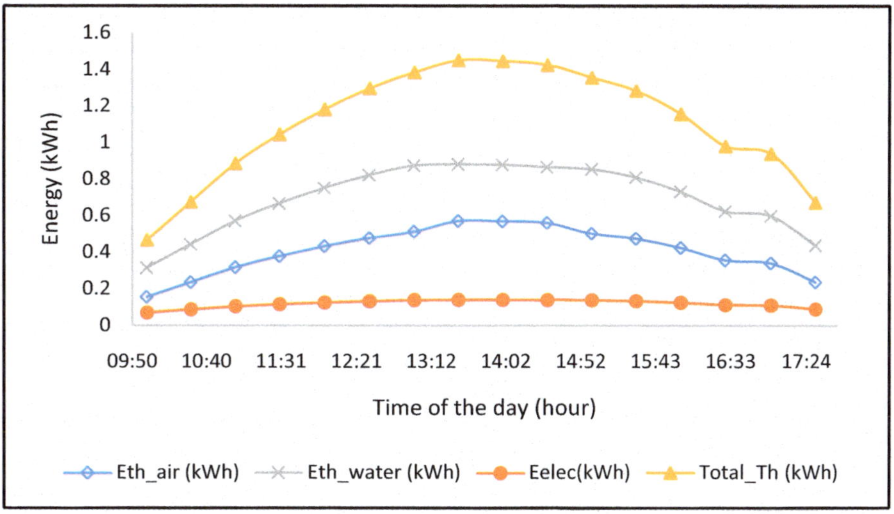

Fig. 64.5 The hourly variation of thermal energy Eth, electrical energyEelec, and overall thermal energy gain Total_Th with time of the day

comparison to water due to the lower heat capacity as the working fluid. The air and water output temperature increases with the increase in solar radiation. At average solar radiation of 977 W/m^2, the output temperature of air and water is 56 °C and 67 °C, respectively; meanwhile during minimum incident solar radiation, the air and water output temperature is 34 °C and 40 °C, respectively. From a psychometric chart, if the initial inlet air temperature is 30 °C, and heated in the solar collector to 40°, the relative humidity will reduce from around 90% to 40%.

Figure 64.5 shows the hourly variation in the air, water, and total thermal and electrical energy output from the PV/T solar collector at air and water mass flow rate of 0.027 kg/s and 0.0066 kg/s, respectively. The calculated thermal energy found to be 4.2 kWh/day and 7.4 kWh/day for air and water output, respectively. Meanwhile, the total electrical energy was found to be 1.2 kWh/day. The produced electricity will be stored in a battery for useful application such as powering the low dc fan. When comparing the simulation results with a greenhouse dryer design studied in [13], such energy produced by the current PV/T bi-fluid solar collector should be sufficient for self-sustainable drying system application.

64.3.2 PVT Bi-fluid Solar Collector Integrated with a Greenhouse Dryer

Considering the air output temperature of the PVT bi-fluid solar collector, the temperature is considered suitable for crop drying application with average maximum allowable temperature of 55 °C [14]. Therefore, in this paper, an integrated PV/T bi-fluid solar greenhouse drying system is introduced. In a typical tray dryer chamber which has been widely used in a hot climate countries, non-uniform distribution of heat transfer causes variation in the moisture level of the drying chamber tray. A tray dryer consists of level of trays placed in an insulated chamber. The heat transfer is either via forced convection of natural convection. The non-uniformity drying throughout the chamber must be avoided to get efficient drying conditions.

In this paper we are proposing two designs of a PV/T bi-fluid solar collector integrated with a greenhouse dryer (see Fig.s 64.6 and 64.7). In the design as shown in Fig. 64.6, instead of using heated air from the PV/T solar collector as the main heat source, crops drying will be performed via heated air from the water to air heat exchanger installed below the bottom tray. Low dc fans are used to circulate the air in the chamber. The heated water for the water to air heat exchanger is obtained from the water output of the PV/T water solar collector. When the heated air is leaving the first tray, the air will increase in its absolute humidity and decrease in temperature.

In a typical tray dryer chamber, this will cause the tray at the top chamber to be low in temperature and high in moisture level. In this drying chamber design, the incoming hot air from the PV/T solar collector will complement the heating process; thus, the temperature and moisture difference between the crops in the bottom and top trays will be reduced. In addition to that, the crop drying will also be assisted by solar heat gained through the glass chamber body. Air with lower humidity and higher temperature will dry the crops in the first tray. The warm water that leaves the water to air heat exchanger configuration can be stored in a water tank. At times of low solar radiation, the valve for the water piping from the solar collector can be closed and hot water from the water tank can be used as the backup heating element via floor heating.

One of the significant disadvantages of solar assisted drying system is the unavailability of solar radiation at night. In design 2 as shown in Fig. 64.7, the PV/T solar dryer is integrated with a low temperature phase change materials (PCM) thermal storage. Although the PCM storage will not be able to provide heating as efficient as during the day, the heat output is considered sufficient to allow continuous drying for at least 3–4 h at times without solar radiation. The working concept of the dryer may be explained as follows: during the day, the heated water charged the PCM thermal storage. With water temperature of approximately 70 °C, efficient heat transfer assumed can be achieved to melt the PCM at 40–50 °C. During the day, crop drying is mainly via hot air output from the solar collector. At night, PCM thermal storage will be discharged of which temperature of up to 55 °C can be achieved. The heat released by the PCM storage from its container will be circulated in the drying chamber using a low dc fan powered by the PV power stored in the battery and an arrangement of fins is introduced to enhance heat transfer rate from the PCM to the air flow.

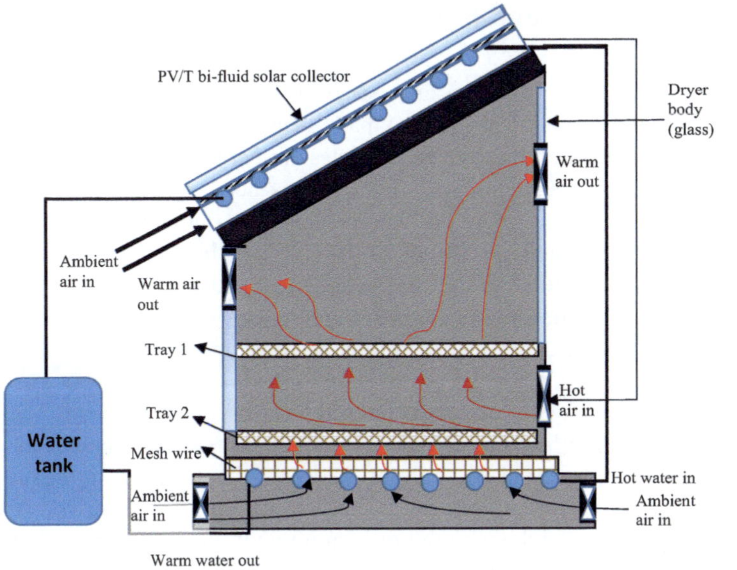

Fig. 64.6 The integrated PV/T bi-fluid solar collector with a greenhouse dryer

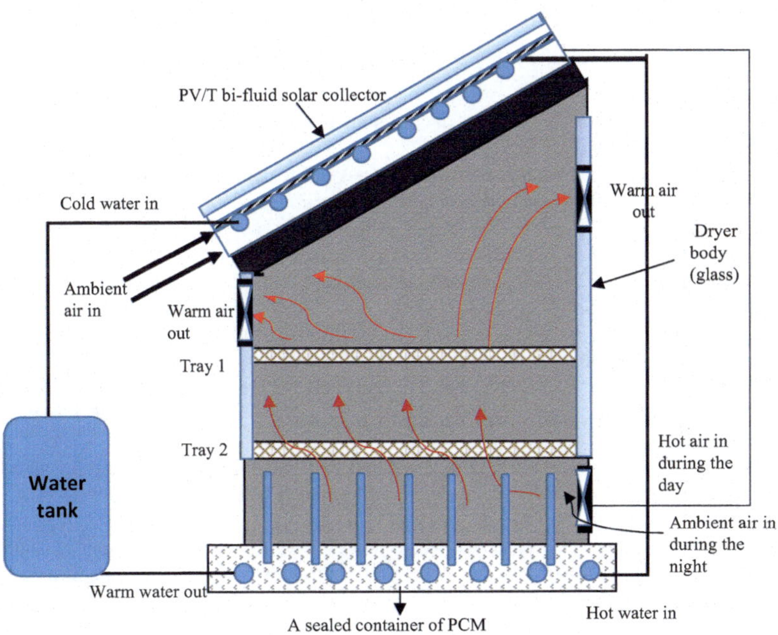

Fig. 64.7 The integrated PV/T bi-fluid solar collector with a greenhouse dryer and PCM thermal storage

64.4 Conclusions and Future Studies

This paper presents a simulation analysis on the thermal performance of a PV/T bi-fluid solar collector using a validated mathematical model. The PV/T solar collector can be operated under three different modes of operation, namely, water mode, air mode, and bi-fluid mode, depending on the energy needs and application. In hot climate, space heating is not required for typical residential and commercial building applications. The potential application of the PV/T bi-fluid solar collector is seen in a solar drying system for fruits, crops, vegetables, and marine products. From the computer simulation, the air temperature range is clearly within the average temperature requirement for solar drying. Therefore, two designs of autonomous solar drying system by integrating the PV/T bi-fluid solar collector with a greenhouse dryer are proposed. The main objective of the proposed solar drying system is to create a drying system which is self-sustainable. This is usually needed at rural area or area with limited access to power. In the current research, the suitable application of the designed PV/T bi-fluid solar collector is made based on the comparison between its thermal performance and temperature requirement in solar drying application. Future studies will involve detail modelling of the crop drying in the chamber and also experimental work to prove the concept.

References

1. Wolf M (1976) Performance analyses of combined heating and photovoltaic power systems for residences. Energy Convers 16:79–90
2. Florschuetz LW (1979) Extension of the Hottel-Whillier model to the analysis of combined photovoltaic/thermal flat plate collectors. Sol Energy 22:361–366
3. Hendrie SD (1982) Photovoltaic/thermal collector development program Rapport Final. Massachusetts Institute of Technology, Cambridge, Mass, USA
4. Tripanagnostopoulos Y (2007) Aspects and improvements of hybrid photovoltaic/thermal solar energy systems. Sol Energy 81:1117–1131
5. Assoa YB, Menezo C, Fraisse G, Yezou R, Brau J (2007) Study of a new concept of photovoltaic–thermal hybrid collector. Sol Energy 81:1132–1143
6. Jarimi H, Abu Bakar MN, Othman M, Din MH (2016) Bi-fluid photovoltaic/thermal (PV/T) solar collector: Experimental validation of a 2-D theoretical model. Renew Energy 85:1052–1067
7. Abu Bakar MN, Othman M, Hj Din M, Manaf NA, Jarimi H (2014) Design concept and mathematical model of a bi-fluid photovoltaic/thermal (PV/T) solar collector. Renew Energy 67:153–164
8. Bakar MNA, Jarimi H (2017) Performance analysis of a bi-fluid type photovoltaic/thermal (PV/T) solar collector under low-cost solar simulator and tropical climate conditions. In: 16th international conference on sustainable energy technologies, Bologna, Italy
9. Othman MY, Hamid SA, Tabook MAS, Sopian K, Roslan MH, Ibarahim Z (2016) Performance analysis of PV/T Combi with water and air heating system: an experimental study. Renew Energy 86:716–722
10. Su D, Jia Y, Huang X, Alva G, Tang Y, Fang G (2016) Dynamic performance analysis of photovoltaic–thermal solar collector with dual channels for different fluids. Energy Convers Manag 120:13–24

11. Baljit SSS, Chan HY, Audwinto VA, Hamid SA, Fudholi A, Zaidi SH et al (2017) Mathematical modelling of a dual-fluid concentrating photovoltaic-thermal (PV-T) solar collector. Renew Energy 114:1258–1271
12. Joshi SS, Dhoble AS (2018) Photovoltaic -thermal systems (PVT): technology review and future trends. Renew Sust Energ Rev 92:848–882
13. Tiwari S, Agrawal S, Tiwari GN (2018) PVT air collector integrated greenhouse dryers. Renew Sust Energ Rev 90:142–159
14. Ekechukwu OV (1999) Review of solar-energy drying systems I: an overview of drying principles and theory. Energy Convers Manag 40:593–613

Chapter 65
Utilization of Biomass Energy in Cement Production: A Pathway Towards Sustainable Infrastructure

Adeyemi Adesina and Paul Awoyera

65.1 Introduction

The need for green energy for different processes is increasing every day as the use of energy from fossil fuel combustion has been concluded to be detrimental to the environment. The cement industry is not an exception as it is currently undergoing serious scrutiny due to its high contribution to the world's human-induced carbon dioxide emissions [1]. In addition, most of the current natural disasters occurring in this century can be attributed to high global warming caused as a result of greenhouse gases which includes carbon dioxide in the atmosphere.

Based on different human activities, the amount of carbon dioxide emitted over the years have increased significantly, and a more exponential increase is expected if alternative sustainable methods are not used for our activities. In order to keep the earth habitable and to prevent a significant increase in temperature, IPCC [2] has stated that the level of carbon dioxide concentration in the atmosphere has to be below 450 ppm. Scientific researchers have indicated an increase in temperature of more than 2 °C will make some parts of our planets inhabitable and several plants and animals facing the threat of exaltation [3]. In the same IPCC assessment report made by United Nation in 2007, increase in the concentration of carbon dioxide in the atmosphere is attributed mainly to the burning of fossil fuel used as a source of energy for different activities including cement production [2]. Figure 65.1 shows that an increase in the global temperature corresponds to an increase in the concentration of carbon dioxide in the atmosphere. Keeping the current carbon concentration below is a hard path that must be threaded carefully, as it is difficult to keep the

A. Adesina (✉)
Concordia University, Montreal, Canada
e-mail: a_adesin@encs.concordia.ca

P. Awoyera
Convenant University, Ota, Nigeria
e-mail: paul.awoyera@covenantuniversity.edu.ng

A. Sayigh (ed.), *Renewable Energy and Sustainable Buildings*, Innovative Renewable Energy, https://doi.org/10.1007/978-3-030-18488-9_65

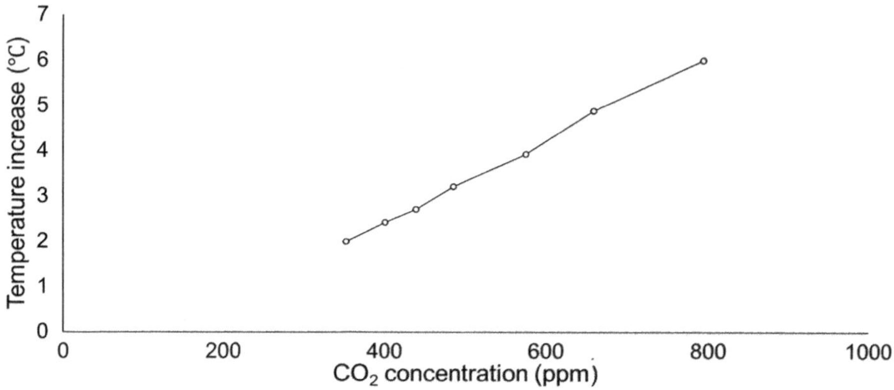

Fig. 65.1 Relationship between temperature increase and carbon dioxide concentration at stabilization (data from [2])

concentration low with rapid developments occurring all over the world. Therefore, it is paramount for all industries especially the major contributors of carbon emission to the atmosphere to find ways to reduce their carbon dioxide contribution significantly in order to save the future of planet earth.

Ordinary Portland cement (OPC) is the main binder in the most used building material in the world (i.e. concrete), and it is the most produced product on earth by mass [4]. Production of OPC is one of the industrial processes responsible for high carbon dioxide emission into the environment. For every mass of OPC produced, an approximate equivalent amount of carbon dioxide is emitted into the environment [5]. About 5–10% of the world's human-induced carbon dioxide emission is from the production of OPC [6, 7]. Carbon dioxide emitted during the production of OPC is from the reaction product formed during the calcination of raw materials and the burning of fossil fuels used to power the production processes. In order to reduce the carbon footprint of cement production, several green initiatives have been developed. The initiatives set by the cement industry include clinker substitution, production process optimization, etc. However, with improvements and developments in these areas, one area that is essential and common to other initiatives is the use of fossil fuels. For example, Even though the advent of special types of cement such as calcium aluminate and calcium sulfoaluminate in which the carbon dioxide produced during calcination of raw materials is reduced, the contribution of the carbon dioxide from the burning fossil fuels still exist. This is also true when OPC production processes are optimized. Therefore, in other to have more reduction in the carbon dioxide emitted during cement production, it is essential to use alternative renewable energy.

As more development and increase in cement production in coming years is expected all over the world especially in developing countries, the availability of a large amount of biomass in several countries can create an opportunity to produce cement in a sustainable way using renewable energy from biomass. Currently, few cement plants in various parts of the world employ the use of biomass energy despite the large production of biomass in some of these countries. For example, Malaysia which has a very large production of palm kernel shell which is an agricultural waste only generates 0.3% of its energy from biomass, while gas and coal still stand at 72.5% and 16.5%, respectively [8].

This paper highlights the one of the green initiative (i.e. use of biomass energy) in which the cement industry can use to reduce the overall concentration of carbon dioxide in the atmosphere. All the carbon dioxide emission described in this paper is related to that associated with cement production and serves as the total greenhouse gases equivalent, except stated otherwise. It is hoped that this overview will help various stakeholders in the cement industry to have more understanding of the benefits and challenges of using renewable energy. It is also hoped that this paper will initiate more research and development in the application of renewable energy for industrial processes.

65.2 Biomass Energy

Generally, biomass energy can be classified as a renewable energy generated from the combustion of different types of biological materials (e.g. grass, coffee husk, wood, cotton residue, livestock manure). Most of these biomass sources are waste materials from the agricultural industry. Though some of these wastes have traditional uses in areas where they are being generated locally, their usage does not maximize the overall benefit of the waste, as it creates additional waste in the environment. For example, most of the agricultural waste can be used for applications such as food for animals or left in the field to be recycled into soil nutrient. These traditional applications are only viable for small-scale agricultural industries where the amount of waste generated is low. But in cases of large agricultural industries, all the wastes generated cannot be fully used for the above-mentioned purposes, so they are disposed of in the environment where they consume large space with the possibility of contaminating resources in the environment. Therefore, use of biomass as a renewable energy source can be deemed a sustainable and effective way to manage these wastes. Fresh plants and trees can also be cultivated and use to biomass for generating energy. However, in other to maximize the benefit of this renewable source of energy, it must be obtained in a sustainable way (i.e. the rate of cultivation should not be greater than the rate of replanting), and by using optimized feeding and combustion systems.

It should be noted that the combustion of some biomass has higher carbon dioxide emission compared to fossil fuels. However, it is referred to as a net zero emission source of energy because it is assumed that the biomass planted to replace the cultivated one will absorb the carbon dioxide from its predecessor's combustion. This is why it's very paramount that a sustainable retrieval method is associated with the use of biomass as a renewable source of energy. Table 65.1 shows the carbon emission of different sources of fuel for cement production, it will be observed when using wood as a biomass for energy generation, about 110 Kg/GJ of carbon dioxide

Table 65.1 Gross and net carbon dioxide emission of fuel sources used for cement production (data from [9])

Fuel source	Gross CO_2 emission factor (Kg/GJ)	Net CO_2 emission factor (Kg/GJ)
Animal food	88	0
Coal	96	96
Liquid hazardous waste	74	74
Petcoke	93	93
Plastic	75	75
Solid hazardous waste	74	74
Waste oil	74	74
Waste wood	110	0
Wood	110	0

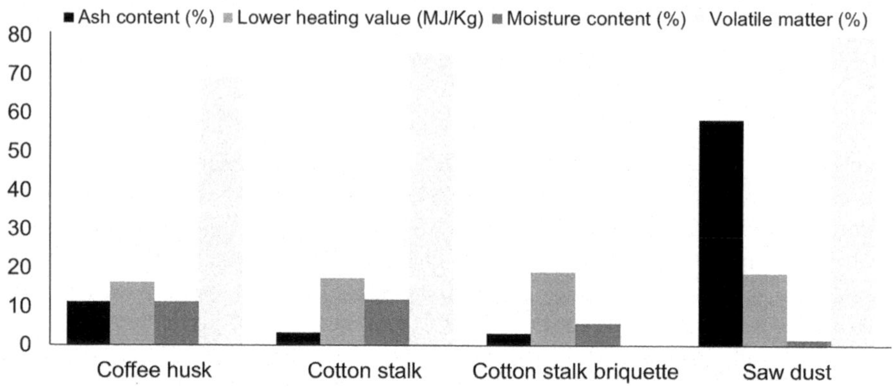

Fig. 65.2 Thermal properties of some types of biomass (data from [10])

emission is associated with it. However, the net emission factor is zero because planting other trees to replace the cultivated one will absorb carbon dioxide emitted during combustion of the preceding biomass. It is worth to mention that different types of biomass have varying thermal properties which affect their performance in terms of being able to generate energy when combusted. Thermal properties will be discussed in another article; however different thermal properties of some biomass are presented in Fig. 65.2. The use of biomass energy as a renewable source of energy is not only a sustainable alternative, but it's also economical in the long run.

65.3 Cement Production

Production of cement is very energy intensive, and fuel is combusted at every level with about 3000–7000 MJ of thermal energy is needed per ton of cement produced [11]. The processes involved in cement production can be summarized in the flow-chart shown in Fig. 65.3. Right from mining or raw materials to packaging of

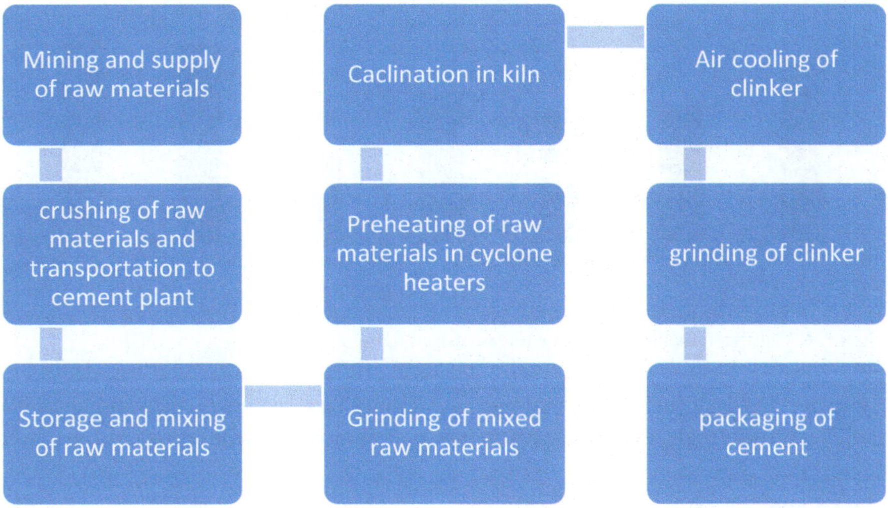

Fig. 65.3 Cement production processes

cement, energy is being used and carbon dioxide is being emitted into the atmosphere. Pyroprocessing is the most energy intensive and process with the highest emission of carbon dioxide in cement production. World Business Council for Sustainable Development [12] reported that about 40% of carbon dioxide emission released is from fossil burning for this process. Pyroprocessing is the operation used to produce clinker which is further grinded to obtain cement.

About 7–38% of the energy used during pyroprocessing is used to eliminate the moisture in the raw materials, and about 5–12% of it is absorbed by the kiln shell in form of heat energy. Though depending on the type of kiln used, energy savings can be made. However, a substantially high amount of energy is required to calcine the raw materials at high temperature.

65.4 Biomass Energy in Cement Production

Renewable energy sourced from biomass can be employed to reduce significantly the fossil fuel consumption by the cement industry. Biomass such as palm kernel shell, coffee pods, rice husk, coconut husks, etc. can be used as biomass in cement production [13]. As the energy from the fossil fuel is mainly used for the heating process of raw materials in cement production, its replacement with biomass energy will lead to a significant reduction in embodied energy and carbon of cement, and concrete in general. In order to replace a type of fuel with another, it is essential to consider its calorific value. This property gives an indication of the amount of energy that can be generated by a specific amount of fuel, and can be measured by the amount of heat generated when an amount of fuel is totally combusted. As shown in Fig. 65.4, the difference in the calorific values of different biomass

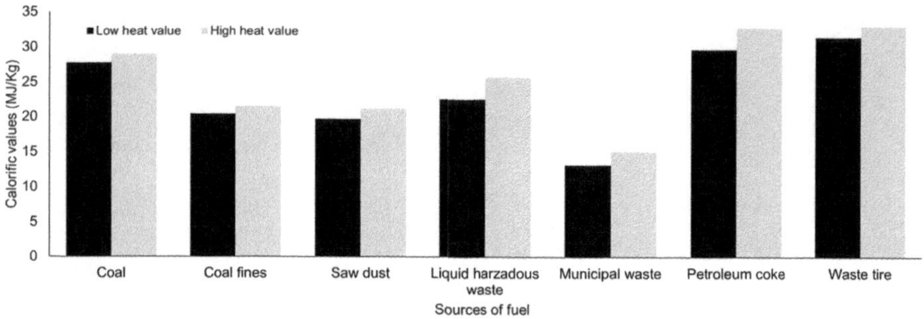

Fig. 65.4 Caloric value for different sources of fuel (data from [15])

compared to fossil fuel shows that biomass cannot be replaced by the same quantity as that of fossil fuel; however, with proper calculations, the right proportion can be determined. Seboka et al. [14] approximated that about 4 kg of biomass can be used as a replacement for 1 L of fuel oil.

Use of renewable energy for cement production is more evident in European countries like the Netherlands and Germany where fossil fuel has been replaced up to 70% with other alternative fuels [16]. But in order to do this, several mechanisms have to be installed and initiated. Large applications of biomass as an energy source are evident in a few places in the world. In Norway, about hundred thousand of biomass is used annually to power the Norcem cement plan, Hima cement plant in Uganda have been able to reduce fossil fuel consumption with biomass (i.e. coffee bean husk) by up to 30% [9, 17]. Malaysia has been able to reduce carbon emission by 60,000 tons annually with the use of palm kernel shell as biomass, and France has been using bone meal as a replacement for fossil fuel since 1998 [17].

Biomass can be used as energy source for cement production by directly feeding it into the preheater or/and pre-calciners in solid form or by converting it to a producer gas to be used in the combustion system.

65.5 Challenges and Possible Solutions

65.5.1 Season

The types of biomass available at a specific location vary with season, making all year supply impossible. For example, biomass such as coffee husk and cotton rusk are only produced during certain months of the year. Therefore, in order to prevent a shortage of supply for renewable energy generation from biomass, proper collection and storage should be done during the production season. Another option will be to alternate different sources of biomass used to generate energy, such that there is a continuous supply of renewable energy regardless of the type of biomass residue that is being generated. However, the non-uniform calorific values of different biomass might pose a challenge as the proper proportion of each type of biomass has to be determined.

65.5.2 Supply Market

In cases where the location where biomass is generated or cultivated is far from the cement production site, a market will have to be put in place such that the demand and supply of materials can be met in a timely manner. However, an increase in the cost of biomass might be associated with a supply market in place has there will be increasing in demand for biomass by cement and other industries. This might, in turn, increase the expenditure incurred by the cement industry, but they will still benefit from the sustainable use of renewable energy. Situating new cement plants that will use biomass energy around agricultural plantations will create an ease of supply of biomass which might lead to significant savings in costs associated with transportation of biomass from rural areas.

65.5.3 Lack of Knowledge

It is evident that the promising advantage of using biomass energy for cement production will be more significant in developing countries. The rapid development expected in coming years means there will lead be a corresponding increase in cement production to meet the demand of the construction industry. However, few of the developing countries currently employ the use of biomass on a large scale for industrial applications. Therefore, for these countries to be able to meet the future demand for energy and cement, it is paramount to get exposure to knowledge about biomass and its applications. Adequate understanding and training of how to maximize the use of biomass energy will propagate more awareness about this renewable source of energy and its applications. Also, creating more awareness alongside more research and development in the field of powering cement production with biomass energy will foster rapid development and understanding among stakeholders.

65.5.4 Capital

Conversion of cement plants from fossil fuel usage to biomass energy requires some initial capital to modify the current infrastructure or construct new ones. Also, the use of biomass requires incorporating pretreatment infrastructures which might lead to a further increase in the capital required. The incorporation of optimized designs into new cement plants to be constructed will help to eliminate future costs that might be incurred as a result of conversion into biomass energy in the future.

65.5.5 Disruption in the Ecosystem

Lack of sustainable retrieval method for biomass might lead to disruption in the ecosystem such as deforestation from using more biomass than being planted. Therefore, as mentioned earlier, it is essential that biomass cultivated for energy generation are replaced accordingly by replanting. Replanting also ensures that the use of biomass energy has a zero net carbon dioxide emission.

65.6 Sustainability of the Use of Biomass Energy for Cement Production

Compared to using other alternative sources of fuel such as rubber tires and plastic as a source of energy in cement production, biomass energy has been found to be more sustainable because the source of biomass is being replenished with replanting. Therefore, the use of biomass energy is not only a renewable source of energy but a net zero carbon emission source of energy. In addition, as some biomass are waste products from the agricultural industry, there use as fuel for cement production creates an avenue to manage these wastes effectively.

65.7 Conclusion

This paper explored the viability of using biomass as a renewable energy source for the production of cement. Based on the overview, the following conclusions can be made.

1. The cement industry is a major contributor of carbon dioxide which has been associated with worldwide temperature variations. High carbon dioxide emission from cement production is from calcination of raw materials and the burning of fossil fuels.
2. Biomass is a renewable energy source due to its viability to be replaced sustainably by planting trees and other vegetation to absorb carbon dioxide generated from biomass combustion.
3. Use of biomass energy in cement production can be used to have more reduction in carbon dioxide emissions and to complement other green initiatives by the cement industry.
4. Though the use of biomass energy to power cement production will lead to a reduction in the carbon dioxide emission of the concrete industry, this initiative alone is not enough to meet the world's industrial goal to keep the carbon dioxide concentration to the minimum as much as possible. Therefore, other ways to have more reduction in carbon dioxide emission has to be developed. One of the promising ways is to develop alkali-activated materials as a suitable replacement for OPC.

5. In addition to the reduction in carbon dioxide emissions, the use of biomass energy will lead to overall reduction in cost and will create an avenue for a nation utilizing biomass energy to conserve foreign exchange that could have been used to purchase and import fossil fuels. In addition, farmers and agricultural industries will be able to make money from the wastes being generated.

References

1. Worrell E, Price L (2000) Evaluating clean development mechanism projects in the cement industry using a process-step benchmarking approach. Ernest Orlando Lawrence Berkeley National Laboratory, Berkeley, CA
2. IPCC (2007) Climate change 2007: synthesis report. Contribution of working groups I, II and III to the fourth assessment report of the Intergovernmental Panel on Climate Change. IPCC, Geneva, Switzerland, p 104
3. Friends of the Earth (2006) The use of palm oil for biofuel and as biomass for energy. http://www.foe.co.uk/resource/briefings/palm_oil_biofuel_position.pdf
4. Imbabi M, Carrigan C, McKenna S (2012) Trends and developments in green cement and concrete technology. Int J Sustain Built Environ 1:194–216
5. Hasanbeigi A, Price L, Lu H, Lan W (2010) Analysis of energy-efficiency opportunities for the cement industry in Shandong Province, China: a case study of 16 cement plants. Energy 35:3461–3473
6. Hienola A, Pietikäinen J, Donnell D, Partanen A, Korhonen H (2017) The role of anthropogenic aerosol emission reduction in achieving the Paris agreement's objective. Geophys Res Abstr 19:EGU2017–EG12544
7. Turner L, Collins F (2013) Carbon dioxide equivalent (CO2-e) emissions: a comparison between geopolymer and OPC cement concrete. Constr Build Mater 43:125–130
8. Shuit S, Tan T, Lee T, Kamaruddin A (2009) Oil palm biomass as a sustainable energy source: a Malaysian case study. Energy 34:1225–1235
9. Tokheim L (2007) Carbon dioxide emission reduction by increased utilization of waste-derived fuels in the cement industry. Telemark University College, Faculty of Technology, Porsgrunn, Norway
10. Da Silva I, Kutty K, Nturanabo F, and Kucel S. 2003 Stand-alone small level power systems based on utilisation of agricultural residue. In: Paper presented at the domestic use of energy conference
11. Szabo L, Hidalgo I, Cıscar J, Soria A, Russ P 2003 Energy consumption and CO_2 emissions from the world cement industry. Institute for prospective technological studies
12. WBCSD (2002) The cement sustainability initiative: our agenda for action. World Business Council for Sustainable Development (WBCSD), Switzerland
13. Murray A, Price L (2008) Use of alternative fuels in cement manufacture: analysis of fuel characteristics and feasibility for use in the chinese cement sector. Ernest Orlando Lawrence Berkeley National Laboratory, Berkeley, CA, LBNL-525E
14. Seboka Y, Getahun M, Haile-Meskel Y (2009) Biomass energy for cement production: opportunities in Ethiopia. CDM capacity development in Eastern and Southern Africa
15. Hansen E (1990) New ways to burn hazard waste
16. Oficemen (2012) Anuario 2011. Madrid: Spain
17. Lafarge (2008) From waste to resource: creating a sustainable industrial system. Lafarge Press Kit, Paris. http://www.lafarge.com

Chapter 66
Blade Optimization of a Small Vertical-Axis Wind Turbine Using the Response Surface Method

Chul-Kyu Kim, Sajid Ali, Sang-Moon Lee, and Choon-Man Jang

66.1 Introduction

Recently, hybrid renewable energy systems (HRES) have emerged as an alternative technology in order to overcome the disadvantages of traditional renewable energy systems by integrating various energy sources to supply a stable and continuous electricity supply, as shown Fig. 66.1 [1]. The main energy sources of such HRES are wind and solar energy. The availability of wind energy greatly differs depending on location, topography, altitude, and season. In the case of South Korea, most regions have very low wind speeds, except for some eastern mountainous areas (refer to Fig. 66.2). So, most sites in the country have favorable wind conditions for small vertical-axis wind turbines, rather than large horizontal-axis wind turbines [2]. The small vertical-axis wind turbine (VAWT) has been attracting attention as a key component of hybrid renewable energy systems due to their many advantages, such as the low cut-in wind speed, low cost, low noise output, and high site flexibility [3]. For these reasons, the research on small vertical wind turbines is increasing.

Danao et al. [4, 5] performed an unsteady-state performance test to discover the features of vertical-axis wind turbines, using both a numerical analysis and experimental data. They conducted experimental study on a vertical-axis wind turbine of a wind tunnel scale with unsteady wind conditions, in addition to performing a

C.-K. Kim · S.-M. Lee
Department of Land, Water and Environment Research, Korea Institute of Civil Engineering and Building Technology, Goyang-si, Republic of Korea

S. Ali · C.-M. Jang (✉)
Department of Land, Water and Environment Research, Korea Institute of Civil Engineering and Building Technology, Goyang-si, Republic of Korea

Construction Environment Engineering, University of Science and Technology, Goyang-si, Republic of Korea
e-mail: jangcm@kict.re.kr

A. Sayigh (ed.), *Renewable Energy and Sustainable Buildings*, Innovative Renewable Energy, https://doi.org/10.1007/978-3-030-18488-9_66

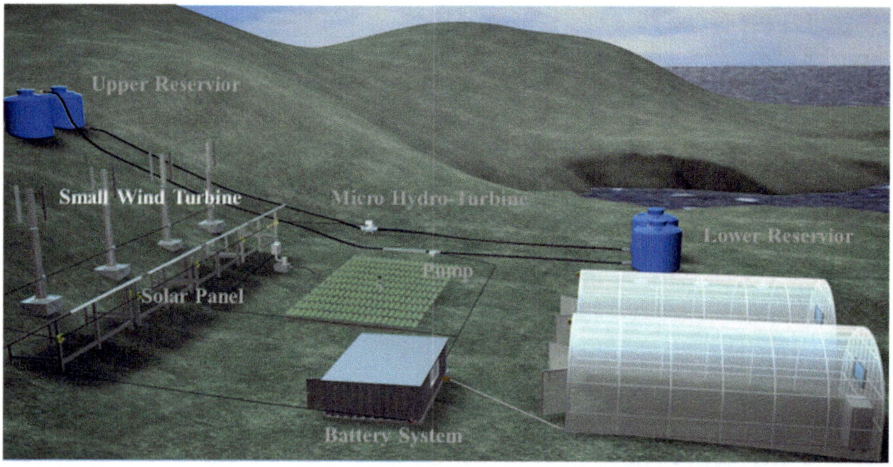

Fig. 66.1 Concept of hybrid renewable energy system for an off-grid island [1]

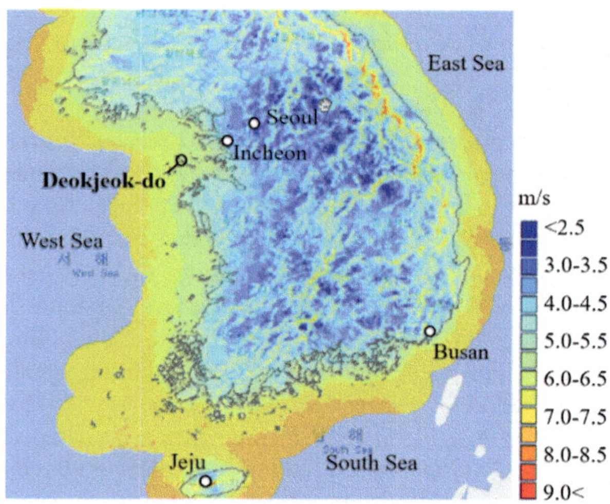

Fig. 66.2 Mean wind speed at 40 m above ground level in South Korea (Sajid et al., 2018)

numerical analysis using Reynolds-averaged Navier-Stokes (RANS) equations to study the effects of steady and the unsteady states on the performance of VAWT. Jaohindy et al. [6] reported the Unsteady Reynolds-averaged Navier-Stokes (URANS) analysis to simulate the unsteady flow behavior of a Savonius rotor with different aspect ratios. Bazilevs et al. [7] reported that a continuously changing angle of attack of wind on the wind turbine rotor drops its aerodynamic efficiency. They performed important function of averaging of real-time data to evaluate the performance of installed wind turbines. Jang et al. (2017) carried out analyzing the effect of different time averaging steps using the real-time-measured wind data of the test site. They also conducted numerical simulations using the URANS analysis and compared 10-, 20-, and 30-min averaged experimental data. In this study, they showed that turbine performance analyzed by using 10-min time-averaged data is in good agreement with the result of the numerical simulation. However, there are very few studies concerning the optimization of the blade of small vertical-axis wind turbines to improve the performance using the RSM as a designing tool.

The main focus of our work is to optimize the blade shape of an H-Darrieus rotor of a small vertical-axis wind turbine to attain the maximum power coefficient using the Response Surface Method for conditions with wind speeds under 4 m/s and TSR 2.6 conditions. A 1.5 kW vertical-axis wind turbine was installed on the Deokjeok-do Island test bed in South Korea and was used as a reference model. The optimization of the blade was the central goal of the Response Surface Method. The maximum thickness location and the chord length were chosen as design variables for the blade optimization in order to enhance the turbine power generation performance. The objective function is the power coefficient, which is a dimensionless performance factor that represents the ratio of turbine power as compared to the theoretical maximum power. The numerical analysis of the URANS was introduced in order to obtain the power coefficient. Consequently, this work showed the optimum model successfully improved the performance as compared to that of the reference model.

66.2 Methodology

66.2.1 Vertical-Axis Wind Turbine and Geographical Location

In the present study, the wind turbine for optimization is a 1.5 kW vertical-axis wind turbine of the H-Darrieus type installed on the Deokjeok-do Island test bed in South Korea. Figure 66.3 shows the shape of the turbine rotor and its performance curve. The design blade section profile is NACA0015, and the blade height and chord length of the turbine rotor are 3 m and 0.2 m, respectively. The rotational diameter of the turbine rotor is 2 m. The cut-in speed of the turbine is 3 m/s, whereas the rated wind speed and rotational speed are 13.5 m/s and 300 rpm, respectively.

Figure 66.4 shows the turbine altitude and an overall view of the test bed. An anemometer and anemoscope are installed at the height of 10 m above the ground.

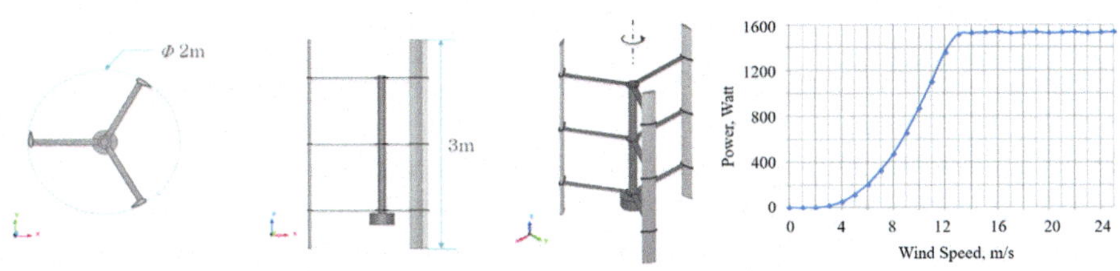

Fig. 66.3 Shape of wind turbine rotor (left) and turbine performance curve (right)

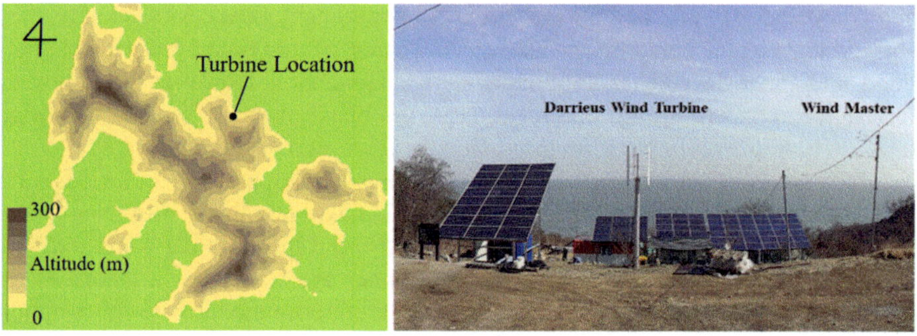

Fig. 66.4 Altitude of the wind turbine (left) and an overall view of system (right)

66.2.2 Wind Environmental Assessment

Analyzing the wind characteristics of a particular site is of immense importance for optimally designing a site-specific wind turbine. The Weibull probability density function (PDF) and cumulative distribution function (CDF) are commonly used tools for assessing the wind potential of a particular site. Many researchers have recommended the use of Weibull functions as the standard tool for wind potential assessment [8–10]. For the wind data analysis on Deokjeok-do Island, eleven years (2005–2015) of measured wind data were utilized, which was recorded by Korean Meteorological Association [11]. Figure 66.5 show the results of the wind data analysis in the forms of a Weibull PDF and CDF plots. As it is clear from Fig. 66.5 that most frequently occurring wind speed is 4 m/s, so that will be used as the reference wind speed for the numerical simulation process. Similarly, Fig. 66.5 also shows the wind rose diagram on Deokjeok-do Island.

Tip speed ratio (TSR) is one of the important parameters for assessing the performance of a wind turbine. The most frequently occurring TSRs and wind speeds will be estimated at the test bed (using experimental data) and will be used as the operating conditions for numerical simulations. As mentioned above, finding out the most probable TSR is of key importance for setting the appropriate value of the rotational speed for numerical simulations. For this purpose, Fig. 66.6 has been prepared showing the distribution of TSRs as a frequency percentage. From the results of Fig. 66.6, it is apparent that the approximate value of the most frequent TSR is 2.6.

66.2.3 Numerical Method

In the numerical simulation, the incompressible three-dimensional unsteady RANS equations were used to analyze the incompressible flow of the computational domain as shown in Fig. 66.7. The grid system of the computational domain was generated using 26 million elements, consisting of tetrahedral, hexahedral, and

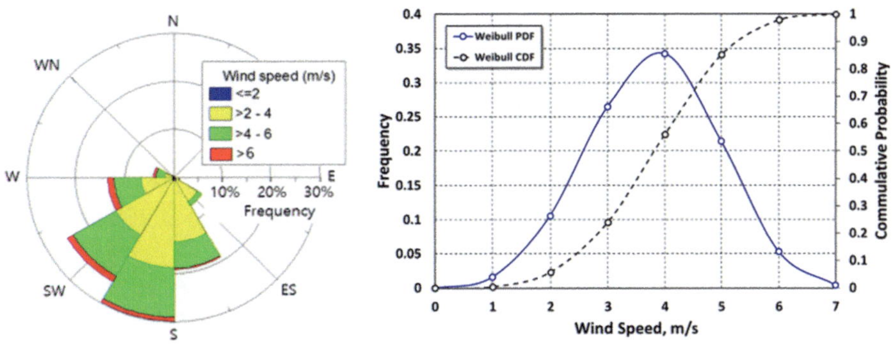

Fig. 66.5 Wind rose (left) and Weibull PDF curve (right) at the Deokjeok-do Island test bed

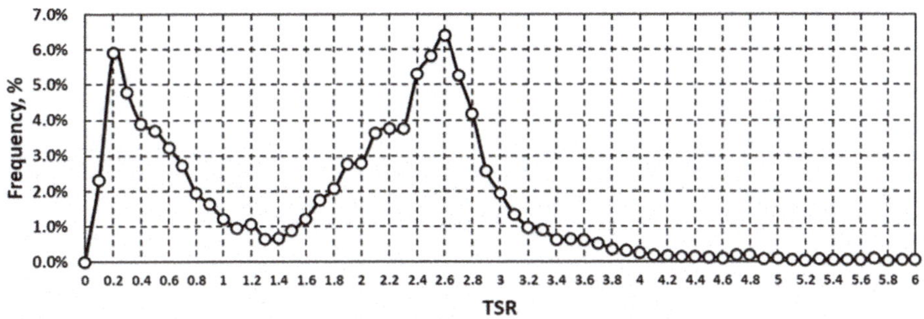

Fig. 66.6 Tip speed ratio (TSR) frequency at the Deokjeok-do Island test bed

prism meshes. The commercial code, SC/Tetra v13, has been employed in the present numerical simulation. The boundary conditions at the inlet and outlet were given to have a velocity of 4 m/s and the atmospheric pressure. The wall surface of the computational the domain was given a slip wall condition allowing the consideration of free flow, whereas the wall surface of the wind turbine was given a no-slip condition. The shear stress transport (SST) model was employed as the turbulence model [12] as it is well known to accurately predict the turbulence in most rotational flows.

66.2.4 Optimization Method

The Response Surface Method (RSM) was used as the optimization method. The RSM is one of the design of experiment (DOE) methods to optimize the response (output variable), which is affected by several independent variables (input variables). The response surface designs are useful for modelling quadratic surfaces. Designed response surface cases can identify points within a design area that have

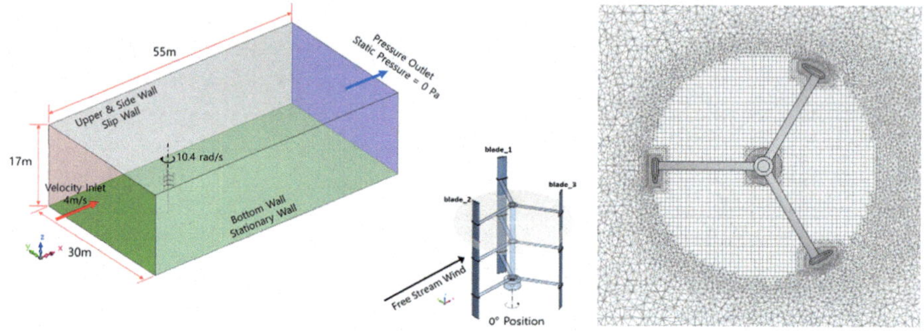

Fig. 66.7 Computational domain (left) and grid system around the turbine rotor (right)

Table 66.1 Cases of response surface design

No.	Max. thickness location, %	Chord length, mm	Remark
Case 1	20	100	
Case 2	20	200	
Case 3	40	200	
Case 4	30	200	Reference model
Case 5	20	300	
Case 6	40	100	
Case 7	30	300	
Case 8	30	100	
Case 9	40	300	

a minimum or maximum response. The optimum design points can also be estimated within the design range through the response surface.

In the present study, the objective function of RSM is considered to be the power coefficient, and the independent variables are considered to be the maximum thickness location and the chord length of the blade. The maximum thickness location can determine the optimal thickness position for the given wind speed condition, and the chord length can determine the optimal length for the TSR condition. According to the Sandia report [13], the maximum thickness location was reported to increase the lift coefficient at a position of 22.9% greater than of a reference model under low wind speed conditions. In the present study, the maximum thickness locations were 20%, 30%, and 40%, and the chord lengths were 100 mm, 200 mm, and 300 mm. The blade section profiles were generated using QBlade [14], an open source program distributed by the Berlin TU. The response surface designs are also shown in Table 66.1 and a total of nine design cases were evaluated using the statistical program, JMP 13 [15].

66.3 Results and Discussion

66.3.1 Verification of Numerical Model

Figure 66.8 shows the results of the numerical model, as compared to the experimental data. The numerical model was compared using the 10-min average wind turbine power [3] with respect to wind speed conditions. The numerical simulation and the experimental data show comparatively good agreement when comparing the results. Therefore, the validated numerical model is applied to the numerical analysis of the response surface design cases.

66.3.2 Determinant of Optimal Design Point Using RSM

Figure 66.9 shows the power coefficient with respect to design cases obtained by numerical simulations. The values of the power coefficient for all nine cases generated by the response surface design were calculated by applying the same numerical method and boundary conditions as the reference numerical model. Using the power coefficient results for each case, a response surface can be created that has the quadratic surface as shown in Fig. 66.10. Here, the optimal design point is the independent variable value at the maximum point of the response surface. Therefore, the optimal design point has a maximum thickness location of 24% and a chord length of 210 mm, within the response surface of the design area.

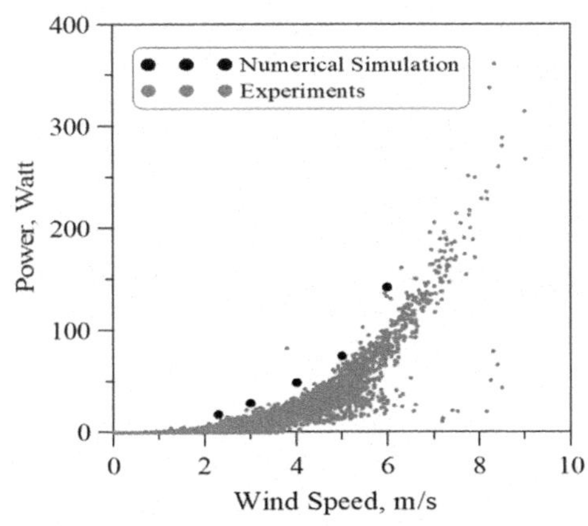

Fig. 66.8 Verification result of numerical model

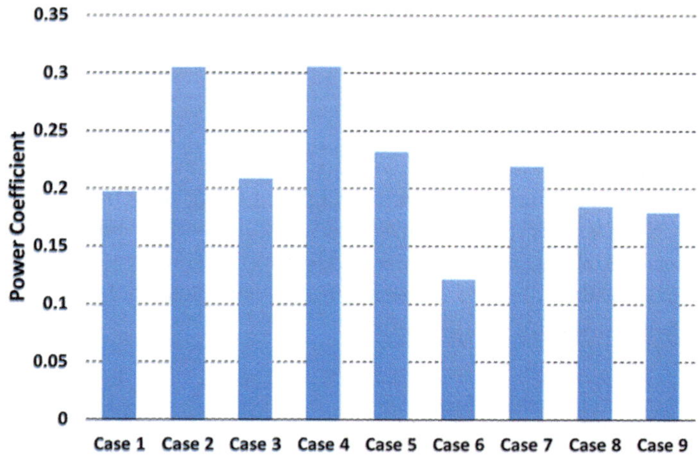

Fig. 66.9 Power coefficient results of nine cases by response surface design

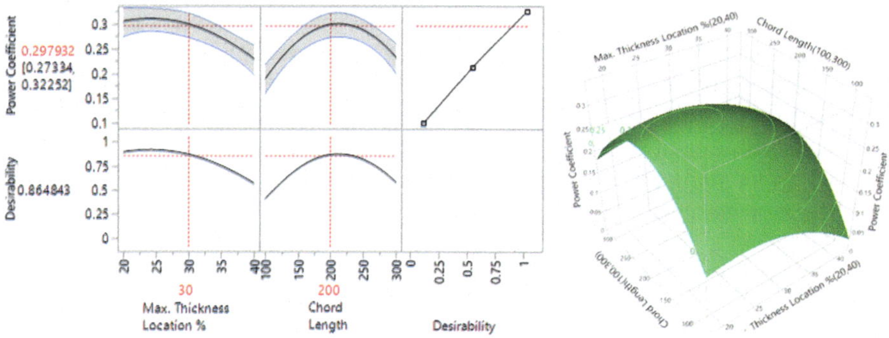

Fig. 66.10 Prediction fit curve of two independent variables (left) and response surface result (right)

66.3.3 Comparison of the Reference Model and Optimum Model

The optimum model was analyzed once again by using design variables determined by the Response Surface Method. Figure 66.11 shows the results of comparing the power coefficient according to the rotational angles of the reference model and the optimum model. The minimum power coefficient of the optimum model increased by 34.9% compared to that of the reference model, where the maximum power coefficient increased by only 18.8%. Overall, the average power coefficient of the reference model was 0.3063; however, the optimum model was 0.3453 which is an improvement of approximately 12.7%. The vertical-axis wind turbine of the optimum design point had a dramatic improvement under TSR 2.6 and 4 m/s wind

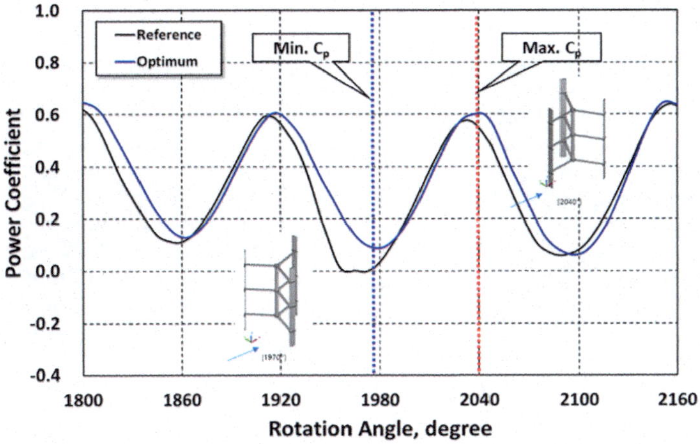

Fig. 66.11 Positions of rotation angle at the min. Power coefficient (1970°) and the max. Power coefficient (2040°)

speed conditions. In other words, the performance improvement of the optimum model can be explained as the increase of the power coefficient by the loss reduction of aerodynamic.

Figure 66.12 shows the velocity magnitude and vorticity around the blade at the position of the minimum power coefficient. In the case of the optimum model, the size of the vorticity flow induced by Blade 2 obstructing the wind flow was greatly reduced compared to that of the reference model. Similarly, the velocity magnitude contour of the optimum model shows a decrease of the vorticity area compared to that of the reference model. Therefore, the power coefficient of the optimum model increased due to the reduction of the turbine loss by a vorticity reduction.

Similarly, Fig. 66.13 shows the velocity magnitude and vorticity around the blade at the position of the maximum power coefficient. For the optimum model, the wake flow of Blade 1 is more stable than that of the reference model. On the other hand, Blade 2 and Blade 3 of the reference and the optimum models show similar flow patterns. In the comparison of the vorticity contours, the optimum model showed a decrease of vorticity flow area around Blade 1 compared to that of the reference model. Therefore, it can be seen that the optimum model has a reduced loss due to the more stable wake flow.

66.4 Conclusions

Using a small vertical-axis wind turbine of the Darrieus-type installed on Deokjeok-do Island in South Korea, turbine power performance has been enhanced by the shape optimization of the turbine blade using the Response Surface Method. The numerical model that analyzed the unsteady-state flow field has also been

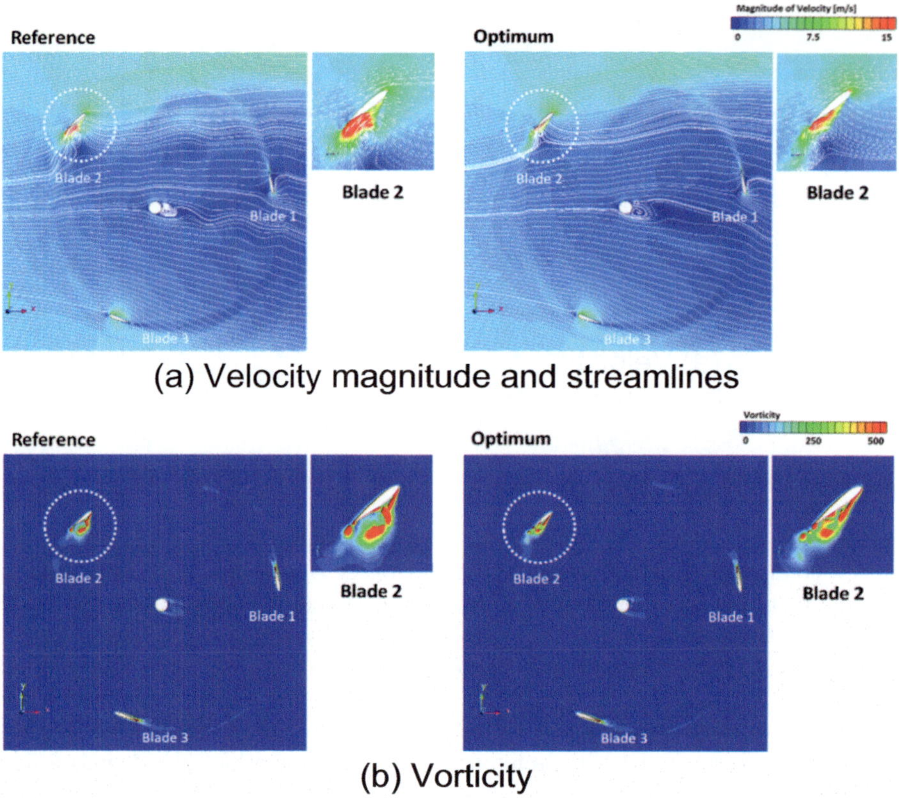

Fig. 66.12 Comparison of velocity magnitude and vorticity at position of minimum power coefficient. (**a**) Velocity magnitude and streamlines. (**b**) Vorticity

validated using the 10-min averaged power-speed experimental data and showed a good agreement. The blade optimization under 4 m/s wind speed and TSR 2.6 conditions has been performed to determine two independent variables of optimum design by RSM that creates the response surface of the output variable. Consequently, the present study can be summarized as follows:

- The maximum blade thickness (%) and chord length (mm) were shortlisted as two independent design variables and considered three different values for each variable. Therefore, a total nine design cases were analyzed at TSR 2.6, and the optimum design point was found on the response surface to improve the performance of small vertical-axis wind turbines.
- The optimum design point had the maximum thickness location of 24% and the chord length of 210 mm as determined by the Response Surface Method. From the simulation results and the Response Surface Method, it successfully improved the power coefficient of wind turbine by 12.7% in comparison to that of the reference model.

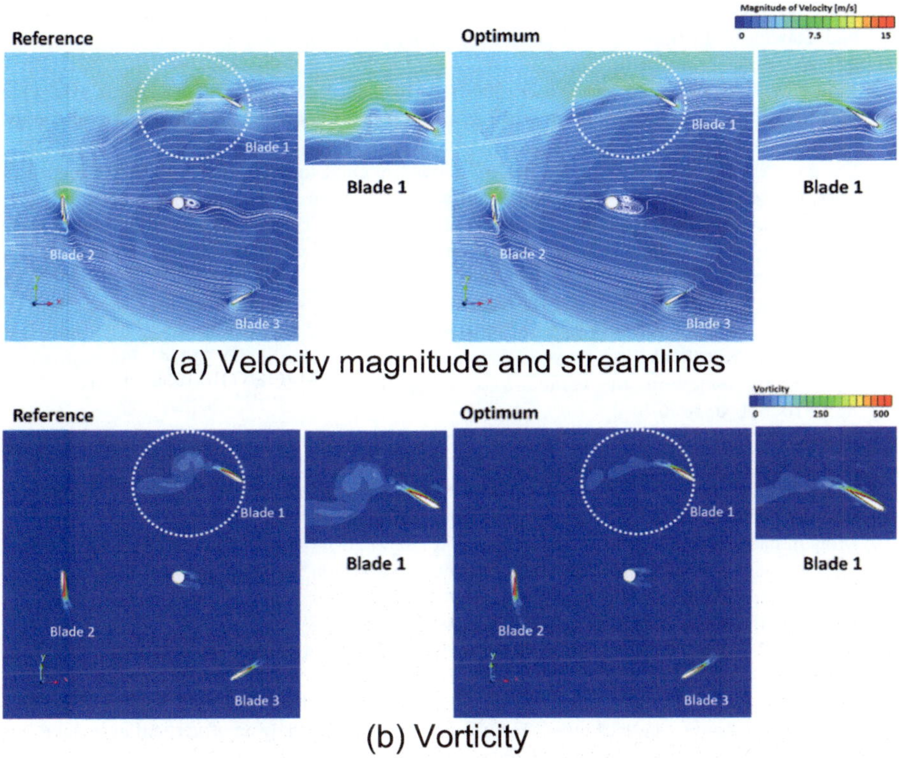

Fig. 66.13 Comparison of velocity magnitude and vorticity at position of maximum power coefficient. (**a**) Velocity magnitude and streamlines. (**b**) Vorticity

- Through a comparison of both, i.e., the power coefficient curves and the flow patterns, the increase of the power coefficient of the optimum model was consistent with the decreasing tendency of vorticity flow and wake flow around the blade. The optimum model has increased the power coefficient of the wind turbine due to the reduction of turbine loss due to vorticity.

Acknowledgments This work was supported by the New and Renewable Energy Core Technology Program of the Korean Institute of Energy Technology Evaluation and Planning (KETEP) and granted financial resources from the Ministry of Trade, Industry and Energy, Republic of Korea (No. 20153010130310).

References

1. Kim CK, Lee SM, Jeon SY, Jang CM (2017) Analysis on the performance of micro hydro turbine used for hybrid renewable energy system. In: 7th Asia-Pacific Forum on Renewable Energy (AFORE) 2017

2. Ali S, Lee SM, Jang CM (2018) Statistical analysis of wind characteristics using WEIBULL and Rayleigh distribution in Deokjoek-do Island-Incheon, South Korea. Renew Energy 123:652–663
3. Jang CM, Kim CK, Lee SM, Ali S (2018) Performance evaluation of a vertical axis wind turbine using real-time measuring wind data. In: Transition towards 100% renewable energy. Springer, Berlin, pp 187–195
4. Danao LA, Eboibi O, Howell R (2013) An experimental investigation into the influence of unsteady wind on the performance of a vertical axis wind turbine. Appl Energy 107:403–411
5. Danao LA, Edwards J, Eboibi O, Howell R (2014) A numerical investigation into the influence of unsteady wind on the performance and aerodynamics of a vertical axis wind turbine. Appl Energy 116:111–124
6. Jaohindy P, McTavish S, Garde F, Bastide A (2013) An analysis of the transient force acting on Savonius rotors with different aspect ratios. Renew Energy 55:286–295
7. Bazilevs Y, Korobenko A, Deng X, Yan J, Kinzel M, Dabiri JO (2014) Fluid-structure interaction modelling of vertical-axis wind turbines. J Appl Mech Trans ASME 81:081006-1–081006-8
8. Deaves D, Lines I (1997) On the fitting of low mean wind speed data to the Weibull distribution. J Wind Eng Ind Aerodynamics 66:169–178
9. Islam M, Saidur R, Rahim N (2011) Assessment of wind energy potentiality at Kudat and Labuan, Malaysia using Weibull distribution function. Energy 36:985–992
10. Mohammadi K, Alavi O, Mostafaeipour A, Goudarzi N, Jalilvand M (2016) Assessing different parameters estimation methods of Weibull distribution to compute wind power density. Energy Convers Manag 108:322–335
11. KMA (2018). http://data.kma.go.kr/. Accessed 23 Mar 2018
12. Menter FR (1994) Two-equation Eddy-viscosity turbulence models for engineering applications. AIAA J 32:1598–1605
13. Robert ES, Paul CK (1981) Aerodynamic characteristics of seven symmetrical airfoil sections through 180-degree angle of attack for use in aerodynamic analysis of vertical axis wind turbines. Sandia National Laboratories Energy Report, California
14. Marten D, Wendler J (2013) QBlade guidelines v0.6
15. SAS Institute Inc (2017) JMP 13 design of experiments guide, 2nd edn. SAS Institute, Cary, NC

Chapter 67
Enhancing Energy Utilisation in Building with Combining Building Integrated PV and Air Source Heat Pump for Underfloor Heating Using Phase Change Materials

Ming Jun Huang and Neil J. Hewitt

67.1 Introduction

Energy use in domestic sector accounted for 29% of the total energy consumption in the UK and 80% of which is used for space heating and hot water. Improving low energy consumption application with heavy thermal mass and utilising renewable energy in buildings in the UK are vital in order to create a sustainable and dependable energy market as well as cutting CO_2 emissions. Therefore installation of renewable energy to building heating systems with cost-effective and affordable ways is becoming an ever-growing factor. Heat pump (HP) is a high energy efficient renewable energy device which can be a renewable-based alternative to fossil fuel boilers in buildings. Heat pumps as a promising solution achieve the carbon reduction target in the domestic sector, but the limitation of electricity consumption and low supply flow temperatures (for most of the standard ASHPs the supplied heat temperature is around 55 °C) will hinder the implementation of ASHP for domestic applications and replace the existing fossil boiler. With 20% of households using heat pumps cause the peak grid demand increases by 7.5 GW (14%) which makes extra burden to the grid. Heat pumps operate intermittently in the on-off mode in order to adjust the heating and cooling capacity to the load required by the building. The issues with the heat pumps for underfloor heating are how to layout the hot water pipes and utilise underfloor materials to achieve cost effective heating effect under an optimised air-source heat pump (ASHP) size and operating mode.

The building integration of photovoltaics (BIPV), where the PV elements actually become an integral part of the building, by simultaneously serving as building envelope material and acting as power generator. There is a growing consensus that

M. J. Huang (✉) · N. J. Hewitt
Centre for Sustainable Technologies, School of Built Environment, Ulster University, Newtownabbey, UK
e-mail: m.huang@ulster.ac.uk

A. Sayigh (ed.), *Renewable Energy and Sustainable Buildings*, Innovative Renewable Energy, https://doi.org/10.1007/978-3-030-18488-9_67

the generated electricity by photovoltaic should provide electricity at the point of use instead of power transmission to the grid. Domestic HPs have great potential to be driven by the solar power on-site. The produced heat/cold by HP can be used for domestic heating or thermal regulation. However, the on-site production of solar electricity is an intermittent source, typically the greatest may not be at or near the time of a building's peak loads for residential demanding. In addition, the electricity generation for BIPV is the elevation of the PV temperature reduces solar to electrical energy conversion efficiency by 0.4–0.5%K^{-1} for crystal silicon PV when it rises above the characteristic power conversion temperature of 25 °C. Maintaining the silicon PV's temperature at a low temperature, preferably lower than or around 25 °C, will retain the maximum conversion efficiency of the PV for practical applications. This specific requirement with PV temperature regulation has limited the utilisation of solar thermal in residential buildings. Passive cooling by natural, wind induced air convection in a duct behind the PV is commonly used in BIPV systems, also some investigations have been carried by using PCMs to regulate the temperature in PVs. Active heat dissipation using water with PCM for the BIPV thermal regulation will be discussed in this paper.

Energy storage plays an important role on building the demand side management. Thermal energy produced by the HP during the off peak time can be stored in the building underfloor structure and released for the later usage. The low temperature water supply by HP can benefit to the underfloor heating in the building during winter and cold in the summer, which provides the most indoor thermal comfort environment. In addition, the high temperature in BIPV need to be regulated in order to avoid the reduction of the solar power conversion efficiency.

Phase change materials (PCMs) as energy storage materials can absorb a large amount of energy as latent heat at a relative constant phase transition temperature and are thus used for heat storage and temperature control. The capacity of a PCM integrated to building for energy storage and temperature control depends on its properties, heat transfer methods and system configuration. The properties of various PCMs, methods of heat transfer enhancement, design configurations of heat storage facilities as part of solar passive and active space heating systems, thermal energy storage systems and building applications have been the subject of recent reviews [1, 2].

PCMs incorporated into BIPV and underfloor heating system in buildings may be suitable for regulating the PV temperature rising and storing the heat from HP for underfloor heating. This may allow the HP to operate in an electricity tariff management mode to store the latent heat for prolonged underfloor heating and keeping the indoor thermal comfort with narrow temperature swing. Optimising the floor heating system design with PCM underfloor heating has been studied [3]. In this paper a two-dimensional finite-volume numerical simulation model used for PCM wall thermal performance study has been modified and validated to examine the dynamic process of the PCM augmented underfloor heating system under steady states heat supply and air-source heat pump on-off heating mode. The studying results provide an insight into the understanding of the structure and thermal performance of the PCM underfloor heating system with optimised design and operations. The previous results show that the PCM can store a great quality of thermal energy and shift the peaks of load. The low thermal conductivity frequently makes an anticipated level of thermal storage untenable within an acceptable time period. Active water flow may be needed to carry on the energy exchange between the PCMs and thermal applications. Floor cover materials and structure and hot water pipe layout along with the ambient conditions play important impact on the indoor environment and energy savings in buildings. A good understanding on the fundamental heat transfer processes involved is essential for accurately predicting the thermal performance of the PCM underfloor heating system driven by HP and avoiding costly system design mistakes.

A schematic diagram of BIPV-PCM-HP-underfloor system is presented in Fig. 67.1. It uses ASHP for supplying heat for PCM layered underfloor heating system or domestic hot water and extracting solar heat from BIPV-PCM for keeping the efficiency of BIPV. The solar energy incidents on the BIPV are converted to electricity and the PCM is used to regulate the temperature rising on PV through the latent heat process. When the temperature regulation capacity of the PCM is out of the range for PV best performance, then stored solar energy in the PCM can be moved away through the ASHP operation. The cycling cupper pipes from evaporator extracts the heat in PCM with stable temperature range for the summer high temperature operation, and the HP provides high temperature heat to the underfloor system during winter. The heat will be stored in the PCM layer in the underfloor construction. The solar thermal energy stored in the BIPV-PCM system can provide constant stable low temperature heat to ASHP without influence from variable ambient influence and provides high temperature hot water for underfloor heating system in the residential buildings. PCMs incorporated into solar energy thermal storage and underfloor construction of heating system in buildings may be suitable for absorbing solar energy

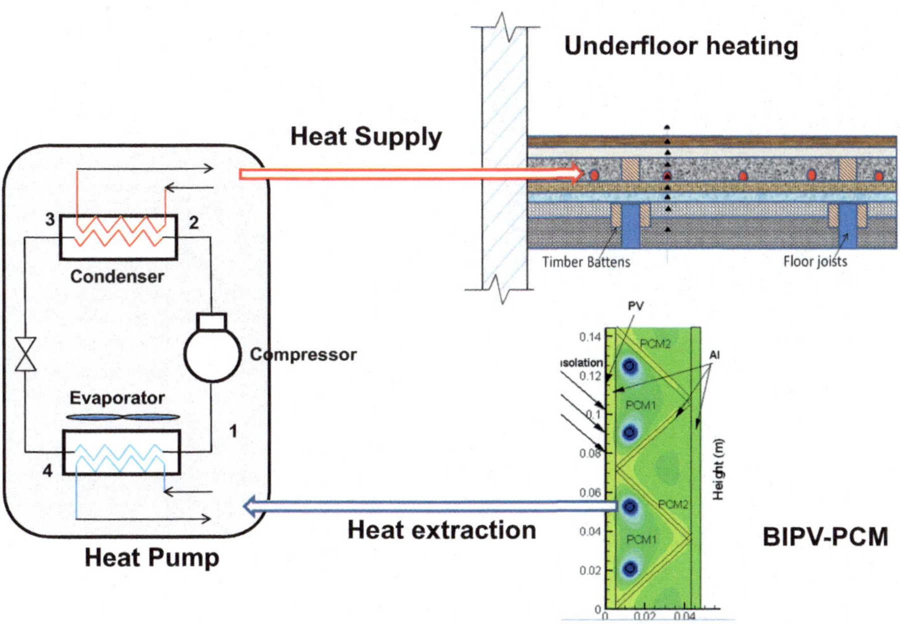

Fig. 67.1 A schematic diagram of using HP for supplying heat for PCM layered underfloor heating system and extracting solar heat from BIPV-PCM through the cycling cupper pipes

directly or store the heat from ASHP during off peak time. Therefore a detailed prediction has been carried on to study the thermal performance of the comprehensive system for PV temperature regulation and store the solar thermal energy and provide high temperature heat for the underfloor heating in the domestic buildings.

A two-dimensional temperature-based finite-volume numerical simulation model has been developed and experimentally validated for PCM energy store and temperature regulation. Using electricity generated by BIPV to drive the ASHP and store the thermal energy to thermal mass of PCM in underfloor heating system has been studied considerably in this paper. It has been found that the effect of using cooling coil of ASHP (evaporator) to augment the temperature regulation in BIPV keeps PV electricity efficiency at a high level. This paper summarises the results of a detailed theoretical investigation and analysis of solar energy production and storage control in buildings. An optimised combination of using solar power to drive ASHP and store the heat in PCM for underfloor heating could improve the energy efficiency of PV through temperature regulation and may achieve a cost-effective and affordable heating system the building.

67.2 Description of the Numerical Simulation Model for BIPV-PCM-HP System

A schematic diagram of BIPV-PCM-HP system is shown in Fig. 67.2. The system will be used for BIPV temperature regulation by PCM latent heat process. When the PCM temperature regulation capacity is out of the PV best performance range, the

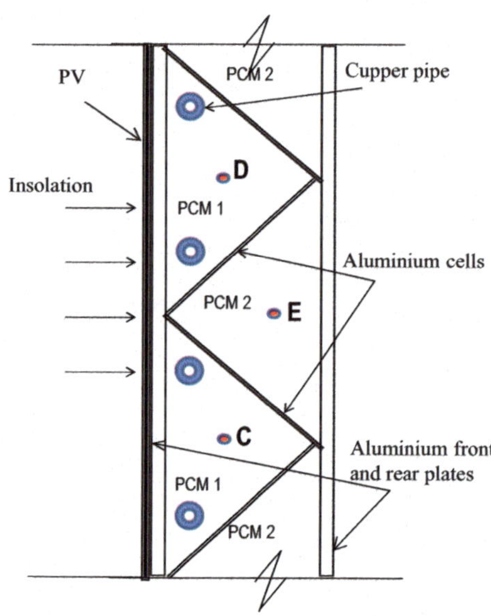

Fig. 67.2 A schematic diagram illustrating the heat transfer in a BIPV-PCM system with cooling pipes from HP

incident solar thermal energy stored in the PCM will be extracted through the cycling copper pipes from heat pump evaporator which are embedded in the PCMs. Therefore the HP can extract low temperature solar thermal energy to keep PV at its best performance when the solar incident energy is high during summer and provides high temperature heat for the underfloor heating system without frosting during extreme winter time. In this work a previous developed two-dimensional numerical simulation model used for BIPV-PCM system thermal regulation [2] has been modified to add cooling effect from HP evaporator.

The thermal performances of the BIPV-PCM-HP system with transient ambient conditions and controlled HP operations under heat charge and discharge processes have been predicted. The effect of the thermal performance of PV with mass material of PCMs under different operation conditions of PCMs with metal cells, PCM metal cells integrating cupper pipes from HP without heat extraction and with heat extracting by 10 °C refrigerant and heat flux have been studied. The effect of insulation with the building integrated system has been studied. The improved energy consumption has been predicted. The following assumptions are made:

1. The end sides at the simulated system are adiabatic.
2. The system construction materials and PCM are homogeneous and isotropic.
3. The convection effect in the molten PCM is neglected.
4. The thermophysical properties of the system materials are constant except the specific heat capacity and density of the PCMs during phase changes.
5. The interfacial resistances between the pipe and PCM material are negligible.
6. The thermal resistance of the pipe wall can be neglected due to the pipe wall is thin and its material has good thermal conductivity. Thus, the model assumes that the heat transfer fluid directly contacts the solid filling PCMs and the pipe thickness can be assumed to be zero.

The energy equation for the BIPV-PCM-HP:

$$\rho c_p \frac{\partial T}{\partial t} + \nabla \cdot \left(k \nabla T\right) = 0 \tag{67.1}$$

where the same equation hold good for all the layers of system integrated material and PCM by incorporating suitable k, ρ, Cp values. The instantaneous continuity of heat flux and temperature at the interfaces of cooling pipes with PCM layer are preserved.

$$k \left.\frac{\partial T}{\partial y}\right|_{y=0} = S_{rad} + h_0 \left(T_{amb,indoor} - T_{y=0}\right) \tag{67.2}$$

$$k \left.\frac{\partial T}{\partial y}\right|_{y=G} = h_G \left(T_{y=G} - T_{amb,ground}\right) \tag{67.3}$$

where h_0 and h_G are the heat transfer coefficients from the PV front surface and rear surface of the BIPV-PCM-HP system to the surroundings. S_{rad} is the heat received by the system on the front surface from the transient incident solar energy. "y" is the position along the horizontal direction. In the front and rear surface boundary, the system is exposed to ambient environment where $y = 0$ and down to rear where "$y = G$," the boundary conditions.

The boundary conditions of the interfaces of pipes with PCM filling materials in the metal cells are:

$$T\left(\left(x,y\right)_{pipe},t\right) = T_w \quad \left(\left(x,y\right) = pipe\ position, t = during\ heating\right) \tag{67.4}$$

$$k\frac{\partial T}{\partial x}\bigg|_{(x,y)=pipeinterface} = S_w \tag{67.5}$$

$$k\frac{\partial T}{\partial y}\bigg|_{(x,y)=pipe\ \ interface} = S_w \tag{67.6}$$

where "T_w" is the average water pipe temperature from HP evaporator and "S_w" is the heat flux supplied by the HP system.

The phase change occurs at a set temperature. In the case of constant specific heat capacities for each phase, the temperature field can be defined as:

$$T = \begin{cases} E / c_S & & T < T_m & (solid\ phase) \\ T_m & 0 < E < H, & T \geq T_m & (melt\ zone) \\ T_m + \left(E - H\right) / c_L & E \geq H, & T > T_m & (liquid\ phase) \end{cases} \tag{67.7}$$

where "E" is the heat stored in the PCMs and "H" is the specified latent heat of PCMs. The heat value "E" of the PCM in the melt zone is modelled as high sensible heat value in each time step and accumulated with time. With time elapsed when the accumulated heat is larger than the specified latent heat "H", then the PCM changes the phase. The governing equation along with the boundary conditions is discretised using the full-implicit finite-volume method. The space grid size is 1 mm and the time step is variable.

Table 67.1 lists the thermophysical properties of RT27 PCM, copper pipe and aluminium plate [4]. The BIPV-PCM-HP system, shown in Fig. 67.2, was simulated to study the effect of metal fins on the BIPV temperature regulation in relation to PCM thermal storage for HP heat extraction. The 5 mm aluminium front and rear walls of the BIPV-PCM-HP system provided a high rate of heat transfer to the PCM. This was enhanced further by a series of embedded metal fins extending into the PCM from the front wall to form different PCM cells. The copper water cycle pipe from the heat exchanger of evaporator in HP is embedded to the PCM cells to extract the stored solar thermal energy. The interior dimensions of the containers were 0.132 m high by

Table 67.1 Thermophysical properties of PCM RT27, copper pipe and aluminium plate [4]

	PCM (Liquid/Solid) and other metals		
Property	Copper pipe	RT27	Aluminium
Density			
Solid (kg m^{-3})	8795	880	2675
Liquid (kg m^{-3})	N/A	760	N/A
Specific heat capacity			
Solid (J m^{-3} K^{-1})	3,386,075	1,566,000	2,415,525
Liquid (J m^{-3} K^{-1})	N/A	1,800,000	N/A
Thermal conductivity			
Solid (W m^{-1} K^{-1})	26	0.2	76
Liquid (W m^{-1} K^{-1})	N/A	0.2	N/A
Melting temperature (°C)	N/A	25–28	N/A
Latent heat of fusion (J kg^{-1})	N/A	184,000	N/A
Kinematic Viscosity (mm^2 s^{-1}) (at 70 °C)	N/A	26.32	N/A
Flash point (°C)	N/A	164	N/A
Volume expansion (%)	N/A	10	2.34 × 10–5

0.04 m depth by 0.050 m thick polystyrene foam, thermal conductivity 0.027 Wm^{-1} K^{-1} provided insulation on the top and bottom of the system.

67.3 Results and Discussions

67.3.1 Effect of Cycle Pipe in the PCM on the BIPV Temperature Regulation

A typical summer day climate conditions in Velari are used as input to predict the thermal performance of the BIPV-PCM-HP. The cold-water temperature for the simulation is from the experimental operation on a residential HP. Figure 67.3 shows the predicted temperatures on the front surfaces of the BIPV-PCM-HP system using the summer weather data from Velari and the cold water from the heat pump evaporator for three hours during the hot mid-day. The predicted results show that the predicted temperatures on the PV can be regulated through the use of PCM along with cold-water cycling. The BIPV temperature regulation can achieve 21–29% improvement comparing with the reference PV response to using PCM only and PCM with 3 h cold-water cycling operations.

Figure 67.4 shows the BIPV-PCM-HP system performance with time evolution. Insolation absorbed on the front surface of the system and conducted through the adjacent layers increased the stored thermal energy in the PCM. The effect of the cold water for PV temperature regulation is clear. All of the cases have demonstrated the effect of temperature regulation in PV, but the copper pipe with HP cooling cyclic can reduce the PV temperature swing with the most regulation capacity.

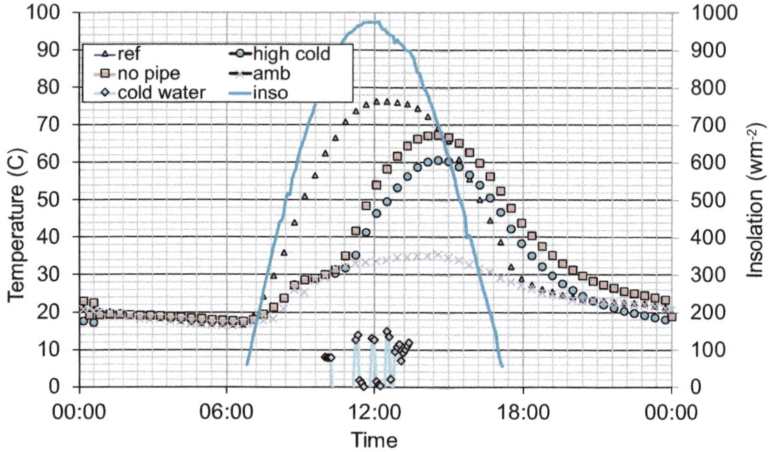

Fig. 67.3 Predicted temperatures on the front surfaces of the BIPV-PCM-HP system under Velari summer conditions with 3 h cold-water cycling from HP evaporator for system temperature regulation

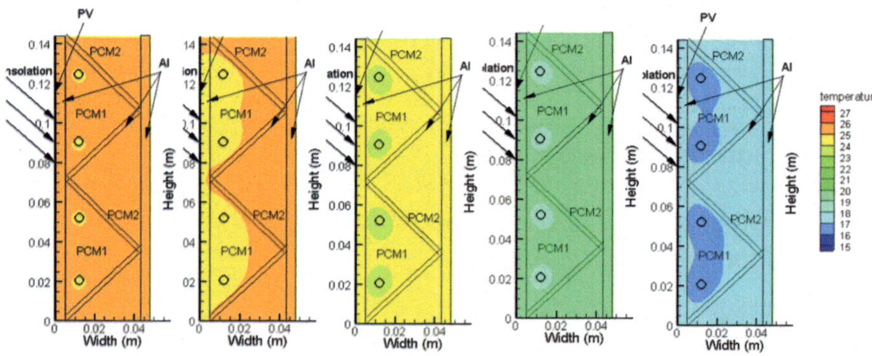

Fig. 67.4 BIPV-PCM-HP system temperature evolution with 3 h cold-water cycling under summer climate in Velari

67.3.2 *Effect of Heat Pump for Underfloor Heating with PCM Layers*

The thermal mass of floor layers plays an important role on the underfloor heating along with floor structure, heating system layout, costs and thermal response. The effect of heating by heat pump for the underfloor heating system is presented in Fig. 67.5. Figure 67.5 presents the predicted thermal isotherms for the cross section of underfloor concrete layer with PCM layer at different time during heating process. The underfloor heating system with PCM surrounding the heating pipe can store more thermal energy during the charging process. The heat loss through the below layer is limited.

Fig. 67.5 Predicted thermal isotherms for the cross section of underfloor concrete structure with PCM layer at different time during heating process

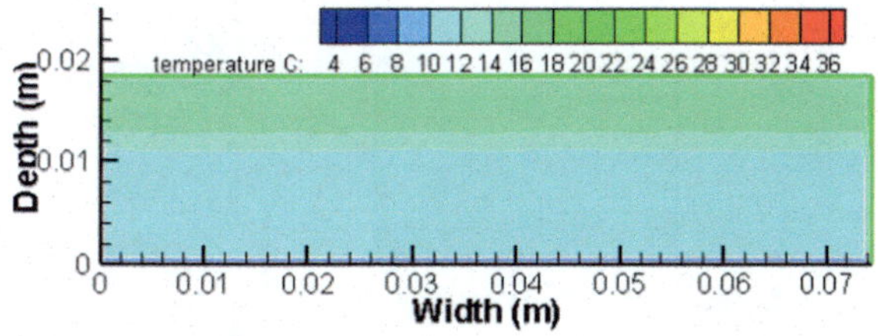

Temperature distribution on the underfloor system before heated by the heat pump

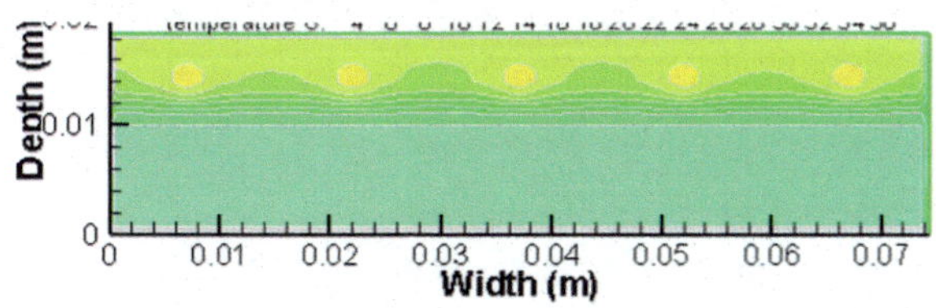

Temperature distribution on the underfloor system after heated by the heat pump for 45 minutes

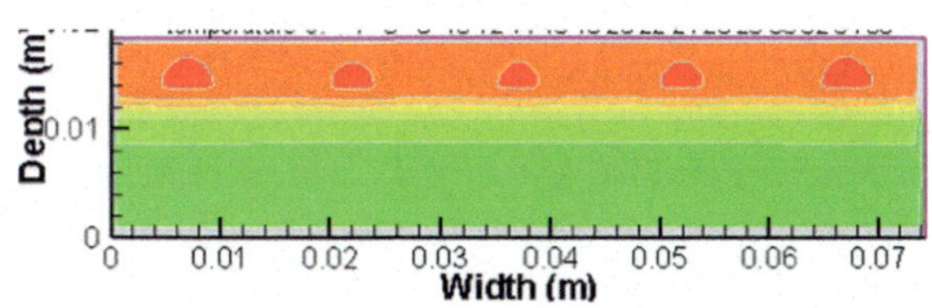

Temperature distribution on the underfloor system after heated by the heat pump for 150 minutes

The predicted temperature on the surface of the floor with hot water from heat pump under winter clime in Belfast, UK, in January is presented on the Fig. 67.6. The heat pump can effectively heat the room through the underfloor heat system. It can charge the floor mass efficiently and quickly rising the floor surface temperature to a steady state. After four hours heating the PCM layer can store the thermal energy and continue release the heat for later use for longer period.

67.4 Conclusions and Future Work

Elevated operating temperatures in the building integrated photovoltaic (PV) reduce the solar to electrical conversion efficiency of crystalline silicon PV cells. A PCM used to passively limit the PV temperature rise and combined with active cold water

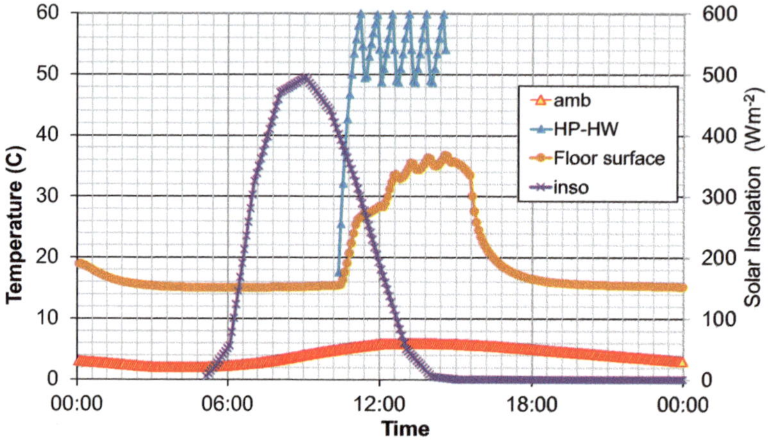

Fig. 67.6 Predicted temperature on the surface of the floor with hot water from heat pump using the weather data for the 21st Jan in Belfast, UK

from heat pump have been studied. The thermal performance of the BIPV-PCM-PV system has been addressed. Effect of using PCM for crystalline silicon PV temperature passive control and active control by heat pump is presented in this paper. It presents an evaluation of the effects of BIPV-PCM-HP system on heat transfer with different system operation conditions. Using RT27 PCM with cyclic cooling pipes, the temperature rise of the BIPV can be reduced when compared to using a single RT27 PCM only under Velari summer conditions. Different layout of the underfloor heating pipes for the concrete floor and PCMs augmented floor were analysed under realistic diurnal temperature boundary conditions and realistic heat operation conditions. The temperature distributions through the floor structure were predicted. The studying results provide an insight into the understanding of the structure and thermal performance of the PCM underfloor heating system with heat pump. Depending on ambient conditions, a PCM underfloor material with an ASHP system may enable the indoor environment to operate near thermal comfort range efficiently and it is believed that it will reduce the electricity cost for a residential home. The simulation results have also indicated that the thermal characteristics of the PCM and the configuration of the PCM floor can result in advantageous control of the heating operation, although more detailed study should carry on with optimised design and operations. In the future, the thermal performance of the PCM floor can be studied through combining with the ASHP different defrosts strategies and innovative PCM structures. The improved fundamental understanding of the processes within the BIPV-PCM-HP system provided by this work can be used to optimise the design of solar energy management systems for buildings to improve the total renewable energy efficiency in buildings.

Acknowledgments The authors would like to acknowledge the support funding from Research Councils UK (RCUK)/Engineering and Physical Sciences Research Council (EPSRC) through Low Temperature Heat Recovery and Distribution Network Technologies (LoT-NET) [EP/R045496/1]. Additionally, the authors thank for the financial contribution from the European Commission via H2020, CHESS-SETUP project [GA680556] and IDEAS project [GA815271].

References

1. Huang MJ, Eames PC, Norton B (2006) Experimental performance of phase change materials for limiting temperature rise building integrated photovoltaics. J Sol Energy 80:1121–1130
2. Huang MJ, Eames PC, Norton B (2004) Thermal regulation of building-integrated photovoltaics using phase change materials. Int J Heat Mass Transf 47:2715–2733
3. Huang MJ, Hewitt NJ (2015) The energy conservation potential of using phase change materials as thermal mass material for air source heat pump driven underfloor heating system in a building. Prog Clean Energy 2:209–227
4. Anon (2016) Rubitherm data sheet. Rubitherm GmbH, Hamburg

Chapter 68
Climate Change Adaptation: Prioritising Districts for Urban Green Coverage to Mitigate High Temperatures and UHIE in Developing Countries

Mohsen Aboulnaga and Mona Mostafa

68.1 Introduction

Heat-related problems have become a global concern due to severe events caused by climate change. Adaptation to climate change has proven significant mean to mitigate such risks, especially in urban areas. The presence of vegetation is significant in urban spaces as it plays a role in enhancing the environmental conditions by shading heat-absorbing surfaces, providing evaporative cooling, and altering wind patterns [1, 2]. Due to the fast urban development, natural surfaces and vegetation are replaced in some cities by materials with characteristics such as high thermal conductivity, high heat absorption, and heat storage capacity as well as low albedo values (reflectivity) that result in absorbing solar radiation during the day and release it at night as sensible heat (Fig. 68.1). Thus, urban areas in cities become warmer than the surrounding countryside: a phenomenon known as urban heat island effect (UHIE) [3].

The scientific term of "UHIE" became well known in 1820 by Luke Howard when compared to two rows of air temperature data measured inside and outside the city, and it was reported that in London in the early 1800s, night temperature is warmer in the city than in the country [4, 5]. Also, UHIE is manifested in almost all cities with few exceptions like tropical deserts such as Sonoran Desert, primarily located in Mexico and America, including Colorado Desert, Denver, and Lechuguilla Desert, Arizona [6].

M. Aboulnaga (✉)
Department of Architecture, Faculty of Engineering, Cairo University, Giza, Egypt

M. Mostafa
Faculty of Engineering, October University for Modern Sciences and Arts, Giza, Egypt

A. Sayigh (ed.), *Renewable Energy and Sustainable Buildings*, Innovative Renewable Energy, https://doi.org/10.1007/978-3-030-18488-9_68

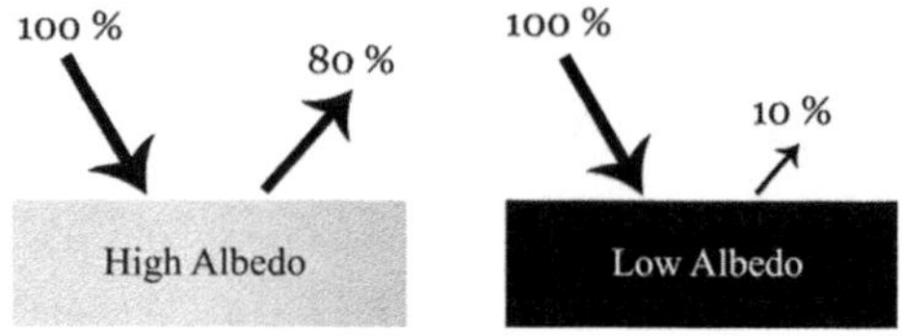

Fig. 68.1 Albedo in terms of reflectivity on black and bright colour horizontal surfaces

68.2 Objectives

The goal of this study is to prioritise districts in urban areas of Cairo Governorate, Egypt, for implementing urban green coverage (UGC) as a method for mitigating high temperatures' impact and contribute to reducing UHIE in these urban areas. The results will furnish landscape architects and urban designers with the study findings to strategically fit UGC that ameliorate the negative effects of UHI and radiant heat absorption in the context of global climate change and in vastly growing cities.

Accordingly, the main research question is as follows: What are the most areas that urgently require UGC implementation in Cairo? This study will furnish answers to such question and will be explored through answering the following sub-questions: (a) What is UHI and how UGC can contribute to mitigating this effect; (b) how to prioritise districts in Cairo Governorate to strategically implement UGC to mitigate high temperatures; and (c) what are the guidelines for UGC applications in developing countries such as Egypt?

68.3 Cities' Challenges and Urban Heat Island Problems

68.3.1 Urban Heat Island Effect and Urban Green Coverage in Egypt

Urban heat island effect (UHIE) refers to a climatic phenomenon expressed when urban areas have higher air temperature than their rural surroundings due to the anthropogenic modifications of land surfaces [6, 7]. However, due to the urban growth, no distinct borderlines exist between urban and rural areas; thus, UHIE can be described in terms of the difference in temperature between the central parts of the city and its surrounding areas [7].

In Cairo city, Egypt, the magnitude of UHIE ranged from 3.11 °C to 5.7 °C [8]. In summer 2012, land surface temperature (LST) varied from 27 °C to 61 °C [7]. The UHI intensity index in Cairo City has five dominant levels, with least intensity in vegetated areas/water bodies while highest in scattered bare land areas and industrial zones [9]. Thus, the above observations make it a necessity to develop means to determine the hot spots in Cairo Governorate, to prioritise and propose strategies for high temperatures mitigation. In addition, the study proposes urban green coverage (UGC) as a method to reduce UHIE [8]. In Egypt, urban areas have expanded on the account of green areas [10]. According to the international standards, in developed countries, areas of green spaces should be between 20 m^2 and 40 m^2 per person, while it should be 12–18 m^2 per person in developing countries [11, 12]. However, in Cairo the most optimistic studies state that the green space per person doesn't exceed 2 m^2. This figure is considered very low when compared to the international norms; nevertheless, more than half of Cairo's population have only 0.5 m^2 per person. Moreover, about a million person in Cairo have green per capita ratio less than 0.1 m^2 per person as green areas are not distributed in spatial justice in Cairo City [12].

68.3.2 Cooling Effect of Urban Green Coverage (UGC)

Urban green coverage (UGC) refers to the use of strategies such as green open spaces, trees, green roofs, and vertical walls. It is important to choose the UGC that fits the place it will be implemented in. The ultimate goals of this step are illustrated in Fig. 68.2.

In cities around the globe, UGC in urban areas can take many forms apart from conventional green urban spaces, such as vertical farming, green walls, and green roofs as illustrated in Fig. 68.3. It is crucial to understand how to

Fig. 68.2 Goals of good selection of urban green coverage. Source: (A. Norton et al., 2015)

Fig. 68.3 Various conventional and innovative set-up of urban green coverage. Photo credits: (Authors 2015, 2016)

Table 68.1 Strategic use of different urban green coverage (UGC)

UGC	Green open spaces	Trees	Green roofs	Vertical greening
Shades canyon surfaces?	Yes, if grass rather than concrete	Yes	Shades roof, not internal canyon surfaces	Yes
Shades people?	Yes, if treed	Yes	No, only very intensive green roofs	No
Increases solar reflectivity?	Yes, when grassed	Yes	Yes, if plants healthy	Yes
Evaporative cooling?	Yes, with water	Yes (unless severe drought)	Yes, with water when hot	Yes, with water when hot
	No, without water		No, without water	No, without water
Priority locations	• Wide streets with low buildings—both sides • Wide streets with tall buildings—sunny side	• Wide streets, low buildings—both sides • Wide streets, tall buildings—sunny side • In green open spaces	• Sun-exposed roofs • Poor insulated buildings. • Low, large buildings. • Dense areas with little available ground space.	• Canyon walls with direct sunlight • Narrow or wide canyons where trees are unviable

Source: (A. Norton et al., 2015)

locate different elements of the UGC in places that would maximise their surfaces' cooling benefits. Table 68.1 presents information on the strategic use of different UGC.

68.4 Methodology

To answer the research question, a study was conducted to investigate the prioritisation of districts on city-scale level, by adopting the Australian model developed by A. Norton et al., in 2015, which suggests that the impact of UHIE is based on a combination of many factors including heat exposure vulnerability to extreme heat and the behavioural exposure occurring (Fig. 68.4a). Thus, a high-priority area can be identified by the intersection of three factors, which are mentioned below.

1. High daytime/night-time surface temperatures (heat exposure)
2. Most vulnerable sections of the society to extreme heat (vulnerability)
3. Zones with many users active outdoor (behavioural exposure)

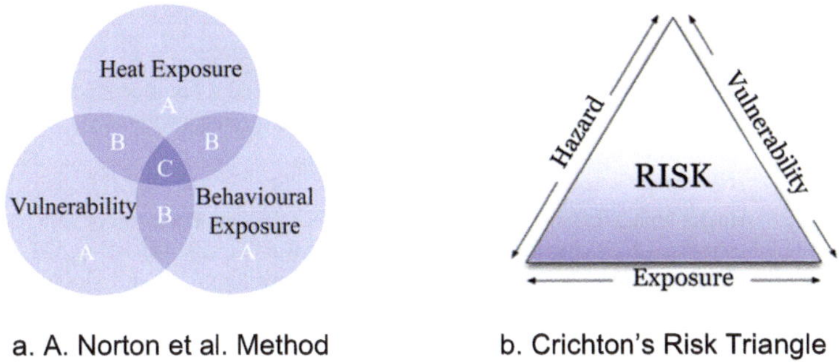

Fig. 68.4 Mapping the relationship between variables that affect the prioritisation of areas. (**a**) A. Norton et al. Method. (**b**) Crichton's Risk Triangle

It is vital to highlight areas, where heat exposure and vulnerability intersect (Area B) that are also considered a priority, as it is strenuously demanding to predict the amount of behavioural exposure in public open spaces [13]. However, in Cairo it was difficult to assess the behaviour of population in outdoor public spaces. Therefore, the study follows "Crichton's Risk Triangle", as conducted by Morabito et al., in 2015, to identify high-risk areas based on the intersection of three layers (Fig. 68.4b). The three components of this triangle are: (1) high daytime/night-time surface temperatures (hazard); (2) total exposed population in a city (exposure); and (3) subpopulations at risk of being harmed during extreme heat (vulnerability).

It is necessary to develop methods to assess the influencing parameters. Firstly, concerning the heat exposure, land surface temperature (LST) for an area provided from satellite or airborne remotely sensed thermal data can be used as an indication for air temperature [13]. Although there is a difference between LST and air temperatures, mitigating areas with high LST will reflect locations, which also have high air temperature [14]. Secondly, pertaining to the vulnerability, the demographic information of a place is needed, as neighbourhoods are categorised depending on social factors [13].

68.4.1 Scope: Study Area, Data, and Period

This study was carried out during summer season (June 2017) and focused on Cairo Governorate that consists of 46 districts. In the simulation, ArcMap version 10.2 was used to handle remote sensing visualisation, manipulation and analysis of satellite images, and geographic data records. Also, Microsoft Excel was used in data processing and analyses. Description of the data is shown in Table 68.2.

Table 68.2 Description of the data used in the study

Data used	Data description	Year	Data source
Cairo district boundaries	Cairo districts geographical boundaries layer in ARC map	2017	Digitised by author based on http://www.cairo.gov.eg/
Daytime land surface temperature (LST)	Raster image of daytime LST covering the study area	(18.06.2017 till 25.06.2017)	NASA satellite remote sensing MODIS (MOD11A2) consisting of LST at 8 days temporal and global 1.00 km spatial resolution Grid V006 both for daytime and night-time conditions during the period, accessed at https://search.earthdata.nasa.gov
Night-time LST	Raster image of night-time LST covering the study area		
Population distribution in Cairo districts	Excel file containing the population distribution in Cairo districts according to age (<5 and >65)	2017	(CAPMAS, 2017)
Deprivation index	Deprivation index table in Cairo districts	2006	(UNDP Egypt and MOPLD, 2006)

68.4.2 Prioritising Districts on City-Scale Level for UGC Implementation

Geospatial technologies, a new opportunity for an integrative system that can assist in hazards' mitigation, are used for mapping and prioritising areas that have severe heat-related impacts. It is important to note that the GIS system should have its inputs' data including urban characteristics, social vulnerabilities, and metrological information in order to derive trends and conclusions [15].

Recent research works attempt to develop heat vulnerability index [16]. In these researches, the methodology adopted for assessing the risk is "Crichton's Risk Triangle" which was applied in many previous studies [15, 17]. This risk definition is specifically adopted in this research work in order to simplify the process of building the GIS database, as it is mainly composed of layering system as the final risk layer is generated from the intersection of three determinants as shown in Fig. 68.5 [17].

The hazard can be accounted by the increase in LST, which its exposure on the population can be determined according to the targeted vulnerable groups. A study conducted by Morabito et al., 2015, took only the elderly as its vulnerability constrain, whereas that conducted by A. Norton et al., 2015, took into account several vulnerability factors such as the elderly and very young population and other affecting socioeconomic factors. Since it is possible to map different vulnerable items using Crichton's Risk Triangle [17], thus this research will take into consideration

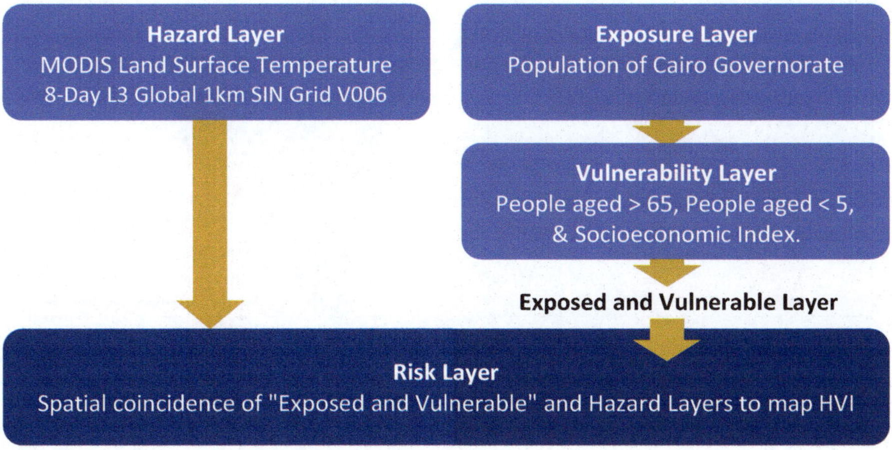

Fig. 68.5 Simplified workflow of the developed geospatial database to map HVI. Source: adopted and developed from: (Tomlinson et al., 2011)

all the vulnerable factors as conducted by A. Norton et al., 2015. Thus, the aim of this section is to develop heat-related vulnerability index (HVI) maps to determine the areas that need UGC intervention. The workflow of the developed geospatial database is shown in Fig. 68.5.

68.4.2.1 Hazard Layer: Daytime and Night-time Summer LST Layers

The daytime and night-time LST layers used in the study are based on NASA satellite imagery, and to cover the study area, two raster images for LST were downloaded. These images were processed using GIS Software "ArcMap 10.2". Finally, the LST values were extracted and scaled using Eq. 68.1, so that LST be displayed in Celsius for easier interpretation.

$$\mathrm{MODIS\_LST2017\_Daytime\_Celsius} = [\mathrm{Value}] * 0.02 - 273.15 \quad (68.1)$$

Then both daytime and night-time LST were normalised using Eq. 68.2, as normalised values will have a standardised scale, which ranges between zero (low) and one (high) as shown in Fig. 68.6.

$$\beta = \frac{x - x_{\mathrm{min}}}{x_{\mathrm{max}} - x_{\mathrm{min}}} \quad (68.2)$$

where β corresponds to the normalised score, x is the original value, while x_{max} and x_{min} are the maximum and minimum values across the complete data set.

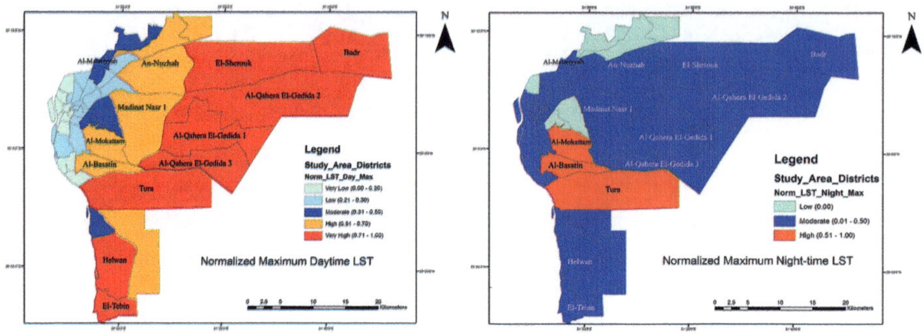

Fig. 68.6 Normalised maximum daytime LST (right) and night-time LST (left) based on MODIS data

68.4.2.2 Exposure Layer

It consists of the population distributed in Cairo Governorate districts by age, obtained from Central Agency for Public Mobilization and Statistics (CAPMAS) according to 2017 census. The data is entered into ArcMap 10.2, and a table is generated, which is merged to the geographic features layer. This table contains deprivation index of districts based on a report by UNDP and MOPLD, 2006 [18].

68.4.2.3 Vulnerability Layers

The vulnerability layer is made up of vulnerable types extracted from the exposure layer which includes number of people aged >65, number of people aged <5, and deprivation index. All layers were normalised using Eq. 68.2 and graphically displayed as illustrated in Fig. 68.7.

68.4.2.4 Final Risk Layer

To obtain the final risk layer (Fig. 68.9), the process followed by Tomlinson et al. (2011) is implemented (Fig. 68.8), which involves the creation of an intermediate layer termed "combined exposed and vulnerable"; thus, the three vulnerability layers were combined into a single layer (each weight of 33.3%) then spatially combined with the hazard layer (each weight of 50%) using Eq. 68.3.

$$\begin{aligned}&[\text{Exposure\_and\_vulnerability\_Norm}]*0.5+[\text{Norm\_Mean\_dayLST}]*0.25\\&+[\text{Norm\_Mean\_nightLST}]*0.25\end{aligned} \tag{68.3}$$

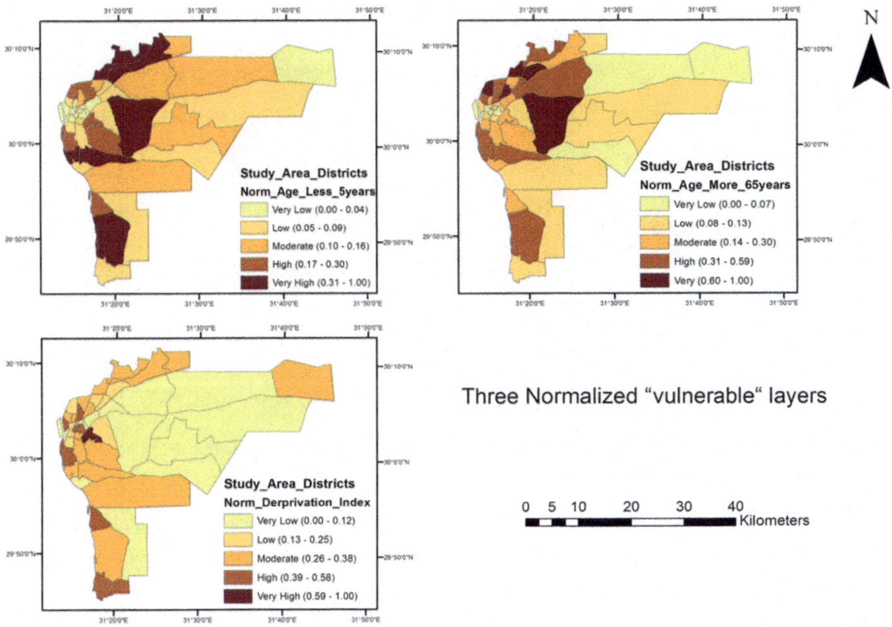

Fig. 68.7 Normalised vulnerability layers of Cairo Governorate districts

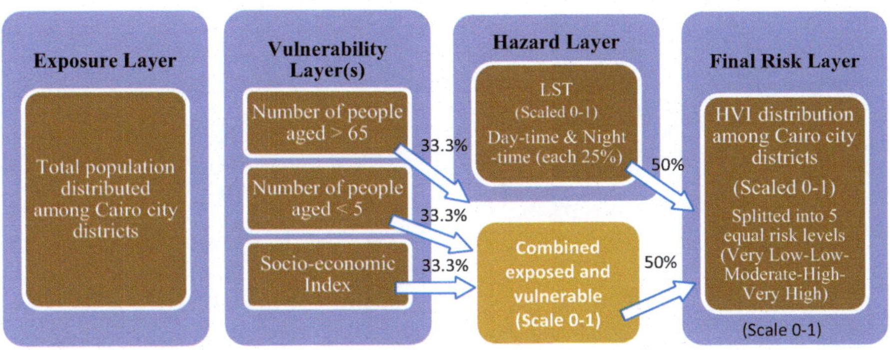

Fig. 68.8 Framework to develop the final mapping of the heat-related vulnerability index (HVI). Source: adopted and developed from: (Tomlinson et al., 2011)

68.5 Results and Discussions

The methodology adopted in this study yields a map depicting the heat-related vulnerability index (HVI) for the 46 districts in Cairo Governorate, Egypt. This HVI is divided into five categories: very high, high, moderate, low, and very low (Fig. 68.9). This enables specialists to determine areas that crucially require UGC intervention.

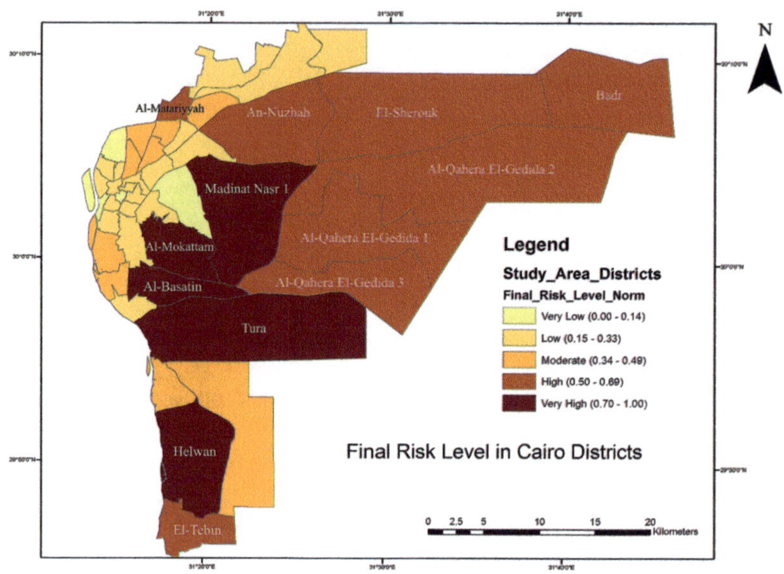

Fig. 68.9 Final risk layer of the HVI's five categories distribution in Cairo Governorate districts

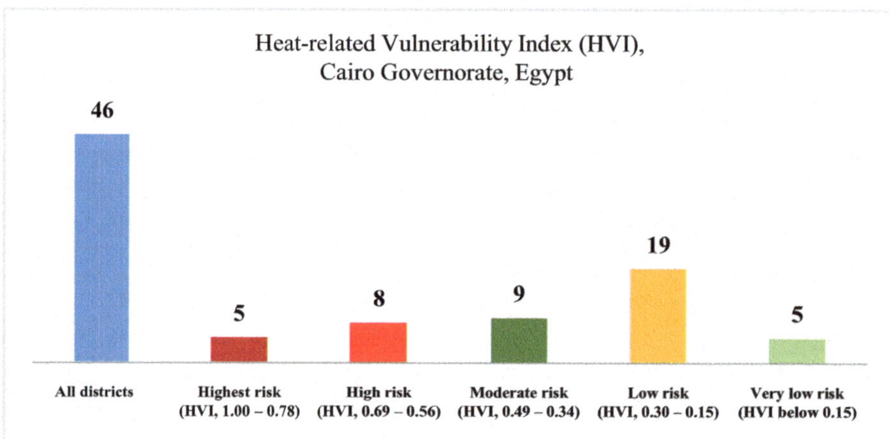

Fig. 68.10 Map of HVI distribution of the 46 districts in Cairo Governorate, Egypt

The highest risk is observed in five districts, namely, Al-Basatin (HDVI = 1), Tura (0.97), Al-Mokattam (0.83), Helwan (0.8), and Madinat Naṣr-1 (0.78) as depicted in Fig. 68.10. Results show that the second category, high risk, is present in eight districts: Badr (0.69), El-Tebin (0.67), Al-Qahera El-Gedida 2 (0.66), Al-Qahera El-Gedida 1 (0.61), Al-Qahera El-Gedida 3 (0.58), An-Nuzhah (0.58), El-Shorouk (0.57), and Al-Matariyyah (0.56) as shown in Fig. 68.10.

The simulation results indicate that nine districts are at moderate risk and have HVI ranging from 0.34 to 0.49. These districts are Ain Shams, 15th of May, Al-Masara, Al-Zawiyah, Al-Ḥamra, Hadaiq Al-Qubbah, Dar Al-Salaam, Miṣr Al-Qadimah, Al-Sharabiyah, and Al-Amireya. It is clear from the HVI map that low-risk districts have the HVI ranging from 0.15 to 0.30. These 19 districts are Al-Zeitoun (0.33), El Marg, Al-Khalifa, Al-Salam-1, Miṣr Al-Gedida, Shubra, Al-Jamaliyah, Bab Al-Shariyah, El-Darb Al-Ahmar, Rud Al-Faraj, Al-Waili, Al-Sayyidah Zaynab, Manshiyat Naser, Al-Zahir, Al-Salam-2, Bulaq, Al-Maadi, Abdin, and Al-Azbakeya (0.17). In addition, only five districts have very low risk. These are Al-Muski, Al-Zamalek, Madinat Naṣr-2, Al-Sahel, and Qaṣr Al-Nil.

Thus, as a first and urgent step, it is important to target areas at the 13 districts with very high risk and high risk, by proposing strategic UGC plans to assist in mitigating high temperatures, UHIE, and play a major role in enhancing the microclimate as well as adaptation to climate change risks. It is also vital to adopt and utilise the following guidelines when planning and implementing UGC in Cairo Governorate and developing countries with similar conditions.

68.5.1 Guidelines for UGC Implementation in Developing Countries: Egypt

This strategy is in line with the 2050 vision, which proposes an increase of the green area per capita in Greater Cairo Region (GCR) to reach 15.00 m^2 [19]. It also attaining to Egypt's Vision 2030 and Sustainable Development Strategy, SDS 2030, mainly the energy and health environment sectors.

The following are opportunities for incorporating new greenery:

(a) Implementing planned green areas that are not planted such as areas localised for green belts, utilising riverbanks as urban green corridor, and hillsides and plateaus greening
(b) Utilising areas that their land use has changed (such as premises and sites of no longer use) or areas that its land use can change (such as the gradual replacement of industrial areas, scattered areas with greenery, informal settlements and deteriorated residential areas)
(c) Doubling the use of spaces such as planting trees in car parking areas, cemeteries' areas, railway roads, streets, airports landscape, etc. or vertical greening [20]

In addition, guidelines regarding sustainable vegetation planting include planting regional native plants, adopting methods to prevent pollution such as (1) decreasing the amount of fertilisers and pesticides used, (2) using integrated pest management techniques, (3) recycling green waste, and (4) reducing runoff and efficient irrigation systems [20].

68.6 Conclusions

Urban heat island effect (UHIE) and urban green coverage (UGC) have been investigated using simulation programmes. Mapping the relationship between land surface temperature (LST) and the peoples' vulnerability is a method that can assist local governments to locate the hot spots that urgently require quick heat prevention and intervention. This study also achieved its goal by reviewing relevant literature on the UHIE and the cooling potential that can be provided by UGC in cities that are characterised with hot and dry summers. In addition, it proposed a framework for prioritising urban areas for UGC implementation. A tool to map the population vulnerability to extreme heat events in Cairo Governorate resulting from UHIE has been developed; it identifies high-priority risk areas that should essentially receive intervention by applying more UGC as a significant action to mitigate UHIE in cities, reduce GHG emission, and assist in adapting to climate change risks. Moreover, guidelines for UGC application to mitigate UHIE in Cairo Governorate have been derived and established. It is important to note that implementing UGC will have a significant environmental impact, knowing that it does not only mitigate high temperatures, but it also reduces air pollution.

As indicated in the latest World Health Organisation (WHO) report [21], air pollution is a major global concern, as it impacts human health severely and leads to about seven million deaths annually worldwide, of which about 500,000 in the MENA region, due to exposure from both outdoor and indoor pollutions. Therefore, implementing UGC extensively in cities, particularly in Cairo Governorate, a mega city, could lessen air pollution and other impacts and assist in reducing mortality rates from air pollution exposure. It is recommended that future research work on the development of UGC in developing countries such as Egypt should be on the street level.

References

1. Oke TR (1989) The micrometeorology of the urban forest. Royal Soc 324(1223):335–349. https://doi.org/10.1098/rstb.1989.0051. Accessed 27 Jun 2018
2. Taha H (1997) Urban climates and heat islands: albedo, evapotranspiration, and anthropogenic heat. Energ Buildings 25(2):99–103
3. Oke TR (1982) The energetic basis of the urban heat island. Q J R Meteorol Soc 108(455):1–24
4. Oke TR (1987) Boundary layer climates, 2nd edn. Methuen Co, London, New York, p 435. http://www.scirp.org/(S(351jmbntvnsjt1aadkposzje))/reference/ReferencesPapers.aspx?ReferenceID=1113473
5. Kolokotroni M, Giannitsaris I, Watkins R (2006) The effect of the London urban heat island on building summer cooling demand and night ventilation strategies. Sol Energy 80(4):383–392
6. Lokoshchenko MA (2014) Urban climate urban "heat island" in Moscow. Urban Climate 10:550–562. https://doi.org/10.1016/j.uclim.2014.01.008. Accessed 30 Jun 2018
7. Abutaleb K, Ngie A, Darwish A, Ahmed M, Arafat S, Ahmed F (2015) Assessment of urban heat island using remotely sensed imagery over greater Cairo, Egypt. Adv Remote Sensing 4(March):35–47

8. Effat HA, Hassan OAK (2014) Change detection of urban heat islands and some related parameters using multi-temporal Landsat images; a case study for Cairo city, Egypt. Urban Climate 10(P1):171–188. https://doi.org/10.1016/j.uclim.2014.10.011
9. El-magd IA, Ismail A, Zanaty N (2016) Spatial variability of urban heat islands in Cairo City, Egypt using time series of landsat satellite images. Int J Adv Remote Sensing GIS 5(3):1618–1638
10. Raymond A (2000) Cairo. Harvard University Press, Cambridge [MA]
11. El-Zafarany AM (2004) Existing green areas in Cairo: comparison with planning criteria and international norms, Cairo, Egypt. http://www.egyptarch.net/research/cairourbandesert.pdf. Accessed 27 Jun 2018
12. Kafafy N, Al-betawi Y (n. d.) Urban green space benefits and the pivotal role of conservation, Cairo's case—Egypt, 1–13. https://core.ac.uk/download/pdf/19396743.pdf. Accessed 28 Jun 2018
13. Norton BA, Coutts AM, Livesley SJ, Harris RJ, Hunter AM, Williams NSG (2015) Planning for cooler cities: a framework to prioritise green infrastructure to mitigate high temperatures in urban landscapes. Landsc Urban Plan 134:127–138. https://doi.org/10.1016/j.landurbplan.2014.10.018. Accessed 30 Jun 2018
14. Matzarakis A, Rutz F, Mayer H (2007) Modelling radiation fluxes in simple and complex environments - application of the RayMan model. Int J Biometeorol 51(4):323
15. Morabito M, Crisci A, Gioli B, Gualtieri G, Toscano P, Di Stefano V et al. (2015) Urban-Hazard risk analysis: mapping of heat-related risks in the elderly in major Italian cities. PLoS ONE 10(5):e0127277. https://doi.org/10.1371/journal.pone.0127277. Accessed 30 Jun 2018
16. Inostroza L, Palme M, De La Barrera F (2016) A heat vulnerability index: spatial patterns of exposure, sensitivity and adaptive capacity for Santiago de Chile. PLoS One 11(9). https://doi.org/10.1371/journal.pone.0162464. Accessed 30 Jun 2018
17. Tomlinson CJ, Chapman L, Thornes JE, Baker CJ (2011) Including the urban heat island in spatial heat health risk assessment strategies: a case study for Birmingham, UK. Int J Health Geogr 10(1):42. https://doi.org/10.1186/1476-072X-10-42
18. UNDP Egypt and MOPLD (2006) Taqareer al-Tanmia Al-basharia li-l-muhafazat Al-Masria: Taqreer Al-Qahira, UNDP, Cairo, [Human Development Reports for Egyptian governorates: Cairo's report]
19. General Office of Physical Planning—GOPP (2009) Cairo future vision 2050: within a national vision of Egypt [Powerpoint presentation], Cairo, Egypt. https://cairofrombelow.files.wordpress.com/2011/08/cairo-2050-vision-v-2009-gopp-12-mb.pdf
20. El-Zafarany AM (2004) Green areas in greater Cairo: the problem and solution opportunities. Cairo, Egypt. http://www.egyptarch.net/research/cairogreanstrategies.pdf
21. World Health Organisation (WHO) 2017. http://www.who.int/airpollution/en/. Accessed 02 Jul 2018

Chapter 69
Operational Strategies for a Large-Scale Horizontal-Axis Wind Turbine During Icing Conditions

D. B. Stoyanov, H. Sarlak, and J. D. Nixon

69.1 Introduction

In 2017, EU invested US$ 29 billion in the wind sector for electrical energy generation [1]. Annual onshore installations marked an increase of 14%, thus adding 12.5 GW to the EU wind power capacity, while offshore annual installation doubled but only added 3.2 GW [2]. As large investments have been made for the past decade, it is of a great importance to ensure optimal operation of wind turbines and minimise losses.

Wind turbine exploitable sites vary in geographical location and weather conditions, which requires different approaches for operation. Special care should be taken for unconventional wind turbine sites such as mountainous and subarctic regions (Cold Climates) [3]. Cold Climate (CC) regions are preferred because of the higher amount of available wind kinetic energy for harvesting [4] and their remoteness from populated areas [3]. However, because of low ambient temperatures and ice buildup, wind turbines operate outside their designed framework. Although icing causes several issues, ice-induced power loss is the one which directly affects the projects' financial viability and investment security. Depending on the geographical location and icing severity, annual power production losses can reach up to 50% [5].

Ice accretion on wind turbine blades and power production in icing conditions have been studied for the past two and a half decades. Both numerically and experimentally, power losses have been investigated for different conditions and wind

D. B. Stoyanov (✉) · J. D. Nixon
Faculty of Engineering, Environment and Computer Science, Coventry University, Coventry, UK
e-mail: stoyanod@uni.coventry.ac.uk; jonathan.nixon@coventry.ac.uk

H. Sarlak
Technical University of Denmark, Lyngby, Denmark
e-mail: hsar@dtu.dk

A. Sayigh (ed.), *Renewable Energy and Sustainable Buildings*, Innovative Renewable Energy, https://doi.org/10.1007/978-3-030-18488-9_69

turbine types [6]. Models have been developed and tested for simulating accretion on blades [7] and mitigating ice buildup [8]. However, only few studies look at operational strategies that can be implemented to ensure continuous and optimal work of wind turbines in icing conditions. Homola et al. [9] conducted a performance losses analysis on a NREL 5-MW reference wind turbine and found that for an ambient temperature of −10 °C and wind speed of 10 ms^{-1}, for one hour of operation, 28% reduction in the turbine power curve can be expected. The authors identified that because of icing the operational point changed and the simulated wind turbine did not operate in its optimal regime, which was prescribed by the wind turbine controller. It was suggested that redesigning the controller can improve the power production for such an icing event. Zanon et al. [10] further investigated the changes in performance due to the implementation of different operational strategies. The realisation of the strategies was achieved by prescribing different rotational speeds (RPM) to the rotor and shifting the operational point of the turbine. The chosen icing conditions included simulations for temperatures of −2 °C and −10 °C, wind speed of 10 ms^{-1} and duration of 8 h. It was reported that by decreasing the rotational speed of the rotor, 6% improvement in performance was met after the events, thus suggesting that changing the rotational speed could be a viable operational strategy. While both studies have reported that the rotational speed modifications can improve on the wind turbine performance in icing conditions, to the authors' knowledge, no study has investigated the uptake and the comparison of different operational strategies during an icing event.

The aim of this paper is to compare three different operational strategies for a large-scale horizontal-axis wind turbine operating in icing conditions. Turbine shutdown, running an electrothermal anti-icing technology and rotational speed modifications have been chosen as operational strategies to mitigate icing losses. By comparing the net power output when each is employed, it is determined which strategy is best suited for the chosen conditions. Ice-induced losses and power output have been simulated numerically. The ice events have been designed considering available icing data for Canada that has been published in the literature.

69.2 Methodology

To examine alternative wind turbine operating strategies during icing events, three modelling approaches have been utilised and combined in this study – scripted blade element momentum theory (BEMT) for calculating power performance, lewINT for simulating ice accretion and JavaFoil for aerodynamic analysis of the iced geometry. Three icing events have been designed by using a selected case study location as a baseline, for which meteorological data is available [11, 12]. Therefore, by comparing the numerically obtained net power output of the wind turbine during icing with and without mitigating strategy, it is determined which strategy is the most suitable choice for operation.

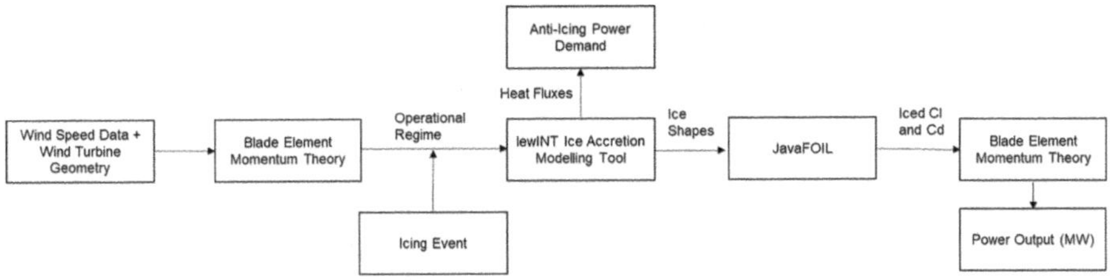

Fig. 69.1 Methodology to obtain the power performance of NREL 5-MW wind turbine in icing conditions

Figure 69.1 provides an overview of the established methodology. Initially, BEMT provides relative velocities and local angles of attack to the blade, so that the icing calculation can be completed with lewINT. The icing parameters (LWC, MVD, ambient temperature and icing event duration (see 3.1)) are next set in lewINT to calculate the 2D iced geometries and the anti-icing heat fluxes according to the prescribed icing events. Then, an aerodynamic analysis is conveyed using JavaFoil, which outputs the iced 2D Cl and Cd tabulated data, needed for the power output estimation. This methodology allows for the calculation of ice-induced power losses and anti-icing power demands for a chosen icing event. Moreover, by changing operational parameters in the BEMT code such as the RPM (revolutions per minute), the rotational modification strategy can be realised.

69.3 Simulation Setup

The simulation setup includes the design of icing events and the setup of simulation tools.

69.3.1 Icing Events

The icing events are defined by their duration and the icing weather parameters—liquid water content in the air (LWC), median volume diameter of the droplets (MVD) and ambient temperature. In the current study they are based on punished icing data, which is available for areas in Canada [11, 13]. Table 69.1 summarises the designed icing events, which are denoted as A, B and C. Lamraoui et al. [13] reported average duration of icing event for LWC of $20 \times 10^{-3}\ kgm^{-3}$ to be 6 h, while for LWC of $36 \times 10^{-3}\ kgm^{-3}$ to be 4 hours. In the current study all events are considered to be 4 hours, thus reducing the number of simulations that is required to compare the ice-induced power losses. The temperature region is assumed between −5 and −20 °C. Those temperatures are suitable for estimating the anti-icing power that is required for protection as it is influenced by the ambient temperature. Simulating in the chosen region will help to set boundary conditions for anti-icing technologies that can be used for the design of anti-icing systems.

Table 69.1 Definition of icing events A, B and C

Icing event	A	B	C
Ambient temperature [°C]	−5	−20	−20
Liquid water content [1×10^{-3} kgm^{-3}]	0.20	0.20	0.36
Median volume diameter [1×10^{-6} m]	36	36	36
Duration [h]	4	4	4

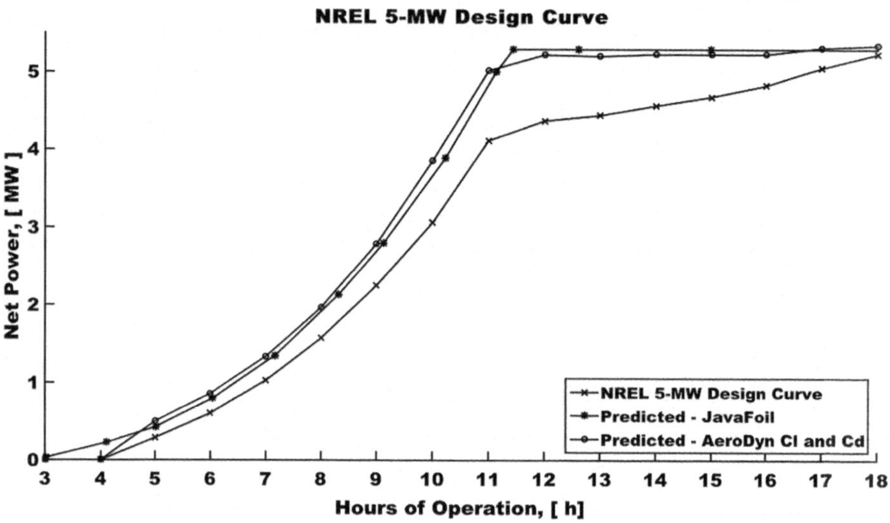

Fig. 69.2 Predictions of NREL 5MW power curve with JavaFoil (graph with cross markers) and AeroDyn against the design curve of the wind turbine

69.3.2 Simulation Tools Setup

The classic BEMT definition of Glauert has been scripted in MATLAB for the NREL 5-MW baseline wind turbine. The blade geometry is discretised as in [14], while the used Cl and Cd data is the one incorporated in the AeroDyn software [15]. The performance of the scripted BEMT can be seen on Figure 69.2. A good agreement is achieved between the scripted BEMT and the design curve of the turbine using the Cl and Cd data from AeroDyn. However, as the aerodynamic analysis in the current study is conducted with JavaFoil, the clean geometry of the wind turbine has been analysed using it and the results are shown on Figure 69.2. As it could be seen, the correlation between the design curve and the predicted one with JavaFoil is within the 18% error margin for velocities between 3 and 18 ms^{-1}. Despite the discrepancies, the tool is still used in the current study because of its robustness and fast computational time. Moreover, the results will be used to determine its suitability for studies on wind turbine performance in icing conditions.

lewINT was used for both the ice accretion modelling and the anti-icing heat flux that is required to keep a prescribed surface temperature. For this study the heat fluxes are calculated for electrothermal technology operating in running-wet regime, thus allowing a film of water to be present on the airfoil surface. Three different surface temperatures are used so that the variation of the required power can be obtained – 1 °C, 3 °C and 5 °C. In the current analysis, the fluxes are calculated assuming that the whole surface of the airfoil is heated, which yields higher than normal energy demand. However, such approach should facilitate the setting of boundary conditions for electrothermal technologies. It is of practical interest to determine under what conditions the electrothermal technologies can be used when considering CCs.

69.4 Results and Discussion

The performance of NREL 5-MW wind turbine for the three icing conditions can be observed on Figure 69.3. There is power drop due to the accumulated ice, immediately after the first hour of the events. The most severe cases of icing are A and C. It is expected as for the former, the produced ice shapes even though being for conditions predominately suitable for rime icing, mixed icing could form (including water layer on the ice surface), which produces more irregular shapes, causing higher aerodynamic penalties. For the latter the LWC is higher, which suggests higher accretion rate on the surface, thus resulting in increased thickness of the ice layer and greater ice deposition. The ice-induced energy losses are summarised in Table 69.2. It can be concluded that for the chosen icing events, the highest losses in terms of energy yield are for icing event C. However, for the operational analysis, icing event A has been chosen as it induces similar

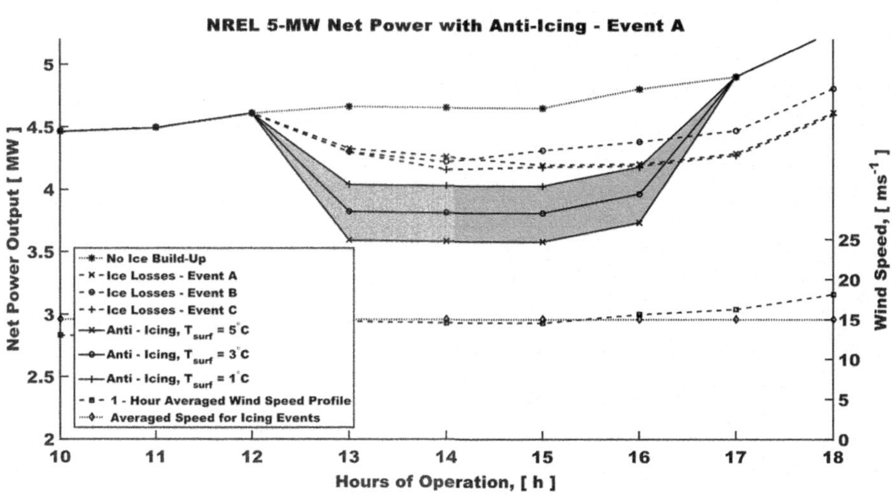

Fig. 69.3 Net Power output for the first operational strategy. Ice losses and wind profile added for clarity

Table 69.2 Ice-induced power losses

Icing event	Net power output first hour, %/MW	Net power output second hour, %/MW	Net power output third hour, %/MW	Net power output fourth hour, %/MW
A	7/4.375	8/4.328	10/4.224	12.5/4.116
B	8/4.328	9/4.281	9/4.281	9/4.281
C	8/4.328	10/4.224	11/4.187	12.8/4.102

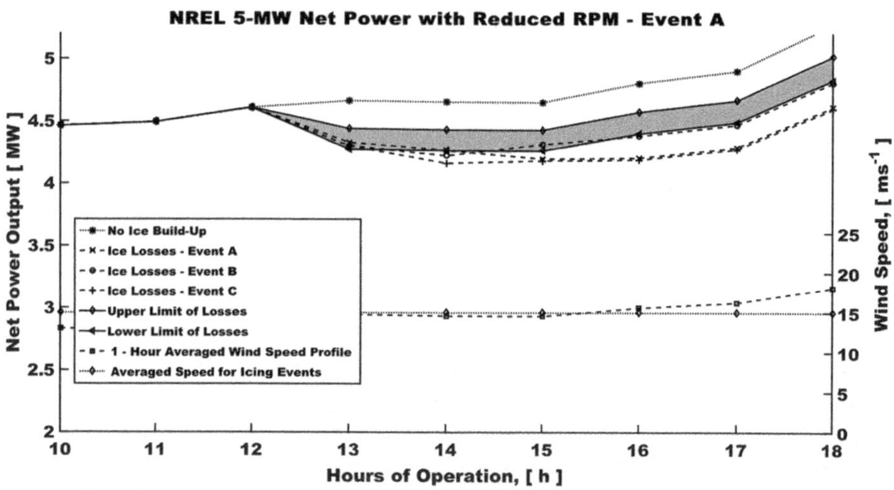

Fig. 69.4 Net Power output for the second operational strategy. Ice losses and wind profile added for clarity

amount of losses but provides more favourable conditions for anti-icing solutions in terms of ambient temperature.

Figure 69.3 depicts the net power output during icing event A, when anti-icing protection is employed for the analysed part of the rotor blade. As it can be seen, the power that is required to keep the surface temperature of the blade, T_{surf}, increases when higher temperatures are assigned. It is the case as an electrothermal technology will demand more power to equilibrate the cooling fluxes from the flow and reach the prescribed T_{surf}. From the results it is apparent that a full geometry protection of the blade is not a viable solution as the net power output is lower than the one that is reduced because of ice buildup. It should be noted that, if an electrothermal design is considered, the required power for anti-icing will drop, which implies that for an optimal T_{surf} and optimal design, an anti-icing can be considered.

The second operational strategy incorporates the reduction of the rotational speed during the icing event, which was proposed in [10] and the results are shown on Figure 69.4. The RPM was reduced to 7 from 12.1, which is the prescribed one from the design curve of the wind turbine. The lower RPM results in lower relative speeds for the blade's leading edge; thus, less ice is accreted because of the reduced ice rate accumulation. The resulting losses are visualised by the shaded region on

Table 69.3 Comparison of the operational strategies to the reference icing losses value for event A

Operational strategy	Net power output during the event, MW	Gained energy during the event, MWh
Anti-icing for $T_{surf} = 1$ °C	4.08	−0.33
RPM reduction	4.31–4.48	0.59
Shutdown	0	0

Figure 69.4, which depicts the net power for this operational strategy. Due to the low fidelity of JavaFoil, only a region can be shown rather than exact graphs. As it could be seen, the net power is higher than the initial icing event (A), suggesting that the operational strategy is a viable solution.

The net power output gained for all operational strategies is summarised in Table 69.3 and compared to the net power output during icing event A. As it could be seen, the optimal strategy for reducing the ice-induced power losses is modifying the rotational speed of the wind turbine during the event. Thus, the ice losses are reduced and overall gain of 3.54% of energy is obtained.

Given the results from the two operational strategies, it can be said that the operational strategies depend on the ice-induced losses. Were the ice losses for event A higher than 25%, the anti-icing technology would be a viable operational strategy. However, if the ambient temperature is lower than −5 °C, then the heating demand would be so high that the only viable operational strategy will be the shutdown of the turbine.

69.5 Conclusion

NREL 5-MW wind turbine operation has been simulated for three icing conditions (A, B and C), each with duration of 4 h. Ice-induced power losses have been estimated and the lowest net power output was found to vary between 4.102 and 4.116 depending on the droplet diameter of the water that is contained in the air. To mitigate the icing losses for conditions at ambient temperature of −5 °C, three operational strategies were considered – anti-icing of the blade surface, reduction in the rotational speed of the rotor and complete shutdown. It has been found that the reduction of the rotational speed is a viable solution for the given conditions and assumptions. Total gain of 3.54% in the net power has been calculated, which accounts for 0.59 MWh of more energy during the event. However, it has been shown that the anti-icing solution for the current assumptions does not improve on the net power output and it depends highly on the prescribed surface temperature of the blade. This can be further addressed if an actual electrothermal anti-icing system design layout is considered. Thus, less surface will be required for heating and reduction in the demand will be seen. Further optimisation will be the finding of optimal surface temperature for the anti-icing technology.

References

1. Louw A (2018) Clean energy investment trends. In: 2017
2. Anon, Global Wind Report, Bloomberg New Energy Finance. (2018) 1–60
3. Battisti L (2015) Chapter 4 icing process, wind turbines in cold climates. Springer, Switzerland, pp 177–248
4. Fortin G, Perron J, Ilinca A (2005) Behavior and modeling of cup anemometers under icing conditions
5. Tammelin B, Cavaliere M, Holttinen H, Morgan C, Seifret H, Santti K (1996-1998) Wind energy production in cold climates (WECO)
6. Makkonen L, Autti M (1991) The effects of icing on wind turbines. Am Soc Mech Eng 575:575–580
7. Myers T (2001) Extension to the messinger model for aircraft icing. AIAA J 36:211–218
8. Dimitrova M, Ramdenee D, Ilinca A (2011) Evaluation and mitigation of ice accretion effects on wind turbine blades, bioenergy technology (BE) World Rene
9. Homola M, Virk M, M Nicklasson PJ (2012) Performance losses due to ice accretion for a 5 MW wind turbine. Wind Energy 15:379–389
10. Zanon A, Gennaro De M, Kuhnelt H (2018) Wind energy harnessing of the NREL 5 MW reference wind turbine in icing conditions under different operational strategies. Renew Energy 115:760–772
11. Pederson MC, Sorenson H, Swytink-Binnema N, Martinez B, Condra T (2018) Measurements from a cold climate site in Canada: boundary conditions and verification methods for CFD icing models for wind turbines. Cold Regions Sci Technol 147:11–21
12. Lamraoui F, Fortin G, Benoit R, Perron J, Masson C (2013) Atmospheric icing severity: quantification and mapping. Atmos Res 128:57–75
13. Lamraoui F, Fortin G, Benoit R, Perron J, Masson C (2014) Atmospheric icing impact on wind turbine production. Cold Regions Sci Technol 100:36–49
14. Jonkman J, Butterfield S, Musial W, Scott G (2009) Definition of a 5-MW reference wind turbine of offshore system development. NREL, Golden, CO
15. Jonkman JM, Hayman GJ, Jonkman BJ, Damiani RR (2009) AeroDyn v15 user's guide and theory manual. NREL, Golden, CO

Chapter 70
Use of Phase Change Materials for Solar Systems Applications

Edward J. Guionneau, Hoda Akbari, and Sarah J. McCormack

70.1 Introduction

The increasing levels of GHG emissions and the energy dependency on other countries have brought governments to develop and improve various ways of producing energy. The sun generates 3.85×10^{24} J of energy per year, meaning that only 1 h of sun exposure is more energy than what is consumed during a year. Of all methods of renewable production of energy, solar photovoltaic has been the most rapidly increasing with a worldwide growth of 60% over 5 years from 2007 to 2012 [1]. This is due to the fact that solar panels are simply designed products that require low maintenance, they are easy to install and are very modular. Indeed, they can be installed anywhere permitting to easy distribution of the energy production among a community or country, thus resolving a lot of grid issues. Furthermore, they also permit to be energy independent on a local or even national scale which have recently been a major geopolitical issue. The major disadvantage of PV electricity production is that it is extremely expensive, with the electricity produced being approximately three times more expensive than that produced by wind and geothermal power, and six times more expensive than electricity produced by coal and gas [2]. This is due to the high cost of PV panels and their low level of efficiency—they only convert 15–20% of solar radiation into energy. Furthermore, the efficiency of solar cells decreases by 0.5% for every degree increase above the solar cells standard operating temperature [3]. Clearly if the trends of increasing PV electricity production are going to continue, the efficiency of PV cells must be increased.

Phase change materials can absorb or release heat while they are changing phase, from solid to liquid for instance. Phase change materials have melting ranges at

E. J. Guionneau · H. Akbari (✉) · S. J. McCormack
Department of Civil, Structural and Environmental Engineering, Trinity College Dublin, University of Dublin, Dublin 2, Ireland
e-mail: guionnee@tcd.ie; hakbari@tcd.ie

A. Sayigh (ed.), *Renewable Energy and Sustainable Buildings*, Innovative Renewable Energy, https://doi.org/10.1007/978-3-030-18488-9_70

which the temperature stays constant until the material is fully liquefied or solidified, this stage is called the latent heat stage. During that stage, PCMs can absorb or release large amount of heat energy. This phase change process results in cooling down system and has proven to be very effective in many applications in the past. Many experiments combining phase change materials and solar panels have been conducted and have shown very encouraging results. However, one of the main issues that has commonly risen, was that the material would not always fully re-solidify overnight, meaning that the PCM would not have its full use the next day.

The main challenge of this project is to use multiple phase change materials to improve the efficiency of PV panels by cooling them and accelerating the re-solidification process of PCMs. Different design concepts have been tested and evaluated to ideally come up with an optimized solution. The objectives of this research are to produce a comprehensive literature review, design novel models that use multiple PCMs using Solidworks, simulate the heat transfer effect on PCMs using ANSYS Fluent, iterate the process to achieve a performant design, test and collect data of the prototype in lab to validate the simulation results.

70.2 Theory

Heat transfer occurs through three different major types of processes: convection, conduction and radiation. Convection is the heat transfer due to the bulk movement of molecules within fluids such as gases and liquids. It therefore cannot take place in most solids since bulk current flow can take place. Conduction occurs by direct contact between two solid materials. The energy goes from an area with high kinetic energy to an area with lower kinetic energy which means that the temperature flow will always occur from hot to cold. Heat transfer will stop once a thermal equilibrium has been reached meaning that the two temperatures are equal. On a molecular scale, conduction happens due to molecular collision, particles with high speed and thus high energy will collide with particles with low speed, as a result their speed will increase, hence also their energy. Conduction depends on various factors such as the material cross section, the material's properties and the temperature gradient which is the quantity that describes the travel path and the rate of heat travel. The greater the size and length of an object the more energy it will require to be heated. Greater surfaces means that more heat is lost, therefore small objects are preferable to minimize heat loss. The most significant material property is the thermal conductivity coefficient, as it dictates how well a material conducts heat.

Convection is a type of heat transfer specific to fluids. When fluids are heated, their molecules expand, the volume of the fluid must increase by the same factor as the temperature increase. This process makes the fluid less dense and therefore rises. While rising, the hot fluid pushes the more dense and cold fluid down. This results in carrying heat away from the source which results in convection. Radiation is generated by random movements of atoms and molecules in matter, the movement of charged protons and electrons results in the emission of electromagnetic waves which carries energy away from the emitting object through a vacuum or any transparent medium. All materials radiate thermal energy which is related to their temperature. The temperature of the object affects the radiation wavelength and frequency, as the temperature of an object increases the wavelength emitted decreases and its frequency increases. Therefore, the hotter an object will be, the more it will radiate.

Solar panels overheat due to these different phenomena. Solar radiation reaches the first layer of the solar panel which is typically a glass cover. The radiation will go through that layer and the second one which is an encapsulant layer, which permits to limit the solar radiation to be reflected. The radiation will then reach the solar cells which will generate energy. That process can generate a lot of heat, the more there is incident solar radiation on the panel, the more heat will be generated. The heat generated is then conducted to the bottom encapsulant layer which will then conduct the heat to the back of the panel. When a container containing phase change material is added to the back of the panel, then heat is also transferred to it by conduction. The hot container will then transfer heat to phase change material inside it by conduction or/and convection depending on the state of the PCM. The heat will then be absorbed by the PCM as long as it is not fully charged. In a similar way, when the PCM discharges heat, it is released and transferred by convection to the container. The heat is then transferred by convection to the outer air and by conduction back to the solar panel system.

70.3 Abbreviations and Acronyms

The aim of this research is to study incorporation of phase change materials to solar systems. It was noticed in previous investigations that one of the main issues of adding a PCM container to a solar panel to cool it down was that the phase change material would tend to not fully solidify overnight. This research will therefore focus on the effect of using multiple PCMs in a same container and evaluate the solidification and melting processes of the systems. Two different types of containers that could contain multiple PCMs were designed using Solidworks to study the effect of combining phase change materials. In both cases, the aim of the designs is to have an optimized heat transfer area for the PCMs contained by having as many interfaces as possible in all directions.

The first design is a honeycomb structured container which is shown in Figs. 70.1, 70.2 and 70.3. Such a structure permits to have multiple interfaces and contacts with the surrounding. The wall thickness between each honeycomb was set to 2 mm to enhance heat transfer. The second design is a circular structured container. Similarly, to the previous design having such a shape permits to create a heat transfer interface in every direction for the PCMs. The minimum wall thickness separating the circular cuts in the container is 2 mm.

It important to take in consideration the fact that there is more solid material (Steel or Aluminium) separating the PCMs in the circular shaped one than in the honeycomb one, meaning that the volume of PCM is greater in the honeycomb container and that contact between each PCM compartment is more effective.

Fig. 70.1 Solar Panel layers

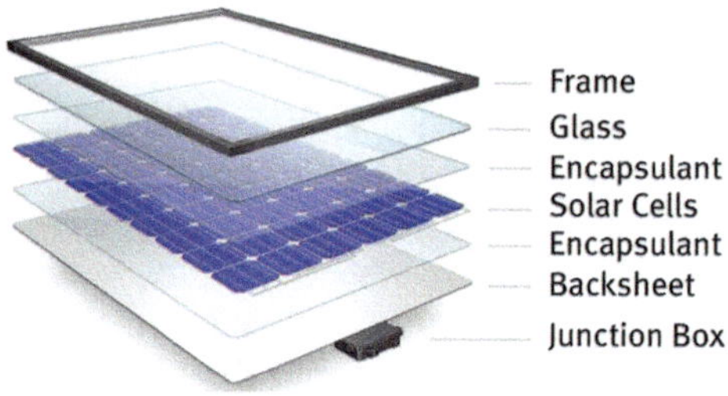

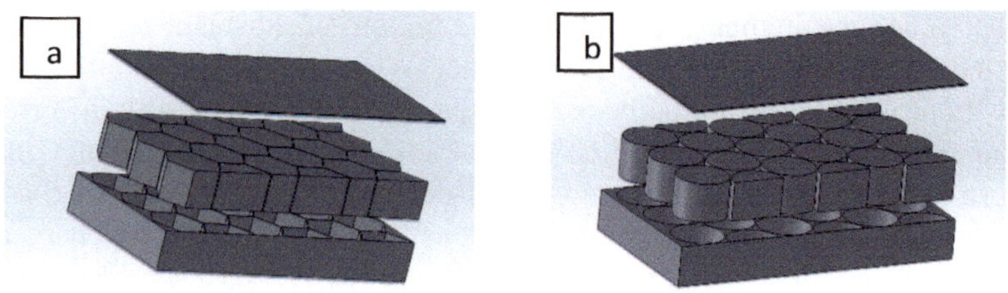

Fig. 70.2 (**a**) Honeycomb container exploded view; (**b**) circular container exploded view

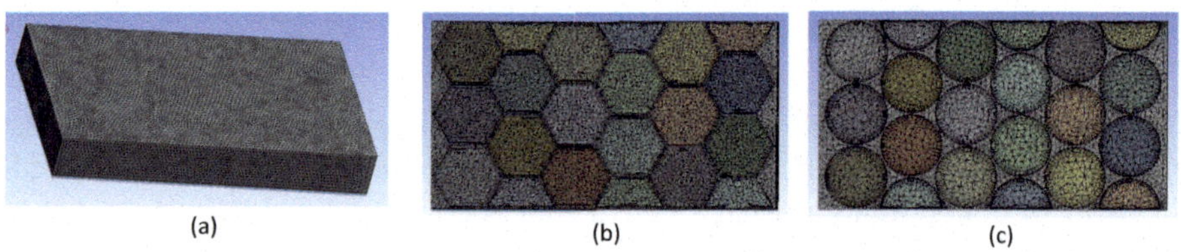

Fig. 70.3 (**a**) Container mesh, (**b**) Honeycomb container inflation layers, (**c**) Circular container inflation layers

70.3.1 Meshing

To obtain accurate results from ANSYS simulation, having an adequate mesh is crucial. An unrefined mesh can give misleading results and make it hard for the solution to converge. An over refined mesh increases the computational load making the solving time extensive and inefficient. An unstructured mesh was created using different body sizing for the different components. Using such a mesh permits to optimize the element distribution across the geometry and can therefore get accurate results. The body sizing of the phase change materials was increased compared to the one of the container as more accuracy is required from the software to compute the phase change process.

It is important to have a more refined mesh at contact areas, as that is where critical heat transfer processes occur.

70.3.2 Boundary Conditions

An investigation was carried out on the roof of the civil engineering in Trinity College Dublin where three PV panels are placed. The first system is a standard solar panel used as a reference, the second system is a standard PV panel on which an empty steel container is added to the back of it. The third system is similar to the second one but the container is filled up with a capric:lauric PCM mixture that suits the Irish weather during summer, as the melting range of the PCM is between 15 °C and 25 °C. The temperature of the three solar panels is acquired using 18 thermocouples in total (six for each panel) each connected to a data logger disposed. Data was acquired from the three panels using six thermocouples for each panel which were connected to a data logger. It was noticed on the data collected that one of the thermocouples placed on the reference panel (channel 316) was giving faulty results. Therefore, only five thermocouples were used for the panel in question. From previous studies it was observed that the overall critical temperature for the reference PV is between 50 and 60 °C. Therefore, we will assume that the temperature use on ANSYS to simulate the heat flux from the Solar Panels will be 55 °C.

70.3.3 Materials

A user-defined function was used to set the properties of the phase change materials. The important values for a numerical transient phase change simulation are the liquid phase temperature, the solid phase temperature, the specific heat capacity, the thermal conductivity and the density. Two different sets of three phase change material were used for this study. The properties of the set PCM used for the first analysis are the following:

For each set of phase change materials, the melting and solidification processes were tested with a steel container and then an aluminium one. The buoyancy of each phase change material was calculated to determine if gravity was required for the simulations. This was done by calculating the ratio between the Grashof number and the Reynolds number and then the Rayleigh number which determines the intensity of the buoyancy flow. Using gravity in such a simulation increases the computational load and can make it hard for the solution to converge. It is therefore critical for transient simulations to determine the buoyancy of the model beforehand otherwise calculations could run for an un-useful amount of time and results can be inaccurate (ANSYS [4]).

From Table 70.3, in all cases, the Grashof Number indicates that the PCMs are buoyancy driven, however the fact that the Rayleigh number is below 10^{-8} means that the flow is laminar and that the buoyancy is not significant enough to have an effect due to gravity.

70.3.4 Fluent Setup

ANSYS Fluent is a commonly used computational fluid dynamic simulation software. The software contains a broad range of physical modelling capabilities which are required to model flow, turbulence, heat transfer and reactions for industrial applications. Fluent covers a broad reach including special models that can simulate cylinder combustion, aero-acoustics or even multiphase systems. In the case of this experiment, a multiphase heat transfer system is being model. For each model simulated in Fluent, settings were done using the same method. The model was set to transient, the energy equation was activated, the flow was set to laminar as determined previously when analysing the buoyancy of the materials and the solidification/melting setting was activated to be able to simulate the phase change. The materials parameter for the phase change materials were inputted using a user-defined function using the characteristics from the Tables 70.1, 70.2, 70.3. The mesh interfaces were manually set to accurately represent the heat transfer contacts between the PCMs and the container. Two interfaces were configured: the first interface coupled the walls of the top of PCMs set and the top of the container with the heat source surface. The second interface coupled the bottom and side walls of the PCMs with the container.

For the melting process of steel containers, the heat source was set to 328.15 K (55 °C) on the top surface of the container, the convection coefficient from the other surfaces of the container to the air were set to 25 W/m K and the free stream temperature (ambient air) was set to 288.15 K (15 °C) as it was assumed it would be the average air temperature in Ireland when a PV panel reaches 55 °C. Both container and PCMs' temperatures were set to 15 °C at the initialization of the simulation as in an ideal scenario it would be at the ambient air temperature. The solution was initialized with the hybrid. The time step size was set to 1 s and the solution was calculated for 10,800 time steps meaning that the solution was simulated over a 3 h period as it was assumed that a solar panel would only reach a temperature of 55 °C during 3 h. For computational load purposes, the number of iterations per time steps was set to 20 as it was found sufficient for the simulation to converge. The solution was recorded every 2 min using the auto save setting.

Table 70.1 Standard PCMs (set 1)

	Density (kg/m^3)	Specific heat (J/kg K)	Thermal conductivity (W/m K)	Viscosity (kg/ms)	Pure solvent melting heat (J/kg)	Solidus temperature (K)	Liquidus temperature (K)
Capric acid	950	2600	0.16	0.00288	150,820	295.95	302.95
Capric–lauric	880	1850	0.142	0.00165	121,020	288.15	293.15
Paraffin–micronal	980	3300	0.2	0.0025	150,000	294.05	299.15

Table 70.2 Improved PCMs (set 2)

	Density (kg/m³)	Specific heat (J/kg K)	Thermal Conductivity (W/m K)	Viscosity (kg/ms)	Pure solvent melting heat (J/kg)	Solidus temperature (K)	Liquidus temperature (K)
Organic—a17	785	2200	0.18	0.0063	150,000	288.15	291.15
Organic—a24	790	2200	0.18	0.0025	150,000	296.50	298.15
Organic—a28	789	2200	0.21	0.00643	150,000	300.15	302.15

Table 70.3 Buoyancy analysis of PCMs

Materials	Thermal diffusivity	Thermal expansion (1/K)	Grashof number	Rayleigh number
Organic—a17	1.04227×10^{-7}	0.0006	108.9	2221.2
Organic—a24	1.03567×10^{-7}	0.0006	691.9	5633.2
Organic—a28	1.20982×10^{-7}	0.0006	104.6	1874.9
Capric acid	6.47773×10^{-8}	0.0006	521.3	7818.1
Capric–lauric	8.72236×10^{-8}	0.00067	1773.6	11316.7
Paraffin–micronal	6.18429×10^{-8}	0.001	1153.1	15723.0

For the solidification process of steel containers, similarly the free stream temperature was set to 10 °C and the convection of the container to the ambient air was set to 25 W/m K for all its surfaces. The solution was also initialized using the hybrid setting and the containers as well as the PCMs' temperatures were set to 55 °C. The time step size was set to 1 s and the simulation was calculated for 25,200 time steps meaning that the results are calculated for a 5 h period. Similarly, to the melting process the number of iteration per time step was set to 20. The solution was recorded every two minutes using the auto save setting.

70.3.5 Method Validation

An experiment was carried out to validate the method used on ANSYS. This experiment consisted in heating up a phase change material, a capric:lauric mixture, in a small aluminium container. Both the container and the phase change material were heated up to 55 °C and then let to cool down in ambient air which was at 23 °C.

The temperatures were recorded using four thermocouples and a thermometer. The first thermocouple was placed on the container itself at the centre point of one of the faces so it could record the temperature change of it. The three other thermocouples were placed inside the container along a bamboo stick, respectively, one quarter up, at mid-point and three quarter up the stick so they could record the temperature change of the PCM. Bamboo was used as it does not conduct heat and thus avoids heat losses. The first step of the experiments was to fill up the container with the capric:lauric PCM mixture. This was done by placing the solid-state material in a beaker and heating it up on a hot plate until it was fully melted. The material was then poured into the container which was then sealed using silicone. The container was then heated on a hot plate until the temperature of both container and PCM reached and stabilized at 55 °C. The container was then taken off the hot plate and let to cool down in ambient air. The temperature change given by the thermocouples were recorded every 2 min for an hour-long period. The experiment was then simulated on ANSYS fluent. The geometry was drawn on Solidworks before being imported to the design modeller. The meshing was done using body sizing functions and inflation layers between the PCM and the container the increase the accuracy of the heat transfer between them. The model was then calculated in fluent, the flow was set as laminar, the energy equation was switched on and the solid/liquid phase change was activated. A user-defined function was set to input the capric:lauric data. The meshing interfaces were manually set. The free stream temperature (ambient air) was set to 296.15 K and the convection coefficient from the container was set to 25 W/m K. The buoyancy was determined as previously described, the characteristics of the capric:lauric PCM required are presented in the table below (Table 70.4):

The ratio between the Grashof and Reynolds number was found to be 318.7 meaning that there is a buoyancy effect induced, however the Rayleigh number was calculated to be equal to 269.7, therefore the effect of buoyancy can be neglected. The effect of gravity was ignored for this experiment. Results from the simulation where collected by creating at data collecting point at the centre point of the PCM and one at the centre of the top surface. Similarly to the experiment, the temperature change were recorded every 2 min over an hour period.

Table 70.4 Buoyancy analysis of validation experiment

Gravitational acceleration (m/s^2)	9.81
Thermal expansion coefficient (1/K)	0.00067
Thermal diffusivity (m^2/s)	8.49×10^{-8}
Temperature gradient (K)	30
Density (kg/m^3)	880
Characteristic length (m)	0.0066
Dynamic viscosity	1.452
Kinematic viscosity	0.00165
Thermal conductivity (Wm^{-1}/K)	0.139
Specific heat capacity (kJ kg^{-1}/K)	1900
Area (m^2)	0.00165
Perimeter (m)	0.25

70.4 Results

In all cases, results depict the temperature and mass fraction variation at a mid-plane cross section. Data points were created at the critical positions for all PCMs which is at the centre of each PCM compartment, at a middle cross section plane of the system, so data could be extracted from the contours. It is therefore important to keep in mind that the following plots only represent characteristics occurring at a certain point among the PCMs and not overall characteristics of the material. The phase change materials used in this experiment can be classified as follows: Low melting range which includes the organic-a17 PCM and the capric–lauric PCM mixture, medium melting range which includes the organic-a24 PCM and the capric-acid PCM and high melting range which includes the organic-b28 PCM and the paraffin–micronal PCM.

70.4.1 Steel Honeycomb Container Standard PCMs

70.4.1.1 Solidification

Figure 70.4 represent the solidified fraction of the PCMs and their temperature at t = 7 h. We can notice that after 7 h the phase change materials are still partly liquefied, the capric–lauric mixture being the most liquid. The half honeycombs on the sides of the container which contain the capric acid PCM have fully solidified. Overall, the most solidified PCM is the paraffin–micronal followed by the capric acid, the capric–lauric mixture is still mostly liquid after 7 h. From the temperature contour, we can observe that even if the paraffin is the most solidified PCM it is also the one with the highest temperature (reaching approximately 302 K in its centre).

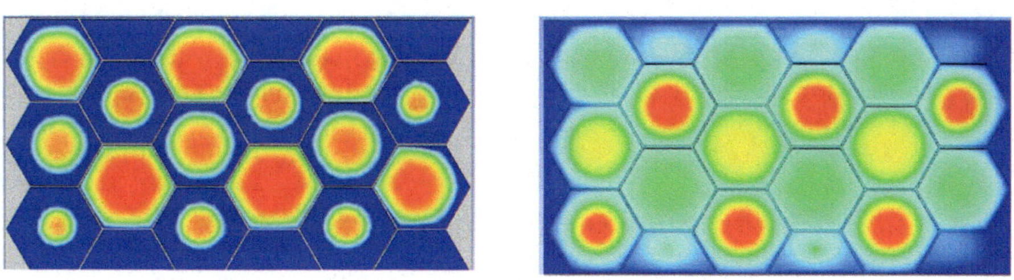

(a) (b)

Fig. 70.4 (**a**) Mass Fraction contour of HC Steel Container Standard PCMs during solidification process at t = 7 h; (b) Temperature contour of HC Steel Container Standard PCMs during solidification process at t = 7 h

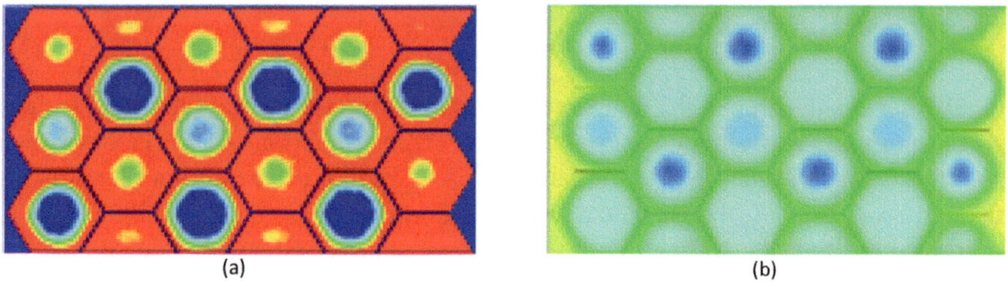

Fig. 70.5 (**a**) Mass Fraction contour of HC Steel Container Standard PCMs during melting process at t = 3 h; (**b**) Temperature contour of HC Steel Container Standard PCMs during melting process at t = 3 h

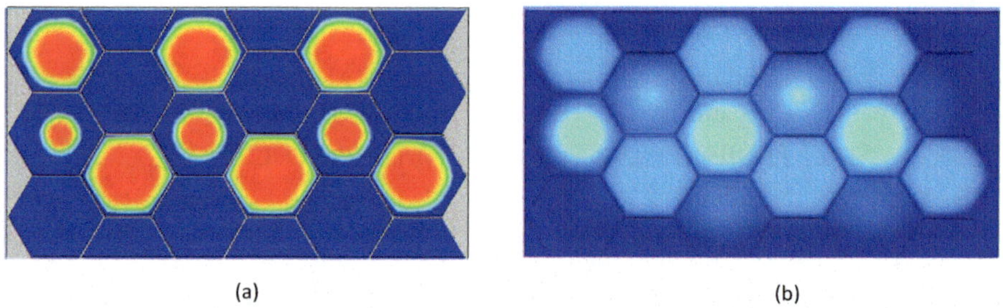

Fig. 70.6 (**a**) Mass Fraction contour of HC Steel Container Improved PCMs during solidification process at t = 7 h; (**b**) Temperature contour of HC Steel Container Improved PCMs during solidification process at t = 7 h

70.4.1.2 Melting

Figure 70.5 shows the melting fraction after 3 h of the following PCMs: capric–lauric, capric acid and paraffin micronal, in a steel honeycomb container. The contour on the left-hand side represents the mass fraction of the PCMs. We can notice that after 3 h, most of the capric–lauric has melted as well as the capric acid contained in the outer compartments. The paraffin–micronal phase change material as well as the capric acid positioned in the centre of the container are mostly in a liquid phase.

70.4.2 Steel Honeycomb Container Improved PCMs

70.4.2.1 Solidification

Figure 70.6 shows the result of the solidification process after 7 h of the organic-a17, organic-a24 and organic-b28 phase change materials. We can notice that all the organic-b28 (high phase change range) material has fully solidified. For the organic-a24, the material located in the half honeycomb compartment also fully solidifies after

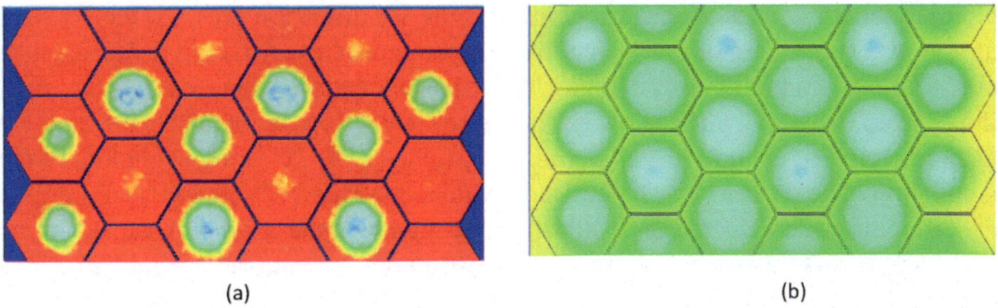

Fig. 70.7 (**a**) Mass Fraction contour of HC Steel Container Improved PCMs during melting process at t = 3 h; (**b**) Temperature contour of HC Steel Container Improved PCMs during melting process at t = 3 h

7 h but when positioned in the centre of the container in a full honeycomb the material only solidifies. The organic-a17 phase change material stays mostly in a liquid phase after 7 h. From the temperature contour, we can notice that the material having the highest temperature is the organic-a24, the honeycombs placed in the centre of the container having a temperature of approximately 298 K.

70.4.2.2 Melting

From Fig. 70.7, the result of the melting process after 3 h of the organic-a17, organic-a24 and organic-b28 phase change materials can be observed. We can observe that at t = 3 h, the phase change material organic-a17 is mostly melted in each of its honeycombs. The organic-a24 PCM is half melted in the honeycombs placed in the centre of the container and fully melted when the honeycombs that are half the size and placed on the outer part of the container. Finally, the organic-b28 PCM which has a higher melting temperature is only partly melted after 3 h.

70.5 Conclusion

In this research, the use of multiple phase change material for the thermal management of solar systems was investigated. Two different geometries were designed in Solidworks, the first one being container with honeycomb compartments and the second a container with circular compartments. The heat transfer was investigated using ANSYS fluent, different boundary conditions and materials were used to carry out the investigations. A first set of the three types following types of PCMs: capric–lauric, capric acid and paraffin–micronal was used in the honeycomb container with the structure being set to steel. A second set of the following PCMs: organic-a17, organic-a24 and organic-b28, was then used in the honeycomb container with the structure being set to steel and then aluminium, as well as in the circular container with a steel structure. For the solidification process, the ambient air temperature was set to 10 °C and the temperature of every material was patched at 55 °C. The simulation was calculated as transient over a 7 h period. For the melting process, a heating surface of 55 °C was set on the top of container, the ambient air was set to 10 °C and every material was patched to 15 °C at the beginning of the simulation. The process was calculated as transient over a 3 h period. A validation experiment was conducted to evaluate the accuracy of the results. A capric–lauric PCM mixture was melted in an aluminium container and let to cool down in ambient air. A simulation was then conducted in ANSYS with the same parameters and boundary conditions. It was found that the difference in results was low enough to validate the model.

It was found that using the second set of PCMs which has similar phase change ranges but better characteristics improves both the solidification process and melting. The best models were found to be the honeycomb structured container made from steel or aluminium. Indeed, after 7 h of solidification most of the honeycomb compartments are solidified for both of these models while for the two other systems, the process is not as advanced. Similarly, for the melting process, the phase change materials melt quicker when encapsulated in these two systems. It is important to consider the fact that aluminium is more susceptible to be corroded by PCMs than steel even if organic PCMs usually have low corrosivity. For future work, it would be interesting to investigate other sets of phase materials with different characteristics and melting ranges. Different boundary conditions to simulate different climates could also be simulated to see the influence of different temperatures on the phase change process and have a prediction model for diverse weather. Finally, a prototype could be built to investigate the models in real conditions.

References

1. Atkin P, Farid MM (2015) Improving the efficiency of photovoltaic cells using PCM infused graphite and aluminium fins. Sol Energy 114:217–228. https://doi.org/10.1016/j.solener.2015.01.037
2. Evans, CJ et al. (2009) G-TRACE: rapid Gal4-based cell lineage analysis in Drosophila. Nat. Methods 6, 603–605
3. Stropnik R, Stritih U (2016) Increasing the efficiency of PV panel with the use of PCM. Renew Energy 97:671–679. https://doi.org/10.1016/j.renene.2016.06.011
4. Ansys (2009) Ansys fluent 12.0 Theory Guide, ANSYS Fluent. https://doi.org/10.1016/0140-3664(87)90311-2

Chapter 71
Quantum Dots Solar Cells in Solar System to Convert Light into Electricity

Amira R. AbouElhamd, Ahmed Hassan, Khaled A. Al-Sallal, and Saleh T. Mahmoud

71.1 Introduction

Several efforts have been studied to benefit from sun radiations and convert it to another form of energy. Solar energy supplies clean and sustainable source of energy for future generations. The challenge, however, relies on developing a technology that utilizes this vast amount of energy and convert it by an economical and efficient way. Whereas solar energy technologies such as silicon photovoltaic cells lead the market, they suffer from the low efficiency of solar energy harvesting [1]. One option to minimize the cost in solar energy generation systems is to concentrate solar radiations in an effective way. Concentrating photovoltaic system (CPV) can be used to concentrate direct radiation only. Limited options are available to concentrate diffused radiations.

The luminescent solar concentrators (LSC) technology was introduced to replace the large area solar cells with an inexpensive, large area of sunlight collectors and hence reduce the cost of module [2]. It consists of a transparent material doped with luminescent species such as organic dye or quantum dotes. The photon incident on a large area are absorbed by the luminescent species on one face of the system, reemitted with larger wavelength, and wave guided with a total internal reflection (TIR) to the edge of the transparent material where they can be collected via solar cells (see Fig. 71.1). The LSC has advantages over other concentrating systems in that it can concentrate not only direct but also diffuse solar radiation and is not limited by a concentration ratio; less losses are associated for solar cells, and they well suit the static nature of buildings and could be used as building-integrated pho-

A. R. AbouElhamd · A. Hassan · K. A. Al-Sallal (✉)
Department of Architectural Engineering,
UAE University, Al Ain, UAE
e-mail: k.sallal@uaeu.ac.ae

S. T. Mahmoud
Department of Physics, UAE University, Al Ain, UAE

A. Sayigh (ed.), *Renewable Energy and Sustainable Buildings*, Innovative Renewable Energy, https://doi.org/10.1007/978-3-030-18488-9_71

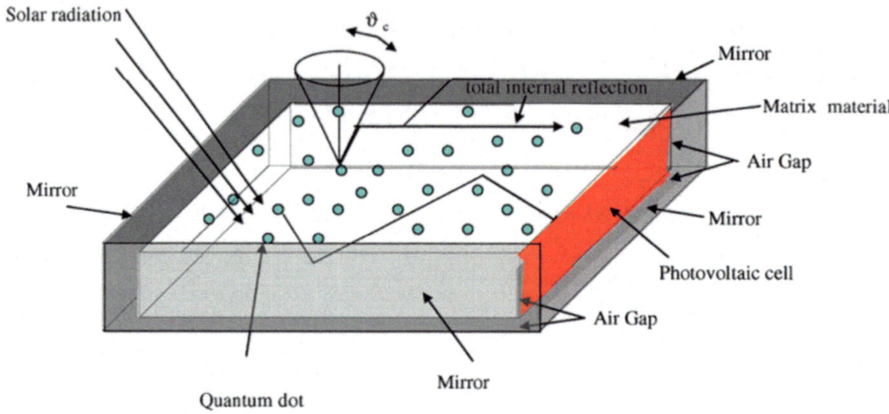

Fig. 71.1 QDSC principle of operation adopted from Gallagher et al. [8]

tovoltaics (BIPV). The organic dyes have some limitations in their properties; they have narrow band absorption and low fluorescence quantum yield (FQY) and are considered less stable compared to quantum dotes [3, 4].

Quantum dots (QDs) are semiconductor material with nanocrystal that has electronic and optical characteristics similar to those found in atoms. They are characterized by their high absorption coefficient, retention time, and reemission in a desired direction with tunable spectrum [5–7]. For these characteristics, they are considered promising to form quantum dot solar concentrators (QDSC) to concentrate both direct and diffused radiation in a desired direction without the need for expensive solar tracking system. QDs have several advantages over the luminescent dyes when used in the fluorescent solar concentrator (FSC) as they are considered more stable, less degradable and have more control on the absorption spectrum by just tuning the dot diameter. The diameter of QDs typically varies from tens to hundreds of nanometers and each size provides distinct electronic and optical properties.

QDs, like any luminescent species, absorb the refracted insolation incident on the device surface from a transparent carrier material which releases photons from QDs at low frequencies. These photons could travel to the edge of the carrier material and leave it or could reflect internally within the device surface and remain inside the concentrator. Higher percentage of the released photons will reach the edge when the carrier material has greater refractive index than that of the surrounding medium, and will follow a total internal reflection. The extent of absorption and reemission of these photons determines the effectiveness of QDs as concentrators. After QDs emit light when they release photons for a short path length within the concentrator, the emitted light within the overlap region began to degenerate and consequentially reduce the efficiency of the solar energy conversion. An effective QDSC should ensure the emitted light to have the shortest path length within the collector as well as absorbing all incident light between the thickness of the collector plate.

QDSC can be integrated to solar cells to improve PV integration and penetration to buildings. Gallagher et al. [8] used a simple principle to illustrates the operation of the QDSC with solar cells integrated to it by using four mirrors on the panels, three of which on the edges and one at the back, to ensure that light will occur either at the top surface or at the forth edge as shown in Fig. 71.1. Air gaps between the QD matrix and the mirror help photons to oppose the total internal reflection in the critical internal incident angle range.

71.2 Aim and Objectives

QDs could be applied to any transparent material such as glazing system of buildings, which helps harvesting the solar energy by PV cells without solar tracking system and without blocking the outdoor visual environment for occupants. This could produce transparent glazing that produced energy to be used in other building applications. This paper aims at reviewing the state of art of QDSC integrated with solar cells that could be applied to buildings (BIPV).

71.3 Types of QDSC

The efficiency of the quantum dots solar concentrators (QDSC) is mainly influenced by the type of QDs used, their shape, size, dot concentration, and optical path length through the device. To design QDSC, the optimum concentration of QDs doping in the QDSC devices is crucial. The following subsections discuss how the type, size, and concentration of QDs affect the efficiency of QDSC systems.

71.3.1 Different Types QDs for QDSCs

CdSe and CdS are the most common types of QDs used in QDSC because they have high potential of light harvesting of visible light [9–11]. Recently, they are used to produce a high efficient QDSC of around 6.33% by doping Mn^{2+} into CdSe [12]. Nevertheless, it has a high band gap (1.7 eV) and may absorb wave length shorter than 700 nm. Therefore, it may not be possible to utilize the full solar spectrum, which limits the performance of the photocurrent density and the power conversion efficiency (PCE) of the solar cells.

$CuInSe_2$ is another type of QDs that has a narrow band gap (1.02 eV), a strong absorbance across the solar spectrum (up to 30%), and high photoluminance quantum yield (η_{PL}) of 65–75%. They also have a large intrinsic stocks shift (Δ_s) between the photoluminescence (PL) band and the onset of strong absorption, which indicates that these DQ has unusual emission mechanism. The PL of these QDs occurs not because of the band-to-band transition, but it happened because of the transition between the conduction-band electron and the localized hole residing in an intragap state attributed to the Cu-related defect, or a self-trapped exciton [13–16]. A crucial advantage that the PL of these QDs has over the normal dyes is that they have high photoluminance quantum yield (η_{pl} > 60percentage) even for high wave lengths (NIR at 800 nm), where common dyes become poor emitters (η_{pl} < 10%). These properties make this type of QD to have a strong performance when they act as LSC fluorophores. Another advantage of using the crises-QD is that they have a neutral-density filter that does not disturbance to perceived colors and thus could be used directly in glass windows. Meinardi et al. developed freestanding, colorless photovoltaic windows that introduce no distortion to perceived colors by incorporating $CuInSe_xS_{2-x}$ quantum dots into photopolymerized poly (lauryl methacrylate) [15]. Additionally, it was proven that controlling the concentration of this type of QDs would allow the control of the degree of shading they can provide for indoor spaces [17] (see Fig. 71.2 adopted from Wu et al. [17]). $CuInSe_xS_2$ is a save element that is toxic free; it reduces the reabsorption of photons and enlarges the coverage of the solar spectrum. Using this type of QDs as QDSC in BIPV was proposed by Berggren et al. [18] by doping them at a loading of 0.4 wt% in a low-iron 10 cm × 10 cm float glass that transmit 40% visible light. The study recorded an optical efficiency of 8.1% while the electrical power conversion efficiency was 2.2% with black background and was 2.9% with reflective one.

The commonly studied Mn^{2+} has a wide gap and is based on zones and ZnS [19]. Nevertheless, they only exhibit a moderate photoluminance quantum yield (maximum of η_{pl} = 50%). $Cd_xZn_{1-x}S$ is a new class of doped Mn^{2+}. The use of alloyed $Cd_xZn_{1-x}S$ makes it more flexible to control the absorption and assists mitigate lattice mismatch with ZnS shell and hence improve the photoluminance quantum yield (η_{pl}) to more than 70% [20, 21]. Wu et al. [17] found that the loses in PL is not due to the reabsorption but rather owing to the weak spectrally independent scattering in the LSC waveguide.

Pubs QDs is a semiconductor that has a very low band gap (near the IR region of 0.41 eV) [22, 23]. It also has high absorption coefficient of 1–5 × 10^5 cm^{-1} and Bohr excitation radius of 18 nm. Hence, PBS could ideally be used as DQSCs. But PbS-

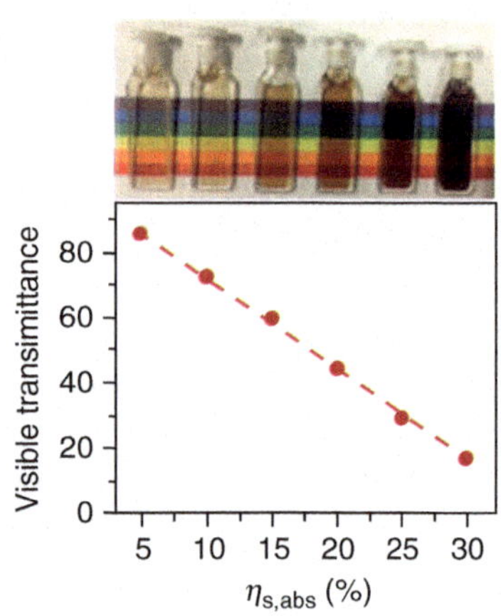

Fig. 71.2 The relation between sunlight absorbance and the visible transmittance for difference concentration of $CuInSe_2$ QDs

sensitized solar cells are much less efficient than the CdS/CdSe. The power conversion efficiency (PCE) of the QDSC with PbS is usually less than 4% without coating or doping. It was found that when the PbS surface is coated with CdS to produce a PbS-CdS cosensitized solar cells, the PCE exceeds 4% [23]. Additionally, Park et al. has produced the highest PCE of 5.6% by doping Hg in PbS [23]. To date, the PCE of the PbS QDSC does not reach expectation owing to many reasons such as the high charge recombination of solar cells and instability of PbS in the electrolyte owing to their slow electron transfer kinetics that affects the charge separation and collection [22, 24, 25]. Tian et al. has increased the PCE of the PbS QDSC to reach 4.01% without doping or cosensitization. They have replaced 30% of deionized water of sulfide electrolyte with methanol which accelerated the redox couple diffusion in the electrolyte and the charge transfer at the interfaces between photoanodes and electrolyte. This also has enhanced the stability of PbS in the electrolyte to reduce the surface charge recombination [26].

71.3.2 QDs Size and Concentration in QDSC System

The different sizes of QDs affect the rate of electron injection in the QD-metal oxide (MO) dynamics. The QDs inject electrons in the MO nanoparticles to enable photocurrent generation in the photoelectrochemical solar cell. Kongkanand et al. [11] assembled CdSe nanoparticles with various diameter sizes, 2.3 nm, 2.6 nm, 3 nm and 3.7 nm, on TiO_2 nanoparticles and nanotubes using a bifunctional surface modifier. Electrons were injected from the CdSe QDs to TiO_2 through the band gap excitation to enable photocurrent generation in the photoelectrochemical solar cells. The energy difference between the conduction bands of CdSe QDs and TiO_2 particles directly affects the injection of electrons from CdSe QDs into TiO_2. Figure 71.3 illustrates a schematic diagram from the study to show the difference of energy levels for the four different-sized CdSe QDs and TiO_2. The study strongly supports that controlling the size of CdSe enables tuning the photoelectrochemical response and photo-conversion efficiency of the cell. It was proven that small size CdSe QDs

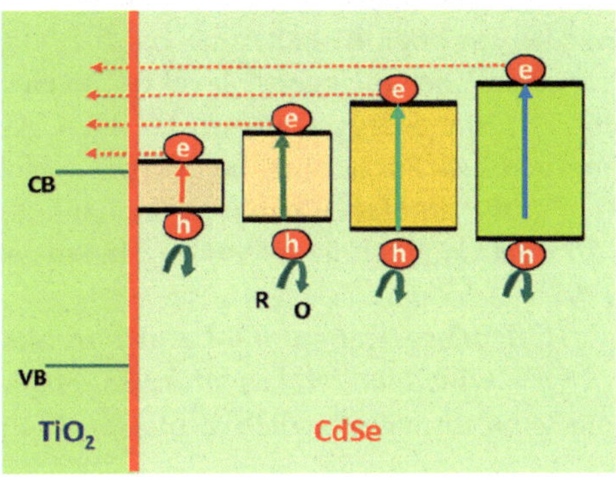

Fig. 71.3 A schematic diagram shows the difference of energy levels for the four different-sized CdSe QDs and TiO_2 [11]

provide greater charge injection rates and higher IPCI (photon-to-charge carrier generation efficiency) while the large particle of CDSE QDs provides better absorption in the visible region. The 3 nm diameter CdSe nanoparticles provided the maximum IPCE with around 35% for particulate TiO_2 and 45% for tubular TiO_2 morphology.

The same results was also found by Robel et al. [27] who investigates the effect of five sizes of CdSe on the band energies and the energy of the charge carriers. The study found that as the particle size decreases from 7.5 to 2.4 nm, the excitonic peak shifts from 645 nm (1.92 eV) to 509 nm (2.44 eV). The study suggests that as the size of QD decreases, a better recovery is suggested through alternate deactivation pathway for the charges separated in CdSe.

Chandra et al. [28] investigated the process of QD size optimization for QDSC. They fabricated five samples on 60 × 40 × 2 plates coated with 0.005, 0.01, 0.03, 0.05, 0.07 wt% Csde/ZnS QDs by a drop casting technique, where the edges were smoothed by polishing technique. The emission transport properties of the QDs were examined using varying optical path length using laser line 405, 488, and 532 nm and white color using cutoff wavelength of 600 nm. It was found that the reabsorption is responsible of reducing the emission for lower QD concentrations while scattering and reabsorption are both responsible for the reduced emissions with high QD concentrations. The study also found that the emission is lost before reaching the edge for 70% for >20 mm optical path length and thereby reduce the efficiency of the system. The optimum QD concentration that gave better optical characteristics for the 60 × 40 × 2 plate was 0.03 wt%, while the optimum concentration for that gave the maximum power output was 0.07 wt%. The study recommended not to use the electrical output results only for optimizing the size and the QD concentration since the optical characteristics were found more reliable as it provides more insights about the emission transportation properties.

71.4 The Challenges Faced by QDSC and Contemporary Solutions

One of the challenged faced in the QDSC system is the voltage losses that occurred as a results of energy level mismatches as well as charge separation. Energy is needed for both the electrons and the holes to separate the excite state into charge carriers. When the energy level of the electron does not match that of band gap for the QD, the energy is lost in between. Another challenge that reduces the performance of this system is the charge recombination. Charge recombination happens due to the difference between the band gap of the absorber and the offset between the electron and hole injection. This means that some losses happen in the system [29].

High concentration of QDs also results in losses such as increased reabsorption of QDs with small stokes shift [23, 30, 31]. When the reabsorption increases, the emission through irradiative relaxation due to the low fluorescent quantum yield

(FQY) of QDs decreases and hence reduces the efficiency of energy conversion of the system. Another drawback of increasing the concentration of QDs is the possibility of cluster formulation that causes scattering losses for the system to increase, and hence reduces the efficacy of the system [32, 33].

Collecting a huge energy using the QDSC on a large surface of glass and concentrating them on a relatively small area of small sheets of photovoltaic cells that could be arrange on the edge of the glass pane is very tricky. Whereas using this system increases the efficiency of the cells as well as reduces the cost by using small sized PV cells, several performance factors should be considered to ensure the optimum performance for the system. Collecting a huge number of electrons from a large glazing area on a small area of PV cells could result in increased temperature of the system. Hence, both electrical and thermal energy are obtained when the QDSC systems is used with photovoltaic cells. This could result in heat generation due to the increase in irradiance and reduction in volume of the system. This generated heat reduces the electric performance of the system and thus needs to be controlled and removed [34].

The tradition methods of cooling the PV cells reply on using either air or water along with circulated air to remove the heat produced by the panel. Nevertheless, this technique of cooling is not the optimum as they consume a huge amount of power by using either pumps or fans while the temperature of the PV could still raise up to 70° [35]. Another method that is used to cool the PV panels is to use photovoltaic phase change materials (PV-PCM), which have been proven to limit the rise in temperature and enhance the performance of PV under one sun [36, 37]. Using nanofluids could further increase the rate of cooling of the PV as found by several studies [38, 39].

71.5 Conclusion

The new technology of large area, high performance LSC-QDs has been reviewed for their use as QDSC-based solar cells to help building designers and system integrators in assessing their use when integrated to buildings to produce transparent glazing that generates energy. This paper analyzed the efficiency, the visibility characteristics, and the environmental impact of several types, sizes, and concentrations of QDs in QDSCs. It also discussed some of the challenged that faces this system at operation and when integrated to buildings. The main highlights of this paper could be summarized as follows:

- Low band gap QDs, such as $CuInSe_2$ and PbS, are preferable over the high band gap QDs, like CdSe and CdS, when used as QDSC as they provide better optical and electrical outcomes.
- $CuInSe_2$ are toxic-free material that has low manufacture cost, neutral-density filter, high coverage of the solar spectrum, and low recombination of photons,

which make them best suited to enhance the power performance, the visibility, and the esthetic appearance QDSCs-based solar cells in BIPV.

- While PbS exhibit a very low band gap, they produce very low photocurrent density compared to the expected value they should produce due to the high percentage of charge recombination and the instability of PbS. More research is needed to enhance their performance when used as QDSCs.
- The unique feature of the ability to tune the size of the QDs provides several advantages. It was found that small size QDs enhances the rate of charge injection and increases the photon-to-charge carrier generation efficiency (IPCI), while the large size QDs exhibit better optical properties in the visible region.
- It is crucial to consider both the optical characteristics and the electrical output when optimizing the size and the concentration of the QDs, since the optical characteristics provide more insights about the emission transportation properties.
- The main challenges faced by the QDSC are the losses in voltage energy, charge recombination, the reduced emission due to reabsorption at both low and high concentrations, and scattering at high concentration.
- Concentrating huge energy on a small sized PV cells reduces the performance of the system as a result of the generated heat. This produced thermal energy should be controlled and removed by using either the traditional cooling methods of PV, using photovoltaic phase change materials, or using nanofluids technologies to convert the excessive heat to thermal energy.

Acknowledgments This work was supported by funds from the Office of the Deputy Vice Chancellor for Research and Graduate Studies at UAE University under fund Number 31R108.

References

1. Bagnall DM, Boreland M (2008) Photovoltaic technologies. Energy Policy 36(12):4390–4396
2. Debije MG, Verbunt PPC (2012) Thirty years of luminescent solar concentrator research: solar energy for the built environment. Adv Energy Mater 2(1):12–35
3. Wittwer V, Heidler K, Zastrow A, Goetzberger A (1981) Theory of fluorescent planar concentrators and experimental results. J Lumin 24:873–876
4. Goetzberger A, Wittwer V (1981) Fluorescent planar collector-concentrators: a review. Solar Cells 4(1):3–23
5. Tang J, Kemp KW, Hoogland S, Jeong KS, Liu H, Levina L et al (2011) Colloidal-quantum-dot photovoltaics using atomic-ligand passivation. Nat Mater 10(10):765
6. Nozik AJ, Beard MC, Luther JM, Law M, Ellingson RJ, Johnson JC (2010) Semiconductor quantum dots and quantum dot arrays and applications of multiple exciton generation to third-generation photovoltaic solar cells. Chem Rev 110(11):6873–6890
7. Sambur JB, Novet T, Parkinson BA (2010) Multiple exciton collection in a sensitized photovoltaic system. Science 330(6000):63–66
8. Gallagher SJ, Rowan BC, Doran J, Norton B (2007) Quantum dot solar concentrator: device optimisation using spectroscopic techniques. Sol Energy 81(4):540–547
9. Zhang Q, Guo X, Huang X, Huang S, Li D, Luo Y et al (2011) Highly efficient CdS/CdSe-sensitized solar cells controlled by the structural properties of compact porous TiO_2 photoelectrodes. Phys Chem Chem Phys 13(10):4659–4667

10. Yu X-Y, Liao J-Y, Qiu K-Q, Kuang D-B, Su C-Y (2011) Dynamic study of highly efficient CdS/CdSe quantum dot-sensitized solar cells fabricated by electrodeposition. ACS Nano 5(12):9494–9500
11. Kongkanand A, Tvrdy K, Takechi K, Kuno M, Kamat PV (2008) Quantum dot solar cells. Tuning photoresponse through size and shape control of CdSe−TiO_2 architecture. J Am Chem Soc 130(12):4007–4015
12. Im J-H, Lee C-R, Lee J-W, Park S-W, Park N-G (2011) 6.5% efficient perovskite quantum-dot-sensitized solar cell. Nanoscale 3(10):4088–4093
13. Berends AC, Rabouw FT, Spoor FCM, Bladt E, Grozema FC, Houtepen AJ et al (2016) Radiative and nonradiative recombination in $CuInS_2$ nanocrystals and $CuInS_2$-based core/shell nanocrystals. J Phys Chem Lett 7(17):3503–3509
14. Rice WD, McDaniel H, Klimov VI, Crooker SA (2014) Magneto-optical properties of $CuInS_2$ nanocrystals. J Phys Chem Lett 5(23):4105–4109
15. Meinardi F, McDaniel H, Carulli F, Colombo A, Velizhanin KA, Makarov NS et al (2015) Highly efficient large-area colorless luminescent solar concentrators using heavy-metal-free colloidal quantum dots. Nat Nanotechnol 10(10):878–885
16. Schuster M, Sisterhenn P, Graf L, Wellmann PJ (2018) Processing and characterization of vacuum-free $CuInSe_2$ thin films from nanoparticle-precursors using novel temperature treatment techniques. J Nano part Res 2:4
17. Wu K, Li H, Klimov VI (2018) Tandem luminescent solar concentrators based on engineered quantum dots. Nat Photonics 12:105–110
18. Bergren MR, Makarov NS, Ramasamy K, Jackson A, Guglielmetti R, McDaniel H (2018) High-performance $CuInS_2$ quantum dot laminated glass luminescent solar concentrators for windows. ACS Energy Lett 3(3):520–525
19. Shao H, Bai X, Cui H, Pan G, Jing P, Qu S et al (2018) White light emission in Bi^{3+}/Mn^{2+} ion co-doped $CsPbCl_3$ perovskite nanocrystals. Nanoscale 10:1023–1029
20. Kim J-U, Lee M-H, Yang H (2008) Synthesis of Zn1−xCdxS: Mn/ZnS quantum dots and their application to light-emitting diodes. Nanotechnology 19(46):465605
21. Levchuk I, Würth C, Krause F, Osvet A, Batentschuk M, Resch-Genger U et al (2016) Industrially scalable and cost-effective Mn^{2+} doped ZnxCd1−xS/ZnS nanocrystals with 70% photoluminescence quantum yield, as efficient down-shifting materials in photovoltaics. Energy Environ Sic 9(3):1083–1094
22. González-Pedro V, Sima C, Marzari G, Boix PP, Giménez S, Shen Q et al (2013) High performance PbS quantum dot sensitized solar cells exceeding 4% efficiency: the role of metal precursors in the electron injection and charge separation. Phys Chem Chem Phys 15(33):13835–13843
23. Lee J-W, Son D-Y, Ahn TK, Shin H-W, Kim IY, Hwang S-J et al (2013) Quantum-dot-sensitized solar cell with unprecedentedly high photocurrent. Sci Rep 3:1050
24. Lai L-H, Protesescu L, Kovalenko MV, Loi MA (2014) Sensitized solar cells with colloidal PbS–CdS core–shell quantum dots. Phys Chem Chem Phys 16(2):736–742
25. Lee H, Leventis HC, Moon SJ, Chen P, Ito S, Haque SA et al (2009) PbS and CdS quantum dot-sensitized solid-state solar cells: "old concepts, new results". Adv Funct Mater 19(17):2735–2742
26. Tian J, Shen T, Liu X, Fei C, Lv L, Cao G (2016) Enhanced performance of PbS-quantum-dot-sensitized solar cells via optimizing precursor solution and electrolytes. Sci Rep 6:23094
27. Robel I, Kuno M, Kamat PV (2007) Size-dependent electron injection from excited CdSe quantum dots into TiO_2 nanoparticles. J Am Chem Soc 129(14):4136–4137
28. Chandra S, McCormack SJ, Kennedy M, Doran J (2015) Quantum dot solar concentrator: optical transportation and doping concentration optimization. Sol Energy 115:552–561
29. Boix PP, Larramona G, Jacob A, Delatouche B, Mora-Seró I, Bisquert J (2011) Hole transport and recombination in all-solid Sb_2S_3-sensitized TiO_2 solar cells using CuSCN as hole transporter. J Phys Chem C 116(1):1579–1587

30. Martí A, Lopez N, Antolin E, Canovas E, Luque A, Stanley CR et al (2007) Emitter degradation in quantum dot intermediate band solar cells. Appl Phys Lett 90(23):233510
31. Coropceanu I, Bawendi MG (2014) Core/shell quantum dot based luminescent solar concentrators with reduced reabsorption and enhanced efficiency. Nano Lett 14(7):4097–4101
32. Cossairt BM, Owen JS (2011) CdSe clusters: at the interface of small molecules and quantum dots. Chem Mater 23(12):3114–3119
33. Koch SW (1993) Semiconductor quantum dots. World Scientific, Singapore
34. Xu Z, Kleinstreuer C (2014) Concentration photovoltaic–thermal energy co-generation system using nanofluids for cooling and heating. Energy Convers Manag 87:504–512
35. Du D, Darkwa J, Kokogiannakis G (2013) Thermal management systems for photovoltaics (PV) installations: a critical review. Sol Energy 97:238–254
36. Kibria MA, Saidur R, Al-Sulaiman FA, Aziz MMA (2016) Development of a thermal model for a hybrid photovoltaic module and phase change materials storage integrated in buildings. Sol Energy 124:114–123
37. Hasan A, Alnoman H, Shah AH (2016) Energy efficiency enhancement of photovoltaics by phase change materials through thermal energy recovery. Energies 9(10):782
38. Elmir M, Mehdaoui R, Mojtabi A (2012) Numerical simulation of cooling a solar cell by forced convection in the presence of a nanofluid. Energy Procedia 18:594–603
39. Wu S, Wang H, Xiao S, Zhu D (2012) Numerical simulation on thermal energy storage behavior of Cu/paraffin nanofluids PCMs. Procedia Eng 31:240–244

Chapter 72
Optical Coupling Sensitivity Study of Luminescent PV Devices Using Monte Carlo Ray-Tracing Model

Mehran Rafiee, Subhash Chandra, Hind Ahmed, Keith Barnham, and Sarah J. McCormack

72.1 Introduction

Luminescent photovoltaic (LPV) devices such as the Luminescent Solar Concentrator (LSC) [1–3] and Luminescent Downshifting Layers (LDS) [4, 5] are cost-effective technologies to optically enhance the performance of PV solar cells. As seen in Figs. 72.1 and 72.2, in both LSC and LDS, luminescent materials are doped in a transparent host polymer layer [6–8]. In LSCs, Fig. 72.1, the incident solar radiation is absorbed by the luminescent material, concentrated and wave-guided to the edges of the device where the PV solar cell is coupled. A thin film LDS layer shown in Fig. 72.2 is deposited on top of the PV solar cell and the absorbed incident solar radiation is red shifted by the LDS layer to the region where the PV cell is more efficient.

LPVs can be fabricated in variety of colors and dimensions [9–14]; therefore, they are attractive choices for building integrated PV (BIPV) systems which brings us closer to the goal of zero carbon energy buildings [6–8].

Absorbance (A) in LPV devices is defined by Beer Lambert law [15] as follows:

$$A = \log_{10} T \tag{1}$$

where T is the value of transmittance. Absorbance is characterized by the concentrations of the luminescent material and the geometry of the device. As shown in Fig. 72.3a, b, the reflection (R) and refraction (T) can be determined from Snell's Law [16]:

M. Rafiee (✉) · S. Chandra · H. Ahmed · S. J. McCormack
Department of Civil, Structural and Environmental Engineering, Trinity College Dublin, Dublin 2, Ireland
e-mail: rafieem@tcd.ie

K. Barnham
Department of Physics, Faculty of Natural Sciences, Imperial College London, London, UK

A. Sayigh (ed.), *Renewable Energy and Sustainable Buildings*, Innovative Renewable Energy, https://doi.org/10.1007/978-3-030-18488-9_72

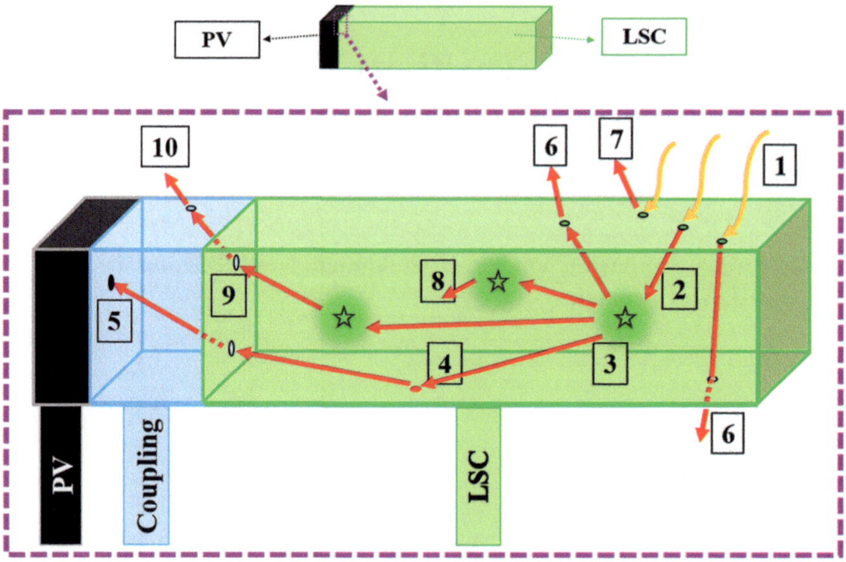

Fig. 72.1 Configuration of an LSC which shows: *1* photon enters the LSC and is *2* absorbed by the luminescent material, then *3* reemitted at longer wavelength and *4* wave-guided by total internal reflection (TIR) and *5* reaches the PV cell through the optical coupling. Losses include: *6* the fraction of light which is lost through the bottom and other surfaces (escape cone loss), *7* front surface reflection, and *8* the emitted photon which is reabsorbed by another luminescent molecules and its energy is decreased. *9* A fraction of radiation striking the optical coupling, might be *10* deviated and lost due to refractive index differences of the optical coupling and luminescent layer. Note that the radiation may also be scattered or attenuated by the host or coupling material which are not shown here

$$R = I - 2(N \cdot I)N$$

$$T = \eta_{\text{it}} I + \left[\eta_{\text{it}} C_{\text{i}} - \sqrt{1 + \eta_{\text{it}}^{2} \left(C_{\text{i}}^{2} - 1 \right)} \right] N \tag{2}$$

where $C_i = \cos(\theta_i)$ and η is the refraction index of the medium. θ_i is the angle of direction between the normal of a surface (N) and the incident ray (I). The ray is reflected by the angle of reflection (θ_r) or refracted by the angle of refraction θ_t. The total internal reflection (TIR) phenomenon can be seen in Fig. 72.3c which is the basic mechanism of trapping and wave-guiding light inside a medium according to solid critical angle (θ_c) of the medium which is calculated as [16]:

$$\theta_{\text{c}} = \sin^{-1} \frac{\eta_2}{\eta_1}. \tag{3}$$

In all aforementioned transmittance mechanisms (reflection, refraction, and TIR), the angle of the ray at the boundaries of the device is characterized by the refraction

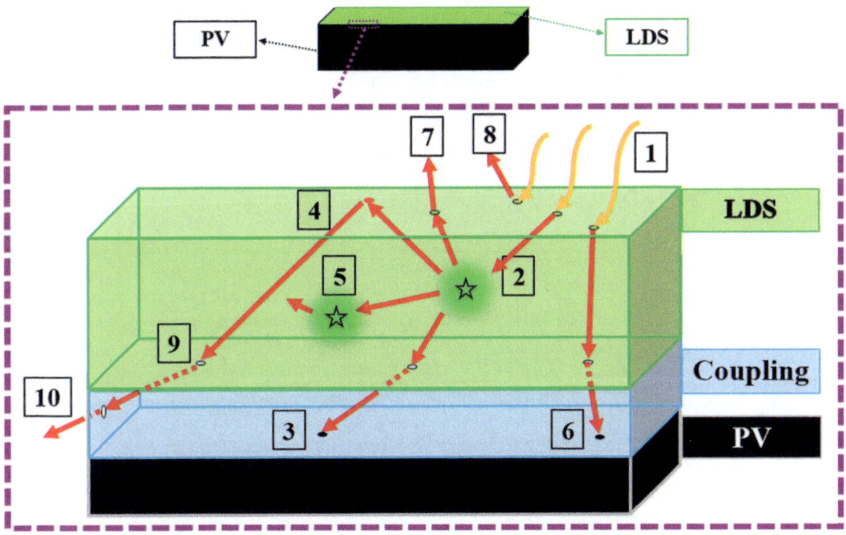

Fig. 72.2 Configuration of an LDS which shows: *1* photon enters the LDS absorbed by the luminescent material. Then, *2* reemitted at longer wavelength and *3* reaches the PV cell through the optical coupling *4* or it is wave-guided by TIR or *5* reabsorbed by another luminescent molecule. *6* Some photons reach the PV cell without red shifting. The other losses include: *7* edge losses and *8* front surface reflection. *9* A fraction of radiation striking the optical coupling, might be *10* deviated and lost due to refractive index differences of the optical coupling and luminescent layer. Note that the radiation may also be scattered or attenuated by the host or coupling material which are not shown here

indices of the inner and outer media. As a result, the PV solar cell (with $\eta \approx 3.5$) and luminescent layers (with $\eta \approx 1.5$) need to be coupled using a compatible index matching fluid [17]. Without using the appropriate index matching coupling material, the total fabrication cost might be decreased; however, an air gap may appear in the coupling between the luminescent device and PV solar cell. The refraction index of air (η_{air}) is $\approx$1; therefore, the radiation at the boundary of the device and the air gap deviates and scatters the light which increases the optical losses. Here, LPV devices have been modeled to investigate their optical losses and performance variations as a function of the optical coupling quality.

72.2 Monte Carlo Ray-Tracing Algorithm (MCRT)

The model used in this study has a statistical nature based on a Monte Carlo algorithm. It is a compatible and robust method to study the behavior of non-linear systems with dramatic uncertainty [18–21]. In MCRT, the inputs were interpreted by weighted probabilities and counting all probable events in the LPV devices such as reflection, refraction, absorption, emission, and TIR. Optical loss mechanisms such

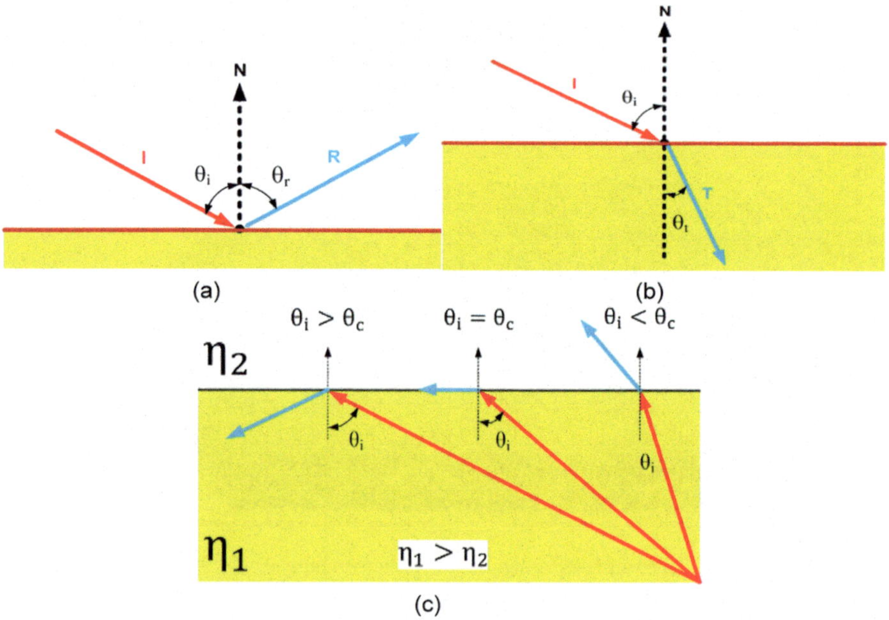

Fig. 72.3 (**a**) Reflection and (**b**) refraction of a ray striking a surface, and (**c**) trapping mechanism of ray inside a medium based on critical angle when the medium is denser than the outer environment ($\eta_1 > \eta_2$) [16]

as escape cone losses, scattering, reabsorption, and attenuation were also considered in the model. The probability of each event was determined according to parameters such as wavelength, angle of the incident photon, characteristics of the host and luminescent material, and dimensions and configuration of the LPV device.

The ray-tracing algorithm was applied based on a set of photon intersection calculation processes in an iterative loop to trace each incident photon. The process continued until the fate of each photon was detected and then the same algorithm was executed for the other incident photons. After applying the iterative loop to all incident photons, the final optical efficiency is calculated by:

$$\eta_{\text{opt}} = \frac{\int_{\lambda_{\min}}^{\lambda_{\max}} E_{P_{\text{OUT}}}(\lambda) \cdot \mathrm{d}(\lambda)}{\int_{\lambda_{\min}}^{\lambda_{\max}} E_{P_{\text{IN}}}(\lambda) \cdot \mathrm{d}(\lambda)} \tag{4}$$

where $E_{P_{\text{OUT}}}(\lambda)$ is the output energy spectrum at the PV cell plane and $E_{P_{\text{IN}}}(\lambda)$ is the energy spectrum of the input solar radiation. Having known the optical efficiency, the solar concentration ratio, C_{P}, can be obtained by:

$$C_{\text{P}} = \Gamma \times \eta_{\text{opt}} \tag{5}$$

where Γ is the geometric gain and it is known as the aperture surface ratio (A_{APR}) to the total PV cell area (A_{PV}) which can be calculated as follows:

$$\Gamma = \frac{A_{APR}}{A_{PV}} \tag{. (6)}$$

The MCRT was used to model a 60 × 60 × 3 mm LSC and 100 × 100 × 0.01 mm LDS with epoxy and PMMA host material, respectively. The specifications of the devices have been summarized in Table 72.1. AM1.5 global solar radiation (Fig. 72.4) was used to generate the incident photon flux for the modelling. Both LPV devices were doped with CdSe/ZnS QDs with quantum yield (QY) of 85%. The QDs emission and absorbance spectra are shown in Fig. 72.5. As it can be seen, the QD absorbed solar radiation below 500 nm and the absorbance peak was at 460 nm. The emission of the QD was a Gaussian spectrum in the region between 430 and 530 nm with a peak at around 485 nm which resulted in a Stokes shift of around 25 nm.

Table 72.1 Configuration of the modeled LSC and LDS devices

	LSC device	LDS thin film
Radiation type	AM1.5G	AM1.5G
Host material type	Epoxy	PMMA
Host material refraction index (η)	1.5	1.5
Length (mm)	60 mm	100 mm
Width (mm)	60 mm	100 mm
Thickness (mm)	3 mm	3 mm
Geometric gain	20	1
CdSe/ZnS QY	0.85	0.85

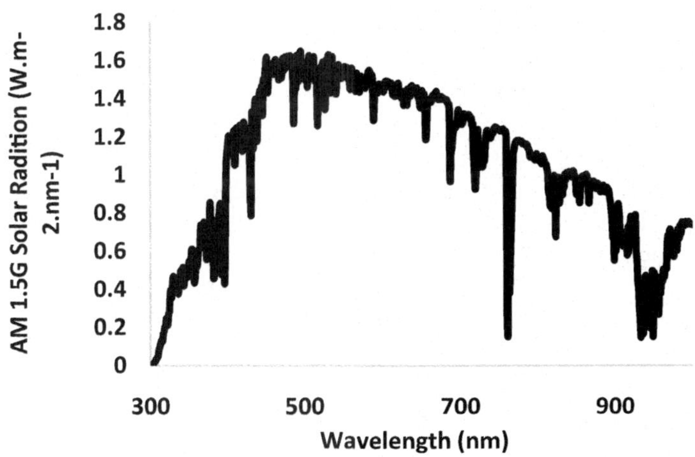

Fig. 72.4 Input AM1.5G solar radiation spectrum

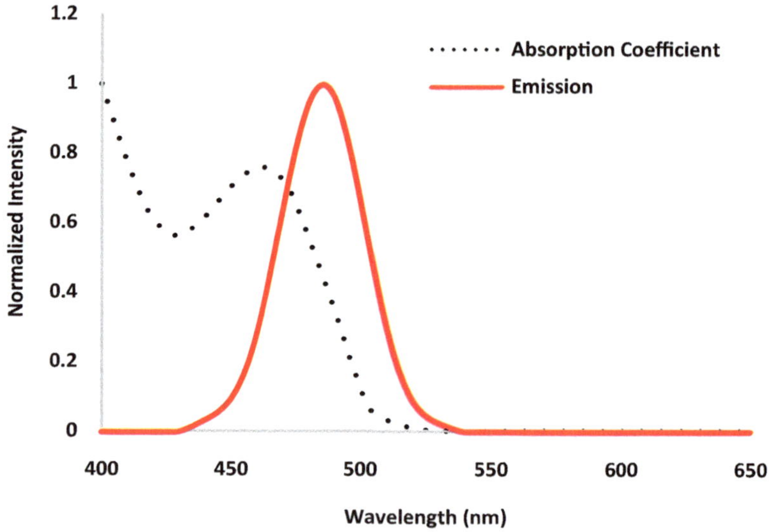

Fig. 72.5 Normalized absorption coefficient and emission spectra for CdSe/ZnS QD

72.3 Results and Discussion

Both LPV devices were modelled under similar circumstances and inputs. Therefore, the amount of top surface reflection was around 17% for both devices due to the close refraction indices of host materials ($\eta \approx 1.5$). The rest of the photons (around 82%) refracted into the layers. The LPV devices were modelled for different air gap thicknesses ($G = 0$–2 mm) between the luminescent layer and PV solar cell.

Figure 72.6a shows the optical efficiency over the air gap variations for the LDS device. The value of η_{opt} significantly decreased from around 82% to 76% when G increased by only 0.1 mm. Further increase in G from 0.1 to 2 mm, resulted in η_{opt} decreasing smoothly to around 66%.

The initial fall in the optical efficiency of the LSC with air gap thickness was significantly less steep for the LSC than the LDS device (Fig. 72.6b), under perfect optical coupling condition ($G = 0$ mm), η_{opt} was around 2.79% which decreased to 2.16% when G increased to 0.5 mm. Above $G = 0.5$ mm, η_{opt} decreased with less slope and reached 1.85 at $G = 2$ mm.

Table 72.2 presents the detailed statistical modelling results. As it can be seen for LDS, by increasing the air gap from 0 to 2 mm, 15.5% increase was found in the rate of the photons exiting the edges. Moreover, thermal losses increased by around 0.32%. Increase in the edge losses was due to the photon deviation in the air gap as a results of refraction index differences. The increase in thermal losses was because of the increase in photon path length resulting in higher attenuation losses inside the air gap. These losses decreased the rate of the photons reaching the bottom-mounted

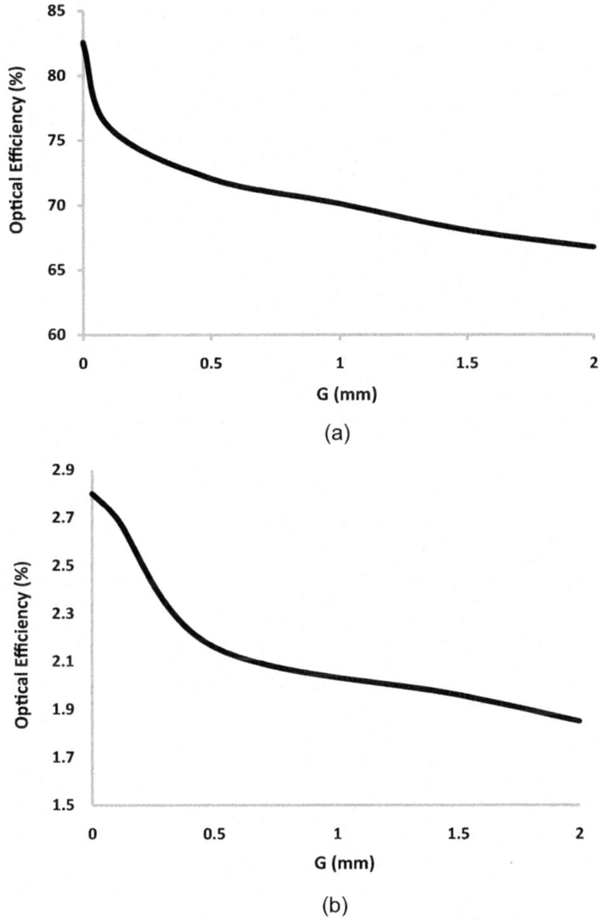

Fig. 72.6 Optical efficiency over air gap variations for the modeled: (**a**) LDS and (**b**) LSC

Table 72.2 Investigating the performance of the modeled LPV devices by changing the air gap thickness

LPV device	LDS		LSC	
G (mm)	0	2	0	2
Reflected (%)	17.42	17.43	17.54	17.53
Refracted (%)	82.58	82.57	82.46	82.47
Thermal loss (%)	0.02	0.34	4.24	4.51
Exited the edges (%)	0.0003	15.50	75.76	76.19
Reached PV (%)	82.55	66.73	2.46	1.77
η_{opt} (%)	82.56	66.74	2.79	1.85
C_p (%)	82.56	66.74	55.80	35.34
Δ_P (%) (between $G = 0$–2 mm)	−19.17%		−36.67%	

PV from around 82% to 66%. As a result, around 19% deduction (Δ_P) was observed in the performance of the device. For the modelled LSC, increasing the air gap resulted in around 0.27% increase in thermal and 0.43% increase in edge losses. Therefore, the rate of the photons reaching the PV cell at the side-edge decreased from 2.46 to 1.77. The total solar concentration ratio decreased from 56% to 36% which resulted in around 37% performance deduction.

72.4 Conclusion

A MCRT algorithm model was developed to model a 60 × 60 × 3 mm LSC and a 100 × 100 × 0.01 mm LDS devices and to investigate their optical coupling sensitivity and related performance. The air gap in both LPV devices was varied between 0 and 2 mm. Increasing the air gap resulted in increasing the attenuation losses and photon deviation through the edges of LPV devices. The performance of LSC decreased by increasing G from 0 to 0.5 mm; while, the performance of LDS was more significantly affected at low air gaps. However, over the full air-gap range studied the LSC loss more efficiency than that the LDS. Around 19% performance deduction was observed for the LDS when G increased from 0 to 2 mm, while the performance of the LSC reduced by around 37% under the same circumstance. This indicated that the LSC was more sensitive to the quality of optical coupling than LDS. The results can be used for further device design and optimization of the fabrication process.

Acknowledgments The authors would like to acknowledge the funding from the European Research Council grant (639760) entitled PEDAL: Plasmonic enhancement of advanced luminescent solar devices and funding from Science Foundation Ireland (SFI).

References

1. Chandra S, Doran J, Mccormack S, Kennedy M, Chatten A (2012) Enhanced quantum dot emission for luminescent solar concentrators using plasmonic interaction. Sol Energy Mater Sol Cells 98:385–390
2. Chatten AJ, Farrell DJ, Bose R, Dixon A, Poelking C, Gödel KC, Mazzer M, Barnham, KW (2011) Luminescent and geometric concentrators for building integrated photovoltaics. Photovoltaic Specialists Conference (PVSC), 2011 37th IEEE. IEEE, 000852-000857
3. Chandra S, Ahmed H, Rafiee M, Mccormack SJ (2017) Plasmonic quantum dot solar concentrator. SPIE Photonics USA, 2017
4. Ahmed H (2014) Materials characterization and plasmonic interaction in enhanced luminescent down-shifting layers for photovoltaic devices. PhD, Dublin Institute of Technology
5. Hovel H, Hodgson R, Woodall J (1979) The effect of fluorescent wavelength shifting on solar cell spectral response. Sol Energy Mater 2:19–29
6. Aste N, Adhikari R, Del Pero C (2011) Photovoltaic technology for renewable electricity production: towards net zero energy buildings. 2011 International Conference on Clean Electrical Power (ICCEP). IEEE, p 446–450

7. Debije MG, Verbunt PP (2012) Thirty years of luminescent solar concentrator research: solar energy for the built environment. Adv Energy Mater 2:12–35
8. Pagliaro M, Ciriminna R, Palmisano G (2010) BIPV: merging the photovoltaic with the construction industry. Prog Photovolt Res Appl 18:61–72
9. Ahmed H, Doran J, Mccormack S (2016) Increased short-circuit current density and external quantum efficiency of silicon and dye sensitised solar cells through plasmonic luminescent down-shifting layers. Sol Energy 126:146–155
10. Ahmed H, Mccormack S, Doran J (2016) External quantum efficiency improvement with luminescent downshifting layers: experimental and modelling. Int J Spectrosc 2016:8543475
11. AHMED H, MCCORMACK S, DORAN J (2017) Plasmonic luminescent down shifting layers for the enhancement of CdTe mini-modules performance. Sol Energy 141:242–248
12. Aste N, Tagliabue LC, Palladino P, Testa D (2015) Integration of a luminescent solar concentrator: effects on daylight, correlated color temperature, illuminance level and color rendering index. Sol Energy 114:174–182
13. Chandra S (2013) Approach to plasmonic luminescent solar concentration. PhD, Dublin Institute of Technology
14. Gajic M, Lisi F, Kirkwood N, Smith TA, Mulvaney P, Rosengarten G (2017) Circular luminescent solar concentrators. Sol Energy 150:30–37
15. Kocsis L, Herman P, Eke A (2006) The modified Beer–Lambert law revisited. Phys Med Biol 51:N91
16. Glassner AS (Ed.) (1989) An introduction to ray tracing. Elsevier.
17. Weber W, Lambe J (1976) Luminescent greenhouse collector for solar radiation. Appl Opt 15:2299–2300
18. Jacques SL, Wang L (1995) Monte Carlo modeling of light transport in tissues. In: Welch AJ, Van Gemert MJC (eds) Optical-thermal response of laser-irradiated tissue. Springer, Boston, MA
19. Joy DC (1995) Monte Carlo modeling for electron microscopy and microanalysis. Oxford University Press, New York
20. Şahin D, Ilan B (2013) Radiative transport theory for light propagation in luminescent media. JOSA A 30:813–820
21. Şahin D, Ilan B, Kelley DF (2011) Monte-Carlo simulations of light propagation in luminescent solar concentrators based on semiconductor nanoparticles. J Appl Phys 110:033108

Chapter 73
Titanium Oxide Film Deposition by Low-Power APS Equipment Using Air Working Gas and Rechargeable Lead Battery

Zine Elabidine Ettayebi, Yoshimasa Noda, Yasutaka Ando, and Mitsumasa Iino

73.1 Introduction

The number of people without access to electricity fell to 1.1 billion in 2016 from 1.7 billion in 2000, and it is on track to decline to 674 million by 2030. Despite widespread global progress, electrification efforts do not keep pace with population growth in sub-Saharan Africa, and those without electricity become increasingly concentrated in that region. By 2030, roughly 600 million of the 674 million people still without access are in sub-Saharan Africa, mostly in rural areas [1]. Despite the continuously declining cost of renewables, mainly PV solar panels, remote areas people and even electrical utilities in developing countries cannot afford them easily due to the considerable capital and maintenance costs required including grid extension. Moreover, some rural areas are hardly linkable to the national grid because of their remote location [2, 3] and are subsequently in need of off-grid power systems, yet fossil fuel is still preferred owing to its low price and usage ease [2–4] such as off-grid diesel generators which are not in favor of a sustainable development. Owing to the rapid exhaustion of fossil fuel energy and the increasing demand for clean energy technologies to reduce greenhouse gases emission, calls for studies of alternatives for powering non-electrified areas have been multiplying [5].

Renewable energy systems, such as mini-hydropower, small-scale wind turbines, bioenergy generators, and solar cells, have become important means for electricity supply within non- electrified rural areas especially in developing countries [2, 6].

Z. E. Ettayebi (✉)
Graduate School of Engineering, Ashikaga University, Tochigi, Japan

Y. Noda · Y. Ando · M. Iino
Department of Mechanical Engineering, Ashikaga University, Tochigi, Japan

A. Sayigh (ed.), *Renewable Energy and Sustainable Buildings*, Innovative Renewable Energy, https://doi.org/10.1007/978-3-030-18488-9_73

Crystalline silicon-based photovoltaic (PV) solar cells standout as a serious competitive of other renewables since widely used as a power source for domestic, commercial, and small- medium industrial demand. However, their manufacturing process is still questionable due its high cost—for example, the high energy required to run the electric arc furnace in order to reduce silicon dioxide—and complexity (which makes them irreparable) and also the involvement of environmentally harmful chemicals such as waste silicon tetrachloride and sulfur hexafluoride.

A promising solar cell commonly known as dye-sensitized solar cell (DSSC) has in contrast shown many assets and stands as an alternative for the expensive silicon-based photovoltaic devices; its composition is not complex and it can be fabricated and maintained even by the end users. The DSSC is composed of four elements, namely, the transparent conducting and counter conducting electrodes, the nanostructured wide bandgap semiconducting layer, the dye molecules (sensitizer), and the electrolyte [7]. In the old generations of photoelectrochemical solar cells (PSC) photoelectrodes were made from bulky semiconductor materials such as Si, gallium arsenide (GaAs), or cadmium sulfide (CdS). However, these kinds of photoelectrodes when exposed to light undergo photo-corrosion that results in poor stability of the photoelectrochemical cell. The use of sensitized wide bandgap semiconductors such as titanium dioxide (TiO_2) or zinc peroxide (ZnO_2) resulted in high chemical stability of the cell due to their resistance to photo-corrosion [7]. As a major component of the DSSC, titanium dioxide (TiO_2), known also as Titania, is indeed known for its photocatalytic properties and for being environmentally friendly [8]. In fact, TiO_2 is among the wide bandgap oxide semiconductors with band energy of 3.0–3.2 eV tolerating to absorb even ultraviolet light [9, 10]. Furthermore, titanium dioxide has attracted great attention in many fields other than nanostructured photovoltaics such as photocatalysts, environmental purification, electronic devices, gas sensors, and photoelectrodes [7].

As for photocatalytic films deposition processes, various techniques have been practically used such as solgel, chemical vapor deposition (CVD), and physical vapor deposition (PVD) including mainly sputtering and evaporation processes. However, many limitations are associated to those techniques such as high capital cost (due to the requirement of a vacuum equipment), reactor's volume capacity that restricts the processable substrate's size and difficulty of attaining large monolithic films due to capillary stresses related to solgel technique. On the other hand, and to overcome those challenging problems, atmospheric thermal plasma spray (APS) has been used on an industrial scale for TiO_2 film deposition. Yet, the conventional thermal spray equipment requires high capital and marginal costs since commercial working gases are required, such as argon, and high electrical input power is necessary. Therefore, it is difficult to use this equipment in developing countries. In this study, in order to fabricate a cost-effective DSSC for rural areas, a 2-kW APS equipment using air working gas was developed and photocatalytic TiO_2 film was deposited.

73.2 Experimental Procedure

73.2.1 Deposition of Photocatalytic TiO_2 Film by APS

Titania film was deposited on a 15 mm^2 × 1 mm^t 304 stainless steel substrate[1]. A grit blasting was necessary so to ensure a roughness and a cleanliness determining the film quality; otherwise, the film would not sufficiently adhere to the substrate and impurities would render the film characterization difficult. Figure 73.1 displays a schematic of the developed equipment. In seek of cutting commercial gases' relative costs, air was used as working gas and a battery-powered air plasma inverter—the battery was connected to a pure sine wave inverter inputting an AC power to the air plasma inverter—was utilized to generate air plasma. A feedstock powder consisting of TiO_2 anatase was used. As for the feeding system, and without using any carrier gas, the feedstock (powder) was injected owing to an enough suction power ensured by the plasma jet and the extension nozzle design. Indeed, this was accomplished due to the high plasma velocity when transferring inside the extension nozzle from ø4 to ø8 (Fig. 73.2) because of the plasma fluid conductance property (the plasma density decreases as electrons and other plasma particles turn less confined and their mean free path increases), thus the surrounding air is carried by the plasma and a negative pressure is created. The purpose was finding a compromise between an enough suction power and an enough feedstock flight time within the plasma so to get more activated (molten) feedstock particles at the extension nozzle outlet. Concentrically to the plasma jet axis, the substrate was set on a substrate holder. The deposition distance, which is the distance between the surface of

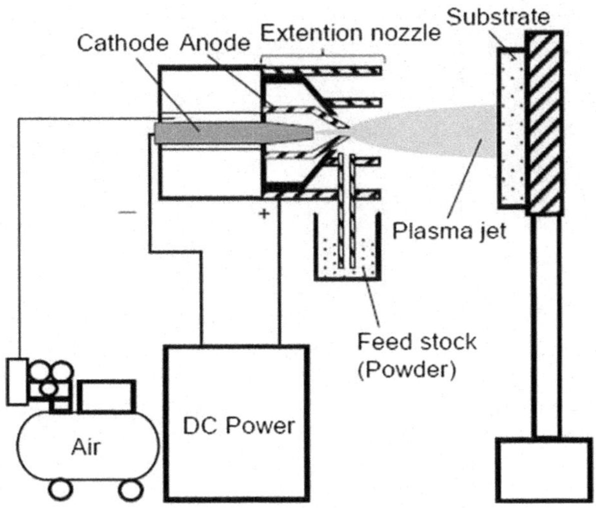

Fig. 73.1 Schematic representation of the developed battery-powered APS equipment

[1] The substrate geometrical characteristics:
- surface area of the substrate: 15mm^2
- thickness of the substrate: 1mm

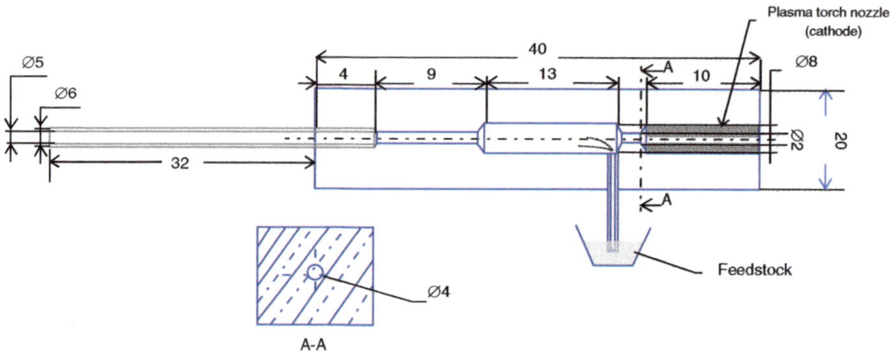

Fig. 73.2 Designed brass plasma torch mounted extension nozzle (the cooling system is not represented)

Table 73.1 Spray conditions

Input current	20 A
Input voltage	100 V
Working gas	Air
Feedstock powder	Anatase (0.2–0.3 μm)
Working gas flow rate	26 L/min
Spray distance	15/40/50 mm
Spray time	15 s

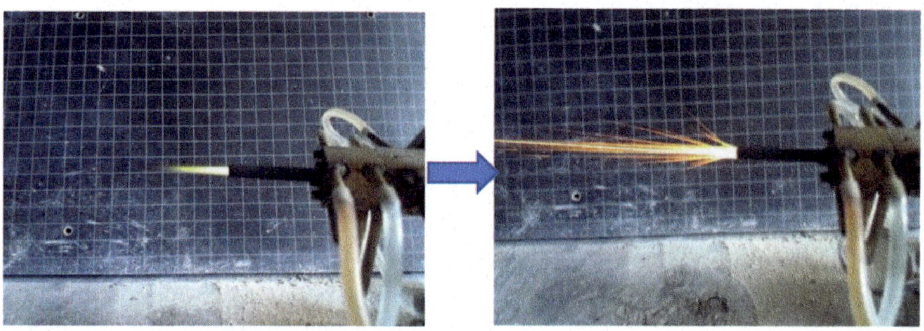

Fig. 73.3 Arc discharge plasma shape and sprayed feedstock

the substrate and the nozzle outlet, was varied from 15 to 50 mm. The deposition time was set at 15 s and the discharge power at 100 V/20 A. In order to maintain the intrinsic electrical conductivity of the extension nozzle's material (brass) once the temperature dramatically increases during the plasma discharge, and also to avoid any melting of the nozzle's inner walls, a nozzle water cooling system was set too. Table 73.1 shows the spray conditions of TiO_2 films deposited by the developed APS equipment. Figure 73.3 shows the plasma jet shape with and without feedstock injection.

73.2.2 Characterization of the Deposited Photocatalytic TiO_2 Films

73.2.2.1 Appearance, X-Ray Diffraction (XRD) and Optical Micrographs of the Cross-Sections of the Samples

Figures 73.4 and 73.5 show the appearance and X-ray diffraction (XRD) patterns of the samples upon titanium oxide films deposition at 15, 40, and 50 mm. At each deposition distance, black-colored films could be deposited and strong crystalline rutile phase peaks were obtained in addition to anatase. However, the anatase peaks of the film deposited at 40 mm were less intense than those deposited at 15 and 50 mm. Table 73.2 roughly presents the crystallinity (anatase-rutile) and crystalline

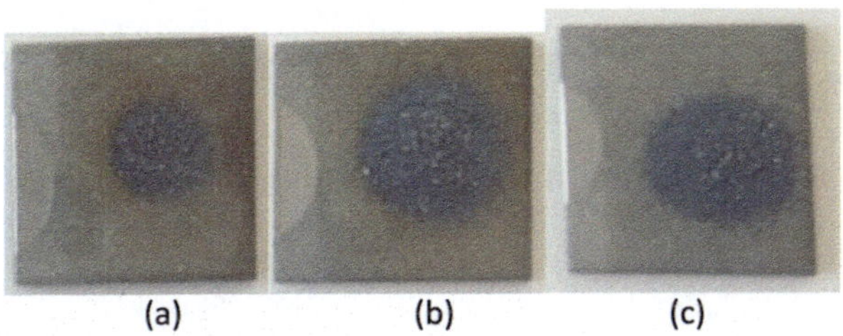

Fig. 73.4 Appearance of the TiO_2-deposited films on stainless steel substrates. (**a**) $d = 15$, (**b**) $d = 40$, and (**c**) $d = 50$ mm

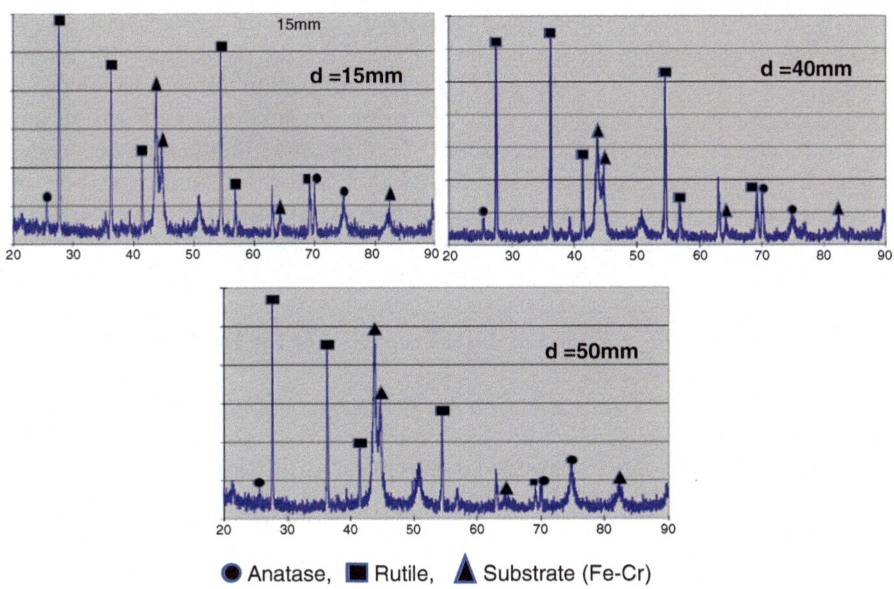

Fig. 73.5 X-ray diffraction patterns of the samples. ● Anatase, ■ rutile, ▲ substrate (Fe-Cr)

Table 73.2 Crystallinity and anatase rates of the deposited Titania films

	Deposition distance (mm)		
	15	40	50
Crystallinity (%)	79.022	82.26	72.09
Anatase (%)	12.74	10.57	11.36

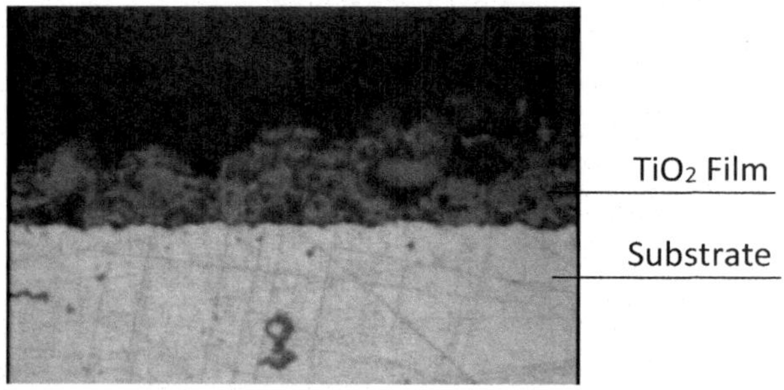

Fig. 73.6 Optical micrograph of the deposited Titania film on the condition of 40 mm as deposition distance

anatase ratios for each deposited film; the films deposited at 15 and 50 mm have the highest anatase ratio values; therefore, higher photocatalytic properties should be exhibited on those two conditions (see Sect. 2.2.2). The film deposited on the condition of 40 mm is more crystalline, however richer in rutile. As for the eventual reason behind that result, the length of the plasma flame is higher than 40 mm but lower than 50 mm, which implies the high rate of crystalline rutile phase on the condition of 40 mm as deposition distance since the feedstock particles have enough flight time to melt within the plasma jet. Furthermore, the deposition efficiency is thought to be lower on the condition of 50 mm as deposition distance which can be evidenced by the narrow values of anatase/rutile ratios between 15 and 50 mm. As for the cross-sections of the deposited TiO_2 films, the film thickness decreased when increasing the deposition distance. Also, the films' porosities were remarkable. Figure 73.6 shows the cross-section pattern of the TiO_2 film deposited on the condition of 40 mm.

73.2.2.2 Assessment of Photocatalytic Properties of the Deposited Films

In order to quantitatively evaluate the photocatalytic properties of the TiO_2-deposited films, a methylene blue solution discoloration test was carried out. As a first step, 0.15 mL of methylene blue was added to 300 mL of distilled water, then using the instrument shown on Fig. 73.7 the light absorption was measured for different concentrations of the prepared solution as depicted on Table 73.3. Therefore, an empirical relation between the concentration (Y) and the light absorption value (X) could be obtained as displayed on Fig. 73.8. A polynomial approximation was utilized for the function Y: $Y = 0.0098X^2 - 2.3527X + 142.38$.

After that, each sample was immersed into a glass container within 30 mL of the prepared solution (100% of concentration) and kept under UV irradiation for 12 h.

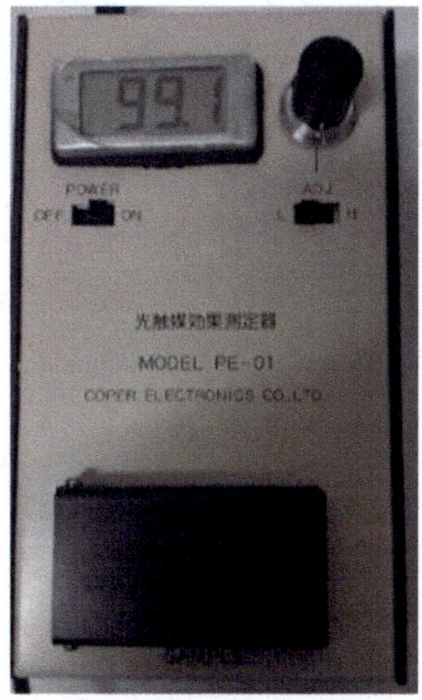

Fig. 73.7 Light absorption meter

Table 73.3 Measured light absorption values for different concentrations of methylene blue solution

Solution_concentration (%)	Light_absorption (*X*)
100	20.2
80	30.5
40	53.8
20	79
10	93.7
0	120

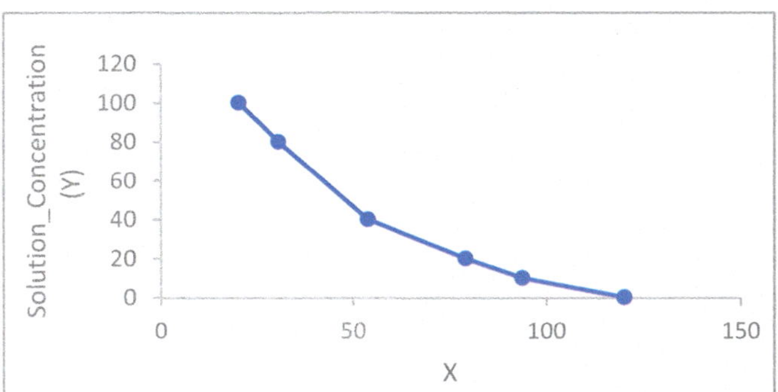

Fig. 73.8 Polynomial interpolation of the solution_concentration (*Y*) function

For each sample, the light absorption value (*X*) was recorded as well as the remaining volume of the solution after 12 h of UV Irradiation. The final concentrations of methylene blue solution for each sample are presented on Table 73.5. Those concentrations were calculated the following relation: Remaining_volume/Initial_volume × *Y*.

As an illustrating example, the final concentration of methylene blue solution where the film deposited at 15 mm was immersed is as follows: $YS_{15} = (13.8/30) \times f(45) = (13.8/30) \times (0.0098 \times 45^2 - 2.3527 \times 45 + 142.38) = 56.3535$.

Table 73.4 Light absorption values and remaining volumes of the methylene blue solutions after 12 h of UV irradiation

Deposition_distance (mm)	Light_absorption (*X*)	Initial_volume (mL)	Remaining_volume (mL)
15	45	30	13.8
40	13.5	30	16.8
50	27.2	30	13

Table 73.5 Final concentrations of methylene blue solutions calculated based on Table.73.4 and the function *Y*

Deposition_distance (mm)	Final_concentration (Y')
15	56.3535
40	112.4046
50	85.636992

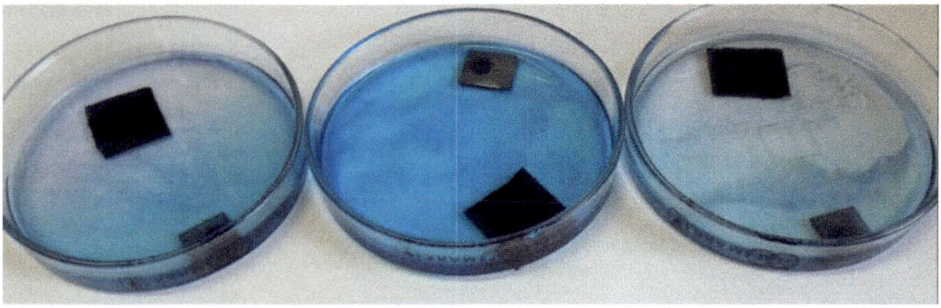

Fig. 73.9 Samples appearance after 12 h of UV irradiation. The corresponding deposition distances are 15, 40, and 50 mm from right to left

In the above case, the solution started to decolor after 6 h of UV irradiation (light_absorption was 27 for a volume decrease of 8.6 mL). Finally, the films deposited on the condition of 15 and 50 mm, and particularly the former, have shown more photocatalytic properties (Table 73.5) as it was expected in Sect. 2.2.1 (see also Fig. 73.9).

73.3 Application to Dye-Sensitized Solar Cells (DSSCs)

Dye-sensitized solar cells were assembled as shown on Fig. 73.10 and their outputs were measured after being irradiated via a solar simulator (1 kW/m^2): open-circuit voltages (OCV) and short-circuit currents (ISC) (Table 73.6). It is important to note that the temperature on the upper plate surface of each DSSC was around 34 °C since the light source was not LED type. Another parameter to take into consideration is the film-coated surface area which was not the same for all samples; the ISC values are thought to be influenced by the aforementioned factor. The lowest ISC

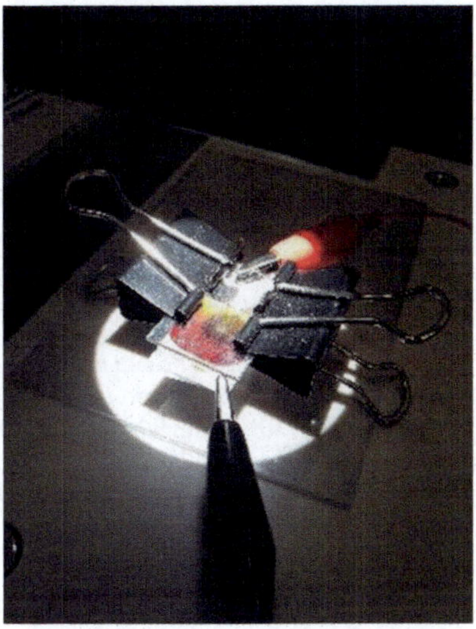

Fig. 73.10 Assembled DSSC

Table 73.6 OCV and ISC of the assembled DSSCs

Deposition_distance (mm)	OCV (V)	ISC (mA)
15	0.8	−0.615
40	0.62	−0.985
50	0.65	−0.992

value corresponds to $d = 15$ mm for which the deposited Titania film surface area is the smallest. The OCV values could be influenced too but the trend was as expected: the highest OCV values are for $d = 15$ mm and $d = 50$ mm.

73.4 Conclusion

Upon the successful deposition of photocatalytic TiO_2 films by 2 kW class APS equipment using air as working gas, the outcome of this study can be summarized as follows:

1. Photocatalytic TiO_2 films could be deposited using an isolated 2 kW APS equipment utilizing air as working gas.
2. Crystalline TiO_2 film could be deposited on all conditions. However, on the condition of 40 mm as deposition distance, the film was more crystalline yet relatively poorer in anatase.
3. The films deposited at $d = 15$ mm and $d = 50$ mm exhibited more photocatalytic properties.
4. The fabricated DSSCs could generate an open-circuit voltage and a theoretical power up to 0.8 V and 645 μW, respectively.

References

1. Daly H, Walton MA (2017) World Energy Outlook Special Report. p 11–12
2. Karakaya E, Sriwannawit P (2015) Barriers to the adoption of photovoltaic systems: the state of the art. Renew Sust Energ Rev 49:60–66
3. Ahlborg H, Hammar L (2014) Drivers and barriers to rural electrification in Tanzania and Mozambique—grid-extension, off-grid, and renewable energy technologies. Renew Energy 61:117–124
4. Palit D (2013) Solar energy programs for rural electrification: experiences and lessons from South Asia. Energy Sustain Dev 17:270–279
5. Ondraczek J (2013) The sun rises in the east (of Africa): a comparison of the development and status of solar energy markets in Kenya and Tanzania. Energy Policy 56:407–417
6. Klunne WJ (2013) Small hydropower in Southern Africa—an overview of five countries in the region. J Energy South Afr 24:14
7. Murni S, Whale J, Urmee T, Davis J, Harries D (2012) The role of micro hydro power systems in remote rural electrification: a case study in The Bawan Valley, Borneo. Procedia Eng 49:189
8. Jasim KE (2011) Dye Sensitized Solar Cells - Working Principles, Challenges and Opportunities, Solar Cells - Dye-Sensitized Devices, Prof. Leonid A. Kosyachenko (Ed.), ISBN: 978-953-307-735-2, InTech, Available from: http://www.intechopen.com/books/solar-cells-dye-sensitized-devices/dye-sensitized-solar-cellsworking-principles-challenges-and-opportunities
9. Mahshid S, Askari M, Ghamsari M, Afshar N, Lahuti S (2009) Mixed-phase TiO_2 nanoparticles preparation using sol–gel method. J Alloy Compd 478:586–589
10. Wang C, Ying J (1999) Sol−gel synthesis and hydrothermal processing of anatase and rutile titania nanocrystals. Chem Mater 11:3113–3120

Chapter 74
Economic and Ecologic Assessment of a Biomass-Based CHP System for a Hotel Resort on the Andaman Islands, India

Moritz Wegener, Yan Zhang, Antonio Isalgue, and Anders Malmquist

74.1 Introduction

Most small islands and island nations rely heavily on fossil fuels to supply their energy demands, which have to be transported to the islands by ships [1]. This implies higher energy costs and less stable fuel supply, while greenhouse gas (GHG) emissions are further increased too. In consequence, the socio-economic development of islands is hindered due to low energy autonomy [2]. Additionally, islands are particularly threatened by rising sea levels because of climate change [3]. Island inhabitants and their administrations are therefore especially motivated to switch from fossil fuel to renewable energy systems based on local resources.

Off-grid, biomass-based combined heat and power (CHP) systems have been identified as a stable, technically feasible and economically viable alternative to conventional energy systems to raise energy autonomy [4]. Kanematsu et al. [5] investigated the long-term sustainability of woody biomass-driven CHP plants on Tanegashima, Japan. They concluded that substantial emission reductions are pos-

M. Wegener (✉)
Architecture and Energy, School of Architecture of Barcelona, Barcelona, Spain

Heat and Power Technology, Department of Energy Technology, KTH Royal Institute of Technology, Stockholm, Sweden
e-mail: mbgw@kth.se

Y. Zhang
Division of Energy Processes, School of Chemical Science and Engineering, KTH-Royal Institute of Technology, Stockholm, Sweden

A. Isalgue
Architecture and Energy, School of Architecture of Barcelona, Barcelona, Spain

A. Malmquist
Heat and Power Technology, Department of Energy Technology, KTH Royal Institute of Technology, Stockholm, Sweden

A. Sayigh (ed.), *Renewable Energy and Sustainable Buildings*, Innovative Renewable Energy, https://doi.org/10.1007/978-3-030-18488-9_74

sible without endangering the harmony of the forest management. Mertzis et al. [6] demonstrated with the sMARt-CHP project that small-scale gasification in combination with internal combustion engines (ICEs) is a flexible alternative to fossil-fuelled systems. A mobile CHP unit was tested in different regions fuelled with different biomass feeds. Most losses occurred during the gasification process, but also potential for optimization of the calibration of the engine fuelled with syngas was identified.

In the tropical regions of Southeast Asia, many countries have recognized the potential of using biomass and are politically supporting the advancement of such off-grid CHP systems [7]. One leading example is India, which plentifully subsidises biomass-based systems, promoting amongst others gasification technology [8].

In this study, an innovative biomass-based CHP system for a hotel resort on the Andaman Islands, India, is presented. In the first step, the conventional diesel generator is substituted by a gas engine fuelled with syngas from a biomass gasifier. The gasifier transforms locally available coconut shells into syngas. In the second step, the system is augmented by heat exchangers to seize the waste heat and by batteries to allow for smaller nominal capacity of the engine. By comparing the systems to the conventional diesel-based reference system, their techno-economic benefits and drawbacks are identified. Additionally, life-cycle-assessment (LCA) reference values are used to determine the ecologic impact of the different systems.

74.2 Methodology

One base-case and two different biomass-based systems have been modelled using the Hybrid Optimization of Multiple Energy Resources (HOMER) software [9]. HOMER allows for a techno-economic optimization of such systems by altering the component sizes within a given set of requirements and/or limitations. For these simulations, HOMER requires different types of input data, which are listed here together with their sources for the hotel resort [10]:

1. The **load profile** and the **technical data** determine the energy demand within the technical possibilities of running such a small-scale CCHP system. The demand data is based on electricity and heat consumption of previous years in the hotel resort. Hourly, daily and monthly fluctuations are considered.
2. The **search space** for sizing the components has been iteratively narrowed down for economic optimization.
3. The **economic data** and **equipment characteristics** of the components are based on values from literature and local suppliers.
4. The **meteorological data** provided by HOMER is based on the surface meteorology and solar energy data base by NASA [11].

HOMER simulates each system from the search space and ranks feasible solutions according to their net present cost (NPC). For calculating the NPC, the following formulae are used:

$$C_{\mathrm{NPC}} = \frac{C_{\mathrm{ann,tot}}}{\mathrm{CRF}\left(i, R_{\mathrm{proj}}\right)} \quad (74.1)$$

$$\mathrm{CRF}\left(i, R_{\mathrm{proj}}\right) = \frac{i(1+i)^R}{(1+i)^R - 1} \quad (74.2)$$

where $C_{\mathrm{ann,\ tot}}$ is the total annualized costs of the entire system, CRF is the capital recovery factor with the interest rate i and R_{proj} is the project lifetime.

In the first step, the costs of each system configuration are calculated based on commercial and scientific values. In the second step, the capital investment costs are lowered by considering government subsidies for such systems [12, 13]. Apart from the economic performance, HOMER also shows energetic performance parameters like fuel consumption, engine efficiency or excess electricity. By altering the cost values of the most expensive assets, a sensitivity analysis is conducted to test the economic robustness of each system. Additionally, by using LCA reference values the ecologic impact of the systems is estimated.

74.3 Case and Component Description

The electricity demand of the hotel resort on Neil Island, India, during the whole year is presented in Fig. 74.1. During peak season from October to March, an average hotel occupancy of 70% is assumed, while during off-season from April to September an average hotel occupancy of 30% is assumed. This leads to peak values in December of up to 152 kW. Often during the daily peak time from 18.00 to 24.00 (when many tourists return to their hotels), the grid is overloaded and breaks down, so that currently a diesel generator is employed for these periods.

For all the economic and technical values chosen, the most conservative estimates of the commercial and scientific sources have been selected. To account for extra shipment and design, 15% extra costs have been added to the source value. The interest rate i has been set to 6% and the project lifetime R_{proj} to 20 years. The possible electricity demand savings due to substituting the electric heaters with water tanks heated by exhaust heat from the engine have been calculated to be 5%. The economic and technical key characteristics of the components are shown in Table 74.1.

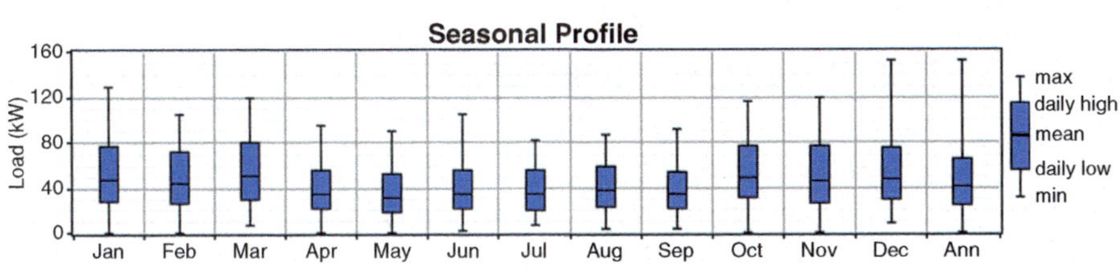

Fig. 74.1 Electricity demand of the hotel resort during the whole year

Table 74.1 Economic and technical key input data of the components

Component	Capital and replacement cost	O&M cost	Technical key characteristics
Diesel engine system	600 USD/kW and 600 USD/kW [14]	0.02 USD/kWh [14]	Lifetime: 20,000 h Minimum load: 40% [15] Maximum elec. efficiency: 38% [15] Emission data from HOMER Diesel price: 1 USD/L [16]
Batteries	1,500 USD/kWh and 1,500 USD/kWh [17]	30USD/year/battery [17]	Model: Surrette 4-KS-25PS 1,900 Ah, 4V deep cycle battery (lead-acid) [17] Minimum battery lifetime: 7 years Lifetime throughput: 10,973 kWh 6 batteries per string (24 V string)
Gasifier and gas engine system	3,000 USD/kW and 3,000 USD/kW	0.03 USD/kWh [18]	Lifetime of combo: 15,000 h [19] Gasifier efficiency: 75% [20] Engine max. elec. efficiency: 24% [21] Syngas price: 0.05 USD/Nm^3 [22] Pre- and post-treatment components included
Converter	750 USD/kW and 750 USD/kW [14]	10 USD/year/kW [23]	Lifetime: 20 years, Inverter efficiency: 90%, Rectifier efficiency 90%
Grid	–	0.42 USD/kWh	Available from 00.00 to 18.00

The base case represents a simple substitution of the diesel generator with 160 kW peak capacity. This capacity is necessary to supply enough electricity during peak demand times. In the first biomass case, the diesel engine is substituted by a gas engine fuelled with syngas from a downdraft gasifier. Any additional components for biomass pre-treatment and syngas post-treatment (e.g. biomass crusher, cyclone, filter) are included in the gasifier and gas engine calculation. The gasifier is fed mainly with coconut shells from surrounding farms. In the second biomass case, the system is enhanced with batteries to allow for load shifting and to operate without grid connection. The system configurations as implemented in HOMER are shown in Fig. 74.2.

74.4 Results

74.4.1 Techno-Economic Performance

The optimal sizes of the components are shown in Table 74.2. In the base case, the diesel engine provides 41% of the electricity needed. There is a decrease in relative electricity generation from the diesel generator to the syngas engine in the on-grid

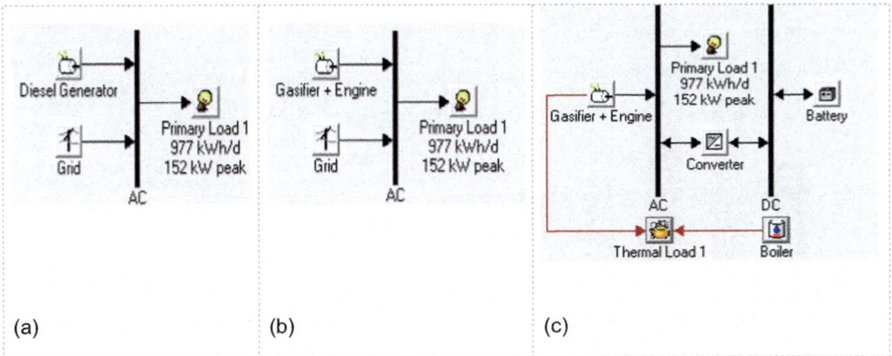

Fig. 74.2 HOMER outlay of the base case and biomass cases: (**a**) a grid-connected, diesel-based system; (**b**) a grid-connected, biomass-based system and (**c**) an off-grid, biomass-based CHP system

Table 74.2 Economically optimized component sizes with absolute and relative electricity generation

Component	Base case	On-grid biomass case	Off-grid biomass case
Grid	233.7 MWh/year 59%	246.6 MWh/year 66%	–
Diesel engine	160 kW 162.3 MWh/year 41%	–	–
Gasifier + gas engine	–	160 kW 125.6 MWh/year 34%	56 kW 367.1 MWh/year 100%
Batteries	–	–	593 kWh (nominal capacity)
Converter	–		36 kW

biomass case, as the biofuel price is higher; hence, it is more cost-effective to use grid electricity. For the battery-assisted CHP case, the gasifier and the gas engine can be downsized substantially to 56 kW while still supplying 100% of the electricity needed. However, this implicates 593 kWh of battery capacity, which would require a battery storage container with adequate spacing and system control.

The average electric efficiency in case 2 is significantly lower with 10.8% compared to 25.8% in case 3. This is due to the lower efficiency in lower partial load operation, which occurs more often in case 2. According to Sinal et al. [22], the total biomass energy available on the island amounts to 2,273 MWh/year. Assuming a gasifier efficiency of 75%, this would mean 1,704 MWh/year of syngas would be available for the engine. Based on the electric efficiencies, a yearly syngas consumption of 1,163 MWh and 1,423 MWh for cases 2 and 3, respectively, can be calculated. Therefore, both systems would not surpass the limit of sustainable biomass consumption on the island.

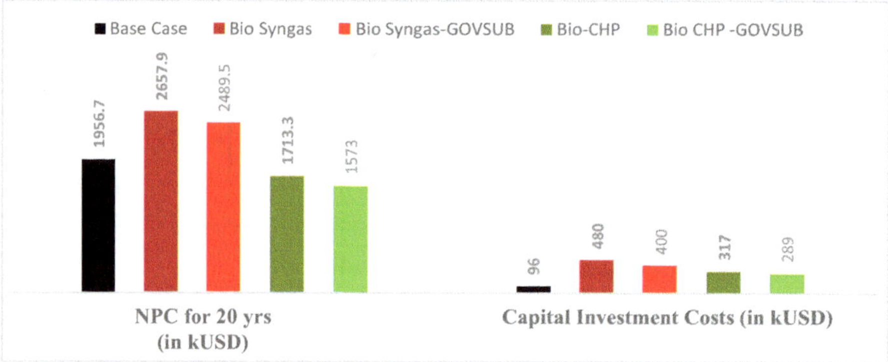

Fig. 74.3 NPC for 20 years and capital investment costs of all cases

Table 74.3 LCA emission reference values

Energy source or storage	Oil	Biomass (gasified)	Battery
Life cycle GHG emissions (g CO_2-eq/kWh)	720[a] [24]	45[a] [24]	200,000[b] [25]

[a]g CO_2-eq per generated unit of electricity ($kWh_{generated}$)
[b]g CO_2-eq per unit of electric storage capacity ($kWh_{storage}$)

According to the optimal system configurations, the NPC and the capital investment costs are shown in Fig. 74.3. The NPC of the first biomass case is worse than the NPC of the base case, even when government subsidies are included. The hotel owner would lose at least 700 kUSD over 20 years. However, by including battery and CHP technology, the second biomass case can lead to substantial savings of more than 200 kUSD and nearly 400 kUSD when including government subsidies. This would mean that the hotel owner could lower his energy expenses by more than 20%. Considering the capital investment costs, the base case outperforms every other case with only 90 kUSD, as it is easy and cheap to replace the diesel generator. The capital investment costs of an off-grid biomass case could be pushed down to 289 kUSD with government subsidies.

74.4.2 LCA Analysis and Environmental Impact

Based on the LCA emission reference values in Table 74.3, the LCA emissions of all three cases have been calculated and are presented in Fig. 74.4. In the base case, more than 285 tonnes CO_2-eq/year are emitted due to emissions from the grid electricity or from the diesel engine employed. In case 2, the emissions are already lowered by more than 100 tonnes CO_2-eq/year by substituting the diesel engine in the hotel resort. In the battery-assisted biomass CHP case, only 26.4 tonnes CO_2-eq/year are emitted, which means a CO_2-eq emission reduction of more than 90% compared to the base case.

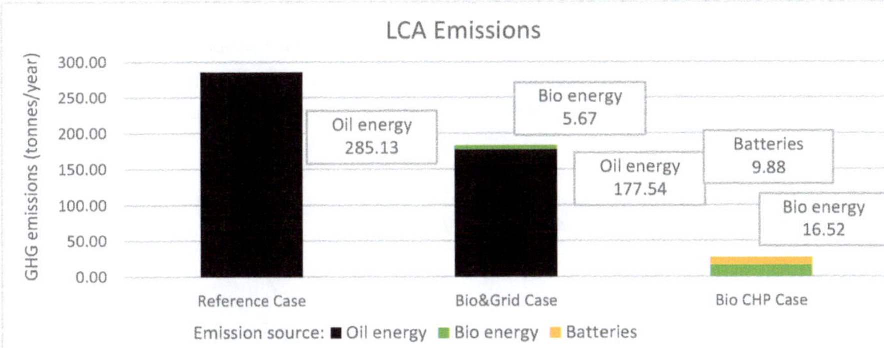

Fig. 74.4 LCA emissions of all three cases

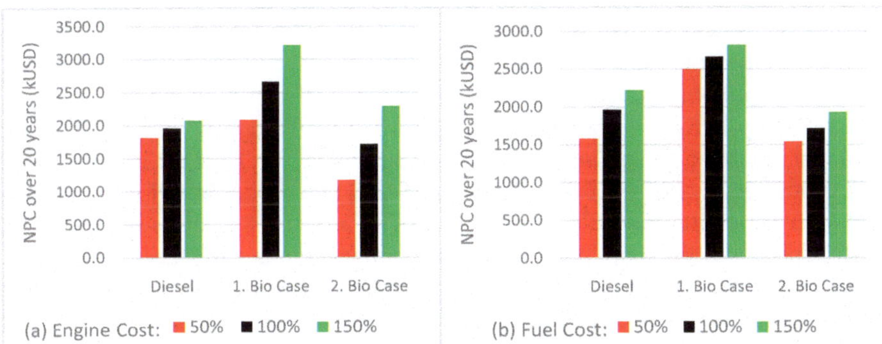

Fig. 74.5 Sensitivity analysis: (**a**) Effect of different engine costs on NPC and (**b**) effect of different fuel costs on NPC

74.4.3 Sensitivity Analysis

The generator and the fuel costs constitute the major costs of the on-grid biomass CHP system. Hence, to test the economic robustness of the system, these values were altered and the effects on the economic system performance of all cases observed. The results are presented in Fig. 74.5. The engine costs only marginally affect the NPC for the base case, but can nearly double the NPC for case 3 when comparing 50% engine cost to 150% engine cost. This is mostly due to the high capital and replacement cost of the gasifier and gas engine system compared to the cheap diesel generator. It should also be noted that with cheaper engine cost, the NPC of the biomass CHP system could drop down to 1167 kUSD. Different fuel costs can raise or reduce the costs of all systems substantially. The NPC of the biomass CHP system would increase by 25% with an increase from 50 to 150% fuel price, while the diesel system would increase by 40% equivalently. The biomass CHP system is therefore more robust. It is remarkable that the first biomass case with a simple substitution of the diesel engine would be outperformed by the base case in all scenarios. It should be noted that oil production and price are outside of the power of the island inhabitants, while biomass production and prices can be determined locally and hence can be kept more stable.

74.5 Discussion and Conclusion

A case study on the benefits and drawbacks of substituting a conventional fossil-fuel-based energy system of a hotel resort on the Andaman Islands, India, with a biomass-based system has been conducted. The calculations are based on recent demand data of the hotel resort, commercial info and scientific references. By simply replacing the diesel engine with a gasifier in connection with a gas engine, the NPC would rise, mostly due to an immense increase in capital investment and fuel costs. However, when smart technologies like CHP and load shifting trough batteries are applied, the peak capacity of the prime mover could be reduced substantially and hence the capital investment costs could be lowered. The NPC of a smart biomass-based CHP system would be lower than that of a conventional system, while at the same time enabling energy independence for the hotel resort from the instable island grid. Additionally, the amount of GHG emissions could be lowered by more than 90% over the entire life cycle of the system. The sensitivity analysis highlights the great dependence on cost-effective gasification technologies for the biomass CHP system to be feasible. It also underlines the greater robustness of a biomass-based system towards fuel price changes compared to a diesel-based system.

The results show that to replace conventional energy system, smart measures are needed to achieve economic superiority. Additionally, initial funding of such a system requires high capital investments, but the costs would be mitigated over the project lifetime. Moreover, the proposed biomass CHP system would enable the hotel resort owner to supply his energy demand independently while reducing his GHG emissions substantially, which could be used for marketing purposes.

The biomass needed for the system could be obtained by local farmers, who in return could receive charcoal as fertilizer. This way, a biomass cycle could be established. Long-term contracts could lead to low and stable biomass costs for the operator as well as an additional source of income for the farmers. In total, the system would improve the socio-economic development of the entire island society. The biomass availability on the island can easily support one biomass CHP system, but with more such systems, sustainable biomass production would be endangered.

Investigations on how to further use exhaust heat (e.g. for air conditioning and/or water purification) and how to seize the solar potential of the island are currently being conducted. A pilot system similar to the system proposed in this study will be installed in India in the near future. The investigated biomass CHP systems could be scaled up and installed on other biomass-rich islands and rural regions; hence, they offer vast entrepreneurial potential. Further development and experience with gasification and gas engine technology are essential for economic and technological feasibility of such systems.

Acknowledgements This research has been conducted in collaboration between UPC (Universitat Politècnica de Catalunya) and KTH Royal Institute of Technology, funded through Erasmus Mundus Joint Doctoral Programme SELECT+, the support of which is gratefully acknowledged. The project STandUP for Energy provided resources for supervision of this study.

A.I. acknowledges Spanish project MOET_BIA2016-77675-R.

The case study has been supported by the industrial partner Pamoja Cleantech AB.

References

1. Blechinger P, Cader C, Bertheau P, Huyskens H, Seguin R, Breyer C (2016) Global analysis of the techno-economic potential of renewable energy hybrid systems on small islands. Energy Policy 98:674–687. https://doi.org/10.1016/J.ENPOL.2016.03.043
2. Liu J, Mei C, Wang H, Shao W, Xiang C (2017) Powering an island system by renewable energy—a feasibility analysis in the Maldives. Appl Energy 227:18–27. https://doi.org/10.1016/J.APENERGY.2017.10.019
3. Pachauri RK, Meyer L, Van Ypersele J-P, Brinkman S, Van Kesteren L, Leprince-Ringuet N et al (2014) IPCC, 2014: climate change 2014: synthesis report. Contribution of working groups I, II and III to the fifth assessment report of the Intergovernmental Panel on Climate Change. IPCC, Geneva
4. Obernberger I, Carlsen H (2003) State-of-the-art and future developments regarding small-scale biomass CHP systems with a special focus on ORC and Stirling engine technologies. Internaional Nordic Bioenergy, Jyvaskyla
5. Kanematsu Y, Oosawa K, Okubo T, Kikuchi Y (2017) Designing the scale of a woody biomass CHP considering local forestry reformation: a case study of Tanegashima, Japan. Appl Energy 198:160–172. https://doi.org/10.1016/J.APENERGY.2017.04.021
6. Mertzis D, Mitsakis P, Tsiakmakis S, Manara P, Zabaniotou A, Samaras Z (2014) Performance analysis of a small-scale combined heat and power system using agricultural biomass residues: the SMARt-CHP demonstration project. Energy 64:367–374. https://doi.org/10.1016/J.ENERGY.2013.11.055
7. Ali Ahmed A, Elizondo Azuela G, Bazilian M, Bertheau P, Cader C, Kempener R, et al. (2015) IRENA – off grid renewable energy systems: status and methodological issue 2015:36. http://www.irena.org/DocumentDownloads/Publications/IRENA_Off-grid_Renewable_Systems_WP_2015.pdf. Accessed 1 Mar 2019
8. Kumar A, Kumar N, Baredar P, Shukla A (2015) A review on biomass energy resources, potential, conversion and policy in India. Renew Sust Energ Rev 45:530–539. https://doi.org/10.1016/j.rser.2015.02.007
9. Homer Energy. HOMER—Hybrid Renewable and Distributed Generation System Design Software 2016. https://www.homerenergy.com/support/docs/3.10/glossary.html. Accessed 7 Mar 2018
10. Bahramara S, Moghaddam MP, Haghifam MR (2016) Optimal planning of hybrid renewable energy systems using HOMER: a review. Renew Sust Energ Rev 62:609–620. https://doi.org/10.1016/j.rser.2016.05.039
11. Surface Meteorology and Solar Energy (n.d.) https://eosweb.larc.nasa.gov/sse/. Accessed 7 Mar 2018
12. Arora J, Khirwar S (2015) 24 × 7 power for all. Gov India. 55 pages
13. Khare DK (2010) Programme on "Biomass Gasifier for industries": F. No. 202/4/2010-BM: 22
14. Ghaem Sigarchian S, Paleta R, Malmquist A, Pina A (2015) Feasibility study of using a biogas engine as backup in a decentralized hybrid (PV/wind/battery) power generation system—case study Kenya. Energy 90:1830–1841. https://doi.org/10.1016/j.energy.2015.07.008
15. Approximate Diesel Generator Fuel Consumption Chart (n.d.). http://www.dieselserviceandsupply.com/Diesel_Fuel_Consumption.aspx Accessed 7 Mar 2018

16. Petrol diesel price in Port Blair, Port-Blair, South-Andaman Today (16/02/18). Current Latest Prices (n.d.). http://www.petroldieselprice.com/petrol-diesel-lpg-cng-fuel-oil-price/Port-Blair/Port-Blair/South-Andaman/Andaman-And-Nicobar-Islands/744101. Accessed 16 Feb 2018
17. Surrette Rolls 4-KS-25PS 1900 Ah, 4 Volt Deep Cycle Battery (n.d.) http://webosolar.com/store/en/deep-cycle-batteries/990-surrette-rolls-4-ks-25ps-1900-ah-4-volt-deep-cycle-battery.html. Accessed 7 Mar 2018
18. Coronado CR, Yoshioka JT, Silveira JL (2011) Electricity, hot water and cold water production from biomass. Energetic and economical analysis of the compact system of cogeneration run with woodgas from a small downdraft gasifier. Renew Energy 36:1861–1868. https://doi.org/10.1016/j.renene.2010.11.021
19. Bhattacharjee S, Dey A (2014) Techno-economic performance evaluation of grid integrated PV-biomass hybrid power generation for rice mill. Sustain Energy Technol Assessments 7:6–16. https://doi.org/10.1016/j.seta.2014.02.005
20. Mamphweli NS, Meyer EL (2010) Evaluation of the conversion efficiency of the 180 Nm^3/h Johansson Biomass Gasifier™. Int J Energy Environ 1:113–120
21. Bates RP, Dölle K, Ravi K (2017) Syngas use in internal combustion engines -a review. Adv Res 10:1–8. https://doi.org/10.9734/AIR/2017/32896
22. Singal SK, Varun G, Singh RP (2007) Rural electrification of a remote island by renewable energy sources. Renew Energy 32:2491–2501. https://doi.org/10.1016/j.renene.2006.12.013
23. Yoo K, Park E, Kim H, Ohm J, Yang T, Kim K et al (2014) Optimized renewable and sustainable electricity generation systems for Ulleungdo Island in South Korea. Sustainability 6:7883–7893. https://doi.org/10.3390/su6117883
24. Chum H, Faaij A, Moreira J, Berndes G, Dhamija P, Dong H, Gabrielle B, Goss Eng A, Lucht W, Mapako M, Cerutti OM, McIntyre T, Minowa T, Pingoud K (2011) Bioenergy. Cambridge University Press, Cambridge
25. Pettinger K-H, Dong W (2017) When does the operation of a battery become environmentally positive? J Electrochem Soc 164:A6274–A6277. https://doi.org/10.1149/2.0401701jes

Chapter 75
Improving the Performance of Cogeneration System in Sugar Factory by the Integration of Superheated Steam Dryer and Parabolic Trough Solar Collector

Somchart Chantasiriwan and Sarocha Charoenvai

75.1 Introduction

A simplified process of raw sugar manufacturing is shown in Fig. 75.1. Inputs to the juice extraction process are 100 t/h of sugar cane and 42 t/h of water. According to Rein [1], the composition of sugar cane is 15% fiber, 16% dissolved solids, and 69% water. It may be assumed that there is no fiber in the extracted juice and all dissolved solids in sugar cane is transferred to the juice. If the bagasse moisture content is 50%, the outputs will be 112 t/h of sugar juice and 30 t/h of bagasse. In order to produce raw sugar, 96 t/h of water must be removed from the juice. According to the recent analysis by Chantasiriwan [2], the amount of water content in juice is three times the amount of low-pressure steam required for the evaporation process. Therefore, the required mass flow rate of steam is 37.33 t/h (10.37 kg/s).

Two main products of raw sugar manufacturing processes are raw sugar and bagasse. The latter can be used as a fuel in a cogeneration system that produces both electrical energy and useful thermal energy. High-pressure steam is produced by a bagasse boiler. It is then fed to a steam turbine that powers the juice extraction process directly. Alternatively, high-pressure steam may be used to generate electrical power in a steam turbine. In either case, low-pressure steam exhausted or extracted from the steam turbine is sent to the juice evaporation process. It should be noted that the quantity of bagasse required by the boiler is less than the quantity of bagasse that is available from the juice extraction process. As a result, there will be excess bagasse that may be used to produce electrical power when the sugar factory is not manufacturing raw sugar or sold for other uses.

S. Chantasiriwan (✉)
Faculty of Engineering, Thammasat University, Pathum Thani, Thailand
e-mail: somchart@engr.tu.ac.th

S. Charoenvai
Faculty of Engineering, Rajamangala University of Technology Thanyaburi, Pathum Thani, Thailand

A. Sayigh (ed.), *Renewable Energy and Sustainable Buildings*, Innovative Renewable Energy, https://doi.org/10.1007/978-3-030-18488-9_75

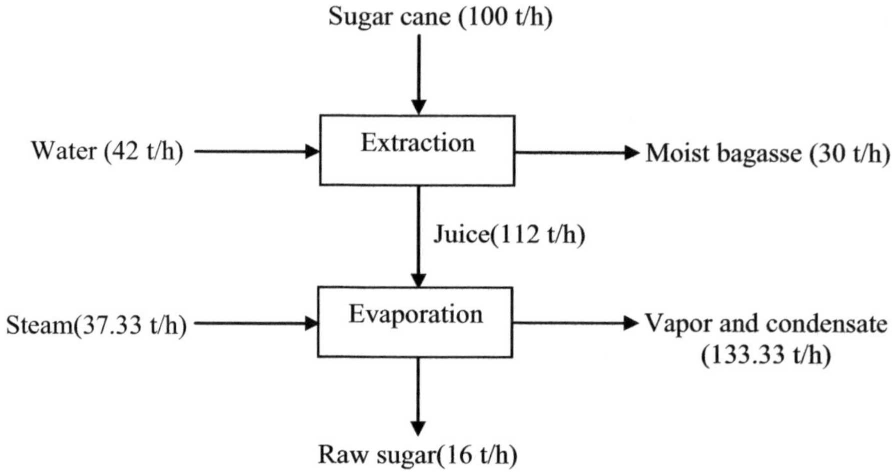

Fig. 75.1 Mass balances in raw sugar manufacturing process

There have been several suggestions of system improvement that will lower the bagasse use in the system. Ensinas et al. [3] used a thermo-economic procedure to optimize the design of sugar juice evaporation system, which resulted in the optimum distribution of heat exchanger surfaces that minimized steam use. Modifications for the cogeneration system improvement suggested by Ensinas et al. [4] include replacing low-efficiency steam turbine used to power the juice extraction process machinery with electrical engine and changes in the juice evaporation process that lower steam demand. Bocci et al. [5] showed that cogeneration systems using gas turbine and fuel cells outperformed typical system that uses steam turbine. Alves et al. [6] compared performances of cogeneration systems using back-pressure turbine and condensing-extraction turbine, and showed that the latter led to larger surplus electrical power generation. Deshmukh et al. [7] suggested that further improved performance could be attained by using biomass integrated gasifier combined cycle. Burin et al. [8] proposed the integration of concentrated solar power in the cogeneration system in order to improve the system performance.

The juice extraction process requires water addition to increase juice extraction. As a consequence, bagasse leaving the juice extraction process has a high moisture content. It is a common practice in many sugar factories to use moist bagasse as a fuel. It is well known that boiler efficiency decreases with increasing bagasse moisture content. Therefore, bagasse drying is a method of cogeneration system improvement. Steam exhausted from back-pressure turbine or extracted from condensing-extraction turbine is usually superheated. Thermal energy released by superheated steam as it becomes condensate may also be used to remove moisture in an indirect dryer, in which the drying medium is air [9, 10]. Alternatively, superheated steam may be mixed with bagasse with high moisture content, which will result in saturated steam and bagasse with lower moisture content.

Recently, Chantasiriwan and Charoenvai [11] developed a model of cogeneration system with superheated steam drying in sugar factory. The model was used to show that the integration of superheated steam dryer in the system resulted in both decreasing bagasse consumption and increasing power generation efficiency. Furthermore, it may be used to show that the performance of the system depends on the degree of superheat, which is the difference between the superheated steam temperature entering the dryer and the saturated steam temperature. The degree of superheat depends on both steam pressure and turbine efficiency. Both parameters are usually fixed values in a cogeneration system. In order to increase the degree of superheat, the integration of parabolic trough solar collector into the system is considered. The use of parabolic trough solar collector to increase steam temperature for drying bagasse is a new proposal that has never been investigated before. Therefore, the objective of this paper is to present a mathematical model of this system, and use it to compare its performance with the performance of the system without parabolic trough solar collector.

75.2 Parabolic Trough Solar Collector

Parabolic trough solar collector consists of a reflector, which is curved as a parabola, and a receiver, which is located at the focus of the parabola. Figure 75.2 shows the cross section of parabolic trough solar collector. Sunlight incident on the reflector is concentrated on the receiver, which consists of two concentric tubes known as the absorber and the cover. The concentrated sunlight heats the absorber, and the fluid flowing inside the absorber is also heated. Heat loss from the absorber to the surroundings is minimized because of the vacuum between the absorber and the cover.

Ideally, the entire solar energy incident on the parabolic trough solar collector is focused onto the receiver. Therefore, the ideal amount of solar energy reaching the receiver is

$$Q_{\text{ideal}} = G_{\text{b}} A_{\text{ap}} \cos\theta \tag{75.1}$$

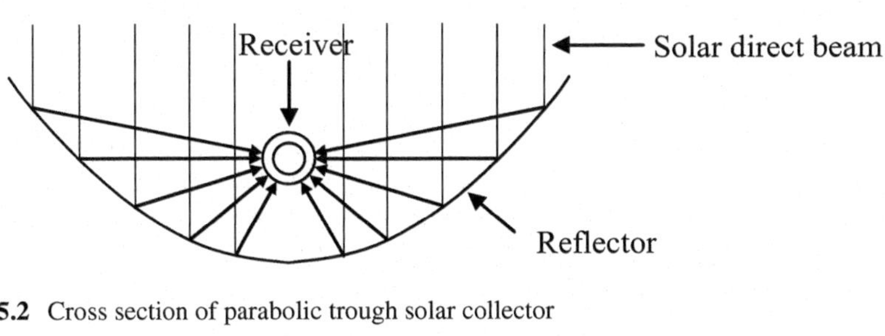

Fig. 75.2 Cross section of parabolic trough solar collector

where G_b is the solar direct beam irradiation, A_{ap} is the aperture area of the solar collector, and θ is the incident angle. It is assumed that sun tracking system is used so that the incident angle is kept at 0°. The actual solar energy reaching the receiver is less than the ideal amount because of losses due to limited reflectance, absorbance, transmittance, and geometrical inaccuracies. The optical efficiency, defined as

$$\eta_{opt} = \frac{Q_{actual}}{Q_{ideal}} \tag{75.2}$$

accounts for all these losses. The actual solar energy (Q_{actual}) reaching the receiver will heat the absorber, which will result in the enthalpy increase of the working fluid. Thermal losses occurring in the absorber are accounted for the thermal efficiency, defined as

$$\eta_{th} = \frac{m_s \Delta h_s}{Q_{actual}} \tag{75.3}$$

where m_s is the mass flow rate of the fluid and Δh_s is the enthalpy increase of the fluid.

Working fluid may be synthetic oil that can reach a high temperature without boiling. The fluid then exchanged heat with water in a heat exchanger to produce superheated steam. Alternatively, superheated steam may be generated directly in the parabolic trough solar collector if water is used as the working fluid. This configuration is known as the direct steam generation. It eliminates the need for an intermediate heat exchanger, which results in a lower cost of the solar plant.

Parabolic trough solar collector is a mature technology that has resulted in construction of solar-thermal power plants in several countries [12]. Most of the previous investigations are concerned with the generation of saturated steam or superheated steam and feed water heating using parabolic trough solar collector. However, the review of the literature reveals that no investigations have been made into the feasibility of using parabolic trough solar collector to generate high-temperature steam for removing moisture in fuels. It is therefore the main objective of this paper to use parabolic trough solar collector for this purpose.

75.3 Integration of Superheated Steam Dryer and Parabolic Trough Solar Collector

Figure 75.3 shows the schematic of the cogeneration system with the integration of superheated steam dryer (SSD) and parabolic trough solar collector (PTSC). In this system, the total amount of bagasse is divided into two parts to allow for the possibility that the available superheated steam is not enough to dry all bagasse. Moist bagasse in the first part with the dry mass flow rate m_{b1} is passed through SSD.

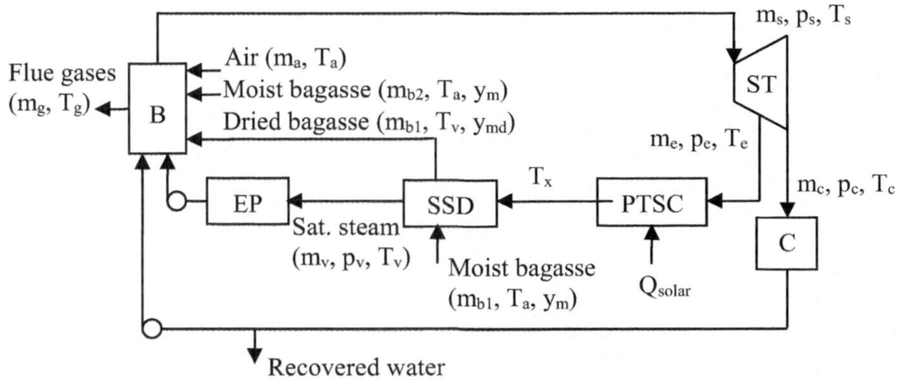

Fig. 75.3 Cogeneration system with superheated steam dryer and parabolic trough solar collector

Mixing between bagasse and superheated steam causes some of the moisture in bagasse to evaporate and the dry-basis moisture content to reduce from y_m to y_{md} before entering the boiler (B). The second part with the dry mass flow rate m_{b2}, which is at ambient temperature, is fed directly to B. Thermal energy released from the combustion of bagasse and ambient air results in the production of high-pressure steam and high-temperature flue gases. The superheated steam leaving the boiler is at pressure p_s and temperature T_s. The flue gases are assumed to be maintained at a fixed temperature T_g. High-pressure steam goes to the condensing-extraction steam turbine (ST). Some of the steam is condensed in the condenser (C), which supplies feed water for B. The remaining steam is extracted at pressure p_e, which is higher than pressure p_v in the evaporation process (EP) to allow for pressure losses in SSD and PTSC. It should be noted that not all of water from C is used as feed water for B. The mass flow rate of recovered water is

$$m_{rw} = \left(y_m - y_{md}\right) m_{b1} \tag{75.4}$$

The extracted steam is superheated, and its temperature (T_e) depends on the isentropic efficiency of ST. In order to increase this temperature, the steam extracted is passed through PTSC before entering SSD. The steam temperature exiting PTSC depends on the rate of heat transfer from PTSC to the steam as follows:

$$T_x = T_e + \frac{\eta_{th}\eta_{opt}G_b A_{ap}}{m_e c_{pv}} \tag{75.5}$$

where Q_{solar} is the rate of heat transfer absorbed by the steam and c_{pv} is the specific heat capacity of the steam.

Mass and energy balances of SSD are

$$m_v = m_e + m_{b1}\left(y_m - y_{md}\right) \tag{75.6}$$

$$(1-x_A)m_{b1}c_{pb}(T_v-T_a)+y_m m_{b1}c_{pw}(T_v-T_a) \\ +x_A m_{b1}c_{pash}(T_v-T_a)+(y_m-y_{md})m_{b1}\Delta h_v = m_e c_{pv}(T_x-T_v) \quad (75.7)$$

where Δh_v is the latent heat of evaporation of water at p_v and x_A is the mass fraction of ash in dry bagasse. The specific heat capacities of dry bagasse (c_{pb}), water (c_{pw}), steam (c_{pv}), and ash (c_{pash}) are, respectively, 0.46, 4.18, 2.00, and 1.00 kJ/kg K.

Assume that the composition of dry bagasse is known. The higher heating value (HHV) can be determined from a formula proposed by Qian et al. [13].

$$\text{HHV} = 8.7352\times10^4\left(\frac{1}{3}x_C + x_H + \frac{1}{8}x_S\right) \quad (75.8)$$

where x_C, x_H, x_O, and x_S are mass fractions of C, H, O, and S, respectively, in dry bagasse. In order to determine the lower heating value (LHV) of bagasse, the amount of water resulting from the combustion of 1 kg of dry bagasse must be known. Since the complete combustion of 1 kg of dry bagasse produces $9x_H$ kg of water, the lower heating value of dry bagasse is

$$\text{LHV} = \text{HHV} - 9x_H\Delta h_r \quad (75.9)$$

where Δh_r is the latent heat of evaporation of water at the standard reference state (2.44×10^3 kJ/kg).

It is assumed that combustion of bagasse is complete. The amount of excess air required for the complete combustion of bagasse depends on bagasse moisture content. According to Rein [1], dry bagasse requires 17% of excess air ($\varphi = 0.17$), and the excess air requirement of moist bagasse is given by

$$\varphi = 47.57x_m^3 + 58.00x_m^2 + 23.99x_m - 3.18 \quad (75.10)$$

where x_m is the wet-basis moisture content of bagasse; $x_m = y_m/(1 + y_m)$. Equation (75.7) is to be used for $0.4 \le x_m \le 0.6$. Once the excess air is known, the mass flow rate of air (m_a) can be computed as follows:

$$m_a = (1+\varphi)\text{AFR}m_b \quad (75.11)$$

where AFR is the stoichiometric air–fuel ratio.

$$\text{AFR} = 11.44x_C + 34.32x_H + 4.29(x_S - x_O) \quad (75.12)$$

The net heat input to the steam generation unit is

$$Q_{in} = (1-\varepsilon)m_b\text{LHV} \quad (75.13)$$

where m_b is the mass flow rate of dry bagasse. The heat loss parameter ε accounts for two sources of heat losses ($\varepsilon = \varepsilon_r + \varepsilon_c$). It is assumed that the heat loss from radiation and convection between the boiler shell and the ambient air is 1.5% of the total heat released by combustion ($\varepsilon_r = 0.015$). In addition, there is unburned carbon that is a result of the difficulty of burning all bagasse completely. It has been found that the amount of unburned carbon depends on bagasse moisture. According to Rein [1],

$$\varepsilon_c = 3.953x_m^3 + 5.000x_m^2 + 2.154x_m - 0.298 \tag{75.14}$$

for $0.4 \le x_m \le 0.6$. If bagasse is dry ($x_m = 0$), $\varepsilon_c = 0.024$.

Energy balance of the boiler requires that heat input is used to (1) evaporate feed water and increase its temperature to T_s; (2) increase the temperature of flue gases to T_g; (3) increase the temperature of the moisture in bagasse to the saturation temperature, evaporate the water, and increase the temperature of the resulting vapor to T_g; and (4) increase the temperature of inert ash to T_g. Therefore, the energy balance equation is

$$Q_{in} + Q_{air} + Q_{bagasse} = Q_{steam} + Q_{flue} + Q_{moisture} + Q_{ash} \tag{75.15}$$

$$Q_{in} = \left[(1-\varepsilon_1)m_{b1} + (1-\varepsilon_1)m_{b2}\right]\mathrm{LHV} \tag{75.16}$$

$$Q_{air} = \left[(1+\varphi_1)m_{b1} + (1+\varphi_2)m_{b2}\right]\mathrm{AFR}c_{pa}(T_a - T_r) \tag{75.17}$$

$$Q_{bagasse} = (1-x_A)m_{b1}c_{pb}(T_v - T_r) + (1-x_A)m_{b2}c_{pb}(T_a - T_r) \tag{75.18}$$

$$Q_{steam} = m_v(h_s - c_{pw}T_v) + (m_s - m_v)(h_s - c_{pw}T_c) \tag{75.19}$$

$$\begin{aligned} Q_{flue} &= \left[1+(1+\varphi_1)\mathrm{AFR} - x_A\right]m_{b1}c_{pg}(T_g - T_r) \\ &+ \left[1+(1+\varphi_2)\mathrm{AFR} - x_A\right]m_{b2}c_{pg}(T_g - T_r) \end{aligned} \tag{75.20}$$

$$\begin{aligned} Q_{moisture} &= y_{md}m_{b1}\left[c_{pw}(T_r - T_v) + \Delta h_r + c_{pw}(T_g - T_r)\right] \\ &+ y_m m_{b2}\left[c_{pw}(T_r - T_a) + \Delta h_r + c_{pw}(T_g - T_r)\right] \end{aligned} \tag{75.21}$$

$$Q_{ash} = x_A m_{b1}c_{pash}(T_g - T_v) + x_A m_{b2}c_{pash}(T_g - T_a) \tag{75.22}$$

where T_r is the reference temperature (25 °C). The flue gases consist of CO_2, H_2O, O_2, N_2, and SO_2. The average heat capacities (c_{pg} and c_{pa}) are determined by taking into account the variation of the heat capacity of each gas with temperature according to Verbanck [14].

The power output of the turbine is found from

$$P = \eta_m \left[m_e \left(h_s - h_e \right) + m_c \left(h_s - h_c \right) \right] \quad (75.23)$$

where η_m is machine efficiency, which is assumed to be 0.95. If isentropic efficiencies (η_{te} and η_{tc}) are known, h_e and h_c are determined from

$$h_e = h_s - \eta_{te} \left(h_s - h_{es} \right) \quad (75.24)$$

$$h_c = h_s - \eta_{tc} \left(h_s - h_{cs} \right) \quad (75.25)$$

where h_{es} is the enthalpy of the extracted steam at pressure p_e and the same entropy as the inlet steam, and h_{cs} is the enthalpy of the exhausted steam at pressure p_c and the same entropy as the inlet steam.

75.4 Results and Discussion

According to Rein [1], the typical composition of dry bagasse is 45.92% of carbon, 43.89% of oxygen, 5.67% of hydrogen, 0.31% of nitrogen, 0.04% of sulfur, and 4.17% of ash. In order to carry out simulation, certain parameters of the systems must be provided. The values of these parameters are p_s = 4.5 MPa, T_s = 440 °C, T_g = 150 °C, p_v = 200 kPa, p_c = 0.01 kPa, and T_a = 30 °C. Furthermore, it is assumed that the power output from the turbine is 6 MW, and the turbine isentropic efficiencies are η_{te} = 0.8 and η_{tc} = 0.75. Once these parameters are specified, simulation results can be obtained for the reference case, in which PTSC is not installed in the system. It can be shown that the rates of dry bagasse consumption ($m_b = m_{b1} + m_{b2}$) and water recovery (m_{rw}) are, respectively, 2.325 kg/s and 0.282 kg/s. The installation of PTSC will result in both the reduction of dry bagasse consumption and the increase in water recovery because of higher superheated steam temperature. The parameters of PTSC are η_{opt} = 0.75, η_{th} = 0.7, and A_{ap} = 1000 m^2. Figure 75.4 shows the effect of the solar direct beam irradiation (G_b) on the reduction of dry bagasse consumption (Δm_b), which is the difference between the rate of bagasse consumption of the reference case and that of the system with PTSC, and the increase of water recovery (Δm_{rw}), which is the difference between the rate of water recovery of the system with PTSC and that of the reference case. It can be seen that both parameters increase linearly with G_b.

Solar direct beam irradiation varies with location. It is interesting to analyze the economic feasibility of the proposed use of PTSC when G_b is 500 W/m^2. Figure 75.4 shows that 1000 m^2 of A_{ap} results in 0.027 kg/s reduction of dry bagasse consumption. It may be assumed that PTSC operates for 8 h each day from 8 am to 4 pm and that the value of 500 W/m^2 is the average solar direct beam irradiation during that period. This means that the amount of dry bagasse that is saved from using PTSC is 777.6 kg each day. For the whole year, the system with PTSC will use 283 tons less dry bagasse that the system without PTSC. It can be shown that 1 kg of dry bagasse

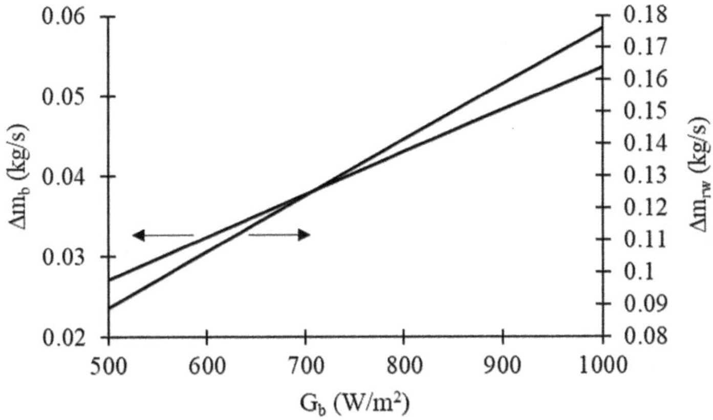

Fig. 75.4 Variations of the reduction of dry bagasse consumption (Δm_b) and the increase in water recovery (Δm_{rw}) with the solar direct beam irradiation (G_b)

can be used to generate 0.69 kW.h of electricity. If the total amount of saved dry bagasse is used to produce electricity, the corresponding electrical power production will be 1.958×10^5 kW.h. Since each kW.h of electricity is valued at \$0.10 in Thailand, the total value of saved dry bagasse is \$28,300. If the unit cost of PTSC is 180 \$/m^2, the simple payback period will be 6.36 years.

75.5 Conclusion

Superheated steam drying may be integrated into the cogeneration system of sugar factory. It will result in both the reduction of dry bagasse consumption and the recovery of water that would be lost with exhausted flue gases. The performance of this system depends on the temperature of superheated steam extracted from the condensing-extraction steam turbine. To improve the system performance, it is proposed that parabolic trough solar collector is used to increase superheated steam temperature. A mathematical model is developed to analyze the proposed system. Simulation results show that for the specified values of system parameters, the annual consumption of dry bagasse in the proposed system is 283 tons less than that in the system without parabolic trough solar collector. The payback period for the investment cost of the proposed system is 6.36 years. However, the recovery of a large amount of water is not taken into account in computing this payback period.

References

1. Rein P (2007) Cane sugar engineering. Verlag, Berlin
2. Chantasiriwan S (2016) Optimum imbibition for cogeneration in sugar factories. Appl Therm Eng 103:1031–1038
3. Ensinas AV, Nebra SA, Lozano MA, Serra L (2007) Design of evaporation systems and heaters networks in sugar cane factories using a thermoeconomic optimization procedure. Int J Thermodyn 10(3):97–105
4. Ensinas AV, Modesto M, Nebra SA, Serra L (2009) Reduction of irreversibility generation in sugar and ethanol production from sugarcane. Energy 34(5):680–688
5. Bocci E, Di Carlo A, Marcelo D (2009) Power plant perspectives for sugarcane mills. Energy 34(5):689–698
6. Alves M, Ponce GHSF, Silva MA, Ensinas AV (2015) Surplus electricity production in sugarcane mills using residual bagasse and straw as fuel. Energy 91:751–757
7. Deshmukh R, Jacobson A, Chamberlin C, Kammen D (2013) Thermal gasification or direct combustion? Comparison of advanced cogeneration systems in the sugarcane industry. Biomass Bioenergy 55:163–174
8. Burin EK, Vogel T, Multhaupt S, Thelen A, Oeljeklaus G, Gorner K, Bazzo E (2016) Thermodynamic and economic evaluation of a solar aided sugarcane bagasse cogeneration power plant. Energy 117:416–428
9. Luk HT, Lam TYG, Oyedun AO, Gebreegziabher T, Hui CW (2013) Drying of biomass for power generation: a case study on power generation from empty fruit bunch. Energy 63:205–215
10. Aziz M, Prawisudha P, Prabowo B, Budiman BA (2015) Integration of energy-efficient empty fruit bunch drying with gasification/combined cycle systems. Appl Energy 139:188–195
11. Chantasiriwan S, Charoenvai S (2018) Using superheated steam dryer for cogeneration system improvement and water recovery (Chapter 6). In: Sayigh A (ed) Transition towards 100% renewable energy: selected papers from the World Renewable Energy Congress WREC 2017. Springer, Cham
12. Baharoon DA, Rahman HA, Omar WZW, Fadhl SO (2015) Historical development of concentrating solar power technologies to generate clean electricity efficiently—a review. Renew Sust Energ Rev 41:996–1027
13. Qian H, Guo X, Fan S, Hagos K, Lu X, Liu C, Huang D (2015) A simple prediction model for higher heat value of biomass. J Chem Eng Data 61:4039–4045
14. Verbanck H (1997) Development of a mathematical model for watertube boiler heat transfer calculations. Proc S Afr Sugar Technol Assoc 71:166–171

Chapter 76
Fermentable Liquid Energy Carriers by Microwave-Assisted Hydrothermal Depolymerisation of Several Biomass Carbohydrates

Javier Remón, Fabio Santomauro, Christopher J. Chuck, Avtar S. Matharu, and James H. Clark

76.1 Introduction

Microwave-assisted hydrothermal depolymerisation allows the conversion of the carbohydrate content of biomass into fermentable sugar-rich aqueous solutions [1]. For the development and commercialisation of this route, it is very important to achieve an optimum control of the process in order to strike a good compromise between sugar production and sugar degradation, as the latter leads to the formation of fermentation inhibitors such as aldehydes, carboxylic acids, ketones and furans [2–5]. Microwave heating is based on the high frequency rotation of polar molecules, which produces a quicker and higher heating of the species with high polarity [6–13]. As lignin has a lower polarity than cellulose and hemicellulose, it is less active during microwave heating. This allows the solubilisation of the carbohydrate content of the biomass without significantly solubilising the lignin fraction [14, 15]. This minimises the presence of lignin-derived inhibitors in the hydrolysate. In addition, as water is highly effective in microwave energy absorption, the combination of hydrothermal conditions together with microwave-assisted heating might result into a very promising technology to achieve a selective and controllable production of saccharide-rich aqueous solution from biomass.

Herein, this work addresses the microwave-assisted hydrothermal treatment of cellulose, hemicellulose and alginic acid (the most representative carbohydrates found in second and third biomass generations) for the production of saccharide-rich solutions. The reactivity of the biomass carbohydrates and the fermentability of the

J. Remón (✉) · A. S. Matharu · J. H. Clark (✉)
Department of Chemistry, Green Chemistry Centre of Excellence, University of York, York, UK
e-mail: javier.remonnunez@york.ac.uk; jremon@icb.csic.es; james.clark@york.ac.uk

F. Santomauro · C. J. Chuck (✉)
Department of Chemical Engineering, University of Bath, Bath, UK
e-mail: c.chuck@bath.ac.uk

A. Sayigh (ed.), *Renewable Energy and Sustainable Buildings*, Innovative Renewable Energy, https://doi.org/10.1007/978-3-030-18488-9_76

liquids produced were thoroughly analysed. In the reactivity analyses, the influence of the temperature, reaction time and solid-to-water ratio was investigated on the yield and chemical composition of the liquids produced. In addition, the effects of the carbohydrates nature and hydrolysates chemical composition on the fermentation results was investigated using the robust yeast *Metschnikowia pulcherrima (MP)*. This yeast was selected due to its ability to metabolise a wide range of mono-/disaccharides and water-soluble oligosaccharides. In addition, it also possesses a high tolerance to inhibitors, such as furfural, 5-HMF and organic acids.

76.2 Experimental

The effects of the temperature (160–210 °C), reaction time (0 and 5 min) and solid/water (S/W) ratio (1/20 and 1/10) were analysed and compared for cellulose, hemicellulose and alginic. The experiments were carried out in a CEM-Mars microwave system using a 70 mL scale PrepPlus® reactor copped with a fibre-optic temperature sensor. For the microwave experiments, different amounts of solid (cellulose, hemicellulose or alginic acid) together with water (40 mL) were loaded into the microwave reactor depending on the solid/water ratio used in each experiment. Then the reactor was closed and placed inside the microwave unit. A ramping time (time to reach the reaction temperature) of 15 min was used for all the experiments. After reaction, the reaction mixture was cooled to room temperature by CEM-Mars default programme, centrifuged, and the solid residue (pellet) was isolated and dried (105 °C). The resultant solid (pellet) and the supernatant (hydrolysate) were stored for further analyses. Inoculum cultures for *M. pulcherrima* were prepared by culturing in 10 mL SPME media (soy peptone: 30 g/L; malt extract: 25 g/L) and 10 mL YPD media (yeast extract: 10 g/L; peptone: 20 g/L; glucose: 20 g/L) and incubated at 20 °C using an agitation speed of 180 rpm for 24 h. The hydrolysates from the microwave processing stage were filtered, adjusted to pH 4 (be means of a 2 mol/L NaOH solution) and diluted 1:2 with a salt solution. This salt solution was made up of KH_2PO_4 2 g/L, $MgSO_4{\cdot}7H_2O$ 0.376 g/L, $MgCl_2{\cdot}6H_2O$ 2.16 g/L, $ZnSO_4{\cdot}7H_2O$ 0.04 g/L, $(NH_4)_2SO_4$ 0.126 g/L, NH_4Cl 0.708 g/L, yeast extract 2.00 g/L, $CaCl_2{\cdot}2H_2O$ 0.300 g/L, tartaric acid 30.0 g/L. Cultures of 1 mL were inoculated with 25 μL of inoculum in 24-well plates and incubated at 180 rpm at 20 °C for 168 h. Throughout the fermentation, the growth of the cultures was estimated for absorbance at 600 nm and the biomass was recovered by centrifugation in 1.5 Eppendorf tubes (10,000 rpm, 5 min). The supernatant and the pellets were stored separately at −20 °C for further analyses.

Table 76.1 lists the response variables used to assess the reactivity of the carbohydrates and the fermentability of the liquids, i.e. the overall carbohydrate conversion, the gas and liquid yields, the liquid carbon efficiency (C-wt.%) and the chemical composition (in carbon basis, C-wt.%) of the hydrolysates. In addition, the biomass growth and 2PE yield produced during fermentation were used to study the suitability of the liquid samples for fermentation. The liquid phase (hydrolysate) was analysed by

Table 76.1 Response variables

Product	Response variable	Analytical method
Solid	$\text{Carbohydrate conversion}(\%) = \frac{\text{initial mass of solid}(g) - \text{final mass of solid}}{\text{initial mass solid}(g)} \times 100$	Gravimetric
	$\text{Residual solid yield}(\%) = \frac{\text{mass of solid after reaction}(g)}{\text{initial mass solid}(g)} \times 100$	Gravimetric
Gas	$\text{Gas yield}(\%) = \frac{\text{mass gas}(g)}{\text{initial mass solid}(g)} \times 100$	Gravimetric (reactor mass balance)
Liquid	$\text{Liquid yield}(\%) = \frac{\text{mass of liquid products}(g)}{\text{initial mass of solid}(g)} \times 100 = 100 - \text{Gas yield}(\%) - \text{Solid yield}(\%)$	Gravimetric (mass difference)
	$\text{Liquid C efficiency}(\%) = \frac{\text{mass of C in the liquid}(g)}{\text{initial mass of C in the solid}(g)} \times 100$	TOC and elemental analysis
	$\text{Composition}(\text{C} - \text{wt.}\%) = \frac{\text{mass of C of each compound}}{\text{total mass of C in solution}} \times 100$	HPLC and TOC
	$\text{Fermentation biomass yield}(\text{wt.}\%) = \frac{\text{biomass produced}(g/L)}{\text{solubilised material}(g/L)} \times 100$	Optical density
	$\text{Fermentation 2PE yield}(\text{wt.}\%) = \frac{\text{2PE produced}(g/L)}{\text{solubilised material}(g/L)} \times 100$	GC-FID

HPLC and total organic carbon (TOC). Optical density (absorbance at 600 nm using a corrector internal factor) was used to estimate the biomass yeast growth. GC-FID was used for the quantification of the 2PE. One-way analysis of variance (one-way ANOVA) with the multiple-range Fisher's least significant difference (LSD) test, both with a significance level of 0.05, were used to evaluate the influence of the carbohydrate type, temperature, solid/water ratio and reaction time on the process.

76.3 Results and Discussion

76.3.1 Carbohydrate Conversion and Gas and Liquid Yields

Figure 76.1 shows the overall carbohydrate conversion, the liquid and gas yields and the liquid C efficiency obtained during the microwave-assisted experiments. The statistical analysis reveals significant differences between the results obtained in the experiments (p-values <0.001). In addition, the Fisher's LSD test shows that both the type of carbohydrate and the operating conditions (temperature, time and S/W ratio) exert a significant influence on the process with 95% confidence.

Cellulose displays a very low reactivity, gas formation is negligible, and the overall carbohydrate conversion and liquid yield and C efficiency are lower than 10% in all the cases. Under these conditions, only the amorphous part of cellulose (around 10 wt.%) is solubilised in the liquid when subjected to microwave-assisted hydrothermal treatment [11]. Conversely, hemicellulose and alginic acid show a much higher reactivity under the same experimental conditions and the operating conditions exert a statistically significant influence on the hydrothermal treatment. For hemicellulose, an increase in the temperature between 160 and 210 °C leads to an increase in the overall carbohydrate conversion, liquid yield and C efficiency regardless of the S/W ratio or reaction time, since an increase in the temperature positively influence the kinetics of the process [1]. The effect of the reaction time depends on the temperature. In particular, an increase in the reaction time from 0 to 5 min increases the overall carbohydrate conversion and the liquid yield when temperatures between 190 and 210 °C are used due to the longer exposure of the material to microwaves. In these cases, the higher the temperature, the greater is the positive effect of the reaction time on the process. The highest gas production takes place when elevated temperatures (>190 °C) along with either long (5 min) reaction

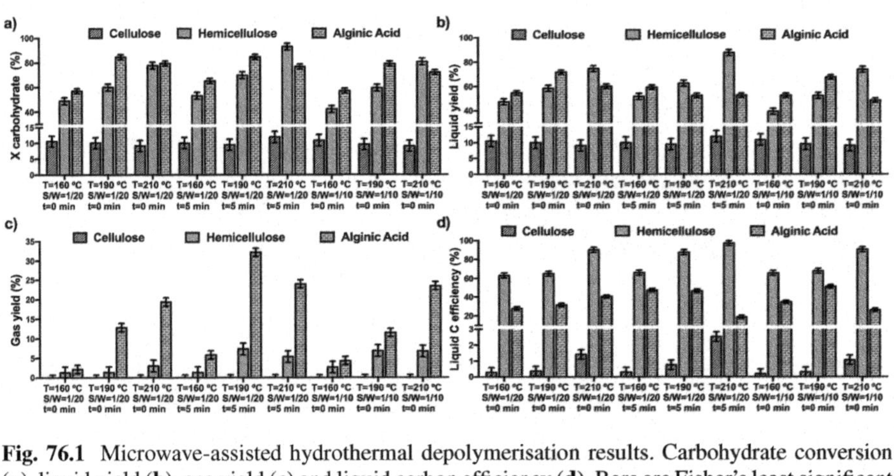

Fig. 76.1 Microwave-assisted hydrothermal depolymerisation results. Carbohydrate conversion (**a**), liquid yield (**b**), gas yield (**c**) and liquid carbon efficiency (**d**). Bars are Fisher's least significant difference (LSD) intervals with 95% confidence

times or low S/W ratio are used. These conditions can promote decarboxylation reactions leading to gas formation [16, 17]. In general, the S/W ratio does not significantly influence the overall carbohydrate conversion. An exception to this occurs when high temperature and low S/W ratios are used. As regards alginic acid depolymerisation, an initial increase in the temperature from 160 to 190 °C increases the carbohydrate conversion; however, a further increase up to 210 °C leads to a decrease in the conversion regardless of the S/W ratio or reaction time. While the initial increase with the temperature is the consequence of the positive kinetic effect of this variable on the process as described earlier [1], the subsequent decrease is accounted for by the formation of solid species, such as char and humins, at high temperature [18–20]. The effect of the S/W ratio and reaction time on the overall conversion depends on the temperature. While an increase in either the time or S/W ratio does not influence the overall conversion at 160 °C, a decrease is observed between 190 and 210 °C due to the formation of humins when high temperature or long reaction times are used [18–20].

The operating conditions also have a significant influence on the liquid yield and liquid C efficiency. Regardless of the S/W ratio, an increase in the temperature when a very short (0 min) reaction time is used has the same consequences for the liquid yield and C efficiency as those described above for the overall conversion, i.e. an initial increase between 160 and 190 °C followed by a subsequent decrease between 190 and 210 °C. However, an increase in the reaction time from 0 to 5 min statistically decreases the liquid yield and C efficiency and modifies the effect of the temperature on this variable. For a 5 min reaction time, an increase in the temperature from 160 to 210 °C leads to a progressive depletion in the liquid yield and C efficiency. Therefore, gas and solid formation results in lower liquid yields. In particular, at medium temperature (190 °C), the decreases in the liquid yield and C efficiency are accounted for by gas formation [1], while at high temperature (210 °C) solid formation is responsible for the decrease observed in the liquid yield [18–20]. The effect of the S/W ratio on the depolymerisation of alginic acid depends on the temperature. At 160 °C, the S/W ratio does not have a significant influence, while an increase in the S/W ratio from 1/20 to 1/10 decreases the overall alginic acid conversion between 190 and 210 °C. This lower overall conversion is the consequence of the greater gas production and solid formation occurring when long reaction times and high temperatures are used. Gas formation is substantial during the hydrothermal treatment of this carbohydrate and gas yields as high as 30% are obtained in some cases. For a 0 min reaction (i.e. only the 15 min ramping without holding), regardless of the S/W ratio, an increase in the temperature from 160 to 210 °C leads to a substantial increase in the gas yield. An increase from 0 to 5 min substantially increases gas formation, especially at medium temperatures. In these cases, an increase in the residence time shifts the reaction towards the formation of low oxygen content liquid products, which are easily converted into gases under the conditions used in this work. This increase occurring when the reaction time increases from 0 to 5 min is less pronounced at 210 °C than at 190 °C. For a 5 min reaction, an increase in the temperature from 190 °C to 210 °C leads to a substantial decrease in gas formation due to the formation of solid products.

With respect to carbohydrates reactivity, it increases as follows: cellulose < hemicellulose < alginic acid. However, and very importantly, the higher reactivity of alginic acid than that of hemicellulose does not produce higher liquid (hydrolysate) yield under some conditions due the substantial formation of gas and solid products during the hydrothermal treatment of alginic acid.

76.3.2 Chemical Composition of the Hydrolysates

The liquid product (hydrolysate) consists of a mixture of oligosaccharides (DP3–6 and DP > 6) and mono-/disaccharides, carboxylic acids, ketones and furans. Mono-/disaccharides include cellobiose, xylose, glucose, fructose, mannose, arabinose, rhamnose and levoglucosan. Carboxylic acids comprise lactic, formic, levulinic, acetic, guluronic and mannuronic acids. Ketones and furans are made of levoglucosenone, 5-HMF and furfural, respectively. The chemical composition in carbon basis (C-wt.%) of the hydrolysates is shown in Fig. 76.2. For cellulose, DP > 6 oligosaccharides are the most abundant species in the liquid phase, followed by monosaccharides and in much lower proportions furans, carboxylic acids and ketones. While the operating conditions exert a statistically significant influence on the

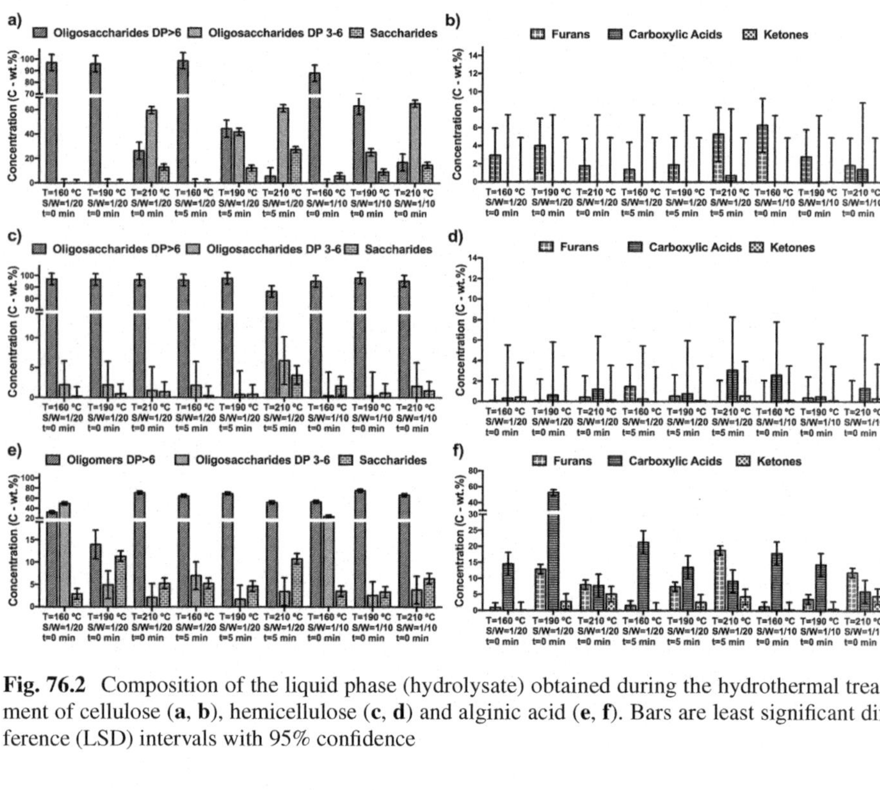

Fig. 76.2 Composition of the liquid phase (hydrolysate) obtained during the hydrothermal treatment of cellulose (**a**, **b**), hemicellulose (**c**, **d**) and alginic acid (**e**, **f**). Bars are least significant difference (LSD) intervals with 95% confidence

proportions of oligo- and monosaccharides, the concentrations of furans, carboxylic acids and ketones are not significantly affected. Regardless of the other conditions, an increase in the reaction time decreases DP > 6 oligosaccharides and increased mono- and disaccharides, as a proportion in the liquid, respectively. An increase in the temperature promotes a further development of the hydrolysis reactions, thus leading to the progressive depolymerisation of the DP > 6 oligosaccharides fraction into DP3–6 oligosaccharides, disaccharides and monosaccharaides [1]. In addition, these reactions are intensified by the reaction time and a greater decrease in the proportions of DP > 6 oligosaccharides occurs when the temperature is increased from 160 to 190 °C for a 5 min reaction than for a 0 min reaction. Furthermore, the increases occurring for the proportions of DP2–6 oligosaccharides and saccharides are also intensified. As a result of this, the liquid has a relatively high proportion of saccharides when high temperatures and long reaction times are used. The S/W ratio has a much lower, although significant, influence on the proportions of oligosaccharides, disaccharides and monosaccharides. Specifically, an increase in the S/W ratio from 1/20 to 1/10 decreases the concentration of DP > 6 oligosaccharides and increases the concentration of DP2–6 and mono-/disaccharides.

With respect to hemicellulose, the liquid phase contains a great proportion of DP > 6 oligosaccharides (>90 C-wt.%). In addition, the operating conditions do not exert a significant influence on the composition of the liquid phase in most of the cases. One exception occurs when a long reaction time (5 min) or high temperatures are used. In this case, an increase in the temperature from 190 to 210 °C slightly decreases the proportion of DP > 6 oligosaccharides and increases the proportion of saccharides. Therefore, at 210 °C, an increase in the reaction time from 0 to 5 min leads to a decrease in the proportions of DP > 6 oligosaccharides and an increase in the proportion of saccharides. The proportions carboxylic acids, furans and ketones are very low (<5 C-wt.%). For these latter compounds, neither the temperature nor the reaction time or the S/W ratio exerts a significant influence on their concentrations.

The liquid phase produced from alginic acid is made up of a complex mixture of several organic compounds. In addition, the operating conditions play a very important role in the chemical composition of the hydrolysates. Regardless of the S/W ratio (1/20 or 1/10), at low temperature and short reaction time (160 °C and 0 min), the liquid phase contains a high proportion of DP > 6 oligomers/oligosaccharides, DP3–6 oligosaccharides, disaccharides and monosaccharides; the total concentration of carboxylic acids, furans and ketones is quite low (<15 C-wt.%). For a low S/W and short time (1/20 and 0 min), the composition of the liquids depends on the temperature with different trends being observed. On the one hand, the relative amounts of DP2–6 saccharides and ketones decrease and increase, respectively, when the temperature increases from 160 to 210 °C. However, the concentration of ketones is very small and this variation is not very important from a practical point of view. On the other hand, maxima and minima occur for the proportions of DP > 6 oligomers/oligosaccharides, di-/monosaccharides, carboxylic acids and furans. Between 160 and 190 °C, the proportion of DP > 6 oligomers/oligosaccharides decreases, while concentrations of monosaccharides, carboxylic acids and furans increase. Conversely, between 190 and 210 °C, the concentration of DP > 6 oligomers/oligosaccharides increases and the proportions of monosaccharides, carboxylic acids and furans decrease. A further increase in the temperature up to 210 °C shifts the process towards the formation of liquid end products such as furans and short-chain carboxylic acids [21–23]. For a S/W ratio of 1/10, the proportion of DP > 6 oligomers/oligosaccharides is not substantially affected by the temperature. Increasing the temperature from 160 to 210 °C decreases the concentrations of DP2–6 saccharides and carboxylic acids and slightly increases the relative amounts of furans and ketones. In addition, the positive effect of the S/W ratio, leading to a greater advancement of the chemical reactions taking place in the liquid phase, correlates well with the increase observed in the proportion of furans. The effect of the reaction time strongly depends on the temperature. At 160 °C, an increase in the reaction time from 0 to 5 min substantially decreases the concentration of DP3–6 oligosaccharides and increases the proportions of DP > 6 oligomers/saccharides and carboxylic acids. The reaction time promotes the transformation of DP2–6 oligosaccharides into smaller saccharides. In addition, the concentration of guluronic and manuronic acids in the liquid also increases due to the transformation of the water-soluble oligomers into these carboxylic acids.

The results described above indicate that the nature of the carbohydrate exerts a very important influence on the chemical composition of the liquid hydrolysates. The liquid phase obtained from alginic acid contains a high amount of products from secondary reactions, such as carboxylic acids, furans and ketones due to the higher reactivity of this carbohydrate under the operating conditions. This is in good agreement with the high overall conversion and liquid and gas yields obtained during the treatment of this carbohydrate. Conversely, cellulose and hemicellulose display a much lower reactivity and the formation of these organic compounds does not take place to a significant extent. This resulted in a liquid phase containing a greater proportion of DP > 6 oligosaccharides, DP3–6 oligosaccharides, disaccharides and monosaccharides.

76.3.3 Fermentability of the Hydrolysates

The hydrolysates produced by the microwave reaction contained mainly oligosaccharides, a range of C5 and C6 sugars and inhibitors. To assess the effect of the MW processing conditions on the fermentability of the hydrolysate from the three carbohydrates, *M. pulcherrima* was cultured in 24 well plates on the hydrolysate with additional N, P and microelements added. The yeast biomass growth is shown in Fig. 76.3.

Overall, the hydrolysates produced from cellulose produce the highest concentration of yeast biomass growth at high temperatures and loadings, and give the most fermentable material over all conditions tested. While at S/W loadings of 1/20 more yeast biomass in total is produced from the hemicellulose fractions, there was

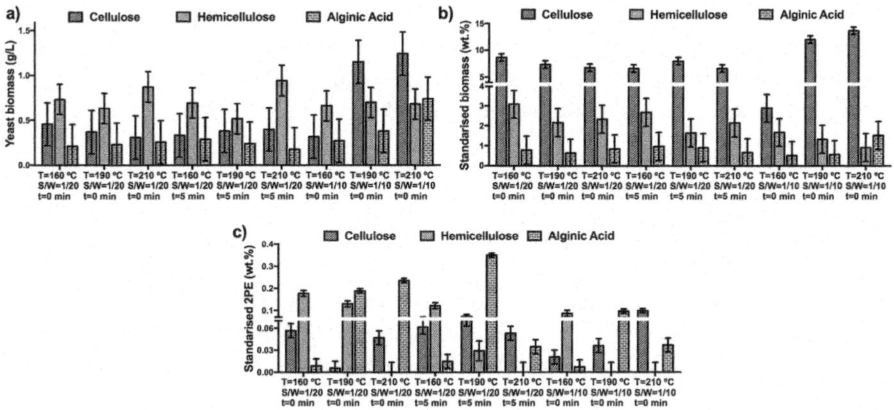

Fig. 76.3 Fermentation results: total yeast biomass concentration (**a**), total yeast biomass as a function of the amount of solubilised material (**b**) and 2PE production as a function of solubilised material (**c**) for the hydrolysates produced from cellulose, hemicellulose and alginic acid. Bars are Fisher's least significant difference (LSD) intervals with 95% confidence

far higher solubilised material available for the yeast, and therefore the conversion is relatively poor. Oligosaccharides from alginic acid were the least fermentable. These differences potentially are the result of the different chemical composition of the liquids obtained from each carbohydrate, with the cellulose producing simple straight chain units as opposed to highly branched oligosaccharides from hemicellulose [24–26], and alginic acid hydrolysates containing high proportions of carboxylic acids and furans. As described earlier, an increase in the S/W ratio decreases the amount of DP > 6 oligosaccharides in the liquids, producing a more accessible hydrolysate. Alternatively, alginic acid hydrolysates contained a particularly high level of inhibitors, whereas there were negligible concentrations of inhibitors such as carboxylic acids and furans in the cellulose hydrolysate. A similar concentration of yeast biomass was produced for every hemicellulose depolymerised material. Though the chemical composition of the liquid produced from hemicellulose is not significantly affected by the operating conditions. *M. pulcherrima* is known to produce high levels of 2PE, with up to 1.5 g/L reported from high concentrations of glucose (100 g/L sugar loading) [27–29]. For all the biomass carbohydrates, the production of 2PE is significantly affected by the hydrothermal conditions used in the microwave experiments. For cellulose, little 2PE is produced from the oligosaccharides, with the highest productivity being observed with the highest yeast biomass and solubilised material present. For hemicellulose, the highest 2PE production occurs during the fermentation of the liquids produced at low temperature (160 °C). Regardless of the reaction time or S/W ratio, an increase in the temperature of the microwave treatment from 160 to 210 °C leads to a substantial decrease in the production of 2PE. In addition, this decrease is more marked when either long reaction times or high S/W ratios are used. Similarly, the highest production of 2PE is observed for alginic acid. Regardless of the S/W ratio or reaction time, an increase in the temperature from 160 to 190 °C significantly increases the production of 2PE, where a maximum is achieved. A further increase up to 210 °C however leads to a decrease in the production of 2PE. This suggests that similar to previous studies a higher concentration of material is necessary to produce high yields of 2PE.

76.4 Conclusions

This work analyses and compares the reactivity of cellulose, hemicellulose and alginic acid when subjected to microwave-assisted hydrothermal treatment for the production of fermentable liquids. The most important conclusions are summarised as follows:

1. When subjected to the same operating conditions, the reactivity of the carbohydrates increased as follows: cellulose < hemicellulose < alginic acid. The overall conversion and liquid and gas yields were not affected by the conditions. Conversely, the temperature and reaction time had a very important influence during the hydrothermal treatment of hemicellulose or alginic acid.
2. The operating conditions did not influence the chemical composition of the hydrolysates produced from hemicellulose and a high concentration of DP > 6 oligosaccharides was obtained in all the experiments. Conversely, the hydrolysates produced from cellulose and alginic acid were influenced by the operating conditions. The former contained high proportions of oligosaccharides and saccharides, the latter comprising high proportions of water-soluble DP > 6 oligomers/oligosaccharides and fermentation inhibitors.
3. The hydrolysates produced from cellulose provided the best results in terms of standardised yeast biomass growth, with lower inhibitors and more accessible oligosaccharides. This did not translate into higher 2PE yields though, with hemicellulose and alginic acid being both suitable materials for the production of this high value compound.

Acknowledgements This research has been funded by the Industrial Biotechnology Catalyst (Innovate UK, BBSRC, EPSRC) to support the translation, development and commercialisation of innovative industrial biotechnology processes (EP/N013522/1). EPSRC for research grant number EP/K014773/1.

References

1. Prado JM, Lachos-Perez D, Forster-Carneiro T, Rostagno MA (2016) Sub- and supercritical water hydrolysis of agricultural and food industry residues for the production of fermentable sugars: a review. Food Bioprod Process 98:95–123
2. Alvira P, Tomas-Pejo E, Ballesteros M, Negro MJ (2010) Pretreatment technologies for an efficient bioethanol production process based on enzymatic hydrolysis: a review. Bioresour Technol 101:4851–4861

3. Huang Y-B, Fu Y (2013) Hydrolysis of cellulose to glucose by solid acid catalysts. Green Chem 15:1095
4. Langan P, Gnanakaran S, Rector KD, Pawley N, Fox DT, Cho DW, Hammel KE (2011) Exploring new strategies for cellulosic biofuels production. Energy Environ Sci 4:3820
5. Tadesse H, Luque R (2011) Advances on biomass pretreatment using ionic liquids: an overview. Energy Environ Sci 4:3913
6. Budarin VL, Clark JH, Lanigan BA, Shuttleworth P, Macquarrie DJ (2010) Microwave assisted decomposition of cellulose: a new thermochemical route for biomass exploitation. Bioresour Technol 101:3776–3779
7. Budarin VL, Shuttleworth PS, Dodson JR, Hunt AJ, Lanigan B, Marriott R, Milkowski KJ, Wilson AJ, Breeden SW, Fan J, Sin EHK, Clark JH (2011) Use of green chemical technologies in an integrated biorefinery. Energy Environ Sci 4:471–479
8. De bruyn M, Fan J, Budarin VL, Macquarrie DJ, Gomez LD, Simister R, Farmer TJ, Raverty WD, McQueen-Mason SJ, Clark JH (2016) A new perspective in bio-refining: levoglucosenone and cleaner lignin from waste biorefinery hydrolysis lignin by selective conversion of residual saccharides. Energy Environ Sci 9:2571–2574
9. de Melo EM, Clark JH, Matharu AS (2017) The Hy-MASS concept: hydrothermal microwave assisted selective scissoring of cellulose for in situ production of (meso)porous nanocellulose fibrils and crystals. Green Chem 19:3408–3417
10. Fan J, De bruyn M, Budarin VL, Gronnow MJ, Shuttleworth PS, Breeden S, Macquarrie DJ, Clark JH (2013) Direct microwave-assisted hydrothermal depolymerization of cellulose. J Am Chem Soc 135:11728–11731
11. Fan J, De bruyn M, Zhu Z, Budarin V, Gronnow M, Gomez LD, Macquarrie D, Clark J (2013) Microwave-enhanced formation of glucose from cellulosic waste. Chem Eng Process Intensif 71:37–42
12. Li T, Remón J, Jiang Z, Budarin VL, Clark JH (2018) Towards the development of a novel "bamboo-refinery" concept: selective bamboo fractionation by means of a microwave-assisted, acid-catalysed, organosolv process. Energy Convers Manag 155:147–160
13. Li T, Remón J, Shuttleworth PS, Jiang Z, Fan J, Clark JH, Budarin VL (2017) Controllable production of liquid and solid biofuels by doping-free, microwave-assisted, pressurised pyrolysis of hemicellulose. Energy Convers Manag 144:104–113
14. Remón J, Matharu AS, Clark JH (2018) Simultaneous production of lignin and polysaccharide rich aqueous solutions by microwave-assisted hydrothermal treatment of rapeseed meal. Energy Convers Manag 165:634–648
15. Zhou L, Budarin V, Fan J, Sloan R, Macquarrie D (2017) Efficient method of lignin isolation using microwave-assisted acidolysis and characterization of the residual lignin. ACS Sustain Chem Eng 5:3768–3774
16. Egües I, Alriols MG, Herseczki Z, Marton G, Labidi J (2010) Hemicelluloses obtaining from rapeseed cake residue generated in the biodiesel production process. J Ind Eng Chem 16:293–298
17. Pińkowska H, Wolak P, Oliveros E (2014) Hydrothermolysis of rapeseed cake in subcritical water. Effect of reaction temperature and holding time on product composition. Biomass Bioenergy 64:50–61
18. Remón J, García L, Arauzo J (2016) Cheese whey management by catalytic steam reforming and aqueous phase reforming. Fuel Process Technol 154:66–81
19. Remón J, Laseca M, García L, Arauzo J (2016) Hydrogen production from cheese whey by catalytic steam reforming: preliminary study using lactose as a model compound. Energy Convers Manag 114:122–141
20. Remón J, Ruiz J, Oliva M, García L, Arauzo J (2016) Cheese whey valorisation: production of valuable gaseous and liquid chemicals from lactose by aqueous phase reforming. Energy Convers Manag 124:453–469
21. Jeon W, Ban C, Kim JE, Woo HC, Kim DH (2016) Production of furfural from macroalgae-derived alginic acid over Amberlyst-15. J Mol Catal A Chem 423:264–269

22. Jeon W, Ban C, Park G, Woo HC, Kim DH (2016) Hydrothermal conversion of alginic acid to furfural catalyzed by Cu(II) ion. Catal Today 265:154–162
23. Wang Y, Delbecq F, Varma RS, Len C (2018) Comprehensive study on expeditious conversion of pre-hydrolyzed alginic acid to furfural in Cu(II) biphasic systems using microwaves. Mol Catal 445:73–79
24. Ding M-Z, Wang X, Liu W, Cheng J-S, Yang Y, Yuan Y-J (2012) Proteomic research reveals the stress response and detoxification of yeast to combined inhibitors. PLoS One 7:e43474
25. Whiffin F, Santomauro F, Chuck CJ (2016) Toward a microbial palm oil substitute: oleaginous yeasts cultured on lignocellulose. Biofuels Bioprod Biorefin 10:316–334
26. Yu X, Zheng Y, Dorgan KM, Chen S (2011) Oil production by oleaginous yeasts using the hydrolysate from pretreatment of wheat straw with dilute sulfuric acid. Bioresour Technol 102:6134–6140
27. Chantasuban T, Santomauro F, Gore-Lloyd D, Parsons S, Henk D, Scott RJ, Chuck C (2018) Elevated production of the aromatic fragrance molecule, 2-phenylethanol, using *Metschnikowia pulcherrima* through both de novo and ex novo conversion in batch and continuous modes. J Chem Technol Biotechnol 93:2118–2130
28. Santamauro F, Whiffin FM, Scott RJ, Chuck CJ (2014) Low-cost lipid production by an oleaginous yeast cultured in non-sterile conditions using model waste resources. Biotechnol Biofuels 7:34
29. Zhou L, Santomauro F, Fan J, Macquarrie D, Clark J, Chuck CJ, Budarin V (2017) Fast microwave-assisted acidolysis: a new biorefinery approach for the zero-waste utilisation of lignocellulosic biomass to produce high quality lignin and fermentable saccharides. Faraday Discuss 202:351–370

Chapter 77
PV Module Temperature Prediction at Any Environmental Conditions and Mounting Configurations

E. Kaplani and S. Kaplanis

77.1 Introduction

The effect of PV temperature on the power output, efficiency and final yield of the PV system is significant [1, 2]. Typical value for the temperature coefficient of maximum power P_m and similarly for the module efficiency is −0.5%/°C. A range of 5–25% reduction in power output is normally expected from the nominal value for 1000 W/m² incident irradiance due to PV temperature alone. The effect is in the higher end of the range for BIPV configurations [3]. PV temperature prediction is essential for forecasting PV power output, PV performance analysis, diagnostic purposes, dynamic predictive management of building integrated photovoltaics (BIPV) in Intelligent Energy Buildings [4], etc.

Several empirical and semi-empirical models have been proposed in the research literature providing an estimation of PV temperature, and taking into account ambient temperature T_a, solar irradiance on the PV plane I_T and wind speed v_w. Some of the well-known models include Eq. (77.1) by King et al. [5], Eq. (77.3) by Faiman [6] while validation of these models' applicability at different climatic regions is presented in [7].

$$T_{pv,b} = T_a + I_T \cdot e^{(a+b \cdot v_w)} \tag{77.1}$$

E. Kaplani (✉)
Engineering, Faculty of Science, University of East Anglia, Norwich, UK
e-mail: e.kaplani@uea.ac.uk

S. Kaplanis
Renewable Energy Systems Lab, Technological Educational Institute of Western Greece, Patra, Greece

A. Sayigh (ed.), *Renewable Energy and Sustainable Buildings*, Innovative Renewable Energy, https://doi.org/10.1007/978-3-030-18488-9_77

where $T_{pv,b}$ corresponds to the temperature at the back of the module. The cell temperature at the centre is then determined from $T_{pv,b}$ according to [5], based on:

$$T_{pv,c} = T_{pv,b} + \Delta T \cdot I_T / 1000 \quad (77.2)$$

where ΔT and the empirically determined coefficients a, b in Eq. (77.1) were defined as:

$a = -3.56$, $b = -0.0750$, $\Delta T = 3$ °C for glass/cell/polymer sheet, open rack module and $a = -2.81$, $b = -0.0455$, $\Delta T = 0$ °C for glass/cell/polymer sheet, insulated back module

In Faiman's model [6] given below, $U_0^{'}, U_1^{'}$ are empirically determined coefficients with average values: $U_o^{'} = 25\ \text{Wm}^{-2}\text{K}^{-1}$ and $U_1^{'} = 6.84\ \text{Wm}^{-3}\text{sK}^{-1}$.

$$T_{pv} = T_a + I_T / \left(U_o^{'} + U_1^{'} \cdot v_w\right) \quad (77.3)$$

In another approach, NN-based models have also been proposed such as Eq. (77.4) proposed by Tamizhmani et al. [8] claiming to be independent of site location and technology type:

$$T_{pv} = 0.943 T_a + 0.028 I_T - 1.528 v_w + 4.3 \quad (77.4)$$

Various theoretical models have also been proposed mainly steady-state models based on the Energy Balance Equation (EBE) such as [9], but so far use empirical models for the determination of the air forced convection coefficient. It has been previously shown by the authors in [10] how the air forced convection coefficient estimated by various empirical and theoretical models differs and the extent to which it influences the T_{pv} prediction. The superiority of theoretical models over empirical for the determination of the air forced convection coefficient was highlighted offering wider applicability independent of latitude.

In [10] a simulation model was developed for the determination of f coefficient in Eq. (77.5) based on theoretical expressions of the heat convection coefficients and the Energy Balance Equation:

$$T_{pv} = T_a + f \cdot I_T \quad (77.5)$$

$$f = \frac{(\tau\alpha) - \eta_{pv}}{U_{L,f} + U_{L,b}} \quad (77.6)$$

The present paper presents a new simulation model developed for the prediction of PV temperature at any environmental conditions and mounting configurations, considering the solar radiation intensity, ambient temperature, wind velocity, wind incidence angle on the PV surface either front or back and mounting geometry.

77.2 PV Temperature Prediction Model

The proposed simulation model is based on the steady-state energy balance equation, shown in Eq. (77.7). It takes into account heat conduction from the cell to the front glass and back polymer sheet, heat transfer by natural convection for both front and back PV sides at any inclination angle, forced convection coefficient at any PV inclination, orientation, wind direction for the windward and leeward side of the PV module based on theoretically derived expressions and, finally, thermal radiation emission for the front and back PV side, taking into account the mounting geometry. The assumptions considered in this model include:

- Temperature of PV module considered uniform
- Thermal capacity of PV module and transient phenomena are neglected. However, the effect of fluctuating wind speed is directly passed onto the temperature of the back PV surface. In these cases the time constant $\tau < 4$ min of the averaging period so these effects are in fact included.

The simplified thermal electrical equivalent used to determine the unknown temperatures $T_{pv,c}$, $T_{pv,f}$, $T_{pv,b}$, is shown in Fig. 77.1(a). Heat conduction from the cell to the front and back polymer sheet is considered through the thermal resistances $R_{c\text{-}f}$ and $R_{c\text{-}b}$ which accounts for the sum of the thermal resistances introduced by the

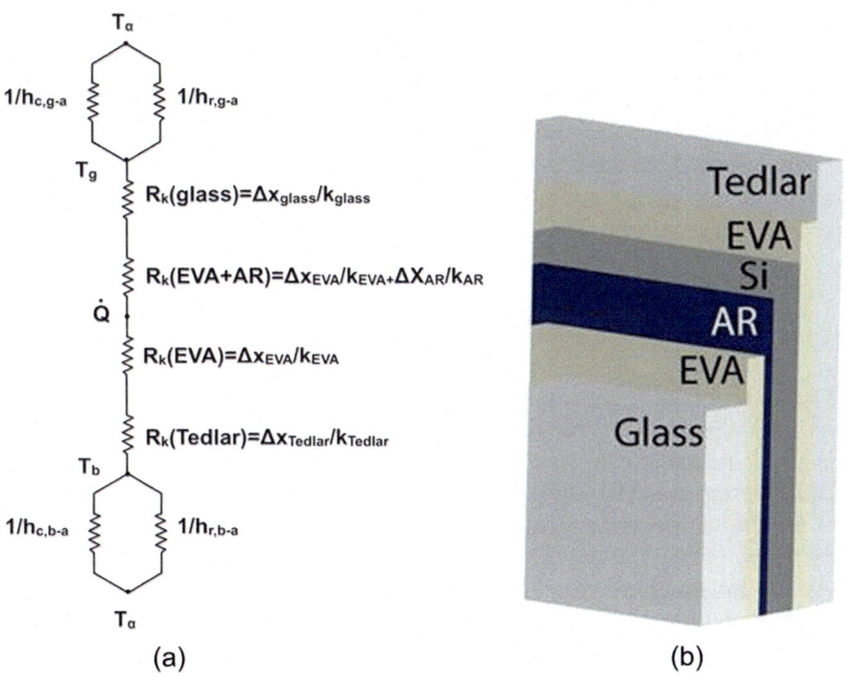

Fig. 77.1 (**a**) Thermal electrical equivalent, (**b**) PV module layers

various PV layers (Fig. 77.1(b)). The heat convection coefficients for the front and back PV sides $h_{c,g\text{-}\alpha}$ and $h_{c,b\text{-}\alpha}$, respectively, and the radiative heat transfer coefficients $h_{r,g\text{-}\alpha}$ and $h_{r,b\text{-}\alpha}$ are taken into account through the overall thermal losses coefficients defined in Eqs. (77.8) and (77.9).

The steady-state energy balance equation becomes:

$$(\tau\alpha)I_{\mathrm{T}} = \eta_{\mathrm{pv}} I_{\mathrm{T}} + \frac{1}{R_{\mathrm{c-f}} + \frac{1}{U_{\mathrm{L,f}}}}\left(T_{\mathrm{pv,c}} - T_{\mathrm{a}}\right) + \frac{1}{R_{\mathrm{c-b}} + \frac{1}{U_{\mathrm{L,b}}}}\left(T_{\mathrm{pv,c}} - T_{\mathrm{a}}\right) \qquad (77.7)$$

where the thermal losses coefficients for the front and back PV side are given by:

$$U_{\mathrm{L,f}} = h_{\mathrm{c,g-a}} + h_{\mathrm{r,g-a}} \qquad (77.8)$$

$$U_{\mathrm{L,b}} = h_{\mathrm{c,b-a}} + h_{\mathrm{r,b-a}} \qquad (77.9)$$

$(\tau\alpha)$ is the transmission-absorptance product. For the sun-tracking PV system $(\tau\alpha)$ is in fact equal to $(\tau\alpha)_n$. For fixed angle PV system or for BIPV, the $(\tau\alpha)$ product is defined through Eq. (77.10) according to [11], considering the beam, diffuse and ground-reflected irradiance components.

$$\frac{(\tau\alpha)}{(\tau\alpha)_n} = \frac{I_{\mathrm{b}} R_{\mathrm{b}} \frac{(\tau\alpha)_{\mathrm{b}}}{(\tau\alpha)_n} + I_{\mathrm{d}} \frac{(\tau\alpha)_{\mathrm{d}}}{(\tau\alpha)_n} \frac{(1+\cos\beta)}{2} + rI \frac{(\tau\alpha)_r}{(\tau\alpha)_n} \frac{(1-\cos\beta)}{2}}{I_{\mathrm{T}}} \qquad (77.10)$$

where the incidence angle modifier (IAM) for the three components is estimated based on Eq. (77.11) with $b_o = 0.136$ using the incidence angle θ for the beam component and the effective angles for the diffuse and ground-reflected radiation according to well-known expressions provided in [11].

$$K_{\tau\alpha} = 1 - b_{\mathrm{o}}\left(\frac{1}{\cos\theta} - 1\right) \qquad (77.11)$$

The natural convection from a PV module is determined through the Nusselt number *Nu* separately for the front PV side and for the back PV side considering the cases of all inclination angles $\beta < 30°$ and $\beta \geq 30°$ according to the equations analytically presented in [10]. In these Prandtl *Pr* and Rayleigh *Ra* numbers are calculated for the temperature of the boundary layer, at the front and back PV side, respectively.

For the forced convection coefficient of the windward side of the module the Sartori's expressions [12] were used, *see* Eqs. (77.12)–(77.14), as demonstrated in [10] to provide very accurate predictions of the *f* coefficient. The distinction between laminar, fully turbulent and mixed flows was made based on conditions $x_c/L \geq 0.95$,

$x_c/L \leq 0.05$ and $x_c/L < 0.95$, respectively. L is the surface length in the wind direction and x_c is the critical length based on the Reynolds number $Re_{x,c}$ and given by: $x_c = Re_{x,c} \cdot \nu / v_w$, where ν the kinematic viscosity of the air (m^2/s).

$$h = 3.83 \cdot v_w^{0.5} \cdot L^{-0.5} \quad \text{for laminar flow} \tag{77.12}$$

$$h = 5.74 \cdot v_w^{0.8} \cdot L^{-0.2} \quad \text{for fully turbulent flow} \tag{77.13}$$

$$h = 5.74 \cdot v_w^{0.8} \cdot L^{-0.2} - 16.46 \cdot L^{-1} \quad \text{for mixed flows} \tag{77.14}$$

For the leeward side of the PV module, the above expressions were considered but with L now given by $4A/S$, where A is the area of the module and S its perimeter.

Combined natural and forced convection was also considered based on the Grashof and Reynolds number comparison according to the following [13, 14]. For the case of combined convection Churchill's expressions Eqs. (77.15) and (77.16) provide a suitable solution with $m = 3$.

- $Gr_L \ll Re_L^2$ forced convection
- $Gr_L \gg Re_L^2$ natural convection
- $0.01 < Gr_L/Re_L^2 < 100$ combined

$$h^m_{c,g-a} = h^m_{c,g-a(\text{natural})} + h^m_{c,g-a(\text{forced})} \tag{77.15}$$

$$h^m_{c,b-a} = h^m_{c,b-a(\text{natural})} \pm h^m_{c,b-a(\text{forced})} \tag{77.16}$$

The radiative heat transfer coefficients for the front PV side and the back PV side were considered normalised to the difference between PV temperature and the ambient to correct for the simplification of the thermal electrical equivalent and EBE which take into account radiative heat transfer with respect to the ambient T_a rather than the sky T_s or ground temperature T_{grd}, respectively. Therefore, the normalised coefficients used in the EBE, Eqs. (77.8) and (77.9) are given by:

$$h_{r,g-a} = \varepsilon_g F_{pv,f-sky} \sigma \frac{T^4_{pv,f} - T^4_s}{T_{pv,f} - T_a} + \varepsilon_g F_{pv,f-grd} \sigma \frac{T^4_{pv,f} - T^4_{grd}}{T_{pv,f} - T_a} \tag{77.17}$$

$$h_{r,b-a} = \varepsilon_b F_{pv,b\ sky} \sigma \frac{T^4_{pv,b} - T^4_s}{T_{pv,b} - T_a} + \varepsilon_b F_{pv,b-grd} \sigma \frac{T^4_{pv,b} - T^4_{grd}}{T_{pv,b} - T_a} \tag{77.18}$$

where the view factors from the PV front or back to the sky and ground, respectively, are:

$$F_{pv,f\text{-}sky} = (1 + \cos\ \beta)/2 \tag{77.19}$$

$$F_{\text{pv,f-grd}} = \left(1 - \cos\ \beta\right)/2 \tag{77.20}$$

$$F_{\text{pv,b-sky}} = \left(1 + \cos\left(\pi - \beta\right)\right)/2 \tag{77.21}$$

$$F_{\text{pv,b-grd}} = \left(1 - \cos\left(\pi - \beta\right)\right)/2 \tag{77.22}$$

For the case of BIPV system, the normalised coefficient for the radiative heat transfer at the back PV side which is facing the indoors environment with temperature T_{ind} is given by:

$$h_{\text{r,b-ind}} = \varepsilon_{\text{b}}\sigma \frac{T_{\text{pv,b}}^4 - T_{\text{ind}}^4}{T_{\text{pv,b}} - T_{\text{a}}} \tag{77.23}$$

Based on the EBE Eq. (77.7) the determination of PV temperature $T_{\text{pv,c}}$ for the free-standing systems is achieved through Eq. (77.24).

$$T_{\text{pv,c}} = \frac{\left(\left(\tau\alpha\right) - \eta_{\text{pv}}\right)I_{\text{T}}}{\dfrac{1}{R_{\text{c-f}} + \dfrac{1}{U_{\text{L,f}}}} + \dfrac{1}{R_{\text{c-b}} + \dfrac{1}{U_{\text{L,b}}}}} + T_{\text{a}} \tag{77.24}$$

For the BIPV configuration the same theoretical model proposed for the free-standing PV system applies for the front PV side considering natural convection, forced convection (windward or leeward) and long-wave radiation, while at the back PV side only natural convection and long-wave radiation apply. In this case, the EBE takes the form of Eq. (77.25) and the $T_{\text{pv,c}}$ may be accurately predicted by Eq. (77.26).

$$\left(\tau\alpha\right)I_{\text{T}} = \eta_{\text{pv}}I_{\text{T}} + \frac{1}{R_{\text{c-f}} + \dfrac{1}{U_{\text{L,f}}}}\left(T_{\text{pv,c}} - T_{\text{a}}\right) + \frac{1}{R_{\text{c-b}} + \dfrac{1}{U_{\text{L,b}}}}\left(T_{\text{pv,c}} - T_{\text{ind}}\right) \tag{77.25}$$

$$T_{\text{pv,c}} = \frac{\left(\left(\tau\alpha\right) - \eta_{\text{pv}}\right)I_{\text{T}} + \dfrac{T_{\text{a}}}{R_{\text{c-f}} + \dfrac{1}{U_{\text{L,f}}}} + \dfrac{T_{\text{ind}}}{R_{\text{c-b}} + \dfrac{1}{U_{\text{L,b}}}}}{\dfrac{1}{R_{\text{c-f}} + \dfrac{1}{U_{\text{L,f}}}} + \dfrac{1}{R_{\text{c-b}} + \dfrac{1}{U_{\text{L,b}}}}} \tag{77.26}$$

In Eqs. (77.7) and (77.24–77.26), the efficiency η_{pv} is considered for the actual conditions of solar irradiance and for the predicted PV temperature according to the following based on [9, 15].

$$\eta_{PV} = \eta_{STC} \cdot \left(1 + \gamma \cdot \left(T_{PV} - T_{PV,STC}\right) + \delta \cdot \ln\left(I_T / I_{T,STC}\right)\right) \quad (77.27)$$

with η_{STC}, $T_{PV,STC}$, and $I_{T,STC}$ the efficiency, PV temperature and solar irradiance at Standard Test Conditions (STC): $T_{PV,STC}$: 25 °C, $I_{T,STC}$: 1000 W/m^2. The temperature coefficient for power γ is taken equal to −0.5%/°C and δ for poly-crystalline PV modules 0.11 and for mono-crystalline 0.085 [15].

Having predicted the PV temperature at the centre of the cell, the proposed simulation algorithm predicts the PV temperature at the front and back of the PV module according to the following continuity principles:

- *thermal losses from cell to glass = thermal losses from cell top to environment*

$$\frac{1}{R_{c-f}}\left(T_{pv,c} - T_{pv,f}\right) = \frac{1}{R_{c-f} + \frac{1}{U_{L,f}}}\left(T_{pv,c} - T_a\right) \quad (77.28)$$

- *Thermal losses cell to back = thermal losses from cell back to environment.*

$$\frac{1}{R_{c-b}}\left(T_{pv,c} - T_{pv,b}\right) = \frac{1}{R_{c-b} + \frac{1}{U_{L,b}}}\left(T_{pv,c} - T_a\right) \quad (77.29)$$

Therefore, the PV temperature at the front glass and at the back PV surface is predicted by:

$$T_{pv,f} = \frac{T_{pv,c} + T_a R_{c-f} U_{L,f}}{R_{c-f} U_{L,f} + 1} \quad (77.30)$$

$$T_{pv,b} = \frac{T_{pv,c} + T_a R_{c-b} U_{L,b}}{R_{c-b} U_{L,b} + 1} \quad (77.31)$$

The simulation algorithm predicts $T_{pv,c}$, $T_{pv,f}$ and $T_{pv,b}$ iteratively with very fast convergence.

77.3 Experimental Data

To test the accuracy of the PV temperature predictions from the proposed simulation model, experimental data were captured from two PV configurations. The first configuration included a two-axis sun-tracking PV system of 4 poly-crystalline Si modules 120 W_p each, *see* Fig. 77.2(a), with PV module parameters monitored on clear-sky days for the duration of a year. These included solar irradiance on the inclined PV plane and module temperature monitored for 4 min every hour. The

(a) (b)

Fig. 77.2 (**a**) two-axis sun-tracking PV system, (**b**) BIPV test cell

modules' inclination and orientation range tested as realised by the sun-tracking system during the year were: $15° \leq \beta \leq 85°$ and $75° \leq \gamma_{PV} \leq 285°$. Additionally the environmental parameters monitored were global and diffuse solar irradiance at horizontal, ambient temperature, wind speed and wind direction in 1 min intervals. The wind speed was corrected to the module height. The two stations were synchronised and 4 min averages of the above parameter values were extracted.

The second configuration included a BIPV experimental test cell comprising of 2 mono-crystalline Si modules 55 W_p each integrated on the rooftop of the test cell at an inclination angle $\beta = 15°$ and South-orientation $\gamma_{PV} = 0°$, *see* Fig. 77.2(b). The same environmental parameters as above were recorded at 1 min intervals, as well as the following BIPV parameters monitored PV module temperature at the back of the modules, solar irradiance on the inclined PV plane and room temperature. The stations were synchronised and data averaged for the same period.

77.4 Results and Analysis

The simulation results from the proposed model with the two-axis sun-tracking PV system are shown in Figs. 77.3 and 77.4 for 3 consecutive days in January and in July. The PV module temperatures at the cell, front and back surface of the PV module were predicted and display higher cell temperature by 0.5–1 °C from that of the back surface which in turn is higher by up to 1 °C from the front PV glass. It is observed that during high wind speed conditions the front and back PV temperatures coincide (see for example Fig. 77.4 for wind speed 6 m/s).

The predicted PV temperature by the simulation algorithm lies very close to the measured values as shown in Figs. 77.5 and 77.6 and outperforms the other well known models [5, 6, 8]. It is noteworthy that Faiman's model shows good predictions particularly at low wind speeds, while larger deviations are observed for mid to high wind speeds. Mani's model shows very good performance in July as shown in Fig. 77.6.

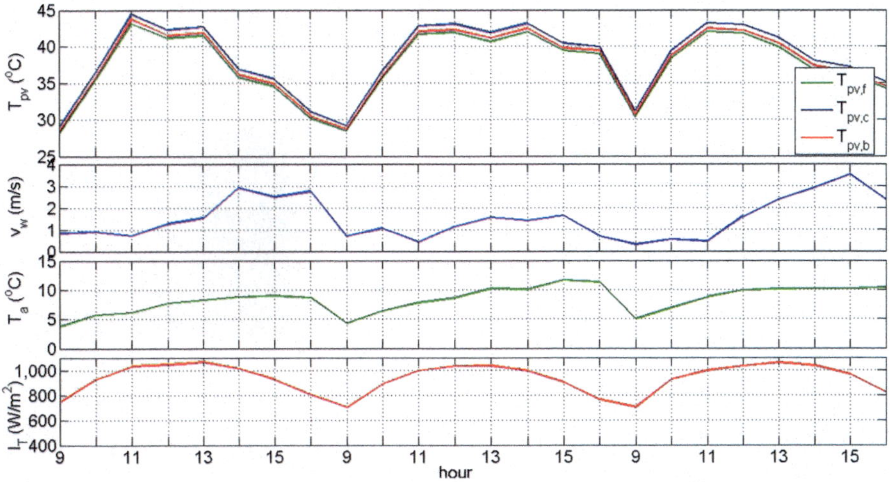

Fig. 77.3 PV temperature at the cell $T_{pv,c}$, back surface $T_{pv,b}$ and front glass $T_{pv,f}$ predicted by the simulation algorithm for the sun-tracking PV system during 3 consecutive days in January. The wind speed v_w, ambient temperature T_a and solar irradiance on PV plane I_T are shown in the subplots

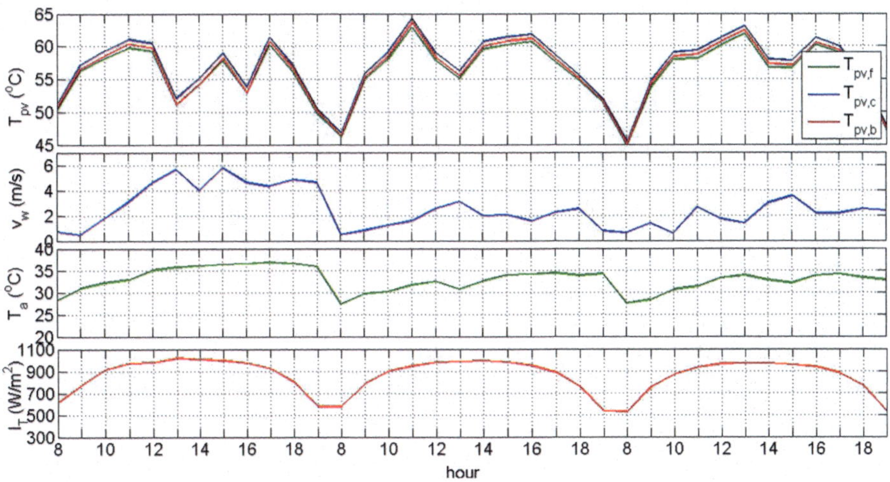

Fig. 77.4 Same as Fig. 77.3 for 3 consecutive days in July

The PV temperature predicted for the BIPV configuration shows the back PV temperature from the cell temperature to be between 0 and 0.5 °C lower and the front glass temperature to be up to 2 °C lower than the cell temperature, with higher differences observed when the front side is windward and lower when leeward (Fig. 77.7). The predicted PV temperature by the simulation algorithm lies very close to the measured values as shown in Fig. 77.8 and outperforms the other models. The other

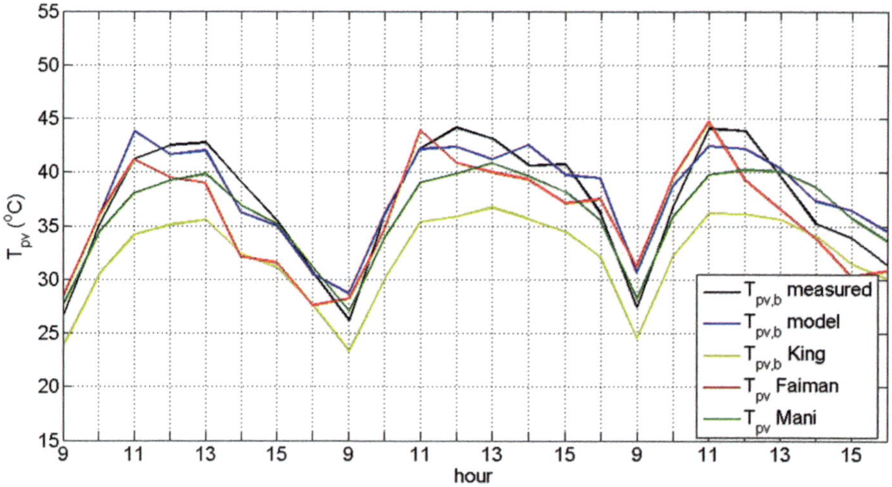

Fig. 77.5 Comparison of the predicted PV temperature by the proposed simulation algorithm with measured data and predictions by other well-known models [5, 6, 8] for the 3 consecutive days in January (*see* Fig. 77.3 for the environmental parameters during this period)

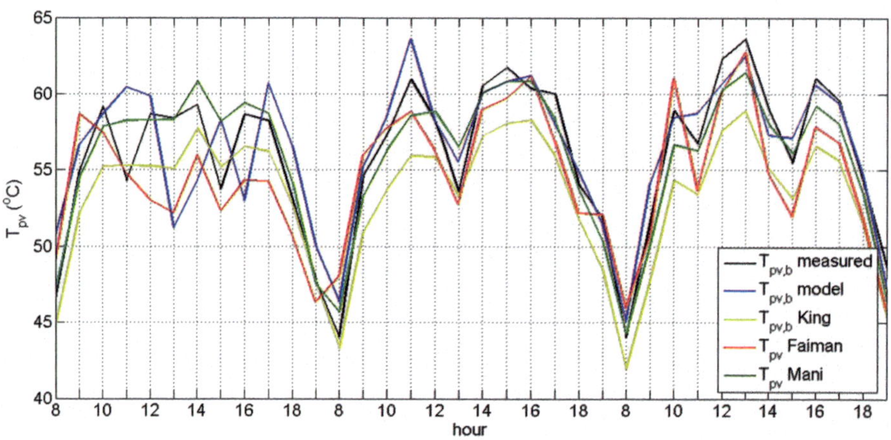

Fig. 77.6 Same as Fig. 77.5 but for the 3 consecutive days in July (*see* Fig. 77.4 for the environmental parameters during this period)

models improve somewhat at very low wind speeds. It is noted that the coefficients a, b in King's model Eq. (77.1) were those corresponding to the insulated back in the case of the BIPV configuration. As shown the PV temperature generally follows the solar irradiance profile, with wind speed playing also a significant role.

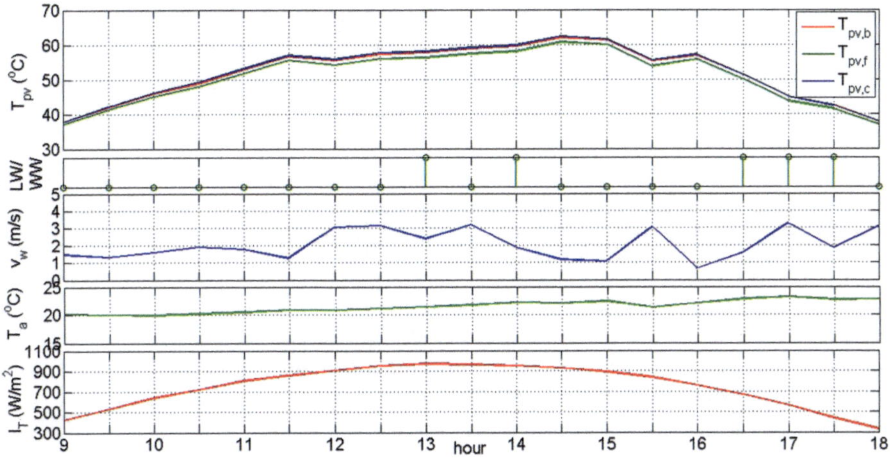

Fig. 77.7 PV temperature at the cell $T_{pv,c}$, back surface $T_{pv,b}$ and front glass $T_{pv,f}$ predicted by the simulation algorithm for the BIPV configuration for a day in May. The wind speed v_w, ambient temperature T_a and solar irradiance on PV plane I_T and whether the front side is leeward (LW) displayed as 0 or windward (WW) as 1 are shown in the subplots

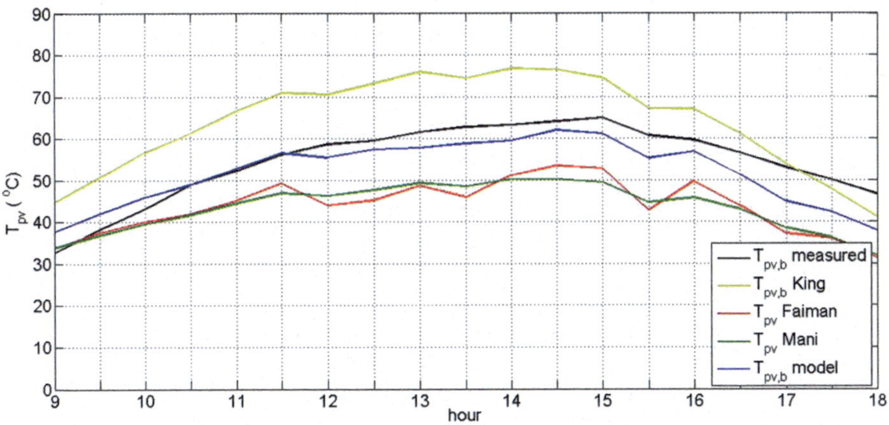

Fig. 77.8 Comparison of the predicted PV temperature by the proposed simulation algorithm with measured data and predictions by other well-known models [5, 6, 8] for a day in May (*see* Fig. 77.7 for the environmental parameters during this period)

77.5 Conclusions

The thermal model developed and presented is theoretical, analytical and predicts PV temperature at the cell, front glass and back surface with high accuracy. The results were validated with measured data captured for the duration of 1 year from a sun-tracking PV system and also from a BIPV system. It was tested for a wide range of inclination angles, orientations, and environmental conditions including solar irradiance, ambient temperature, wind speed and wind direction. The model results were also compared with those produced by some of the well-known models empirical, semi-empirical, NN-based, and was shown in general to outperform them. Empirical models are very easy to apply but in most cases are fine-tuned to the experimental set-up, location, technology type, mounting configuration, at which the empirical parameters were derived. Theoretical models give an advantage over the empirical models and may be successfully applied for any PV module orientation, inclination, environmental conditions, and mounting configurations. The estimated air forced convection coefficient significantly affects the prediction of PV temperature and therefore it is important the coefficient is theoretically derived. Sartori's equations used in the simulation model provided an excellent estimate of the air forced convection coefficient. The thermal radiation emission from the front and the back of the PV module is another critical component that needs to be included in the prediction models, whereas several existing models neglect this.

The robustness of the simulation algorithm developed in the prediction of PV module temperature was presented and its clear advantage over empirical models was illustrated. Further work is currently carried out in including transient phenomena related to fluctuations in the solar irradiance.

References

1. King DL, Kratochvil JA, Boyson WE (1997) Temperature coeffficients for PV modules and arrays: Measurement, methods, difficulties and results. In: Proceeding of 26th IEEE Photovoltaic Specialists Conference Anaheim, California, Sept 29–Oct 3
2. Skoplaki E, Palyvos JA (2009) On the temperature dependence of photovoltaic module electrical performance: a review of efficiency/power correlations. Sol Energy 83:614–624
3. Kaplanis S, Kaplani E (2014) On the relationship factor between the PV module temperature and the solar radiation on it for various BIPV configurations. AIP Conf Proc 1618:341 . (2014). https://doi.org/10.1063/1.4897744
4. Kaplanis S, Kaplani E (2015) Intelligent energy buildings based on RES and nanotechnology. AIP Conf Proc 1702:150005 . (2015). https://doi.org/10.1063/1.4938926
5. King DL, Boyson WE, Kratochvil JA (2004) Photovoltaic array performance model, SANDIA Report, SAND2004–3535. Sandia National Laboratories, New Mexico
6. Faiman D (2008) Assessing the outdoor operating temperature of photovoltaic modules. Prog Photovol 16:307–315
7. Koehl M, Heck M, Wiesmeier S, Wirth J (2011) Modeling of the nominal operating cell temperature based on outdoor weathering. Sol Energy Mater Sol Cell 95:1638–1646
8. Tamizhmani G, Ji L, Tang Y, Petacci L, Osterwald C (2003) Photovoltaic module thermal/wind performance: Long term monitoring and model development for energy rating. NREL/CD-520-33586, 936-939
9. Mattei M, Notton G, Cristofari C, Muselli M, Poggi P (2006) Calculation of the polycrystalline PV module temperature using a simple method of energy balance. Renew Energy 31:553–567
10. Kaplani E, Kaplanis S (2014) Thermal modelling and experimental assessment of the dependence of PV module temperature on wind velocity and direction, module orientation and inclination. Sol Energy 107:443–460

11. Duffie JA, Beckman WA (2013) Solar engineering of thermal processes, 4th edn. Wiley, Hoboken
12. Sartori E (2006) Convection coefficient equations for forced air flow over flat surfaces. Sol Energy 80:1063–1071
13. White FM (1988) Heat and mass transfer. Addison-Wesley Publishing, Boston
14. Churchill SW (1977) A comprehensive correlating equation for laminar, assisting, forced and free convection. AICHE J 23(1):10–16
15. Anderson AJ (1996) Photovoltaic translation equations: A new approach. In: Final Subcontract Report NREL/TP-411-20279

Chapter 78
Potential of Adsorption Refrigeration System for Off-Grid Cooling Applications

Michael John, Cuthbert Z. M. Kimambo, Trygve M. Eikevik, Ole J. Nydal, and Joseph Kihedu

78.1 Introduction

Significant implications to the environment such as ozone layer depletion, global warming and huge electrical energy consumption associated with the use of conventional halogen-based refrigeration systems have made scientists search for environmental friendly refrigerating technologies [1]. Many rural areas of developing countries lack access to grid electricity. The International Energy Agency (IEA) reported that more than 20% of the world's population lacked access to electricity in the year 2010. Among them, 57% lived in rural areas of sub-Saharan Africa and had no hope of being connected to the centralised power supply soon [2].

Among the important application for refrigeration in the health sector is to keep vaccines at appropriate +2 °C to +8 °C temperatures in intermediate vaccine stores and in health facilities [3]. Water-ammonia kerosene and gas-driven absorption refrigerators have been used to store vaccines. However, they do not meet the standards established by the WHO on Performance, Quality and Safety System [4]. PV-powered cooling systems preserve vaccines more efficiently and in an environmental friendly manner. However, batteries are needed [5]. Batteries live shorter than refrigerators, implying extra costs. Also, PV systems have low possibility of being manufactured in most developing countries [6, 7]. Preservation of food is also an important application as FAO estimated that 32% of all food produced in the world was wasted in the year 2009 [8]. Adsorption refrigeration systems can utilise low-temperature waste heat or renewable energy sources like solar thermal energy to produce a cooling effect [9]. Also, the use of heat generated by burning agricultural

M. John (✉) · C. Z. M. Kimambo · J. Kihedu
Department of Mechanical and Industrial Engineering, University of Dar es Salaam, Dar es Salaam, Tanzania

T. M. Eikevik · O. J. Nydal
Department of Energy and Process Engineering, Norwegian University of Science and Technology, Trondheim, Norway

A. Sayigh (ed.), *Renewable Energy and Sustainable Buildings*, Innovative Renewable Energy, https://doi.org/10.1007/978-3-030-18488-9_78

waste or biomass in general is possible, in the remote parts of developing countries or islands where conventional cooling is difficult [10, 11]. Adsorption refrigerators use clean and renewable energy resource, operate with environmentally harmless refrigerants and can be manufactured with locally available resources [10, 12].

78.2 The Basic Adsorption Refrigeration Cycle

The basic adsorption refrigeration cycle was shown in Fig. 78.1. It involves four basic steps: (a), (b), (c) and (d), as explained below [13].

(a) Heating and pressurising: At this stage, the adsorbent contains a large concentration of refrigerant. The adsorbent bed receives heat which causes an increase of temperature and pressure as well. The mass in the generator is considered to remain constant until the condensing pressure is reached.
(b) Heating and desorption: The adsorbent bed continues receiving heat while it is connected to the condenser; thus, the condenser pressure prevails within the system. Further increase in the adsorbent temperature induces desorption of the refrigerant vapour and the vapour condenses in a condenser. The concentration of refrigerant decreases as the temperature increases from the lower to the upper generating temperatures.
(c) Cooling and depressurisation: The adsorbent bed is cooled, which induces a change in pressure from condensing to evaporation pressure. The mass in the generator is considered to remain constant until the evaporating pressure is reached.
(d) Cooling and adsorption: The adsorption bed is cooled while connected to the evaporator; thus, the evaporator pressure prevails within the system. The adsorbent temperature continues decreasing, which induces adsorption of the refrigerant vapour that is adsorbed in the adsorption bed. This adsorbed vapour is vaporised in the evaporator with the evaporator heat being supplied by the heat source at lower temperature. The concentration increases as the temperature decreases from initial adsorption to final adsorption temperature.

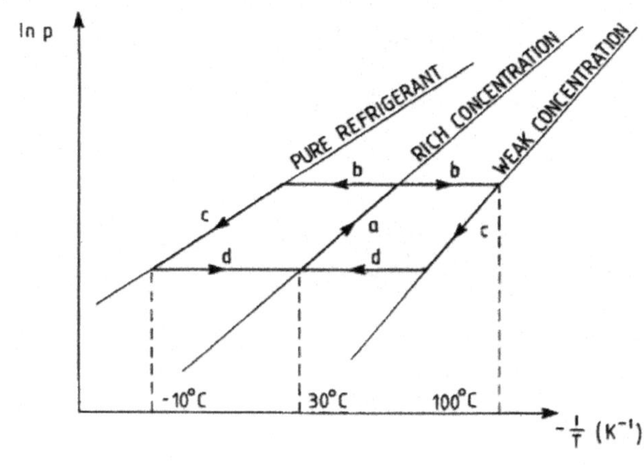

Fig. 78.1 The basic adsorption refrigeration cycle [14]

78.3 Development on Adsorption Prototypes

Several adsorption refrigeration prototypes have been reported to have been developed for different applications ranging from ice making, cooling of food products and vaccines, air conditions and for hybrid system of hot water and ice making. Investigated working pairs for adsorption ice maker prototypes include activated carbon-methanol pair [15, 16], monolithic carbon-ammonia pair [17] and compound adsorbent (activated carbon-$CaCl_2$) and ammonia pair [18]. In food products and vaccines storages, silica gel-water [19–21], activated carbon-ethanol [22] and activated carbon-methanol [2, 7, 23] were among the reported adsorption pairs. The AQSOA-FAM-Z02-water [24], silica gel-water [25–28] and carbon-ammonia [29] were used in air-condition applications. Wang et al. [30] used activated carbon-methanol pair for hybrid system of solar-powered water heater for hot water supply and ice maker. Figures 78.2, 78.3 and 78.4 show various types of adsorption system prototypes.

Adsorption refrigeration systems have been reported to have low coefficient of performance (COP) as compared with the conventional refrigerators [26, 31]. Maximum-cycle COP of 0.83 was realised by system-employing composites of silica gel and chloride-water pair [32]. Boubakri et al. [15] presented a solar COP from 0.08 to 0.12 for solar ice makers producing 5.2 kg of ice with temperature between −15 °C and −5 °C a day. The COP of 0.120 and SCP of 60 W/kg-carbon were reported for improved ice-making machine [17]. Lu et al. [18] reported improved performance from 161.2 W/kg and 0.12 to 770.4 W/kg and 0.39 for SCP and COP, respectively, by using compound adsorbent of activated carbon and $CaCl_2$-ammonia for ice maker for fishing boat driven by waste heat from exhaust and solar ice maker driven by a solar water heater. Hildbrand et al. [19] reported a gross solar cooling COP of 0.10–0.25 by the cylindrical tubes adsorber and the solar flat plate, 2 m^2 double glazed collector with air-cooled condenser prototype shown in Fig. 78.4. Frazzica et al. [22] reported a power density of 95 W/kg and 50 W/kg and COPs of

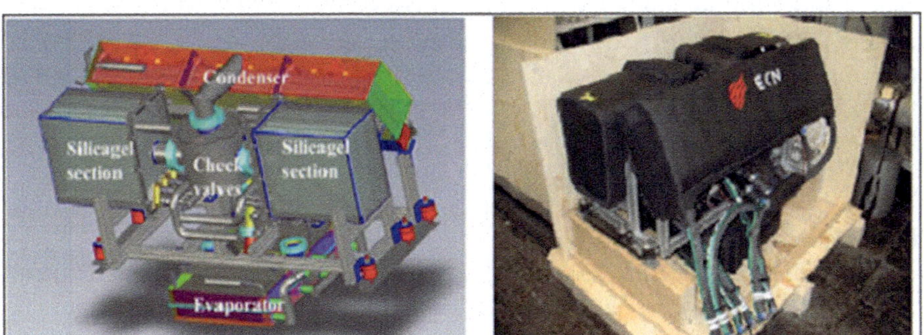

(a) (b)

Fig. 78.2 Prototype adsorption chiller (**a**) drawing and (**b**) prototype [25]

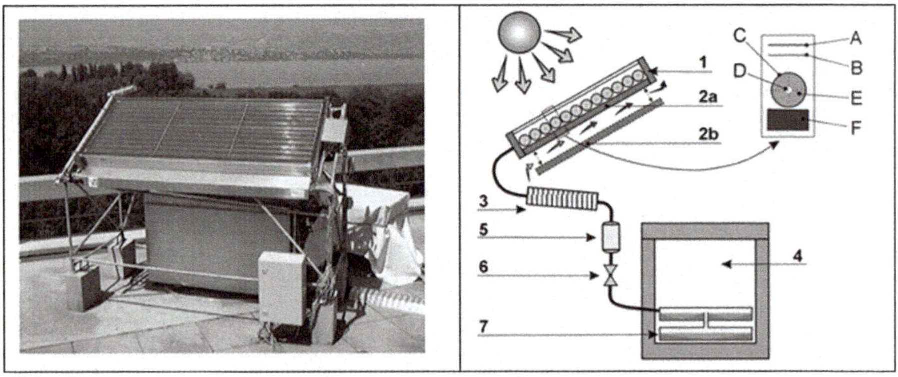

Fig. 78.3 Photograph and plan of adsorptive solar refrigerator [19]

Fig. 78.4 Photograph of experimental solar adsorption refrigerator [20]

0.09–0.11 for air conditioning and refrigeration cycles, respectively. Bouzeffour and Khelidj [20] presented a solar COP of 0.083–0.09, with a daily global irradiance of 773–837 W/m^2 and maximum adsorbent bed temperatures of 95–117 °C from total energy received by solar collector of 18 MJ/m^2 to provide +5 °C to +8 °C as evaporator temperatures; this prototype is shown in Fig. 78.5. Anyanwu and Ezekwe [7] presented adsorption cycle with the overall COPs ranging between 0.056–0.093 cycle and 0.007–0.015 solar, respectively, for evaporator temperatures 1.0–8.5 °C which is recommended for drugs, fruits and vegetables with preservation temperatures in the ranges of 4–16 °C.

Lemmini and Errougani [23] presented the solar-powered adsorption refrigerator using activated carbon AC35-methanol which produced cold air of 0 °C to 5 °C with COP of 0.05–0.08 for an irradiation of 12,000–27,000 kJ/m^2 and a daily mean ambient temperature of 14–18 °C. Magnetto et al. [25] demonstrated the concept of a waste heat driven adsorption cooling system using silica gel-water pair for comfort cooling purposes in vehicles, which produced 2 kW of chilling power with a COP

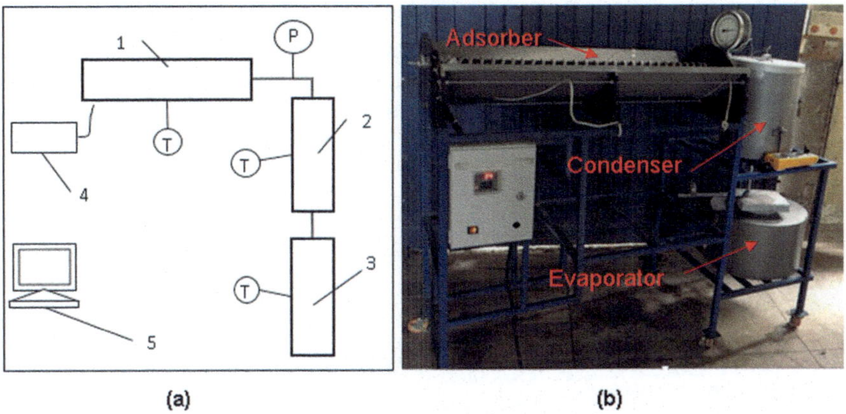

Fig. 78.5 Experimental adsorption refrigeration unit (**a**) Schematic; *1* adsorber, *2* condenser, *3* evaporator, *4* temperature control system, *5* data collection system, *P* pressure gauge and *T* thermocouples. (**b**) Photograph

of 0.4; Fig. 78.3 shows the prototype. Xia et al. [26] presented an adsorption chiller driven by hot water of 60–90 °C which produced cooling power of 8.70 kW and COP of 0.39 for the heat source of 82.5 °C, cooling water of 30.4 °C and chilled water outlet temperature of 12 °C. Critoph et al. [29] presented a laboratory prototype refrigerator with heating power input of 1.130 kW, cooling power of 0.5 kW, COP of 0.44 and SCP of 0.180 kW/kg-carbon.

78.4 Laboratory Experiments

The adsorption refrigerator laboratory prototype (Fig. 78.5) consists of the adsorber which contains the adsorbent-adsorbate pair of calcium chloride-ammonia, the water-cooled condenser and the evaporator which contain the cold chamber, an electric tape heater and controller, connecting pipeline and affiliated parts. The refrigeration rate was tested at different desorption temperatures and desorption times. The adsorbent bed was heated by temperature-controlled electric tape heater thermocoax isopad IT-20. Water-cooled condenser of 36 litres of water was used while cold chamber was filled up with 0.5 L of water. Experiments have been conducted by varying the desorption temperatures from 75 °C to 100 °C and desorption time from 1 to 4 h. Adsorber temperature, condenser water temperature, cold chamber water temperature and ambient temperature have been recorded using Pico TC-08 USB Thermocouple Data Logger with PicoLog Data Logging Software.

The adsorption refrigeration cycle includes two processes [33]: adsorption refrigeration process and heating desorption process. At low temperatures, adsorbent adsorbs refrigerant evaporates in the evaporator. Liquid refrigerant evaporates by adsorbing heat from the refrigeration cold box. Thus, the refrigeration occurs until the adsorbent is saturated. When the adsorber rich with adsorbed refrigerant is heated, temperature and pressure increase. When the system pressure increases to condensation pressure, the desorbed refrigerant gas condenses in the condenser. The adsorbent will cool down by natural convection which will also decrease the pressure. Then, the next adsorption refrigeration cycle begins when the pressure is reduced to evaporation pressure.

78.5 Results and Discussions

78.5.1 Adsorption Refrigeration Rate

Figures 78.6 and 78.7 show the trends of adsorption refrigeration rate of $CaCl_2$ and ammonia at different desorption temperatures. Figure 78.6 shows the adsorption rate after heating desorption of 2 h for various desorption temperatures while Fig. 78.7 shows the adsorption rate after the heating desorption of 4 h. The evaporation temperature decreases rapidly at the beginning period of the adsorption refrigeration process then the rate decreases. Low temperatures at the cold chamber was attained by higher desorption temperatures and longer desorption time.

Adsorption rate trend for CaI_2-ammonia for this experiment followed similar trends for CaI_2-ammonia adsorption rate as reported by Wang et al. [33] and shown in Fig. 78.8. However, in the present work, the initial adsorption temperature was observed to be higher due to the high ambient temperature. It is also important to note that the adsorption refrigeration system deployed in present work has no valve installed between adsorber and evaporator, and therefore the adsorption process was self-started.

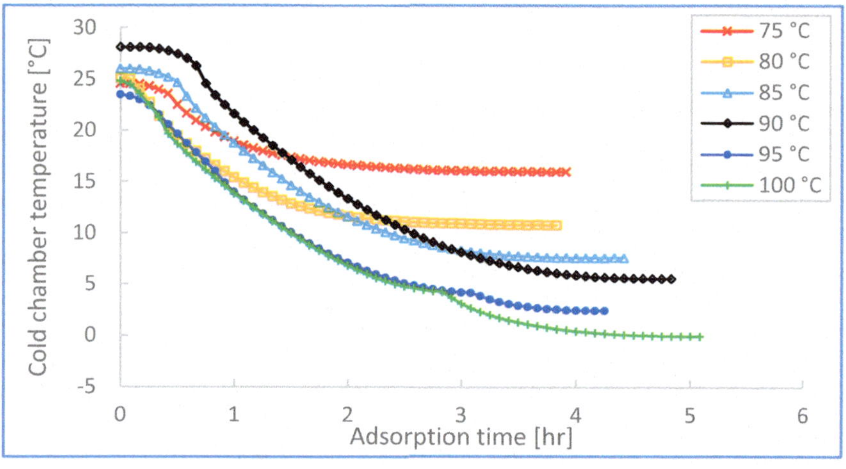

Fig. 78.6 Adsorption refrigeration rate for 2-h desorption time

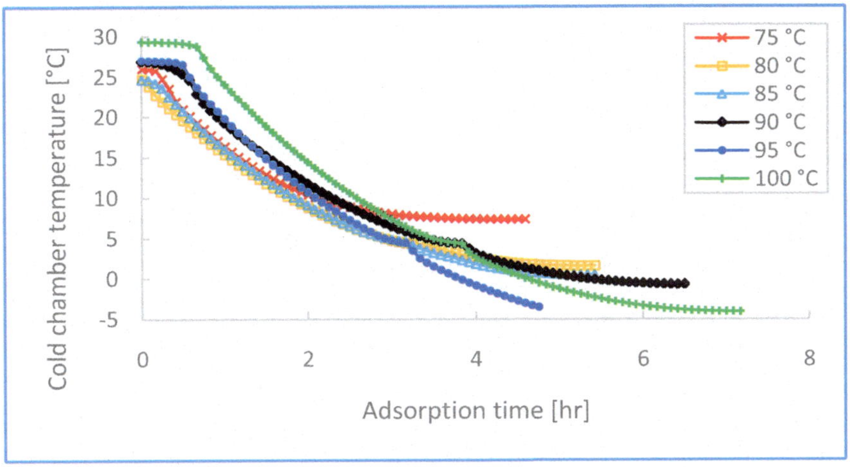

Fig. 78.7 Adsorption refrigeration rate for 4-h desorption time

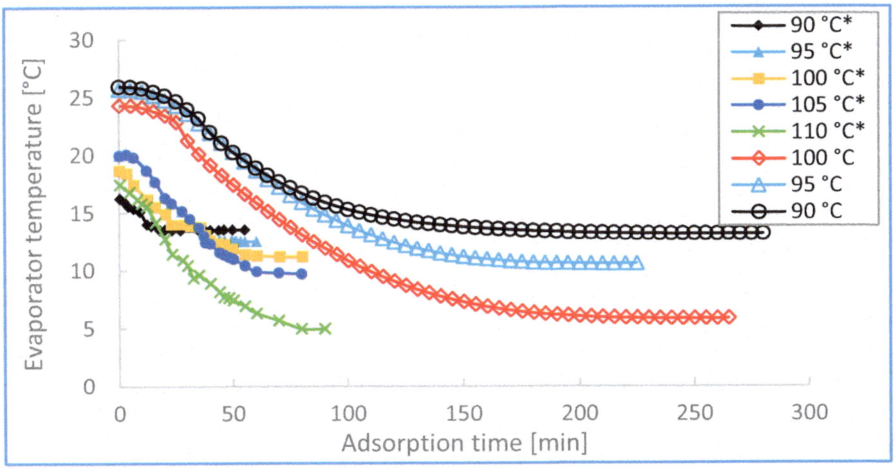

Fig. 78.8 Adsorption refrigeration rate of $CaCl_2$ compared to Wang et al. [33]

78.5.2 Cold Chamber Temperature

Figure 78.9 show the minimum attained cold chamber temperatures at different heating desorption times for various heating desorption temperatures. For heating desorption temperatures from 85 °C and heating desorption time greater than 1 h, the cold chamber of the prototype obtained temperature between 2 °C and 8 °C which is suitable for storage of food products and vaccines [3]. Temperature below 0 °C which is suitable for ice making was obtained for heating desorption temperature greater than 95 °C with heating desorption time from 2 h. Heating desorption temperature below 85 °C can be used for air-condition application as they have provided cold

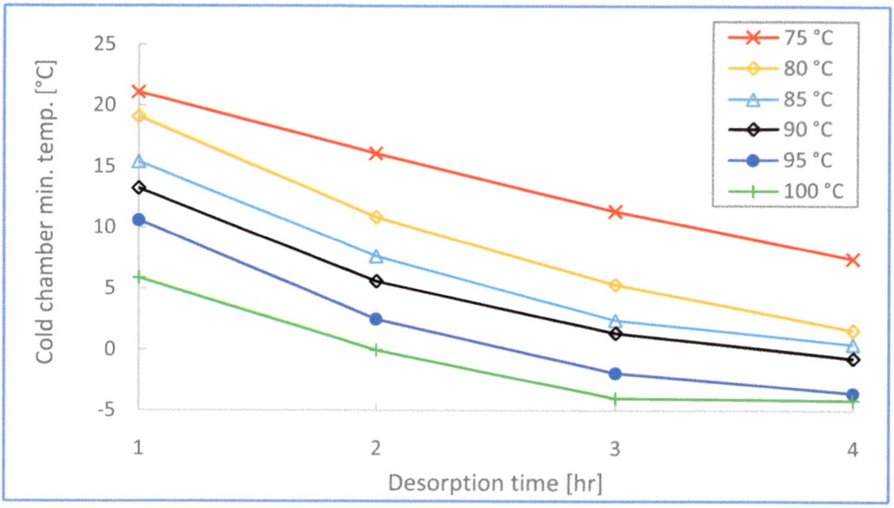

Fig. 78.9 Cold chamber minimum attained temperature

chamber temperature below 15 °C for heating desorption time greater than 1 h. The tested desorption temperatures are common temperatures which can be obtained by flat plate and evacuated tube solar collectors.

78.6 Conclusion

Adsorption refrigeration systems have shown great potential to meet cooling demands in off-grid areas. The tested laboratory adsorption prototype attained temperatures potential for storage of vaccines and foods products, producing ice or in air conditioning by using low-temperature heat source of 75–100 °C. The tested temperature can be obtained by flat plate and evacuated tube solar collectors. Also, the use of heat generated by burning agricultural waste or biomass in general is possible, in the remote parts of developing countries or islands where conventional cooling is difficult. Besides, adsorption refrigerators use clean and renewable energy resource, operate with environmentally harmless refrigerants and can be manufactured with locally available resources.

References

1. Anupam K, Palodkar AV, Halder G (2016) Experimental study on activated carbon-nitrogen pair in a prototype pressure swing adsorption refrigeration system. Heat Mass Transf 52(4):753–761
2. Allouhi A, Kousksou T, Jamil A, Agrouaz Y, Bouhal T, Saidur R, Benbassou A (2016) Performance evaluation of solar adsorption cooling systems for vaccine preservation in sub-Saharan Africa. Appl Energy 170:232–241

3. WHO (2006) Temperature sensitivity of vaccines. WHO, Geneva, pp 1–62
4. McCarney S, Robertson J, Arnaud J, Lorenson K, Lloyd J (2013) Using solar-powered refrigeration for vaccine storage where other sources of reliable electricity are inadequate or costly. Vaccine 31(51):6050–6057
5. Tina GM, Grasso AD (2014) Remote monitoring system for stand-alone photovoltaic power plants: the case study of a PV-powered outdoor refrigerator. Energy Convers Manag 78:862–871
6. Axaopoulos PJ, Theodoridis MP (2009) Design and experimental performance of a PV icemaker without battery. Sol Energy 83(8):1360–1369
7. Anyanwu EE, Ezekwe CI (2003) Design, construction and test run of a solid adsorption solar refrigerator using activated carbon/methanol, as adsorbent/adsorbate pair. Energy Convers Manag 44(18):2879–2892
8. Lipinski B, Hanson C, Lomax J, Kitinoja L, Waite R, Searchinger T (2013) Reducing food loss and waste: working paper. World Resources Institute, Washington, DC
9. El-Sharkawy II, Saha BB, Koyama S, He J, Ng KC, Yap C (2008) Experimental investigation on activated carbon–ethanol pair for solar powered adsorption cooling applications. Int J Refrig 31(8):1407–1413
10. Ullah KR, Saidur R, Ping HW, Akikur RK, Shuvo NH (2013) A review of solar thermal refrigeration and cooling methods. Renew Sust Energ Rev 24:499–513
11. Tamainot-Telto Z, Metcalf SJ, Critoph RE, Zhong Y, Thorpe R (2009) Carbon-ammonia pairs for adsorption refrigeration applications: ice making, air conditioning and heat pumping. Int J Refrig 32(6):1212–1229
12. Allouhi A, Kousksou T, Jamil A, Bruel P, Mourad Y, Zeraouli Y (2015) Solar driven cooling systems: an updated review. Renew Sust Energ Rev 44:159–181
13. Zhong Y (2006) Studies on equilibrium and dynamic characteristics of new adsorption pairs. PhD thesis, University of Warwick
14. Critoph RE (1989) Activated carbon adsorption cycles for refrigeration and heat pumping. Carbon 27(1):63–70
15. Boubakri A, Arsalane M, Yous B, Ali-Moussa L, Pons M, Meunier F, Guilleminot J (1992) Experimental study of adsorptive solar-powered ice makers in Agadir (Morocco)-1. Performance in actual site. Renew Energy 2(1):7–13
16. Li M, Wang RZ, Xu Y, Wu J, Dieng A (2002) Experimental study on dynamic performance analysis of a flat-plate solar solid-adsorption refrigeration for ice maker. Renew Energy 27(2):211–221
17. Tamainot-Telto Z, Critoph R (1997) Adsorption refrigerator using monolithic carbon-ammonia pair. Int J Refrig 20(2):146–155
18. Lu Z, Wang RZ, Wang LW, Chen C (2006) Performance analysis of an adsorption refrigerator using activated carbon in a compound adsorbent. Carbon 44(4):747–752
19. Hildbrand C, Dind P, Pons M, Buchter F (2004) A new solar powered adsorption refrigerator with high performance. Sol Energy 77(3):311–318
20. Bouzeffour F, Khelidj B (2016) Experimental investigation of a solar adsorption refrigeration system working with silicagel/water pair: a case study for Bou-Ismail solar data. Sol Energy 131:165–175
21. Brites GJVN, Costa JJ, Costa VAF (2016) Influence of the design parameters on the overall performance of a solar adsorption refrigerator. Renew Energy 86:238–250
22. Frazzica A, Palomba V, Dawoud B, Gullì G, Brancato V, Sapienza A, Vasta S, Freni A, Costa F, Restuccia G (2016) Design, realization and testing of an adsorption refrigerator based on activated carbon/ethanol working pair. Appl Energy 174:15–24
23. Lemmini F, Errougani A (2005) Building and experimentation of a solar powered adsorption refrigerator. Renew Energy 30(13):1989–2003
24. Vasta S, Freni A, Sapienza A, Costa F, Restuccia G (2012) Development and lab-test of a mobile adsorption air-conditioner. Int J Refrig 35(3):701–708
25. Magnetto D, de Boer R, Taklanti A (2011) A mobile air conditioning system operated by the engine waste heat. SAE Technical Paper, No. 2011-01-0135

26. Xia ZZ, Wang RZ, Wang DC, Liu YL, Wu JY, Chen CJ (2009) Development and comparison of two-bed silica gel–water adsorption chillers driven by low-grade heat source. Int J Therm Sci 48(5):1017–1025
27. Kubota M, Ueda T, Fujisawa R, Kobayashi J, Watanabe F, Kobayashi N, Hasatani M (2008) Cooling output performance of a prototype adsorption heat pump with fin-type silica gel tube module. Appl Therm Eng 28(2):87–93
28. Alahmer A, Wang X, Al-Rbaihat R, Alam KA, Saha BB (2016) Performance evaluation of a solar adsorption chiller under different climatic conditions. Appl Energy 175:293–304
29. Critoph RE, Tamainot-Telto Z, Davies GL (2000) A prototype of a fast cycle adsorption refrigerator utilizing a novel carbon-aluminium laminate. Proc Inst Mech Eng A 214:439–448
30. Wang RZ, Li M, Xu YX, Wu JY (2000) An energy efficient hybrid system of solar powered water heater and adsorption ice maker. Sol Energy 68(2):189–195
31. Li TX, Wang RZ, Li H (2014) Progress in the development of solid-gas sorption refrigeration thermodynamic cycle driven by low-grade thermal energy. Prog Energy Combust Sci 40:1–58
32. Askalany AA, Salem M, Ismael IM, Ali AHH, Morsy MG, Saha BB (2013) An overview on adsorption pairs for cooling. Renew Sust Energ Rev 19:565–572
33. Wang L, Chen L, Wang HL, Liao DL (2009) The adsorption refrigeration characteristics of alkaline-earth metal chlorides and its composite adsorbents. Renew Energy 34(4):1016–1023

Chapter 79
Deposition of Porous Photocatalytic TiO_2 Film for Dye-Sensitized Solar Cells by Low-Powered Atmospheric Plasma Spray Equipment

Alabi Kelvin Oluwafunmilade, Zine Abidine, Yasutaka Ando, Yoshimasa Noda, and Mitsumasa Iino

79.1 Introduction

The utilization of solar energy becomes increasingly important, as the fossil and mineral sources are limited and also the main sources of environmental pollution. Dye-sensitized solar cells (DSCs) based on nanostructured semiconductor films as a type of photovoltaic device have been considered as promising alternatives to silicon-based solar cells due to their low cost of materials and manufacturing. However, improving their relatively low power conversion efficiency remains a big challenge confronting us because of the highest efficiency of up to 13% compared with ~20% of silicon-based solar cells. A photoelectrode, when loaded with dyes, as a component of DSCs is important due to its role as a light harvester, a charge generator as well as an electron collector. The pore size, specific surface area, surface chemistry and defects, charge mobility and charge recombination at the interface are all critical in achieving high-power conversion efficiency [1].

Sodium chloride, also known as salt, is an ionic compound with the chemical formula NaCl representing a 1:1 ratio of sodium and chloride ions. With molar masses of 22.99 and 35.45 g/mol, 100 g of NaCl contains 39.34 g Na and 60.66 g Cl, density of 2.16 g/mL and melting point of 801 °C, respectively. Sodium chloride is the salt most responsible for the salinity of seawater and of the extracellular fluid of many multicellular organisms. In its edible form of table salt, it is commonly used as a condiment and

A. K. Oluwafunmilade (✉)
Graduate School of Engineering, Ashikaga University, Tochigi Prefecture, Japan

Z. Abidine
Faculty of Engineering, Ashikaga University, Tochigi Prefecture, Japan

Y. Ando
Collaborative Research Center, Ashikaga University, Tochigi Prefecture, Japan

Y. Noda · M. Iino
Department of Mechanical Engineering, Ashikaga University, Tochigi Prefecture, Japan

A. Sayigh (ed.), *Renewable Energy and Sustainable Buildings*, Innovative Renewable Energy, https://doi.org/10.1007/978-3-030-18488-9_79

food preservative. Large quantities of sodium chloride are used in many industrial processes, and it is a major source of sodium and chlorine compounds used as feedstock for further chemical syntheses. A second major application of sodium chloride is de-icing of roadways in sub-freezing weather. Sodium chloride is readily soluble in water and other polar solvents. It is a stable solid which makes it suitable for this experiment.

Consequently, it was confirmed that the porous photocatalytic TiO_2 film could be deposited on the condition of Ar and N_2 has the working gases (on the condition of 1.0 L/min and 0.5 L/min in Ar and N_2 working gas flow rates, respectively). Nevertheless, since Ar is a very expensive gas, low-price working gas is required for practical use of the APS equipment. In this study, to develop a deposition process of porous TiO_2 film for DSSC using 1 kW class APS equipment on the condition of high N_2 concentration, Ar/N_2 working gas was carried out.

79.2 Method

This experiment was conducted in two phases using TiO_2 and TiO_2/NaCl as the feeding stock. Figure 79.1 shows the schematic diagram of the thermal spray equipment used in this study. This equipment consists of plasma torch, DC power source, feedstock supplying system and working gas supplying the system. Except the feedstock supplying system, the constitution of this equipment was the same as the APS and the ASPPS equipment used in our previous studies and conventional high-power thermal plasma spray equipment. In the case of the conventional thermal spray equipment, the mechanically and electrically driven type powder feeder is generally used. On the other hand, since the powder feeder is very expensive, a suction-type powder feeder, which can feed the powder into plasma jet by negative pressure generated by thermal plasma jet, was developed in this study. Table 79.1 shows the film deposition conditions. As the feedstock, commercial TiO_2 (anatase) powder was used. Figure 79.2 shows the appearance of the air plasma jet and after feedstock has been injected.

Fig. 79.1 Schematic diagram of the atmospheric thermal plasma spray equipment

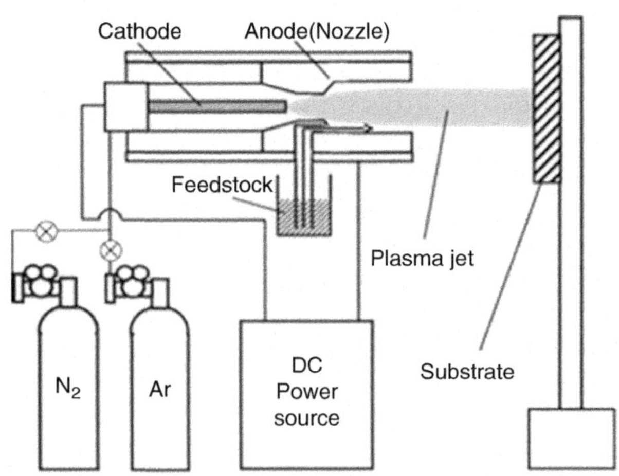

Table 79.1 Film deposition conditions

Plasma working gas	Ar/N_2
Nitrogen gas flow rate	0.5 L/min
Argon gas flow rate	1.0 L/min
Operating discharge power	50 A/20 V
Deposition distance	50–70 mm
Deposition time	1 min
Feedstock materials	TiO_2 (anatase) powder, salt (NaCl)
Substrate	304 stainless steel plate

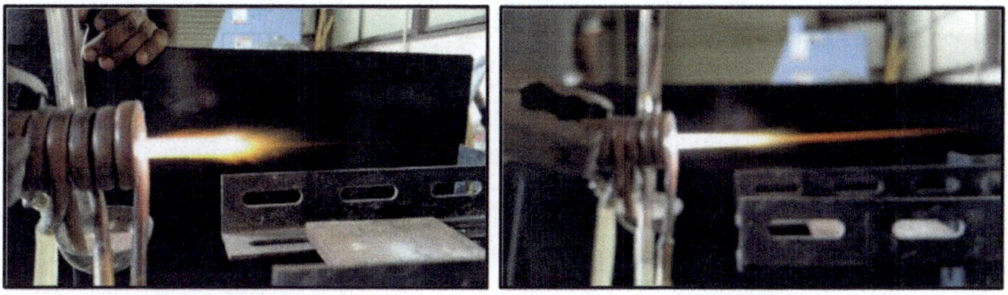

(a) (b)

Fig. 79.2 Appearances of the air plasma jets. (**a**) Air plasma jet. (**b**) Feedstock injected

Fig. 79.3 Ultrasonic cleaner used for NaCl residual removal

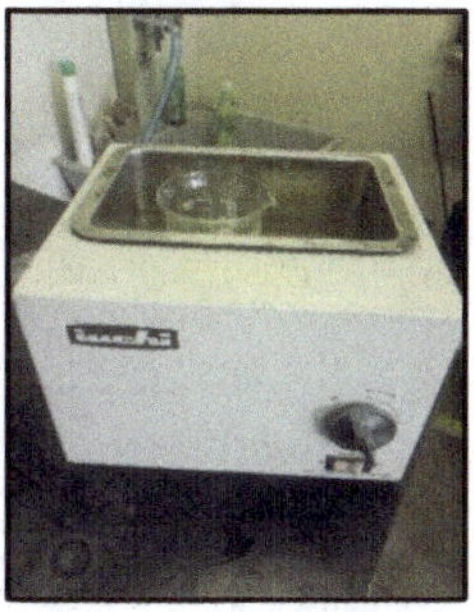

Figure 79.3 shows TiO_2/NaCl sample submerged in a solvent (H_2O) using an ultrasonic cleaner to remove NaCl residual; the sample was cleaned for about 10 min after which X-ray diffraction was conducted. Figures 79.4 and 79.5 show the 15 mm × 15 mm × 1 mm 304 stainless steel plate with grit blasted surface used as substrate before and after deposition has occurred.

The substrate was horizontally set on the substrate holder and the central area of the sample was placed perpendicular to the axial centre of the plasma jet. The input power for discharge was fixed at 50 A. After TiO_2 film deposition, the microstructure of the film was investigated by optical microscope and X-ray diffraction with CuKα at 40 kV and 100 mA.

Fig. 79.4 Substrate before spraying

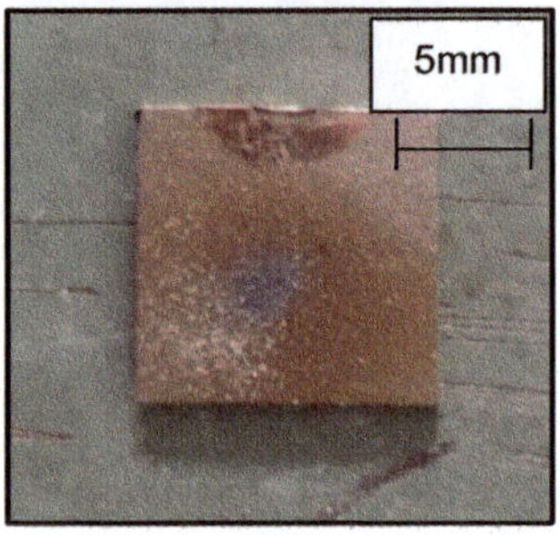

Fig. 79.5 Substrate after spraying

79.3 Results and Discussion

From the experimental results in our previous study [2], the addition of N_2 was proved to be effective in increasing the thermal energy of Ar plasma jet. However, the following are the disadvantages on Ar working gas as well as its cost.

- In the case of Ar plasma jet, thermal energy of the plasma jet was increased with increasing Ar working gas flow rate. On the other hand, thermal energy of the plasma jet was decreased with increasing Ar working gas flow rate on the conditions of N_2 working gas flow rate fixed Ar/N_2 plasma jets because thermal energy of Ar plasma jet is much lower than that of N_2 plasma jet.
- Besides, since feedstock powder feed rate was increased with increasing working gas flow rate in the case of suction-type feedstock feeder, non-melted particle content in the deposited film was increased with increasing Ar working gas flow rate on the conditions of the N_2 flow rate fixed Ar/N_2 plasma jets.
- The NaCl was able to create a porous surface on the TiO_2 after removal with H_2O solution.

Figures 79.6, 79.7, 79.8, 79.9, 79.10 and 79.11 show the XRD patterns of the TiO_2 films and TiO_2/NaCl porous films deposited samples used in this study. Although the colour of the feedstock TiO_2 (anatase) powder was white, the colour of the particles turned into grey during flight in the plasma jet and the grey colour film was deposited, and anatase dominant film was deposited in both cases.

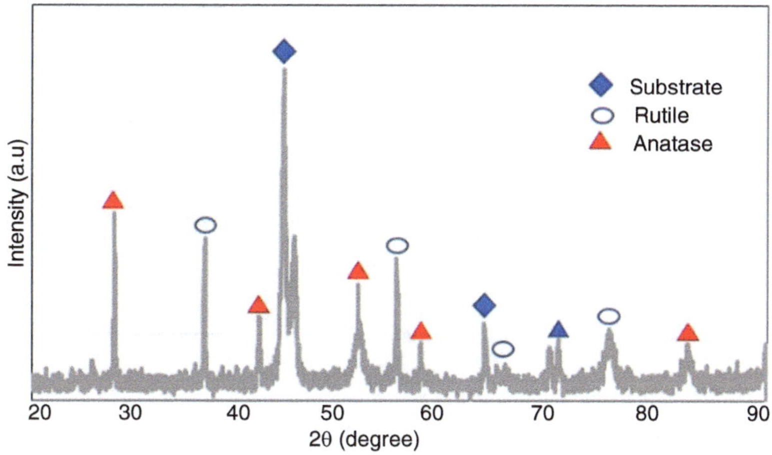

Fig. 79.6 XRD pattern of the TiO_2 film deposited sample (50 mm)

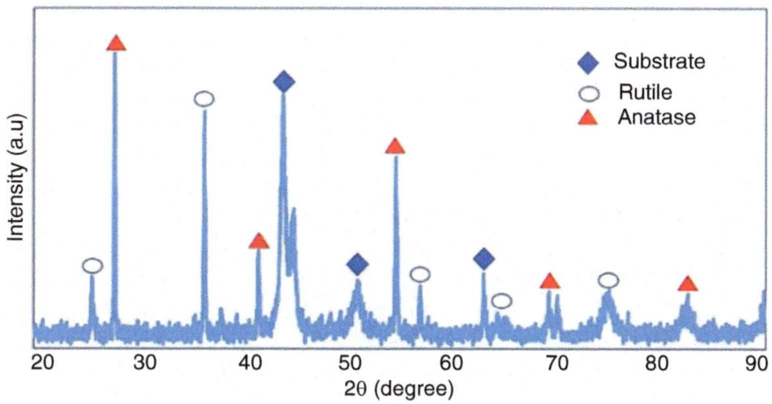

Fig. 79.7 XRD pattern of the TiO_2 film deposited sample (60 mm)

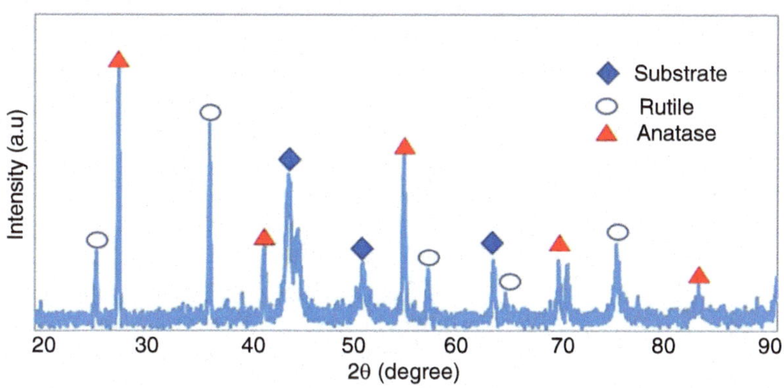

Fig. 79.8 XRD pattern of the TiO_2 film deposited sample (70 mm)

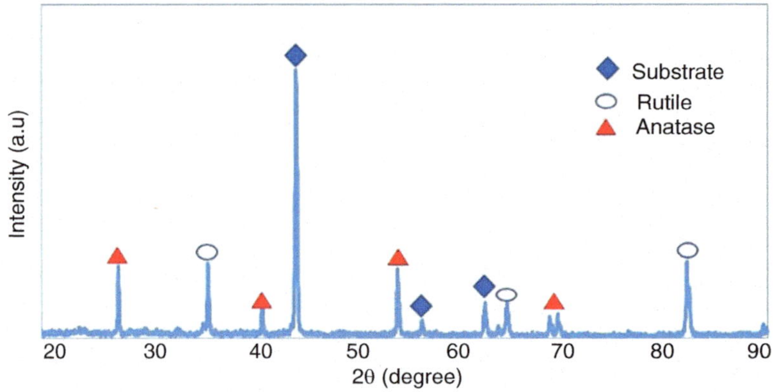

Fig. 79.9 XRD pattern of the TiO_2/NaCl porous film deposited sample (50 mm)

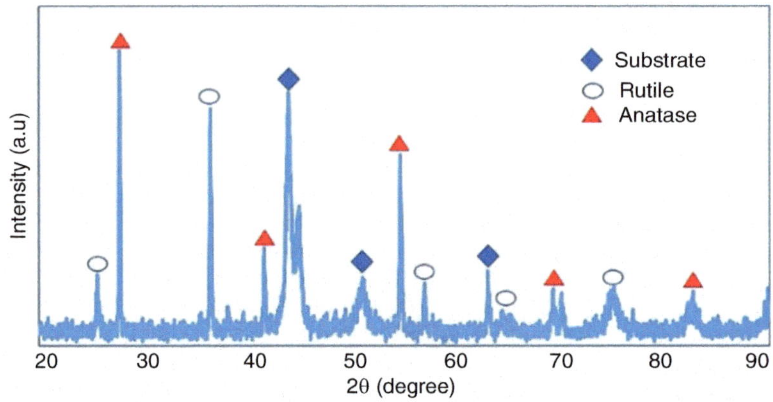

Fig. 79.10 XRD pattern of the TiO_2/NaCl porous film deposited sample (60 mm)

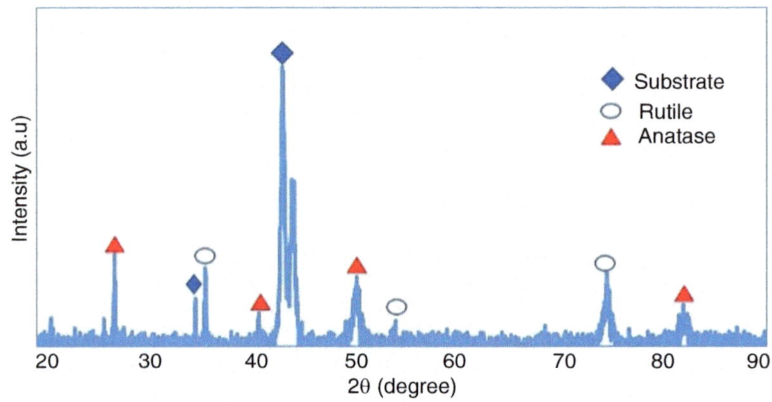

Fig. 79.11 XRD pattern of the TiO_2/NaCl porous film deposited sample (70 mm)

As for photocatalytic properties detection method, methylene blue solution test was carried out. In this experiment, this test was carried out in order to confirm the photocatalytic properties of both oxide films. Table 79.2 shows the result of methylene blue solution test for the TiO_2 film. Firstly, 0.15 mL of methylene blue was added to 300 mL water solution (H_2O); its light absorption value was measured at different concentrations of water (H_2O) to test its photocatalytic properties. Afterwards, each substrate was submerged in a 30 mL glass plate of H_2O/methylene blue solution and kept under UV light for 10 h. The initial and final light absorption values (*X*) were recorded using a light absorption meter; the final volumes of each substrate were also recorded respectively (Figs. 79.12 and 79.13)

Table 79.2 Shows the relationship between H_2O/methylene blue concentration to light absorption value (*X*)

Concentration (H_2O to methylene blue)	Value (*X*)
100	11.7
80	15
40	44
20	62.3
10	90.3
0	108.7

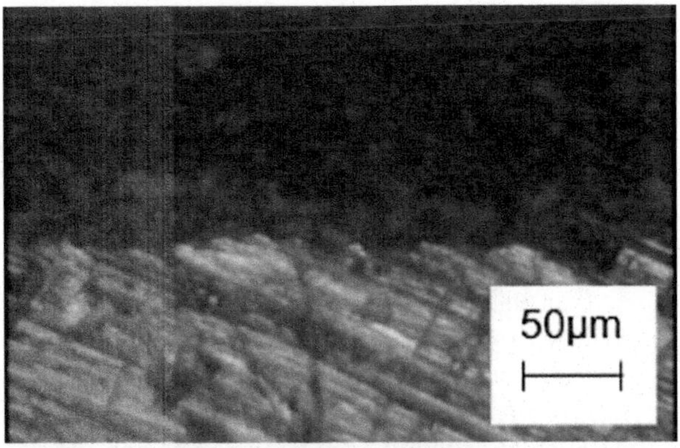

Fig. 79.12 Cross-sectional optical micrograph of s of the TiO_2 film deposited sample (deposition distance = 50 mm)

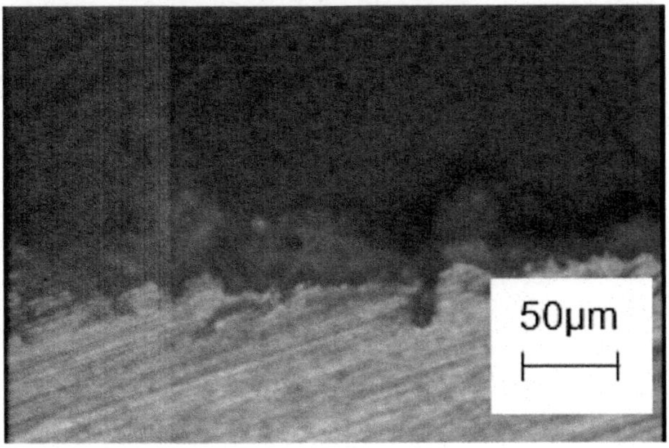

Fig. 79.13 Cross-sectional optical micrograph of the TiO_2/NaCl film deposited sample (deposition distance = 50 mm)

79.3.1 Performance of DSSCs

Equation of graph $(Y) = -0.9972X + 95.511$ = Tentative Concentration
Equation of graph $(Y) = -0.9972X + 95.511$, $Y = f(X)$
For dense TiO_2 film
(***50 mm***) $Y = f(X) = f(12) = -0.9722(12) + 95.511 = 83.84$, Actual Concentration = $(8/30) \times 83.84 = 22.36$
(***60 mm***) $Y = f(X) = f(11.5) = -0.9722(11.5) + 95.511 = 84.33$, Actual Concentratio n = $(9/30) \times 84.33 = 25.30$
(***70 mm***) $Y = f(X) = f(12) = -0.9722(12) + 95.511 = 83.84$, Actual Concentration = $(15/30) \times 83.84 = 41.92$
For porous TiO_2 film
(***50 mm***) $Y = f(X) = f(15.9) = -0.9722(15.9) + 95.511 = 80.05$, Actual Concentratio n = $(5/30) \times 80.05 = 13.34$
(***60 mm***) $Y = f(X) = f(19) = -0.9722(19) + 95.511 = 77.04$, Actual Concentration = $(6/30) \times 77.04 = 15.41$
(***70 mm***) $Y = f(X) = f(12) = -0.9722(10.8) + 95.511 = 85.01$, Actual Concentration = $(13/30) \times 85.01 = 36.84$

79.4 Conclusion

In order to develop a low-cost photovoltaic device manufacturing process, deposition of porous titanium oxide film for photovoltaic device for DSSC was conducted by 1 kW class APS equipment. Consequently, the following results were obtained:

1. Even in the case of 1 kW class APS equipment, titanium oxide film could be deposited on the stainless-steel substrate on the condition of non-cooled substrate holder.
2. The results of X-RD patterns and optical micrograph of the films proved that particle dispensability can be varied by varying deposition distances.
3. From these results, this technique was found to have high potential for high rate, simple and low-cost titanium oxide film deposition.
4. Tables 79.3 and 79.4 show that the deposition distances vary with the light absorption value (X); a lower deposition distance yields a better concentration.

Table 79.3 Shows the results of dense TiO_2 after 10 h exposure to UV light

Deposition distance (mm)	Initial volume (mL)	Final volume (mL)	Light absorption value (X)
50	30	8	12.0
60	30	9	11.5
70	30	15	12.0

Table 79.4 Shows the results of porous TiO_2/NaCl after 10 h exposure to UV light

Deposition distance (mm)	Initial volume (mL)	Final volume (mL)	Light absorption value (*X*)
50	30	5	15.9
60	30	6	19
70	30	13	10.8

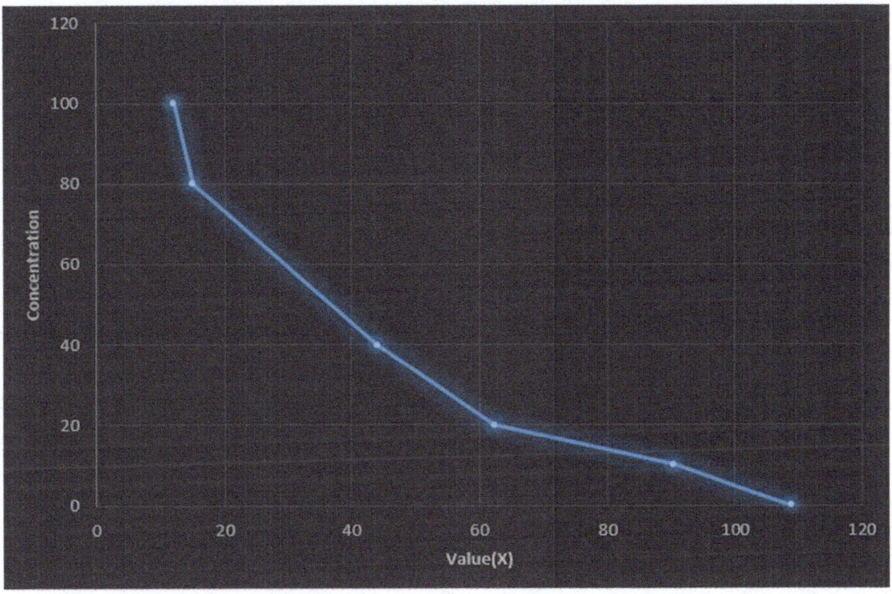

Fig. 79.14 Shows the relationship between the H_2O/methylene blue concentration and light absorption value (*X*), equation of graph $(Y) = -0.9972X + 95.511$

5. The graph results in Fig. 79.14 indicate the lower the value (*X*), the higher the concentration of methylene blue which results to better DSSC performance.
6. The actual concentrations of the porous TiO_2/NaCl films were higher compared to the dense TiO_2 film which makes it more suitable for DSSC.

References

1. Zhao Z et al (2015) Dye-sensitized solar cells based on hierarchically structured porous TiO_2 filled with nanoparticles. J Mater Chem A 3:11320–11329
2. Ando Y, Noda Y, Kobayashi A (2014) Al_2O_3 film deposition by atmospheric thermal plasma spray using 1 kW class electric power source. Front Appl Plasma Technol 7(2):75–80

Chapter 80
Achievement of Low-Energy Buildings in High-Latitude Countries Through Passive Solar Systems

Dorota Chwieduk

80.1 Introduction

The architecture and civil engineering design of buildings are crucial for their energy needs. To create low-energy buildings, it is necessary to take into account geographical and climatic conditions, including availability of solar energy. Energy consumption of every building depends on its location: local and more general geographical, its architecture and civil engineering concept, structure, materials, heating installations, interior arrangement and usage patterns. A building in a city has different environmental conditions than one located in the suburbs or in the countryside. The energy needs of a building are influenced strongly by the specific local environment. In cities, access to heat or electricity sources is easy. However, it is determined by the presence of centralized energy sources, district heating systems and power grids. Outside cities, it is easier to make independent decisions on the architecture of a building, its location and systems of energy supply, including type of fuels. It does not mean that any citizen cannot have an influence on the place where they live. However, the possibility of selection of energy media is limited, because some are not available in cities. But solar radiation reaches everywhere. It is important to know how solar energy can be used in both a passive and active way. Architecture, shape, structure and materials of a building can help to use solar energy when its availability is small or to protect a building against overheating when there is too much solar radiation gain. Well-thought-out design and construction of a new building is the simplest way of reducing space heating/cooling energy needs. Any activity to reduce energy consumption in a building should start from the proper design of a building including application of solar passive systems.

D. Chwieduk (✉)
Institute of Heat Engineering, Faculty of Power and Aeronautical Engineering,
Warsaw University of Technology, Warsaw, Poland
e-mail: dorota.chwieduk@itc.pw.edu.pl

A. Sayigh (ed.), *Renewable Energy and Sustainable Buildings*, Innovative Renewable Energy, https://doi.org/10.1007/978-3-030-18488-9_80

80.2 Main Features of Solar Passive Systems in High-Latitude Countries

Solar radiation availability depends of course on geographical conditions, which are connected to climate. However, there are some common issues. Solar energy used in a passive way by a building, no matter what are geographical conditions, should take into account the following features:

- Location and shape of the building with regard to availability of solar energy appropriate to the season of the year
- Location of rooms with regard to their different functions in accordance to their exposure to solar radiation
- Possibility of adaptation of a building's structure to use solar energy in a passive way by applying solar passive systems in the form of buffer spaces, solar collector walls and storage walls depending on climatic conditions
- Selection of building's structural elements and their materials to use solar energy or to provide protection against too much solar radiation depending on a season of the year
- Utilization of a building's surroundings to enhance or reduce solar energy availability, including planning of vegetation
- Utilization of a building's surroundings for natural cooling and ventilation, precooling or preheating of ventilation air in structural elements of a building and underground channels
- Adaptation of structural, material and installation solutions for waste energy utilization, including energy recovery from ventilation air systems, wastewater, etc.

Passive solutions utilize the building structure to absorb, store and distribute the collected solar energy [1–6]. A building should be exposed to solar radiation impact on the southern side in winter time, and from the north, it should be opaque to environment influence, i.e. well insulated and tight. This means that transparent elements of the envelope, especially windows, should not be used at that side or their use should be minimized only to fulfil daylighting requirements. Using passive solutions decreases energy consumption for space heating and cooling. In many cases, the use of passive solutions can be combined with daylighting systems. It is particularly important in the case of large building structures: residential, industrial and especially in public utility buildings.

The use of solar passive systems leads often to the use of large glazing surfaces, glazed verandas and atria [7]. As a result, a closer link between rooms and the neighbouring environment is achieved. Application of solar passive systems also utilizes the building's structure for the heat storage and distribution of heat gained. Utilization of passive solutions also requires designing appropriate shading elements of the building envelope, e.g. overhangs, wing walls and its surrounding such as trees and other greenery. Introduction of passive solutions requires careful analysis of energy demand and conditions for collecting solar energy for specific climate conditions.

In higher-latitude countries, direct solar gain systems can be recommended for winter. Large-area windows and glazed façades should be applied on the southern elevation. Such windows should also be characterized by low U-values (heat transfer coefficient). Large-area windows in climate conditions observed in higher latitudes often result in large indoor temperature variations, in both summer and winter. This can lead to overheating of rooms in summer and overcooling of rooms in winter. Because of this, especially designed shading devices should be used in summer and specially insulating extra covers can be used for winter.

80.3 Solar Buffer Spaces as a Recommended Solution for High-Latitude Countries

One of the well-known forms of the solar passive systems is a buffer space, also known as a greenhouse. Some examples are presented in Fig. 80.1.

Passive systems in the form of solar buffer spaces can be recommended in high-latitude countries. However, the type of a buffer space depends on the specific climatic conditions.

Regular solar buffer spaces in the form of greenhouses can be attached to buildings at the south side in a moderate climate [8]. Buffer spaces presented in Fig. 80.1 are of that type. They can be recommended for moderate climate or specific climatic conditions like those in the United Kingdom. In the United Kingdom, even in high-latitude regions, winters are not cold because of the influence of the Gulf Stream. Usually winter climate is mild, so the large glazed surfaces do not cause high heat losses and greenhouses constitute very good buffer zones protecting interiors against sudden changes of weather conditions.

In high-latitude countries, the climatic conditions can be severe in winter, with very low ambient air temperature and solar irradiance. However, in some northern countries with a continental climate, as in Poland, in summer ambient air temperature and solar irradiance can be high. Therefore, a specific type of the buffer space is necessary. The buffer spaces should be incorporated into the interior of a building,

Fig. 80.1 Examples of solar buffer spaces

not attached to it. No glazed horizontal or inclined roofs should be used. Horizontal or small inclination glazed surfaces are characterized by high heat losses in winter. Such surfaces can see only the sky, and in winter nights there can be a significant heat transfer through thermal radiation between the horizontal roof and sky. The heat transfer is very intensive in cold nights with a clear sky. Then usually the temperature difference between the glazed roof and the sky is large, and the difference of ambient air temperature and the sky can be about 30 K, so between the glazed roof and the sky the temperature difference is especially large. Conversely, in summer, horizontal surfaces or surfaces of small inclination are especially exposed to solar radiation. In Poland, solar irradiance at noon can be even as high as 800 W/m^2. At that time of the day, the incident angle of solar radiation on horizontal surface is small and transmissivity of the glass is at its highest level. In addition, such glazed surfaces can see the Sun for nearly the whole day. In Poland, sunrise can be about 3 o'clock in the morning and sunset after 9 o'clock in the evening. Figure 80.2 presents solar buffer spaces recommended in high-latitude countries with the influence of a continental climate, such as Poland.

The solar buffer space presented in Fig. 80.2 can improve much thermal energy balance of a building. It assures good protection against strong winter conditions and hot summer. The energy needed for space heating is much reduced and there is no need for summer cooling. The building presented in Fig. 80.2 is characterized by very low energy consumption. A solar buffer space incorporated into the structure of a house is the main solar passive system in the building. A large glazed south façade accounts for 35 m^2, which is nearly 70% of the total south façade area. There are two overhangs: external and internal. The buffer space contains two cuboid subspaces with no partition between them. The external subspace is higher and the internal lower. The overhangs are beneficial in reducing solar energy impact in summer, but they allow penetration of solar radiation in winter. The internal overhang is a part of the internal building's horizontal partition, being a part of a ceiling for the first floor and at the same time being a part of a floor on the second floor. The internal overhang was designed to allow the solar radiation to enter fully into the interior space of the house for 3 months: November to January. The building does not need

Fig. 80.2 Solar buffer space incorporated into the interior of a building, view from inside (in summer) and from outside (in winter)

much energy for space heating. In winter, the buffer space acts as a natural passive solar heater significantly reducing consumption of energy for space heating, and in summer the buffer space acts as a natural passive barrier significantly reducing the influence of solar radiation. For space heating and domestic hot water, the indices of final and primary energy consumption are equal to 22.56 kWh/(m^2a) and 25.58 kWh/(m^2a), accordingly. This house is a really low-energy building.

80.4 Other Recommendations for Solar Passive Solutions

Apart from the solar buffer space idea, which seems a very good solution for the solar passive systems in high-latitude countries, there are some basic ones, which always should be fulfilled.

A building must have a shape suitable to climatic conditions and to the given location of a building. Any building of cuboid shape should have a main façade exposed to the south, the extension in direction east–west to direction south–north should be in the range from 1.5:1 to 2:1 [9]. At the southern part, a main glazing façade can be planned, but direct solar gain systems are not recommended. For summer, special shading devices must be planned, and for winters some external covers or blinds, especially for their use at night.

Quite often, a cylindrical shape is recommended for low-energy buildings in high-latitude countries. Usually the cylindrical shape is extended from the east to the south, and the north wall of a building is a regular vertical one. In the case of glazed semi-cylindrical façades, the extension of the glazed façade should be limited by the range of azimuth angles of the sunrise and sunset in winter (in December) in a given geographical location. It means that in Poland it should be extended from the south-east to the south-west (in December the azimuth angle of the sunrise is about −50° and the sunset is about +50°). The southern glazed façade must be fully vertical. In case of the extension of the main façade to the west, a vertical partition should be only partly transparent. This is because of high solar irradiation of west vertical surfaces in summer in countries of similar climatic conditions to Polish ones.

When energy needs are really reduced significantly thanks to solar passive solutions, then innovative active energy systems can be introduced.

80.5 Conclusion

The aim of using passive systems is the reduction of energy demand for room heating and cooling. Each building exchanges heat with its surrounding. The intensity of this heat transfer depends of course on weather (climatic) conditions, but also on the structure and materials used for opaque partitions and windows. Utilization of advanced and innovative materials, design solutions and installations make it possible to reduce the amount of energy needed for heating and cooling to maintain thermal comfort inside a building at the required level. Application of solar passive systems is the first step in assuring a building to be low energy. Passive solutions in architecture improve the energy balance of a building. At the same time, they affect the external look of a building and create esthetical and functional solutions. For instance, the introduction of solar buffer spaces brings the inhabitants closer to nature and their environment. It is always good to look at the features of traditional architecture in the region. Passive solar energy solutions, such as overhangs, extended roofs or wingwalls, have been used in architecture for many years. They allow the limiting of the excessive influence of solar radiation during summer without disturbing the access of solar energy in winter. Modern solar architecture, not only in the case of small residential houses but also in large city buildings, should combine the ability to use energy available in the environment with the possibility of restricting unwanted energy interactions, when required. At the same time, it should be based on proven solutions and traditions in planning and constructing buildings according to the local geographic and climatic conditions. Modern solar passive technologies help us to realize ideas that have already been discovered in the past by traditional architecture and civil engineering methods.

Nowadays, a solar passive system can be an important element of a total design solution including ventilation systems, solar active thermal systems, PV systems and daylighting solutions of a building. Planned solar energy use in buildings can be done through application of the building integrated solar technologies, such as BIPV, BIST and BIPV/T [10]. Summarizing, the achievement of low-energy buildings in high-latitude countries can be through the common use of the passive solar systems, which nowadays are integrated with active solar energy technologies. Thus, low-energy buildings should utilize solar energy in a planned and comprehensive way through the application of

- Solar passive solutions
- Building integrated solar thermal systems (BISTs)
- Building integrated photovoltaic systems (BIPVs)
- Building integrated solar thermal/photovoltaic (BIPV/Ts)

Thanks to these building integrated solar energy technologies, the collection, storage and distribution of energy take place on site for the self-use of the energy producer, who is focused on reduction on energy consumption. In this way, the achievement of low-energy buildings becomes easier and more understandable for everybody.

References

1. Anderson B (1975) Solar energy: fundamentals in building design. Total Environmental Action, Inc., Harrisville, New Hampshire
2. Athienitis AK, Santamouris M (2002) Thermal analysis and design of passive solar buildings. James & James Ltd, Hong Kong

3. Balcomb JD, Jones RW, McFarland RD, Wray WO (1984) Passive solar heating analysis—a design manual. ASHRAE, Atlanta
4. Balcomb JD (1992) Passive solar buildings. The MIT Press, Cambridge, MA
5. Kolokotsa D, Santamouris A, Karlessi T (2014) Passive solar architecture. In: Comprehensive renewable energy, vol 3. Elsevier, Amsterdam, pp 637–665
6. Hastings R (ed) (1995) Solar low energy houses of IEA Task 13, Solar Heating & Cooling Programme, IEA. James & James, London
7. Leftheriotis G, Yianoulis P (2012) Glazing and coatings. In: Comprehensive renewable energy, vol 3. Elsevier, Oxford, pp 313–355
8. Porterous C, MacGregor K (2012) Solar architecture in cool climates. EARTHSCAN, London, Sterling, VA
9. Chwieduk D (2014) Solar energy in buildings. Thermal energy balance for efficient heating and cooling. Elsevier, Academic Press, San Diego, CA
10. Tripanagnastopoulos Y, Chwieduk D, Farkas I (2017) New systems and application options. In: Kalogirou S (ed) Building integrated solar thermal systems. Design and application handbook. p 263–280. COST Action TU1205 (BIST), COST Office

Chapter 81
Advancements and Challenges Affecting Wind Turbine Implementation in the GCC Countries

Abdul Salam Darwish

81.1 Introduction

GCC countries are rapidly growing, driving the Middle East power demand to be doubled in 2040. The demand has been growing from 472 TWh in the year 2000 to 1132 TWh with an average growth of 5.6% a year [1]. This is due to the increase in population, growing economy and high living standard. Renewable energy has come recently of interest and expected to contribute a share of more than 10% by 2040. The population of the GCC countries is 07% of the world population in 2014 [2]. The GCC contributes to a higher percentage of the global greenhouse gas emission than the European Union [2]. The total energy consumption is shown in Fig. 81.1.

Renewable energy was always the first-choice option for a modern power system. In the GCC area, wind and solar are now competitive with conventional sources and commanded 90% of the investments in renewable power. The reasons behind this are the potential of these resources and the technological advancement that has made it easier to use and cheaper to invest in [4]. The cost of wind turbines has fallen by nearly 1/3 since 2009 and that of solar photovoltaic (PV) modules by 80% [5]. Increased number of countries that held auctions to deploy renewables [40]. Global investment in renewables raised from less than $ 50 billion in 2004 to more than $ 360 billion in 2018 [6]. Figure 81.2 shows renewable power capacity installations have exceeded non-renewables by a rising margin.

Wind energy had a very good trend over the years and reached a worldwide capacity of 540 GW in 2017 [7]. It is expected to reach more than 610 GW (10% increase) by the end of 2018 as shown in Fig. 81.3.

Wind energy utilization in GCC countries have a moderate potential for the capacity of wind power generation. It can be considered as low speed region. However, the wind potential varies substantially from one country to another.

A. S. Darwish (✉)
Phoenix Renewable Energy Centre, Manchester, UK
e-mail: Darwish@phintc.com

A. Sayigh (ed.), *Renewable Energy and Sustainable Buildings*, Innovative Renewable Energy, https://doi.org/10.1007/978-3-030-18488-9_81

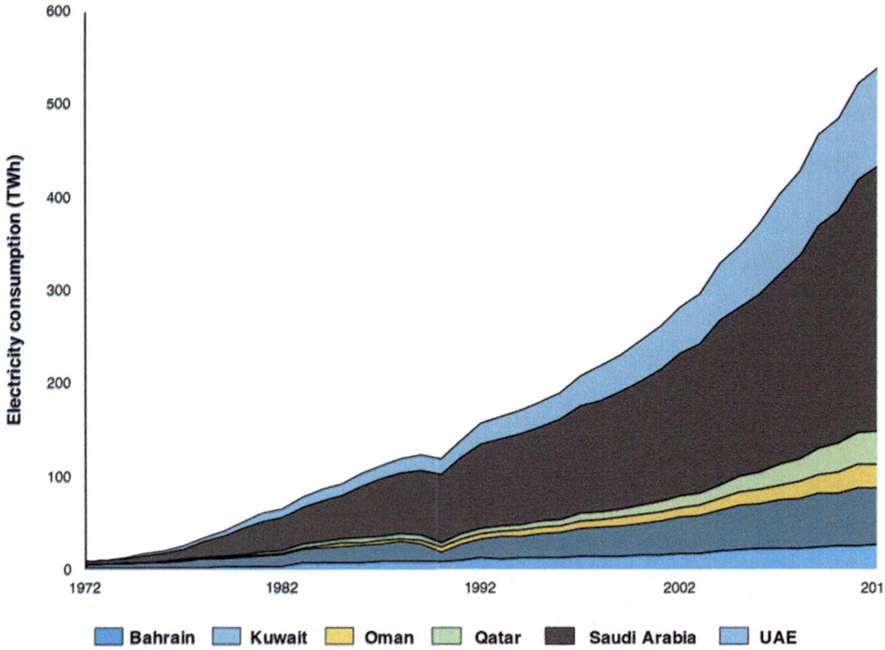

Fig. 81.1 Total primary energy consumption from 1972 to 2012, million tons of oil equivalent (Mtoe) [3]

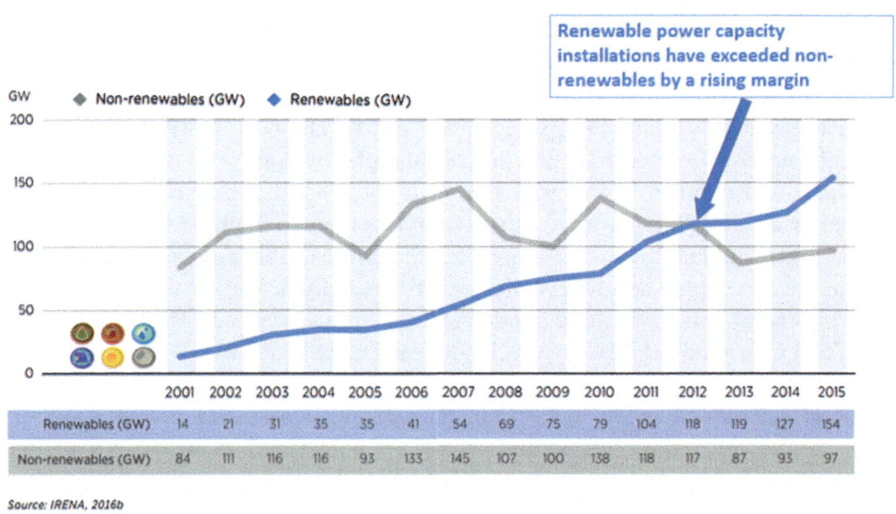

	2001	2002	2003	2004	2005	2006	2007	2008	2009	2010	2011	2012	2013	2014	2015
Renewables (GW)	14	21	31	35	35	41	54	69	75	79	104	118	119	127	154
Non-renewables (GW)	84	111	116	116	93	133	145	107	100	138	118	117	87	93	97

Fig. 81.2 Renewable and non-renewable power capacity additions, 2001–2015 [5]

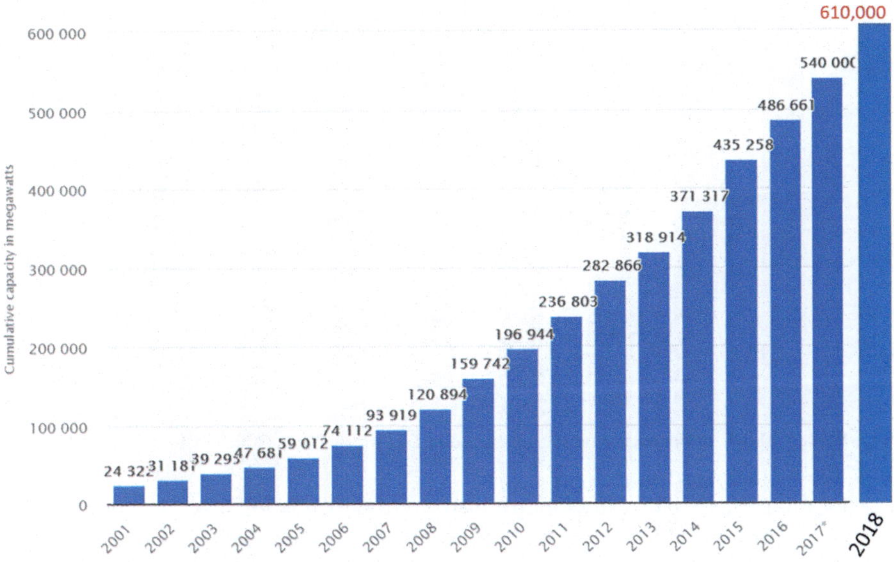

Fig. 81.3 Cumulative installation capacity

Table 81.1 Wind speeds at the GCC [8]

Country	Wind speed (m/s)	Hours of full load of wind per year
Bahrain	5–6	1360
Kuwait	5–5.5	1605
Oman	4–6	1463
Qatar	5–7	1421
Saudi Arabia	2.5–4.5	1789
United Arab Emirates	3.5–4.5	1176

Table 81.1 [8] shows the recorded hours of the full load of wind per year for the GCC.

It can be seen that Saudi Arabia has the highest with 1789 h and UAE has the lowest with 1176 h. Higher speeds can be found at mountain terrain, coastal and higher heights. According to the RES, countries with more than 1400 h per year are considered to have economically viable wind energy utility [9]. The greatest potential is on some of the sites on the 2000 km coastline for offshore wind. An example of the average mean wind speed (AMWS) and capacity factor for Duqm, Oman, is seen in Fig. 81.4 [10].

The GCC region is in need for this source of energy as the solar can cover the daily demand to a certain extent but the wind can substitute the lack of supply

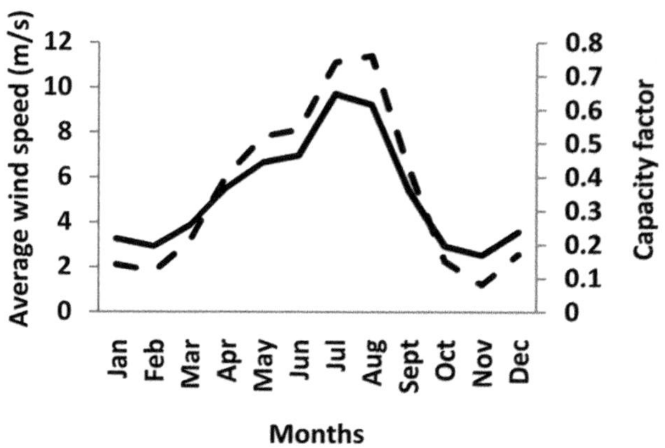

Fig. 81.4 AMWS and capacity factor measured at Duqm for 2004–2008 [10]

during night-time. The wind energy utilization depends on several factors. These factors are the available wind speed frequency distribution over several years, social acceptance, economic benefits, technical support, distance from the main electricity grids and the impact on the environment. Matching between the site suitability and the suitable wind turbine is an important issue before starting installing wind turbines in any country. Several studies were done to specify the most suitable sites and showed the feasibility of implementation, e.g. [11]. The lack of public awareness and detailed researches has slowed down the implementation. This paper aims to raise awareness and presents advancements in the technology and data analysis that could make the implementation possible.

Wind energy has become an important energy resource for many developed and developing countries. Technological advancements have supported the efficient harnessing of available wind energy and how to integrate it into energy transmission networks but is also due to growing recognition of the importance of renewable energy sources in climate change mitigation strategies [12, 13]. Many countries have relied on this source of energy to meet the total demand for electricity in a considerable amount. For example, Denmark is depending for 40% of the domestic energy consumption [14] and hoping to achieve 50% by 2020. In 2015, China reaches a total capacity of 148 GW [15].

High wind energy potential is available at different sites in GCC. For instance, recent study showed that Duqm, Oman, is a high wind speed region. Simulating a wind farm at that area showed the possibility of having a site with 25 MW with wind power density ranging between 200 and 800 W/m^2 at 80 m above the ground and can generate a total of about 75 GWh net AEP [16]. Wind availability across GCC is encouraging (e.g. KSA-central). It peaks during evening hours and decreases during daylight hours while it has an opposite behaviour at the coastal area (e.g. KSA-east, Abu Dhabi) [17]. The highest wind speed is nearly 8 (m/s) at winter time and 6.7 m/s at summer time. It is at higher values at higher heights.

Barriers to the implementation of wind power at the GCC region is merging within the barriers for overall renewable energy implementation which slows down all efforts for energy utilization, such as [18–20]:

(a) Lack of renewable market. This is due to the subsidies of fossil fuel and energy use, and non-liberalized market entry resulted in no renewable system that can compete with the existing energy prices [19].
(b) Lack of related policies and regulations.
(c) Lack of appointed financial support.
(d) Reduced public awareness for implementing such schemes.
(e) No clear view on how to integrate the renewable energy systems with the national grid.

81.2 The Advancements of Wind Turbine Technology

Wind turbine technology had to fight for legitimacy among other renewable systems, which had an impact on the development of its technical improvements including aerodynamic parts, control, size and other facilities. Advancements in those areas have also generated pains against the barriers that have always acted as obstacles. The recent improvements have enabled access to wind power and reduced the overall cost in spite of the existing industrial challenges [21]. The following advancements are enhancing the wind energy implementation:

- *Larger turbines improve efficiency and performance*
  Any percentage increase in average rotor diameter resulted in a large amount in energy generation and considered by many investors as attracted technology. The largest wind turbine ready for testing in the UK is the Haliade-X 12 MW [22]. Engineering giant GE Renewable Energy and the UK's Offshore Renewable Energy (ORE) signed a 5-year research to test the world's largest offshore wind turbine at a facility in northeast England. The turbine designed with 220 m rotor diameter, 107 m blade length, 260 height with 63% capacity factor [23]. Dong Energy has installed the world's largest wind turbines, Burbo Bank wind farm in the Irish Sea [24]. The turbine generates 8 megawatts (MW), with the largest rotor diameter of 164 m.
- *Clean energy meets battery storage*
  Most recent advancement in wind technology is the batteries. In this respect, Tesla has made recently a surprising effort on the global stage to combine wind generation with battery storage in a way the world has not seen before. Batteries will support the power supply smoothness and help level out any trouble spots between the whims of the weather and the more predictable cycles of daily life [21].
- *Stable way of installation with confidence*
  Nowadays, wind turbine installations have been done with confidence as the whole process has been well modernized and experience has been built up to its highest level.
- *Advanced and efficient site selection methodologies*
  Site selection has been well formulated with latest advanced models and programs that have the ability to selecting and matching the right technology for the

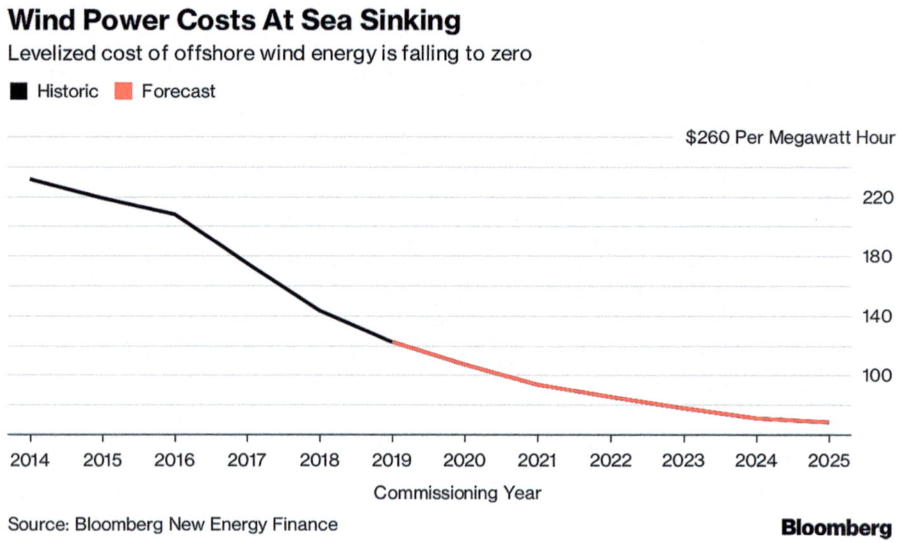

Fig. 81.5 Wind power costs falling steeply [26]

right site. Site selection is a critical part of the whole process and could have a direct effect on the success of a wind project. Advanced monitoring systems and surveying techniques have contributed to this selection process.

- *Specialized blades are developed to generate more energy from low wind areas*
  The need for specialized wind rotor blades is essential particularly for the low speed regions to maximize the power extracted from the turbine over the hours of operation at any available wind speed [25].
- *Prices of offshore wind is falling steeply and will keep doing so*
  Figure 81.5 [26] shows that the offshore wind cost is falling steeply. This reflects on all types of wind technology and indicates a bright future for wind energy implementation, which will have an impact on the decision for investment.

81.3 Why Do We Need Wind Energy in GCC?

Several drivers are initiating the need for wind energy at the GCC countries in spite of the wealth of sunshine compared with the available wind energy potential. Those are the following:

- To secure power generation at night-time or through any unexpected climate change periods.
- Wind turbine normal costs dropped by more than 37%.

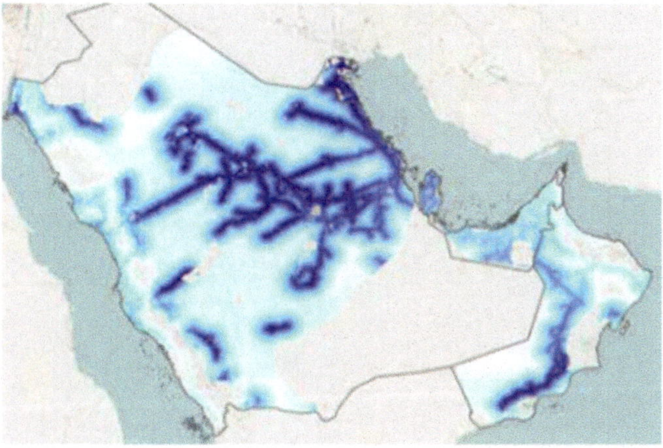

Fig. 81.6 Suitability scores for grid-connected wind up to 75 km from the grid [28]

- By 2025, average electricity costs generated by the sun and wind could drop by as much as 59% [27].
- Technology to produce power from lower wind speeds is improving [25].
- Desalination is cheaper with the wind [27].
- More than 56% of the GCC's surface area has significant potential for wind deployment.
- Covering just 1% of this area could translate into an equivalent 60 GW of capacity.
- Figure 81.6 shows more than 70% (light blue) and up to 100% suitability (dark blue). Higher scores represent increased suitability [28].

81.4 Wind Energy Resources in the GCC Region Versus Oil

GCC regions exhibit significant untapped potential for wind power, driven by abundant wind resources and increasing power consumption [29]. The average wind speeds in GCC countries is comparable to Germany (3–7 m/s); wind potential has remained largely untapped so far with GCC cumulative capacities in 2012 being less than 2 MW (Germany: more than 31 GW) [29]. Wind power generation is significantly cheaper than oil-based power generation. Effectively, wind power frees up more locally produced oil available for export for all countries in this region. The average LCOE (Levelized Energy Cost) in favourable wind regions (i.e. >5 m/s) varies between 9 and 16 USD cents/kWh. This is well below the present cost of oil-fired electricity generation of around 19 USD cents/kWh at an oil price of 100 USD/

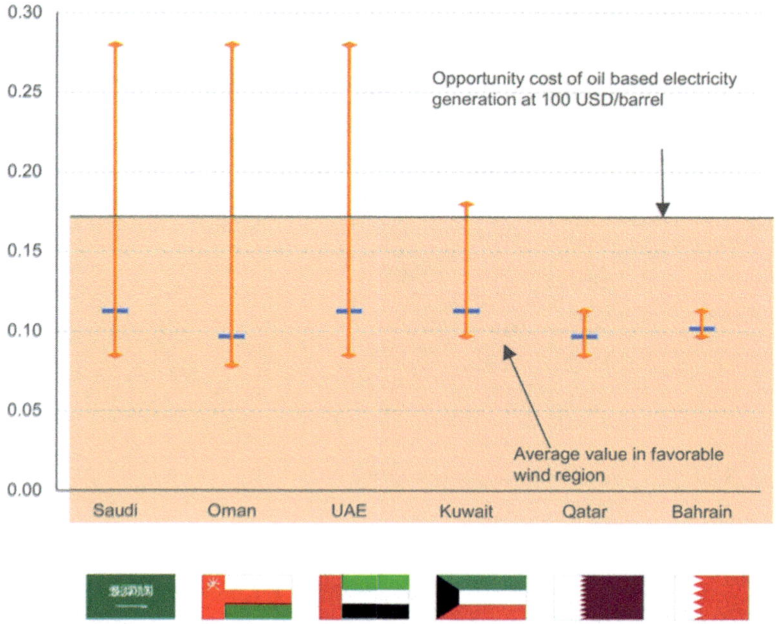

Fig. 81.7 Levelized cost of energy (inflation adjusted) based on full load hours for different sites employing Vestas V100 turbine [29]

barrel. Figure 81.7 shows the levelized cost of energy (inflation adjusted) based on full load hours for different sites employing Vestas V100 turbine [29].

81.5 Regional Suitability for Onshore Wind Power Generation

Report by [30] indicated that site suitability is with a minimum wind speed of around 3.7 m/s at a height of 10 m in order for wind generation to be competitive with the LCOE of oil-fired electricity generation. Accordingly, Fig. 81.8 shows that many suitable sites are available in GCC. For instance, in Saudi Arabia, Yanbu in the West and Juaymah and Dammam/Al Khobar/Dhahran in the East have a considerable potential for utilization and promising sites.

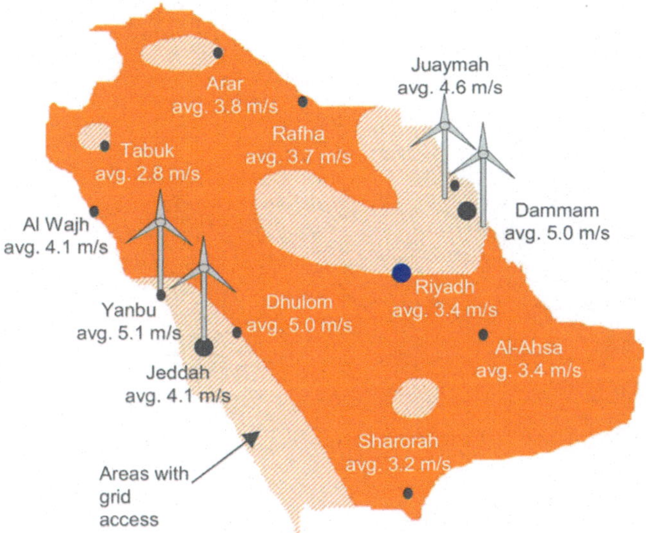

Fig. 81.8 Apricum analysis. Wind speeds measured at 10 m height [30]

81.6 Wind Energy Availability Overview in the GCC Region

The GCC countries have started to raise awareness on the real need for alternative energy supply in spite of the wealth of fossil fuel. The following sections provide a brief overview for each country in this aspect.

81.6.1 Oman

This country enjoys a promising amount of wind potential and started to implement the project for utilization. Wind speed is more than 8 m/s during the summer months (June to September) (e.g. Duqm area) and is shown in many other areas between 5 and 7 m/s [16]. Figure 81.9 shows wind power density at 80 m above the ground for the whole country and monthly distribution of wind speed at 80 m above the ground [16]. The 50 MW project is to be installed by Masdar in Abu Dhabi [31]. This is to supply 16,000 homes, which can offset 110,000 tonnes of CO_2 emissions/year.

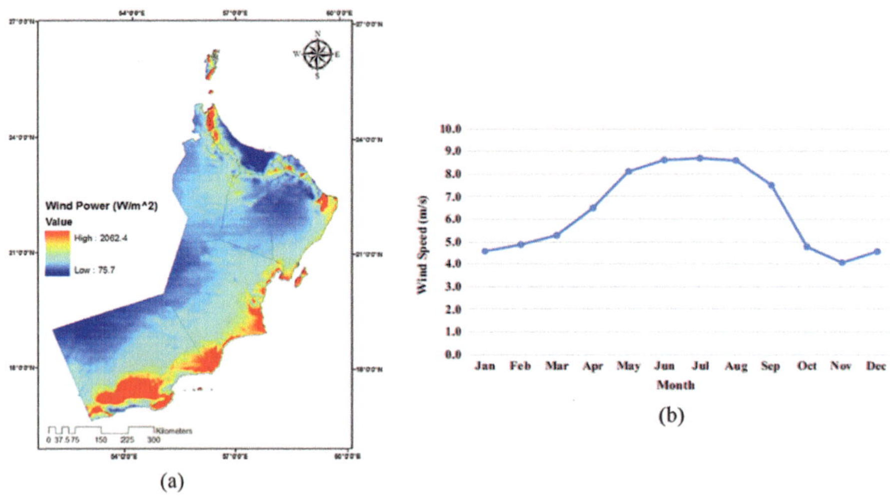

Fig. 81.9 (**a**) Wind power density at 80 m above the ground for the whole country. (**b**) Monthly distribution of wind speed at 80 m above the ground [16]

81.6.2 Bahrain

Bahrain has been utilizing wind energy starting with the World Trade Center (1950 and renewed in 2007), generating 0.66 MW which supply 15% of the building consumption. The country enjoys a good potential, which could attract investors. Assessment study by [32] reveals a wind speed of 4.8 m/s at 10 m height and mean Weibull scale and shape parameters C and K of 4.8 m/s and 1.74, respectively. This indicates that the wind potential is high and of good quality (strong enough winds of long duration). Several projects with a capacity of 200 MW are possible to be invested at several locations in the unpopulated central and southern parts of the country in spite of the fact that central and northern parts are with higher wind potential, Fig. 81.10 [32]. Since wind energy can be used for reverse osmosis (RO) desalination plants [27], it can, therefore, create opportunities for investment in many areas in Bahrain.

81.6.3 Saudi Arabia

Within their energy vision on 2030, Saudi Arabia targets to generate 54 GW of renewable energy. The 9 GW target is from wind in 2040 which contributes 6.3% of the total power capacity. Figure 81.11 shows the wind map of Saudi Arabia for February and August 2015 [33] and also shows three regions of wind availability in the coastal region, 7.5–8 m/s at Hafar Al-Batin, 7–7.5 m/s at Yanbu and 5.5–6 m/s at Riyadh in the central region [33]. Saudi Aramco commissioned the first wind turbine in Turaif [34]; 2.75 MW can supply power to 250 households that saved 19,000

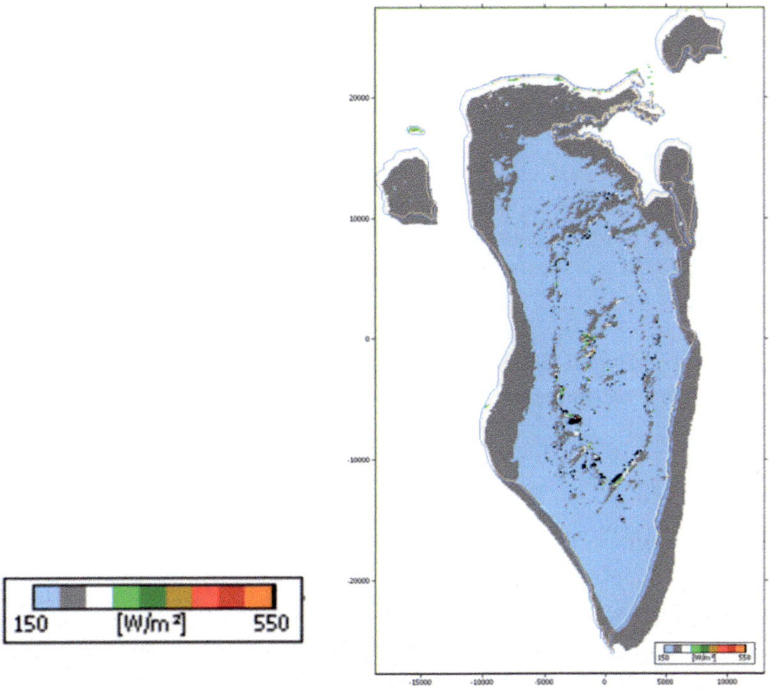

Fig. 81.10 The wind power geographical distribution at 10 m height [32]

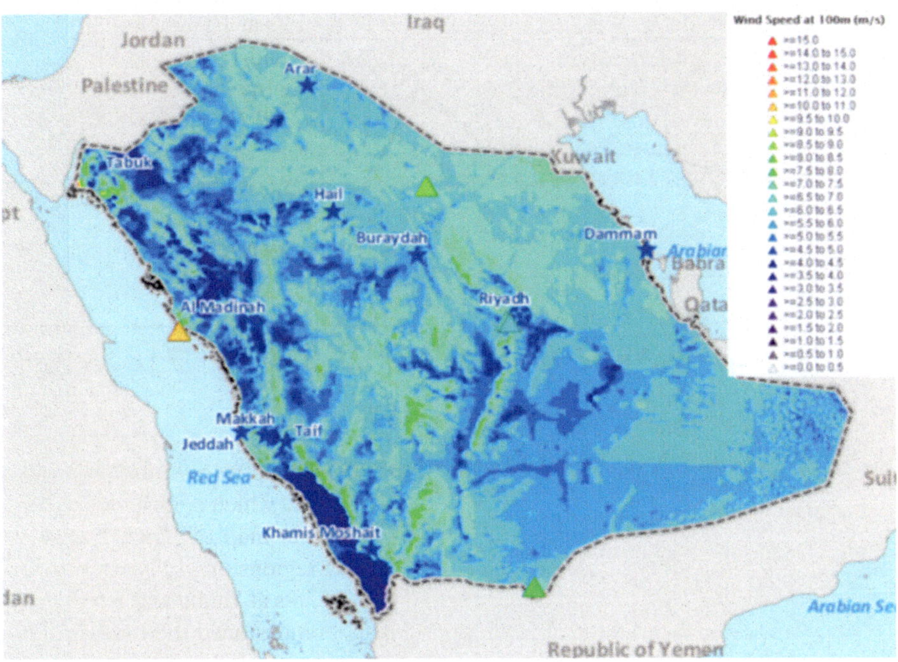

Fig. 81.11 Average wind speed in m/s for February 2015

barrels of oil equivalent. Keeping a similar investment will obviously enable the country to achieve their targets earlier.

81.6.4 *Qatar*

Wind energy potential in Qatar has also been studied [35]. Results showed that the average wind speed is 4.4 m/s. A good potential can be seen over the year between March and July (see Fig. 81.12). Utilization of available wind energy is possible through small wind energy systems, which can be both efficient and competitive. Wind pumping and small-scale electricity generation can be implemented. However, the present work recommends the offshore wind farms as the nearby areas are promising.

81.6.5 *Kuwait*

Annual average wind velocity at the standard height of 10 m was found to be between 3.7 and 5.5 m/s [36]. This indicates a promising potential and many projects can generate a reasonable amount of energy. The Salmi Mini-wind Farm of 2.4 MW was established to test and measure the performance of the equipment and

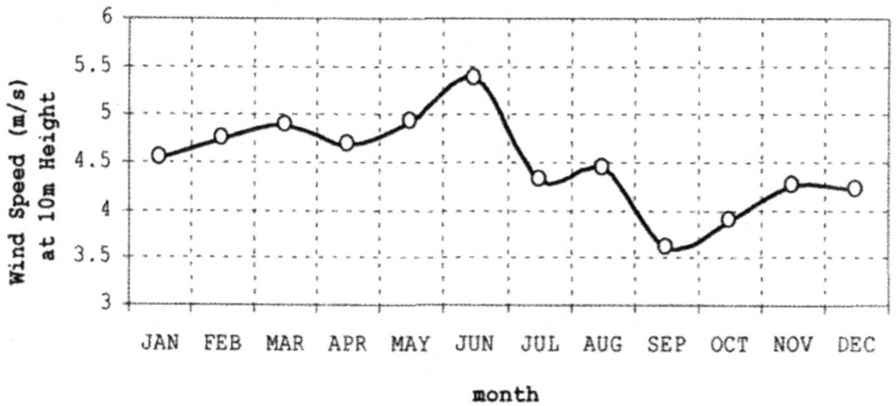

Fig. 81.12 Mean monthly variation of wind speed in Doha (1976–1989) [35]

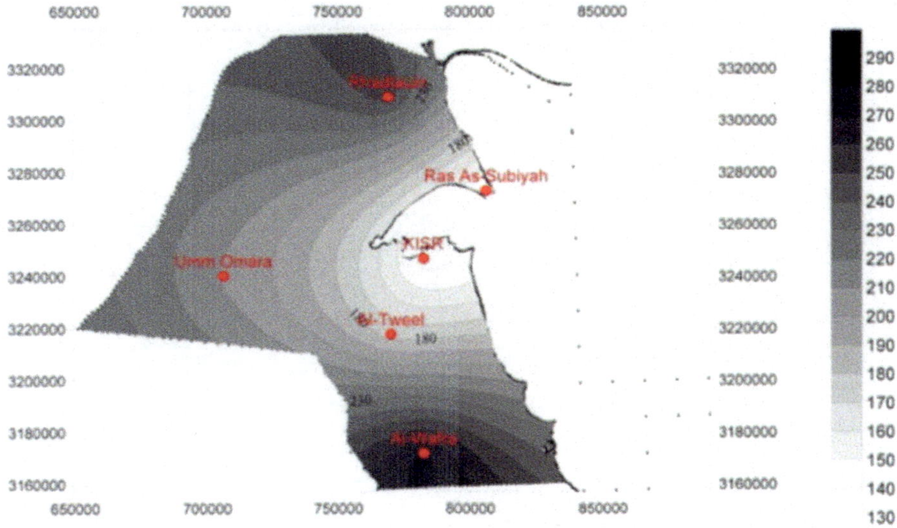

Fig. 81.13 Distribution of wind power density over Kuwait at 30 m height [37]

was located near the site of the first 70 MW, Shagaya Renewable Energy Park [37]. The wind will have the capacity to generate 10 MW. The monthly variation analysis shows high WPD during the high electricity demand summer season than other months with maximum WPD of 555 at Al-Wafra (see Fig. 81.13).

81.6.6 United Arab Emirates

Studies have shown [38] that the possible appearance of wind speeds greater than >5 m/s is monitored onshore and exceed >7.0 m/s nearshore. Moreover, it is possible for energy utilization from the wind at a height of 80 m to achieve a full load hour of 1176 h per year which refers to low capacity factor. Wind farms with capacity of 1–2 MW are possible with this available wind speeds if careful attention is taken to the recent advances. Other low speed turbines can well be implemented in small to medium size. The maximum wind speed is generated in the north region in the UAE, whereas the maximum wind speed potential is mainly derived from coastal areas as seen in Fig. 81.14 [39]. High potential is also visible offshore.

81.7 Challenges to Implementation

The following are some challenges that might form an obstacle to the implementation of wind energy systems at the GCC regions:

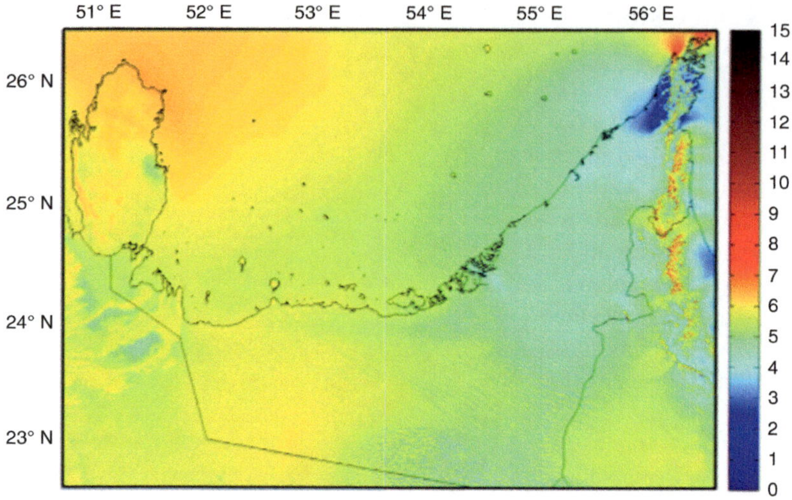

Fig. 81.14 The UAE wind resource map using GIS, WindPRO and WAsP programs [39]

- *Suitable wind sites are with low average wind speed*
  - Low speed rotors have been developed for low wind speed regions [25]
- *High initial investment*
  - Establishing a viable wind turbine site just cannot compete with electricity generated from more conventional means.
  - This is reduced over the years.
- *Geography*
  - Sites that best suit wind generation are not close to population centres and will require to build lengthy and expensive transmission lines.
- *Transportation*
- *Noise solved by*
  - Design changes (more air converted to power and less goes to acoustic).
  - Proper siting.
  - Insulating materials can be used to minimize noise impacts.
- *Visual impacts* (proper siting decisions can help avoid any aesthetic impacts to the landscape and using today's larger and more efficient models of wind turbines).
- *Avian/bat mortality* (is minimized by proper site selection and the use of latest technology to deter birds away from the turbine blades).
- *Primary health and safety considerations*
- Interference with radar and telecommunication facilities

81.8 Conclusion

The present wind energy business case is prevented by a number of barriers, but our survey results clearly show that the respondents believed that policies can make a major difference and allow for an expansion of wind system installations at the GCC region. This work has reviewed the main advances that happened with the designs, implementation and installation of the wind energy systems. It aimed to highlight the fact that all barriers that stand in front of the attempts to use this wind technology in this region is a myth considering the available wind potential in each country and the advancement in technology. Hurdles still hold back the deployment of more wind systems across many promising locations in the GCC. These hurdles disappeared as the technology has developed for the low wind speed areas and with the sharp reduction in cost. The advancement in the data analysis techniques enabled the accurate assessment of the wind availability in these countries and showed a wider range of promising sites for utilization, which formed a competitive advantage against the other forms of energy generation. Companies all over the world are much experienced nowadays to select, install and maintain, which provides confidence to investors. This study highlights these advancements and the good potential of wind energy at the GCC region and recommends more investments in utilizing that potential in order to secure power generation during night-time or through any unexpected climate change periods. Wind energy resource for those countries is to be the second source of clean energy after the solar if implemented in the right way.

References

1. Al Bawaba (2017) Business. https://www.albawaba.com. Accessed from albawaba
2. Indraganti M, Boussaa D (2018) An adaptive relationship of thermal comfort for the Gulf Cooperation Council (GCC) countries: the case of offices in Qatar. Energ Buildings 159:201–212
3. Kapsarc.org (2018). https://www.kapsarc.org/wp-content/uploads/2017/11/KS-2017-MP04-GCC-Energy-Overview-2017.pdf. Accessed 10 Jul 2018
4. Darwish A (2016) Sustainable green smart buildings: future energy survivor. J Sci Technol 12(21):35–42
5. IRENA (2017) REthinking Energy 2017: accelerating the global energy transformation. International Renewable Energy Agency, Abu Dhabi
6. Centre/BNEF, r. S.-U (2018) Global trends in renewable energy investment 2018. Frankfurt School of Finance & Management gGmbH. https://cleanenergysolutions.org/sites/default/files/···/christine-2018_e_short.pdf
7. Rehman S, Alam M (2018) A review of wind-turbine structural stability, failure and alleviation. Advances in Civil, Environmental, & Materials Research (ACEM18). Songdo Convensia, Incheon, Korea
8. Ferroukhi R, Ghazal-Aswad N, Androulaki S, Hawila D, Mezher T (2013) Renewable energy in the GCC: status and challenges. Int J Energy Sector Manag 7(1):84–112. https://doi.org/10.1108/17506221311316498
9. Alnaser WE, Alnaser NW (2009) Solar and wind energy potential in GCC countries and some related projects. J Renew Sustainable Energy 1(2):1–28

10. Shawon MJ, Chaar LE, Lamont LA (2011) The GCC: wind technology deployment potential. IEEE GCC Conference and Exhibition (GCC), Dubai. p 174–177
11. Eltamaly AM (2013) Design and implementation of wind energy system in Saudi Arabia. Renew Energy 60:42–52
12. Bruckner, T., Bashmakov, I., Mulugetta, Y., Chum, H., de la Vega Navarro A., Edmonds J., Faaij A., Fungtammasan B., Garg A., Hertwich E., et al (2014) Climate change 2014: mitigation of climate change. In: Contribution of Working Group III to the Fifth Assessment Report of the Intergovernmental Panel on Climate Change. Cambridge University Press, Cambridge/ New York, NY
13. Genton MG (2004) Skew-elliptical distributions and their applications: a journey beyond normality. CRC Press, Boca Raton
14. DEA (2016) The Danish energy model: innovative, efficient and sustainable. Danish Energy Agency, Copenhagen
15. WWEA (2016) WWEA Quarterly Bulleting Issue 1–206: Community Wind Special. World Wind Energy Association, Bonn
16. Al-Yahyai S, Charabi Y (2015a) Assessment of large-scale wind energy potential in the emerging city of Duqm (Oman). Renew Sust Energ Rev 47:438–447
17. Alotaibi S (2011) Energy consumption in Kuwait: prospects and future approaches. Energy Policy 39:637–643
18. Mokri A, Ali MA, Emziane M (2013) Solar energy in the United Arab Emirates: a review. Renew Sust Energ Rev 28:340–375
19. Mondal A, Khalil HS (2012) Renewable energy readiness assessment report: the GCC countries. Masdar Institute, Masdar
20. Lilliestam J, Anthony P (2015) Barriers, risks and policies for renewables in the gulf states. Energies 8:8263–8285. https://pdfs.semanticscholar.org/2c2d/0606dc6582592fd1f43dd8ddefd9562e9921.pdf
21. Matthews K (2017 BLOG POST. www.planetizen.com.Accessed 7 Apr 2018 from planetizen. Inc. (US)
22. Murray J (2018) Haliade-X: world's largest wind turbine coming to the UK. https://www.businessgreen.com/bg/news/3030821/haliade-x-worlds-largest-wind-turbine-coming-to-the-uk. Accessed 27 Apr 2018 from BusinessGreen
23. Energy GR (n.d.) HALIADE-X offshore wind turbine platform. https://www.ge.com/renewableenergy/wind-energy/turbines/haliade-x-offshore-turbine. Accessed 24 June 2018 from ge.com
24. Environment (n.d.) Dong installs world's largest wind turbines off UK coast. https://www.reuters.com/article/us-dong-energy-windfarm-britain-idUSKCN11E1F9. Accessed 27 Jun 2018 from Reuters
25. Salam Darwish A et al (2019) A methodology for improving wind energy production in low wind speed regions, with a case study application in Iraq. Comput Indust Eng 127:89–102
26. Shankleman J (2017) Subsidy-free wind farms risk ruining the industry's reputation. https://about.bnef.com/blog/subsidy-free-wind-farms-risk-ruining-the-industrys-reputation/. Accessed from Bloomberg NEF
27. Dodd J (2017). https://www.windpowermonthly.com/article/1436396/uae-desalination-cheaper-wind. Accessed 25 Apr 2018
28. IRENA (2016) Renewable energy market analysis: the GCC region. IRENA, Abu Dhabi
29. Mayr F (2013) Vast untapped growth potential exists for wind players in the Gulf region. https://www.apricum-group.com/hidden-champions-of-the-emerging-wind-markets/. Accessed from Apricum
30. Eltamaly et al. 2012, Rehman et al. 2003, Apricum analysis, Apricum. www.apricum-group.com
31. grupotsk (n.d.) TSK to build the largest wind farm in the Gulf. http://en.grupotsk.com/noticias/tsk-construira-el-mayor-parque-eolico-del-golfo. Accessed 25 Apr 2018 from en.grupotsk

32. Hanan Albuflasa, Murray Thomson, David Infield, Simon Watson (2007) Feasibility study of wind power in the kingdom of Bahrain. European Wind Energy Conference and Exhibition 2007, EWEC 2007. p 3
33. Ramli MAM et al (2017) Analyzing the potential and progress of distributed generation applications in Saudi Arabia: the case of solar and wind resources. Renew Sust Energ Rev 70:287–297
34. Damoom MM et al (2018) Adding sustainable sources to the Saudi Arabian electricity sector. Electr J 31(4):20–28
35. Marafia AH, Ashour HA (2002) Feasibility of wind energy utilization in Qatar. Eng J Univ Qatar 15:241–251
36. Al-Nassar W, Alhajraf S, Al-Enizi A, Al-Awadhi L (2005) Potential wind power generation in the state of Kuwait. Renew Energy 30:2149–2161
37. Fathi SA (2017) All Arab states should follow renewable energy possibilities. https://gulf-news.com/business/analysis/all-arab-states-should-follow-renewable-energy-possibilities-1.2026865. Accessed from Gulf News
38. Redha AM, Dincer I, Gadalla M (2011) Thermodynamic performance assessment of wind energy systems: an application. Energy 36(7):4002–4010
39. Sgouridis S et al (2016) RE-mapping the UAE's energy transition: an economy-wide assessment of renewable energy options and their policy implications. Renew Sust Energ Rev 55:1166–1180
40. IRENA (2018) Renewable energy auctions. http://www.irena.org: http://www.irena.org/policy/Renewable-Energy-Auctions

Conclusions

So we come to the conclusion of another exciting and successful world renewable energy congress. We enjoyed the presentations from our opening and plenary speakers who reminded us of the pressing urgency of our mission to ensure that renewable energy takes its rightful role as world leading source of energy and equally important is the need to make our built environment safe and sustainable.

We heard papers from more than 90 countries around the world gaining a truly global perspective on the energy and construction industries. All economies and societies are facing unique circumstances in which to solve their energy and environmental demands. Coming together allows us to share their experience, gain knowledge and new approaches from them. Each paper included here adds to that construct of expanding technology and policy.

Ali Sayigh

A. Sayigh (ed.), *Renewable Energy and Sustainable Buildings*, Innovative Renewable Energy, https://doi.org/10.1007/978-3-030-18488-9

Index

A. Sayigh (ed.), *Renewable Energy and Sustainable Buildings*, Innovative Renewable Energy, https://doi.org/10.1007/978-3-030-18488-9

D

E

M

U

V

Printed by Printforce, the Netherlands